ESSENTIALS of
LIVING WORLD

seventh edition

George
Johnson

Joel
Bergh

McGraw Hill

ESSENTIALS OF THE LIVING WORLD, SEVENTH EDITION

Published by McGraw Hill LLC, 1325 Avenue of the Americas, New York, NY 10019. Copyright ©2023 by McGraw Hill LLC. All rights reserved. Printed in the United States of America. Previous editions ©2020, 2017, and 2013. No part of this publication may be reproduced or distributed in any form or by any means, or stored in a database or retrieval system, without the prior written consent of McGraw Hill LLC, including, but not limited to, in any network or other electronic storage or transmission, or broadcast for distance learning.

Some ancillaries, including electronic and print components, may not be available to customers outside the United States.

This book is printed on acid-free paper.

1 2 3 4 5 6 7 8 9 LWI 27 26 25 24 23 22

ISBN 978-1-264-15652-8 (bound edition)
MHID 1-264-15652-9 (bound edition)
ISBN 978-1-264-40783-5 (loose-leaf edition)
MHID 1-264-40783-1 (loose-leaf edition)

Portfolio Manager: *Ian Townsend*
Product Developer: *Melisa Seegmiller*
Marketing Manager: *Britney Ross*
Content Project Managers: *Maria McGreal, Tammy Juran*
Buyer: *Sandy Ludovissy*
Content Licensing Specialist: *Lori Hancock*
Cover Image: *Seregraff/Shutterstock*
Compositor: *MPS Limited*

All credits appearing on page or at the end of the book are considered to be an extension of the copyright page.

Library of Congress Cataloging-in-Publication Data

Names: Johnson, George B. (George Brooks), 1942- author. | Bergh, Joel, author.
Title: Essentials of the living world / George Johnson, Joel Bergh.
Description: Seventh edition. | New York : McGraw Hill LLC, [2023] | Includes index.
Identifiers: LCCN 2021017835 | ISBN 9781264156528 (hardcover) | ISBN 9781264407835 (spiral bound)
Subjects: LCSH: Biology–Textbooks.
Classification: LCC QH308.2 .J6199 2023 | DDC 570–dc23
LC record available at https://lccn.loc.gov/2021017835

The Internet addresses listed in the text were accurate at the time of publication. The inclusion of a website does not indicate an endorsement by the authors or McGraw Hill LLC, and McGraw Hill LLC does not guarantee the accuracy of the information presented at these sites.

mheducation.com/highered

Brief Contents

Chapter 0 Studying Biology 2

Part 1 The Study of Life

Chapter 1 The Science of Biology 16

Part 2 The Living Cell

Chapter 2 The Chemistry of Life 34
Chapter 3 Molecules of Life 48
Chapter 4 Cells 64
Chapter 5 Energy and Life 90
Chapter 6 Photosynthesis: Acquiring Energy from the Sun 104
Chapter 7 How Cells Harvest Energy from Food 118

Part 3 The Continuity of Life

Chapter 8 Mitosis 134
Chapter 9 Meiosis 152
Chapter 10 Foundations of Genetics 166
Chapter 11 DNA: The Genetic Material 190
Chapter 12 How Genes Work 208
Chapter 13 The New Biology 226

Part 4 The Evolution and Diversity of Life

Chapter 14 Evolution and Natural Selection 250
Chapter 15 Exploring Biological Diversity 278
Chapter 16 Evolution of Microbial Life 294
Chapter 17 Evolution of Plants 322
Chapter 18 Evolution of Animals 342

Part 5 The Living Environment

Chapter 19 Populations and Communities 382
Chapter 20 Ecosystems 402
Chapter 21 Behavior and the Environment 428
Chapter 22 Human Influences on the Living World 442

Part 6 Animal Life

Chapter 23 The Animal Body and How It Moves 460
Chapter 24 Circulation and Respiration 476
Chapter 25 The Path of Food Through the Animal Body 496
Chapter 26 Maintaining the Internal Environment 514
Chapter 27 How the Animal Body Defends Itself 524
Chapter 28 The Nervous System 546
Chapter 29 Chemical Signaling Within the Animal Body 572
Chapter 30 Reproduction and Development 586

Part 7 Plant Life

Chapter 31 Plant Form and Function 606
Chapter 32 Plant Reproduction and Growth 626

Glossary 640
Index 654

About the Authors

George Johnson

Dr. George B. Johnson taught biology and genetics at Washington University in Saint Louis for over three decades and now serves as a Professor Emeritus of Biology. An English major at Dartmouth College before obtaining his PhD in Biology at Stanford University, Dr Johnson has long been involved in innovative efforts to incorporate interactive learning into our nation's classrooms. He continues to be at the forefront of adapting new modalities to learning biology. Over three million students have been taught from biology textbooks authored by Dr. Johnson. His college textbooks include *Biology, Understanding Biology, The Living World,* and this text, *Essentials of the Living World*. He has also authored two widely used high school biology texts. For many years, he has written a weekly column "On Science" for the Saint Louis Post Dispatch (you can read them at his home page *www.biologywriter.com*). He has chronicled the recent coronavirus pandemic in a series of letters you can read at *www.biologywriter.com* Dr. Johnson lives in St. Louis, Missouri with his wife Barbara and small dog Paddington.

©George B. Johnson

Joel Bergh

Dr. Joel Bergh is a Senior Lecturer in the Department of Biology at Texas State University in San Marcos, Texas. Dr. Bergh earned his BS in Biology from the University of Houston-Clear Lake. In 2003, he earned a PhD from the University of Delaware. His dissertation was focused on the role of vitamin D_3 on bone formation rates. Before coming to Texas State University, he taught at St. Edwards University in Austin, Texas. Dr. Bergh's teaching focuses on majors and nonmajors introductory biology, genetics, and cancer biology. When not in the classroom, he works on innovative new ways to make connections between the events students have gone through in their lives and advances in the scientific community. Dr. Bergh lives in Austin, Texas with his wife Errin and their two children.

©Joel Bergh

Preface

Science Is All Around Us

Day in and day out, every person is making observations, forming hypotheses, and drawing conclusions. The global pandemic has highlighted the importance of making the observations and being scientifically literate. Does wearing a mask reduce your risk of COVID-19 infection? Is COVID-19 effectively treated with hydroxychloroquine? With news coming at us from around the globe, on our television, our phone, and social media, how can you identify what is accurate and correct? The first step is for you to grasp clearly how science is done. This book highlights the essential biological concepts which will let you do this.

pixelfit/Getty Images

Today, Biology Hits Close to Home

While every student may not need to learn in detail how COVID-19 infects human cells and causes sickness and death, there is no denying the fact that this virus has impacted your life, and that of every student, and that it will continue to do so in the future. In this text, we have tried to give you a useful tool for explaining the many ways the pandemic is affecting everyone's life. We have deliberately sought to include many pages about COVID-19 in this edition. We provide details on how the coronavirus pandemic began, the life cycle of the virus, and its many impacts on society. We give sharp focus to the technology that will end the pandemic: development of a vaccine. Writing this edition was exciting, as it allowed us to present up-to-date biological information in as real time as possible, on a subject that is impacting the lives of every student, even as they read this text.

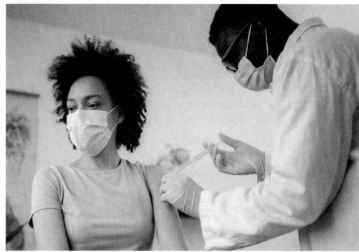

Ivan Pantic/E+/Getty Images

Making Learning Biology Relevant

The use of real-world examples to demonstrate the importance of biology in your lives not only makes learning biology more fun, but is also widely recognized as an effective teaching strategy for the introductory biology classroom. Every student wants to learn about things happening in their life, the posts they encounter on social media, and the topics their friends are interested in and talk about. To give students and their instructors tools to better address the relevancy of what is taught in the classroom, this edition offers several relevancy-based resources:

Relevancy Readings

In many chapters, new *Answering Your Questions About* . . . readings directly address matters of immediate interest to students. Topics include vaping and e-cigarettes, wearing masks in a pandemic, energy drinks, how to read a food label, global warming, cloning your dog, LGBTQ, the opioid epidemic, and many others. The *A Closer Look* feature "A Day in the Life of Your Body" lets students consider how often their heart beats and their lungs inhale, how fast their hair and fingernails grow, and other fascinating events.

Relevancy Modules

New Relevancy Modules correspond with each unit in *Essentials of the Living World, 7e*. These modules demonstrate the connections between biological content and topics that are of interest to society as a whole. Each module consists of an overview of basic scientific concepts and then a closer look at the application of these concepts to the topic. Assessment questions, specific to the module, are also available. These modules are available as a supplementary e-book to the existing text within Connect and may be assigned by the instructor for use in a variety of ways in the classroom. Examples of topics covered include cancer biology, fermentation science, weed evolution, antibiotic resistance, mega crops, the biology of weight gain, and climate change. New topics are planned for launch each year to keep this resource current.

Relevancy Videos: BioNow

Like the *Inquiry and Analysis* feature at the end of each chapter of *Essentials of the Living World*, BioNow videos, narrated and produced by educator Jason Carlson, provide a relevant, applied approach that allows students to feel they can actually do and learn biology themselves. While tying directly to the content of your course, the series of videos helps students relate their daily lives to the biology you teach, and then connect what they learn back to their lives. Each video provides an engaging and entertaining story about applying the science of biology to a real situation or problem. Attention is given to using tools and techniques that the average person would have access to, so students see the science as something they can do and understand.

McGraw Hill

Relevancy Readings

Answering Your Questions About...

Medical Marijuana 114
Energy Drinks 127
Vaping 149
Cloning Your Dog 241
Pandemics 303
Weather and Climate 416
Global Warming 446
Vaping 493
Food Labels 500
Plastic Microbeads 527
The Opioid Crisis 556

A Closer Look

Beer and Wine—Products of Fermentation 101
The Biology of Sexual Orientation 163
CRISPR-Edited Human Babies 205
Darwin and Moby Dick 258
Microbial Bartenders 304
The Oldest Living Things 337
Metabolic Efficiency and the Length of Food Chains 408
The Dance Language of the Honeybee 437
A Day in the Life of Your Body 472
On Being a Whale 489
How Hormones Control Your Kidney's Functions 520
A Sense of Where You Are 562
How the Platypus Sees with Its Eyes Shut 569
A Closer Look at the Science Behind Sexual Orientation 590

Today's Biology

The Polar Ice Caps Are Disappearing 27
Acid Rain 45
Prions and Mad Cow Disease 58
When Membranes Don't Work Right 86
Babies with Three Parents 156
Tracing Your Family History with DNA 182
The Father's Age Affects the Risk of Mutation 197
Testing for COVID-19 233
DNA and the Innocence Project 236
Race and Medicine 287
Has Life Evolved Elsewhere? 298
SARS-CoV-2 Life Cycle 310
Answering Your Questions About Wearing Masks To Prevent The Spread of Coronavirus 311
Marijuana—A Cash Crop? 339
Fish Out of Water 370
The War Against Urban Deer 387
How to Stop Cars from Hitting Deer? Add Wolves! 399
The Global Decline in Amphibians 450
Test-Tube Hamburgers? 511
Inventing a Vaccine for COVID-19 538

Biology and Staying Healthy

Can Lactose-Intolerant People Ever Enjoy Ice Cream? 98
The Paleo Diet 130
Curing Cancer 144
Cancer and COVID 148
Protecting Your Genes 199
Drugs from Dirt 223
Bird and Swine Flu 309
Why Don't Men Get Breast Cancer? 577

New to This Edition

The Coronavirus Pandemic. Over the span of our lifetime, nothing has changed the world we live in quite like the global COVID-19 pandemic. The opening essay of chapter 1 on pages 16–17 describes how the virus that causes COVID-19 grew into a global pandemic, while the closely related SARS and MERS coronaviruses that appeared in the last decade were more contained. COVID-19 originated in China, but quickly spread to Europe and then the United States. Within the first eighteen months of the pandemic, 206 million people were known to be infected.

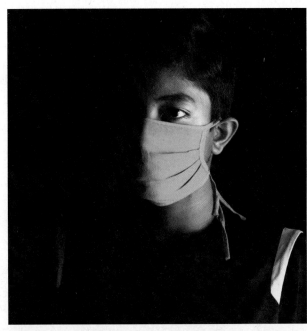
Anubhab Roy/Shutterstock

Earth's Polar Icecaps Are Disappearing. Scientists have been fearfully watching the amount of ice at the planet's poles. With atmospheric CO_2 levels at an all-time high and the world struggling to reduce emissions, the planet is warming. As described in chapter 1 on page 27, Antarctica is feeling the brunt of this warming. The ice in Antarctica took millions of years to form, but with summer temperatures reaching 70°F, it is becoming more unstable. In 2021, an iceberg larger than the state of Rhode Island broke off. Our main hope for reversing global warming lies with every country's active participation in the Paris Climate Accord.

Curing Cancer. Because nearly half of all Americans will be diagnosed with cancer at some point in their life, doctors and scientists are working hard to come up with better cancer treatments. To help develop these treatments, you have to identify the proper targets to give the treatments a chance. As described in chapter 8 on page 144, doctors are focused on what signals are given to cells that cause them to become cancerous and what is necessary to allow cancer cells to spread to other parts of the body. Identifying and understanding the critical steps in the process give targets for potential anticancer therapies.

Cancer and COVID-19. The health of a person is never more fragile than it is when someone is undergoing cancer treatments. When the COVID-19 pandemic began, doctors saw a dramatic decrease in common cancer screening appointments. As described in chapter 8 on page 148, the delay in cancer screenings because of the pandemic is expected to increase the number of cancer deaths. The delay in screening gives any tumor that may be present more time to grow and spread. Patients receiving chemotherapy are already in an immune-compromised state and their body will have a harder time getting rid of any infection. This highlights why it is so important to try to prevent the spread of the virus.

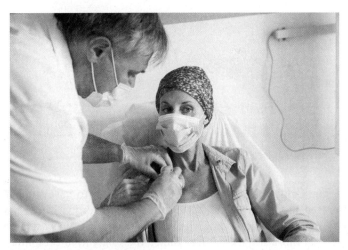
AMELIE-BENOIST/BSIP/Alamy Stock Photo

Human Sexual Orientation. Adult sexual behavior is determined by the activation, early in development, of fetal genes located on the Y chromosome that lead to secretion of the hormone testosterone. This testosterone acts on the brain of the fetus in a way that determines individual's future sexual behavior. Depending upon its action, the sexual orientation of the individual will be to persons of the opposite gender (heterosexual), to the same gender (homosexual), or to both genders (bisexual). Non-heterosexual individuals commonly refer to themselves as LGBTQ (lesbian, gay, bisexual, transgender, queer). The CDC estimates the number of these individuals as 5% of the total U.S. population—some 13 million LGBTQ individuals. As discussed in chapter 9 on page 163, sexual orientation of LGBTQ individuals is determined before birth, and is not subject to later changes.

CRISPR-Edited Human Babies. The invention of gene editing with CRISPR has led quickly to an ethical nightmare. As recounted in chapter 11 on page 205, a Chinese researcher in November 2018 announced he had used CRISPR to edit the genomes of human babies! Do you see the problem? In the CRISPR applications, DNA was edited in the somatic (body) tissues of adults, changes that could not be passed on to future generations. As described on page 205, the Chinese researcher edited the DNA of a single-cell embryo. This alters all the tissues that derive from it, germ-line as well as somatic—the changes he created with CRISPR will be passed on in the germ line to future human generations.

Testing for COVID-19. When the 2020 coronavirus pandemic began, it was a guessing game as to who had COVID and who was sick for some other reason. It was important to develop a sensitive and accurate test to determine if someone had the virus. Chapter 13, page 233, explores the different types of tests that have become available. PCR and antigen tests can determine if a person actively has COVID, while the antibody tests can determine if they already had an infection.

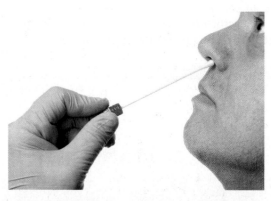
Henrik Dolle/Shutterstock

The Origin of COVID-19. COVID-19 seems to have suddenly appeared in China in late 2019 but, in truth, coronaviruses have been around for a long time. The start of the pandemic was caused by one of those strains that had gained the ability to infect humans. But where did the virus come from and how did it gain the ability to infect us? These questions, explored in chapter 16 on page 309, have led scientists to answers that may help prevent future pandemics. COVID has been passed between animals for centuries, but a colony of bats in southwest China carried a virus that was 96% identical to the virus that infected humans. The origin of the outbreak was traced to some wet markets, where live animals are often sold, in Wuhan, China, but there were no bats sold at the market. There virus must have gotten to the market through a different host, but which one? We still do not know the intermediary, but what we do know is once COVID infected humans, it spread rapidly and is continuing to evolve.

The SARS-CoV-2 Life Cycle. Researchers have been hard at work to identify the genetic code and the mechanism through which COVID infects humans. As described in chapter 16 on page 309, they have been able to identify how COVID gets into our cells and what happens once it is in. The spike protein of COVID is needed to bind to receptors on the surface of our cells. Once inside, the viral RNA uses your own ribosomes to produce more viral proteins. It even codes for an RNA polymerase that will be used to generate thousands of copies of the viral RNA. Together, the viral proteins and viral RNA, when packaged, fill the cytoplasm with more coronavirus particles.

Masks Prevent the Spread of Coronavirus. To slow the spread of the pandemic, health care experts recommended that we all wear facemasks. But are masks effective at preventing the spread of the coronavirus? This question is explored in chapter 16 on page 310. When you cough or sneeze, water droplets can carry viral particles up to 10-feet away. Wearing a simple cloth mask reduces the spread to less than a foot, keeping more of your water droplets to yourself. Wearing a mask also reduces the likelihood that you will inhale the virus that was just exhaled by someone else.

Cannabis as a Cash Crop. Marijuana is derived from the Cannabis plant. In 1996, California was the first state to legalize Cannabis and, since then, more than 32 states have followed suit. The sale and cultivation of Cannabis have become close to $1 billion a year industry. As described in chapter 17 on page 339, there is more to Cannabis than just the psychoactive effect. There are two groups of Cannabis plants, marijuana and hemp, and they are distinguished by the amount of cannabinoid compounds in each.

Sergio Azenha/Alamy Stock Photo

A Vaccine for COVID-19. COVID-19 easily caused a global pandemic because we do not have a naturally occurring defense to the virus, nor were their any successful treatments. To end the pandemic, the race was on to develop a vaccine. As described in chapter 27 on page 539, companies were using both traditional methods, like live attenuated virus vaccines and inactivated virus vaccines, as well as new approaches like nucleic acid vaccines. The first two vaccines to make it to the public were both RNA vaccines. The RNA for the spike protein of SARS-COV2 was placed inside a lipid carrier and injected into patients. As COVID-19 vaccines become more available worldwide, there is hope the pandemic can be ended.

Acknowledgments

Every author knows that he or she labors on the shoulders of many others. The text you see is the result of hard work by an army of "behind-the-scenes" editors, spelling and grammar checkers, photo researchers, and artists who perform their magic on our manuscript, plus an even larger army of production managers and staff who then transform this manuscript into a bound book. We cannot thank them all enough.

Portfolio Manager Ian Townsend and Product Developer Melisa Seegmiller were our editorial team, with whom we worked every day. They provided valuable advice and support to a sometimes querulous and always anxious pair of authors.

Content Project Manager Maria McGreal spearheaded our production team. The photo program was carried out by Senior Content Licensing Specialist Lori Hancock. David Hash did a great job with the cover design and was unbelievably tolerant of the author's many "creative" changes. The book was produced and the interior design was modified by MPS Limited.

The marketing of Essentials of the Living World, 7e has been planned and supervised by Marketing Manager Britney Ross. Her enthusiasm and commitment to this book has been a major contributor to its success. No author could wish for a better, more fiercely competitive person to market their book.

No text goes through seven editions without the strong support of its editors, past and present. Dr. Johnson would like to extend special thanks to Pat Reidy, who got him over many rough bumps in early editions, and particularly to Michael Lange for his early and continued strong support of this project.

Reviewers of the Seventh Edition

Judy Bluemer
Morton College

Kristin Y. Bridge
Motlow State Community College

Gregory Burchett
Riverside City College, Riverside California

Mark Chiappone
Miami Dade College, Homestead Campus

Andrea L. Corbett
Cleveland State University

Stacy Dowd
Motlow State Community College

Michelle C. Farrell
North Carolina A&T State University

Gregory D. Frederick
LeTourneau University

Carrie L. Geisbauer
Moorpark College

Meshagae Hunte-Brown
Drexel University

Ragupathy Kannan
University of Arkansas, Fort Smith

Todd Kostman
University of Wisconsin, Oshkosh

Andre Kulisz
Richard J. Daley College

Jerry Lasnik
Moorpark College

Jake Marquess
Arkansas State University, Beebe

Tiffany B. McFalls-Smith
Elizabethtown Community & Technical College

Melissa Meador
Arkansas State University, Beebe

Pele Eve Rich
North Hennepin Community College

Avodotun Sodipe
Texas Southern University, Houston

Kimberly Taugher
Diablo Valley College

Encarni Trueba
The Community College of Baltimore County

Brian Weaver
Arkansas State University, Beebe

Heather Wilson-Ashworth
Utah Valley University

Aaron Woodyatt
South College

Shawn Xiong
Pennsylvania State University

Shirley Zajdel
Housatonic Community College

ReadAnywhere

Read or study when it's convenient for you with McGraw Hill's free ReadAnywhere app. Available for iOS or Android smartphones or tablets, ReadAnywhere gives users access to McGraw Hill tools including the eBook and SmartBook 2.0 or Adaptive Learning "RAssignments in Connect. Take notes, highlight, and complete assignments offline – all of your work will sync when you open the app with WiFi access. Log in with your McGraw Hill Connect username and password to start learning – anytime, anywhere!

Neustockimages/Getty Images

Minerva Studio/Shutterstock

Remote Proctoring & Browser-Locking Capabilities

New remote proctoring and browser-locking capabilities, hosted by Proctorio within Connect, provide control of the assessment environment by enabling security options and verifying the identity of the student.

Seamlessly integrated within Connect, these services allow instructors to control students' assessment experience by restricting browser activity, recording students' activity, and verifying students are doing their own work.

Instant and detailed reporting gives instructors an at-a-glance view of potential academic integrity concerns, thereby avoiding personal bias and supporting evidence-based claims.

Instructors: Student Success Starts with You

Tools to enhance your unique voice

Want to build your own course? No problem. Prefer to use an OLC-aligned, prebuilt course? Easy. Want to make changes throughout the semester? Sure. And you'll save time with Connect's auto-grading too.

65%
Less Time Grading

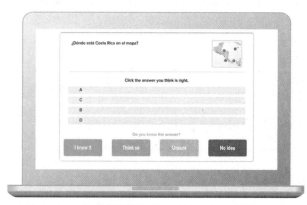

Laptop: McGraw Hill Education

Study made personal

Study resources in Connect help your students be better prepared in less time. You can transform your class time from reviewing the basics to dynamic discussion. Hear from your peers about the benefits of Connect at **www.mheducation.com/highered/connect**

Affordable solutions, added value

Make technology work for you with LMS integration for single sign-on access, mobile access to the digital textbook, and reports to quickly show you how each of your students is doing. And with our Inclusive Access program you can provide all these tools at a discount to your students. Ask your McGraw Hill representative for more information.

Padlock: Jobalou/Getty Images

Solutions for your challenges

A product isn't a solution. Real solutions are affordable, reliable, and come with training and ongoing support when you need it and how you want it. Visit **www.supportateverystep.com** for videos and resources both you and your students can use throughout the semester.

Checkmark: Jobalou/Getty Images

Students: Get Learning that Fits You

Effective tools for efficient studying

Connect is designed to help you be more productive with simple, flexible, intuitive tools that maximize your study time and meet your individual learning needs. Get learning that works for you with Connect.

Study anytime, anywhere

Download the free ReadAnywhere app and access your online eBook when it's convenient, even if you're offline. And since the app automatically syncs with your eBook in Connect, all of your notes are available every time you open it. Find out more at **www.mheducation.com/readanywhere**

> *"I really liked this app—it made it easy to study when you don't have your textbook in front of you."*
>
> - Jordan Cunningham, Eastern Washington University

Calendar: owattaphotos/Getty Images

Everything you need in one place

Your Connect course has everything you need—whether reading on your digital eBook or completing assignments for class, Connect makes it easy to get your work done.

Learning for everyone

McGraw Hill works directly with Accessibility Services Departments and faculty to meet the learning needs of all students. Please contact your Accessibility Services Office and ask them to email accessibility@mheducation.com, or visit **www.mheducation.com/about/accessibility** for more information.

Top: Jenner Images/Getty Images, Left: Hero Images/Getty Images, Right: Hero Images/Getty Images

Contents

Preface v

0 Studying Biology 2

Learning 4
- 0.1 How to Study 4
- 0.2 Using Your Textbook 7
- 0.3 Using Your Textbook's Internet Resources 9
- 0.4 Online Labs 10

Putting What You Learn to Work 11
- 0.5 Science Is a Way of Thinking 11
- 0.6 How to Read a Graph 13

Part 1
The Study of Life

1 The Science of Biology 16

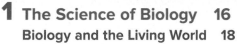
Anubhab Roy/Shutterstock

Biology and the Living World 18
- 1.1 The Diversity of Life 18
- 1.2 Properties of Life 19
- 1.3 The Organization of Life 20
- 1.4 Biological Themes 22

The Scientific Process 24
- 1.5 Stages of a Scientific Investigation 24
- 1.6 Theory and Certainty 26

Core Ideas of Biology 28
- 1.7 Four Theories Unify Biology as a Science 28

Part 2
The Living Cell

2 The Chemistry of Life 34

Herbert Spichtinger/Image Source

Some Simple Chemistry 36
- 2.1 Atoms 36
- 2.2 Ions and Isotopes 38
- 2.3 Molecules 39

Water: Cradle of Life 42
- 2.4 Unique Properties of Water 42
- 2.5 Water Ionizes 44

3 Molecules of Life 48

Forming Macromolecules 50
- 3.1 Building Big Molecules 50

Types of Macromolecules 52
- 3.2 Proteins 52
- 3.3 Nucleic Acids 56
- 3.4 Carbohydrates 59
- 3.5 Lipids 61

4 Cells 64

The World of Cells 66
- 4.1 Cells 66

Kinds of Cells 69
- 4.2 Prokaryotic Cells 69
- 4.3 Eukaryotic Cells 70

Tour of a Eukaryotic Cell 72
- 4.4 The Plasma Membrane 72
- 4.5 The Nucleus: The Cell's Control Center 74
- 4.6 The Endomembrane System 76
- 4.7 Organelles that Harvest Energy 78
- 4.8 The Cytoskeleton: Interior Framework of the Cell 80

Transport Across Plasma Membranes 82
- 4.9 Diffusion and Osmosis 82
- 4.10 Bulk Passage into and out of Cells 84
- 4.11 Transport through Proteins 85

5 Energy and Life 90

Cells and Energy 92
- 5.1 The Flow of Energy in Living Things 92
- 5.2 The Laws of Energy 92

Cell Chemistry 94
- 5.3 Chemical Reactions 94

xiv

Enzymes 95
- 5.4 How Enzymes Work 95
- 5.5 How Cells Regulate Enzymes 96

How Cells Use Energy 99
- 5.6 ATP: The Energy Currency of the Cell 99

6 Photosynthesis: Acquiring Energy from the Sun 104

Photosynthesis 106
- 6.1 An Overview of Photosynthesis 106
- 6.2 How Plants Convert Sunlight to Chemical Energy 110
- 6.3 Building New Molecules 113

Photorespiration 115
- 6.4 Photorespiration: Putting the Brakes on Photosynthesis 115

7 How Cells Harvest Energy from Food 118

An Overview of Cellular Respiration 120
- 7.1 Where Is the Energy in Food? 120

Respiration Without Oxygen: Glycolysis 122
- 7.2 Using Coupled Reactions to Make ATP 122

Respiration with Oxygen: The Krebs Cycle 123
- 7.3 Harvesting Electrons from Chemical Bonds 123
- 7.4 Using the Electrons to Make ATP 125

Harvesting Electrons Without Oxygen: Fermentation 128
- 7.5 Cells Can Metabolize Food Without Oxygen 128

Other Sources of Energy 129
- 7.6 Glucose Is Not the Only Food Molecule 129

Part 3
The Continuity of Life

8 Mitosis 134
Cell Division 136
- 8.1 Prokaryotes Have a Simple Cell Cycle 136
- 8.2 Eukaryotic Cell Cycle 137
- 8.3 Chromosomes 138
- 8.4 Cell Division 140

Cancer and the Cell Cycle 143
- 8.5 What Is Cancer? 143

9 Meiosis 152
Meiosis 154
- 9.1 Discovery of Meiosis 154
- 9.2 The Sexual Life Cycle 155
- 9.3 The Stages of Meiosis 157

Comparing Meiosis and Mitosis 160
- 9.4 How Meiosis Differs from Mitosis 160
- 9.5 Evolutionary Consequences of Sex 162

10 Foundations of Genetics 166
Mendel 168
- 10.1 Mendel and the Garden Pea 168
- 10.2 What Mendel Observed 170
- 10.3 Mendel Proposes a Theory 172
- 10.4 Mendel's Laws 175

From Genotype to Phenotype 176
- 10.5 How Genes Influence Traits 176
- 10.6 Why Some Traits Don't Show Mendelian Inheritance 178

Chromosomes and Heredity 181
- 10.7 Human Chromosomes 181
- 10.8 The Role of Mutations in Human Heredity 184

11 DNA: The Genetic Material 190
Genes Are Made of DNA 192
- 11.1 Discovering the Structure of DNA 192

DNA Replication 194
- 11.2 How DNA Copies Itself 194

Altering the Genetic Message 196
- 11.3 Mutation 196

Editing Genes 200
- 11.4 Gene Editing with CRISPR 200
- 11.5 Gene Drives 203

12 How Genes Work 208
From Gene to Protein 210
- 12.1 The Central Dogma 210
- 12.2 From DNA to RNA: Transcription 211
- 12.3 From RNA to Protein: Translation 212
- 12.4 Gene Expression 215

Regulating Gene Expression 218
- 12.5 Transcriptional Control in Prokaryotes 218
- 12.6 Transcriptional Control in Eukaryotes 219
- 12.7 RNA-Level Control: Silencing Genes 221

13 The New Biology 226

Genomics 228
- 13.1 The Human Genome 228

Genetic Engineering 230
- 13.2 A Scientific Revolution 230
- 13.3 Genetic Engineering and Medicine 234
- 13.4 Genetic Engineering and Agriculture 238

The Revolution in Cell Technology 242
- 13.5 Reproductive Cloning 242
- 13.6 Stem Cell Therapy 244
- 13.7 Gene Therapy 246

Part 4
The Evolution and Diversity of Life

14 Evolution and Natural Selection 250

Evolution 252
- 14.1 Darwin's Voyage on HMS *Beagle* 252
- 14.2 Darwin's Evidence 254
- 14.3 The Theory of Natural Selection 255

The Theory of Evolution 257
- 14.4 The Evidence for Evolution 257
- 14.5 Evolution's Critics 261

How Populations Evolve 263
- 14.6 Genetic Change in Populations 263
- 14.7 Agents of Evolution 264

Adaptation Within Populations 269
- 14.8 Sickle-Cell Disease 269
- 14.9 Peppered Moths and Industrial Melanism 271

How Species Form 273
- 14.10 The Biological Species Concept 273
- 14.11 Isolating Mechanisms 274

15 Exploring Biological Diversity 278

The Classification of Organisms 280
- 15.1 The Invention of the Linnaean System 280
- 15.2 Species Names 281
- 15.3 Higher Categories 282
- 15.4 What Is a Species? 283

Inferring Phylogeny 284
- 15.5 How to Build a Family Tree 284

Kingdoms and Domains 288
- 15.6 The Kingdoms of Life 288
- 15.7 Domain: A Higher Level of Classification 290

16 Evolution of Microbial Life 294

Origin of Life 296
- 16.1 How Cells Arose 296

Prokaryotes 300
- 16.2 The Simplest Organisms 300

Viruses 305
- 16.3 Structure of Viruses 305
- 16.4 How Viruses Infect Organisms 308

Protists 311
- 16.5 General Biology of Protists 312
- 16.6 Kinds of Protists 314

Fungi 316
- 16.7 A Fungus Is Not a Plant 316
- 16.8 Kinds of Fungi 318

17 Evolution of Plants 322

Plants 324
- 17.1 Adapting to Terrestrial Living 324
- 17.2 Plant Evolution 326

Seedless Plants 328
- 17.3 Nonvascular Plants 328
- 17.4 The Evolution of Vascular Tissue 329
- 17.5 Seedless Vascular Plants 330

The Advent of Seeds 332
- 17.6 Evolution of Seed Plants 332
- 17.7 Gymnosperms 334

The Evolution of Flowers 336
- 17.8 Rise of the Angiosperms 336

18 Evolution of Animals 342

Introduction to the Animals 344
- 18.1 General Features of Animals 344
- 18.2 Six Key Transitions in Body Plan 346

Evolution of the Animal Phyla 348
- 18.3 The Simplest Animals 348
- 18.4 Advent of Bilateral Symmetry 351
- 18.5 Changes in the Body Cavity 354
- 18.6 Redesigning the Embryo 363

The Parade of Vertebrates 368
- 18.7 Overview of Vertebrate Evolution 368
- 18.8 Fishes Dominate the Sea 371
- 18.9 Amphibians and Reptiles Invade the Land 373
- 18.10 Birds Master the Air 376
- 18.11 Mammals Adapt to Colder Times 377
- 18.12 Human Evolution 378

Part 5
The Living Environment

19 Populations and Communities 382
Ecology 384
19.1 What Is Ecology? 384

Populations 385
19.2 Population Growth 385
19.3 The Influence of Population Density 388
19.4 Life History Adaptations 389
19.5 Population Demography 390

How Competition Shapes Communities 391
19.6 Communities 391
19.7 The Niche and Competition 392

Species Interact in Many Ways 394
19.8 Coevolution and Symbiosis 394
19.9 Predation 396

Community Stability 398
19.10 Ecological Succession 398

20 Ecosystems 402
The Energy in Ecosystems 404
20.1 Ecosystems 404
20.2 Ecological Pyramids 409

Materials Cycle Within Ecosystems 410
20.3 The Water Cycle 410
20.4 The Carbon Cycle 412
20.5 The Nitrogen and Phosphorus Cycles 413

How Weather Shapes Ecosystems 415
20.6 The Sun and Atmospheric Circulation 415
20.7 Latitude and Elevation 417

Major Kinds of Ecosystems 418
20.8 Ocean Ecosystems 418
20.9 Freshwater Ecosystems 420
20.10 Land Ecosystems 422

21 Behavior and the Environment 428
The Study of Behavior 430
21.1 Instinctive Behavioral Patterns 430
21.2 Genetic Effects on Behavior 431

Behavior Can Be Influenced by Learning 432
21.3 How Animals Learn 432
21.4 Animal Cognition 433

Evolutionary Forces Shape Behavior 434
21.5 A Cost-Benefit Analysis of Behavior 434
21.6 Migratory Behavior 436

Social Behavior 438
21.7 Animal Societies 438
21.8 Human Social Behavior 439

22 Human Influences on the Living World 442
Global Change 444
22.1 Pollution 444
22.2 Global Warming 445
22.3 Loss of Biodiversity 449

Saving Our Environment 451
22.4 Preserving Nonreplaceable Resources 451
22.5 Curbing Population Growth 453

Solving Environmental Problems 454
22.6 Preserving Endangered Species 454
22.7 Individuals Can Make the Difference 457

Part 6
Animal Life

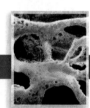

23 The Animal Body and How It Moves 460
The Animal Body Plan 462
23.1 Organization of the Vertebrate Body 462

Tissues of the Vertebrate Body 464
23.2 Epithelium Is Protective Tissue 464
23.3 Connective Tissue Carries Out Various Functions 465
23.4 Muscle Tissue Lets the Body Move 467
23.5 Nerve Tissue Conducts Signals Rapidly 468

The Skeletal and Muscular Systems 469
23.6 Types of Skeletons 469
23.7 Muscles and How They Work 471

24 Circulation and Respiration 476
Circulation 478
24.1 Open and Closed Circulatory Systems 478
24.2 Architecture of the Vertebrate Circulatory System 480
24.3 Blood 483
24.4 Human Circulatory System 485

Respiration 488
24.5 Types of Respiratory Systems 488
24.6 The Human Respiratory System 490
24.7 How Respiration Works: Gas Exchange 492

25 The Path of Food Through the Animal Body 496

Food Energy and Essential Nutrients 498
25.1 Food for Energy and Growth 498

Digestion 501
25.2 Types of Digestive Systems 501
25.3 The Human Digestive System 502
25.4 The Mouth and Teeth 503
25.5 The Esophagus and Stomach 505
25.6 The Small and Large Intestines 507
25.7 Accessory Digestive Organs 509

26 Maintaining the Internal Environment 514

Homeostasis 516
26.1 How the Animal Body Maintains Homeostasis 516

Osmoregulation 518
26.2 Regulating the Body's Water Content 518
26.3 Eliminating Nitrogenous Wastes 521

27 How the Animal Body Defends Itself 524

Three Lines of Defense 526
27.1 Skin: The First Line of Defense 526
27.2 Cellular Counterattack: The Second Line of Defense 529
27.3 Specific Immunity: The Third Line of Defense 531

The Immune Response 533
27.4 Initiating the Immune Response 533
27.5 T Cells: The Cellular Response 533
27.6 B Cells: The Humoral Response 535
27.7 Active Immunity Through Clonal Selection 537
27.8 Vaccination 540
27.9 Antibodies in Medical Diagnosis 541

Defeat of the Immune System 542
27.10 Overactive Immune System 542
27.11 AIDS: Immune System Collapse 543

28 The Nervous System 546

Neurons and How They Work 548
28.1 The Nervous System 548
28.2 Nerve Impulses 549
28.3 The Synapse 551

The Central Nervous System 553
28.4 How the Brain Works 553
28.5 The Spinal Cord 557

The Peripheral Nervous System 558
28.6 The Voluntary and Autonomic Nervous Systems 558

The Sensory Nervous System 560
28.7 Sensing the Internal Environment 560
28.8 Sensing Gravity and Motion 561
28.9 Sensing Chemicals: Taste and Smell 564
28.10 Sensing Sounds: Hearing 565
28.11 Sensing Light: Vision 566

29 Chemical Signaling Within the Animal Body 572

The Endocrine System 574
29.1 Hormones 574
29.2 How Hormones Target Cells 576

The Major Endocrine Glands 579
29.3 The Pituitary 579
29.4 The Pancreas 581
29.5 The Thyroid, Parathyroid, and Adrenal Glands 582

30 Reproduction and Development 586

Modes of Reproduction 588
30.1 Asexual and Sexual Reproduction 588

The Human Reproductive System 591
30.2 Males 591
30.3 Females 593
30.4 Hormones Coordinate the Reproductive Cycle 595

The Course of Development 597
30.5 Embryonic Development 597
30.6 Fetal Development 600

Birth Control and Sexually Transmitted Diseases 602
30.7 Contraception and Sexually Transmitted Diseases 602

Part 7
Plant Life

31 Plant Form and Function 606

Structure and Function of Plant Tissues 608
31.1 Organization of a Vascular Plant 608
31.2 Plant Tissue Types 609

The Plant Body 613
31.3 Roots 613
31.4 Stems 615
31.5 Leaves 617

Judy Freilicher/Alamy Stock Photo

Plant Transport and Nutrition 620
- **31.6** Water Movement 620
- **31.7** Carbohydrate Transport 623

32 Plant Reproduction and Growth 626
Flowering Plant Reproduction 628
- **32.1** Angiosperm Reproduction 628
- **32.2** Seeds 631
- **32.3** Fruit 632
- **32.4** Germination 633

Regulating Plant Growth 634
- **32.5** Plant Hormones 634
- **32.6** Auxin 635

Plant Responses to Environmental Stimuli 636
- **32.7** Photoperiodism and Dormancy 636
- **32.8** Tropisms 637

Glossary 640
Index 654

0 Studying Biology

LEARNING PATH ▼

Learning
1. How to Study
2. Using Your Textbook
3. Using Your Textbook's Internet Resources
4. Online Labs

Putting What You Learn to Work
5. Science Is a Way of Thinking
6. How to Read a Graph

SUCCESS WILL depend not only on how much you study, but also on when.
Randy Faris/Getty Images

Pulling an All-Nighter

At some point in the next months, you will face that scary rite: the first exam in this course. Many students face the challenge of exams by cramming. They live and die by the all-nighter, black coffee becomes their closest friend during exam week, and sleep is a luxury they can't afford. Trying to cram enough in to meet any possible question, they feel they can't waste time sleeping.

Sleeping

If you take this approach, you won't have much luck. Why doesn't the hard work of cramming give good grades? Because of how humans learn. Researchers have demonstrated that memory of newly learned information improves only after hours of sleeping. If you wanted to do well in an exam, you could not have chosen a poorer way to prepare than an all-nighter.

Learning is, in its most basic sense, a matter of forming memories. Research shows that a person trying to learn something does not improve his or her knowledge until after they have had more than six hours of sleep (preferably eight). It seems the brain needs time to file new information away in the proper slots so it can be retrieved later. Without enough sleep to do this filing, new information does not get properly encoded into the brain's memory circuits.

A Closer Look at Learning

To sort out the role of sleep in learning, Harvard Medical School researchers used undergrads as guinea pigs. The undergraduates were trained to look for particular visual targets on a computer screen and to push a button as soon as they were sure they had seen one. At first, responses were relatively sluggish—it typically took 400 milliseconds for a target to reach a student's conscious awareness. With an hour's training, however, many students were hitting the button correctly in 75 milliseconds.

How well had they learned? When retested from 3 to 12 hours later on the same day, there was no further improvement past a student's best time in the training session. The researchers let a student get a little sleep, but less than six hours, then retested the next day, the student still showed no improvement in performing the target identification.

Steve Hix/Fuse/Getty Images

For students who slept more than six hours, the story was very different. Sleep greatly improved the performance. Students who achieved 75 milliseconds in the training session would reliably perform the target identification in 62 milliseconds after a good night's sleep! After several nights of ample sleep, they often got even more proficient.

Moving Your Short-Term Experiences to Long-Term Memory

Why six or eight hours and not four or five? The sort of sleeping you do at the beginning of a night's sleep and the sort you do at the end are different, and both, it appears, are required for efficient learning.

The First Two Hours. The first two hours of sleeping are spent in deep sleep, what psychiatrists call slow-wave sleep. During this time, certain brain chemicals become used up, which allows information that has been gathered during the day to flow out of the memory center of the brain, the hippocampus, and into the cortex, the outer covering of the brain where long-term memories are stored. Like moving information in a computer from active memory to the hard drive, this process preserves experience for future reference. Without it, long-term learning cannot occur.

The Next Hours. Over the next hours, the cortex sorts through the information it has received, distributing it to various locations and networks. Particular connections between nerve cells become strengthened as memories are preserved, a process that is thought to require the time-consuming manufacturing of new proteins. If you halt this process before it is complete, the day's memories do not get fully "transcribed," and you don't remember all that you would have, had you allowed the process to continue to completion. A few hours are just not enough time to get the job done. Four hours, the Harvard researchers estimate, is a minimum requirement.

The Last Two Hours. The last two hours of a night's uninterrupted sleep are spent in rapid-eye-movement (rem) sleep. This is when dreams occur. The brain shuts down the connection to the hippocampus and runs through the data it has stored over the previous hours. This process is also important to learning, as it reinforces and strengthens the many connections between nerve cells that make up the new memory. Like a child repeating a refrain to memorize it, the brain goes over what it has learned, until practice makes perfect.

That's why getting by on three or four hours of sleep during exam week and crashing for 12 hours on weekends don't work. After a few days, all of the facts memorized during "all-nighters" fade away, never given a chance to integrate properly into memory circuits.

Learning

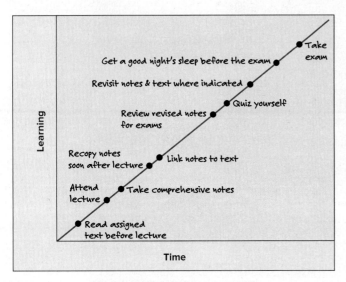

Figure 0.1 A learning timeline.

0.1 How to Study

Studying and learning are not the same thing. Just putting in time is not enough—it is not how long you study, but how effectively.

Taking Notes

> **LEARNING OBJECTIVE 0.1.1** Explain why it is important to recopy your lecture notes promptly.

Listening to lectures and reading the text are only the first steps in learning enough to do well in a biology course. The key to mastering the mountain of information and concepts you are about to encounter is to take careful notes. Studying from poor-quality notes that are sparse, disorganized, and barely intelligible is not a productive way to approach preparing for an exam.

There are three simple ways to improve the quality of your notes:

1. **Take many notes.** Always attempt to take the most complete notes possible during class. If you miss class, take notes yourself from a tape of the lecture, if at all possible. It is the process of taking notes that promotes learning. Using someone else's notes is but a poor substitute. When someone else takes the notes, that person tends to do most of the learning as well.
2. **Take paraphrased notes.** Develop a legible style of abbreviated note-taking. Obviously, there are some things that cannot be easily paraphrased (referred to in a simpler way), but using abbreviations and paraphrasing will permit more comprehensive notes. Attempting to write complete organized sentences in note-taking is frustrating and too time-consuming—people just talk too fast!
3. **Revise your notes.** As soon as possible after lecture, you should decipher and revise your notes. Nothing else in the learning process is more important because this is where most of your learning will take place. By revising your notes, you meld the information together and put it into a context that is understandable to you. As you revise your notes, organize the material into major blocks of information with simple "heads" to identify each block. Add ideas from your reading of the text, and note links to material in other lectures. Clarify terms and concepts that might be confusing with short notes and definitions. Thinking through the ideas of the lecture in this organized way will crystallize them for you, which is the key step in learning. Also, simply rewriting your notes to make them legible, neat, and tidy can be a tremendous improvement that will further enhance your ease of learning (**figure 0.1**).

Remembering and Forgetting

> **LEARNING OBJECTIVE 0.1.2** Name two things you can do to slow down the forgetting process.

Learning is the process of placing information in your memory. Just as in your computer, there are two sorts of memory. The first, *short-term memory*, is analogous to the RAM (random access memory) of a computer, holding information for only a short period of time. As in your computer, this memory is constantly being "written over" as new information comes in. The second kind of memory, *long-term memory*, consists of information that you have stored in your memory banks for future retrieval, like storing files on your computer's hard drive. In its simplest context, learning is the process of transferring information to your hard drive.

BIOLOGY & YOU

Improving Memory. There is an active market on the Internet for commercial products that claim to improve your memory. Many of these products involve repetitive games or other gimmicks; few have any lasting impact on memory. Psychologists have carried out considerable research on this subject and have found that the best way to improve memory seems to be to increase the supply of oxygen to the brain. How do you do this? These researchers recommend aerobic exercise. Walking for three hours each week significantly increases brain oxygen levels, as does swimming or cycling. One study found that chewing gum while studying will supply the brain with enough oxygen to improve memorizing items simply because of the muscle movement.

Forgetting is the loss of information stored in memory. Most of what we forget when taking exams is the natural consequence of short-term memories not being effectively transferred to long-term memory. Forgetting occurs very rapidly, dropping to below 50% retention within one hour after learning and leveling off at about 20% retention after 24 hours.

There are many things you can do to slow down the forgetting process. Here are two important ones:

1. **Recopy your notes as soon as possible after lecture.** Remember, there is about a 50% memory loss in the first hour. You should use your textbook as well when recopying your notes.
2. **Establish a purpose for reading.** When you sit down to study your textbook, have a definite goal to learn a particular concept. Each chapter begins with a preview of its key concepts—let them be your guides. Do not try and learn the entire contents of a chapter in one session; break it up into small pieces that are "easily digested."

Learning

> **LEARNING OBJECTIVE 0.1.3** List three general means of rehearsal.

Learning may be viewed as the efficient transfer of information from your short-term memory to your long-term memory. Learning strategists refer to this transfer as *rehearsal*. As its name implies, rehearsal always involves some form of repetition. There are four general means of rehearsal in the jargon of education called "critical thinking skills" (**figure 0.2**).

Repeating. The most obvious form of rehearsal is repetition. To learn facts, the sequence of events in a process, or the names of a group of things, you write them down, say them aloud, and mentally repeat them over and over until you have "memorized" them. This often is a first step on the road to learning. Many students mistake this as the only step. It is not, as it involves only rote memory instead of understanding. If all you do in this course is memorize facts, you will not succeed.

Organizing. It is important to organize the information you are attempting to learn because the process of sorting and ordering increases retention. For example, if you place a sequence of events in order, such as the stages of cell division, you will be able to recall the entire sequence if you can remember what gets the sequence started.

Linking. Biology has a natural hierarchy of information, with terms and concepts nested within other terms and concepts. You will learn facts and concepts more easily if you attempt to connect them with something you have already already learned or a related concept you will be learning in a later chapter. Throughout this textbook, you will see arrows, like the ones in **figure 0.3**, indicating such links. Use them to cross-check concepts and processes you have already learned. You will be surprised how much doing this will help you learn the new material.

Connecting. You will learn biology much more effectively if you relate what you are learning to the world around you. The many challenges of living in today's world are often related to the information presented in this course, and understanding these relationships will help you learn. In each chapter of this textbook, you will encounter several Apps (Application dialogs) in the outer margins that allow you to briefly explore a "real-world" topic related to what you are learning. Read them. You may not be tested on these Apps, but reading them will provide you with another "hook" to help you learn the material on which you will be tested.

Figure 0.2 Learning requires work.

Learning is something you do, not something that happens to you.

Image Source/Getty Images

IMPLICATION FOR YOU If you are honest with yourself, how many of the four rehearsal techniques (critical thinking skills) do you use when you take a science course like this one? Do you think they are as important in nonscience classes such as English or history? Why?

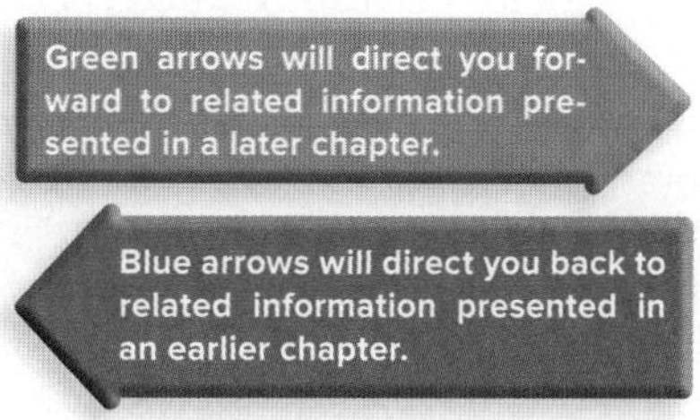

Green arrows will direct you forward to related information presented in a later chapter.

Blue arrows will direct you back to related information presented in an earlier chapter.

Figure 0.3 Linking concepts.

These linking arrows, found throughout the text, will help you to form connections between related topics covered in different chapters of the text.

Studying to Learn

> **LEARNING OBJECTIVE 0.1.4** Describe three strategies to improve studying efficiency.

How Long Should I Study? If I have heard it once, I have heard it a thousand times, "Gee, Professor Johnson, I studied for 20 hours straight and I still got a D." By now, you should be getting the idea that just throwing time at the material does not ensure a favorable outcome. Many students treat studying for biology like penance: If you do it, you will be rewarded for having done so. Not always.

The length of time spent studying and the spacing between study or reading sessions directly affect how much you learn. If you had 10 hours to spend studying, you would be better off if you broke it up into 10 one-hour sessions than to spend it all in one or two sessions. There are two good reasons for this:

First, we know from formal cognition research (as well as from our everyday life experiences) that we remember "beginnings" and "endings" but tend to forget "middles." Thus, the learning process can benefit from many "beginnings" and "endings."

Second, unless you are unusual, after 30 minutes or an hour, your ability to concentrate is diminished. Concentration is a critical component of studying to learn. Many short, topic-focused study sessions maximize your ability to concentrate effectively. For most of us, effective concentration also means a comfortable, quiet environment with no outside distractions, such as loud music or conversations.

It is important to realize that learning biology is not something you can do passively. Many students think that simply possessing a lecture video or a set of class notes will get them through. In and of themselves, videos and notes are no more important than the Nautilus machine an athlete works out on. It is not the machine *per se* but what happens when you use it effectively that is of importance.

Figure 0.4 Critical learning occurs in the classroom.

Learning occurs in at least four distinct stages: doing assigned textbook readings before lecture; attending class; listening and taking notes during lecture; and recopying notes shortly after lecture.
Fuse/Getty Images

Four Keys to Success Common sense will have a great deal to do with your success in learning biology, as it does in most of life's endeavors. Your success in this biology course will depend on simple, obvious things (figure 0.4):

- *Attend class.* Go to all the lectures and be on time.
- *Read the assigned readings before lecture.* If you have done so, you will hear things in lecture that will be familiar to you, a recognition that is a vital form of learning reinforcement. Later you can go back to the text to check details.
- *Take comprehensive notes.* Recognizing and writing down lecture points is another form of recognition and reinforcement. Later, studying for an exam, you will have already forgotten lecture material you did not record, and so even if you study hard, you will miss exam questions on this material.
- *Revise your notes soon after lecture.* Actively interacting with your class notes while you still hold much of the lecture in short-term memory provides perhaps the most powerful form of reinforcement and will be a key to your success.

As you proceed through this textbook, you will encounter a blizzard of terms and concepts. Biology is a field rich with ideas and the technical jargon needed to describe them. What you discover reading this textbook is intended to support the lectures that provide the core of your biology course. Integrating what you learn here with what you learn in lecture will provide you with the strongest possible tool for successfully mastering the basics of biology. The rest is just hard work.

0.2 Using Your Textbook

Your text is more than a list of what you are supposed to know. It is a tool that helps you understand: This is the key to true learning.

A Textbook Is a Tool

> **LEARNING OBJECTIVE 0.2.1** Describe how you can use your text to reinforce what you learn in lecture.

A student enrolled in an introductory biology course, as you are, almost never learns everything from the textbook. Your text is a tool to explain and amplify what you learn in lecture. No textbook is a substitute for attending lectures, taking notes, and studying them. Success in your biology course is like a stool with three legs: lectures, class notes, and text reading—all three are necessary. Used together, they will take you a long way toward success in the course.

Seregraff/Shutterstock

When to Use Your Text. While you can glance at your text at any time to refresh your memory or answer a question that pops into your mind, your use of your text as a learning tool should focus on providing support for the other two "legs" of course success: lectures and class notes.

Do the Assigned Reading. Many instructors assign reading from the text, reading that is supposed to be done before lecture. The timing here is very important: If you already have a general idea of what is being discussed in lecture, it is much easier to follow the discussion and take better notes.

Link the Text to Your Lecture Notes. Your text provides you with a powerful tool to reinforce ideas and information you encounter in lecture. Text illustrations and detailed explanations can pound home an idea quickly grasped in lecture and answer any questions that might occur to you as you sort through the logic of an argument. Thus it is absolutely essential that you follow along with your text as you recopy your lecture notes, keying your notes to the textbook as you go. Annotating your notes in this way will make them far better learning tools as you study for exams later.

Review for Exams. It goes without saying that you should review your recopied lecture notes to prepare for an exam. But that is not enough. What is often missed in gearing up for an exam is the need to also review that part of the text that covers the same material. Reading the chapter again, one last time, helps place your lecture notes in perspective, so that it will be easier to remember key points when a topic explodes at you off the page of your exam.

How to Use Your Text. The single most important way to use your text is to read it. As your biology course proceeds and you move through the text, read each assigned chapter all the way through at one sitting. This will give you valuable perspective. Then, guided by your lecture notes, go back through the chapter, one *learning objective* at a time, and focus on the concepts in each learning objective as you recopy your notes. Pay attention to the "linking" arrows in the text, as using them will reinforce what you are learning. As discussed earlier, building a bridge between text and lecture notes is a very powerful way to learn. Remember, your notes don't take the exam, and neither does the textbook; you do, and the learning that occurs as you integrate text pages and lecture notes in your mind will go a long way toward your taking it well.

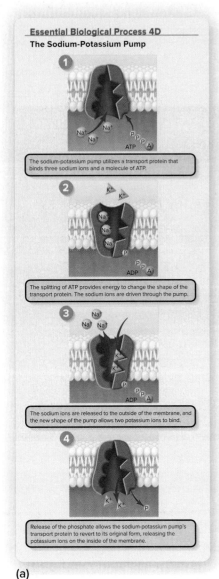

Figure 0.5 Visual learning tools.
(a) An example of an *Essential Biological Process* illustration.
(b) An example of a Phylum Facts illustration.

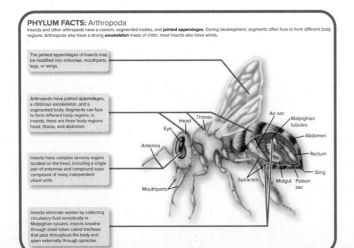

Learning Tools at Your Disposal

LEARNING OBJECTIVE 0.2.2 Review the assessment tools that your text provides to help you master the material.

A textbook is more than just words. What do you see when you flip through the pages of this text? Pictures, lots of them. And questions, scattered through each chapter and clustered at chapter's end. The pictures and quiz questions you will encounter within each chapter can be an important part of your learning experience.

Let the Illustrations Teach You. All introductory biology texts are rich with colorful photographs and diagrams. They are there not to decorate but to aid your comprehension of ideas and concepts. When the text refers you to a specific figure, look at it—the visual link will help you remember the idea much better than restricting yourself to cold words on a page.

Three sorts of illustrations offer particularly strong reinforcement:

Essential Biological Process Illustrations. While you will be asked to learn many technical terms in this course, learning the names of things is not your key goal. Your goal is to master a small set of concepts. There are several essential biological processes that explain how organisms (living individuals) work the way they do. When you have understood these processes, much of the heavy lifting in learning biology is done. Every time you encounter one of these essential biological processes in the text, you will be provided with an illustration to help you better understand. These *Essential Biological Process* illustrations break the process down into easily understood stages, so that you can grasp how the overall process works without being lost in a forest of details (**figure 0.5a**).

Bubble Links. Illustrations teach best when they are simple. Unfortunately, some of the structures and processes being illustrated just aren't simple. Every time you encounter a complex diagram in the text, it will be "predigested" for you—the individual components of the diagram will each be identified with a number in a colored circle, or bubble. This same number is also placed in the text narrative right where that component is discussed. These bubble links allow the text to step you through the illustration, explaining what is going on at each stage—the illustration is a feast you devour one bite at a time.

Phylum Facts. In chapter 18, you will encounter a train of animal phyla (a phylum is a major category of organisms). In such a sea of information, what should you learn? Every time you encounter a phylum in chapter 18, you will be provided with a *Phylum Facts* illustration that selects the key bits of information about the body and lifestyle of that kind of animal (**figure 0.5b**). If you learned and understood only the items highlighted there, you would have mastered much of what you need to know.

Check What You Know. As you move through a chapter, addressing first one topic and then another, it will be important that you monitor your progress—not only what you have read, but also how well you have understood it.

Putting Concepts to Work. At the end of each learning objective within a chapter, you will encounter a "putting the concept to work" question you can use to reinforce what you are learning. If you can't answer it, you should go back and have another look at that section.

LearnSmart Questions. When you complete a chapter, you can gauge how well you have learned the material by answering the *LearnSmart* questions provided by *Connect,* if your class utilizes this program (discussed in the next section).

0.3 Using Your Textbook's Internet Resources

Some of the most powerful learning tools this text provides are delivered over the Internet.

Connect

> **LEARNING OBJECTIVE 0.3.1** Describe the three benefits of using *Connect*.

It probably came as no surprise to you that you were instructed in section 0.2 to read your text in order to learn the material on which you will be tested, using its illustrations to fortify your understanding. It thus came as something of a surprise to education researchers when they found that most successful students do exactly the opposite. Watching how college students actually use their textbooks, they repeatedly observed students going first to the illustrations, then to the captions beneath them, and only later to the words of the text, using the text to clarify their understanding of the illustrations! Said simply, successful students are often visual learners.

A visual learner is best tested with visual and interactive questions. If your class is utilizing an instructor-guided learning program called *Connect*, just such an approach is available to you. As a platform for tackling such interactive assessment of how you are doing, *Connect* provides you with a fully interactive *eBook* version of this text, with embedded animations, as well as notes and highlights added by your instructor. For each class assignment, the instructor then assigns you a series of interactive questions. *Connect* grades each answer for you. If you have trouble with a question, the program connects that question to the learning objective in the *eBook* where the question is answered.

How *Connect* Helps You to Learn. *Connect* is not simply a testing machine used by the instructor to look over your shoulder and spy on how you are studying. Far from it. It is a powerful learning platform you can use to help understand instructor-assigned material. By the time you have successfully navigated the series of questions assigned by your instructor, you will be well on the way to mastering the assignment. *Connect* is no substitute for reading your text and linking it to your lecture notes. Make no mistake about it—your text and lecture notes are the only sure road to success in this course. The great utility of *Connect* is that it provides a way for you to check how you are doing. The visual and interactive questions you access through *Connect* are self-study questions fully integrated with the text. They provide you with a powerful—and fun—way to identify holes in your understanding of an assignment and the means to fill them in. Why wait until an exam to find out what you don't know? Your course grade will be far superior if you find and solve these problems before the exam.

Adaptive Learning

> **LEARNING OBJECTIVE 0.3.2** Describe how *SmartBook* tests how well you have learned.

Not all students come to a biology class with the same level of preparation or remember equally well what they learned in high school. *SmartBook*, delivered through *Connect*, addresses this problem in a direct way, tailoring a learning plan to each student individually.

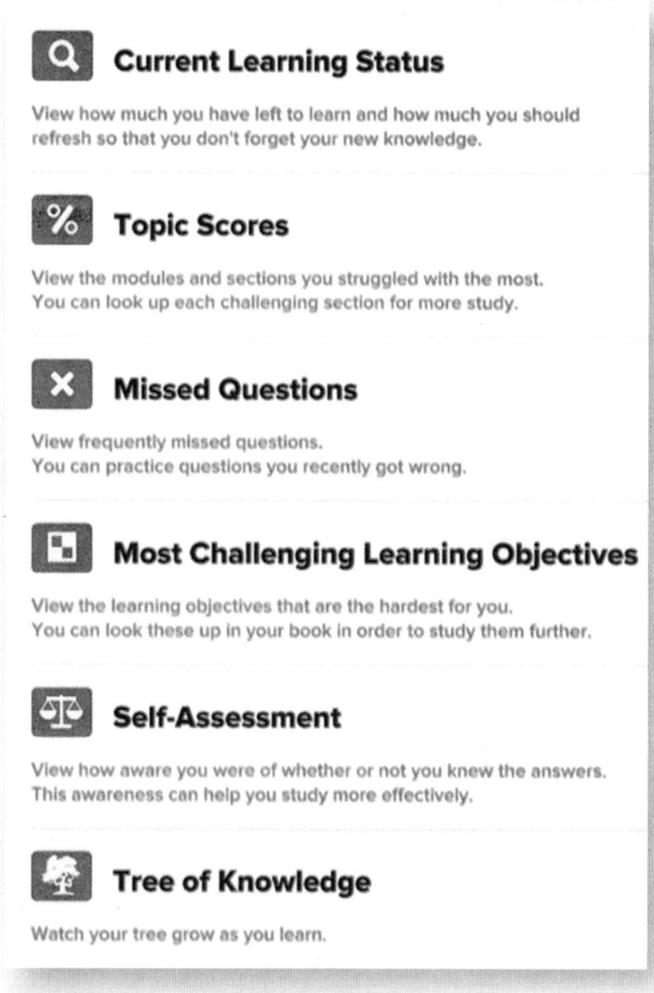

Current Learning Status
View how much you have left to learn and how much you should refresh so that you don't forget your new knowledge.

Topic Scores
View the modules and sections you struggled with the most. You can look up each challenging section for more study.

Missed Questions
View frequently missed questions. You can practice questions you recently got wrong.

Most Challenging Learning Objectives
View the learning objectives that are the hardest for you. You can look these up in your book in order to study them further.

Self-Assessment
View how aware you were of whether or not you knew the answers. This awareness can help you study more effectively.

Tree of Knowledge
Watch your tree grow as you learn.

Figure 0.6 Using *SmartBook* Reports.
SmartBook reports provide you with six tools to help you identify what you do not yet understand, so you can better focus your available study time.

Figure 0.7 Recharge to retain knowledge longer.
SmartBook predicts when you are in danger of forgetting previously learned content and adds a reminder in your To Do list so you can make time to refresh your knowledge.
McGraw Hill

SmartBook adapts to each learner. The power of *SmartBook* is that it adapts directly to you. As you read, yellow highlights focus your attention on the concepts you need to learn at that moment in time. After reading a while, you are prompted to *Practice* what you have learned. As you answer questions correctly, highlights turn green to indicate mastery and new content is highlighted. If you answer incorrectly or if you've been in *Practice* mode for a while, you'll be directed to go back to *Read* mode. In both *Read* and *Practice* modes, *Learning Resources* offer videos and other tools to clarify concepts you are struggling to master. This process of going back and forth between reading and quizzing will help you learn more effectively.

Reports show your progress and identify knowledge gaps. Use *SmartBook* reports like the one you see in figure 0.6 to track your progress and see where you need to focus your study time.

- **Current Learning Status** shows how many items you have left to learn in a chapter and the approximate time you will need to take to learn them, as well as items you need to revisit from earlier chapters.
- **Missed Questions and Most Challenging LOs** can help make your study time more efficient by focusing your attention on the questions and learning outcomes that were difficult for you to master. These reports can be used before a test or exam for additional practice.
- **Self-Assessment** provides insight into how aware you are of what you know and what you don't know. Being able to identify your knowledge gaps helps you focus your study where you need it.
- **Tree of Knowledge** illustrates what portions of the chapter's web of concepts you have mastered and which others still need attention.

Recharge to retain knowledge throughout the semester. *SmartBook* uses your response data to predict when you are in danger of forgetting previously learned content. These *Recharge* reminders appear in the "To Do" section of *Connect* to prompt you to plan time to review (figure 0.7).

0.4 Online Labs

One of the best ways to retain newly-learned information is to put it into practice.

Virtual Labs and Lab Simulations

> **LEARNING OBJECTIVE 0.4.1** Describe how virtual labs improve student learning.

Biology is a hands-on discipline, and many instructors will choose to augment your course with laboratory experience. At the end of every chapter of this text you will find an Inquiry & Analysis feature that walks you through a real biology experiment, so you can learn how an experiment is designed and analyzed.

Even better, Connect provides you with Virtual Lab Simulations that allow you to carry out an experiment yourself, virtually. As you proceed, the simulation will check your understanding and provide feedback. In addition, your instructor can customize each lab assignment to you using adaptive pre-lab and post-lab assessments.

These lab simulations provide a valuable connection between the lecture environment and your own hands-on experience, helping you to visualize complex scientific processes by putting them to work.

Figure 0.8 Virtual Lab Simulations.
Connect Virtual Labs allow you to put your knowledge to work in real-world settings, helping you deepen your understanding.
McGraw Hill

Putting What You Learn to Work

0.5 Science Is a Way of Thinking

In the study of biology, you will encounter a great deal of information, a forest of new terms and definitions, and a lot of descriptions of how things work—body processes, evolutionary relationships, interactions within ecosystems, and many others. In all of this, you will be asked to accept that what you are being taught is "true," that it accurately reflects reality. In fact, what it accurately reflects is what we know about reality. Of the things that you learn in this course, some will be altered by future scientists as they learn more. As we will discuss in chapter 1, our knowledge of science is always incomplete, the picture of reality we construct always a rough draft.

One of the most important things you can learn in a biology course is how these adjustments are made. Long after this class is completed, you will be making decisions that involve biology, and they will be better, more informed decisions if you have acquired the skill to evaluate scientific claims for yourself. Because it is printed in the newspaper or cited on a Web site doesn't make a scientific claim valid. Figure 0.9 illustrates the sorts of biology you encounter today in the news—and this is just a small sample. They are, all of them, important issues that will affect your own life. How do you reach informed opinions about them?

You do it by asking the question, "How do we know this?" Science is a way of thinking that demands to see the evidence, that challenges the validity of every claim. If you can learn to do this, to apply this skill in the future to personal decisions about biology as it impacts your life, you will have taken from this course a valuable lesson.

Figure 0.9 Biology in the news.

How Do We Know What We Know?

> **LEARNING OBJECTIVE 0.5.1** Analyze one example of how biological scientists have come to a conclusion when confronted with problems of major public importance.

A useful way to learn how scientists think, how they constantly check and question what they know, is to look at real cases. What follows are four instances in which biologists have come to a conclusion. All four of these cases will be treated at length in later chapters—here they serve only to introduce you to the process of scientific questioning.

Does Cigarette Smoking Cause Lung Cancer? According to the American Cancer Society, 606,520 Americans died of cancer in 2020. Fully one in four of the students using this textbook can be expected to die of cancer, almost a third of them of lung cancer.

As you might imagine, something that kills so many of us has been the subject of much research. The first step biologists took was to ask a simple question: "Who gets lung cancer?" The answer came back loud and clear: Fully 87% of those who die from lung cancer are cigarette smokers. Delving into this more closely, researchers looked to see if the incidence of lung cancer (that is, how many people contract it per 100,000 people) can be predicted by how many cigarettes a person smokes each day. As you can see in the graph in figure 0.10, it can. The more cigarettes smoked, the higher the occurrence of lung cancer. Based on this study and lots of others like it, biologists concluded that smoking cigarettes causes lung cancer.

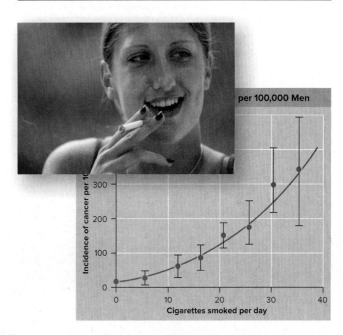

Figure 0.10 Does smoking cause lung cancer?

Grantly Lynch /UK Stock Images Ltd./Alamy Stock Photo

Does Carbon Dioxide Cause Global Warming? Our world is getting warmer—a lot warmer. The great ice caps that cover Antarctica and Greenland are melting, and sea levels are rising. Looking for the cause, atmospheric scientists soon began to suspect what might at first seem an unlikely culprit: carbon dioxide (CO_2), a gas that is a minor component (0.03%) of the air we breathe. As you will learn in this course, burning coal and other fossil fuels releases CO_2 into the atmosphere. Problems arise because CO_2 traps heat. As the modern world industrializes, more and more CO_2 is released. Does this lead to a hotter earth? To find out, researchers looked to see if the rise in global temperature reflected a rise in the atmosphere's CO_2. As you can see in the graph in **figure 0.11**, it does. After these and other careful studies, which we will explore in detail in chapter 22, scientists concluded that rising CO_2 levels are indeed the cause of global warming.

Figure 0.11   Does carbon dioxide cause global warming?

Kelly Cheng/Moment/Getty Images

Does Obesity Lead to Type 2 Diabetes? The United States is in the midst of an obesity epidemic. From 1999 to 2017, the percentage of Americans who are obese has more than tripled, from 13% to 42.4%. Coincidentally (or was it a coincidence?), the number of Americans suffering from type 2 diabetes (a disorder in which the body loses its ability to regulate glucose levels in the blood, often leading to blindness and amputation of limbs) more than tripled over the same period, from 7 million to more than 23 million (that's one in every 14 Americans!).

What is going on here? When researchers compared obesity levels with type 2 diabetes levels, they found a marked correlation, clearly visible in the graph in **figure 0.12**. Investigating more closely, the researchers found that an estimated 80% of people who develop type 2 diabetes are obese. Detailed investigations described in chapter 29 have now confirmed the relationship that these early studies hinted at: Overeating triggers changes in the body that lead to type 2 diabetes.

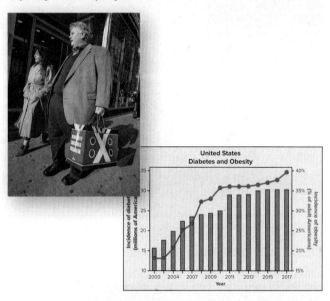

Figure 0.12   Does obesity lead to type 2 diabetes?

Lars A. Niki

What Causes the Ozone Hole? Forty years ago, atmospheric scientists first reported a loss of ozone (O_3 gas) high in the atmosphere over Antarctica. This was scary because ozone in the upper atmosphere absorbs ultraviolet radiation from the sun, protecting the earth's surface from these harmful rays. Trying to understand the reason for this "ozone hole," researchers considered many possibilities, and one of them seemed a good candidate: chlorofluorocarbons, or CFCs. CFCs are supposedly inert chemicals that are widely used as heat exchangers in air conditioners. However, further studies indicated that CFCs are not inert after all—in the intense cold temperatures high over Antarctica, they cause O_3 to be converted to O_2. Scientists concluded that CFCs were indeed causing the ozone hole over Antarctica. This finding led to international treaties beginning in 1990 blocking the further manufacture of CFC chemicals. As you can see in the graph in **figure 0.13**, the size of the ozone hole seems to have stopped expanding.

Looking at the Evidence

One thing these four cases have in common is that in each, scientists reached their conclusion not by applying established rules but rather by looking in detail at what was going on and then testing possible explanations. In short, they gathered data and analyzed it. If you are going to think independently about scientific issues in the future, then you will need to learn how to analyze data and understand what it is telling you. In each case above, the data is presented in the form of a graph. Said simply, you will need to learn to read a graph.

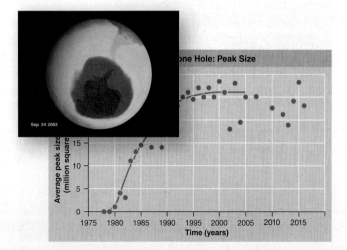

Figure 0.13   What causes the ozone hole?

SVS/TOMS/NASA

0.6 How to Read a Graph

The most basic of all tools you will need to study and understand science is the ability to evaluate data. Said simply, you need to learn to read a graph.

Variables and Graphs

> **LEARNING OBJECTIVE 0.6.1** Explain why correlation of dependent variables does not prove causation.

In section 0.5, you encountered four graphs illustrating what happened to variables, such as global temperature, size of the ozone hole, incidence of obesity, and incidence of lung cancer, when other variables changed. A **variable**, as its name implies, is something that can change. Variables are the tools of science, and you will encounter many different kinds as you proceed through this text. Many of the variables biologists study are examined in graphs like the ones you saw in the previous section. A **graph** shows what happens to one variable when another one changes.

There are two types of variables. The first kind, an **independent variable**, is one that a researcher deliberately changes—for example, the concentration of a chemical in a solution or the number of cigarettes smoked per day. The second kind, a **dependent variable**, is what happens in response to the changes in the independent variable—for example, the intensity of a solution's color or the incidence of lung cancer. Importantly, the change in a dependent variable that is measured in an experiment is not predetermined by the investigator.

In science, all graphs are presented in a consistent way. The independent variable is always presented and labeled across the bottom, called the *x axis*. The dependent variable is always presented and labeled along the side (usually the left side), called the *y axis* (**figure 0.14**).

Some research involves examining correlations between sets of variables, rather than the deliberate manipulation of a variable. For example, a researcher who measures both diabetes and obesity levels (as described in section 0.5) is actually comparing two dependent variables. While such a comparison can reveal correlations and so suggest potential relationships, *correlation does not prove causation*. What is happening to one variable may actually have nothing to do with what happens to the other variable. Only by manipulating a variable (making it an independent variable) can you test for causality.

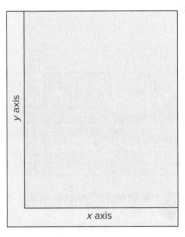

Figure 0.14 The two axes of a graph.

The independent variable is almost always presented along the *x* axis, and the dependent variable is usually shown along the *y* axis.

Using the Appropriate Scale and Units

> **LEARNING OBJECTIVE 0.6.2** Discriminate between arithmetic and logarithmic scales.

A key aspect of presenting data in a graph is the selection of proper scale. Data presented in a table can utilize many scales, from seconds to centuries, with no problems. A graph, however, typically has a single scale on the *x* axis and a single scale on the *y* axis, which might consist of molecular units (for example, nanometers, microliters, micrograms) or macroscopic units (for example, feet, inches, liters, days, milligrams). In each instance, a scale must be chosen that fits what is being measured. Changes in centimeters would not be obvious in a graph scaled in kilometers. Also, if a variable changes a great deal over the course of the experiment, it is often useful to use an expanding scale. A **log** or **logarithmic scale** is a series of numbers plotted as powers of 10 (1, 10, 100, 1,000,...) rather than in the linear progression seen on most graphs (2,000, 4,000, 6,000,...). Consider the two graphs in **figure 0.15**, where the *y* axis is plotted on a linear scale

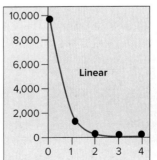 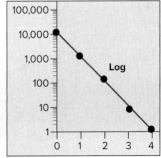

Figure 0.15 Linear and log scale: two ways of presenting the same data.

on the left and on a log scale on the right. Notice that the interval *between* each y axis number is not linear either—the interval between each number is itself subdivided on a log scale. Thus, 50 (the fourth tick mark between 10 and 100) is plotted much closer to 100 than to 10.

Individual graphs use different units of measurement, each chosen to best display the experimental data. By international convention, scientific data are presented in **metric units**, a system of units expressed as powers of 10. For example, weight is expressed in units called *grams*. Ten grams make up a decagram, and 1,000 grams is a kilogram. Smaller weights are expressed as a portion of a gram—for example, a centigram is a hundredth of a gram, and a milligram is a thousandth of a gram. The units of measurement employed in a graph are by convention indicated in parentheses next to the variable label.

Drawing a Line

> **LEARNING OBJECTIVE 0.6.3** Explain how a regression line is drawn.

Most of the graphs that you will find in this text are **line graphs,** which are graphs composed of data points and one or more lines. Line graphs are typically used to present *continuous data*—that is, data that are discrete samples of a continuous process. An example might be data measuring how quickly the ozone hole develops over Antarctica in August and September of each year. You could in principle measure the area of the ozone hole every day, but to make the project more manageable in time and resources, you might actually take a measurement only once a week. Measurements reveal that the ozone hole increases in area rapidly for about six weeks before shrinking, yielding six data points during its expansion. These six data points are like individual frames from a movie—frozen moments in time. The six data points might indicate a very consistent pattern, or they might not.

Consider the hypothetical data in the graphs of figure 0.16. The data points on the left graph are changing in a very consistent way, with little variation from what a straight line (drawn in red) would predict. The graph in the middle shows more experimental variation, but a straight line still does a good job of revealing the overall pattern of how the data are changing. Such a straight "best-fit line" is called a **regression line** and is calculated by estimating the distance of each point to possible lines, adding the values, and selecting the line with the lowest sum. The data points in the graph on the right are randomly distributed and show no overall pattern, indicating that there is no relationship between the dependent and the independent variables.

Other Graphical Presentations of Data

Sometimes the independent variable for a data set is not continuous but rather represents discrete sets of data. A line graph, with its assumption of continuity, cannot accurately represent the variation occurring in discrete sets of data, where the data sets are being compared with one another. In these cases, the preferred presentation is that of a **histogram,** a kind of bar graph. For example, if you were surveying the heights of pine trees in a park, you might group their heights (the independent variable) into discrete "categories" such as 0 to 5 meters tall, 5 to 10 meters, and so on. These categories are placed on the *x* axis. You would then count the number of trees in each category and present that dependent variable on the *y* axis, as shown in figure 0.17.

Some data represent proportions of a whole data set, for example the different types of trees in the park as a percentage of all the trees. This type of data is often presented in a **pie chart** (figure 0.18).

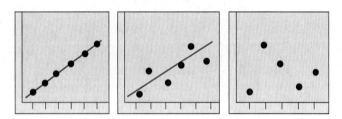

Figure 0.16 **Line graphs: hypothetical growth in size of the ozone hole.**

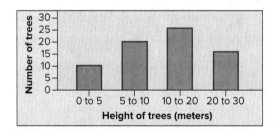

Figure 0.17 **Histogram: the frequency of tall trees.**

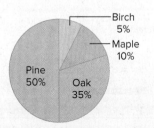

Figure 0.18 **Pie chart: the composition of a forest.**

Putting Your Graph-Reading Skills to Work: Inquiry & Analysis

> **LEARNING OBJECTIVE 0.6.4** Discuss the four distinct steps scientists use to analyze a graph.

The sorts of graphs you have encountered here in this brief introduction are all used frequently by scientists in analyzing and presenting their experimental results, and you will encounter them often as you proceed through this text.

Learning to read a graph and understanding what it does and does not tell you is one of the most important things you can take away from a biology course. To help you develop this skill, every chapter of this text ends with an *Inquiry & Analysis* feature. Each of these end-of-chapter features describes a real scientific investigation. You will be introduced to a question and then given a hypothesis posed by a researcher (a hypothesis is a kind of explanation) to answer that question. The feature will then tell you how the researcher set about evaluating his or her hypothesis with an experiment and will present a graph of the data the researcher obtained. You are then challenged to analyze the data and reach a conclusion about the validity of the hypothesis.

As an example, consider research on the ozone hole. What sort of graph might you expect to see? A *line graph* can be used to present data on how the size of the ozone hole changes over the course of one year. Because the dependent variable is the size of the ozone hole measured continuously over a single season, a smooth curve accurately portrays what is actually going on. In this case, the regression line is not a straight line but rather a curve (figure 0.19).

Can line graphs and histograms be used to present the same data? In some cases, yes. The mode of its presentation does not alter the data; it only serves to emphasize the point being investigated. The *histogram* in figure 0.20 presents data on how the peak size of the ozone hole changed in two-year intervals over 26 years. However, this same data could also have been presented as a line graph (see figure 0.13).

Presented with a graph of the data obtained in the investigation of the *Inquiry & Analysis*, it will be your job to analyze it. Every analysis of a graph involves four distinct steps, some more complex than others but all essential to the process.

Applying Concepts. Your first task is to make sure you understand the nature of the variables and the scale at which they are being presented in the graph. As a self-test, it is always a good idea to ask yourself to identify the dependent variable.

Interpreting Data. Look at the graph. What is changing? How much? How quickly? Is the change continuous? Progressive? What in fact has happened?

Making Inferences. Looking at what has happened, can you logically infer that the independent variable has caused the change you see in the dependent variable?

Drawing Conclusions. Does the inference you were able to make support the hypothesis that the experiment set out to test?

This process of inquiry and analysis is the nuts and bolts of science, and by mastering it, you will go a long way toward learning how a scientist thinks.

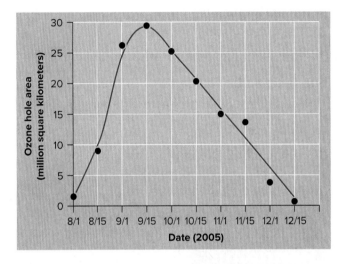

Figure 0.19 Changes in the size of the ozone hole over one year.

This graph shows how the size of the ozone hole changes over time, first expanding in size and then getting smaller.

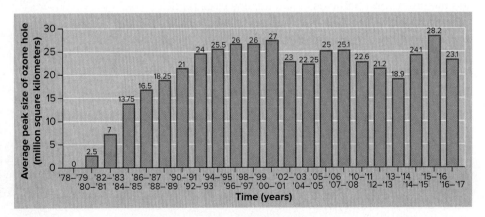

Figure 0.20 The peak sizes of the ozone hole.

This histogram shows how the size of the ozone hole increased in size (determined by its peak size) for 20 years before it stabilized.

PART 1 The Study of Life

1 The Science of Biology

LEARNING PATH ▼

Biology and the Living World
1. The Diversity of Life
2. Properties of Life
3. The Organization of Life
4. Biological Themes

The Scientific Process
5. Stages of a Scientific Investigation
6. Theory and Certainty

Core Ideas of Biology
7. Four Theories Unify Biology as a Science

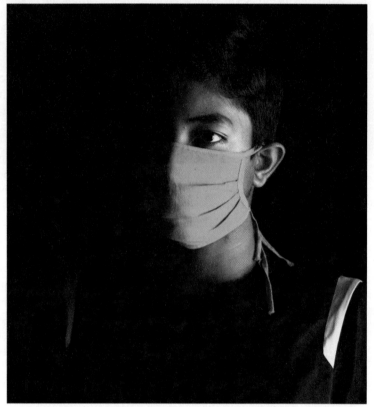
Anubhab Roy/Shutterstock

Pandemic

The student in this photo is wearing a mask. He is not disguising himself or robbing a bank. He is protecting himself and others from infection by a dangerous virus. Many of you have worn the same sort of facemask, for the same reason. The mask helps block your inhaling virus particles in the air around you. Floating in the air like the minute carbon particles in smoke, tiny virus particles are very good at passing from one human to another. This is particularly true of so-called respiratory viruses, which pass from person to person via tiny airborne water droplets. When we breathe the droplets in, the viruses infect the lungs. The 1918 pandemic strain of the respiratory influenza virus killed millions of people in just two years. Strains of flu continue to arise in Asia that infect lots of Americans today. A total of 34,200 Americans died of flu in 2018–2019.

Appearance of a Deadly Coronavirus

Among the respiratory viruses is a distinctive group called the coronaviruses, named for their crown (corona) of surface proteins which you can see in the photo. The virus uses these surface proteins to attach to and penetrate animal

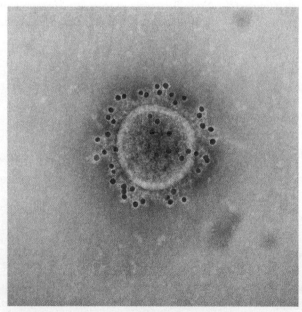
National Institute of Allergy and Infectious Diseases (NIAID)

host cells. Some 20% to 30% of common colds are caused by coronaviruses, but until recently coronaviruses have not caused serious illnesses in people, and so nobody has paid much attention to them.

That all changed in 2002 with the appearance of a deadly coronavirus in China. Called **S**evere **A**cute **R**espiratory **S**yndrome, or **SARS**, the virus infected 8,098, killing just under 10% of those it infected. The outbreak was contained quickly because the virus didn't spread before the symptoms appeared. No known transmission of SARS has occurred since 2004.

Having made this abrupt appearance on the human stage, the coronaviruses have not stayed away. In Saudi Arabia, another deadly respiratory coronavirus appeared in 2012 called **M**iddle **E**ast **R**espiratory **S**yndrome, or **MERS**. Over 2,000 cases of MERS were reported over two years before the outbreak subsided. Fully 36% of those diagnosed with the disease died of it, a very high number indeed!

COVID-19

It is a third coronavirus that we are dealing with today, a virus closely related to SARS that appeared on our radar screens for the first time in Wuhan, the 9th largest city in China (its population of 11 million far exceeds New York City's population of 8 million). The first cases of this new virus were reported to the World Health Organization (WHO) on December 31, 2019. When in a few weeks reports emerged from China of large numbers of people sick from this virus, the WHO on January 23 declared a Public Health Emergency, assigning the formal name of the virus as SARS-CoV2 (that is, **SARS Co**rona **Vi**rus, version **2**) because it is so like the SARS virus. The infection caused by SARS-CoV2 is called **Co**rona**vi**rus **D**isease-2019—the name is usually shortened to COVID-19, with the 19 referring to the year 2019 when the disease first appeared.

Unlike SARS, people infected with COVID-19 do not immediately show symptoms. The virus first colonizes the nose and throat, only infecting the lungs after several days. The person is most contagious two to three days before they are aware that they are ill. Some 35% of infected individuals never exhibit symptoms, although fully contagious. As you can imagine, COVID-19 infections spread rapidly, particularly among people living in densely populated cities, the bulk of transmission of the virus by people not sick enough to get the attention of doctors.

Pandemic: The Outbreak Spreads

Passing from one person to another wherever people congregate, often between people with no symptoms, COVID-19 has spread out from China like a wild fire, passing through Europe and the United States, then to Brazil and all around the globe—a pandemic. The virus began killing Americans in March. Starting in New York City, it soon spread throughout the country, infecting over 33 million people and killing 606,570 of them in the first seventeen months. In one day in January, the United States had 300,969 cases—in a single day. Although most severe in the United States, this same pattern of spread by person-to-person contact was seen all over the world. In these same seventeen months, India had 30 million cases and Brazil 19 million. How many people overall, worldwide? 187 million!

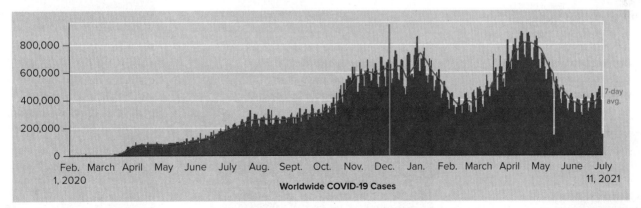

Putting Science to Work

How can we beat the virus? The virus spreads in tiny water droplets in the air. To stop its transmission, it is necessary to keep people at least six feet apart, so the virus cannot bridge the distance between them. In particular, you should avoid bars, sporting events, and other places where people congregate in large numbers. The key to blocking the transmission of the virus is to wear face masks to prevent you from spreading it to others. To control the outbreak in communities where the virus is already common, it is important to identify infected individuals so that they can be isolated away from anyone else. For the first months of the pandemic, testing of large numbers of individuals was difficult, as a complicated machine was needed to carry out the analysis. The development by researchers of simple saliva tests solved that problem. Meanwhile scientists worldwide have been furiously working to develop and improve vaccines to protect all of us from COVID-19. As you proceed through this text, you will hear a lot more about viruses, testing, and vaccines.

Biology and the Living World

What are these? Throughout this text you will encounter blue and green linking arrows. These "Flash-back" and "Flash-forward" concept links will direct you to related material in earlier chapters or in later chapters. Using them will help you master the interconnected nature of concepts in biology.

Figure 1.1 The six kingdoms of life.

Biologists assign all living things to six major categories called *kingdoms*. Each kingdom is profoundly different from the others.

Archaea: Power and Syred/Science Source; **Bacteria:** Alfred Pasieka/Science Source; **Giant Green:** Tammy616/Island Effects/E+/Getty Images; **Fungi:** Russell Illig/Getty Images; **tinctoria blooms:** Kirk Gulden/Shutterstock; **Animalia:** Alan and Sandy Carey/Getty Images

1.1 The Diversity of Life

> **LEARNING OBJECTIVE 1.1.1** List the six kingdoms of life.

In its broadest sense, biology is the study of living things—the science of life. The living world is teeming with a breathtaking variety of creatures—whales, algae, mushrooms, bacteria, pine trees—all of which can be categorized into six groups, or **kingdoms,** of organisms, living things (figure 1.1). Organisms that are placed into a kingdom possess similar characteristics with all other organisms in that same kingdom and are very different from organisms in the other kingdoms. Biologists study the diversity of life in many different ways. They live with gorillas, collect fossils, isolate bacteria, and study molecules of heredity. In the midst of all this diversity, it is easy to lose sight of the key lesson of biology, which is that all living things have much in common.

> **Putting the Concept to Work**
> Identify the kingdom to which you belong.

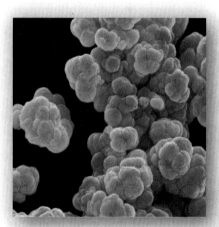

Archaea. This kingdom of prokaryotes (simple cells that do not have nuclei) includes this methanogen, which manufactures methane.

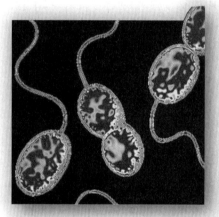

Bacteria. This group is the second of the two prokaryotic kingdoms. Shown here are purple sulfur bacteria, which are able to convert light energy into chemical energy.

Protista. Most of the unicellular eukaryotes (those whose cells contain a nucleus) are grouped into this kingdom, and so are the multicellular algae pictured here.

Fungi. This kingdom contains nonphotosynthetic organisms, mostly multicellular, that digest their food externally, such as these mushrooms.

Plantae. This kingdom contains photosynthetic multicellular organisms that are primarily terrestrial, such as the flowering plant pictured here.

Animalia. Organisms in this kingdom are nonphotosynthetic multicellular organisms that digest their food internally, such as this ram.

1.2 Properties of Life

LEARNING OBJECTIVE 1.2.1 Name and describe the five basic properties shared by all living things.

Biology is the study of life—but what does it mean to be alive? What are the properties that define a living organism? This is not as simple a question as it seems because some of the most obvious properties of living organisms are also properties of many nonliving things—for example, *complexity* (a computer is complex), *movement* (clouds move in the sky), and *response to stimulation* (a soap bubble pops if you touch it). To appreciate why these three properties, so common among living things, do not help us define life, imagine a mushroom standing next to a television: The television seems more complex than the mushroom, the picture on the television screen moves while the mushroom just stands there, and the television responds to a remote control device while the mushroom continues to just stand there—yet it is the mushroom that is alive. All living things share five basic properties:

1. **Cellular organization.** All living things are composed of one (**figure 1.2**) or more cells. A cell is a tiny compartment with a thin covering called a *membrane*. Some cells have simple interiors, while others are complexly organized, but all are able to grow and reproduce. A human body contains about 10 to 100 trillion cells (depending on how big you are). That is a lot—a string 100 trillion centimeters long could wrap around the world 1,600 times!
2. **Metabolism.** All living things use energy. Moving, growing, thinking—everything you do requires energy. Where does all this energy come from? It is captured from sunlight by plants, algae, and certain bacteria and used to build sugars in a process called photosynthesis. To get the energy that powers our lives, we extract it from plants or from animals that eat plants or that eat plant-eating animals (**figure 1.3**). Metabolism refers to all the chemical reactions that occur within cells and involves the transfer of energy from one form to another.
3. **Homeostasis.** While the environment often varies a lot, organisms act to keep their interior conditions relatively constant in a process called *homeostasis*. For example, your body acts to maintain an internal temperature of about 37°C (98.6°F), regardless of how hot or cold the weather might be. Stable internal conditions allow other complex body processes to be better coordinated.
4. **Growth and reproduction.** All living things grow and reproduce. Bacteria increase in size and simply split in two, as often as every 15 minutes. More complex organisms grow by increasing the number of cells and reproduce sexually by producing gametes (sperm or eggs) that combine, giving rise to offspring.
5. **Heredity.** All organisms possess a genetic system that is based on a long molecule called *DNA* (*deoxyribonucleic acid*). The information that determines what an individual organism will be like is contained in a code that is dictated by the order of the subunits making up the DNA molecule, just as the order of letters on this page determines the sense of what you are reading. Each set of instructions within the DNA is called a *gene*. DNA is faithfully copied from one generation to the next, and so any change in a gene is preserved and passed on to future generations. The transmission of characteristics from parent to offspring is a process called *heredity*.

Putting the Concept to Work
Explain why complexity, movement, and response to stimuli are not properties that define life.

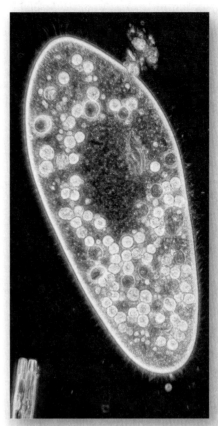

Figure 1.2 Cellular organization.

These paramecia are complex single-celled protists that have just ingested several yeast cells.

MELBA PHOTO AGENCY/Alamy Stock Photo

IMPLICATION FOR YOU In your body, bacterial cells outnumber the body's own cells by a 10 to 1 ratio. What do you imagine these billions of bacteria are doing?

Figure 1.3 Metabolism.

This heron obtains the energy it needs to move, grow, and carry out its body processes by eating fish that eat algae. The bird metabolizes this food using chemical processes that occur within cells.

MZPHOTO.CZ/Shutterstock

1.3 The Organization of Life

Perhaps the most important property of life is organization. Living things are very complex but in a very orderly way.

A Hierarchy of Increasing Complexity

> **LEARNING OBJECTIVE 1.3.1** List the 13 hierarchical levels of the organization of life.

A key to understanding the organization of living things is to appreciate that they exhibit different levels of complexity, layered one on top of another like layers of paint. We will examine the complexity of life at three levels: cellular, organismal, and populational.

> This text follows this hierarchy of increasing complexity. The cellular level is discussed in chapters 2 through 13; the organismal level is discussed in chapters 23 through 32; and the population level is discussed in chapters 14 through 22.

Cellular Level. Following down the first section of **figure 1.4**, you can see that structures within cells get more and more complex—that there is a *hierarchy* of increasing complexity within cells.

1. **Atoms.** The fundamental elements of matter are atoms.
2. **Molecules.** Atoms are joined together into complex clusters called molecules.
3. **Macromolecules.** Large complex molecules are called macromolecules, such as DNA, which stores hereditary information.
4. **Organelles.** Complex biological molecules are assembled into tiny compartments within cells called organelles, such as the nucleus within which the cell's DNA is stored.
5. **Cells.** Organelles and other elements are assembled into membrane-bounded units we call cells. Cells are the smallest level of organization that can be considered alive.

Figure 1.4 Levels of organization.

A traditional and very useful method of sorting through the many ways in which the organisms of the living world interact is to arrange them in terms of levels of organization, proceeding from the very small and simple to the very large and complex.

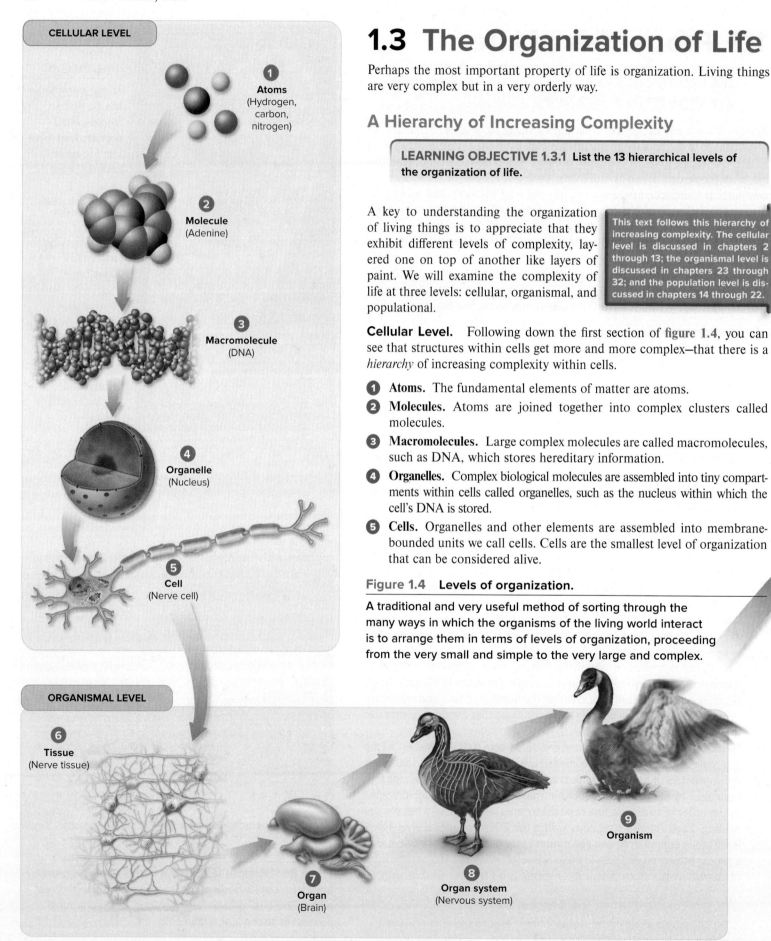

Organismal Level. At the organismal level, in the second section of figure 1.4, cells are organized into four levels of complexity.

⑥ **Tissues.** The most basic level is that of tissues, which are groups of similar cells that act as a functional unit. Nerve tissue is one kind of tissue, composed of cells called neurons that carry electrical signals.

⑦ **Organs.** Tissues, in turn, are grouped into organs, which are body structures composed of several different tissues that form a structural and functional unit. Your brain is an organ composed of nerve cells and a variety of tissues that form protective coverings and distribute blood.

⑧ **Organ systems.** At the third level of organization, organs are grouped into organ systems. The nervous system, for example, consists of sensory organs, the brain and spinal cord, neurons that convey signals throughout the body, and supporting cells.

⑨ **Organism.** Organ systems function together to form an organism.

Populational Level. Organisms are further organized into several hierarchical levels within the living world, as you can see to the right.

⑩ **Population.** The most basic of these is the population, which is a group of organisms of the same species living in the same place. A flock of geese living together on a pond is a population.

⑪ **Species.** All the populations of a particular kind of organism together form a species, its members similar in appearance and able to interbreed. All Canada geese are members of the species *Branta canadensis*. Sandhill cranes are a different species.

⑫ **Community.** At a higher level of biological organization, a community consists of all the populations of different species living together in one place. Geese, for example, may share their pond with ducks, fish, grasses, and many kinds of insects.

⑬ **Ecosystem.** At the highest tier of biological organization, a biological community and the soil and water within which it lives together constitute an ecological system, or ecosystem.

> **Putting the Concept to Work**
> The nucleus of one of your skin cells is bounded by a membrane and contains DNA. Why is it not considered to be alive?

Emergent Properties

> **LEARNING OBJECTIVE 1.3.2** Explain the origin of emergent properties.

At each higher level in the living hierarchy, novel properties emerge, properties that were not present at simpler levels. These **emergent properties** result from the way in which components interact and are the natural consequence of the hierarchy and structural organization of life. Functional properties emerge from more complex organization. Metabolism is an emergent property of life. The chemical reactions within a cell arise from interactions between molecules that are orchestrated by the orderly environment of the cell's interior. Consciousness is an emergent property of the brain that results from the interactions of many neurons in different parts of the brain.

> **Putting the Concept to Work**
> Describe an emergent property for each of the three general levels of life's complexity.

POPULATIONAL LEVEL

⑩ Population

⑪ Species

⑫ Community

⑬ Ecosystem

Canada geese: Ron Rowan Photography/Shutterstock; Canadian goose: Karl Blessing/Getty Images; Canada geese: Ron Rowan Photography/Shutterstock; Sandhill cranes/Canada geese: Cal Sport Media/Alamy Stock Photo; Sandhill crane: Saddako/Getty Images; Sandhill cranes/Canada geese: Cal Sport Media/Alamy Stock Photo; Sandhill cranes/Canada geese: Cal Sport Media/Alamy Stock Photo

1.4 Biological Themes

> **LEARNING OBJECTIVE 1.4.1** Explain the five general themes that define biology as a science.

Just as a house is organized into thematic areas such as bedroom, kitchen, and bathroom, the living world is organized by major *themes,* such as how energy flows within the living world from one part to another. As you study biology in this text, five general themes will emerge repeatedly, themes that serve to both unify and explain biology as a science.

Evolution

Evolution is genetic change in a species over time. Charles Darwin was an English naturalist who, in 1859, proposed the idea that this change is a result of a process called **natural selection.** Simply stated, those organisms whose characteristics make them better able to survive the challenges of their environment live to reproduce, passing their favorable characteristics on to their offspring. Darwin was thoroughly familiar with variation in domesticated animals (in addition to many nondomesticated organisms). He knew that varieties of pigeons could be selected by breeders to exhibit exaggerated characteristics, a process called *artificial selection.* You can see some of these extreme-looking pigeons pictured here. We now know that the characteristics selected are passed on through generations because DNA is transmitted from parent to offspring. Darwin visualized how selection in nature could be similar to that which had produced the different varieties of pigeons. Thus, the many forms of life we see about us on earth today and the way we ourselves are constructed and function reflect a long history of natural selection.

The Flow of Energy

All organisms require energy to carry out the activities of living—to build bodies and do work and think thoughts. All of the energy used by most organisms comes from the sun and is passed in one direction through ecosystems. The simplest way to understand the flow of energy through the living world is to look at who uses it. The first stage of energy's journey is its capture by green plants, algae, and some bacteria by the process of photosynthesis. This process uses energy from the sun to synthesize sugars that photosynthetic organisms like plants store in their bodies. Plants then serve as a source of life-driving energy for animals that eat them. Other animals, like the eagle shown here, may then eat the plant eaters. At each stage, some energy is used for the processes of living, some is transferred, and much is lost primarily as heat. The flow of energy is a key factor in shaping ecosystems, affecting how many and what kinds of organisms live in a community.

Rock pigeon: David Thyberg/iStockphoto/Getty Images; **Swallow Pigeon:** Custom Life Science Images/Alamy Stock Photo; **Fantail pigeon:** Tom McHugh/Science Source; **Bald eagle:** Jeff Vanuga/Corbis

Cooperation

The ants cooperating in the photo on the right protect the plant on which they live from predators and shading by other plants, while the plant returns the favor by providing the ants with nutrients (the yellow structures at the tips of the leaves). This type of cooperation has played a critical role in the evolution of life on earth. For example, organisms of two different species that live in direct contact, such as the ants and the plant on which they live, form a type of relationship called **symbiosis**. Many types of cells possess organelles that are the descendants of symbiotic bacteria, and symbiotic fungi helped plants first invade land from the sea. The coevolution of flowering plants and insects—in which changes in flowers influenced insect evolution and, in turn, changes in insects influenced flower evolution—has been responsible for much of life's great diversity.

Dreedphotography/Getty Images

Structure Determines Function

One of the most obvious lessons of biology is that biological structures are very well suited to their functions. You will see this at every level of organization: Enzymes, which are macromolecules that cells use to carry out chemical reactions, are precisely structured to match the shapes of the chemicals the enzymes must manipulate. Within the many kinds of organisms in the living world, body structures seem carefully suited to carry out their functions—the long tongue with which the moth to the right sucks nectar from deep inside a flower is one example. The superb fit of structure to function in the living world is no accident. Life has existed on earth for over 2 billion years, a long time for evolution to favor changes that better suit organisms to meet the challenges of living. It should come as no surprise to you that after all this honing and adjustment, biological structures carry out their functions well.

Steve Byland/Getty Images

Homeostasis

The high degree of specialization we see among complex organisms is possible only because these organisms act to maintain a relatively stable internal environment, a process introduced earlier called **homeostasis**. Without this constancy, many of the complex interactions that need to take place within organisms would be impossible. Maintaining homeostasis in a body as complex as yours requires a great deal of signaling back-and-forth between cells. For example, homeostasis often involves water balance to maintain proper blood chemistry. All complex organisms need water—some, like this hippo, luxuriate in it. Others, like the kangaroo rat that lives in arid conditions where water is scarce, obtain water from food and never actually drink.

You will encounter these five biological themes repeatedly in this text. But just as a budding architect must learn more than the parts of buildings, so your study of biology should teach you more than a list of themes, concepts, and parts of organisms. Biology is a dynamic science that will affect your life in many ways, and that lesson is in a very fundamental respect the most important you will learn.

Peter Johnson/Corbis/VCG/Getty Images

> **Putting the Concept to Work**
> What key ability allows a complex body like yours to maintain homeostasis?

The Scientific Process

1.5 Stages of a Scientific Investigation

> **LEARNING OBJECTIVE 1.5.1** Explain how the six stages of a scientific investigation allow biologists to discover general principles by careful examination of specific cases.

Science is not simply a matter of asking questions—the sort of questions that are asked and *how* they are asked are critical to any scientific investigation.

How Science Is Done

Scientists establish general principles as a way to explain the world around us, but how do scientists determine which general principles are true from among the many that might be? They do this by systematically testing alternative proposals. If these proposals prove inconsistent with experimental observations, they are rejected as untrue. After making careful observations concerning a particular area of science, scientists construct a **hypothesis**, which is a suggested explanation that accounts for those observations: an "educated guess." A hypothesis is a proposition that might be true. Those hypotheses that have not yet been disproved are retained. They are useful because they fit the known facts, but they are always subject to future rejection in the light of new information.

We call the test of a hypothesis an experiment. To evaluate alternative hypotheses, you would conduct an experiment designed to eliminate one or more of the hypotheses. Note that this does not prove that any of the other hypotheses are true; it merely demonstrates that one of them is not. A successful experiment is one in which one or more of the alternative hypotheses is demonstrated to be inconsistent with the results and is thus rejected.

In this text, you will encounter a great deal of information, often accompanied by explanations. These explanations are hypotheses that have withstood the test of experiment. Many will continue to do so; others will be revised as new observations are made. Biology, like all science, is in a constant state of change, with new ideas appearing and replacing the old ones.

The Scientific Process

Scientific investigations can be said to have six stages, as illustrated in **figure 1.5**: ❶ observing what is going on; ❷ forming a set of hypotheses; ❸ making predictions; ❹ testing them; ❺ carrying out controls, until one or more of the hypotheses have been eliminated; and ❻ forming conclusions based on the remaining hypothesis. To better understand how a scientist progresses through these stages, let's examine an actual scientific investigation: the discovery and analysis of the ozone hole.

1. **Observation.** The key to any successful scientific investigation is careful *observation*. Scientists had studied the skies over the Antarctic for many years, noting a thousand details about temperature, light, and levels of chemicals. Had these scientists not kept careful records of what they observed, they might not have noticed that levels of the atmospheric gas ozone were dropping.
2. **Hypothesis.** When the unexpected drop in ozone was reported, scientists made a guess why—perhaps something was destroying the ozone; maybe the culprit was chlorofluorocarbons (CFCs), an industrial chemical used

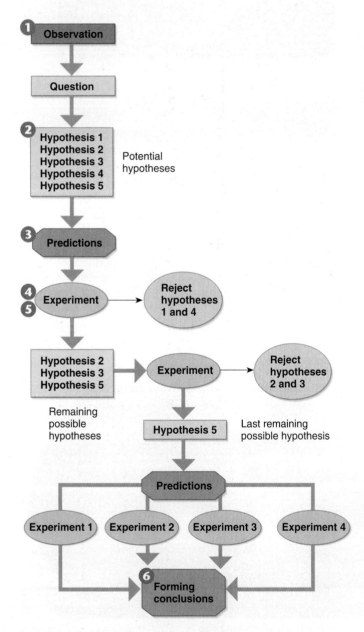

Figure 1.5 The scientific process.
This diagram illustrates the stages of a scientific investigation. Further predictions can be made based on an accepted hypothesis and tested with more experiments. If a hypothesis is validated by numerous experiments and stands the test of time, it may eventually become a theory.

as a coolant in air conditioners, propellants in aerosols, and foaming agents in making Styrofoam. Of course, this was not a guess in the true sense; scientists had some working knowledge of CFCs and what they might be doing in the upper atmosphere. We call such a guess a ***hypothesis.*** What the scientists guessed was that chlorine from the breakdown of CFCs was reacting chemically with ozone over the Antarctic (figure 1.6), converting ozone (O_3) into oxygen gas (O_2) and removing the ozone shield from our earth's atmosphere. Often, scientists will form *alternative hypotheses* if they have more than one guess about what they observe. In this case, there were several other hypotheses advanced to explain the ozone hole. One suggestion explained it as a normal consequence of the spinning of the earth, the ozone spinning away from the polar regions much as water spins away from the center as a clothes washer moves through its spin cycle. Another hypothesis was that the ozone hole was simply due to sunspots and would soon disappear.

3. **Predictions.** If the CFC hypothesis is correct, then several consequences can reasonably be expected. We call these expected consequences *predictions.* A prediction is what you expect to happen if a hypothesis is true. The CFC hypothesis predicts that if CFCs are responsible for producing the ozone hole, then it should be possible to detect CFCs in the upper Antarctic atmosphere, as well as the chlorine released from CFCs that attack the ozone.

4. **Testing.** Scientists set out to test the CFC hypothesis by attempting to verify some of its predictions. We call the test of a hypothesis an *experiment.* To test the hypothesis, atmospheric samples were collected from the stratosphere over six miles up by a high-altitude balloon. Analysis of the samples revealed the presence of CFCs, as predicted. Were the CFCs interacting with the ozone? The samples contained free chlorine and fluorine, confirming the breakdown of CFC molecules. The results of the experiment thus support the hypothesis.

5. **Controls.** Events in the upper atmosphere can be influenced by many factors. We call each factor that might influence a process a **variable.** To evaluate alternative hypotheses about one variable, all the other variables must be kept constant so that we do not get misled or confused by these other influences. This is done by carrying out two experiments in parallel: In the first experimental test, we alter one variable in a known way to test a particular hypothesis; in the second, called a *control experiment,* we do *not* alter that variable. In all other respects, the two experiments are the same. To further test the CFC hypothesis, scientists carried out control laboratory experiments in which they detected no drop in ozone levels in the absence of CFCs. The result of the control was consistent with the predictions of the hypothesis.

6. **Conclusion.** A hypothesis that has been tested and not rejected is tentatively accepted. The hypothesis that CFCs released into the atmosphere are destroying the earth's protective ozone shield is now supported by a great deal of experimental evidence and is widely accepted. A collection of related hypotheses that have been tested many times and not rejected is called a **theory.** A theory indicates a higher degree of certainty; however, in science, nothing is "certain." For example, the theory of the ozone shield—that ozone in the upper atmosphere shields the earth's surface from harmful UV rays by absorbing them—is supported by a wealth of observation and experimentation and is widely accepted.

Figure 1.6 **Living under the ozone hole.**

The sky above these Adelie penguins shields them from the sun's harmful UV radiation, just as the sky above you shields you. Not in the summer, however. In the Antarctic summer an "ozone hole" appears, depleting the ozone above these penguins and exposing them to the danger of UV radiation. It appears that human activities far to their north are having a serious impact on the environment of these penguins.

Joel Simon/Digital Vision/Getty Images

IMPLICATION FOR YOU If the ozone hole above these penguins extends only over Antarctica (and sometimes the tip of South America), why should people living in the United States worry about it?

Putting the Concept to Work

If no theory is certain, can you propose a future discovery that if confirmed would disprove the theory of the ozone shield?

1.6 Theory and Certainty

> **LEARNING OBJECTIVE 1.6.1** Distinguish between hypothesis and theory.

A theory is a unifying explanation for a broad range of observations. Thus, we speak of the theory of gravity, the theory of evolution, and the theory of the atom. Theories are the solid ground of science, that of which we are the most certain. However, there is no absolute truth in science, only varying degrees of uncertainty. A scientist's acceptance of a theory is always provisional because the possibility always remains that future evidence will cause a theory to be revised; some "once-accepted" theories are discussed in figure 1.7.

The word *theory* is thus used very differently by scientists than by the general public. To a scientist, a theory represents that of which he or she is most certain; to the general public, the word *theory* implies a *lack* of knowledge or a guess. How often have you heard someone say, "It's only a theory!"? As you can imagine, confusion often results. In this text, the word *theory* will always be used in its scientific sense, in reference to a generally accepted scientific principle.

The Scientific "Method"

It was once fashionable to claim that scientific progress is the result of applying a series of steps called the *scientific method;* that is, a series of logical "either/or" predictions tested by experiments to reject one alternative. The assumption was that trial-and-error testing would inevitably lead one through the maze of uncertainty that always slows scientific progress. If this were indeed true, a computer would make a good scientist—but science is not done this way! If you ask successful scientists how they do their work, you will discover that without exception they design their experiments with a pretty fair idea of how they will come out. Environmental scientists understood the chemistry of chlorine and ozone when they formulated the CFC hypothesis, and they could imagine how the chlorine in CFCs would attack ozone molecules. A hypothesis that a successful scientist tests is not just any hypothesis. Rather, it is a "hunch" or educated guess in which the scientist integrates all that he or she knows in an attempt to get a sense of what *might* be true. It is because insight and imagination play such a large role that some scientists are so much better at science than others—just as Beethoven and Mozart stand out among composers.

The Limitations of Science

Scientific study is limited to organisms and processes that we are able to observe and measure. Supernatural and religious hypotheses are beyond the realm of scientific analysis because they cannot be scientifically studied, analyzed, or explained. Supernatural hypotheses can be used to explain any result and cannot be disproven by experiment or observation. Scientists in their work are limited to objective interpretations of observable phenomena.

> A nonscientific argument called Intelligent Design has been suggested as an alternative to evolution. The ability to study and test this argument versus a scientific one is discussed in chapter 14.

> **Putting the Concept to Work**
> Why don't computers, which are good at trial-and-error testing, make good scientists?

The theory that the earth was flat—the idea that the earth was flat was generally accepted until about 330 B.C., when Aristotle provided observational evidence for a spherical earth using constellations. About 100 years later, Eratosthenes provided quantitative evidence by estimating the circumference of the earth.

The theory that the sun circles earth—this geocentric view of the universe was accepted as fact until Copernicus first proposed a sun-centered solar system in the 16th century, which was supported by quantitative evidence in the early 1600s by Galileo.

The theory of creationism—this explanation of the origin and diversity of life on earth following the account presented in the Bible was widely accepted, although other explanations that proposed that life evolved on earth were also put forth. The theory of creationism was questioned and then dispelled in 1859, when Charles Darwin proposed an explanation for evolution, a process called natural selection, which is the widely accepted scientific explanation of the origin and diversity of life on earth.

The theory of acquired traits—Jean-Baptiste Lamarck, a predecessor of Darwin, proposed an explanation of evolution. He stated that traits were acquired during one's life and were then passed on to offspring. Darwin later proposed the explanation of natural selection as the driving force of evolution.

The hypothesis of cold nuclear fusion—the experimental results that led to this hypothesis, that nuclear reactions could be performed at near room temperature and pressure, were presented in 1989. The results could have revolutionized the energy industry, but the results could not be duplicated, and widespread support for the hypothesis never developed.

Figure 1.7 Rejected theories.
To illustrate how experimentation changes scientific thought, consider these scientific theories that were once accepted as true but have since been rejected.
Land surface: Pixtal/Stocktrek/age fotostock; Sunset: Tom Pepeira/Iconotec/Alamy Stock Photo; Nuclear power station: Andrew Holt/Digital Vision/Getty Images

Today's Biology

The Polar Ice Caps Are Disappearing

Michelle Valberg/All Canada Photos/Alamy Stock Photo

ESA/ZUMA Press/Newscom

Antarctica, the southernmost continent, has 5.4 million square miles of land mass covered, like icing on a cake, by more than 7.2 million square miles of ice. The ice is as much as a 1 mile thick in places, making it home to over 70% of the world's fresh water. It took some 30 millions of years of snow falling during the winter and not melting in the summer to form this giant layer of ice. As the layer of ice forms in the mountainous interior and becomes thicker, it slides downhill toward the Southern Ocean. One such ice flow, extending far out from the land mass into the ocean, is called the West Antarctic Ice Shelf. This lip of ice, 530,000 cubic feet in size, reaches a depth of 1.5 miles! Despite its massive size, the West Antarctic Ice Shelf has remained stable for centuries, the yearly amount of ice melting into the ocean being equaled by the amount of new ice flowing downward from the continent's ice-forming interior.

If the Antarctic ice shelves, like the West Antarctic Ice Shelf, were to melt completely, the water level in the world's oceans would increase by more than 8 feet. What would it take to melt that much ice? We are on our way to finding out. In 2020, for the first time ever in recorded history, Antarctica had temperature recordings close to 70°F. The average temperature in the region has increased by 5°F over the last 50 years. As the temperature increases, the ocean water temperature also increases, melting more submerged ice from the ice shelves. This, in turn, causes the ice shelves to sink lower into the water and further melt. This cycle will continue to happen until parts of the ice shelf become unstable enough that they calve, breaking off a large chunk of ice. In 2019, the largest ice chunk ever to calve off Antarctica was recorded. At 631 square miles, the D28 iceberg covers more area than the entire city of Houston, Texas. Similar observations are seen with the Arctic ice cap. In September 2020, the amount of Arctic sea ice covered just 1.7 square miles, making it the second smallest record of Arctic sea ice. Satellite images show that the top of Earth that used to be covered in white ice is now turning blue because of all of the exposed water.

How can we stop this? The Paris Climate Accord is an international agreement between most countries in the world with a mission to enact stronger environmental policies to reduce climate change to within 3.5°F of the temperatures recorded around the 1900s. The plan intentionally sets out to reduce greenhouse gas emissions, which are directly responsible for the warming of our oceans and the decay of our polar ice caps. Increasing greenhouse gas accumulation in the atmosphere leads to more absorption of solar radiation, trapping, and holding heat in the atmosphere. By reducing greenhouse gas emissions, more heat will be allowed to escape, and we can slow the heating of the oceans and prevent the collapse of the ice caps. However, time is getting late. Even if the goals of the Paris Climate Accord are met, a 3.5°F increase in average temperature will cause enough ice to melt that some of the world's smaller island nations will be completely submerged.

Core Ideas of Biology

1.7 Four Theories Unify Biology as a Science

You will learn a great deal about biology in this text, but much of it will come down to four central theories.

The Cell Theory: Organization of Life

> **LEARNING OBJECTIVE 1.7.1** State the cell theory.

All organisms are composed of cells, life's basic units. Cells were discovered by Robert Hooke in England in 1665. Hooke was using one of the first microscopes, one that magnified 30 times. Looking through a thin slice of cork, he observed many tiny chambers, which reminded him of monks' cells in a monastery. Not long after that, the Dutch scientist Anton van Leeuwenhoek used microscopes capable of magnifying 300 times and discovered an amazing world of single-celled life in a drop of pond water like you see in figure 1.8. In 1839, the German biologists Matthias Schleiden and Theodor Schwann, summarizing a large number of observations by themselves and others, concluded that all living organisms consist of cells. Their conclusion forms the basis of what has come to be known as the *cell theory*. Later, biologists added the idea that all cells come from other cells. The cell theory, one of the basic ideas in biology, is the foundation for understanding the reproduction and growth of all organisms. You will learn much more about cells in chapter 4.

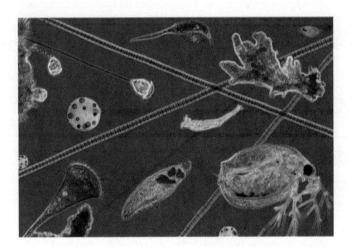

Figure 1.8 Life in a drop of pond water.

All organisms are composed of cells. Some organisms, including protists, are single celled, while others, such as plants, animals, and fungi, consist of many cells.

Mary Martin/Science Source

> **Putting the Concept to Work**
> Why was it not possible to develop the cell theory until the end of the 17th century?

The Gene Theory: Molecules and Inheritance

> **LEARNING OBJECTIVE 1.7.2** Define the term *gene*.

Even the simplest cell is incredibly complex, more intricate than a computer. The information that specifies what a cell is like—its detailed plan—is encoded in a long cablelike molecule called **DNA** (**deoxyribonucleic acid**). Each DNA molecule is formed from two long chains of building blocks, called nucleotides, wound around each other. You can see in figure 1.9 that the two chains face each other, like two lines of people holding hands. The chains contain information in the same way this sentence does—as a sequence of letters. There are four different nucleotides in DNA (symbolized as A, T, C, and G in the figure), and the sequence in which they occur encodes the information. Specific sequences of several hundred to many thousand nucleotides make up a *gene*, a discrete unit of hereditary information. All organisms encode their genes in strands of DNA. The *gene theory*, illustrated in figure 1.10, states that the proteins encoded by an organism's genes determine what it will be like. Genes are the subject of chapters 12 and 13.

> How DNA functions in the cell is directly related to its structure and is discussed in detail in chapter 3 and again in chapter 11.

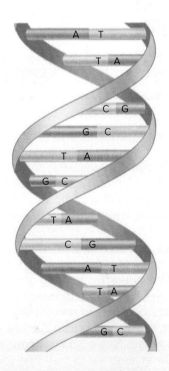

Figure 1.9 Genes are made of DNA.

Winding around each other like the rails of a spiral staircase, the two strands of a DNA molecule make a double helix.

> **Putting the Concept to Work**
> Do you imagine that all the cells of an organism contain the same genes? If so, how can its body have different kinds of cells?

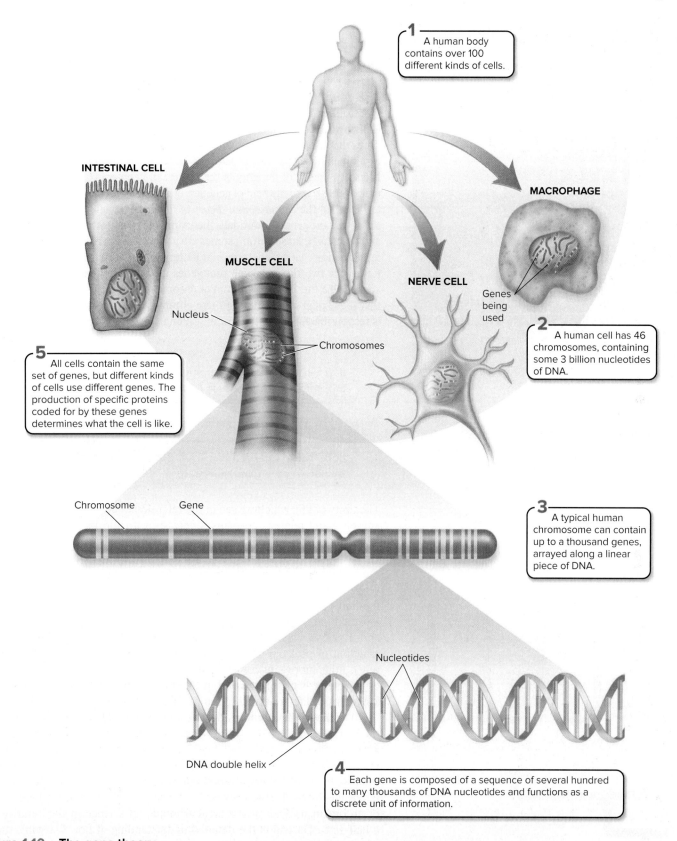

Figure 1.10 **The gene theory.**

The gene theory states that what an organism is like is determined in large measure by its genes. Here you see how the many kinds of cells in the body of each of us are determined by which genes are used in making each particular kind of cell.

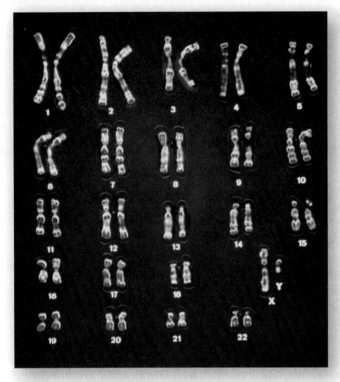

Figure 1.11 Human chromosomes.

The chromosomal theory of inheritance states that genes are located on chromosomes. This human karyotype (an ordering of chromosomes) shows banding patterns on chromosomes that represent clusters of genes.
CNRI/Science Source

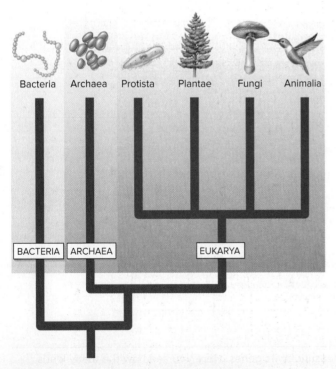

Figure 1.12 The three domains of life.

Biologists categorize all living things into three overarching groups called domains: Bacteria, Archaea, and Eukarya.

The Theory of Heredity: Unity of Life

> **LEARNING OBJECTIVE 1.7.3** State how the theory of heredity is related to the chromosomal theory of inheritance.

The storage of hereditary information in genes composed of DNA is common to all living things. The *theory of heredity* first advanced by Gregor Mendel in 1865 states that the genes of an organism are inherited as discrete units. A triumph of experimental science developed long before genes and DNA were understood, Mendel's theory of heredity is the subject of chapter 10. Soon after Mendel's theory gave rise to the field of genetics, other biologists proposed what has come to be called the *chromosomal theory of inheritance*, which in its simplest form states that the genes of Mendel's theory are physically located on chromosomes and that it is because chromosomes are parceled out in a regular manner during reproduction that Mendel's regular patterns of inheritance are seen. In modern terms, the two theories state that genes are a component of a cell's chromosomes (like the 23 pairs of human chromosomes you see in figure 1.11) and that the regular duplication of these chromosomes during sexual reproduction is responsible for the pattern of inheritance we call Mendelian segregation.

> **Putting the Concept to Work**
> If genes are located on chromosomes, does it necessarily follow that chromosomes contain DNA?

The Theory of Evolution: Diversity of Life

> **LEARNING OBJECTIVE 1.7.4** State how the theory of evolution is related to the gene theory.

The unity of life, which we see in the retention of certain key characteristics among many related life-forms, contrasts with the incredible diversity of living things that have evolved. These diverse organisms are sorted by biologists into six kingdoms, as you learned in section 1.1. Organisms placed in the same kingdom have in common some general characteristics. In recent years, biologists have added a classification level above kingdoms, based on fundamental differences in cell structure. The six kingdoms are each now assigned into one of three great groups called *domains:* Bacteria, Archaea, and Eukarya (figure 1.12). Bacteria and Archaea each consist of one kingdom of prokaryotes (single-celled organisms with little internal structure). Four more kingdoms composed of organisms with more complexly organized cells are placed within the domain Eukarya, the eukaryotes. The kingdoms of life are discussed in detail in chapter 15.

The **theory of evolution,** advanced by Charles Darwin in 1859, attributes the diversity of the living world to natural selection. Those organisms best able to respond to the challenges of living will leave more offspring, he argued, and thus their traits become more common in the population. It is because the world offers diverse opportunities that it contains so many different life-forms. Today, scientists can decipher all the genes (the genome) of an organism. One of the great triumphs of science in the century and a half since Darwin is the detailed understanding of how Darwin's theory of evolution is related to the gene theory—of how changes in life's diversity result from changes in individual genes (figure 1.13).

> **Putting the Concept to Work**
> Of what domain are you a member?

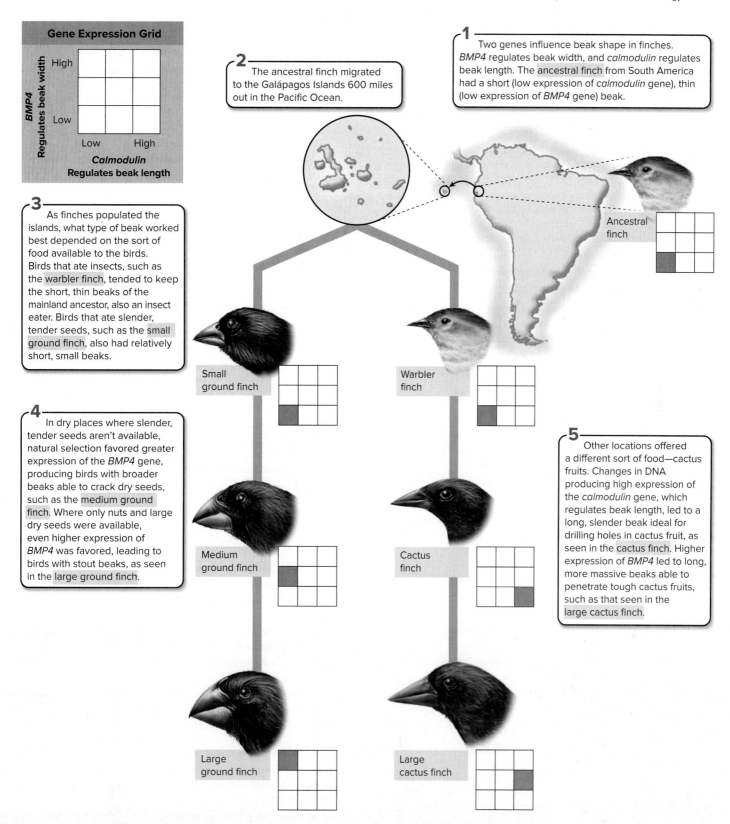

Figure 1.13 **The theory of evolution.**

Darwin's theory of evolution proposes that many forms of a gene may exist among members of a population and that those members with a form better suited to their particular habitat will tend to reproduce more successfully, and so their traits become more common in the population, a process Darwin dubbed "natural selection." Here you see how this process is thought to have worked on two pivotal genes that helped generate the diversity of finches on the Galápagos Islands, visited by Darwin in 1831 on his round-the-world voyage on **HMS** *Beagle.*

Putting the Chapter to Work

1 Mitochondria are compartments within cells that specialize in energy production.

Based on this description of mitochondria, at what level, in the five hierarchy levels of the cell, would you place mitochondria?

2 During a sleep apnea study, researchers have three groups of participants who are given different concentrations of medicine through an inhaler mask while they sleep. Because of previous studies on the medicine, the researchers know the amount of medicine administered can vary the results the participants will achieve. However, there is still the question of which amount achieves the best results. To determine the best administration dosage, one of the three groups will act as the control group.

Describe the amount of medicine administered to the control group through the participants' inhaler mask.

3 The cells that line your stomach have different characteristics than the cells that make up your skin.

Even though all cells have the same DNA, why do cells, like those described here, have different characteristics?

Retracing the Learning Path

Biology and the Living World

1.1 The Diversity of Life

1. Biology is the study of life. While the living world is astonishingly diverse, all living organisms share common characteristics. Organisms are categorized by biologists into six groups called kingdoms. The six kingdoms are Bacteria, Archaea, Protista, Fungi, Plantae, and Animalia.

1.2 Properties of Life

1. All living organisms share five basic properties:

   Cellular organization: All living organisms are composed of cells.
   Metabolism: All living organisms use energy.
   Homeostasis: All living organisms maintain stable internal conditions.
   Growth and reproduction: All living organisms grow in size and reproduce.
   Heredity: All living organisms possess genetic information in DNA that determines how each organism looks and functions, and this information is passed on to future generations.

1.3 The Organization of Life

1. Living organisms exhibit increasing levels of complexity within their cells (cellular level), within their bodies (organismal level), and within ecosystems (populational level).

2. Life's hierarchical organization is responsible for novel emergent properties that characterize the living world. These properties are the natural consequences of ever more complex structural organization.

1.4 Biological Themes

1. Five themes emerge from the study of biology: (1) evolution, (2) the flow of energy, (3) cooperation, (4) structure determines function, and (5) homeostasis.

The Scientific Process

1.5 Stages of a Scientific Investigation

1. In their studies, scientists systematically eliminate hypotheses that are not consistent with observation. Hypotheses are possible explanations that are used to form predictions. These predictions are tested experimentally. Some hypotheses are rejected on the basis of experimental results, whereas others are tentatively accepted.

   - Scientific investigations often use a series of stages, called the scientific process, to study a scientific question. These stages are observations, forming hypotheses, making predictions, testing, establishing controls, and drawing conclusions.

   - The discovery of the hole in the ozone required careful observations of data collected from the atmosphere. Based on these observations, scientists proposed an explanation of what caused a decrease in the levels of ozone over the Antarctic. This explanation is called a hypothesis. They then formed predictions and tested the hypothesis against controls. The hypothesis that CFCs released into the atmosphere were causing the breakdown of ozone to oxygen gas was supported by the data. Further experimentation allowed scientists to form the conclusion that CFCs were responsible for the loss of ozone over the Antarctic.

1.6 Theory and Certainty

1. Hypotheses that hold up to testing over time are combined into statements called theories. Theories carry a higher degree of certainty, although no theory in science is absolute.

   - The process of science was once viewed as a series of "either/or" predictions that were tested experimentally. Now, this discrete series of steps, which used to be referred to as the "scientific method," is often modified to take into account the role of judgment and intuition.

   - Science can only study what can be tested experimentally. A hypothesis can be established through science only if it can be tested and potentially disproven.

Core Ideas of Biology

1.7 Four Theories Unify Biology as a Science

1. The cell theory states that all living organisms are composed of cells, which grow and reproduce to form other cells.

2. The gene theory states that long molecules of DNA carry instructions for producing cellular components. These instructions are encoded in the nucleotide sequences in the strands of DNA. The nucleotides are organized into discrete units called genes, and the genes determine how an organism looks and functions.

3. The theory of heredity states that the genes of an organism are passed as discrete units from parent to offspring.

4. Organisms are organized into kingdoms based on similar characteristics. The organisms within a kingdom show similarities but exhibit differences from those of other kingdoms. The kingdoms are further organized into three major groups called domains based on their cellular characteristics. The three domains are Bacteria, Archaea, and Eukarya.

   - The theory of evolution states that modifications in genes that are passed from parent to offspring result in changes in future generations. Today's biological diversity is the product of a long evolutionary journey.

Inquiry and Analysis

Do "Stay-At-Home" Orders Slow the Spread of Coronavirus?

The novel coronavirus, SARS-CoV-2, was first detected in the United States on January 19, 2020. The spread of the virus was initially quite slow, with only 30 cases confirmed by March 1, but by March 10, the number of cases swelled to 937 and by March 20, the United States had 18,747 confirmed cases. No one then dreamed the number of cases would grow to over 30 million worldwide within the year.

Soon many were dying. Infection with the SARS-2 virus leads to a respiratory illness called COVID-19 (<u>co</u>rona <u>vi</u>rus <u>d</u>isease 20<u>19</u>). About a third of those infected don't get sick, but those that do have trouble breathing, and need hospitalization and supplemental oxygen. Two of every hundred die.

In these early months, New York State was the epicenter of SARS-CoV-2 infection, with 7,845 cases on March 20, 2020, and was desperate to stop the spread of the virus. Scientists did not yet know a lot about the SARS-CoV-2 coronavirus and how it caused the often-fatal COVID-19 disease, so public health officials simply recommended good hygiene, like washing your hands often and reducing facial touching, as the best way to slow the spread of the virus. However, these simple things were not enough to reduce the spread, and the cases continued to climb. On March 20, the Governor of New York signed an executive order entitled, **"New York State on PAUSE,"** that mandated a closure of all nonessential businesses and temporarily banned nonessential gatherings of groups of any size for any reason. While shutting nonessential businesses down is bad for the economy, is this sort of broad "stay-at-home" order an effective strategy to reduce the rampant spread of COVID-19?

To determine if "stay-at-home" orders are successful in slowing the spread of COVID-19, doctors and scientists examined the number of new cases diagnosed following the issuance of the Governor's order. The graph shows the number of COVID-19 cases (*y* axis) in New York (blue line) and the cumulative number of cases in the United States (red line) from March 8 through May 3 (*x* axis). COVID-19 has an incubation time of 1 to 14 days, with most cases developing four to six days following exposure.

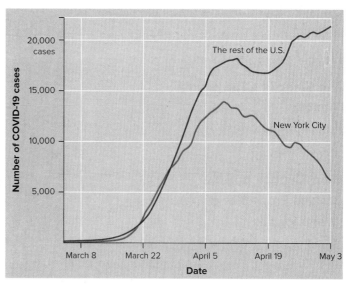

New York Times; https://www.nytimes.com/interactive/2020/05/06/opinion/coronavirus-deaths-statistics.html

Analysis

1. **Applying Concepts**
   Variable. In the graph, what is the dependent variable?
2. **Interpreting Data**
   a. On March 22, what was the total number of diagnosed COVID-19 cases in the entire United States?
   b. When was the peak of diagnosed cases in New York observed? How many cases were identified at that time?
3. **Making Inferences**
   a. From March 8 through the 22, what percentage of total cases in the United States were diagnosed in New York?
   b. If New York waited one more week before issuing the "stay-at-home" orders, how would that change the curve?
4. **Drawing Conclusions**
   a. Knowing that the incubation time for COVID-19 is 1 to 14 days, was the "stay-at-home" order issued by Governor Cuomo effective at reducing the spread of the virus?
5. **Further Analysis**
   a. Examine the data for the rest of the United States; why do you think there was a continued increase in the number of cases followed by a decrease around April 10?
   b. For the data for the rest of the United States, if you ignore the brief drop in cases, from April 10 through April 22, what do you observe about the trend in total cases? What do you think is responsible for this observation?

PART 2 The Living Cell

2 The Chemistry of Life

LEARNING PATH ▼

Some Simple Chemistry
1. Atoms
2. Ions and Isotopes
3. Molecules

Water: Cradle of Life
4. Unique Properties of Water
5. Water Ionizes

Death in the Meadow

Chemistry often plays a role in fiction, particularly detective stories involving deadly poisons such as cyanide. In real life, it is sometimes difficult to tell the difference between scientific investigation and good detective work. On a warm spring day a week before the Kentucky Derby, biologists were called in to solve a mystery that Sherlock Holmes would have enjoyed.

A Wave of Death

Our mystery starts in April in Kentucky, the Bluegrass State. At this time of year, the lush pastures of Kentucky's many horse farms become filled with thoroughbred mares and their leggy foals. Kentucky is the acknowledged center of thoroughbred horse breeding in the United States, a $1.2 billion industry. Normally, about 10,000 thoroughbred foals are born in Kentucky each year.

On this April day, in horse farms all over Kentucky, newly pregnant mares began to abort unborn fetuses in the first trimester of pregnancy. Foals also began to die unexpectedly soon after birth, some surviving only a few days. A certain number of stillborn or weak foals is to be expected, but not this wave of death.

There were few clues as to the cause of the problem. It seemed to affect all horse breeds and bloodlines and so

PRANCING IN ITS meadow, this Kentucky colt is threatened by a chemical enemy.
Herbert Spichtinger/Image Source

probably wasn't due to bad genes. A similar problem arose years ago in 1980–81, when many unexplained horse deaths occurred. That mystery has never been solved.

Veterinarians dubbed the illness causing hundreds of Kentucky mares to deliver stillborn or weak foals "mare reproductive loss syndrome."

How widespread was this syndrome? An alarming 497 of the year's foals had died by May. The number of lost fetuses was unknown, but much larger—roughly 1 out of 20 foals due to be born that spring had aborted or been stillborn. During May, the University of Kentucky Diagnostic Laboratory was receiving more than 20 dead foals/fetuses a day. Financial losses were estimated to be at least $225 million. By the end of the summer, the wave of foal death subsided.

Search for a Cause

Breeders, after eliminating heredity and veterinary procedures as possible causes, focused on pastures. The

Kevin R. Morris/Corbis/VCG/Getty Images

Kevin R. Morris/Corbis/VCG/Getty Images

weather that spring in Kentucky had been abnormal, with warm weather followed by a hard freeze and unusually dry conditions. This suggested a working hypothesis: Perhaps the unusual weather had encouraged mold to grow on the grass the horses had been eating. Mold can produce powerful mycotoxins, poisons that could easily harm the reproduction of animals eating the toxins.

Testing this hypothesis, University of Kentucky Agriculture Department investigators looked for mold and mycotoxins on the grass of pastures where foals were dying. They didn't find any. The mycotoxin hypothesis was not supported.

The investigators did find something else, however—something unexpected: cyanide.

The Culprit: A Deadly Poison

Cyanide is a deadly poison to all vertebrates (animals with backbones), as it interferes with metabolism. The low cyanide concentrations found in Kentucky pastures were not enough to kill an adult horse, but were high enough to kill fetuses and weaken foals.

How did cyanide get onto Kentucky's pastures? Sherlock Holmes would have loved the answer: cherry trees and caterpillar feces! Here is the train of events that transpired, as the University of Kentucky biologists pieced it together. Black cherry trees are common all over Kentucky, and their leaves naturally contain substantial amounts of cyanide. Analyzing the leaves, the scientists discovered that the warm, dry weather followed by a frost had caused the cyanide levels in the tree leaves to spike sky high, far exceeding the usual high levels. Seedlings had even more. Eastern tent caterpillars eat the leaves of black cherry trees, unaffected by the cyanide, which does not harm insect metabolism. They are voracious eaters and can defoliate a tree in days, eating every leaf. Agronomists had reported a heavy infestation of Eastern tent caterpillars in Kentucky that spring. Laden with cyanide from the leaves they had eaten, the caterpillars left cyanide-laced feces on the surrounding grass. Even worse, small black cherry seedlings grew among the grass of pastures, exposing grazing horses to the unusually high levels of cyanide they contained that spring. The pasture poisoned the mares who ate it and killed their foals. The spike in cyanide levels in black cherry leaves and the caterpillar infestation were both over by summer's end, and so was the problem. Case solved.

Some Simple Chemistry

Figure 2.1 Replacing electrolytes.

During extreme exercise, athletes will often consume drinks that contain electrolytes, chemicals such as calcium, potassium, and sodium that play an important role in muscle contraction.

Ron Chapple/Thinkstock Images/Getty Images

2.1 Atoms

> **LEARNING OBJECTIVE 2.1.1** Describe the basic structure of an atom in terms of three subatomic particles.

Biology is the science of life, and all life, in fact even all nonlife, is made of substances. Chemistry is the study of the properties of these substances. So, while it may seem tedious or unrelated to examine chemistry in a biology text, it is essential. Organisms are chemical machines (figure 2.1), and to understand them, we must learn a little chemistry.

Any substance in the universe that has mass and occupies space is defined as matter. All matter is composed of extremely small particles called **atoms**. An atom is the smallest particle into which a substance can be divided and still retain its chemical properties.

Subatomic Particles

Every atom has the same basic structure you see in figure 2.2. At the center of every atom is a small, very dense nucleus formed of two types of subatomic particles, **protons** (illustrated by purple balls) and **neutrons** (the pink balls in the figure). Whizzing around the nucleus is an orbiting cloud of a third kind of subatomic particle, the **electron** (depicted by yellow balls on concentric rings). Neutrons have no electrical charge, whereas protons have a positive charge and electrons have a negative one. In a typical atom, there is an orbiting electron for every proton in the nucleus. The electron's negative charge balances the proton's positive charge so that the atom is electrically neutral.

An atom is typically described by the number of protons in its nucleus or by the overall mass of the atom. The terms *mass* and *weight* are often used interchangeably, but they have slightly different meanings. Mass refers to the amount of a substance, whereas weight refers to the force gravity exerts on a substance. Hence, an object has the same mass whether it is on the earth or the moon, but its weight will be greater on the earth, because the earth's gravitational force is greater than the moon's. For example, an astronaut weighing 180 pounds on earth will weigh about 30 pounds on the moon. He didn't lose any significant mass during his flight to the moon; there is just less gravitational pull on his mass.

Elements

The number of protons in the nucleus of an atom is called the **atomic number**. For example, the atomic number of carbon is 6 because it has six protons. Atoms with the same atomic number (that is, the same number of protons) have the same chemical properties and are said to belong to the same **element**. Formally speaking, an element is any substance that cannot be broken down into any other substance by ordinary chemical means.

Neutrons are similar to protons in mass, and the number of protons and neutrons in the nucleus of an atom is called the **mass number**. A carbon atom that has six protons and six neutrons has a mass number of 12. An electron's contribution to the overall mass of an atom is negligible. The atomic numbers and mass numbers of some of the most common elements in living organisms are shown in table 2.1.

> **Putting the Concept to Work**
> Each of the elements in table 2.1 has a mass number that is about double its atomic number. Why?

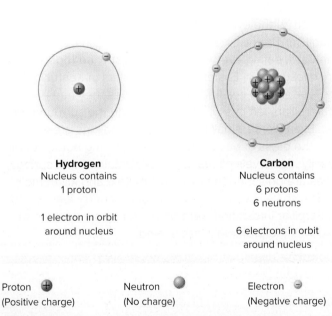

Figure 2.2 Basic structure of an atom.

All atoms have a nucleus consisting of protons and neutrons, except hydrogen, the smallest atom, which has only one proton and no neutrons in its nucleus. Electrons spin around the nucleus in orbitals a far distance away from the nucleus.

Electrons Determine What Atoms Are Like

LEARNING OBJECTIVE 2.1.2 Explain why electrons determine the chemical behavior of atoms.

Electrons have very little mass (only about 1/1,840 the mass of a proton). Of all the mass contributing to your weight, the portion that is contributed by electrons is less than the mass of your eyelashes. And yet electrons determine the chemical behavior of atoms because they are the parts of atoms that come close enough to each other in nature to interact. Almost all the volume of an atom is empty space. Protons and neutrons lie at the core of this space, while orbiting electrons are very far from the nucleus. If the nucleus of an atom were the size of an apple, the orbit of the nearest electron would be more than a mile out!

Putting the Concept to Work
How much of the mass of the earth (6×10^{24} kg) is electrons?

Electrons Carry Energy

LEARNING OBJECTIVE 2.1.3 Explain how electrons carry energy.

Because electrons are negatively charged, they are attracted to the positively charged nucleus, but they also repel each other's negative charge. It takes work to keep them in orbit, just as it takes work to hold an apple in your hand when gravity is pulling the apple down toward the ground. The apple in your hand is said to possess **energy**, the ability to do work, because of its position—if you were to release it, the apple would fall. Similarly, electrons have energy of position, called *potential energy*. It takes work to oppose the attraction of the nucleus, so moving the electron farther out away from the nucleus requires an input of energy and results in an electron with greater potential energy. Cells use the potential energy of atoms to drive chemical reactions.

> Potential energy is discussed in more detail in chapter 5, and chapters 6 and 7 discuss how the potential energy from electrons is used in biological systems.

The volume of space around a nucleus where an electron is most likely to be found is called the *orbital* of that electron. Each energy level of an atom, called an *electron shell*, has a specific number of orbitals, and each orbital can hold up to two electrons. The first shell in any atom contains one orbital. Helium, shown in figure 2.3a, has one electron shell with one orbital that corresponds to the lowest energy level. The orbital contains two electrons, shown above and below the nucleus. In atoms with more than one electron shell, the second shell contains four orbitals and holds up to eight electrons. Nitrogen, shown in figure 2.3b, has two electron shells; the first one is completely filled with two electrons, but three of the four orbitals in the second electron shell are not filled because nitrogen's second shell contains only five electrons (openings in orbitals are indicated with dotted circles). In atoms with more than two electron shells, subsequent shells also contain up to four orbitals with a maximum of eight electrons. Atoms with unfilled electron orbitals tend to be more reactive because they lose, gain, or share electrons in order to fill their outermost electron shell. An atom with a completely filled outermost shell is more stable.

Putting the Concept to Work
Why don't electrons, which carry a negative charge, simply crash into the positively charged nucleus?

TABLE 2.1	Elements Common in Living Organisms		
Element	Symbol	Atomic Number	Mass Number
Hydrogen	H	1	1.008
Carbon	C	6	12.011
Nitrogen	N	7	14.007
Oxygen	O	8	15.999
Sodium	Na	11	22.989
Phosphorus	P	15	30.974
Sulfur	S	16	32.064
Chlorine	Cl	17	35.453
Potassium	K	19	39.098
Calcium	Ca	20	40.080
Iron	Fe	26	55.847

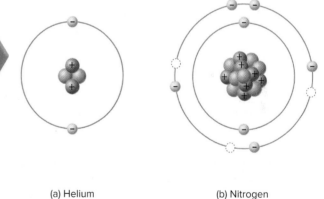

(a) Helium   (b) Nitrogen

Figure 2.3 Electrons in electron shells.
(a) An atom of helium has two protons, two neutrons, and two electrons. The electrons fill the one orbital in its one electron shell, the lowest energy level. (b) An atom of nitrogen has seven protons, seven neutrons, and seven electrons. Two electrons fill the orbital in the innermost electron shell, and five electrons occupy orbitals in the second electron shell (the second energy level). The orbitals in the second electron shell can hold up to eight electrons; therefore, there are three vacancies in the outer electron shell of a nitrogen atom.

IMPLICATION FOR YOU Helium gas is often used to inflate party balloons, and if you release these balloons, they rise up into the air. If you fill the party balloons with nitrogen gas, they don't rise. Why this difference in party balloon behavior?

2.2 Ions and Isotopes

Not all atoms of an element are the same. They sometimes differ in the number of their electrons or neutrons.

Ions

> **LEARNING OBJECTIVE 2.2.1** Differentiate between a cation and an anion.

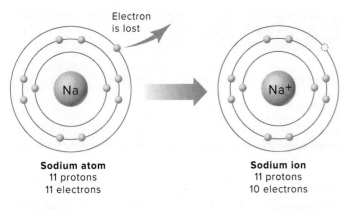

Figure 2.4 Making a sodium ion.

An electrically neutral sodium atom has 11 protons and 11 electrons. Sodium ions bear a positive charge when they ionize and lose one electron. Sodium ions have 11 protons and only 10 electrons.

Sometimes an atom may gain or lose an electron from its outer shell (we will look at why this happens in the next section). Atoms in which the number of electrons does not equal the number of protons because they have gained or lost one or more electrons are called **ions**. All ions are electrically charged. For example, an atom of sodium (on the left in **figure 2.4**) becomes a positively charged ion (Na^+), called a *cation*, when it loses an electron (on the right); one proton in the nucleus is left with an unbalanced charge (11 positively charged protons and only 10 negatively charged electrons). Negatively charged ions, called *anions*, form when an atom gains one or more electrons from another atom.

> **Putting the Concept to Work**
> Does table salt contain more cations or anions?

Isotopes

> **LEARNING OBJECTIVE 2.2.2** Differentiate between an ion and an isotope.

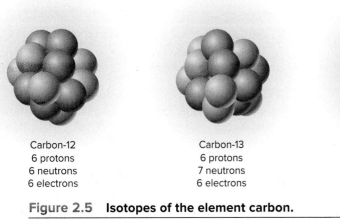

Figure 2.5 Isotopes of the element carbon.

The three most abundant isotopes of carbon are carbon-12, carbon-13, and carbon-14. The yellow "clouds" in the diagrams represent the orbiting electrons, whose numbers are the same for all three isotopes. Protons are shown in purple, and neutrons are shown in pink.

The number of neutrons in an atom of a particular element can vary without changing the chemical properties of the element. Atoms that have the same number of protons and electrons, but different number of neutrons, are called **isotopes**. Isotopes of an atom have the same atomic number but differ in their mass number. Most elements in nature exist as mixtures of different isotopes. For example, there are three isotopes of the element carbon, all of which possess six protons (the purple balls in **figure 2.5**). The most common isotope of carbon (99% of all carbon) has six neutrons (the pink balls). Because its mass number is 12 (six protons plus six neutrons), it is referred to as carbon-12 (on the left). The isotope carbon-14 (on the right) is rare (1 in 1 trillion atoms of carbon) and unstable, such that its nucleus tends to break up into particles with lower atomic numbers, a process called **radioactive decay**. Radioactive isotopes are used in dating fossils, as discussed in the "Inquiry and Analysis" feature at the end of the chapter, and in medicine.

Medical Uses of Radioactive Isotopes. Radioactive isotopes are used in many medical procedures. Short-lived isotopes, those that decay fairly rapidly and produce harmless products, are commonly used as *tracers,* a radioactive substance that is taken up and used by the body. Emissions from the radioactive isotope tracer are detected using special laboratory equipment and can reveal key diagnostic information about the functioning of the body. For example, PET and PET/CT (positron emission tomography/computerized tomography) imaging procedures can be used to identify a cancerous area in the body (**figure 2.6**).

Figure 2.6 Using a radioactive tracer to identify cancer.

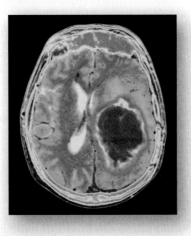

This patient has been injected intravenously with a radioactive tracer that is absorbed in greater amounts by cancer cells. The tracer emits radioactivity that is detected using PET and PET/CT equipment. Cancerous areas in the neck are seen as bright red glowing areas.

ALFRED PASIEKA/SCIENCE PHOTO LIBRARY/Getty Images

> **Putting the Concept to Work**
> Why do the three abundant isotopes of carbon all form the same types of chemical bonds?

2.3 Molecules

Molecules are collections of atoms linked together by chemical bonds.

Nature of the Chemical Bond

> **LEARNING OBJECTIVE 2.3.1** Define a chemical bond, and describe the three principal kinds.

A **molecule** is a group of atoms held together by energy. The energy acts as "glue," ensuring that the various atoms stick to one another. The energy or force holding two atoms together is called a **chemical bond.** There are three principal kinds of chemical bonds: ionic bonds, where the force is generated by the attraction of oppositely charged ions; covalent bonds, where the force results from the sharing of electrons; and hydrogen bonds, where the force is generated by the attraction of opposite partial electrical charges.

> **Putting the Concept to Work**
> Can you name an atom that is *not* part of a molecule?

Ionic Bonds

> **LEARNING OBJECTIVE 2.3.2** Explain how ionic bonds promote crystal formation.

Chemical bonds called **ionic bonds** form when atoms are attracted to each other by opposite electrical charges. Just as the positive pole of a magnet is attracted to the negative pole of another, so an atom can form a strong link with another atom if they have opposite electrical charges. Because an atom with an electrical charge is an ion, these bonds are called ionic bonds.

Everyday table salt is built of ionic bonds. The sodium and chlorine atoms that make up table salt are ions. The sodium you see in the yellow panels of figure 2.7a gives up the sole electron in its outermost shell (leaving a filled outer shell), and chlorine, in the light green panels, gains an electron to complete its outermost shell. Recall from section 2.1 that an atom is more stable when its outermost electron shell is filled. As a result of this electron hopping, sodium atoms in table salt are positive sodium ions and chlorine atoms are negative chloride ions. Because each ion is electrically attracted to all surrounding ions of opposite charge, they form an elaborate matrix of ionic bonds between alternating sodium and chloride ions—a crystal (figure 2.7b). That is why table salt is composed of tiny crystals and is not a powder.

The two key properties of ionic bonds that make them form crystals are that they are strong (although not as strong as covalent bonds) and that they are *not* directional. An ion is attracted to the electrical field contributed by all nearby ions of opposite charge. Ionic bonds do not play an important part in most biological molecules because of this lack of directionality. Biological molecules require more specific associations made possible by directional bonds.

> **Putting the Concept to Work**
> In figure 2.7b, how many sodium ions interact with an individual chloride ion?

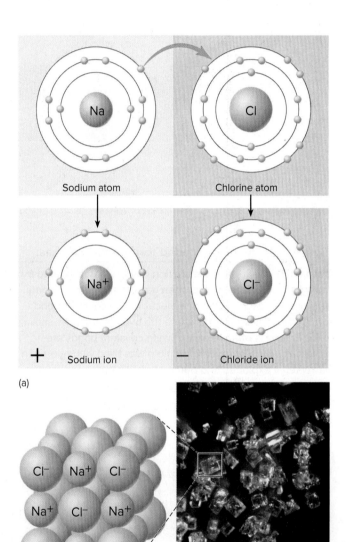

Figure 2.7 The formation of ionic bonds in table salt.

(a) When a sodium atom donates an electron to a chlorine atom, the sodium atom, lacking that electron, becomes a positively charged sodium ion. The chlorine atom, having gained an extra electron, becomes a negatively charged chloride ion. (b) Sodium chloride forms a highly regular lattice of alternating sodium ions and chloride ions. You are familiar with these crystals as everyday table salt.

(b) Evelyn Jo Johnson/McGraw Hill

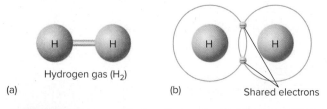

Figure 2.8 A covalent bond forms hydrogen gas.
Covalent bonds involve the sharing of electrons, indicated by the light blue bar in (a). Hydrogen gas consists of two atoms of hydrogen, each having one electron. The covalent bond forms when the two atoms share the two electrons, as shown in (b). Energy is often released when covalent bonds are broken. (c) The *Hindenberg* dirigible was filled with hydrogen gas when it exploded and burned in 1937; the energy of the inferno came from the breaking of H_2 covalent bonds.

(c) Bettmann/Getty Images

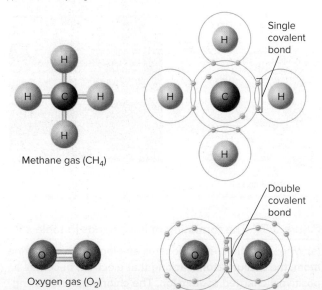

Figure 2.9 Single and double covalent bonds.
Single covalent bonds involve the sharing of one pair of electrons, as in methane (CH_4). In double covalent bonds, two pairs of electrons are shared, as in oxygen gas.

Covalent Bonds

> **LEARNING OBJECTIVE 2.3.3** Distinguish between polar and nonpolar covalent bonds.

Strong chemical bonds called **covalent bonds** form between two atoms when they share electrons (figure 2.8). Most of the atoms in your body are linked to other atoms by covalent bonds. Why do atoms in molecules share electrons? Remember, all atoms seek to fill up their outermost shell of orbiting electrons, which in all atoms (except tiny hydrogen and helium) takes eight electrons. For example, an atom with six outer shell electrons seeks to share them with an atom that has two outer shell electrons or with two atoms that have single outer shell electrons. The carbon atom has four electrons in its outermost shell, and so carbon can form as many as four covalent bonds in its attempt to fully populate its outermost shell of electrons. Because there are many ways that four covalent bonds can form, carbon atoms participate in many different kinds of molecules. The strength of covalent bonds increases as more electrons are shared. If only one pair of electrons is shared between two atoms, the covalent bond is called a *single covalent bond*. If two pairs of electrons are shared, it is a stronger bond, called a *double covalent bond,* and if three pairs of electrons are shared, it is a very strong *triple covalent bond*.

> The atoms in nitrogen gas (N_2) are held together by a very strong triple covalent bond. As described in section 20.5, only a few kinds of bacteria are able to break this bond and make the nitrogen available to other organisms.

The two key properties of covalent bonds that make them ideal for their molecule-building role in living systems are that (1) they are strong, involving the sharing of lots of energy and (2) they are very directional—allowing bonds to form between two specific atoms, rather than creating a generalized attraction of one atom for its neighbors. For example, compare the covalent bonds of the methane molecule in figure 2.9 with the nondirectional ionic bonds of the NaCl crystal in figure 2.7.

Polar Covalent Bonds

When a covalent bond forms between two atoms, one nucleus may be much better at attracting the shared electrons than the other. In water, for example, the shared electrons are much more strongly attracted to the oxygen atom than to the hydrogen atoms. When this happens, shared electrons spend more time in the vicinity of the oxygen atom, making it somewhat negative in charge; they spend less time in the vicinity of the hydrogens, and these become somewhat positive in charge. The charges are not full electrical charges, as in ions, but rather tiny *partial charges*. What you end up with is a sort of molecular magnet, with positive and negative ends, or "poles." Molecules like this are said to be **polar molecules** (see figure 2.10).

> **Putting the Concept to Work**
> Why don't covalent bonds promote crystal formation?

Hydrogen Bonds

> **LEARNING OBJECTIVE 2.3.4** Predict which molecules will form hydrogen bonds with each other.

Polar molecules like water are attracted to one another through a special type of weak chemical bond called a **hydrogen bond**. Hydrogen bonds occur when the positive end of one polar molecule is attracted to the negative end of another, like two magnets drawn to each other. In a hydrogen bond, an electropositive hydrogen from one polar molecule is attracted to an electronegative atom, often oxygen (O) or nitrogen (N), from another polar molecule.

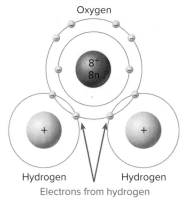

(a) Electron shells in a water molecule

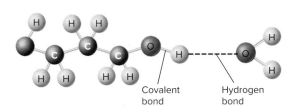

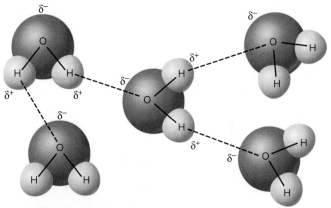

(b) Hydrogen bonding in water molecules

Water is very polar, because the oxygen atoms in water molecules are more electronegative (attract electrons more strongly) than the hydrogen atoms (**figure 2.10**). In a water molecule, each oxygen has a partial negative charge (δ^-) and each hydrogen has a partial positive charge (δ^+). Hydrogen bonds (shown as dashed lines in **figure 2.10**) form between the positive end of one polar water molecule and the negative end of another polar water molecule.

Two key properties of hydrogen bonds cause them to play an important role in the molecules found in organisms. First, hydrogen bonds are highly directional. Second, they are weak and so are not effective over long distances like more powerful covalent and ionic bonds. Too weak to actually form stable molecules, they act more like Velcro, forming a tight bond by the additive effects of many weak interactions.

> **Putting the Concept to Work**
>
> Would you expect a sodium ion to form hydrogen bonds with surrounding water molecules?

Figure 2.10 Hydrogen bonding in water molecules.

(a) A water molecule is composed of one oxygen atom (red) and two hydrogen atoms (blue). (b) Because electrons are more attracted to oxygen atoms than to hydrogen atoms, the shared electrons spend more time near the oxygen atom, giving it a partial negative charge (δ^-) and each hydrogen atom a partial positive charge (δ^+). Hydrogen bonds (dashed lines) form between the partial charges. (c) Water molecules form strong hydrogen bonds with each other, giving liquid water many unique properties.

(c) Digital Archive Japan/Alamy Stock Photo

Water: Cradle of Life

TABLE 2.2	The Properties of Water
Property	Explanation
Heat storage	Hydrogen bonds require considerable heat before they break, minimizing temperature changes.
Ice formation	Water molecules in an ice crystal are spaced relatively far apart because of hydrogen bonding.
High heat of vaporization	Many hydrogen bonds must be broken for water to evaporate, which requires considerable heat.
Cohesion	Hydrogen bonds hold molecules of water together.
High polarity	Water molecules are attracted to ions and other polar compounds.

2.4 Unique Properties of Water

Three-fourths of the earth's surface is covered by liquid water. About two-thirds of your body is water, and you cannot exist long without it. All other organisms also require water. The chemistry of life, then, is water chemistry.

Water's ability to form hydrogen bonds is responsible for much of the organization of living chemistry, from the structure of membranes that encase every cell to how large molecules fold. The weak hydrogen bonds that form between a hydrogen atom of one water molecule and the oxygen atom of another produce a lattice of hydrogen bonds within liquid water. Each of these bonds is individually very weak and short-lived—a single bond lasts only 1/100,000,000,000 of a second. However, the cumulative effect of large numbers of these bonds is enormous and is responsible for many of the important physical properties of water (table 2.2).

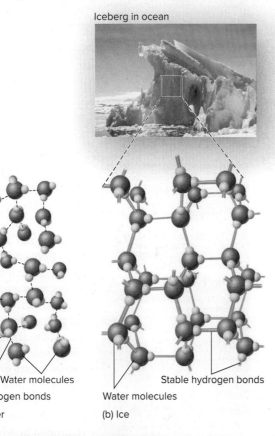

Water Stores Heat

LEARNING OBJECTIVE 2.4.1 Explain why water heats up so slowly.

The temperature of any substance is a measure of how rapidly its individual molecules are moving. Due to the many hydrogen bonds that water molecules form with one another, a large input of thermal energy is required to disrupt the organization of liquid water and raise its temperature. Because of this, water heats up more slowly than almost any other compound and holds its temperature longer. That is a major reason why your body, which is mostly water, is able to maintain a relatively constant internal temperature.

Putting the Concept to Work
Would adding table salt to a pan of water cause the water to come to a boil faster on a stove?

Water Forms Ice

LEARNING OBJECTIVE 2.4.2 Explain why ice floats.

If the temperature is low enough, very few hydrogen bonds between water molecules will break. Instead, the lattice of these bonds assumes a crystal-like structure, forming a solid we call ice. Interestingly, ice is less dense than water—that is why icebergs and ice cubes float. Why is ice less dense? This is best understood by comparing the molecular structures of liquid water and ice that you see in figure 2.11. At temperatures above freezing (0°C or 32°F), water molecules in figure 2.11a move around each other with hydrogen bonds breaking and forming. As temperatures drop, the movement of water molecules decreases, allowing hydrogen bonds to stabilize and hold individual molecules farther apart, as in figure 2.11b. This stabilized lattice makes the ice structure less dense.

Figure 2.11 Ice formation.
When water (a) cools below 0°C, it forms a regular crystal structure (b) that floats. The individual water molecules are spaced apart and held in position by hydrogen bonds.
(c) Kelly Cheng/Moment/Getty Images

IMPLICATION FOR YOU Will the melting of the polar ice cap (ice floating in the ocean) due to global warming raise sea level? [Hint: If you leave a glass of ice water on the table, does the level of water in the glass change as the ice melts?]

Putting the Concept to Work
Would adding table salt to a bottle of water cause the water to freeze to ice faster when you place the bottle in the freezer?

Water Evaporates

LEARNING OBJECTIVE 2.4.3 Explain why sweating cools you.

If the temperature is high enough, many hydrogen bonds between water molecules will break, with the result that the liquid is changed into vapor (a gas). A considerable amount of heat energy is required to do this—every gram of water that evaporates from your skin removes 2,452 joules of heat from your body, which is equal to the energy released by lowering the temperature of 586 grams of water 1°C (which is quite a lot of heat). That is why sweating cools you off; as the sweat evaporates (vaporizes), it takes energy with it, in the form of heat, cooling the body.

> **Putting the Concept to Work**
> Does an athlete wearing a sweat suit cool off as quickly as one wearing only shorts?

Water Clings to Polar Molecules

LEARNING OBJECTIVE 2.4.4 Distinguish cohesion from adhesion.

Because water molecules are very polar, they are attracted to other polar molecules—hydrogen bonds bind polar molecules to each other. When the other polar molecule is another water molecule, the attraction is called **cohesion**. The surface tension of water is created by cohesion. Surface tension is the force that causes water to bead or supports the weight of an insect (figure 2.12). When the other polar molecule is a different substance, the attraction is called **adhesion**. Water clings to any substance with which it can form hydrogen bonds. Adhesion is why things get "wet" when they are dipped in water and why waxy substances do not—they are composed of nonpolar molecules that don't form hydrogen bonds with water molecules.

> Cohesion and adhesion are properties of water that are necessary for the movement of water in plants from the roots to the leaves, as described in section 31.6.

> **Putting the Concept to Work**
> Explain why an insect can walk on water, whereas you cannot.

Water Is Very Polar

LEARNING OBJECTIVE 2.4.5 Explain why oil will not dissolve in water.

Water molecules in solution tend to form the maximum number of hydrogen bonds possible and gather around polar molecules or molecules with an electrical charge. For example, when a salt crystal dissolves in water, what really happens is that individual ions break off from the crystal and become surrounded by water molecules (figure 2.13). Water molecules orient around each ion and form a *hydration shell*, preventing the ions from reassociating with the crystal. Polar molecules that dissolve in water in this way are said to be **soluble** in water. In contrast, when nonpolar molecules are placed in water, water molecules shy away. The nonpolar molecules are forced into association with one another and are referred to as **hydrophobic** (Greek *hydros*, water, and *phobos*, fearing). Polar molecules, on the other hand, are called **hydrophilic** (Greek *hydros*, water, and *philic*, loving).

> **Putting the Concept to Work**
> What molecules would you expect to dissolve in vegetable oil?

Figure 2.12 Cohesion.

(a) Cohesion allows water molecules to stick together and form droplets. (b) Surface tension is a property derived from cohesion—that is, water has a "strong" surface due to the force of its hydrogen bonds. Some insects, such as this water strider, literally walk on water.

(a) Tony Sweet/Digital Vision/Getty Images; (b) Jan Miko/Shutterstock

> **IMPLICATION FOR YOU** If you were to devise very large footpads made of lightweight film, would you be able to walk on water like this water strider? Why would you have to wear footpads and not just bare feet?

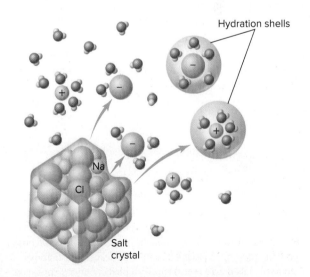

Figure 2.13 How salt dissolves in water.

Salt is soluble in water because the partial charges on water molecules are attracted to the charged sodium and chloride ions. The water molecules surround the ions, forming what are called hydration shells. When all of the ions have been separated from the crystal, the salt is said to be dissolved.

2.5 Water Ionizes

LEARNING OBJECTIVE 2.5.1 Predict the change in hydrogen ion concentration represented by a difference of 1 on the pH scale.

One of the most important properties of water arises from the ease with which its chemical bonds break.

Ionization

The covalent bonds within a water molecule sometimes break spontaneously. When it happens, a proton (hydrogen atom nucleus) dissociates from the molecule as a positively charged ion, *hydrogen ion* (H^+). The rest of the dissociated water molecule, which has retained the shared electron from the covalent bond, is a negatively charged *hydroxide ion* (OH^-).

$$H_2O \leftrightarrow OH^- + H^+$$
$$\text{water} \quad \text{hydroxide ion} \quad \text{hydrogen ion}$$

pH

A convenient way to express the hydrogen ion concentration of a solution is to use the **pH scale** (figure 2.14). This scale ranges from 0 (highest hydrogen ion concentration) to 14 (lowest hydrogen ion concentration). Pure water has a pH of 7. Each pH unit represents a 10-fold change in hydrogen ion concentration. This means that a solution with a pH of 4 has *10 times* the H^+ concentration of a solution with a pH of 5 and *100 times* the H^+ concentration of a solution with a pH of 6.

Acids. Any substance that dissociates in water to increase the concentration of H^+ is called an acid. Acidic solutions have pH values below 7. The stronger an acid, the more H^+ and so the lower its pH. For example, hydrochloric acid (HCl), which is abundant in your stomach, ionizes completely in water, giving the solution a pH of 1.

Bases. A substance that combines with H^+ when dissolved in water is called a base. By combining with H^+, a base lowers the H^+ concentration in the solution. Basic (or alkaline) solutions, therefore, have pH values above 7. Very strong bases, such as sodium hydroxide (NaOH), have pH values of 12 or more.

Buffers

The pH inside almost all living cells, and of the fluid surrounding cells in multicellular organisms, is fairly close to 7. The many proteins that govern metabolism are all extremely sensitive to pH, and slight alterations in pH can cause the molecules to take on different shapes that disrupt their activities. For this reason, it is important that a cell maintain a constant pH level. The pH of your blood, for example, is 7.4, and you would survive only a few minutes if it were to fall to 7.0 or rise to 7.8.

What keeps an organism's pH constant? Cells contain chemical substances called **buffers** that minimize changes in concentrations of H^+ and OH^- by taking up or releasing hydrogen ions into solution as the hydrogen ion concentration of the solution changes. As the "Today's Biology" feature *Acid Rain* explains, when acid in rain or snow exceeds the buffering capacity of a tree or other organism, death may result.

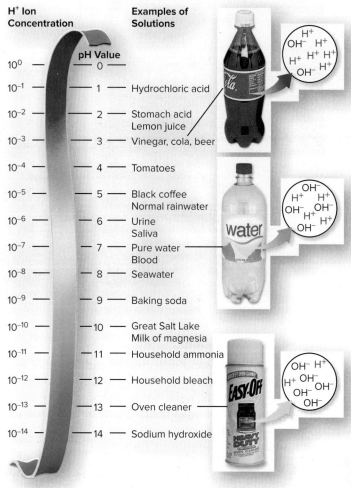

Figure 2.14 The pH scale.

A fluid is assigned a value according to the number of hydrogen ions present in a liter of that fluid. The scale is logarithmic, so that a change of only 1 means a 10-fold change in the concentration of hydrogen ions; thus lemon juice with a pH of 2 is 100 times more acidic than tomatoes with a pH of 4, and seawater is 10 times more basic than pure water.

Top: Bob Coyle/McGraw Hill; **Middle:** Jacques Cornell/McGraw Hill; **Bottom:** Jill Braaten/McGraw Hill

BIOLOGY & YOU

Heartburn. Your stomach contains large amounts of hydrochloric acid, used to digest food. Sometimes this acid backs up from the stomach into the esophagus (the food pipe) stretching up from the stomach to the throat. This escape of acid from the stomach, called acid reflux, causes a painful burning sensation as the acid attacks the inner lining of the esophagus. Because the esophagus lies just behind the heart, the burning sensation is informally referred to as "heartburn." Nearly one-third of the adult population of the United States experiences acid reflux to some degree at least once a month. For minor heartburn, you might take an antacid, such as TUMS, which is a base that counteracts stomach acidity.

Putting the Concept to Work

When you drink a cola, you are consuming an acid. Why doesn't your body's pH go down as a result?

Today's Biology

Acid Rain

As you study biology, you will learn that hydrogen ions play many roles in the chemistry of life. When conditions become overly acidic—too many hydrogen ions—serious damage to organisms often results. One important example of this is acid precipitation, more informally called **acid rain.**

The Problem

Acid precipitation is just what it sounds like, the presence of acid in rain or snow. Where does the acid come from? Coal-burning power plants send smoke high into the atmosphere through tall smokestacks, each of which is over 65 meters high. The smoke the stacks belch out contains high concentrations of sulfur dioxide (SO_2) because the coal that the plants burn is rich in sulfur. The sulfur-rich smoke is dispersed and diluted by winds and air currents. Since the 1950s, such tall stacks have become popular in the United States and Europe—there are now over 800 of them in the United States alone. Eighty are very tall indeed, with stacks extending 700 feet up.

In the 1970s, 20 years after the stacks were introduced, ecologists began to report evidence that the tall stacks were not eliminating the problems associated with the sulfur, just exporting the ill effects elsewhere. The lakes and forests of the Northeast suffered drastic drops in biodiversity, with forests dying and lakes becoming devoid of life. It turned out that the SO_2 introduced into the upper atmosphere by high smokestacks combines with water vapor to produce sulfuric acid (H_2SO_4). When this water later falls back to earth as rain or snow, it carries the sulfuric acid with it. When schoolchildren measured the pH of natural rainwater as part of a nationwide project, rain and snow in the Northeast often had a pH as low as 2 or 3—more acidic than vinegar.

The Impact

After accumulating in soils for over 60 years, the effects of acid rain are now only too evident. The impact of acid rain on forests first became apparent in the Northeast. Some 15% of the lakes in New England have become chronically acidic and are dying biologically as their pH levels fall to below 5.0. Many of the forests of the northeastern United States and Canada have also been seriously damaged. The trees in this photo show the ill effects of acid precipitation. In the last decades, acid added to forest soils has caused the loss from these soils of over half the essential plant nutrients calcium and magnesium. Researchers blame excess acids for dissolving Ca^{++} and Mg^{++} ions into drainage waters much faster than weathering rocks can replenish them. Without them, trees stop growing and die. Serious damage is also done to the mycorrhizae, fungi growing within the cells of the tree roots. Trees need mycorrhizae in order to extract Ca^{++} and Mg^{++} ions from the soil.

Yujnyj/Shutterstock

Acid rain effects became apparent in the Southeast decades later than in New England. Researchers suggest the reason for the delay is that southern soils are generally thicker than northern ones and thus able to sponge up far more acid. But now that southern forest soils are becoming saturated, they too are beginning to die. In a third of the southeastern streams studied, fish are declining or already gone.

The Solution

The solution is straightforward: capture and remove the emissions instead of releasing them into the atmosphere. Progressively tougher pollution laws over the past four decades have reduced U.S. emissions of sulfur dioxide by about 40% from its 1973 peak of 28.8 metric tons a year. Despite this significant progress, much remains to be done. Researchers predict that unless levels are cut further, forests may not recover for centuries.

An informed public will be essential. While textbook treatments have in the past tended to minimize the impact of this issue on students ("the vast majority of North American forests are not suffering substantially from acid precipitation"), it is important that we face the issue squarely and support continued efforts to address this serious problem.

Putting the Chapter to Work

1 To demonstrate your understanding of the mass of an atom, calculate the mass number of an isotope that contains 6 protons, 8 neutrons, and 6 electrons.

2 Water contains polar covalent bonds that result in "poles" forming around the oxygen and hydrogen atoms. When HCl ($H^+ + Cl^- =$ HCl) is dissolved in water, the ions associate around the poles of the oxygen and hydrogens and form hydrogen bonds.
In a mixture of water and HCl, with which atoms of water would the H^+ of HCl form hydrogen bonds? The Cl^- ion?

3 If you were to heat water in a pot on the stove, the transfer of heat from the stove coils to the water may, if hot enough, cause steam to rise from the pot.
Which type of chemical bond is breaking to release this steam?

Retracing the Learning Path

Some Simple Chemistry

2.1 Atoms

1. An atom is the smallest particle that retains the chemical properties of its substance. Atoms contain a core nucleus of protons and neutrons; electrons spin around the nucleus. The number of electrons equals the number of protons in a typical atom. The number of protons in an atom is called its atomic number.

2. Protons are positively charged particles, and neutron particles carry no charge. Electrons are negatively charged particles that orbit around the nucleus at different energy levels. Electrons determine the chemical behavior of an atom because they are the subatomic particles that interact with other atoms.

3. It takes energy to hold the electrons in their orbits; this energy of position is called potential energy. The amount of potential energy of an electron is based on its distance from the nucleus. Most electron shells hold up to eight electrons, and atoms will undergo chemical reactions in order to fill the outermost electron shell, either by gaining, losing, or sharing electrons.

2.2 Ions and Isotopes

1. Ions are atoms that have either gained one or more electrons (negative ions called anions) or lost one or more electrons (positive ions called cations).

2. Isotopes are atoms that have the same number of protons but differing numbers of neutrons. Isotopes tend to be unstable and break up into other elements through a process called radioactive decay. All isotopes of an atom have the same chemical properties.

2.3 Molecules

1. Molecules are atoms linked together by chemical bonds. There are three main types of chemical bonds.

2. Ionic bonds form when ions of opposite electrical charge are attracted to each other. Table salt is formed by ionic bonds between positive sodium ions and negative chloride ions.

3. Most biological molecules are held together by covalent bonds. Covalent bonds form when two atoms share electrons, attempting to fill empty electron orbitals. Covalent bonds are stronger when more electrons are shared.

4. Polar molecules are held together by covalent bonds in which the shared electrons are unevenly distributed around their nuclei, giving the molecule a slightly positive end and a slightly negative end. Hydrogen bonds form when the positive end of one molecule is attracted to the negative end of another.

Water: Cradle of Life

2.4 Unique Properties of Water

1. Water molecules form a network of hydrogen bonds with each other in liquid, and dissolve other polar molecules. Many of the key properties of water arise because it takes considerable energy to break liquid water's many hydrogen bonds.

2. The hydrogen bonds that hold water molecules together become more stable at lower temperatures, and as a result, they lock water molecules into place in solid crystal structures called ice.

3. In order for water to vaporize into a gas, a significant input of heat energy is needed to break the hydrogen bonds.

4. Because water molecules are polar molecules, they will form hydrogen bonds with other polar molecules. If the other polar molecules are water molecules, the process is called cohesion. If the other polar molecules are some other substance, the process is called adhesion.

5. Water molecules tend to surround other polar molecules, forming a barrier around them called a hydration shell. Nonpolar molecules do not form hydrogen bonds, are water-insoluble, and will cluster together when placed in water.

2.5 Water Ionizes

1. A tiny fraction of water molecules spontaneously ionize at any moment, forming H^+ and OH^- ions. The pH of a solution is a measure of its H^+ concentration. Low pH values indicate high H^+ ion concentrations (acidic solutions), and high pH values indicate low H^+ ion concentrations (basic solutions).

- An acid has a pH below 7. A base has a pH above 7. A buffer is a chemical substance that minimizes changes in pH by taking up excess H^+ in acidic solution, or releasing H^+ in basic solutions.

Inquiry and Analysis

Using Radioactive Decay to Date the Iceman

In the fall of 1991, two German hikers found a corpse sticking out of the melting snow on the crest of a high pass near the mountainous border between Italy and Austria. Right away it was clear the body was very old, frozen in an icy trench where he had sought shelter long ago and only now released as the ice melted. In the years since this startling find, scientists have learned a great deal about the dead man, who they named Ötzi. They know his age, his health, the clothing he wore, what he ate, and that he died from an arrow that ripped through his back. Its tip is still embedded in the back of his left shoulder. From the distribution of chemicals in his teeth and bones, we know he lived his life within 60 kilometers of where he died.

When did this Iceman die? Scientists answered this key question by measuring the degree of decay of the short-lived carbon isotope ^{14}C in Ötzi's body. While most carbon atoms are the stable isotope ^{12}C, a tiny proportion is the unstable radioactive isotope ^{14}C, created by the bombardment of nitrogen-14 (^{14}N) atoms with cosmic rays. This proportion of ^{14}C is captured by plants in photosynthesis, the process plants use to capture sunlight to make food, and is present in the carbon molecules of the animal's body that eats the plant. After the plant or animal dies, it no longer accumulates any more carbon, and the ^{14}C present at the time of death decays over time back to ^{14}N. Thus, over time the ratio of ^{14}C to ^{12}C decreases. It takes 5,730 years for half of the ^{14}C present to decay, a length of time called the **half-life** of the ^{14}C isotope. Because the half-life is a constant that never changes, the extent of radioactive decay allows you to date a sample. Thus a sample that had one-quarter of its original proportion of ^{14}C remaining would be approximately 11,460 years old (two half-lives).

The graph to the right displays the radioactive decay curve of the carbon isotope ^{14}C. Scientists know it takes 5,730 years for half of the ^{14}C present in a sample to decay to nitrogen-14 (^{14}N). When Ötzi's carbon isotopes were analyzed, researchers determined that the ratio of ^{14}C to ^{12}C (a ratio is the size of one variable relative to another), also written as the fraction $^{14}C/^{12}C$, in Ötzi's body was 0.435 of the fraction found in tissues of a person who has recently died.

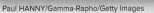
Paul HANNY/Gamma-Rapho/Getty Images

Gerhard Zwerger-Schoner/imageBROKER/Alamy Stock Photo

Analysis

1. **Applying Concepts** What proportion (a proportion is the size of a variable relative to the whole) of the ^{14}C present in Ötzi's body when he died is still there today? When he died, it would have been 1.0.
2. **Interpreting Data** Plot this proportion on the ^{14}C radioactive decay curve above. How many half-lives does this point represent?
3. **Making Inferences** If Ötzi were indeed a recent corpse, made to look old by the harsh weather conditions found on the high mountain pass, what would you expect the ratio of ^{14}C to ^{12}C to be, relative to that in your own body?
4. **Drawing Conclusions** When did Ötzi the Iceman die?

3 Molecules of Life

LEARNING PATH ▼

Forming Macromolecules
1. Building Big Molecules

Types of Macromolecules
2. Proteins
3. Nucleic Acids
4. Carbohydrates
5. Lipids

Buffing Up with Anabolic Steroids

Among the most notorious of lipids (one of the main types of macromolecules found in organisms) in recent years has been the class of synthetic hormones known as anabolic steroids. Since the 1950s, some athletes have been taking these chemicals to build muscle and boost athletic performance. Both because of the intrinsic unfairness of this and health risks, the use of anabolic steroids has been banned in sports for decades. Controversy over their use in professional baseball has put anabolic steroids on the nation's front pages.

What Are Anabolic Steroids?

Anabolic steroids were developed in the 1930s to treat hypogonadism, a condition in which the male testes do not produce sufficient amounts of the hormone testosterone for normal growth and sexual development. Scientists soon discovered that by slightly altering the chemical structure of testosterone, they could produce synthetic versions that facilitated the growth of skeletal muscle in laboratory animals. The word anabolic means growing or building. Further tweaking reduced the added impact of these new chemicals on sexual development.

BASEBALL SUPERSTAR Alex Rodriguez was suspended from playing by Major League Baseball for using steroids.
Ron Antonelli/NY Daily News Archive/Getty Images

More than 100 different anabolic steroids have been developed, most of which have to be injected to be effective. All are banned in professional, college, and high school sports.

Another way to increase the body's level of testosterone is to use a chemical that is not itself anabolic but that the body converts to testosterone. One such chemical is 4-androstenedione, more commonly called "andro." It was first developed in the 1970s by East German scientists trying to enhance their athletes' Olympic performances. Because andro does not have the same side effects as anabolic steroids, it was legally available until 2004. It was used by baseball star Mark McGwire, but it is now banned in all sports, and possession of andro is a federal crime.

How Anabolic Steroids Work

Anabolic steroids work by signaling muscle cells to make more protein. They bind to special "androgenic receptor" proteins within the cells of muscle tissue. Like jabbing these proteins with a poker, the binding prods the receptors into action, causing them to activate genes on the cell's chromosomes that produce muscle tissue proteins, triggering an increase in protein synthesis. At the same time, the anabolic steroid molecules bind to so-called "cortisol receptor" proteins in the cell, preventing these receptors from doing their job of causing protein breakdown, the muscle cell's way of suppressing inflammation and promoting the use of proteins for fuel during exercise. By increasing protein production and inhibiting the breakdown of proteins in muscle cells after workouts, anabolic steroids significantly increase the mass of an athlete's muscle tissue.

What's Wrong with Using Steroids?

If the only effect of anabolic steroids on your body was to enhance your athletic performance by increasing your muscle mass, using them would still be wrong, for one very simple and important reason: fairness. To gain advantage in competition by concealed use of anabolic steroids—"doping"—is simply cheating. That is why these drugs are banned in sports.

The use of anabolic steroids by athletes and others is not only wrong, but also illegal because increased muscle mass is not the only effect of using these chemicals. Among adolescents, anabolic steroids can also lead to premature termination of the adolescent growth spurt, so that for the rest of their lives, users remain shorter than they would have been without the drugs. Adolescents and adults are also affected by steroids in the following ways. Anabolic steroids can lead to potentially fatal liver cysts and liver cancer (the liver is the organ of the body that attempts to detoxify the blood), hypertension (which can promote heart attack and stroke), and acne.

Because anabolic steroids are not legal in most countries, they are often produced not in pharmaceutical laboratories but in small home-made "underground" labs, using raw substances imported from abroad. Black market importation of commercially manufactured anabolic steroids also occurs from Mexico, Thailand, and other countries where steroids are illegal.

Cheating in Sports Has Been Widespread

All Russian track and field athletes were banned by the International Olympic Committee from competing in the 2016 Summer Olympics in Rio de Janeiro, Brazil, and the 2018 Winter Olympics in PyeongChang, South Korea, because of steroid use. The Russian Ministry of Sport was found to have operated an extensive "doping" program and to have covered up any positive results from Olympic Committee antidoping tests. New antidoping tests have been used to catch well-known American sports figures as well. Tour de France champion Lance Armstrong, Olympic athlete Marion Jones, and baseball sluggers Alex Rodriguez, Barry Bonds, and Mark McGwire have all been involved in steroid use.

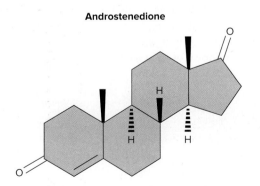

Androstenedione

Forming Macromolecules

3.1 Building Big Molecules

> **LEARNING OBJECTIVE 3.1.1** Distinguish between a polymer and a monomer.

The bodies of organisms contain thousands of different kinds of molecules and atoms. Organisms obtain many of these molecules from their surroundings and from what they consume. You might be familiar with some of the substances listed on nutritional labels. But what do the words on these labels mean? Some of them are names of minerals, such as calcium and iron (discussed in chapters 23, 25, 32, and others). Others are vitamins, which are discussed in chapter 26. Still others are the subject of this chapter: large molecules that are found in our food and that make up the bodies of organisms, such as proteins, carbohydrates (including sugars), and lipids (including fats, trans fats, saturated fats, and cholesterol, like that in popcorn, shown in **figure 3.1**). These molecules are called *organic molecules*.

The bodies of organisms contain thousands of different kinds of organic molecules, but much of the body is made of just four kinds: *proteins, nucleic acids, carbohydrates,* and *lipids*. Called **macromolecules** because they can be very large, these four are the building materials of cells, the "bricks and mortar" that make up the body of a cell and the machinery that runs within it.

Macromolecules are assembled by sticking smaller bits, called **monomers,** together much as a train is built by linking railcars together. A molecule built up of long chains of similar subunits is called a **polymer.**

> **Putting the Concept to Work**
> Which macromolecule is not present in popcorn?

Figure 3.1 What's in the food you eat?

Fats, cholesterol, carbohydrates, and proteins—all molecules found in popcorn—are discussed in this chapter.

Burke/Triolo Productions/Getty Images

Making (and Breaking) Macromolecules

> **LEARNING OBJECTIVE 3.1.2** Contrast hydrolysis with dehydration synthesis.

The four different kinds of macromolecules (proteins, nucleic acids, carbohydrates, and lipids) are built from different monomers, as shown in **figure 3.2**, but all have their subunits put together in the same way. A covalent bond is formed between two subunits in which a hydroxyl group (OH) is removed from one subunit and a hydrogen (H) is removed from the other. This process, illustrated in **figure 3.3a**, is called *dehydration synthesis* because, in effect, the removal of the OH and H groups (highlighted by the

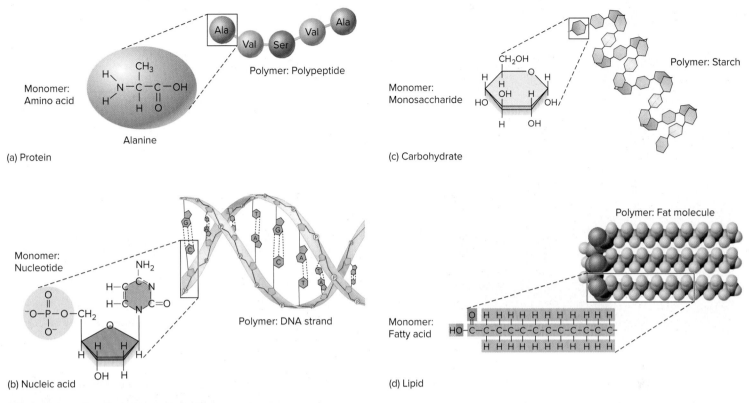

Figure 3.2 Polymers are built from monomers.

Each macromolecule polymer is built from different monomers. (a) A protein polymer, called a polypeptide, is built from amino acid monomers. (b) A nucleic acid polymer, such as a strand of DNA, is built from nucleotide monomers. (c) A carbohydrate polymer, such as a starch molecule, is built from monosaccharide monomers. (d) A lipid polymer, such as a fat molecule, is built from fatty acids.

blue oval) constitutes removal of a molecule of water—the word *dehydration* means "taking away water." This process requires the help of a special class of proteins called **enzymes** to facilitate the positioning of the molecules so that the correct chemical bonds are stressed and broken. The process of tearing down a molecule such as the protein or fat contained in the food you eat is essentially the reverse of dehydration synthesis: instead of removing a water molecule, one is added. When a water molecule comes in, as shown in figure 3.3b, a hydrogen becomes attached to one subunit and a hydroxyl to another, and the covalent bond is broken. The breaking up of a polymer in this way is called **hydrolysis**.

> **Putting the Concept to Work**
>
> Do you think hydrolysis of a protein requires an enzyme like dehydration synthesis does? Explain.

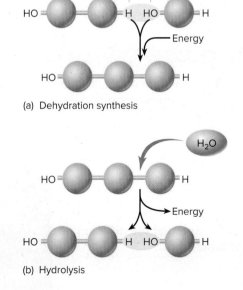

Figure 3.3 Dehydration and hydrolysis.

(a) Biological molecules are formed by linking subunits with a covalent bond in a dehydration synthesis, during which a water molecule is released. (b) Breaking such a bond requires the addition of a water molecule, a reaction called hydrolysis.

Types of Macromolecules

3.2 Proteins

> **LEARNING OBJECTIVE 3.2.1** Explain what proteins do by listing five functional groupings of proteins.

Complex macromolecules called **proteins** are important biological macromolecules within the bodies of all organisms. One of the most important types of proteins are *enzymes,* which have the key role in cells of helping to carry out particular chemical reactions. Other proteins play structural roles. Cartilage, bones, and tendons all contain a structural protein called collagen. Keratin, another structural protein, forms hair, the horns of a rhinoceros, and feathers. Still other proteins act as chemical messengers within the brain and throughout the body. Figure 3.4 presents an overview of the wide-ranging functions of proteins.

Putting the Concept to Work
Of the five functional groupings of proteins shown in figure 3.4, are any of them not present in your little finger?

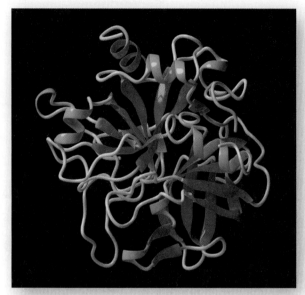

(a) Enzymes: Globular proteins called enzymes play a key role in many chemical reactions. This is a computer model of an enzyme.
MOLEKUUL/SCIENCE PHOTO LIBRARY/Getty Images

(b) Structural proteins (keratin): Keratin forms hair, nails, feathers, and components of horns.
Con Tanasiuk/Design Pics

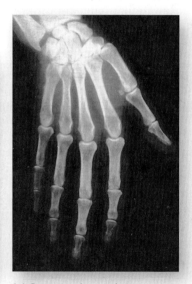

(c) Structural proteins (collagen): Collagen is present in bones, tendons, and cartilage.
Steve Allen/Getty Images

(d) Contractile proteins: Proteins called actin and myosin are present in muscles.
Ruslan Semichev/Shutterstock

Figure 3.4 Some of the different types of proteins.

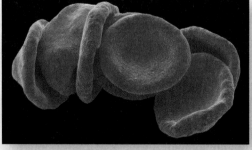

(e) Transport proteins: Red blood cells contain the protein hemoglobin, which transports oxygen in the body.
STEVE GSCHMEISSNER/Science Photo Library RF/Getty Images

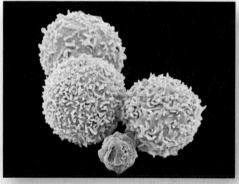

(f) Defensive proteins: White blood cells destroy foreign cells in the body and make antibody proteins that attack invaders.
Steve Gschmeissner/Getty Images

Amino Acids

> **LEARNING OBJECTIVE 3.2.2** Diagram the structure of a peptide bond.

Despite their diverse functions, all proteins have the same basic structure: a long polymer chain made of subunits called amino acids. **Amino acids** are small molecules with a simple basic structure: a central carbon atom attached to an amino group (—NH_2), a carboxyl group (—COOH), a hydrogen atom (H), and a functional group, designated "R."

There are 20 common amino acids that differ from one another by the identity of their functional R group. The R group of an amino acid largely determines its chemical properties. Some amino acid R groups are polar, interacting with water; some are nonpolar, shying away from water; and others have special chemical groups that are important in forming links between protein chains or in forming kinks in their shapes.

Linking Amino Acids. An individual protein is made by linking specific amino acids together in a particular order, just as a word is made by putting letters of the alphabet together in a particular order. The covalent bond linking two amino acids together is called a **peptide bond** (figure 3.5). You can see in the figure below that a water molecule is released as the peptide bond forms. Long chains of amino acids linked by peptide bonds are called **polypeptides**.

> **Putting the Concept to Work**
> Why don't the amino groups of amino acids make all proteins polar?

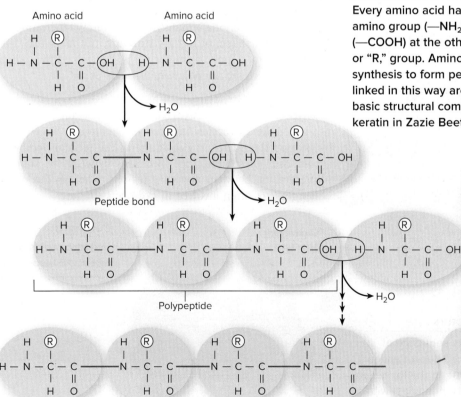

Figure 3.5 The formation of a peptide bond.

Every amino acid has the same basic structure, with an amino group (—NH_2) at one end and a carboxyl group (—COOH) at the other. The only variable is the functional, or "R," group. Amino acids are linked by dehydration synthesis to form peptide bonds. Chains of amino acids linked in this way are called polypeptides and are the basic structural components of proteins, such as the keratin in Zazie Beetz's hair.

Andrea Raffin/Alamy Stock Photo

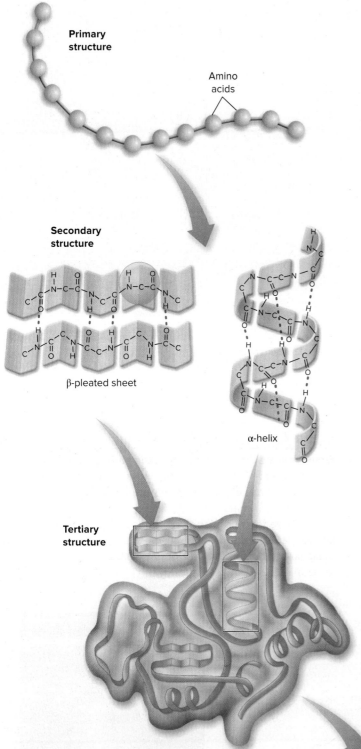

Protein Structure

LEARNING OBJECTIVE 3.2.3 Describe the four general levels of protein structure and how the polar nature of water influences them.

Functional polypeptides are more commonly called proteins. Some proteins, such as keratin in human hair, form long, thin fibers, whereas others are globular, their strands coiled up and folded back on themselves. The shape of a protein is very important because it determines the protein's function. There are four general levels of protein structure: primary, secondary, tertiary, and quaternary (figure 3.6); all are ultimately determined by the sequence of amino acids.

Primary Structure. The sequence of amino acids of a polypeptide chain is termed the polypeptide's primary structure. The amino acids are linked together by peptide bonds, forming long chains like a "beaded strand." The primary structure of a protein, the sequence of its amino acids, determines all other levels of protein structure. Because amino acids can be assembled in any sequence, a great diversity of proteins is possible.

Secondary Structure. Hydrogen bonds forming between different parts of the polypeptide chain stabilize the folding of the polypeptide. As you can see, these stabilizing hydrogen bonds, indicated by red dotted lines, do not involve the R groups themselves, rather the polypeptide backbone. This initial folding is called the secondary structure of a protein. Hydrogen bonding within this secondary structure can fold the polypeptide into coils, called α-helices, and sheets, called β-pleated sheets.

Tertiary Structure. Because some of the amino acids are nonpolar, a polypeptide chain folds up in water, which is very polar, pushing nonpolar amino acid functional groups from the watery environment. The final three-dimensional shape, or tertiary structure, of the protein, folded and twisted in the case of a globular molecule, is determined in a polypeptide chain by exactly where the nonpolar amino acids occur.

Quaternary Structure. When a protein is composed of more than one polypeptide chain, the spatial arrangement of the several component chains is called the quaternary structure of the protein. For example, four subunits make up the quaternary structure of the protein hemoglobin. In proteins composed of subunits, the interfaces between the subunits often involve interactions between nonpolar regions, whereas the polar regions of the subunits are often involved in hydrogen bonding.

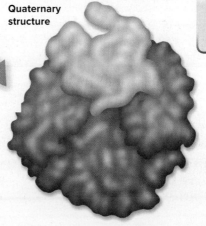

Putting the Concept to Work
What force holds the subunits of a protein like hemoglobin together?

Figure 3.6 Levels of protein structure.

The *primary structure* of a protein is its sequence of amino acids. Twisting or pleating of the chain of amino acids, called *secondary structure*, is due to the formation of localized hydrogen bonds (the *red* dotted lines) within the chain. More complex folding of the chain is referred to as *tertiary structure*. Two or more polypeptide chains associated together form a *quaternary structure*.

Protein Folding and Denaturation

> **LEARNING OBJECTIVE 3.2.4** Explain the forces that cause a protein to denature.

The polar nature of the watery environment in the cell influences how the polypeptide folds into the functional protein. The protein in figure 3.7 is folded in such a way that allows it to carry out its function. If the polar nature of the protein's environment changes by either increasing temperature or lowering pH, both of which alter hydrogen bonding, the protein may unfold, as in the lower right of the figure. When this happens the protein is said to be *denatured*. When proteins are denatured, they usually lose their ability to function properly. When the polar nature of the solvent is reestablished, some proteins may spontaneously refold, but most don't. Cooking an egg is an example of denaturing proteins that do not refold. The egg proteins denature as temperature increases but do not refold as the egg cools down—they are permanently denatured. Protein denaturation is also the rationale behind traditional methods of preserving food. Prior to the ready availability of refrigerators and freezers, a practical way to keep microorganisms from growing in food was to keep the food in a solution containing a high concentration of vinegar, a treatment called pickling. The low pH of the vinegar denatures proteins in microorganisms and keeps them from growing on the food.

> **Putting the Concept to Work**
> Why do you think most heat-denatured proteins don't spontaneously refold when they cool back to room temperature?

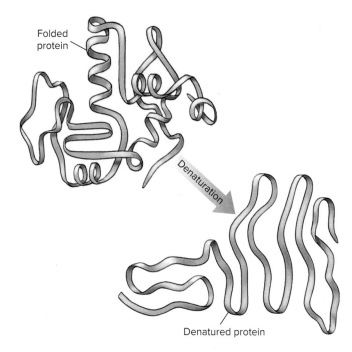

Figure 3.7 Protein denaturation.
Changes in a protein's environment, such as variations in temperature or pH, can cause a protein to unfold and lose its shape in a process called denaturation. In this denatured state, proteins are biologically inactive.

> **IMPLICATION FOR YOU** A tiger can eat raw meat with relish, but most of us humans prefer our steak cooked. Is this just a matter of taste, or does cooking meat before eating it have any benefit?

Protein Structure Determines Function

> **LEARNING OBJECTIVE 3.2.5** Describe how the structure of an enzyme enables it to catalyze a chemical reaction.

The three-dimensional shape of a protein determines its function. For example, many structural proteins assume long cablelike shapes that let them play architectural roles within cells (figure 3.8a). *Enzymes* are globular proteins that help particular chemical reactions to occur in the cell. When the polypeptide folds correctly, the enzyme surface has a groove or depression that precisely fits a particular molecule. For example, the red molecule binding to the groove on the surface of the enzyme in figure 3.8b is a sugar. Once within the groove, the molecule is induced to undergo a chemical reaction, such as the formation or breaking of one of its covalent bonds. By bringing two atoms close together, an enzyme can make it easier for them to share electrons and form covalent bonds. Other enzymes function by positioning a molecule so that there is stress on a particular bond so it breaks.

Because the primary structure of a protein (its sequence of amino acids) determines how the protein folds into its functional shape, a change in the identity of even one amino acid can have profound effects on a protein's shape and so on its ability to function properly.

> **Putting the Concept to Work**
> What sorts of chemicals might act to block the action of a particular enzyme and not others?

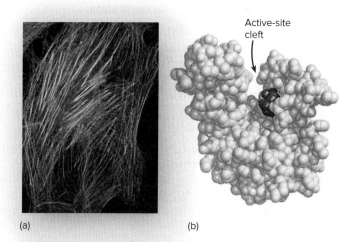

Figure 3.8 Protein structure determines function.
(a) Fluorescently labeled structural proteins within a cell.
(b) Enzymes are globular proteins that aid chemical reactions in the cell. This enzyme (*blue*) has a deep groove that binds a specific chemical (*red*) at a site on the enzyme called the active site.

(a) Dr. Gopal Murti/Science Source

3.3 Nucleic Acids

Your body stores hereditary information and puts it to work by utilizing nucleic acids.

Nucleotides

> **LEARNING OBJECTIVE 3.3.1** Name the three parts of a nucleotide.

Very long polymers called **nucleic acids** serve as the genetic information storage devices of cells, just as hard drives store the information that computers use. Nucleic acids are long polymers of repeating subunits called **nucleotides.** Each nucleotide is a complex organic molecule composed of three parts shown in **figure 3.9a**: a 5-carbon sugar (in blue), a phosphate group (in yellow, PO_4), and an organic nitrogen-containing base (in orange). In the formation of a nucleic acid, the individual sugars with their attached nitrogenous bases are linked through dehydration reactions in a line by the phosphate groups in very long **polynucleotide chains** (shown to the right). The nitrogenous bases extend out from the backbone of the chain.

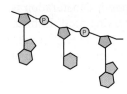

How does the long, chainlike structure of a nucleic acid permit it to store the information necessary to specify what an organism is like? If nucleic acids were simply monotonous repeating polymers, they could not encode the message of life. Imagine trying to write a story using only the letter *E* and no spaces or punctuation. All you could ever say is "EEEEEEE. . . ." You need more than one letter to communicate—the English alphabet uses 26 letters. Nucleic acids can encode information because they contain more than one kind of nucleotide. There are five different nucleotides found in nucleic acids: two larger ones that contain the nitrogenous bases adenine and guanine (shown in the top row of **figure 3.9b**), and three smaller ones that contain the nitrogenous bases cytosine, thymine, and uracil (in the bottom row). Nucleic acids encode information by varying the identity of the nucleotide at each position in the polymer.

> **Putting the Concept to Work**
> Try writing a 10-word sentence using only five letters.

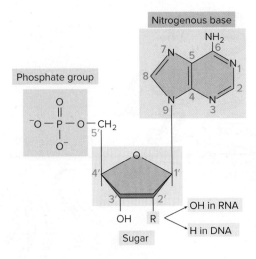

Figure 3.9 The structure of a nucleotide.

(a) Nucleotides are composed of three parts: a 5-carbon sugar, a phosphate group, and an organic nitrogenous base. The nitrogenous base can be one of five, shown in (b).

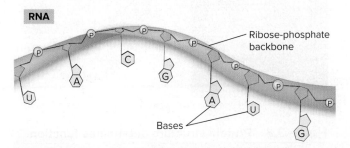

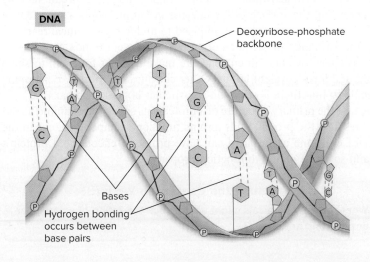

Figure 3.10 How DNA structure differs from RNA.

(*left*) RNA is a single strand of ribose nucleotides containing uracil. (*right*) DNA is a double strand of deoxyribose nucleotides that contain thymine and that are wrapped around each other, A pairing with T and G with C.

DNA and RNA

> **LEARNING OBJECTIVE 3.3.2** State the two major chemical differences between DNA and RNA.

Nucleic acids come in two varieties, **deoxyribonucleic acid (DNA)** and **ribonucleic acid (RNA)**. Both are polymers of nucleotides, but they have different functions in the cell and they differ in their structures. RNA is similar to DNA but with two major chemical differences. First, RNA molecules contain the sugar ribose, in which the 2' carbon (this is the carbon labeled 2' in **figure 3.9a**) is bonded to a hydroxyl group (—OH). In DNA, this hydroxyl group is replaced with a hydrogen atom, forming the sugar to deoxyribose. Second, DNA contains the thymine nucleotide, RNA molecules do not; they contain the uracil nucleotide instead. Structurally, RNA is also different. RNA is a long, single strand of nucleotides (**figure 3.10**), whereas DNA consists of *two* polynucleotide chains wound around each other in a **double helix,** like strands of a pearl necklace twisted together (**figures 3.10** and **3.11**). RNA and DNA also have different roles to play in the cell. DNA stores the genetic information (**figure 3.12**). The sequence of nucleotides in DNA determines the order of amino acids in the primary structure of the protein. Small RNA molecules regulate how the information in DNA is used, while large RNA molecules carry this information from the DNA to the protein-making machinery in the cell.

> **Putting the Concept to Work**
> What prevents RNA from forming a double helix like DNA does?

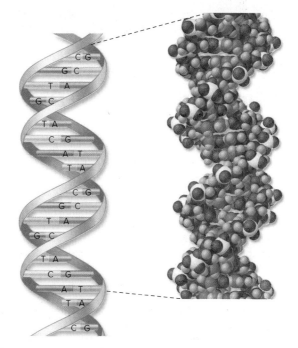

Figure 3.11 The DNA double helix.

The DNA molecule is composed of two polynucleotide chains twisted together to form a double helix. The two chains of the double helix are joined by hydrogen bonds between the A–T and G–C base pairs. The section of DNA on the right is a space-filling model of DNA, where atoms are indicated by colored balls.

The Double Helix

> **LEARNING OBJECTIVE 3.3.3** Explain why four of the six potential base pairings do not occur in DNA molecules.

How does DNA form a double helix? Look carefully at the structure of DNA in **figure 3.11**: the bases of each chain point inward toward the other. The bases of the two chains are linked in the middle of the molecule by hydrogen bonds, like two columns of people holding hands across (see also **figure 3.10**). The key to understanding the double helix structure of DNA is revealed by looking at the nucleotide bases: *only two base pairs are possible.* Because the distance between the two strands is consistent, this suggests that two big bases cannot pair together—the combination is simply too bulky to fit; similarly, two little ones cannot pair, as they would pinch the helix inward too much. To form a double helix, it is necessary to pair a big base with a little one. *In every DNA double helix, adenine (A) pairs with thymine (T), and guanine (G) pairs with cytosine (C).* The reason A doesn't pair with C, and G doesn't pair with T, is that these base pairs cannot form proper hydrogen bonds—the electron-sharing atoms are not aligned with each other.

The simple A–T, G–C base pairs within the DNA double helix allow the cell to copy the information in a very simple way. It just unzips the helix and adds the nucleotides with complementary bases to each strand! That is the great advantage of a double helix—it actually contains two copies of the information, one the mirror image of the other. If the sequence of one chain is ATTGCAT, the sequence of its helix partner *must* be TAACGTA.

> **Putting the Concept to Work**
> If one chain of a DNA strand has the sequence ACCTGGAAT, what is the sequence of the other strand?

Figure 3.12 Extracting DNA.

Working under fluorescent light, this investigator is inserting a syringe into a glowing band of DNA, which can then be extracted and studied.

Maximilian Stock/Age fotostock

Today's Biology

Prions and Mad Cow Disease

For decades, scientists have been fascinated by a peculiar group of fatal brain diseases. These diseases have the unusual property that they are transmissible from one individual to another, but it is years and often decades before the disease is detected in infected individuals. The brains of these individuals develop numerous small cavities as neurons die, producing a marked spongy appearance, as shown here in the photo. Called **transmissible spongiform encephalopathies (TSEs),** these diseases include scrapie in sheep, "mad cow" disease in cattle, and kuru and Creutzfeldt-Jakob disease in humans.

Discovery of the Disease

TSEs can be transmitted between individuals of a species by injecting infected brain tissue into a recipient animal's brain. TSEs can also spread via tissue transplants and, apparently, food. Kuru was common in the Fore people of Papua New Guinea when they practiced ritual cannibalism, literally eating the brains of infected individuals. Mad cow disease spread widely among the cattle herds of England in the 1990s because cows were fed bone meal prepared from sheep carcasses to increase the protein content of their diet. Like the Fore, the British cattle were literally eating the tissue of sheep that had died of scrapie.

Identifying the Infectious Agent

In the 1960s, British researchers T. Alper and J. Griffith noted that infectious TSE preparations remained infectious even after exposure to radiation that would destroy DNA or RNA. They suggested that the infectious agent was a protein. Perhaps, they speculated, the protein usually preferred one folding pattern but could sometimes misfold and then catalyze other proteins to do the same, the misfolding spreading like a chain reaction. This heretical suggestion was not accepted by the scientific community, as it violates a key tenet of molecular biology: only DNA or RNA act as hereditary material, transmitting information from one generation to the next.

Prusiner's Prions

In the early 1970s, physician Stanley Prusiner, moved by the death of a patient from Creutzfeldt-Jakob disease, began to study TSEs. Prusiner became fascinated with Alper and Griffith's hypothesis. Try as he might, Prusiner could find no evidence of nucleic acids, bacteria, or viruses in the infectious TSE preparation. He concluded, as Alper and Griffith had, that the infectious agent was a *protein,* which he named a **prion,** for "proteinaceous infectious particle."

Prusiner went on to isolate a distinctive prion protein, and for two decades he continued to amass evidence that prions play a key role in triggering TSEs. The scientific community resisted Prusiner's renegade conclusions, but eventually experiments done in Prusiner's and other laboratories began to convince many. For example, when Prusiner injected prions of different abnormal conformations into several different hosts, these hosts developed prions with the same abnormal conformations as the parent prions. In another important experiment, Charles Weissmann showed that mice genetically engineered to lack Prusiner's prion protein are immune to TSE infection.

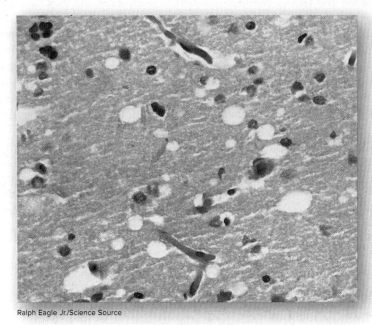

Ralph Eagle Jr./Science Source

However, if brain tissue with the prion protein is grafted into the mice, the grafted tissue—but not the rest of the brain—can then be infected with TSE. In 1997, Prusiner was awarded the Nobel Prize in Physiology or Medicine for his work on prions.

Humans at Risk

Can humans catch mad cow disease by eating infected meat? Many scientists are becoming worried that prions may be transmitting such diseases to humans. Specifically, they worry that prion-caused bovine spongiform encephalopathy (BSE), a brain disease in cows commonly known as mad cow disease, may infect humans and produce a similar fatal disorder, variant Creutzfeldt-Jakob disease (vCJD). In 1996, an outbreak of mad cow disease in Britain, with over 80,000 cows infected, created widespread concern. BSE, a degeneration of the brain caused by prions, appears to have entered the British cattle herds from sheep! Sheep are subject to a prion disease called *scrapie,* and the disease is thought to have passed from sheep to cows through protein-supplemented feed pellets containing ground-up sheep brains. The passage of prions from one species to another has a scary consequence: Prions can also pass from cows to people! One-hundred fifty British consumers of BSE-infected beef have subsequently died of vCJD. Tissue from the brains of the dead Britons and from BSE cows induce the same brain lesions in mice, whereas classic CJD produces quite different lesions—clearly the form of vCJD that killed them was caused by the same agent that caused BSE.

The occasional appearance of BSE-infected cows in United States herds led Europe, Russia, Brazil, China, and many other countries to ban U.S. beef imports in 2004, a ban that was only lifted in China in 2017 after heightened scrutiny of commercial beef in the United States in an attempt to eliminate any possibility of BSE contamination.

3.4 Carbohydrates

> **LEARNING OBJECTIVE 3.4.1** Distinguish between monosaccharides and polysaccharides.

Polymers called **carbohydrates** make up the structural framework of certain cells and play a critical role in energy storage. A carbohydrate is any molecule that contains carbon, hydrogen, and oxygen in the ratio 1:2:1. Some carbohydrates are small monomers or dimers and are called simple carbohydrates. Others are long polymers and are called complex carbohydrates. Because they contain many carbon-hydrogen (C—H) bonds, carbohydrates are well-suited for energy storage. Such C—H bonds are the ones most often broken by organisms to obtain energy. Table 3.1 shows some examples of carbohydrates.

Simple Carbohydrates

The simplest carbohydrates are the *simple sugars* or **monosaccharides** (from the Greek *monos,* single, and *saccharon,* sweet). These molecules consist of one subunit. For example, glucose, the sugar that carries energy to the cells of your body, is made of six carbons and has the chemical formula $C_6H_{12}O_6$. A molecule of glucose is pictured in several ways in figure 3.13. Another type of simple carbohydrate is a **disaccharide,** which forms when two monosaccharides link together through a dehydration reaction. Table sugar is a disaccharide, sucrose, made by linking two 6-carbon sugars together: a glucose and a fructose (see table 3.1).

Complex Carbohydrates

Organisms store their metabolic energy by converting sugars, which are water-soluble, into insoluble forms that can be deposited in specific storage areas in the body. This trick is achieved by linking the sugars together into long polymer chains called **polysaccharides.** Plants and animals store energy in polysaccharides formed from glucose. The glucose polysaccharide that plants use to store energy is called *starch*—that is why potatoes are referred to as "starchy" food. In animals, energy is stored in *glycogen,* a highly insoluble macromolecule formed of glucose polysaccharides that are very long and, unlike starch, highly branched.

Plants and animals also use glucose chains as building materials, linking the subunits together in different orientations not recognized by most enzymes. These structural polysaccharides are *chitin* in animals and *cellulose* in plants. The cellulose deposited in the cell walls of plant cells, like the cellulose strand shown in figure 3.14, cannot be digested by humans and makes up the fiber in our diets. Microbes in the digestive tracts of cows and horses, however, have the cellulose-digesting enzymes we humans lack. The microbial enzyme breaks the bonds holding the glucose molecules together so that they can be used by the animal cells for energy. These animals can thrive on a diet of grass, but you can't. Termites have these enzymes too, which is why a termite can eat wood while you would starve on a diet of tree limbs or lumber.

> As discussed in section 16.7, fungi possess enzymes that can break down cellulose, which is why fungi often grow on dead trees.

> **Putting the Concept to Work**
> Why can you digest starch but not cellulose?

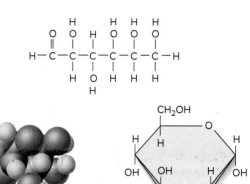

Figure 3.13 The structure of glucose.

Glucose is a monosaccharide and consists of a linear 6-carbon molecule that forms a ring when placed in water.

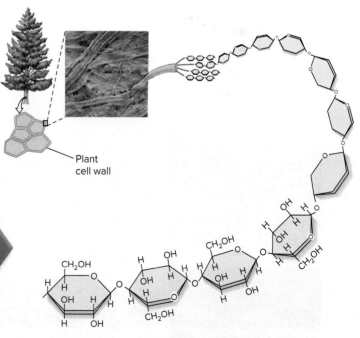

Figure 3.14 A polysaccharide: cellulose.

The polysaccharide cellulose is found in the cell walls of plant cells and is composed of glucose subunits.

Scimat/Science Source

TABLE 3.1 Carbohydrates and Their Functions

Carbohydrate	Example	Description
Transport Disaccharides		
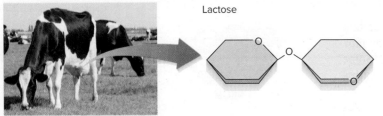 Lactose, Sucrose	Glucose is transported within some organisms as a disaccharide. In this form, it is less readily metabolized because the normal glucose-utilizing enzymes of the organism cannot break the bond linking the two monosaccharide subunits. One type of disaccharide is called lactose. Many mammals supply energy to their young in the form of lactose, which is found in milk. Another transport disaccharide is sucrose. Many plants transport glucose throughout the plant in the form of sucrose, which is harvested from sugarcane to make granulated sugar.
Storage Polysaccharides		
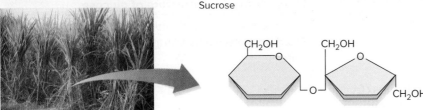 Starch	Organisms store energy in long chains of glucose molecules called polysaccharides. The chains tend to coil up in water, making them insoluble and ideal for storage. The storage polysaccharides found in plants are called starches, which can be branched or unbranched. Starch is found in potatoes and in grains, such as corn and wheat.
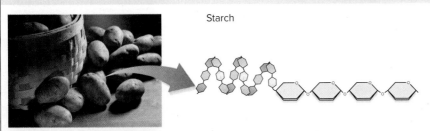 Glycogen	In animals, glucose is stored as glycogen. Glycogen is similar to starch in that it consists of long chains of glucose that coil up in water and are insoluble. But glycogen chains are much longer and highly branched. Glycogen can be stored in muscles and the liver.
Structural Polysaccharides		
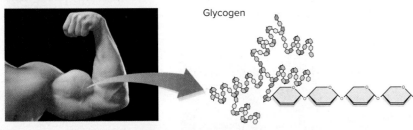 Cellulose	Cellulose is a structural polysaccharide found in the cell walls of plants; its glucose subunits are joined in a way that cannot be broken down readily. Cleavage of the links between the glucose subunits in cellulose requires an enzyme most organisms lack. Some animals, such as cows, are able to digest cellulose by means of bacteria and protists they harbor in their digestive tract, which provide the necessary enzymes.
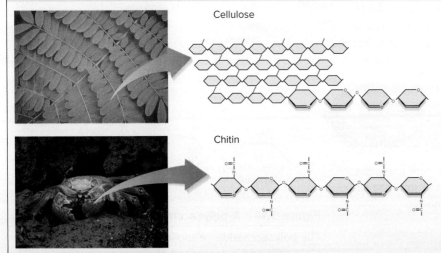 Chitin	Chitin is a type of structural polysaccharide found in the external skeletons of many invertebrates, including insects and crustaceans, and in the cell walls of fungi. Chitin is a modified form of cellulose with a nitrogen group added to the glucose units. When cross-linked by proteins, it forms a tough, resistant surface material.

(a) **Dairy Cows:** jvdwolf/123RF; (b) **Sugarcane:** Dinodia/Pixtal/age fotostock; (c) **Potatoes:** Purestock/SuperStock; (d) **Hand of athlete:** Ruslan Semichev/Shutterstock; (e) **Fern leaves:** PictureNet/Corbis/Getty Images; (f) **Crab:** Lieutenant Elizabeth Crapo/NOAA

3.5 Lipids

> **LEARNING OBJECTIVE 3.5.1** Distinguish between saturated and unsaturated fats, and explain why one is a solid and the other a liquid at room temperature.

For long-term energy storage, organisms usually convert glucose into fats, another kind of storage molecule that contains more energy-rich C—H bonds than carbohydrates. Fats and all other biological molecules that are not soluble in water are called **lipids**. Lipids are nonpolar; in water, fat molecules cluster together because they cannot form hydrogen bonds with water molecules. This is why oil forms into a layer on top of water when the two substances are mixed. A special type of lipid called a phospholipid is important because it forms boundary layers in cells called membranes (**figure 3.15**). They will play a major role in chapter 4!

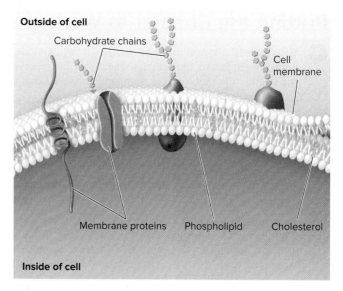

Figure 3.15 Lipids are a key component of biological membranes.

Lipids are one of the most common molecules in the human body because the membranes of all cells are composed of phospholipids. Membranes also contain cholesterol, another type of lipid.

Fats

Fat molecules are lipids composed of two kinds of subunits: fatty acids and glycerol. A *fatty acid* is a long chain of carbon and hydrogen atoms. Glycerol contains three carbons and forms the backbone to which three fatty acids are attached through dehydration reactions.

The chemical composition of the fatty acids that make up a fat molecule can affect its physical properties. Fats whose fatty acid chains are composed of the maximum number of hydrogen atoms are said to be *saturated* (**figure 3.16a**). Saturated fats are solid at room temperature. Animal fats are often saturated and occur as hard fats. On the other hand, fats composed of fatty acids with double bonds between one or more pairs of carbon atoms contain fewer than the maximum number of hydrogen atoms and are called *unsaturated* (**figure 3.16b**). Unsaturated fats are liquid at room temperature. Many plant fats are unsaturated and occur in oils. Unsaturated fats in food products may be artificially *hydrogenated* (industrial addition of hydrogens), extending the shelf life of products such as peanut butter. In some cases, the hydrogenation creates *trans fats,* a type of unsaturated fat linked to heart disease.

Other Types of Lipids

Other types of lipids include phospholipids and cholesterol, which play key roles in the membranes that encase all cells of your body. Cholesterol is a type of lipid called a *steroid*. Cholesterol plays a key role in biological membranes, helping them stay flexible. The male and female sex hormones testosterone and estradiol are also steroids. The chemical structure of steroids consists of multiple carbon rings and looks somewhat like a section of chicken wire. Rubber, waxes, and light-absorbing pigments are other important biological lipids.

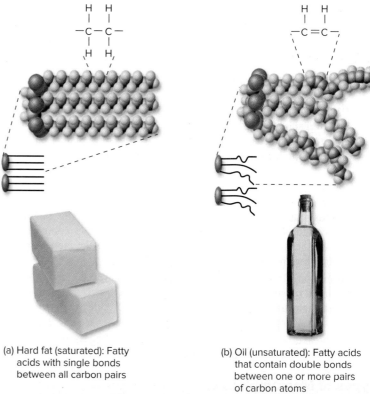

(a) Hard fat (saturated): Fatty acids with single bonds between all carbon pairs

(b) Oil (unsaturated): Fatty acids that contain double bonds between one or more pairs of carbon atoms

Figure 3.16 Saturated and unsaturated fats.

Fat molecules each contain a 3-carbon glycerol to which is attached three fatty acid tails. (a) Most animal fats are "saturated" (every carbon atom carries the maximum load of hydrogens). Their fatty acid chains fit closely together and form immobile arrays called hard fats. (b) Most plant fats are unsaturated, which prevents close association between chains and so results in oils.

(a) Two bricks of butter: Photodisc/Getty Images; (b) Bottle of olive oil: C Squared Studios/Getty Images

> **Putting the Concept to Work**
> If you hydrogenate an unsaturated fat, does the treatment make the fat more solid? Explain.

Putting the Chapter to Work

1 Myoglobin is a protein found in muscle tissue that binds iron and stores oxygen for working muscles. The myoglobin protein has only one subunit: the amino acid chain wound into an α-helix.

Based on this information, at what level of protein structure does myoglobin differ from hemoglobin?

2 Adenosine triphosphate (ATP) is used as energy by the cells of living organisms. This molecule is made of the nitrogen-containing base, adenine, ribose, and phosphate groups.

This molecule resembles nucleotides found in which nucleic acid?

3 Some bacteria can create a layer outside of their cells called a capsule. Capsules provide a protective coating around the cell that is difficult to break down. They are composed of many saccharides linked together through dehydration synthesis and contain cross-linkages.

Based on this information, would a capsule be classified as a transport disaccharide, storage polysaccharide, or a structural polysaccharide?

4 Grapeseed oil, avocado oil, and flaxseed oil are examples of healthy fats that can be incorporated into a diet. These fats do not have the maximum number of hydrogen atoms on the fatty acid chains.

Would these fats be liquid or solid at room temperature?

Retracing the Learning Path

Forming Macromolecules

3.1 Building Big Molecules

1. Living organisms produce organic macromolecules, which are large carbon-based molecules. The chemical properties of these molecules are due to unique functional groups that are attached to the carbon core.
2. Macromolecules are formed by the linking together of subunits, called monomers, to form long chains called polymers. Amino acids are the subunits that link together to form polypeptides. Nucleotide monomers link together to form nucleic acids. Monosaccharide monomers link together to form carbohydrates. Fatty acids are the monomers that link together to form a type of lipid called fats.
- Polymers are formed by dehydration reactions. Dehydration reactions produce covalent bonds that link monomers together. The reaction is called a dehydration reaction because a water molecule is removed as each link is formed.
- The breakdown of macromolecules involves hydrolysis reactions, where adding a water molecule causes the bond to break.

Types of Macromolecules

3.2 Proteins

1. Proteins are long chains of amino acids that fold into complex shapes.
2. There are 20 different amino acids found in proteins. All amino acids have the same basic core structure. They differ in the type of functional group attached to the core. The functional groups are referred to as R groups. Some functional groups are polar, some are nonpolar, and still others give the amino acid unique chemical properties. Amino acids are linked together with covalent bonds referred to as peptide bonds.
3. The sequence of amino acids within the polypeptide is the primary structure of the protein. The chain of amino acids can twist into a secondary structure, where hydrogen bonding holds portions of the polypeptide in a coiled shape called an α-helix or in sheets called β-pleated sheets. Further bending and folding of the polypeptide results in its tertiary structure. When more polypeptides are present in a protein, the interaction of these polypeptide subunits is its quaternary structure.
4. Changes in environmental conditions that disrupt hydrogen bonding can cause a protein to unfold, a process called denaturation. Proteins cannot function if they are denatured. Some denatured proteins can refold back into their functional shapes.
5. The location of nonpolar amino acids in a sequence determines how the strand folds into its three-dimensional shape, forcing these amino acids into the protein interior.

3.3 Nucleic Acids

1. Nucleic acids, such as DNA and RNA, are long chains of nucleotides. Nucleotides contain three parts: a 5-carbon sugar, a phosphate group, and a nitrogenous base. DNA and RNA function in information storage and retrieval in the cell, carrying the information needed to build proteins. The information is stored as different sequences of nucleotides that determine the order of amino acids in proteins.
2. DNA and RNA differ chemically in that the sugar found in DNA is deoxyribose and in RNA it is ribose. The nitrogenous bases in DNA are cytosine, adenine, guanine, and thymine, and the same are found in RNA except for thymine, which is substituted in RNA with uracil.
- DNA and RNA also differ structurally. DNA contains two strands of nucleotides wound around each other, a double helix. RNA is typically a single strand of nucleotides.
3. The two strands of the DNA double helix are held together through hydrogen bonding between their bases: adenine (A) pairs with thymine (T), and cytosine (C) pairs with guanine (G).

3.4 Carbohydrates

1. Carbohydrates are macromolecules that contain C, H, and O atoms in the ratio 1:2:1. They serve two primary functions in the cell: structural framework and energy storage.
- Carbohydrates that consist of only one or two monomers are called simple carbohydrates, such as the glucose monosaccharide and the disaccharide sucrose. Carbohydrates that consist of long chains of monomers are called complex carbohydrates or polysaccharides.
- Polysaccharides such as starch and glycogen provide a means of storing energy in the cell. They are broken down in the cells when energy is needed. Carbohydrates such as cellulose and chitin provide structural integrity and are not broken down by animals because they lack the enzyme necessary. Microbes in the guts of some animals are able to break down cellulose.

3.5 Lipids

1. Lipids are large nonpolar molecules that are insoluble in water. Lipids called phospholipids are components of biological membranes. Lipids called fats function in long-term energy storage. Other lipids are steroids (including sex steroids and cholesterol), rubber, and pigments.

Inquiry and Analysis

Is Excess Fat or Sugar Worse for Your Health?

Excess weight is one of the leading causes of poor health in the United States, with close to two-thirds of the American population falling in the category of overweight or obese. Thus, it comes as no surprise that an estimated 45 million people go on a weight-loss diet each year. There are many diets out there, some very low in carbohydrates, others low in fat or having a reduced overall calorie count.

Why is excess weight so deadly? One major reason is its association with heart disease. The American Heart Association reports that nearly one-third of Americans have at least one form of cardiovascular disease, that most of these Americans are overweight, and that losing excess weight produces substantial cardiovascular benefits. However, few comparative studies have examined whether different diets are equally effective at improving your cardiovascular health. This raises an interesting question: Are low-carbohydrate and low-fat diets equally effective at reducing weight and improving overall cardiovascular health?

This hypothesis was tested in a group of 148 men and women without clinical cardiovascular disease. These subjects were randomly distributed into two test groups. The 73 participants assigned to the low-fat diet group were instructed to keep their total fat consumption to less than 30% of their daily energy intake. The 75 participants in the low-carbohydrate diet group were instructed to keep their intake of digestible carbohydrates to less than 40 grams per day. At the onset of the study, all participants were weighed and had blood tests performed to establish baseline levels of two key body chemicals associated with cardiovascular health: HDL cholesterol and LDL cholesterol. For each patient, a Framingham coronary heart disease risk score, based on a patient's age, sex, LDL cholesterol, HDL cholesterol, and blood pressure, was calculated by physicians. Study participants were followed for a period of one year, with intermediate weight checks and bloodwork performed at 3 months, 6 months, and 12 months after the beginning of the study. The graphs shown here represent the data that were collected, with the results of the low-fat diet indicated in blue and the low-carbohydrate diet in red. The graph on the left is the change in body weight (y axis) over time (x axis), while the graph on the right indicated changes in the average cardiovascular risk score (y axis) over time (x axis). The lower the cardiovascular risk score, the lower the risk of cardiovascular disease.

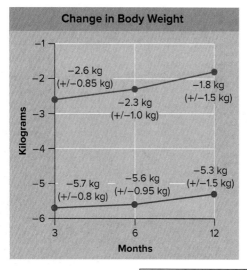
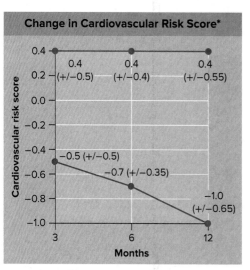

*Estimated 10-year risk for coronary heart disease, represented by Framingham risk score
Source: Annals of Internal Medicine

Analysis

1. **Applying Concepts Variable** In the weight loss graph, which is the dependent variable? In the other graph?
2. **Interpreting Data**
   a. On the red line, at what time point is the maximal amount of weight loss observed?
   b. Which diet resulted in an improvement in cardiovascular health? For this diet, what was the change in the cardiovascular score at three months? At six months?
3. **Making Inferences**
   a. What general statement can be made about individuals being placed on a low-carbohydrate diet versus a low-fat diet?
   b. Patients switching to a low-fat diet ended up with a higher cardiovascular risk score, indicating a slightly higher chance of having a cardiovascular attack. Why do you think that result was observed?
   c. Drawing Conclusions Does this data support the hypothesis that low-carbohydrate diets and low-fat diets are equally effective at decreasing weight and improving overall cardiovascular health?
4. **Further Analysis**
   a. In this experiment, participants on the low-fat diet put weight back on faster than the low-carbohydrate participants. Provide a hypothesis for why the low-fat diet group put weight on faster than the low-carbohydrate diet participants.
   b. Examine the low-carbohydrate diet data. At the beginning of the study, the participants in this group had a very large decrease in their body weight, followed by a slow weight gain over the next nine months. Even with the weight gain, this group still saw decreases in their cardiovascular risk score over the entire study. Why do you think this persistent improvement in cardiovascular health is observed even when some weight is put back on?

4 Cells

LEARNING PATH ▼

The World of Cells
1. Cells

Kinds of Cells
2. Prokaryotic Cells
3. Eukaryotic Cells

Tour of a Eukaryotic Cell
4. The Plasma Membrane
5. The Nucleus: The Cell's Control Center
6. The Endomembrane System
7. Organelles that Harvest Energy
8. The Cytoskeleton: Interior Framework of the Cell

Transport Across Plasma Membranes
9. Diffusion and Osmosis
10. Bulk Passage into and out of Cells
11. Transport through Proteins

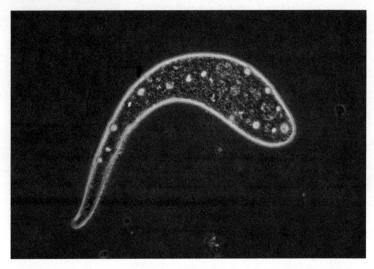

THIS CELL IS a ferocious predator in the microscopic world of pond scum.

Eric V. Grave/Science Source

Zombie Apocalypse in Pond Scum

Few natural communities are more complex, diverse, and hated than pond scum. Small freshwater ponds, when well supplied with nutrients, develop complex populations of microscopic algae and animals on their surface. This is a beauty only a biologist would love, however. What someone else may see when walking by such a pond is a dense layer of green scum covering the surface of the water. Rich with cellular life, the pond may seem dead, icky, and perhaps a little smelly.

A Closer Look

On a microscopic scale, things are a lot more interesting. Imagine for a moment that you could shrink yourself down to the size of the creatures living in that world. You would find yourself in a jungle of green filaments and spirals, often slicing through dense masses of green goo. The filaments are cyanobacteria, the spirals blue-green algae, and the masses single-celled green algae called desmids. All of these creatures have bodies packed chock-full of tiny chlorophyll-rich chloroplasts, and on sunny days they are photosynthesizing and growing like mad.

A Dangerous World

This is not a peaceful world, however. Danger lurks in this green world, in the form of aggressive single-celled killers called *Dileptus*, too small to see with the naked eye.

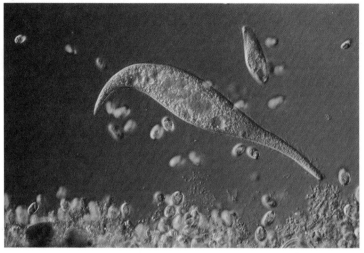

Michael Abbey/Science Source

septemberlegs/Getty Images

An innocent microscopic organism of the pond can at any moment find itself attacked by crowds of these predators, pale and lethal, like an advancing row of zombies in some cheap horror movie—but a lot more deadly. The alarming creature shown in this essay and in the chapter opening photo is a *Dileptus*, magnified a thousand times. At its front end, which you can see better in the photo at top right, is a long, mobile noselike extrusion lined with toxic sacs with which it stuns organisms before consuming them.

Living Small

We are accustomed to thinking of living as something that big animals do: complicated and demanding. Yet everything you as a human must do to survive and prosper, *Dileptus* must do with only the equipment its one tiny cell provides. Just as you move about using legs to walk, so *Dileptus* uses rows of hairlike projections (called cilia) that cover its body to propel itself through the water. Like many oars rowing a boat, they can shoot the body forward through the water with a burst of speed to catch fleeing prey.

Just as your brain is the control center of your body, the compartment called the macronucleus, deep within the interior of *Dileptus*, controls the activities of this complex and very active cell. *Dileptus* has no mouth to take in food as you do. Instead, it takes in food particles and other molecules through its surface, in ways you will learn about later in this chapter.

In a larger sense, this versatile microscopic creature is capable of leading a complex life because the interior of *Dileptus* is subdivided into compartments, each of which carries out different activities. Your body is organized this way too, with digestion going on in your stomach, breathing in your lungs, and thinking in your brain. Functional specialization is the hallmark of the *Dileptus* cell interior, a powerful approach to cellular organization that is shared by all eukaryotes (creatures with a nucleus). In this chapter, we will explore these cell compartments in detail.

The World of Cells

4.1 Cells

> **LEARNING OBJECTIVE 4.1.1** State the three principles of the cell theory.

Hold your finger up and look at it closely. What do you see? Skin. It looks solid and smooth, creased with lines and flexible to the touch. But if you were able to remove a bit and examine it under a microscope, it would look very different. **Figure 4.1** takes you on a journey into your fingertip. The crammed bodies you see in panels ❸ and ❹ are skin cells, laid out like a tiled floor. As your journey continues, you travel inside one of the cells and see organelles, structures in the cell that perform specific functions. Proceeding even farther inward, you encounter the molecules of which the structures are made, and finally the atoms shown in panels ❽ and ❾. While some organisms are composed of a single cell, your body is composed of many cells. A human body has as many cells as there are stars in a galaxy, 10 trillion to 100 trillion, depending on your size. All of your body cells, however, are small. In this chapter, we look more closely at cells and learn something of their internal structure and how they communicate with their environment.

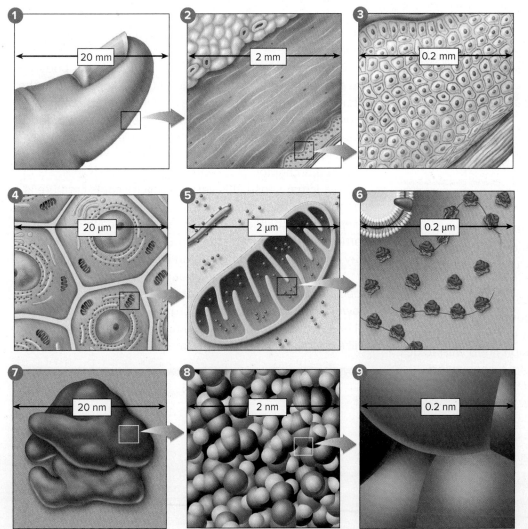

The Cell Theory

Cells are small, so small that no one observed them until microscopes were invented in the mid-17th century. Robert Hooke first described cells in 1665, when he used a microscope he had built to examine a thin slice of nonliving plant tissue called cork. Hooke observed a honeycomb of tiny, empty (because the cells were dead) compartments. He called the compartments in the cork *cellulae* (Latin, small rooms), and the term has come down to us as **cells.** For another century and a half, however, biologists failed to recognize the importance of cells. In 1838, botanist Matthias Schleiden made a careful study of plant tissues and developed the first statement of the cell theory. He stated that all plants "are aggregates of fully individualized, independent, separate beings, namely the cells themselves." In 1839, Theodor Schwann reported that all animal tissues also consist of individual cells.

Figure 4.1 The size of cells and their contents.

This diagram shows the size of human skin cells, organelles, and molecules. In general, the diameter of a human skin cell is a little less than 20 micrometers (μm), of a mitochondrion is 2 μm, of a ribosome is 20 nanometers (nm), of a protein molecule is 2 nm, and of an atom is 0.2 nm.

The idea that all organisms are composed of cells is called the **cell theory**. In its modern form, the cell theory includes three principles:

1. All organisms are composed of one or more cells within which the processes of life occur.
2. Cells are the smallest living things. Nothing smaller than a cell is considered alive.
3. Cells arise only by division of a previously existing cell. Although life likely evolved spontaneously in the environment of the early earth, biologists have concluded that no additional cells are originating spontaneously at present. Rather, life on earth represents a continuous line of descent from those early cells.

> **Putting the Concept to Work**
> Some viruses are larger than a small cell. Are they alive?

Most Cells Are Very Small

> **LEARNING OBJECTIVE 4.1.2** Explain why most cells are so small.

Cells are not all the same size. Individual marine alga cells, for example, can be up to 5 centimeters long—as long as your little finger. In contrast, the cells of your body are typically from 5 to 20 micrometers (μm) in diameter, too small to see with the naked eye. It would take anywhere from 100 to 400 human cells to span the diameter of the head of a pin. The cells of bacteria are even smaller than your cells, only a few micrometers thick (**figure 4.2**).

Why are most cells so tiny? Most cells are small because larger cells do not function as efficiently. In the center of every cell is a command center that must issue orders to all parts of the cell, directing the synthesis of certain enzymes, the entry of ions and molecules from the exterior, and the assembly of new cell parts. These orders must pass from the core to all parts of the cell, and it takes a single molecule a long time to reach the periphery of a large cell. For this reason, an organism composed of relatively small cells has an advantage over one made of large cells.

The Importance of Surface Area. Another reason cells are not larger is the advantage of having a greater surface area. A cell's surface provides the interior's only opportunity to interact with the environment, as its surface provides the only way for substances to pass into and out of the cell. As cells grow larger, their interior volume increases much more than their surface area, and as a result there is far less surface available to service each unit of volume. In the same way, before airplanes and trains were invented, the size of cities was limited because the surrounding countryside could not support all the people living in a big city—only so many roads could be built into the city, only so many farms were close enough to the city to use them.

Some larger cells still function quite efficiently because they have structural features that increase surface area. Cells of the nervous system have long threadlike extensions called axons extending more than a meter in length, so thin that their interior regions are not far from the surface at any point. For the same reason, many body cells are flat and plate-shaped. Another structural feature that increases surface area are small fingerlike projections called microvilli that dramatically increase a cell's surface area.

> **Putting the Concept to Work**
> Why is it advantageous for your body to be made up of so many flat, plate-shaped cells?

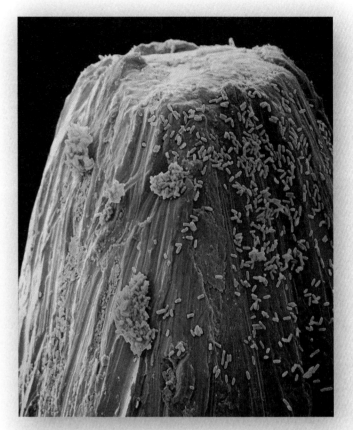

Figure 4.2 **Bacteria on the point of a pin (175×).**
Dr. Tony Brain & David Parker/Science Source

Visualizing Cells

> **LEARNING OBJECTIVE 4.1.3** Explain how biologists are able to visualize cells too small to see with the naked eye.

The reason we can't see small objects like cells is the limited resolution of the human eye. *Resolution* is defined as the minimum distance two points can be apart and still be distinguished as two separated points. The limit of resolution of the human eye is about 100 micrometers. When two objects are closer together than about 100 micrometers, the light reflected from each strikes the same "detector" cell at the rear of the eye.

Microscopes. One way to increase resolution is to increase magnification, so that small objects appear larger. Modern *light microscopes* use two magnifying lenses (and a variety of correcting lenses) to achieve very high magnification and clarity by magnifying the image and focusing it on the receptor cells inside the back of the eye. Microscopes that magnify in stages using several lenses are called compound microscopes. They can resolve structures that are separated by as little as 200 nanometers (nm).

Increasing Resolution. Light microscopes, even compound ones, are not powerful enough to resolve many structures within cells. Why not just add another magnifying lens to the microscope and so increase its resolving power? Because when two objects are closer than a few hundred nanometers, the light beams reflecting from the two images start to overlap. One way to avoid overlap is by using a beam of electrons rather than a beam of light. Electrons have a much shorter wavelength, and a microscope employing electron beams has 1,000 times the resolving power of a light microscope. A transmission electron microscope is capable of resolving objects only 0.2 nanometers apart—just twice the diameter of a hydrogen atom (**figure 4.3**)! A second kind of electron microscope, the scanning electron microscope, beams the electrons onto the surface of the specimen. The beam knocks electrons off of atoms on the surface, these electrons are amplified, and the image created is transmitted to a screen and photographed, producing an often striking three-dimensional picture.

> **Putting the Concept to Work**
> Why can't you see a ribosome with a light microscope?

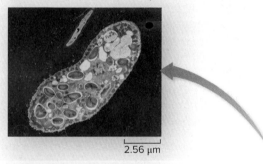

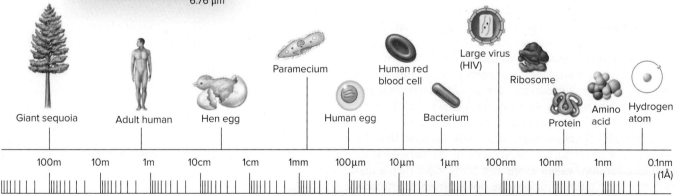

Figure 4.3 A scale of visibility.

Most cells are microscopic in size, although some animal gametes can be seen with the unaided eye. Bacterial cells are generally 1 to 2 μm in diameter.

Top: Michael Abbey/Science Source; **Middle:** Microworks/Phototake; **Bottom:** SPL/Science Source

Kinds of Cells

4.2 Prokaryotic Cells

> **LEARNING OBJECTIVE 4.2.1** Describe the interior of a prokaryotic cell.

There are two major kinds of cells: prokaryotes and eukaryotes. **Prokaryotes** have a relatively uniform interior that is not subdivided by internal membranes into separate compartments. They do not, for example, have special membrane-bounded compartments, called *organelles,* or a *nucleus* (a membrane-bounded compartment that holds hereditary information). As discussed in chapter 1, figure 1.1, the two main groups of prokaryotes are *bacteria* and *archaea;* all other organisms are eukaryotes.

Single-Celled Organisms

Prokaryotes are the simplest cellular organisms. Over 5,000 species are recognized, but doubtless many times that number actually exist and have not yet been described. Although these species are diverse in form, their organization is fundamentally similar: They are single-celled organisms; the cells are small (typically about 1 to 10 micrometers thick); the cells are enclosed by a plasma membrane; and there are no distinct interior compartments. Outside of almost all bacteria and archaea is a *cell wall,* composed of different molecules in different groups (see table 15.1 and section 16.2). In some bacteria, another layer called the *capsule* encloses the cell wall. Archaea are extremely diverse and inhabit diverse environments. Bacteria are abundant and play critical roles in many biological processes. Bacteria assume many shapes, such as the sausage or spiral shapes shown in figure 4.4a,b. They can also adhere in masses or chains, as shown in figure 4.4c, but in these cases the individual cells remain functionally separate from one another.

The Cell Interior

If you were able to peer into a prokaryotic cell, you would be struck by its simple organization. The entire interior of the cell is one unit, with no internal compartments bounded by membranes and little or no internal support structure (the rigid wall, the purple layer surrounding the cell in figure 4.5, supports the cell's shape). Scattered throughout the cytoplasm of prokaryotic cells are small structures called *ribosomes,* the small spherical structures you see inside the cell in figure 4.5. Ribosomes, the sites where proteins are made, lack a membrane boundary. Prokaryotic DNA is found in a region of the cytoplasm called the *nucleoid region.* Although the DNA is localized in this region of the cytoplasm, it is not considered a nucleus because, as you can see in figure 4.5, the nucleoid region and its associated DNA are not enclosed within an internal membrane.

Flagella

Some prokaryotes use a **flagellum** (plural, **flagella**) to move. Flagella are long, threadlike structures made of protein fibers that project from the surface of a cell. They are used in locomotion and feeding. Bacteria can swim at speeds of up to 20 cell diameters per second, rotating their flagella like screws.

Some prokaryotic cells contain **pili** (singular, **pilus**), which are short flagella (only several micrometers long, and about 7.5 to 10 nanometers thick). Pili help the prokaryotic cell attach to appropriate substrates and aid in the exchange of genetic information between cells.

> **Putting the Concept to Work**
> Are there membrane-bounded organelles within a bacterial cell?

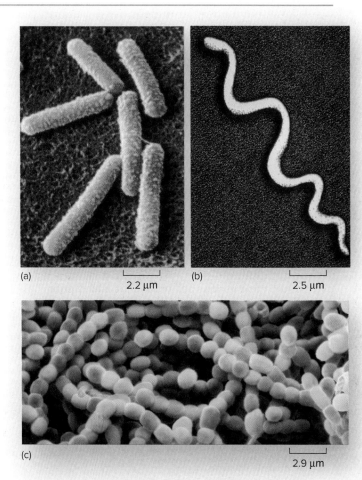

Figure 4.4 Bacterial cells have different shapes.

(a) *Bacillus* is a rod-shaped bacterium. (b) *Treponema* is a coil-shaped bacterium; rotation of internal filaments produces a corkscrew movement. (c) *Streptomyces* is a more or less spherical bacterium in which the individuals adhere in chains.

(A): Andrew Syred/Science Source; (B): Alfred Pasieka/Science Source; (C): Microfield Scientific Ltd/Science Source

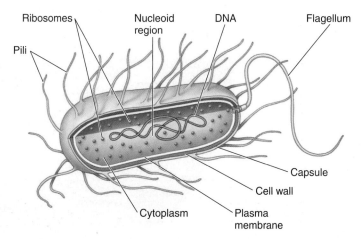

Figure 4.5 Organization of a prokaryotic cell.

Prokaryotic cells lack internal compartments. Not all prokaryotic cells have a flagellum or a capsule like the one illustrated here, but all have a nucleoid region, ribosomes, a plasma membrane, cytoplasm, and a cell wall.

4.3 Eukaryotic Cells

> **LEARNING OBJECTIVE 4.3.1** List the organelles unique to eukaryotic cells, and state which of them are not present in plant cells.

Eukaryotic cells are much larger and profoundly different from prokaryotic cells, with a complex interior organization. Figures 4.6 and 4.7 present cross-sectional diagrams of idealized animal and plant cells. As you can see, the interior of a eukaryotic cell is much more complex than that of the prokaryotic cell you encountered in figure 4.5. The **plasma membrane** ❶ encases a semifluid matrix called the **cytoplasm** ❷, which contains within it the nucleus and various cell structures called organelles. An **organelle** is a specialized structure within which particular cell processes occur. Each organelle, such as a **mitochondrion** ❸, has a specific function in the eukaryotic cell. The organelles are anchored at specific locations in the cytoplasm by an interior scaffold of protein fibers, the **cytoskeleton** ❹.

One of the organelles is very visible when these cells are examined with a microscope, filling the center of the cell like the pit of a peach: the **nucleus** ❺ (plural, *nuclei*), from the Latin word for "kernel." Inside the nucleus, the DNA is wound tightly around proteins and packaged into compact units called chromosomes. It is the nucleus that gives **eukaryotes** their name, from the Greek words *eu*, true, and *karyon*, nut; by way of contrast, the earlier-evolving bacteria and archaea are called prokaryotes ("before the nut").

Figure 4.6 Structure of an animal cell.

In this generalized diagram of an animal cell, the plasma membrane encases the cell, which contains the cytoskeleton and various cell organelles and interior structures suspended in a semifluid matrix called the cytoplasm. Some kinds of animal cells possess fingerlike projections called microvilli. Other types of eukaryotic cells—for example, many protist cells—may possess flagella, which aid in movement, or cilia, which can have many different functions.

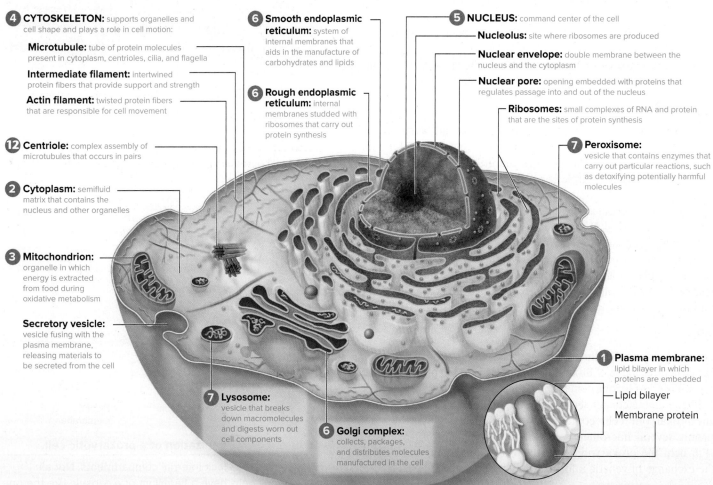

❹ **CYTOSKELETON:** supports organelles and cell shape and plays a role in cell motion:

- **Microtubule:** tube of protein molecules present in cytoplasm, centrioles, cilia, and flagella
- **Intermediate filament:** intertwined protein fibers that provide support and strength
- **Actin filament:** twisted protein fibers that are responsible for cell movement

❶❷ **Centriole:** complex assembly of microtubules that occurs in pairs

❷ **Cytoplasm:** semifluid matrix that contains the nucleus and other organelles

❸ **Mitochondrion:** organelle in which energy is extracted from food during oxidative metabolism

Secretory vesicle: vesicle fusing with the plasma membrane, releasing materials to be secreted from the cell

❻ **Smooth endoplasmic reticulum:** system of internal membranes that aids in the manufacture of carbohydrates and lipids

❻ **Rough endoplasmic reticulum:** internal membranes studded with ribosomes that carry out protein synthesis

❼ **Lysosome:** vesicle that breaks down macromolecules and digests worn out cell components

❻ **Golgi complex:** collects, packages, and distributes molecules manufactured in the cell

❺ **NUCLEUS:** command center of the cell

- **Nucleolus:** site where ribosomes are produced
- **Nuclear envelope:** double membrane between the nucleus and the cytoplasm
- **Nuclear pore:** opening embedded with proteins that regulates passage into and out of the nucleus
- **Ribosomes:** small complexes of RNA and protein that are the sites of protein synthesis

❼ **Peroxisome:** vesicle that contains enzymes that carry out particular reactions, such as detoxifying potentially harmful molecules

❶ **Plasma membrane:** lipid bilayer in which proteins are embedded

- Lipid bilayer
- Membrane protein

If you examine the organelles in figures 4.6 and 4.7, you can see that most of them form separate compartments within the cytoplasm, bounded by their own membranes. *The hallmark of the eukaryotic cell is this compartmentalization.* This internal compartmentalization is achieved by an extensive **endomembrane system** ❻ that weaves through the cell interior.

Vesicles ❼ (small membrane-bounded sacs that store and transport materials) form closed-off compartments in the cell, allowing different processes to proceed simultaneously without interfering with one another, just as rooms do in a house. For example, organelles called *lysosomes* are recycling centers that have acidic interiors in which old organelles are broken down and their component molecules recycled. This acid would be very destructive if released into the cytoplasm.

Comparing figure 4.6 with figure 4.7, you will see the same set of organelles, with a few interesting exceptions. For example, the cells of plants, fungi, and many protists have strong exterior **cell walls** ❽ composed of cellulose or chitin fibers, while the cells of animals lack cell walls. All plants and many kinds of protists have **chloroplasts** ❾, within which photosynthesis occurs. No animal or fungal cells contain chloroplasts. Plant cells also contain a large **central vacuole** ❿ that stores water and **plasmodesmata** ⓫, which are openings in the cell wall that create cytoplasmic connections between cells. **Centrioles** ⓬, which will be described later, are present in animal cells but absent in plant and fungal cells.

> **Putting the Concept to Work**
> If they work so well, why don't animal cells use central vacuoles?

Figure 4.7 Structure of a plant cell.

Most mature plant cells contain large central vacuoles, which occupy a major portion of the internal volume of the cell, and organelles called chloroplasts, within which photosynthesis takes place. The cells of plants, fungi, and some protists have cell walls, although the composition of the walls varies among the groups. Plant cells have cytoplasmic connections through openings in the cell wall called plasmodesmata. Flagella occur in sperm of a few plant species but are otherwise absent in plant and fungal cells. Centrioles are also absent in plant and fungal cells.

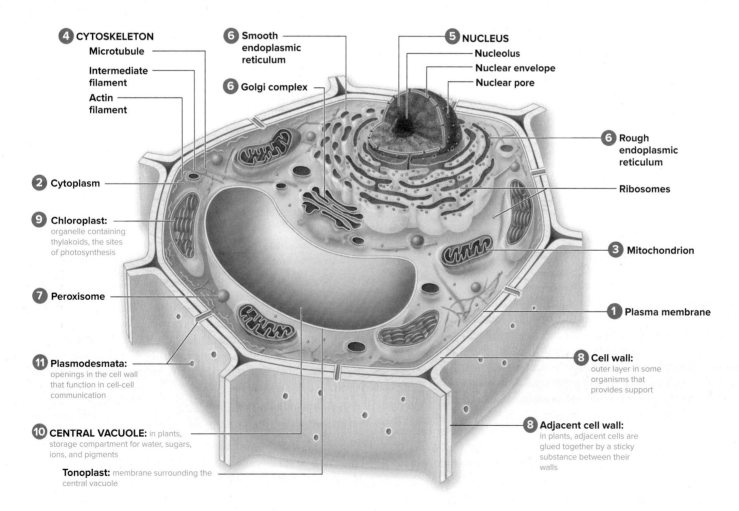

Tour of a Eukaryotic Cell

4.4 The Plasma Membrane

LEARNING OBJECTIVE 4.4.1 Explain how proteins are anchored within the lipid bilayer of a membrane.

Encasing all living cells is a delicate sheet of molecules called the **plasma membrane**. It would take more than 10,000 of these molecular sheets, which are about 5 nanometers thick, piled on top of one another to equal the thickness of this sheet of paper. However, the sheets are not simple in structure, like a soap bubble's skin. Rather, they are made up of a diverse collection of proteins floating within a lipid framework like small boats bobbing on the surface of a pond.

The Fluid Mosaic Model

Regardless of the kind of cell they enclose, all plasma membranes have the same basic structure of proteins embedded in a sheet of lipids, called the **fluid mosaic model**. The fluid mosaic model refers to the assortment of molecules (proteins and lipids) that are constantly moving while still maintaining a barrier between the outside environment and the inside of the cell.

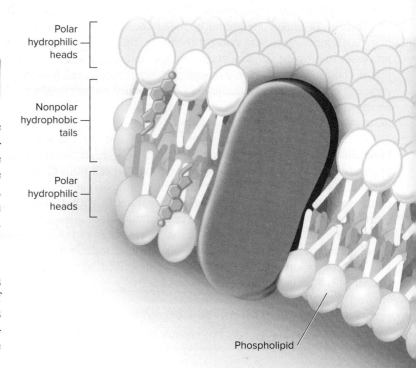

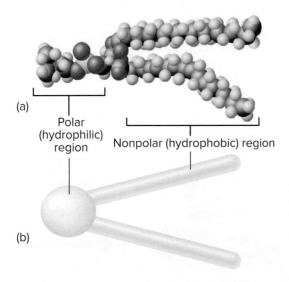

Phospholipids. The lipid layer that forms the foundation of a plasma membrane is composed of modified fat molecules called **phospholipids**. A phospholipid molecule can be thought of as a polar head with two nonpolar tails attached to it, as shown above. The head of a phospholipid molecule has a phosphate chemical group linked to it, making it extremely polar (and thus water-soluble). The other end of the phospholipid molecule is composed of two long fatty acid chains. Recall from chapter 3 that fatty acids are long chains of carbon atoms with attached hydrogen atoms. The carbon atoms are the gray spheres you see above. The fatty acid tails are strongly nonpolar and thus water-insoluble. The phospholipid is often depicted diagrammatically as a ball with two tails.

Lipid Bilayers. Imagine what happens when a collection of phospholipid molecules is placed in water. A structure called a **lipid bilayer** forms spontaneously. How can this happen? The long nonpolar tails of the phospholipid molecules are pushed away by the water molecules that surround them, shouldered aside as the water molecules seek partners that can form hydrogen bonds. After much shoving and jostling, every phospholipid molecule ends up with its polar head facing water and its nonpolar tail facing away from water. The phospholipid molecules form a *double* layer, called a bilayer. As you can see in the figure above, the watery environments inside and outside the plasma membrane push the nonpolar tails to the interior of the bilayer. Because there are two layers with the tails facing each other, no tails are ever in contact with water. Thus, the interior of a lipid bilayer is completely nonpolar, and it repels any water-soluble molecules that attempt to pass through it, just as a layer of oil stops the passage of a drop of water (that's why ducks do not get wet).

Cholesterol, another nonpolar lipid molecule, resides in the interior portion of the bilayer. Cholesterol is a multiringed molecule that affects the fluid nature of the membrane. Although cholesterol is important in maintaining the integrity of the plasma membrane, it can accumulate in blood vessels, forming plaques that lead to cardiovascular disease.

Proteins Within the Membrane

The second major component of every biological membrane is a collection of **membrane proteins** that float within the lipid bilayer. Membrane proteins function as transporters, receptors, and cell

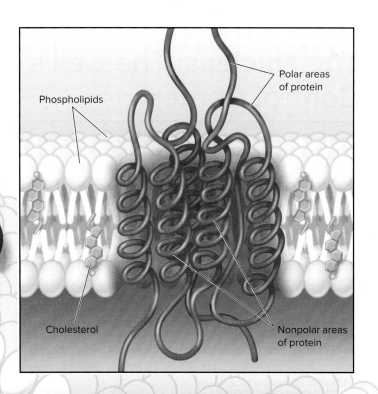

surface markers. As you can see here in the figure, some proteins (the purple structures) pass through the lipid bilayer, providing channels through which molecules and information pass. While some membrane proteins are fixed into position, others move about freely.

Cell Surface Proteins. Many membrane proteins project up from the surface of the plasma membrane like buoys, often with carbohydrate chains or lipids attached to their tips like flags. These *cell surface proteins* act as markers to identify particular types of cells or as beacons to bind specific hormones or proteins to the cell.

Transmembrane Proteins. Proteins that extend all the way across the bilayer can provide passageways for ions and polar molecules like water so they can pass into and out of the cell. How do these *transmembrane proteins* manage to span the membrane, rather than just floating on the surface in the way that a drop of water floats on oil? The part of the protein that actually traverses the lipid bilayer is a specially constructed spiral helix of nonpolar amino acids—the red coiled areas of the transmembrane protein you see in the inset. Water responds to these nonpolar amino acids much as it does to nonpolar lipid chains, and as a result the helical spiral is held within the lipid interior of the bilayer, anchored there by the strong tendency of water to avoid contact with these nonpolar amino acids.

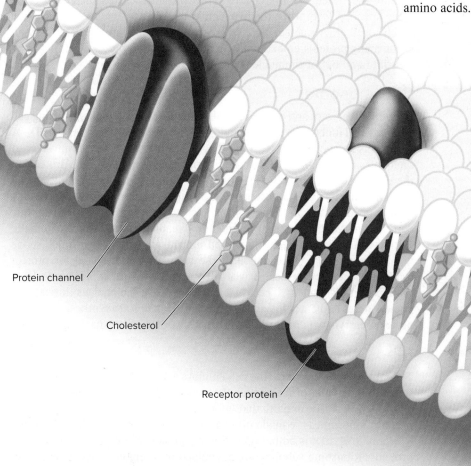

Putting the Concept to Work
How many different proteins are required for a transmembrane channel such as illustrated above?

4.5 The Nucleus: The Cell's Control Center

> **LEARNING OBJECTIVE 4.5.1** Recount two functions of the cell nucleus.

Eukaryotic cells have many structural components and organelles in common (table 4.1). If you were to journey far into the interior of one of your cells, you would eventually reach the center of the cell. There you would find, cradled within a network of fine filaments like a ball in a basket, the **nucleus** ❶ (figure 4.8). The nucleus is the command and control center of the cell, directing all of its activities. It is also the genetic library where the hereditary information is stored.

Nuclear Membrane

The surface of the nucleus is bounded by a special kind of membrane called the **nuclear envelope** ❷. The nuclear envelope is actually *two* membranes, one outside the other, like a sweater over a shirt. The nuclear envelope acts as a barrier between the nucleus and the cytoplasm, but substances need to pass through the envelope. The exchange of materials occurs through openings scattered over the surface of this envelope. Called **nuclear pores** ❸, these openings form when the two membrane layers of the nuclear envelope pinch together. A nuclear pore is not an empty opening, however; rather, it has many proteins embedded within it that permit proteins and RNA to pass into and out of the nucleus ❹.

Chromosomes

In both prokaryotes and eukaryotes, all hereditary information specifying cell structure and function is encoded in DNA. However, unlike prokaryotic DNA that forms an enclosed circle, the DNA of eukaryotes is divided into several segments and associated with protein, forming **chromosomes.** The proteins in the chromosome permit the DNA to wind tightly and condense during cell division. Under a light microscope, these condensed chromosomes are readily seen in dividing cells as densely staining rods. After cell division, eukaryotic chromosomes uncoil and fully extend into threadlike strands called **chromatin** ❺ that can no longer be distinguished individually with a light microscope within the nucleus. Once uncoiled, the chromatin is available for protein synthesis. The process begins when RNA copies of genes are made from the DNA in the nucleus. The RNA molecules leave the nucleus through the nuclear pores and enter the cytoplasm where proteins are synthesized. These proteins, carrying out many different functions, determine what the cell is like.

Ribosomes

To make its many proteins, the cell employs a special structure called a **ribosome,** a kind of platform on which the proteins are built. Ribosomes read the RNA copy of a gene and use that information to direct the construction of a protein. Ribosomes are made up of a special form of RNA called *ribosomal RNA,* or *rRNA,* that is bound up within a complex of several dozen different proteins. Ribosome subunits are assembled in a region within the nucleus called the **nucleolus** ❻.

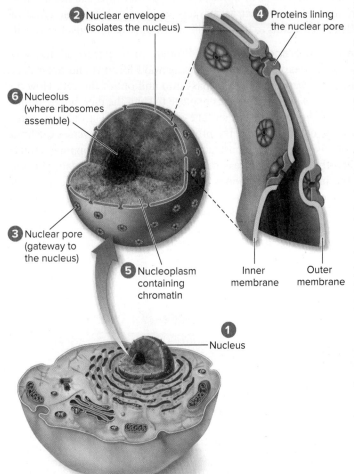

Figure 4.8 The nucleus.

The nucleus is composed of a double membrane, called a nuclear envelope, enclosing a fluid-filled interior containing the chromosomes. In cross section, the individual nuclear pores are seen to extend through the two membrane layers of the envelope. A pore is lined with protein, which acts to control access through the pore.

> **Putting the Concept to Work**
> Where in the cell are ribosomes assembled, and where do they function?

TABLE 4.1 Eukaryotic Cell Organelles and Their Functions

Structure	Description	Function
Structural Elements		
Cell wall	Outer layer of cellulose or chitin; absent in animal cells	Protection; support
Cytoskeleton	Network of protein filaments	Structural support; cell movement
Flagella and cilia	Cellular extensions with 9 + 2 arrangement of pairs of microtubules	Motility or moving fluids over surfaces
Plasma Membrane and Endomembrane System Organelles		
Plasma membrane	Lipid bilayer in which proteins are embedded	Regulates what passes into and out of cell; cell-to-cell recognition
Endoplasmic reticulum	Network of internal membranes	Forms compartments and vesicles; participates in protein and lipid synthesis
Nucleus	Structure (usually spherical) that contains chromosomes; surrounded by double membrane	Control center of cell; directs protein synthesis and cell reproduction
Golgi complex	Stacks of flattened vesicles	Packages proteins for export from the cell; forms secretory vesicles
Lysosomes	Vesicles derived from Golgi complex that contain hydrolytic digestive enzymes	Digest worn-out organelles and cell debris; play role in cell death
Energy-Producing Organelles		
Mitochondria	Bacteria-like elements with double membrane	Sites of oxidative metabolism; provide ATP for cellular energy
Chloroplasts	Bacteria-like organelles found in plants and algae; complex inner membrane consists of stacked vesicles	Sites of photosynthesis
Organelles of Gene Expression		
Chromosomes	Long threads of DNA that form a complex with protein	Contain hereditary information
Nucleolus	Site of genes for rRNA synthesis	Assembles ribosomes
Ribosomes	Small, complex assemblies of protein and RNA, often bound to endoplasmic reticulum	Sites of protein synthesis

4.6 The Endomembrane System

Surrounding the nucleus within the interior of the eukaryotic cell is a tightly packed mass of membranes. They fill the cell, dividing it into compartments, channeling the transport of molecules through the interior of the cell and providing the surfaces on which enzymes act. The system of internal compartments created by these membranes in eukaryotic cells constitutes the most fundamental distinction between the cells of eukaryotes and prokaryotes.

Endoplasmic Reticulum: Transportation System

> **LEARNING OBJECTIVE 4.6.1** Distinguish between rough ER and smooth ER.

The extensive system of internal membranes is called the **endoplasmic reticulum,** often abbreviated **ER**. The term *endoplasmic* means "within the cytoplasm," and the term *reticulum* is a Latin word meaning "little net." Looking at the sheets of membrane weaving through the interior of the cell in **figure 4.9**, you can see how the ER got its name. The ER creates a series of channels and interconnections, and it also isolates some spaces as membrane-enclosed sacs called **vesicles**.

The surface of the ER is the place where the cell makes proteins intended for export (such as enzymes secreted from the cell surface). The surface of those regions of the ER devoted to the synthesis of such transported proteins is heavily studded with ribosomes and appears pebbly, like the surface of sandpaper, when seen through an electron microscope. For this reason, these regions are called **rough ER**. Regions in which ER-bound ribosomes are relatively scarce are correspondingly called **smooth ER**. The surface of the smooth ER is embedded with enzymes that aid in the manufacture of carbohydrates and lipids.

> **Putting the Concept to Work**
> Explain why proteins intended for export are made at rough ER, whereas proteins intended for the cytoplasm are not.

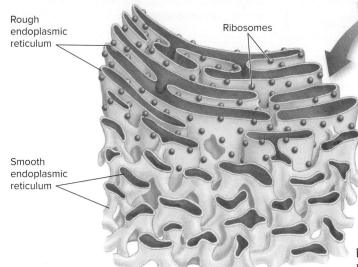

Figure 4.9 The endoplasmic reticulum.

The endoplasmic reticulum provides the cell with an extensive system of internal membranes for the synthesis and transport of materials. Ribosomes are associated with only one side of the rough ER; the other side is the boundary of a separate compartment within the cell into which the ribosomes extrude newly made proteins destined for secretion. Smooth endoplasmic reticulum has few to no bound ribosomes.

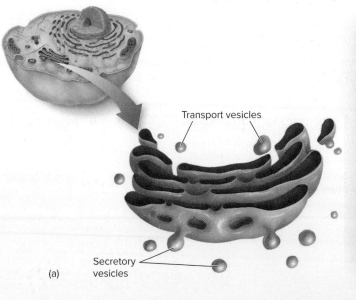

Figure 4.10 Golgi complex.

This vesicle-forming system, called the Golgi complex after its discoverer, is an integral part of the cell's internal membrane system. The Golgi complex processes and packages materials for transport to another region within the cell and/or for export from the cell. (a) Diagram of a Golgi complex. (b) Micrograph of a Golgi complex showing vesicles.

(B): Biophoto Associates/Science Source

The Golgi Complex: Delivery System

> **LEARNING OBJECTIVE 4.6.2** Explain how Golgi bodies ensure the correct delivery of substances made in the ER.

As new molecules are made on the surface of the ER, they are passed from the ER to flattened stacks of membranes called Golgi bodies (figure 4.10), which look like pancakes stacked one on top of the other. Golgi bodies function in the collection, packaging, and distribution of molecules manufactured in the cell. Scattered through the cytoplasm, Golgi bodies are collectively referred to as the **Golgi complex.**

The rough ER, smooth ER, and Golgi work together as an endomembrane transport system in the cell. Figure 4.11 walks you through the path that molecules take from the ER through the Golgi and out to their final destinations. Proteins and lipids that are manufactured on the ER membranes are transported through the channels of the ER and are packaged into transport vesicles that bud off from the ER ❶. The vesicles fuse with the membrane of the Golgi bodies, dumping their contents into the Golgi ❷. Within the Golgi bodies, the molecules may take one of many paths, indicated by the branching arrows in figure 4.11. Many of these molecules become tagged with carbohydrates. The molecules collect at the ends of the membranous folds of the Golgi bodies; these folds are given the special name *cisternae* (Latin, collecting vessels). Vesicles that pinch off from the cisternae carry the molecules to the different compartments of the cell ❸ and ❹, or to the inner surface of the plasma membrane, where molecules to be secreted are released to the outside ❺.

Lysosomes. Other organelles called lysosomes arise from the Golgi complex (the light orange vesicle budding at ❸) and contain a concentrated mix of the powerful enzymes manufactured in the rough ER that break down macromolecules. Lysosomes are the recycling centers of the cell, breaking down failing organelles and other structures within cells, digesting worn-out cell components, and recycling the proteins and other materials of the old parts.

> **Putting the Concept to Work**
> How do vesicles pinched off from the Golgi "know" where to go?

Vacuoles: Storage Compartments

> **LEARNING OBJECTIVE 4.6.3** Describe the function of plant vacuoles.

The interiors of plant and many protist cells contain membrane-bounded storage compartments called **vacuoles.** The center of the plant cell shown in figure 4.12, as in all plant cells, contains a large, apparently empty space, called the *central vacuole*. This vacuole is not really empty; it contains large amounts of water and other materials, such as sugars, ions, and pigments. The central vacuole functions as a storage center for these important substances.

> **Putting the Concept to Work**
> What do you think the function of the vacuole membrane is? Explain.

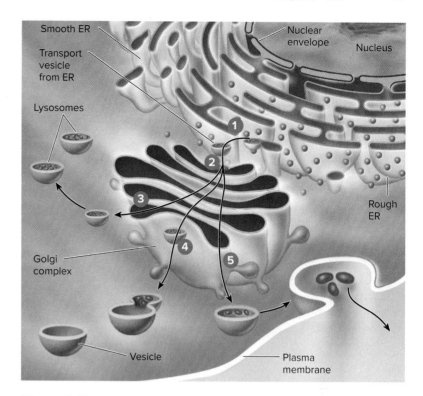

Figure 4.11 How the endomembrane system works.

A highly efficient highway system within the cell, the endomembrane system transports material from the ER to the Golgi and from there to other destinations.

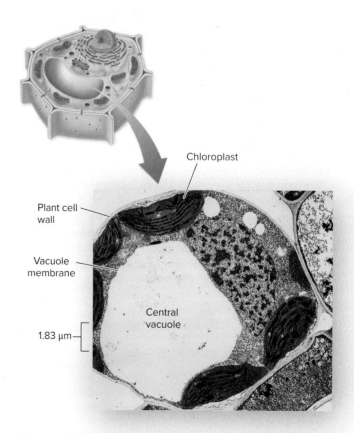

Figure 4.12 A plant's central vacuole.

A plant's central vacuole stores dissolved substances and can grow in size to grow the surface area of a plant cell.

Biophoto Associates/Science Source

4.7 Organelles that Harvest Energy

> **LEARNING OBJECTIVE 4.7.1** Differentiate between mitochondria and chloroplasts.

Eukaryotic cells contain several kinds of complex, energy-harvesting organelles that contain their own DNA and appear to have been derived from ancient bacteria. These organelles include mitochondria (which occur in the cells of all but a very few eukaryotes) and chloroplasts (which occur only in algae and plant cells—they do not occur in animal or fungal cells).

Mitochondria: Powerhouses of the Cell

Eukaryotic organisms extract energy from organic molecules ("food") in a complex series of chemical reactions called **cellular respiration,** which takes place only in their mitochondria. **Mitochondria** (singular, **mitochondrion**) are sausage-shaped organelles about the size of a bacterial cell. Mitochondria are bounded by two membranes. The outer membrane, shown partially cut away in **figure 4.13**, is smooth and apparently derives from the plasma membrane of the host cell that first took up the bacterium long ago. The inner membrane, apparently the plasma membrane of the bacterium that gave rise to the mitochondrion, is bent into numerous folds called **cristae** (singular, **crista**) that resemble the folded plasma membranes in various groups of bacteria. The cutaway view of the figure shows how the cristae partition the mitochondrion into two compartments, an inner **matrix** and an outer compartment, called the **intermembrane space.** As you will learn in chapter 7, this architecture is critical to successfully carrying out cellular respiration. Mitochondria have a circular molecule of DNA, called mitochondrial DNA, that closely resembles the circular DNA molecule of a bacterium. On this mtDNA are several genes that produce some of the proteins essential for cellular respiration.

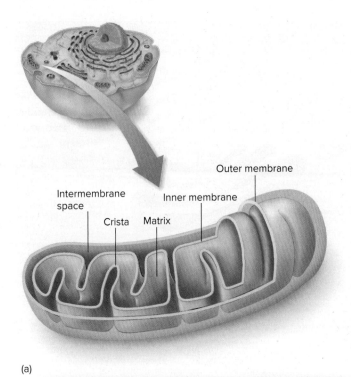

(a)

(b)

Figure 4.13 Mitochondria.
The mitochondria of a cell are sausage-shaped organelles within which oxidative metabolism takes place, and energy is extracted from food using oxygen. (a) A mitochondrion has a double membrane. The inner membrane is shaped into folds called cristae. The space within the cristae is called the matrix. The cristae greatly increase the surface area for oxidative metabolism. (b) Micrograph of mitochondrion, cut lengthwise.
(B): BSIP SA/Alamy Stock Photo

Chloroplasts: Energy-Capturing Centers

Some organisms are able to capture energy from sunlight and use it to power the production of sugars in a process called **photosynthesis.** All photosynthesis in plants and algae takes place within another bacteria-like organelle, the **chloroplast** (figure 4.14). There is strong evidence that chloroplasts, like mitochondria, were derived from bacteria by a mutually sharing arrangement called a symbiosis. A chloroplast is bounded, like a mitochondrion, by two membranes, the inner derived from the original bacterium and the outer resembling the host cell's ER. Chloroplasts are larger than mitochondria and have a more complex organization. Inside the chloroplast, another series of membranes is fused to form stacks of closed vesicles called **thylakoids.** The energy-harvesting reactions of photosynthesis take place within the thylakoids. The thylakoids are stacked on top of one another to form a column called a **granum** (plural, **grana**). The interior of a chloroplast is bathed with a semiliquid substance called the **stroma.**

Like mitochondria, chloroplasts have a circular DNA molecule. This DNA contains many of the genes coding for the proteins necessary to carry out photosynthesis. Plant cells can contain from one to several hundred chloroplasts, depending on the species. Neither mitochondria nor chloroplasts can be grown in a cell-free culture; they are totally dependent on the cells within which they occur.

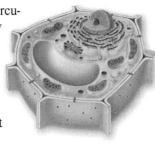

> **Putting the Concept to Work**
> What exactly does a mitochondrion need to get from a living cell?

Theory of Endosymbiosis

> **LEARNING OBJECTIVE 4.7.2** Describe the evidence that mitochondria evolved from ancient bacteria.

Symbiosis is a close, integrated relationship between organisms of different species that live together. The theory of **endosymbiosis** proposes that some of today's eukaryotic organelles evolved by a symbiosis in which one cell of a prokaryotic species was engulfed by and lived inside the cell of another species of prokaryote that was a precursor to eukaryotes. **Figure 4.15** shows how this is thought to have occurred. Many cells take up food or other substances through endocytosis, a process whereby the plasma membrane of a cell wraps around the substance, enclosing it within a vesicle inside the cell. According to the endosymbiont theory, the engulfed prokaryotes provided their hosts with certain advantages associated with their special metabolic abilities. Two key eukaryotic organelles just described are believed to be the descendants of these endosymbiotic prokaryotes: mitochondria, which are thought to have originated as bacteria capable of carrying out oxidative metabolism; and chloroplasts, which apparently arose from photosynthetic bacteria.

The endosymbiont theory is supported by a wealth of evidence. Both mitochondria and chloroplasts are surrounded by two membranes; the inner membrane probably evolved from the plasma membrane of the engulfed bacterium, while the outer membrane is probably derived from the plasma membrane or ER of the host cell. Mitochondria are about the same size as most bacteria, and the cristae formed by their inner membranes resemble the folded membranes in various groups of bacteria. Mitochondrial ribosomes are also similar to bacterial ribosomes in size and structure. Both mitochondria and chloroplasts contain circular molecules of DNA similar to those in bacteria. Finally, mitochondria divide by simple fission, splitting in two just as bacterial cells do, and they apparently replicate and partition their DNA in much the same way as bacteria do.

During the 1.5 billion years in which chloroplasts and mitochondria have existed in eukaryotic cells, most of their genes have been transferred to the chromosomes of the host cells. But chloroplasts and mitochondria still have some of their original genes.

> **Putting the Concept to Work**
> What is the principal difference between mitochondria and chloroplasts?

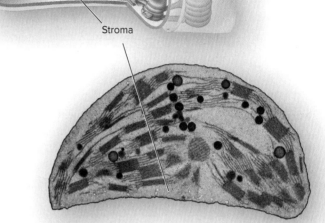

Figure 4.14 A chloroplast.

Bacteria-like organelles called chloroplasts are the sites of photosynthesis in photosynthetic eukaryotes. Like mitochondria, they have a complex system of internal membranes on which chemical reactions take place.

Photo Researchers/Science History Images/Alamy Stock Photo

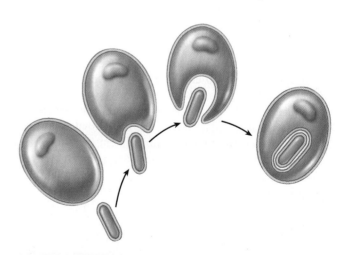

Figure 4.15 Endosymbiosis.

This figure shows how a double membrane may have been created during the symbiotic origin of mitochondria or chloroplasts.

4.8 The Cytoskeleton: Interior Framework of the Cell

LEARNING OBJECTIVE 4.8.1 Describe the protein fibers of the cytoskeleton.

If you were to shrink down and enter into the interior of a eukaryotic cell, your view would be similar to what you see in the illustration shown here: a dense network of protein fibers called the **cytoskeleton** provides a framework that supports the shape of the cell. The cytoskeleton also anchors organelles such as the mitochondria to fixed locations within the cell interior. The protein fibers of the cytoskeleton are a dynamic system, constantly being formed and disassembled. There are three different kinds of protein fibers that make up the cytoskeleton, shown as enlargements below.

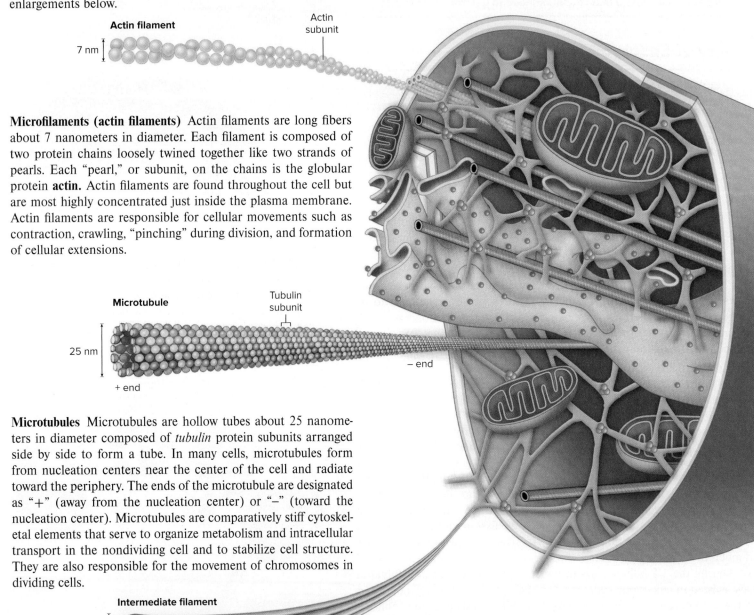

Microfilaments (actin filaments) Actin filaments are long fibers about 7 nanometers in diameter. Each filament is composed of two protein chains loosely twined together like two strands of pearls. Each "pearl," or subunit, on the chains is the globular protein **actin**. Actin filaments are found throughout the cell but are most highly concentrated just inside the plasma membrane. Actin filaments are responsible for cellular movements such as contraction, crawling, "pinching" during division, and formation of cellular extensions.

Microtubules Microtubules are hollow tubes about 25 nanometers in diameter composed of *tubulin* protein subunits arranged side by side to form a tube. In many cells, microtubules form from nucleation centers near the center of the cell and radiate toward the periphery. The ends of the microtubule are designated as "+" (away from the nucleation center) or "−" (toward the nucleation center). Microtubules are comparatively stiff cytoskeletal elements that serve to organize metabolism and intracellular transport in the nondividing cell and to stabilize cell structure. They are also responsible for the movement of chromosomes in dividing cells.

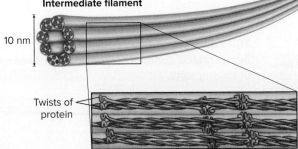

Intermediate filaments Intermediate filaments are composed of overlapping staggered twists of protein. These twists are then bundled into cables. This molecular arrangement allows for a ropelike structure that imparts tremendous mechanical strength to the cell. Intermediate filaments are intermediate in size between actin filaments and microtubules. Once formed, intermediate filaments are stable and usually do not break down. They provide structural reinforcement to the cell and organelles.

> **Putting the Concept to Work**
> Which of the three fibers of the cytoskeleton is the thickest?

Flagella and Cilia

> **LEARNING OBJECTIVE 4.8.2** Explain how animal cells move.

Some eukaryotic cells contain flagella (singular, flagellum), fine, long, threadlike organelles protruding from the cell surface. **Figure 4.16a** shows how a flagellum arises from a microtubular structure called a basal body, with groups of microtubules arranged in rows of three, shown in the cross-sectional view. Some of these microtubules extend up into the flagellum, which consists of a circle of nine microtubule pairs surrounding two central microtubules. This 9 + 2 arrangement is a fundamental feature of eukaryotes and apparently evolved early in their history. In humans, we find a single long flagellum on each sperm cell that propels the cell in a swimming motion. If flagella are numerous and organized in dense rows, they are called cilia (**figure 4.16b**). Cilia do not differ from flagella in their structure, but cilia are usually shorter. In humans, dense mats of cilia project from cells that line our breathing tube, the trachea, to move mucus and dust particles out of the respiratory tract into the throat (where we can expel these contaminants by spitting or swallowing). Eukaryotic flagella serve a similar function as the bacterial flagella discussed in section 4.2, but are very different structurally.

> **Putting the Concept to Work**
> Which element of the cytoskeleton is active in cell movement?

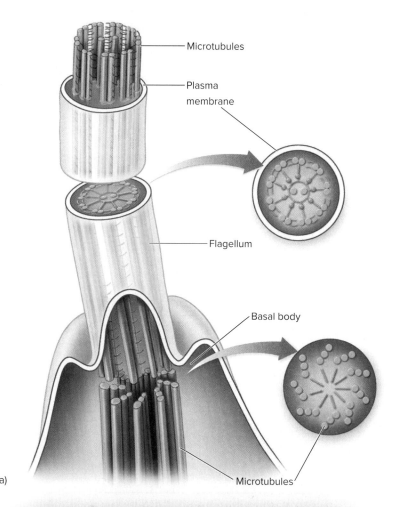

(a)

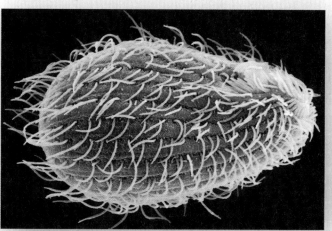

(b)

Figure 4.16 **Flagella and cilia.**

(a) A eukaryotic flagellum springs directly from a basal body and is composed of a ring of nine pairs of microtubules with two microtubules in its core. (b) The surface of this microscopic *Tetrahymena* is covered with a dense forest of cilia, which it uses to propel itself.

(B): Aaron J. Bell/Science Source

Transport Across Plasma Membranes

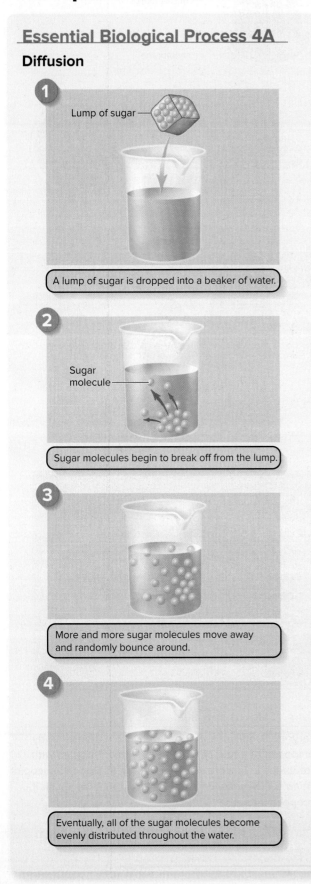

Essential Biological Process 4A

Diffusion

1. A lump of sugar is dropped into a beaker of water.
2. Sugar molecules begin to break off from the lump.
3. More and more sugar molecules move away and randomly bounce around.
4. Eventually, all of the sugar molecules become evenly distributed throughout the water.

4.9 Diffusion and Osmosis

For cells to survive, nutrients, water, and other materials must pass into the cell, and waste materials must be eliminated. All of this moving back and forth across the cell's plasma membrane occurs in one of three ways: (1) water and other substances diffuse through the membrane; (2) food particles are engulfed by the plasma membrane folding around them; or (3) proteins in the membrane act as doors that admit certain molecules only.

Diffusion

> **LEARNING OBJECTIVE 4.9.1** Explain why diffusion occurs down a concentration gradient, rather than up.

How a molecule moves—just where it goes—is totally random, like shaking marbles in a cup, so if two kinds of molecules are added together, they soon mix. The random motion of molecules always tends to produce uniform mixtures because a substance moves from regions where its concentration is high to regions where its concentration is lower (that is, *down* the **concentration gradient**). How does a molecule "know" in what direction to move? It doesn't—molecules don't "know" anything. A molecule is equally likely to move in any direction and is constantly changing course in random ways. There are simply more molecules able to move from where they are common than from where they are scarce. This mixing process is called **diffusion**. Diffusion (*Essential Biological Process 4A*) is the net movement of molecules down a concentration gradient toward regions of lower concentration (that is, where there are relatively fewer of them) as a result of random motion. For example, a lump of sugar dropped into a beaker of water will break apart into individual sugar molecules that will move about randomly. However, they will tend to travel away from the area of high concentration (the sugar cube) to an area of lower concentration (the rest of the beaker). Eventually, the substance will achieve a state of *equilibrium*, where there is no net movement toward any particular direction (as shown in panel 4). The individual molecules of the substance are still in motion, but there is no net change in direction.

> **Putting the Concept to Work**
> Would you expect polar molecules to diffuse into water more completely than nonpolar ones? Explain.

Osmosis

> **LEARNING OBJECTIVE 4.9.2** Discuss how solute concentration affects the movement of water by osmosis.

Water moves into and out of cells by a form of diffusion called **osmosis**.

Free Water. As in diffusion, water passes across a cell membrane down its concentration gradient, a process called osmosis (*Essential Biological Process 4B*). To understand how water moves into and out of a cell, let's first focus on the water molecules already present inside a cell. What are they doing? Many of them are interacting with the sugars, proteins, and other polar molecules inside. Remember, water is very polar itself and readily interacts with other polar molecules. Instead of freely moving about, a shell of water molecules remains clustered around each polar molecule inside the cell. As a result, a water molecule coming into the cell by random motion may not be free to come out again. For example, in *Essential Biological Process 4B,* the addition of a polar solute reduces the number of free water molecules on the right side of the beaker, which can be thought of as the inside of a cell. Because the "outside" of the cell (on the left) has more unbound water molecules, water moves by diffusion into the cell (to the right).

Aquaporins. Ions and polar molecules cannot cross the very nonpolar environment found in the lipid core of the membrane bilayer. However, the movement of water molecules, which are very polar, is not blocked—water diffuses freely across the plasma membrane. How is this possible? Water molecules pass through small channels, called **aquaporins,** that traverse the membrane. While water can cross the membrane very slowly on its own, aquaporins allow water to move much quicker. These water channels are very selective, even blocking the passage of protons (hydrogen ions), which are smaller than water molecules—a cluster of positively charged amino acids that line the pore repel protons, which are also positively charged.

Osmotic Concentration. The concentration of *all* molecules dissolved in a solution (the solutes) is called the osmotic concentration of the solution. If the osmotic concentrations of two solutions are equal, the solutions are isotonic (Greek *iso,* the same). If two solutions have unequal osmotic concentrations, the solution with the higher solute concentration is said to be hypertonic (Greek *hyper,* more than), and the solution with the lower one is hypotonic (Greek *hypo,* less than).

Movement of water into a cell by osmosis creates pressure, called osmotic pressure, which can cause a cell to swell and burst (**figure 4.17**). Most animal cells cannot withstand osmotic pressure unless their plasma membranes are braced to resist the swelling. If placed in pure water, they soon burst like overinflated balloons. That is why the cells of so many kinds of organisms have cell walls to stiffen their exteriors. In fact, this osmotic pressure, called turgor pressure in plants, is important for plant cells to maintain their shape. Without adequate water inside the cells, the plants wilt. In animals, the fluids bathing the cells have as many polar molecules dissolved in them as the cells do, making them isotonic, so the problem doesn't arise.

> **Putting the Concept to Work**
> Explain how water diffuses across the plasma membrane.

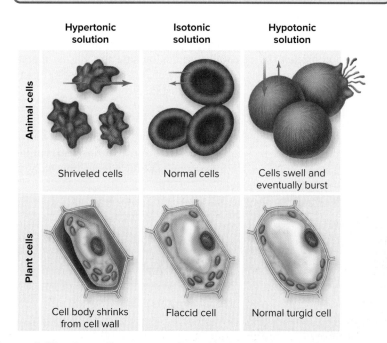

Figure 4.17 Osmotic pressure in animal and plant cells.

Essential Biological Process 4B
Osmosis

1 Semipermeable membrane / Water molecules / Isotonic

Diffusion causes water molecules to distribute themselves equally on both sides of a semipermeable membrane.

2 Hypotonic / Hypertonic / Urea

Addition of solute molecules that cannot cross the membrane reduces the number of free water molecules on that side, as they bind to the solute.

3

Diffusion then causes free water molecules to move from the side where their concentration is higher to the solute side, where their concentration is lower.

4.10 Bulk Passage into and out of Cells

> **LEARNING OBJECTIVE 4.10.1** Distinguish phagocytosis from pinocytosis.

Not all movement of materials into and out of cells passes through channels.

Endocytosis

The cells of many eukaryotes take in food and liquids by extending their plasma membranes outward toward food particles. The membrane engulfs the particle and forms a vesicle—a membrane-bounded sac—around it. This process is called **endocytosis** (figure 4.18).

If the material the cell takes in is particulate (made up of discrete particles), such as an organism, like the red bacterium in figure 4.18a, or some other fragment of organic matter, the process is called **phagocytosis** (Greek *phagein,* to eat, and *cytos,* cell). If the material the cell takes in is liquid or substances dissolved in a liquid, like the small particles in figure 4.18b, it is called **pinocytosis** (Greek *pinein,* to drink). Pinocytosis is common among animal cells. Mammalian egg cells, for example, "nurse" from surrounding cells; the nearby cells secrete nutrients that the maturing egg cell takes up by pinocytosis. Virtually all eukaryotic cells constantly carry out these kinds of endocytosis, trapping particles and extracellular fluid in vesicles and ingesting them. Endocytosis rates vary from one cell type to another. They can be surprisingly high: Some types of white blood cells ingest 25% of their cell volume each hour!

Exocytosis

The reverse of endocytosis is **exocytosis,** the discharge of material from vesicles at the cell surface. The vesicle in figure 4.19 contains a substance to be discharged, or released, from the cell. The purple particles remain suspended in the vesicle as it fuses with the plasma membrane. The membrane that forms the vesicle is made of phospholipids, and as it comes in contact with the plasma membrane, the phospholipids of both membranes interact, forming a pore through which the contents leave the vesicle to the outside. In plant cells, exocytosis is an important means of exporting the materials needed to construct the cell wall that lies outside the plasma membrane. In animal cells, exocytosis provides a mechanism for secreting many hormones, digestive enzymes, and other substances.

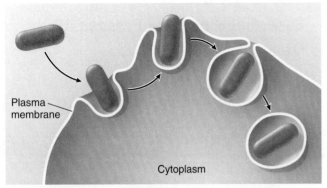

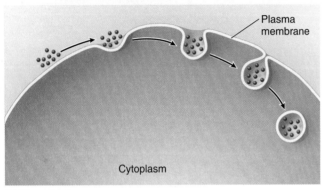

Figure 4.18 Endocytosis.

Endocytosis is the process of engulfing material by folding the plasma membrane around it, forming a vesicle. (a) When the material is an organism or some other relatively large fragment of organic matter, the process is called phagocytosis. (b) When the material is a liquid, the process is called pinocytosis.

Figure 4.19 Exocytosis.

Exocytosis is the discharge of material from vesicles at the cell surface. Proteins and other molecules are secreted from cells in small pockets called secretory vesicles, whose membranes fuse with the plasma membrane, thereby allowing the secretory vesicles to release their contents to the cell surface.

©Dr. Birgit Satir, Albert Einstein College of Medicine

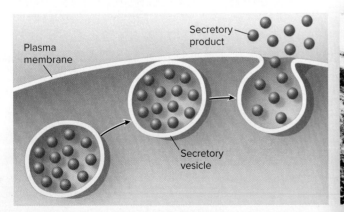

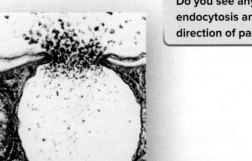

> **Putting the Concept to Work**
> Do you see any basic difference between endocytosis and exocytosis, except the direction of particle movement?

4.11 Transport through Proteins

> **LEARNING OBJECTIVE 4.11.1** Distinguish between selective diffusion and facilitated diffusion.

From the point of view of efficiency, the problem with endocytosis is that it is expensive to carry out—the cell must make and move a lot of membrane. Also, endocytosis is not picky—in pinocytosis particularly, engulfing liquid does not allow the cell to choose which molecules come in. Cells solve this problem by using proteins in the plasma membrane as channels and carrier proteins to pass molecules into and out of the cell. Because each kind of transport allows passage of only a certain kind of molecule, the cell can control what enters and leaves, an ability called **selective permeability.**

Selective Diffusion

Some transport proteins are channels that act like open doors. As long as a molecule fits the channel, it is free to pass through in either direction. Diffusion tends to equalize the concentration of such molecules on both sides of the membrane, with the molecules moving toward the side where they are scarcest. This mechanism of transport is called **selective diffusion.** One class of selectively open channels consists of ion channels, which are pores that span the membrane. Ions that fit the pore can diffuse through it in either direction. Such ion channels play an essential role in signaling by the nervous system.

Facilitated Diffusion

Most diffusion occurs through use of special carrier proteins. These proteins bind only certain kinds of molecules, such as a particular sugar, amino acid, or ion. The molecule physically binds to the carrier on one side of the membrane and is released to the other side. The direction of the molecule's net movement depends on its concentration gradient across the membrane. If the concentration is greater outside the cell, the molecule is more likely to bind to the carrier on the extracellular side of the membrane, as shown in **panel 1** of *Essential Biological Process 4C*, and be released on the cytoplasmic side, as in **panel 3**. If the concentration of the molecule is greater inside the cell, the net movement will be from inside to outside. Thus the net movement always occurs from high concentration to low, just as it does in simple diffusion, but the process is facilitated by the carriers. For this reason, this mechanism of transport is given a special name, **facilitated diffusion.**

A characteristic feature of transport by carrier proteins is that its rate can be saturated. If the concentration of a substance is progressively increased, the rate of transport of the substance increases up to a certain point and then levels off. There are a limited number of carrier proteins in the membrane, and when the concentration of the transported substance is raised high enough, all the carriers will be in use. The transport system is then said to be "saturated." When an investigator wishes to know if a particular substance is being transported across a membrane by a carrier protein, or is diffusing across, he or she conducts experiments to see if the transport system can be saturated. If it can be saturated, it is carrier-mediated; if it cannot be saturated, it is not.

> **Putting the Concept to Work**
> Explain why saturation indicates the action of a carrier protein.

Essential Biological Process 4C
Facilitated Diffusion

1 Particular molecules can bind to special protein carriers in the plasma membrane.

2 The protein carrier helps (facilitates) the diffusion process and does not require energy.

3 The molecule is released on the far side of the membrane. Protein carriers transport only certain molecules across the membrane but will take them in either direction down their concentration gradients.

Today's Biology

When Membranes Don't Work Right

Cystic fibrosis is a fatal disease in which the body cells of affected individuals secrete a thick mucus that clogs the airways of the lungs. The cystic fibrosis patient in the photograph is breathing into a Vitalograph, a device that measures lung function. Cystic fibrosis is usually thought of as a children's disease because until recently few affected individuals lived long enough to become adults. Even today, half die before their mid-twenties. There is no known cure.

Cystic Fibrosis Is a Gene Defect

Cystic fibrosis results from a defect in a single gene that is passed down from parent to child. It is the most common fatal genetic disease of Caucasians. One in 20 individuals possesses at least one copy of the defective gene. Most of these individuals are not afflicted with the disease; only those children who inherit a copy of the defective gene from each parent succumb to cystic fibrosis—about 1 in 2,500 infants.

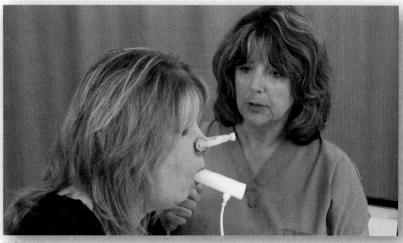
McGraw Hill

What Exactly Is the Problem?

Cystic fibrosis has proven difficult to study, as many organs are affected by the disease. What is the core problem? The first clear clue was obtained when an investigator seized on a commonly observed characteristic of cystic fibrosis patients, that their sweat is abnormally salty. They isolated a sweat duct from a small piece of skin and placed it in a solution of salt (NaCl) that was three times as concentrated as the NaCl inside the duct. They then monitored the movement of ions. Diffusion tends to drive both the sodium (Na^+) and the chloride (Cl^-) ions into the duct because of the higher outside ion concentrations. In skin isolated from normal individuals, Na^+ and Cl^- both entered the duct, as expected. In skin isolated from cystic fibrosis individuals, however, only Na^+ entered the duct—no Cl^- entered. For the first time, the molecular nature of cystic fibrosis became clear. Water accompanies chloride and was not entering the ducts because chloride was not, thus creating thick mucus. Cystic fibrosis is a defect in a plasma membrane protein called CFTR (*c*ystic *f*ibrosis *t*ransmembrane *c*onductance *r*egulator) that normally regulates passage of Cl^- into and out of the body's cells.

Repairing cf in Rats

The defective *cf* gene was soon isolated, and its position on a particular human chromosome (chromosome 7) was pinpointed. Experiments were immediately begun to see if it would be possible to cure cystic fibrosis by gene therapy—that is, by transferring healthy *cf* genes into the cells with defective ones. First results were promising: a working *cf* gene was successfully transferred into the lung cells of a living animal, a rat. The *cf* gene was first inserted into the DNA of a cold virus called adenovirus that easily infects lung cells of living animals. The treated virus was then inhaled by the rat. Carried piggyback, the *cf* gene entered the rat's lung cells and began producing the normal human CFTR protein within these cells!

These results were very encouraging, and at first, the future for all cystic fibrosis patients seemed bright. Clinical tests using adenovirus to introduce healthy *cf* genes into cystic fibrosis patients were begun with much fanfare.

A Human Is Not a Rat

They were not successful. As described in detail in chapter 13, there were insurmountable problems with the adenovirus being used to transport the *cf* gene into human cystic fibrosis patients. Administered to humans, the adenovirus frequently led to cancer! It turned out that sometimes adenovirus enters into the human chromosome—right in the middle of a key cancer-protecting gene. This disables the gene, blocking its ability to ward off cancer. The cure proved more deadly than the disease.

Enter CRISPR

Recently, investigators have begun attempting to use a molecular tool called CRISPR to edit out the defect in the *cf* gene directly, eliminating the need for a virus. As explained in chapter 12, CRISPR allows a researcher to snip out a particular DNA sequence and replace it with another. In this case, the defective *cf* sequence is removed, and the healthy one inserted in its place. While figuring out exactly how to do this efficiently is no simple task, this approach shows great promise. The steady persistence of researchers has taken us a long way, and again the future for cystic fibrosis patients seems bright.

Active Transport

LEARNING OBJECTIVE 4.11.2 Describe the operation of the sodium-potassium pump, including the role of ATP.

Other carrier proteins through the plasma membrane are closed doors. These proteins open only when energy is provided. They are designed to enable the cell to maintain high or low concentrations of certain molecules, much more or less than exists outside the cell. Like motor-driven turnstiles, these transport proteins operate to move a certain substance *up* its concentration gradient. The operation of these one-way, energy-requiring proteins results in **active transport,** the movement of molecules across a membrane to a region of higher concentration by the expenditure of energy.

You might think that the plasma membrane possesses all sorts of active transport proteins for the transport of sugars, amino acids, and other molecules, but in fact, most of the active transport in cells is carried out by one kind of transporter, the sodium-potassium pump.

The Sodium-Potassium Pump. The most important active transport protein is the *sodium-potassium (Na^+-K^+) pump,* which expends metabolic energy to actively pump sodium ions (Na^+) out of cells and potassium ions (K^+) into cells (*Essential Biological Process 4D*). More than one-third of all the energy expended by your body's cells is spent driving Na^+-K^+ pump carrier proteins. This energy is derived from *adenosine triphosphate (ATP),* a molecule we will learn more about in chapter 5. The transportation of two different ions in opposite directions happens because energy causes a change in the shape of the protein carrier. The panels to the right walk you through one cycle of the pump. Each transport protein can move over 300 sodium ions per second when working full tilt. As a result of all this pumping, there are far fewer sodium ions in the cell.

Coupled Transport. The plasma membranes of many cells are studded with facilitated diffusion transport proteins, which offer a path for sodium ions that have been pumped out by the Na^+-K^+ pump to diffuse back in. There is a catch, however; these transport proteins require that the sodium ions have a partner in order to pass through which is why these are called *coupled* transport proteins. Coupled transport proteins won't let sodium ions across unless another molecule tags along, crossing hand in hand with the sodium ion. In some cases, the partner molecule is a sugar; in others, an amino acid or other molecule. Because the concentration gradient for sodium is so large, many sodium ions are trying to get back in, and this diffusion pressure drags in the partner molecules as well, even if they are already in high concentration within the cell. In this way, sugars and other actively transported molecules enter the cell—via special coupled transport protein channels.

Putting the Concept to Work
Explain how cells are able to take up sugar even when they already have a high concentration of that molecule in their cytoplasm.

Essential Biological Process 4D
The Sodium-Potassium Pump

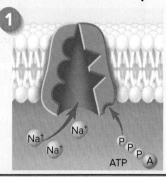

The sodium-potassium pump utilizes a transport protein that binds three sodium ions and a molecule of ATP.

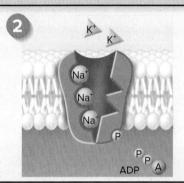

The splitting of ATP provides energy to change the shape of the transport protein. The sodium ions are driven through the pump.

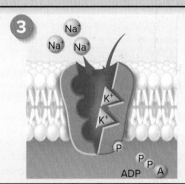

The sodium ions are released to the outside of the membrane, and the new shape of the pump allows two potassium ions to bind.

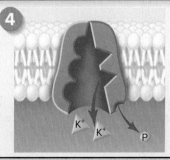

Release of the phosphate allows the sodium-potassium pump's transport protein to revert to its original form, releasing the potassium ions on the inside of the membrane.

Putting the Chapter to Work

1 The phospholipid bilayer of the plasma membrane is essential in regulating cellular activities and contents.

Which organelle would be involved in aiding in the synthesis and replacement of damaged phospholipids?

2 Plant cells have organelles specific to the needs of plants. List and provide the function of the organelles found in plant cells but not in animal or fungal cells.

3 Insulin is classified as a secretory protein because after it is produced within the cell, it is secreted via exocytosis. Demonstrate your understanding of the endomembrane transport system by stating which organelle creates the vesicle for the secretion of insulin from the cell.

4 If a person has a potassium deficiency, their sodium-potassium pump will not be as active and sodium ions can accumulate outside of the cell.

If sodium builds up outside of the cell, creating a hypertonic environment, will the cell shrink or swell?

Retracing the Learning Path

The World of Cells

4.1 Cells

1. Cells are the smallest living structure. They consist of cytoplasm enclosed in a plasma membrane. All living things are composed of one or more cells.
2. A smaller cell can function more efficiently, from transporting materials throughout the cell to transporting materials across the plasma membrane. A smaller cell has a larger surface-to-volume ratio, which increases the area through which materials may pass.
3. Most cells and their components are so small they can only be seen using microscopes. There are many different types of microscopes, each providing a slightly different view.

Kinds of Cells

4.2 Prokaryotic Cells

1. Prokaryotic cells are simple unicellular organisms that lack nuclei or other internal organelles and are usually encased in a rigid cell wall.

4.3 Eukaryotic Cells

1. Eukaryotic cells are larger and more structurally complex compared with prokaryotic cells. Eukaryotic cells have a system of internal membranes and membrane-bounded organelles such as the nucleus that subdivide the cell interior into functional compartments.

Tour of a Eukaryotic Cell

4.4 The Plasma Membrane

1. All cells are encased within a delicate double layer of lipids, called the plasma membrane, in which proteins are embedded that act as cell markers or transports through the membrane. The lipid bilayer of the plasma membrane is made up of lipid molecules called phospholipids, which have a polar (water-soluble) end and a nonpolar (water-insoluble) end. The bilayer forms because the nonpolar ends move away from the watery surroundings, forming the two layers. Transmembrane proteins are held in place by the interaction of nonpolar sections of amino acids with the interior lipid portion of the bilayer.

4.5 The Nucleus: The Cell's Control Center

1. The nucleus is the command and control center of the cell. It contains the cell's DNA, which encodes the hereditary information that runs the cell. An RNA copy of the DNA guides the production of the proteins that carry out cell activities.

4.6 The Endomembrane System

1. The endomembrane system is an extensive system of interior membranes that organize and divide the cell's interior into functional compartments: The endoplasmic reticulum (ER) is a transport system that modifies and moves proteins and other molecules produced in the ER to the Golgi complex.
2. The Golgi complex is a delivery system that carries molecules to the surface of the cell where they are released to the outside.
3. Lysosomes contain enzymes that digest worn out organelles. Vacuoles are storage compartments.

4.7 Organelles that Harvest Energy

1. The mitochondrion is the site of oxidative metabolism, an energy-extracting process. Chloroplasts are the site of photosynthesis and are present in plant and algal cells.
2. Mitochondria and chloroplasts are cell-like organelles that have their own DNA and appear to be ancient bacteria that formed endosymbiotic relationships with early eukaryotic cells.

4.8 The Cytoskeleton: Interior Framework of the Cell

1. The interior of the cell contains a latticework of protein fibers called the cytoskeleton that determines the shape of the cell and anchors organelles to particular locations within the cytoplasm.
2. Cilia and flagella propel the cell through its environment.

Transport Across Plasma Membranes

4.9 Diffusion and Osmosis

1. Random movements of molecules cause them to move to areas of lower concentration, a process called diffusion.
2. Water molecules associated with polar solutes are not free to diffuse, causing a net movement of water across a membrane toward the side with less "free" water, a process called osmosis.

4.10 Bulk Passage into and out of Cells

1. The plasma membrane can engulf materials by endocytosis, folding the membrane around the material to encase it within a vesicle. Exocytosis is essentially this process in reverse, using vesicles to expel substances from the cell.

4.11 Transport through Proteins

1. Selective transport of materials across the membrane is accomplished by facilitated diffusion and active transport. In both, substances must bind to a membrane transporter, called a carrier, in order to pass across the membrane.
2. Active transport involves the input of energy to transport substances against (or up) their concentration gradients. An example is the sodium-potassium pump, which pumps sodium ions out of the cell and potassium ions into the cell.

Inquiry and Analysis

Are Microplastics Being Consumed by Ocean Animals?

Each year, close to 8 million tons of plastic will find its way into our oceans. The plastic, being less dense than water, floats near the surface of the ocean where it is exposed to ultraviolet light from the sun and the mechanical forces of the waves. These forces lead to the degradation of the plastic into smaller and smaller pieces termed *microplastics*. Microplastics are small pieces of plastic less than 5 mm in length. Being inherently non-biodegradable, microplastics persist in the environment and can be further degraded into nanoparticles (size up to 1 μm). Realizing that these microplastics are of similar size to plankton and krill, scientists hypothesized that small marine animals may be unintentionally consuming the plastic.

To test this hypothesis, scientists conducted an experiment in which groups of the marine blue mussel (*Mytilus edulis*) were placed in microplastic-containing solutions. Other groups of mussels were housed in the same water, but without the microplastic, to serve as controls. In one experiment, 36 tanks, each containing three mussels, were filled with 2 liters of filtered ocean water. Microplastic was added to half the tanks (2.5 g of microplastic per liter of ocean water). The other half of the tanks contained only ocean water. The mussels in all 36 tanks were fed every 12 hours. At regular post-treatment intervals (3 hours, 6 hours, 12 hours, 24 hours, 48 hours, and 96 hours), the mussels from three treated tanks and three control tanks were analyzed for the effects of the microplastic on the mussel. Using polarized light microscopy, scientists were able to make a qualitative determination for the relative amount of microplastic found in the gills and in the lysosomal membranes of the mussels. Qualitative data are information that cannot be precisely measured but can be compared to what is observed in other samples. Microplastic accumulation in mussel lysosomal membranes was measured periodically for four days and the results are plotted in the graph. Green circles indicate the relative abundance of microplastic-treated samples; red circles indicate the amount of microplastic found in mussels left untreated.

Paulo de Oliveira/NHPA/Avalon/Newscom

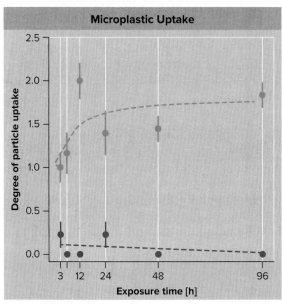

Analysis

1. **Applying Concepts**
   a. Variable: In the graph, which is the dependent variable?
   b. Control: What substance(s) are absent in the control group?
2. **Interpreting Data**
   a. On the green line, at what time point has the animal accumulated about 50% of the total amount of microplastic?
   b. Why is there no additional accumulation of microplastic in the red line?
   c. At what time do you see a slowing of microplastic accumulation in the treatment group? In the control group?
3. **Making Inferences**
   a. Based on the data presented here, what general statement can be made about mussels living in water that has microplastic in it?
   b. Microplastics were still detected in some samples in the control group. Why do you think that result was observed?
4. **Drawing Conclusions** Does this data support the hypothesis that microplastics in the environment are being consumed by animals?
5. **Further Analysis**
   a. In this experiment, none of the mussels had a degree of particle intake greater than 2.0. Provide a hypothesis for why there seems to be an upper limit of 2.0.
   b. Knowing that the lysosome functions as the digestive tool of the cell, what impact would you expect the microplastic accumulation to have on the ability of the lysosome to degrade other particles?

5 Energy and Life

LEARNING PATH ▼

Cells and Energy
1. The Flow of Energy in Living Things
2. The Laws of Energy

Cell Chemistry
3. Chemical Reactions

Enzymes
4. How Enzymes Work
5. How Cells Regulate Enzymes

How Cells Use Energy
6. ATP: The Energy Currency of the Cell

Vegetarians and Vegans

We humans are omnivores, meaning we can eat a broad range of plant and animal tissues—but not all of us choose to do so. Each of us makes choices about what we eat. Some people, for example, don't like spinach, but love steak (bovine-atarians?), while others select vegetables and choose not to eat meat. People that make this particular dietary choice, often claiming it mirrors that of our early human ancestors, are called vegetarians. Some become vegetarians because they judge it to be a more healthy diet—plants are lower in saturated fats, which are linked to heart disease and obesity, and high in vitamins, minerals, and fiber. Others make the choice for ethical reasons, sensitive to the animal rights issues associated with livestock agriculture. Still others simply don't like meat.

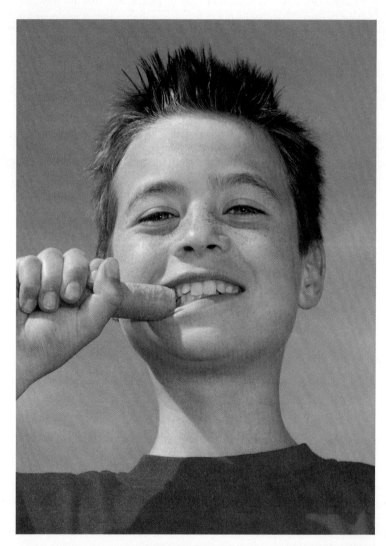

IT'S CRUNCHY, but is it a healthy diet?
Claudia Rehm/Getty Images

Vegetarians

The diets of vegetarians are as varied as the reasons for becoming one. Some vegetarians are selective in what animal products are eaten, eliminating red meat and/or poultry from their diets while still eating fish. Others avoid all meat, including poultry and fish, but still eat animal "products," such as eggs, and/or dairy products, such as yogurt, milk, cheese, and ice cream.

Vegans

The vegan diet is the most extreme form of a vegetarian diet. Vegans avoid all animal proteins, eating neither meat nor animal by-products. They don't eat red meat, poultry,

Mitch Hrdlicka/Photodisc/Getty Images

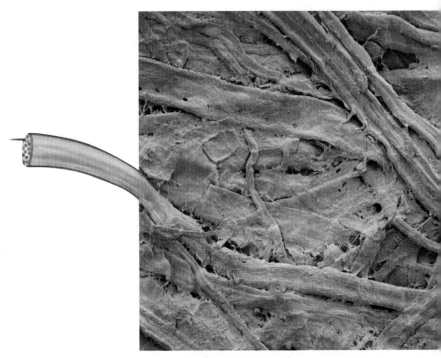
Scimat/Science Source

fish, eggs, or milk. Instead, they obtain all nutrients from grains, vegetables, fruits, legumes, nuts, and seeds. The vegan diet is very challenging, because no single fruit, vegetable, or grain contains all of the essential amino acids that humans require in their diet. Vegetal foods must be eaten in particular combinations to provide this necessary balance. For example, beans and rice eaten together provide a balanced diet, but neither food does when eaten alone. For calcium not obtained from milk, vegans must eat green leafy vegetables such as broccoli or spinach.

Nutritional Guidelines

Nutritional guidelines recommend that a person should eat 0.8 grams of protein per kilogram (2.2 pounds) of body weight—multiply your ideal weight in pounds by 0.8, then divide that number by 2.2. A typical omnivore human diet contains about 16% protein. It is important to understand that not all of this protein has to come from hamburgers. A vegetarian diet can be almost as rich in protein. A vegetarian diet that includes eggs and dairy contains about 13% protein, while a typical vegan diet contains a little less, between 11 and 12% protein, adequate for most people.

The Importance of Fiber

Regardless of how you feel about eating meat, an extremely beneficial aspect of a vegetarian diet, one that even meat-eaters would do well to incorporate into their own diets, is the high fiber content of most plant foods. The human digestive system cannot break down and digest the cellulose fibers found in plants, with the consequence that they pass completely through the human digestive system, greatly facilitating its work.

In fact, only a limited number of microorganisms and one group of animals possess the enzymes necessary to digest cellulose. Plant-eating animals such as cows and horses are only able to live on a diet of grass because they harbor large colonies of microorganisms that have these enzymes in their guts. Vegetarians, and in fact all humans, could obtain vastly more energy from the plants they eat, and indeed could live on a diet of grass like a wild horse does, if only they too could digest cellulose.

Cells and Energy

5.1 The Flow of Energy in Living Things

> **LEARNING OBJECTIVE 5.1.1** Differentiate between kinetic and potential energy.

All life is driven by energy. The concepts and processes discussed in the next three chapters are key to life. We are chemical machines, powered by chemical energy, and for the same reason that a successful race car driver must learn how the engine of a car works, we must look at cell chemistry. Indeed, if we are to understand ourselves, we must "look under the hood" at the chemical machinery of our cells and see how it operates.

As described in chapter 2, **energy** is defined as the ability to do work. It can be considered to exist in two states: kinetic energy and potential energy. **Kinetic energy** is the energy of motion. Objects that are not in the process of moving but have the capacity to move are said to possess **potential energy,** or stored energy (figure 5.1a). A boulder perched on a hilltop (figure 5.1b) has potential energy; after the boulder is pushed and it begins to roll downhill (figure 5.1c), some of the boulder's potential energy is converted into kinetic energy. All of the work carried out by organisms also involves the transformation of potential energy to kinetic energy.

Thermodynamics

Energy exists in many forms: mechanical energy, heat, sound, electric current, light, or radiation. Because it can exist in so many forms, there are many ways to measure energy. The most convenient is in terms of heat, because all other forms of energy can be converted to heat. Thus the study of energy is sometimes called *thermodynamics,* meaning "heat changes."

Energy flows into the biological world from the sun, which shines a constant beam of light on the earth. It is estimated that the sun provides the earth with more than 13×10^{23} calories per year, or 40 million billion calories per second! Plants, algae, and certain kinds of bacteria capture a fraction of this energy through photosynthesis. In photosynthesis, energy garnered from sunlight is used to combine small molecules (water and carbon dioxide) into more complex molecules (sugars). These complex sugar molecules have potential energy due to the arrangement of their atoms. This potential energy, in the form of chemical energy, will eventually be used by the cell to do its work.

> **Putting the Concept to Work**
> Is the energy in a "high-energy" food bar kinetic or potential?

(a) Potential energy

(b) Potential energy

(c) Kinetic energy

Figure 5.1 Potential and kinetic energy.
Objects that have the capacity to move but are not moving have potential energy, whereas objects that are in motion have kinetic energy. (a) The kinetic energy of this skateboarder becomes potential energy at the peak of this slope. (b) The energy required to move the ball up the hill is stored as potential energy. (c) This stored energy is released as kinetic energy as the ball rolls down the hill.
Top: Fuse/Corbis/Getty Images

5.2 The Laws of Energy

Running, thinking, singing, reading these words—all activities of living organisms involve changes in energy. A set of universal energy laws we call the laws of thermodynamics govern these and all other energy changes in the universe.

The Law of Energy Conservation

> **LEARNING OBJECTIVE 5.2.1** State the law of energy conservation.

The first of these universal energy laws, sometimes called the **first law of thermodynamics,** concerns the amount of energy in the universe. It states that energy can change from one state to another (from potential to kinetic, for example), but it can never be destroyed, nor can new energy be made. The total amount of energy in the universe remains constant.

The field mouse eating a blackberry in **figure 5.2** is in the process of acquiring energy. The mouse isn't creating new energy; rather it is merely transferring some of the potential energy stored in the tissues of the blackberry to its own body. Within any living organism, this chemical potential energy can be shifted to other molecules and stored in chemical bonds, or it can be converted into kinetic energy or into other forms of energy. During each conversion, some of the energy dissipates into the environment as heat energy, a measure of the random motions of molecules (and, hence, a measure of one form of kinetic energy). Energy continuously flows through the biological world in one direction, with new energy from the sun constantly entering the system to replace the energy dissipated as heat.

Figure 5.2 Acquiring energy.

Rudmer Zwerver/iStockphoto/Getty Images

> **Putting the Concept to Work**
> When Albert Pujols's bat strikes a baseball, is what happens to the baseball a chemical reaction?

The Law of Increasing Disorder

> **LEARNING OBJECTIVE 5.2.2** State the law of increasing disorder.

The second universal law of energy, sometimes called the **second law of thermodynamics,** concerns this transformation of potential energy into heat, or random molecular motion. The law of increasing disorder states that the disorder in a closed system like the universe is continuously increasing. Put simply, disorder is more likely than order. For example, it is much more likely that a column of bricks will tumble over than that a pile of bricks will arrange themselves spontaneously to form a column. Also, without an input of energy from the teenager (or a parent), the ordered room in **figure 5.3** falls into disorder. When the input of energy is localized, one area can become far more organized than its disordered surroundings, like cleaning one room of a messy house. It is in just this way that a cell uses energy to keep more organized than its surroundings.

Entropy is a measure of the degree of disorder of a system, so the second law of thermodynamics can also be stated simply as "entropy increases."

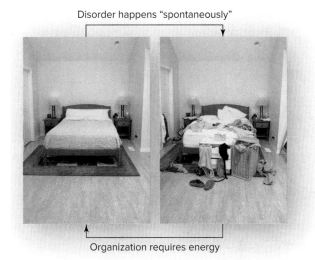

Figure 5.3 Entropy in action.

As time elapses, a teenager's room becomes more disorganized. It takes energy to clean it up.

Keith Eng, 2008

BIOLOGY & YOU

Creationists and the Law. One objection often made by creationists to the theory of evolution is that it violates the law of increasing disorder, the second law of thermodynamics. If disorder is continually increasing in the universe, making disorder more likely than order, then they assert that a more ordered system (cells) could not arise spontaneously without direction. Evolutionary biologists respond to this criticism by pointing out that this creationist objection misstates the second law of thermodynamics by omitting a key provision: "*in a closed system.*" The living world is by no means a closed system, as energy is continually being supplied to it by the sun. It is this energy, captured in photosynthesis, that powers the organized system we call life. In no sense does this in any way violate the second law.

> **Putting the Concept to Work**
> When an organism dies, what happens to its entropy?

Cell Chemistry

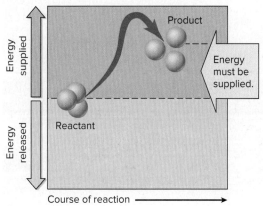

❶ Endergonic reaction

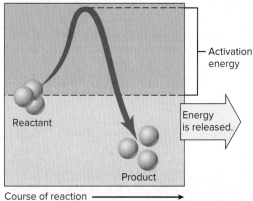

❷ Exergonic reaction

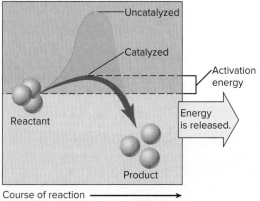

❸ Catalyzed reaction

Figure 5.4 Chemical reactions and catalysis.

❶ The products of endergonic reactions contain more energy than the reactants. ❷ The products of exergonic reactions contain less energy than the reactants, but exergonic reactions do not necessarily proceed rapidly because it takes energy to get them going. The "hill" in this energy diagram represents energy that must be supplied to destabilize existing chemical bonds. ❸ Catalyzed reactions occur faster because the amount of activation energy required to initiate the reaction—the height of the energy hill that must be overcome—is lowered.

5.3 Chemical Reactions

> **LEARNING OBJECTIVE 5.3.1** Differentiate between endergonic and exergonic reactions.

A **chemical reaction** is a process that changes molecules to form different molecules. In a chemical reaction, the original molecules before the chemical reaction occurs are called **reactants,** or sometimes **substrates,** whereas the molecules that result after the reaction has taken place are called the **products.** Not all chemical reactions are equally likely to occur. Just as a boulder is more likely to roll downhill than uphill, so a reaction is more likely to occur if it releases energy than if it needs to have energy supplied. Consider how the chemical reaction proceeds in figure 5.4 ❶. Like rolling a boulder uphill, energy needs to be supplied because the product of the reaction contains more energy than the reactant. This type of chemical reaction, called **endergonic,** does not occur spontaneously. By contrast, an **exergonic** reaction, shown in ❷, tends to occur spontaneously because the product has less energy than the reactant, like a boulder that has rolled downhill.

> **Putting the Concept to Work**
> If exergonic reactions tend to occur spontaneously, why haven't they all done so? What stops the world's gasoline from burning?

Activation Energy

> **LEARNING OBJECTIVE 5.3.2** Define activation energy and catalysis.

If all chemical reactions that release energy tend to occur spontaneously, it is fair to ask, "Why haven't all exergonic reactions occurred already?" They haven't because almost all chemical reactions require an input of energy to get started—it is first necessary to break existing chemical bonds in the reactants, and this takes energy. The extra energy required to destabilize existing chemical bonds and initiate a chemical reaction is called **activation energy,** indicated by brackets in figure 5.4 ❷ and ❸. You must first nudge a boulder out of the hole it sits in before it can roll downhill. Activation energy is simply a chemical nudge.

Catalysis

One way to make a reaction more likely to happen is to lower the necessary activation energy. Like digging away the ground below your boulder, lowering activation energy reduces the nudge needed to get things started. The process of lowering the activation energy of a reaction is called **catalysis.** Catalysis cannot make an endergonic reaction occur spontaneously—you cannot avoid the need to supply energy—but it can make a reaction, endergonic or exergonic, proceed much faster. Compare the activation energy levels (the red arched arrows) in the second and third panels to the left: The catalyzed reaction ❸ has a lower barrier to overcome.

> **Putting the Concept to Work**
> Why is the speed of a chemical reaction affected by the amount of activation energy required to initiate it?

Enzymes

5.4 How Enzymes Work

> **LEARNING OBJECTIVE 5.4.1** Differentiate between active site and substrate binding site.

Macromolecules called **enzymes** are the protein catalysts used by cells to touch off particular chemical reactions. By controlling which enzymes are present and when they are active, cells are able to control what happens within themselves, just as a conductor controls the music an orchestra produces by dictating which instruments play when.

Active Sites

An enzyme works by binding to a specific molecule in such a way as to make a particular reaction more likely (*Essential Biological Process 5A*). The key to this activity is the shape of the enzyme. An enzyme is specific for a particular reactant, or substrate, because the enzyme surface provides a mold that very closely fits the shape of the desired reactant. For example, the blue-colored lysozyme enzyme in figure 5.5 is contoured to fit a specific sugar molecule (the yellow reactant). Other molecules that fit less perfectly simply don't adhere to the enzyme's surface. The site on the enzyme surface where the reactant fits is called the **active site**. The site on the reactant that binds to an enzyme is called its **binding site**. In figure 5.5b, the edges of the lysozyme hug the sugar molecule, leading to an "induced fit" between the enzyme and its reactant, like a hand wrapping around a baseball. The enzyme is not affected by the chemical reaction and is available to be used again.

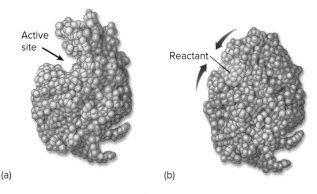

Figure 5.5 An enzyme's shape determines its activity.
(a) A groove runs through the lysozyme enzyme that fits the shape of the reactant—in this case, a chain of sugars. (b) When such a chain of sugars slides into the groove, it induces the protein to change its shape slightly and embrace the substrate more intimately. This induced fit causes a chemical bond between two sugar molecules within the chain to break.

Biochemical Pathways

Every organism contains thousands of different kinds of enzymes that together catalyze a bewildering variety of reactions. Often, several of these reactions occur in a fixed sequence called a **biochemical pathway,** the product of one

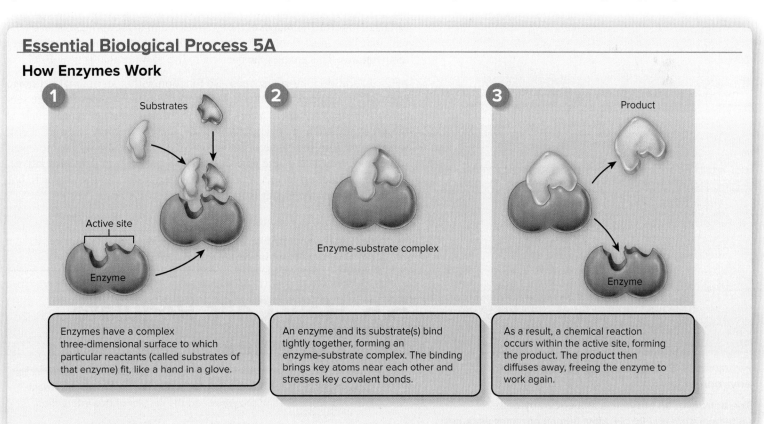

Essential Biological Process 5A

How Enzymes Work

1. Enzymes have a complex three-dimensional surface to which particular reactants (called substrates of that enzyme) fit, like a hand in a glove.

2. An enzyme and its substrate(s) bind tightly together, forming an enzyme-substrate complex. The binding brings key atoms near each other and stresses key covalent bonds.

3. As a result, a chemical reaction occurs within the active site, forming the product. The product then diffuses away, freeing the enzyme to work again.

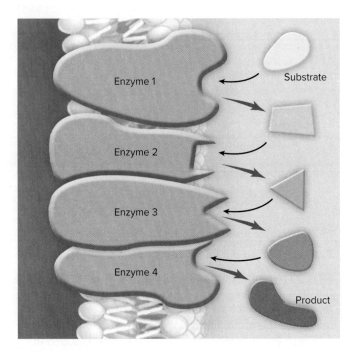

Figure 5.6 A biochemical pathway.

The original substrate is acted on by Enzyme 1, changing the substrate to a new form recognized by Enzyme 2. Each enzyme in the pathway acts on the product of the previous stage.

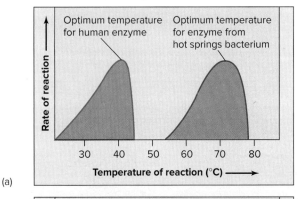

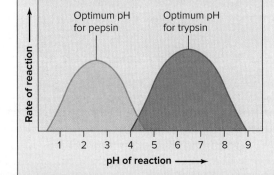

Figure 5.7 Enzymes are sensitive to their environment.

The activity of an enzyme is influenced by both (a) temperature and (b) pH. Most human enzymes work best at temperatures near 37°C and within a pH range of 6 to 8.

reaction becoming the substrate for the next. You can see in figure 5.6 how the initial substrate is altered by Enzyme 1 so that it now fits into the active site of another enzyme, becoming the substrate for Enzyme 2, and so on until the final product is produced. Because these reactions occur in sequence, the enzymes involved are often positioned near each other in the cell. For example, the enzymes involved in this biochemical pathway are all embedded in a membrane near each other. The close proximity of the enzymes allows the reactions of a biochemical pathway to proceed faster. Biochemical pathways are the organizational units of metabolism.

> **Putting the Concept to Work**
> Describe how an enzyme lowers the activation energy of a chemical reaction.

Factors Affecting Enzyme Activity

> **LEARNING OBJECTIVE 5.4.2** Explain the effects of temperature and pH on enzyme-catalyzed reactions.

Enzyme activity is affected by any change in condition that alters the enzyme's three-dimensional shape. For this reason, temperature and pH can have a major influence on the action of enzymes.

Temperature. When the temperature increases, the bonds that determine enzyme shape are too weak to hold it in the proper position, and the enzyme denatures. As a result, enzymes function best within an optimum temperature range, which is relatively narrow for most human enzymes. In the human body, enzymes work best at temperatures near the normal body temperature of 37°C, as shown by the brown curve in figure 5.7a. Also notice that the rates of enzyme reactions tend to drop quickly at higher temperatures, when the enzyme begins to unfold. This is why an extremely high fever in humans can be fatal. However, the shapes of the enzymes found in hot springs bacteria (the red curve) are more stable, allowing the enzymes to function at much higher temperatures. This enables the bacteria to live in water that is near 70°C.

pH. In addition, most enzymes also function within an optimal pH range because the shape-determining polar interactions of enzymes are quite sensitive to hydrogen ion (H^+) concentration. Most human enzymes, such as the protein-degrading enzyme trypsin (the dark blue curve in figure 5.7b), work best within the range of pH 6 to 8. Blood has a pH of 7.4. However, some enzymes, such as the digestive enzyme pepsin (the light blue curve), are able to function in very acidic environments such as the stomach, but can't function at the higher pH where trypsin works best.

> **Putting the Concept to Work**
> Explain how hot springs bacteria can live in near-boiling water (70°C = 158°F) that would kill a human bather.

5.5 How Cells Regulate Enzymes

> **LEARNING OBJECTIVE 5.5.1** Describe how repressors interact with allosteric sites of enzymes and the results of this interaction.

Enzymes can be turned *on* and *off* by altering their shape.

Essential Biological Process 5B

Regulating Enzyme Activity

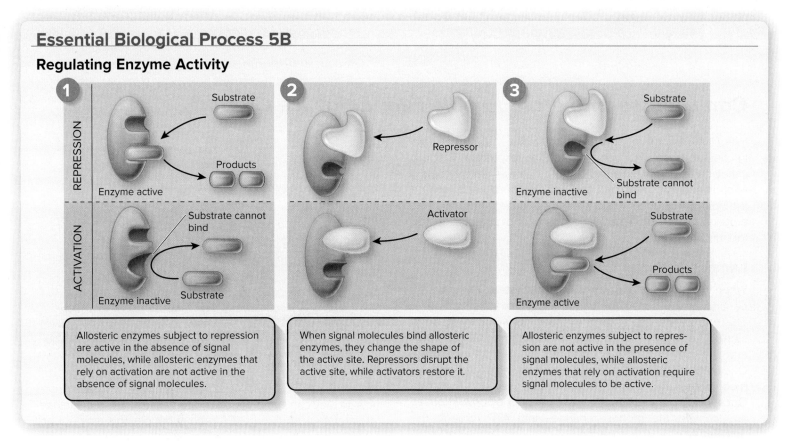

Allosteric enzymes subject to repression are active in the absence of signal molecules, while allosteric enzymes that rely on activation are not active in the absence of signal molecules.	When signal molecules bind allosteric enzymes, they change the shape of the active site. Repressors disrupt the active site, while activators restore it.

Allosteric enzymes subject to repression are not active in the presence of signal molecules, while allosteric enzymes that rely on activation require signal molecules to be active.

Changing an Enzyme's Shape

Because an enzyme must have a precise shape to work correctly, it is possible for the cell to control when an enzyme is active by altering its shape. Many enzymes have shapes that can be altered by the binding of "signal" molecules to their surfaces, making them work better (activation) or worse (inhibition). These are called *allosteric enzymes*. For example, the upper panels of *Essential Biological Process 5B* show an enzyme that is inhibited. The binding of a signal molecule called a **repressor** (*panel 2*) alters the shape of the enzyme's active site such that it cannot bind the substrate. In other cases, the enzyme may not be able to bind the reactants *unless* the signal molecule is bound to the enzyme. The lower set of panels shows a signal molecule serving as an **activator.** The substrate cannot bind to the enzyme's active site unless the activator is first in place.

Feedback Inhibition

Enzymes are often regulated by a mechanism called **feedback inhibition,** where the product of the reaction acts as a repressor. Feedback inhibition can occur in two ways: *competitive* and, much more commonly, *noncompetitive*. The blue molecule in **figure 5.8a** functions as a competitive inhibitor, blocking the active site so that the substrate cannot bind. The yellow molecule in **figure 5.8b** functions as a noncompetitive inhibitor. It binds to an allosteric site, changing the shape of the enzyme such that it is unable to bind to the substrate.

> **Putting the Concept to Work**
> Can a single type of signal molecule activate one enzyme and repress another?

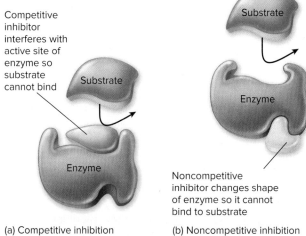

(a) Competitive inhibition (b) Noncompetitive inhibition

Figure 5.8 How enzymes can be inhibited.

(a) In competitive inhibition, the inhibitor interferes with the active site of the enzyme. (b) In noncompetitive inhibition, the inhibitor binds to the enzyme at a place away from the active site, affecting a conformational change in the enzyme so that it can no longer bind to its substrate.

> **IMPLICATION FOR YOU** Many antibiotics work by inhibiting enzymes. The antibiotic penicillin inhibits an enzyme bacteria used in making cell walls. Imagine that, as a confused patient, you mistakenly took two to three times the number of prescribed penicillin pills at one time. Is it likely that you would be seriously harmed? Explain.

Biology and Staying Healthy

Can Lactose-Intolerant People Ever Enjoy Ice Cream?

Every living organism requires energy to live. As you will learn in chapter 7, you get yours from cellular respiration. Eating a candy bar and drinking a soda before class are ways to obtain enough energy to make it through a 2-hour lecture. So, what's the best human fuel? While we can get energy from fats and proteins, carbohydrates are a better source of potential energy for humans.

Lactose Is Sugar from Milk

When we are looking for a quick snack, instead of the candy bar and soda, we may opt for a milkshake or ice cream cone. The energy we get from dairy products such as these is stored in a sugar called lactose, a disaccharide found in milk and other dairy products that are produced by mammals. Once digestion occurs, the monosaccharides diffuse into the bloodstream and eventually make their way into our cells. Once inside our cells, the mitochondria act as powerhouses and convert the monosaccharides into ATP through the pathway of cellular respiration.

Not everybody can do this effectively, however. A problem for some folks is that they are lactose intolerant—they are not able to digest ice cream or other dairy products that contain lactose. If they eat foods that contain high levels of lactose, they often suffer from diarrhea, nausea, abdominal cramping, and gas.

Lactose Intolerance

It is estimated that 33% or approximately 40 million people in the United States are lactose intolerant. Lactose intolerance can come in different forms. Primary lactose intolerance is the most common form. This is when the production of lactase declines dramatically as we grow older. This form of lactose intolerance has a genetic basis to it. Individuals who have an African, Asian, or Hispanic ancestry are more likely to develop primary lactose intolerance as they get older.

Secondary lactose intolerance occurs when the small intestine decreases production of lactase after an injury or illness. Crohn's disease and celiac disease are associated with secondary lactose intolerance. The third type of lactose intolerance is a rare, autosomal recessive genetic disorder called developmental lactose intolerance. If children are homozygous for this recessive mutation, they will be unable to produce lactase when they are born.

Solutions for Lactose Intolerance

The good news is that folks who are lactose intolerant have a variety of options available to them. The easiest solution to

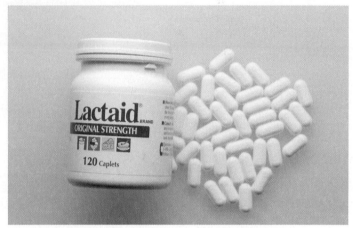

Will & Deni McIntyre/Getty Images

their problem is to avoid large servings of milk and dairy products so they can avoid the discomfort associated with lactose intolerance. Unfortunately, for many people, this may not be a realistic solution. Ice cream, milkshakes, and other dairy products are very tempting. A different option is to consume dairy products in small quantities. Each lactose-intolerant individual needs to figure out their individual threshold level, so they have a gauge on how much dairy they can consume before they experience the discomfort of lactose intolerance.

Fortunately, science and the food industry have come up with a variety of options for lactose-intolerant folks who still want to enjoy dairy. Lactose-reduced ice cream and milk products are available for people who still want a milkshake but don't want to deal with the symptoms of lactose intolerance. The idea is that the levels of lactose in these foods will be low enough that they won't exceed an individual's lactose threshold. Unfortunately, these foods don't work for everyone. An additional option for people is to take an artificial enzyme that will mimic the function of lactase. This enzyme needs to be eaten just before the individual indulges in dairy products. Artificial lactase comes in liquid or tablet form and is available as an over-the-counter product. Celebrities such as Rachel Hunter, Jessica Simpson, Miley Cyrus, and Billy Bob Thornton are all reported to suffer from different forms of lactose intolerance.

So the Answer Is Yes

Unfortunately, there is no cure for lactose intolerance. With enough awareness of their sensitivity to lactose, a shift in their eating habits, and the possible use of an artificial lactase enzyme, lactose-intolerant folks can enjoy ice cream for dessert.

How Cells Use Energy

5.6 ATP: The Energy Currency of the Cell

> **LEARNING OBJECTIVE 5.6.1** Explain how the phosphate groups of ATP store potential energy and how organisms use this energy to power endergonic reactions.

Cells use energy to do all those things that require work, but how does the cell use energy from the sun or the potential energy stored in molecules to power its activities? These energy sources cannot be used directly to run a cell, any more than money invested in stocks and bonds or real estate can be used to buy a candy bar at a store. To be useful, the energy from the sun or food molecules must first be converted to a source of energy that a cell can use, like someone converting stocks and bonds to ready cash. The "cash" molecule in the body is **adenosine triphosphate (ATP)**.

Structure of the ATP Molecule

Each ATP molecule is composed of three parts (figure 5.9): (1) a sugar (colored blue) that serves as the backbone to which the other two parts are attached, (2) adenine (colored peach), which is also one of the four nitrogenous bases in DNA and RNA, and (3) a chain of three phosphates (colored yellow) that contain high-energy bonds. As you can see in the figure, the phosphates carry negative electrical charges, so it takes considerable chemical energy to hold the line of three phosphates next to one another at the end of ATP. Like a compressed spring, the phosphates are poised to push apart. It is for this reason that the chemical bonds linking the phosphates are such chemically reactive bonds. When the endmost phosphate is broken off an ATP molecule, a sizable packet of energy is released. The reaction converts ATP to adenosine diphosphate (ADP) and P_i, inorganic phosphate:

$$ATP \longleftrightarrow ADP + P_i + energy$$

Chemical reactions require activation energy, and endergonic reactions require the input of even more energy, so these reactions in the cell are usually coupled with the breaking of the phosphate bond in ATP. In **coupled reactions**, the energy released from an exergonic reaction, such as the breakdown of ATP, drives an endergonic reaction. Because almost all chemical reactions in cells require less energy than is released by this reaction, ATP is able to power many of the cell's activities. Table 5.1 introduces you to some of the key cell activities powered by the breakdown of ATP. ATP is continually recycled from ADP and P_i via the ATP-ADP cycle (figure 5.10).

The Energy Cycle

Cells use two different but complementary processes to convert energy from the sun and food molecules into potential energy stored in the chemical bonds of ATP. Some cells convert energy from the sun into molecules of ATP through the process of **photosynthesis,** the subject of chapter 6. This ATP is then used to manufacture sugar molecules, converting the energy from ATP into potential energy stored in the bonds that hold the atoms in the sugar molecule together. All cells convert the potential energy found in food molecules such as sugar into ATP through **cellular respiration,** the subject of chapter 7.

> **Putting the Concept to Work**
>
> Compare and contrast ATP with the nucleotides found in DNA and RNA. [Hint: See figure 3.10.]

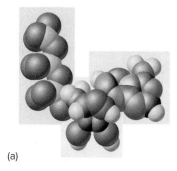

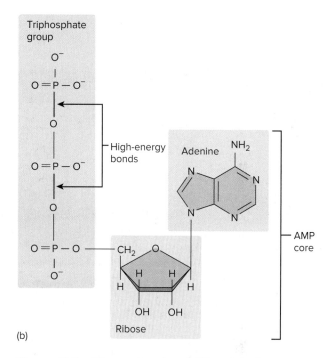

Figure 5.9 The parts of an ATP molecule.

The model (a) and structural diagram (b) both show that ATP consists of three phosphate groups attached to a ribose (5-carbon sugar) molecule. The ribose molecule is also attached to an adenine molecule (also one of the nitrogenous bases of DNA and RNA). When the endmost phosphate group is split off from the ATP molecule, considerable energy is released.

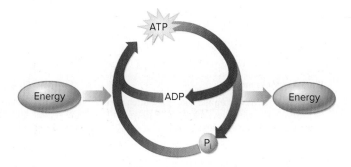

Figure 5.10 The ATP-ADP cycle.

TABLE 5.1 How Cells Use ATP Energy to Power Cellular Work

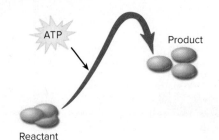

Biosynthesis

Cells use the energy released from the exergonic hydrolysis of ATP to drive endergonic reactions like those of protein synthesis, an approach called energy coupling.

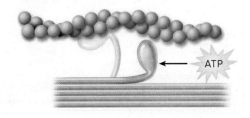

Contraction

In muscle cells, filaments of protein repeatedly slide past each other to achieve contraction of the cell. An input of ATP is required for the filaments to reset and slide again.

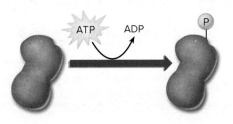

Chemical Activation

Proteins can become activated when a high-energy phosphate from ATP attaches to the protein, activating it. Other types of molecules can also become phosphorylated by transfer of a phosphate from ATP.

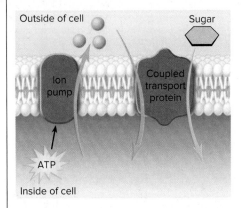

Importing Metabolites

Metabolite molecules such as amino acids and sugars can be transported into cells against their concentration gradients by coupling the intake of the metabolite to the inward movement of an ion moving down its concentration gradient, this ion gradient being established using ATP.

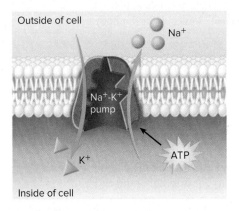

Active Transport: Na^+-K^+ Pump

Most animal cells maintain a low internal concentration of Na^+ relative to their surroundings, and a high internal concentration of K^+. This is achieved using a protein called the sodium-potassium pump, which actively pumps Na^+ out of the cell and K^+ in, using energy from ATP.

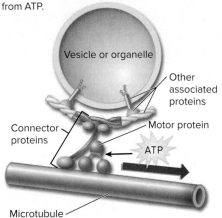

Cytoplasmic Transport

Within a cell's cytoplasm, vesicles or organelles can be dragged along microtubular tracks using molecular motor proteins, which are attached to the vesicle or organelle with connector proteins. The motor proteins use ATP to power their movement.

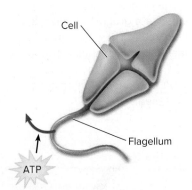

Flagellar Movements

Microtubules within flagella slide past each other to produce flagellar movements. ATP powers the sliding of the microtubules.

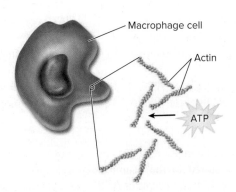

Cell Crawling

Actin filaments in a cell's cytoskeleton continually assemble and disassemble to achieve changes in cell shape and to allow cells to crawl over substrates or engulf materials. The dynamic character of actin is controlled by ATP molecules bound to actin filaments.

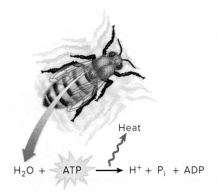

Heat Production

The hydrolysis of the ATP molecule releases heat. Reactions that hydrolyze ATP often take place in mitochondria or in contracting muscle cells and may be coupled to other reactions. The heat generated by these reactions can be used to maintain an organism's temperature.

A Closer Look

Beer and Wine—Products of Fermentation

Alcoholic fermentation, the anaerobic conversion of sugars into alcohol, predates human history. Like many other natural processes, humans seem to have first learned to control this process by stumbling across its benefit—the pleasures of beer and wine. Artifacts from the third millennium B.C. contain wine residue, and we know from the pollen record that domesticated grapes first became abundant at that time. The production of beer started even earlier, back in the fifth millennium B.C., making beer possibly one of the oldest beverages produced by humans.

Beer fermentors.

ClarkandCompany/Getty Images

The process of making beer is called brewing and involves the fermentation of cereal grains. Grains are rich in starches, which you will recall from chapter 3 are long chains of glucose molecules linked together. First, the cereal grains are crushed and soaked in warm water, creating an extract called *mash*. Enzymes called amylases are added to the mash to break down the starch within the grains into free sugars. The mash is filtered, yielding a darker, sugary liquid called wort. The wort is boiled in large tanks, like the ones shown in the photo here, called fermentors. This helps break down the starch and kills any bacteria or other microorganisms that might be present before they begin to feast on all the sugar. Hops, another plant product, is added during the boiling stage. Hops gives the solution a bitter taste that complements the sweet flavor of the sugar in the wort.

Now we are ready to make beer. The temperature of the wort is brought down and yeast is added. This begins the fermentation process. It can be done in either of two ways, yielding the two basic kinds of beer.

Lager

The yeast *Saccharomyces uvarum* is a bottom-fermenting yeast that settles on the bottom of the vat during fermentation. This yeast produces a lighter, pale beer called lager. Bottom fermentation is the most widespread method of brewing. Bottom fermentation occurs at low temperatures (5–8°C). After fermentation is complete, the beer is cooled further to 0°C, allowing the beer to mature before the yeast is filtered out.

Ale

The yeast *S. cerevisiae,* often called "brewer's yeast," is a top-fermenting yeast. It rises to the top during fermentation and produces a darker, more aromatic beer called ale (also called porter or stout). In top fermentation, less yeast is used. The temperature is higher, around 15–25°C, and so fermentation occurs more quickly but less sugar is converted into alcohol, giving the beer a sweeter taste.

Most yeasts used in beer brewing are alcohol-tolerant only up to about 5% alcohol—alcohol levels above that kill the yeast. That is why commercial beer is typically 5% alcohol.

Carbon dioxide is also a product of fermentation and gives beer the bubbles that form the head of the beer. However, because only a little CO_2 forms naturally, the last step in most brewing involves artificial carbonation, with CO_2 gas being injected into the beer.

Wine

Wine fermentation is similar to beer fermentation in that yeasts are used to ferment sugars into alcohol. Wine fermentation, however, uses grapes. Rather than the starches found in cereal grains, grapes are fruits rich in the sugar sucrose, a disaccharide combination of glucose and fructose. Grapes are crushed and placed into barrels. There is no need for boiling to break down the starch, and grapes have their own amylase enzymes to cleave sucrose into free glucose and fructose sugars. To start the fermentation, yeast is added directly to the crush. *Saccharomyces cerevisiae* and *S. bayanus* are common wine yeasts. They differ from the strains of yeast used in beer fermentation in that they have higher alcohol tolerances—up to 12% alcohol (some wine yeasts have even higher alcohol tolerances).

A common misconception about wine is that the color of the wine results from the color of the grape juice used, red wine using juice from red grapes and white wines using juice from white grapes. This is not true. The juice of all grapes is similar in color, usually light. The color of red wine is produced by leaving the skins of red or black grapes in the crush during the fermentation process. White wines can be made from any color of grapes and are white simply because the skins that contribute red coloring are removed before fermentation.

White wines are typically fermented at low temperatures (8–19°C). Red wines are fermented at higher temperatures of 25–32°C with yeasts that have a higher heat tolerance. During wine fermentation, the carbon dioxide is vented out, leaving no carbonation in the wine. Champagnes have some natural carbonation that results from a two-step fermentation process: In the first fermentation step, the carbon dioxide is allowed to escape; in a second fermentation step, the container is sealed, trapping the carbon dioxide. But, as in beer brewing, the natural carbonation is often supplemented with artificial carbonation.

Putting the Chapter to Work

1 Read through the scenario below.

A young boy climbs into a chair and waits patiently for his guardian to read him a story. His guardian approaches and hands him a bag of chips to munch on while they enjoy story time. The child excitedly takes the chips and begins eating. As his guardian reads to him, he quickly falls asleep.

Identify the selected activities, from the scenario, as examples of kinetic energy or potential energy.
 a. The child waiting patiently
 b. The guardian handing the child a bag of chips
 c. The child sleeping

2 In a chemical reaction, starch (a long chain of glucose molecules) is broken down into pairs of glucose molecules called maltose. The process of removing a maltose molecule from starch is catalyzed by the enzyme amylase.

What is the substrate in this reaction?

3 During cellular respiration, sugar is broken down to produce more ATP than is consumed in the reaction. The energy produced can then be used by your cells for work. Using this scenario, identify the following activities as endergonic or exergonic.
 a. Cellular respiration
 b. Cells using the ATP for work

Retracing the Learning Path

Cells and Energy

5.1 The Flow of Energy in Living Things

1. Energy is the ability to do work, either actively (kinetic energy) or stored for later possible use (potential energy).

- Kinetic energy is the energy of motion. Potential energy is stored energy, which exists in objects that aren't in motion but have the capacity to move, like a ball poised at the top of a hill. Work carried out by living organisms involves the transformation of potential energy into kinetic energy.

- Energy flows from the sun to the earth, where it is trapped by photosynthesis and stored in carbohydrates as potential energy. This energy is transferred during chemical reactions when the covalent bonds linking atoms together are formed or broken.

5.2 The Laws of Energy

1. The laws of energy describe changes in energy in our universe. The first law explains that energy cannot be created or destroyed, only changed from one form to another. The total amount of energy in the universe remains constant.

2. The second law of energy states that disorder in the universe tends to increase—a conversion of potential energy into random molecular motion. This conversion of energy progresses from an ordered but less stable form to a disordered but stable form. Entropy, which is a measure of disorder in a system, is thus constantly increasing. Energy must be used to oppose this tendency in order to maintain order.

Cell Chemistry

5.3 Chemical Reactions

1. Chemical reactions involve the breaking or formation of covalent bonds. The starting molecules are called the reactants, and the molecules produced by the reaction are called the products. Chemical reactions in which the products contain more potential energy than the reactants are called endergonic reactions. Chemical reactions that release energy are called exergonic reactions and are more likely to occur.

2. All chemical reactions require an input of energy. The energy required to start a reaction is called activation energy.

- A chemical reaction proceeds faster when its activation energy is lowered, a process called catalysis.

Enzymes

5.4 How Enzymes Work

1. Enzymes are molecules that act as cellular catalysts by lowering the activation energy of chemical reactions in the cell.

- An enzyme binds to the reactant, or substrate, at its active site. The enzyme molds around the reactant and acts to stress covalent bonds or bring atoms into closer proximity. The actions of the enzyme increase the likelihood that chemical bonds will break or form. An enzyme lowers the activation energy of the reaction. The enzyme is not affected by the reaction and can be used over and over again.

2. In the cell, chemical reactions are regulated by controlling which enzymes are active. Other factors, such as temperature and pH, also affect enzyme shape and activity. Most enzymes have an optimal temperature and pH range.

- Higher temperatures can disrupt the bonds that hold the enzyme in its proper shape, decreasing its ability to catalyze a chemical reaction. The bonds that hold the enzyme's shape are also affected by hydrogen ion concentrations, and so increasing or decreasing the pH can disrupt the enzyme's function.

5.5 How Cells Regulate Enzymes

1. An enzyme can be inhibited or activated in the cell by temporarily altering the enzyme's shape. A repressor molecule can bind to the active site of the enzyme, blocking it, or bind to a different site elsewhere on the enzyme, altering the shape of the protein so that its active site can no longer bind the substrate.

How Cells Use Energy

5.6 ATP: The Energy Currency of the Cell

1. Cells store chemical energy in ATP molecules. ATP contains a sugar, an adenine, and a chain of three phosphates held together with high-energy bonds. When the endmost phosphate bond breaks, considerable energy is released. A cell uses this energy to drive reactions in the cell by coupling the breakdown of ATP with other chemical reactions in the cell.

Inquiry and Analysis

Do All Your Body's Cells Make the Same Amount of Catalase Enzyme?

Hydrogen peroxide (H_2O_2) has been used as an antiseptic since the early part of the 20th century. H_2O_2 is a strong oxidizing agent (it is the oxidant used in rocket fuel!) capable of breaking down most organic molecules, including the proteins and carbohydrates found in the cell walls of bacteria. The antiseptic properties of H_2O_2 are the result of its destruction of the microbial cell wall, thus killing the bacterial cell and cleaning the wound.

When H_2O_2 is poured onto a human cut, bubbles appear where the solution contacts the wound. This bubbling is the result of the breakdown of H_2O_2 into water (H_2O) and oxygen gas (O_2). Infecting bacteria are killed, while the surrounding tissue is undamaged. What protects these surrounding body cells from oxidation powerful enough to kill bacteria? Answer: Most of your body's cells produce H_2O_2 as a natural by-product of the electron transport chain during cell respiration. (You will learn more about this in chapter 7.) With H_2O_2 being potentially so dangerous, your cells synthesize a protective enzyme called catalase that breaks down H_2O_2 into water and oxygen. One of the fastest-working of all enzymes, one catalase molecule can convert millions of molecules of H_2O_2 to water and oxygen gas each second! That is why when you apply H_2O_2 to a cut, you only observe bubbles (the release of O_2 gas) at the sight of the wound and not at the adjacent, undamaged tissue.

This leads to an interesting question: Does H_2O_2 work better as an antiseptic for some tissues than others? This might be true if some tissues of the body produce more catalase than others. To address this question, investigators measured the rate of H_2O_2 decomposition in various tissues. Experiments were conducted by incubating H_2O_2 with tissue samples (or a tissue-free control sample), measuring for each sample the amount of oxygen produced over time. In each tissue, the greater the breakdown of the substrate, the more oxygen gas produced. In the graph, time is represented on the *x* axis, while the amount of oxygen produced is presented on the *y* axis. H_2O_2 decomposition curves were obtained for equally weighted tissue samples from liver tissue, skeletal muscle tissue, and from a tissue-free control sample. These particular tissues were chosen because they have very different metabolic rates, with liver being highly metabolically active and skeletal muscle being less active.

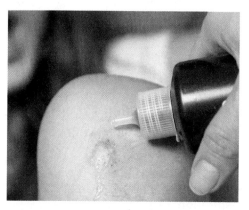

Ekaterina Vidyasova/Shutterstock

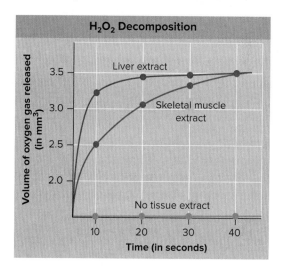

Analysis

1. **Applying Concepts**
   a. **Variable.** In the graph, what is the dependent variable?
   b. **Concentration.** After 20 seconds, which reaction has produced the most oxygen?
2. **Interpreting Data**
   a. At what timepoint is saturation reached in the experiment containing the liver extract? The skeletal muscle extract?
   b. What is the reason for including a sample that did not contain any tissue extract?
   c. How much oxygen was produced in each reaction at 10 seconds? At 20 seconds? At 30 seconds?
   d. What would you conclude about the effect of catalase concentration on oxygen production?
3. **Making Inferences**
   a. For both tissues, the experiments eventually reach the same level of saturation, but at different rates. Why do you observe one sample reaching saturation before another sample?
   b. The experiment containing no catalase still produced a small amount of oxygen. Why might you observe this in the absence of any tissue being present?
4. **Drawing Conclusions** Does the type of tissue affect the rate of H_2O_2 decomposition? Why?
5. **Further Analysis** The outer layers of your skin do not actively divide. Why might this have a role in the lack of bubbles observed when H_2O_2 is applied to healthy skin?

6 Photosynthesis: Acquiring Energy from the Sun

LEARNING PATH ▼

Photosynthesis
1. An Overview of Photosynthesis
2. How Plants Convert Sunlight to Chemical Energy
3. Building New Molecules

Photorespiration
4. Photorespiration: Putting the Brakes on Photosynthesis

Eco-Warriors Attack Global Warming at Sea

THESE RESEARCHERS ARE fertilizing the Southern Ocean with iron, hoping this may help cool the earth's climate.
Thomas Pickard/Getty Images

Atmospheric carbon dioxide (CO_2) levels are at a 2-million-year high, with most of the rise occurring since 1900. In response, global temperatures are rising rapidly. If nothing is done to reverse this trend, earth's oceans are going to rise as polar ice melts and flood coastal cities such as Miami. Drought and weather extremes will become commonplace.

Few Feasible Solutions

What should we do? There are only two feasible solutions to global warming induced by the CO_2 released by humans into earth's atmosphere. One is to reduce the amount of CO_2 humans release. While attempts are being made to reduce car and power plant emissions, no serious impact has been achieved. The other potential solution is to remove the excess CO_2 from the atmosphere. Attempts to store (sequester) atmospheric CO_2 in deep wells do not seem practical on a large scale. Where else might one put it? The most promising answer, oddly enough, is: back where it came from.

Fertilizing the Oceans

In 1988, ocean ecologist John Martin famously said, "Give me half a tanker of iron and I'll give you an ice age." He was pointing out that earth's oceans are rich in marine algae, their growth limited primarily by lack of iron (Fe). Adding iron to the upper ocean could trigger an algal "bloom," its intense photosynthesis sequestering massive amounts of CO_2 in organic matter that would then fall to the ocean's bottom, where it could no longer react with earth's atmosphere. In essence, fertilizing the oceans with Fe would reverse human-caused global warming, returning CO_2 to the state from which the burning of fossil fuels had released it.

The Idea Seems Promising

In the laboratory, experiments had suggested that every pound of iron added to ocean water could remove as much as 100,000 pounds of carbon from the air, so Martin's quip was taken seriously. However, while exciting, Martin's iron dump hypothesis is not an easy idea to test. In 12 small-scale ocean experiments carried out since 1988 (white dots on the map), adding iron to sea waters always succeeded in drawing carbon from the atmosphere into the oceans, although not as efficiently as in the laboratory.

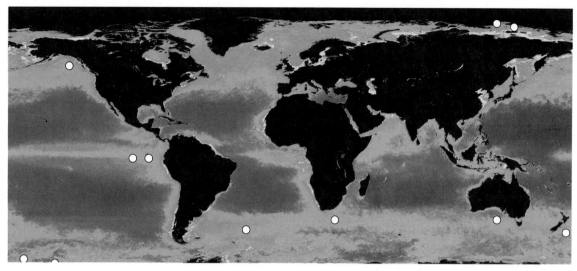
NASA

But Do the Algae Sink?

However, two potential problems were revealed by these experiments: (1) Tiny zooplankton that live on the ocean's surface eat much of the bloom of algae as soon as it forms, returning the CO_2 back to the atmosphere; (2) Surface bacteria consume dead algae, and in doing so remove life-giving oxygen from the water. These two processes both assume the algae do not sink beneath the ocean surface, as Martin had claimed they would. Do they? In these studies, ocean currents have confused our ability to assess this critical question, but researchers in 2012 found a solution. In the Southern Ocean near Antarctica, currents swirl, forming stable eddies within which the necessary measurements could be made. If Martin is right, adding Fe to the ocean waters should produce an algal bloom within the eddy that would then fall downward into the deep ocean. To test this prediction, a businessman-turned-eco-warrior named Russ George added 100 tons of iron sulfate from a fishing boat into the ocean eddy, and then for a month measured organic carbon and chlorophyll in a 100-meter-deep column of ocean water.

You can see in the graph what happened. Red dots are from within the Fe-fertilized patch; blue dots are from outside the eddy where no iron was added. Because of the eddy barrier, waters from the two locations do not mix. The results were clear-cut and exciting: A week after fertilization, an algal bloom exploded within the eddy waters, reflected in the rapidly rising levels of organic carbon. After another week, the bloom reached its maximum size, and then began to fall as the algal mass sank to the ocean bottom. Martin's iron dump hypothesis was confirmed.

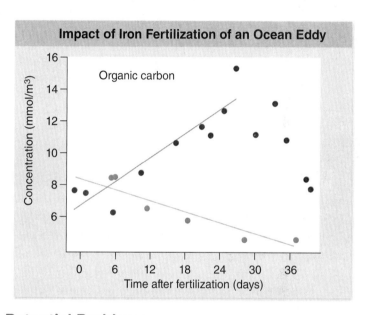

Potential Problems

Ocean fertilization remains a highly controversial approach. Legal and ethical issues arise from the fact that no one country owns the earth's oceans (for example, George's experiment, done without obtaining any international approvals, was greeted with outrage by many countries). The key problem is that the potential biological impact of extensive and prolonged Fe fertilization is poorly characterized and may be significant even if the Fe-induced algal blooms sink below 1,000 m. It seems likely, for example, that Fe fertilization will promote growth of the bacteria responsible for nitrification in the oceans, and N_2O is a much more powerful greenhouse gas than CO_2. Said simply, there is a great deal we don't know, making iron fertilization of earth's oceans ecologically risky until matters are more clearly understood. But the problem we face is massive, and this approach is promising.

Photosynthesis

6.1 An Overview of Photosynthesis

> **LEARNING OBJECTIVE 6.1.1** Name the three layers of a leaf through which light must pass to reach chloroplasts.

Life is powered by sunshine. All of the energy used by almost all living cells comes ultimately from the sun, captured by plants, algae, and some bacteria through the process of **photosynthesis.** Every oxygen atom in the air we breathe was once part of a water molecule, liberated by photosynthesis, as you will discover in this chapter. Life as we know it is possible only because our earth is awash in energy streaming inward from the sun. Each day, the radiant energy that reaches the earth is equal to that of about 1 million Hiroshima-sized atomic bombs. About 1% of it is captured by photosynthesis and provides the energy needed to synthesize carbohydrates that drives almost all life on earth. Use the arrows between the drawings in this section to follow the path of energy from the sun through photosynthesis.

Trees. Many kinds of organisms carry out photosynthesis, not only the diversity of plants that make our world green, but also bacteria and algae. Photosynthesis is somewhat different in bacteria, but we will focus our attention on photosynthesis in plants, starting with this maple tree crowned with green leaves. Later, we will look at the grass growing beneath the maple tree—it turns out that grasses and other related plants sometimes take a different approach to photosynthesis, depending on the conditions.

Leaves. To learn how this maple tree captures energy from sunlight, follow the light. It comes beaming in from the sun, down through earth's atmosphere, bathing the top of the tree in light. What part of the maple tree is actually being struck by this light? The green leaves. Each branch at the top of the tree ends in a spread of these leaves, each leaf flat and thin like the page of a book. Within these green leaves is where photosynthesis occurs. No photosynthesis occurs within this tree's stem, covered with bark, and none in the roots, buried within the soil—little to no light reaches these parts of the plant. The tree has a very efficient internal plumbing system that transports the products of photosynthesis to the stem, roots, and other parts of the plant so that they too may benefit from the capture of the sun's energy.

The Leaf Surface. Now follow the light as it passes into a leaf. The beam of light first encounters a waxy protective layer called the cuticle. The cuticle acts a bit like a layer of clear fingernail polish, providing a thin, watertight, and surprisingly strong layer of protection. Light passes right through this transparent wax and then proceeds to pass right on through a layer of cells immediately beneath the cuticle called the epidermis. Only one cell layer thick, this epidermis acts as the "skin" of the leaf, providing more protection from damage and, very importantly, controlling how gases and water enter and leave the leaf. Very little of the light is absorbed by the cuticle or the epidermis.

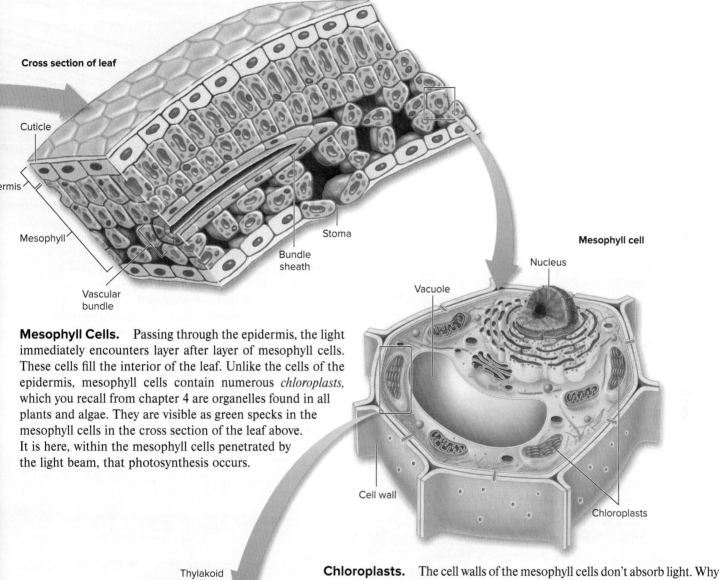

Mesophyll Cells. Passing through the epidermis, the light immediately encounters layer after layer of mesophyll cells. These cells fill the interior of the leaf. Unlike the cells of the epidermis, mesophyll cells contain numerous *chloroplasts*, which you recall from chapter 4 are organelles found in all plants and algae. They are visible as green specks in the mesophyll cells in the cross section of the leaf above. It is here, within the mesophyll cells penetrated by the light beam, that photosynthesis occurs.

Chloroplasts. The cell walls of the mesophyll cells don't absorb light. Why not? Because the cell walls contain few, if any, molecules that absorb visible light. If chloroplasts were not also present in these cells, most of this light would pass right through, just as it passed through the epidermis. But chloroplasts are present, lots of them. One chloroplast is highlighted by a box in the mesophyll cell above. Light passes into the cell and when it reaches the chloroplast, it passes through the outer and inner membranes to reach the thylakoid structures within the chloroplast, clearly seen as the green disks in the cutaway chloroplast shown here.

> **Putting the Concept to Work**
> Why is there no photosynthesis within an oak tree's stem?

Inside the Chloroplast

LEARNING OBJECTIVE 6.1.2 Diagram the structure of a chloroplast, and contrast the light-dependent and light-independent reactions that occur there.

All the important events of photosynthesis happen inside the chloroplast. The journey of light into the chloroplasts ends when the light beam encounters a series of internal membranes within the chloroplast organized into flattened sacs called *thylakoids*. Often, numerous thylakoids are stacked on top of one another in columns called *grana*. In the drawing below, the grana look not unlike piles of dishes. Although each thylakoid is a separate compartment that functions more or less independently, the membranes of the individual thylakoids are all connected, part of a single continuous membrane system. Occupying much of the interior of the chloroplast, this thylakoid membrane system is submerged within a semiliquid substance called *stroma*, which fills the interior of the chloroplast in much the same way that cytoplasm fills the interior of a cell. Suspended within the stroma are many enzymes and other proteins, including the enzymes that act later in photosynthesis to assemble organic molecules from CO_2 in reactions that do not require light and that are discussed later.

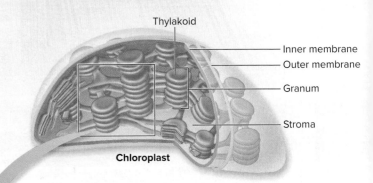

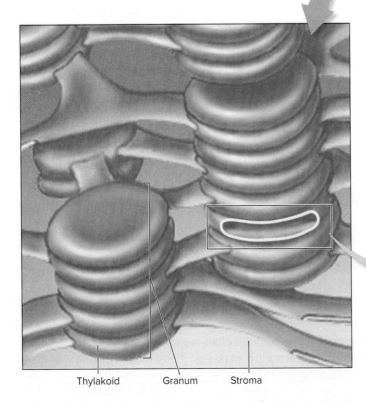

Penetrating the Thylakoid Surface. The first key event of photosynthesis occurs when a beam of sunlight strikes the surface membrane of a thylakoid. Embedded within this membrane, like icebergs on an ocean, are clusters of light-absorbing pigments. A pigment molecule is a molecule that absorbs light energy. The primary pigment molecule in most photosystems is **chlorophyll,** an organic molecule that absorbs red and blue light but does not absorb green wavelengths of light. The green light is instead reflected, giving the thylakoid and the chloroplast that contains it an intense green color. Plants are green because they are rich in green chloroplasts. Except for some alternative pigments also present in thylakoids, no other parts of the plant absorb visible light with such intensity.

Striking the Photosystem. Within each pigment cluster, the chlorophyll molecules are arranged in a network called a *photosystem*. The light-absorbing chlorophyll molecules of a photosystem act together as an antenna to capture photons (units of light energy). A lattice of structural proteins, indicated by the purple element inserted into the thylakoid membrane in the diagram at the end of the green arrow, anchors each of the chlorophyll molecules of a photosystem into a precise position, such that every chlorophyll molecule is touching several others. Wherever light strikes the photosystem, some chlorophyll molecule will be in position to receive it.

Energy Absorption. When sunlight strikes any chlorophyll molecule in the photosystem, the chlorophyll molecule absorbs energy. The energy becomes part of the chlorophyll molecule, boosting some of its electrons to higher energy levels. Possessing these more energetic electrons, the chlorophyll molecule is said to now be "excited." With this key event, the biological world has captured energy from the sun.

Excitation of the Photosystem. The excitation that the absorption of light creates is then passed from the chlorophyll molecule that was hit to another chlorophyll molecule and then to another, like a hot potato being passed down a line of people. This shuttling of excitation is not a chemical reaction, in which an electron physically passes between atoms. Rather, it is energy that passes from one chlorophyll molecule to its neighbor. A crude analogy to this form of energy transfer is the initial "break" in a game of pool.

yanukit/iStockphoto/Getty Images

If the cue ball squarely hits the point of the triangular array of 15 billiard balls, the two balls at the far corners of the triangle fly off, and none of the central balls moves at all. The kinetic energy is transferred through the central balls to the most distant ones. In much the same way, the sun's excitation energy moves through the photosystem from one chlorophyll to the next.

Energy Capture. As the energy shuttles from one chlorophyll molecule to another within the photosystem network, it eventually arrives at a key chlorophyll molecule, the only one that is touching a membrane-bound protein. Like shaking a marble in a box with a walnut-sized hole in it, the excitation energy will find its way to this special chlorophyll just as sure as the marble will eventually find its way to and through the hole in the box. The special chlorophyll then transfers an excited (high-energy) electron to the acceptor molecule it is touching.

The Light-Dependent Reactions. Like a baton being passed from one runner to another in a relay race, the electron is then passed from that acceptor protein to a series of other proteins in the membrane that put the energy of the electron to work making ATP and the electron carrier NADPH (nicotinamide adenine dinucleotide phosphate). In a way you will explore later in this chapter, the energy is used to power the movement of protons across the thylakoid membrane to make ATP and to synthesize NADPH. So far, photosynthesis has consisted of two stages, indicated by numbers in the diagram to the lower left: ❶ capturing energy from sunlight—accomplished by the photosystem and ❷ using the energy to make ATP and NADPH. These first two stages of photosynthesis take place only in the presence of light, and together are traditionally called the **light-dependent reactions**. ATP and NADPH are important energy-rich chemicals, and after this, the rest of photosynthesis becomes a chemical process.

The Light-Independent Reactions. The ATP and NADPH molecules generated by the light-dependent reactions are then used to power a series of chemical reactions in the stroma of the chloroplast, each catalyzed by an enzyme present there. Acting together like the many stages of a manufacturing assembly line, these reactions accomplish the synthesis of carbohydrates from CO_2 in the air ❸. This third stage of photosynthesis, the formation of organic molecules such as glucose from atmospheric CO_2, is called the **Calvin cycle** but is also referred to as the **light-independent reactions** because it doesn't require light directly. We will examine the Calvin cycle in detail later in this chapter.

This completes our brief overview of photosynthesis. In the rest of the chapter, we will revisit each stage and consider its elements in more detail. For now, the overall process may be summarized by the following simple equation:

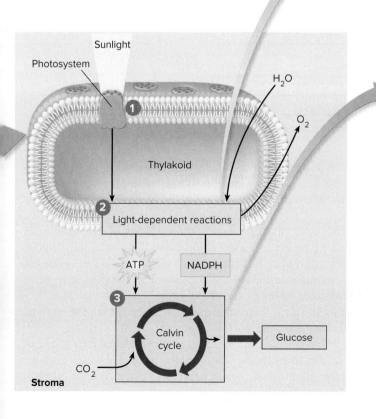

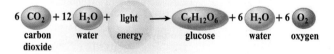

$$6\ CO_2 + 12\ H_2O + \text{light energy} \longrightarrow C_6H_{12}O_6 + 6\ H_2O + 6\ O_2$$
carbon dioxide + water + light energy → glucose + water + oxygen

---

Putting the Concept to Work
Why does a photosystem capture photons better than a random mixture of chlorophyll molecules?

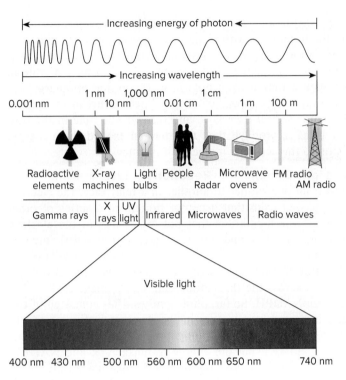

Figure 6.1 Photons of different energy: the electromagnetic spectrum.

Light is composed of packets of energy called photons. Some of the photons in light carry more energy than others. The shorter the wavelength of light, the greater the energy of its photons. Visible light represents only a small part of the electromagnetic spectrum, that with wavelengths between about 400 and 740 nanometers.

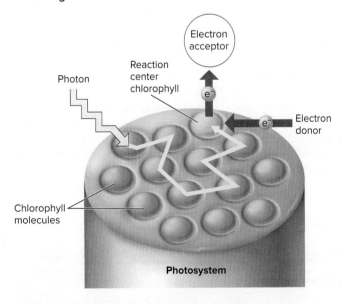

Figure 6.2 How a photosystem works.

When light of the proper wavelength strikes any pigment molecule within a photosystem, the light is absorbed, and its excitation energy is then transferred from one molecule to another within the cluster of pigment molecules (the antenna complex) until it encounters the reaction center, which exports the energy as high-energy electrons to an acceptor molecule.

6.2 How Plants Convert Sunlight to Chemical Energy

> **LEARNING OBJECTIVE 6.2.1** State in what way a photon's energy is related to its wavelength.

Where is the energy in light? Light actually consists of tiny packets of energy called **photons**, which have properties both of particles and of waves. When light shines on your hand, your skin is being bombarded by a stream of these photons.

Sunlight contains photons of many energy levels, only some of which we "see." We call the full range of these photons the **electromagnetic spectrum.** As you can see in figure 6.1, some of the photons in sunlight have shorter wavelengths (toward the left side of the spectrum) and carry a great deal of energy—for example, gamma rays and ultraviolet (UV) light. Others, such as radio waves, have longer wavelengths (hundreds to thousands of meters long) and carry very little energy. Molecules that absorb light energy are called **pigments**. When we speak of visible light, we refer to those wavelengths that the pigment within human eyes, called *retinal,* can absorb— roughly with wavelengths from 400 nanometers (violet) to 740 nanometers (red). Plants are even more picky, absorbing mainly blue and red light and reflecting back what is left of the visible light. Plants are perceived by our eyes as green simply because only the green wavelengths of light are reflected off the plant leaves.

> **Putting the Concept to Work**
> What color would your body be if you reflected all wavelengths of visible light? if you absorbed only green light?

Pigments and Photosystems

> **LEARNING OBJECTIVE 6.2.2** List the five stages of the light-dependent reactions.

The main pigment in plants that absorbs light is chlorophyll, present in two versions: chlorophyll *a* and chlorophyll *b*. While chlorophyll absorbs fewer kinds of photons than our eye pigment retinal, it is much more efficient at capturing them.

In plants and algae, the light-dependent reactions of photosynthesis occur in specialized organelles called chloroplasts. The chlorophyll molecules and proteins involved in the light-dependent reactions are embedded in the thylakoid membranes inside the chloroplasts. This complex of proteins and pigment molecules makes up the **photosystem.**

Like a magnifying glass focusing light on a precise point, the *antenna complex* of a photosystem, the portion that contains the light-harvesting pigment molecules, channels the excitation energy gathered by any one of its pigment molecules to a specific chlorophyll *a* molecule, which is called the reaction center chlorophyll. For example, in figure 6.2, a chlorophyll

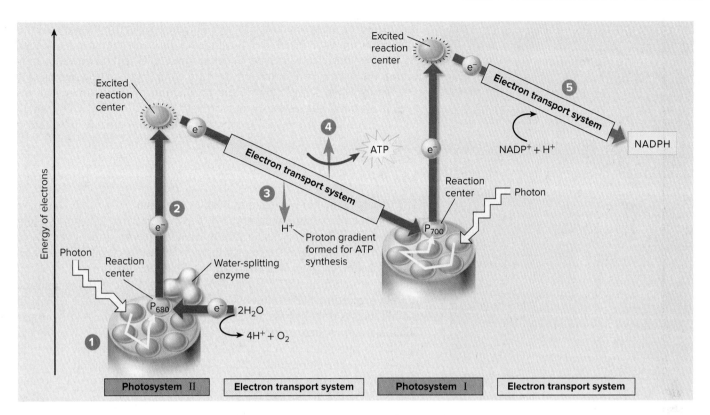

Figure 6.3 Plants use two photosystems in the light-dependent reactions.

In stage ❶, a photon excites pigment molecules in photosystem II. In stage ❷, a high-energy electron is transferred to the electron transport system. In stage ❸, the excited electron is used to pump a proton across the membrane. In stage ❹, the concentration gradient of protons is used to produce a molecule of ATP. In stage ❺, the ejected electron then passes to photosystem I to power the forming of NADPH.

molecule on the outer edge of the photosystem is excited by the photon, and this energy passes through the antenna complex from one chlorophyll molecule to another, indicated by the yellow zigzag arrow, until it reaches the reaction center molecule. This molecule then passes the energy, in the form of an excited electron, out of the photosystem to drive the synthesis of ATP and organic molecules.

Using Two Photosystems

In the light-dependent reactions, plants and algae use two photosystems, photosystems I and II (figures 6.3 and 6.4). Photosystem II captures the energy that is used to produce the ATP needed to build sugar molecules. The light energy that it captures is used to transfer the energy of a photon of light ❶ to an excited electron ❷, which is then shuttled along a series of electron-carrier molecules embedded in the membrane called the **electron transport system** ❸. The energy of this electron is then used to produce ATP ❹, which will be used in the Calvin cycle.

The electron then enters photosystem I, where it is "re-energized" by the absorption of another photon of light. Photosystem I passes this energized electron to a second electron transport system, where the electron, carried by a hydrogen ion (a proton), is used to form NADPH from NADP$^+$ ❺. NADPH shuttles hydrogens to the Calvin cycle, where sugars are made. This whole process produces oxygen gas because the

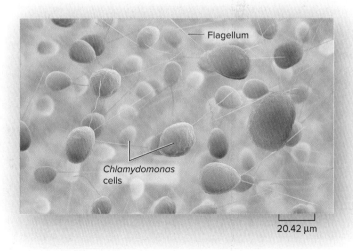

Figure 6.4 Plants aren't the only ones that carry out photosynthesis.

Each of these green balls is a single-celled photosynthetic organism, a green alga called *Chlamydomonas* that is common in pond water.

Andrew Syred/Science Source

Figure 6.5 Capturing energy from sunlight.
Sunlight beams down in this forest glade. Although only a few of the plants you see are bathed in direct sunlight, all of the trees and shrubs and grasses are receiving an ample amount of sunlight, their photosystems running full bore even in dim light.
Adrian_am13/Shutterstock

water-splitting enzyme acting at stage ❷ removes electrons from water to replace the light-energized ones flowing into the electron transport system. As soon as four electrons have been removed as H⁺ from two water molecules, the two oxygen atoms left over are combined and released as O_2.

Photosystem I thus powers the production of the hydrogen atoms needed to build sugars and other organic molecules from CO_2, which has no hydrogen atoms, while photosystem II powers the production of ATP.

Making ATP

The thylakoid membrane is impermeable to protons, so protons that have been pumped across the membrane by the first electron transport system (see figure 6.3) build up inside the thylakoid space, creating a very large concentration gradient. These protons now diffuse back out of the thylakoid space, down their concentration gradient, passing through a special channel protein called *ATP synthase*. The ATP synthase is the membrane protein protruding like a knob out of the external surface of the thylakoid membrane. As protons pass out of the thylakoid through the ATP synthase channels, a phosphate group is added onto ADP to form ATP in a process called phosphorylation. ATP is then released into the stroma (the fluid matrix inside the chloroplast). Because the chemical formation of ATP is driven by a diffusion process similar to osmosis, this type of ATP formation is called **chemiosmosis**.

Making NADPH

Now, with ATP formed, let's return to where photosystem I accepts an electron from the electron transport system. Energy is fed to photosystem I by an antenna complex of chlorophyll molecules. The electron arriving from the first electron transport system has by no means lost all of its light-excited energy; almost half remains. Thus, the absorption of another photon of light energy by photosystem I boosts the electron leaving its reaction center to a very high energy level (see figures 6.3 and 6.5).

Like photosystem II, photosystem I passes electrons to an electron transport system. When two of these electrons reach the end of this electron transport system, they are then donated to a molecule of NADP⁺ along with a proton (a hydrogen ion) to form NADPH. Because the reaction occurs on the stromal side of the membrane and involves the uptake of a proton, it contributes further to the proton concentration gradient established during photosynthetic electron transport.

Products of the Light-Dependent Reactions

The ATP and NADPH produced in the light-dependent reactions end up being passed on to the Calvin cycle in the stroma of the chloroplast. The stroma contains the enzymes that catalyze the light-independent reactions. There, ATP is used to power chemical reactions that build carbohydrates, with NADPH providing the necessary hydrogens and electrons.

> **Putting the Concept to Work**
> What would happen to photosynthesis if you were to puncture the thylakoid membrane?

6.3 Building New Molecules

The Calvin Cycle

> **LEARNING OBJECTIVE 6.3.1** Explain why the Calvin cycle requires NADPH, as well as ATP.

Stated very simply, photosynthesis is a way of making organic molecules from carbon dioxide (CO_2). The actual assembly of new molecules employs a complex battery of enzymes in what is called the **Calvin cycle,** or **C_3 photosynthesis** (C_3 because the first molecule produced in the process is a 3-carbon molecule). The process takes place in three stages, indicated by the three panels in Essential Biological Process 6A. Three turns of the cycle are needed to produce one molecule of glyceraldehyde 3-phosphate. In any *one* turn of the cycle, a carbon atom from a carbon dioxide molecule is first added to a 5-carbon sugar, producing two 3-carbon sugars. This process is called **carbon fixation** because it attaches a carbon atom that was in a gas to an organic molecule. The cycle has to "turn" six times in order to form a new glucose molecule. The cycle is driven by energy from ATP, and hydrogen atoms are supplied by NADPH, both produced in the light-dependent reactions.

> **Putting the Concept to Work**
> What three substances are needed by the Calvin cycle to produce a molecule of glucose?

Essential Biological Process 6A

The Calvin Cycle

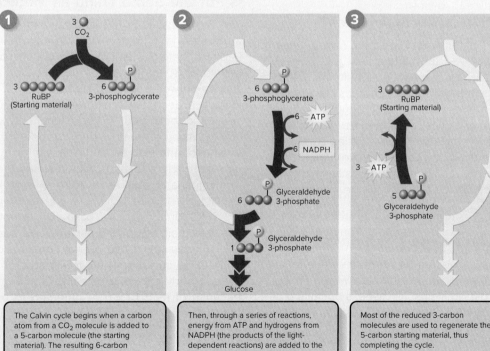

1. The Calvin cycle begins when a carbon atom from a CO_2 molecule is added to a 5-carbon molecule (the starting material). The resulting 6-carbon molecule is unstable and immediately splits into 3-carbon molecules. (Three "turns" of the cycle are indicated here with three molecules of CO_2 entering the cycle.)

2. Then, through a series of reactions, energy from ATP and hydrogens from NADPH (the products of the light-dependent reactions) are added to the 3-carbon molecules. The now-reduced 3-carbon molecules either combine to make glucose or are used to make other molecules.

3. Most of the reduced 3-carbon molecules are used to regenerate the 5-carbon starting material, thus completing the cycle.

Answering Your Questions About Medical Marijuana

Is Marijuana a Vegetable?

In common usage, a *vegetable* is "a plant grown for food," so while marijuana is indeed composed of vegetable matter and is often consumed by people, it is not a food crop and so not a vegetable. However, the marijuana plant *Cannabis sativa* and two close relatives are cultivated throughout the world. Why this popularity, if it is not a food? For two reasons: (1) The flowers and leaves of *Cannabis* plants contain a psychoactive drug called tetrahydrocannabinol (THC), which has both medical and recreational uses. (2) The stem makes great rope.

What Are the Medical Uses of Marijuana?

For thousands of years, marijuana has been used to treat pain. Around A.D. 200, Chinese physicians were using marijuana as an anesthetic when performing surgery. They would also boil the leaves and use the oils to treat ailments such as blood clots, tapeworms, constipation, and baldness. Today, marijuana is widely used to treat schizophrenia, stop seizures, heal broken bones, alleviate the nausea that is often a side effect of chemotherapy, and restore the appetites of AIDS patients. In England, *Cannabis* was approved in 2010 as a drug to treat multiple sclerosis and neuropathic pain.

Is Marijuana Native to North America?

Cannabis is native to Central Asia. During the Middle Ages, marijuana found its way into the Middle East and Europe, where it was used to treat a variety of different health problems. The Spanish introduced the *Cannabis* plant to the Americas in the 1500s. Interestingly, only South American colonies got THC-rich varieties of *Cannabis*. The plants introduced into North America were of varieties cultivated for the fiber portion of the plant, used to make hemp rope. While this variety had little THC, hemp rope was an important commercial crop in the United States until the invention of nylon ropes in the 1940s.

Is Medical Marijuana Legal?

As of January 2021, 36 states and 4 territories have legalized the use of medical marijuana, although with a variety of restrictions on this use. In almost all cases, a doctor's prescription is required, and in most states, amounts are strictly limited. In marked contrast to this widespread acceptance of marijuana's medical usefulness, the federal government has, since 1970, designated marijuana to be a Schedule 1 drug and listed it as having no medicinal value. In 2016, a majority of Americans (51%) had used cannabis, about 12% in the last year and 7% in the last month. This makes marijuana the most commonly used illegal drug in the United States.

What Is the THC in Marijuana, and How Does It Get You High?

THC is a lipid manufactured by the *Cannabis* plant as a defense against insect predators. In the human brain, THC binds to special receptors on brain neurons. Among animals, only chordates have these receptors, which are normally used to modulate pain. It is only an accident of nature that a plant chemical "fits this lock" on our brain cells. Triggered by THC binding, these brain receptors proceed to inhibit the enzyme adenylate cyclase and so decrease the concentration in the brain of a "second messenger" molecule called cAMP. Normally, cAMP is a tool the brain uses to keep our mood calm, so lowering its level in the brain leads to a feeling of euphoria.

Is Pot the Same Thing as Medical Marijuana?

Yes. Since the early 1900s, marijuana has been smoked as a recreational drug, the THC promoting euphoria, a high or "stoned" feeling. Onset of effects is within minutes if smoked, and about 30 to 60 minutes when cooked and eaten. Effects last for between two and six hours. Short-term side effects include a decrease in short-term memory, impaired motor skills, and feelings of anxiety. Commonly known as "pot" or "weed," marijuana has for decades been considered the drug of stoners and hippies, but in recent years, its use has seen far more social acceptance. In 18 states, the recreational use of marijuana is legal as of 2021—in each instance, in direct conflict with federal law.

THC-containing oil ("hash oil") is not to be confused with hemp oil, made by crushing hemp seeds from varieties of *Cannabis sativa* that do not contain significant amounts of THC. Refined hemp oil, totally free of THC, is primarily used in body care products and is of high nutritional value because of its 3:1 ratio of omega-6 to omega-3 essential fatty acids.

Photorespiration

6.4 Photorespiration: Putting the Brakes on Photosynthesis

> **LEARNING OBJECTIVE 6.4.1** Contrast C_3, C_4, and CAM photosynthesis.

Many plants have trouble carrying out C_3 photosynthesis when the weather is hot. As temperatures increase in hot, arid weather, plants partially close their **stomata** (singular, **stoma**) to conserve water. The stomata are openings in the epidermis through which water vapor and O_2 pass out of the leaf and CO_2 passes in. As a result, when stomata close, CO_2 and O_2 are not able to enter and exit the leaves (figure 6.6). The concentration of CO_2 in the leaves falls, while the concentration of O_2 in the leaves rises. Under these conditions, the active site of the enzyme that carries out the first step of the Calvin cycle (called rubisco) binds O_2 instead of CO_2. When this occurs, CO_2 is ultimately released as a by-product of an alternate reaction, effectively short-circuiting the Calvin cycle. This failure of photosynthesis is called **photorespiration**.

C_4 Photosynthesis

Some plants are able to adapt to climates with higher temperatures by performing **C_4 photosynthesis**. In this process, plants such as sugarcane, corn, and many grasses are able to fix carbon using different types of cells and chemical reactions within their leaves, thereby avoiding a reduction in photosynthesis due to higher temperatures.

A cross section of a leaf from a C_4 plant is shown in figure 6.7. Examining it, you can see how these plants solve the problem of photorespiration. In the enlargement, you see two cell types: The green cell is a mesophyll cell and the tan cell is a bundle-sheath cell. In the mesophyll cell, CO_2 combines with a 3-carbon molecule instead of RuBP (the 5-carbon starting molecule in the Calvin cycle). This reaction produces a 4-carbon molecule, oxaloacetate (hence the name, C_4 photosynthesis), rather than the 3-carbon molecule, phosphoglycerate, shown in panel 1 of *Essential Biological Process 6A*. C_4 plants carry out this process using a different enzyme. The oxaloacetate is then converted to malate, which is transferred to the bundle-sheath cells of the leaf. In the tan bundle-sheath cell, malate is broken down to regenerate CO_2, which enters the Calvin cycle and sugars are synthesized. Why go to all this trouble? Because the bundle-sheath cells are impermeable to CO_2 and so the concentration of CO_2 increases within them, which substantially lowers the rate of photorespiration.

CAM Photosynthesis

A second strategy to decrease photorespiration is used by many succulent (water-storing) plants such as cacti and pineapples. Called **crassulacean acid metabolism (CAM)** plants after the plant family Crassulaceae in which it was first discovered, these plants open their stomata and fix CO_2 into organic compounds during the night, when it's cooler, and then close the stomata during the day.

> **Putting the Concept to Work**
> Are C_4 plants more likely to be common in Arizona or Maine? Why?

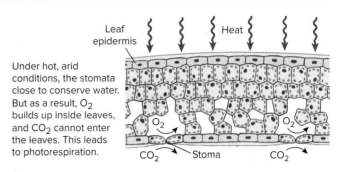

Figure 6.6 **Photorespiration.**

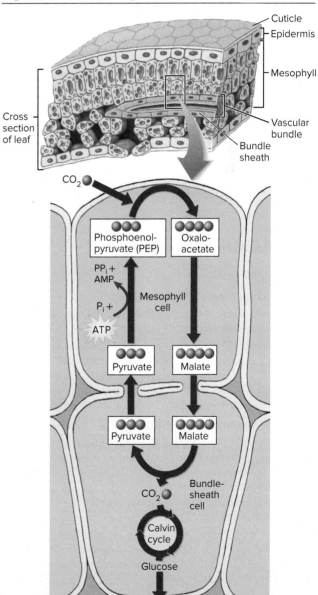

Figure 6.7 **Carbon fixation in C_4 plants.**

In mesophyll cells, oxaloacetate is converted into malate that is transported into bundle-sheath cells. Once there, malate undergoes a chemical reaction, producing carbon dioxide. The carbon dioxide is trapped in the bundle-sheath cell, where it enters the Calvin cycle.

Putting the Chapter to Work

1 The electromagnetic spectrum encompasses a broad range of energy. Some of this energy we can see as photons of visible light, while other portions of the energy spectrum are invisible to our eyes. Gamma rays are an example of energy we cannot see.

How do gamma rays compare in wavelength and energy to the photons we can see?

2 Photosynthesis is the ultimate source of the oxygen in earth's atmosphere. The production of oxygen occurs in photosystem II of the light-dependent reaction.

What molecule is split by a photon of light in photosystem II to produce oxygen?

3 Photosynthesis is the process that uses sunlight to convert atmospheric CO_2 into organic molecules for cells to use. It consists of two reactions: (1) the light-dependent reaction and (2) the light-independent reaction.

What does the light-dependent reaction provide that is necessary for the light-independent reaction to take place?

4 The light-dependent reaction has two photosystems that generate different reactants used in the light-independent reaction.

If photosystem I is not excited by another photon of light, which molecule will not be available for the light-independent reaction?

Retracing the Learning Path

Photosynthesis

6.1 An Overview of Photosynthesis

1. Photosynthesis uses energy from sunlight to power the synthesis of organic molecules from CO_2 in the air.
2. Photosynthesis consists of a series of chemical reactions that occurs in two stages: the light-dependent reactions that produce ATP and NADPH occur on the thylakoid membranes of chloroplasts in plants, and the light-independent reactions (the Calvin cycle) that synthesize carbohydrates occur in the stroma of the chloroplasts.

6.2 How Plants Convert Sunlight to Chemical Energy

1. Sunlight contains packets of energy called photons, which contain varying amounts of energy. As light wavelengths increase in size, the amount of energy in the photons decreases. Visible light consists of wavelengths absorbed by pigments in the human eye (between 400 and 740 nanometers). Pigments are molecules that capture light energy. Plants use the pigment chlorophyll to absorb light energy.
2. Plants appear green because of their chlorophyll pigments. Chlorophyll absorbs wavelengths in the far ends of the visual spectrum (the blue and red wavelengths) and reflects the green wavelengths, which is why leaves appear green.
- The light-dependent reactions occur on the thylakoid membranes of chloroplasts in plants. The chlorophyll molecules and other pigments involved in photosynthesis are embedded in a complex of proteins within the membrane called a photosystem.
- The energy from a photon of light is absorbed by a chlorophyll molecule and is transferred between chlorophyll molecules in the photosystem. Once the energy is passed to the reaction center, it excites an electron, which is transferred to the electron transport system.
- The energized electron is used to generate ATP and NADPH. ATP powers the Calvin cycle, and NADPH donates hydrogen atoms toward the building of carbohydrate molecules. Plants utilize two photosystems that occur in series. Photosystem II leads to the formation of ATP, and photosystem I leads to the formation of NADPH.
- The excited electron that leaves the reaction center of photosystem II is replenished with an electron captured from the breakdown of a water molecule. Oxygen gas is released as a by-product of this reaction.
- The excited electron is passed from one protein to another in the electron transport system, where energy from the electron is used to operate a proton pump that pumps hydrogen ions across the membrane against a concentration gradient.
- The hydrogen ion concentration gradient is used as a source of energy to generate molecules of ATP. As the concentration of H^+ ions inside the thylakoid increases, H^+ ions diffuse back across the membrane through a specialized channel protein called ATP synthase, which catalyzes the formation of ATP, a process called chemiosmosis.
- After the electron passes along the first electron transport system, it is then transferred to a second photosystem, photosystem I, where it gets an energy boost from the capture of another photon of light. This reenergized electron is passed along another electron transport system to an ultimate electron acceptor, $NADP^+$. $NADP^+$ binds electrons and an H^+ ion to produce NADPH, which is shuttled to the Calvin cycle.

6.3 Building New Molecules

1. ATP and NADPH from the light-dependent reactions are shuttled to the stroma, where they are used in the Calvin cycle.
- The Calvin cycle is carried out by a series of enzymes that use the energy from ATP, and electrons and hydrogen ions from NADPH, to build molecules of carbohydrates by reducing CO_2 from the air.

Photorespiration

6.4 Photorespiration: Putting the Brakes on Photosynthesis

1. Photorespiration occurs as a response to the buildup of oxygen within photosynthetic cells: a product of photosynthesis, oxygen tends to push photosynthesis backward. In hot, dry weather, plants will close the stomata in their leaves to conserve water. As a result, the levels of O_2 increase in the leaves and CO_2 levels drop. Under these conditions, the Calvin cycle, also called C_3 photosynthesis, is disrupted. When there is a higher internal concentration of oxygen, O_2 rather than CO_2 enters the Calvin cycle, a process called photorespiration. In this case, the first enzyme in the Calvin cycle, rubisco, binds oxygen instead of carbon dioxide.
- C_4 plants reduce the effects of photorespiration by modifying the carbon-fixation step, splitting it into two steps that take place in different cells. The C_4 pathway produces malate in mesophyll cells. Malate is then transferred to bundle-sheath cells, where it breaks down to produce carbon dioxide. This CO_2 then enters the Calvin cycle in the bundle-sheath cells.

Inquiry and Analysis

Does Increased Carbon Dioxide Affect the Growth of All Plants Equally?

Climate change has its immediate impact on local temperature and rainfall. A broad range of scientific observations tell us the primary driver of climate change is the increase in atmospheric carbon dioxide (CO_2) produced by the burning of fossil fuels. If the temperature and rainfall patterns for a given region change too much, you might expect there would be an observable shift in plant biodiversity in that region, as less successful plants shift their ranges and plants better able to cope with the changing condition migrate in. Over the last 100 years, atmospheric CO_2 levels have increased from 302 parts per million (PPM) to 411 PPM, an increase of 109 PPM or 36%. Just in the last 20 years, we have increased atmospheric CO_2 levels 49 PPM! Clearly the amount of CO_2 in our atmosphere is rapidly increasing.

It has been argued that the increase in atmospheric CO_2 creating climate change may be beneficial to some plants: If higher levels of CO_2 were to increase the rates of photosynthesis, this would be expected to lead to increased plant growth and yield. Why should higher CO_2 levels affect photosynthesis? Because when the weather is hot, C_3 plants have trouble carrying out photosynthesis. Heat increases the rate of photorespiration, which undoes the work of photosynthesis. C_4 plants, more common in warmer climates, utilize a different process to fix CO_2 that avoids this reduction. In laboratory experiments, C_3 plants do indeed have decreased rates of photosynthesis as CO_2 levels increase, while at the same CO_2 levels, C_4 plant photosynthesis is relatively constant.

But these studies were performed over a relatively short time frame of just a few years, in a greenhouse. Out in the real world, do C_3 and C_4 plants respond the same way to the long-term elevations in atmospheric CO_2 associated with climate change?

Scientists tested this question by examining the amount of total plant growth (biomass) in eighty-eight experimental plots over a twenty-year period. In each of the plots, scientists planted equal amounts of four different C_3 grasses and four different C_4 grasses. During the late fall of each growing season, the aboveground and below-ground biomass were measured and recorded. The graphs shown here represent the data that were collected. Each data point represents a moving 3-year average, averaging the year before, the current year, and the next year. The total biomass for C_3 grasses is indicated in red and the total biomass in C_4 grasses is indicated in blue. The graph is plotted as the total biomass, measured in gram weight per square meter (y-axis) over time (x-axis).

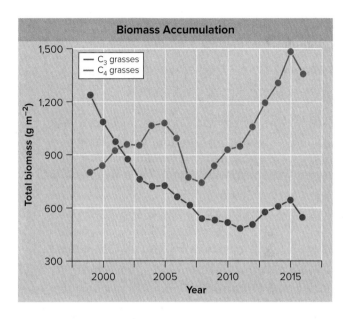

Analysis

1. **Applying Concepts**
   Variable. In the graph, what is the dependent variable?
2. **Interpreting Data**
   a. On the red line, what year did the C_3 grasses have the smallest biomass? What about the blue line representing the C_4 grasses?
   b. Which type of grass had an increase in total biomass over time? What was the total biomass in 2000? In 2015?
3. **Making Inferences**
   In the data from 2000, what was the ratio of the total biomass of C_3 grasses compared to C_4 grasses?
4. **Drawing Conclusions**
   Does this data support the hypothesis that, over long periods of time, C_3 grasses are able to use higher levels of atmospheric CO_2 to increase in their biomass?
5. **Further Analysis**
   a. In this experiment, the C_3 grasses had higher total biomass measurements early in the study but decreased in biomass in subsequent years. Provide a hypothesis for why the C_3 plants lost biomass early in the experiment.
   b. Examine both data sets and you will notice that around 2012, both C_3 and C_4 grasses experienced a decrease in their respective biomasses. What do you think would cause a decrease in biomass in both types of grass at the same time?

7 How Cells Harvest Energy from Food

LEARNING PATH ▼

An Overview of Cellular Respiration
1. Where Is the Energy in Food?

Respiration Without Oxygen: Glycolysis
2. Using Coupled Reactions to Make ATP

Respiration with Oxygen: The Krebs Cycle
3. Harvesting Electrons from Chemical Bonds
4. Using the Electrons to Make ATP

Harvesting Electrons Without Oxygen: Fermentation
5. Cells Can Metabolize Food Without Oxygen

Other Sources of Energy
6. Glucose Is Not the Only Food Molecule

THEY ARE HARVESTING ENERGY from a cheeseburger—some 500 calories, enough to fuel their body for many hours.
©Corbis/VCG/Getty Images

Fad Diets and Impossible Dreams

There are very few of us who at one time or another don't seek to lose weight. Even those skinny in their teens tend to put on the pounds as they age. How to lose the excess pounds? The classic American answer to this perennial question: Go on a diet. You will probably not be surprised to learn that diet books are the most popular best sellers, year in and year out. Most present fad diets, quick solutions based on "imaginative" science that melt away the pounds—painlessly, of course. Chief among these are the Atkins diet and, more recently, the paleo diet. The reason these diets don't deliver on their promise of pain-free weight loss is well understood by science but not by the general public. Only hope and hype make them best sellers.

Avoiding Carbs

The secret of the Atkins and paleo diets, stated simply, is to avoid carbohydrates. The Atkins basic proposition is that your body, if it does not detect blood glucose (from metabolizing carbohydrates), will think it is starving and start to burn body fat, even if there is lots of fat already circulating in your bloodstream. You may eat all the fat and protein you want, all the steak and eggs and butter

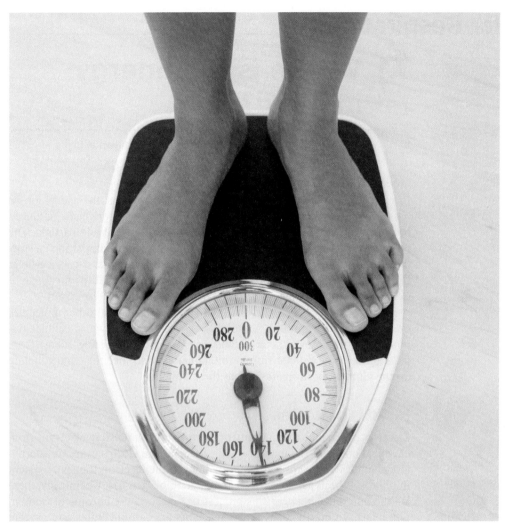
Stockbyte/Getty Images

The Atkins and paleo diets are the sort of diets the American Heart Association tells us to avoid (all those saturated fats and cholesterol promote the formation of plaque on arterial walls and so greatly increase the chance of heart disease and fatty liver disease). While it is difficult to stay on a low-carb diet, if you do hang in there, you will lose weight—simply because you eat less. Studies of this diet by federal laboratories confirm both the caloric restriction and the fact that the weight loss is rarely permanent.

The Basic Laws of Dieting

There are two basic laws that no diet can successfully violate:

1. All calories are equal.
2. (calories in) − (calories out) = fat

The fundamental fallacy of the Atkins and paleo diets, and indeed of all fad diets, is the idea that somehow carbohydrate calories are different from fat and protein calories. This is scientific foolishness. Every calorie you eat contributes equally to your eventual weight, whether it comes from carbohydrate, fat, or protein.

and cheese, and you will still burn fat and lose weight—just don't eat more than 6% carbohydrates (very little). No bread or pasta or potatoes. The paleo diet assumes ancient humans ate a diet high in protein and animal fat, and low in carbohydrates—similar to the Atkins diet.

People who try these diets often lose up to 10 pounds in a few weeks. In three months, it is usually all back and then some. So what happened? Where did the pounds go, and why did they come back? The temporary weight loss turns out to have a simple explanation: because carbohydrates act as water sponges in your body, forcing your body to become depleted of carbohydrates causes your body to lose water. The 10 pounds initially lost on this diet was not fat weight but water, quickly regained with the first starchy foods eaten.

Successful Dieting

To the extent these diets work at all, they do so because they obey the second law. By reducing calories in, they reduce fat. Studies of successful weight losers—those who have shed more than 30 pounds and kept them off for at least a year—reveal that they limit their food consumption to only about 1,400 calories a day. Keeping the weight off had nothing to do with the type of diet—low fat, low-carb—with which it was originally lost.

This doesn't mean that we should give up and learn to love our fat. Rather, we must accept the hard fact that we cannot beat the requirements of the two diet laws. The real trick is not to give up. Eat less and exercise more, and keep at it. In one year, or two, or three, your body will readjust its metabolism to reflect the new reality you have imposed by constant struggle. There simply isn't any easy way to lose weight.

An Overview of Cellular Respiration

7.1 Where Is the Energy in Food?

> **LEARNING OBJECTIVE 7.1.1** Distinguish between oxidation and reduction.

In plants and animals, and in fact in almost all organisms, the energy for living is obtained by breaking down the organic molecules originally produced by photosynthetic organisms, such as plants, algae, and certain bacteria. The energy invested in building the organic molecules is retrieved by stripping away the energetic electrons and using them to make ATP, a process called **cellular respiration.** Do not confuse the term *cellular respiration* with the breathing of oxygen gas that your lungs carry out, which is called simply respiration.

The cells of plants fuel their activities with sugars and other molecules that they produce through photosynthesis and break down in cellular respiration. Nonphotosynthetic organisms eat plants, extracting energy from plant tissue using cellular respiration. Other animals, like the child gobbling up the hamburger in figure 7.1, eat these animals.

Redox Reactions

Eukaryotes produce the majority of their ATP by harvesting electrons from chemical bonds of the food molecule glucose. The electrons are transferred along an electron transport chain (similar to the electron transport system in photosynthesis) and eventually donated to oxygen gas. Chemically, there is little difference between this process in a cell and the burning of wood in a fireplace. In both instances, the reactants are carbohydrates and oxygen, and the products are carbon dioxide, water, and energy:

$$C_6H_{12}O_6 + 6\ O_2 \rightarrow 6\ CO_2 + 6\ H_2O + \text{energy (heat or ATP)}$$

In a chemical reaction, when an atom or molecule loses an electron, it is said to be *oxidized,* and the process by which this occurs is called **oxidation.** The name reflects the fact that in biological systems, oxygen, which attracts electrons strongly, is the most common electron acceptor. This is certainly the case in cellular respiration, where oxygen is the final electron acceptor. Conversely, when an atom or molecule gains an electron, it is said to be *reduced,* and the process is called **reduction.** Oxidation and reduction always take place together because every electron that is lost by an atom through oxidation is gained by some other atom through reduction. Therefore, chemical reactions of this sort are called **oxidation-reduction (redox) reactions.** In redox reactions, energy follows the electron, as shown in figure 7.2.

Cellular Respiration

Cellular respiration is carried out in two stages, illustrated in figure 7.3. The first stage uses coupled reactions to make ATP. This stage, *glycolysis,* takes place in the cell's cytoplasm (the blue area in figure 7.3). Importantly, it is anaerobic (that is, it does not require oxygen). This ancient energy-extracting process is thought to have evolved over 2 billion years ago, when there was no oxygen in the earth's atmosphere.

The second stage is aerobic (requires oxygen) and takes place within the mitochondrion (the tan sausage-shaped structure in figure 7.3). The focal point of this stage is the *Krebs cycle,* a cycle of chemical reactions that harvests

Figure 7.1 A human acquiring energy.

Energy that this child extracts from the hamburger will be used to power their thinking, fuel their running, and build a bigger body.

Ingram Publishing/Getty Images

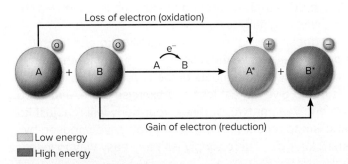

Figure 7.2 Redox reactions.

Oxidation is the loss of an electron; reduction is the gain of an electron. Here the charges of Molecules A and B are shown in small circles to the upper right of each molecule. Molecule A loses energy as it loses an electron, while Molecule B gains energy as it gains an electron.

electrons from C—H chemical bonds and passes the energy-rich electrons to the electron-carrier molecules called NADH and FADH$_2$. These molecules deliver the electrons to an electron transport chain, which uses their energy to power the production of ATP. The harvesting of electrons, a form of *oxidation,* is far more powerful than glycolysis at recovering energy from food molecules and is how the bulk of the energy used by eukaryotic cells is extracted from food molecules.

> **Putting the Concept to Work**
> The Krebs cycle harvests electrons from what kind of chemical bond? Are the carbon atoms of the bond being oxidized or reduced?

Figure 7.3 An overview of cellular respiration.
Electrons harvested from C—H chemical bonds are first transferred to NADH and FADH$_2$. These then carry the electrons to the electron transport chain, as indicated by the long red arrow on the left. The energy-depleted electron is finally donated with a proton to oxygen, forming a molecule of water.

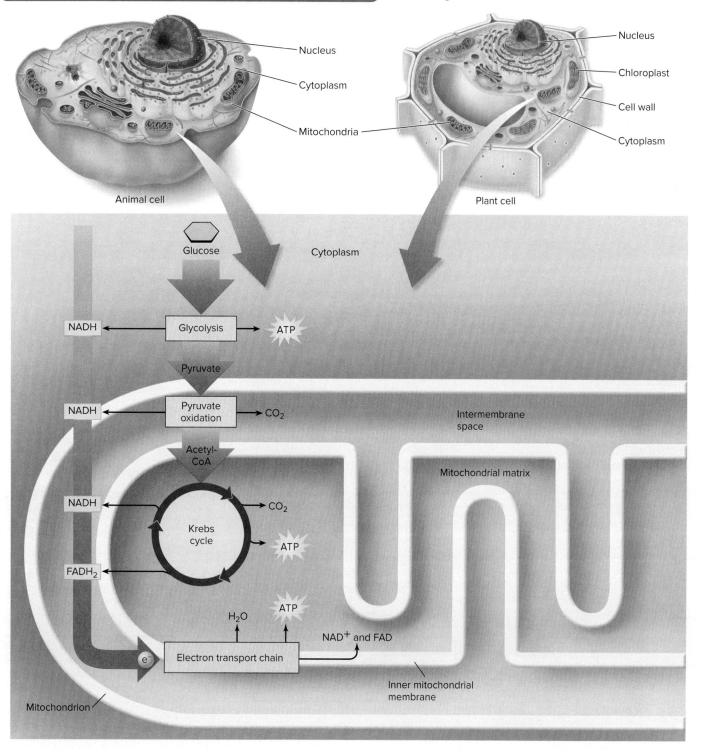

Respiration Without Oxygen: Glycolysis

7.2 Using Coupled Reactions to Make ATP

LEARNING OBJECTIVE 7.2.1 Explain how glycolysis uses coupled reactions to produce ATP from glucose.

The first stage in cellular respiration is a series of 10 reactions called **glycolysis** (*Essential Biological Process 7A*) in which the 6-carbon sugar glucose is cleaved into two 3-carbon molecules of pyruvate. Where is the energy extracted? In each of two coupled reactions, the breaking of a chemical bond releases enough energy to drive the formation of an ATP molecule from ADP (an endergonic reaction). This transfer of a high-energy phosphate group from a substrate to ADP is called **substrate-level phosphorylation.** In the absence of oxygen, this is the only way organisms can get energy from food.

Putting the Concept to Work
How many ATPs are made from glucose in the absence of oxygen?

Essential Biological Process 7A

Glycolysis

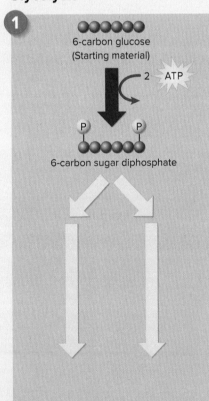

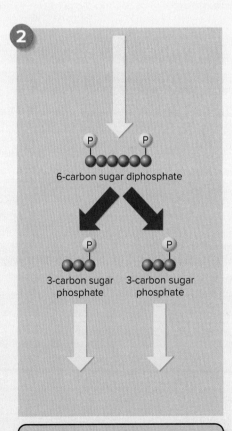

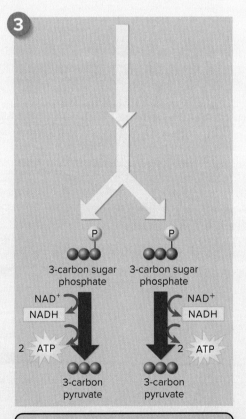

Priming reactions. Glycolysis begins with the addition of energy. Two high-energy phosphates from two molecules of ATP are added to the 6-carbon molecule glucose, producing a 6-carbon molecule with two phosphates.

Cleavage reactions. Then, the phosphorylated 6-carbon molecule is split in two, forming two 3-carbon sugar phosphates.

Energy-harvesting reactions. Finally, in a series of reactions, each of the two 3-carbon sugar phosphates is converted to pyruvate. In the process, an energy-rich hydrogen is harvested as NADH, and two ATP molecules are formed for each pyruvate.

Respiration with Oxygen: The Krebs Cycle

7.3 Harvesting Electrons from Chemical Bonds

> **LEARNING OBJECTIVE 7.3.1** Name and describe the enzyme that removes CO_2 from pyruvate.

The first step of oxidative respiration in the mitochondrion is the oxidation of the 3-carbon molecule called pyruvate, which is the end product of glycolysis. The cell harvests electrons from pyruvate in two steps: first, by oxidizing pyruvate to form acetyl-CoA and then by oxidizing acetyl-CoA in the Krebs cycle.

Step One: Producing Acetyl-CoA

Pyruvate is oxidized in a single reaction that cleaves off one of pyruvate's three carbons. Pyruvate dehydrogenase, the complex of enzymes that removes CO_2 from pyruvate, is one of the largest enzymes known. It contains 60 subunits! In the course of the reaction, a hydrogen and electrons are removed from pyruvate and donated to NAD^+ to form NADH. Now focus on **figure 7.4**. The 2-carbon fragment (called an acetyl group) that remains after removing CO_2 from pyruvate still contains a great deal of energy. It is joined to a cofactor called coenzyme A (CoA) by pyruvate dehydrogenase, forming a compound known as **acetyl-CoA**. If the cell has a plentiful supply of ATP, acetyl-CoA is funneled into fat synthesis, with its energetic electrons preserved for later needs. If the cell needs ATP now, the fragment is directed instead into the Krebs cycle, where its energy is harvested to make ATP.

> **Putting the Concept to Work**
> Does the oxidation of pyruvate harvest any energy? If so, how?

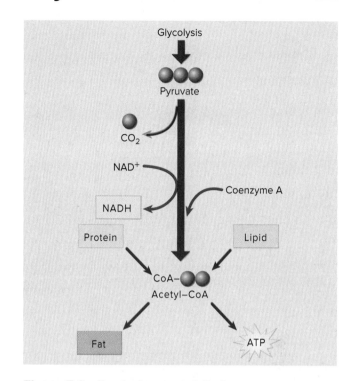

Figure 7.4 Producing acetyl-CoA.

Pyruvate, the 3-carbon product of glycolysis, is oxidized to the 2-carbon molecule acetyl-CoA and in the process loses one carbon atom as CO_2 and an electron (donated to NAD^+ to form NADH). Almost all the molecules you use as foodstuffs are converted to acetyl-CoA, which is then channeled into fat synthesis or into ATP production, depending on your body's needs.

Step Two: The Krebs Cycle

> **LEARNING OBJECTIVE 7.3.2** Identify the overall substrates for the nine-reaction Krebs cycle and the overall products.

The next stage in oxidative respiration, called the **Krebs cycle** after the scientist who discovered it, harvests the energy of acetyl-CoA fragments. The Krebs cycle reactions take place within the mitochondrion.

Stage 1. The cycle starts when the 2-carbon acetyl-CoA fragment produced from pyruvate is stuck onto a 4-carbon sugar, producing a 6-carbon molecule (*Essential Biological Process 7B*).

Stage 2. Then, in rapid-fire order, two carbons are removed as CO_2, their electrons donated to NAD^+, and a 4-carbon molecule is left. A molecule of ATP is also produced.

Stage 3. When it is all over, two carbon atoms have been expelled as CO_2, more energetic electrons are extracted and taken away as NADH or $FADH_2$, and we are left with the same 4-carbon sugar we started with.

Essential Biological Process 7B

The Krebs Cycle

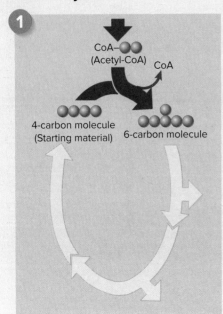
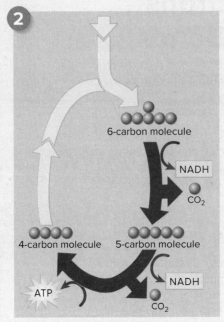
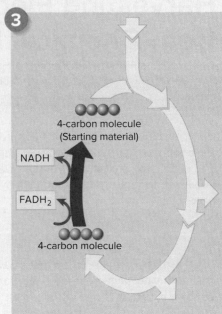

1. The Krebs cycle begins when a 2-carbon fragment is transferred from acetyl-CoA to a 4-carbon molecule (the starting material).

2. Then, the resulting 6-carbon molecule is oxidized (a hydrogen removed to form NADH) and decarboxylated (a carbon removed to form CO_2). Next, the 5-carbon molecule is oxidized and decarboxylated again, and a coupled reaction generates ATP.

3. Finally, the resulting 4-carbon molecule is further oxidized (hydrogens removed to form $FADH_2$ and NADH). This regenerates the 4-carbon starting material, completing the cycle.

The process is a cycle—a circle of nine reactions. In each turn of the cycle, a new acetyl group replaces the two CO_2 molecules lost, and more electrons are extracted. In the process of cellular respiration, glucose is entirely consumed. All that is left to mark the passing of the glucose molecule into six CO_2 molecules is its energy (figure 7.5), preserved in four ATP molecules (two from glycolysis and two from the Krebs cycle) and electrons carried by 10 NADH and two $FADH_2$ carriers.

> **Putting the Concept to Work**
> Why can't the Krebs cycle function in the absence of oxygen?

Figure 7.5 Putting food to work.

This chipmunk has cheeks full of leaves. Recently part of a plant, the leaves are soon destined to become part of this chipmunk's life. Climbing trees, chewing on leaves, and seeing, smelling, and hearing its surroundings, thinking the thoughts that chipmunks think—all are powered by **ATP** made from its food.

Isabella Piroyan/EyeEm/Getty Images

7.4 Using the Electrons to Make ATP

Electrons carry energy from one part of a cell to another.

Moving Electrons Through the Electron Transport Chain

> **LEARNING OBJECTIVE 7.4.1** Describe the components of the electron transport chain.

In eukaryotes, aerobic respiration takes place within the mitochondria present in virtually all cells. The internal compartment, or **matrix,** of a mitochondrion contains the enzymes that carry out the reactions of the Krebs cycle. The electrons harvested by oxidative respiration in the matrix are then passed along the electron transport chain, and the energy they release transports protons out of the matrix and into the **intermembrane space.**

The NADH and $FADH_2$ molecules formed during the first stages of aerobic respiration each contain electrons and hydrogens that were gained when NAD^+ and FAD were reduced (refer back to figure 7.3). The NADH and $FADH_2$ molecules carry their electrons to the inner mitochondrial membrane, where they transfer the electrons to a series of membrane-associated molecules collectively called the **electron transport chain** (figure 7.6). The electron transport chain works much like the electron transport system you encountered in studying photosynthesis.

Harvesting Electrons.

❶ A protein complex (the light pink structure in figure 7.6) receives the electrons and, using a mobile carrier, passes these electrons to a second protein complex (the purple structure).

Pumping Protons Out.

❷ This protein complex, along with others in the chain, operates as a proton pump, using the energy of the electrons to drive a proton out across the membrane into the intermembrane space. The arrows indicate the transport of the protons into the top half of the figure, which represents the intermembrane space.

Forming Water.

❸ The electron is then shuttled by another carrier to a third protein complex (the light blue structure). This complex uses electrons such as this one to link oxygen atoms with hydrogen ions to form molecules of water. It is the availability of a plentiful supply of oxygen, the electron acceptor molecule, that makes oxidative respiration possible.

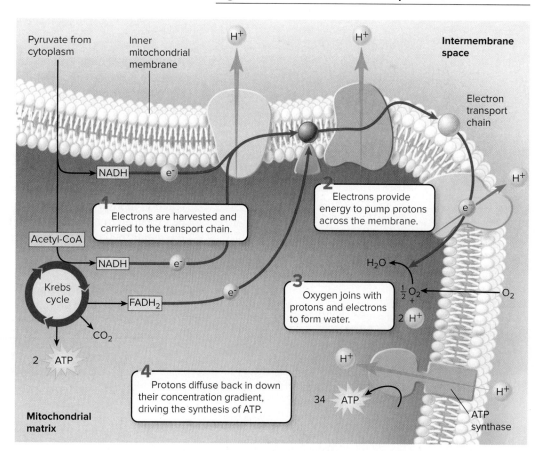

Figure 7.6 The electron transport chain.

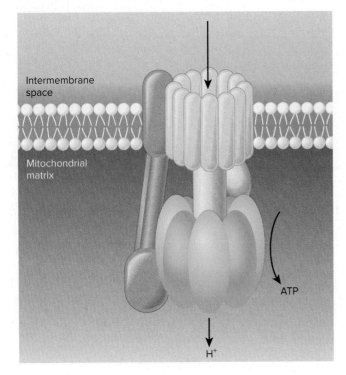

Figure 7.7 The ATP synthase is a highly conserved enzyme and is one of the few true rotary motors known in living organisms.

The electron transport chain used in aerobic respiration is similar to, and may well have evolved from, the electron transport system employed in photosynthesis. Photosynthesis is thought to have preceded cellular respiration in the evolution of biochemical pathways, generating the oxygen that is necessary as the electron acceptor in cellular respiration. Natural selection didn't start from scratch and design a new biochemical pathway for cellular respiration; instead, it built on the photosynthetic pathway that already existed, which uses many of the same reactions.

Making ATP.

❹ As the proton concentration in the intermembrane space rises above that in the matrix, the concentration gradient induces the protons to reenter the matrix by diffusion through a special proton channel called **ATP synthase**. ATP synthase channels are embedded in the inner mitochondrial membrane, as shown in the illustration. As the protons pass through, these channels synthesize ATP from ADP and P_i within the matrix. The ATP is then transported by facilitated diffusion out of the mitochondrion and into the cell's cytoplasm. This ATP synthesizing process is the same chemiosmosis process that you encountered in studying photosynthesis in chapter 6.

> **Putting the Concept to Work**
> What does NADH do that $FADH_2$ cannot? Why can't $FADH_2$ do this?

How Many ATPs Can One Glucose Molecule Yield?

> **LEARNING OBJECTIVE 7.4.2** Calculate how many ATP molecules a cell can harvest from a glucose molecule in the presence of oxygen and in its absence.

Although we have discussed electron transport and chemiosmosis as separate processes, in a cell they are integrated, as shown in figure 7.6. The electron transport chain uses two electrons harvested in glycolysis, two harvested in pyruvate oxidation, and eight harvested in aerobic respiration (red arrows in figure 7.6) to pump a large number of protons out across the inner mitochondrial membrane (shown in the upper right of the panel in figure 7.6). Their subsequent reentry back into the mitochondrial matrix drives the synthesis of 34 ATP molecules by chemiosmosis (shown in the lower right in figure 7.6). Two additional ATPs were harvested by a coupled reaction in glycolysis and two more in the Krebs cycle. As two ATPs must be expended to transport NADH into the mitochondria by active transport, the grand total of ATPs harvested is thus 36 molecules.

> **Putting the Concept to Work**
> If the electron transport chain uses the energy harvested from C–H bonds to drive protons out of the matrix, how is it that the ATP molecules formed as a consequence are *within* the matrix?

Answering Your Questions About Energy Drinks

What Is an Energy Drink?
Energy drinks are nonalcoholic drinks that contain caffeine, taurine (an amino acid), and B vitamins, in addition to sugar and other brand-specific ingredients. They are big business, marketed in your local supermarket and drugstore as stimulants that improve performance and increase energy. Global sales of Red Bull, Monster, 5-Hour Energy, and other energy drinks in 2017 exceeded $55 billion. Over 6 billion cans of Red Bull were sold last year.

Where Does the "Pick Up" Come From?
Drinking a can of energy drink floods your body with caffeine, a strong stimulant, raising blood pressure and blood glucose levels. How much caffeine? In a single can of Red Bull, 80 mg, about as much as in a freshly brewed cup of coffee.

Jeffrey Blackler/Alamy Stock Photo

What's the Problem?
The World Health Organization has issued a warning about the health risks of energy drinks, primarily related to their caffeine content. Consuming more than 400 mg of caffeine in a day's time leads to headache, insomnia, irritability—you get jittery. For this reason, it's important not to overdo Red Bull or Monster. If you are an adolescent, a Type 2 diabetic, pregnant, or obese, far more serious health problems may result.

Is It Safe to Mix Energy Drinks with Alcohol?
No. No. No. Many bars and clubs sell "goldfish bowls" of Red Bull and vodka. The rush of caffeine reduces drowsiness without diminishing the effects of alcohol, resulting in "wide-awake drunkenness." Mixing caffeine and alcohol in this way, by leading to a loss of inhibition, increases the risk that you will engage in risky and dangerous behavior.

What About the Other Stuff in Energy Drinks?
Each brand of energy drink contains its own "energy blend" of stimulants such as guarana, carnitine, ephedrine, and ginseng. Not approved by the U.S. Food and Drug Administration as safe in the food supply, their health risks have been recently assessed as suspicious. Research published in the *Journal of the American Heart Association* in 2017 compared healthy men and women after drinking energy drinks and after drinking another concoction with the same amount of caffeine but none of the other ingredients. What did they find? Electrocardiograms indicated that the energy drinks but not the caffeine-only drinks produced irregular heartbeats sometimes associated with life-threatening heart disorders.

Do Energy Drinks Cause Heart Attacks?
Yes. The excessive levels of caffeine and other stimulants reduce the blood flow in coronary blood vessels (those supplying freshly oxygenated blood to the heart). In someone with an unsuspected heart condition in which the coronary arteries are partially occluded (blocked) by plaque buildup, this can produce a heart attack. Researchers regard this as the single greatest health danger of energy drinks.

Harvesting Electrons Without Oxygen: Fermentation

7.5 Cells Can Metabolize Food Without Oxygen

> **LEARNING OBJECTIVE 7.5.1** Distinguish between ethanol fermentation and lactic acid fermentation.

Electrons on the move always find a home.

Fermentation

In the absence of oxygen, aerobic metabolism (the Krebs cycle and electron transport chain) cannot occur, and cells must rely exclusively on glycolysis to produce ATP. Under these conditions, the hydrogen atoms and electrons that were involved in the oxidation of NAD$^+$ to NADH in glycolysis are donated to organic molecules instead of the electron transport chain, in a process called **fermentation**. Fermentation recycles NAD$^+$ so that glycolysis can continue.

Ethanol Fermentation. Bacteria carry out more than a dozen kinds of fermentations, all using some form of organic molecule to accept the hydrogen atom from NADH. By contrast, eukaryotic cells are capable of only a few types of fermentation. In one type, which occurs in single-celled fungi called yeast, the molecule that accepts hydrogen from NADH is derived from pyruvate, the end product of glycolysis itself. Yeast enzymes remove a CO$_2$ group from pyruvate through decarboxylation, producing a 2-carbon molecule called acetaldehyde. The CO$_2$ released causes bread made with yeast to rise, whereas bread made without yeast (unleavened bread) does not. The acetaldehyde accepts a hydrogen atom from NADH, producing NAD$^+$ and ethanol (**figure 7.8**, *upper panel*). This particular type of fermentation is of great interest to humans because it is the source of the ethanol in wine and beer. Ethanol is a by-product of fermentation that is actually toxic to yeast; as it approaches a concentration of about 12%, it begins to kill the yeast. That is why naturally fermented wine contains only about 12% ethanol.

Lactic Acid Fermentation. Most animal cells regenerate NAD$^+$ by a second type of fermentation, using an enzyme called lactate dehydrogenase to transfer a hydrogen atom from NADH back to the pyruvate that is produced by glycolysis. This reaction converts pyruvate into lactic acid and regenerates NAD$^+$ from NADH (**figure 7.8**, *lower panel*). It therefore closes the metabolic circle, allowing glycolysis to continue as long as glucose is available. Circulating blood removes excess lactate (the ionized form of lactic acid) from muscles. It was once thought that during strenuous exercise, when the removal of lactic acid cannot keep pace with its production, the accumulation induces muscle fatigue. However, scientists now believe that lactic acid is actually used by muscles as another source of fuel.

> **Putting the Concept to Work**
> What would happen to glycolysis if NAD$^+$ wasn't recycled?

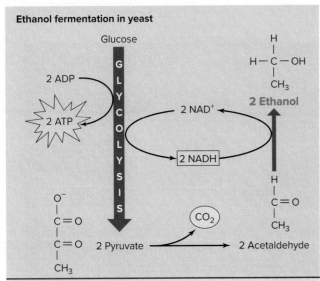
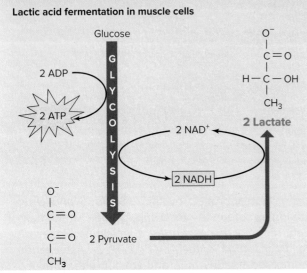

Figure 7.8 Fermentation.
Yeasts carry out the conversion of pyruvate to ethanol. Muscle cells convert pyruvate into lactate, which is less toxic than ethanol. In both cases, NAD$^+$ is regenerated to allow glycolysis to continue.

> **IMPLICATION FOR YOU** Despite our best intentions, once in a while most of us consume a little more alcohol than we should and wake up the next morning with a hangover—a pounding headache, nausea, shakiness, and often a very dry mouth. Many of these symptoms are those of dehydration. What sort of hangover prevention does this "dehydration" hypothesis suggest?

Other Sources of Energy

7.6 Glucose Is Not the Only Food Molecule

> **LEARNING OBJECTIVE 7.6.1** Describe how cells garner energy from proteins and from fats.

We have considered in detail the fate of a molecule of glucose, a simple sugar, in cellular respiration. But how much of what you eat is sugar? As a more realistic example of the food you eat, consider the fate of a fast-food hamburger. The hamburger is composed primarily of carbohydrates, fats, and proteins. This diverse collection of complex molecules is broken down by the process of digestion in your stomach and intestines into simpler molecules. Carbohydrates are broken down into simple sugars, fats into fatty acids, and proteins into amino acids. Nucleic acids are also present in the food you eat, but these macromolecules store little energy that the body actually uses.

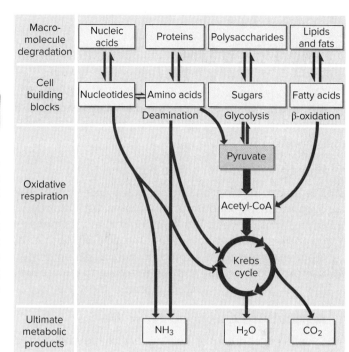

Figure 7.9 How cells obtain energy from foods.
Most organisms extract energy from organic molecules by oxidizing them. First, the subunits of macromolecules other than carbohydrates, such as proteins and fats, must be converted into products that can enter the biochemical pathways found in oxidative respiration.

Cellular Respiration of Protein

Proteins (the second category in **figure 7.9**) are first broken down into their individual amino acids. A series of *deamination* reactions removes the nitrogen side groups (called amino groups) and converts the rest of the amino acid into a molecule that takes part in the Krebs cycle. For example, alanine is converted into pyruvate, glutamate into α-ketoglutarate, and aspartate into oxaloacetate. The reactions of the Krebs cycle then extract the high-energy electrons from these molecules and put them to work making ATP.

Cellular Respiration of Fat

Lipids and fats (the fourth category in **figure 7.9**) are first broken down into fatty acids. A fatty acid typically has a long tail of 16 or more —CH$_2$ links, and the many C—H bonds in these long tails provide a rich harvest of energy. Enzymes in the matrix of the mitochondrion first remove one 2-carbon acetyl group from the end of a fatty acid tail, and then another, and then another, in effect chewing down the length of the tail in 2-carbon bites. Eventually, the entire fatty acid tail is converted into acetyl groups. Each acetyl group then combines with coenzyme A to form acetyl-CoA, which feeds into the Krebs cycle. This process is known as *β-oxidation*.

> Recall from the discussion of lipids in Chapter 3 that fats are composed of a 3-carbon glycerol backbone attached to three fatty acid tails. The breakdown of fats, as described here, occurs with the removal of two carbons at a time from the fatty acid tails.

> **Putting the Concept to Work**
> What part of a hamburger would yield energy in the absence of oxygen?

Biology and Staying Healthy

The Paleo Diet

The trendy diet of the last few years is the so-called "paleo diet" that purports to match the diet of our ancestors in the Paleolithic, before agriculture was developed. In 2016, the paleo diet was Google's most searched weight-loss topic.

Hunter-Gatherers Ate Simply

The basic assumption of the paleo diet is that humans evolved eating food very unlike what we eat today. Over several million years, as hunter-gatherers, humans of the Paleolithic became adapted to the sort of foods available then: wild plants and hunted animals. The human genome was molded then, and this diet is based on the presumption that our genes have altered relatively little since.

Agriculture Changed How We Eat

Everything about what humans eat changed ten thousand years ago, with the advent of agriculture and the domestication of animals. Humans began for the first time to consume cereal grains (what we know in modern times as wheat, corn, and rice), legumes (things the farmer picks, such as peas, beans, and peanuts), and dairy products (milk, butter, yogurt, cheese). By modern times, the human diet had increasingly become centered on high-calorie processed foods, rich in corn syrup, sugars, and fat. The human genome has not been able to keep up with this radical alteration of the human diet, paleo diet proponents claim, leading directly to the modern-day "epidemic" of obesity, diabetes, and heart disease. Surely, they claim, eating the way we are designed to eat by our genes will lead to longer, healthier lives.

A Pretty Severe Diet

Now that's a great diet story. Deferring for a moment the question of whether the paleo diet story is valid, let's first look at where it takes us. The diet that this story claims to be best for us—eating more whole foods and less processed foods—is one recommended by physicians and dietitians for years, but in an extreme form. The paleo diet, taken seriously, restricts all grains, dairy, legumes, beans, sugar, and salt, focusing instead

Mauricio Anton/Science Source

on organic fruits and vegetables, wild-caught fish, grass-fed beef, and cage-free poultry. Make no bones about it—this is a seriously restrictive diet. Imagine living with the people of a Paleolithic village, you eating what they ate. Of course, there are not very many mastodons around today, but you get the point. No white bread, peanut butter, or ice cream.

The Evolutionary Discordance Hypothesis

So how right is the paleo diet story? Not very, according to research carried out by geneticists, evolutionists, and paleontologists. The rationale for the diet, first articulated in 1985 as the evolutionary discordance hypothesis, states that "*many chronic diseases and degenerative conditions evident in modern Western populations have arisen because of a mismatch between Stone Age genes and recently adopted lifestyles*" (Eaton & Konner, 1985). This is a testable scientific hypothesis.

So, what does science tell us? First, we don't in fact know much about what Paleolithic peoples ate, except that they evolved to be flexible eaters. Most researchers find the idea that Paleolithic diets were based largely on plant foods at best hypothetical. We just don't know what they ate. It is often argued that Paleolithic hunter-gatherers did not suffer from the diseases of affluence simply because they did not live long enough to develop them. The health concerns of today's world indeed reflect our modern diet, these scientists conclude, but they stem not from a particular diet but from an imbalance between the energy we take in and the energy we expend. The basic rule of fad diets still applies: calories in minus calories out equals fat (and obesity-related disease).

Nor is the human genome as unchanging as the evolutionary discordance hypothesis assumes. The 10,000 years since the Pleistocene allows time for a lot of evolution, as the many medical differences among human races attest. The evolution of lactose intolerance among dairy eaters is but one example.

Our Dietary Partners

Importantly, our genomes are not the only ones that play a key role in how we digest what we eat. As you will discover on page 502, the bacteria in our intestines play a major role in determining what foods our bodies absorb. In future, the best way to modulate how many calories we take in may be to alter the microbial gut community with which we share our food.

So, how seriously should we take the paleo diet? It will certainly reduce the number of calories you consume, if you are able to follow it for any period of time. And it is certainly healthy to consume more whole foods and fewer processed foods. But doing without peanut butter?

Christian Jegou/Publiphoto/Science Source

Putting the Chapter to Work

1 There are two main stages in cellular respiration. Each stage has reactions in which molecules are oxidized and reduced. For the following, indicate whether the event describes oxidation or reduction.

   a. Removing the electrons from glucose in glycolysis
   b. The cleaving of pyruvate that results in the removal of a hydrogen and electrons during the production of acetyl-CoA
   c. Electrons being received by NAD^+ in the Krebs cycle

2 The enzyme ATP synthase, located in the inner mitochondrial membrane, catalyzes the formation of ATP. This enzyme utilizes a proton gradient to drive the ATP-forming reaction via facilitated diffusion.

What two molecules directly provide the protons for the proton gradient?

3 Energy derived from the sugar molecule glucose is used to synthesize ATP.

If there is enough ATP available within a cell, the cell will instead convert this excess energy to fat, first directing that pyruvate be converted to what molecule that leads to fat biosynthesis?

4 In fermentation, NAD^+ is reduced to NADH during glycolysis and then oxidized back to NAD^+.

What type of molecule is reduced when NADH is oxidized?

Retracing the Learning Path

An Overview of Cellular Respiration

7.1 Where Is the Energy in Food?

1. Nonphotosynthetic organisms acquire energy from the breakdown of food, either by eating plants that store the food or by eating animals that have eaten plants. Energy stored in carbohydrate molecules is extracted through the process of cellular respiration and is stored in the cell as ATP.
- Coupled reactions, called oxidation-reduction or redox reactions, involve the transfer of electrons from one atom or molecule to another. The atom or molecule that loses an electron is said to be oxidized and loses energy. The atom or molecule that gains the electron is said to be reduced and gains energy.
- Cellular respiration is carried out in two stages: glycolysis occurring in the cytoplasm and oxidation occurring in the mitochondria.

Respiration Without Oxygen: Glycolysis

7.2 Using Coupled Reactions to Make ATP

1. Glycolysis is an energy-extracting series of 10 chemical reactions that shuffle around the chemical bonds of glucose to produce two net molecules of ATP by substrate-level phosphorylation.
- Electrons extracted from glucose are donated to a carrier molecule, NAD^+, which becomes NADH. NADH carries electrons and hydrogen atoms to be used in a later stage of oxidative respiration.

Respiration with Oxygen: The Krebs Cycle

7.3 Harvesting Electrons from Chemical Bonds

1. The two molecules of pyruvate formed in glycolysis are passed into the mitochondrion, where they are converted into two molecules of acetyl-CoA. In the process, another molecule of NADH is formed. What the cell does with acetyl-CoA depends on the needs of the cell. If the cell has enough ATP, acetyl-CoA is used in synthesizing fat molecules. If the cell needs energy, acetyl-CoA is directed to the Krebs cycle.
- The formation of NADH is an enzyme-catalyzed reaction. The enzyme brings the substrate and NAD^+ into close proximity. Through a redox reaction, a hydrogen atom and an electron are transferred to NAD^+, reducing it to NADH. NADH then carries the electrons and hydrogen to a later step in oxidative respiration.

2. Acetyl-CoA enters a series of chemical reactions called the Krebs cycle, where one molecule of ATP is produced in a coupled reaction. Energy is also harvested in the form of electrons that are transferred to molecules of NAD^+ and FAD to produce NADH and $FADH_2$, respectively. The Krebs cycle makes two turns for every molecule of glucose that is oxidized.

7.4 Using the Electrons to Make ATP

1. The electrons harvested by oxidizing food molecules are used to power proton pumps that chemiosmotically drive the production of ATP. The molecules of NADH and $FADH_2$ that were produced during glycolysis and the Krebs cycle carry electrons to the inner mitochondrial membrane, where they drive proton pumps that move H^+ across the membrane from the matrix to the intermembrane space, creating an H^+ concentration gradient.
- When electrons reach the end of the electron transport chain, they bind with oxygen and hydrogen to form water molecules.

2. ATP is produced in the mitochondrion through chemiosmosis. The H^+ concentration gradient in the intermembrane space drives H^+ back across the membrane through ATP synthase channels. The energy from the movement of H^+ through the channel is transferred to the chemical bonds in ATP.

Harvesting Electrons Without Oxygen: Fermentation

7.5 Cells Can Metabolize Food Without Oxygen

1. In the absence of oxygen, other molecules can be used as electron acceptors. When the electron acceptor is an organic molecule, the process is called fermentation. Depending on what type of organic molecule accepts the electrons, either ethanol or lactic acid, in the form of lactate, is formed.

Other Sources of Energy

7.6 Glucose Is Not the Only Food Molecule

1. Food sources other than glucose are also used in oxidative respiration. Macromolecules, such as proteins, lipids, and nucleic acids, are broken down into intermediate products that feed into cellular respiration in different reaction steps.

Inquiry and Analysis

How Do Swimming Fish Avoid Low Blood pH?

Animals that live in oxygen-poor environments, such as worms living in the oxygen-free mud at the bottom of lakes, are not able to obtain the energy required for muscle movement from the Krebs cycle. Their cells lack the oxygen needed to accept the electrons stripped from food molecules. Instead, these animals rely on glycolysis to obtain ATP, donating the electron to pyruvate, forming lactic acid. While much less efficient than the Krebs cycle, glycolysis does not require oxygen. Even when oxygen is plentiful, the muscles of an active animal may use up oxygen more quickly than it can be supplied by the bloodstream and so be forced to temporarily rely on glycolysis to generate the ATP for continued contraction.

This presents a particular problem for fish. Fish blood is much lower in carbon dioxide than yours, and as a consequence, the amount of sodium bicarbonate acting as a buffer in fish blood is also quite low. Now imagine you are a trout and need to suddenly swim very fast to catch a mayfly for dinner. The vigorous swimming will cause your muscles to release large amounts of lactic acid into your poorly buffered blood; this could severely disturb the blood's acid-base balance and so impede contraction of your swimming muscles before the prey is captured.

The graph to the right presents the results of an experiment designed to explore how a trout solves this dilemma. In the experiment, the trout was made to swim vigorously for 15 minutes in a laboratory tank and then allowed a day's recovery. The lactic acid concentration in its blood was monitored periodically during swimming and recovery phases.

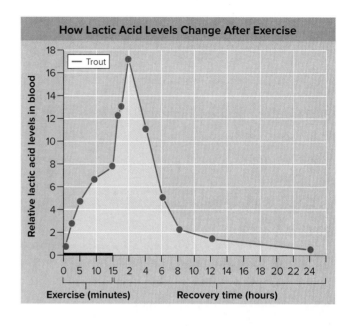

Analysis

1. **Applying Concepts** Lactic acid levels are presented for both swimming and recovery periods. In what time units are the swimming data presented? The recovery data?

2. **Interpreting Data** What is the effect of exercise on the level of lactic acid in the trout's blood? How does the level of lactic acid change after exercise stops?

3. **Making Inferences** About how much of the total lactic acid created by vigorous swimming is released after this exercise stops? [Hint: Notice that the x axis scale changes from minutes to hours.]

4. **Drawing Conclusions** Is this result consistent with the hypothesis that fish maintain blood pH levels by delaying the release of lactic acid from muscles? Why might this be beneficial to the fish?

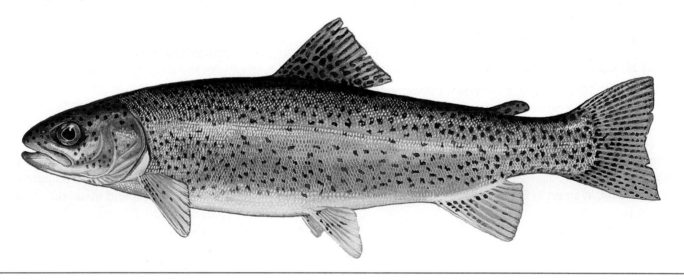

PART 3 The Continuity of Life

8 Mitosis

LEARNING PATH ▼

Cell Division
1. Prokaryotes Have a Simple Cell Cycle
2. Eukaryotic Cell Cycle
3. Chromosomes
4. Cell Division

Cancer and the Cell Cycle
5. What Is Cancer?

Tanning Your Way to Skin Cancer

Almost 9,000 Americans will die this year from melanoma, a virulent form of skin cancer. And even more will die next year. The death rate has doubled, and doubled again, since 1973. What is going on here? Unexpectedly, the problem seems to be suntans. When you get a suntan, the sunlight does more than change the tone of your skin. When cells on your body's surface are badly damaged by the sun—what we call a sunburn—the damaged cells slough off. Recall the peeling that you experience after a bad sunburn. Other damage is more subtle—and deadly.

Tanning Becomes Fashionable

Up until the early 20th century, a tan was a condition that people went to great lengths to avoid. Before the Industrial Revolution, a tanned body was a sign of the working class, people who had to work in the sun. The wealthy elite avoided the sun with pale skin being in fashion. The Greeks and Romans would use chalks and lead paints to whiten their skin. Women in Elizabethan England would even paint blue lines on their skin to make their skin appear translucent.

THIS TANNING BED may shorten the journey to their death bed.
Andrew Fox/Alamy Stock Photo

All of this changed in the 1920s, when tans became a status symbol, with the wealthy able to travel to warm, sunny destinations, even in the middle of winter. That tanned, bronzed glow that people would sit in the sun for hours to achieve was thought to be both healthy and attractive.

...at a Price

During the 1970s, doctors started to see an uptick in the number of cases of melanoma and other skin cancers. New cases were increasing about 6% each year, an alarming rate. Researchers proposed that ultraviolet (UV) rays from

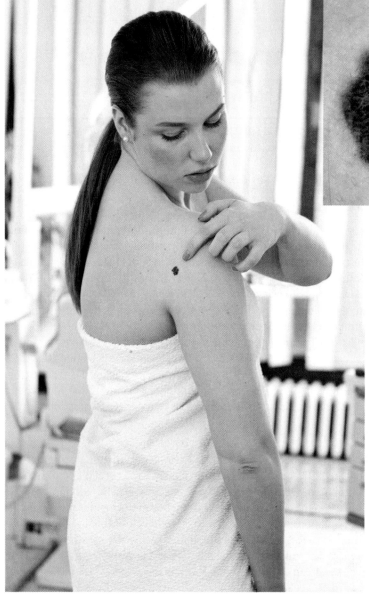

dean bertoncelj/Shutterstock; (Inset) Australis Photography/Shutterstock

the sun were the underlying cause of this epidemic of skin cancer and warned people to avoid the sun when possible and protect themselves with sunscreen.

Malignant Melanoma

Malignant melanoma is the most deadly of skin cancers, although treatable if caught early. Melanoma is cancer of melanocyte cells. Melanoma lesions usually appear as shades of tan, brown, and black and can begin in or near a mole, and so changes in a mole is a symptom of melanoma. Melanoma is most prevalent in fair-skinned people, but unlike the other forms of skin cancer, it can also affect people with darker complexions.

But the public has been slow to respond to this warning, perhaps because the cosmetic benefits of tanning are immediate, while the health hazards are much delayed. The desire to achieve that tanned, bronzed body is as strong as ever.

Indoor Tanning—A Deadly Fad

A good tan requires regular exposure to the sun to maintain it, so indoor tanning salons have become popular. Tanning booths emit concentrated UV rays from two sides, allowing a person to tan in less time and in sun, rain, or snow. Nearly a third of white women ages 16 to 25 use an indoor tanning bed each year, as the person in the chapter's opening photo is doing. An estimated 28 million Americans tan this way annually; hence, the increase in melanoma.

Actually, people had thought that building up a tan through the use of tanning booths would protect a person's skin from burning and would reduce the time exposed to the UV radiation, both leading to a reduced risk of skin cancer. However, recent research does not support these assumptions. A 2003 study of 106,000 Scandinavian women showed that exposure to UV rays in a tanning booth as little as once a month can increase your risk of melanoma by 55%, especially when the exposure is during early adulthood. Those women who were in their 20s and used sun lamps to tan were at the highest risk, about 150% higher than those who didn't use a tanning bed. As with other studies, fair-skinned women were at the greatest risk. In fact, tanning booths, even for those people who tan more easily, heighten the risk for skin cancer because people use the tanning booths year-round, increasing their cumulative exposure.

Cell Division

8.1 Prokaryotes Have a Simple Cell Cycle

LEARNING OBJECTIVE 8.1.1 Diagram the prokaryotic cell cycle, identifying DNA replication, DNA partitioning, and cell fission.

In prokaryotes, the DNA divides first, then the cell divides.

Copying the DNA

All species reproduce, passing their hereditary information on to their offspring. In prokaryotes, the hereditary information—that is, the genes that specify the prokaryote—is encoded in a single circle of DNA, called a prokaryotic chromosome. Before a prokaryotic cell divides to reproduce, the DNA circle makes a copy of itself, a process called *replication*. Starting at one point, the origin of replication, the double helix of DNA begins to unzip, exposing the two strands. The enlargement on the right of **figure 8.1a** shows how the DNA replicates. The new double helix is formed from each naked strand by placing on each exposed nucleotide its complementary nucleotide (that is, A with T, G with C, as discussed in chapter 3). DNA replication is discussed in more detail in chapter 11. When the unzipping has gone all the way around the circle, the cell possesses two copies of its hereditary information.

Dividing the Cell

When the DNA has been copied, the cell grows, resulting in elongation. The newly replicated DNA molecules are partitioned toward each end of the cell. This partitioning process involves DNA sequences near the origin of replication and results in these sequences being attached to the membrane. When the cell reaches an appropriate size, the prokaryotic cell begins to split into two equal halves, a process called **binary fission**. New plasma membrane and cell wall are added at a point between where the two DNA copies are partitioned, indicated by the green divider in **figure 8.1a**. As the growing plasma membrane pushes inward, the cell is constricted in two, eventually forming two *daughter cells*. Each contains one prokaryotic chromosome that is genetically identical to the parent cell's, and each is a complete living cell in its own right.

> **Putting the Concept to Work**
> How does a dividing bacterial cell ensure that one copy of its chromosome goes to each daughter cell?

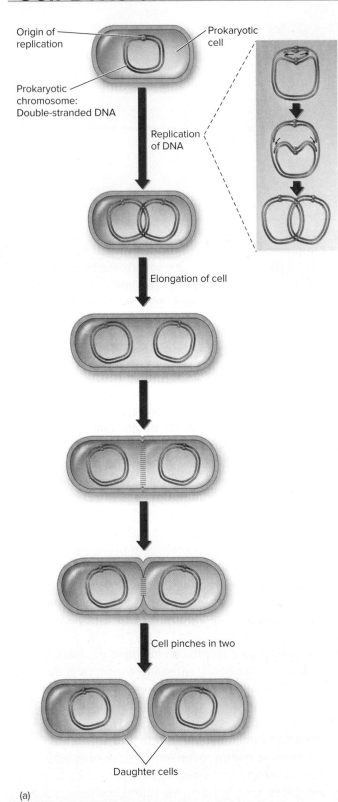

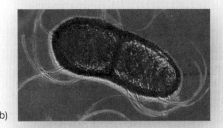

(a)
(b)

Figure 8.1 The prokaryotic cell cycle.

Prokaryotic cells divide by a process of binary fission. (a) Before the cell splits, the circular DNA molecule of a prokaryote initiates replication at a single site, called the origin of replication, moving out in both directions. When the two moving replication points meet on the far side of the molecule, its replication is complete. The cell then undergoes binary fission, in which the cell divides into two daughter cells. (b) Here, a prokaryotic cell has divided in two and is about to be pinched apart by the growing plasma membrane.

(b) Science Photo Library - CNRI/SPL/Getty Images

8.2 Eukaryotic Cell Cycle

> **LEARNING OBJECTIVE 8.2.1** Describe the phases of the eukaryotic cell cycle, including the three phases of interphase.

Cell division in eukaryotes is more complex than in prokaryotes. Eukaryotes contain far more DNA than prokaryotes, and eukaryotic DNA is wound tightly around proteins that condense into a compact shape, the eukaryotic **chromosome**. The cells of eukaryotic organisms either undergo mitosis or meiosis to divide up their DNA. **Mitosis** is the mechanism of cell division that occurs in an organism's nonreproductive cells, called *somatic cells*. **Meiosis** divides the DNA in sexually reproductive *germ-line cells*.

Essential Biological Process 8A outlines the eukaryotic cell cycle:

Interphase. Interphase is composed of three phases:
- **G_1 phase.** This "first gap" phase is the cell's primary growth phase. For most organisms, this phase occupies much of the cell's life span.
- **S phase.** In this "synthesis" phase, the DNA replicates, producing two copies of each chromosome.
- **G_2 phase.** Cell division preparation begins in the "second gap" phase with the replication of mitochondria, chromosome condensation, and the synthesis of microtubules.

M phase. In mitosis, which is continuous but can be thought of as occurring in four phases, a microtubular apparatus binds to the chromosomes and moves them apart. The four phases of mitosis are prophase, metaphase, anaphase, and telophase.

C phase. In cytokinesis, the cytoplasm divides, creating two daughter cells.

> **Putting the Concept to Work**
> Do eukaryotic chromosomes condense before they replicate?

Essential Biological Process 8A

The Cell Cycle

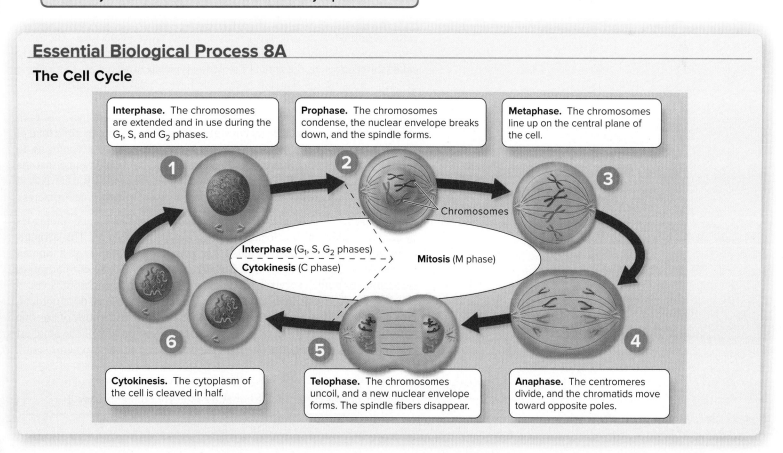

1. **Interphase.** The chromosomes are extended and in use during the G_1, S, and G_2 phases.
2. **Prophase.** The chromosomes condense, the nuclear envelope breaks down, and the spindle forms.
3. **Metaphase.** The chromosomes line up on the central plane of the cell.
4. **Anaphase.** The centromeres divide, and the chromatids move toward opposite poles.
5. **Telophase.** The chromosomes uncoil, and a new nuclear envelope forms. The spindle fibers disappear.
6. **Cytokinesis.** The cytoplasm of the cell is cleaved in half.

Interphase (G_1, S, G_2 phases) / Cytokinesis (C phase) — Mitosis (M phase)

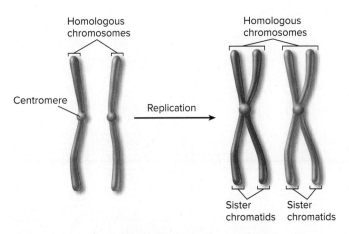

Figure 8.2 The difference between homologous chromosomes and sister chromatids.

Homologous chromosomes are a pair of the same chromosome—say, chromosome number 16. Sister chromatids are the two replicas of a single chromosome held together by the centromere after DNA replication. A duplicated chromosome looks somewhat like an X.

8.3 Chromosomes

> **LEARNING OBJECTIVE 8.3.1** Distinguish between a homologue and a sister chromatid.

Chromosomes were first observed by the German embryologist Walther Flemming in 1879, while he was examining the rapidly dividing cells of salamander larvae. When Flemming looked at the cells through what would now be a rather primitive light microscope, he saw minute threads within their nuclei that appeared to be dividing lengthwise. Flemming called their division *mitosis,* based on the Greek word *mitos,* meaning "thread."

Chromosome Number

Since their initial discovery, chromosomes have been found in the cells of all eukaryotes examined. Their number may vary enormously from one species to another. A few kinds of organisms—such as the Australian ant *Myrmecia* spp.; the plant *Haplopappus gracilis,* a relative of the sunflower that grows in North American deserts; and the fungus *Penicillium*—have only 1 pair of chromosomes, while some ferns have more than 500 pairs. Most eukaryotes have between 10 and 50 chromosomes in their body cells.

Homologous Chromosomes

Chromosomes exist in somatic cells as pairs, called **homologous chromosomes,** or **homologues.** Homologues carry information about the same traits at the same locations on each chromosome, but the information can vary between homologues, which will be discussed in chapter 10. Cells that have two of each type of chromosome are called **diploid cells.** One chromosome of each pair is inherited from the mother (colored green in figure 8.2) and the other from the father (colored purple).

Chromatids. Before cell division, each homologous chromosome replicates, resulting in two identical copies, called **sister chromatids**. The sister chromatids remain joined together after replication at a special linkage site called the **centromere**, the knoblike structure typically observed near the middle of each chromosome. Human body cells have a total of 46 chromosomes, which are actually 23 pairs of homologous chromosomes. In their duplicated state, before mitosis, there are still only 23 pairs of chromosomes, but each chromosome has duplicated and consists of two sister chromatids, for a total of 92 chromatids. The duplicated sister chromatids can make it confusing to count the number of chromosomes in an organism, but keep in mind that the number of centromeres doesn't increase with replication, and so you can always determine the number of chromosomes simply by counting the centromeres.

Karyotypes. The 46 human chromosomes can be paired as homologues by comparing size, shape, location of centromeres, and so on. This arrangement of chromosomes is called a **karyotype.** An example of a human karyotype is shown in figure 8.3. Possession of all 46 of the chromosomes is essential to human survival. Individuals missing even one chromosome, a condition called monosomy, do not usually survive embryonic development. Nor does the human embryo develop properly with an extra copy of any one chromosome, a condition called trisomy. For all but a few of the smallest chromosomes, trisomy is fatal; even in those cases, serious problems result.

> **Putting the Concept to Work**
> How many chromosomes does a cell of your finger possess?

Chromosome Structure

> **LEARNING OBJECTIVE 8.3.2** Discuss the function of a nucleosome.

Chromosomes are composed of **chromatin,** a complex of DNA and protein; most are about 40% DNA and 60% protein. A significant amount of RNA is also associated with chromosomes because chromosomes are the sites of RNA synthesis.

A Single DNA Fiber. The DNA of a chromosome is one very long, double-stranded fiber that extends unbroken through the entire length of the chromosome. A typical human chromosome contains about 140 million (1.4×10^8) nucleotides in its DNA. Furthermore, if the strand of DNA from a single chromosome were laid out in a straight line, it would be about 5 centimeters (2 inches) long. The amount of information in one human chromosome would fill about 2,000 printed books of 1,000 pages each! Fitting such a strand into a nucleus is like cramming a string the length of a football field into a baseball—and that's only 1 of 46 chromosomes! In the cell, however, the DNA is coiled, allowing it to fit into a much smaller space than would otherwise be possible.

Chromosome Coiling. Eukaryotic DNA is formed into chromosomes, such as the duplicated sister chromatids seen in figure 8.3, by winding and twisting the long DNA strands into much more compact forms. Winding up DNA presents an interesting challenge. Because the phosphate groups of DNA molecules have negative charges, it is impossible to just tightly wind up DNA—all the negative charges would simply repel one another. As you can see in figure 8.4, the DNA helix wraps around proteins with positive charges called **histones.** The positive charges of the histones counteract the negative charges of the DNA, so that the complex has no net charge.

Nucleosomes. Every 200 nucleotides, the DNA duplex is coiled around a core of eight histone proteins, forming a complex known as a **nucleosome.** The nucleosomes are further coiled into a solenoid. This solenoid is then organized into looped domains. The final organization of the chromosome is not known, but it appears to involve further radial looping into rosettes around a preexisting scaffolding of protein. This complex of DNA and histone proteins, coiled tightly, forms a compact chromosome.

> **Putting the Concept to Work**
> Why does DNA coiling require positively charged histone proteins?

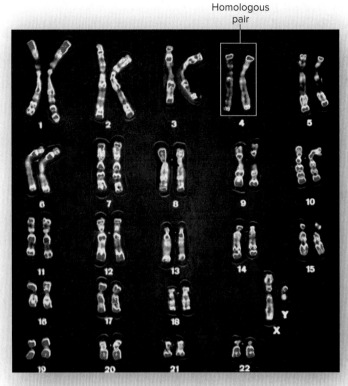

Figure 8.3 The 46 chromosomes of a human.
In this photograph, the individual chromosomes of a human male have been paired with their homologues, creating an organized display called a karyotype.
CNRI/Science Source

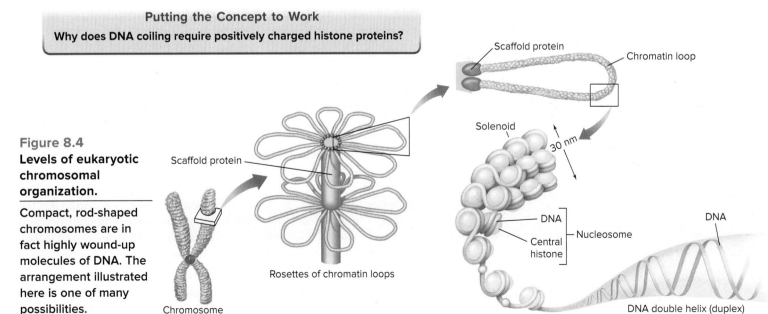

Figure 8.4
Levels of eukaryotic chromosomal organization.

Compact, rod-shaped chromosomes are in fact highly wound-up molecules of DNA. The arrangement illustrated here is one of many possibilities.

Essential Biological Process 8B

Cell Division

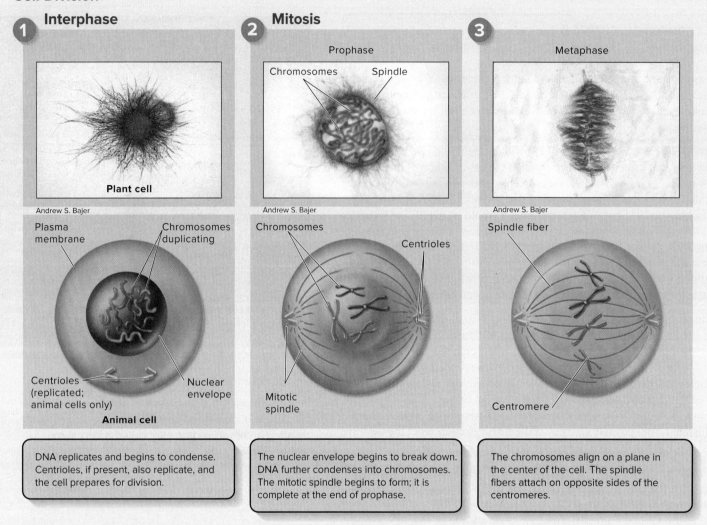

8.4 Cell Division

> **LEARNING OBJECTIVE 8.4.1** Describe the stages of mitosis.

Interphase

When cell division begins in interphase (**panel 1** of *Essential Biological Process 8B*), chromosomes first replicate and then begin to wind up tightly, a process called **condensation.** Chromosomes are not usually visible under the microscope during interphase.

Mitosis

Interphase is followed by nuclear division, called *mitosis*. Although the process of mitosis is continuous, with the stages flowing smoothly one into another, for ease of study, mitosis is traditionally subdivided into four stages: prophase, metaphase, anaphase, and telophase.

Prophase: Mitosis Begins. In **prophase** of mitosis (**panel 2**), the individual condensed chromosomes first become visible with a light microscope. As the replicated chromosomes condense, the cell dismantles the nuclear

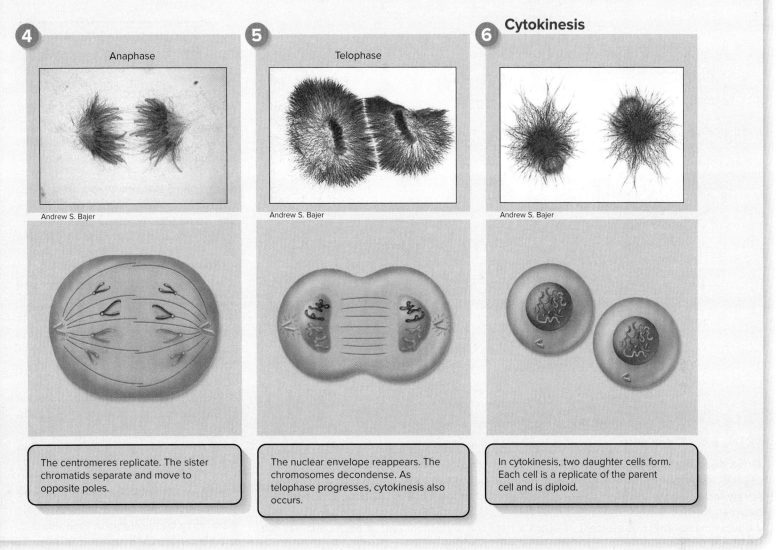

envelope, and two centrosomes (centrioles in animal cells) begin to assemble the apparatus it will use to pull the replicated sister chromatids to opposite ends ("poles") of the cell. In the center of an animal cell, the pairs of centrioles separate and move apart toward opposite poles of the cell, forming between them as they move apart a network of protein cables called the **spindle.** Each cable is called a *spindle fiber* and is made of microtubules, which are long, hollow tubes of protein. Plant cells lack centrioles and instead brace the ends of the spindle toward the poles. The spindle fibers attach to the chromosomes, and when the process is complete, one sister chromatid of each pair is attached by microtubules to one pole and the other sister chromatid to the other pole.

Metaphase: Alignment of the Chromosomes. The second phase of mitosis, **metaphase,** begins when the chromosomes, each consisting of a pair of sister chromatids, align in the center of the cell along an imaginary plane that divides the cell in half, referred to as the equatorial plane. Spindle fibers attached to the centromeres extend back toward the opposite poles of the cell (**figure 8.5**).

Anaphase: Separation of the Chromatids. In **anaphase,** the centromeres split, and the sister chromatids are freed from each other.

Figure 8.5 A cell's chromosomes in metaphase.

The cell you see above is a dividing cell of the Oregon newt *Taricha granulosa,* a kind of salamander. The image captures the cell in metaphase, when all the blue-stained chromosomes are lined up on the equatorial plane. Soon the red-stained spindle fibers will draw duplicates of the homologous chromosomes to opposite poles of the cell.

©Andrew S. Bajer

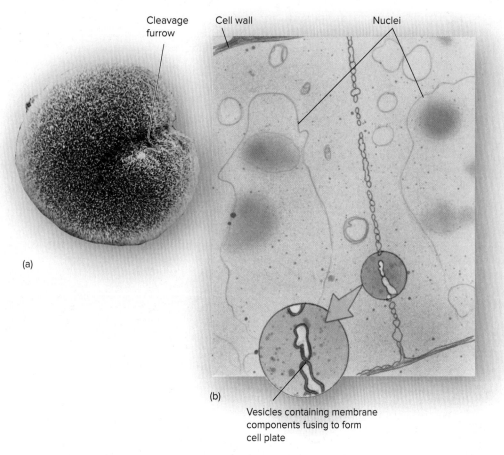

Figure 8.6 Cytokinesis.

The division of cytoplasm that occurs after mitosis is called cytokinesis and cleaves the cell into roughly equal halves. (a) In an animal cell, such as this sea urchin egg, a cleavage furrow forms around the dividing cell. (b) In this dividing plant cell, a cell plate is forming between the two newly forming daughter cells. Note that the flattened vesicles are forming a double membrane, one element of which is destined to become part of each daughter cell.

(a) David M. Phillips/Science Source

Cytokinesis

> **LEARNING OBJECTIVE 8.4.2**
> Contrast cytokinesis in plant and animal cells.

At the end of telophase, mitosis is complete. The cell has divided its replicated chromosomes into two nuclei, which are positioned at opposite ends of the cell. As mitosis ends, the division of the cytoplasm, called **cytokinesis,** occurs, and the cell is cleaved into roughly equal halves. The formation of these two daughter cells, shown in the last panel of *Essential Biological Process 8B,* signals the end of cell division.

In animal cells, which lack cell walls, cytokinesis is achieved by pinching the cell in two with a contracting belt of actin filaments (**figure 8.6a**). As contraction proceeds, a *cleavage furrow* becomes evident around the cell's circumference, where the cytoplasm is being progressively pinched inward by the decreasing diameter of the actin belt.

Plant cells have rigid walls that are far too strong to be deformed by actin filament contraction. A different approach to cytokinesis has therefore evolved in plants. Plant cells assemble membrane components in their interior, at right angles to the mitotic spindle. In **figure 8.6b**, you can see how membrane

Cell division is now simply a matter of reeling in the spindle fibers, dragging the sister chromatids (now referred to as daughter chromosomes) to the poles.

Telophase: Re-formation of the Nuclei. In **telophase,** the mitotic spindle disassembles, and a nuclear envelope forms around each set of chromosomes while they begin to uncoil. The nucleolus also reappears.

is deposited between the daughter cells by vesicles that fuse together. This expanding partition, called a *cell plate,* grows outward until it reaches the interior surface of the plasma membrane and fuses with it, at which point it has effectively divided the cell in two. Cellulose is then laid down over the new membranes, forming the cell walls of the two new cells.

> **Putting the Concept to Work**
> After interphase, how many chromatids does a cell contain?

> **Putting the Concept to Work**
> How do you suppose the cell plate forming in figure 8.6b divides to form *two* cell walls, one for each of the two daughter cells?

Cancer and the Cell Cycle

8.5 What Is Cancer?

LEARNING OBJECTIVE 8.5.1 Explain how mutation is linked to cancer, noting the genes most often involved.

Cancer is a growth disorder of cells. It starts when an apparently normal cell begins to divide in an uncontrolled way. The result is a cluster of cells, called a tumor, that constantly expands in size. The cluster of pink lung cells in the photo in figure 8.7 have begun to form a malignant tumor called a *carcinoma*. Malignant tumors are invasive. Their cells are able to break away from the tumor, enter the bloodstream, and spread to other areas of the body (figure 8.8), forming new tumors at distant sites called **metastases**.

A Deadly Disease

Cancer is perhaps the most devastating and deadly disease. Most of us have had family or friends affected by the disease. In 2020, over 1 million American men and women were diagnosed with cancer; in that same year, over half a million Americans died of cancer. One in every two Americans born will be diagnosed with some form of cancer during their lifetime. In the United States, the three deadliest human cancers are lung cancer, cancer of the colon and rectum, and breast cancer. Lung cancer, responsible for the most cancer deaths, is largely preventable; most cases result from smoking cigarettes. Colorectal cancers appear to be fostered by the high-meat diets so favored in the United States. The cause of breast cancer is still a mystery.

Mutations Cause Cancer

Not surprisingly, researchers are expending a great deal of effort to learn the cause of cancer. Scientists have made considerable progress in the last 30 years using molecular biological techniques, and the rough outlines of understanding are now emerging. We now know that cancer is a gene disorder of somatic tissue, in which damaged genes fail to properly control cell growth and division. The cell division cycle is regulated by a sophisticated group of proteins called growth factors. Cancer results from damage to the genes encoding these proteins. Damage to DNA, such as damage to these genes, is called **mutation.** Cancer can be caused by chemicals that alter DNA such as the tars in cigarette smoke (we examine the link between smoking and lung cancer in considerable detail in the following text), by environmental factors such as UV rays that damage DNA (discussed in chapter 11), or in some instances by viruses that circumvent the cell's normal growth and division controls (viruses are discussed in chapter 16).

There are two general classes of growth factor genes that are usually involved in cancer: proto-oncogenes and tumor-suppressor genes.

Stepping on the Accelerator. Genes known as **proto-oncogenes** encode proteins that stimulate cell division. Mutations that activate or improve the functioning of these genes "step on the accelerator" of cell division, causing mutated cells to divide excessively. Mutated proto-oncogenes become cancer-causing genes called **oncogenes.**

Releasing the Brakes. The second class of cancer-causing genes are called **tumor-suppressor genes.** Cell division is normally turned off in healthy

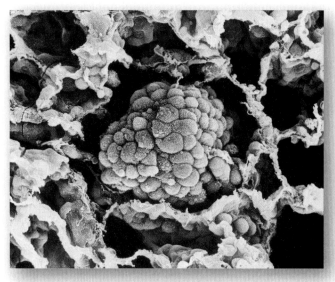

Figure 8.7 Lung cancer cells (300X).

These cells are from a tumor located in the alveolus (air sac) of a human lung.

Moredun Scientific/Science Source

IMPLICATION FOR YOU 140,730 people died of lung cancer in the United States in 2020, almost all of them cigarette smokers. Fully 7.5% of pack-a-day smokers will die of lung cancer within 30 years of their first cigarette. That's 1 in 13. Do you smoke? Do you vape? Do any of your friends? Can you think of a reason that would justify the risk?

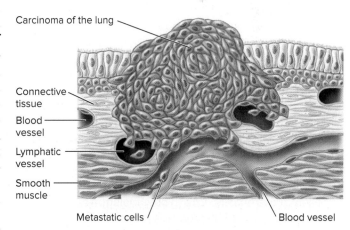

Figure 8.8 Portrait of a tumor.

This ball of cells is a carcinoma (cancer tumor) developing from epithelial cells that line the interior surface of a human lung. As the mass of cells grows, it invades surrounding tissues, eventually penetrating lymphatic and blood vessels, both of which are plentiful within the lung. These vessels carry metastatic cancer cells throughout the body, where they lodge and grow, forming new masses of cancerous tissue.

Biology and Staying Healthy

Curing Cancer

Half of all Americans will face cancer at some point in their lives. Potential cancer therapies are being developed on many fronts. Some act to prevent the start of cancer within cells. Others act outside cancer cells, preventing tumors from growing and spreading. The figure on the right indicates targeted areas for the development of cancer treatments. The following discussion will examine each of these areas.

Preventing the Start of Cancer

Many promising cancer therapies act within potential cancer cells, focusing on different stages of the cell's "Shall I divide?" decision-making process.

1 Receiving the Signal to Divide The first step in the decision process is receiving a "divide" signal, usually a small protein called a growth factor released from a neighboring cell. The growth factor, the red ball at #1 in the figure, is received by a protein receptor on the cell surface. Like banging on a door, its arrival signals that it's time to divide. Mutations that increase the number of receptors on the cell surface amplify the division signal and so lead to cancer. Over 20% of breast cancer tumors prove to overproduce a protein called HER2 associated with the receptor for epidermal growth factor (EGF).

Therapies directed at this stage of the decision process utilize the human immune system to attack cancer cells. Special protein molecules called *monoclonal antibodies,* created by genetic engineering, are the therapeutic agents. These monoclonal antibodies are designed to seek out and stick to HER2. Like waving a red flag, the presence of the monoclonal antibody calls down attack by the immune system on the HER2 cell. Because breast cancer cells overproduce HER2, they are killed preferentially. The biotechnology research company Genentech's recently approved monoclonal antibody, called herceptin, has given promising results in clinical tests.

Up to 70% of colon, prostate, lung, and head/neck cancers have excess copies of a related receptor, epidermal growth factor 1 (HER1). The monoclonal antibody C225, directed against HER1, has succeeded in shrinking 22% of advanced, previously incurable colon cancers in early clinical trials. Apparently blocking HER1 interferes with the ability of tumor cells to recover from chemotherapy or radiation.

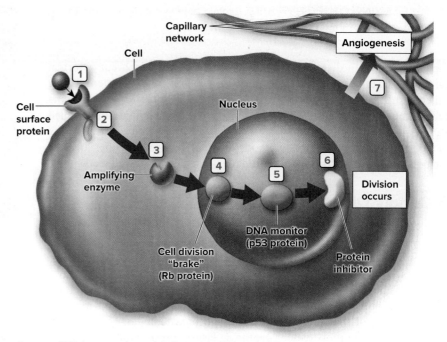

Seven different stages in the cancer process.

(1) On the cell surface, a growth factor's signal to divide is increased. (2) Just inside the cell, a protein relay switch that passes on the divide signal gets stuck in the "ON" position. (3) In the cytoplasm, enzymes that amplify the signal are amplified even more. In the nucleus, (4) a "brake" preventing DNA replication is inoperable, (5) proteins that check for damage in the DNA are inactivated, and (6) other proteins that inhibit the elongation of chromosome tips are destroyed. (7) The new tumor promotes angiogenesis, the formation of new blood vessels that promote growth.

2 Passing the Signal via a Relay Switch The second step in the decision process is the passage of the signal into the cell's interior, the cytoplasm. This is carried out in normal cells by a protein called Ras that acts as a relay switch, #2 in the figure. When growth factor binds to a receptor like EGF, the adjacent Ras protein acts like it has been "goosed," contorting into a new shape. This new shape is chemically active and initiates a chain of reactions that passes the "divide" signal inward toward the nucleus. Mutated forms of the Ras protein behave like a relay switch stuck in the "ON" position, continually instructing the cell to divide when it should not. Thirty percent of all cancers have a mutant form of Ras. So far, no effective therapies have been developed targeting this step.

3 Amplifying the Signal The third step in the decision process is the amplification of the signal within the cytoplasm. Just as a TV signal needs to be amplified in order to be received at a distance, so a "divide" signal must be amplified if it is to reach the nucleus at the interior of the cell, a very long journey at a molecular scale. To get a signal all the way into the nucleus, the cell employs a sort of pony express. The "ponies" in this case are enzymes called *tyrosine kinases*, #3 in the figure. These enzymes add phosphate groups to proteins but only at a particular amino acid, tyrosine. No other enzymes in the cell do this, so the tyrosine kinases form an elite core of signal carriers not confused by myriad other molecular activities going on around them.

Cells use an ingenious trick to amplify the signal as it moves toward the nucleus. Ras, when "ON," activates the initial protein kinase. This protein kinase activates other protein kinases that in their turn activate still others. The trick is that once a protein kinase enzyme is activated, it goes to work like a demon, activating hordes of others every second! And each and every one it activates behaves the same way too, activating still more, in a cascade of ever-widening effect. At each stage of the relay, the signal is amplified a thousandfold.

Mutations stimulating any of the protein kinases can dangerously increase the already amplified signal and lead to cancer. Some 15 of the cell's 32 internal tyrosine kinases have been implicated in cancer. Five percent of all cancers, for example, have a mutant hyperactive form of the protein kinase Src. The trouble begins when a mutation causes one of the tyrosine kinases to become locked into the "ON" position, sort of like a stuck doorbell that keeps ringing and ringing.

To cure the cancer, you have to find a way to shut the bell off. Each of the signal carriers presents a different problem, as you must quiet it without knocking out all the other signal pathways the cell needs. The cancer therapy drug Gleevec, a monoclonal antibody, has just the right shape to fit into a groove on the surface of the tyrosine kinase called "abl." Mutations locking abl "ON" are responsible for chronic myelogenous leukemia, a lethal form of white blood cell cancer. Gleevec totally disables abl. In clinical trials, blood counts revert to normal in more than 90% of cases.

[4] **Releasing the Brake** The fourth step in the decision process is the removal of the "brake" the cell uses to restrain cell division. In healthy cells, this brake, a tumor-suppressor protein called Rb, blocks the activity of a protein called E2F, #4 in the figure. When free, E2F enables the cell to copy its DNA. Normal cell division is triggered to begin when Rb is inhibited, unleashing E2F. Mutations that destroy Rb release E2F from its control completely, leading to ceaseless cell division. Forty percent of all cancers have a defective form of Rb.

Therapies directed at this stage of the decision process are only now being attempted. They focus on drugs able to inhibit E2F, which should halt the growth of tumors arising from inactive Rb. Experiments in mice in which the E2F genes have been destroyed provide a model system to study such drugs, which are being actively investigated.

[5] **Checking That Everything Is Ready** The fifth step in the decision process is the mechanism used by the cell to ensure that its DNA is undamaged and ready to divide. This job is carried out in healthy cells by the tumor-suppressor protein p53, which inspects the integrity of the DNA, #5 in the figure. When it detects damaged or foreign DNA, p53 stops cell division and activates the cell's DNA repair systems. If the damage doesn't get repaired in a reasonable time, p53 pulls the plug, triggering events that kill the cell. In this way, mutations such as those that cause cancer are either repaired or the cells containing them eliminated. If p53 is itself destroyed by mutation, future damage accumulates unrepaired. Among this damage are mutations that lead to cancer. Fifty percent of all cancers have a disabled p53. Fully 70% to 80% of lung cancers have a mutant inactive p53—the chemical benzo[*a*]pyrene in cigarette smoke is a potent mutagen of p53.

[6] **Stepping on the Gas** Cell division starts with replication of the DNA. In healthy cells, another tumor suppressor "keeps the gas tank nearly empty" for the DNA replication process by inhibiting production of an enzyme called *telomerase*. Without this enzyme, a cell's chromosomes lose material from their tips, called *telomeres*. Every time a chromosome is copied, more tip material is lost. After some 30 divisions, so much is lost that copying is no longer possible. Cells in the tissues of an adult human have typically undergone 25 or more divisions. Cancer can't get very far with only the five remaining cell divisions, so inhibiting telomerase is a very effective natural brake on the cancer process, #6 in the figure. It is thought that almost all cancers involve a mutation that destroys the telomerase inhibitor, releasing this brake and making cancer possible. It should be possible to block cancer by reapplying this inhibition. Cancer therapies that inhibit telomerase are just beginning clinical trials.

Preventing the Spread of Cancer

[7] **Stopping Tumor Growth** Once a cell begins cancerous growth, it forms an expanding tumor. As the tumor grows ever-larger, it requires an increasing supply of food and nutrients, obtained from the body's blood supply. To facilitate this necessary grocery shopping, tumors leak out substances into the surrounding tissues that encourage the formation of small blood vessels, a process called angiogenesis, #7 in the figure. Chemicals that inhibit this process are called *angiogenesis inhibitors*. Two such natural angiogenesis inhibitors, angiostatin and endostatin, caused tumors to regress to microscopic size in mice, but initial human trials were disappointing.

Laboratory drugs are more promising. A monoclonal antibody drug called Avastin, targeted against a blood vessel growth-promoting substance called vascular endothelial growth factor (VEGF), destroys the ability of VEGF to carry out its blood-vessel-forming job. Given to hundreds of advanced colon cancer patients as part of a large clinical trial, Avastin improved colon cancer patients' chance of survival by 50% over chemotherapy.

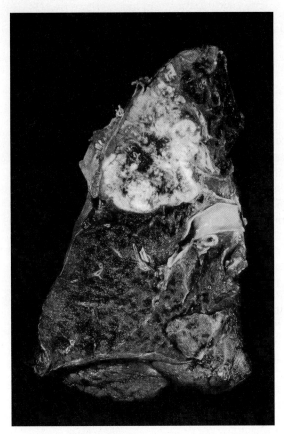

Figure 8.9 Lung cancer.

A cancerous tumor has almost completely taken over the top half of this lung, and black cigarette tars have damaged most of the lung tissue.
Science Source

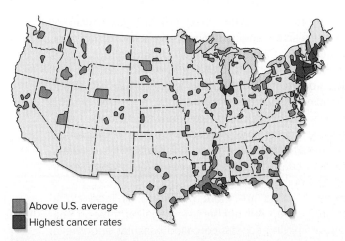

■ Above U.S. average
■ Highest cancer rates

Figure 8.10 Cancer in the United States.

The incidence of cancer per 1,000 people is not uniform throughout the United States. It is centered in cities where chemical manufacturing is common and in the Mississippi Delta.

cells by proteins encoded by tumor-suppressor genes. Mutations to these genes essentially "release the brakes" of cell division, allowing the cell containing the mutated gene to divide uncontrolled. The cell cycle never stops in a cancerous line of cells.

> **Putting the Concept to Work**
> What things might lead to the DNA damage that initiates cancer?

The Nature of Lung Cancer

> **LEARNING OBJECTIVE 8.5.2** Evaluate the evidence that cigarette smoking causes lung cancer.

Of all human diseases, none is more feared than cancer. Nearly one in every six deaths in the United States was caused by cancer in 2020. The American Cancer Society estimates that 606,520 people died of cancer in the United States in 2020. About 22% of these—135,720 people—died of **lung cancer** (figure 8.9). About 135,000 cases of lung cancer are diagnosed each year and 90% of these persons died within three years. What has caused lung cancer to become a major killer of Americans?

The search for a cause of cancers, such as lung cancer, has uncovered a host of environmental factors that appear to be associated with cancer. For example, the incidence of cancer per 1,000 people is not uniform throughout the United States. Rather, it is centered in cities, such as the heavily populated Northeast, indicated by the red areas in figure 8.10, and in the Mississippi Delta, indicated by the red and brown areas, suggesting that environmental factors such as pollution and pesticide runoff may contribute to cancer. When the many environmental factors associated with cancer are analyzed, a clear pattern emerges: Most cancer-causing agents, or carcinogens, share the property of being potent mutagens. A mutagen is a chemical that damages DNA, destroying or changing genes (a change in DNA is called a mutation). That cancer is caused by mutation is now supported by an overwhelming body of evidence.

What sort of genes are being mutated? In the last several years, researchers have found that mutation of only a few genes is all that is needed to transform normally dividing cells into cancerous ones. In identifying and isolating these cancer-causing genes, investigators have learned that all are involved with regulating cell proliferation (how fast cells grow and divide). A key element in this regulation are tumor suppressors, genes that actively prevent tumors from forming.

How do tumor suppressor genes work? The *p53 protein,* a tumor suppressor sometimes called the "guardian angel" of the cell, inspects the DNA before the cell divides. When p53 detects damaged or foreign DNA, it stops cell division and activates the cell's DNA repair systems. If the damage isn't repaired in a certain amount of time, p53 pulls the plug, triggering events that kill the cell. In this way, mutations such as those that cause cancer are either repaired or the cells containing them eliminated. If the gene that produces the p53 protein is itself destroyed by mutation, future damage accumulates unrepaired, including mutations that lead to cancer. Fully 50% of all cancers have a disabled *p53* gene.

Smoking Causes Lung Cancer

If cancer is caused by damage to growth-regulating genes, what then has led to the rapid increase in lung cancer in the United States? Two lines of evidence are particularly telling. The first consists of detailed information about cancer rates among smokers. The annual incidence of lung cancer among nonsmokers is only a few per 100,000 but increases with the number of cigarettes smoked per day to a staggering 300 per 100,000 for 30-cigarettes-a-day smokers.

A second line of evidence consists of changes in the incidence of lung cancer that mirror changes in smoking habits. Look carefully at the data presented in figure 8.11. The upper graph is compiled from data on men and shows the incidence of smoking (blue line) and of lung cancer (red line) in the United States since 1900. As late as 1920, lung cancer was a rare disease. About 30 years after the incidence of smoking began to increase among men, lung cancer also started to become more common. Now look at the lower graph, which presents data on women. Because of social mores, a significant number of women in the United States did not smoke until after World War II (see blue line), when many social conventions changed. As late as 1963, only 6,588 women had died of lung cancer. But as women's frequency of smoking has increased, so has their incidence of lung cancer (red line), again with a lag of about 30 years. Women today have achieved equality with men in the number of cigarettes they smoke, and their lung cancer death rates are now rapidly approaching those for men. In 2014, 72,330 American women died of lung cancer.

How does smoking cause cancer? Cigarette smoke contains many powerful mutagens, among them benzo[*a*]pyrene, and smoking introduces these mutagens to the lung tissues. Benzo[*a*]pyrene, for example, binds to three sites on the *p53* gene and causes mutations at these sites that inactivate the gene. In 1997, scientists studying this tumor-suppressor gene demonstrated a direct link between cigarettes and lung cancer. They found that the *p53* gene is inactivated in 70% of all lung cancers. When these inactivated *p53* genes are examined, they prove to have mutations at just the three sites where benzo[*a*]pyrene binds. Clearly, the presence of the chemical benzo[a]pyrene in cigarette smoke was responsible for the lung cancer.

The most effective way to avoid lung cancer is not to smoke. On a statistical basis, smoking a single cigarette lowers your life expectancy 10.7 minutes (more than the time it takes to smoke the cigarette!). Every pack of 20 cigarettes bears an unwritten label: *The price of smoking this pack of cigarettes is 3 1/2 hours of your life.* Smoking a cigarette is very much like going into a totally dark room with a person who has a gun and standing still. The person with the gun cannot see you, does not know where you are, and shoots once in a random direction. A hit is unlikely, and most shots miss. As the person keeps shooting, however, the chance of eventually scoring a hit becomes more likely. Similarly, every time an individual smokes a cigarette, mutagens are being shot at his or her genes. The more cigarettes smoked, the more likely that genes will be damaged.

In 2020, 13.5% of Americans were cigarette smokers. In an attempt to lower that number, the federal government has proposed new regulations strictly limiting the amount of nicotine in cigarettes to nonaddictive levels.

> **Putting the Concept to Work**
>
> In 2010, nearly as many women died of lung cancer as men, while 50 years earlier women accounted for only one out of every 10 lung cancer deaths. What happened in those 50 years?

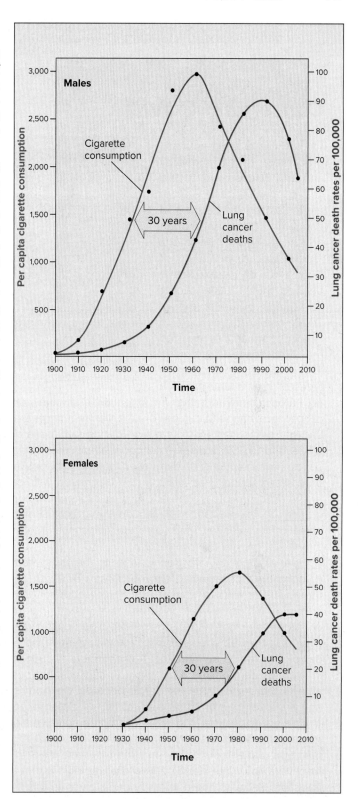

Figure 8.11 Incidence of lung cancer in men and women.
Lung cancer was a rare disease a century ago. As men in the United States increased smoking in the early 1900s, the incidence of lung cancer also increased. Women followed suit years later, and in 2008, more than 71,000 women died of lung cancer, a rate of 40 per 100,000. In that year, about 20% of women smoked and about 25% of men.

Biology and Staying Healthy

Cancer and COVID

The COVID-19 pandemic has presented cancer patients and their doctors with a new set of challenges. Cancer never stops; yet during a pandemic, treatments often do. As COVID-19 infections rapidly spread, hundreds of thousands of people have delayed their routine cancer screening. In the spring of 2020, one of the leaders in cancer diagnostics reported a 46% decrease in screening for the top six most common types of cancer and nearly a 90% drop in mammograms. The longer the delay, the more the cancers grow and the number of potential treatments that can be used decreases. Doctors are predicting that these delays will lead to 10,000 additional cancer deaths in the future. The cancers will eventually be diagnosed, but the tumors will have progressed further. Patients that are diagnosed with cancer face a difficult choice: If they delay their cancer treatments to reduce the risk of COVID-19 infection, they risk the continued spread of their cancer; if they continue their cancer treatments, they are at greater risk of infection.

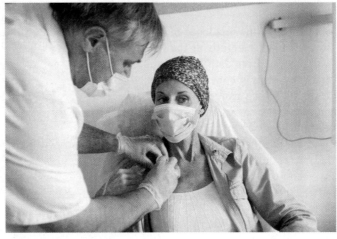
AMELIE-BENOIST/BSIP/Alamy Stock Photo

Developing an infection is one of the most common complications of receiving cancer therapy. Because chemotherapy treatments for cancer typically target rapidly dividing cells, the immune system is often hit very hard. Patients on chemotherapy lose a lot of their white blood cells, the major players in the immune system, which puts them at increased risk of developing an infection. A chemotherapy-suppressed immune system makes a patient with cancer more susceptible to the coronavirus pandemic because the body's skin and mucosal membranes—the first line of defense against a virus—are weakened due to chemotherapies, making it easier for COVID-19 to infect the patient. If COVID-19 does infect the patient, his or her immune response kicks in neutrophils (a key white blood cell) dividing rapidly to combat the infection. Because chemotherapy drugs given to this cancer patient greatly impede the rapid increase of neutrophils in the blood, this makes the patient even more susceptible to COVID-19 infection.

Nothing in the world can ever prepare you to hear the words, "You have cancer." Finding out you have COVID-19 on top of cancer can be even more frightening. Nearly 16% of cancer patients who contract COVID-19 die from the virus infection, not the cancer. This is more than triple the mortality rate we see from COVID-19 patients who do not have cancer. These numbers show why cancer patients are considered to be at particularly high risk to contract COVID-19, and have led changes in how cancer is treated: When possible, treatments are now given orally instead of intravenously, radiation treatments have been shortened, and there have been modifications to immunotherapy usage. However, while bloodwork, mammograms, non-urgent surgeries and preventative screenings can all be delayed, they cannot be put off indefinitely. Recognizing the importance of early detection highlights the necessity of avoiding the danger of exposure to COVID-19. That is why it is particularly important that cancer patients and their caregivers maintain proper social distancing, practice good hand washing, and wear a mask when they are out in public.

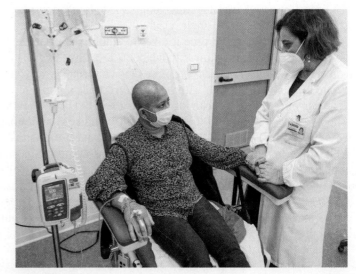
Fabrizio Villa/Getty Images

Answering Your Questions About Vaping

What Is an E-Cigarette?

Cigarettes have been an immense commercial success because even though the tars in tobacco may lead to lung cancer, the nicotine in tobacco leads to addiction. Said simply, smokers keep buying cigarettes, even if tobacco is dangerous, because they quickly become addicted to the nicotine. Eighty-five percent of people who try to quit smoking on their own relapse. However, as the evidence connecting cigarette smoking to lung cancer became impossible to avoid, public acceptance of smoking has faded. As you might expect, cigarette sales rapidly fell as fewer people took up smoking. While 40% of Americans smoked in 1965, only 16% did in 2017. The commercial answer? Sell the addictive nicotine without the cancer-causing tars! Rather than burning tobacco to deliver nicotine in smoke, simply deliver the nicotine to a customer directly—an e-cigarette. First patented back in the 1960s, e-cigarettes really didn't come onto the scene in the United States until 2007. In the ten years since, they have become as widely used as tobacco cigarettes.

What Is Vaping?

E-cigarettes do not burn tobacco products. Rather they contain nicotine solutions that are heated to make a vapor (hence the term *vape* or *vaping*). Typically, the nicotine in an e-cigarette is combined with chemicals to vaporize the nicotine, chiefly propylene glycol (closely related to ethylene glycol, a key ingredient in antifreeze) and glycerin, which together comprise about 95% of the e-liquid. Flavorings may also be added, either natural or artificial. Stored in a tube that looks like a cigarette or small flashlight, the e-liquid is heated by a battery to about 250°C, producing a nicotine-laced aerosol vapor. The amount of nicotine delivered by an e-cigarette into a smoker's bloodstream is comparable to that delivered by smoking a tobacco cigarette.

What Is a Juul Vape Pen?

The Juul (pronounced "jewel") vape pen is an e-cigarette with a lot of nicotine. Smoking a Juul delivers far more nicotine than most commercially available e-cigarettes. How does it do this? Each Juul vaporizes, rather than heats, a liquid that is 5% nicotine. Packaged in a sleek, small, flat stick, a Juul looks a bit like a USB computer thumb drive. Juuls are not particularly expensive, at $35 for the pen and $16 for a four-pack of prefilled cartridges (a "Juul pod"). The hit of nicotine is powerful: one pod of Juuls contains as much nicotine as a pack of cigarettes. A Juul is thus an ideal commercial product to get its users addicted to its use, in just the same way tobacco hooked so many cigarette smokers. Juuls represent half of the $2 billion e-cigarettes sold annually today and are particularly trendy among high school students and young adults.

How Widespread Is Vaping Among Young Adults?

While manufacturers claim that e-cigarettes are not aimed at kids, there are over 7,000 flavors on the market, including gummy bear, bubble gum, and cotton candy. Is the marketing working? Government statistics found millions of teens have tried vaping: nearly 36% of 12th graders had tried some form of e-cigarette in 2017, 17% within the last 30 days (that's more than smoked cigarettes). Because they are tobacco-free, e-cigarettes are currently unregulated in the United States, although that may change. Many states are considering restrictions.

Is Vaping Safer than Smoking Cigarettes?

Yes. The DNA-mutating tars responsible for lung cancer are not present in e-cigarette vapor. Nor do Juul pens produce second-hand smoke. If e-cigarettes were used only as replacement products for tobacco smokers who have been unable to quit smoking, they would certainly be a harm-reducing product. However, e-cigarette use among young adults is not primarily for tobacco use cessation, for few of the high school kids being introduced to vaping are tobacco users. The safety of vaping is much disputed. Manufacturers tout their safety, while public health officials point out that while e-cigarettes do not produce carcinogenic smoke, there is accumulating evidence that vaping may carry negative health effects that should not be ignored.

Are E-Cigarettes a Gateway Drug?

This is, of course, the key question. Does vaping lead to cigarette smoking? Researchers report that teens who vape are six times more likely as those who have never vaped to later begin smoking. Once addicted to nicotine, a vaping young adult can as easily satisfy a craving for nicotine with a cigarette as a Juul. The danger is very real.

Putting the Chapter to Work

1 Prokaryotic cells undergo a much simpler division than eukaryotic cells.

For both types of division, binary fission and mitosis, how many daughter cells are produced, and what is the genetic similarity to the parent cell?

2 Sister chromatids result from the replication of homologous chromosomes.

In which phase of Interphase does the replication of homologous chromosomes occur?

3 During a study of cellular mitotic division, scientists notice that by creating a certain mutation, the cells will divide without replicating mitochondria.

Which part of the cell cycle has the mutation affected?

4 *If scientists were developing a study to understand how sister chromatids separate during eukaryotic cell division to the opposite ends of the dividing cell, what phase of mitosis would be their main interest?*

Retracing the Learning Path

Cell Division

8.1 Prokaryotes Have a Simple Cell Cycle

1. Prokaryotic cells divide in a two-step process: DNA replication followed by binary fission. The genetic information in a prokaryotic cell is present as a single loop of DNA. The DNA begins replication at a site called the origin of replication. The DNA double strand unzips, and new strands form along the original strands, producing two circular prokaryotic chromosomes that separate to the ends of the cell. New plasma membrane and cell wall are added down the middle of the cell, splitting the cell in two. This cell division, called binary fission, produces two daughter cells that are genetically identical to the parent cell.

8.2 Eukaryotic Cell Cycle

1. Cell division in eukaryotes is more complex than in prokaryotes because eukaryotic cells contain more DNA and their DNA is packaged into linear chromosomes.

- Eukaryotic cells divide by one of two methods: mitosis or meiosis. Mitosis occurs in nonreproductive cells, called somatic cells. Meiosis occurs in cells that are involved in sexual reproduction, forming germ-line cells, such as sperm and eggs.

- Interphase, the first portion of the cell cycle, is broken down into three phases: The G_1 phase is the growing phase and takes up the major portion of the cell's life cycle. The S phase is the synthesis phase, when the DNA is replicated. The G_2 phase involves the final preparations for cell division with the replication of mitochondria, chromosome condensation, and synthesis of microtubules.

- During the following M phase, the chromosomes are distributed into opposite sides of the cell. During the C phase, the cell divides its cytoplasm into two separate daughter cells.

8.3 Chromosomes

1. All eukaryotic cells store their hereditary information in chromosomes. Coiling of the DNA into chromosomes allows it to fit into the nucleus. The number of chromosomes varies greatly between species.

- Chromosomes exist in cells as pairs called homologous chromosomes. Two chromosomes that carry copies of the same genes are homologous chromosomes.

- Before cells divide, the DNA replicates, forming two identical copies of each chromosome, called sister chromatids. Sister chromatids stay connected at an area called the centromere. Human somatic cells have 46 chromosomes, and after DNA replication, there are 92 sister chromatids.

- Chromosomes are not uniform; they vary in size, shape, and placement of centromeres. These variations allow researchers to match up homologues making an array, called a karyotype, where homologues are positioned next to each other.

2. The DNA in a chromosome is one long double-stranded fiber. After the DNA is replicated, it associates with histone proteins, forming chromatin. There are several levels of chromosomal organization. The DNA wraps around a histone complex forming a nucleosome and then further folds and loops on itself forming a compact chromosome.

8.4 Cell Division

1. In interphase, the chromosomes replicate and then begin to condense. In the mitosis that follows, they will be drawn by microtubules to opposite ends of the cell.

- Prophase signals the beginning of mitosis. The DNA that was replicated during interphase condenses into chromosomes. The sister chromatids stay attached at the centromeres. The nuclear envelope disappears. Centrioles, when present, migrate to opposite sides of the cell, called the poles, and begin forming the spindle. Microtubules that form the spindle extend from the poles and attach to the chromosomes at the centromeres, anchoring sister chromatids to opposite poles.

- Metaphase involves the alignment of sister chromatids along the equatorial plane.

- During anaphase, the centromeres split, freeing the sister chromatids. The microtubules shorten, pulling the sister chromatids apart and toward opposite poles.

- Telophase signals the completion of nuclear division. The microtubule spindle is dismantled, the chromosomes begin to uncoil; nuclear envelopes form.

2. Following mitosis, the cell separates into two daughter cells in a process called cytokinesis. Cytokinesis in animal cells involves a pinching in of the cell around its equatorial plane until the cell eventually splits into two cells.

Cancer and the Cell Cycle

8.5 What Is Cancer?

1. Cancer is a growth disorder of cells caused by damage to one or more of the genes that regulate the cell cycle. Cells begin to divide in an uncontrolled way, forming a mass of cells called a tumor. Metastases occur when cells from a tumor break away from the mass and spread to other tissues.

Inquiry and Analysis

Why Do Human Cells Age?

For many years, we thought growing older was under hormonal control. Not so. It appears human cells have built-in life spans. In 1961, cell biologist Leonard Hayflick reported the startling result that skin cells growing in tissue culture, such as those growing in culture flasks in the photo below, will divide only a certain number of times. After about 50 population doublings, cell division stops (a **doubling** is a round of cell division producing two daughter cells for each dividing cell; for example, going from a population of 30 cells to 60 cells). If a cell sample is taken after 20 doublings and frozen, when thawed it resumes growth for 30 more doublings and then stops. An explanation of the "Hayflick limit" was suggested in 1986, when researchers first glimpsed an extra length of DNA at the end of chromosomes. Dubbed *telomeres*, these lengths proved to be composed of the simple DNA sequence TTAGGG, repeated nearly a thousand times. Importantly, telomeres were found to be substantially shorter in the cells of older body tissues. This led to the hypothesis that a run of some 16 TTAGGGs was where the DNA replicating enzyme, called polymerase, first sat down on the DNA (16 TTAGGGs being the size of the enzyme's "footprint"), and because of being its docking spot, the polymerase was unable to copy that bit. Thus, a 100-base portion of the telomere was lost

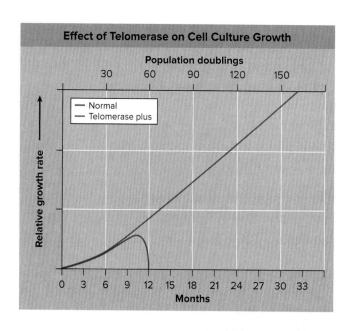

TTAGGG TTAGGG TTAGGG TTAGGG TTAGGG—

by a chromosome during each doubling as DNA replicated. Eventually, after some 50 doubling cycles, each with a round of DNA replication, the telomere would be used up and there would be no place for the DNA replication enzyme to sit. The cell line would then enter senescence, no longer able to proliferate.

This hypothesis was tested in 1998. Using genetic engineering, researchers transferred into newly established human cell cultures a gene that leads to expression of an enzyme called *telomerase* that all cells possess but no body cell uses. This enzyme adds TTAGGG sequences back to the end of telomeres, in effect rebuilding the lost portions of the telomere. Laboratory cultures of cell lines with (telomerase plus) and without (normal) this gene were then monitored for many generations. The graph above displays the results.

Analysis

1. **Applying Concepts** Comparing continuous processes, how do normal skin cells (blue line) differ in their growth history from telomerase plus cells with the telomerase gene (red line)?
2. **Interpreting Data** After how many doublings do the normal cells cease to divide? The telomerase plus cells?
3. **Making Inferences** After nine population doublings, would the rate of cell division be different between the two cultures? After 15? Why?
4. **Drawing Conclusions** How does the addition of the telomerase gene affect the senescence (death by old age) of skin cells growing in culture? Does this result confirm the telomerase hypothesis this experiment had set out to test?

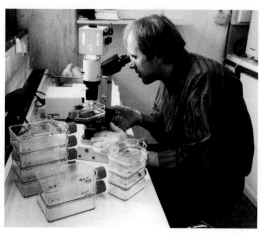

Courtesy Priv. Doz. Dr. Roland Zell

9 Meiosis

LEARNING PATH ▼

Meiosis
1. Discovery of Meiosis
2. The Sexual Life Cycle
3. The Stages of Meiosis

Comparing Meiosis and Mitosis
4. How Meiosis Differs from Mitosis
5. Evolutionary Consequences of Sex

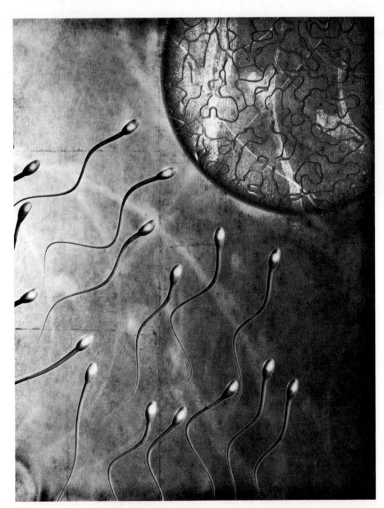

THESE SPERM ARE about to fertilize a human egg, both cells the products of meiosis. But is this complex process necessary?
Colin Anderson/Brand X Pictures/Getty Images

Why Sex?

Very few reading this text are ignorant of the basic *"facts of life"* — we humans reproduce when a sperm cell from a male fertilizes an egg cell from a female, a process we call sex. We learn about sex as teenagers and live with it most of the rest of our lives. Most novel, film, and television plots revolve around sex, and so do key events in each of our lives. This chapter is devoted to describing how sex cells—sperm and egg—are made. As you will see, the process is complex. It might come as a surprise, then, to learn that sex isn't necessary for animals to reproduce. In many animals, the budding off of a localized bunch of cells can grow by mitosis to form a new individual identical in every way to its parent. So why sex? Why go to all the trouble of meiosis?

Sex Is Not Necessary to Reproduce

It isn't just some special benefit coming from the meiosis process that makes sex cells. Even when meiosis and the production of sperm and eggs occur, there may still be reproduction without sex. The development of an adult from an unfertilized egg, called parthenogenesis, is a common form of reproduction in arthropods. Among bees, for example, fertilized eggs develop into diploid females, but unfertilized eggs develop into haploid males. Parthenogenesis even occurs among the vertebrates (animals with backbones). Populations of the checkered whiptail lizard, seen on the following page, are all female. There are no males. A female lizard's unfertilized egg undergoes a mitotic nuclear division without cell cleavage to produce a diploid cell, which then develops into an adult.

...So Why Do It?

If a lizard can do it, why not we? If reproduction in vertebrates can occur without sex, why does sex occur at all?

James Gerholdt/Photolibrary/Getty Images

retroimages/Getty Images

This question has generated considerable discussion. Sex is of great evolutionary advantage for species, which, as you will learn in this chapter, benefit from the genetic variability generated by meiosis. However, evolution occurs because of changes at the level of *individual* survival and reproduction, and no obvious advantage is gained by the progeny of an individual that engages in sexual reproduction. In fact, the segregation of chromosomes during meiosis tends to disrupt advantageous combinations of genes. It is, therefore, a puzzle to know what a well-adapted individual gains from participating in sexual reproduction, as *all* of its progeny could maintain its successful gene combinations if that individual simply reproduced asexually.

The DNA Repair Hypothesis

Several geneticists have suggested that sex occurs because only a diploid cell can effectively repair certain kinds of chromosome damage, particularly double-strand breaks in DNA. Both radiation and chemical events within cells can induce such breaks. As organisms became larger and longer-lived, it must have become increasingly important for them to be able to repair such damage. As you will see, in early stages of meiosis, pairs of homologous chromosomes line up precisely. This arrangement may well have evolved originally as a mechanism for repairing double-strand damage to DNA: The undamaged homologous chromosome could be used as a template to repair the damaged chromosome. Sex may just be a way to repair DNA.

Muller's Ratchet

The geneticist Herman Muller pointed out in 1965 that asexual populations tend to accumulate mutations: once harmful mutations arise, asexual populations have no way of eliminating them, and they accumulate over time, like turning a ratchet. Sexual populations, on the other hand, can employ recombination to generate individuals that do not carry the mutation, which selection can then favor. Sex may just be a way to keep the mutational load down.

The Red Queen Hypothesis

One evolutionary advantage of sex may be that it allows populations to "store" recessive forms of a trait that are currently bad but have promise for reuse at some time in the future, when conditions are different. In all species, selection is constantly acting against detrimental traits when they are expressed. But in sexual species, selection can never entirely remove recessive detrimental variants, as only the dominant forms of a trait are expressed in heterozygote individuals. The evolution of most sexual species, most of the time, thus has "variation in the bank" to keep pace with ever-changing, and often rapidly changing, environments. This "treadmill evolution" is sometimes called the "Red Queen hypothesis," after the Queen of Hearts in Lewis Carroll's *Through the Looking Glass,* who tells Alice, "Now, here, you see, it takes all the running you can do, to keep in the same place." Sex may just be a way to keep up in a rapidly changing world.

Meiosis

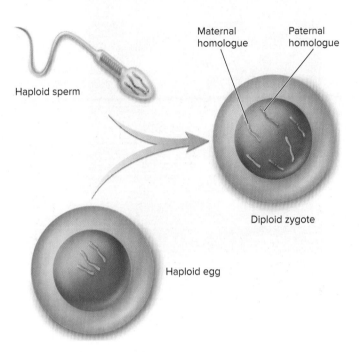

Figure 9.1 Diploid cells carry chromosomes from two parents.

A diploid cell contains two versions of each chromosome, a maternal homologue contributed by the haploid egg of the mother and a paternal homologue contributed by the haploid sperm of the father.

Figure 9.2 Sexual and asexual reproduction.

Reproduction in an organism is not always either sexual or asexual. The strawberry reproduces both asexually (runners) and sexually (flowers).

MIXA/Getty Images

9.1 Discovery of Meiosis

LEARNING OBJECTIVE 9.1.1 Distinguish sexual from asexual reproduction, haploid from diploid, gamete from zygote, and meiosis from mitosis.

Only a few years after Walther Flemming's discovery of chromosomes in 1879, Belgian scientist Pierre-Joseph van Beneden was surprised to find different numbers of chromosomes in different types of cells in the roundworm *Ascaris*. Specifically, he observed that the **gametes** (eggs and sperm) each contained two chromosomes, whereas the somatic (nonreproductive) cells of embryos and mature individuals each contained four. From his observations, van Beneden proposed in 1887 that an egg and a sperm, each containing half the complement of chromosomes found in other cells, fuse to produce a single cell called a **zygote**. The zygote, like all of the somatic cells ultimately derived from it, contains two copies of each chromosome. The fusion of gametes to form a new cell is called **fertilization,** or **syngamy**.

The Problem Solved by Meiosis

It was clear even to early investigators that gamete formation must involve some mechanism that reduces the number of chromosomes to half the number found in other cells. If it did not, the chromosome number would double with each fertilization, and after only a few generations, the number of chromosomes in each cell would become impossibly large. For example, after one generation, the 46 chromosomes present in human cells would increase to 92 (46×2^1) chromosomes; after two generations, it would increase to 184; and by 10 generations, the 46 chromosomes present in human cells would increase to over 47,000 (46×2^{10}) chromosomes.

The number of chromosomes does not explode in this way because of a special reduction division that occurs during gamete formation, producing cells with half the normal number of chromosomes. The subsequent fusion of two of these cells ensures a consistent chromosome number from one generation to the next. This reduction division process is known as **meiosis**.

Sexual Reproduction

Meiosis and fertilization together constitute a cycle of reproduction. Two sets of chromosomes are present in the somatic cells of adult individuals, making them **diploid** cells (Greek, *di,* two and often indicated by $2n$, where "n" is the number of sets of chromosomes), but only one set is present in the gametes, which are thus **haploid** (Greek, *haploos,* one and often indicated by $1n$). For example, if a sperm cell containing three chromosomes from the father fused with an egg cell containing three chromosomes from the mother, a diploid zygote with six chromosomes (three pairs of homologous chromosomes) would result (figure 9.1). Reproduction that involves this alternation of meiosis and fertilization is called **sexual reproduction**. Some organisms, however, reproduce by mitotic division and don't involve the fusion of gametes. Reproduction in these organisms is referred to as **asexual reproduction**. Many plants and some animals are able to reproduce both asexually and sexually (figure 9.2).

Putting the Concept to Work
How many chromatids are present in a human gamete?

9.2 The Sexual Life Cycle

> **LEARNING OBJECTIVE 9.2.1** Compare the life cycles of plants and animals, distinguishing between somatic and germ-line cells.

Having two parents is basic to being an animal or plant.

Alternation of Generations

The life cycles of all sexually reproducing organisms follow the same basic pattern of alternation between diploid chromosome numbers and haploid ones (figure 9.3). In unicellular eukaryotic organisms like the protist shown in figure 9.4a, individuals are haploid for most of their lives. When they encounter environmental stress, haploid cells sometimes fuse with other haploid cells to form a diploid cell. Later, this cell undergoes meiosis to re-form the haploid phase.

In most animals, like the frog you see in figure 9.4b, meiosis forms gametes, and fertilization results in the formation of a diploid zygote. This single diploid cell divides by mitosis, eventually giving rise to the adult frog shown in the photo. Almost all animals are diploid for the multicellular stage of their life cycle.

In plants like the fern you see in figure 9.4c, haploid and diploid individuals alternate. Certain cells of a diploid individual undergo meiosis, making haploid sex cells (gametes) that divide repeatedly by mitosis to form a multicellular haploid individual. Some cells of this haploid individual eventually differentiate into eggs or sperm, which fuse to form a diploid zygote. Dividing by mitosis, the zygote forms a diploid individual.

Germ-Line Tissues

In animals, the cells that will eventually undergo meiosis to produce gametes are set aside from other cells early in the course of development. The cells of the body are called **somatic cells**, from the Latin word for "body," while gamete-forming cells, located in the reproductive organs of males and females, are referred to as **germ-line cells**. Both the somatic cells and the gamete-producing germ-line cells are diploid. Somatic cells undergo mitosis to form genetically identical, diploid daughter cells. The germ-line cells undergo meiosis, producing haploid gametes.

> **Putting the Concept to Work**
> How is the plant sexual life cycle different from the animal sexual life cycle?

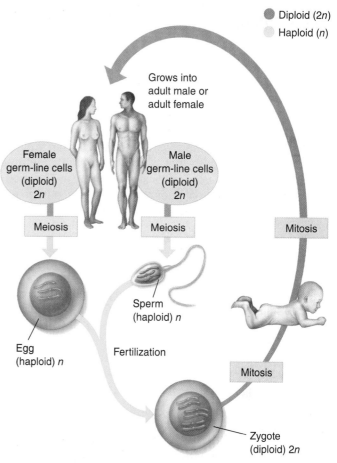

Figure 9.3 The sexual life cycle in animals.

In animals, the completion of meiosis is followed soon by fertilization. Thus, the vast majority of the life cycle is spent in the diploid stage. In this text, n stands for haploid and $2n$ stands for diploid. Germ-line cells are set aside early in development and undergo meiosis to form haploid gametes (eggs or sperm). The rest of the body cells are called somatic cells.

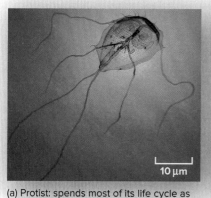

(a) Protist: spends most of its life cycle as a haploid individual

(b) Animal: spends most of its life cycle as a diploid individual

(c) Plant: spends significant portions of its life cycle as haploid and diploid individuals

Figure 9.4 Three types of sexual life cycles.

In sexual reproduction, haploid cells or organisms alternate with diploid cells or organisms.

(a) corbis/MedicalRF.com; (b) Craig K. Lorenz/Science Source; (c) Don Paulson Photography/Purestock/SuperStock.

Today's Biology

Babies with Three Parents

Like everyone you know, you have two parents. A sperm cell from your father joined with an egg cell from your mother to create the zygote that became you, the zygote getting 23 chromosomes from dad and 23 from mom. These 46 chromosomes make up your genetic instructions, determining the traits that you, their baby, will express.

Problems with Making a Baby

Unfortunately, some couples have trouble making a baby the old-fashioned way. An alarming 15.7% of American couples in 2017 are having problems getting pregnant (up from 5.4% in 1984!). As a loose rule of thumb, a couple is said to have impaired fertility if they are unable to get pregnant after one year of trying.

In about 40% of these cases, daddy is the basic problem: about 7% of men are effectively infertile. In many instances, their bodies are producing antibodies that attack their own sperm cells.

In 60% of cases of impaired fertility, the egg is the issue: 12.1% of U.S. women aged 15 to 44 are effectively infertile, with the effect far more pronounced among older women.

In Vitro Fertilization

What, then, to do? For years, couples have been able to turn to *in vitro* fertilization as a method to help them make a baby. The procedure is not complicated. The mother's egg is removed and fertilized with the father's sperm in the laboratory. The fertilized egg, now a zygote, is then implanted into the uterus of the mother, where it can grow into a baby. This *in vitro* procedure does not always lead to the birth of a baby. The success rate for women under 35 is about 40%. Sadly, then, there are many infertile couples that still strike out when trying to make a baby with *in vitro* fertilization.

Mitochondrial Problems

What went wrong? One scenario in which normal *in vitro* fertilization may not work is when dad or mom has a genetic defect in their mitochondria. The mitochondria act as powerhouses in a living cell, producing the ATP that powers its growth and development. Sometimes the cell producing sperm or egg has deficient mitochondria. A mutation in the mitochondrial DNA may damage a gene that encodes a Krebs cycle enzyme, for example, injuring metabolism.

If a sperm-making cell has such mutant mitochondria, the sperm it produces will not fully develop, leading to male infertility. This problem is not passed on to a zygote, of course, as sperm don't carry mitochondria along with them during fertilization. However, if it is the egg cell rather than the sperm-producing cell that contains such defective mitochondria, then the zygote inherits the problem too and will not have a functional source of power. Since the zygote is unable to complete development, this will lead to termination of the pregnancy.

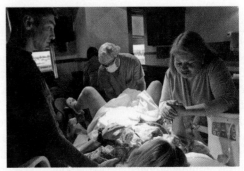

Image Bank/Alamy Stock Photo

A Baby with Two Moms

A new form of *in vitro* fertilization can help couples struggling with such mitochondrial issues. To accomplish this type of *in vitro* fertilization, an egg is harvested from a third "parent," a healthy donor woman. This egg will then have its nucleus removed, leaving behind the mitochondria, each with 37 undamaged and healthy mitochondrial genes. At the same time this is going on, the nucleus is removed from one of mom's eggs, leaving behind its bad mitochondria. Mom's nucleus, along with dad's sperm, are placed inside the donor woman's egg cell, forming a zygote.

This zygote has three parents! Twenty-three chromosomes come from mom's nucleus, 23 come from dad's sperm, and the mitochondria DNA with its 37 genes were already in the donor's egg. The happy result is a functional zygote free of mitochondrial defects. While some folks view this baby as having three parents, others disagree, pointing out that the human genome consists of approximately 25,000 genes. Inheriting 37 genes out of 25,000 is not viewed as enough genetic contribution to consider the egg cell donor to be a parent to the baby.

Future of Three-Parent Babies

With the rapid growth of gene editing technologies, the future of *in vitro* fertilization is wide open. New technologies may allow parents to remove chromosome bits that contain a bad gene or to rewrite the problem sequence to repair the defect. Depending on the number and sort of changes involved in the formation of the zygote, we may see any number of "parents" contributing to the making of a baby in the near future.

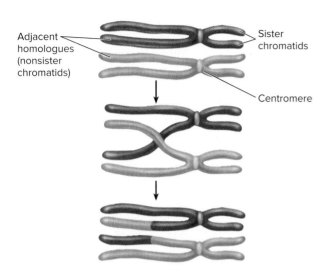

Figure 9.5 Crossing over.

In crossing over, the two homologues of a chromosome exchange portions. During the crossing-over process, nonsister chromatids that are next to each other exchange chromosome arms or segments.

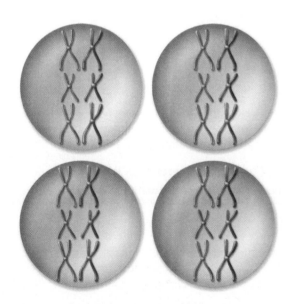

Figure 9.6 Independent assortment.

Independent assortment occurs because the orientation of chromosomes on the metaphase plate is random. Shown here are four possible orientations of chromosomes in a hypothetical cell. Each of the many possible orientations results in gametes with different combinations of parental chromosomes.

9.3 The Stages of Meiosis

Now, let's look more closely at the process of meiosis. Just as in mitosis, the chromosomes have replicated before meiosis begins, during a period called interphase. The first of the two divisions of meiosis, called **meiosis I**, serves to separate the homologous chromosomes (or homologues); the second division, **meiosis II**, serves to separate the *sister chromatids*. Thus, when meiosis is complete, what started out as one diploid cell ends up as four haploid cells. Because there was one replication of DNA but *two* cell divisions, the process reduces the number of chromosomes by half.

Meiosis I

> **LEARNING OBJECTIVE 9.3.1** Outline what happens in the four stages of meiosis I.

Meiosis I is traditionally divided into four stages (see the left side of *Essential Biological Process 9A*):

1. Prophase I. The two versions of each chromosome (the two homologues) pair up and exchange segments.
2. Metaphase I. The chromosomes align on a central plane.
3. Anaphase I. One homologue with its two sister chromatids still attached moves to a pole of the cell, and the other homologue moves to the opposite pole.
4. Telophase I. Individual chromosomes gather together at the two poles.

Prophase I. In **prophase I,** individual chromosomes first become visible, when viewed with a light microscope, as their DNA coils more and more tightly. Because the DNA replicates before the onset of meiosis, each of the threadlike chromosomes actually consists of two sister chromatids joined at their centromeres, just as in mitosis. However, now meiosis begins to differ from mitosis. During prophase I, the two homologous chromosomes line up side by side, physically touching one another, as you see in **figure 9.5**. At this point, a process called **crossing over** is initiated, in which the chromosomes actually break in the same place on both nonsister chromatids, and sections of chromosomes are swapped between the homologous chromosomes, producing a hybrid chromosome that is part maternal chromosome (the green sections) and part paternal chromosome (the purple sections).

Metaphase I. In **metaphase I,** the spindle apparatus forms, but because homologues are held close together by crossovers, spindle fibers can attach to only the outward-facing portion of each centromere. For each pair of homologues, the orientation on the metaphase plate is random. Each orientation of homologues results in gametes with different combinations of parental chromosomes. This process is called **independent assortment** (figure 9.6).

Anaphase I. In **anaphase I,** the spindle fibers shorten and the homologues are pulled apart and move toward opposite poles. At the end of anaphase I, each pole has half as many chromosomes as were present in the cell when meiosis began. Remember that the chromosomes replicated and thus contained two sister chromatids before the start of meiosis, but sister chromatids are not counted as separate chromosomes. As in mitosis, count the number of centromeres to determine the number of chromosomes.

Essential Biological Process 9A

Meiosis

Meiosis I

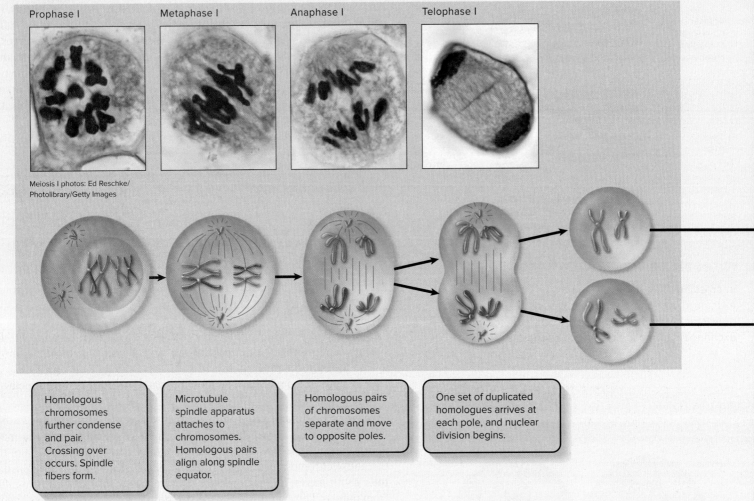

Meiosis I photos: Ed Reschke/Photolibrary/Getty Images

- Homologous chromosomes further condense and pair. Crossing over occurs. Spindle fibers form.
- Microtubule spindle apparatus attaches to chromosomes. Homologous pairs align along spindle equator.
- Homologous pairs of chromosomes separate and move to opposite poles.
- One set of duplicated homologues arrives at each pole, and nuclear division begins.

Telophase I. In telophase I, the chromosomes gather at their respective poles to form two chromosome clusters. After an interval, meiosis II occurs, in which the sister chromatids are separated as in mitosis.

> **Putting the Concept to Work**
> At the end of meiosis I, are sister chromatids still attached?

Meiosis II

> **LEARNING OBJECTIVE 9.3.2** Outline the events of meiosis II, stating how many copies of each chromosome are present at the end.

Following meiosis I, the cells enter a brief interphase, in which no DNA synthesis occurs, and then the second meiotic division begins. Meiosis II is simply a mitotic division involving the products of meiosis I (*Essential Biological Process 9A*). At the end of telophase I, each pole has a haploid complement of chromosomes, each of which is still composed of two sister chromatids attached at the centromere. Like meiosis I, meiosis II is divided into four stages:

1. **Prophase II.** At the two poles of the cell, the clusters of chromosomes enter a brief prophase II, where a new spindle forms.

BIOLOGY & YOU

Bloom's Syndrome. A rare inherited disorder called Bloom's syndrome has long puzzled investigators. Affected people have severe growth deficiencies, are predisposed to a wide variety of cancers, and are often sterile—a very unhappy state of affairs. What sort of gene disorder could lead to such a broad array of defects? Recent reports implicate events early in meiosis. It turns out that the gene defective in Bloom's syndrome patients encodes a protein dubbed BLM that binds to homologues early in meiotic prophase I. A properly functioning BLM protein is required to release homologues that are interlocked during crossover, in preparation for homologous chromosome separation during anaphase I. In Bloom's syndrome individuals, the BLM protein is defective and cannot remove cross-links that are created during crossover, resulting in a high incidence of gaps and breaks in homologous chromosomes as they separate.

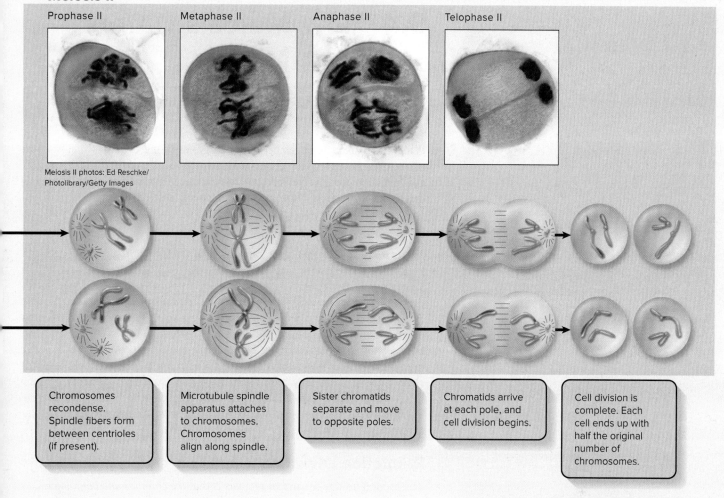

2. **Metaphase II.** Spindle fibers bind to both sides of the centromeres, and the chromosomes line up along a central plane (figure 9.7).
3. **Anaphase II.** The spindle fibers shorten, splitting the centromeres and moving the sister chromatids to opposite poles.
4. **Telophase II.** Finally, the nuclear envelope re-forms around the four sets of daughter chromosomes.

The main outcome of the four stages of meiosis II is to separate the sister chromatids. The final result of this division is four cells containing haploid sets of chromosomes. No two are alike because of the crossing over in prophase I.

If you think about it, the key to meiosis is that the sister chromatids of each chromosome do not separate from each other in the first division, prevented by the crossing over that occurred in prophase I. Imagine two people dancing closely. You can tie a rope to the back of each person's belt, but you cannot tie a second rope to their belt buckles in front because the two dancers are facing each other and are very close. In just the same way, microtubules cannot attach to the inner sides of the centromeres to pull the two sister chromatids apart because crossing over holds the homologous chromosomes together like dancing partners.

Putting the Concept to Work

Do sister chromatids separate in meiosis I or II?

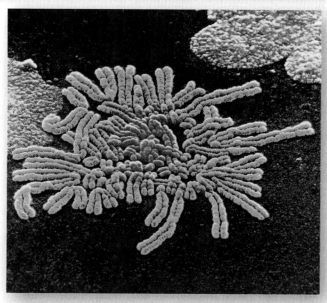

Figure 9.7 Getting ready to divide.

These human metaphase chromosomes are lining up to divide. Note that the sister chromatids have not yet been pulled apart.

Adrian T Sumner/Science Source

Comparing Meiosis and Mitosis

9.4 How Meiosis Differs from Mitosis

> **LEARNING OBJECTIVE 9.4.1** Explain the two unique features that distinguish meiosis from mitosis.

While there are differences between eukaryotes in the details of meiosis, two consistent features are seen in the meiotic processes of every eukaryote: synapsis and reduction division. Indeed, these two unique features are the key differences that distinguish meiosis from mitosis, which you studied in chapter 8.

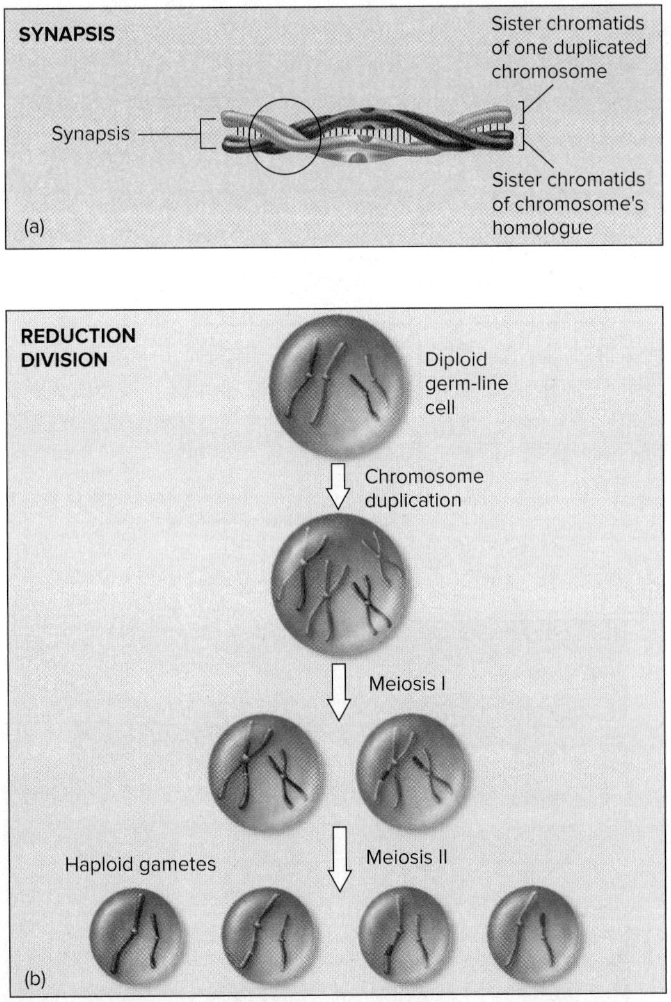

Synapsis

The first of these two features happens early during the first nuclear division. Following chromosome replication, homologous chromosomes or homologues *pair all along their lengths,* with sister chromatids being held together by proteins called cohesin. While homologues are thus physically joined, *genetic exchange occurs at one or more points between them.* The process of forming these complexes of homologous chromosomes is called **synapsis,** and the exchange process between paired homologues, as described earlier, is crossing over. Figure 9.8*a* shows how the homologous chromosomes are held together closely enough that they are able to physically exchange segments of their DNA. Sister chromatids do not separate from each other in the first nuclear division, so each homologue is still composed of two chromatids joined at the centromere and still considered one chromosome.

Reduction Division

The second unique feature of meiosis is that *the chromosome homologues do not replicate between the two nuclear divisions,* so that chromosome assortment in the second division separates sister chromatids of each chromosome into different daughter cells.

In most respects, the second meiotic division is identical to a normal mitotic division. However, because of the crossing over that occurred during the first division, the sister chromatids in meiosis II are not identical to each other. Also, there are only half the number of chromosomes in each cell at the beginning of meiosis II because only one of each homologue is present. Figure 9.8*b* shows how reduction division occurs. The diploid cell in the figure contains four chromosomes (two homologous pairs). After meiosis I, the cells contain just two chromosomes (remember to count the number of *centromeres,* because sister chromatids are not considered separate chromosomes). During meiosis II, the sister chromatids separate, but each gamete still only contains two chromosomes, half as many of the germ-line cell.

Because mitosis and meiosis use similar terminology, it is easy to confuse the two processes. Figure 9.9 compares the two processes side-by-side. Both processes start with a diploid cell, but you can see that early during meiosis I, crossing over occurs and that as a consequence homologous *pairs,* not individual centromeres, line up along the meiosis I metaphase plate. These two differences result in haploid cells in meiosis and diploid cells in mitosis.

> **Putting the Concept to Work**
> How do two sister chromatids entering meiosis II differ?

Figure 9.8 Unique features of meiosis.

(a) Synapsis draws homologous chromosomes together, all along their lengths, creating a situation (indicated by the circle) in which two homologues can physically exchange portions of arms, a process called crossing over. (b) Reduction division, omitting a chromosome duplication before meiosis II, produces haploid gametes, thus ensuring that the chromosome number remains the same as that of the parents following fertilization.

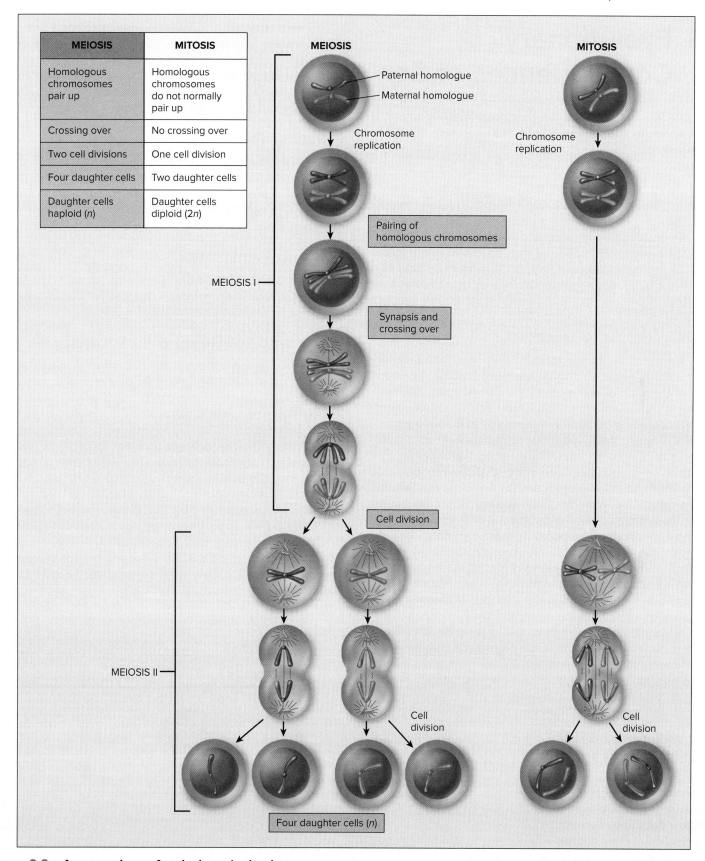

Figure 9.9 **A comparison of meiosis and mitosis.**

Meiosis differs from mitosis in several key ways, highlighted by the orange boxes. Meiosis involves two nuclear divisions with no DNA replication between them. It thus produces four daughter cells, each with half the original number of chromosomes. Also, crossing over occurs in prophase I of meiosis. Mitosis involves a single nuclear division after DNA replication. Thus, it produces two daughter cells, each containing the original number of chromosomes, which are genetically identical to those in the parent cell.

9.5 Evolutionary Consequences of Sex

> **LEARNING OBJECTIVE 9.5.1** List the principal mechanisms that generate new genetic combinations in meiosis.

Meiosis is a lot more complicated than mitosis. Why has evolution gone to so much trouble? While our knowledge of how meiosis and sex evolved is still growing, it is abundantly clear that meiosis and sexual reproduction have an enormous impact on how species continue to evolve today because of their ability to rapidly generate new genetic combinations. Three mechanisms each make key contributions: independent assortment, crossing over, and random fertilization.

Independent Assortment

The reassortment of genetic material that takes place during meiosis is the principal factor that has made possible the evolution of eukaryotic organisms, in all their bewildering diversity, over the past 1.5 billion years. Sexual reproduction represents an enormous advance in the ability of organisms to generate genetic variability. To understand, recall that most organisms have more than one pair of chromosomes. For example, the organism represented in **figure 9.10** has three pairs of chromosomes, each offspring receiving three homologues from each parent, purple from the father and green from the mother. The offspring in turn produces gametes, but the distribution of homologues into the gametes is completely random. A gamete could receive all homologues that are paternal in origin, as on the far left; or it could receive all maternal homologues, as on the far right; or any combination. Independent assortment alone leads to eight possible gamete combinations in this example. In humans, each gamete receives one homologue of each of the 23 chromosomes, but which homologue of a particular chromosome it receives is determined randomly. Each of the 23 pairs of chromosomes migrates independently, so there are 2^{23} (more than 8 million) different possible kinds of gametes that can be produced.

To make this point to his class, one professor offers an "A" course grade to any student who can write down all the possible combinations of heads and tails (an "either/or" choice, like that of a chromosome migrating to one pole or the other) with flipping a coin 23 times (like 23 chromosomes moving independently). No student has ever won an "A"; there are over 8 million possibilities.

Crossing Over

The DNA exchange that occurs when the arms of nonsister chromatids cross over adds even more recombination to the independent assortment of chromosomes that occurs later in meiosis. Thus, the number of possible genetic combinations that can occur among gametes is virtually unlimited.

Random Fertilization

Also, the zygote that forms a new individual is created by the fusion of two gametes, each produced independently, so fertilization squares the number of possible outcomes ($2^{23} \times 2^{23} = 70$ trillion).

Importance of Generating Diversity

Paradoxically, the evolutionary process is both revolutionary and conservative. It is revolutionary in that the pace of evolutionary change is quickened by genetic recombination, much of which results from sexual reproduction. It is conservative in that change is not always favored by selection, which may instead preserve existing combinations of genes. These conservative pressures appear to be greatest in some asexually reproducing organisms that do not move around freely and that live in especially demanding habitats. In vertebrates, on the other hand, the evolutionary premium appears to have been on versatility, and sexual reproduction is the predominant mode of reproduction.

Whatever the forces that led to sexual reproduction, its evolutionary consequences have been profound. No genetic process generates diversity more quickly; and as you will see in chapter 10, genetic diversity is the raw material of evolution, the fuel that drives it and determines its potential directions.

> **Putting the Concept to Work**
> You only need one set of instructions for your body to do all the jobs it needs to carry out. So why aren't humans simply haploid all their lives?

Figure 9.10 Independent assortment increases genetic variability.

Independent assortment contributes new gene combinations to the next generation because the orientation of chromosomes on the metaphase plate is random. In the cell shown here with three chromosome pairs, there are eight different gametes that can result, each with different combinations of parental chromosomes.

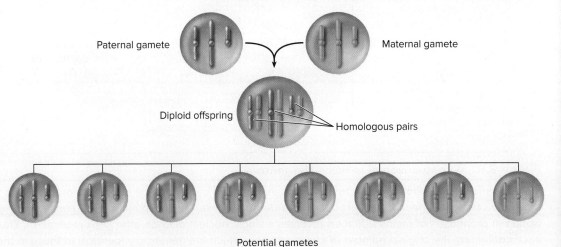

A Closer Look

The Biology of Sexual Orientation

Classic experiments in the 1950s and '60s showed us how the sex chromosomes of a fetus determine whether that individual will grow up behaving as a male or a female.

The Key Impact of Testosterone

The key to adult sexual behavior, it turns out, is the activation early in development of fetal genes located on the Y chromosome that lead the developing testis to secrete the hormone testosterone. This testosterone acts on the brain of the fetus in a way that determines that individual's future sexual behavior.

Early experiments were carried out on rats. In a baby rat, the release of testosterone does not occur until after birth. When experimenters gave testosterone to newborn female rats, their sexual behavior was malelike when they grew up. Conversely, when experimenters removed the testes of newborn males, the result was femalelike behavior. Similar experiments have since been carried out in species whose babies are born much more mature than rats, such as monkeys. The same results were obtained as with rats, indicating that the same processes occur in the brain of these species, but earlier, during pregnancy.

Epigenetics Is the Tool

But a monkey is not a human. Does this process of determining sexuality during pregnancy occur in humans, too? Researchers think so. Male human embryos form testes very early in pregnancy (at about 10 weeks), and these testes immediately start to secrete testosterone. So the male brain is exposed to testosterone early in development. What does the testosterone do in the male brain? Recent experiments give a hint. In the hypothalamus (a portion of the brain that plays a major role in sexual behavior), a female fetus suppresses malelike behavior by chemically modifying certain key genes. The process is an example of epigenetics (see section 12.6), the female fetus adding methyl groups to the genes to silence them. Giving females testosterone reduces this, removing some of the methyl groups and releasing these genes to become active. Such females behave more like males. In much the same fashion, giving a drug to newborn males that blocks demethylation causes these males to behave more like females. It seems that the default human sexual behavior is female, with testosterone in the male fetus releasing genes that determine future malelike behavior.

So, in a nutshell, how we behave sexually—our sexual orientation—is determined by what **genes** we have, by **hormones** that influence these genes in developing males, and by **environmental factors** that may modify the rate at which these genes are epigenetically activated.

LGBTQ

As a rough measure, your sexual orientation (your enduring pattern of sexual or romantic attraction) can be to a person or persons of the opposite gender (heterosexual), to the same gender (homosexual), or to both genders (bisexual). Nonheterosexual individuals commonly refer to themselves as **LGBTQ**, a term that stands for **l**esbian, **g**ay, **b**isexual, **t**ransgender, and **q**ueer. The CDC estimates the number of nonheterosexual individuals as 5% of the total U.S. population in 2014—some 13 million LGBTQ individuals.

In the CDC study, 1.5% of adult American women self-identify as lesbian, or homosexual women, while about 1.8% of adult men self-identify as gay, or homosexual men. About 1.3% of adult Americans self-identify as bisexual.

There is another possibility, of course—individuals whose gender identity differs from their anatomical sex at birth. These people are said to be transgender. The CDC estimates that transgender individuals, an often-not-discussed but not rare minority, represent 0.6% of the U.S. population.

So what about the Q in LGBTQ? While for decades a demeaning label for any LGBT individual, the term *queer* has been embraced by younger members of the LGBTQ community as signifying "questioning"—someone who is in the process of figuring out their sexual identity.

Conversion Therapy

Can the sexual orientation of an adult be changed? No. No major mental health professional organization sanctions efforts to change a person's sexual orientation (a discredited process its advocates call "conversion therapy"), and virtually all of these organizations have adopted policy statements cautioning against conversion therapy. The underlying assumption of conversion therapy—that homosexuality is the result of troubled family dynamics and so can be reversed by intensive psychotherapy—is now uniformly understood to have been based on misinformation and prejudice. There is no substantive evidence that the nature of parenting or early childhood experience plays any role in the formation of a person's sexual orientation.

The developmental decisions that determine sexual orientation act at the gene level during pregnancy. There is much we do not yet know about which genes are involved and the factors influencing the timing of their epigenetic activation, but this much is clear: The events determining sexual orientation, once made in a fetus, are irrevocable and are not subject to later change. Just as left-handed children, about 10% of all kids, can be trained to write with their right hand but continue for all their lives to be left-handed, so it is with the 5% of children born LGBTQ. Their sexual orientation, like left-handedness, is not voluntary or "curable," although the overt expression of their sexuality is certainly subject to conscious choice. You are born with your sexuality but choose how to express it.

Putting the Chapter to Work

If sex is so advantageous, why is parthenogenesis common in nature?

1. Mice reproduce via sexual reproduction. The gamete-producing germ-line cells, in both males and females, consist of 40 chromosomes.

   What will be the number of chromosomes in the gametes that both males and females produce?

2. During meiosis I, genetic variation occurs because of the events that take place in prophase 1 and in metaphase 1.

   If the variation-causing event in prophase 1 was skipped, what is the name of the variation-causing process that would not occur?

3. There have been scientific reports of meiosis events in which the second phase, meiosis II, does not occur.

   If this second cellular division does not occur, would the gametes contain a haploid or diploid number of chromosomes?

4. Independent assortment creates genetic variation among gametes. It results from the random orientation of the pairs of homologues arrayed along the central axis of the cell.

   During which phase of meiosis does this event occur?

5. The products of asexual reproduction are different from the gametes formed by meiosis. State how the end results of meiosis differ.

Retracing the Learning Path

Meiosis

9.1 Discovery of Meiosis

1. Meiosis is a form of cell division in which the number of chromosomes is halved during gamete formation. In sexually reproducing organisms, a gamete from the male fuses with a gamete from the female to form a cell called the zygote. This process is called fertilization. The number of chromosomes in gametes must be halved to maintain the correct number of chromosomes in offspring of sexually reproducing organisms.

- A cell that contains a full set of chromosomes, two copies of each chromosome, is called a diploid cell. Cells such as gametes that contain only one copy of each chromosome are haploid.

- Sexual reproduction involves meiosis, but some organisms also undergo asexual reproduction, which is reproducing by mitosis or binary fission.

9.2 The Sexual Life Cycle

1. In the sexual life cycle, there is an alternation between diploid and haploid stages. Three types of sexual life cycles exist: In many protists, the majority of the life cycle is devoted to the haploid stage; in most animals, the majority of the life cycle is devoted to the diploid stage; and in plants, the life cycle is split more equally between a haploid stage and a diploid stage.

- Germ-line cells of an organism are diploid but produce haploid gametes through meiosis.

9.3 The Stages of Meiosis

1. Meiosis involves two nuclear divisions, meiosis I and meiosis II, each containing a prophase, metaphase, anaphase, and telophase. Like mitosis, the DNA replicates during interphase, before meiosis begins. Because there are two nuclear divisions but only one round of DNA replication, the four daughter cells contain half the number of chromosomes as the parent cell.

- During meiosis I, homologous chromosomes move to opposite poles of the cell. Meiosis I is divided into four stages: prophase I, metaphase I, anaphase I, and telophase I. Prophase I is distinguished by the exchange of genetic material between homologous chromosomes, a process called crossing over. In this process, homologous chromosomes align with each other along their lengths, and sections of nonsister chromatids are physically exchanged. This recombines the genetic information contained in the chromosomes.

- During metaphase I, microtubules in the spindle apparatus attach to homologous chromosomes, and chromosome pairs align along the metaphase plate. The alignment of the chromosomes is random, leading to the independent assortment of chromosomes into the gametes.

- The homologous chromosomes separate during anaphase I, being pulled apart by the spindle apparatus toward their respective poles. This differs from mitosis and later in meiosis II, in which sister chromatids separate in anaphase.

- In telophase I, the chromosomes cluster at the poles. This leads to the next phase of meiosis, called meiosis II.

2. Meiosis II mirrors mitosis in that it involves the separation of sister chromatids through the phases of prophase II, metaphase II, anaphase II, and telophase II. Meiosis II differs from mitosis in that there is no DNA replication before meiosis II. Because homologous pairs were separated during meiosis I, each daughter cell has only one-half the number of chromosomes. Also, these are not genetically identical because of crossing over.

Comparing Meiosis and Mitosis

9.4 How Meiosis Differs from Mitosis

1. Two processes distinguish meiosis from mitosis: crossing over through synapsis and reduction division.

- When homologous chromosomes come together during prophase I, they associate with each other along their lengths, a process called synapsis. During synapsis, sections of homologous chromosomes are physically exchanged in crossing over, resulting in daughter cells that are not genetically identical to the parent cell or to each other.

- Meiosis also differs from mitosis in reduction division, in which the daughter cells contain half the number of chromosomes as the parent cell. Reduction division occurs because meiosis contains two nuclear divisions but only one round of DNA replication during interphase.

- The primary reasons for the differences in meiosis and mitosis stem from the synapsis of homologous chromosomes in prophase I. The close association of the homologous chromosomes in synapsis blocks the inner centromeres from attaching to the spindle. As a result, sister chromatids do not separate during meiosis I, resulting in reduction division.

9.5 Evolutionary Consequences of Sex

1. Independent assortment and crossing over, like shuffling a deck of cards, give a species the ability to rapidly generate new genetic combinations, a key to evolutionary success.

Inquiry and Analysis

Are New Microtubules Made When the Spindle Forms?

During interphase, before the beginning of meiosis, only a few long microtubules extend from the so-called *centrosome* (a zone around the centrioles of animal cells where microtubules are organized) to the cell periphery. Like most microtubules, they are refreshed at a low rate with resynthesis. Late in prophase, however, a dramatic change can be seen: the centrosome appears to divide into two, and a large increase is seen in the number of microtubules radiating from each of the two daughter centrosomes. The two clusters of new microtubules are easily seen as the green fibers connecting to the two sets of purple chromosomes in the micrograph of early prophase below (a **micrograph** is a photo taken through a microscope). This burst of microtubule assembly marks the beginning of the formation of the spindle characteristic of prophase and metaphase. When these clusters of new microtubules first became known to cell biologists, they asked whether these were existing microtubules being repositioned in the spindle or newly synthesized microtubules.

The graph to the upper right displays the results of an experiment designed to answer this question. Mammalian cells in culture (cells **in culture** are growing in the laboratory on artificial medium) were injected with microtubule subunits (tubulin) to which a fluorescent dye had been attached (a **fluorescent dye** is one that glows when exposed to ultraviolet or short-wavelength visual light). After the fluorescent subunits had become incorporated into the cells' microtubules, all the fluorescence in a small region of a cell was bleached by an intense laser beam, destroying the microtubules there. Any subsequent rebuilding of microtubules in the bleached region would have to employ the fluorescent subunits present in the cell, causing recovery of fluorescence in the bleached region. The graph reports this recovery as a function of time for interphase and metaphase cells. The dotted line represents the time for 50% recovery of fluorescence ($t_{1/2}$) (that is, $t_{1/2}$ is the time required for half of the microtubules in the region to be resynthesized).

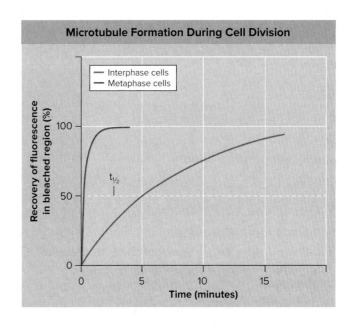

Analysis

1. **Applying Concepts** Are new microtubules synthesized during interphase? (Hint: Does the blue line in the graph rise?) What is the $t_{1/2}$ of this replacement synthesis? (Hint: Where does the dotted line in the graph cross the blue line?) Are new microtubules synthesized during metaphase? (Hint: Does the red line rise?) What is the $t_{1/2}$ of this replacement synthesis?
2. **Interpreting Data** Is there a difference in the rate at which microtubules are synthesized during interphase and metaphase? How big is the difference? What might account for it?
3. **Making Inferences**
   a. What general statement can be made regarding the relative rates of microtubule production before and during meiosis?
   b. Is there any difference in the final amount of microtubule synthesis that would occur if this experiment were to be continued for an additional 15 minutes?
4. **Drawing Conclusions** When are the microtubules of the spindle assembled?

Dr. Alexey Khodjakov/Science Source

10 Foundations of Genetics

LEARNING PATH ▼

Mendel
1. Mendel and the Garden Pea
2. What Mendel Observed
3. Mendel Proposes a Theory
4. Mendel's Laws

From Genotype to Phenotype
5. How Genes Influence Traits
6. Why Some Traits Don't Show Mendelian Inheritance

Chromosomes and Heredity
7. Human Chromosomes
8. The Role of Mutations in Human Heredity

IDENTICAL TWINS. How identical are they?
Image Source/Getty Images

Does Environment Affect IQ?

In a very general way, everybody knows what IQ is—it's how smart you are, right? A person's IQ score, their "intelligence quotient," has been believed for some time to be determined largely by that person's genes. How did science come to that conclusion? Scientists measure the degree to which genes influence traits such as IQ by examining where variation in IQ comes from.

What Is IQ?

What factors contribute to the variance of IQ scores? There are three: (1) The first is variation at the gene level, some gene combinations leading to higher IQ scores than others. (2) The second is variation at the environmental level, some environments leading to higher IQ scores than others. (3) The third is what statisticians call covariance, the degree to which environment affects genes.

The degree to which genes influence a trait such as IQ, the *heritability* of IQ, is given the symbol H and is defined simply as the fraction of the total variance that is genetic.

How Important Are Genes to IQ?

So how heritable is IQ? Geneticists estimate the heritability of IQ by measuring the environmental and genetic contributions to the total variance of IQ scores. The environmental contributions can be measured by comparing the IQ scores of identical twins reared together with those reared apart (any differences should reflect environmental influences). The genetic contributions can be measured by comparing identical twins reared together (who are 100% genetically identical) with fraternal twins reared together (who are 50% genetically identical). Any differences should reflect genes, as twins share identical prenatal conditions and are raised in virtually identical environmental circumstances. When traits are more commonly shared between identical twins than fraternal twins, the difference is likely genetic.

When these sorts of "twin studies" have been done in the past, researchers have uniformly reported that IQ is highly heritable, with values of H typically reported as being around 0.7 (a very high value). While it didn't seem significant at the time, almost all the twins available for study had come from middle-class or wealthy families.

Pete Collins/iStock/Getty Images

Are Some Groups Genetically Inferior?

The study of IQ has proven controversial, because IQ scores are often different when social and racial groups are compared. What is one to make of the observation that IQ scores of poor children measure lower as a group than do scores of children of middle-class and wealthy families? This difference has led to the controversial suggestion that the poor are genetically inferior.

A Faulty Assumption

What should we make of such a harsh conclusion? These studies have all made a critical assumption, that environment does not affect gene expression, so that covariance makes no contribution to the total variance in IQ scores.

Studies have allowed a direct assessment of this assumption. Importantly, it proves to be flat wrong. In November 2003, researchers reported an analysis of twin data from a study carried out in the late 1960s. The National Collaborative Prenatal Project, funded by the National Institutes of Health, enrolled nearly 50,000 pregnant women, most of them Black and quite poor, in several major U.S. cities. Researchers collected abundant data and gave the children IQ tests seven years later. Although not designed to study twins, this study was so big that many twins were born, 623 births. Seven years later, 320 of these pairs were located and given IQ tests. This thus constitutes a huge "twin study," the first ever conducted of IQ among the poor.

Poverty Lessens IQ Scores

When the data were analyzed, the results were unlike any ever reported. The heritability of IQ was different in different environments! Most notably, the influence of genes on IQ was far less in conditions of poverty, where environmental limitations seem to block the expression of genetic potential. Specifically, for children of high socioeconomic status, $H = 0.72$, much as reported in previous studies, but for children raised in poverty, $H = 0.10$, a very low value, indicating genes were making little contribution to observed IQ scores. The lower a child's socioeconomic status, the less impact genes had on IQ. These data say, with crystal clarity, that the genetic contributions to IQ don't mean much in an impoverished environment. How does poverty in early childhood affect the brain? Neuroscientists report that many children growing up in very poor families experience poor nutrition, which impairs their neural development. This affects language development and memory the rest of their lives.

Clearly, improvements in the growing and learning environments of poor children can be expected to have a major impact on their IQ scores. Additionally, these data argue that the controversial differences reported in mean IQ scores between racial groups may well reflect no more than poverty and are no more inevitable.

Mendel

Figure 10.1 **Families look alike.**
These two are parent and child. It is no accident that they look so much alike, as the child shares half their parent's genes.
Nancy Ney/Getty Images

10.1 Mendel and the Garden Pea

> **LEARNING OBJECTIVE 10.1.1** Contrast the experiments of Gregor Mendel with earlier classic experiments on pea plants.

When you were born, many things about you resembled your mother or father (figure 10.1). This tendency for traits to be passed from parent to offspring is called **heredity**. *Traits* are the expressions of a character or a heritable feature. How does heredity happen? Before DNA and chromosomes were discovered, this puzzle was one of the greatest mysteries of science. The key to understanding the puzzle of heredity was found in the garden of an Austrian monastery over a century ago by a monk named Gregor Mendel. Mendel's solution to the puzzle of heredity was the first step on our journey to understanding heredity and one of the greatest intellectual accomplishments in the history of science.

Mendel's Experiments

Gregor Mendel was born in 1822 to peasant parents and was educated in a monastery. He became a monk and was sent to the University of Vienna to study science and mathematics. Although he aspired to become a scientist and teacher, he failed his university exams and returned to the monastery, where he spent the rest of his life, eventually becoming abbot. Upon his return, Mendel joined an informal neighborhood science club. Under the patronage of a local nobleman, each member set out to undertake scientific investigations, which were then discussed at meetings and published in the club's own journal. Mendel undertook to repeat some classic crosses with pea plants done in the 1790s, but unlike earlier plant breeders, he intended to count the numbers of each kind of offspring in the hope that the numbers would give some hint of what was going on. Quantitative approaches to science—measuring and counting—were just becoming fashionable in Europe.

Gregor Mendel

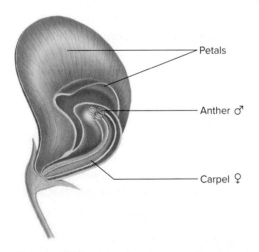

Figure 10.2 **The garden pea flower.**
Because it is easy to cultivate and because there are many distinctive varieties, the garden pea, *Pisum sativum*, was a popular choice as an experimental subject in investigations of heredity for as long as a century before Mendel's studies. The flower shown here is partially cut away to show internal structures.
Gregor (Johann) Mendel: Pixtal/age fotostock

> **Putting the Concept to Work**
> Do you think it fair that earlier classic experiments with peas are not mentioned alongside Mendel today? Explain your response.

Mendel's Experimental System: The Garden Pea

> **LEARNING OBJECTIVE 10.1.2** List four characteristics that made the garden pea easy for Mendel to study.

Mendel chose to study the garden pea because several of its characteristics made it easy to work with:

1. Many varieties were available. Mendel selected seven pairs of lines that differed in easily distinguished traits (including the white versus purple flowers that earlier classic experiments had reported 60 years earlier).
2. Mendel knew from the work of earlier experimenters that he could expect the infrequent version of a character to disappear in one generation and reappear in the next. He knew, in other words, that he would have something to count.
3. Pea plants are small, easy to grow, produce large numbers of offspring, and mature quickly.
4. The reproductive organs of peas are enclosed within their flowers (see figure 10.2). Left alone, the flowers do not open. They simply fertilize themselves with their own pollen (male gametes). To carry out a cross, Mendel had only to pry the petals apart, reach in with a pair of scissors, and snip off the male organs (anthers); he could then dust the female organs (the tip of the carpel) with pollen from another plant to make the cross.

> Flowers are the reproductive structures in a group of plants called angiosperms. Angiosperm reproduction, through the pollination and fertilization of flowers, is discussed in more detail in section 32.1.

Mendel's Experimental Design

Mendel's experimental design was the same as that of experimenters 60 years earlier, only Mendel counted his plants. The crosses were carried out in three steps that are presented in **figure 10.3**:

① Mendel began by letting each variety self-fertilize for several generations. This ensured that each variety was true-breeding, meaning that it contained no other varieties of the trait and so would produce only offspring of the same variety when it self-pollinated. The white flower variety, for example, produced only white flowers and no purple ones in each generation. Mendel called these lines the P generation (P for parental).

② Mendel then conducted his experiment: He crossed two pea varieties exhibiting alternative traits, such as white versus purple flowers. The offspring that resulted he called the F_1 generation (F_1 for "first filial" generation, from the Latin word for "son" or "daughter").

③ Finally, Mendel allowed the plants produced in the crosses of step 2 to self-fertilize, and he counted the numbers of each kind of offspring that resulted in this F_2 ("second filial") generation. As reported 60 years earlier and shown in step 3, the white flower trait reappeared in the F_2 generation, although not as frequently as the purple flower trait.

> **Putting the Concept to Work**
> In Mendel's crosses, the offspring of the F_1 generation self-fertilized to produce the F_2 generation. Why didn't Mendel allow the parent plants to self-fertilize like this in making the F_1 generation?

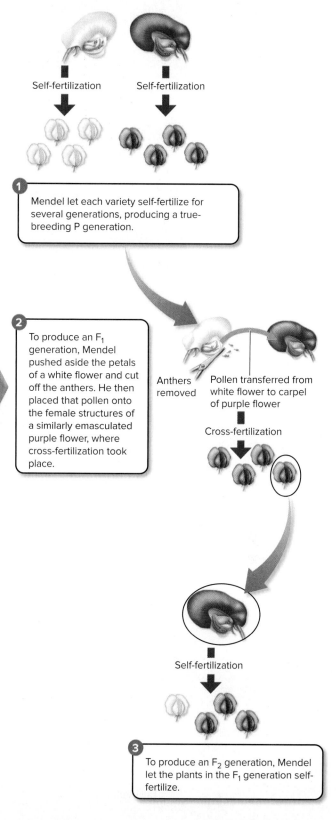

Figure 10.3 How Mendel conducted his experiments.

10.2 What Mendel Observed

LEARNING OBJECTIVE 10.2.1 Describe what Mendel observed when crossing two contrasting traits.

Mendel experimented with a variety of traits in the garden pea (figure 10.4) and repeatedly made similar observations. In all, Mendel examined seven pairs of contrasting traits, as shown in table 10.1. For each pair of contrasting traits that Mendel crossed, he observed the same result, shown in figure 10.3, where a trait disappeared in the F_1 generation only to reappear in the F_2 generation. We will examine in detail Mendel's crosses with flower color.

The F_1 Generation

In the case of flower color, when Mendel crossed purple and white flowers, all the F_1 generation plants he observed were purple; he did not see the contrasting trait, white flowers. Mendel called the trait expressed in the F_1 plants **dominant** and the trait not expressed **recessive**. In this case, purple flower

Figure 10.4 The pod of a garden pea plant.
Richard Gross/McGraw Hill

TABLE 10.1 Seven Characters Mendel Studied in His Experiments

Dominant Form	Character ×	Recessive Form	F_2 Generation Dominant: Recessive	Ratio
Purple flowers	×	White flowers	705:224	3.15:1 (3/4:1/4)
Yellow seeds	×	Green seeds	6022:2001	3.01:1 (3/4:1/4)
Round seeds	×	Wrinkled seeds	5474:1850	2.96:1 (3/4:1/4)
Green pods	×	Yellow pods	428:152	2.82:1 (3/4:1/4)
Inflated pods	×	Constricted pods	882:299	2.95:1 (3/4:1/4)
Axial flowers	×	Terminal flowers	651:207	3.14:1 (3/4:1/4)
Tall plants	×	Dwarf plants	787:277	2.84:1 (3/4:1/4)

color was dominant and white flower color recessive. Mendel studied several other characters in addition to flower color, and for every pair of contrasting traits Mendel examined, one proved to be dominant and the other recessive. The dominant and recessive traits for each character he studied are indicated in table 10.1.

The F_2 Generation

After allowing individual F_1 plants to mature and self-fertilize, Mendel collected and planted the seeds from each plant to see what the offspring in the F_2 generation would look like. Mendel found (as others had earlier) that some F_2 plants exhibited white flowers, the recessive trait. The recessive trait had disappeared in the F_1 generation, only to reappear in the F_2 generation. It must somehow have been present in the F_1 individuals but unexpressed!

At this stage, Mendel instituted his radical change in experimental design. He *counted* the number of each type among the F_2 offspring. He believed the proportions of the F_2 types would provide some clue about the mechanism of heredity. In the cross between the purple-flowered F_1 plants, he counted a total of 929 F_2 individuals (see table 10.1). Of these, 705 (75.9%) had purple flowers and 224 (24.1%) had white flowers. Approximately one-fourth of the F_2 individuals exhibited the recessive form of the trait. Mendel carried out similar experiments with other traits, such as round versus wrinkled seeds (figure 10.5), and obtained the same result: Three-fourths of the F_2 individuals exhibited the dominant form of the character, and one-fourth displayed the recessive form. In other words, the dominant:recessive ratio among the F_2 plants was always approximately 3:1.

Figure 10.5 Round versus wrinkled seeds. One of the differences among varieties of pea plants that Mendel studied was the shape of the seed. In some varieties the seeds were round, whereas in others they were wrinkled.
MShieldsPhotos/Alamy Stock Photo

A Disguised 1:2:1 Ratio

Mendel let the F_2 plants self-fertilize for another generation and found that the one-fourth that were recessive were true-breeding—future generations showed nothing but the recessive trait. Thus, the white F_2 individuals described previously showed only white flowers in the F_3 generation (as shown on the right in figure 10.6). Among the three-fourths of the plants that had shown the dominant trait in the F_2 generation, only one-third of the individuals were true-breeding in the F_3 generation (as shown on the left in figure 10.6). The others showed both traits in the F_3 generation (as shown in the center)—and when Mendel counted their numbers, he found the ratio of dominant to recessive to again be 3:1! From these results, Mendel concluded that the 3:1 ratio he had observed in the F_2 generation was in fact a disguised 1:2:1 ratio:

1	2	1
true-breeding :	not-true-breeding :	true-breeding
dominant	dominant	recessive

> **Putting the Concept to Work**
> If you were to pick an F_2 individual at random, what is the probability that the individual you pick will be not-true-breeding?

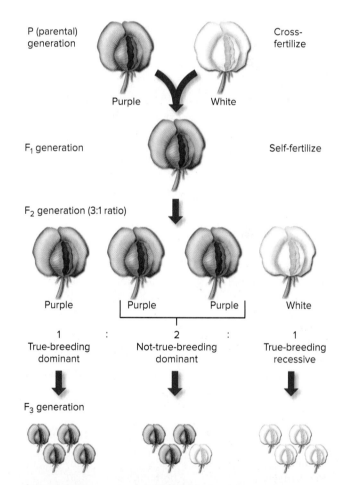

Figure 10.6 The F_2 generation is a disguised 1:2:1 ratio.

By allowing the F_2 generation to self-fertilize, Mendel found from the offspring (F_3) that the ratio of F_2 plants was one true-breeding dominant, two not-true-breeding dominant, and one true-breeding recessive.

10.3 Mendel Proposes a Theory

LEARNING OBJECTIVE 10.3.1 State the five hypotheses of Mendel's theory, distinguishing between gene and allele, and between genotype and phenotype.

To explain his results, Mendel proposed a simple set of five hypotheses that would faithfully predict the results he had observed. Now called Mendel's theory of heredity, it has become one of the most famous theories in the history of science.

Hypothesis 1: Parents do not transmit traits directly to their offspring. Rather, they transmit information about the traits, what Mendel called *merkmal* (the German word for "factor"). These factors act later, in the offspring, to produce the trait. In modern terminology, we call Mendel's factors **genes**.

Hypothesis 2: Each parent contains two copies of the factor governing each trait. The two copies may or may not be the same. If the two copies of the factor are the same (both encoding purple or both white flowers, for example), the individual is said to be homozygous. If the two copies of the factor are different (one encoding purple, the other white, for example), the individual is said to be **heterozygous**.

Hypothesis 3: Alternative forms of a factor lead to alternative traits. Alternative forms of a factor are called **alleles**. Mendel used lowercase letters to represent recessive alleles and uppercase letters to represent dominant ones. In modern terms, we call the appearance of an individual its **phenotype**. Appearance is determined by which alleles of a gene an individual receives from its parents, and we call those particular alleles the individual's **genotype**. Thus a pea plant might have the phenotype "white flower" and the genotype *pp*.

Hypothesis 4: The two alleles that an individual possesses do not affect each other, any more than two letters in a mailbox alter each other's contents. Each allele is passed on unchanged when the individual matures and produces its own gametes (egg or sperm). At the time, Mendel did not know that his factors were carried from parent to offspring on chromosomes.

Hypothesis 5: The presence of an allele does not ensure that a trait will be expressed in the individual that carries it. In heterozygous individuals, only the dominant allele achieves expression; the recessive allele is present but unexpressed.

These five hypotheses, taken together, constitute Mendel's model of the hereditary process. Many traits in humans exhibit dominant or recessive inheritance similar to the traits Mendel studied in peas (table 10.2).

Putting the Concept to Work
Can you name a recessive allele displayed by a member of your family?

BIOLOGY & YOU

Look-alike Meter. Have you ever wondered which parent you resemble more? Using state-of-the-art face recognition technology, Internet services now allow you to answer this question in a very objective way. You upload a photo of your face, looking straight into the camera and similar photos of your parents, and the service analyzes them to provide you with your own personal "look-alike meter." One parent's photo is shown on the left end of the meter scale, the other parent's photo on the right. If you resemble both parents equally, the meter's needle points to the middle (straight up). If, on the other hand, you look more like one parent than the other, then the needle will swing over toward the side of that parent. The more you resemble them, the closer to that side of the meter the needle points. When the baby of Tom Cruise and Katie Holmes was analyzed in this way with the look-alike meter shown below, Suri proved to look more like Tom, by 15%. Why do you think it is possible for Suri to look more like their father than their mother? Didn't they get half their genes from each parent?

Katie Holmes: Dave M. Benett/Getty Images;
Suri Cruise: Marcel Thomas/FilmMagic/Getty Images;
Tom Cruise: Dave M. Benett/Getty Images

TABLE 10.2 Some Dominant and Recessive Traits in Humans

Recessive Traits	Phenotypes
Common baldness	M-shaped hairline receding with age
Albinism	Lack of melanin pigmentation
Alkaptonuria	Inability to metabolize homogentisic acid
Red-green color blindness	Inability to distinguish red and green wavelengths of light

Dominant Traits	Phenotypes
Mid-digital hair	Presence of hair on middle segment of finger
Brachydactyly	Short fingers
Phenylthiocarbamide (PTC) sensitivity	Ability to taste PTC as bitter
Camptodactyly	Inability to straighten the little finger
Polydactyly	Extra fingers and toes

Analyzing Mendel's Results

> **LEARNING OBJECTIVE 10.3.2** Using a Punnett square analysis, explain the basis of the 3:1 Mendelian ratio.

To analyze Mendel's results, it is important to remember that each trait is determined by the inheritance of alleles from the parents, one allele from the mother and the other from the father. These alleles, present on chromosomes, are distributed to gametes during meiosis. Each gamete receives one copy of each chromosome and therefore one copy of an allele.

Consider again Mendel's cross of purple-flowered with white-flowered plants. Like Mendel, we will assign the symbol *P*, written in uppercase, to the dominant allele associated with the production of purple flowers and the symbol *p*, written in lowercase, to the recessive allele associated with the production of white flowers.

In this system, the genotype of an individual true-breeding for the recessive white-flowered trait would be designated *pp*, as both copies of the allele specify the white phenotype. Similarly, the genotype of a true-breeding purple-flowered individual would be designated *PP*, and a heterozygote would be designated *Pp* (dominant allele first). Using these conventions and denoting a cross between two strains with ×, we can symbolize Mendel's original cross as *pp* × *PP*.

Punnett Squares

The possible results from a cross between a true-breeding, white-flowered plant (*pp*) and a true-breeding, purple-flowered plant (*PP*) can be visualized with a **Punnett square**. In a Punnett square, the possible gametes of one individual are listed along the horizontal side of the square, while the possible gametes of the other individual are listed along the vertical side. The genotypes of potential offspring are represented by the cells within the square. Figure 10.7 walks you through the set-up of a Punnett square crossing two plants that are heterozygous for flower color (*Pp* × *Pp*).

The frequency that these genotypes occur in the offspring is usually expressed by a probability. For example, in a cross between a homozygous white-flowered plant (*pp*) and a homozygous purple-flowered plant (*PP*), *Pp* is the only possible genotype for all individuals in the F$_1$ generation, as shown by the Punnett square on the left of figure 10.8. Because *P* is dominant to *p*, all individuals in the F$_1$ generation have purple flowers. When individuals from the F$_1$ generation are crossed, as shown by the Punnett square on the right, the probability of obtaining a homozygous dominant (*PP*) individual in the F$_2$ is 25% because one-fourth of the possible genotypes are *PP*. Similarly, the probability of an individual in the F$_2$ generation being homozygous recessive (*pp*) is 25%. Because the heterozygous genotype has two possible ways of occurring (*Pp* and *pP*, but both written as *Pp*), it occurs in half of the cells within the square; the probability of obtaining a heterozygous (*Pp*) individual is 50% (25% + 25%).

> **Putting the Concept to Work**
> What is the probability of NOT obtaining a heterozygous F$_2$ individual?

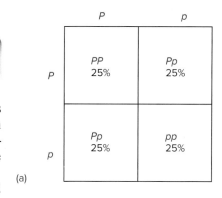

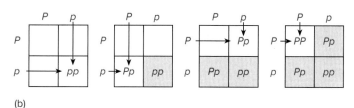

(a)

(b)

Figure 10.7 A Punnett square analysis.
(a) Each square represents 1/4 or 25% of the offspring from the cross. The squares in (b) show how the square is used to predict the genotypes of all potential offspring.

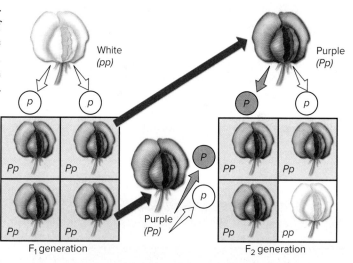

Figure 10.8 How Mendel analyzed flower color.
The only possible offspring of the first cross are *Pp* heterozygotes, purple in color. These individuals are known as the F$_1$ generation. When two heterozygous F$_1$ individuals cross, three kinds of offspring are possible: *PP* homozygotes (purple flowers); *Pp* heterozygotes (also purple flowers), which may form two ways; and *pp* homozygotes (white flowers). Among these individuals, known as the F$_2$ generation, the ratio of dominant phenotype to recessive phenotype is 3:1.

The Testcross

> **LEARNING OBJECTIVE 10.3.3** Diagram how a testcross reveals the genotype of an individual exhibiting a dominant phenotype.

How did Mendel know which of the purple-flowered individuals in the F_2 generation (or the F_1 generation) were homozygous (*PP*) and which were heterozygous (*Pp*)? It is not possible to tell simply by looking at them. For this reason, Mendel devised a simple and powerful procedure called the **testcross** to determine an individual's actual genetic composition. Consider a purple-flowered plant. It is impossible to determine such a plant's genotype simply by looking at its phenotype. To learn its genotype, you must cross it with some other plant. What kind of cross would provide the answer? If you cross it with a homozygous dominant individual, all of the progeny will show the dominant phenotype whether the test plant is homozygous or heterozygous. It is also difficult (but not impossible) to distinguish between the two possible test plant genotypes by crossing with a heterozygous individual. However, if you cross the test plant with a homozygous recessive individual, the two possible test plant genotypes will give totally different results.

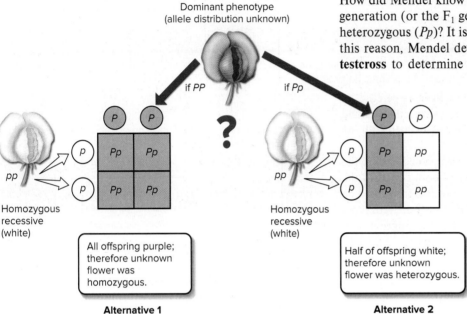

Figure 10.9 How Mendel used the testcross to detect heterozygotes.

To determine whether an individual exhibiting a dominant phenotype, such as purple flowers, was homozygous (*PP*) or heterozygous (*Pp*) for the dominant allele, Mendel devised the testcross. He crossed the individual in question with a known homozygous recessive (*pp*)—in this case, a plant with white flowers.

Analyzing a Test Cross. To see how this works, step through a testcross of a purple-flowered plant with a white-flowered plant. **Figure 10.9** shows you the two possible alternatives:

Alternative 1 (on left): Unknown plant is homozygous (*PP*). *PP* × *pp*: All offspring have purple flowers (*Pp*), as shown by the four purple squares.

Alternative 2 (on right): Unknown plant is heterozygous (*Pp*). *Pp* × *pp*: One-half of offspring have white flowers (*pp*) and one-half have purple flowers (*Pp*), as shown by the two white and two purple squares.

In one of his testcrosses, Mendel crossed F_1 individuals exhibiting the dominant trait back to the parent homozygous for the recessive trait. He predicted that the dominant and recessive traits would appear in a 1:1 ratio, and that is what he observed, as you can see illustrated in Alternative 2 above.

Test Crosses Using Two Genes. Testcrosses can also be used to determine the genotype of an individual when two genes are involved. Mendel carried out many two-gene crosses, some of which we will soon discuss. He often used testcrosses to verify the genotypes of particular dominant-appearing F_2 individuals. Thus an F_2 individual showing both dominant traits (*A_ B_*) might have any of the following genotypes: *AABB*, *AaBB*, *AABb*, or *AaBb*. By crossing dominant-appearing F_2 individuals with homozygous recessive individuals (that is, *A_ B_* × *aabb*), Mendel was able to determine if either or both of the traits bred true among the progeny—any that did could not be heterozygous—and so determine the genotype of the F_2 parent.

> **Putting the Concept to Work**
> How did Mendel know which F_2-generation peas were heterozygous?

BIOLOGY & YOU

Is Your Black Labrador Puppy Homozygous? In Labrador retrievers, the gene for black coat color is dominant over the gene for brown coat color. A puppy with a black-colored coat may be either true-breeding, with two alleles for black coat color, or heterozygous, with one allele for black coat color and one for brown coat color. To determine which genetic endowment your black lab puppy has, you can mate it (when it grows up) to a so-called chocolate lab, which has brown fur and two alleles for brown coat color. If all the offspring of this mating are black, your dog must be true-breeding and have both alleles for black coat color. If some of the young are chocolate brown, your dog must be heterozygous and have one allele for black coat color and one for brown coat color. Yellow labs, on the other hand, lack the allele for pigmentation, so they have light-colored coats.

10.4 Mendel's Laws

> **LEARNING OBJECTIVE 10.4.1** State Mendel's first and second laws, showing how they account for his experimental results.

Mendel's understanding of heredity can be summarized in two simple laws.

Mendel's First Law: Segregation

Mendel's model brilliantly predicts the results of his crosses, accounting in a neat and satisfying way for the ratios he observed. Similar patterns of heredity have since been observed in countless other organisms. Traits exhibiting this pattern of heredity are called *Mendelian traits*. Because of its overwhelming importance, Mendel's theory is often referred to as Mendel's first law, or the **law of segregation.** In modern terms, Mendel's first law states that *the two alleles of a trait separate during the formation of gametes, so that half of the gametes will carry one copy and half will carry the other copy.*

Mendel's Second Law: Independent Assortment

Mendel went on to ask if the inheritance of one factor, such as flower color, influences the inheritance of other factors, such as plant height. To investigate this question, he followed the inheritance of two separate traits, called a **dihybrid** cross (*di* meaning two; a *monohybrid* cross examines one trait). He first established a series of true-breeding lines of peas that differed from one another with respect to two of the seven pairs of characteristics and then crossed them. Figure 10.10 shows an experiment in which the P generation consists of homozygous individuals with round, yellow seeds (*RRYY* in the figure) that are crossed with individuals that are homozygous for wrinkled, green seeds (*rryy*). This cross produces F_1 individuals that have round, yellow seeds and are heterozygous for both of these traits (*RrYy*). The chromosomes then are allocated to the gametes during meiosis such that there are four types of gametes for these two traits.

Mendel then allowed the dihybrid individuals to self-fertilize. If the segregation of alleles affecting seed shape and alleles affecting seed color was independent, the probability that a particular pair of seed-shape alleles would occur together with a particular pair of seed-color alleles would simply be a product of the two individual probabilities that each pair would occur separately. For example, the probability of an individual with wrinkled, green seeds appearing in the F_2 generation would be equal to the probability of an individual with wrinkled seeds (1 in 4) multiplied by the probability of an individual with green seeds (1 in 4), or 1 in 16.

In his dihybrid crosses, Mendel found that the frequency of phenotypes in the F_2 offspring closely matched the 9:3:3:1 ratio predicted by the Punnett square analysis shown in figure 10.10. He concluded that for the pairs of traits he studied, the inheritance of one trait does not influence the inheritance of the other trait, a result often referred to as Mendel's second law, or the **law of independent assortment.** We now know that this result is valid only for genes not located near one another on the same chromosome. Thus, in modern terms, Mendel's second law is often stated as follows: *Genes located on different chromosomes are inherited independently of one another.*

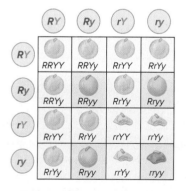

Figure 10.10 Analysis of a dihybrid cross.

This dihybrid cross shows round (*R*) versus wrinkled (*r*) seeds and yellow (*Y*) versus green (*y*) seeds. The ratio of the four possible phenotypes in the F_2 generation is predicted to be 9:3:3:1.

> **Putting the Concept to Work**
>
> If two genes *are* located near each other on a chromosome, would recessive alleles of each of these genes still exhibit 3:1 Mendelian ratios in the F_2 generation of a dihybrid cross?

From Genotype to Phenotype

10.5 How Genes Influence Traits

> **LEARNING OBJECTIVE 10.5.1** Describe how genotype determines phenotype.

It is useful, before considering Mendelian genetics further, to gain a brief understanding of how genes work. With this in mind, we will sketch, in broad strokes, a picture of how a Mendelian trait is influenced by a particular gene. We will use the protein hemoglobin, which is found in human red blood cells, as our example—you can follow along using figure 10.11.

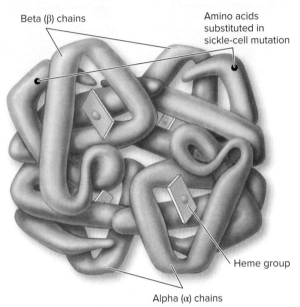

Structure of hemoglobin

Beta (β) chains
Amino acids substituted in sickle-cell mutation
Heme group
Alpha (α) chains

From DNA to Protein

Each body cell of an individual contains the same set of DNA molecules, called the genome of that individual. The human genome contains about 20,000 to 25,000 genes, parceled out into 23 pairs of chromosomes. You can see in figure 10.11 that the hemoglobin gene (*Hb*) is located on chromosome 11.

Individual genes are "read" from the chromosomal DNA by enzymes that create an RNA transcript of the gene sequence. After editing out unnecessary bits, this RNA transcript of the hemoglobin gene leaves the nucleus as messenger RNA (mRNA) and is delivered to ribosomes in the cytoplasm. Each ribosome is a tiny protein-assembly plant and uses the nucleotide sequence of the mRNA to determine the amino acid sequence of a particular polypeptide. In the case of beta-hemoglobin, the mRNA encodes a strand of 146 amino acids.

How Proteins Determine the Phenotype

The beta-hemoglobin amino acid chain, which resembles beads on a string in the figure, spontaneously folds into a complex three-dimensional shape. This beta-hemoglobin associates with another beta chain and two alpha chains to form the active four-subunit hemoglobin protein (shown to the left) that binds oxygen in red blood cells. As a general rule, genes influence the phenotype by specifying the kind of proteins present in the body, which determines in large measure how that body looks and functions.

How Mutation Alters Phenotype

A change in the identity of a single nucleotide within a gene, called a mutation, can have a profound effect if the new version of the protein folds differently, as this may alter or destroy its function. For example, how well hemoglobin performs its oxygen transport duties depends on the precise shape that the protein subunits assume when they fold. A change in the sixth amino acid of beta-hemoglobin from glutamic acid to valine causes the hemoglobin molecules to aggregate into long chains, which deform blood cells into a sickle shape that can no longer carry oxygen efficiently. The resulting sickle-cell disease can be fatal, as discussed further in section 10.8.

> **Putting the Concept to Work**
> Explain how a single nucleotide change in the gene for beta-hemoglobin can lead to organ damage and death.

BIOLOGY & YOU

Sickle-Cell Disease. The inherited blood disorder sickle-cell disease occurs among African Americans in 3 of every 1,000 (or about 1 in 375) live births. In its severe form, it affects more than 50,000 people. The first account of what was then called "sickle-cell anemia" was in 1910, when James Herrick, a Chicago physician, described the symptoms of a 20-year-old Black male student from the West Indies. The patient reported shortness of breath and had severe anemia (low red blood cell count). In a 1904 examination of the patient's blood cells under a microscope, Dr. Ernest Iron, Herrick's intern, described "elongated and irregular, sickle-shaped red blood cells." Because these deformed sickle cells can clog blood vessels, organ damage often eventually results. Rap artist Prodigy of the legendary New York rap duo Mobb Deep died of complications arising from sickle-cell disease. Open about battling the disease for years, he was hospitalized for treatment in June 2017 and choked on an egg when hit with a sickle-cell disease crisis.

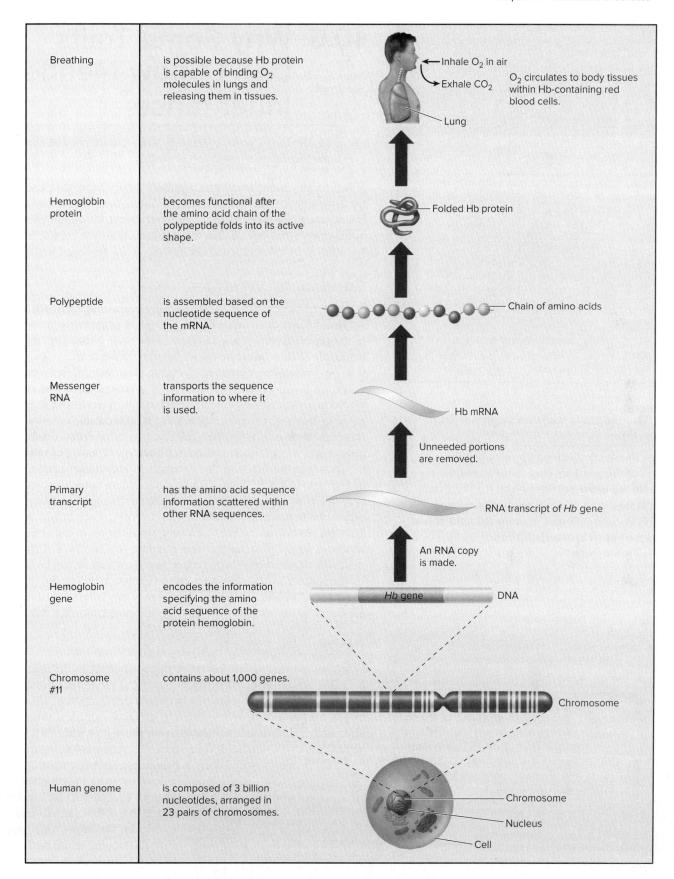

Figure 10.11 The journey from DNA to phenotype.

What an organism is like is determined in large measure by its genes. Here you see how one gene of the 20,000 to 25,000 in the human genome plays a key role in allowing oxygen to be carried throughout your body. The many steps on the journey from gene to trait are the subject of chapters 11 and 12.

10.6 Why Some Traits Don't Show Mendelian Inheritance

> **LEARNING OBJECTIVE 10.6.1** List five factors that can disguise the Mendelian segregation of alleles.

Scientists attempting to confirm Mendel's theory often had trouble obtaining the same simple ratios he had reported. Often, the expression of the genotype is not straightforward. Most phenotypes reflect the action of many genes. Some phenotypes can be affected by alleles that lack complete dominance, are affected by environmental conditions, or are expressed together.

Continuous Variation

When multiple genes act jointly to influence a character such as height or weight, the character often shows a range of small differences. Because all of the genes that play a role in determining these phenotypes segregate independently of each other, we see a gradation in the degree of difference when many individuals are examined. A classic illustration of this sort of variation is seen in **figure 10.12**, a photograph of a college class. The students were placed in rows according to their heights, under 5 feet toward the left and over 6 feet to the right. You can see that there is considerable variation in height in this population of students. We call this type of inheritance **polygenic** (many genes), and we call this gradation in phenotypes **continuous variation.**

How can we describe the variation in a character such as the height of the individuals in **figure 10.12a**? Individuals range from quite short to very tall, with average heights more common than either extreme. What we often do is to group the variation into categories. Each height, in inches, is a separate phenotypic category. Plotting the numbers in each height category produces a histogram, such as that in **figure 10.12b**. The histogram approximates an idealized bell-shaped curve, and the variation can be characterized by the mean and spread of that curve. Compare this to the inheritance of plant height in Mendel's peas: They were either tall or dwarf; no intermediate height plants existed because only one gene controlled that trait.

(a)

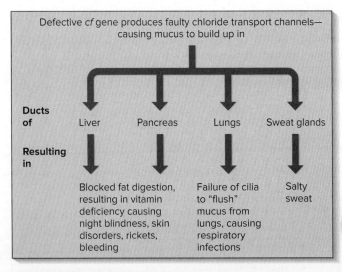

(b)

Figure 10.12 Height is a continuously varying character in humans.

(a) A college professor teaching heredity had the 82 members of the genetics class array themselves on the lawn in the order of their height. (b) This graph illustrates the bell-shaped distribution they created that day. This is quite different from the 3:1 ratio of two alternative types seen in Mendel's peas.

(a) David Hyde/Wayne Falda/McGraw Hill

Pleiotropic Effects

Often, an individual allele has more than one effect on the phenotype. Such an allele is said to be **pleiotropic.** When the pioneering French geneticist Lucien Cuenot studied yellow fur in mice, a dominant trait, he was unable to obtain a true-breeding yellow strain by crossing individual yellow mice with one another. Individuals homozygous for the yellow allele died, because the yellow allele was pleiotropic: one effect was yellow color, but another was a lethal developmental defect. A pleiotropic gene alteration may be dominant with respect to one phenotypic consequence (yellow fur) and recessive with respect to another (lethal developmental defect). In pleiotropy, one gene affects many characters, in marked contrast to polygeny, where many genes affect one character. Pleiotropic effects are difficult to predict, because the genes that affect a character often perform other functions we may know nothing about.

Cystic fibrosis, discussed in chapter 4's "Today's Biology" section *When Membranes Don't Work Right*, is caused by a mutation in a single gene that disrupts the functions of a chloride channel in the plasma membrane. This allele has pleiotropic effects because the chloride channel is found in many different types of cells in the body.

Figure 10.13 Pleiotropic effects of the cystic fibrosis gene, *cf*.

Figure 10.14 **Incomplete dominance.**

In a cross between a red-flowered Japanese four o'clock (genotype $C^R C^R$), and a white-flowered one ($C^W C^W$), neither allele is dominant. The heterozygous progeny have pink flowers and the genotype $C^R C^W$. If two of these heterozygotes are crossed, the phenotypes of their progeny occur in a ratio of 1:2:1 (red:pink:white).

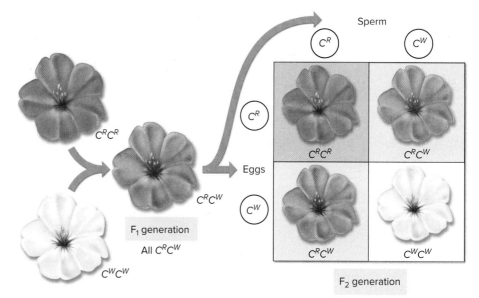

Pleiotropic effects are characteristic of many inherited disorders, such as cystic fibrosis and sickle-cell disease, discussed later in this chapter. In these disorders, multiple symptoms can be traced back to a single gene defect. As shown in figure 10.13, cystic fibrosis patients exhibit overly sticky mucus, salty sweat, liver and pancreas failure, and a battery of other symptoms. All are pleiotropic effects of a single defect, a mutation in a gene that encodes a chloride ion transmembrane channel. In sickle-cell disease, a defect in the oxygen-carrying hemoglobin molecule causes anemia, heart failure, kidney failure, enlargement of the spleen, and many other symptoms.

Incomplete Dominance

Not all alternative alleles are fully dominant or fully recessive in heterozygotes. Some pairs of alleles exhibit **incomplete dominance** and produce a heterozygous phenotype that is intermediate between those of the parents. For example, the cross of red- and white-flowered Japanese four o'clocks described in figure 10.14 produced red-, pink-, and white-flowered F_2 plants in a 1:2:1 ratio—heterozygotes are intermediate in color. This is different than in Mendel's pea plants that didn't exhibit incomplete dominance; the heterozygotes expressed the dominant phenotype.

Epigenetics and Environmental Effects

The degree to which many alleles are expressed depends on the environment. Some alleles are heat-sensitive, for example. Traits influenced by such alleles are more sensitive to temperature or light than are the products of other alleles. The Arctic foxes in figure 10.15, for example, make fur pigment only when the weather is warm. Can you see why this trait would be an advantage for the fox? Imagine a fox that didn't possess this trait and was white all year round. It would be very visible to predators in the summer, standing out against its darker surroundings.

Sometimes the environment can affect genes directly, by adding or removing chemical groups to DNA or the proteins involved in packaging DNA into chromosomes. Called **epigenetic modification,** these changes can accumulate as a person ages and can even be passed from one generation to the next. While most genes are not epigenetically modified, some that influence human behaviors seem to be.

Figure 10.15 **Environmental effects on an allele.**

(a) An Arctic fox in winter has a coat that is almost white, so it is difficult to see the fox against a snowy background. (b) In summer, the same fox's fur darkens to a reddish brown, so that it resembles the color of the surrounding tundra.

(a) Dmitry Deshevykh/E+/Getty Images; (b) BMJ/Shutterstock

Codominance

A gene may have more than two alleles in a population, and in fact most genes possess several different alleles. Often in heterozygotes there isn't a dominant allele; instead, the effects of both alleles are expressed. In these cases, the alleles are said to be **codominant**. Codominance is seen in the color patterning of some animals, such as the "roan" pattern exhibited by some varieties of horses and cattle. A roan animal expresses both white and colored hairs because both alleles are being expressed. The horse in figure 10.16 is exhibiting the roan pattern. It looks like it has gray hairs, but if you were able to examine its coat closely, you would see both white hairs and dark brown hairs.

The ABO Blood Groups. A human gene that exhibits more than one dominant allele is the gene that determines ABO blood type. This gene encodes an enzyme that adds sugar molecules to lipids on the surface of red blood cells. These sugars act as recognition markers for cells in the immune system and are called cell surface antigens. The gene that encodes the enzyme, designated I, has three common alleles: (1) I^B, whose product adds galactose; (2) I^A, whose product adds galactosamine; and (3) i, whose product does not add a sugar.

Different combinations of the three I gene alleles occur in different individuals because each person may be homozygous for any allele or heterozygous for any two. An individual heterozygous for the I^A and I^B alleles produces both forms of the enzyme and adds both galactose and galactosamine to the surfaces of red blood cells. Because both alleles are expressed simultaneously in heterozygotes, the I^A and I^B alleles are codominant. Both I^A and I^B are dominant over the i allele because both I^A or I^B alleles lead to sugar addition and the i allele does not. The different combinations of the three alleles produce four different phenotypes (figure 10.17):

1. Type A individuals add only galactosamine. They are either I^AI^A homozygotes or I^Ai heterozygotes (the three darkest boxes).
2. Type B individuals add only galactose. They are either I^BI^B homozygotes or I^Bi heterozygotes (the three lightest-colored boxes).
3. Type AB individuals add both sugars and are I^AI^B heterozygotes (the two intermediate-colored boxes).
4. Type O individuals add neither sugar and are ii homozygotes (the one white box).

These four different cell surface phenotypes are called the **ABO blood groups**. A person's immune system can distinguish between these four phenotypes. If a type A individual receives a transfusion of type B blood, the recipient's immune system recognizes that the type B blood cells possess a "foreign" antigen (galactose) and attacks the donated blood cells, causing the cells to clump or agglutinate. This also happens if the donated blood is type AB. However, if the donated blood is type O, it contains no galactose or galactosamine antigens on the surfaces of its blood cells and so elicits no immune response to these antigens. For this reason, the type O individual is often referred to as a "universal donor." Because neither galactose nor galactosamine is foreign to type AB individuals (whose red blood cells have both sugars), those individuals ("universal recipients") may receive any type of blood.

Figure 10.16 Codominance in color patterning.

This roan horse is heterozygous for coat color. The offspring of a cross between a white homozygote and a black homozygote, it expresses both phenotypes. Some of the hairs on its body are white and some are black (their color modified in this horse by other genes to a dark brown).

McGraw Hill

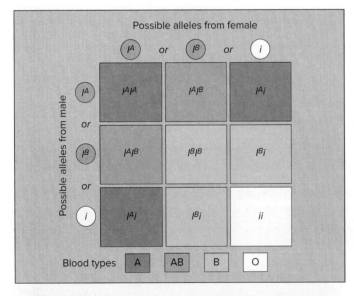

Figure 10.17 Multiple alleles controlling the ABO blood groups.

Three common alleles control the ABO blood groups. The different combinations of the three alleles result in four different blood type phenotypes: type A (either I^AI^A homozygotes or I^Ai heterozygotes), type B (either I^BI^B homozygotes or I^Bi heterozygotes), type AB (I^AI^B heterozygotes), and type O (ii homozygotes).

> **Putting the Concept to Work**
> Is human blood type A dominant over blood type B? Explain.

Chromosomes and Heredity

10.7 Human Chromosomes

LEARNING OBJECTIVE 10.7.1 Contrast aneuploidy with nondisjunction, monosomic with trisomic, and Klinefelter syndrome with Turner syndrome.

Each human somatic cell normally has 46 chromosomes, which in meiosis form 23 pairs. Homologous chromosomes can be identified according to size, shape, and appearance. Of the 23 pairs of human chromosomes, 22 are perfectly matched in both males and females and are called **autosomes.** The remaining pair, the **sex chromosomes,** consist of two similar chromosomes in females and two dissimilar chromosomes in males. In humans, females are designated XX and males are XY. The genes present on the Y chromosome determine "maleness," and therefore humans who inherit the Y chromosome develop into males.

> Recent evidence suggests that a gene called *SRY* for "Sex-determining Region of the *Y* chromosome" may be responsible for the development of "maleness." This is discussed in more detail in section 30.1.

Nondisjunction

Sometimes during meiosis, homologous chromosomes or sister chromatids that paired up during metaphase remain stuck together instead of separating. The failure of chromosomes to separate correctly during either meiosis I or II is called **nondisjunction.** Nondisjunction leads to **aneuploidy,** an abnormal number of chromosomes. The nondisjunction you see in figure 10.18 occurs because the homologous pair of larger chromosomes failed to separate in anaphase I. The gametes that result from this division have unequal numbers of chromosomes. Under normal meiosis (refer back to *Essential Biological Process 9A*), all gametes in figure 10.18 would be expected to have two chromosomes, but as you can see, two of these gametes have three chromosomes, and the others have just one.

Almost all humans of the same sex have the same karyotype (refer back to figure 8.4) simply because other arrangements don't work well. Humans who have lost even one copy of an autosome (called **monosomics**) do not survive development. In all but a few cases, humans who have gained an extra autosome (called **trisomics**) also do not survive. However, five of the smallest chromosomes—those numbered 13, 15, 18, 21, and 22—can be present in humans as three copies and still allow the individual to survive for a time. The presence of an extra chromosome 13, 15, or 18 causes severe developmental defects, and infants with such a genetic makeup die within a few months. In contrast, individuals who have an extra copy of chromosome 21 or, more rarely, chromosome 22, usually survive to adulthood. In such individuals, the maturation of the skeletal system is delayed, so they generally are short and have poor muscle tone. Their mental development is also affected, and children with trisomy 21 or trisomy 22 are always mentally impaired.

Down Syndrome. The developmental defect produced by trisomy 21, an extra copy of chromosome 21 seen in the karyotype in figure 10.19, was first described in 1866 by J. Langdon Down; for this reason, it is called Down syndrome. About 1 in every 750 children exhibits Down syndrome, and the frequency is similar in all racial groups. It is much more

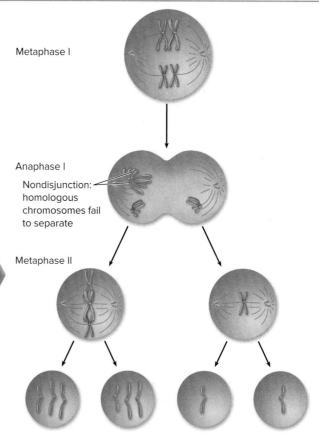

Results in four gametes: two are $n+1$ and two are $n-1$

Figure 10.18 Nondisjunction in anaphase I.

In nondisjunction that occurs during meiosis I, one pair of homologous chromosomes fails to separate in anaphase I, and the gametes that result have one too many or one too few chromosomes. Nondisjunction can also occur in meiosis II, when sister chromatids fail to separate during anaphase II.

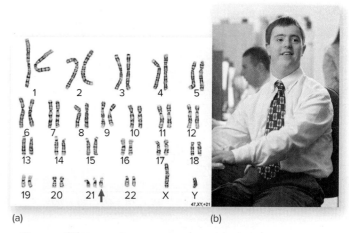

Figure 10.19 Down syndrome.

(a) In this karyotype of a male individual with Down syndrome, the trisomy at position 21 can be clearly seen. (b) A young man with Down syndrome.

(a) ©Director, Karen Swisshelm, PhD, FACMG/The Colorado Genetics Laboratory, Denver, CO; (b) George Doyle/Stockbyte/Veer/Getty Images

Today's Biology

Tracing Your Family History with DNA

For most of us, our family history is a bit uncertain. We can often go back a couple of generations in old family letters and records to figure out where our ancestors came from, but beyond that, most of us don't have a clue. Do you know in what country your great, great grandfather was born? His great, great grandfather? Our unrecorded ancestry is a curtain through which little could be seen—until recently.

What changed was the Human Genome Project, which in 2003 worked out the complete DNA sequence of the human genome. Now any human's DNA sequence can be compared to this reference and the differences noted. The chemistry is very straightforward and simple. A scrap of tissue from inside your mouth yields cells from which DNA can be extracted, amplified, and sequenced. When compared to the reference human genome sequence, any differences reveal mutations that occurred in your ancestors. How do you sort out your past ancestry? The principle is very simple: people with the same ancestors tend to have inherited the same collections of ancient mutations. If your ancestors are from Ireland, then your DNA will contain the same array of differences that other people from Ireland do. Once a lot of people from all over the world have been sampled, any one person can be located on the map with ease.

Autosomal DNA Testing

As you may recall from chapter 4, your cells actually contain three different sources of DNA: 22 pairs of chromosomes called autosomes, one pair of sex chromosomes (usually XX or XY), and a small amount of DNA stored in the cell's mitochondria. As most of your DNA is found in your autosomes, that is where it is easiest to look for any differences from the reference sequence. It can provide information about your ethnicity and reveal the various regions of the world where your ancestors lived over the past several hundred years. Autosomal tests screen for the presence in your DNA of any of a large collection of known single-base DNA mutations (called SNPs, or single-nucleotide polymorphisms). The collection of SNPs that you contain will match some ancestral groups and not others and in this way reveal which are your ancestors. The drawback of the autosomal test is that genetic recombination and crossing over mixes up the combinations of autosomes each generation like shuffling a deck of cards, so it can accurately determine your relatedness to an ancestor only as far back as four to five generations. But within five generations it can be powerful. One study of 200,000 SNPs among Europeans allowed researchers to pinpoint where a new European person was from to within a few hundred kilometers.

Y Chromosome Testing

To look further, we must look to the other two sources of cellular DNA. Y-DNA testing, because it is only available for men, looks exclusively at the paternal side of your family tree. The Y chromosome contains some 200 genes, about 1% of the cell's DNA. Because there is only one Y chromosome in a cell, there is no crossing over in meiosis to mix up combinations of SNPs. Also, the Y chromosome contains

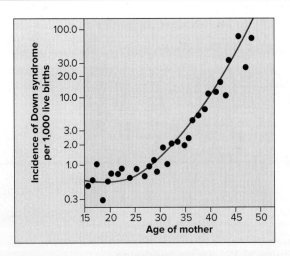

Figure 10.20 Correlation between maternal age and the incidence of Down syndrome.

As women age, the chances they will bear a child with Down syndrome increase. After a woman reaches age 35, the frequency of Down syndrome increases rapidly.

common in children of older mothers. The graph in figure 10.20 shows the increasing incidence in older mothers. In mothers under 30 years old, the incidence is only about 0.6 per 1,000 (or 1 in 1,500 births), whereas in mothers 30 to 35 years old, the incidence doubles to about 1.3 per 1,000 births (or 1 in 750 births). In mothers over 45, the risk is as high as 63 per 1,000 births (or 1 in 16 births). The reason that older mothers are more prone to Down syndrome babies is that all the eggs that a woman will ever produce are present in her ovaries by the time she is born, and as she gets older, they may accumulate damage that can result in nondisjunction.

Nondisjunction Involving the Sex Chromosomes

As noted, 22 of the 23 pairs of human chromosomes are perfectly matched in both males and females and are called autosomes. The remaining pair are the sex chromosomes: X and Y. In humans, as in *Drosophila* (but by no means in all diploid species), females are XX and males XY; any individual with at least one Y chromosome is male. The Y chromosome is highly condensed and bears few functional genes in most organisms. Some of the active genes the Y chromosome does possess are responsible for the features

(a) NetPhotos/Alamy Stock Photo; (b) Martin Shields/Alamy Stock Photo; (c) pictoKraft/Alamy Stock Photo

short segments that have been duplicated many, many times, like a DNA-repeating stutter. Called short tandem repeats, the hundreds of repeats in a segment can gain or lose elements, but this happens only very rarely, a very slow evolutionary change that can help identify ancestral connections that go back many generations.

mtDNA Testing

mtDNA tests ignore the DNA in your nucleus and focus instead on the DNA found in your mitochondria. Because you inherit your mitochondria from your mother, this test explores only the maternal side of your family tree, looking at the approximately 16,500 genetic base pairs of DNA that are found within the mitochondria. Like Y-DNA, the DNA of the mitochondria is not subject to recombination during meiosis and so can provide very precise and accurate information about your ancestry going back a very long time, sometimes 10,000 years or more.

Which Test Is Right for You?

So, how do you go about testing your DNA? These days you can have a commercial company do the entire analysis for you! A wide variety of companies will do the entire analysis for about $100, including MyHeritage, 23andMe, AncestryDNA, and LivingDNA. The companies differ in emphasis, some stressing relatedness to ethnic groups or locating geographic areas where your ancestors came from, whereas others focus on health issues. All of them attempt to provide accurate and objective analysis, and state that they guarantee privacy of your information. Should you take them up on their offer, you will soon find yourself gazing deep into your family's past.

---

associated with "maleness." Individuals who gain or lose a sex chromosome do not generally experience the severe developmental abnormalities caused by changes in autosomes. Such individuals may reach maturity, but with somewhat abnormal features.

Nondisjunction of the X Chromosome. When X chromosomes fail to separate during meiosis, some of the gametes that are produced possess both X chromosomes and so are XX gametes; the other gametes that result from such an event have no sex chromosome and are designated "O."

Figure 10.21 shows what happens if gametes from X chromosome nondisjunction combine with sperm. If an XX egg combines with an X sperm, the resulting XXX zygote develops into a female who is taller than average, but other symptoms can vary greatly. Some are normal in most respects, others may have lower reading and verbal skills, and still others are mentally retarded. If an XX egg combines with a Y sperm, the XXY zygote develops into a sterile

Figure 10.21 Nondisjunction of the X chromosome.

Nondisjunction of the X chromosome can produce sex chromosome aneuploidy—that is, abnormalities in the number of sex chromosomes.

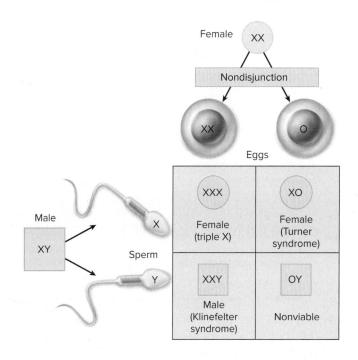

male who has many female body characteristics and, in some cases, diminished mental capacity. This condition, called *Klinefelter syndrome,* occurs in about 1 in 500 male births.

If an O egg fuses with a Y sperm, the OY zygote is nonviable and fails to develop further because humans cannot survive when they lack the genes on the X chromosome. If an O egg fuses with an X sperm, the XO zygote develops into a sterile female of short stature, with a webbed neck and immature sex organs that do not undergo changes during puberty. The mental abilities of XO individuals are normal in verbal learning but lower in nonverbal/math-based problem solving. This condition, called *Turner syndrome,* occurs roughly once in every 5,000 female births.

Nondisjunction of the Y Chromosome. The Y chromosome can also fail to separate in meiosis, leading to the formation of YY sperm. When these sperm combine with X eggs, the XYY zygotes develop into fertile males of normal appearance. The frequency of the XYY genotype is about 1 per 1,000 newborn males.

> **Putting the Concept to Work**
> Is it possible for a female to have Klinefelter syndrome? Explain.

10.8 The Role of Mutations in Human Heredity

> **LEARNING OBJECTIVE 10.8.1** Contrast the inheritance of hemophilia and sickle-cell disease.

Human heredity is a matter of inheriting gene mutations.

Mutations

The proteins encoded by most of your genes must function in a very precise fashion for you to develop properly and for the many complex processes of your body to function correctly. Unfortunately, genes sometimes sustain damage or are copied incorrectly. We call these accidental changes in genes **mutations.** Mutations occur only rarely because your cells monitor your genes and attempt to correct any damage they encounter. Still, some mutations get through. Many of them are bad for you in one way or another. It is easy to see why. Mutations hit genes at random—imagine that you randomly

> Mutations can occur in somatic cells or in germ-line cells. Both can have dramatic effects, but as discussed in section 11.3, only mutations in germ-line tissues are passed on to offspring and affect future generations.

TABLE 10.3	Some Important Genetic Disorders				
Disorder	**Symptom**	**Defect**	**Dominant/ Recessive**	**Frequency Among Human Births**	
Cystic fibrosis	Mucus clogs lungs, liver, and pancreas	Failure of chloride ion transport mechanism	Recessive	1/2,500 (Caucasians)	
Sickle-cell disease	Poor blood circulation	Abnormal hemoglobin molecules	Recessive	1/625 (African Americans)	
Tay-Sachs disease	Deterioration of central nervous system in infancy	Defective enzyme (hexosaminidase A)	Recessive	1/3,500 (Ashkenazi Jews)	
Phenylketonuria (PKU)	Brain fails to develop in infancy	Defective enzyme (phenylalanine hydroxylase)	Recessive	1/12,000	
Hemophilia	Blood fails to clot	Defective blood-clotting factor VIII	Sex-linked recessive	1/10,000 (Caucasian males)	
Huntington's disease	Brain tissue gradually deteriorates in middle age	Production of an inhibitor of brain cell metabolism	Dominant	1/24,000	
Muscular dystrophy (Duchenne)	Muscles waste away	Degradation of myelin coating of nerves stimulating muscles	Sex-linked recessive	1/3,700 (males)	
Congenital hypothyroidism	Increased birth weight, puffy face, constipation, lethargy	Failure of proper thyroid development	Recessive	1/1,000 (Hispanics) 1/700 (Native Americans)	
Hypercholesterolemia	Excessive cholesterol levels in blood, leading to heart disease	Abnormal form of cholesterol cell surface receptor	Dominant	1/500	

changed the number of a part on the design of a jet fighter. Sometimes it won't matter critically—a seat belt becomes a radio, say. But what if a key rivet in the wing becomes a roll of toilet paper? The chance of a random mutation in a gene improving the performance of its protein is about the same as that of a randomly selected part making the jet fly faster.

Human Hereditary Disorders. Most mutations are rare in human populations. Almost all result in recessive alleles, and so they are not eliminated from the population by evolutionary forces—because they are not expressed in most individuals (heterozygotes) in which they occur. Do you see why they occur mostly in heterozygotes? Because mutant alleles are rare, it is unlikely that a person carrying a copy of the mutant allele will marry someone who also carries it. Instead, he or she will typically marry someone homozygous dominant, and so none of their children would be homozygous for the mutant allele. While most mutations are harmful to normal functions and are usually recessive, that is not to say all mutations are undesirable; some mutations can lead to enhanced function. Nor are all mutations recessive; rarely, they can occur as dominant alleles.

In some cases, particular mutant alleles have become more common in human populations. In these cases, the harmful effects that they produce are called *genetic disorders*. Some of the most common genetic disorders are listed in **table 10.3**.

Analyzing Pedigrees. To study human heredity, scientists look at the results of crosses that have already been made. They study family trees, or **pedigrees,** to identify which relatives exhibit a trait. Then they can often determine whether the expression of the gene is dominant or recessive (**figure 10.22**), and whether the gene producing the trait is **sex-linked** (that is, located on the X chromosome) or autosomal (**figure 10.23**). Frequently, the pedigree will also help an investigator infer which individuals in a family are homozygous and which are heterozygous for the allele specifying the trait.

In a pedigree, females are indicated with circles and males with squares. The lines connect the parents and display the offspring. Solid shapes indicate an individual affected by a disorder; half-filled shapes indicate a carrier (someone who is heterozygous); open shapes indicate an individual not affected.

Hemophilia: A Sex-Linked Trait

Blood in a cut clots as a result of the polymerization of protein fibers circulating in the blood. A dozen proteins are involved in this process, and all must function properly for a blood clot to form. A mutation causing any of these proteins to lose their activity leads to a form of **hemophilia**, a hereditary condition in which the blood clots slowly or not at all.

Hemophilias are recessive disorders, expressed only when an individual does not possess any copy of the normal allele and so cannot produce one of the proteins necessary for clotting. Most of the genes that encode the blood-clotting proteins are on autosomes, but two (designated *VIII* and *IX*) are on the X chromosome. These two genes are thus sex-linked. Any male who inherits a mutant allele will develop hemophilia because his other sex chromosome is a Y chromosome that lacks any alleles of those genes.

The most famous instance of hemophilia, often called the Royal hemophilia, is a sex-linked form that arose in the royal family of England. This hemophilia was caused by a mutation in gene *IX* that occurred in one of the parents of Queen Victoria of England (1819–1901). The

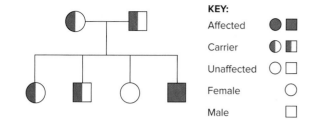

Figure 10.22   A general pedigree.

This pedigree is consistent with the inheritance of a recessive trait.

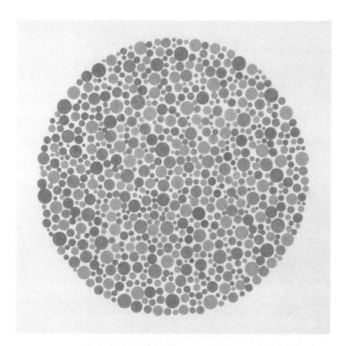

Figure 10.23   A sex-linked trait: Color blindness.

Test samples called Ishihara plates are used to determine if a person is color blind. The test plates contain different colored dots arranged to reveal a shape, often a number. Individuals who are red-green color blind cannot see the number in the test plate shown above, as all the dots appear the same color.

PRISMA ARCHIVO/Alamy Stock Photo

Figure 10.24 The Royal hemophilia pedigree.

Queen Victoria's daughter Alice introduced hemophilia into the Russian and Prussian royal houses, and her daughter Beatrice introduced it into the Spanish royal house. Victoria's son Leopold, himself a victim, also transmitted the disorder in a third line of descent. In this photo, Queen Victoria of England is surrounded by some of her descendants in 1894. Standing behind Victoria and wearing feathered boas are two of Victoria's granddaughters, Alice's daughters: Princess Irene of Prussia *(right)* and Alexandra *(left)*, who would soon become czarina of Russia. Both Irene and Alexandra were also carriers of hemophilia.

Bettmann/Getty Images

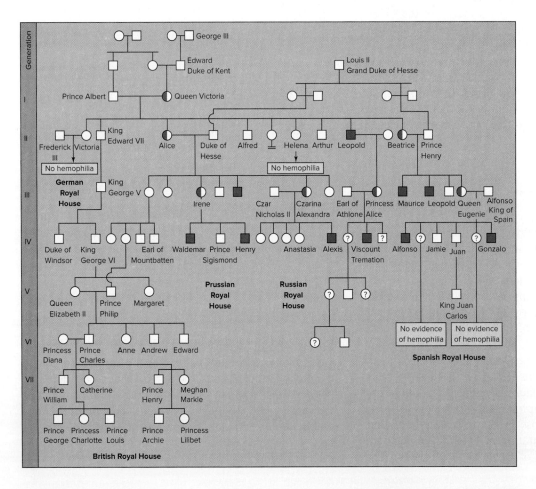

pedigree in **figure 10.24** shows that in the six generations since Queen Victoria, 10 of her male descendants have had hemophilia (the solid squares). The present British royal family has escaped the disorder because Queen Victoria's son, King Edward VII, did not inherit the defective allele, and all the subsequent rulers of England are his descendants. Three of Victoria's nine children did receive the defective allele, however, and they carried it by marriage into many of the other royal families of Europe. As you will learn in chapter 13, this type of hemophilia has been successfully treated using gene therapy.

Sickle-Cell Disease: A Recessive Trait

Sickle-cell disease is a recessive hereditary disorder. Its inheritance is shown in the pedigree in **figure 10.25**, in which affected individuals are homozygous, carrying two copies of the mutated gene. Affected individuals have defective molecules of hemoglobin, the protein within red blood cells that carries oxygen. Consequently, these individuals are unable to properly transport oxygen to their tissues.

Sickle-Cell Hemoglobin. The hemoglobin in the defective red blood cells differs from that in normal red blood cells in only one of beta-hemoglobin's 146 amino acid subunits. In the defective hemoglobin, the amino acid valine replaces a glutamic acid at a single position in each of the two beta-subunit proteins. Interestingly, the position of the change is far from the active site of hemoglobin where the iron-bearing heme group binds oxygen. Instead, the change occurs on the outer corners of the 4-subunit hemoglobin molecule. Why then is the result so catastrophic? The sickle-cell mutation puts a very nonpolar amino acid (valine) on the surface of the hemoglobin protein, creating a "sticky patch" that sticks to other such patches—nonpolar amino acids tend to associate with one another in polar environments such as water. The defective hemoglobin molecules adhere to one another in chains. These hemoglobin chains form stiff, rodlike structures that result in sickle-shaped red blood cells (see **figure 10.25**). As a result of their stiffness and irregular shape, these cells have difficulty moving through the smallest blood vessels; they tend to accumulate in those vessels and form clots. People who have large proportions of sickle-shaped red blood cells tend to have intermittent illness and a shortened life span.

Resistance to Malaria. Individuals heterozygous for the sickle-cell allele are generally indistinguishable from normal persons. However, some of their red blood cells show the sickling characteristic when they are exposed to low levels of oxygen. The allele responsible for sickle-cell disease is particularly common among people of African descent because the sickle-cell allele is more common in Africa. About 9% of African Americans are heterozygous for this allele, and about 0.2% are homozygous and therefore have the disorder. In some groups of people in Africa, up to 45% of all individuals are heterozygous for this allele, and fully 6% are homozygous and express the disorder. What factors determine the high frequency of sickle-cell disease in Africa? It turns out that heterozygosity for the sickle-cell allele increases resistance to malaria, a common and serious disease in central Africa. Comparing the two maps shown in **figure 10.26**, you can see that the area of the sickle-cell trait matches well with the incidence of malaria. The interactions of sickle-cell disease and malaria are discussed further in chapter 14.

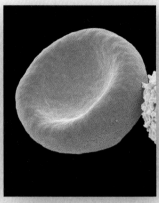

Sickled red blood cell | Normal red blood cell

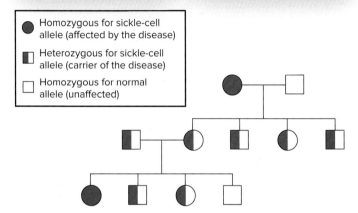

Figure 10.25 Inheritance of sickle-cell disease.

(left) Stocktrek Images/Getty Images; (right) STEVE GSCHMEISSNER/Science Photo Library/Alamy Stock Photo

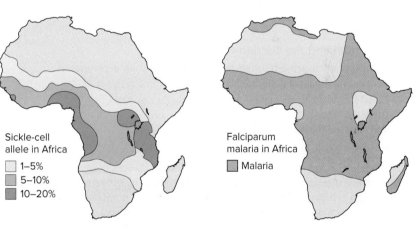

Figure 10.26 The sickle-cell allele confers resistance to malaria.

The distribution of sickle-cell disease closely matches the occurrence of malaria in central Africa. This is not a coincidence. The sickle-cell allele, when heterozygous, confers resistance to malaria, a very serious disease.

Putting the Chapter to Work

1 Mendel used a testcross to determine if plants were homozygous or heterozygous.

If two purple plants (purple being the dominant phenotype over white) were crossed, resulting in the F_1 and F_2 generations of all purple plants, would you conclude the parent plants from the F_1 generation were homozygous or heterozygous?

2 Mice have two hair lengths: short and long (short, S, being the dominant trait).
   a. Demonstrate your understanding of the following terms by providing the allele representation of the following:
      Heterozygous
      Homozygous recessive
      Homozygous dominant
   b. Perform a monohybrid cross of the homozygous recessive mouse with the homozygous dominant mouse and provide the phenotypes and genotypes of the offspring in the F_1 generation.

3 The Chesapeake Bay retriever is a large dog found in the sporting group. The dominant fur type of this dog breed is dark brown (*B*) and wavy (*W*); the recessive fur type is tan (b) and straight (w). Demonstrate your understanding of dihybrid cross by providing the ratios and phenotypes of the F_2 generation if a brown wavy-haired (homozygous dominant) was mated with a tan straight-haired retriever (homozygous recessive) to create the F_1 generation.

Retracing the Learning Path

Mendel

10.1 Mendel and the Garden Pea
1. Mendel studied heredity using the garden pea and an experimental system that included counting his results.
2. Mendel crossed plants that were truebreeding for easily scored alternative traits, then allowed the offspring of these crosses to self-fertilize and counted the numbers of each type of offspring.

10.2 What Mendel Observed
1. In Mendel's experiments, the first (F_1) generation plants all expressed the same alternative form, called the dominant trait. In the second (F_2) generation, 3/4 of the offspring expressed the dominant trait and 1/4 expressed the other form, called the recessive trait. Mendel found this 3:1 ratio in the F_2 generation in all the seven traits he studied. Mendel then found that this 3:1 ratio was actually a 1:2:1 ratio—1 true-breeding dominant: 2 not-true-breeding dominant: 1 true-breeding recessive.

10.3 Mendel Proposes a Theory
1. Mendel's theory of heredity proposes that characteristics are passed from parent to offspring, one version (called an allele) inherited from each parent. If both of the alleles are the same, the individual is homozygous for the trait. If the individual has one dominant and one recessive allele, it is heterozygous for the trait. An individual's alleles are its genotype, and the expression of those alleles is its phenotype.
2. A Punnett square can be used to predict the probabilities of inheriting certain genotypes and phenotypes in the offspring of a cross. The alleles are assigned letters: an uppercase letter for the dominant allele and a lowercase letter for the recessive allele.
3. A testcross determines the genotype of an individual exhibiting the phenotype of a dominant trait by mating the individual of unknown genotype with an individual that is homozygous recessive and seeing if there are any offspring that do not exhibit the dominant trait.

10.4 Mendel's Laws
1. Mendel's law of segregation states that alleles are distributed into gametes so that half of the gametes will carry one copy of a trait and the remaining gametes carry the other copy of the trait. Mendel's law of independent assortment states that the inheritance of one trait does not influence the inheritance of other traits. Genes located on different chromosomes are inherited independent of each other.

From Genotype to Phenotype

10.5 How Genes Influence Traits
1. Genes determine phenotype because DNA encodes the amino acid sequences of proteins, and proteins are the outward expression of genes.

10.6 Why Some Traits Don't Show Mendelian Inheritance
1. Not all traits exhibit the inheritance patterns outlined by Mendel. Continuous variation results when more than one gene contributes in a cumulative way to a phenotype, resulting in a continuous array of phenotypes. This pattern of inheritance is called polygenic. Pleiotropic effects result when one gene influences more than one trait. Incomplete dominance results when heterozygous individuals express a phenotype that is intermediate between the dominant and recessive phenotypes. The expression of some genes is influenced by environmental factors, such as the changing of fur color triggered by heat-sensitive alleles. Codominance occurs when there isn't a dominant allele—both alleles are expressed equally in heterozygotes.

Chromosomes and Heredity

10.7 Human Chromosomes
1. Humans have 23 pairs of homologous chromosomes, for a total of 46 chromosomes. They have 22 pairs of autosomes and one pair of sex chromosomes. Nondisjunction occurs when sister chromatids or homologous pairs fail to separate during meiosis, resulting in gametes with too many or too few chromosomes. Nondisjunction of autosomes is usually fatal, Down syndrome being an exception, but the effects of nondisjunction of sex chromosomes are less severe.

10.8 The Role of Mutations in Human Heredity
1. Mutations can lead to genetic disorders such as hemophilia.

Inquiry and Analysis

Why Woolly Hair Runs in Families

The boy in the photo on the right does not cut his hair. His hair breaks off naturally as it grows, keeping it from getting long. Other members of his family have the same sort of hair, suggesting it is a hereditary trait. Because of its curly, fuzzy texture, this trait has been given the name "woolly hair."

While the woolly hair trait is rare, it flares up in certain families. The extensive pedigree below (drawn curved to fit the large families produced by the second and subsequent generations) records the incidence of woolly hair in five generations (indicated by the Roman numerals on the left) of a Norwegian family. As is the convention, affected individuals are indicated by solid symbols, with circles indicating females and squares indicating males. The pedigree will provide you with all the information you need to discover how this trait is inherited within human families.

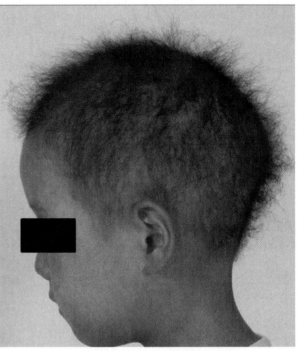

Courtesy Yutaka Shimomura, MD, PhD Biochemistry, Genetics and Molecular Biology, "Current Genetics in Dermatology", book edited by Naoki Oiso, ISBN 978-953-51-0971-6, Published: February 6, 2013. http://www.intechopen.com/books/current-genetics-in-dermat

Analysis

1. **Applying Concepts** In the diagram below, how many individuals are documented? Are all of them related?
2. **Interpreting Data**
   a. Does the woolly hair trait appear in both sexes equally?
   b. Does every woolly hair child have a woolly hair parent?
   c. What percentage of the offspring born to a woolly haired parent are also woolly haired?
3. **Making Inferences**
   a. Is woolly hair sex-linked or autosomal?
   b. Is woolly hair dominant or recessive?
   c. Is the woolly hair trait determined by a single gene or by several?
4. **Drawing Conclusions**
   a. How many copies of the woolly hair allele are necessary to produce a detectable change in a person's hair?
   b. Are there any woolly hair homozygous individuals in the pedigree? Explain.

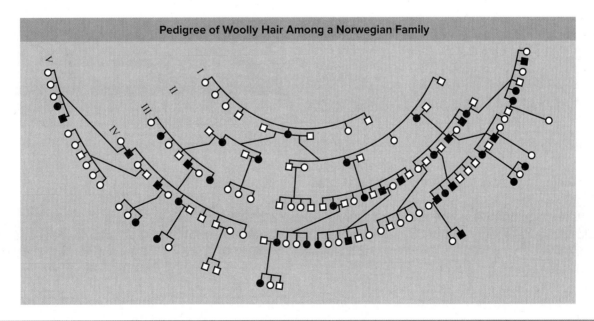

Pedigree of Woolly Hair Among a Norwegian Family

11 DNA: The Genetic Material

LEARNING PATH ▼

Genes Are Made of DNA
1. Discovering the Structure of DNA

DNA Replication
2. How DNA Copies Itself

Altering the Genetic Message
3. Mutation

Editing Genes
4. Gene Editing with CRISPR
5. Gene Drives

Using DNA: Crime Scene Investigation

Television programs marketed as entertainment are sometimes surprisingly educational, none more than the long-popular CSI programs. The letters CSI stand for Crime Scene Investigation. In the spirit of this sort of drama-enhanced science education, we will begin this chapter as detectives, tagging along on a CSI investigation, one in which DNA—the subject of this chapter—plays a central role.

It's Murder

Our episode begins with a murder scene. Ten-month-old baby Joey is crying in the playpen, their mother dead on the floor with a bullet in her right temple. A blood-splattered revolver lies by the body. A sad suicide, it would seem—but the CSI investigators note that the dead woman's hands are clean. Spattered with her blood, the revolver could not have been held by her hand without it becoming spattered as well, so clearly she was not holding the gun when it was fired. Not suicide, then, but murder.

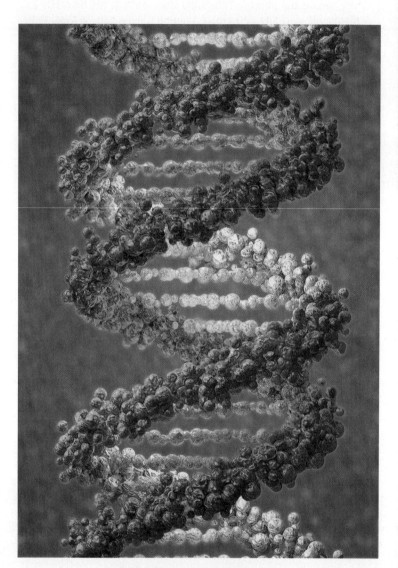

THE DNA MOLECULE often provides a key tool in criminal investigations.
Michael Dunning/Getty Images

A Virgin Birth?

So far, the crime has seemed unexceptional, but the medical examiner soon changes that. The autopsy reveals the dead woman is a virgin! Her hymen is still

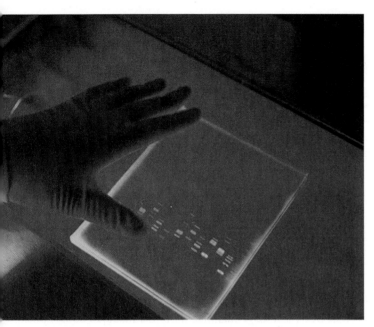
Eurelios/Science Source

Andrew Brookes/Getty Images

intact, which means she never had intercourse. The scar on her abdomen confirms that she gave birth to Joey by Caesarean section. Was this a miraculous conception? No. The hymen presents no barrier to semen, and she could easily have gotten pregnant from "playing around" without full penetration.

So Who's the Father?

The CSI investigators sample the child's and victim's blood, and request a cheek swab from the boyfriend, so that they can compare DNAs to confirm that he is indeed the father of the dead woman's child. Your DNA is like a molecular fingerprint, a calling card that reveals your true identity and your relationships with other people.

As for the boyfriend's DNA: It's not a match! Whoever fathered the dead woman's child, it wasn't her boyfriend. You have to admit, this sure is an unusual little family: Mom's a virgin, and Daddy isn't the daddy.

Next Question: Who's the Mother?

Time to step back and look at the DNA evidence with a more jaundiced eye. Who hasn't the CSI team checked? The dead woman. When the murder victim's DNA profile is examined, all bets are off. Her DNA doesn't match the DNA of her son!

The dead woman was in fact a surrogate mother—the embryo that was to become baby Joey began life as a leftover fertilized egg from a fertility clinic. His murdered birth mother had "adopted" him nine months before he was born.

A New Suspect

Why then the murder? The CSI team collected the clothing worn recently by the biological parents of the embryo to check for bloodstains. They are deeply offended to be considered suspects. They had gone to the clinic because they wanted a child, they said, but not enough to commit murder. Now, of course, they will take custody of Joey as his biological parents.

The CSI team collects all the clothing in the laundry basket and finds a size 14 blouse—much larger than the dead woman wore—with GSR (gun shot residue) adhering to it. This blouse belongs to the biological mother's own mother. The long journey of crime scene investigation is over, this murder solved with the help of DNA.

Genes Are Made of DNA

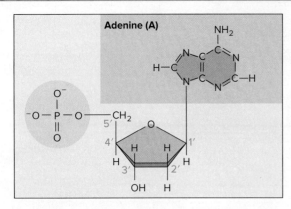
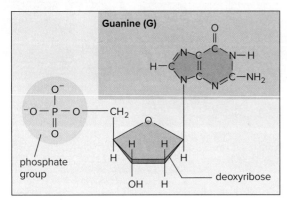
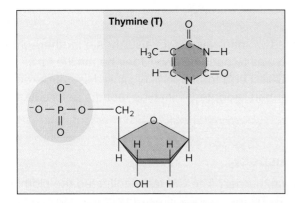
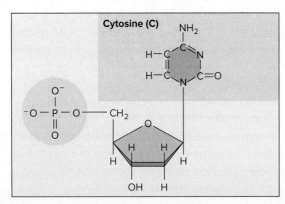

Figure 11.1 The four nucleotide subunits that make up DNA.

The nucleotide subunits of DNA are composed of three parts: a central 5-carbon sugar called deoxyribose, a phosphate group, and an organic, nitrogen-containing base.

11.1 Discovering the Structure of DNA

> **LEARNING OBJECTIVE 11.1.1** Explain how Watson and Crick's proposed structure of DNA explained Chargaff's rule.

After Mendel's work in the 1800s, a variety of different experiments in the first half of the 1900s showed that genes are composed of DNA.

As it became clear that DNA stored the hereditary information, researchers began to question how a molecule like DNA could carry out the complex function of inheritance. Scientists at that time did not know what the DNA molecule looked like.

The Chemical Structure of DNA

We now know that DNA is a long, chainlike molecule made up of subunits called **nucleotides**. As you can see in figure 11.1, each nucleotide has three parts: a central sugar called deoxyribose to which a phosphate (PO_4) group and an organic nitrogen-containing base are attached. The sugar (lavender pentagon structure) and the phosphate group (yellow-circled structure) are the same in every nucleotide of DNA. However, there are four different kinds of bases: two large ones with double-ring structures and two small ones with single rings. The large bases, called **purines**, are **A** (adenine) and **G** (guanine). The small bases, called **pyrimidines**, are **C** (cytosine) and **T** (thymine). The carbon atoms that make up the sugar of the deoxyribose are numbered from 1' to 5', as shown in the top panel. The phosphate group binds to the 5' carbon, and the organic base binds to the 1' carbon.

Early in the analysis of DNA, a key observation was made by Erwin Chargaff. He noted that DNA molecules always had equal amounts of purines and pyrimidines. In fact, with slight variations due to imprecision of measurement, the amount of A always equals the amount of T, and the amount of G always equals the amount of C. This observation (A = T, G = C), known as Chargaff's rule, suggested that DNA had a regular structure.

Discovering the Shape of the DNA Molecule

In 1950, the British chemist Maurice Wilkins carried out the first X-ray diffraction experiments on DNA. In his experiments, DNA fibers bombarded with X-ray beams created a pattern on photographic film that looked like the ripples created by tossing a rock into a smooth lake (figure 11.2a). In 1951, Wilkins concluded from his research that the DNA molecule was composed of two strands, wound around one another in the shape of a coiled spring or a corkscrew, a form called a helix.

In January 1953, Wilkins shared a particularly clear X-ray diffraction photograph taken in 1952 in his lab by post-doc Rosalind Franklin with two researchers at Cambridge University: Francis Crick and James Watson. Using Tinkertoy-like models of the bases, Watson and Crick deduced the structure of DNA (figure 11.2b): The DNA molecule is a **double helix,** a structure that resembles a winding staircase (figure 11.2c). The sugar and phosphate groups form the stringers of

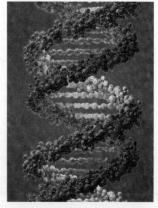

Model of DNA double helix
Michael Dunning/Getty Images

the staircase, and the bases of the nucleotides form the steps. The significance of Chargaff's rule is now clear—every bulky purine on one strand is paired with a slender pyrimidine on the other strand. Specifically, A pairs with T, and G pairs with C.

> **Putting the Concept to Work**
> In a DNA double helix, why doesn't the purine adenine pair with the pyrimidine cytosine?

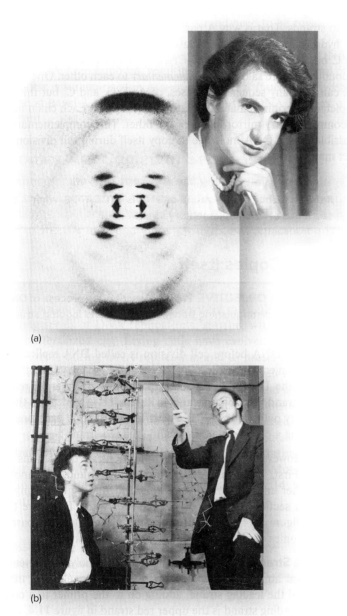

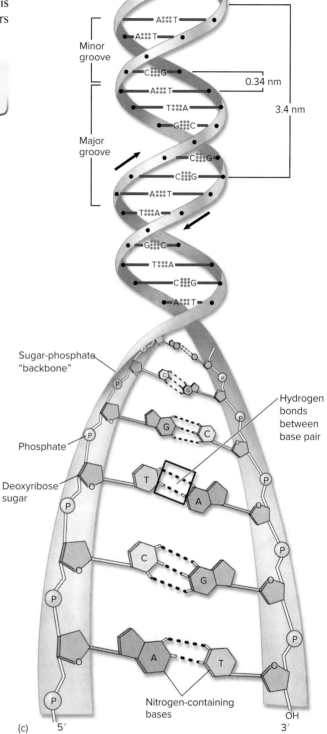

Figure 11.2 The DNA double helix.

(a) This X-ray diffraction photograph was made in 1952 by Rosalind Franklin (inset) in the laboratory of Maurice Wilkins. (b) Using it, Watson and Crick deduced in 1953 that the two strands of a DNA molecule pair with one another in a highly specific way determined by hydrogen bonding. James Watson (seated and peering up at their homemade model of the DNA molecule) was a young American postdoctoral student, and Francis Crick (pointing) was an English post-doc. For their discovery of how the strands of the DNA molecule pair with one another, Watson and Crick were awarded the Nobel Prize in 1962; the other portion of this Nobel Prize was awarded to Maurice Wilkins for his use of X-ray diffraction to discover in 1951 that the DNA molecule was a double helix. (c) The dimensions of the double helix were revealed by the X-ray diffraction studies. In a DNA duplex molecule, only two base pairs are possible: adenine (A) with thymine (T) and guanine (G) with cytosine (C). A G–C base pair has three hydrogen bonds; an A–T base pair has only two.

(a—left) ©Omikron/Science Source; (a—right) National Library of Medicine/Science Source; (b) A. Barrington Brown/Science Source

Editing Genes

11.4 Gene Editing with CRISPR

> **LEARNING OBJECTIVE 11.4.1** Explain how CRISPR is used to edit genes.

Discovery of CRISPR

The cute fellow you see in the photo had his genes edited by researchers, an ability to reach in and adjust the gene message of primates undreamed of until recent advances. This new gene editing technology is having a profound impact on society. Here, briefly, is how it was discovered.

Discovery of a Very Odd DNA Sequence. The story of this remarkable new advance starts quite a while back, over 30 years ago in 1987. A Japanese biologist studying the DNA sequences of bacteria observed an odd repeating pattern within the DNA sequence at one end of a bacterial gene: A sequence of several dozen DNA bases was followed by the same sequence in reverse, then 30 seemingly random bases of "spacer" DNA. This three-part pattern was repeated, with different sequences of the random spacer DNA, again and again. RNA molecules copied from a DNA sequence repeated in reverse (called palindromes) fold back on themselves. This looped RNA can then bind to proteins, the loop's sequence determining to which protein it attaches.

The Sequence Binds a DNA-Cutting Enzyme. When the RNA loops produced by the palindromes discovered by the Japanese researcher were examined by other researchers, it turned out these RNA loops bind to a DNA-cutting enzyme (technically called an endonuclease). There are several varieties of these endonucleases, with uninspired names such as Cas9 and Cpf1; each variety cuts DNA in a slightly different way, but all bind to palindromic RNA loops.

The Sequence Targets Viral DNA for Cutting. The next step in this rapid journey of discovery happened in 2005, when genome sequence comparisons made from databases stored on the Internet revealed that the 30-base spacer DNA sequences originally identified by the Japanese researcher, thought to be random, actually were not random at all—they matched DNA sequences found in the genomes of viruses that infect and kill bacteria. The researcher had, without intending to, discovered a weapon that bacteria use to fight viruses! How does this work? First, the "spacer" RNA binds invading virus DNA of matching sequence, then the DNA-cutting enzyme attached to the loop cuts up the virus DNA.

Re-aiming at a New Target. Now, quickly, came the key advance that will affect all of us. It turns out that, using now-standard gene engineering approaches, any 30-base sequence can be substituted for the spacer sequence. Why is this the key advance? Because it allows the researcher to target ANY gene for modification or destruction! This tool, called **CRISPR** (the initials of a very awkward name: "**C**lustered **R**egularly **I**nterspersed **S**hort **P**alindromic **R**epeats"), powerful and easy to use, is rapidly altering the scientific landscape. Within a few years, a wide variety of animal cells have been CRISPR-modified, including human ones.

Courtesy Yongchang Chen, PhD of the following study: Niu Y, Shen B, Cui Y, Chen Y, Wang J, Wang L, Kang Y, Zhao X, Si W, Li W, Xiang AP, Zhou J, Guo X, Bi Y, Si C, Hu B, Dong G, Wang H, Zhou Z, Li T, Tan T, Pu X, Wang F, Ji S, Zhou Q, Huang X, Ji W, Sha J. "Generation of gene-modified cynomolgus monkey via Cas9/RNA-mediated gene targeting in one-cell embryos." Cell. 2014 Feb 13; 156(4):836–43.

Using CRISPR to Edit Genes

CRISPR allows an investigator to "nick" a particular gene, but how does this then allow the investigator to edit the gene, changing what it says? When CRISPR-associated Cas9 or Cpf1 cuts a target gene sequence, this creates a gap that the cell quickly fills in, using its DNA repair machinery to insert a new sequence of nucleotides into the gap and so restoring the genome's integrity. In a diploid organism like a human, the alternative homolog chromosome usually provides a template for the repair machinery to copy. However, if the investigator floods the cell with many, many copies of a DNA sequence of the investigator's choosing, the cell's DNA repair enzymes will then be induced to select and insert one of these copies into the CRISPR-created gap. The result? The target gene has been "edited," now containing the DNA sequence provided by the investigator. In the few years since gene editing and base editing approaches have been developed, investigators have used CRISPR systems to treat many diseases and disorders.

In mice, which are easily studied in the laboratory, investigators have already succeeded in using CRISPR to "correct" the single-base gene mutations responsible for sickle-cell disease, muscular dystrophy, and cystic fibrosis. Work on these and other disorders in humans is proceeding at a rapid pace. In coming years, this novel new tool is certain to have a major impact on human health.

Using CRISPR Gene Editing to Treat Human Disease

Curing AIDS. Researchers have used CRISPR to target a gene needed for successful infection of AIDS patients. As you will learn in chapter 16, proteins called receptors on the surface of human cells are required for viruses to enter, acting a bit like doors to which the virus is a key. In the case of the HIV virus that causes AIDS, the entry port is a protein called the CCR_5 receptor. Protein spikes protruding from the HIV virus (photo) fit into the CCR_5 receptor, unlocking the door. In a highly controversial experiment, a Chinese doctor used CRISPR to disable CCR_5 in human embryos whose father had HIV. Used in *in vitro* fertilization, these CRISPR-modified embryos developed into healthy twin girls born in November 2018, the first gene-edited humans.

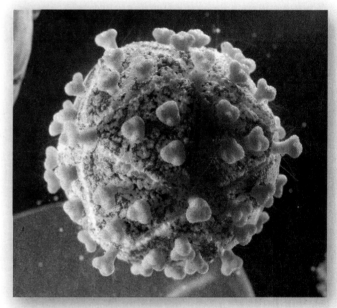

Ian Cuming/Alamy Stock Photo

Organ Transplants from Pigs. A particularly powerful aspect of CRISPR is that it can be used to alter ALL copies of a particular gene. Biomedical researchers have taken advantage of CRISPR's wide reach to solve a long-standing problem facing patients requiring organ transplants: There are three times as many patients needing transplants than there are available human organs. Oddly enough, pig organs would work well for many transplants—except that pigs contain a host of RNA viruses embedded within their chromosomes (called "porcine endogenous retroviruses," or *PERVS*) that may be harmful to humans. How many *PERVS* in a pig's cells? Fully 62, too many to remove using standard approaches. But using CRISPR, researchers in 2017 succeeded in eradicating all 62 in one step. Working with pig embryos, the researchers used CRISPR to search out each copy of the *PERV* gene in embryos and, using the Cas9 endonuclease, snipped them out. Transferred to surrogate sows, these embryos were born as PERV-inactivated piglets. As adults, these pigs will be a suitable source of organs for human transplant.

Luhan Yang Xinhua News Agency/Newscom

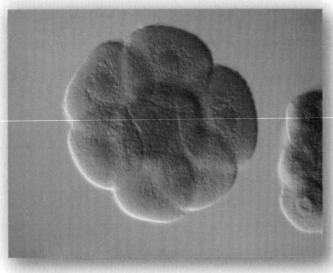

Manfred Kage/Science Source

Correcting Disease-Causing Mutations. Recently, scientists for the first time have edited genes in human embryos to repair a disease-causing mutation. The mutation, called *MYBPC3,* causes the heart muscle to thicken and is the leading cause of sudden death in young athletes. The mutation is dominant, which as you will recall from chapter 10 means that a child need only inherit one copy of the mutated gene to be affected. In the 2017 experiment, sperm carrying the *MYBPC3* mutation were allowed to fertilize human eggs in culture, and the resulting embryos were immediately treated with CRISPR carrying the Cas9 DNA-cutting machinery. Of 58 human embryos treated, 42 were successfully edited to contain two normal copies of the *MYBPC3* gene. In each case, CRISPR filled in the blank created by Cas9, using for a template the healthy copy of the *MYBPC3* gene inherited from the egg donor. In the photograph, you can see some of the resulting edited embryos at the eight-cell stage. While these embryos were not implanted to continue development, any adult who were to develop from such an embryo would be free of the disease and would not transmit it to children.

Repairing disease-causing mutations with CRISPR need not be limited to embryos. In Paris in 2017, doctors succeeded in curing an adult patient of sickle-cell disease by extracting bone marrow from the patient, harvesting blood stem cells (stem cells are cells that form a particular kind of tissue—you will encounter them in chapter 13), and using CRISPR to edit the DNA of these cells so they would make normal hemoglobin—basically, correcting the sickle-cell mutation. The edited stem cells, when placed back in the patient's bone, led to the production of normal hemoglobin.

Gene Therapy for Cancer. In August 2017, the Food and Drug Administration approved the first-ever treatment that genetically modifies a patient's own cells to fight cancer. The therapy was approved for children and young adults with an aggressive form of leukemia, a usually fatal cancer of white blood cells. In clinical trials, 83% of treated children achieved remission within three months! The first patient to be treated was Emily Whitehead, who was six years old and near death. Now 12 years old (that's her in the photo), she has been free of leukemia for more than five years. In 2018 researchers reported using CRISPR to repair a gene mutation in dogs with Duchenne muscular dystrophy. The gene-editing system restored production of dystrophin, a protein crucial for healthy muscle function, in the treated dog's muscles. About 20,000 children are diagnosed every year with this fatal inherited disorder, in which the body's muscles slowly deteriorate. Because of the possibility of unwanted other DNA changes, the long-term health of the CRISPR-treated dogs must be assessed before human clinical trials can begin.

MCT/Contributor/Getty Images

11.5 Gene Drives

> **LEARNING OBJECTIVE 11.5.1** Explain how a gene drive might be used to eliminate malaria.

What about other applications of CRISPR? For example, could CRISPR be used to attack and eliminate malaria? The approach has not looked promising, for while a mosquito gene could potentially be modified in the lab to promote resistance to malaria, the effect would be limited to lab mosquitos. Released outdoors, CRISPR-modified mosquitos would not take over wild populations for the simple reason that making a mosquito unable to transmit the malaria parasite does not improve the mosquito's ability to survive and reproduce. In a sexually reproducing diploid species such as mosquitos (or us), a gene has a 50% chance of being inherited by each parent, so that without improved survival, its frequency doesn't change from one generation to the next. Every beginning biology student learns this as the Hardy-Weinberg rule (described in chapter 14).

The Idea of a Gene Drive

But what if you stacked the deck? If the gene is passed on MORE than half the time, it could quickly increase in frequency within a population, couldn't it? This sort of bias, called "gene drive" by professional biologists, doesn't happen often in nature—but might it be possible to find a way to do it?

Yep. Years ago, British geneticist Austin Burt suggested a thought experiment: use target-specific DNA-cutting enzymes (endonucleases) to attack mosquito genes necessary for transmitting the malarial parasite. Because the DNA-cutting enzyme would act on both chromosomes, the effect would be to change the 50% chance of the change being inherited to 100%. In effect, Burt said, the targeted nature of the nuclease attack would DRIVE the gene change through the population.

In 2015, Burt's theoretical idea of an endonuclease drive was put to the test in Panama and succeeded in reducing wild populations of dengue fever-carrying *Aedes aegypti* mosquitos 93%. Gene drives really work, just as Burt said they would.

Using CRISPR to Power a Gene Drive

Enter CRISPR, which allows a particular gene sequence to be replaced by a new one devised in the laboratory. What if that new sequence was a two-part cassette that included not only the gene to be replaced, but also a copy of the CRISPR sequence? Do you see? A gamete with this cassette in its DNA will act within a fertilized egg on the DNA of the gamete contributed by the other parent so that after CRISPR did its thing on that other gamete, BOTH of an offspring's chromosome sets now contain the cassette, and all of that individual's offspring will bear the "new gene plus CRISPR" cassette. In the future, every individual mated by any of these offspring will suffer the same fate, as will all of their offspring—a chain reaction!

Using CRISPR to Eliminate Malaria

Now imagine again Burt's proposal of a gene drive for mosquitos. DNA researchers in the laboratory could insert into a mosquito a DNA cassette containing both CRISPR and a gene preventing transmission of the malarial parasite, directing CRISPR to cut the original but not the edited version of the gene. Released outdoors, this mosquito would mate with a wild individual, their offspring inheriting one wild and one laboratory

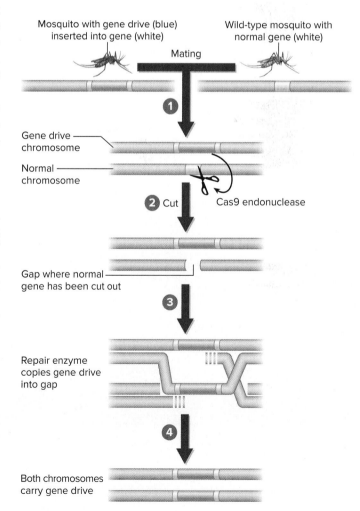

James Gathany/CDC

copy of the parasite transmission gene. Fifty percent transmission, right? But now CRISPR attacks the normal wild copy, inserting in its place the edited version + CRISPR. And so it begins: A CRISPR-driven chain reaction!

When tested in fruit flies in early 2015, this sort of CRISPR-driven chain reaction proved to be more than just an interesting possibility—it really works! In fact, it is shockingly efficient as a gene drive. Using a CRISPR drive cassette, researchers at the University of California, San Diego "drove" a recessive mutation that blocks pigmentation of fruit flies from a 50% chance of being inherited to 97%!

Geneticists are already busy at work building and testing a "Burt" laboratory mosquito with a CRISPR cassette containing the genes necessary to block parasite transmission. Their first job is to defeat a chromosomal repair program in the mosquito DNA that acts to counter-edit CRISPR. Then, using the CRISPR gene drive, geneticists hope to be able to drive the parasite resistance genes into and through wild mosquito populations and so eliminate malaria. Field tests are scheduled to begin in Africa as soon as 2024. Over half a million people die every year from malaria, so what these researchers are trying to do is no small thing—and think of the range of other diseases that could be conquered using a similar approach.

The Need for Care

It's a little scary, though. CRISPR gene drives would allow any scientist to spread nearly any gene alteration through nearly any sexually reproducing population. Humans don't reproduce as rapidly as mosquitos, of course, so it would take hundreds of years for a CRISPR-driven change to spread through human populations. And yet we already hear of researchers using CRISPR to modify the genomes of nonviable human embryos.

Surely some sorts of safeguards are in order. In 2017, the U.S. National Academy of Science and the National Academy of Medicine convened researchers and other experts "to explore the scientific, ethical, and policy issues associated with human gene-editing research." Their conclusion: While the potential for future good is great, decisions about when and how to use CRISPR gene drives should be made collectively. If we are going to fool with Mother Nature, in other words, we should do it carefully.

In October 2020, Dr. Emmanuelle Charpentier and Dr. Jennifer A. Doudna were awarded the Nobel Prize in Chemistry for their pioneering work on the CRISPR/Cas9 system. Their work in the early development and characterizing of this gene-editing tool system made them pioneers in the onslaught of applications in the medical, agricultural, and biotechnological fields.

A Closer Look

CRISPR-Edited Human Babies

The invention of gene editing with CRISPR has led quickly to an ethical nightmare. In the six years since its introduction in 2012, laboratories have begun to explore the use of CRISPR to treat a variety of human diseases. Clinical trials have been approved for treatments of AIDS, cancer, muscular dystrophy, sickle-cell anemia, and an inherited form of blindness.

A Dangerous Step

So: A promising approach, very real progress, laboratories all over the world excited, working feverishly . . . Then, on November 26, 2018, a shock. A little-known Chinese researcher posted YouTube videos announcing that he had created CRISPR-edited human babies! His research team, in experiments done without publicity and thus coming as a complete surprise, had guided the birth of twin baby girls with CRISPR-crippled CCR_5, a modification the twin girls would pass on to their descendants.

The researcher's approach was not complicated. Working with eight couples in which the men were HIV-positive and the women were HIV-negative, his team first washed the men's sperm to ensure that HIV virus particles were not present, then injected a mixture of the sperm and CRISPR-Cas9 enzymes into unfertilized eggs from the men's partners. The CRISPR-Cas9 was targeted at CCR_5, hoping to disable this gene. The procedure produced 16 viable embryos. Did any of them exhibit disabled CCR_5? Yes. Two of the embryos from one couple did. In one of these embryos, both chromosomes expressed disabled CCR_5; in the other embryo, one copy of CCR_5 was disabled, the other intact. The research team implanted both embryos into the female partner, who carried the twin girls successfully to term, giving birth to healthy girls.

So What's the Problem?

Do you see the problem? In previous experiments, CCR_5 was CRISPR-disabled in <u>adults.</u> Called *somatic gene engineering,* changes created using this approach cannot be passed on to future generations, because it is somatic tissue that is altered. The Chinese experiment edited the DNA of a single-cell *embryo,* and so altered all the tissues that derive from that zygote, germ line as well as somatic. In *germ-line engineering,* any changes created in CCR_5 will be passed on to future generations. And THAT is an ethical nightmare.

Here's why. Said simply, the future individuals are as yet unborn, and so do not get a say in gene changes that will directly affect their bodies. Nor will their children. Or their children's children. So such choices, made for them, must be made VERY carefully. There is an entire field of ethics devoted to making these difficult ethical decisions. Four principles are universally accepted (1). It goes without saying that the parents have to be very carefully informed of all potential risks, and give their full, informed consent; (2). Neither their health nor the health of the babies may be put at risk; (3). The process must be carefully monitored and well followed up; and (4). The procedure should only be done to address serious unmet needs in medical treatment.

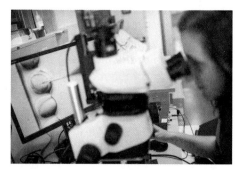
Gregor Fischer/dpa/Alamy Stock Photo

Failing the Test

The Chinese CRISPR-editing of human babies fails all of these tests:

1. **Consent.** The parents were counseled by the investigator, not by an independent authority who would place the needs of parents and children above the needs of the experiment.
2. **Safety.** Loss of CCR_5 is thought to make a patient more susceptible to flu infection. Indeed, the long-term consequences of lacking a functional CCR_5 gene in humans are simply unknown.
3. **Monitoring.** The results of the experiment—full genome DNA sequencing of the parents and twins, for example, to ensure that CRISPR has not unexpectedly altered other genes—have not been published, nor have any details of how the twins' later lives will be monitored for any unanticipated CRISPR-caused future problems.
4. **Medical necessity.** There is simply no substantial medical need being met by this gene edit. For one thing, disabling CCR_5 doesn't actually make the twins resistant to AIDS, as some strains of HIV use a different port of entry called $CXCR_4$. Only one of the twins would be protected from AIDS in the best of cases, because the other child has an unaltered CCR_5 gene and thus assumes all of the risks with no potential benefits.

Gene editing of humans is illegal in the United States, Canada, Australia, and England, but not in China. The uproar over the CRISPR-editing of human babies there has started a series of wide-ranging discussions among scientists all over the world about the need for clear ethical guidelines. No one wants "designer babies" with genes selected by their parents to express traits the parents judge will benefit their child. On the other hand, everyone can see that the world would be better off without gene disorders like cystic fibrosis. But where do you draw the line? This is a problem that even a few years ago would have been thought of as futuristic, something our children might have to deal with. But for gene editing, the future has arrived, and the ethics of editing human babies needs to be sorted out now.

Putting the Chapter to Work

1 Demonstrate your understanding of DNA structure by providing the sequence of the strand in a DNA duplex molecule that would be paired with:

3′-ATTGCCACGATCCAGGTAACGCAAT-5′

2 Erwin Chargaff provided evidence that there are equal amounts of purines and pyrimidines in a DNA molecule.

In DNA replication, which enzyme facilitates the addition of the matching purine to its pyrimidine in the growing complementary DNA strand?

3 DNA replication requires primers to be put down to begin DNA replication.

Which strand, the leading or lagging strand, would require more primers during its synthesis?

4 DNA replication occurs at a surprisingly fast rate. Occasionally, mistakes in selecting the proper complementary base can cause mutations to occur. Observe the following DNA parent strand (in blue) and daughter strand (in red), and identify what type of sequence mutation has occurred.

5′-AACGTCCAGGCAATTCGGA-3′
3′-TTCAGGTCCGTTAAGCCT-5′

5 Scientists study the roles of enzymes by causing mutations in the genes that encode them. Mutations that destroy the enzyme's ability to function are referred to as knock-out mutations.

If they were studying the role of the ligase enzyme through a knock-out mutation experiment, what would be the impact of the mutation on the lagging strand as DNA replication proceeded?

Retracing the Learning Path

Genes Are Made of DNA

11.1 Discovering the Structure of DNA

1. While the shape of the DNA molecule was not known, its chemical components were known to be nucleotides. Each nucleotide has a similar structure: a deoxyribose sugar attached to a phosphate group and one of four organic bases.
- Erwin Chargaff observed that two sets of bases are always present in equal amounts in a molecule of DNA (the amount of A nucleotides equals the amount of T nucleotides, and C nucleotides equals G nucleotides).
- Using X-ray diffraction, Maurice Wilkins was able to form a "picture" of DNA. These first crude X-ray images suggested to him that the DNA molecule was coiled, a form called a helix.
- Using Chargaff's and Wilkin's research and an improved X-ray image taken by Rosalind Franklin, Watson and Crick argued that these data could best be explained if DNA is a *double* helix, two strands that are connected by base pairing between the nucleotide bases. An A nucleotide on one strand pairs with T on the other, and similarly G pairs with C.

DNA Replication

11.2 How DNA Copies Itself

1. The basis for the great accuracy of DNA replication is *complementarity* (A pairs only with T, and C only with G): If you know the sequence of one strand, the other strand's sequence is fully determined.
2. In replication, the DNA molecule first unwinds. Each DNA strand is then copied by the actions of an enzyme called DNA polymerase. The two original strands serve as templates to the new DNA strands. DNA polymerase adds complementary nucleotides to a growing strand but can only add on to the end of an existing strand, so a primer is first added to the start of the sequence by a different enzyme. Nucleotides are added to this growing strand in a 5′ to 3′ direction.
- The point where the DNA separates is called the replication fork. Because nucleotides can only be added onto the 3′ end of the growing DNA strand, DNA copies in a continuous manner on one of the strands, called the leading strand, and in a discontinuous manner on the other strand, called the lagging strand. On the lagging strand, primers are inserted at the replication fork, and nucleotides are added in sections. Before the new DNA strands rewind, the primers are removed and the DNA segments are linked together with another enzyme called DNA ligase.
- Errors can occur during the replication of DNA. The cell has proofreading mechanisms to correct mistakes or damage, comparing one strand against its complementary strand to correct errors, but this system is not foolproof.

Altering the Genetic Message

11.3 Mutation

1. A mutation is a change in the nucleotide sequence of the genetic message. Mutations that change one or only a few nucleotides are called point mutations. Some mutations cause only minor changes, whereas others can have dramatic consequences.

Editing Genes

11.4 Gene Editing with CRISPR

1. Using a DNA-snipping tool called CRISPR, it is possible to replace one gene with another.

11.5 Gene Drives

1. The CRISPR tool can be incorporated within a CRISPR-targeted gene sequence, creating a chain reaction of edits that spreads throughout a population.

Inquiry and Analysis

Are Mutations Random or Directed by the Environment?

Once biologists appreciated that Mendelian traits were in fact alternative versions of DNA sequences, which resulted from mutations, a very important question arose and needed to be answered: Are mutations random events that might happen anywhere on a DNA chromosome, or are they directed to some degree by the environment? Do the mutagens in cigarettes, for example, damage DNA at random locations, or do they preferentially seek out and alter specific sites such as those regulating the cell cycle?

This key question was addressed and answered in an elegant and deceptively simple experiment carried out in 1943 by two of the pioneers of molecular genetics: Salvadore Luria and Max Delbruck. They chose to examine a particular mutation that occurs in laboratory strains of the bacterium *Escherichia coli*. These bacterial cells are susceptible to T1 viruses, tiny chemical parasites that infect, multiply within, and kill the bacteria. If 10^5 bacterial cells are exposed to 10^{10} T1 viruses and the mixture spread on a culture dish, not one cell grows—every single *E. coli* cell is infected and killed. However, if you repeat the experiment using 10^9 bacterial cells, lots of cells survive! When tested, these surviving cells prove to be mutants, resistant to T1 infection. The question is, did the T1 virus cause the mutations, or were the mutations present all along, too rare to be present in a sample of only 10^5 cells but common enough to be present in 10^9 cells?

To answer this question, Luria and Delbruck devised a simple experiment they called a "fluctuation test," illustrated here. Five cell generations are shown for each of four independent bacterial cultures, all tested for resistance in the fifth generation. If the T1 virus causes the mutations (top row), then each culture will have more or less the same number of resistant cells, with only a little fluctuation (that is, variation among the four). If, on the other hand, mutations are spontaneous and so equally likely to occur in any generation, then bacterial cultures in which the T1 resistance mutation occurs in earlier generations will possess far more resistant cells by the fifth generation than cultures in which the mutation occurs in later generations, resulting in wide fluctuation among the four cultures. The table presents the data they obtained for 20 individual cultures.

Number of Bacteria Resistant to T1 Virus

Culture number	Resistant colonies found	Culture number	Resistant colonies found
1	1	11	107
2	0	12	0
3	3	13	0
4	0	14	0
5	0	15	1
6	5	16	0
7	0	17	0
8	5	18	64
9	0	19	0
10	6	20	35

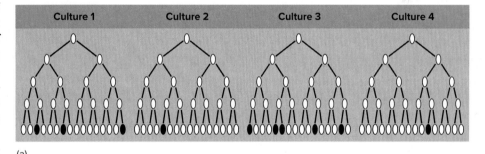

(a)

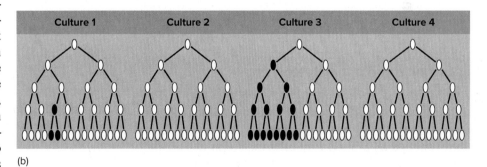

(b)

Analysis

1. **Interpreting Data** What is the mean number of T1-resistant bacteria found in the 20 individual cultures?

2. **Making Inferences**
   a. Comparing the 20 individual cultures, do the cultures exhibit similar numbers of T1-resistant bacterial cells?
   b. Which of the two alternative outcomes illustrated above, (*a*) or (*b*), is more similar to the outcome obtained by Luria and Delbruck in this experiment?

3. **Drawing Conclusions** Are these data consistent with the hypothesis that the mutation for T1 resistance among *E. coli* bacteria is caused by exposure to the T1 virus? Explain.

12 How Genes Work

LEARNING PATH ▼

From Gene to Protein
1. The Central Dogma
2. From DNA to RNA: Transcription
3. From RNA to Protein: Translation
4. Gene Expression

Regulating Gene Expression
5. Transcriptional Control in Prokaryotes
6. Transcriptional Control in Eukaryotes
7. RNA-Level Control: Silencing Genes

Build-a-Baby

Did you ever wonder what life would be like for you if you could change just one thing about yourself? Some of us desperately want to be healthier. Others would like to be smarter or stronger or more slender. Well, presto! With a wave of almost magic technology, laboratory scientists are moving a long way toward granting our wish.

Recipe for Building a Better You

The first step on the journey to becoming the new, improved you is to write down a full description of the person you are now. An impossible task? Not at all. Modern biology teaches us that each of us is built a bit like a laptop computer, with a set of basic instructions that determine what happens when the keyboard is used. The basic instructions that drive our bodies are written in molecular code on a long, long molecule called DNA. These instructions determine what we are like—our hair and skin color, our sex, how big we are, all the aspects of the newborn person that life molds into who we are. Your DNA is the recipe for building you. These days, it is no problem to obtain that recipe, called your "genome." You can order it up for about $100. To build a better you, you need to change your recipe. You need to rewrite your DNA.

A STRING OF NUCLEOTIDES encodes a baby.
Don Farrall/Getty Images

Altering the Recipe

The second step on the journey to becoming the new you is where the magic technology comes in. Using a sort of molecular snipper with the unusual name of CRISPR that you encountered in chapter 11, scientists search out the key DNA affecting the trait you wish to change, nip it out, and fill in the blank space with whatever new DNA instructions you desire. Impossible only a few years ago, this slight-of-hand technology is now being done all the time.

Designing a Better You

So, what gene are you going to snip? Not an easy question to answer. For most of the traits we are interested in changing, the DNA instructions are hard to put your fingers on—like looking for a particular sentence in your city public library. You know it's there, but not where. The third step on your journey to a better

Tanya Little/Shutterstock

you, then, is to be precise about your change. For many traits, that's a tall order, but not for all. Some traits such as sickle-cell disease or cystic fibrosis are caused by a particular DNA defect, and we know just what the DNA code of the defect is. Scientists are curing diseases like these using CRISPR as you read this.

Building a Better Baby

Well, it looks like it will be difficult to change yourself in a lot of ways you might like, not knowing your recipe well enough. But your journey need not end here. This same DNA-changing technology, called gene editing, can also be targeted at your offspring. When a new person is a single cell just after conception, any editing of the genome of that cell will affect every cell of the individual-to-be. With emerging technologies, it will soon be possible to survey that cell's entire genome for defects and, using CRISPR, correct them.

But *should* you? Is it moral to alter a baby who has no vote in what you do but must live with it for a lifetime? Should family planning be a build-a-baby process, with the parents selecting a certain version of each DNA instruction available? Is it moral to assemble a healthier baby than nature would give you? Where is the harm in avoiding heritable disease? Not easy questions.

From Gene to Protein

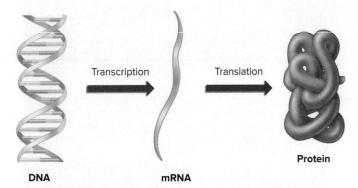

Figure 12.1 **The central dogma: DNA to RNA to protein.**

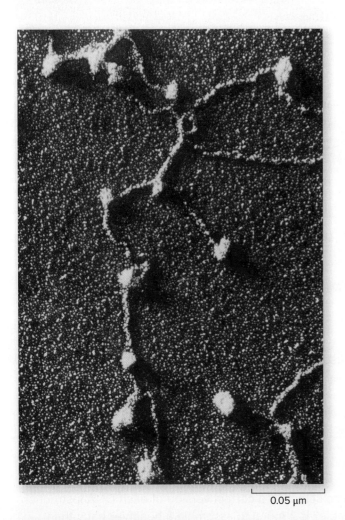

Figure 12.2 **RNA polymerase.**
In this electron micrograph, the dark circles are RNA polymerase molecules synthesizing RNA from a DNA template.
Omikron/Science Source

12.1 The Central Dogma

LEARNING OBJECTIVE 12.1.1 State the "central dogma."

The discovery that genes are made of DNA, discussed in chapter 11, left unanswered the question of how the information in DNA is used. How does a string of nucleotides in a spiral molecule determine if you have red hair? We now know that the information in DNA is arrayed in little blocks, like entries in a dictionary, and each block is a gene that specifies the sequence of amino acids for a polypeptide. These polypeptides form the proteins that determine what a particular cell will be like.

All organisms, from the simplest bacteria to ourselves, use the same basic mechanism of reading and expressing genes, so fundamental to life as we know it that it is often referred to as the "central dogma": Information passes from the genes (DNA) to an RNA copy of the gene, and the RNA copy directs the sequential assembly of a chain of amino acids (figure 12.1). Said briefly, **DNA → RNA → protein**.

A cell uses four kinds of RNA in the synthesis of proteins: messenger RNA (mRNA), microRNA (miRNA), ribosomal RNA (rRNA), and transfer RNA (tRNA). These types of RNA are described in more detail later in this chapter.

Transcription: An Overview

The first step of the central dogma is the transfer of information from DNA to RNA, which occurs when an mRNA copy of the gene is produced. Because the DNA sequence in the gene is transcribed into an RNA sequence, this stage is called *transcription*. Transcription is initiated when the enzyme *RNA polymerase* binds to a special nucleotide sequence called a *promoter* located at the beginning of a gene. Starting there, the RNA polymerase moves along the strand into the gene (figure 12.2). As it encounters each DNA nucleotide, it adds the corresponding complementary RNA nucleotide to a growing mRNA strand. This chain is a complementary transcript of the gene from which it was copied.

Translation: An Overview

The second step of the central dogma is the transfer of information from RNA to protein, which occurs when the information contained in the mRNA transcript is used to direct the sequence of amino acids during the synthesis of polypeptides by ribosomes. This process is called *translation* because the nucleotide sequence of the mRNA transcript is translated into an amino acid sequence in the polypeptide. Translation begins when an rRNA molecule within the ribosome recognizes and binds to a "start" sequence on the mRNA. The ribosome then moves along the mRNA molecule, three nucleotides at a time. Each group of three nucleotides is a code word that specifies which amino acid will be added to the growing polypeptide chain and is recognized by a specific tRNA molecule. MicroRNA molecules help the cell fine-tune gene use by influencing which transcripts are translated. The use of information in DNA to direct the production of particular polypeptides is called **gene expression**.

> **Putting the Concept to Work**
> What four roles does RNA play in gene expression?

12.2 From DNA to RNA: Transcription

> **LEARNING OBJECTIVE 12.2.1** Indicate the chemical direction in which an mRNA chain is assembled during transcription.

Just as an architect protects building plans from loss or damage by keeping them safe in a central place and issuing only blueprint copies to on-site workers, your cells protect their DNA instructions by keeping them safe within a central DNA storage area, the nucleus. The DNA never leaves the nucleus. Instead, the process of **transcription** creates "blueprint" copies of particular genes that are sent out into the cell to direct the assembly of proteins (figure 12.3). These working copies of genes are made of ribonucleic acid (RNA) rather than DNA (see figures 3.10 and 3.11).

The Transcription Process

The RNA copy of a gene used in the cell to produce a polypeptide is called **messenger RNA (mRNA)**—it is the messenger that conveys the information from the nucleus to the cytoplasm. The copying process that makes the mRNA is called transcription—just as monks in monasteries used to make copies of manuscripts by faithfully transcribing each letter, so enzymes within the nuclei of your cells make mRNA copies of your genes by faithfully complementing each nucleotide.

In your cells, the transcriber is a large and very sophisticated protein called **RNA polymerase**. It binds to one strand of a DNA double helix at the promoter site and then moves along the DNA strand like a train engine on a track. Although DNA is double-stranded, the two strands have complementary rather than identical sequences, so RNA polymerase is able to bind only one of the two DNA strands (the one with the promoter-site sequence it recognizes). As RNA polymerase goes along the DNA strand it is copying, it pairs each nucleotide with its complementary RNA version (G with C, A with U—recall that RNA uses a nucleotide with uracil, U, in place of thymine). The mRNA chain is built in the 5′ to 3′ direction (figure 12.4).

> **Putting the Concept to Work**
> Why is an mRNA chain assembled in only one direction?

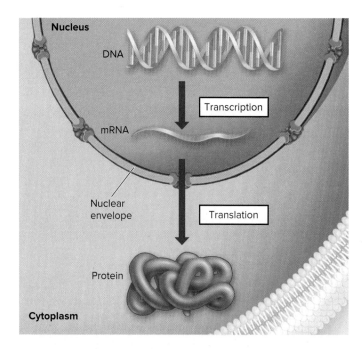

Figure 12.3 Overview of gene expression in a eukaryotic cell.

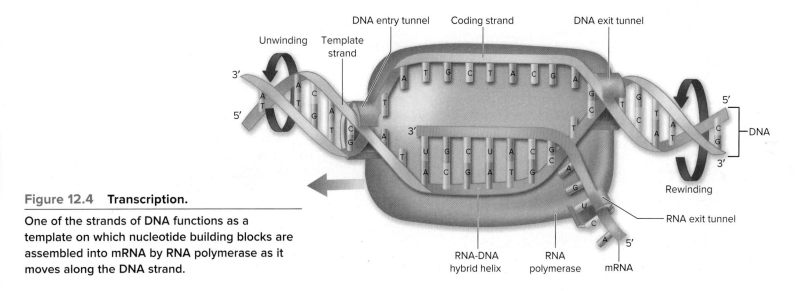

Figure 12.4 Transcription.
One of the strands of DNA functions as a template on which nucleotide building blocks are assembled into mRNA by RNA polymerase as it moves along the DNA strand.

12.3 From RNA to Protein: Translation

The Genetic Code

> **LEARNING OBJECTIVE 12.3.1** Define codon.

The essential lesson of genetics is that the information determining hereditary traits is encoded information, written within the chromosomes in blocks called genes. To read a gene, a cell must translate the information encoded in DNA into the language of proteins—that is, it must convert the order of the gene's nucleotides into the order of amino acids in a polypeptide, a process called **translation**. The rules that govern this translation are called the **genetic code**.

The mRNA nucleotide sequence is "read" by a ribosome in three nucleotide units called **codons**. Biologists worked out which codon corresponds to which amino acid by trial-and-error experiments carried out in test tubes. In these experiments, investigators used artificial mRNAs to direct the synthesis of polypeptides in the tube and then looked to see the sequence of amino acids in the newly formed polypeptides. An mRNA that was a string of UUUUUU . . . , for example, produced a polypeptide that was a string of phenylalanine (Phe) amino acids, telling investigators that the codon UUU corresponded to the amino acid Phe. Most amino acids are specified by more than one codon. The entire genetic code dictionary is presented in **figure 12.5**.

The genetic code is universal, the same in practically all organisms. GUC codes for valine in bacteria, fruit flies, eagles, and in your own cells. The only exception biologists have ever found to this rule is in how cell organelles that contain DNA (mitochondria and chloroplasts) and a few microscopic protists read the "stop" codons. In every other instance, the same genetic code is employed by all living things.

> **Putting the Concept to Work**
> How many codons do not code for an amino acid?

Figure 12.5 The genetic code (RNA codons).

A codon consists of three nucleotides read in sequence. For example, ACU codes for threonine. To determine which amino acid corresponds to a codon, you read the chart first down the left side, then across the top, then down the right side. For the codon ACU, find the first letter, A, in the First Letter column; then follow the row over to the second letter, C, in the Second Letter row across the top; and then in that column trace down to the third letter, U, in the Third Letter column. Most amino acids are specified by more than one codon. For example, threonine is specified by four codons, which differ only in the third nucleotide (ACU, ACC, ACA, and ACG).

The Genetic Code

First Letter	Second Letter: U	Second Letter: C	Second Letter: A	Second Letter: G	Third Letter
U	UUU Phenylalanine UUC Phenylalanine UUA Leucine UUG Leucine	UCU Serine UCC Serine UCA Serine UCG Serine	UAU Tyrosine UAC Tyrosine UAA Stop UAG Stop	UGU Cysteine UGC Cysteine UGA Stop UGG Tryptophan	U C A G
C	CUU Leucine CUC Leucine CUA Leucine CUG Leucine	CCU Proline CCC Proline CCA Proline CCG Proline	CAU Histidine CAC Histidine CAA Glutamine CAG Glutamine	CGU Arginine CGC Arginine CGA Arginine CGG Arginine	U C A G
A	AUU Isoleucine AUC Isoleucine AUA Isoleucine AUG Methionine; Start	ACU Threonine ACC Threonine ACA Threonine ACG Threonine	AAU Asparagine AAC Asparagine AAA Lysine AAG Lysine	AGU Serine AGC Serine AGA Arginine AGG Arginine	U C A G
G	GUU Valine GUC Valine GUA Valine GUG Valine	GCU Alanine GCC Alanine GCA Alanine GCG Alanine	GAU Aspartate GAC Aspartate GAA Glutamate GAG Glutamate	GGU Glycine GGC Glycine GGA Glycine GGG Glycine	U C A G

Translating the RNA Message

> **LEARNING OBJECTIVE 12.3.2** Contrast the roles of tRNA and ribosomes in reading the genetic code.

The final result of the transcription process is the production of an mRNA copy of a gene. Like a photocopy, the mRNA can be used without damage or wear and tear on the original. After transcription of a gene is finished, the mRNA passes out of the nucleus into the cytoplasm through pores in the nuclear envelope. There, translation of the genetic message occurs. In translation, organelles called **ribosomes** use the mRNA produced by transcription to direct the synthesis of a polypeptide, following the genetic code.

The Protein-Making Factory. Ribosomes are the polypeptide-making factories of the cell. Ribosomes are very complex, containing over 50 different proteins (shown in purple in figure 12.6, top) and three chains of **ribosomal RNA (rRNA)** of some 3,000 nucleotides (shown in tan). It had been traditionally assumed that the proteins in a ribosome act as enzymes to catalyze the assembly process, with the RNA acting as a scaffold to position the proteins. Powerful atomic-resolution X-ray diffraction studies have unexpectedly revealed the many proteins of a ribosome to be scattered over its surface like decorations on a Christmas tree. The role of these proteins seems to be to stabilize the many bends and twists of the RNA chains. Importantly, there are no proteins on the inside of the ribosome where the chemistry of protein synthesis takes place—just twists of RNA. It is the ribosome's RNA, not its proteins, that catalyzes the synthesis of proteins!

Ribosomes are composed of two parts, or subunits, one nested into the other like a fist in the palm of your hand. The "fist" is the smaller of the two subunits, the pink structure in figure 12.6. Its rRNA has a short nucleotide sequence exposed on the surface of the subunit. This exposed sequence is identical to a sequence called the leader region that occurs at the beginning of all genes. Because of this, an mRNA molecule binds to the exposed rRNA of the small subunit like a fly sticking to flypaper.

The Key Role of tRNA. Directly adjacent to the exposed rRNA sequence are three small pockets or dents, designated the A, P, and E sites, in the surface of the ribosome shown in figure 12.6. These sites have just the right shape to bind yet a third kind of RNA molecule, **transfer RNA (tRNA)**. It is tRNA molecules that bring amino acids to the ribosome used in making proteins. tRNA molecules are chains about 80 nucleotides long. The string of nucleotides folds back on itself, forming a three-looped structure shown in figure 12.7a. The looped structure further folds into a compact shape shown in figure 12.7b, with a three-nucleotide sequence at the bottom (the pink loop) and an amino acid attachment site at the top (the 3′ end).

The three-nucleotide sequence on the tRNA, called the **anticodon**, is very important: It is the complementary sequence to 1 of the 64 codons of the genetic code! Special enzymes, called *activating enzymes,* match amino acids in the cytoplasm with their proper tRNAs. The anticodon determines which amino acid will attach to a particular tRNA. Because the first dent in the ribosome, called the A site (the attachment site where amino-acid-bearing tRNAs will bind), is directly adjacent to where the mRNA binds to the rRNA, three nucleotides of the mRNA are positioned directly facing the anticodon of the tRNA. Like the address on a letter, the anticodon ensures that an amino acid is delivered to its correct "address" on the mRNA where the ribosome is assembling the polypeptide.

> **Putting the Concept to Work**
> How many different activating enzymes does a cell require?

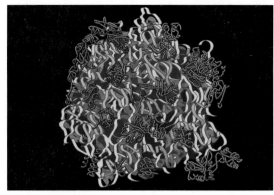

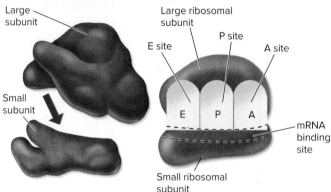

Figure 12.6 A ribosome is composed of two subunits.
The smaller subunit fits into a depression on the surface of the larger one. The A, P, and E sites on the ribosome play key roles in protein synthesis.

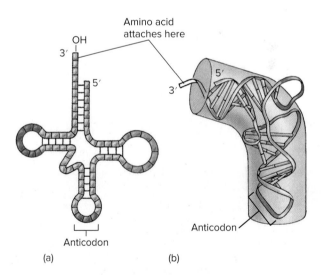

Figure 12.7 The structure of tRNA.

tRNA, like mRNA, is a long strand of nucleotides. However, unlike mRNA, hydrogen bonding occurs between its nucleotides, causing the strand to form hairpin loops, as seen in (a). The loops then fold up on each other to create the compact, three-dimensional shape seen in (b). Amino acids attach to the free, single-stranded —OH end of a tRNA molecule. A three-nucleotide sequence called the anticodon in the lower loop of tRNA interacts with a complementary codon on the mRNA.

Essential Biological Process 12A

Control of Eukaryotic Gene Expression

1 The initial tRNA occupies the P site on the ribosome. Subsequent tRNAs with bound amino acids first enter the ribosome at the A site.

2 The tRNA that binds to the A site has an anticodon complementary to the codon on the mRNA.

3 The ribosome moves three nucleotides to the right as the initial amino acid is transferred to the second amino acid at the P site.

4 The initiating tRNA leaves the ribosome at the E site, and the next tRNA enters at the A site.

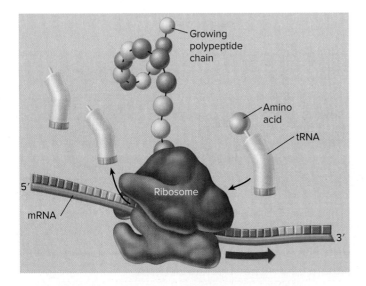

Figure 12.8 Ribosomes guide the translation process.

tRNA binds to an amino acid as determined by the anticodon sequence. Ribosomes bind the loaded tRNAs to their complementary sequences on the strand of mRNA. tRNA adds its amino acid to the growing polypeptide chain, which is released as the completed protein.

Making the Polypeptide

> **LEARNING OBJECTIVE 12.3.3** Describe in detail how ribosomes guide the translation process.

Once an mRNA molecule has bound to the small ribosomal subunit, the other larger ribosomal subunit binds as well, forming a complete ribosome. The ribosome then begins the process of translation, shown in *Essential Biological Process 12A*. The mRNA begins to thread through the ribosome like a string passing through the hole in a doughnut, moving from left to right, as shown in **figure 12.8**. The mRNA passes through in short spurts, three nucleotides at a time, and at each burst of movement, a new three-nucleotide codon on the mRNA is positioned opposite the site in the ribosome where a tRNA molecule binds.

As each new tRNA brings in an amino acid to each new codon, the old tRNA paired with the previous codon is passed along to an adjacent site where a peptide bond forms between the incoming amino acid and the growing polypeptide chain. The tRNA eventually exits, and the amino acid it carried is attached to the end of a growing amino acid chain. So as the ribosome proceeds down the mRNA, one tRNA after another is selected to match the sequence of mRNA codons, and the growing polypeptide chain extending out from the ribosome exits (**figure 12.8**). Translation continues until a "stop" codon is encountered, which signals the end of the polypeptide. The ribosome complex falls apart, and the newly made polypeptide is released into the cell.

> **Putting the Concept to Work**
> In what order are the sites of a ribosome occupied by a particular amino acid?

12.4 Gene Expression

The central dogma, discussed in section 12.1, is the same in all organisms. **Figure 12.9** presents an overview of the components needed for the key processes of DNA replication, transcription, and translation and the products that are formed in each. In general, the components are the same, the processes are the same, and the products are the same whether in prokaryotes or eukaryotes. However, there are some differences in gene expression between the two types of cells.

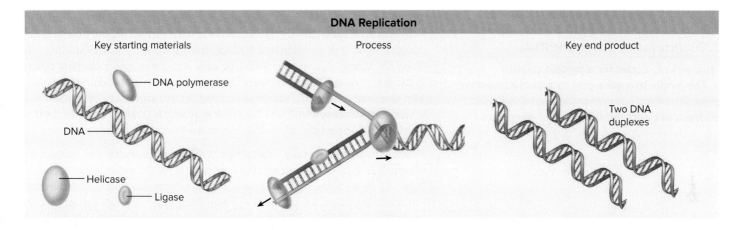

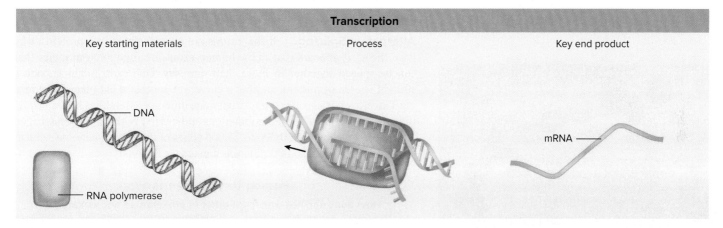

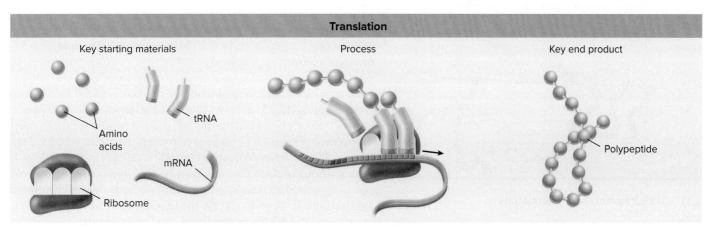

Figure 12.9 The processes of DNA replication, transcription, and translation.

These processes are generally the same in prokaryotes and eukaryotes.

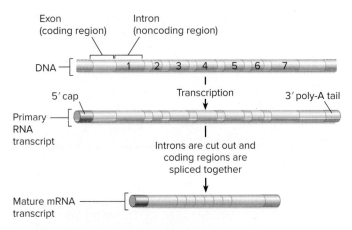

Figure 12.10 Processing eukaryotic RNA.
The gene shown here codes for a protein called ovalbumin. The ovalbumin gene and its primary transcript contain seven segments not present in the mRNA used by the ribosomes to direct the synthesis of the protein.

Architecture of the Gene

> **LEARNING OBJECTIVE 12.4.1** Compare the architecture of prokaryotic and eukaryotic genes, explaining how introns allow alternative splicing.

Introns and Exons. In prokaryotes, a gene is an uninterrupted stretch of DNA nucleotides whose transcript is read three nucleotides at a time to make a chain of amino acids. In eukaryotes, by contrast, genes are fragmented. In these more complex genes, the DNA nucleotide sequences encoding the amino acid sequence of a polypeptide, called **exons,** are interrupted frequently by extraneous nucleotides, "extra stuff" called **introns**. In the segment of DNA illustrated in figure 12.10, the exons are the blue areas and the introns are the orange areas. Imagine looking at an interstate highway from a satellite. Scattered randomly along the thread of concrete would be cars, some moving in clusters, others individually; most of the road would be bare. That is what a eukaryotic gene is like: scattered exons embedded within much longer sequences of introns. In a typical human gene, only 5% of the DNA sequence is devoted to the exons that encode the polypeptide, while 95% is devoted to noncoding introns!

Processing the Primary Transcript. When a eukaryotic cell transcribes a gene, it first produces a *primary RNA transcript* of the entire gene, shown in figure 12.10 with the exons in green and the introns in orange. Enzymes add modifications called a *5′ cap* and a *3′ poly-A tail,* which protect the RNA transcript from degradation. The primary transcript is then processed: Enzyme-RNA complexes excise out the introns and join together the exons to form the shorter mature mRNA transcript that is actually translated into an amino acid chain.

Alternative Splicing. If the introns are not translated into proteins, why have them? It appears that many human exons are functional modules that can be spliced together in more than one way. One exon might encode a straight stretch of protein, another a curve, yet another a flat place. Like mixing Tinkertoy parts, you can construct quite different assemblies by employing the same exons in different combinations and orders. With this process, called **alternative splicing,** the 20,000 to 25,000 genes of the human genome seem to encode as many as 120,000 different messenger RNAs.

> **Putting the Concept to Work**
> How does an RNA transcript differ in prokaryotes and eukaryotes?

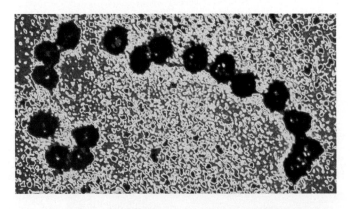

Protein Synthesis

> **LEARNING OBJECTIVE 12.4.2** Describe the six stages of eukaryotic protein synthesis.

Prokaryotic cells lack a nucleus, and so a gene can be translated as it is being transcribed (figure 12.11). In eukaryotic cells, a nuclear membrane separates the process of transcription from translation, making protein synthesis much more complicated. Figure 12.12 walks you through the entire process. Transcription (step ❶) and RNA processing (step ❷) occur within the nucleus. In step ❸, the mRNA travels to the cytoplasm where it binds to the ribosome. In step ❹, tRNAs bind to their appropriate amino acids, which correspond to their anticodons. In steps ❺ and ❻, the tRNAs bring the amino acids to the ribosome, and the mRNA is translated into a polypeptide.

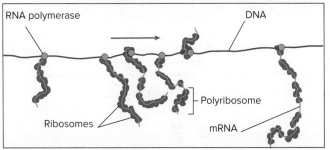

Figure 12.11 Transcription and translation in prokaryotes.
Ribosomes attach to an mRNA as it is formed, producing polyribosomes that translate the gene soon after it is transcribed.

Top: Omikron/Science Source

> **Putting the Concept to Work**
> Why is protein synthesis more complex in eukaryotes?

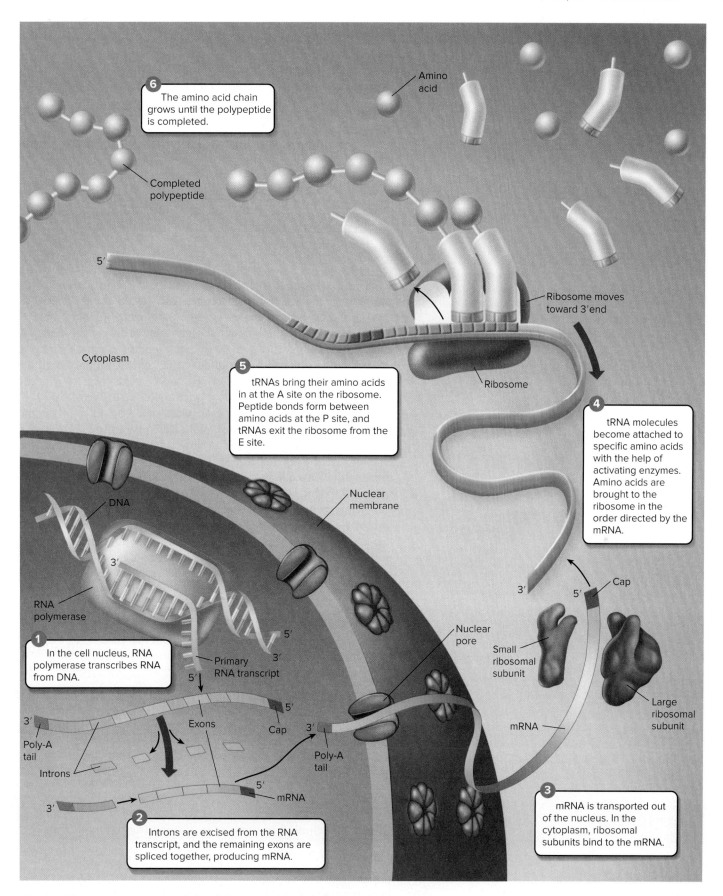

Figure 12.12 **How protein synthesis works in eukaryotes.**

Regulating Gene Expression

12.5 Transcriptional Control in Prokaryotes

Prokaryotes control when their genes are used by controlling immediate access to them.

The Need for Gene Controls

> **LEARNING OBJECTIVE 12.5.1** Describe the fundamental goal of transcriptional control in prokaryotes.

Being able to translate a gene into a polypeptide is only part of gene expression. Every cell must also be able to regulate when particular genes are used. Imagine if every instrument in a symphony played at full volume all the time, all the horns blowing full blast and each drum beating as fast and loud as it could! No symphony plays that way because music is more than noise—it is the controlled expression of sound. In the same way, growth and development are due to the controlled expression of genes, each brought into play at the proper moment to achieve precise and delicate effects.

Control of gene expression is accomplished very differently in prokaryotes than in the cells of complex multicellular organisms. Prokaryotic cells have been shaped by evolution to grow and divide as rapidly as possible, enabling them to exploit transient resources. Proteins in prokaryotes turn over rapidly. This allows them to respond quickly to changes in their external environment by changing patterns of gene expression. In prokaryotes, the primary function of gene control is to adjust the cell's activities to its immediate environment. Changes in gene expression alter which enzymes are present in the cell in response to the quantity and type of available nutrients and the amount of oxygen present. Almost all of these changes are fully reversible, allowing the cell to adjust its enzyme levels up or down as the environment changes.

> **Putting the Concept to Work**
> Your body maintains quite constant internal conditions. Why might prokaryotic gene regulation not be suitable for you?

How Prokaryotes Turn Genes Off and On

> **LEARNING OBJECTIVE 12.5.2** Explain how activators and repressors work together to control transcription.

Prokaryotes control the expression of their genes largely by saying *when* individual genes are to be transcribed. At the beginning of each gene are special regulatory sites that act as points of control. Regulatory proteins within the cell bind to these sites, turning transcription of the gene off or on.

For a gene to be transcribed, the RNA polymerase has to bind to a **promoter,** a specific sequence of nucleotides on the DNA that signals the beginning of a gene. In prokaryotes, the expression of a gene is regulated by controlling the access of RNA polymerase to the gene's promoter.

A gene is turned off by the binding of a **repressor** protein to the DNA, which blocks access to the promoter. Genes are turned on by the binding of an **activator,** a protein that makes the promoter more accessible to RNA polymerase.

Repressors. Most prokaryotic genes are "negatively" controlled: They are turned off except when needed. In these genes, the regulatory site is located between the place where the RNA polymerase binds to the DNA (the promoter site) and the beginning edge of the gene. When a regulatory protein called a repressor is bound to its regulatory site, called the *operator,* its presence blocks the movement of the polymerase toward the gene (figure 12.13). Imagine if you went to sit down to eat dinner and someone was already sitting in your chair—you could not begin your meal until this person was removed from your chair. In the same way, the polymerase cannot begin transcribing the gene until the repressor protein is removed.

To turn on a gene whose transcription is blocked by a repressor, cells use special "signal" molecules. The binding of the signal molecules to the repressor causes it to contort into a shape that doesn't fit DNA; the repressor falls off, removing the barrier to transcription.

A specific example demonstrating how repressor proteins work is the set of genes called the *lac* operon in the bacterium *Escherichia coli.* An **operon** is a segment of DNA containing a cluster of genes that are transcribed as a unit. The *lac* operon, shown in figure 12.13, consists of both protein-encoding genes (labeled genes 1, 2, and 3, which code for enzymes involved in breaking down the sugar lactose) and associated regulatory elements—the operator, the promoter, and other binding sites. When *E. coli* encounters the sugar lactose, a form of the sugar called allolactose binds to the repressor protein and induces a twist in its shape that causes the repressor to fall from the DNA (figure 12.13*b*); RNA polymerase, no longer blocked, starts to transcribe the genes needed to break down the lactose to get energy.

> **Putting the Concept to Work**
> If lactose is added to their growth medium, bacterial cells that had contained no lactose-metabolizing enzymes abruptly possess them. How does this "enzyme induction" occur?

12.6 Transcriptional Control in Eukaryotes

> **LEARNING OBJECTIVE 12.6.1** Describe the regulatory role of DNA methylation in eukaryotes.

Eukaryotes also control the long-term use of genes.

Chromatin Structure

In multicellular organisms with relatively constant internal environments, the primary function of gene control in a cell is to participate in regulating the body as a whole. Some of these changes in gene expression compensate for changes in the physiological condition of the body. Others mediate the decisions that ultimately produce the body, ensuring that the right genes are expressed in the right cells at the right time during development.

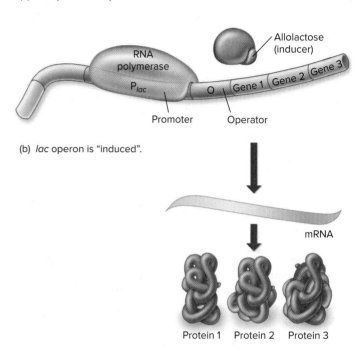

Figure 12.13 How the *lac* operon works.

(a) The *lac* operon is shut down ("repressed") when the repressor protein is bound to the operator site. Because promoter and operator sites overlap, RNA polymerase and the repressor cannot bind at the same time. (b) The *lac* operon is transcribed ("induced") when allolactose binds to the repressor protein, changing its shape so that it can no longer sit on the operator site and block polymerase binding.

The first hurdle faced by RNA polymerase in transcribing a eukaryotic gene is gaining access to it. As described in chapter 8, eukaryotic chromosomes consist of chromatin, a complex of DNA and protein. In chromatin, the DNA is packaged with histone proteins into nucleosomes (figure 12.14). During mitosis, nucleosomes wind, twist, and coil into chromosomes, but even during interphase when transcription is occurring, some sections of DNA remain as part of nucleosomes. These nucleosomes seem to block the binding of RNA polymerase and other proteins called transcription factors to the promoter site. Histones can be chemically modified to result in a greater condensation of the chromatin, making it even less accessible. Another chemical modification of the DNA—adding a methyl group ($-CH_3$) to cytosine nucleotides—blocks accidental transcription of "turned-off" genes. DNA methylation ensures that once a gene is turned off, it stays off.

> **Putting the Concept to Work**
> Do you think methylated genes would be found in nucleosomes?

Transcription Factors and Enhancers

> **LEARNING OBJECTIVE 12.6.2** Describe how enhancers regulate gene expression in eukaryotes.

Eukaryotic transcription requires not only the RNA polymerase molecule, but also a variety of other proteins, called *transcription factors,* that interact with the polymerase to form an *initiation complex.* Transcription factors are necessary for the assembly of the initiation complex and recruitment of RNA polymerase to a promoter (figure 12.15).

There are up to 1,600 different transcription factors known to act in the human genome. Different cell types in the body's tissues often employ different transcription factors, and one cell type may activate different transcription factors at different times.

Although prokaryotic gene control sites are positioned immediately upstream of the coding region, eukaryotes can have regulatory sites located far from the gene. Eukaryotic activators can bind at distant sites called **enhancers,** and DNA bending to form a loop brings the activator in contact with the RNA polymerase/initiation complex (see figure 12.15).

Complex transcriptional control is only one aspect of gene regulation in eukaryotes. Other processes associated with translation, such as gene silencing, protein synthesis, and post-translational modification, are described in *Essential Biological Process 12A* and provide a cell the ability to produce finely graded responses to environmental and developmental signals.

> **Putting the Concept to Work**
> If an enhancer is located far from a gene, how can it control the gene's transcription?

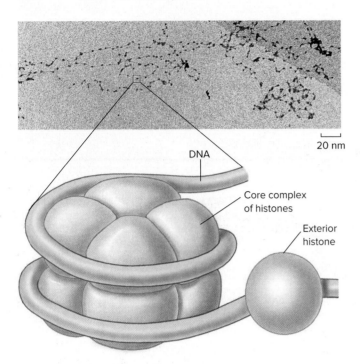

Figure 12.14 DNA coils around histones.

In chromatin, DNA is packaged into nucleosomes. In the electron micrograph (*top*), the DNA is partially unwound, and individual nucleosomes can be seen. In a nucleosome, the DNA double helix coils around a core complex of eight histones; one additional histone binds to the outside of the nucleosome.

Top: Gopal Murti/Medical Images

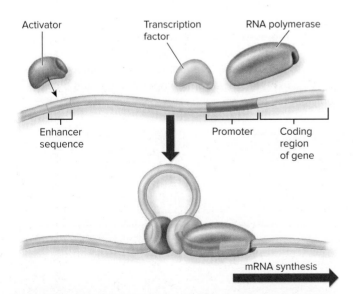

Figure 12.15 Transcription factors and enhancers.

The activator binding site, or enhancer, is often located far from the gene. The binding of an activator protein to the transcription factor is required for polymerase initiation.

12.7 RNA-Level Control: Silencing Genes

Discovery of RNA Interference

> **LEARNING OBJECTIVE 12.7.1** Define RNA interference.

As will be discussed in detail in chapter 13, the bulk of the eukaryotic genome is not translated into proteins. This finding was puzzling at first but began to make sense in 1998, when a simple experiment was carried out for which Americans Andrew Fire and Craig Mello later won the Nobel Prize in Physiology or Medicine in 2006. These investigators injected double-stranded RNA molecules into the nematode worm *Caenorhabditis elegans*. This resulted in the silencing of the gene whose sequence was complementary to the double-stranded RNA while silencing no other gene. The investigators called this very specific effect *gene silencing*, or **RNA interference.** What is going on here? A group of viruses called RNA viruses (those that contain RNA as their hereditary storage molecule rather than DNA) replicate themselves through double-stranded intermediates. RNA interference may have evolved as a cellular defense mechanism against these viruses, with double-stranded viral RNAs being targeted for destruction by RNA interference machinery. Without intending to do so, the nematode researchers had stumbled across this defense.

> **Putting the Concept to Work**
> How can attacking double-stranded DNA provide a defense against RNA viruses?

How RNA Interference Works

> **LEARNING OBJECTIVE 12.7.2** Explain how small RNA molecules carry out RNA interference.

Investigating interference, researchers noted that in the process of silencing a gene, plants produced short RNA molecules (ranging in length from 21 to 28 nucleotides) that matched the gene being silenced. Earlier researchers focusing on far larger messenger RNA (mRNA), transfer RNA (tRNA), and ribosomal RNA (rRNA) had not noticed these smaller bits, tossing them out during experiments. These bits of RNA are called microRNA or miRNA. How do these small fragments of RNA silence the activity of specific genes? In a complex way we are just beginning to understand clearly, the small RNA fragments bind to any mRNA molecules in the cell that have a complementary sequence. Silencing of the gene that produced this mRNA is achieved in one of two ways (figure 12.16): Either the expression of the mRNA is inhibited by blocking its translation into protein, or the mRNA is simply destroyed. In either case, the specific gene that produced that mRNA fails to be expressed—it is silenced.

> **Putting the Concept to Work**
> If RNA is not double-stranded, how can microRNA bind to mRNA?

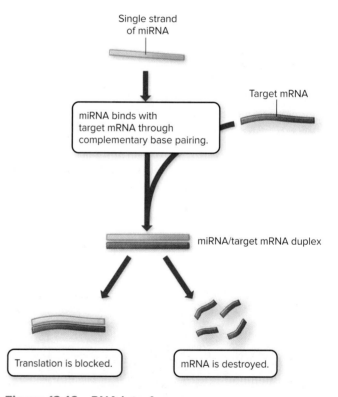

Figure 12.16 RNA interference.

RNA interference stops gene expression either by blocking the translation of the gene (on the left) or by targeting the mRNA for destruction before it can be translated.

Essential Biological Process 12B

Control of Eukaryotic Gene Expression

1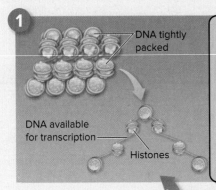

Chromatin structure
Access to many genes is affected by the packaging of DNA and by chemically altering the histone proteins.

2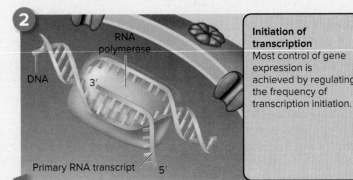

Initiation of transcription
Most control of gene expression is achieved by regulating the frequency of transcription initiation.

3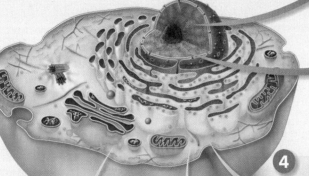

RNA splicing
Gene expression can be controlled by altering the rate of splicing in eukaryotes. Alternative splicing can produce multiple mRNAs from one gene.

4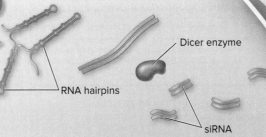

Gene silencing
Cells can silence genes with siRNAs, which are cut from inverted sequences that fold into double-stranded loops. siRNAs bind to mRNAs and block their translation.

6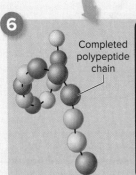

Post-translational modification
Phosphorylation or other chemical modifications can alter the activity of a protein after it is produced.

5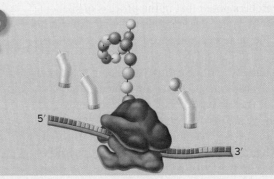

Protein synthesis
Many proteins take part in the translation process, and regulation of the availability of any of them alters the rate of gene expression by speeding or slowing protein synthesis.

Biology and Staying Healthy

Drugs from Dirt

In 1928, researcher Alexander Fleming noticed that mold had inadvertently drifted onto the surface of a lab culture dish on which he was growing bacteria. Surprisingly, the bacteria on the dish would not grow in a zone on the culture dish surrounding where the mold had landed—something was diffusing through the growth media from the mold that was killing any nearby bacteria. Fleming had accidentally discovered penicillin, the world's first antibiotic. When a person takes penicillin, it works to disrupt the formation of a strong bacterial cell wall, causes any infecting bacterial cell to explode when that cell attempts to make copies of itself. Widely administered to people sick from bacterial infections in the years after World War II, penicillin saved countless lives.

Drug-Resistant Superbugs

Then in the mid-1960s, penicillin stopped helping many patients—bacterial strains had evolved resistance to the drug. Luckily, other antibiotics had been discovered by then that could be used to cure these patients. Over a dozen different antibiotics have been introduced since 1943, when penicillin was first widely used. But resistance to these other antibiotics has developed too, hastened by their overuse in people and animals. All but one of these antibiotics have by now been reported to have generated resistant strains. Even worse, the resistant bacteria are very efficient at passing their resistance from one strain to another, leading to the emergence of "superbugs" resistant to many and sometimes all available antibiotics. As a consequence, deadly bacterial infections are not simply specters of our past, conquered by modern medical technology. They kill today, and often. Drug-resistant infections helped cause an estimated 2 million illnesses and 23,000 deaths in the United States in 2017.

Seeking New Antibiotics in Dirt

The emergence of superbugs resistant to all available antibiotics has left scientists scrambling to find new antibiotics. The problem is that they have already looked in all the likely places. Practically any organism that can be grown in a laboratory has been checked. This leaves us with organisms that cannot be grown in the lab. Stand outside in nature and look around. Where are the most bacteria? Under your feet. Countless bacteria grow in soil, unknown to science because almost none of them can be cultured in the lab. *There* is where scientists are now looking for new antibiotics!

While soil bacteria can't be grown in the lab, it is possible to simply skip this step and extract the DNA of all bacteria present in a soil sample. Each cell present in the soil sample contributes only a tiny smidgen, of course, but that is all modern cell biologists need, as today's technology allows a single DNA strand to be amplified to billions of copies. The trick is to find the right smidgen. What researchers at Rockefeller University did in 2018 was to screen the DNA from a soil sample for bits of DNA with a sequence that encoded proteins able to bind calcium. Why select calcium-binding ability as their target? Because many of the enzymes of bacterial metabolism only work properly when bound to a calcium ion. If you are looking for something to disrupt bacterial metabolism, that's a good place to start.

Discovery of Malacidin

The Ca^{++}-protein-encoding DNA bits the researchers found in dirt were then each amplified and sequenced. In each case, comparing the result with a database of known bacteria gene sequences allowed the researchers to identify likely candidates, DNA-encoding proteins associated with key metabolic processes. Putting these DNA segments into bacteria easily grown in the lab allowed the researchers to test if any of these soil DNA sequences made proteins that hurt the bacteria. One did. It blocked the synthesis of stuff needed to build the bacteria's cell wall. Nor were bacteria able to easily generate resistance to the protein encoded by this DNA; in 21 days of continuous exposure, no resistance to this drug appeared among the bacteria exposed to it. The researchers dubbed their newly discovered antibiotic malacidin.

DrugsFromDirt.org

The discovery of this dirt-derived antibiotic is very promising, as malacidin kills a variety of superbugs, including resistant strains of MRSA. Most exciting of all is the success of the general approach to discovering new antibiotics—drugs from dirt. The researchers have set up an online citizens' science project (DrugsFromDirt.org) to solicit soil donations from around the world. Indeed, the sandy soil from which malacidin was extracted was shipped to the Rockefeller University lab in New York City by the parents of one of the researchers who lived in the Southwestern United States.

Comstock/Getty Images

Putting the Chapter to Work

1 Transcription is the first step in "gene to protein" outlined in the central dogma. The main enzyme involved is the RNA polymerase. Demonstrate your understanding of the transcription process by providing the template DNA sequence to the gene used by RNA polymerase to produce the mRNA:

5'-AUG ACU AAU GAU AGU CGC UGA-3'

2 Each amino acid genetic code is complementary to an anticodon on a tRNA molecule.

If the codon UCU is read by a tRNA containing the anticodon AGT, would the amino acid added to the growing polypeptide be changed?

Explain your answer.

3 Cancers are caused by the expression of oncogenes (cancer-causing genes). Recent studies have linked the expression of some oncogenes to a change in chromatin configuration. When this occurs, the oncogene region of the DNA is released from the nucleosome.

How would this release from the nucleosome result in the expression of the oncogenes?

4 Tetracycline is a common antimicrobial drug used to treat many bacterial infections. It works by binding to the A site in the large portion of the prokaryotic ribosome. Discuss how blocking this site would interfere with protein synthesis, which would eventually result in the death of prokaryotic cells.

Retracing the Learning Path

From Gene to Protein

12.1 The Central Dogma

1. DNA is the storage site of genetic information in the cell. The process of gene expression, DNA to RNA to protein, is called the central dogma.
   - Gene expression occurs in two stages using different types of RNA: transcription, where an mRNA copy is made from the DNA, and translation, where the information on the mRNA is translated into a protein using rRNA and tRNA.

12.2 From DNA to RNA: Transcription

1. Transcription is the process of reading the DNA message. In transcription, DNA serves as a template for mRNA synthesis by an enzyme called RNA polymerase. RNA polymerase binds to one strand of the DNA at a site called the promoter. RNA polymerase adds complementary nucleotides onto the growing mRNA.

12.3 From RNA to Protein: Translation

1. The message within the mRNA is coded in its sequence of nucleotides, which are read in three nucleotide units called codons. Codons correspond to particular amino acids. The rules that govern the translation of codons on the mRNA into amino acids are called the genetic code.
2. In translation, the mRNA carries the message to the cytoplasm. rRNA combines with proteins to form a structure called a ribosome, the platform on which proteins are assembled.
   - A ribosome contains a small and a large subunit. Translation begins when the mRNA binds to the small ribosomal subunit, which triggers the binding of the large subunit to the small subunit, forming a complete ribosome. The ribosome has three sites, called A, P, and E, to which a third type of RNA, called transfer RNA (tRNA), binds.
   - The tRNA molecules carry amino acids to the ribosome to build the polypeptide chain. The amino acid that attaches to the tRNA is determined by a three-nucleotide sequence on the tRNA called the anticodon.
3. A ribosome moves along mRNA, while the tRNA molecules that contain anticodon sequences that are complementary to the codons bring the appropriate amino acids to the ribosome. An incoming tRNA first attaches to an entry site, then moves on as the ribosome moves, leaving the entry site open for the next tRNA. The amino acids add to the growing polypeptide chain. When a "stop" codon is reached, the ribosome dissociates and releases the polypeptide into the cell.

12.4 Gene Expression

1. Eukaryotic genes are fragmented, containing coding regions called exons and noncoding regions called introns. The entire eukaryotic gene is transcribed into RNA, but the introns are spliced out before translation. Exons can be spliced together in different ways, a process called alternative splicing.
2. In prokaryotes, ribosomes attach to the mRNA as it is transcribed, beginning translation before transcription is completed. In eukaryotes, the RNA transcript is produced and its introns spliced out in the nucleus. The mRNA then travels to the cytoplasm, where it is translated into a polypeptide.

Regulating Gene Expression

12.5 Transcriptional Control in Prokaryotes

1. Prokaryotes control the expression of genes by determining when they are transcribed.
2. In prokaryotic cells, genes are turned off when a repressor protein binds to a site called the operator and blocks the promoter.
   - The *lac* operon contains a cluster of genes that are involved in the breakdown of the sugar lactose. When the proteins produced by the *lac* operon genes are needed, an inducer molecule will bind to the repressor protein so it can't attach to the DNA, thereby freeing the promoter so that RNA polymerase can bind.

12.6 Transcriptional Control in Eukaryotes

1. The coiling of DNA around histones restricts RNA polymerase's access to the DNA. Controlling gene expression may involve chemical modification of histones or methylation of the DNA itself that keeps a gene turned off.
2. In eukaryotic cells, transcription requires the binding of transcription factors before RNA polymerase can bind to the promoter. Eukaryotic genes are controlled from distant locations called enhancers. A regulatory protein binds to the enhancer region far from the gene, and the DNA forms a loop, bringing the distant enhancer region into contact with the transcription factors. Eukaryotic cells have several levels of gene expression.

12.7 RNA-Level Control: Silencing Genes

1. RNA interference is gene silencing by blocking translation.
2. In RNA interference, small bits of RNA, called microRNA, bind to mRNA in the cytoplasm, blocking translation or destroying mRNA.

Inquiry and Analysis

Building Proteins in a Test Tube

The complex mechanisms used by cells to build proteins were not discovered all at once. Our understanding came slowly, accumulating through a long series of experiments, each telling us a little bit more. To gain some sense of the incremental nature of this experimental journey and to appreciate the excitement that each step gave, it is useful to step into the shoes of an investigator back when little was known and the way forward was not clear.

The shoes we will step into are those of Paul Zamecnik, an early pioneer in protein synthesis research. Working with colleagues at Massachusetts General Hospital in the early 1950s, Zamecnik first asked the most direct of questions: Where in the cell are proteins synthesized? To find out, they injected radioactive amino acids into rats. After a few hours, the labeled amino acids could be found as part of newly made proteins in the livers of the rats. And, if the livers were removed and checked only minutes after injection, radioactive-labeled proteins were found only associated with small particles in the cytoplasm. Composed of protein and RNA, these particles, later named ribosomes, had been discovered years earlier by electron microscope studies of cell components. This experiment identified them as the sites of protein synthesis in the cell.

After several years of trial-and-error tinkering, Zamecnik and his colleagues had worked out a "cell-free" protein-synthesis system that would lead to the synthesis of proteins in a test tube. It included ribosomes, mRNA, and ATP to provide energy. It also included a collection of required soluble "factors" isolated from homogenized rat cells that somehow worked with the ribosome to get the job done. When Zamecnik's team characterized these required factors, they found most of them to be proteins, as expected, but also present in the mix was a small RNA, very unexpected.

To see what this small RNA was doing, they performed the following experiment. In a test tube, they added various amounts of ^{14}C-leucine (that is, the radioactively labeled amino acid leucine) to the cell-free system containing the soluble factors, ribosomes, and ATP. After waiting a bit, they then isolated the small RNA from the mixture and checked it for radioactivity. You can see the results in graph (a) above.

In a follow-up experiment, they mixed the radioactive leucine–small RNA complex that this experiment had generated with cell extracts containing intact endoplasmic reticulum (that is, a cell system of ribosomes on membranes quite capable of making protein). Looking to see where the radioactive label now went, they then isolated the newly made protein [red in graph (b)], as well as the small RNA [blue in graph (b)].

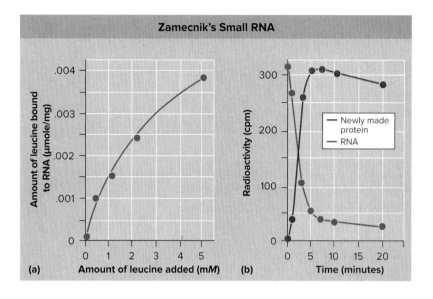

Analysis

EXPERIMENT A, shown in graph (a)

1. **Interpreting Data** Does the amount of leucine added to the test tube have an effect on the amount of leucine found bound to the small RNA?
2. **Making Inferences** Is the amount of leucine bound to small RNA proportional to the amount of leucine added to the mixture?
3. **Drawing Conclusions** Can you reasonably conclude from this result that the amino acid leucine is binding to the small RNA?

EXPERIMENT B, shown in graph (b)

1. **Interpreting Data**
   a. Monitoring radioactivity for 20 minutes after the addition of the radioactive leucine-small RNA complex to the cell extract, what happens to the level of radioactivity in the small RNA (blue)?
   b. Over the same period, what happens to the level of radioactivity in the newly made protein (red)?
2. **Making Inferences** Is the same amount of radioactivity being lost from the small RNA that is being gained by the newly made protein?
3. **Drawing Conclusions**
   a. Is it reasonable to conclude that the small RNA is donating its amino acid to the growing protein? (Hint: As a result of this experiment, the small RNA was called transfer RNA.)
   b. If you were to isolate the protein from this experiment made after 20 minutes, which amino acids would be radioactively labeled? Explain.

13 The New Biology

LEARNING PATH ▼

Genomics
1. The Human Genome

Genetic Engineering
2. A Scientific Revolution
3. Genetic Engineering and Medicine
4. Genetic Engineering and Agriculture

The Revolution in Cell Technology
5. Reproductive Cloning
6. Stem Cell Therapy
7. Gene Therapy

THIS WATERHEMP WEED growing in a soybean field has survived the field's spraying with Roundup.

Tim Barker/Post-Dispatch

Frankenfoods

We are living in a world of molecular agriculture, with genetically engineered foods now a common part of our daily lives. Over 93% of soybean, corn, and cotton crops being raised in the United States have been genetically modified. Many people, including influential activists and members of the scientific community, have expressed concern that genetic engineers are "playing God" by tampering with genetic material. What kind of unforeseen impact on the ecosystem might "improved" crops have? Should we be eating "Frankenfood"?

GM Foods Are Controversial

With genetically modified (GM) foods so much in evidence, these questions have led to considerable controversy and protest. Many states now mandate that food items produced from GM crops be so labeled in order that grocery shoppers be able to make an informed choice. In this chapter, you will learn how we measure the risks associated with the genetic engineering of plants and evaluate concerns about the use of GM crops.

An Herbicide Arms Race

Farmers in the American Midwest use chemical pesticides to fight a superweed called Palmer amaranth (waterhemp). Untreated, amaranth can quickly overrun soybean fields. GM soybeans carry a gene making them resistant to the herbicide Roundup, so fields can be sprayed with Roundup to kill amaranth weeds without harming the soybean crop. However, the amaranth weed eventually becomes resistant to Roundup, just as many pests have become resistant to the high levels of chemical pesticide we sprayed on crops. With years of spraying the herbicide, a farmer will eventually run across an individual weed with natural immunity to glyphosate. This individual will survive the spraying, and its offspring will carry the same natural resistance. Soon the field will be overrun with Roundup-resistant weeds. To combat this, new GM soybeans were developed to be resistant to both glyphosate and a second pesticide, dicamba. Any plant naturally resistant to one will not escape the other.

Dicamba Drift

Unfortunately, the powerful herbicide dicambra sometimes volatizes, turning from liquid to vapor. Carried by wind, dicamba spreads across adjacent fields, killing all crops

fotokostic/iStock/Getty Images

there that are not gene-modified to be dicamba-resistant. Farmers raising soybeans that are not gene-modified lost 3.6 million acres in this way in 2017 and are urging restrictions on the use of these new double-barreled GM crops.

Gene Escape

How about the possibility that introduced genes will pass from GM crops to their wild or weedy relatives? For the first round of major GM crops, there is usually no potential relative around to receive the modified gene from the GM crop. There are no wild relatives of soybeans in Europe, for example. Thus, there can be no gene escape from GM soybeans in Europe, any more than genes can flow from you to your pet dog or cat. For secondary crops only now being gene-modified, the risks are more tangible. There is mounting evidence that passage of modified genes to other plants can be significant, and the possibility that it might occur must be carefully evaluated. Frankenfoods, very much part of our lives, bear watching.

Jill Braaten/McGraw Hill

Genomics

13.1 The Human Genome

LEARNING OBJECTIVE 13.1.1 Describe the three major surprises discovered with the sequencing of the human genome.

The full complement of genetic information of an organism—all of its genes and other DNA—is called its **genome**. On June 26, 2000, geneticists announced that the entire human genome had been sequenced. This effort presented no small challenge, as the human genome is huge—more than 3 billion base pairs, which was the largest genome sequenced to date. To get an idea of the magnitude of the task, consider that if all 3.05 billion base pairs were written down on the pages of this book, the book would be 500,000 pages long, and it would take you about 60 years, working eight hours a day, every day, at five bases a second, to read it all.

The sequencing of the human genome—The Human Genome Project—was initiated in September 1990. One of the largest collaborative projects in the history of biological science, portions of the genome were sequenced in 20 universities in the United States, England, Japan, France, Germany, Spain, and China, and then the portions stitched together by analyzing overlaps. A parallel project was launched in 1998 by the private company Celera Genomics, using a "shotgun" approach which first fragmented the entire DNA, sequenced all the fragments, then used raw computer power to detect overlaps, and ultimately deduced the full sequence. Both efforts were completed simultaneously and announced to the world on February 12, 2001 to great acclaim.

Reading the human genome for the first time, geneticists encountered three big surprises.

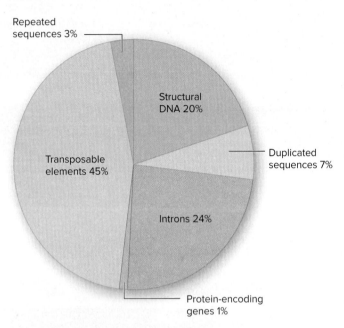

Figure 13.1 The human genome.
Very little of the human genome is devoted to protein-encoding genes, indicated by the light-blue section in this pie chart.

The Number of Genes Is Surprisingly Low

The human genome sequence contains only 19,969 protein-encoding genes, only 1% of the genome (**figure 13.1**). This is scarcely more genes than in a nematode worm (21,000 genes) while far more than the number in a fruit fly (13,000 genes). Researchers had confidently anticipated at least four times as many genes because over 100,000 unique messenger RNA (mRNA) molecules can be found in human cells—surely, they argued, it would take as many genes to make them.

How can human cells contain more mRNAs than genes? Recall from chapter 12 that in a typical human gene, the sequence of DNA nucleotides that specifies a protein is broken into many bits called exons, scattered among much longer segments of nontranslated DNA called introns. Imagine this paragraph as a human gene: all the occurrences of the letter "e" could be considered exons, while the rest would be noncoding introns, which make up 24% of the human genome.

When a cell uses a human gene to make a protein, it first manufactures mRNA copies of the gene, then splices the exons together, getting rid of the intron sequences in the process. Now here's the turn of events researchers had not anticipated: The exon portions of human gene transcripts are often spliced together in different ways, called *alternative splicing*. As we discussed in chapter 12, each exon is actually a module; one exon may code for one part of a protein, another for a different part of a protein. When the exon transcripts are mixed in different ways, very different protein shapes can be built.

With alternative splicing, it is easy to see how 19,969 genes can encode four times as many proteins. The added complexity of human proteins occurs because the gene parts are put together in new ways. Great music is made from simple tunes in much the same way.

Some Chromosomes Have Very Few Genes

In addition to the fragmenting of genes by the scattering of exons throughout the genome, there is another interesting "organizational" aspect of the genome. Genes are not distributed evenly over the genome. The small chromosome number 19 is packed densely with genes, transcription factors, and other functional elements. The much larger chromosome numbers 4 and 8, by contrast, have few genes, scattered like isolated hamlets in a desert. On most chromosomes, vast stretches of seemingly barren DNA fill the chromosomes between clusters rich in genes.

Most of the Genome Is Noncoding DNA

The third notable characteristic of the human genome is the startling amount of noncoding DNA it possesses. Only 1% to 1.5% of the human genome is coding DNA, devoted to genes encoding proteins. Each of your cells has about 6 feet of DNA stuffed into it, but of that, less than 1 inch is devoted to genes! Nearly 99% of the DNA in your cells seems to have little or nothing to do with the instructions that make you who you are (table 13.1).

There are four major types of noncoding human DNA:

Noncoding DNA Within Genes. As discussed earlier, a human gene is made up of numerous fragments of protein-encoding information (exons) embedded within a much larger matrix of noncoding DNA (introns). Introns make up 24% of the human genome, exons only 1%!

Structural DNA. About 20% of the chromosomes remain tightly coiled and untranscribed throughout the cell cycle. These portions tend to be localized around the centromere or near the ends of the chromosome.

Repeated and Duplicated Sequences. Scattered about chromosomes (figure 13.2) are short sequences, some only a few bases long, repeated like a broken record thousands and thousands of times. These make up about 10% of the human genome.

Transposable Elements. Fully 45% of the human genome consists of mobile parasitic bits of DNA called transposable elements, bits of DNA that are able to jump from one location on a chromosome to another, like tiny molecular versions of jumping beans. Nested within the human genome are over half a million copies of an ancient transposable element called Alu, composing fully 10% of the entire human genome. Often jumping right into genes, Alu transpositions cause many harmful mutations.

> **Putting the Concept to Work**
> The noncoding portions of human and mouse genomes have been conserved more than the coding portions. How might you explain this?

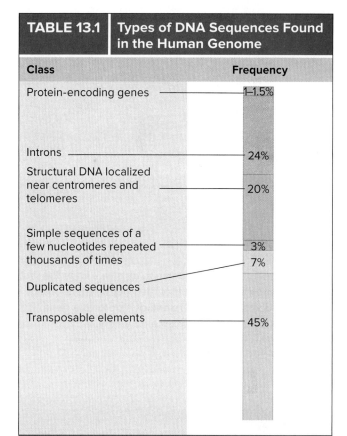

TABLE 13.1 Types of DNA Sequences Found in the Human Genome

Class	Frequency
Protein-encoding genes	1–1.5%
Introns	24%
Structural DNA localized near centromeres and telomeres	20%
Simple sequences of a few nucleotides repeated thousands of times	3%
Duplicated sequences	7%
Transposable elements	45%

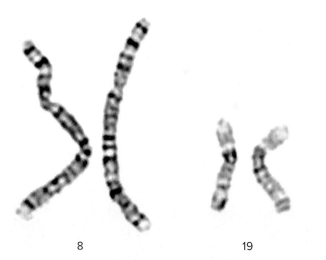

8   19

Figure 13.2 Banding on karyotyped chromosomes.
The dark bands are areas rich in repeated sequences of A and T nucleotides, which indicate nontranslated areas. Lighter, gene-heavy areas of the chromosome are rich in repetitive sequences containing C and G nucleotides.

Courtesy The Colorado Genetics Laboratory, Denver, CO; Director, Karen Swisshelm, PhD, FACMG

Genetic Engineering

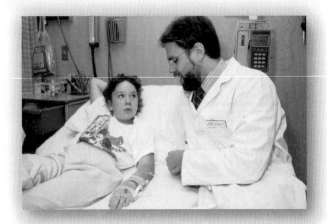

Curing disease. One of two young girls who were the first humans "cured" of a hereditary disorder by transferring into their bodies healthy versions of a defective gene. Twenty years later, the girls remain healthy.

Courtesy Dr. Ken Culver, Photo by John Crawford, National Institutes of Health

Increasing yields. The genetically engineered salmon on the *right* have shortened production cycles and are heavier than the nontransgenic salmon of the same age on the *left*.

Courtesy R. Devlin, Fisheries and Oceans Canada

Pest-proofing plants. The genetically engineered cotton plants on the *left* have a gene that inhibits feeding by weevils; the cotton plants on the *right* lack this gene and produce far fewer cotton bolls.

Left: image BROKER/Alamy Stock Photo; Right: SK Hasan Ali/Alamy Stock Photo

Figure 13.3 Examples of genetic engineering.

13.2 A Scientific Revolution

In recent years, **genetic engineering**—the ability to manipulate genes and move them from one organism to another—has led to great advances in medicine and agriculture (figure 13.3).

Restriction Enzymes

> **LEARNING OBJECTIVE 13.2.1** Explain how restriction enzymes are used to transfer genes between organisms.

The first stage in any genetic engineering experiment is to chop up the "source" DNA to get a copy of the gene you wish to transfer. This first stage is the key to successful transfer of the gene, and learning how to do it is what has led to the genetic revolution. The trick is in how the DNA molecules are cut. The cutting must be done in such a way that the resulting DNA fragments have "sticky ends" that can later be joined with another molecule of DNA.

This special form of molecular surgery is carried out by **restriction enzymes,** also called *restriction endonucleases,* which are special enzymes that bind to specific short sequences (typically four to six nucleotides long) on the DNA. These sequences are very unusual in that they are symmetrical—the two strands of the DNA duplex have the same nucleotide sequence, running in opposite directions! The sequence in figure 13.4, for example, is GAATTC. Try writing down the sequence of the opposite strand: it is CTTAAG—the same sequence, written backward. This sequence is recognized by the restriction enzyme *Eco*RI.

What makes the DNA fragments "sticky" is that most restriction enzymes do not make their incision in the center of the sequence; rather, the cut is made to one side. In the sequence in figure 13.4 ❶, the cut is made on both strands between the G and A nucleotides, G/AATTC. This produces a break, with short, single strands of DNA dangling from each end. Because the two single-stranded ends are complementary in sequence, they could pair up and heal the break, with the aid of a sealing enzyme—or they could pair with *any other DNA fragment cut by the same enzyme* because all would have the same single-stranded sticky ends. Figure 13.4 ❷ shows how DNA from another source (the orange DNA), also cut with *Eco*RI, has the same sticky ends as the original source DNA. Any gene in any organism cut by the enzyme that attacks GAATTC sequences will have the same sticky ends and can be joined to any other with the aid of a sealing enzyme called *DNA ligase* ❸.

> **Putting the Concept to Work**
> Would a restriction enzyme that cuts the sequence GAATTC between the A and the T be as useful in genetic engineering as one that cuts the sequence between the G and the A? Explain.

Formation of cDNA

> **LEARNING OBJECTIVE 13.2.2** Explain why cDNA is necessary to produce eukaryotic proteins in bacteria.

As we have discussed, eukaryotic genes are encoded in segments called exons separated from one another by numerous nontranslated sequences called introns. The entire gene is transcribed by RNA polymerase, producing what is called the primary RNA transcript (figure 13.5). Before a eukaryotic gene can be translated into a protein, the introns must be cut out of this primary transcript. The fragments that remain are then spliced together to form the mRNA, which is eventually translated in the cytoplasm. When transferring eukaryotic genes into bacteria (discussed in section 13.3), it is necessary to transfer DNA that has had the intron information removed because bacteria lack the enzymes to carry out this processing. Bacterial genes do not contain introns. To produce eukaryotic DNA without introns, genetic engineers first isolate from the cytoplasm the processed mRNA corresponding to a particular gene. The cytoplasmic mRNA has *only* exons, properly spliced together. An enzyme called *reverse transcriptase* is then used to make a complementary DNA strand of the mRNA, from which there forms a double-stranded DNA molecule called **complementary DNA, or cDNA**.

> **Putting the Concept to Work**
> If you wished to produce a bacterial protein in a human cell culture, would you need to employ cDNA? Explain.

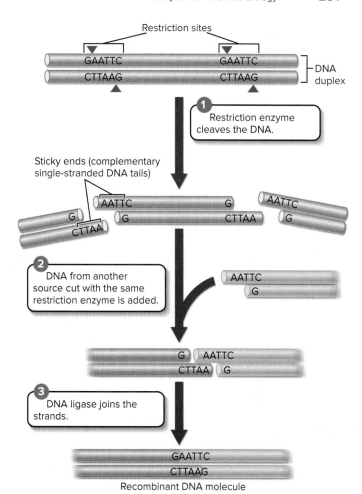

Figure 13.4 **How restriction enzymes produce DNA fragments with sticky ends.**

The restriction enzyme *Eco*RI always cleaves the sequence GAATTC between G and A. Because the same sequence occurs on both strands, both are cut. However, the two sequences run in opposite directions on the two strands. As a result, single-stranded tails are produced that are complementary to each other, or "sticky."

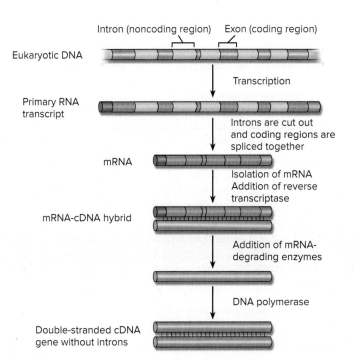

Figure 13.5 **cDNA: Producing an intron-free version of a eukaryotic gene for genetic engineering.**

In eukaryotic cells, a primary RNA transcript is processed into the mRNA, which is translated into a protein. To obtain a gene that bacteria will translate into the protein, the mRNA is isolated from eukaryotic cells and converted into cDNA. The cDNA is then transferred to bacterial cells, which can use it to direct the synthesis of the desired protein.

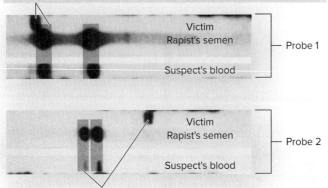

Probe 1 tags the same fragments of DNA in the rapist and the suspect but tags a different fragment of DNA in the victim. This probe indicates a match between the rapist and the suspect.

Like probe 1, probe 2 tags the same fragments of DNA in the rapist and the suspect but tags a different fragment of DNA in the victim. This probe indicates another match between the rapist and the suspect.

Figure 13.6 DNA fingerprints that led to conviction.

The two DNA probes seen here were used to characterize DNA isolated from the victim, the semen left by the rapist, and the suspect's blood. There is a clear match between the suspect's DNA and the DNA of the rapist's semen.

Hologic

BIOLOGY & YOU

Plant Witness for the Prosecution. The first case in which a murderer was convicted on plant DNA evidence was in Phoenix, Arizona. A pager left at the scene of the murder of a young woman led the police to a prime suspect. He admitted picking up the victim at a bar but claimed she had robbed him of his wallet and pager. The crime scene squad examined the suspect's pickup truck and collected pods wedged into the front fender that were later identified as the fruits of the palo verde tree (*Cercidium spp.*). One squad member went back to the murder scene and found several palo verde trees, one recently damaged by a car. If the pods on the suspect's car could be linked to the damaged tree at the crime scene, they had their man. The detective heading the crime scene squad contacted a geneticist at the University of Arizona to establish evidence that would stand up in court that the pod indeed came from this very tree and no other. First, crucially, it was necessary to establish that individual palo verde trees have unique patterns of DNA. Sampling different palo verde trees at the murder scene and elsewhere quickly established that each palo verde tree is unique in its DNA pattern. Next, the detectives compared the DNA pattern of a pod found on the suspect's truck to that of the damaged palo verde tree at the murder scene. They were the same. This placed the suspect at the scene of the murder beyond any doubt, and the suspect was convicted.

DNA Fingerprinting and Forensic Science

LEARNING OBJECTIVE 13.2.3 Explain how DNA variation is used in forensic science to identify culprits in criminal investigations.

DNA fingerprinting is a process used to compare samples of DNA. Because each person differs genetically from others in many ways, comparing DNA samples from two people can be used as a tool to tell them apart, much as detectives use fingerprints.

The process of DNA fingerprinting uses probes to fish out particular sequences to be compared from the thousands of other sequences in the human genome. Fragments of the subject's DNA are exposed to a "probe," a DNA fragment of a short, defined sequence that has been radioactively tagged so that it can be viewed. DNA fragments that have sequences complementary to the probes will bind the probes, which are then visible on autoradiographic film as dark bands. The resulting pattern of parallel bars on X-ray film resembles the line patterns of the universal price codes found on groceries, in essence a DNA "fingerprint" that can be used in criminal investigations and other identification applications.

Figure 13.6 shows the DNA fingerprints a prosecuting attorney presented in a rape trial in 1987. This trial was the first time DNA evidence was used in a U.S. court of law. Usually six to eight probes are used to identify the source of a DNA sample; two such probes are shown in figure 13.6. The probes are unique DNA sequences found in noncoding regions of human DNA that vary much more frequently from one individual to the next than do coding regions of the DNA. The chances of any two individuals, other than identical twins, having the same restriction pattern for these noncoding sequences vary from 1 in 800,000 to 1 in 1 billion, depending on the number of probes used. Since 1987, DNA fingerprinting has been admitted as evidence in thousands of court cases.

> **Putting the Concept to Work**
> If the suspect in the 1987 rape trial was innocent, what differences would you expect to see in the probes of figure 13.6?

PCR Amplification

LEARNING OBJECTIVE 13.2.4 Explain how the tiny amount of DNA in a single hair can be enough for forensic analysis.

Tiny DNA samples, as little as that found on a single human hair, can be magnified to many millions of copies by a process called **PCR** (for **polymerase chain reaction**). The use of DNA in forensic analysis often depends critically on PCR technology. In PCR, a double-stranded DNA fragment is heated so it becomes single-stranded; each of the strands is then copied by DNA polymerase to produce two double-stranded fragments. The fragments are heated again and copied again to produce four double-stranded fragments. This cycle is repeated many times, each time doubling the number of copies, creating enough DNA copies for analysis.

PCR is so powerful that a single hair will do. Biologists used to think that DNA was present only in the cells at the root of a hair. But we now know that hair follicle cells become incorporated into the growing shaft, and their DNA is sealed in by the protein keratin.

> **Putting the Concept to Work**
> If a PCR cycle took 20 minutes, how long would it take to convert a single copy of a DNA fragment into a million copies?

Today's Biology

Testing for COVID-19

In the year 2020, the coronavirus pandemic raced across the world, infecting nearly 100 million people with the COVID-19 virus. Face masks and social distancing slowed the spread by making it more difficult for a virus particle to pass from one individual to another, but did not halt the wave of infection. What to do? In many countries, the solution that has worked best has been identification and contact tracing: test individuals to identify who has the virus, quarantine those who are infected, and then find and test those who came in contact with the infected individuals, and quarantine any of those who test positive. As some 40% of infected individuals have no symptoms, widespread testing provides the only key to quarantining the infected and thereby controlling the outbreak.

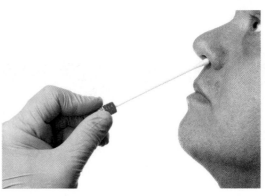

Henrik Dolle/Shutterstock

Devising a PCR Test

COVID-19 particles are tiny, so any test to detect their presence must be very powerful. It did not take scientists very long to devise one. Within months, they had devised a test that relies on a chemical procedure you have encountered in this chapter: the polymerase chain reaction (PCR).

1. A sample is collected from the patient (typically from far up the nose!) and any genetic material in the sample isolated.
2. The tiny amount of COVID-19 RNA, if present, is then added to a PCR machine, which converts the minute amount of single-stranded RNA in the sample to the corresponding DNA sequence using an enzyme called reverse transcriptase.
3. The machine then heats the sample to 95°C. This heating dissociates the two strands of the DNA duplex, as they are held together only by weak hydrogen bonds.
4. The sample is cooled to about 60°C. At this temperature, tiny molecular "probes" can bind to any complementary sequence in the tube. The probes are themselves fragments of COVID-19 genes, like the one encoding virus surface molecules, called spike proteins.
5. The machine then warms the sample to 72°C and activates the enzyme DNA polymerase. If the COVID-19 probe finds its target, matching a sequence on the sample single strand, the polymerase promptly replicates that DNA strand into a duplex. You now have two duplex molecules where before you only had one. The machine then reheats the mixture and repeats the cycle, again and again and again.
6. There is a fluorescent dye in the mix that lights up when it binds to DNA. If a COVID-19 gene is being amplified in the test, the sample will begin to glow as more copies of DNA are produced. After a few dozen cycles, the glow will become intense, indicating the patient is infected with COVID-19.

Typically, the test is carried out twice, using probes directed at two COVID-19 genes. If the first test is positive, the test is repeated on the second gene. Only if this is also positive is the result considered conclusive.

Need for a Better Test

Versions of this test have been used successfully throughout the world. It is both sensitive and accurate. But there have been problems. The PCR machines and reagents are expensive and require trained technicians to carry out the tests. In addition, the tests are slow. With many individuals needing to be tested, this can delay results for days and even weeks, a disaster when contact tracing is being attempted. Finally, many people avoid being tested because of the uncomfortable nature of the way-up-your-nose sample collection.

An Antigen Test for COVID-19

A slightly less accurate COVID-19 test is cheap, rapid and avoids sending samples to a laboratory for analysis. It uses a credit card–sized device to detect COVID-19 antigens. After a health care professional conducts a nasal swab of a patient, the swab is inserted into the device along with a few drops of solution. The solution causes the sample's material to flow across a strip containing antibodies to the COVID-19 spike protein antigen. If the COVID-19 virus is present in the swab sample, it will bind to the antibodies and create a color change to signal the presence of the virus. This antigen test was reported to correctly identify seriously ill patients infected with the COVID-19 virus 97.1% of the time, and people without the virus 98.5% of the time. However, its ability to detect small amounts of the virus in people not seriously ill is much less certain—as many as 40% may be missed.

13.3 Genetic Engineering and Medicine

> **LEARNING OBJECTIVE 13.3.1** Explain how genetic engineering permits production of medically important proteins.

Much of the excitement about genetic engineering has focused on its potential to improve medicine to aid in curing and preventing illness. Major advances have been made in the production of proteins used to treat illness and in the creation of new vaccines to combat infections.

Making Magic Medicine

Many genetic defects occur because our bodies fail to make critical proteins. Juvenile diabetes is such an illness. The body is unable to control levels of sugar in the blood because a critical protein, insulin, cannot be made. This failure can be overcome if the body can be supplied with the protein it lacks. The donated protein is in a very real sense a "magic medicine" to combat the body's inability to regulate itself.

Until recently, the principal problem with using regulatory proteins as drugs was in manufacturing the protein. Proteins that regulate the body's functions are typically present in the body in very low amounts, and this makes them difficult and expensive to obtain in quantity. With genetic engineering techniques, the problem of obtaining large amounts of rare proteins has been largely overcome. The cDNA of genes encoding medically important proteins is now introduced into bacteria. Because the host bacteria can be grown cheaply, large amounts of the desired protein can be easily isolated. In 1982, the U.S. Food and Drug Administration approved the use of human insulin produced from genetically engineered bacteria, the first commercial product of genetic engineering.

The use of genetic engineering techniques in bacteria has provided ample sources of therapeutic proteins, but the application extends beyond bacteria. Today, hundreds of pharmaceutical companies around the world are busy producing other medically important proteins, expanding the use of these genetic engineering techniques. A human gene added to the DNA of the mouse on the right in figure 13.7 produces human growth hormone, allowing the mouse to grow larger than its twin.

Figure 13.7 Genetically engineered human growth hormone.

These two mice are genetically identical, but the large one has one extra gene: the gene encoding human growth hormone. The gene was added to the mouse's genome by genetic engineers and is now a stable part of the mouse's genetic makeup. In humans, growth hormone is used to treat various forms of dwarfism.

Chicago Tribune/KRT/Newscom

> **Putting the Concept to Work**
> When a human gene is added to the DNA of a mouse, is it added as a DNA segment containing the gene or as cDNA? Explain.

Powerful New Vaccines

> **LEARNING OBJECTIVE 13.3.2** Describe how a piggyback vaccine is constructed.

Another area of potential significance involves the use of genetic engineering to produce **subunit vaccines** against disease-causing viruses. Genes encoding part of the protein-polysaccharide coat of a virus such as herpes simplex or hepatitis B are spliced into a fragment of the vaccinia (cowpox) virus genome (figure 13.8). The vaccinia virus, which is essentially harmless to humans, was used by British physician Edward Jenner more than 200 years ago in his pioneering vaccinations against smallpox. Vaccines produced in this way, also known as **piggyback vaccines,** elicit an immune response in the recipient against the coat of the disease-causing virus.

Recently, clinical trials have begun for a new kind of vaccine, called a **DNA vaccine.** DNA containing a viral gene is injected into and taken up by the body, where the gene is expressed. The infected cells trigger a cellular immune response in which blood cells known as killer T cells attack the infected cells. The first DNA vaccines spliced an influenza virus gene into a plasmid, which was then injected into mice. The mice developed strong cellular immune responses against influenza. The approach offers great promise.

In 2010, the first effective cancer vaccines were announced. A cancer vaccine is therapeutic rather than preventive, stimulating the immune system to attack a tumor in the same way invading microbes are attacked. The first cancer vaccine approved for clinical use employs proteins from prostate cancer cells to induce the immune system to attack prostate cancer tumors.

> **Putting the Concept to Work**
> Is a cancer vaccine also a piggyback vaccine?

Figure 13.8 Constructing a subunit, or piggyback, vaccine for the herpes simplex virus.

The harmless vaccinia virus can be used as a vector to carry a herpes simplex viral coat gene in a subunit, or piggyback, vaccine. Constructing this subunit vaccine begins with ❶ extracting the herpes simplex viral DNA and ❷ isolating a gene that codes for a protein on the surface of the virus. The cowpox viral DNA is extracted and cleaved ❸, and the herpes gene is combined with the cowpox DNA ❹. The recombinant DNA is inserted into a cowpox virus. Many copies of the recombinant virus, which have the outside coat of a herpes virus, are produced. When this recombinant virus is injected into a human ❺, the immune system produces antibodies directed against the coat of the recombinant virus ❻. The person therefore develops an immunity to the virus.

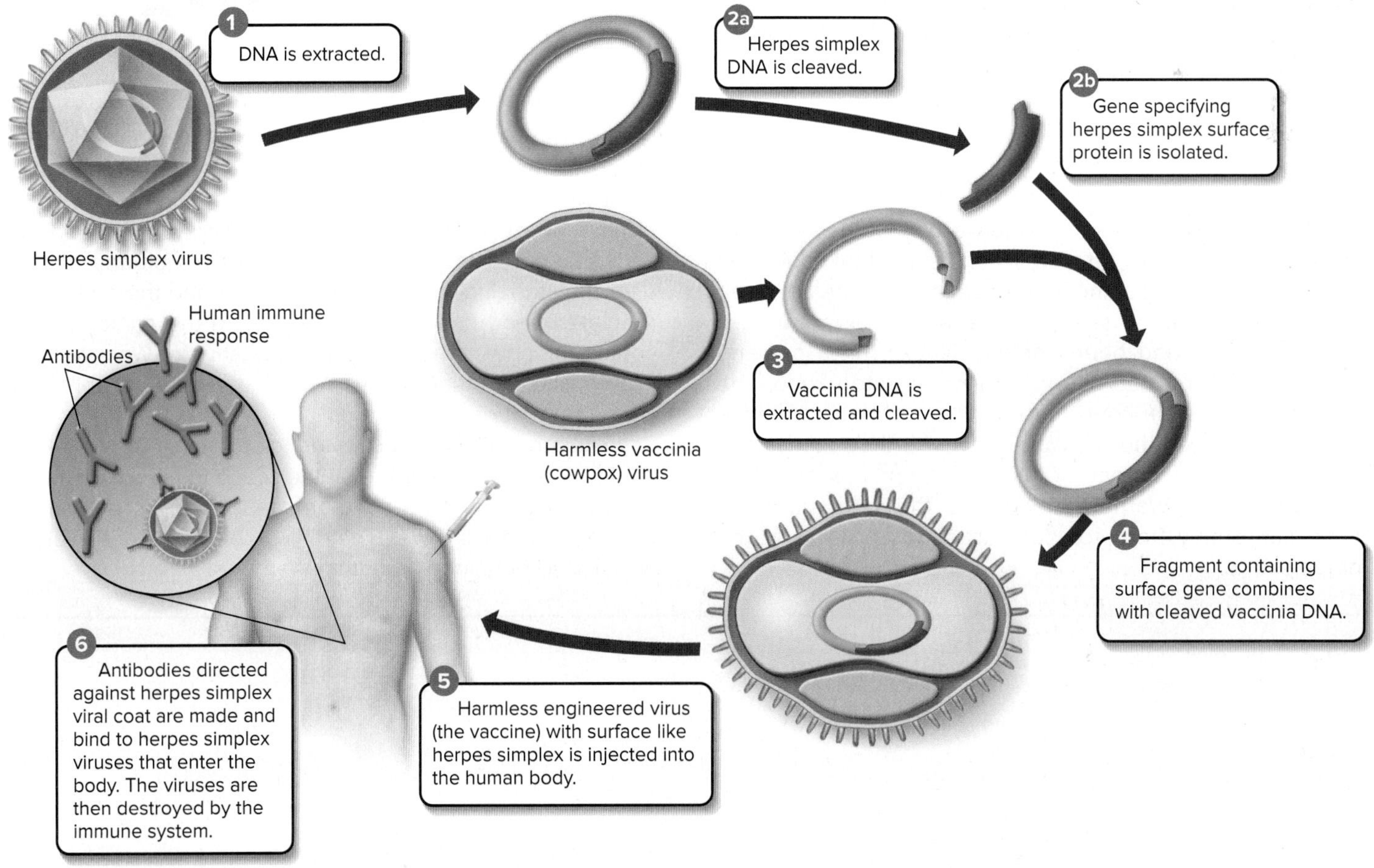

Today's Biology

DNA and the Innocence Project

Every person's DNA is uniquely their own, a sequence of nucleotides found in no other person. Like a molecular Social Security number a billion digits long, the nucleotide sequence of an individual's genes can provide proof-positive identification of a rapist or murderer from DNA left at the scene of a crime—proof more reliable than fingerprints, more reliable than an eyewitness, even more reliable than a confession by the suspect.

The Warney Case

Never was this demonstrated more clearly than on May 16, 2006, in a Rochester, New York, courtroom. There, a judge freed convicted murderer Douglas Warney after 10 years in prison.

The murder occurred on New Year's Day in 1996. The bloody body of William Beason, a prominent community activist, was found in his bed. Police called in all the usual suspects, in this case everyone known to be an acquaintance of the victim. Warney, an unemployed 34-year-old who had dropped out of school in the eighth grade, committed robberies, and worked as a male hustler, learned that detectives wanted to speak to him about the killing and went to the police station for questioning. Within hours, he was charged with murder.

A Signed Confession

Warney's interrogation was not recorded, but it resulted in a signed confession based on the words that the detective sergeant said Warney uttered, a confession that contained accurate details about the murder scene that had not been made public. Warney said that the victim was wearing a nightgown and had been cooking chicken in a pot and that the murderer had used a 12-inch serrated knife. The case against him in court rested almost entirely on these vivid details. Even though Warney recanted his confession at his trial, the accuracy of his confession was damning. The details that Warney provided to the sergeant could have come only from someone who was present at the crime scene.

Digital Vision/Getty Images

Problems with the Confession

Issues with Warney's confession emerged during the trial. Three elements of the signed statement's account of that night were clearly not true. It said Warney had driven to the victim's house in his brother's car, but his brother did not own a car; it said Warney disposed of his bloody clothes after the stabbing in the garbage can behind the house, but the can, buried in snow from the day of the crime, did not contain bloody clothes; it named a relative of Warney as an accomplice, but that relative was in a secure rehabilitation center on that day.

The most difficult bit of evidence to match to the written confession was blood found at the scene that was not that of the victim. Drops of a second person's blood were found on the floor and on a towel. The difficult bit was that the blood was a different blood type than Warney's.

Douglas Arthur Warney holds hands with his mother, Sheila, as he speaks to reporters at the Monroe County Courthouse in Rochester, N.Y., on Tuesday, May 16, 2006. Warney, who has been behind bars since January 1996, left the courtroom a free man after a judge vacated his conviction on a charge of second-degree homicide.

Michael J. Okoniewski/The New York Times/Redux

Trial and Conviction

At trial, prosecutors pointed out that the blood could have come from the accomplice mentioned in the confession, and they hammered home the point that the details in the confession could only have come from first-hand knowledge of the crime.

After a short trial, Douglas Warney was convicted of the murder of William Beason and sentenced to 25 years.

Calling in the Innocence Project

When appeals failed, Warney in 2004 sought help from the Innocence Project, a nonprofit legal clinic that helps identify wrongly convicted individuals and secure their freedom. Set up at the Benjamin N. Cardozo School of Law in New York by lawyers Barry Scheck and Peter Neufeld, the Innocence Project specializes in using DNA technology to establish innocence.

The Innocence Project staff petitioned the court for additional DNA testing using new sensitive DNA probes, arguing that this might disclose evidence that would have resulted in a different verdict. The judge refused, ruling that the possibility that the blood found at the scene of the crime might match that of a criminal already in the state databank was "too speculative and improbable" to warrant the new tests.

New DNA Tests

Someone in the prosecutor's office must have been persuaded, however. Without notifying Warney's legal team, Project Innocence, or the court, this good Samaritan arranged for new DNA tests on the blood drops found at the crime scene. When compared to the New York State criminal DNA database, they hit a strong match. The blood was that of Eldred Johnson, in prison for slitting his landlady's throat in Utica two weeks before Beason was killed.

When confronted, the prisoner Johnson readily admitted to the stabbing of Beason. He said he was the sole killer and had never met Douglas Warney.

Another Look at Warney's Confession

This leaves the interesting question of how Warney's signed confession came to include such accurate information about the crime scene. It now appears he may have been fed critical details about the crime scene by the homicide detective leading the investigation, Sergeant John Gropp. It seems Gropp's partner had stepped out to get some papers when the confession was obtained. Sergeant Gropp died in March 2006.

DNA testing has become a pillar of the American criminal justice system. It has provided key evidence that has established the guilt of thousands of suspects beyond any reasonable doubt—and, as you see here, also provided the evidence that our criminal justice system sometimes convicts and sentences innocent people. Over more than two decades, the Innocence Project and other similar efforts have cleared hundreds of convicted people, strong proof that wrongful convictions are not isolated or rare events. DNA testing opens a window of hope for the wrongly convicted.

13.4 Genetic Engineering and Agriculture

> **LEARNING OBJECTIVE 13.4.1** Explain how genetic engineering is being used to increase food production.

Pest Resistance

An important effort of genetic engineers in agriculture has involved making crops resistant to insect pests without spraying with pesticides, a great saving to the environment. Consider cotton. Its fibers are a major source of raw material for clothing throughout the world, yet the plant itself can hardly survive in a field because many insects attack it. Over 40% of the chemical insecticides used today are employed to kill insects that eat cotton plants. The world's environment would greatly benefit if these thousands of tons of insecticide were not needed.

One successful alternative uses a kind of soil bacterium, *Bacillus thuringiensis* (Bt), that produces a protein that is toxic when eaten by crop pests. When the gene producing the Bt protein was inserted into the chromosomes of tomatoes, the plants began to manufacture Bt protein. While not harmful to humans, it makes the tomatoes highly toxic to hornworms, one of the most serious pests of commercial tomato crops; in cotton, it has the same effect against cotton borers, the chief pests of the plant.

Herbicide Resistance

A major advance has been the creation of crop plants that are resistant to the herbicide *glyphosate*, a powerful biodegradable herbicide that kills most actively growing plants. Glyphosate is used in orchards and agricultural fields to control weeds. Growing plants need to make a lot of protein, and glyphosate stops them from making protein by destroying an enzyme necessary for the manufacture of so-called aromatic amino acids (that is, amino acids that contain a ring structure). Humans are unaffected by glyphosate because we don't make aromatic amino acids—we obtain them from plants we eat! To make crop plants resistant to this powerful plant killer, genetic engineers screened thousands of organisms until they found a species of bacteria that could make aromatic amino acids in the presence of glyphosate. They then isolated the gene encoding the resistant enzyme and successfully introduced the gene into plants. They inserted the gene into the plants using DNA particle guns, also called gene guns. You can see in figure 13.9 how a DNA particle gun works. Small tungsten or gold pellets are coated with DNA (red in the figure) that contains the gene of interest and placed in the DNA particle gun. The DNA gun literally shoots the gene into plant cells in culture where the gene can be incorporated in the plant genome and then expressed. Plants that have been genetically engineered in this way are shown in figure 13.10. The two plants on top were genetically engineered to be resistant to glyphosate, the herbicide that killed the two plants at the bottom of the photo.

> **Putting the Concept to Work**
> What sort of bacterial gene would enable bacterial cells to make aromatic amino acids in the presence of glyphosate?

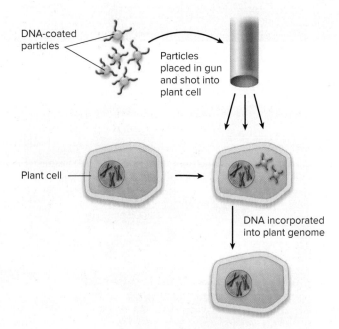

Figure 13.9 Shooting genes into cells.

A DNA particle gun, also called a gene gun, fires tungsten or gold particles coated with DNA into plant cells. The DNA-coated particles pass through the cell wall and into the cell, where the DNA is incorporated into the plant cell's DNA. The gene encoded by the DNA is expressed.

Figure 13.10 Genetically engineered herbicide resistance.

All four of these petunia plants were exposed to equal doses of an herbicide. The two on *top* were genetically engineered to be resistant to glyphosate, the active ingredient in the herbicide, whereas the two dead ones on the *bottom* were not.

Courtesy Monsanto Company

TABLE 13.2	Genetically Modified Crops
Rice	Genes have been added to commercial rice from daffodils for vitamin A, and from beans, fungi, and wild rice to supply dietary iron; transgenic strains that are cold-tolerant are under development.
Wheat 	New strains of wheat, resistant to the herbicide glyphosate, greatly reduce the need for tilling and so reduce loss of topsoil.
Soybean 	A major animal feed crop, soybeans tolerant of the herbicide glyphosate were used in over 94% of U.S. soybean acreage in 2014. Varieties are being developed that contain the *Bt* gene to protect the crop from insect pests without chemical pesticides.
Corn	Corn varieties resistant to insect pests (Bt corn) are widely planted (81% of U.S. acreage in 2016; fully 89% of corn acreage was planted with Roundup-resistant seeds in 2017). Varieties that are drought-resistant are being developed, as well as nutritionally improved lines.
Cotton 	Cotton crops are attacked by cotton bollworm, budworm, and other lepidopteran insects; more than 40% of all chemical pesticide tonnage worldwide is applied to cotton. A form of the *Bt* gene toxic to all lepidopterans but harmless to other insects has transformed cotton to a crop that requires few chemical pesticides. Over 85% of U.S. acreage is Bt cotton.
Peanut	The lesser cornstalk borer causes serious damage to peanut crops. An insect-resistant variety is under development by gene engineers to control this pest.
Potato	Verticillium wilt (a fungal disease) infects the water-conducting tissues of potatoes, reducing crop yields 40%. An antifungal gene from alfalfa reduces infections sixfold.

More Nutritious Crops

> **LEARNING OBJECTIVE 13.4.2** Describe how researchers genetically engineered a more nutritious type of rice.

The cultivation of genetically modified (GM) crops of corn, cotton, soybeans, and other plants (table 13.2) has become commonplace in the United States. In 2010, 90% of soybeans in the United States were planted with seeds genetically modified to be herbicide resistant. The result has been that less tillage is needed, and as a consequence, soil erosion is greatly reduced. Pest-resistant GM corn in 2010 comprised over 86% of all corn planted in the United States, and pest-resistant GM cotton comprised 93% of all cotton.

The real promise of plant genetic engineering is to produce genetically modified plants with desirable traits that directly benefit the consumer. One recent advance, nutritionally improved "golden" rice, gives us a hint of what is to come. In developing countries, large numbers of people live on simple diets that are poor sources of iron and vitamin A. Iron deficiency affects about 30% of the world population, and an estimated 250 million children are deficient in vitamin A. These deficiencies are especially severe in developing countries where the major staple food is rice.

To solve the problem of dietary iron deficiency among rice eaters, gene engineers first asked why rice is such a poor source of dietary iron. The problem, and the answer, proved to have three parts:

1. *Too little iron.* The proteins of rice endosperm have unusually low amounts of iron. To solve this problem, a ferritin gene (Fe in figure 13.11) was transferred into rice from beans. Ferritin is a protein with an extraordinarily high iron content, and so it greatly increased the iron content of the rice.
2. *Inhibition of iron absorption by the intestine.* Rice contains an unusually high concentration of a chemical called phytate, which inhibits iron absorption in the intestine—it stops your body from taking up the iron in the rice. To solve this problem, a gene encoding an enzyme called phytase (Pt) that destroys phytate was transferred into rice from a fungus.
3. *Too little sulfur for efficient iron absorption.* The human body requires sulfur for the uptake of iron, and rice has very little of it. To solve this problem, a gene encoding a sulfur-rich protein (S) was transferred into rice from wild rice.

To solve the problem of vitamin A deficiency, the same approach was taken. First, the problem was identified. It turns out rice goes only partway toward making vitamin A; there are no enzymes in rice to catalyze the last four steps. To solve the problem, genes encoding these four enzymes (abbreviated A_1 A_2 A_3 A_4) were added to rice from a flower, the daffodil.

Is Eating Genetically Modified Food Dangerous? Many consumers worry that when genetic engineers introduce novel genes into GM crops like rice, there may be dangerous consequences. Nor is rice the only

Figure 13.11 Transgenic "golden" rice.

Golden rice goes a long way toward addressing serious nutritional deficiencies in one of the world's major food crops. The development of this transgenic rice by researchers is the first key step toward real progress. Many years will be required to breed the golden rice genes into lines adapted to local conditions.

Courtesy Golden Rice Humanitarian Board: www.goldenrice.org

> **IMPLICATION FOR YOU** The "Free Rice Game" on the Internet (www.freerice.com) asks you quiz questions. For each answer you get right, 20 grains of rice are donated to the UN World Food Program to help end hunger. Since its inception in October 2007, almost 98 billion grains of rice have been donated, about 20,000,000 grains a day. After each answer you get right, you are given a harder question. How many questions in a row do you think you can get right? (The average is four.)

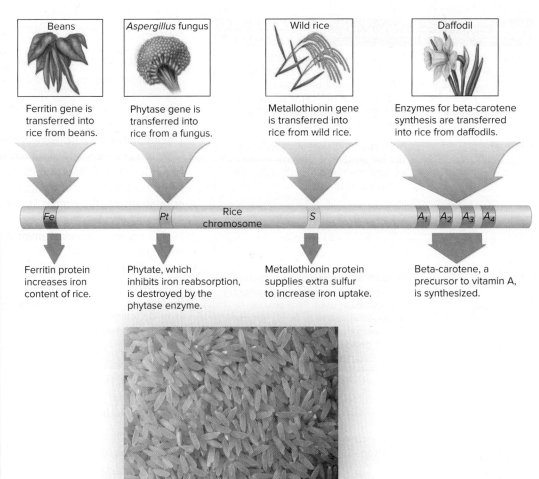

food that has been genetically modified (**table 13.2**). The introduction of glyphosate resistance into soybeans is a typical example. Is the soybean that results nutritionally different? No. But could introduced proteins such as the enzyme making the GM soybeans glyphosate tolerant cause a fatal immune reaction in some people? Because the potential danger of allergic reactions is quite real, every time a protein-encoding gene is introduced into a GM crop, it is necessary to carry out extensive tests of the introduced protein's allergen potential. No GM crop currently being produced in the United States contains a protein that acts as an allergen to humans. On this score, then, the risk of genetic engineering to the food supply seems to be slight.

Putting the Concept to Work
Seven genes were transferred into GM golden rice. How could researchers ensure that the genes were expressed in the rice cells?

Answering Your Questions About Cloning Your Dog

Did Barbra Streisand Really Clone Her Puppy?

Yes. When Streisand's beloved dog Samantha died at age 14, two puppies—Miss Violet and Miss Scarlett—were cloned from cells taken from Sammie's mouth and stomach. This was not ground-breaking biology: Since Snuppy the Afghan hound was cloned in 2005, hundreds of dogs have been cloned. The process is straightforward: A cell from the dog is fused with a canine egg whose nucleus has been removed. After the egg develops into an embryo, the embryo is transferred to a surrogate mother, which bears the puppy. The process works only about 33% to 40% of the time, however, and is expensive. It cost Barbra Streisand $50,000 a pup.

Can You Clone Your Cat? Your Hamster?

Sure. Pet cats are being cloned in much the same way as dogs. Indeed, any mammal can potentially be cloned—the first animal to be cloned was Dolly, a sheep. Most reproductive cloning of animals is done on the farm. The first Jersey cow was cloned in 2010, and thousands of cattle are now cloned each year. Horses were first cloned in 2003, and many of today's prized show horses and polo ponies are clones. Even camels are now being cloned.

How About a Dinosaur?

Since *Jurassic Park* was written by Michael Crichton in 1990, people have speculated about the possibility of bringing the great extinct dinosaurs back to life—of cloning *T. rex*. Unfortunately, cloning a dinosaur would require dinosaur DNA, and none has yet been found—not even in dino-biting mosquitoes encased in amber, as Crichton had proposed. The closest scientists have come is the discovery in 2016 of a fossil *T. rex* that was pregnant when it died. Fragmentary DNA was recovered from its bones, but not enough to sequence. DNA is just not stable enough to survive 66 million years. Scientists are trying to clone a woolly mammoth from 4,000-year-old tissue long frozen in the Arctic, but so far with no success.

Have People Been Cloned?

No, not yet. In 2018, scientists successfully cloned a macaque monkey, with the birth of two live and healthy females. As monkeys are primates and thus closely related to humans, there is every reason to believe that humans could be successfully cloned using the same procedures. The ethical problems with such an attempt are enormous, however. To state but one issue: The very low success rate of reproductive cloning means an attempt to clone a human would involve the loss of large numbers of human embryos and fetuses. In the United States, the reproductive cloning of a human is universally condemned but is not specifically illegal.

Would Human Clones Be Identical?

Clones are in essence identical twins. Like identical twins, they share the same DNA genome and so have identical potential. But no two people experience exactly the same things as they grow up. Differences will develop between clones for the same reason they do between identical twins. For example, the realization of highly heritable traits such as intelligence depends critically on childhood nutrition and mental stimulation. So no, human clones would not be identical to one another. But they sure would be a lot alike.

Would Human Clones Be Persons?

No, not in the sense that you are. The most distinguishing characteristic of humans is their individuality. No two people are alike. Meiosis sees to that. Not so with reproductive cloning. Dispensing with genetic recombination eliminates the variation that is the core of the human condition. So while a human clone might possess the same physical body as you or I and so satisfy the dictionary definition of a person, it lacks the potential uniqueness that every person possesses.

Might Human Clones Be Superior to Us?

The recent discovery of CRISPR and other techniques for editing genes raises the possibility of altering the genome of humans. While gene drives would take many generations to spread any gene edit through the world's humans, all human clones would possess any edits made to the human genome, raising the scary possibility of our redesigning who and what we are. Such an attempt to improve on what evolution has produced is far beyond our ability today but is not beyond possibility. It is a future we should avoid.

The Revolution in Cell Technology

13.5 Reproductive Cloning

LEARNING OBJECTIVE 13.5.1 Describe experiments that have demonstrated the possibility of cloning animals from adult tissue.

One of the most active and exciting areas of biology involves recently developed approaches to manipulating animal cells. In this section, you will encounter three areas where landmark progress is being made in cell technology: (1) reproductive cloning of farm animals, (2) stem cell research, (3) and gene therapy. Advances in cell technology hold the promise of revolutionizing our lives.

The idea of cloning animals was first suggested in 1938 by German embryologist Hans Spemann (called the "father of modern embryology"), who proposed what he called a "fantastical experiment": remove the nucleus from an egg cell (creating an enucleated egg) and put in its place a nucleus from another cell. When attempted many years later (figure 13.12), this experiment actually succeeded in frogs, sheep, monkeys, and many other animals. However, only donor nuclei extracted from early embryos seemed to work. After repeated failures using nuclei from adult cells, many researchers became convinced that the nuclei of animal cells become irreversibly committed to a developmental pathway after the first few cell divisions of the developing embryo.

Wilmut's Lamb

Then, in the 1990s, a key insight was made in Scotland by geneticist Keith Campbell, a specialist in studying the cell cycle of agricultural animals. Recall from chapter 8 that the division cycle of eukaryotic cells progresses in several stages. Campbell reasoned, "Maybe the egg and the donated nucleus need to be at the same stage in the cell cycle." This proved to be a key insight. In 1994, researchers succeeded in cloning farm animals from advanced embryos by first starving the cells, so that they paused at the beginning of the cell cycle. Two starved cells are thus synchronized at the same point in the cell cycle.

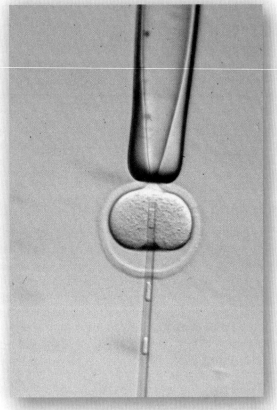

Figure 13.12 A cloning experiment.

In this photo, a nucleus is being injected from a micropipette *(bottom)* into an enucleated egg cell held in place by a pipette.

Handout/Getty Images

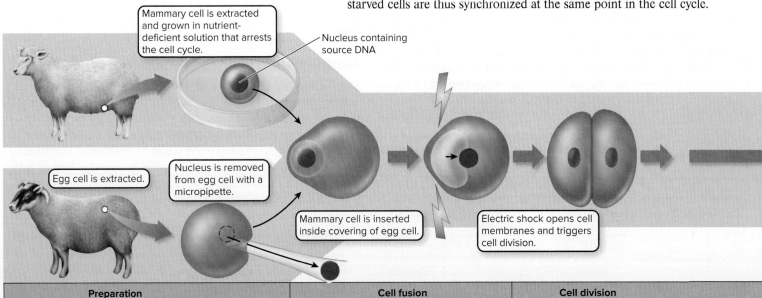

Figure 13.13 Wilmut's animal cloning experiment.

AFP PHOTO/FILES/ALESSANDRO ABBONIZIO/COLIN MCPHERSON/Getty Images

Campbell's colleague Ian Wilmut then attempted the key breakthrough, the experiment that had been eluding researchers: He set out to transfer the nucleus from an adult differentiated cell into an enucleated egg and to allow the resulting embryo to grow and develop in a surrogate mother, hopefully producing a healthy animal (figure 13.13). Approximately five months later, on July 5, 1996, the mother gave birth to a lamb. This lamb, "Dolly," was the first successful clone generated from an adult animal cell. Dolly grew into a healthy adult and went on to have healthy offspring normal in every respect.

Progress with Reproductive Cloning

Since Dolly's birth in 1996, scientists have successfully cloned a wide variety of farm animals with desired characteristics, including cows, pigs, goats, horses, and donkeys, as well as pets such as cats and dogs. Snuppy, the puppy in figure 13.14, was the first dog to be cloned. For most farm animals, cloning procedures have become increasingly efficient since Dolly was cloned. However, the development of clones into adults tends to go unexpectedly haywire. Almost none survives to live a normal life span.

The Importance of Gene Reprogramming

What is going wrong? It turns out that as mammalian eggs and sperm mature, their DNA is conditioned by the parent female or male, a process called reprogramming. Chemical changes are made to the DNA that alter when particular genes are expressed without changing the nucleotide sequences. In the years since Dolly, scientists have learned a lot about gene reprogramming, also called **epigenetics.** Epigenetics works by blocking the cell's ability to read certain genes. A gene is locked in the off position by adding a —CH_3 (methyl) group to some of its cytosine nucleotides. After a gene has been altered like this, the polymerase protein that is supposed to "read" the gene can no longer recognize it. The gene has been shut off.

We are only beginning to learn how to reprogram human DNA, so any attempt to clone a human is simply throwing stones in the dark, hoping to hit a target we cannot see. For this and many other reasons, human reproductive cloning is regarded as highly unethical.

Figure 13.14 Cloning the family pet.

This puppy named "Snuppy" is the first dog cloned. Beside him, to the left, is the adult male dog who provided the skin cell from which Snuppy was cloned. The dog in the photo on the right was Snuppy's surrogate mother.

Seoul National University/Getty Images

IMPLICATION FOR YOU Do you have a dog or cat? Do you think it would be ethical for you to have your dog or cat cloned when it grows old, so you can continue to enjoy its company for many more years? Then, in later years, do you clone the clone? What, if any, limits would you place on this process? Discuss.

Putting the Concept to Work
Why didn't gene reprogramming of gamete DNA prevent the successful cloning of Dolly?

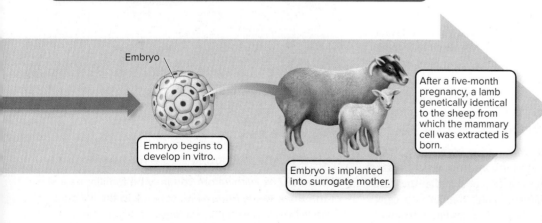

©AFP/Getty Images

13.6 Stem Cell Therapy

> **LEARNING OBJECTIVE 13.6.1** Describe how stem cells offer the possibility of replacing damaged human tissues.

You can see a mass of human embryonic stem cells in figure 13.15. Many are **totipotent**—able to form any body tissue and even an entire adult animal. What is an embryonic stem cell, and why is it totipotent? To answer this question, we need to consider for a moment where an embryo comes from. At the dawn of a human life, a sperm fertilizes an egg to create a single cell destined to become a child. As development commences, that cell begins to divide, producing after four divisions a small mass of 16 **embryonic stem cells.** Each of these embryonic stem cells has all of the genes needed to produce a normal individual.

As development proceeds, some of these embryonic stem cells become committed to forming specific types of tissues, such as nerve tissues and, after this step is taken, cannot ever produce any other kind of cell. For example, in nerve tissue, they are then called *nerve stem cells.* Others become specialized to produce blood cells, others to produce muscle tissue, and still others to form the other tissues of the body. Each major tissue is formed from its own kind of tissue-specific **adult stem cell.** Because an adult stem cell forms only that one kind of tissue, it is not totipotent.

Using Stem Cells to Repair Damaged Tissues

Stem cells offer the exciting possibility of restoring damaged tissues. To understand how embryonic stem cells could be used to repair damaged tissue, follow along in figure 13.17. A few days after fertilization, an embryonic stage called the *blastocyst* forms ❶. Embryonic stem cells are harvested from its *inner cell mass* or from cells of the embryo at a later stage ❷. These embryonic stem cells can be grown in tissue culture (see figure 13.15) and in principle be induced to form any type of tissue in the body ❸. The resulting healthy tissue can then be injected into the patient, where it will grow and replace damaged tissue ❹. Alternatively, where possible, adult stem cells can be isolated and, when injected back into the body, can form certain types of tissue cells.

Success in Mice. Both adult and embryonic stem cell transfer experiments have been carried out successfully in mice. Adult blood stem cells have been used to cure leukemia. Heart muscle cells grown from mouse embryonic stem cells have successfully replaced the damaged heart tissue of a living mouse. In other experiments, damaged spinal neurons have been partially repaired. In mouse brains, DOPA-producing neurons, whose progressive loss is responsible for Parkinson's disease, have been successfully replaced with embryonic stem cells, as have the islet cells of the pancreas, whose loss leads to juvenile diabetes. Because the course of development is broadly similar in all mammals, these experiments in mice suggest exciting possibilities for stem cell therapy in humans. The hope is that individuals with conditions such as Parkinson's disease (figure 13.16) might be partially or fully cured with stem cell therapy. As you might imagine, work proceeds intensively in this field of research.

Ethical Issues. There are ethical objections to using embryonic stems cells, but new experimental results hint at ways around this ethical maze. In 2007, researchers in two independent laboratories reported that they had engineered embryonic stemlike cells from normal adult human skin cells. The cells they created were pluripotent—they could differentiate into many different cell types. Whether pluripotency extends to totipotency is still being investigated. How were these cells transformed? The essential clue came six years earlier, when fusing adult cells with embryonic stem cells transformed the adult cells into pluripotent cells, as if factors had been transmitted to the adult cells that conferred pluripotency. Then, in a crucial

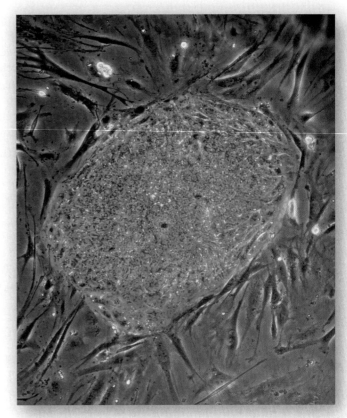

Figure 13.15 Human embryonic stem cells (×20).
This mass is a colony of undifferentiated human embryonic stem cells growing in tissue culture and surrounded by fibroblasts (elongated cells) that serve as a "feeder layer."
Mary Martin/Science Source

Figure 13.16 Promoting a cure for Parkinson's.
Michael J. Fox, with whom you may be familiar as a star of the *Back to the Future* film series and the TV show *Family Ties,* is a victim of Parkinson's disease and a prominent spokesman for those who suffer from it. Here you see him testifying before the U.S. Senate (along with fellow advocate Mary Tyler Moore) on the need for vigorous efforts to support research seeking a cure.
Alex Wong/Staff/Getty Images

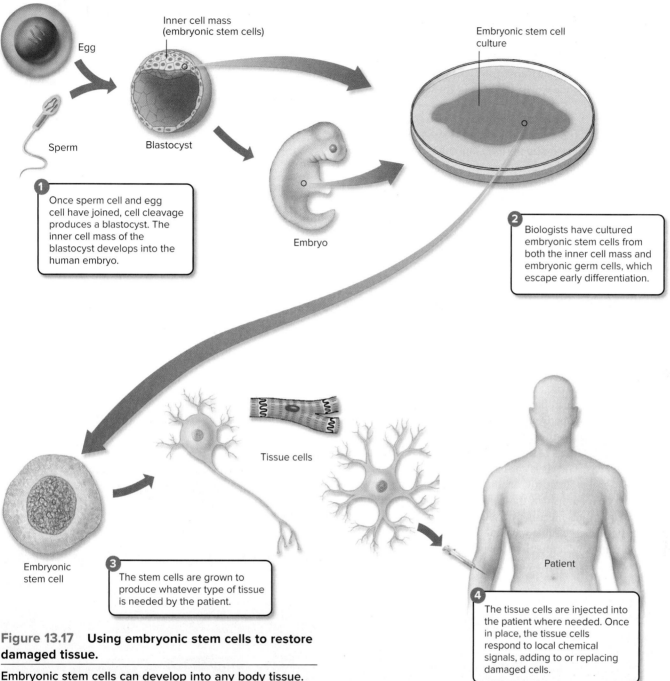

Figure 13.17 Using embryonic stem cells to restore damaged tissue.

Embryonic stem cells can develop into any body tissue. Methods are being developed for growing the tissue and using it in adults to repair damaged tissue, such as the brain cells of multiple sclerosis patients, heart muscle, and spinal nerves.

advance in 2006, Japanese cell biologist Shinya Yamanaka introduced into adult mammalian skin cells not the entire contents of an embryonic stem cell but just the genes for four transcription factors. The genes were transferred into the skin cells carried piggyback on viruses. Once inside, these four factors induced a series of events that converted the adult skin cells to pluripotency. In effect, he had found a way to reprogram the adult cells to be embryonic stem cells. For his discovery, Yamanaka was awarded the Nobel Prize for Medicine in 2012. From proof of principle in a laboratory culture dish to actual medical application is still a leap, but the possibility is exciting.

Putting the Concept to Work

Can you see any ethical objection to treating Michael J. Fox's Parkinson's with pluripotent cells obtained from his skin?

13.7 Gene Therapy

> **LEARNING OBJECTIVE 13.7.1** Explain attempts to cure hereditary disorders such as cystic fibrosis by transferring a healthy gene into the cells of affected tissues and why they have not yet succeeded.

The third major advance in cell technology involves introducing "healthy" genes into cells that lack them. For decades, scientists have sought to cure often-fatal genetic disorders such as cystic fibrosis, muscular dystrophy, and multiple sclerosis by replacing the defective gene with a functional one.

Early Success

A successful **gene transfer therapy** procedure was first demonstrated in 1990 (see section 13.2). Two girls were cured of a rare blood disorder due to a defective gene for the enzyme adenosine deaminase. Scientists isolated working copies of this gene and introduced them into bone marrow cells taken from the girls. The gene-modified bone marrow cells were allowed to proliferate, then were injected back into the girls. The girls recovered and stayed healthy. For the first time, a genetic disorder was cured by gene therapy.

The Rush to Cure Cystic Fibrosis

Researchers quickly set out to apply the new approach to one of the big killers, cystic fibrosis. The defective gene, labeled *cf*, had been isolated in 1989. Five years later, in 1994, researchers successfully transferred a healthy *cf* gene into a mouse that had a defective one—they in effect had cured cystic fibrosis in a mouse. They achieved this remarkable result by adding the *cf* gene to a virus that infected the lungs of the mouse, carrying the gene with it "piggyback" into the lung cells. The virus chosen as the "vector" was adenovirus (the red viruses in **figure 13.18**), a virus that causes colds and is very infective of lung cells. Very encouraged by these preliminary trials with mice, several labs set out to cure cystic fibrosis by transferring healthy copies of the *cf* gene into human patients the same way. But the gene-modified cells in the patients' lungs soon came under attack by the patients' own immune systems. The "healthy" *cf* genes were lost and with them any chance of a cure.

Problems with the Vector

In retrospect, although it was not obvious then, the problem with these early attempts seems predictable. Adenovirus causes colds. Do you know anyone who has never had a cold? When you get a cold, your body produces antibodies to fight off the infection, and so all of us have antibodies directed against adenovirus. We were introducing genes in a vector our bodies are primed to destroy.

A second serious problem is that when the adenovirus infects a cell, it inserts its DNA into the human chromosome. Unfortunately, it does so at a random location. This means that the insertion events could cause mutations—if the viral DNA inserts into the middle of a gene, it could inactivate that gene. Because the spot where the adenovirus inserts is random, some of the mutations that result can be expected to cause cancer, an unacceptable consequence first reported in 1999.

In 2003, gene therapy clinical trials attempting to cure severe combined immune deficiency (SCID) were halted when 5 of the 20 patients in the trial developed leukemia. Apparently, the vector had contained a small segment of DNA homologous to a leukemia-causing human gene. When the vector inserted there, the leukemia-causing genes were activated.

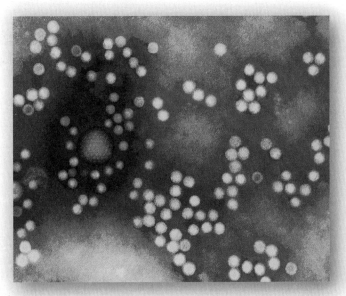

Figure 13.18 Adenovirus and AAV vectors (×200,000).

Adenovirus, the *red* virus particles above, has been used to carry healthy genes in clinical trials of gene therapy. Its use as a vector is problematic, however. AAV, the much smaller *bluish-green* virus particles seen in association with adenovirus here, lacks the many problems of adenovirus and is a much more promising gene transfer vector.

Dr. Linda M. Stannard, University of Cape Town/Science Source

Success with New Vectors

Researchers are now using a much more promising vector, a tiny virus called *adeno-associated virus* (AAV—the smaller bluish-green viruses in figure 13.18) that has only two genes. To create a vector for gene transfer, researchers remove both of the AAV genes. The shell that remains is still quite infective and can carry genes into patients (figure 13.19). AAV does not elicit a strong immune response—cells infected with AAV are not eliminated by a patient's immune system.

In 2011, researchers using AAV as a vector succeeded in attempts to treat hemophilia, which you will recall from chapter 10 is a blood clotting disorder due to an X-linked recessive mutation. The researchers stripped out the AAV genes of the vector, replacing them with the relatively small *factor IX* gene, and then injected the recombinant vector into patients. After a single injection, four of the six patients were able to stop the usual injections of factor IX, costing $300,000 a year, and make adequate amounts of factor IX on their own for over 22 months (the other two patients continued to need injections, but less frequently). Trials using AAV as a vector are under way for a wide variety of disorders.

The AAV vector is being used by researchers to produce an AIDS vaccine that is very effective in rhesus monkeys. The antibody protein produced by the gene researchers spliced into AAV works by simultaneously grabbing onto both of the two surface spikes of HIV that allow the virus to dock onto and enter human cells (section 16.4). Human trials are under way.

> **Putting the Concept to Work**
> Do you think the improved vector from the AAV clinical trials will work as a means of curing cystic fibrosis? Explain.

Figure 13.19 Using gene therapy to cure a retinal degenerative disease in dogs.

Researchers were able to use genes from healthy dogs to restore vision in dogs blinded by an inherited retinal degenerative disease. This disease also occurs in human infants and is caused by a defective gene that leads to early vision loss, degeneration of the retinas, and blindness. In the gene therapy experiments, genes from dogs without the disease ❶ were placed in an AAV vector ❷ and inserted into three-month-old dogs that were known to carry the defective gene ❸. All of the dogs had been blind since birth. Six weeks after the treatment, the dogs' eyes were producing the normal form of the gene's protein product, and by three months ❹, tests showed that the dogs' vision was restored. In 2017, this treatment was approved for humans suffering this disorder.

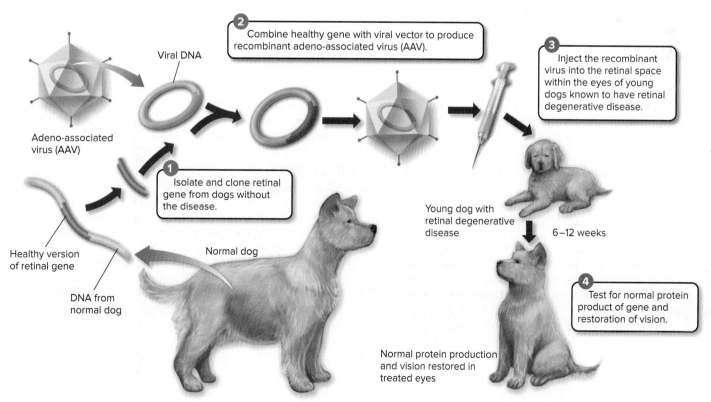

Putting the Chapter to Work

1. Restriction enzymes are important tools used by genetic engineers to cut DNA at a particular site. Below is a list of three restriction enzymes with their corresponding cut sites.

Which of these restriction enzymes would cut the provided DNA sequence at the highlighted region of interest?

ApaI cuts site G/GGCC
BseYI cuts site C/CCAG
XhoI cuts site C/TCGA

5'-AACCTCGG**CCCAG**TTCATC-3'

2. Genetically modified foods have been proven to be safe in all but one respect.

If you were to create a new genetically modified food, what is the one danger you would need to assess before releasing it to the public for consumption?

3. This chapter discusses three major advances in cell technology. Read the example below and identify the type of cell technology used.

In 2001, a young lady incurred a spinal cord injury in a car accident that left her paralyzed from the waist down. Surgeons opted to used stem cells to regenerate the damaged tissue. By 2004, she had regained the ability to use her legs and began walking.

Retracing the Learning Path

Genomics

13.1 The Human Genome

1. The genetic information of an organism, its genes and other DNA, is called its genome. The sequencing and study of genomes is an area of biology called genomics.
- The 3.2-billion-base-pair human genome contains about 25,000 genes, organized in different ways in the genome, with nearly 99% of the human genome containing noncoding segments.

Genetic Engineering

13.2 A Scientific Revolution

1. Genetic engineering is the process of moving genes from one organism to another. The key tools are so-called restriction enzymes, a special kind of enzyme that binds to specific short sequences of DNA and cuts them. The cut made is often a staggered cut producing "sticky ends," which are sections of single-stranded DNA. The significance of these sticky ends is that any two molecules of DNA cut with the same restriction enzyme will have the same sticky end sequences. This allows the two sources of DNA to combine through base pairing of their sticky ends.

2. Before transferring a eukaryotic gene into a bacterial cell, its intron regions must be removed. This is accomplished with the use of cDNA, a copy of the gene made from the processed mRNA that doesn't contain the introns.

3. DNA fingerprinting is a process using probes to compare samples of DNA. The probes bind to the DNA samples, creating restriction patterns that can be compared. Two DNA samples with the same restriction patterns are likely from the same source.

4. PCR technology greatly amplifies the amount of DNA in a tiny sample.

13.3 Genetic Engineering and Medicine

1. Genes encoding medically important proteins can be inserted into bacteria, which then produce large quantities of the proteins that can be administered to patients.

2. A gene that encodes a viral protein of a pathogenic virus can be inserted into the DNA of a harmless virus that serves as a vector. The vector carrying the recombinant DNA is injected into a human. The body elicits an immune response against the vector-born viral proteins, protecting the person from an infection by that pathogenic virus in the future.

13.4 Genetic Engineering and Agriculture

1. Genetic engineering has been used in crop plants to make them more cost-effective to grow or more nutritious.

2. GM rice has been altered to greatly improve its nutrient value.

3. So far, GM foods are safe to eat, but in some instances, they may create unintended hybrids in the field.

The Revolution in Cell Technology

13.5 Reproductive Cloning

1. The idea of cloning animals by placing the nucleus from an adult cell into an enucleated egg cell is not new, but early cloning attempts had mixed results. In 1996, Ian Wilmut succeeded in cloning a sheep by synchronizing the donated nucleus and the egg cell to the same stage of the cell cycle.

- Other animals have been successfully cloned, but problems and complications arise if modifications are not made to the DNA, which turns certain genes on or off, a process called epigenetic reprogramming.

13.6 Stem Cell Therapy

1. Embryonic stem cells are totipotent, able to divide and develop into any type of cell in the body or develop into an entire individual. Because of the totipotent nature of embryonic stem cells, they could be used to replace lost or damaged tissues.

13.7 Gene Therapy

1. Gene therapy attempts to cure a patient with a genetic disorder by replacing a defective gene with a "healthy" version. In theory, this should work, but early attempts to cure cystic fibrosis failed because of immunological reactions to the adenovirus vector used to carry the healthy genes into the patient.

- The recent focus of gene therapy has been to identify new vectors. Promising results in experiments using a virus called adeno-associated virus (AAV) has scientists hopeful that new vectors will eliminate the problems seen with adenovirus.

Inquiry and Analysis

Can Modified Genes Escape from GM Crops?

In section 13.4, the question of whether gene flow of GM crops posed a problem to the environment was discussed. A field experiment conducted in 2004 by the Environmental Protection Agency assessed the possibility that introduced genes could pass from genetically modified golf course grass to other plants. Investigators introduced a gene conferring herbicide resistance (the EPSP synthetase gene for resistance to glyphosate) into golf course bentgrass, *Agrostis stolonifera,* and then looked to see if the gene passed from the GM grass to other plants of the same species and if it passed to other related species.

The map at the bottom displays the setup of this elaborate field study. A total of 178 *A. stolonifera* plants were placed outside the golf course, many of them downwind. An additional 69 bentgrass plants were found to be already growing downwind, most of them the related species *A. gigantea.* Seeds were collected from each of these plants, and the DNA of resulting seedlings tested for the presence of the gene introduced into the GM golf course grass. In the graph, the upper red histogram (a **histogram** is a "bar graph" that sorts data into a series of discontinuous categories, the value of each bar representing the number of individuals in a category, or, as in this case, the average value of entries in that category) presents the relative frequency with which the gene was found in *A. stolonifera* plants located at various distances from the golf course. The lower blue histogram does the same for *A. gigantea* plants.

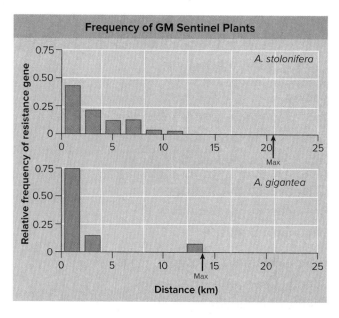

Analysis

1. **Applying Concepts**
   a. **Reading a Histogram.** Does the gene conferring resistance to herbicide pass to other plants of this species, *A. stolonifera*? To individuals of the related species *A. gigantea*?
   b. What is the maximal distance over which the herbicide resistance gene is transferred to other plants of this species? Of the related species? What are these distances, expressed in miles (km × 0.62 = mile)?
2. **Interpreting Data**
   a. What general statement can be made about the effect of distance on the likelihood that the herbicide resistance gene will pass to another plant?
   b. Are there any significant differences in the gene flow to individuals of *A. stolonifera* and to individuals of the related species *A. gigantea*?
3. **Making Inferences** What mechanism do you propose to account for this gene flow?
4. **Drawing Conclusions** Is it fair to conclude that genetically modified traits can pass from crops to other plants? What qualifications would you place on your conclusion?

Jack Hollingsworth/Getty Images

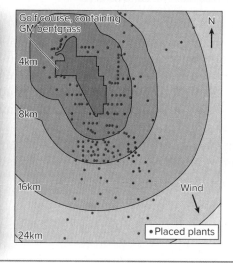

PART 4 The Evolution and Diversity of Life

14 Evolution and Natural Selection

LEARNING PATH ▼

Evolution
1. Darwin's Voyage on HMS *Beagle*
2. Darwin's Evidence
3. The Theory of Natural Selection

The Theory of Evolution
4. The Evidence for Evolution
5. Evolution's Critics

How Populations Evolve
6. Genetic Change in Populations
7. Agents of Evolution

Adaptation Within Populations
8. Sickle-Cell Disease
9. Peppered Moths and Industrial Melanism

How Species Form
10. The Biological Species Concept
11. Isolating Mechanisms

FIDO, A FAMILY DOG, has the spirit—and perhaps the genes—of a wolf.

Nick Ridley/Oxford Scientific/Getty Images

Your Dog's Inner Wolf

The dog family, the canids, is a diverse group of 34 species, including wolves, coyotes, dogs, and jackals, and is found in almost every environment—in forest, savanna, desert, and tundra. Both fossil and DNA evidence tell us the canids first appeared roughly 40 million years ago, early in the evolution of carnivores. The wolves and jackals of genus *Canis* are its oldest members. The coyotes are newcomers among the canids, having a common North American grey wolf ancestor only 2 million years ago.

Wolf, Coyote, or Jackal?

What of the domestic dog, man's best friend? Most of the world's estimated 400 million dogs are pets. The earliest fossils of the domestic dog date from 33,000 to 36,000 years ago. From which of the canids did the domestic dog arise? Darwin suggested that wolves, coyotes, and jackals—each of which has the same number of chromosomes as dogs (39 pairs, compared with 23 pairs in humans) and can interbreed and produce fertile offspring—may all have played a role, producing a complex dog ancestry that might prove

Dieter Hopf/Glow Images

Wolf

The molecular hunt for the first Fido began in 2002, when investigators sequencing small segments of mitochondrial DNA found the greatest diversity in southern China and concluded that Chinese village dogs evolved from now-extinct Chinese wolves about 32,000 years ago. These results finger Chinese hunter-gatherers as the dogs' first friends.

. . . But from Where?

A very different result was obtained in 2012 by an international team of scientists that focused its attention on one small portion of the mitochondrial DNA called the control region because it was known to vary a lot among mammals. The team collected blood, tissue, or hair from 140 dogs of 67 breeds and from 162 wolves, coyotes, and jackals living in North America, Europe, Asia, and Arabia.

What did they find? The domestic dog would appear to be an extremely close relative of the grey wolf, differing from it by at most 0.2% of the mitochondrial DNA sequence. Coyotes and jackals were a lot more different from dogs than wolves were. That settles the core issue. Dogs are indeed domesticated wolves. But did dogs evolve in south China 32,000 years ago, as the earlier study suggested? There is another possibility. The early Chinese dogs may have had wolflike DNA simply because they were dog-wolf hybrids. To settle the matter, it was necessary to look at ancient DNA.

Ancient DNA Settles the Matter: Europe

Recent advances pioneered by Svante Paabo in sequencing the Neanderthal genome allow the extraction of DNA from fossils—and, although rare, there are fossil dogs to look at. Comparing mitochondrial DNA of 27 ancient dog genomes, the investigators found that living dogs are most closely related to extinct wolves from Europe! When? Between 18,000 and 30,000 years ago. At that time, Northern Europe was covered in glaciers, and the southern portion was a grassland where our hunter-gatherer ancestors hunted mammoths and other large game. Dogs would have been great hunting companions.

impossible to unravel. In the 1950s, Nobel Prize–winning behaviorist Konrad Lorenz suggested an alternative that did not involve interbreeding—that some dog breeds derive from jackals, others from wolves.

Based on anatomy, most biologists put their money on the wolf, a judgment which has been confirmed repeatedly in recent years by comparisons of canid DNA. But *when* did dogs evolve from wolves and *where*? The DNA evidence seemed to be telling two different stories, only one of which could be correct.

Evolution

Figure 14.1 The theory of evolution by natural selection was proposed by Charles Darwin.

This rediscovered photograph appears to be the last ever taken of the great biologist. It was taken in 1881, the year before Darwin died.

Library of Congress Prints and Photographs Division [LC-DIG-ggbain-03485]

14.1 Darwin's Voyage on HMS *Beagle*

LEARNING OBJECTIVE 14.1.1 Recount the story of Darwin's voyage on the *Beagle*.

The great diversity of life on earth—ranging from bacteria to elephants and roses—is the result of a long process of **evolution,** the change that occurs in organisms' characteristics through time. In 1859, the English naturalist Charles Darwin (1809–82; figure 14.1) first suggested an explanation for why evolution occurs, a process he called *natural selection*. Biologists soon became convinced Darwin was right and now consider evolution one of the central concepts of the science of biology. In this chapter, we examine Darwin and evolution in detail, as the concepts we encounter will provide a solid foundation for your exploration of the living world.

The theory of evolution proposes that a population can change over time, sometimes forming a new species. A **species** is a group of populations that possess similar characteristics and whose members can interbreed and produce fertile offspring. This famous theory provides a good example of how a scientist develops a hypothesis—in this case, a hypothesis of how evolution occurs—and how, after much testing, the hypothesis is eventually accepted as a theory.

Darwin and Evolution

Charles Robert Darwin was an English naturalist who, after years of study and observation, wrote in 1859 one of the most famous and influential books of all time. This book, *On the Origin of Species by Means of Natural Selection, or The Preservation of Favoured Races in the Struggle for Life,* created a sensation when it was published, and the ideas Darwin expressed in it have played a central role in the development of human thought ever since.

In Darwin's time, most people believed that the various kinds of organisms and their individual structures resulted from direct actions of the Creator. Species were thought to be specially created and unchangeable over the course of time. In contrast to these views, a number of earlier philosophers had presented the view that living things must have changed during the history of life on earth. Darwin proposed a concept he called natural selection as a coherent, logical explanation for this process. Darwin's book, as its title indicates, presented a conclusion that differed sharply from conventional wisdom. Although his theory did not challenge the existence of a Divine Creator, Darwin argued that this Creator did not simply create things and then leave them forever unchanged. Instead, Darwin's God expressed Himself through the operation of natural laws that produced change over time—evolution.

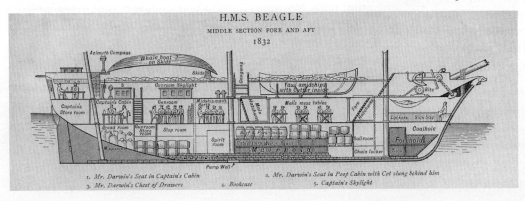

Figure 14.2 Cross section of HMS *Beagle*.

HMS *Beagle,* a 10-gun brig of 242 tons, only 90 feet in length, had a crew of 74 people! After he first saw the ship, Darwin wrote to his college professor Henslow: "The absolute want of room is an evil that nothing can surmount."

Chronicle/Alamy Stock Photo

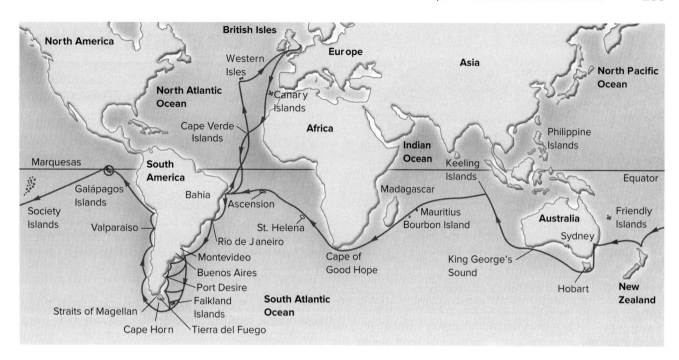

Figure 14.3 The five-year voyage of HMS *Beagle*.

Although the ship sailed around the world, most of the time was spent exploring the coasts and coastal islands of South America, such as the Galápagos Islands. Darwin's studies of the animals of these islands played a key role in the eventual development of his theory of evolution by means of natural selection.

Darwin's Voyage. The story of Darwin and his theory begins in 1831, when he was 22 years old. The small British naval vessel HMS *Beagle* (figure 14.2) was about to set sail on a five-year navigational mapping expedition around the coasts of South America. The red arrows in figure 14.3 indicate the route taken by HMS *Beagle*. The young (26-year-old) captain of HMS *Beagle*, unable by British naval tradition to have social contact with his crew and anticipating a voyage that would last many years, wanted a gentleman companion, someone to talk to. Indeed, the *Beagle*'s previous skipper had broken down and shot himself to death after three solitary years away from home.

On the recommendation of John Stevens Henslow, one of his professors at Cambridge University, Darwin, the son of a wealthy doctor and very much a gentleman, was selected to serve as the captain's companion, primarily to share his table at mealtime during every shipboard dinner of the long voyage. Darwin paid his own expenses and even brought along a manservant.

Darwin took on the role of ship's naturalist (the official naturalist, a man named Robert McKormick, left the ship before the first year was out). During this long voyage, Darwin had the chance to study a wide variety of plants and animals on continents and islands and in distant seas. He was able to explore the biological richness of the tropical forests, examine extraordinary fossils at the southern tip of South America, and observe the remarkable series of related forms of life on the Galápagos Islands.

A Life of Study. When Darwin returned from the voyage at the age of 27, he began a long period of study and contemplation. During the next 10 years, he published important books on several different subjects, including the formation of oceanic islands from coral reefs and the geology of South America. He then devoted eight years to a detailed study of barnacles, a group of small shelled marine animals that inhabit rocks and pilings, eventually writing a four-volume work on their classification and natural history. In 1842, Darwin and his family moved out of London to a country home at Down, in the county of Kent. In these pleasant surroundings, Darwin lived, studied, and wrote for the next 40 years.

> **Putting the Concept to Work**
> How long after his original voyage on HMS *Beagle* did Darwin publish *On the Origin of Species*?

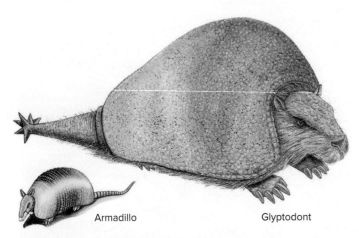

Armadillo | Glyptodont

Figure 14.4 Fossil evidence of evolution.

The now-extinct glyptodont was a large 2,000-kilogram South American armadillo (about the size of a small car), much larger than the modern armadillo, which weighs an average of about 4.5 kilograms and is about the size of a house cat. The similarity of fossils such as the glyptodonts to living organisms found in the same regions suggested to Darwin that evolution had taken place.

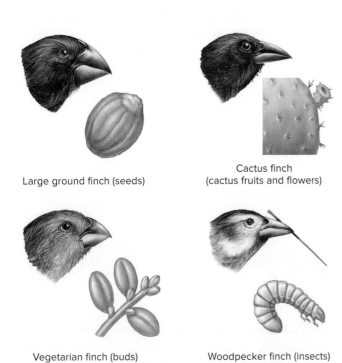

Large ground finch (seeds) | Cactus finch (cactus fruits and flowers)

Vegetarian finch (buds) | Woodpecker finch (insects)

Figure 14.5 Four Galápagos finches and what they eat.

Darwin observed 14 different species of finches on the Galápagos Islands, differing mainly in their beaks and feeding habits. These four finches eat very different food items, and Darwin surmised that the very different shapes of their beaks represented evolutionary adaptations improving their ability to do so.

14.2 Darwin's Evidence

> **LEARNING OBJECTIVE 14.2.1** Describe the fossils and patterns of life Darwin observed on the voyage of HMS *Beagle*.

One of the obstacles that had blocked the acceptance of any theory of evolution in Darwin's day was the incorrect notion, widely believed at that time, that the earth was only a few thousand years old. The discovery of thick layers of rocks, evidence of extensive and prolonged erosion, and the increasing numbers of diverse and unfamiliar fossils discovered during Darwin's time made this assertion seem less and less likely. The great geologist Charles Lyell, whose *Principles of Geology* (1830) Darwin read eagerly as he sailed on HMS *Beagle,* outlined for the first time the story of an ancient world of plants and animals in flux, species constantly becoming extinct while others were emerging.

What Darwin Saw

When HMS *Beagle* set sail, Darwin was fully convinced that species were immutable, meaning that they were not subject to being changed. Indeed, it was not until two or three years after his return that he began to seriously consider the possibility that they could change. Nevertheless, during his five years on the ship, Darwin observed a number of phenomena that were of central importance to him in reaching his ultimate conclusion.

Fossils In South America

In the rich fossil beds of southern South America, he observed fossils of the extinct armadillo shown on the right in figure 14.4. They were surprisingly similar in form to the armadillos that still lived in the same area, shown on the left. Why would similar living and fossil organisms be in the same area unless the earlier form had given rise to the other? Later, Darwin's observations would be strengthened by the discovery of other fossils that show intermediate characteristics, pointing to successive change.

The Galápagos Islands. Repeatedly, Darwin saw that the characteristics of similar species varied somewhat from place to place. These geographical patterns suggested to him that organismal lineages change gradually as individuals move into new habitats. On the Galápagos Islands, 900 kilometers (540 miles) off the coast of Ecuador, Darwin encountered a variety of different finches on the islands. The 14 species, although related, differed slightly in appearance. Darwin felt it most reasonable to assume all these birds had descended from a common ancestor blown by winds from the South American mainland several million years ago. Eating different foods, on different islands, the species had changed in different ways, most notably in the size of their beaks. The larger beak of the ground finch in the upper left of figure 14.5 is better suited to crack open the large seeds it eats. As the generations descended from the common ancestor, these ground finches changed and adapted, what Darwin referred to as "descent with modification".

In a more general sense, Darwin was struck by the fact that the plants and animals on these relatively young volcanic islands resembled those on the nearby coast of South America. If each one of these plants and animals had been created independently and simply placed on the Galápagos Islands, why didn't they resemble the plants and animals of islands with similar climates, such as those off the coast of Africa? Why did they resemble those of the adjacent South American coast instead?

> **Putting the Concept to Work**
> What did Darwin see on the Galápagos Islands that hinted at evolution?

14.3 The Theory of Natural Selection

> **LEARNING OBJECTIVE 14.3.1** Explain how Malthus's proposition that nature acts to limit population numbers leads to the process Darwin called natural selection.

It is one thing to observe the results of evolution, but quite another to understand how it happens. Darwin's great achievement lies in his formulation of the hypothesis that evolution occurs because of natural selection.

Darwin and Malthus

Of key importance to the development of Darwin's insight was his study of Thomas Malthus's *Essay on the Principle of Population* (1798). In his book, Malthus pointed out that populations of plants and animals (including human beings) tend to increase geometrically, while their food supply increases only arithmetically. In a geometric progression, the elements increase by a constant factor; the blue line in **figure 14.6** shows the progression 2, 6, 18, 54, . . . where each number is three times the preceding one. In an arithmetic progression, in contrast, the elements increase by a constant difference; the red line shows the progression 2, 4, 6, 8, . . . where each number is two greater than the preceding one.

Because populations increase geometrically, virtually any kind of animal or plant would cover the entire surface of the world within a surprisingly short time, if it could reproduce unchecked. Instead, population sizes of species remain fairly constant year after year, because death limits population numbers. Malthus's conclusion provided the key ingredient for Darwin to develop the hypothesis that evolution occurs by natural selection.

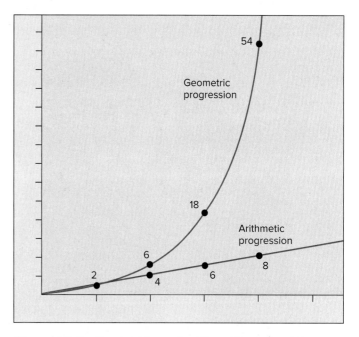

Figure 14.6 Geometric and arithmetic progressions.
An arithmetic progression increases by a constant difference (for example, units of 1 or 2 or 3), while a geometric progression increases by a constant factor (for example, by 2 or by 3 or by 4).

Natural Selection

Sparked by Malthus's ideas, Darwin saw that although every organism has the potential to produce more offspring than can survive, only a limited number actually do survive and produce further offspring. Many examples appear in nature. Sea turtles, for instance, will return to the beaches where they hatched to lay their eggs. Each female will lay about 100 eggs. The beach could be covered with thousands of hatchlings, like in **figure 14.7**, trying to make it to the water's edge. Less than 10% will actually reach adulthood and return to this beach to reproduce. Darwin combined his observation with what he had seen on the voyage of HMS *Beagle*, as well as with his own experiences in breeding domestic animals, and made an important association: Those individuals that possess physical, behavioral, or other attributes that help them live in their environment are more likely to survive than those that do not have these characteristics. Darwin called this process **natural selection.** By surviving, they gain the opportunity to pass on their favorable characteristics to their offspring. As the frequency of these characteristics will increase in the population, the nature of the population as a whole will gradually change, or evolve.

Figure 14.7 Sea turtle hatchlings.

These newly hatched sea turtles make their way to the ocean from their nests on the beach. Thousands of eggs may be laid on a beach during a spawning, but less than 10% of hatchlings will survive to adulthood. Natural predators, human egg poachers, and environmental challenges prevent the majority of offspring from surviving.

Rene Frederick/Digital Vision/Getty Images

Figure 14.8 Darwin greets his monkey ancestor.
In his time, Darwin was often portrayed unsympathetically, as in this drawing from an 1874 publication.
Science Source

The driving force of evolution that Darwin identified has often been referred to as survival of the fittest. However, this is not to say the biggest or the strongest always survive. These characteristics may be favorable in one environment but less favorable in another. The organisms that are "best suited" to their particular environment survive more often and therefore produce more offspring than others in the population and in this sense are the "fittest."

Darwin's theory that evolution is caused by natural selection provides a simple and direct explanation of biological diversity, or why animals are different in different places—because habitats differ in their requirements and opportunities, the organisms with characteristics favored locally by natural selection will tend to vary in different places. As we will discuss in section 14.7, there are five evolutionary forces that can affect biological diversity, although natural selection is the only evolutionary force that produces *adaptive* changes.

Darwin Drafts His Argument

Darwin drafted the overall argument for evolution by natural selection in a preliminary manuscript in 1842. After showing the manuscript to a few of his closest scientific friends, however, Darwin put it in a drawer and for 16 years turned to other research. No one knows for sure why Darwin did not publish his initial manuscript—it is very thorough and outlines his ideas in detail.

Wallace Has the Same Idea

The stimulus that finally brought Darwin's theory into print was an essay he received in 1858. A young English naturalist named Alfred Russel Wallace (1823-1913) sent the essay to Darwin from Malaysia; it concisely set forth the theory of evolution by means of natural selection, a theory Wallace had developed independently of Darwin. Like Darwin, Wallace had been greatly influenced by Malthus's 1798 book. After receiving Wallace's essay, Darwin arranged for a joint presentation of their ideas at a seminar in London. Neither attended the seminar: Wallace was still in Malaysia, and Darwin's young daughter drowned the day before the seminar took place. Darwin then expanded the 1842 manuscript he had written so long ago and submitted it for publication the following year as *On the Origin of Species by Means of Natural Selection*.

Publication of Darwin's Theory

Darwin's book appeared in November 1859 and caused an immediate sensation. Although people had long accepted that humans closely resembled apes in many characteristics, the possibility that there might be a direct evolutionary relationship was unacceptable to many. Darwin did not actually discuss this idea in his book, but it followed directly from the principles he outlined. In a subsequent book, *The Descent of Man*, Darwin presented the argument directly, building a powerful case that humans and living apes have common ancestors. Many people were deeply disturbed with the suggestion that human beings were descended from the same ancestor as apes, and his book on evolution caused Darwin to become a victim of the satirists of his day—the cartoon in **figure 14.8** is a vivid example. Darwin's arguments for the theory of evolution by natural selection were so compelling, however, that his views were almost completely accepted among scientists after the 1860s.

> **IMPLICATION FOR YOU** In 2008, the Spanish parliament approved resolutions granting to gorillas, chimpanzees, and orangutans statutory rights currently applicable only to humans. This was the first time a country had taken such action. The resolutions were based on the Great Ape Project, a framework designed by scientists and philosophers to provide humans' closest relatives with the right to life, liberty, and protection from torture. Zoos can still legally hold apes, but living conditions must be "optimal." Using apes in performances is illegal. The law also bans using apes in potentially useful research if it might hurt the ape in any way. Do you think this resolution, particularly the last condition, should be adopted in the United States? Explain.

> **Putting the Concept to Work**
> How did Malthus influence both Darwin and Wallace?

The Theory of Evolution

14.4 The Evidence for Evolution

> **LEARNING OBJECTIVE 14.4.1** Outline a four-step procedure to test the theory of evolution using fossil evidence.

The evidence that Darwin presented in *The Origin of Species* to support his theory of evolution was strong. We will now examine other lines of evidence supporting Darwin's theory, including information revealed by examining fossils, anatomical features, and molecules such as DNA and proteins.

The Fossil Record

The most direct evidence of evolution is found in the fossil record. **Fossils** are the preserved remains, tracks, or traces of once-living organisms. Fossils are created when organisms become buried in sediment. The calcium in bone or other hard tissue mineralizes, and the surrounding sediment eventually hardens to form rock. Most fossils are, in effect, skeletons, as shown by the dinosaur fossil in figure 14.9. In the rare cases when fossils form in very fine sediment, feathers may also be preserved. When remains are frozen or become suspended in amber (fossilized plant resin), however, the entire body may be preserved. The fossils contained in layers of sedimentary rock reveal a history of life on earth.

By dating the rock in which a fossil occurs, we can get an accurate idea of how old the fossil is. Rocks are dated by measuring the amount of certain radioactive isotopes, also called radioisotopes, in the rock (see Chapter 2). A radioisotope will break down, or decay, into other isotopes or elements. This occurs at a constant rate, and so the amount of a radioisotope present in the rock is an indication of the rock's age.

Using Fossils to Test the Theory of Evolution

If the theory of evolution is correct, then the fossils we see preserved in rock should represent a history of evolutionary change. The theory makes the clear prediction that a parade of successive changes should be seen, as first one change occurs and then another. If the theory of evolution is not correct, on the other hand, then such orderly change is not expected.

To test this prediction, biologists follow a very simple procedure:

1. *Assemble a collection of fossils of a particular group of organisms.* You might, for example, gather together a collection of fossil titanotheres, a hoofed mammal that lived about 50–35 million years ago.
2. *Date each of the fossils.* In dating the fossils, it is important to make no reference to what the fossil looks like. Imagine it as being concealed in a black box of rock, with only the box being dated.
3. *Order the fossils by their age.* Without looking in the "black boxes," place them in a series, from oldest to youngest.
4. *Now examine the fossils.* Do the differences between the fossils appear jumbled, or is there evidence of successive change as evolution predicts? You can judge for yourself in figure 14.10, which traces the fossils through time from the oldest at the bottom to the more recent at the top. During the 15 million years spanned by this collection of titanothere fossils, the small, bony protuberance located above the nose 50 million years ago evolved in a series of continuous changes into relatively large, blunt horns.

Figure 14.9 Dinosaur fossil of *Parasaurolophus*.

Andy Crawford/Getty Images

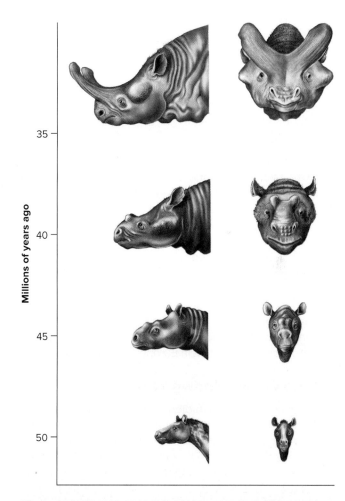

Figure 14.10 Testing the theory of evolution with fossil titanotheres.

Here you can trace the changes in a group of hoofed mammals known as titanotheres from about 50 million years ago (at the bottom) through 35 million years ago (at the top). During this time, the small, bony protuberance located above the nose 50 million years ago evolved into relatively large, blunt horns.

A Closer Look

Darwin and Moby Dick

Moby Dick, the white whale hunted by Captain Ahab in Herman Melville's novel, was a sperm whale. One of the ocean's great predators, a large sperm whale is a voracious meat-eater that may span over 60 feet and weigh 50 tons. A sperm whale is not a fish, though. A sperm whale is a mammal, just as you are! This raises an interesting question. If Darwin is right about the fossil record reflecting life's evolutionary past, then fossils tell us mammals evolved from reptiles on land at about the time of the dinosaurs. How did they end up back in the water?

Whales Once Lived on Land

Only in recent years have fossils been discovered that reveal the answer to this intriguing question. Whales, it turns out, are the descendants of four-legged land mammals that reinvaded the sea some 50 million years ago, much as seals and walruses are doing today.

. . . As Hippos

From what group of land mammals did whales arise? Researchers had long speculated that it might be a hoofed meat-eater with three toes known as a mesonychid, related to rhinoceroses. But, surprisingly, ankle bones from two newly described 50 million-year-old whale species are those of an artiodactyl, a four-toed mammal related to hippos, cattle, and pigs. Even more recently, Japanese researchers studying DNA have discovered unique genetic markers shared today only by whales and hippos.

Biologists now conclude that whales, like hippos, are descended from a group of early four-hoofed mammals called anthracotheres, modest-sized grazing animals with a piggish appearance abundant in Europe and Asia 50 million years ago.

The Oldest Known Whale

In Pakistan in 1994, biologists discovered its descendant, the oldest known whale. The fossil was 49 million years old, had four legs, each with four-toed feet and a little hoof at the tip of each toe. Dubbed *Ambulocetus* (walking whale), it was sharp-toothed and about the size of a large sea lion. Analysis of the minerals in its teeth reveal it drank fresh water, so like a seal it was not yet completely a marine animal. Its nostrils were on the end of the snout, like a dog's.

Appearing in the fossil record a few million years later is *Rodhocetus,* also seal-like but with smaller hind limbs and the teeth of an ocean water drinker. Its nostrils are shifted higher on the skull, halfway toward the top of the head.

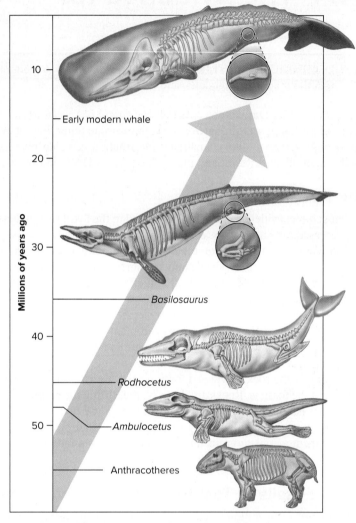

Almost 10 million years later, about 37 million years ago, we see the first representatives of *Basilosaurus,* a giant 60-foot-long serpentlike whale with shrunken hind legs still complete down to jointed knees and toes.

Modern Whales Appear

The earliest modern whales appear in the fossil record 15 million years ago. The nostrils are now in the top of the head, a "blowhole" that allows it to break the surface, inhale, and resubmerge without having to stop or tilt the head up. The hind legs are gone, with vestigial tiny bones remaining that are unattached to the pelvis. Still, today's whales retain all the genes used to code for legs—occasionally a whale is born having sprouted a leg or two.

So it seems to have taken 35 million years for a whale to evolve from the piglike ancestor of a hippopotamus—intermediate steps preserved in the fossil record for us to see. Darwin, who always believed that gaps in the vertebrate fossil record would eventually be filled in, would have been delighted.

It is important not to miss the key point of the result you see illustrated in figure 14.10: Evolution is an observation, not a conclusion. Because the dating of the samples is independent of what the samples are like, *successive change through time is a data statement*. While the statement that evolution is the result of natural selection is a theory advanced by Darwin, the statement that evolution has occurred is a factual observation.

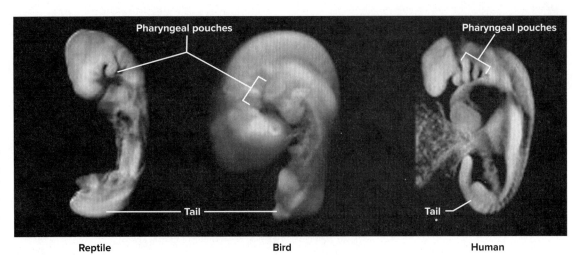

Figure 14.11 **Embryos show our early evolutionary history.**

These embryos, representing various vertebrate animals, show the primitive features that all vertebrates share early in their development, such as pharyngeal pouches and a tail.

Courtesy Michael Richardson and Ronan O'Rahilly

> **Putting the Concept to Work**
> If titanothere horns had *not* evolved continuously, how would you expect the fossil data analysis to have come out?

The Anatomical Record

> **LEARNING OBJECTIVE 14.4.2** Discriminate between analogous and homologous structures.

Much of the evolutionary history of vertebrates, which are animals with backbones, can be seen in the way in which their embryos develop. **Figure 14.11** shows three different embryos early in development, and as you can see, all vertebrate embryos have pharyngeal pouches (in fish, these develop into gill slits); also every vertebrate embryo has a long bony tail, even if the tail is not present in the fully developed animal. These relict developmental forms strongly suggest that all vertebrates share a basic set of developmental instructions.

As vertebrates have evolved, the same bones are sometimes still there but put to different uses, their presence betraying their evolutionary past. For example, the forelimbs of vertebrates are all **homologous structures;** that is, although the structures and functions of the bones have diverged, they are derived from the same body part present in a common ancestor. You can see in **figure 14.12** how the bones of the forelimb have been modified to make up the human forearm, wrist, and fingers, the wing of the bat, the full leg of the horse, and the paddle in the fin of the porpoise. Sometimes homologous structures are put to no use at all, such as the human appendix. The great apes, our closest relatives, have an appendix much larger than ours, attached to their gut tube. This appendix acts as a pouch to hold bacteria used in the ape gut to digest the cellulose walls of the plants it eats. Your appendix is a vestigial structure that now serves no function, as the human gut cannot digest cellulose.

Not all similar features are homologous. Sometimes features found in different lineages come to resemble each other as a result of parallel evolutionary adaptations to similar environments. This form of evolutionary change is referred to as *convergent evolution,* and these similar-looking features are called analogous structures. For example, the wings of birds, pterosaurs, and bats are **analogous structures,** with different bones modified through natural selection to serve the same function and therefore produce wings that look the same.

> **Putting the Concept to Work**
> If similar features are either homologous or analogous, how can both be used as arguments supporting evolution?

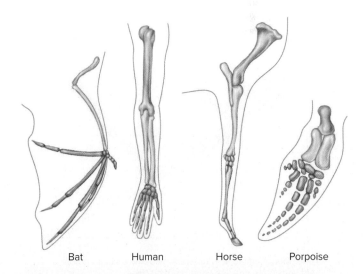

Figure 14.12 **Homology among vertebrate limbs.**

Homologies among the forelimbs of four mammals show the ways in which the proportions of the bones have changed in relation to the particular way of life of each organism. Although considerable differences can be seen in form and function, the same basic bones are present in each forelimb.

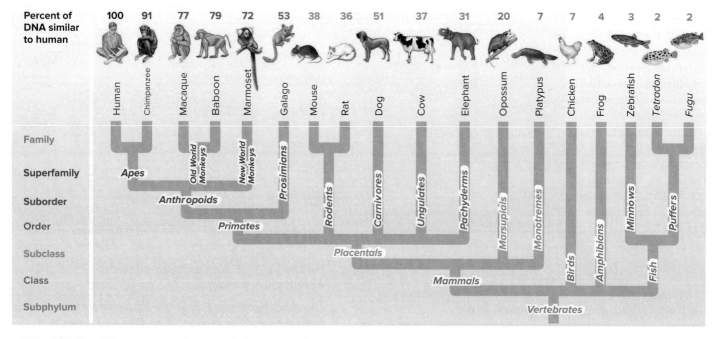

Figure 14.13 The more closely you are related, the more similar your DNA.

The number above each kind of vertebrate is the percent of DNA nucleotides in that animal's genome that match those of the human genome. The result is striking: As you proceed from very distant human relatives on the right to very close ones on the left, the genomes become more and more similar. Over their 300-million-year history, vertebrates have accumulated more and more genetic change in their DNA.

The Molecular Record

LEARNING OBJECTIVE 14.4.3 Describe the DNA evidence supporting the hypothesis of evolutionary divergence.

Traces of our evolutionary past are also evident at the molecular level. We possess the same set of color vision genes as our ancestors, only more complex, and during early development, we employ pattern formation genes that all animals share. Indeed, if you think about it, the fact that organisms have evolved from a series of simpler ancestors implies that a record of evolutionary change is present in the cells of each of us, in our DNA. According to evolutionary theory, new alleles arise from older ones by mutation and come to predominance through favorable selection. A series of evolutionary changes thus implies a continual accumulation of genetic changes in the DNA. Thus, evolutionary theory makes a clear prediction: Organisms that are more distantly related should have accumulated a greater number of evolutionary differences than two species that are more closely related.

This prediction is now subject to direct test. Recent DNA research allows us to directly compare the genomes of different organisms. The result is clear: For a broad array of vertebrates, the more distantly related two organisms are, the greater their genomic difference (figure 14.13).

This same pattern of divergence can be clearly seen at the protein level. Comparing the hemoglobin amino acid sequence of different species with the human sequence in figure 14.14, you can see that species more closely related to humans have fewer differences in the amino acid structure of their hemoglobin. Macaques, primates closely related to humans, have fewer differences from humans (only eight different amino acids) than do more distantly related mammals such as dogs (which have 32 different amino acids). Nonmammalian terrestrial vertebrates differ from us even more, and marine vertebrates are the most different of all.

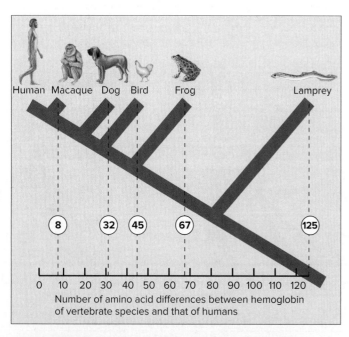

Figure 14.14 Molecules reflect evolutionary divergence.

The greater the evolutionary distance from humans (as revealed by the *blue* evolutionary tree based on the fossil record), the greater the number of amino acid differences in the vertebrate hemoglobin polypeptide.

Putting the Concept to Work

Can you propose a nonevolutionary explanation of the constant rate at which gene-encoding proteins such as hemoglobin accumulate mutations?

14.5 Evolution's Critics

LEARNING OBJECTIVE 14.5.1 Evaluate the scientific merit of common criticisms of Darwin's theory of evolution.

Critics of evolution have raised a variety of objections to Darwin's theory of evolution by natural selection:

1. **Evolution is not solidly demonstrated.** *"Evolution is just a theory,"* critics point out, as if theory meant lack of knowledge, some kind of guess. Scientists, however, use the word *theory* in a very different sense than the general public does (see section 1.6). Theories are the solid ground of science, supported with much experimental evidence and that of which we are most certain. Few of us doubt the theory of gravity because it is "just a theory."

2. **There are no fossil intermediates.** *"No one ever saw a fin on the way to becoming a leg,"* critics claim, pointing to the many gaps in the fossil record in Darwin's day. Since then, however, most fossil intermediates in vertebrate evolution have indeed been found. A clear line of fossils now traces the transition between whales and hoofed mammals, between reptiles and mammals, and between apes and humans. The fossil evidence of evolution between major forms is compelling (figure 14.15).

3. **The intelligent design argument.** *"The organs of living creatures are too complex for a random process to have produced them."* This classic "argument from design" was first proposed nearly 200 years ago by William Paley in his book *Natural Theology*—the existence of a clock is evidence of the existence of a clockmaker, Paley argues. Similarly, Darwin's critics argue that organs such as the mammalian ear are too complex to be due to blind evolution. There must have been a designer. Biologists do not agree. Complex structures like the mammalian ear evolved as a progression of slight improvements. The intermediates in the evolution of the mammalian ear are well documented in the fossil record, each favored by natural selection because they each had value—being able to amplify sound a little is better than not being able to amplify it at all. Nor is the solution always optimal, as your own eyes attest. As you can see in the blown-up image in figure 14.16, the receptor cells in the human eye are actually facing backward to the stimulus (light). No intelligent designer would design an eye backward!

4. **Evolution violates the second law of thermodynamics.** *"A jumble of soda cans doesn't by itself jump neatly into a stack. Things become more disorganized due to random events, not more organized."* Biologists point out that this argument ignores what the second law really says: Disorder increases *in a closed system*, which the earth most certainly is not. Energy enters the biosphere from the sun, fueling life and all the processes that organize it.

5. **Proteins are too improbable.** *"Hemoglobin has 141 amino acids. The probability that the first one would be leucine is 1/20 and that all 141 would be*

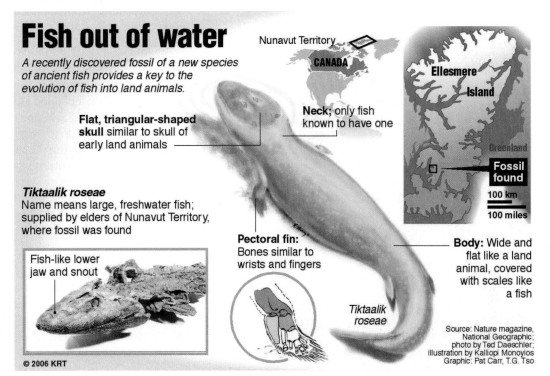

Figure 14.15 An intermediate fossil.

The animal that produced this fossil is an extinct lobe-finned fish (genus *Tiktaalik*) that lived approximately 375 million years ago. Coined by its discoverer as a "fishopod," it clearly has some characteristics that are fishlike, similar to fish that lived about 380 million years ago, and others that are more like early tetrapods, which lived about 365 million years ago. *Tiktaalik* appears to be a transitional animal, between fish and amphibians.

Tso KRT/Newscom

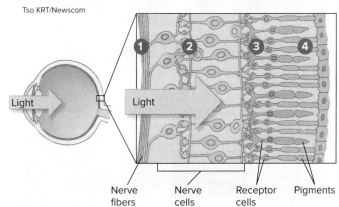

Figure 14.16 The vertebrate eye is poorly designed.

The visual pigments in a vertebrate eye that are stimulated by light are embedded in the retinal tissue, facing backward to the direction of the light. The light has to pass through nerve fibers ❶, nerve cells ❷, and receptor cells ❸ before reaching the pigments ❹.

the ones they are by chance is $(1/20)^{141}$, an impossibly rare event." You cannot use probability to argue backward. The probability that a student in a classroom has a particular birthday is 1/365; arguing this way, the probability that everyone in a class of 50 would have the birthdays they do is $(1/365)^{50}$—and yet there the class sits.

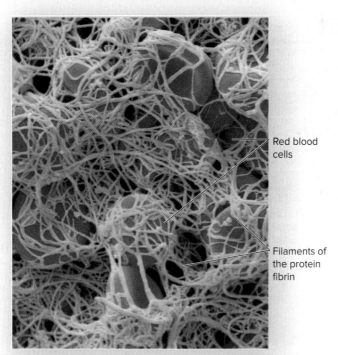

(a) A blood clot

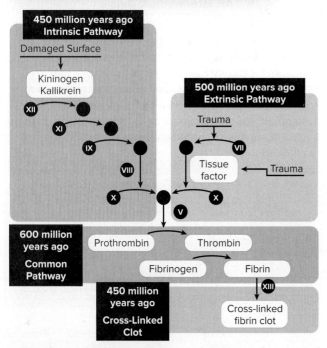

(b) The blood clotting system

Figure 14.17 How blood clotting evolved.
The blood clotting system evolved in steps, with new proteins adding on to the preceding step.

Steve Gschmeissner/Science Photo Library/Getty Images

The Irreducible Complexity Fallacy

The century-and-a-half-old "intelligent design" argument of William Paley has been recently articulated in a new molecular guise. Today's proponents of intelligent design now argue that the intricate molecular machinery of our cells is so elaborate that it cannot be explained by evolution from simpler stages—it is "irreducibly complex." Each part of a molecular machine plays a vital role. Remove just one, they claim, and cell molecular machinery cannot function.

Is Blood Clotting Irreducibly Complex? As an example of such an irreducibly complex system, intelligent design advocates point to the series of more than a dozen blood clotting proteins that act in our body to cause blood to clot around a wound. Take out any step in the complex cascade of reactions that leads to coagulation of blood, they say, and your body's blood would leak out from a cut like water from a ruptured pipe. If dozens of different proteins all must work correctly to clot blood, how could natural selection act to fashion any one of the individual proteins? No one protein does anything on its own, just as a portion of a watch doesn't tell time. Like Paley's watch, the blood clotting system must have been designed all at once, as a single functioning machine.

A Flawed Argument. What's wrong with this argument, as evolutionary scientists have been quick to point out, is that evolution acts on the system, not its parts. Natural selection can evolve a complex system because at every stage of its evolution, the system functions. Parts that improve function are added and, because of later changes, eventually become essential, in the same way that the second rung of a ladder becomes essential once you have added a third.

Blood Clotting Evolved a Bit at a Time. The mammalian blood clotting system, for example, has evolved in stages from much simpler systems (**figure 14.17**). The core of the vertebrate clotting system, called the "common pathway" (highlighted in blue), evolved at the dawn of the vertebrates approximately 600 million years ago and is found today in lampreys, the most primitive fish. As vertebrates evolved, proteins were added to the clotting system, improving its efficiency. The so-called extrinsic pathway (highlighted in pink), triggered by substances released from damaged tissues, was added 500 million years ago. Each step in the pathway amplifies what goes before, so adding the extrinsic pathway greatly increases the amplification and thus the sensitivity of the system. Fifty million years later, a third component was added, the so-called intrinsic pathway (highlighted in tan). It is triggered by contact with the jagged surfaces produced by injury. Again, amplification and sensitivity were increased to ultimately end up with blood clots formed by the cross linking of a protein called fibrin (highlighted in green). At each stage as the clotting system evolved to become more complex, its overall performance came to depend on the added elements. Mammalian clotting, which utilizes all three pathways, no longer functions if any one of them is disabled. Blood clotting has become "irreducibly complex"—as the result of Darwinian evolution. Intelligent design proponents claim that complex cellular and molecular processes can't be explained by Darwinism. Indeed, examination of the human genome reveals that the cluster of blood clotting genes arose through duplication of genes, with increasing amounts of change. The evolution of the blood clotting system is an observation, not a surmise. Its irreducible complexity is a fallacy.

> **Putting the Concept to Work**
> Is the theory of intelligent design a scientific theory? Explain.

How Populations Evolve

14.6 Genetic Change in Populations

> **LEARNING OBJECTIVE 14.6.1** State the Hardy–Weinberg rule.

Darwin was puzzled why dominant **alleles** (alternative forms of a gene) do not drive recessive alleles out of populations. The solution to this puzzle was explained in 1908 by G. H. Hardy and W. Weinberg who studied **allele frequencies** (the proportion of alleles of a particular type in a population). Hardy and Weinberg pointed out that in a large population, in the absence of forces that change allele frequencies, the original genotype proportions remain constant from generation to generation. Dominant alleles do not, in fact, replace recessive ones. Because their proportions do not change, the genotypes are said to be in **Hardy–Weinberg equilibrium**.

Figure 14.18 Hardy–Weinberg equilibrium.
In the absence of factors that alter them, the frequencies of gametes, genotypes, and phenotypes remain constant generation after generation. The example shown here involves a population of 1,000 cats in which 160 are white and 840 are black. White cats are *bb*, and black cats are *BB* or *Bb*. The potential crosses in this cat population can be determined using a Punnett square analysis.

Allele Frequencies

The Hardy–Weinberg rule is viewed as a baseline to which the frequencies of alleles in a population can be compared. If the allele frequencies are not changing (they are in Hardy–Weinberg equilibrium), the population is not evolving. If, however, allele frequencies differ greatly from what would be expected under Hardy–Weinberg equilibrium, then the population is undergoing evolutionary change.

Hardy and Weinberg came to their conclusion by analyzing the frequencies of alleles in successive generations. The **frequency** of something is defined as the proportion of individuals with a certain characteristic, compared to the entire population. Thus, in the population of 1,000 cats shown in figure 14.18, there are 840 black cats and 160 white cats. To determine the frequency of black cats, divide 840 by 1,000 (840/1,000), which is 0.84. The frequency of white cats is 160/1,000 = 0.16.

Knowing the frequency of the phenotype, one can calculate the frequency of the genotypes and alleles in the population. By convention, the frequency of the dominant (and usually more common) of two alleles (in this case, *B* for the black allele) is designated by the letter *p* and that of the recessive allele (*b* for the white allele) by the letter *q*. Because there are only two alleles, the sum of *p* and *q* must always equal 1 ($p + q = 1$).

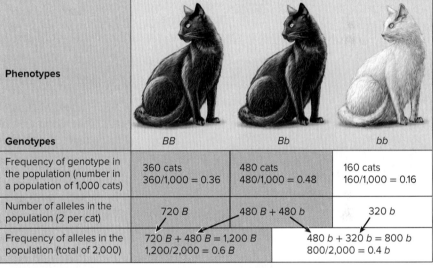

Phenotypes			
Genotypes	BB	Bb	bb
Frequency of genotype in the population (number in a population of 1,000 cats)	360 cats 360/1,000 = 0.36	480 cats 480/1,000 = 0.48	160 cats 160/1,000 = 0.16
Number of alleles in the population (2 per cat)	720 B	480 B + 480 b	320 b
Frequency of alleles in the population (total of 2,000)	720 B + 480 B = 1,200 B 1,200/2,000 = 0.6 B		480 b + 320 b = 800 b 800/2,000 = 0.4 b

The Hardy–Weinberg Equation

In algebraic terms, the Hardy–Weinberg equilibrium is written as an equation. For a gene with two alternative alleles, *B* (frequency *p*) and *b* (frequency *q*), the equation looks like this:

$$p^2 + 2pq + q^2 = 1$$

- p^2: Individuals homozygous for allele *B*
- $2pq$: Individuals heterozygous for alleles *B* and *b*
- q^2: Individuals homozygous for allele *b*

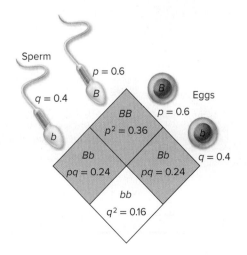

> **Putting the Concept to Work**
> Does the sum of allele frequencies always equal the sum of genotype frequencies? Discuss.

TABLE 14.1 Agents of Evolution

Factor		Description
Mutation	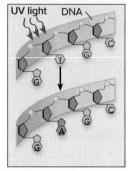	Individual mutations occur so rarely that mutation alone does not change allele frequency much.
Migration		Migration acts to promote evolutionary change by enabling populations that exchange members to converge toward one another.
Genetic drift		Chance events may result in the loss of individuals and therefore the loss of alleles in a population. Usually occurs only in very small populations.
Nonrandom mating	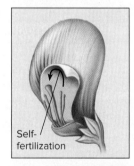	Inbreeding is the most common form of nonrandom mating. It does not alter allele frequency but decreases the proportion of heterozygotes ($2pq$).
Selection		The only form that produces *adaptive* evolutionary changes. Only rapid for allele frequency greater than .01

14.7 Agents of Evolution

> **LEARNING OBJECTIVE 14.7.1** Discuss five evolutionary forces that have the potential to significantly alter allele and genotype frequencies in populations.

Many factors can alter allele frequencies. But only five alter the proportions of homozygotes and heterozygotes enough to produce significant deviations from the proportions predicted by the Hardy-Weinberg rule (**table 14.1**):

1. Mutation
2. Migration
3. Genetic drift
4. Nonrandom mating
5. Selection

Mutation

A **mutation** is a change in a nucleotide sequence in DNA. For example, a T nucleotide could undergo a mutation and be replaced with an A nucleotide. Mutation from one allele to another obviously can change the proportions of particular alleles in a population. But mutation rates are generally too low to significantly alter Hardy-Weinberg proportions of common alleles. Many genes mutate 1 to 10 times per 100,000 cell divisions. Some of these mutations are harmful, whereas others are neutral or, even rarer, beneficial. Also, the mutations must affect the DNA of the germ cells (egg and sperm), or the mutation will not be passed on to offspring. The mutation rate is so slow that few populations are around long enough to accumulate significant numbers of mutations. However, no matter how rare, mutation is the ultimate source of genetic variation in a population.

Migration

Migration is defined in genetic terms as the movement of individuals between populations. It can be a powerful force, upsetting the genetic stability of natural populations. Migration includes movement of individuals into a population, called *immigration,* or the movement of individuals out of a population, called *emigration.* If the characteristics of the newly arrived individuals differ from those already there and if the newly arrived individuals adapt to survive in the new area and mate successfully, then the genetic composition of the receiving population may be altered.

Sometimes migration is not obvious. Subtle movements include the drifting of gametes of plants or of the immature stages of marine organisms from one place to another. For example, a bee can carry pollen (structures that contain male gametes) from a flower in one population to a flower in another population. By doing this, the bee may be introducing new alleles into a population. However it occurs, migration can alter the genetic characteristics of populations and cause a population to be out of Hardy-Weinberg equilibrium. Thus, migration can cause evolutionary change. The magnitude of effects of migration is based on two factors: (1) the proportion of migrants in the population and (2) the difference in allele frequencies between the migrants and the original population. The actual evolutionary impact of migration is difficult to assess and depends heavily on the selective forces prevailing at the different places where the populations occur.

Genetic Drift

In small populations, the frequencies of particular alleles may be changed drastically by chance alone. In an extreme case, individual alleles of a given gene may all be represented in few individuals and may be accidentally lost if those individuals fail to reproduce or die. This loss of individuals and their alleles is due to random events rather than the fitness of the individuals carrying those alleles. This is not to say that alleles are always lost, but allele frequencies appear to change randomly, as if the frequencies were drifting; thus, random changes in allele frequencies is known as **genetic drift.** A series of small populations that are isolated from one another may come to differ strongly as a result of genetic drift.

When one or a few individuals migrate and become the founders of a new, isolated population at some distance from their place of origin, the alleles that they carry, even if rare in the source population, will become a significant fraction of the new population's genetic endowment. This is called the **founder effect.** As a result of the founder effect, rare alleles and combinations often become more common in new, isolated populations. The founder effect is particularly important in the evolution of organisms that occur on oceanic islands, such as the Galápagos Islands that Darwin visited. Most of the kinds of organisms that occur in such areas were probably derived from one or a few initial founders. In a similar way, isolated human populations are often dominated by the genetic features that were characteristic of their founders, particularly if only a few individuals were involved initially (figure 14.19).

Nonrandom Mating

Individuals with certain genotypes sometimes mate with one another either more or less frequently than would be expected on a random basis, a phenomenon known as **nonrandom mating.** One type of nonrandom mating is **inbreeding,** or mating with relatives, such as in the self-fertilization of a flower. Inbreeding increases the proportions of individuals that are homozygous because no individuals mate with any genotype but their own. As a result, inbred populations contain more homozygous individuals than predicted by the Hardy-Weinberg rule. This is why populations of self-fertilizing plants contain so many more homozygous individuals than outcrossing plants, which have a higher proportion of heterozygous individuals. Nonrandom mating alters genotype frequencies but not allele frequencies—the alleles are just distributed differently among the offspring.

Selection

As Darwin pointed out, some individuals leave behind more progeny than others, and the likelihood they will do so is affected by their inherited characteristics. The result of this process is called **selection** and was familiar even in Darwin's day to breeders of domestic animals. In so-called **artificial selection,** the breeder selects for the desired characteristics. For example, mating larger animals with each other produces offspring that are larger. In **natural selection,** Darwin suggested the environment plays this role, with conditions in nature determining which kinds of individuals in a population are the most fit (best suited to their environment) and so affecting the proportions of genes among individuals of future populations. The environment imposes the conditions that determine the results of selection and, thus, the direction of evolution.

> **Putting the Concept to Work**
> If a population is in Hardy–Weinberg equilibrium, is it evolving?

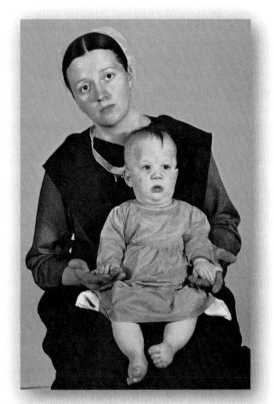

Figure 14.19 The founder effect.

This Amish woman is holding her child, who has Ellis-van Creveld syndrome. The characteristic symptoms are short limbs, dwarfed stature, and extra fingers. This disorder was introduced in the Amish community by one of its founders in the eighteenth century and persists to this day because of reproductive isolation.

Courtesy Dr. Victor A. McKusick, Johns Hopkins University

Kinds of Selection

> **LEARNING OBJECTIVE 14.7.2** Compare the operations of stabilizing, disruptive, and directional selection.

Selection operates in natural populations of a species as skill does in a football game. In any individual game, it can be difficult to predict the winner because chance can play an important role in the outcome. But over a long season, the teams with the most skillful players usually win the most games. In nature, those individuals best suited to their environments tend to win the evolutionary game by leaving the most offspring, although chance can play a major role in the life of any one individual. While you cannot predict the fate of any one individual, just as you cannot predict "heads" or "tails" in any one coin toss, it is possible to predict which trait will tend to become more common in populations of a species. For example, it is possible to predict the proportion of heads after many coin tosses.

In nature, many traits, perhaps most, are affected by more than one gene. The interactions between genes are typically complex, as you saw in chapter 10. For example, alleles of many different genes play a role in determining human height (see figure 10.12). In such cases, selection operates on all the genes, influencing most strongly those that make the greatest contribution to the phenotype. How selection changes the population depends on which genotypes are favored. Three types of natural selection have been identified: (1) stabilizing selection, (2) disruptive selection, and (3) directional selection. Figure 14.20 shows the results of these three types of selection on body size.

Figure 14.20 Three kinds of natural selection.
In the top panels, the *blue* areas indicate the phenotypes that are being selected for, and the *red* areas are the phenotypes that are not being preferentially selected for. The bottom panels show the phenotypic results of the selection. (a) In *stabilizing selection,* individuals with midrange phenotypes are favored, with selection acting against both ends of the range of phenotypes.
(b) In *disruptive selection,* individuals in the middle of the range of phenotypes of a certain trait are selected against, and the extreme forms of the trait are favored.
(c) In *directional selection,* individuals concentrated toward one extreme of the array of phenotypes are favored.

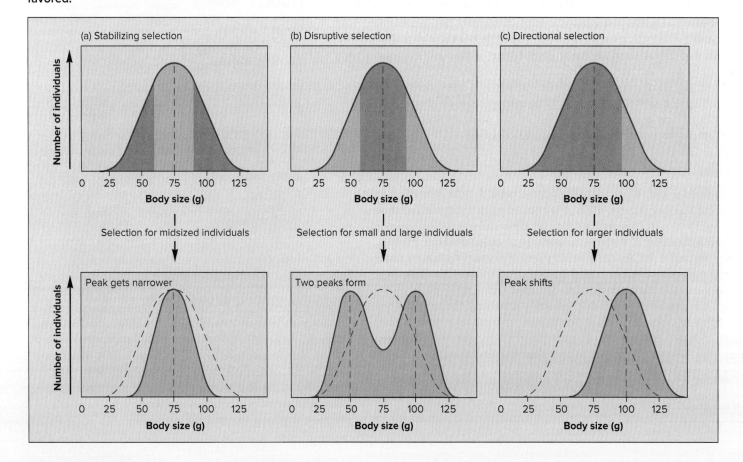

Stabilizing Selection

When selection acts to eliminate both extremes from an array of phenotypes, the result is an increase in the frequency of the already common intermediate phenotype. This is called **stabilizing selection.** In effect, selection is operating to prevent change away from the middle range of values. In a classic study carried out after an "uncommonly severe storm of snow, rain, and sleet" on February 1, 1898, a total of 136 starving English sparrows were collected and brought to the laboratory of H. C. Bumpus at Brown University in Providence, Rhode Island. Of these, 64 died and 72 survived. Bumpus took standard measurements on all the birds. He found that among the female birds that perished were many more individuals that had extreme measurements, either very large or very small. Selection had acted most strongly against these "extreme-sized" birds. Stabilizing selection does not change which phenotype is the most common of the population—the average-sized birds were already the most common phenotype—but rather makes it even more common by eliminating extremes. Many examples similar to Bumpus's female sparrows are known. In humans, infants with intermediate weight at birth have the highest survival rate (figure 14.21a).

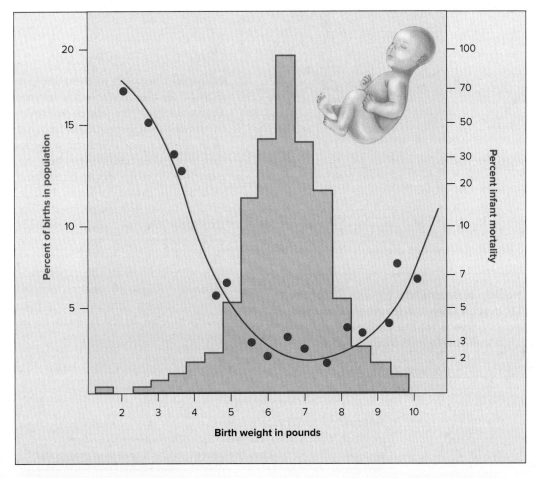

(a) **Stabilizing selection for birth weight.** The death rate among human babies is lowest at an intermediate birth weight between 7 and 8 pounds, indicated by the *red* line. The intermediate weights are also the most common in the population, indicated by the *blue* area. Larger and smaller babies both occur less frequently and have a greater tendency to die at or near birth.

Figure 14.21 Examples of selection.

(b) **Disruptive selection for large and small beaks.** Differences in beak size in the black-bellied seedcracker finch of West Africa are the result of disruptive selection for two distinct food sources.

Disruptive Selection

In some situations, selection acts to eliminate the intermediate type (figure 14.20b), resulting in the two more extreme phenotypes becoming more common in the population. This type of selection is called **disruptive selection**. A clear example is the different beak sizes of the African black-bellied seedcracker finch *Pyrenestes ostrinus* (figure 14.21b). Populations of these birds contain individuals with large and small beaks but very few individuals with intermediate-sized beaks. As their name implies, these birds feed on seeds, and the available seeds fall into two size categories: large and small. Only large-beaked birds, like the one on the left in the figure, can open the tough shells of large seeds, whereas birds with the smallest beaks, like the one on the right, are more adept at handling small seeds. Birds with intermediate-sized beaks are at a disadvantage with both seed types: unable to open large seeds and too clumsy to efficiently process small seeds. Consequently, selection acts to eliminate the intermediate phenotypes, partitioning the population into two distinct groups.

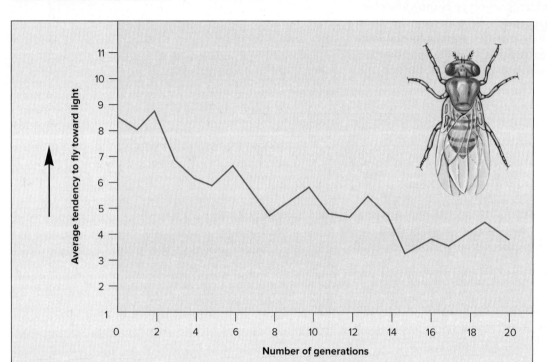

(c) **Directional selection for negative phototropism in Drosophila.** Individuals of the fly *Drosophila* were selectively bred. Flies that moved toward light were discarded; only flies that moved away from light were used as parents for the next generation. After 20 generations, the offspring had an ever greater tendency to avoid light.

Figure 14.21 Examples of selection (*cont'd*)

Directional Selection

When selection acts to eliminate one extreme from an array of phenotypes (figure 14.20c), the alleles determining this extreme become less frequent in the population. This form of selection is called **directional selection**. An experiment set up to test directional selection is shown in figure 14.21c. In the *Drosophila* population, flies that flew toward light, a behavior called phototropism, were eliminated from the population. The remaining flies were mated and the experiment repeated. After 20 generations of selected mating, flies exhibiting phototropism were far less frequent in the population.

> **Putting the Concept to Work**
> Selection for faster racehorses is no longer very successful, although it used to be. Can you think of a reason why and what might be done to increase the success of such breeding programs?

Adaptation Within Populations

14.8 Sickle-Cell Disease

> **LEARNING OBJECTIVE 14.8.1** Explain how stabilizing selection maintains sickle-cell disease in Central Africa.

In the time since Darwin suggested the pivotal role of natural selection in evolution, many examples have been found in which natural selection is clearly acting to change the genetic makeup of species, just as Darwin predicted. Here we will examine two examples: (1) sickle-cell disease (a defect in human hemoglobin proteins) and (2) industrial melanism in European moths.

A Mutation in the Hemoglobin Gene

Sickle-cell disease (which used to be called sickle-cell anemia) is a hereditary disease affecting hemoglobin molecules in the blood. It was first detected in 1904 in Chicago in a blood examination of an individual complaining of tiredness. You can see the original doctor's report in figure 14.22. The disorder arises as a result of a single nucleotide change in the gene encoding β-hemoglobin, one of the key proteins used by red blood cells to transport oxygen. The sickle-cell mutation changes the sixth amino acid in the β-hemoglobin chain (position B6) from glutamic acid (very polar) to valine (nonpolar). The unhappy result of this change is that the nonpolar *valine* at position B6, protruding from a corner of the hemoglobin molecule, fits nicely into a nonpolar pocket on the opposite side of another hemoglobin molecule; the nonpolar regions associate with each other. As the two-molecule unit that forms still has both a B6 valine and an opposite nonpolar pocket, other hemoglobins clump on, and long chains form as in figure 14.23a. The result is the deformed "sickle-shaped" red blood cell you see in figure 14.23b. In normal hemoglobin, by contrast, the polar amino acid *glutamic acid* occurs at position B6. This polar amino acid is not attracted to the nonpolar pocket, so no hemoglobin clumping occurs, and the cells are normal in shape, as in figure 14.23c.

Persons homozygous for the sickle-cell genetic mutation (referred to as the *s* allele) in the β-hemoglobin gene often have a reduced life span. This is because red blood cells that are sickled do not flow smoothly through the tiny capillaries but instead jam up and block blood flow, which interferes with oxygen transport. Heterozygous individuals, who have both a defective and a normal form of the gene, make enough functional hemoglobin to keep their red blood cells healthy.

The Puzzle: Why So Common?

The disorder is now known to have originated in Central Africa, where the frequency of the sickle-cell allele is about 0.12. One in 100 people is homozygous for the defective allele and develops the fatal disorder. Sickle-cell disease affects roughly two African Americans out of every thousand but is almost unknown among other racial groups.

If Darwin is right and natural selection drives evolution, then why has natural selection not acted against the defective allele in Africa and eliminated it from the human population there? Why is this potentially fatal allele instead relatively common there?

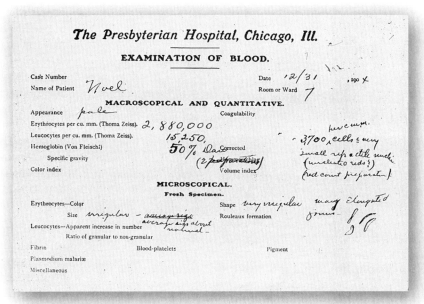

Figure 14.22 The first known patient with sickle-cell disease.
Dr. Ernest Irons's blood examination report on his patient Walter Clement Noel, December 31, 1904, described his oddly shaped red blood cells.
Herrick, James Brian. Papers, Crerar Ms 44, Box 10, Folder 3, Special Collections Research Center, University of Chicago Library

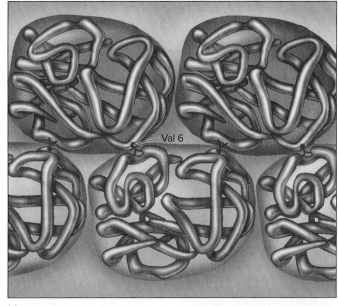

(a)

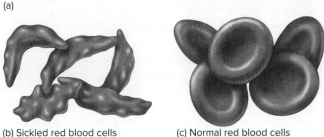

(b) Sickled red blood cells   (c) Normal red blood cells

Figure 14.23 Why the sickle-cell mutation causes hemoglobin to clump.

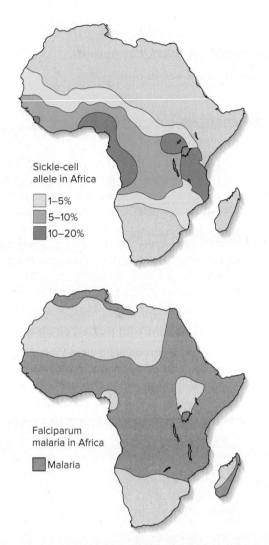

Figure 14.24 How stabilizing selection maintains sickle-cell disease.

The diagrams show the frequency of the sickle-cell allele (*top*) and the distribution of falciparum malaria (*bottom*). Falciparum malaria is one of the most devastating forms of the often fatal disease. As you can see, its distribution in Africa is closely correlated with that of the allele of the sickle-cell characteristic.

IMPLICATION FOR YOU One in 500 African Americans has sickle-cell disease (SCD). Although there is no cure, the antitumor drug hydroxyurea can be used to block expression of the disorder. Administration of hydroxyurea reverses the course of hemoglobin development, inducing the reading of the fetal hemoglobin gene in place of the adult hemoglobin gene, and in so doing preventing sickling in SCD patients. Can you suggest a reason why the substitution of fetal for adult hemoglobin might prevent sickling and so block expression of the disorder?

The Answer: Stabilizing Selection

The defective *s* allele has not been eliminated from Central Africa because people who are heterozygous for the sickle-cell allele are much less susceptible to malaria, one of the leading causes of death in Central Africa. Examine the maps in figure 14.24, and you will see the relationship between sickle-cell disease and malaria clearly. The upper map shows the frequency of the sickle-cell allele, the darker green areas indicating a 10% to 20% frequency of the allele. The map on the bottom indicates the distribution of malaria in dark orange. Clearly, the areas that are colored in darker green on the upper map overlap many of the dark orange areas in the map at the bottom. Although the population pays a high price—the many individuals in each generation who are homozygous for the sickle-cell allele die—the deaths are far fewer than would occur due to malaria if the heterozygous individuals were not malaria resistant. One in 5 individuals (20%) is heterozygous and survives malaria, while only 1 in 100 (1%) is homozygous and dies of sickle-cell disease.

Heterozygote Advantage. Similar inheritance patterns of the sickle-cell allele are found in other areas frequently exposed to malaria, such as around the Mediterranean, India, and Indonesia. Natural selection has favored the sickle-cell allele in Central Africa and other areas hit by malaria because the payoff in survival of heterozygotes more than makes up for the price in death of homozygotes. This phenomenon is an example of **heterozygote advantage.**

Stabilizing selection (also called *balancing selection*) is thus acting on the sickle-cell allele: (1) selection tends to eliminate the sickle-cell allele because of its lethal effects on homozygous individuals and (2) selection tends to favor the sickle-cell allele because it protects heterozygotes from malaria. Like a manager balancing a store's inventory, natural selection increases the frequency of an allele as long as there is something to be gained by it, until the cost balances the benefit.

Removing Heterozygote Advantage. Stabilizing selection occurs because malarial resistance counterbalances lethal sickle-cell disease. Malaria is a tropical disease that has essentially been eradicated in the United States since the early 1950s, and stabilizing selection has not favored the sickle-cell allele here. Africans brought to America several centuries ago have not gained any evolutionary advantage in all that time from being heterozygous for the sickle-cell allele. There is no benefit to being resistant to malaria if there is no danger of getting malaria anyway. As a result, the selection against the sickle-cell allele in America is not counterbalanced by any advantage, and the allele has become less common among African Americans than among native Africans living in Central Africa.

Recent advances promise a long-time cure for sickle-cell disease. To fix the defective *s* allele, researchers are turning to the gene-editing technique CRISPR (see chapter 11). In mice with a mutated form of the hemoglobin gene giving them sickled red blood cells, researchers have collected immature red blood cells, "corrected" the damaged gene in these cells, and then transplanted the CRISPR-modified cells back into the same mice. The treated mice show no signs of sickle-cell disorder. This same approach using CRISPR has succeeded in "correcting" the *s* allele mutation in human blood cells in the lab. Clinical trials are underway to see if the approach works on human patients.

Putting the Concept to Work

Do you think natural selection will ever completely eliminate the *s* allele in the United States? What factors will be at play?

14.9 Peppered Moths and Industrial Melanism

> **LEARNING OBJECTIVE 14.9.1** Assess the evidence that natural selection has led to melanism in moths.

The peppered moth, *Biston betularia,* is a European moth that rests on tree trunks during the day. Until the mid-nineteenth century, almost every captured individual of this species had light-colored wings. From that time on, individuals with dark-colored wings increased in frequency in the moth populations near industrialized centers, until they made up almost 100% of these populations. Dark individuals had a dominant allele that was present but very rare in populations before 1850. Biologists soon noticed that in industrialized regions where the dark moths were common, the tree trunks were darkened almost black by the soot of pollution. Dark moths were much less conspicuous resting on them than light moths were. In addition, air pollution that was spreading in the industrialized regions had killed many of the light-colored lichens on tree trunks, making the trunks darker.

Selection for Melanism

Can Darwin's theory explain the increase in the frequency of the dark allele? Why did dark moths gain a survival advantage around 1850? An amateur moth collector named J. W. Tutt proposed in 1896 what became the most commonly accepted hypothesis explaining the decline of the light-colored moths. He suggested that light forms were more visible to predators on sooty trees that had lost their lichens. Consequently, birds ate the light moths resting on the trunks of darkened trees during the day. The dark forms, in contrast, were at an advantage because they were camouflaged (**figure 14.25**). Although Tutt initially had no evidence, British ecologist Bernard Kettlewell tested the hypothesis in the 1950s by rearing populations of peppered moths with equal numbers of dark and light individuals. Kettlewell then released these populations into two sets of woods: one near heavily polluted Birmingham, and the other in unpolluted Dorset. Kettlewell set up traps in the woods to see how many of both kinds of moths survived. To evaluate his results, he had marked the released moths with a dot of paint on the underside of their wings, where birds could not see it.

In the polluted area near Birmingham, Kettlewell trapped 19% of the light moths but 40% of the dark ones. This indicated that dark moths had a far better chance of surviving in these polluted woods, where the tree trunks were dark. In the relatively unpolluted Dorset woods, Kettlewell recovered 12.5% of the light moths but only 6% of the dark ones. This indicated that where the tree trunks were still light-colored, light moths had a much better chance of survival. Kettlewell later solidified his argument by placing dead moths on trees and filming birds looking for food. Sometimes the birds actually passed right over a moth that was the same color as its background.

Industrial Melanism

Industrial melanism is a term used to describe the evolutionary process in which darker individuals come to predominate over lighter individuals since the Industrial Revolution as a result of natural selection. The process is widely believed to have taken place because the dark organisms are better concealed from their predators in habitats that have been darkened by soot and other forms of industrial pollution, as suggested by Kettlewell.

Figure 14.25 Tutt's hypothesis explaining industrial melanism.

These photographs show color variants of the peppered moth (*Biston betularia*). Tutt proposed that the dark moth is more visible to predators on unpolluted trees (*top*), whereas the light moth is more visible to predators on bark blackened by industrial pollution (*bottom*).

Top: Perennou Nuridsany/Science Source; Bottom: Bill Coster IN/Alamy Stock Photo

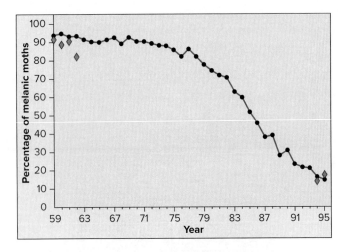

Figure 14.26 Selection against melanism.
The dots indicate the frequency of melanic *Biston betularia* moths at Caldy Common in England, sampled continuously from 1959 to 1995. Red diamonds indicate frequencies of melanic *B. betularia* in Michigan from 1959 to 1962 and from 1994 to 1995.

Dozens of other species of moths have changed in the same way as the peppered moth in industrialized areas throughout Eurasia and North America, with dark forms becoming more common from the mid-nineteenth century onward as industrialization spread.

Selection Against Melanism

As of the second half of the twentieth century, with the widespread implementation of pollution controls, these trends were reversing, not only for the peppered moth in many areas in England, but also for many other species of moths throughout the northern continents. These examples provide some of the best-documented instances of changes in allelic frequencies of natural populations as a result of natural selection due to specific factors in the environment.

In England, the air pollution promoting industrial melanism began to reverse following enactment of Clean Air legislation in 1956. Beginning in 1959, the *Biston* population at Caldy Common outside Liverpool has been sampled each year. The frequency of the melanic (dark) form dropped from a high of 94% in 1960 to a low of 19% in 1995 (**figure 14.26**). Similar reversals have been documented at numerous other locations throughout England. The drop correlates well with a drop in air pollution, particularly with tree-darkening sulfur dioxide and suspended particulates.

Interestingly, the same reversal of industrial melanism appears to have occurred in America during the same time that it was happening in England. Industrial melanism in the American subspecies of the peppered moth was not as widespread as in England, but it has been well documented at a rural field station near Detroit. Of 576 peppered moths collected there from 1959 to 1961, a total of 515 were melanic, a frequency of 89%. The American Clean Air Act, passed in 1963, led to significant reductions in air pollution. Resampled in 1994, the Detroit field station peppered moth population had only 15% melanic moths! The moths in Liverpool and Detroit, both part of the same natural experiment, exhibit strong evidence of natural selection.

Reconsidering the Target of Natural Selection

Tutt's hypothesis, widely accepted in the light of Kettlewell's studies, is currently being reevaluated. The problem is that the recent selection against melanism does not appear to correlate with changes in tree lichens. At Caldy Common, the light form of the peppered moth began its increase in frequency long before lichens began to reappear on the trees. At the Detroit field station, the lichens never changed significantly as the dark moths first became dominant and then declined over the last 30 years. In fact, investigators have not been able to find peppered moths on Detroit trees at all, whether covered with lichens or not. Wherever the moths rest during the day, it does not appear to be on tree bark. Some evidence suggests they rest on leaves on the treetops, but no one is sure.

The action of selection may depend on other differences between light and dark forms of the peppered moth as well as their wing coloration. Researchers report, for example, a clear difference in their ability to survive as caterpillars under a variety of conditions. Perhaps natural selection is also targeting the caterpillars rather than the adults. While we can't yet say exactly what the targets of selection are, the evidence is very clear that selection is at work here—it is indeed one of the best-documented instances of natural selection in action.

Putting the Concept to Work
If birds aren't eating noncryptic moths, is there evidence of selection? What is it?

BIOLOGY & YOU

Natural Selection for Melanism in Mice. Melanism isn't restricted to just insects. Many mammals have melanic forms that are subject to natural selection in much the same way as moths. The coat color of desert pocket mice that live on differently colored rock habitats provides a clear-cut example of natural selection acting on melanism. In Arizona and New Mexico, these small, wild pocket mice live in isolated black volcanic lava beds and the pale soils between them. Melanin synthesis during hair development of pocket mice is regulated by the receptor gene *MC1R*. Mutations that disable *MC1R* lead to melanism. Such mutations are dominant alleles, so whenever they are present in a population, dark pocket mice are seen. When wild populations of pocket mice were surveyed by biologists from the University of Arizona, there was a striking correlation between coat color and the color of the rock on which the population of pocket mice lived. As you can see in the upper photographs, the close match between coat color and background color gives the mice cryptic protection from avian predators, particularly owls. These mice are very visible when placed in the opposite habitats (lower photos).

Michael W. Nachman, Hopi E. Hoekstra, and Susan L. D'Agostino, "The genetic basis of adaptive melanism in pocket mice," PNAS, Vol. 100, No. 9, April 29, 2003, pp. 5268-5273. ©National Academy of Sciences, U.S.A. Used with permission.

How Species Form

14.10 The Biological Species Concept

> **LEARNING OBJECTIVE 14.10.1** Define the biological species concept.

A key aspect of Darwin's theory of evolution is his proposal that small-scale evolutionary changes or adaptations (what is called microevolution) lead ultimately to large-scale changes leading to species formation and higher taxonomic groups (macroevolution). The way natural selection leads to the formation of new species has been thoroughly documented by biologists, who have observed the stages of the species-forming process, or **speciation**, in many different plants and animals. Speciation usually involves successive change: first, local populations become increasingly specialized; then, if they become different enough, natural selection may act to keep them that way.

The Importance of Reproductive Isolation

Before we can discuss how one species gives rise to another, we need to understand exactly what a species is. The **biological species concept** defines species as "groups of actually or potentially interbreeding natural populations which are reproductively isolated from other such groups" (Mayr, 1942). In other words, the biological species concept says that a species is composed of populations whose members mate with each other and produce fertile offspring—or would do so if they came into contact. Conversely, populations whose members do not mate with each other or cannot produce fertile offspring are said to be *reproductively isolated* and, thus, members of different species.

Isolating Mechanisms

What causes reproductive isolation? If organisms cannot interbreed or cannot produce fertile offspring, they clearly belong to different species. However, some populations that are considered to be separate species can interbreed and produce fertile offspring, but they ordinarily do not do so under natural conditions. They are still considered to be reproductively isolated in that genes from one species generally will not be able to enter the gene pool of the other species. Table 14.2 summarizes the steps at which barriers to successful reproduction may occur. Examine this table carefully. We will return to it throughout our discussion of species formation. Such barriers are termed **reproductive isolating mechanisms** because they prevent genetic exchange between species. We will first discuss *prezygotic isolating mechanisms,* those that prevent the formation of zygotes. Then we will examine *postzygotic isolating mechanisms,* those that prevent the proper functioning of zygotes after they have formed.

Even though the definition of what constitutes a species is of fundamental importance to evolutionary biology, this issue has still not been completely settled and is currently the subject of considerable research and debate. For example, the biological species concept has had a number of problems. Plants of different species cross-fertilize and produce fertile hybrids at much higher frequencies than first thought. Hybridization also occurs in animal populations. This is not to say that hybridization is rampant, but it is common enough to cast doubt about whether reproductive isolation is the only force maintaining the integrity of species.

TABLE 14.2	Isolating Mechanisms
Mechanism	**Description**
Prezygotic Isolating Mechanisms	
Geographic isolation	Species occur in different areas, which are often separated by a physical barrier such as a river or mountain range.
Ecological isolation	Species occur in the same area, but they occupy different habitats. Survival of hybrids is low because they are not adapted to either environment of their parents.
Temporal isolation	Species reproduce in different seasons or at different times of the day.
Behavioral isolation	Species differ in their mating rituals.
Mechanical isolation	Structural differences between species prevent mating.
Prevention of gamete fusion	Gametes of one species function poorly with the gametes of another species or within the reproductive tract of another species.
Postzygotic Isolating Mechanisms	
Hybrid inviability or infertility	Hybrid embryos do not develop properly, hybrid adults do not survive in nature, or hybrid adults are sterile or have reduced fertility.

Putting the Concept to Work

Studies of DNA reveal that modern humans interbred with Neanderthals. Does this mean the two were the same species?

14.11 Isolating Mechanisms

LEARNING OBJECTIVE 14.11.1 Describe five prezygotic isolating mechanisms, comparing them to postzygotic mechanisms.

Prezygotic Isolating Mechanisms

Geographical Isolation. This mechanism is perhaps the easiest to understand; species that exist in different areas are not able to interbreed. The two populations of flowers in the first panel of table 14.2 are separated by a mountain range and so would not be capable of interbreeding.

Ecological Isolation. Even if two species occur in the same area, they may utilize different portions of the environment and thus not hybridize because they do not encounter each other, like the lizards in the second panel of table 14.2. One lives on the ground and the other in the trees. Another example in nature is the ranges of lions and tigers in India. Their ranges overlapped until about 150 years ago. Even when they did overlap, however, there were no records of natural hybrids. Lions stayed mainly in the open grassland and hunted in groups called prides; tigers tended to be solitary creatures of the forest. Because of their ecological and behavioral differences, lions and tigers rarely came into direct contact with each other, even though their ranges overlapped thousands of square kilometers. Figure 14.27 shows that hybrids are possible; the liger shown in figure 14.27c is a hybrid of a male lion and a female tiger (a tigon is the hybrid of a male tiger and a female lion). These matings do not occur in the wild but can happen in artificial environments such as zoos.

Temporal Isolation. *Lactuca graminifolia* and *L. canadensis,* two species of wild lettuce, grow together along roadsides throughout the southeastern United States. Hybrids between these two species are easily made experimentally and are completely fertile. But such hybrids are rare in nature because *L. graminifolia* flowers in early spring and *L. canadensis* flowers in summer. This is called temporal isolation and is shown in the third panel in table 14.2. When the blooming periods of these two species overlap, as they do occasionally, the two species do form hybrids, which may become locally abundant.

Behavioral Isolation. In chapter 21, we will consider the often elaborate courtship and mating rituals of some groups of animals, which tend to keep these species distinct in nature even if they inhabit the same places. This behavioral isolation is pictured in the fourth panel of table 14.2. For example, mallard and pintail ducks are perhaps the two most common freshwater ducks in North America. In captivity, they produce completely fertile offspring, but in nature they nest side-by-side and rarely hybridize.

Figure 14.27 Lions and tigers are ecologically isolated.

The ranges of lions and tigers used to overlap in India. However, lions and tigers do not hybridize in the wild because they utilize different portions of the habitat. (a) Tigers are solitary animals that live in the forest, whereas (b) lions live in open grassland. (c) Hybrids, such as this liger, have been successfully produced in captivity, but hybridization does not occur in the wild.

(a): Terje Håheim/Moment Open/Getty Images; (b): Carl D. Walsh/Getty Images; (c): Porterfield/Chickering/Science Source

Mechanical Isolation. The prevention of mating due to structural differences between related species of animals and plants is called mechanical isolation and is shown in panel five of table 14.2. Flowers of related species of plants often differ significantly in their proportions and structures. Some of these differences limit the transfer of pollen from one plant species to another. For example, bees may pick up the pollen of one species on a certain place on their bodies; if this area does not come into contact with the receptive structures of the flowers of another plant species, the pollen is not transferred.

Prevention of Gamete Fusion. In animals that shed their gametes directly into water, eggs and sperm derived from different species may not attract one another. Many land animals may not hybridize successfully because the sperm of one species may function so poorly within the reproductive tract of another that fertilization never takes place. In plants, the growth of pollen tubes may be impeded in species hybrids. In both plants and animals, the operation of such isolating mechanisms prevents the union of gametes even after mating.

Postzygotic Isolating Mechanisms

If hybrid matings do occur, and zygotes are produced, many factors may still prevent those zygotes from developing into normally functioning, fertile individuals. In hybrids, the genetic complements of two species may be so different that they cannot function together normally in embryonic development. For example, hybridization between sheep and goats usually produces embryos that die in the earliest developmental stages.

Figure 14.28 shows four species of leopard frogs (genus *Rana*) and their ranges throughout North America. It was assumed for a long time that they constituted a single species. However, careful examination revealed that although the frogs appear similar, successful mating between them is rare because of problems that occur as the fertilized eggs develop. Many of the hybrid combinations cannot be produced even in the laboratory.

Even if hybrids survive the embryo stage, however, they may not develop normally. If the hybrids are weaker than their parents, they will almost certainly be eliminated in nature. Even if they are vigorous and strong, as in the case of the mule, a hybrid between a female horse and a male donkey, they may still be sterile and thus incapable of contributing to succeeding generations. Sterility may result in hybrids because the development of sex organs may be abnormal, because the chromosomes derived from the respective parents may not pair properly, or from a variety of other causes.

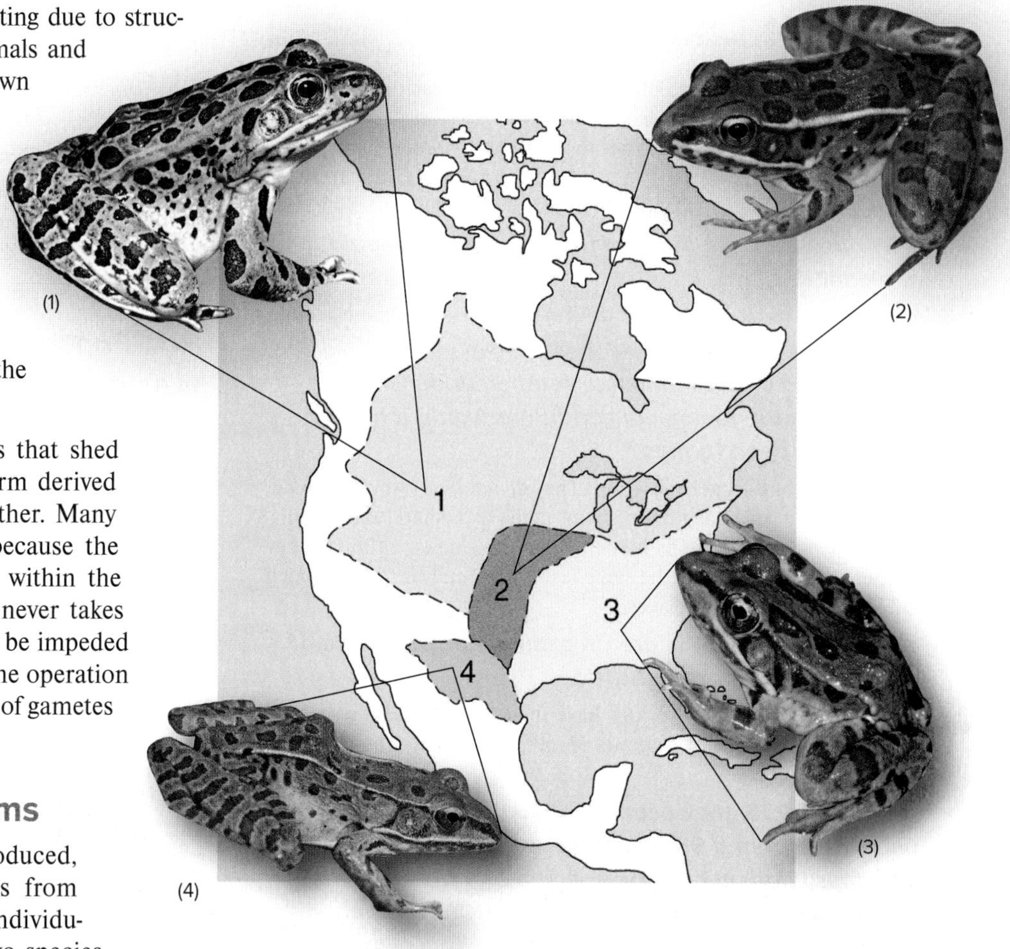

Figure 14.28 Postzygotic isolation in leopard frogs.
Numbers indicate the following species in the geographic ranges shown: **(1)** *Rana pipiens*; **(2)** *Rana blairi*; **(3)** *Rana sphenocephala*; **(4)** *Rana berlandieri*. These four species resemble one another closely in their external features. Their status as separate species was first suspected when hybrids between them were found to produce defective embryos in the laboratory. Subsequent research revealed that the mating calls of the four species differ substantially, indicating that the species have both pre- and postzygotic isolating mechanisms.

(1): James Gerholdt/Photolibrary/Getty Images; **(2):** Kenneth M. Highfill/Science Source; **(3):** Rick & Nora Bowers/Alamy Stock Photo; **(4):** Michael Hare/Shutterstock

> **Putting the Concept to Work**
> Why do ligers and tigons not occur in the wild?

Putting the Chapter to Work

1 Most frogs lay between 1,000 and 4,000 eggs in each clutch—figure 2,000 eggs in an average clutch. However, due to eggs and tadpoles being eaten by predators, the survival rate to adulthood is approximately 5%. A female frog will typically lay two clutches a year and live 10 years. How many adult offspring can that frog be expected to produce in her lifetime?

Why doesn't the number of frogs in natural populations explode?

2 Some animal tissues such as bone form fossils readily, whereas other tissues such as feathers form fossils only rarely.

If you found a fossil dinosaur feather, under what environmental/geological conditions would it most likely have been formed?

3 Penguins are birds that have wings. They do not fly, however; instead, they use their wings as fins to swim, much as a shark uses its fins. Thus, the swimming limbs of penguins and sharks, although they look alike and are used in the same fashion, actually formed through different evolutionary lineages.

Their swimming limbs are an example of what type of evolutionary structures?

4 There are five objections critics have made to evolution by natural selection. Read the following and identify which objection to evolution is refuted.

In March 2000, the second limbed snake fossil was uncovered from a 95-million-year-old excavation site. This fossil had much shorter limbs that the first snake fossil discovered. More recently, in 2015, researchers discovered a four-limbed snake fossil. This allows scientists to trace the evolution of snake limbs from four limbs, to two limbs, to no limbs.

Retracing the Learning Path

Evolution

14.1 Darwin's Voyage on HMS *Beagle*
1. Darwin's theory that evolution is a consequence of natural selection is overwhelmingly accepted by scientists and is considered to be a central concept of the science of biology.

14.2 Darwin's Evidence
1. The fossils and patterns of life that Darwin observed on his voyage eventually convinced him that evolution had taken place.

14.3 The Theory of Natural Selection
1. Key to Darwin's hypothesis was the observation by Malthus that the food supply limits population growth. Using Malthus's observations and his own, Darwin proposed that individuals that are better suited to their environments survive to produce offspring, gaining the opportunity to pass their characteristics on to future generations, what Darwin called natural selection.

The Theory of Evolution

14.4 The Evidence for Evolution
1. The fossil record provides a clear record of successive evolutionary change. The fossil record in many cases reveals organisms that are intermediate in form.
2. The evidence for evolution includes the anatomical record, which reveals similarities in structures between species.
3. The molecular record traces changes in the genomes and proteins of species.

14.5 Evolution's Critics
1. Darwin's theory of evolution through natural selection has always had its critics. Their criticisms of the theory of evolution, however, are without scientific merit.

How Populations Evolve

14.6 Genetic Change in Populations
1. If a population is small, has selective mating, experiences mutations or migrations, or is under the influence of natural selection, allele frequencies will be different from those predicted by the Hardy-Weinberg rule.

14.7 Agents of Evolution
1. Five evolutionary forces act on populations to change their allele and genotype frequencies: (1) Mutations are changes in DNA. (2) Migrations are the movements of individuals into or out of a population. (3) Genetic drift is the random loss of alleles in a population due to chance occurrences. (4) Nonrandom mating occurs when individuals seek out mates based on certain traits. (5) Selection occurs when individuals with certain traits leave more offspring because their traits allow them to better respond to the challenges of their environment.
2. Stabilizing selection tends to reduce extreme phenotypes, disruptive selection tends to reduce intermediate phenotypes, and directional selection tends to reduce one extreme phenotype.

Adaptation Within Populations

14.8 Sickle-Cell Disease
1. Sickle-cell disease is an example of a disease maintained in human populations by natural selection. In areas with malaria, people who are heterozygous for the sickle-cell trait survive better than individuals with either of the two homozygous phenotypes.

14.9 Peppered Moths and Industrial Melanism
1. Natural selection favors the dark form of the peppered moth in areas of heavy pollution.

How Species Form

14.10 The Biological Species Concept
1. The biological species concept defines a species as a group of organisms that are capable of mating with each other and producing fertile offspring. If they cannot mate, or can mate but produce only sterile offspring, they are said to be reproductively isolated.

14.11 Isolating Mechanisms
1. There are two types of isolating mechanisms: prezygotic mechanisms prevent the formation of a hybrid zygote, and postzygotic mechanisms prevent development of a hybrid zygote or result in sterile offspring.

Inquiry and Analysis

Does Natural Selection Act on Enzyme Polymorphism?

The essence of Darwin's theory of evolution is that, in nature, selection favors some gene alternatives over others. Many studies of natural selection have focused on genes encoding enzymes because populations in nature tend to possess many alternative alleles of their enzymes (a phenomenon called *enzyme polymorphism*). Often, investigators have looked to see if weather influences which alleles are more common in natural populations. A particularly nice example of such a study was carried out on a fish, the mummichog (*Fundulus heteroclitus*), which ranges along the East Coast of North America. Researchers studied allele frequencies of the gene encoding the enzyme lactate dehydrogenase, which catalyzes the conversion of pyruvate to lactate. As you learned in chapter 7, this reaction is a key step in energy metabolism, particularly when oxygen is in short supply. There are two common alleles of lactate dehydrogenase in these fish populations, with allele *a* being a better catalyst at lower temperatures than allele *b*.

In an experiment, investigators sampled the frequency of allele *a* in 41 fish populations located over 14 degrees of latitude, from Jacksonville, Florida (31° North), to Bar Harbor, Maine (44° North). Annual mean water temperatures change 1°C per degree change in latitude. The survey is designed to test a prediction of the hypothesis that natural selection acts on this enzyme polymorphism. If it does, then you would expect that allele *a*, producing a better "low-temperature" enzyme, would be more common in the colder waters of the more northern latitudes. The graph on the right presents the results of this survey. The points on the graph are derived from pie chart data such as shown for 20 populations in the map (a **pie chart diagram** assigns a slice of the pie to each variable; the size of the slice is proportional to the contribution made by that variable to the total). The blue line on the graph is the line that best fits the data (a **"best-fit" line,** also called a regression line, is determined statistically by a process called *regression analysis*).

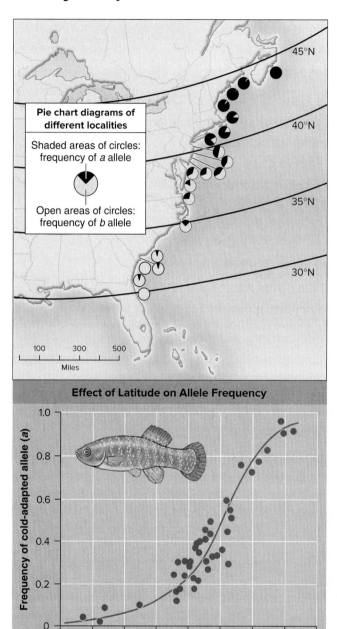

Analysis

1. **Applying Concepts**
   a. In the fish population located at 35°N latitude, what is the frequency of the *a* allele? Locate this point on the graph.
   b. Compare the frequency of allele *a* among fish captured in waters at 44°N latitude with the frequency among fish captured at 31°N latitude. Is there a pattern? Describe it.
2. **Interpreting Data** At what latitude do fish populations exhibit the greatest variability in allele *a* frequency?
3. **Making Inferences**
   a. Are fish populations in cold waters at 44°N latitude more or less likely to contain heterozygous individuals than fish populations in warm waters at 31°N latitude?
   b. Where along this latitudinal gradient in the frequency of allele *a* would you expect to find the highest frequency of heterozygous individuals? Why?
4. **Drawing Conclusions** Are the differences in population frequencies of allele *a* consistent with the hypothesis that natural selection is acting on the alleles encoding this enzyme? Explain.

15 Exploring Biological Diversity

LEARNING PATH ▼

The Classification of Organisms
1. The Invention of the Linnaean System
2. Species Names
3. Higher Categories
4. What Is a Species?

Inferring Phylogeny
5. How to Build a Family Tree

Kingdoms and Domains
6. The Kingdoms of Life
7. Domain: A Higher Level of Classification

HUNTING BLIND
Dave Watts/Alamy Stock Photo

How the Platypus Sees with Its Eyes Shut

In 1799, the skin of a most unusual animal was sent to England by Captain John Hunter, governor of the British penal colony in New South Wales (Australia). Covered in soft fur, it was less than 2-feet long. As it had mammary glands with which to suckle its young, it was clearly a mammal, but in other ways, it seemed very like a reptile. It had a shared urinary and reproductive tract opening called a cloaca, and it laid eggs, as reptiles do. It thus seemed a confusing mixture of mammalian and reptilian traits. Adding to this impression was its appearance: It has a tail not unlike that of a beaver, a bill not unlike that of a duck, and webbed feet! It was as if a child had mixed together body parts at random—a most unusual animal. Individuals like it, pictured here, are abundant in freshwater streams of eastern Australia today. What does one call such a beast? *Ornithorhynchus anatinus* (ducklike animal with a bird's snout)—informally, the duckbill platypus.

Platypus Hunting Behavior Is Unusual . . .

It turns out that platypuses also have some very unique behaviors. Until recently, few scientists had studied the platypus in its natural habitat—it is elusive, spending its days in burrows it constructs on the banks of waterways. Also, a platypus is active mostly at night, diving in streams and lagoons to capture bottom-dwelling invertebrates, such as shrimps and insect larvae. Interestingly, unlike whales and other marine mammals, a platypus cannot stay under water long. Its dives typically last a minute and a half. (Try holding your breath that long!)

. . . It Hunts with Its Eyes Shut!

When scientists began to study platypuses' diving behavior, they soon observed a curious fact: The eyes and ears of a platypus are located within a muscular groove, and when a platypus dives, the sides of these grooves close over tightly. Imagine pulling your eyebrows down to your cheeks—effectively blindfolded, you wouldn't be able to see a thing! To complete its isolation, the nostrils at the end of the snout also close. So how in the world does the animal find its prey?

JohnCarnemolla/Getty Images

The Bill Is the Key

For over a century, biologists have known that the soft surface of the platypus bill is pierced by hundreds of tiny openings. In recent years, Australian neuroscientists (scientists who study the brain and nervous system) have learned that these pores contain sensitive nerve endings. Nestled in an interior cavity, the nerves are protected from damage by the bill but are linked to the outside streamwater via the pore. The nerve endings act as sensory receptors, communicating to the brain information about the animal's surroundings. These pores in the platypus bill are its diving "eyes."

Using Electricity to See at a Distance

But how does the platypus locate its prey at a distance in murky water with its eyes shut? When a platypus feeds, it swims along steadily wagging its bill from side to side, two or three sweeps per second, until it detects and homes in on prey. How is the platypus able to sense that a prey individual is nearby and orient itself to it? The sensory receptors in its bill detect the tiny electrical currents generated by the muscle movements of its prey as the shrimp or insect larva moves to evade the approaching platypus!

It is easy to demonstrate this, once you know what is going on. Just drop a small 1.5-volt battery into the stream. A platypus will immediately orient to it and attack it, from as far away as 30 centimeters. Some sharks and fishes have the same sort of sensory system. In muddy, murky waters sensing the muscle movements of a prey individual is far superior to trying to see its body or hear it move—which is why the platypus that you see in the photo above is hunting with its eyes shut.

The Classification of Organisms

15.1 The Invention of the Linnaean System

> **LEARNING OBJECTIVE 15.1.1** Compare the classification systems of the Romans with those of Linnaeus.

It is estimated that our world is populated by some 10 to 100 million different kinds of organisms. To talk about them and study them, it is necessary to give them names, just as it is necessary that people have names. Of course, no one can remember the name of every kind of organism, so biologists use a kind of multilevel grouping of individuals called **classification**.

Organisms were first classified more than 2,000 years ago by the Greek philosopher Aristotle, who categorized living things as either plants or animals. He classified animals as either land, water, or air dwellers, and he divided plants into three kinds based on stem differences. This simple classification system was expanded by the Greeks and Romans, who grouped animals and plants into basic units such as cats, horses, and oaks. Eventually, these units began to be called **genera** (singular, **genus**), the Latin word for "group." Starting in the Middle Ages, these names began to be systematically written down, using Latin, the language used by scholars at that time. Thus, cats were assigned to the genus *Felis*, horses to *Equus*, and oaks to *Quercus*—names that the Romans had applied to these groups. For genera that were not known to the Romans, new names were invented.

The Binomial System

Today's universal system of naming animals, plants, and other organisms stems from the work of the Swedish biologist Carolus Linnaeus (1707-78). Linnaeus (figure 15.1) devoted his life to a challenge that had defeated many biologists before him—cataloging all the different kinds of organisms. As a kind of shorthand in classifying plants and animals, Linnaeus in the 1750s assigned a two-part name for each species. These two-part names, or **binomials** (*bi* is the Latin prefix for "two"), have become our standard way of designating species. For example, he designated the willow oak (shown in figure 15.2a with its smaller, unlobed leaves) *Quercus phellos* and the red oak (with the larger deeply lobed leaves in figure 15.2b) *Quercus rubra*. We also use binomial names for ourselves, our so-called given and family names.

Figure 15.1 Carolus Linnaeus (1707–78).
This 18th-century Swedish professor, physician, and naturalist devised the binomial system for naming species of organisms and established the major categories that are used in the hierarchical system of biological classification.
Time & Life Pictures/Getty Images

(a) *Quercus phellos* (Willow oak) (b) *Quercus rubra* (Red oak)

Figure 15.2 How Linnaeus named two species of oaks.
(a) Willow oak, *Quercus phellos*. (b) Red oak, *Quercus rubra*. Although they are clearly oaks (members of the genus *Quercus*), these two species differ sharply in the shapes and sizes of their leaves and in many other features, including their overall geographical distributions.

Linnaeus took the naming of organisms a step further, grouping similar organisms into higher-level categories based on similar characteristics (discussed in section 15.3). Although not intended to show evolutionary connections between different organisms, this hierarchical system acknowledged that there were broad similarities shared by groups of species that distinguished them from other groups.

> **Putting the Concept to Work**
> Can you construct a descriptive name for yourself that would distinguish you from anyone else? Do it.

15.2 Species Names

> **LEARNING OBJECTIVE 15.2.1** Describe the two parts of a scientific name.

A group of organisms at a particular level in a classification system is called a **taxon** (plural, **taxa**), and the branch of biology that identifies and names such groups of organisms is called **taxonomy**. Taxonomists are in a real sense detectives, biologists who must use clues of appearance and behavior to identify and assign names to organisms.

A Universal Naming System

By formal agreement among taxonomists throughout the world, no two organisms can have the same name. So that no one country is favored, a language spoken by no country—Latin—is used for the names. Because the scientific name of an organism is the same anywhere in the world, this system provides a standard and precise way of communicating, whether the language of a particular biologist is Chinese, Arabic, Spanish, or English. This is a great improvement over the use of common names, which often vary from one place to the next. As you can see in figure 15.3, in America corn refers to the plant in the upper left photo, but in Europe it refers to the plant Americans call wheat, the lower left photo. A bear is a large placental omnivore in the United States (the upper right photo), but in Australia it is a koala, a vegetarian marsupial (the lower right photo).

By convention, the first word of the binomial name is the genus to which the organism belongs. This word is always capitalized. The second word, called the *specific epithet,* refers to the particular species and is not capitalized. The two words together are called the **scientific name,** or species name, and are written in italics. The system of naming animals, plants, and other organisms established by Linnaeus has served the science of biology well for over 270 years.

> **Putting the Concept to Work**
> Do you think a system of scientific names based on numbers would be as valid as one based on Latin names? Explain.

(a) (b)

Figure 15.3 Common names make poor labels.

The common names corn (a) and bear (b) bring clear images to our minds (photos on *top*), but the images would be very different to someone living in Europe or Australia (photos on *bottom*). There, the same common names are used to label very different species.

Top Left: stevanovicigor/iStock/Getty Images; **Bottom Left:** GarethPriceGFX/Getty Images; **Top Right:** Comstock/PunchStock; **Bottom Right:** Andras Deak/Getty Images

15.3 Higher Categories

> **LEARNING OBJECTIVE 15.3.1** List the eight categories used to classify organisms.

A biologist needs more than two categories to classify all the world's living things.

A Hierarchical Naming System

Taxonomists group the genera with similar properties into a cluster called a **family**. For example, the eastern gray squirrel at the bottom in figure 15.4 is placed in a family with other squirrel-like animals including prairie dogs, marmots, and chipmunks. Similarly, families that share major characteristics are placed into the same **order** (for example, squirrels placed in with other rodents). Orders with common properties are placed into the same **class** (squirrels in the class Mammalia), and classes with similar characteristics into the same **phylum** (plural, **phyla**) such as the Chordata. Botanists (scientists who study plants) also call plant phyla "divisions." Finally, the phyla are assigned to one of several gigantic groups, the **kingdoms.** Biologists currently recognize six kingdoms: two kinds of prokaryotes (Archaea and Bacteria), a largely unicellular group of eukaryotes (Protista), and three multicellular groups of eukaryotes (Fungi, Plantae, and Animalia). To remember the seven categories in their proper order, it may prove useful to memorize a phrase such as "**k**indly **p**ay **c**ash **o**r **f**urnish **g**ood **s**ecurity" or "**K**ing **P**hilip **c**ame **o**ver **f**or **g**reen **s**paghetti" (**k**ingdom-**p**hylum-**c**lass-**o**rder-**f**amily-**g**enus-**s**pecies).

In addition, an eighth level of classification, called *domains,* is sometimes used. Domains are the broadest and most inclusive taxa, and biologists recognize three of them, Bacteria, Archaea, and Eukarya—which are discussed later in this chapter.

For example, consider a squirrel:

Level 1: Its **species** name, *Sciurus carolinensis,* tells you it is the eastern gray squirrel, with gray fur and white-tipped hairs on the tail.
Level 2: Its **genus** name, *Sciurus,* tells you it is a tree squirrel.
Level 3: Its **family**, *Sciuridae,* tells you it has four front toes and five back toes.
Level 4: Its **order,** *Rodentia,* is distinguished by its gnawing teeth.
Level 5: Its **class,** *Mammalia,* tells you it has fur and nipples.
Level 6: Its **phylum**, *Chordata*, **Subphylum** *Vertebrata*, tells us it has a backbone.
Level 7: Its **kingdom,** *Animalia,* says that it is a multicellular heterotroph whose cells lack cell walls.
Level 8: An addition to the Linnaean system, its **domain**, *Eukarya*, says that its cells contain membrane-bounded organelles.

Figure 15.4 The hierarchical system used to classify a gray squirrel.

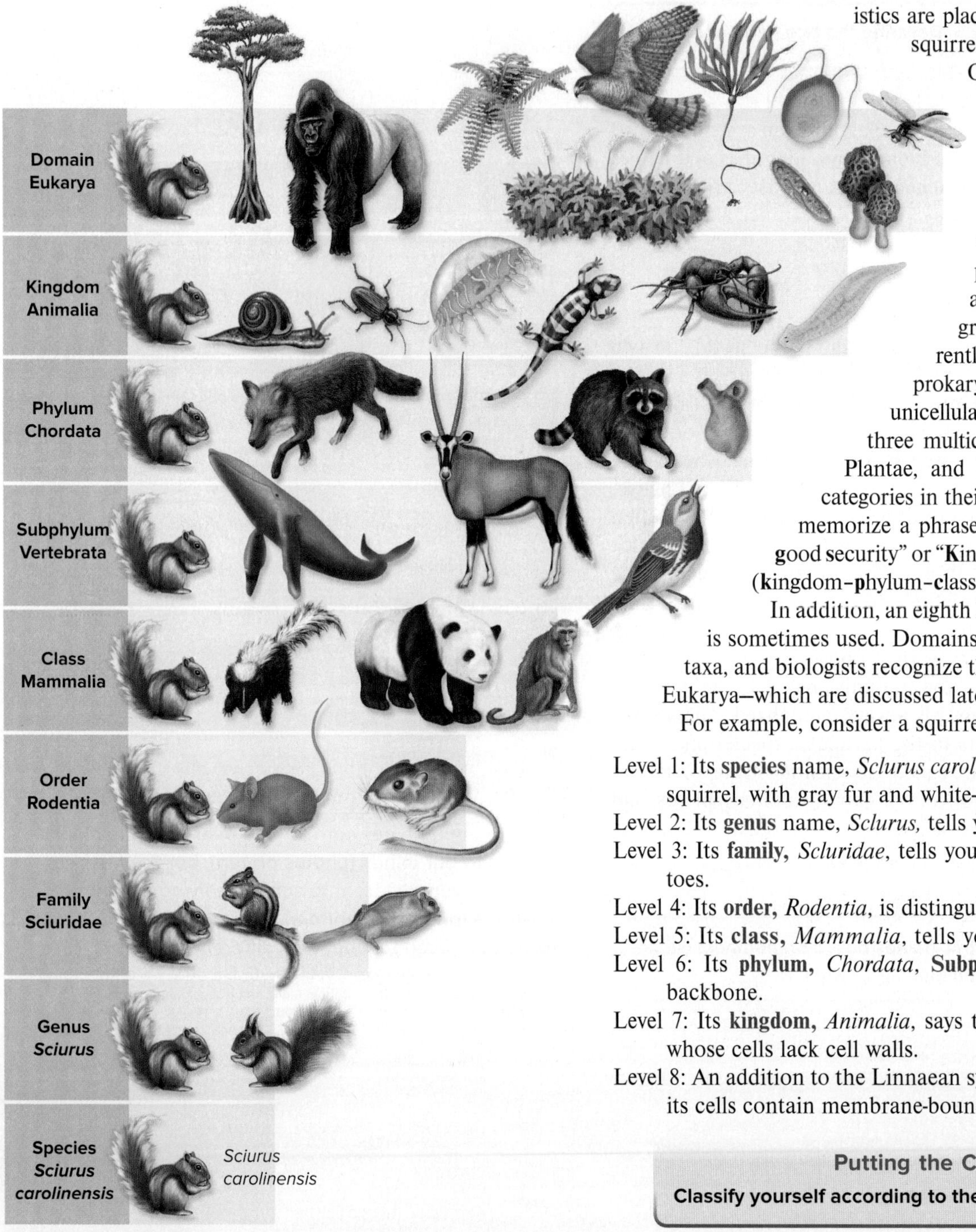

> **Putting the Concept to Work**
> Classify yourself according to the eight-stage system above.

15.4 What Is a Species?

> **LEARNING OBJECTIVE 15.4.1** Contrast Ray's definition of species with the biological species concept.

The basic biological unit in the Linnaean system of classification is the species. John Ray (1627-1705), an English clergyperson and scientist, was one of the first to propose a general definition of species. In about 1700, he suggested a simple way to recognize a species: all the individuals that belong to it can breed with one another and produce fertile offspring. By Ray's definition, the offspring of a single mating were all considered to belong to the same species, even if they contained different-looking individuals, as long as these individuals could interbreed. All domestic cats are one species (they can all interbreed), whereas carp are not the same species as goldfish (they cannot interbreed). The donkey you see in figure 15.5 is not the same species as the horse because when they interbreed, the offspring—mules—are sterile.

The Biological Species Concept

When the evolutionary ideas of Darwin were joined to the genetic ideas of Mendel in the 1920s, it became desirable to define the category of species more precisely. The definition that emerged, the *biological species concept*, defines species as groups that are reproductively isolated. Hybrids (offspring of different species that mate) are rare in nature, whereas individuals that belong to the same species are able to interbreed freely.

A Problem: Asexual Reproduction Is Common. The biological species concept works fairly well for animals, where strong barriers to hybridization between species exist, but very poorly for members of the other kingdoms. The problem is that in prokaryotes and many protists, fungi, and plants, asexual reproduction (reproduction without sex) predominates. These species clearly cannot be characterized in the same way as animals—they do not breed with one another, much less with individuals of other species.

Another Problem: Reproductive Barriers Are Not Common. Complicating matters further, the reproductive barriers that are key to the biological species concept, although common among animal species, are not typical of other kinds of organisms. In fact, there are essentially no barriers to hybridization between the species in many groups of trees, such as oaks, and other plants, such as orchids. Even among animals, fish species are able to form fertile hybrids with one another, though they may not do so in nature.

In practice, biologists today recognize species in different groups in much the way they always did, as groups that differ from one another in their visible features. However, molecular data are causing a reevaluation of traditional classification systems and are changing the way scientists classify plants, protists, fungi, prokaryotes, and even animals.

How Many Kinds of Species Are There?

Since the time of Linnaeus, about 1.5 million species have been named. But the actual number of species in the world is undoubtedly much greater, judging from the very large numbers that are still being discovered. Some scientists estimate that at least 10 million species exist on earth, two-thirds in the tropics.

> **Putting the Concept to Work**
> Why does the biological species concept work so much better for animals than for plants?

Horse

Donkey

Mule

Figure 15.5 Ray's definition of a species.

According to Ray, donkeys and horses are not the same species. Even though they produce very hardy offspring (mules) when they mate, the mules are sterile, meaning that two mules that mate cannot produce offspring. The sterility arises because the two species have different numbers of chromosomes: donkeys have 62 chromosomes, whereas horses have 64.

Top: Juniors Bildarchiv/Alamy Stock Photo; **Center:** perseomedusa/123RF; **Bottom:** Steve Taylor/Alamy Stock Photo

Inferring Phylogeny

15.5 How to Build a Family Tree

> **LEARNING OBJECTIVE 15.5.1** Explain the difference between cladistics and traditional taxonomy.

By looking at the differences and similarities between organisms, biologists attempt to reconstruct the tree of life, inferring which organisms evolved from which other ones, in what order, and when. The evolutionary history of an organism and its relationship to other species is called **phylogeny.**

Cladistics

A simple and objective way to construct an evolutionary history, or **phylogenetic tree,** is to focus on key characters that some organisms share because they have inherited them from a common ancestor. A **clade** is a group of organisms related by descent, and this approach to constructing a phylogeny is called **cladistics.** Cladistics infers phylogeny (that is, builds family trees) according to similarities derived from a common ancestor, so-called **derived characters.** Derived characters are defined as characters that are present in a group of organisms that arose from a common ancestor that lacked the character. The key to the approach is being able to identify morphological, physiological, or behavioral traits that differ among the organisms being studied and can be attributed to a common ancestor. By examining the distribution of these traits among the organisms, it is possible to construct a **cladogram,** a branching diagram that represents the phylogeny. A cladogram of the vertebrates is shown in figure 15.6.

Cladograms Compare Groups. Cladograms are not true family trees, derived directly from data that document ancestors and descendants like the fossil record does. Instead, cladograms convey comparative information about *relative* relationships. Organisms that are closer together on a cladogram simply share a more recent common ancestor than those that are farther apart. Because the analysis is comparative, it is necessary to have something to anchor the comparison to, some solid ground against which the comparisons can be made. To achieve this, each cladogram must contain an *outgroup*, a rather different organism (but not too different) to serve as a baseline for comparisons among the other organisms being evaluated, called the *ingroup*. For example, in figure 15.6, the lamprey is the outgroup to the clade of animals that have jaws. Comparisons are then made up the cladogram, beginning with lampreys and sharks, figure 15.6, based on the emergence of derived characters. For example, the shark differs from the lamprey in that it has jaws, the derived character missing in the lamprey. The derived characters are in the colored boxes along the main line of the cladogram. Salamanders differ from sharks in that they have lungs, and so on up the cladogram.

"Weighting" Cladograms. Sometimes cladograms are adjusted to "weight" characters, or take into account the variation in the "strength" (importance) of a character—the size or location of a fin, the effectiveness of a lung. For example, let's say that the following are five unique events that occurred on September 11, 2001: (1) my cat was declawed, (2) I had a wisdom tooth pulled, (3) I sold my first car, (4) terrorists attacked the United States using commercial airplanes, and (5) I passed physics.

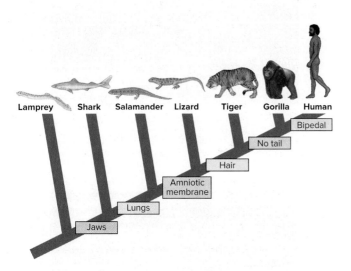

Figure 15.6 A cladogram of vertebrate animals.

The derived characters between the branch points are shared by all the animals to the right of each character and are not present in any organisms to the left of it.

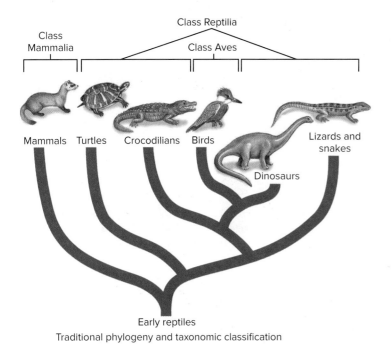
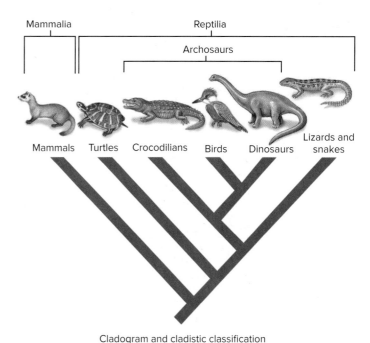

Figure 15.7 Two ways to classify terrestrial vertebrates.

Traditional taxonomic analyses place birds in their own class (Aves) because birds have evolved several unique adaptations that separate them from the reptiles. Cladistic analyses, however, place crocodiles, dinosaurs, and birds together (as archosaurs) because they share many derived characters, indicating a recent shared ancestry. In practice, most biologists adopt the traditional approach and consider birds as members of the class Aves rather than Reptilia.

Without weighting the events, each one is assigned equal importance. In a nonweighted cladistic sense, they are equal (all happened only once, and on that day), but in a practical, real-world sense, they certainly are not. One event, the terrorist attack, had a far greater impact and importance than the others. Because evolutionary success depends so critically on just such high-impact events, these weighted cladograms attempt to assign extra weight to the evolutionary significance of key characters.

Traditional Taxonomy

Weighting characters lies at the core of **traditional taxonomy**. In this approach, phylogenies are constructed based on all available information about the morphology and biology of the organism. The large amount of information permits a knowledgeable weighting of characters according to their biological significance. For example, in classifying the terrestrial vertebrates, traditional taxonomists, shown by the phylogeny on the left in **figure 15.7**, place birds in their own class (Aves), giving great weight to the characters that made powered flight possible, such as feathers. However, a cladogram of vertebrate evolution, as shown on the right, lumps birds in among the reptiles with crocodiles and dinosaurs. This accurately reflects their ancestry but ignores the immense evolutionary impact of a derived character such as feathers.

Overall, phylogenetic trees based on traditional taxonomy are information-rich, whereas cladograms often do a better job of deciphering evolutionary histories. Traditional taxonomy is the better approach when a great deal of information is available to guide character weighting. However, cladistics is the preferred approach when less information is available about how the character affects the life of the organism. For example, the cat family tree in **figure 15.8** is based on thousands of DNA differences among the different groups of felines, few of which affect the life of the cats in ways we know and understand.

> **Putting the Concept to Work**
> Identify a derived character that discriminates yourself from a monkey.

Figure 15.8 The cat family tree.

Studies of DNA similarities have allowed biologists to construct this feline family tree of the eight major cat lineages and their individual species. Among the oldest of all cats are the four big panthers: tiger, lion, leopard, and jaguar. The other big cats, the cheetah and mountain lion, are members of a much younger lineage and are not close relatives of the big four. Domestic cats *(Felix silvestris)* evolved most recently.

Kitten in straw: Markus Botzek/Corbis; **Mountain lion (Felis concolor), close-up:** Alan and Sandy Carey/Getty Images; **Cheetah (Acinonyx jubatus) running:** Alan and Sandy Carey/Getty Images; **Bobcat, Lynx rufus, Uinta National Forest, Utah, USA:** Kevin Schafer/Getty Images; **Canadian lynx, Lynx canadensis, single cat in snow:** Erni/Shutterstock; **Ocelot:** James Gritz/Getty Images; **Close-up of Caracal head and shoulders:** Stockbyte/Getty Images; **Snow Leopard (Panthera uncia). Woodland Park Zoo. Seattle. Washington. USA:** Rene Frederic/Pixtal/age fotostock; **Bengal Tiger Snarling:** Chase Swift/Getty Images; **Face of African Lion:** Jeff Vanuga/Getty Images; **Head of Leopard Panthera Pardus Resting in Shade:** Peter Johnson/Corbis /VCG/Getty Images; **Portrait of jaguar in jungle:** Alan and Sandy Carey/Getty Images

Domestic cat — Domestic cat, African wildcat, black-footed cat, Chinese mountain cat, European wildcat, jungle cat, sand cat

Leopard cat, flat-headed cat, fishing cat

Mountain lion (also called cougar or puma)

Cheetah

Bobcat

Canada lynx — Bobcat, Canada lynx, Eurasian lynx, Spanish lynx

Ocelot — Ocelot, Andean mountain cat, Geoffroy's cat, kodkod, margay, oncilla, pampas cat

Caracal — Caracal, African golden cat

Bay cat, Asiatic golden cat

Snow leopard — Snow leopard and clouded leopard

Tiger

Lion

Leopard

Jaguar

Millions of years ago

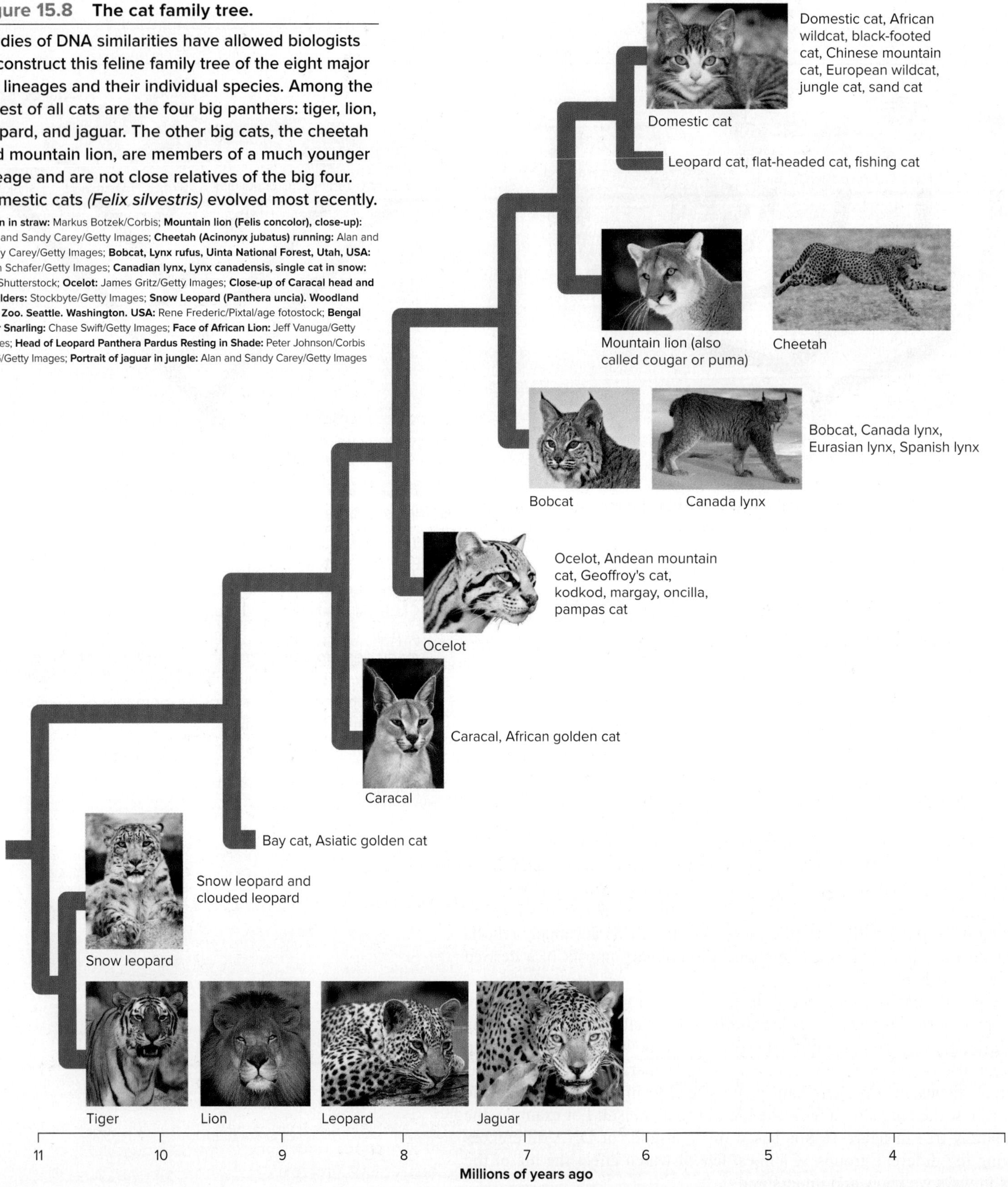

Today's Biology

Race and Medicine

Few issues in biology have stirred more social controversy than race. *Race* has a deceptively simple definition, referring to groups of individuals related by ancestry that differ from other groups, but not enough to constitute separate species. The controversy arises because of the way people have used the concept of race to justify the abuse of humans. The African slave trade is but one obvious example. Geneticists point out that if one looked at genes rather than faces, the differences between the genes of an African and a European would be hardly greater than the difference between those of any two Europeans. For 40 years, gene data has continually reinforced the validity of this observation, and the concept of genetic races has been largely abandoned.

Image Source/Getty Images

Haplotypes Define Our Ancestry

Recent detailed comparisons of the genomes of people around the world reveal that particular alleles defining human features often occur in clusters on chromosomes. Because they are close together, the genes experience little recombination over the centuries. The descendants of a person who has a particular combination of alleles will almost always have that same combination. The set of alleles, technically called a "haplotype," reflects the common ancestry of these descendants from that ancestor.

Ignoring skin color and eye shape, and instead comparing the DNA sequences of hundreds of regions of the human genome, investigators at the University of Southern California showed that when a large sample of people from around the world are compared, the computer sorts them into five large groups containing similar clusters of variation: (1) Europe, (2) East Asia, (3) Africa, (4) America, and (5) Australasia. That these are more or less the major races of traditional anthropology is not the point. The point is that humanity evolved in these five regions in isolation from one another, that today each of these groups is composed of individuals with a shared ancestry, and that by analyzing the DNA of an individual, we can deduce that ancestry. You can submit your DNA to any of a number of services available on the Internet to map out your ancestry by analyzing DNA haplotypes.

Many Diseases Correlate with Haplotypes

Analysis of the human genome is revealing that many diseases are influenced by alleles that have arisen since the five major branches of the human family tree separated from one another and are much more common in the human ancestral group within which the DNA mutation causing the disease first occurred. For example, a mutation causing hemochromatosis, a disorder of iron metabolism, is rare or absent among Indians or Chinese but very common among northern Europeans (it occurs in 7.5% of Swedes), who also commonly possess an allele leading to adult lactose intolerance (inability to digest lactose) not common in many other groups. Similarly, the hemoglobins mutation causing sickle-cell disease is common among Africans of Bantu ancestry but seems to have arisen only there.

This is a pattern we see again and again as we compare genomes of people living in different parts of the world—and it has a very important consequence. Because of common ancestry, genetic diseases (disorders that are inherited) have a lot to do with geography. The risk that an African American man will be afflicted with hypertensive heart disease or prostate cancer is nearly three times greater than that for a European American man, whereas the European American is far more likely to develop cystic fibrosis or multiple sclerosis.

These differences in medically important DNA variants carry over to genetic differences in how individuals respond to treatment. African Americans, for example, respond poorly to some of the main drugs used to treat heart conditions, such as beta-blockers and angiotensin enzyme inhibitors.

Haplotype Is Not the Same as Race

It is important to keep clearly in mind the goal of sorting out individual human ancestry, which is to identify common lines of descent that share a common response to potential therapies. It is NOT to assign people to overarching racial categories. This point is being made with great clarity in a course being taught at Pennsylvania State University, where each year students have their DNA sampled and compared with four of the five major human groups. Many of these students had thought of themselves as "100% white," but only a few are. One "white" student learned that 14% of his DNA came from Africa, and 6% from East Asia. Similarly, "black" students found that as much as half of their genetic material came from Europe and a significant amount from Asia as well.

The point is, rigid ideas about the biological basis of identity are wrong. Humanity is much more complex than indicated by a few genes affecting skin color and eye shape. The more clearly we can understand that complexity, the better we can deal with the medical consequences and the great potential that human diversity provides all of us.

Kingdoms and Domains

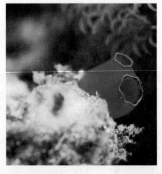
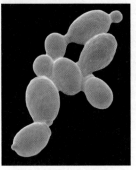

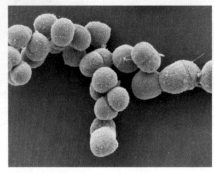

Figure 15.9 **Pick a kingdom. To which kingdom would you assign these interesting creatures?**

Top Left: Gerry Pearce/Alamy Stock Photo; **Top Right:** Jung Hsuan/Shutterstock; **Bottom Left:** Steve Gschmeissner/Science Photo Library/Getty Images; **Bottom Center:** Dennis Kunkel Microscopy/Science Source; **Bottom Right:** Eye of Science/Science Source

15.6 The Kingdoms of Life

> **LEARNING OBJECTIVE 15.6.1** Name the six kingdoms of life, assigning each to a domain.

Classification systems have gone through their own evolution of sorts, as illustrated in **figure 15.9**. The earliest classification systems recognized only two kingdoms of living things: animals and plants. But as biologists discovered microorganisms and learned more about other organisms such as the protists and the fungi, they added kingdoms in recognition of fundamental differences. Most biologists now use a six-kingdom system (**table 15.1**).

Four Eukaryotic Kingdoms

In this system, four kingdoms consist of eukaryotic organisms. The two most familiar kingdoms, **Animalia** and **Plantae**, contain organisms that are multicellular during most of their life cycle. These groups of animals and plants are no doubt familiar to you. The kingdom **Fungi** contains multicellular forms, such as mushrooms and molds, and single-celled yeasts, which are thought to have multicellular ancestors. Fundamental differences divide these three kingdoms. Plants are mainly stationary, but some have motile sperm; fungi have no motile cells; animals are mainly motile. Animals ingest their food, plants manufacture it, and fungi digest it by means of secreted extracellular enzymes. Each of these kingdoms probably evolved from a different single-celled ancestor.

The large number of unicellular eukaryotes are arbitrarily grouped into a single kingdom called **Protista**. They include the algae and many kinds of microscopic aquatic organisms.

Two Prokaryotic Kingdoms

The remaining two kingdoms, **Archaea** and **Bacteria,** consist of prokaryotic organisms, which are vastly different from all other living things. The prokaryotes with which you are most familiar, those that cause disease or are used in industry, are members of the kingdom Bacteria. Archaea are a diverse group including some that live in physically extreme conditions. They differ greatly from bacteria in many ways. The characteristics of these six kingdoms are presented in **table 15.1**.

Domains

As biologists have learned more about the archaea, it has become increasingly clear that this ancient group is very different from all other organisms. When the full genomic DNA sequences of an archaean and a bacterium

---

BIOLOGY & YOU

How Biodiversity Benefits You—Direct Economic Value. Many threatened species have direct value to you as sources of food, medicine, clothing, energy, and shelter:

1. **Gene variation.** Most of the world's food crops, such as corn, wheat, and rice, are derived from a small number of domesticated plants and so contain relatively little genetic variation; on the other hand, their wild relatives have great diversity. In the future, genetic variation from wild strains will be needed if we are to improve yields or find a way to breed resistance to new pests.

2. **Medicine.** About 70% of the world's population depends directly on wild plants as their source of medicine. In addition, some 40% of the prescription drugs used today have active ingredients extracted from animals or plants. Aspirin was first extracted from the leaves of a tropical willow tree. Vinblastine, the most effective drug for combating childhood leukemia, is extracted from the rosy periwinkle vine. The cancer-fighting drug taxol is produced from the bark of the Pacific yew tree.

3. **Ecosystem services.** Diversity is of vital importance to the health of ecosystems (communities of organisms and the places where they live). Diverse biological communities preserve water quality, soil richness, and local climates, while absorbing pollution and recycling wastes. The world's stability and productivity depend critically on these services, which in turn depend critically on species richness.

TABLE 15.1 Characteristics of the Six Kingdoms

Domain	Bacteria	Archaea	Eukarya			
Kingdom	Bacteria	Archaea	Protista	Plantae	Fungi	Animalia
Cell type	Prokaryotic	Prokaryotic	Eukaryotic	Eukaryotic	Eukaryotic	Eukaryotic
Nuclear envelope	Absent	Absent	Present	Present	Present	Present
Mitochondria	Absent	Absent	Present or absent	Present	Present or absent	Present
Chloroplasts	None (photosynthetic membranes in some types)	None (bacteriorhodopsin in one species)	Present in some forms	Present	Absent	Absent
Cell wall	Present in most; peptidoglycan	Present in most; polysaccharide, glycoprotein, or protein	Present in some forms; various types	Cellulose and other polysaccharides	Chitin and other noncellulose polysaccharides	Absent
Means of genetic recombination, if present	Conjugation, transduction, transformation	Conjugation, transduction, transformation	Fertilization and meiosis	Fertilization and meiosis	Fertilization and meiosis	Fertilization and meiosis
Mode of nutrition	Autotrophic (chemosynthetic, photosynthetic) or heterotrophic	Autotrophic (photosynthesis in one species) or heterotrophic	Photosynthetic or heterotrophic or combination of both	Photosynthetic, chlorophylls a and b	Absorption	Digestion
Motility	Bacterial flagella, gliding, or nonmotile	Unique flagella in some	9 + 2 cilia and flagella; amoeboid, contractile fibrils	None in most forms, 9 + 2 cilia and flagella in gametes of some forms	Nonmotile	9 + 2 cilia and flagella, contractile fibrils
Multicellularity	Absent	Absent	Absent in most forms	Present in all forms	Present in most forms	Present in all forms

were first compared in 1996, the differences proved striking. Based on comparing DNA and cell structures, archaea are as different from bacteria as bacteria are from eukaryotes. Recognizing this, biologists have in recent years adopted a taxonomic level higher than kingdom that recognizes three **domains** (figure 15.10). Archaea are in one domain, bacteria in a second, and eukaryotes in the third. While the domain Eukarya contains four kingdoms of organisms, the domains Bacteria and Archaea contain only one kingdom in each.

> **Putting the Concept to Work**
> Under Linnaeus's two-kingdom system, which kinds of protists do you think would have been classified as plants? Why?

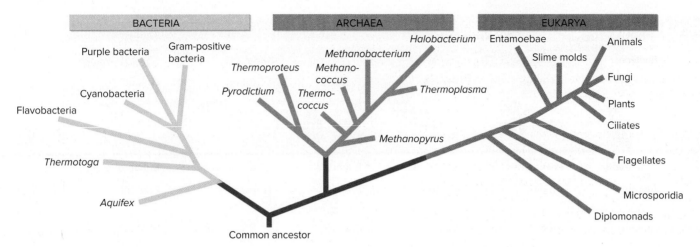

Figure 15.10 A tree of life.
This phylogeny, prepared from rRNA analyses, shows the evolutionary relationships among the three domains. The root of the tree is within the bacterial domain. Archaea and eukaryotes diverged later and are more closely related to each other than either is to bacteria.

15.7 Domain: A Higher Level of Classification

> **LEARNING OBJECTIVE 15.7.1** Explain how organisms in the three domains differ in basic cellular structure.

Domain Bacteria

The domain Bacteria contains one kingdom of the same name, Bacteria. Bacteria are the most abundant organisms on earth. There are more living bacteria in your mouth than there are mammals living on earth. Although too tiny to see with the unaided eye, bacteria play critical roles throughout the biosphere. For example, they extract from the air all the nitrogen used by organisms, and they play key roles in cycling carbon and sulfur.

> All organisms need nitrogen, but only a few kinds of bacteria can transform the nitrogen in the air into a form that can be used by other organisms. This process, called nitrogen fixation, is discussed in section 20.5.

There are many different kinds of bacteria, and the evolutionary links between them are not well understood. The archaea and eukaryotes are more closely related to each other than to bacteria and are on a separate evolutionary branch of the phylogenetic tree in **figure 15.10**.

Domain Archaea

The domain Archaea contains one kingdom by the same name, the Archaea. The term *archaea* (Greek, *archaio,* ancient) refers to the ancient origin of this group of prokaryotes, which most likely diverged very early from the bacteria. Notice in **figure 15.10** that the Archaea, in red, branched off from a line of prokaryotic ancestors that led to the evolution of eukaryotes. Though a diverse group, all archaea share certain key characteristics: They possess very unusual cell walls, lipids, and ribosomal RNA (rRNA) sequences. Also, some of their genes possess introns, unlike those of bacteria.

As the genomes of archaea have become better known, microbiologists have been able to identify signature sequences of DNA present in all archaea and in no other organisms. When samples from soil or seawater are tested for genes matching these signature sequences, many of the prokaryotes living there prove to be archaea. Clearly, archaea are not restricted to extreme habitats, as microbiologists used to think.

Domain Eukarya

For at least 1 billion years, prokaryotes ruled the earth. No other organisms existed to eat them or compete with them, and their tiny cells formed the world's oldest fossils. The third great domain of life, the eukaryotes, appear in the fossil record much later, only about 1.5 billion years ago.

Three Largely Multicellular Kingdoms. Fungi, plants, and animals are well-defined evolutionary groups. They are largely multicellular, each group clearly stemming from a different single-celled eukaryotic ancestor. Within each multicellular kingdom, there is astonishing diversity. Some fungi are multicellular mushrooms, others single-celled yeasts; some plants are thin blades of grass, others towering trees; some animals are too small to see, others are as large as elephants (figure 15.11), and still others exhibit a bizarre collection of traits usually seen in quite separate groups. The ancestor of each of these three distinct multicellular lines is a member of the kingdom Protista.

A Fourth Very Diverse Kingdom. When multicellularity evolved, the diverse kinds of single-celled organisms that existed at that time did not simply become extinct. A wide variety of unicellular, as well as multicellular, eukaryotes exists today in the kingdom Protista. Protists are a fascinating group containing many organisms of intense interest and great biological significance.

Figure 15.11 An African elephant.

ac productions/Blend Images LLC

Symbiosis and the Origin of Eukaryotes. The hallmark of eukaryotes is complex cellular organization, highlighted by an extensive endomembrane system that subdivides the eukaryotic cell into functional compartments called organelles (see chapter 4). Not all of these organelles, however, are derived from the endomembrane system. Mitochondria and chloroplasts are both believed to have entered early eukaryotic cells by a process called endosymbiosis (*endo,* inside) in which an organism such as a bacterium is taken into the cell and remains functional inside the cell.

With few exceptions, all modern eukaryotic cells possess energy-producing organelles, the mitochondria. Mitochondria are about the size of bacteria and contain DNA. Comparison of the nucleotide sequence of this mitochondrial DNA with that of a variety of organisms indicates clearly that mitochondria are the descendants of purple bacteria that were incorporated into eukaryotic cells early in the history of the group. Some protist phyla have also acquired chloroplasts during the course of their evolution and thus are photosynthetic. These chloroplasts are derived from cyanobacteria that became symbiotic in several groups of protists early in their history. Figure 15.12 shows how this could have happened, with the green cyanobacterium being engulfed by an early protist. Some of these photosynthetic protists gave rise to land plants. We discussed the endosymbiotic origin of mitochondria and chloroplasts in chapter 4.

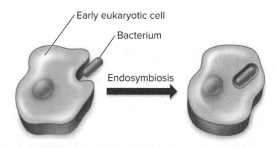

Figure 15.12 Endosymbiosis.

Mitochondria and chloroplasts are thought to have arisen in early eukaryotic cells through endosymbiosis: a bacterium is taken into the cell through a process similar to endocytosis but remains functional inside the host cell.

> **Putting the Concept to Work**
> What precisely is the difference between archaea and bacteria?

Putting the Chapter to Work

1 Carolus Linnaeus devised the binomial system to assign uniform names to animals and plants. The horse is a member of the taxonomic genus *Equus*, which are large grazing mammals that have an odd number of toes—horses, donkeys, and zebras. Among these, only the domesticated horse has a single toe. Linnaeus assigned this single-toed *Equus* group the name *caballus*. Demonstrate your understanding of the writing of correct species names by providing the binomial name for the domesticated horse.

2 Organisms are classified by many characteristics.

If you were given the following description of an organism, which domain and kingdom would you place it in?

The organism has mitochondria, its cell walls contain chitin, it obtains nutrients through absorption, and it is nonmotile.

3 Prokaryotes are found in the domains Bacteria and Archaea. While these two groups of prokaryotes have many similarities, there are major differences at the genetic level.

What physical difference in cell wall structure could be used to differentiate prokaryotes in the two domains?

4 One of the characteristics of the six kingdoms is the mode of nutrition.

In which kingdom would you place an organism that has a digestive system?

Retracing the Learning Path

The Classification of Organisms

15.1 The Invention of the Linnaean System

1. Scientists use classification to group similar organisms together. Latin, a language no longer spoken, is used to avoid confusion among scientists speaking many different languages.

- The polynomial system of classification named an organism by using a list of adjectives that described the organism. The binomial system, using a two-part name, was originally developed by Linnaeus as a "shorthand" reference to the polynomial name.

15.2 Species Names

1. Taxonomy is the area of biology involved in identifying, naming, and grouping organisms. Since Linnaeus, scientific names consist of two parts. The first part identifies the genus to which a species belongs, and the second part distinguishes that particular species from other species in the genus.

15.3 Higher Categories

1. A hierarchical system is used to classify organisms, in which the higher categories convey more general information about the organisms in a particular group. The most general category, domain, is the largest grouping, followed by ever-increasingly specific information that is used to group organisms into a kingdom, phylum, class, order, family, genus, and species.

15.4 What Is a Species?

1. The biological species concept states that a species is a group of organisms that is reproductively isolated, meaning that the individuals mate and produce fertile offspring with each other but not with other organisms.

- This concept works well to define animal species because animals regularly outcross (mate with other individuals) as do many kinds of plants. However, the concept does not apply to other organisms (fungi, protists, prokaryotes, and some plants) that regularly reproduce without mating through asexual reproduction. The classification of these organisms relies more on physical, behavioral, and molecular characteristics.

Inferring Phylogeny

15.5 How to Build a Family Tree

1. The study of taxonomy gives us a glimpse of the evolutionary history of life on earth, because organisms with similar characteristics are more likely to be related to each other. The evolutionary history of an organism and its relationship to other species is called phylogeny, and relationships are often mapped out using phylogenetic trees.

- Phylogenetic trees are created by assuming that key characteristics are shared by organisms with a common ancestor. A group of organisms related by descent is called a clade, and a phylogenetic tree organized in this manner is called a cladogram.

- Phylogenies can sometimes be misleading when key characteristics turn out to be less important than first thought. Cladograms in which all characteristics are weighted equally provide a more accurate picture of when changes occurred.

- Traditional taxonomy focuses more on the significance or evolutionary impact of a characteristic and not just on the commonality of the characteristic. For example, in traditional taxonomy, birds are placed in their own class, even though, cladistically, birds fall within the reptile group. Traditional taxonomy is used when more information is available to weight characteristics that seem more significant, whereas cladistics places more emphasis on the order or timing in which unique, or derived, characteristics appear.

Kingdoms and Domains

15.6 The Kingdoms of Life

1. The designation of kingdoms, the second-highest category used in classification, has changed over the years as more and more information about organisms has been uncovered. Currently, six kingdoms have been identified: Bacteria, Archaea, Protista, Fungi, Plantae, and Animalia.

- The domain level of classification was added in the mid-1990s, recognizing three fundamentally different types of cells: Eukarya (eukaryotic cells), Archaea (prokaryotic archaea), and Bacteria (prokaryotic bacteria).

15.7 Domain: A Higher Level of Classification

1. The domain Bacteria contains prokaryotic organisms in the kingdom Bacteria. These single-celled organisms are the most abundant organisms on earth and play key roles in ecology.

- The domain Archaea contains prokaryotic organisms in the kingdom Archaea. Although they are prokaryotes, archaea are as different from bacteria as they are from eukaryotes. These single-celled organisms are found in diverse environments, including very extreme environments.

- The domain Eukarya contains very diverse organisms from four kingdoms that are similar in that they are all eukaryotes. Fungi, plants, and animals are multicellular organisms, and the protists are primarily single-celled but very diverse. Eukaryotes contain cellular organelles, some that were most likely acquired through endosymbiosis.

Inquiry and Analysis

Do Zombie Ants Prevent Colony-Wide Fungal Infections?

Sometimes zombies are the good guys. Consider the ant *Camponotus castaneus*. Occasionally it becomes infected by the parasitic fungus *Ophiocordyceps unilateralis*. Reproductive cells of the fungus called spores can attach to the exterior body of the ant and enzymatically degrade parts of its exoskeleton. Once the fungus has penetrated the exoskeleton, fungal filaments called hyphae can grow throughout the ant's body and produce compounds that result in the ant climbing up the stems of nearby plants and using its mandibles to clamp to a leaf vein. Once attached to the leaf, the ant dies and the fungal body cells (called stroma) grow out from the head of the ant, which is what you see happening in the photo. Because of their peculiar behavior, the infected ants are called zombie ants.

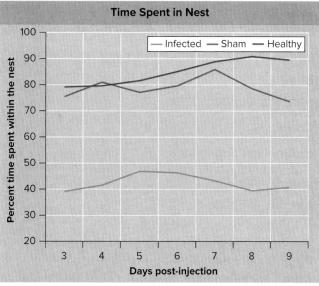

Dr Morley Read/Science Source

These ants live in colonies that can contain thousands, if not millions, of individuals. However, when ants infected with *O. unilateralis* are found, scientists have discovered to their surprise that the infection is limited to a few individuals instead of an entire colony. What is going on here?

To investigate, scientists collected *C. castaneus* ants, divided them into treatment groups, and placed them in a lab-based colony. Seventy-six ants were collected and divided into three treatment groups, "healthy," "infected," and "sham" treated. The "infected" ants were injected under the thorax with 1 microliter of a hyphae/media mixture. The hyphae/media mixture was generated by homogenizing the hyphae from a single fungal colony with a growth medium. The "sham"-treated ants were injected with 1 microliter of the growth medium alone. "Healthy" ants remained untreated. Colonies with all three treatment groups were housed in 15 cm^2 wooden boxes placed on 452 cm^2 sandy foraging arena. Ants were kept on 12 hour day/night cycles. A camera was affixed on top of the colony chamber and recordings were performed 24 hours a day. Using the data produced by the camera, scientists measured the time that each ant spent inside the nest. The results are plotted in the graph. The red line is the data for the "healthy" ants, the blue line is for the "sham"-treated ants, and the green line is for the "infected" ants.

Analysis

1. **Applying Concepts**
   a. **Variable**   In the graph, which is the dependent variable?
   b. **Control**   What is the difference between the "healthy" group and the "sham" group?
2. **Interpreting Data**
   a. On Day 4, what is the difference in the amount of time that the "healthy" ants spend in the nest versus the "infected" ants?
   b. Compare the "healthy" to the "sham." What is the difference?
3. **Making Inferences**
   a. Based on the data presented here, what general statement can be made about the "sham" treatment compared to the untreated "healthy" ants?
   b. Starting on Day 5, the "healthy" ant populations start spending more time inside the nest. Why?
4. **Drawing Conclusions**
   Does this data support the hypothesis that infected ants will be isolated from the rest of the colony?
5. **Further Analysis**
   a. Why was it important to include the "sham" treatment in this experiment?
   b. Ants are very social animals and participate in the mutual exchange of regurgitated food and liquid. With the data presented here, why do you think infected ants spend more time outside of the nest than healthy or sham-treated ants?

16 Evolution of Microbial Life

LEARNING PATH ▼

Origin of Life
1. How Cells Arose

Prokaryotes
2. The Simplest Organisms

Viruses
3. Structure of Viruses
4. How Viruses Infect Organisms

Protists
5. General Biology of Protists
6. Kinds of Protists

Fungi
7. A Fungus Is Not a Plant
8. Kinds of Fungi

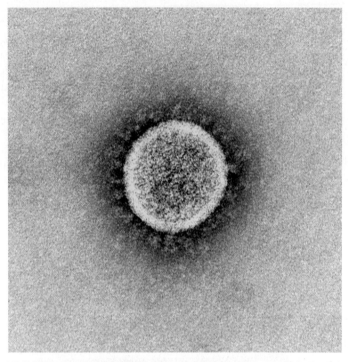

ELECTRON MICROGRAPH IMAGE of the SARS-CoV-2 virus that caused the COVID-19 pandemic.
NIAID/NIH/Science Source

Evolution of a Global Pandemic

All living organisms evolve as changes in their DNA either help the organism thrive or hurt its chances for survival. Viruses, while not alive as cells are, do contain DNA or RNA, and so evolve too, changing from one generation to the next. Sometimes virus evolution has a major impact on humans. You receive a flu vaccine every year because the virus that causes flu evolves very rapidly, with changes in the genetic composition of the flu virus making last year's vaccine less effective.

Recently, virus evolution has had a major impact on all our lives. A virus native to bats in Northern China evolved a new form which you will have heard of—SARS-CoV-2. While coronaviruses have been around in animals for years, they usually are not harmful to humans, at worst causing common colds. In 2019, however, a coronavirus evolved to a point that when it infected humans, it could cause serious illnesses and death.

This virus became a killer. In 18 months, more than 200 million people became infected, and over 4 million of them died.

The Pandemic Begins

In early December of 2019, several patients in the Hubei Province of China went to the doctor for pneumonia-like symptoms: difficulty breathing, fevers, and coughs. Pneumonia can be caused by both bacteria and viruses, but, regardless of the source, patients typically respond to the standard treatment regimens fairly well and recover within a few weeks. The patients in China, however, did not respond to any of the treatments and continued to get worse. Patients were struggling to get enough oxygen into their system and they had to be placed on ventilators. Patients were dying and the doctors were stumped. It was not until January 12, 2020, that the cause of the mystery illness was identified. It was a new member of the coronavirus family and the disease was named COVID-19

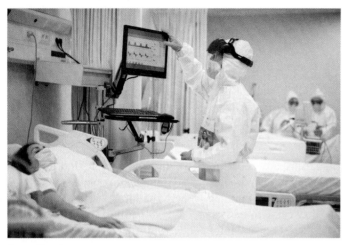
Tempura/Getty Images

HECTOR RETAMAL/AFP/Getty Images

(**co**rona **vi**rus **d**isease of 20**19**). But by the time the virus was found, it had already spread beyond the borders of China. Patients in Thailand and Japan were sick with the same symptoms. The virus had left the country and was spreading around the world.

A Killer Lurks

If the virus that causes COVID-19 first appeared in humans in late 2019, where was it before that? With approximately 75% of all new human diseases having origins in other animals, scientists started looking there. Coronaviruses are a large family of viruses that have likely been around for centuries and are quite common in a vast number of animals, including humans, dogs, cats, camels, and, especially important, bats. Bats have an amazingly strong and robust immune system that allows them to be perfect hosts for a multitude of viruses while not themselves becoming sick. A species of bat, the *Rhinolophus affinis* bat, is quite common and widespread throughout Asia. A small colony captured in Pu'er City in China was found to be a reservoir for a coronavirus that was 96.1% identical to the virus that causes COVID-19. This high level of similarity allowed scientists to conclude that the two viruses share a common ancestor and that COVID-19 is a bat virus.

Zoonotic Transfer

Zoonotic transfer is the spread of a disease from an animal to a human. It seems that the first COVID-19-infected human must have somehow picked up the virus from a bat. The origin of the outbreak has been traced back to a "wet" market in Wuhan, a large city in China. Wet markets are collections of open-air stalls that sell a wide range of fresh seafood, meat, fruit, and even live animals.

But among all the animals being sold at this market where the pandemic began, no one was selling bats. So how did the virus transfer from bats to humans? There must have been an intermediary animal! Scientists have not determined what animal served as this zoonotic stepping stone, but they do know this process of moving between species is extremely difficult. To be able to infect a new species, the proteins on the surface of the virus (called spike proteins) must precisely fit into receptor proteins located on the surface of the new host's cells. Like a key unlocking a door, that is how the virus enters the cell it is infecting. When it gains entry, it must also be able to take over the replication machinery of the cell so as to produce more viruses. It is no easy feat, evolving enough to jump between species.

The Virus Keeps Evolving

People from all over the world have been infected by COVID-19, but they are not all being infected by the same identical virus. When the virus that started the pandemic in China was first identified, it was designated as an "S" isolate. The version that was found to have spread all over China and into other countries was a slightly different version, named the "L" isolate. While it is unclear if these early mutational differences had any effect on the infection rates or disease severity, it is clear that the virus has been changing. There were different isolates of the virus spreading across the world, each just a little different than the others. The "L" isolate was disappearing from China and being replaced by isolates like the more infectious Alpha variant that first appeared in Europe in September, 2020, and the even more infectious Delta variant that appeared in India a month later. The new mutations are clustered around four small regions of the viral genome which code for part of the spike proteins and for RNA polymerase; the rest of the virus is unchanged. By targeting vaccines at the regions that are not rapidly evolving, researchers hope to be able to combat any mutant variants of COVID-19 that might evolve in the future.

Origin of Life

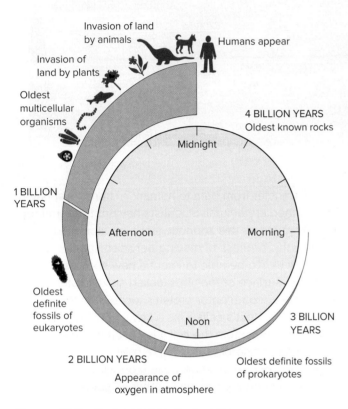

Figure 16.1 A clock of biological time.
A billion seconds ago, most students using this text had not yet been born. A billion minutes ago, the year was 20 A.D. A billion hours ago, the first modern humans were beginning to appear. A billion days ago, the ancestors of humans were beginning to use tools. A billion months ago, the last dinosaurs had not yet been hatched. A billion years ago, no creature had ever walked on the surface of the earth.

16.1 How Cells Arose

LEARNING OBJECTIVE 16.1.1 Describe the Miller–Urey experiment, assessing the significance of its results.

All living organisms are constructed of the same four kinds of macromolecules discussed in chapter 3. They are the building blocks of a cell, like the bricks and mortar of a building. Where the first macromolecules came from and how they came to be assembled together into cells are among the least understood questions in biology—questions that address the very origin of life itself.

No one knows for sure where the first organisms (thought to be like today's bacteria) came from. It is not possible to go back in time and watch how life originated, nor are there any witnesses.

In this chapter, we will attempt to understand whether the forces of evolution could have led to the origin of life and, if so, how the process might have occurred. This is not to say that evolution is definitely the source of life on earth. Nothing rules out the possibility that life did not originate on earth at all but instead was carried to it, perhaps as an extraterrestrial infection of spores originating on a planet of a distant star. Or life-forms may have been put on earth by supernatural or divine forces. In biology, we limit the scope of our inquiry to scientific matters. Of these possibilities, only evolution permits testable hypotheses to be constructed and so provides the only scientific explanation—that is, one that could potentially be disproved by experiment.

Forming Life's Building Blocks

If we look at the development of living organisms as a 24-hour clock of biological time shown in figure 16.1, with the formation of the earth 4.5 billion years ago being midnight, the first cells appear just before noon, but humans do not appear until the day is almost all over, only minutes before its end. How then can we learn about the origin of the first cells? One way is to try to reconstruct what the earth was like when life originated 2.5 billion years ago. We know from rocks that were forming at that time that there was little or no oxygen in the earth's atmosphere then and more of the hydrogen-rich gases hydrogen sulfide (H_2S), ammonia (NH_3), and methane (CH_4). Electrons in these gases would have been frequently pushed to higher energy levels by photons crashing into them from the sun or by electrical energy in lightning. Today, high-energy electrons are quickly soaked up by the oxygen in earth's atmosphere (the atmosphere or air is 21% oxygen, all of it contributed by photosynthesis) because oxygen atoms have a great "thirst" for such electrons. But in the absence of oxygen, high-energy electrons would have been free to help form biological molecules.

The Miller–Urey Experiment. When the scientists Stanley Miller and Harold Urey reconstructed the oxygen-free atmosphere of the early earth in their laboratory and subjected it to the lightning and UV radiation it would have experienced then, they found that many of the building blocks of organisms, such as amino acids and nucleotides, formed spontaneously. They concluded that life may have evolved in a "primordial soup" of biological molecules formed in the ancient earth's oceans.

Problems with the Primordial Soup Hypothesis. Recently, concerns have been raised regarding the primordial soup hypothesis as the origin of life on earth. If the earth's atmosphere had no oxygen soon after it was formed, as Miller and Urey assumed (and most evidence supports this assumption), then there would have been no protective layer of ozone to shield the earth's surface from the sun's damaging UV radiation. Without an ozone layer, scientists think UV light would have destroyed any ammonia and methane in the atmosphere. When these gases are missing, the Miller–Urey experiment does not produce key biological molecules such as amino acids. If the necessary ammonia and methane were not in the atmosphere, where were they?

An Alternative Hypothesis. The bubble model, shown in figure 16.2, proposes that the key chemical processes generating the building blocks of life took place not in a primordial soup but rather within bubbles on the ocean's surface. Inside the bubbles, methane and ammonia would have been protected from destruction by UV radiation.

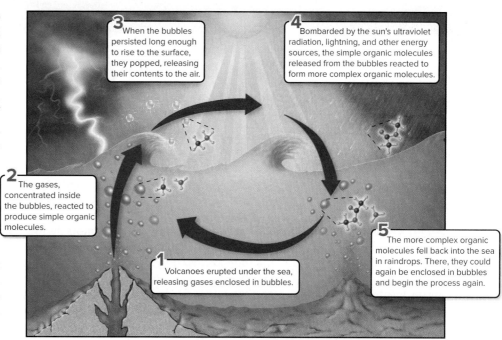

Figure 16.2 **A chemical process involving bubbles may have preceded the origin of life.**

In 1986, geophysicist Louis Lerman proposed that the chemical processes leading to the evolution of life took place within bubbles on the ocean's surface.

> **Putting the Concept to Work**
> Why can't biological macromolecules such as amino acids form spontaneously today, if they did 2.5 billion years ago?

The First Cells

> **LEARNING OBJECTIVE 16.1.2** Explain why scientists feel that the first macromolecules to form were RNA rather than protein.

We don't know how the first cells formed, but most scientists suspect they aggregated spontaneously. When complex carbon-containing macromolecules are present in water, they tend to gather together, sometimes forming aggregations big enough to see without a microscope. Try vigorously shaking a bottle of oil-and-vinegar salad dressing—tiny bubbles called *microspheres* form spontaneously, suspended in the vinegar. Similar microspheres might have represented the first step in the evolution of cellular organization. Such microspheres have many cell-like properties—their outer boundary resembles the membranes of a cell in that it has two layers (see section 4.3), and the microspheres can increase in size and divide.

Scientists suspect that the first macromolecules to form were RNA molecules, which can behave as enzymes and catalyze their own assembly. Eventually, DNA may have taken the place of RNA as the storage molecule for genetic information because the double-stranded DNA would have been more stable than single-stranded RNA.

As you can see, the scientific vision of life's origin is at best a hazy outline. How life might have originated naturally and spontaneously remains a subject of intense interest, research, and discussion among scientists.

> **Putting the Concept to Work**
> Is there anything about a spontaneous RNA model of the origin of life on earth that would rule out life originating on other planets?

Today's Biology

Has Life Evolved Elsewhere?

Are we alone? Perhaps the most fundamental question of biology is whether humanity is unique. Has life evolved elsewhere in the universe? We have been looking.

Our Solar System

We found no life on the moon when we reached it in 1969. Mars seemed a better bet, but the space probes that landed on the surface of Mars in the mid-1970s found no evidence of life in the soil samples they analyzed. The two trusty little rovers NASA has had sniffing around the Martian landscape in recent years have revealed only the teasing suggestion of ancient watery seas. A salty frigid lake lies a mile beneath the South Pole, tantalizing but impossible to explore.

Europa, a large moon of Jupiter, is a more promising candidate. Photos taken in close orbit in 1998 reveal that Europa is covered with seas of liquid water beneath a thin skin of ice. Larger than earth's oceans, Europa's are warmed by the push and pull of the gravitational attraction of Jupiter's many large satellite moons. Conditions are far less hostile to life than the conditions that existed in the oceans of the primitive earth. NASA is sending the *Europa Clipper* satellite mission, scheduled to launch in the 2020s, to look more closely.

Europa

JPL-Caltech/SETI Institute/NASA

Exoplanets

Might life exist on other solar systems in distant galaxies? The universe contains 10^{20} (100,000,000,000,000,000,000) stars similar to our sun. Do many of them have planets? In a word, yes. Too far away to be seen directly by telescope, an exoplanet (a planet orbiting another sun) can be detected by looking very precisely at the star it circles. Each time the exoplanet rotates across its face, the star will appear to "blink" or dim. The size of the dip in light can tell you how big the exoplanet is, whereas the length of time between blinks tells you how many days long its year is and thus how far out it is from its star and so how warm it is likely to be.

The first exoplanet was discovered in 1995 orbiting the star 51 Pegasi, about 50 light-years from earth. The *Kepler* spacecraft was launched in 2009 to look for more. Its telescope stares at some 160,000 stars in a patch of sky in the constellation Cygnus, about 1/400th of the entire sky. As of late 2017, *Kepler* had detected many more—over 5,000 exoplanet candidates and more than 2,500 confirmed exoplanets orbiting distant stars.

Goldilocks Planets

Is it likely that any of these far planets harbor life? The evolution of a carbon-based life form is probably possible only within the narrow range of temperatures that exist on earth, which is directly related to its distance from the star it orbits, the sun. To harbor life, a planet would also have to be big enough for its gravitational pull to harbor an atmosphere, but not so big as to create a superdense atmosphere through which light could not pass to the planet's surface. To harbor life, an exoplanet would have to be what is referred to as a "Goldilocks" planet (from the fairy tale of Goldilocks and the three bears: "Not too hot, not too cold. . .").

The first Goldilocks exoplanet was found orbiting a star 1,400 light-years from earth. Labeled Kepler 452b, it circles a star very much like our sun, taking only 20 days longer to get around than earth does. Temperatures on Kepler 452b would be similar to lukewarm water. Far away from earth, though. Other Goldilocks exoplanets are much closer. One named LHS 1140b orbits a faint red dwarf star only 41 light-years away and is just earth's size. Also orbiting a red dwarf is the earthlike exoplanet Ross 128b, only 11 light-years away. An even more likely candidate, the earth-sized exoplanet Proxima b, orbits a sunlike star only 4.2 light-years away from earth, practically a next-door neighbor.

Over 30 of the sunlike stars in the Cygnus galaxy surveyed by the *Kepler* space telescope have what appear to be habitable earth-sized exoplanets. In our own small spiral galaxy called the Milky Way, there are many millions of stars, and the universe contains over 200 billion such galaxies. Thus, there could be many billions of earth-sized exoplanets basking in lukewarm conditions suitable for liquid water and, perhaps, for the evolution of life. In April of 2018, a new satellite, *TESS* (for *Transiting Exoplanet Survey Satellite*), was launched to expand the search and take a closer look. TESS will even be able to determine the types of atmospheres around these exoplanets. In 2020, TESS completed its primary mission and discovered 66 new Goldilocks exoplanets and identified nearly 2,100 other exoplanet candidates that astronomers are evaluating.

So Where Are They?

It does not seem likely that we are alone. Looking up on a clear night, it is impossible not to ask yourself, *"On planets orbiting how many of these stars is someone studying the stars and speculating on my existence?"*

Of course, with so many worlds on which life might have emerged, we should have heard from someone by now. . . . In a wonderful Calvin and Hobbes cartoon, Calvin says, *"I was reading about how countless species are being pushed toward extinction by man's destruction of forests."* He goes on in the next frame to say, *"Sometimes I think the surest sign that intelligent life exists elsewhere in the universe is that none of it has tried to contact us."* As our rapidly warming world looks to the stars, we can only hope that Calvin is being too cynical, that we will wise up and stop destroying our planet, so we will have something worthwhile to say to whoever out there might be trying to communicate with us.

Prokaryotes

16.2 The Simplest Organisms

> **LEARNING OBJECTIVE 16.2.1** Distinguish between gram-negative and gram-positive bacteria.

For over a billion years, prokaryotes were the only living things.

Our Earliest Ancestors

Judging from fossils in ancient rocks, prokaryotes have been plentiful on earth for over 2.5 billion years. The diverse array of early living forms gave rise to three sorts of organisms that survive today: Eukaryotes like you, and two kinds of prokaryotes named Bacteria and Archaea. The fossil record indicates that eukaryotic cells, being much larger than prokaryotes and exhibiting elaborate shapes in some cases, did not appear until about 1.5 billion years ago. For at least 1 billion years, prokaryotes were the only organisms that existed.

Today, prokaryotes are the simplest and most abundant form of life on earth. In a spoonful of farmland soil, 2.5 billion bacteria may be present. In one hectare (about 2.5 acres) of wheat land in England, the weight of bacteria in the soil is approximately equal to that of 100 sheep!

It is not surprising, then, that prokaryotes occupy a very important place in the web of life on earth. They play a key role in cycling minerals within the earth's ecosystems. In fact, photosynthetic bacteria were in large measure responsible for the introduction of oxygen into the earth's atmosphere. Pathogenic bacteria are responsible for some of the most deadly animal and plant diseases, including many human diseases. Bacteria and archaea are our constant companions, present in everything we eat and on everything we touch.

The Structure of a Prokaryote

Prokaryotes are single-celled organisms too tiny to see with the naked eye. Each cell has a simple interior with a single circle of DNA. Importantly, this DNA is not confined by a nuclear membrane in a nucleus, as in the cells of eukaryotes. Table 16.1 illustrates the seven key ways in which prokaryotes differ from eukaryotes.

A prokaryotic cell's plasma membrane is surrounded by a cell wall that is chemically quite different from those found in eukaryotic cells (protists, fungi, and plants). Those with thin cell walls have an outer membrane surrounding the cell that appears to protect them from attack by chemicals, such as the antibiotic penicillin. Those with thick cell walls lack an outer membrane. Many bacteria also have a gelatinous layer outside the cell wall, called a capsule.

Bacterial cell wall thickness and complexity was first discovered by a Danish researcher, Hans Gram, who developed a staining process that stains cell types differently. Bacteria that possess a thinner cell wall stain red and are called gram-negative bacteria. Those with thicker cell walls stain purple and are called gram-positive bacteria. Today, scientists and medical doctors alike still refer to bacteria as either gram-positive or gram-negative, which indicates the type of antibiotic that might be an effective treatment.

TABLE 16.1 Prokaryotes Compared to Eukaryotes

Internal compartmentalization. Unlike eukaryotic cells, prokaryotic cells contain no internal compartments, no internal membrane system, and no cell nucleus.

Cell size. Most prokaryotic cells are only about 1 micrometer in diameter, whereas most eukaryotic cells are well over 10 times that size.

Prokaryotic cell   Eukaryotic cell

Unicellularity. All prokaryotes are fundamentally single-celled.

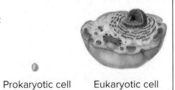
Unicellular bacteria

Chromosomes. Prokaryotes do not possess chromosomes as eukaryotes do. Instead, their DNA exists as a single circle in the cytoplasm.

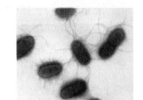
Prokaryotic chromosome   Eukaryotic chromosomes

Cell division. Cell division in prokaryotes takes place by binary fission. In eukaryotes, microtubules pull chromosomes to opposite poles during mitosis.

Binary fission in prokaryotes   Mitosis in eukaryotes

Flagella. Prokaryotic flagella are composed of a single fiber of protein that spins like a propeller. Eukaryotic flagella are more complex structures, with a 9 + 2 arrangement of microtubules.

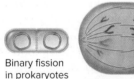
Simple bacterial flagellum

Metabolic diversity. Prokaryotes possess many metabolic abilities that eukaryotes do not, including anaerobic photosynthesis and the ability to fix atmospheric nitrogen.

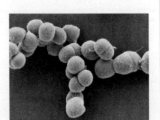
Chemoautotrophs

Unicellular bacteria: Kwangshin Kim/Science Source; Chemoautotrophs: Eye of Science/Science Source

Biology and Your Health

Drugs from Dirt

In 1928, researcher Alexander Fleming noticed that mold had inadvertently drifted onto the surface of a lab culture dish on which he was growing bacteria. Surprisingly, the bacteria on the dish would not grow in a zone on the culture dish surrounding where the mold had landed—something was diffusing through the growth media from the mold that was killing any nearby bacteria. Fleming had accidentally discovered penicillin, the world's first antibiotic. When a person takes penicillin, it works to disrupt the formation of a strong bacterial cell wall, and causes any infecting bacterial cell to explode when that cell attempts to make copies of itself. Widely administered to people sick from bacterial infections in the years after World War II, penicillin saved countless lives.

Drug-Resistant Superbugs

Then, in the mid-1960s, penicillin stopped helping many patients—bacterial strains had evolved resistance to the drug. Luckily, other antibiotics had been discovered by then that could be used to cure these patients. Over a dozen different antibiotics have been introduced since 1943, when penicillin was first widely used. But resistance to these other antibiotics has developed too, hastened by their overuse in people and animals. All but one of these antibiotics have by now been reported to have generated resistant strains. Even worse, the resistant bacteria are very efficient at passing their resistance from one strain to another, leading to the emergence of "superbugs" resistant to many and sometimes all available antibiotics. As a consequence, deadly bacterial infections are not simply specters of our past, conquered by modern medical technology. They kill today, and often. Drug-resistant infections helped cause an estimated 2 million illnesses and 23,000 deaths in the United States in 2017.

Seeking New Antibiotics in Dirt

The emergence of superbugs resistant to all available antibiotics has left scientists scrambling to find new antibiotics. The problem is that they have already looked in all the likely places. Practically any organism that can be grown in a laboratory has been checked. This leaves us with organisms that cannot be grown in the lab. Stand outside on your lawn and look around. Where are the most bacteria? Under your feet. Countless bacteria grow in soil, unknown to science because almost none of them can be cultured in the lab. *There* is where scientists are now looking for new antibiotics!

While soil bacteria can't be grown in the lab, it is possible to simply skip this step and extract the DNA of all bacteria present in a soil sample. Each cell present in the soil sample contributes only a tiny smidgen, of course, but that is all modern cell biologists need, because today's technology allows a single DNA strand to be amplified to billions of copies. The trick is to find the right smidgen. What researchers at Rockefeller University did in 2018 was to screen the DNA from a soil sample for bits of DNA with a sequence that encoded proteins able to bind calcium. Why select calcium-binding ability as their target? Because many of the enzymes of bacterial metabolism only work properly when bound to a calcium ion. If you are looking for something to gum up bacterial metabolism, that's a good place to start.

Discovery of Malacidin

The Ca^{++}-protein-encoding DNA bits the researchers found in dirt were then each amplified and sequenced. In each case, comparing the result with a database of known bacteria gene sequences allowed the researchers to identify likely candidates, DNA-encoding proteins associated with key metabolic processes. Putting these DNA segments into bacteria easily grown in the lab allowed the researchers to test if any of these soil DNA sequences made proteins that hurt the bacteria. One did. It blocked the synthesis of stuff needed to build the bacteria's cell wall. Nor were bacteria able to easily generate resistance to the protein encoded by this DNA; in 21 days of continuous exposure, no resistance to this drug appeared among the bacteria exposed to it. The researchers dubbed their newly discovered antibiotic malacidin.

DrugsFromDirt.org

The discovery of this dirt-derived antibiotic is very promising, as malacidin kills a variety of superbugs, including resistant strains of MRSA. Most exciting of all is the success of the general approach to discovering new antibiotics—drugs from dirt. The researchers have set up an online citizens' science project (DrugsFromDirt.org) to solicit soil donations from around the world. Indeed, the sandy soil from which malacidin was extracted was shipped to the Rockefeller University lab in New York City by the parents of one of the researchers who lived in the Southwest United States.

Comstock/Getty Images

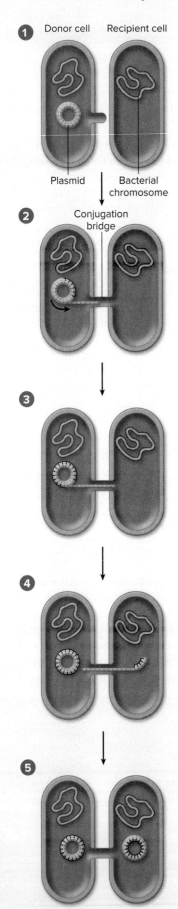

Figure 16.3 Bacterial conjugation.

How Prokaryotes Move

Many kinds of bacteria possess threadlike **flagella,** long strands of protein that may extend out several times the length of the cell body. Bacteria swim by twisting these flagella in a corkscrew motion, as shown below. Some bacteria also possess shorter outgrowths called **pili** (singular, **pilus**), which act as docking cables, helping the cell attach to surfaces or to other cells.

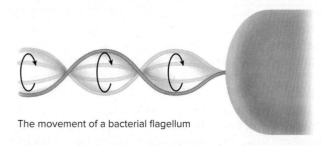

The movement of a bacterial flagellum

When exposed to harsh conditions (dryness or high temperature), some bacteria form thick-walled **endospores** around their DNA and a small bit of cytoplasm. These endospores are highly resistant to environmental stress and may germinate to form new active bacteria even after centuries.

> **Putting the Concept to Work**
> Name seven ways prokaryotes differ from eukaryotes.

How Prokaryotes Reproduce

> **LEARNING OBJECTIVE 16.2.2** Contrast the process of bacterial conjugation with binary fission.

Prokaryotes, like all other living cells, grow and divide. Prokaryotes reproduce using a process called **binary fission** in which an individual cell simply increases in size and divides in two. Following replication of the cell's DNA, the plasma membrane and cell wall grow inward and eventually divide the cell by forming a new wall from the outside toward the center of the old cell (see section 8.1 and **figure 8.1**).

Some bacteria can exchange genetic information by passing plasmids from one cell to another in a process called **conjugation.** A plasmid is a small, circular fragment of DNA that replicates outside the main bacterial chromosome. In bacterial conjugation, seen in **figure 16.3**, the pilus of a donor cell contacts a recipient cell ❶; the pilus draws the two cells close together. A passageway called a *conjugation bridge* forms between the two cells. The plasmid in the donor cell begins to replicate its DNA ❷, passing the replicated copy out across the bridge into the recipient cell ❸, where a complementary strand is synthesized ❹. The remaining strand in the donor cell serves as a template for the synthesis of its complement. Both the recipient cell and the donor cell now contain a complete copy of the genetic material found in the plasmid of the donor cell ❺.

> **Putting the Concept to Work**
> Are recipient and donor copies of the plasmid identical after conjugation?

Answering Your Questions About Pandemics

What Is a Pandemic?

A pandemic is a disease outbreak that occurs on a global scale. For a virus to cause a pandemic, three things must happen: (1) a never-before-seen virus must appear, (2) the virus must spread rapidly, and (3) there is no natural human defense to the virus. All three of these happened in 2020 with the COVID-19 virus. There is a lot we still do not know about COVID-19, but we do know a lot about one of the last pandemics, the Ebola pandemic that broke out in 2013.

How Do Pandemics Begin?

It is not uncommon to see new viruses arise and disappear fairly quickly. The Ebola virus first made an appearance in 1976. People in a few small central African villages were found dead, with blood running from their bodies. While it was not evident what the root cause was at the time, health workers were initially able to contain the spread of the Ebola virus to these local villages. The Ebola virus, which spreads only by contact with bodily fluids, killed more than 90% of those infected, making it one of the most lethal viruses known.

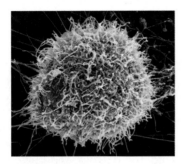
Callista Images/Getty Images

How Do Pandemics Spread Across the Globe?

The pattern of Ebola changed on December 6, 2013. A 2-year-old boy named Emile died of fever in Meliandou, a small rain forest village in Guinea, West Africa, near the borders of the nations of Sierra Leone and Liberia. These are three of the poorest nations on earth, racked by years of civil war. Emile's mother then died, suffering extensive bleeding, on December 13. His 3-year-old sister died on December 29, and his grandmother on January 1. Several people from neighboring villages attending the grandmother's funeral died bloodily in the following weeks. By the time the outbreak in Guinea was recognized as Ebola, dozens more people had died in eight communities, and cases were cropping up in Sierra Leone and Liberia. Never before had an Ebola outbreak affected so many different locations. Modern transportation was the culprit, with infected individuals able to travel quite far in the 21 days before symptoms revealed the infection.

How Long Do Pandemics Last?

Pandemics last until there is a vaccine that builds up immunity in people. For Ebola, there is no vaccine yet, and research into treatment of infected individuals, while promising, is still at the laboratory stage. The only way to combat the disease is to isolate infected individuals. This is no simple task in countries with poor public health systems and deep-rooted traditions of family burial. It came as no surprise that over the course of the spring months of 2014, the West Africa Ebola outbreak did not abate, but rather gained in intensity, despite the best efforts of international health organizations to isolate patients. By August, there were 3,039 Ebola cases, with 1,522 deaths. This 52% lethality, while mercifully less than the 90% of earlier outbreaks, is a grim reaper. By the end of the outbreak in mid-2015, over 24,000 people had caught the virus, and recorded deaths had climbed past 10,000.

Nor is this the last time Ebola will appear in Africa. The fruit bats aren't going anywhere. With research on vaccines and treatments being given priority, we can hope to be better prepared for the next outbreak.

Why Has Ebola Not Broken Out in America?

The simplest answer to this question is that fruit bats, the likely natural reservoir of Ebola virus, do not inhabit North America, nor South America. Known as "flying foxes," they live in dense forests in Africa, Asia, and Australia, where they exist in huge colonies and are active mostly at night. To date, outbreaks of Ebola have never occurred in any continent but Africa.

The more interesting answer is the question "Or has it?" In late 1989 in Reston, Virginia (located near the Washington, DC, metropolitan area), a lab population of several hundred Philippine macaque monkeys became sick and many died, the infection passing from cage to cage through the air. Tissue samples revealed the presence of antibodies to Ebola! All animals in the facility were promptly euthanized, in a quiet, quick, and massive operation involving 178 people.

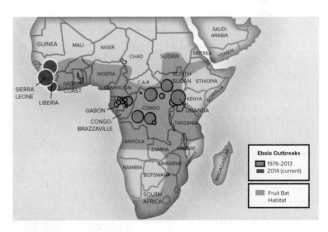

Six of the animal handlers in that operation subsequently developed antibodies to Ebola virus—but none got sick. Apparently, this strain of Ebola virus, now called Reston Ebolavirus, is not pathogenic in humans. It does infect and kill pigs, often used as models of human infectivity. Apparently, we dodged a bullet: There are five kinds of Ebola virus, four very fatal in humans; this was the fifth.

A Closer Look

Microbial Bartenders

Alcoholic fermentation, the anaerobic conversion of sugars into alcohol (see section 7.5), predates human history. As with many other natural processes, humans seem to have first learned to control this process by stumbling across its benefit—the pleasures of beer and wine. Artifacts from the third millennium b.c. contain wine residue, and we know from the pollen record that domesticated grapes first became abundant at that time. The production of beer started even earlier, back in the fifth millennium b.c., making beer possibly one of the oldest beverages produced by humans.

Making Beer. The process of making beer is called brewing and involves the fermentation of cereal grains. Grains are rich in starches, which you will recall from chapter 3 are long chains of glucose molecules linked together. First, the cereal grains are crushed and soaked in warm water, creating an extract called *mash*. Enzymes called amylases are added to the mash to break down the starch within the grains into free sugars. The mash is filtered, yielding a darker, sugary liquid called wort. The wort is boiled in large tanks, like the ones shown in the photo here, called fermentors. This helps break down the starch and kills any bacteria or other microorganisms that might be present before they begin to feast on all the sugar. Hops, another plant product, is added during the boiling stage. Hops gives the solution a bitter taste that complements the sweet flavor of the sugar in the wort.

Now we are ready to make beer. The temperature of the wort is brought down, and yeast is added. This begins the fermentation process. It can be done in either of two ways, yielding the two basic kinds of beer.

Lager. The yeast *Saccharomyces uvarum* is a bottom-fermenting yeast that settles on the bottom of the vat during fermentation. This yeast produces a lighter, pale beer called lager. Bottom fermentation is the most widespread method of brewing. Bottom fermentation occurs at low temperatures (5–8°C). After fermentation is complete, the beer is cooled further to 0°C, allowing for the beer to mature before the yeast is filtered out.

Alan Levenson/Getty Images

Ale. The yeast *S. cerevisiae*, often called "brewer's yeast," is a top-fermenting yeast. It rises to the top during fermentation and produces a darker, more aromatic beer called ale (also called porter or stout). In top fermentation, less yeast is used. The temperature is higher, around 15–25°C, and so fermentation occurs more quickly, but less sugar is converted into alcohol, giving the beer a sweeter taste.

Most yeasts used in beer brewing are alcohol-tolerant only up to about 5% alcohol—alcohol levels above that kill the yeast. That is why commercial beer is typically 5% alcohol.

Carbon dioxide is also a product of fermentation and gives beer the bubbles that form the head of the beer. However, because only a little CO_2 forms naturally, the last step in most brewing involves artificial carbonation, CO_2 gas being injected into the beer.

Making Wine. Wine fermentation is similar to beer fermentation in that yeasts are used to ferment sugars into alcohol. Wine fermentation, however, uses grapes. Rather than the starches found in cereal grains, grapes are fruits rich in the sugar sucrose, a disaccharide combination of glucose and fructose. Grapes are crushed and placed into barrels. There is no need for boiling to break down the starch, and grapes have their own amylase enzymes to cleave sucrose into free glucose and fructose sugars. To start the fermentation, yeast is added directly to the crush. *S. cerevisiae* and *S. bayanus* are common wine yeasts. They differ from the yeast used in beer fermentation in that they have higher alcohol tolerances—up to 12% alcohol (some wine yeasts have even higher alcohol tolerances).

A common misconception about wine is that the color of the wine results from the color of the grape juice used, red wine using juice from red grapes and white wines using juice from white grapes. This is not true. The juice of all grapes is similar in color, usually light. The color of red wine is produced by leaving the skins of red or black grapes in the crush during the fermentation process. White wines can be made from any color of grapes and are white simply because the skins that contribute red coloring are removed before fermentation.

Corbis/VCG/Getty Images

Viruses

16.3 Structure of Viruses

> **LEARNING OBJECTIVE 16.3.1** Describe the two biological macromolecules of which a virus is made and where each is found in a virus.

The border between the living and the nonliving is very clear to a biologist. Living organisms are cellular and able to grow and reproduce independently, guided by information encoded within DNA. As discussed in section 16.2, the simplest creatures living on earth today that satisfy these criteria are prokaryotes. Viruses, on the other hand, do not satisfy the criteria for "living" because they possess only a portion of the properties of organisms. **Viruses** are literally "parasitic" chemicals, segments of DNA (or sometimes RNA) wrapped in a protein coat. They cannot reproduce on their own, and for this reason, they are not considered alive by biologists. They can, however, replicate within cells, using the host cell's DNA enzymes and ribosomes.

Size

Viruses are very small. The smallest viruses are only about 17 nanometers in diameter; the largest ones are long filamentous viruses that measure up to 1,000 nanometers in length. Viruses are so small that they are smaller than many of the molecules in a cell. Most viruses can be detected only by using the higher resolution of an electron microscope.

Chemical Makeup

Each virus is in fact a mixture of two chemicals: nucleic acid and protein. The tobacco mosaic virus (TMV) pictured in **figure 16.4a** has the structure of a Twinkie, a tube made of an RNA core (the green springlike structure) surrounded by a coat of protein (the purple structures that encircle the RNA). Researchers were able to separate the RNA from the protein and purify and store each chemical. Later, when they mixed the protein and RNA, the two components reassembled into an infectious unit fully able to infect healthy tobacco plants. Because it cannot reproduce on its own, a virus is best regarded as a chemical assembly rather than a living organism.

Structure

Viruses infect all organisms, from bacteria to humans, and in every case, the basic structure of the virus is the same, a core of nucleic acid surrounded by protein. All viruses reproduce by first disintegrating into their component nucleic acid and protein molecules within a cell. Viruses then use the cell's machinery to manufacture new components and re-form them into a new generation of virus particles.

There is considerable difference, however, in the details of virus structure. In **figure 16.4**, you can compare the structure of plant, bacterial, and animal viruses—they are clearly quite different from one another, and there is even a wide variety of shapes and structures within each group of viruses. Bacterial viruses, called *bacteriophages*, can have elaborate structures like the bacteriophage in **figure 16.4b**, which looks more like a lunar module than a virus. Many plant viruses such as tobacco mosaic virus (TMV) have

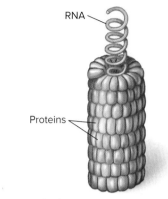

(a) Tobacco mosaic virus

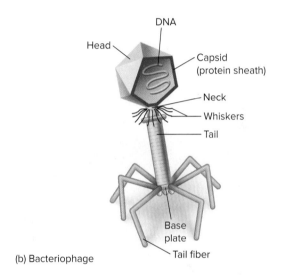

(b) Bacteriophage

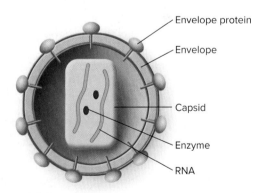

(c) Human immunodeficiency virus

Figure 16.4 The structure of plant, bacterial, and animal viruses.

(a) TMV infects plants and consists of 2,130 identical protein molecules (*purple*) that form a cylindrical coat around the single strand of RNA (*green*), protecting it. (b) Bacterial viruses, called bacteriophages, often have a complex structure. (c) In the human immunodeficiency virus (HIV), the RNA core is held within a capsid that is encased by a protein envelope.

a core of RNA, and some animal viruses such as HIV (figure 16.4c) and SARS-CoV-2 do too. One or more different segments of DNA may be present in other animal virus particles, along with many different kinds of protein. Like TMV, most viruses form a protein sheath, or *capsid,* around their nucleic acid core. In addition, many viruses form a membranelike *envelope,* rich in proteins and lipids, around the capsid.

> **Putting the Concept to Work**
> Is a virus alive? Can it evolve? Defend your answers.

The Origin of Viral Diseases

> **LEARNING OBJECTIVE 16.3.2** Name and describe six emerging viruses.

Sometimes viruses that originate in one organism pass to another, causing a disease in the new host. New pathogens arising in this way, called *emerging viruses,* represent a greater threat today than in the past, as air travel and world trade allow infected individuals to move about the world quickly.

Robert Burton/USFWS

Influenza. One of the most lethal virus in human history has been the influenza virus. Between 40 and 100 million people worldwide died of flu within 19 months in 1918 and 1919—an astonishing number. The natural reservoir of influenza virus is in ducks, chickens, and pigs in central Asia. Major flu pandemics (that is, worldwide epidemics) have arisen in Asian ducks through recombination within multiple infected individuals, putting together novel combinations of virus surface proteins unrecognizable by human immune defenses. The Asian flu of 1957 killed over 100,000 Americans. The Hong Kong flu of 1968 infected 50 million people in the United States alone, of which 70,000 died. The swine flu epidemic of 2009 killed thousands of American children.

Digital Vision/Getty Images

HIV, human immunodeficiency virus. The virus that causes AIDS first entered humans from chimpanzees somewhere in Central Africa, probably between 1910 and 1950. The chimpanzee virus, called simian immunodeficiency virus, or SIV (now known as HIV), spread to human populations worldwide. AIDS was first reported as a disease in 1981. In the following years, 25 million people have died of AIDS, and 33 million more are currently infected. Where did chimpanzees acquire SIV? SIV viruses are rampant in African monkeys, and chimpanzees eat monkeys. Study of the nucleotide sequences of monkey SIVs reveal that one end of the chimp virus RNA closely resembles the SIV found in red-capped mangabey monkeys, while the other end resembles the virus from the greater spot-nosed monkey. It thus seems certain that chimpanzees acquired SIV from the monkeys they ate.

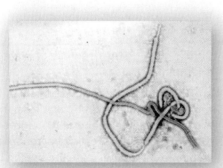
Courtesy of Dr. Frederick A. Murphy, Centers for Disease Control and Prevention

Ebola virus. Among the most lethal of emerging viruses are the filamentous Ebola and Marburg viruses. First arising in Central Africa, they attack the endothelial cells that line human blood vessels. With lethality rates that can exceed 90%, these so-called filoviruses cause some of the most lethal infectious diseases known. In the summer of 2014, an outbreak of Ebola in West Africa spread through three densely populated countries, killing over 10,000 people. As mentioned in the "Answering Your Questions About Pandemics" feature *What Is a Pandemic?,* researchers have reported evidence implicating fruit bats as Ebola hosts. These large bats are eaten for food everywhere in Central Africa where outbreaks have occurred.

Zika virus. A sudden outbreak of microcephaly (incomplete head and brain development in newborns) in Brazil in 2016 was soon traced to the mosquito-transmitted Zika virus. First discovered in Uganda, Africa, in the 1940s, Zika virus is now common in the tropics, but until recently caused only mild fever in infected humans. By 2018, more than 22,000 cases were reported in the Americas.

James Gathany/CDC

SARS. A recently emerged strain of coronavirus was responsible for a worldwide outbreak in 2003 of *severe acute respiratory syndrome* (SARS), a respiratory infection with pneumonia-like symptoms that in over 8% of cases is fatal. When the 29,751-nucleotide RNA genome of the SARS virus was sequenced, it proved to be a completely new form of coronavirus, not closely related to any of the three previously described forms. Virologists in 2005 identified the Chinese horseshoe bat as the natural host of the SARS virus. Because these bats are healthy carriers not sickened by the virus and occur commonly throughout Asia, it will be difficult to prevent future outbreaks.

Marko König/imageBROKER/Alamy Stock Photo

West Nile Virus. A mosquito-borne virus, West Nile virus first infected people in North America in 1999. Carried by infected crows and other birds, the virus proceeded to spread across the country, with 4,156 cases at its peak in 2002, a total of 284 of which were fatal. By 2005, the wave of infection had greatly lessened. The virus is thought to be transmitted to humans by mosquitoes that have previously bitten infected birds. An earlier spread of the virus through Europe also abated after several years.

Tim Zurowski/Getty Images

SARS-CoV-2. SARS-CoV-2, the virus that causes COVID-19, appeared in China in late 2019 and quickly spread around the world. A close relative of SARS, COVID-19 has led to a world-wide pandemic. COVID-19 attacks the upper and lower respiratory tracks. Within a year, there were more than 120 million confirmed COVID-19 cases worldwide and over 2.6 million deaths. When the RNA genome of this new virus was sequenced, it was 96.1% identical to a bat virus, making it likely that, like SARS, COVID-19 originated in the bats that are common across Asia.

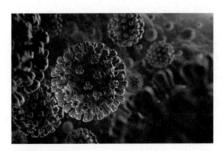
fpm/E+/Getty Images

> **Putting the Concept to Work**
> What do these six emerging viruses all have in common?

16.4 How Viruses Infect Organisms

> **LEARNING OBJECTIVE 16.4.1** Compare how viruses infect bacterial and animal cells.

All viruses require a means of getting into living cells.

Infecting Bacteria

Bacteriophages are viruses that infect bacteria. They are diverse both structurally and functionally, and are united solely by their occurrence in bacterial hosts. Bacteriophages with double-stranded DNA have played a key role in molecular biology. Many of these bacteriophages are large and complex, with relatively large amounts of DNA and proteins. Some of them have been named as members of a "T" series (T1, T2, and so forth); others have been given different kinds of names. To illustrate the diversity of these viruses, T3 and T7 bacteriophages are icosahedral and have short tails. In contrast, the so-called T-even bacteriophages (T2, T4, and T6) are more complex, as shown by the T4 bacteriophage in figure 16.4b. T-even phages have an icosahedral head that holds the DNA (shown in the cutaway view), a capsid that consists primarily of three proteins, a connecting neck with a collar and "whiskers," a long tail, and a complex base plate. Many of these structures are also visible in the photograph of a T4 bacteriophage in figure 16.5a.

During the process of bacterial infection by a T4 bacteriophage, the tail fibers set the bacteriophage perpendicular to the surface of the bacterium and bring the base plate into contact with the cell surface, as seen on the left side in figure 16.5b. After the bacteriophage is in place, the tail contracts, and the tail tube passes through an opening that appears in the base plate, piercing the bacterial cell wall (as shown on the right side of figure 16.5b). The contents of the head, mostly DNA, are then injected into the host cytoplasm. The viral DNA that is injected in the cell is transcribed and translated by the host cell into viral components that are assembled in the host cell. Eventually, the host cell ruptures and the new phages are released, ready to infect more cells.

Infecting Animal Cells

Unlike bacterial viruses, which punch a hole in the bacterial cell wall and inject their DNA inside, animal viruses typically enter their host cells by membrane fusion or endocytosis, the cell's plasma membrane dimpling inward, surrounding and engulfing the virus particle.

How does an animal virus recognize and enter an animal cell? Studding the surface of each animal virus are spikes that bang into any cell the virus encounters. When a spike happens onto an animal cell surface marker that matches its shape, the binding of spike proteins to cell surface proteins triggers membrane fusion.

Once inside the animal cell, the animal virus particle sheds its protective coat. This leaves the virus nucleic acid floating in the cell's cytoplasm, free to replicate and direct the production of new viruses, which eventually exit the animal cell by bursting through the plasma membrane. This rupturing destroys the animal cell's physical integrity and kills it.

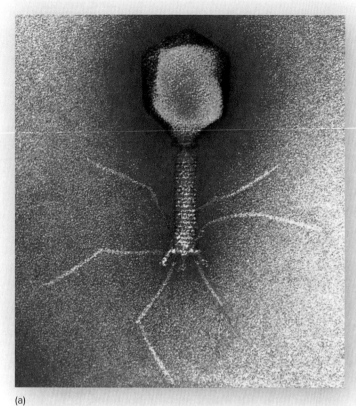

(a)

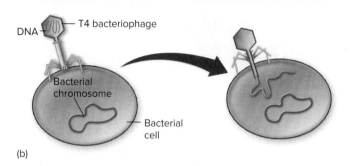

(b)

Figure 16.5 A T4 bacteriophage.

(a) Electron micrograph of T4 and (b) diagram of a T4 bacteriophage infecting a bacterial cell.

(a) Department of Microbiology, Biozentrum, University of Basel/Science Source

> **Putting the Concept to Work**
> Why can you not be infected by a bacterial virus?

Biology and Staying Healthy

Bird and Swine Flu

The influenza virus has been one of the most lethal viruses in human history. Flu viruses are animal RNA viruses containing 11 genes. An individual flu virus resembles a sphere studded with spikes composed of two kinds of protein. Different strains of flu virus, called subtypes, differ in their protein spikes. One of these proteins, hemagglutinin (H), aids the virus in gaining access to the cell interior. The other, neuraminidase (N), helps the daughter virus break free of the host cell once virus replication has been completed. Flu viruses are currently classified into 13 distinct H subtypes and 9 distinct N subtypes, each of which requires a different vaccine to protect against infection. Thus, the virus that caused the Hong Kong flu epidemic of 1968 has Type 3 H molecules and Type 2 N molecules, and is called H3N2.

How New Flu Strains Arise

Worldwide epidemics of the flu in the last century have been caused by shifts in flu virus H-N combinations. The "killer flu" of 1918, H1N1, thought to have passed directly from birds to humans, killed between 40 and 100 million people worldwide. The Asian flu of 1957, H2N2, killed over 100,000 Americans, and the Hong Kong flu of 1968, H3N2, killed 70,000 Americans. Every year, some 36,000 Americans die of flu, most often H3N2.

It is no accident that new strains of flu usually originate in the Far East. The most common hosts of influenza virus are ducks, chickens, and pigs, which in Asia often live in close proximity to each other and to humans. Pigs are subject to infection by both bird and human strains of the virus, and individual animals are often simultaneously infected with multiple strains. This creates conditions favoring genetic recombination between strains, as illustrated above, sometimes putting together novel combinations of H and N spikes unrecognizable by human immune defenses specific for the old configuration. The Hong Kong flu, for example, arose from recombination between H3N8 from ducks and H2N2 from humans. The new strain of influenza, in this case H3N2, then passed back to humans, creating an epidemic because the human population had never experienced that H-N combination before.

Conditions for a Pandemic

Not every new strain of influenza creates a worldwide flu epidemic. Three conditions are necessary: (1) The new strain must contain a novel combination of H and N spikes so that the human population has no significant immunity from infection; (2) the new strain must be able to replicate in humans and cause death—many bird influenza viruses are harmless to people because they cannot multiply in human cells; (3) the new strain must be efficiently transmitted between humans. The H1N1 killer flu of 1918 spread in water droplets exhaled from infected individuals and subsequently inhaled into the lower respiratory tract of nearby people.

The new strain need not be deadly to every infected person in order to produce a pandemic—the H1N1 flu of 1918 had an

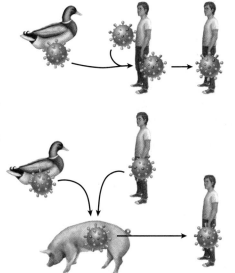

Recombination within humans
A person infected with a flu virus can become infected with another type of flu virus by direct contact with birds. The two viruses can undergo genetic recombination to produce a third type of virus, which can spread from human to human.

Recombination within pigs
Pigs can contract flu viruses from both birds and humans. The flu viruses can undergo genetic recombination in the pig to produce a new kind of flu virus, which can spread from pigs to humans.

overall mortality rate of only 2% and yet killed 40 to 100 million people. Why did so many die? Because so much of the world's population was infected.

Bird Flu

A potentially deadly new strain of flu virus emerged in Hong Kong in 1997: H5N1. Like the 1918 pandemic strain, H5N1 passes to humans directly from infected birds, usually chickens or ducks, and for this reason has been dubbed "bird flu." Bird flu satisfies the first two conditions of a pandemic: H5N1 is a novel combination of H and N spikes for which humans have little immunity, and the resulting strain is particularly deadly, with a mortality of 59% (much higher than the 2% mortality of the 1918 H1N1 strain). Fortunately, the third condition for a pandemic is not yet met: The H5N1 strain of flu virus does not spread easily from person to person, and the number of human infections remains small.

Swine Flu

A second potentially pandemic form of flu virus, H1N1, emerged in Mexico in 2009, passing to humans from infected pigs. It seems to have arisen by multiple genetic recombination events between humans, birds, and pigs. Like the 1918 H1N1 virus, this flu (dubbed "swine flu") passes easily from person to person, and within a year it spread around the world. Also like the 1918 virus, the initial wave of swine flu infection triggered only mild symptoms in most people. The 1918 H1N1 virus only became deadly in a second wave of infection, after the virus had better adapted to living in the human body. As with bird flu, public health officials continue to watch swine flu carefully for fear that a subsequent wave of infection may also become more lethal.

Today's Biology

SARS-CoV-2 Life Cycle

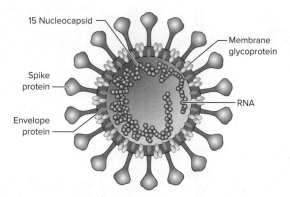

The human body is a fine-tuned machine, capable of defending itself against a large range of threats, but every so often, a threat emerges that our bodies are not equipped to fight. COVID-19 is an illness caused by severe acute respiratory syndrome-coronavirus type 2 (SARS-CoV-2) virus. Humans do not have much protection from this virus, so many people are falling ill and a large number are dying. Scientists are racing to study how this virus gets into our cells and then, once inside, how the virus replicates. Knowing this, we will have some important information needed to fight the virus.

events. First, once the virus has inserted its RNA genome into your cell, your very own ribosomes translate the RNA, making viral proteins. The newly synthesized viral proteins are not structural in nature, rather they are necessary and play a critical role in RNA synthesis. Two of these nonstructural proteins (nsp's), nsp13 and nsp14, have helicase and proofreading functions, respectively, comparable to human genes. One of the most critical elements to get translated is nsp12, which codes for the RNA-dependent RNA polymerase (RdRP), the protein that will make thousands of copies of the viral genome. Increasing the amount of viral RNA in the cell is not enough. To complete the process of viral replication, there must be an increase in the abundance of the structural proteins. There will also be an increase in translation of the four major structural proteins which will form the envelope, membrane, nucleocapsid, and spike proteins. Finally, with all of the structural components made and the RNA copied, the new viruses are assembled and fill the cytoplasm of the host cell.

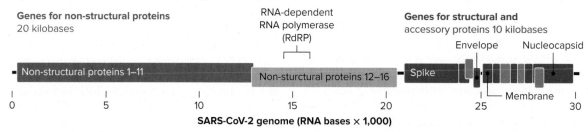

Getting into the Cell

Being exposed to a virus is one thing, being infected is another. Coronaviruses can bind to a cell surface protein called angiotensin converting enzyme 2 (ACE2), which is a cell surface protein involved in the regulation of blood pressure. The structure of the coronavirus is key to infection. The virus has spike proteins around the outside. These spike proteins are what promote the attachment of the virus to the cell by binding to ACE2. After binding of one type of spike protein to the host cell, another subunit of the spike protein facilitates fusion of the viral and host cell membranes. Once fused, the viral genome, which is RNA based, is free to seize control of the now infected cell.

Replicating inside Your Cells

Viruses are not living; they lack the ability to replicate their genome on their own. To replicate, viruses depend on the activity of proteins inside of your cells. To accomplish the hijacking of your cell, SARS-CoV-2 utilizes a very ordered sequence of

What Can We Do?

One of the factors that makes treating COVID-19 infections difficult and challenging is the broad range of disease symptoms, ranging from no symptoms to a mild-flu like state to respiratory distress and death. Knowing how SARS-CoV-2 enters into the cell and replicates gives scientists great targets to develop therapeutics. There are currently drugs that block the binding of molecules to ACE2. If we could find the right drug that would prevent the binding of the spike protein of SARS-CoV-2, we could prevent the cell from ever being infected. This would be a great first step to slow the spread of the virus. Another area for therapeutic development targets the activity of the viral proteins. Remdesivir, an antiviral agent first developed for the treatment of the Ebola virus and further found useful in the SARS outbreak in 2003, was an early drug used in the treatment of COVID-19. Remdesivir interferes with the activity of RdRP, decreasing viral RNA production. While not effective as a solo therapy, it has been demonstrated to reduce the severity of the infection.

Today's Biology

Answering Your Questions About Wearing Masks To Prevent The Spread of Coronavirus

Nick Gregory/Alamy Stock Photo

Why Do We Need to Wear a Mask?

With the outbreak of COVID-19 beginning in China in late 2019, public health officials have been grappling with policy decisions that will best protect the public. COVID-19 is most commonly spread through water droplets released when infected patients speak, sneeze, and cough. Preventing spread is key to slowing the infection rates and health officials are asking us to wear masks. But is a mask effective to reducing the spread of the virus? An uncovered sneeze easily could travel over 10 feet. Sneezing into your elbow can reduce the distance that the sneeze travels to a foot or less. Clearly, covering the mouth and nose when sneezing is critical to limit the spread. Masks provide a hands-free method of reducing the transmission while allowing for most activities to normally proceed.

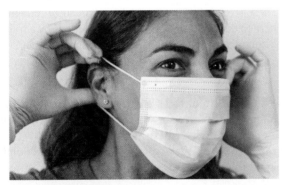
Alessandro Biascioli/Shutterstock

Do Masks Prevent You from Exposing Others to Coronavirus?

COVID-19 has a surprisingly long latency period, 5 to 14 days. This means that it may be up to 2 weeks from when someone was exposed and infected to when they developed symptoms. More than 35% who have COVID-19 are asymptomatic, which means they are infected but are not showing any signs of infection. Think of all the people that an asymptomatic person could have encountered in those 2 weeks. Just when talking, we expel droplets around 1 µM in diameter. The size of these droplets increases to 100 µM when we sneeze and to a whopping 250 µM when we cough. These droplets do not include all the fine aerosols that are produced. Studies have demonstrated that surgical masks, like the N-95, can reduce the release of these droplets to the environment by 95%, while homemade cloth masks show a reduction in the range of 50% to 80%. By wearing a mask, the data demonstrates that you will keep more of your droplets to yourself. This is of paramount importance in the long asymptomatic phase of a COVID-19 infection.

Does Wearing a Mask Reduce Your Chance of being Exposed to Coronavirus?

You cannot control whether the people you encounter wear masks, but does wearing a mask keep you safe? Data have demonstrated that masks made from household materials like tightly woven cotton tea towels can filter out 60% of airborne particles that are less than 1 µM in size. This level of filtration is better than you might think since surgical masks filter about 75% of the same particles. Surgical masks cost money and are designed to be disposable. With simple materials you already have around the house, you can make multiple masks and wash and reuse them after every use. Not only do simple cloth masks reduce your risk of exposure, they do so in a cost-effective manner.

Can Wearing a Mask Increase Risk of Exposure?

There is some concern from public health officials that the wearing of a mask, no matter the type, may increase the chance of exposure to COVID-19. There are two main areas of concern: (1) the safe handling of used masks and (2) an increase in risk behavior. When you wear a mask in public, what do you do with the mask when you take it off? If you touch the mask itself and not the ear loops, you have just risked getting whatever is on your mask onto your hands. If you then touch your face, you may have just infected yourself. How does wearing a mask change your behavior while in public? In some studies, it has been noted that individuals wearing masks in public demonstrate less social distancing and decreased frequency of washing their hands. Wearing a mask has given some people a false sense of security. It is important to maintain, in the time of a pandemic and regardless of whether you are wearing a mask, proper social distancing, lots of hand washing, and resistance to the urge to touch your face.

Protists

Figure 16.6 A unicellular protist.

The protist kingdom is a catch-all kingdom for many different groups of unicellular organisms, such as this *Vorticella* (phylum Ciliophora), which is heterotrophic, feeding on bacteria, and has a contractible stalk.

Dennis Kunkel Microscopy/Science Source

BIOLOGY & YOU

Traveler Beware. If you travel to an under-developed country, particularly one where sanitation is poor, you would be well-advised to drink only bottled water. Over 10 million people each year get "traveler's diarrhea" from drinking fecally contaminated water. Usually this means a few days of cramps and frequent trips to the bathroom, and 80% of the time antibiotics quickly eliminate the bacterial infection. In some 20% of the cases, however, the problem is not so simply solved because the infecting agent is a protist, not a bacterium, and a lot harder to kill. Giardia attaches to the inside of the small intestine, whereas cryptosporidium (a relative of malaria) and amoebic parasites each form dormant cysts. All three protists leave the body in feces, which, when contaminating drinking water, spread the infection to others. All three protists cause serious diseases that are more difficult to treat than bacterial infections. Cysts make treatment particularly difficult because dormant cysts are resistant to antibiotics (most of which work by interfering with metabolism or genome replication). In a sign of the medical importance of amoebic dysentery, which infects 50 million people worldwide, the amoebic parasite (*Entamoeba histolytica*) was one of the first protists to have its genome completely sequenced.

16.5 General Biology of Protists

LEARNING OBJECTIVE 16.5.1 Describe the wide range of forms, locomotion, nutrition, reproduction, and cellular aggregation among the protists.

As discussed in chapter 15, animals, plants, fungi, and protists are all eukaryotes. **Protists** are the most ancient eukaryotes and are united on the basis of a single negative characteristic: They are not fungi, plants, or animals. In all other respects, they are highly variable with no uniting features. Many are unicellular, like the *Vorticella* you see in figure 16.6 with its contractible stalk, but there are numerous colonial and multicellular groups. Most are microscopic, but some are as large as trees.

The Cell Surface

Protists possess varied types of cell surfaces. All protists have plasma membranes. Some protists, such as algae and molds, are additionally encased within strong cell walls. Other protists, such as diatoms and radiolarians, secrete glassy shells of silica.

Locomotor Organelles

Movement in protists is also accomplished by diverse mechanisms. Protists move by flagella, cilia, pseudopods, or gliding mechanisms. Many protists wave one or more flagella to propel themselves through the water, whereas others use banks of short, flagella-like structures called cilia to create water currents for their feeding or propulsion. Among amoeba, large, blunt extensions of the cell body called *pseudopodia* are the chief means of locomotion. Other related protists extend thin, branching protrusions, sometimes supported by axial rods of microtubules. These so-called *axopodia* can be extended or retracted. Because the tips can adhere to adjacent surfaces, the cell can move by a rolling motion, shortening the axopodia in front and extending those in the rear.

Cyst Formation

Many protists with delicate surfaces are successful in quite harsh habitats. How do they manage to survive so well? They survive inhospitable conditions by forming *cysts*. A cyst is a dormant form of a cell with a resistant outer covering in which cell metabolism is more or less completely shut down. Amoebic parasites form cysts that are resistant to gastric acidity.

Nutrition

Protists employ every form of nutritional acquisition except chemoautotrophy (the ability of an organism to use energy from chemical reactions to make its own food), which has so far been observed only in prokaryotes. Some protists are photosynthetic autotrophs (using energy from the sun to make food as plants do) and are called *phototrophs*. Others are *heterotrophs* that obtain energy from organic molecules synthesized by other organisms. Among heterotrophic protists, those that ingest visible particles of food are called phagotrophs. Those ingesting food in soluble form are called saprozoic feeders.

Phagotrophs ingest food particles into intracellular vesicles called *food vacuoles,* or phagosomes. Lysosomes fuse with the food vacuoles, introducing enzymes that digest the food particles within. The food vacuoles shrink as the digested molecules are absorbed across their membranes.

Reproduction

Protists typically reproduce asexually, reproducing sexually only in times of stress. Asexual reproduction involves mitosis, but the process is often somewhat different from the mitosis that occurs in multicellular animals, discussed in chapter 8. The nuclear membrane, for example, often persists throughout mitosis, with the microtubular spindle forming within it. In some groups, asexual reproduction involves fission (figure 16.7a), and in others, it involves the formation of **spores,** which are haploid reproductive cells capable of developing without uniting with another cell. Sexual reproduction occurs only rarely, typically by exchanging nuclei (figure 16.7b).

Multicellularity

A single cell has its limits. As a cell becomes larger, there is too little surface area for so much volume. The evolution of multicellular individuals composed of many cells solved this problem. **Multicellularity** is a condition in which an organism is composed of many cells, permanently associated with one another, that integrate their activities. The key advantage of multicellularity is that it allows specialization—distinct types of cells, tissues, and organs can be differentiated within an individual's body, each with a different function. With such functional "division of labor" within its body, a multicellular organism can possess cells devoted specifically to protecting the body, others to moving it about, still others to seeking mates and prey, and yet others to carry on a host of other activities. In just this way, a small city of 50,000 inhabitants is vastly more complex and capable than a crowd of 50,000 people in a football stadium—each city dweller is specialized in a particular activity that is interrelated to everyone else's, rather than just being another body in a crowd.

Colonies. A colonial organism is a collection of cells that are permanently associated but in which little or no integration of cell activities occurs. Many protists form colonial assemblies, consisting of many cells with little differentiation or integration. In some protists, the distinction between colonial and multicellular is blurred. For example, in the green algae *Volvox* shown in figure 16.8, individual motile cells aggregate into a hollow ball of cells that moves by a coordinated beating of the flagella of the individual cells—like scores of rowers all pulling their oars in concert. A few cells near the rear of the moving colony are reproductive cells, but most are relatively undifferentiated.

Multicellular Individuals. True multicellularity, in which the activities of the individual cells are coordinated and the cells themselves are in contact, occurs only in eukaryotes and is one of their major characteristics. Three groups of protists have independently attained true but simple multicellularity—the brown algae (phylum Phaeophyta), green algae (phylum Chlorophyta), and red algae (phylum Rhodophyta). In multicellular organisms, individuals are composed of many cells that interact with one another and coordinate their activities.

Putting the Concept to Work

With modern advanced communications, do you think a country has all of the properties of a multicellular organism?

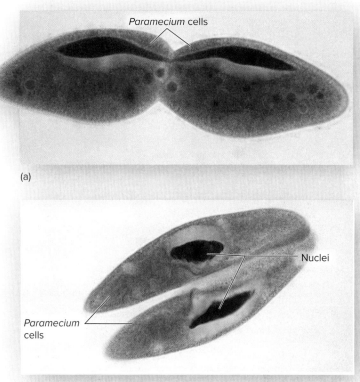

Figure 16.7 Reproduction among paramecia.

(a) When *Paramecium* reproduces asexually, a mature individual divides, and two genetically identical individuals result. (b) In sexual reproduction, two mature cells fuse in a process called conjugation (×100) and exchange haploid nuclei.

(a) Ed Reschke/Peter Arnold/Getty Images; (b) Ed Reschke/Photolibrary/Getty Images

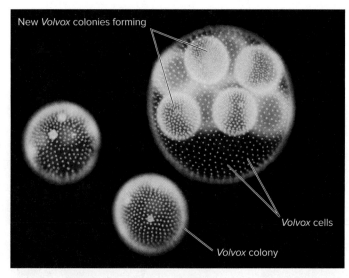

Figure 16.8 A colonial protist.

Individual, motile, unicellular green algae are united in the protist *Volvox* as a hollow colony of cells that moves by the beating of the flagella of its individual cells. Some species of *Volvox* have cytoplasmic connections between the cells that help coordinate colony activities. The *Volvox* colony is a highly complex form that has many of the properties of multicellular life.

Stephen Durr

16.6 Kinds of Protists

> **LEARNING OBJECTIVE 16.6.1** Identify the protist group thought to be ancestral to animals, the one ancestral to fungi, and the one ancestral to plants.

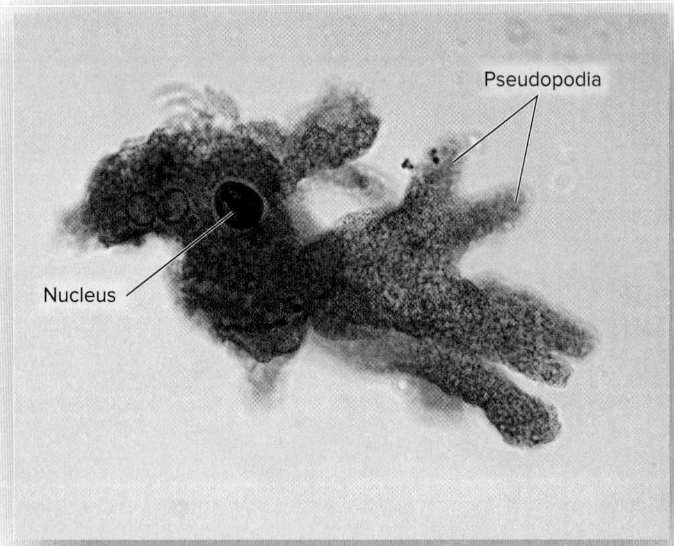

Protists are the most diverse of the four kingdoms in the domain Eukarya. The 200,000 different forms in the kingdom Protista include many unicellular (**figure 16.9**), colonial, and multicellular groups (**table 16.2**).

Recently, with the advent of DNA technology, it has become possible to compare the genomes of the 11 distantly related protist phyla and for the first time generate a rough outline of the phylogenetic tree of this kingdom, identifying the kinds of protists from which animals, fungi, and plants originated (**figure 16.10**).

How Protists Impact Your Life

Protists have major impacts on humanity. Some of humanity's most important diseases are caused by protists. Sleeping sickness, which causes extreme fatigue, is caused by trypanosomes and affects millions of people each year in tropical areas, and makes it impossible to raise domestic cattle in a large portion of Africa. Malaria, caused by the *Plasmodium* parasite, is one of humanity's greatest scourges, killing over a million people each year.

Protists have great economic impact, both negative and positive. The poisonous and destructive "red tides" in coastal areas are caused by population explosions, or "blooms," of red-pigmented dinoflagellates. Diatoms, which are photosynthetic protists with shells of silica, have many industrial applications as grinding materials. Algae are particularly useful. Massive marine algae called kelp that occur in shallow ocean waters throughout the world provide commercial fiber. Sushi rolls are wrapped in nori, a red algae. Polysaccharides from red algae are used to thicken ice cream and cosmetics.

> **Putting the Concept to Work**
> Which protists are multicellular? What do they have in common?

Figure 16.9 An amoeba.

Amoeba proteus is a relatively large amoeba (×45). The projections are pseudopodia; an amoeba moves simply by flowing cytoplasm into them. The nucleus is plainly visible.

De Agostini Picture Library/Science Source

Figure 16.10 A phylogenetic tree for the protists.

This tree shows the 11 major groups of protists and a tentative array of their relationship to each other.

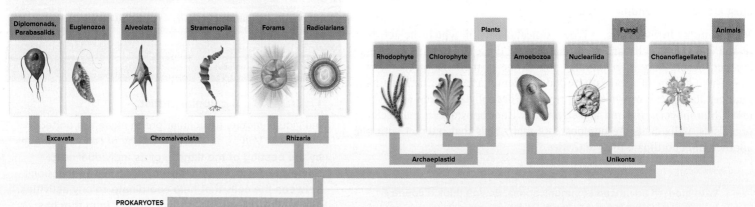

TABLE 16.2 Kinds of Protists

Supergroup	Phylum or Group	Key Characteristics
Excavata		
	Diplomonads (*Giardia*)	Move by flagella; have two nuclei; lack mitochondria
	Parabasalids (*Trichomonas*)	Undulating membrane; some are pathogens, others digest cellulose in gut of termites
	Euglenozoa (*Euglena*)	Unicellular; some photosynthetic
	Kinetoplastids (*Trypanosoma*)	Heterotrophic; mitochondrial DNA in two circles
Chromalveolata		
	Alveolata	
	Dinoflagellates (red tides)	Unicellular, with two flagella
	Apicomplexans (malaria)	Unicellular, nonmotile; spores contain a complex mass of organelles
	Ciliates (*Paramecium*)	Heterotrophic; unicellular; cells possess two nuclei and many cilia
	Phaeophyta (Brown algae or kelp)	Multicellular; contain chlorophylls a and c
	Chrysophyta (Diatoms)	Unicellular; double shells of silica
	Oomycetes (Water molds)	Terrestrial and freshwater parasites; motile spores bear two unequal flagella
Rhizaria		
	Actinopoda (Radiolarians)	Glassy skeletons; needlelike pseudopods
	Foraminifera (Forams)	Rigid shells called tests that are often multichambered
Archaeplastida		
	Rhodophyta (Red algae)	Lack flagella and centrioles; reproduce sexually; multicellular
	Chlorophyta (*Volvox*)	Unicellular, colonial, and multicellular forms; contain chlorophylls a and b
	Streptophytes (Charophytes)	Plasmodesmata and flagellated sperm; the ancestors of plants
Unikonta		
	Amoebozoa (Plasmodial and cellular slime molds)	Amoeboid individuals; plasmodial slime molds form huge multinucleate plasmodium for feeding
	Nucleariida	Single-celled heterotrophic amoeba; the ancestors of fungi
	Choanoflagellida	Single-celled, with single flagellum surrounded by funnel-shaped collar; the ancestors of animals

Fungi

Figure 16.11 Mushrooms.

Most of the body of a fungus is below ground, a network of fine threads that often spread over great distances. Mushrooms are aboveground reproductive structures that release spores into the air, allowing the fungus to invade new habitats. (a) Yellow morel. (b) *Amanita* mushroom.

(a) Robert Marien/Corbis/Getty Images; (b) Ro-ma Stock Photography/Corbis/Getty Images

16.7 A Fungus Is Not a Plant

> **LEARNING OBJECTIVE 16.7.1** List five significant differences between fungi and plants.

The **fungi** are a distinct kingdom of organisms, comprising about 74,000 named species, some microbial, most multicellular. Mycologists, scientists who study fungi, believe there may be many more species in existence.

How Fungi Differ from Plants

Although fungi were at one time included in the plant kingdom, they lack chlorophyll and resemble plants only in their general appearance and lack of mobility. Significant differences between fungi and plants include the following:

Fungi are heterotrophs. Perhaps most obviously, a mushroom is not green because it does not contain chlorophyll. Virtually all plants are photosynthesizers, whereas no fungi carry out photosynthesis. Instead, fungi obtain their food by secreting digestive enzymes onto whatever they are attached to and then absorbing into their bodies the organic molecules that are released by the enzymes.

Fungi have filamentous bodies. A plant is built of groups of functionally different cells called tissues, with different parts typically made of several different tissues. Fungi by contrast are basically filamentous in their growth form (that is, their bodies consist entirely of cells organized into long, slender filaments called *hyphae*), even though these filaments may be packed together to form a mass, called a *mycelium*.

Fungi have nonmotile sperm. Some plants have motile sperm with flagella. The majority of fungi do not.

Fungi have cell walls made of chitin. The cell walls of fungi contain chitin, the same tough material that a crab shell is made of. The cell walls of plants are made of cellulose, also a strong building material. Chitin, however, is far more resistant to microbial degradation than is cellulose.

> The cellulose that makes up plant cell walls and the chitin that makes up fungal cell walls are both polysaccharides, made of glucose subunits. Refer back to table 3.1 in section 3.4 to see the structural difference between cellulose and chitin.

Fungi have nuclear mitosis. Mitosis in fungi is different from mitosis in plants, animals, and most protists in one key aspect: The nuclear envelope does not break down and re-form; instead, all of mitosis takes place *within* the nucleus. A spindle apparatus forms there, dragging chromosomes to opposite poles of the *nucleus* (not the cell).

You could build a much longer list, but already the take-home lesson is clear: Fungi are not like plants at all! Their many unique features are strong evidence that fungi are not closely related to any other group of organisms.

The Body of a Fungus

Fungi exist mainly in the form of slender filaments, barely visible with the naked eye, called **hyphae** (singular, **hypha**). A hypha is basically a long string of cells. Many hyphae associate with each other to form a much larger structure, called a **mycelium** (plural, **mycelia**), like the mushrooms in **figure 16.11**. The main body of a fungus, however, is not the mushroom, which is a temporary reproductive structure, but rather the extensive network of fine hyphae that penetrate the soil, wood, or flesh in which the fungus is growing. A mycelium may contain many meters of individual hyphae.

Cytoplasmic Streaming. Fungal cells are able to exhibit a high degree of communication within such structures because although most cells of fungal hyphae are separated by cross-walls called *septa* (singular, *septum*), these septa rarely form a complete barrier. Openings in the septa allow cytoplasm to flow throughout the hyphae from one cell to another.

Because of such cytoplasmic streaming, proteins synthesized throughout the hyphae can be carried to the hyphal tips. *This novel body plan is perhaps the most important innovation of the fungal kingdom.* As a result of it, fungi can respond quickly to environmental changes, growing very rapidly when food and water are plentiful and the temperature is optimal. Also due to cytoplasmic streaming, many nuclei may be connected by the shared cytoplasm of a fungal mycelium. None of them (except for reproductive cells) is isolated in any one cell; all of them are linked cytoplasmically with every cell of the mycelium. Indeed, the entire concept of multicellularity takes on a new meaning among the fungi, the ultimate communal sharers among the multicellular organisms.

How Fungi Reproduce

Fungi reproduce both asexually and sexually. All fungal nuclei except for the zygote are haploid. Often in the sexual reproduction of fungi, individuals of genetically different "mating types" must participate, much as two sexes are required for human reproduction. Sexual reproduction is initiated when two hyphae of different mating types (indicated by "+" and "−") come in contact, and the hyphae fuse (figure 16.12). What happens next? In animals and plants, when the two haploid gametes fuse, the two haploid nuclei immediately fuse to form the diploid nucleus of the zygote. As you might by now expect, fungi handle things differently. In most fungi, the two nuclei do not fuse immediately. Instead, they remain unmarried inhabitants of the same house, coexisting in a common cytoplasm for most of the life of the fungus! A fungal hypha that has two nuclei is called **dikaryotic.** The nuclei in certain cells will fuse, forming a zygote.

Spores are a common means of reproduction among the fungi. The puffball fungus in figure 16.13 is releasing spores in a somewhat explosive manner. Spores are well suited to the needs of an organism anchored to one place. They are so small and light that they may remain suspended in the air for long periods of time and may be carried great distances. When a spore lands in a suitable place, it germinates and begins to divide, giving rise to a new fungal hypha.

How Fungi Obtain Nutrients

All fungi obtain food by secreting digestive enzymes into their surroundings and then absorbing back into the fungus the organic molecules produced by this *external digestion*. Many fungi are able to break down the cellulose in wood, cleaving the linkages between glucose subunits and then absorbing the glucose molecules as food. That is why fungi are so often seen growing on trees (figure 16.14).

> **Putting the Concept to Work**
> When two haploid hyphae of different mating types fuse, is the resulting cell diploid? Explain.

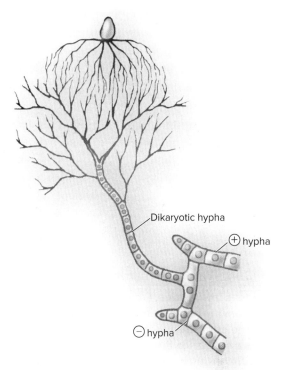

Figure 16.12 Sexual reproduction in fungi.

In some fungi, different mating strains ("+" and "−") fuse and form a dikaryotic hypha that grows into a mycelium. The nuclei in certain cells will eventually fuse, forming a diploid zygote.

Figure 16.13 Many fungi produce spores.

Spores explode from the surface of a puffball fungus.

RF Company/Alamy Stock Photo

16.8 Kinds of Fungi

> **LEARNING OBJECTIVE 16.8.1** Distinguish the major fungal phyla by their mode of reproduction.

Figure 16.14 The oyster mushroom.
Not only does this species, *Pleurotus ostreatus,* use cellulose from the tree as a food source, but is also a predator and attracts tiny roundworms called nematodes. It secretes a substance that immobilizes nematodes, which the fungus then uses as a source of food.
Westend61/Getty Images

(a) (b)

Figure 16.15 Edible and poisonous mushrooms.
Edible mushrooms include (a) button mushrooms (*Agaricus bisporus*). Poisonous mushrooms, including (b) *Amanita muscaria,* can cause a range of symptoms, from slight allergic and digestive reactions, to organ failure and death.
(a) Image Source/Getty Images; (b) Gallo Images-Nigel Dennis/Digital Vision/Getty Images

Fungi are an ancient group of organisms at least 400 million years old. There are nearly 74,000 described species, and many more awaiting discovery. Many fungi are harmful because they decay, rot, and spoil many different materials as they obtain food and because they cause serious diseases in animals and particularly in plants. Some mushrooms are edible, but others are poisonous (figure 16.15). Other fungi are used for commercial purposes. The manufacture of both bread and beer depends on the biochemical activities of yeasts, single-celled fungi that produce abundant quantities of carbon dioxide and ethanol. Fungi are used in industry to convert one complex organic molecule into another; many commercially important steroids are synthesized in this way.

DNA analysis is making major contributions to our understanding of fungal diversity. Traditionally, fungi have been classified according to their mode of sexual reproduction; the 17,000 species for which sexual reproduction had not been observed were dubbed "imperfect fungi." With a vast amount of genome sequence data now available, fungal biologists (called mycologists) have recently agreed on eight phyla (table 16.3). The microsporidia are sisters to all other fungi, but there remains some disagreement as to whether or not they are true fungi.

Ecological Roles as Decomposers

Fungi, together with bacteria, are the principal **decomposers** earth's ecosystems. They break down organic materials and return the substances that had been locked in those molecules to circulation in the ecosystem. Fungi are virtually the only organisms capable of breaking down lignin, one of the major constituents of wood. By breaking down such substances, fungi release critical building blocks, such as carbon, nitrogen, and phosphorus, from the bodies of dead organisms and make them available to other organisms.

In breaking down organic matter, some fungi attack living plants and animals, acting as disease-causing organisms. Pathogenic fungi are responsible for billions of dollars in agricultural losses every year. Not only are fungi the most harmful pests of living plants, but they also attack food products once they have been harvested and stored. In addition, fungi often secrete substances into the foods they are attacking that make these foods unpalatable or poisonous.

Commercial Uses

The same aggressive metabolism that makes fungi ecologically important has been put to commercial use in many ways. The manufacture of both bread and beer depends on the biochemical activities of yeasts, single-celled fungi mentioned in the "A Closer Look" feature in this chapter. Cheese and wine achieve their delicate flavors because of the metabolic processes of certain fungi, and others make possible the manufacture of soy sauce. Vast industries depend on the biochemical manufacture of organic substances

TABLE 16.3 Fungi

Phylum	Key Characteristics	Number of Species	Phylum	Key Characteristics	Number of Species
Microsporidia	Spore-forming intracellular parasites of animals; among smallest known eukaryotes	1,500	Zygomycota	Multinucleate hyphae lack septa except for reproductive structures; fusion of hyphae leads directly to formation of a zygote	1,050
Blastocladiomycota	Flagellated gametes (zoospores); alternation of haploid and diploid generations	140	Glomeromycota	Possess large nonflagellated multi-nuclear asexual spores; form symbiotic mycorrhizal associations with plants	150
Neocallimastigomycota	Zoospores with numerous flagella; lack mitochondria and so are obligate anaerobes; found in digestive tracts of herbivores	10	Basidiomycota	Reproduce by sexual means; basidiospores are borne on club-shaped structures called basidia	22,000
Chytridiomycota	Produce flagellated gametes (zoospores); predominately aquatic, some freshwater and some marine	1,500	Ascomycota	Reproduce by sexual means; ascospores are formed inside a sac called an ascus; asexual reproduction is also common	32,000

Microsporidia: Eye of Science/Science Source; **Blastocladiomycota:** Biology Pics/Science Source; **Neocallimastigomycota:** Dr. Daniel A. Wubah

such as citric acid by fungi in culture, and yeasts are now used on a large scale to produce protein for the enrichment of animal food. Many antibiotics, including the first one that was used on a wide scale, penicillin, are derived from fungi.

Fungal Associations

Two kinds of mutualistic associations between fungi and autotrophic organisms are ecologically important: **Mycorrhizae** and **lichens.** In each case, a photosynthetic organism fixes atmospheric carbon dioxide and thus makes organic material available to the fungi. The metabolic activities of the fungi, in turn, enhance the overall ability of the symbiotic association to exist in a particular habitat. Mycorrhizae are symbiotic associations between fungi and the roots of plants in which the fungal partner expedites the plant's absorption of essential nutrients such as phosphorus (figure 16.16). Lichens are symbiotic associations between fungi and either green algae or cyanobacteria. They are prominent nearly everywhere in the world, especially in unusually harsh habitats such as bare rock (figure 16.17).

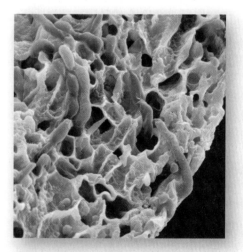

Figure 16.16 Mycorrhizae on the roots of pines.

The mycorrhizae act to expand the surface area of the root, increasing water and nutrient absorption in the root.

Eye of Science/Science Source

> **Putting the Concept to Work**
> No fungi are photosynthetic, and none has a mouth or stomach. From where do fungi get their food?

Figure 16.17 Lichens growing on a rock.

Perry Mastrovito/Corbis/Getty Images

Putting the Chapter to Work

1 In 2012, people attending a private party became ill with botulism after eating beets. The cause was later proved to be improper canning procedures.

What bacterial structure of the organism, **Clostridium botulinum,** *allowed it to survive within the beets after canning had occurred?*

2 Methicillin-resistant *Staphylococcus aureus* is a strain of bacteria that is not affected by the antimicrobial drug methicillin. Scientists believe this organism evolved its resistance by obtaining a plasmid containing the methicillin-resistance gene.

If this process of genetic exchange happened through conjugation, in what way did the resistant donor cell pass the plasmid to **Staphylococcus aureus?**

3 Viruses are too small to be viewed using a compound light microscope; the use of an electron microscope is required.

If you were viewing a virus with an electron microscope and you observed a head, tail, and tail fibers, which type of organism would you conclude that it would infect?

4 Bread mold is a type of fungus. If you were to look at a sample of the mold under the microscope, you would observe multinucleated hyphae and fusion of septated hyphae that result from reproduction.

Based on this observation, in which phylum would you place bread mold?

Retracing the Learning Path

Origin of Life

16.1 How Cells Arose

1. Life appeared on earth 2.5 billion years ago. It may have arisen spontaneously, although the nature of the process is not clearly understood.

2. Life on earth may have arisen in bubbles in the ocean. The "bubble model" suggests that biological molecules were captured in bubbles where they underwent chemical reactions, leading to the origin of life. The first cells may have formed from bubble-enclosed molecules, such as RNA.

Prokaryotes

16.2 The Simplest Organisms

1. Prokaryotes are the smallest and simplest organisms, single cells with no internal compartments or organelles. The plasma membrane of prokaryotes is encased in a cell wall chemically different from the cell wall of archaea.
   - Many bacteria have flagella or pili and may form endospores.

2. Bacteria reproduce by splitting in two, called binary fission, and may exchange genetic information through conjugation. In conjugation, bacterial cells exchange genetic information by transferring copies of plasmids.

Viruses

16.3 Structure of Viruses

1. Viruses are not living organisms but are parasitic genomes of DNA or RNA, encased in a protein shell, that can infect cells and replicate within them. They are chemical assemblies, not cells, and are not alive.

2. Viruses are responsible for some of the most lethal diseases of humans, including influenza and AIDS.

16.4 How Viruses Infect Organisms

1. Bacterial viruses inject their genes into prokaryotic cells, whereas animal viruses typically bind to specific cell surface receptors that trigger endocytosis.

Protists

16.5 General Biology of Protists

1. Protists are a very diverse group of eukaryotes, having diverse locomotion, nutrition, and reproduction. Most protists reproduce asexually except during times of stress, when they engage in sexual reproduction. Some exist as single cells, whereas others form colonies or aggregates.

16.6 Kinds of Protists

1. The classification of organisms within the kingdom Protista is now based on comparing the DNA sequences of their genomes. Some forms of protists are as different from one another as they are from plants or animals.

Fungi

16.7 A Fungus Is Not a Plant

1. Although anchored in place, a fungus is not like a plant. Fungi are nonmobile heterotrophs. The body of a fungus is composed of long, slender filaments called hyphae that pack together to form a mycelium. Fungi have cell walls made of chitin, which is different from the cellulose found in plant cell walls. Most fungal cells are separated by an incomplete wall called a septum, which allows cytoplasm to pass between cells for intercellular communication.

   - Fungi reproduce both sexually and asexually. Sexual reproduction takes place when two hyphae of different mating types fuse; their haploid nuclei may stay separate in the cell.
   - Many fungi form spores. Spores are released and may be carried by wind considerable distances away to where they will land and germinate in new environments.
   - Fungi obtain nutrients through external digestion. They grow on their food, releasing enzymes that digest the food. The products of digestion are then absorbed by the fungus.

16.8 Kinds of Fungi

1. Fungi have traditionally been classified according to their mode of sexual reproduction but now are grouped according to their DNA sequence similarity into eight phyla.

   - Fungi play key roles in the environment as decomposers. They break down the bodies of dead organisms, returning substances to the ecosystem. Certain fungi have commercial uses. Some are edible, whereas others are poisonous. The manufacturing of bread, beer, wine, and cheese uses single-celled yeast.
   - Fungi are involved in symbiotic associations with the roots of some plants, called mycorrhizae, and with cyanobacteria or algae, called lichens.

Inquiry and Analysis

Did COVID-19 Become Less Lethal?

The first COVID-19 death in New York City occurred on March 1, 2020. The virus spread rapidly in the city, and soon was causing deaths in other cities throughout the United States. By April, the United States was seeing over 20,000 new cases each day. In response, cities and states began to issue "stay-at-home" orders, and schools and businesses began closing their doors. As the public became more aware, and more scared, some very basic social distancing practices like trying to stay 6 feet away from others and wearing a mask became part of the normal routine. By keeping people in their homes and away from others, the rate of daily new COVID infections peaked in early April and began to go down.

As the number of daily new infections and deaths began to drop, there was hope that the pandemic was stopped. It did not take long for people to grow weary of the lockdown and safety measures. Businesses began to reopen, and most states cancelled their "shelter in place" orders. COVID-19 cases continued to fall in May, but only slowly—by the end of June, the United States still had close to 20,000 new cases a day. With summer weather, beaches were crowded, community-wide gatherings common—and the number of new COVID-19 cases began to climb. By mid-July, this surge had risen to more than 70,000 new cases each day. The United States braced for a huge increase in the number of dying patients. Makeshift hospitals were set up in convention centers and in parks.

To determine if COVID-19 cases had the same fatality rate in the surge as was observed in the first few months of the pandemic, doctors and scientists kept daily records of the number of new infections and the number of COVID-19-related deaths. The graph on the left shows the number of new cases that are diagnosed (y axis) over time (x axis) while the graph on the left plots the number COVID-19-related fatalities (y axis) per day (x axis).

Analysis

1. **Applying Concepts**
   Variable. In the graph on the left, what is the dependent variable? What is the dependent variable on the right graph?
2. **Interpreting Data** On the graph for the number of daily new infections, what was the peak number of infections in the first wave? What about the second wave?
3. **Making Inferences**
   a. At the highest number of daily infections in the first wave, what was the case fatality rate?
   b. What was the case fatality rate for COVID-19 at the peak of new daily cases during the second wave?
4. **Drawing Conclusions** Do these data support the hypothesis that COVID-19 was becoming less lethal as the pandemic spread?
5. **Further Analysis**
   a. During the first and second waves of the spread of COVID-19, there were no significant improvements made in medical treatments, yet the number of deaths did not climb at the same rate as new infections. Why do you think the peak number of deaths in the first wave is higher than the peak number in the second wave?
   b. If you examine both sets of data, you will notice the number of deaths in the second wave peaks about two weeks after the number of daily infections peaks. What do you think caused the delay?

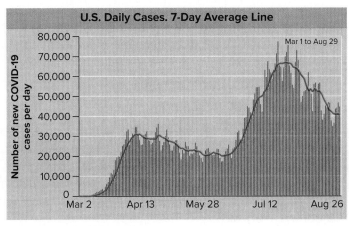

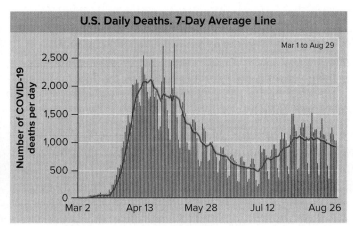

Source: COVID tracking project

17 Evolution of Plants

LEARNING PATH ▼

Plants
1. Adapting to Terrestrial Living
2. Plant Evolution

Seedless Plants
3. Nonvascular Plants
4. The Evolution of Vascular Tissue
5. Seedless Vascular Plants

The Advent of Seeds
6. Evolution of Seed Plants
7. Gymnosperms

The Evolution of Flowers
8. Rise of the Angiosperms

THE ISLAND OF Surtsey forms as an undersea volcano erupts in the North Atlantic.
NGDC/NOAA/Howell Williams

Life Comes to Surtsey

On the morning of November 14, 1963, the watchman on a fishing vessel sailing off the southern coast of Iceland saw a plume of smoke that appeared to be rising from the ocean. Assuming it was a ship on fire, he roused the captain, and they notified the coast guard. Upon closer inspection, they realized it wasn't a ship in trouble but instead an explosive eruption underwater, releasing a column of black smoke and ash. An underwater volcano was erupting.

Birth of an Island

The explosions continued for several weeks, but after only a few days, an island began to form, shown in the photo on the facing page. The watchman and the crew had seen what very few people have seen, the birth of a new island. Seen here on the right shortly after the eruption finished, the volcano top that protrudes from the sea is not a huge island, about 1.4 square kilometers. Its cooling rock contained no life—the glowing lava that bubbled up through the earth's crust is 900 to 3,000°C, far too hot to contain life. The island was given the name Surtsey after the fire giant Surtur from Norse mythology. Surtsey isn't far from the mainland, 33 kilometers (less than 20 miles) away. Newly born in 1963, it was a dead place, bare rock that only served as a perch for tired birds.

A Barren Rock

So, how does life begin on a barren mound of rock? The difficulty of the challenge this first invasion poses cannot be overstated. Animals could not be the first to invade. Imagine you were a seagull flying over from the mainland for a visit—what could you eat? There are no plants. There is not even any dirt!

Ragnar Th Sigurdsson Agency/Arctic Images/Alamy stock photo

Yann Arthus-Bertrand/Science Source

Lichens Colonize

The evolutionary answer to this challenge was a sort of "You scratch my back, I'll scratch yours" cooperation between two unlikely partners, fungi and photosynthetic algae. Together they formed a partnership called a lichen that let them grow on bare rock. You can see the orange associations in the photo here, growing on grey, wind-swept rock. Working together, each member of the partnership does something the other cannot: The algae within the lichen harvest energy from sunlight, while the fungal cells within it extract minerals from the rock. Together they grow where before there was no life.

Perry Mastrovito/Corbis/Getty Images

Plants Soon Follow

Now the island begins to change. The acidic secretions of lichen and the decomposition of dying lichen and bacteria change the surface. Finally, dirt! Soon, mosses are able to take up residence on the island. These pioneers are the first plants to venture onto the new lands but are not alone for long. Vascular plants, their seeds carried by strong winds from the mainland, quickly established themselves in the soil now rich with nutrients made available by the growth of mosses. Scientists were amazed at the speed with which these colonizing events occurred on Surtsey—on the order of months rather than years.

And Then Come Animals

Not such a lonely place any more, Surtsey. Attracted by the plant life, seagulls have reason to stay around and soon set up living on the island. Like a row of falling dominos, things begin to change quickly. The nutrient content of the soil on Surtsey improved with increasing numbers of gulls, allowing more nutrient-demanding plants to survive. By the summer of 2000, a total of 55 species of vascular plants had been discovered by scientists visiting the island. Earthworms, spiders, and insects have arrived as passengers on drifting bits of storm-tossed material and in 2014 thrive on the island.

Life has come to Surtsey.

Plants

Figure 17.1 The importance of plants.

Plants are key components in the biosphere. Photosynthesis conducted by plants and algae creates the majority of oxygen in the atmosphere. Also, through photosynthesis, plants are the main source of food for humans and many other organisms.

Charles Smith/Corbis/VCG/Getty Images

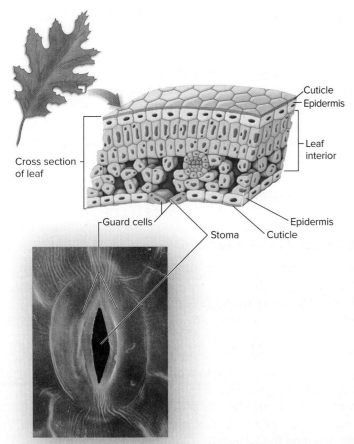

Figure 17.2 A stoma.

A stoma is a passage through the cuticle covering the epidermis of a leaf (×400). Water and oxygen pass out through the stoma, and carbon dioxide enters by the same portal. The cells flanking the stoma are called guard cells and control the opening and closing of the stoma.

David M. Phillips/Science Source

17.1 Adapting to Terrestrial Living

> **LEARNING OBJECTIVE 17.1.1** List three environmental challenges overcome by plants as they adapted to life on land.

Plants are complex multicellular organisms that are terrestrial **autotrophs**—that is, they occur almost exclusively on land and feed themselves by photosynthesis. The name *autotroph* comes from the Greek, *autos,* self, and *trophos,* feeder. Today, plants are the dominant organisms on the surface of the earth and are key components of ecosystems (figure 17.1). Nearly 300,000 species are now in existence, covering almost every part of the terrestrial landscape. In this chapter, we see how plants adapted to life on land.

The green algae that were probably the ancestors of today's plants are aquatic organisms that are not well adapted to living on land. Before their descendants could live on land, they had to overcome many environmental challenges. For example, they had to absorb minerals from the rocky surface. They had to find a means of conserving water. They had to develop a way to reproduce on land.

Absorbing Minerals

Plants require relatively large amounts of six inorganic minerals: nitrogen, potassium, calcium, phosphorus, magnesium, and sulfur. Each of these minerals constitutes 1% or more of a plant's dry weight. Algae absorb these minerals from water, but where is a plant on land to get them? From the soil. Soil is the weathered outer layer of the earth's crust. It is composed of a mixture of ingredients, which may include sand, rocks, clay, silt, humus (partly decayed organic material), and various other forms of mineral and organic matter. The soil is also rich in microorganisms that break down and recycle organic debris. Plants absorb these materials, along with water, through their *roots*. Most roots are found in topsoil, which is a mixture of mineral particles, living organisms, and humus. When topsoil is lost due to erosion, the soil loses its ability to hold water and nutrients.

> Topsoil is a nonreplaceable resource, as discussed in section 22.7. The success of the U.S. agricultural industry depends on the quality of the topsoil, and so preserving the topsoil is of major concern for the U.S. and world economies.

The first plants seem to have developed a special relationship with fungi, which was a key factor in their ability to absorb minerals in terrestrial habitats. Within the roots or underground stems of many early fossil plants such as *Cooksonia* (see figure 17.7) and *Rhynia,* fungi can be seen living intimately within and among the plant's cells. These kinds of symbiotic associations are called **mycorrhizae** (discussed in chapter 16). In plants with mycorrhizae, the fungi enable the plant to take up phosphorus and other nutrients from rocky soil, while the plant supplies organic molecules to the fungus.

Conserving Water

One of the key challenges to living on land is to avoid drying out. To solve this problem, plants have a watertight outer covering called a **cuticle.** The covering is formed from a waxy substance that is impermeable to water. Like the wax on a shiny car, the cuticle prevents water from entering or leaving the stem or leaves. Water enters the plant only through the roots, while the cuticle prevents water loss to the air. Passages do exist through the cuticle, in the form of specialized pores called **stomata** (singular, **stoma**) in the leaves and sometimes the green portions of the stems. **Figure 17.2** shows a stoma on the underside of a leaf. The cutaway view allows you to see the placement of the stoma in relation to other cells in the leaf. Stomata, which occur on at least some portions of all plants except liverworts, allow carbon dioxide to pass into the plant bodies (by diffusion) for photosynthesis and allow water and oxygen gas to pass out of them. The two *guard cells* that border the stoma swell (opening the stoma) and shrink (closing the stoma) when water moves into and out of them by osmosis. This controls the loss of water from the leaf while allowing the entrance of carbon dioxide (see also figure 32.22). In most plants, water enters through the roots as a liquid and exits through the stomata as water vapor. Water movement within a plant is discussed in more detail in chapter 32, section 32.6.

Reproducing on Land

To reproduce sexually on land, it is necessary to pass gametes from one individual to another, which is a challenge because plants cannot move about. In the first plants, the eggs were surrounded by a jacket of cells, and a film of water was required for the sperm to swim to the egg and fertilize it. In later plants, pollen evolved, providing a means of transferring gametes without drying out. The pollen grain is protected from drying and allows plants to transfer gametes by wind or animals.

Changing the Life Cycle. Among many algae, haploid cells occupy the major portion of the life cycle. The zygote formed by the fusion of gametes is the only diploid cell, and it immediately undergoes meiosis to form haploid cells again. In early plants, by contrast, meiosis was delayed and the cells of the zygote divided to produce a multicellular diploid structure. Thus, for a significant portion of the life cycle, the cells were diploid. This change resulted in an **alternation of generations,** in which a diploid generation alternates with a haploid one. **Figure 17.3** shows a life cycle that exhibits an alternation of generations. Botanists call the haploid generation (indicated by the upper yellow area) the **gametophyte** because it forms haploid *gametes* by mitosis ❶. The diploid generation (indicated by the lower blue area) is called the **sporophyte** because it forms haploid *spores* by meiosis ❹.

When you look at primitive plants such as liverworts and mosses, you see mostly gametophyte tissue, the green leafy structures in figure 17.4a—the sporophytes are smaller brown structures attached to or enclosed within the tissues of the larger gametophyte. When you look at plants that evolved later, such as a pine tree, you see mostly sporophyte tissue. The gametophytes of these plants are always much smaller than the sporophytes and are often enclosed within sporophyte tissues. Figure 17.4b shows the male gametophyte (pollen grains) of a pine tree. They are not photosynthetic cells and are dependent upon the sporophyte tissue in which they are usually enclosed.

> **Putting the Concept to Work**
> Is a sporophyte haploid or diploid?

Figure 17.4 Two types of gametophytes.

(a) This gametophyte of a moss (a more primitive plant) is green and free-living. (b) Male gametophytes (pollen) of a pine tree (a more advanced plant) are barely large enough to see with the naked eye (×200).

(a) Steven P. Lynch; (b) Richard Gross/McGraw Hill

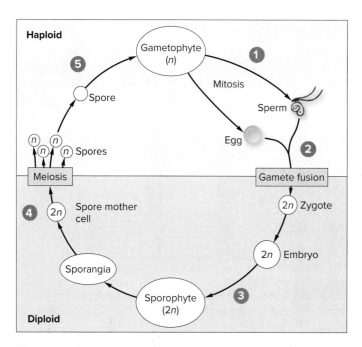

Figure 17.3 Generalized plant life cycle.

In a plant life cycle, a diploid generation alternates with a haploid one. Gametophytes, which are haploid (n), alternate with sporophytes, which are diploid ($2n$). ❶ Gametophytes give rise by mitosis to sperm and eggs. ❷ The sperm and egg ultimately come together to produce the first diploid cell of the sporophyte generation, the zygote. ❸ The zygote undergoes cell division, ultimately forming the sporophyte. ❹ Meiosis takes place within the sporangia, the spore-producing organs of the sporophyte, resulting in the production of the spores, which are haploid and are the first cells ❺ of gametophyte generations.

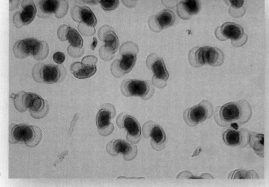

Figure 17.5 The evolution of plants.

17.2 Plant Evolution

> **LEARNING OBJECTIVE 17.2.1** List four key innovations in the evolution of plants.

Once plants became established on land, they gradually developed many other features that aided their evolutionary success in this new, demanding habitat. For example, among the first plants there was no fundamental difference between the aboveground and the belowground parts. Later, roots and shoots with specialized structures evolved, each suited to its particular environment below or above ground.

Key Innovations

As we explore plant diversity, we will examine several key evolutionary innovations that have given rise to the wide variety of plants we see today. **Table 17.1** provides you with an overview of the plant phyla and their characteristics.

Four key evolutionary innovations serve to trace the evolution of plants:

1. **Alternation of generations.** Although algae exhibit a haploid and diploid phase, the diploid phase is not a significant portion of their life cycle. By contrast, even in early plants (the nonvascular plants indicated by the first vertical bar in figure 17.5), the diploid sporophyte is a larger structure and offers protection for the egg and developing embryo. The dominance of the sporophyte, both in size and the proportion of time devoted to it in the life cycle, becomes greater throughout the evolutionary history of plants.
2. **Vascular tissue.** A second key innovation was the emergence of vascular tissue. Vascular tissue transports water and nutrients throughout the plant body and provides structural support. With the evolution of vascular tissue, plants were able to more efficiently supply the upper portions of their bodies with water absorbed from the soil and had some rigidity, allowing the plants to grow larger and in drier conditions. The first vascular plants were the seedless vascular plants, the second vertical bar shown in figure 17.5.
3. **Seeds.** The evolution of seeds was a key innovation that allowed plants to dominate their terrestrial environments. Seeds provide nutrients and a tough, durable cover that protects the embryo until it encounters favorable growing conditions. The first plants with seeds were the gymnosperms, the third vertical bar in figure 17.5.
4. **Flowers and fruits.** The evolution of flowers and fruits were key innovations that improved the chances of successful mating in sedentary organisms and facilitated the dispersal of their seeds. Flowers protected the egg and improved the odds of its fertilization, allowing plants that were located at considerable distances to mate successfully. Fruit, which surrounds the seed and aids in its dispersal, allowed plant species better opportunities to invade new and possibly more favorable environments. The angiosperms, the fourth vertical bar in figure 17.5, are the only plants to produce flowers and fruits.

> **Putting the Concept to Work**
> A pine tree is a gymnosperm. Do pine trees have flowers at any point in their life cycle? If so, when?

BIOLOGY & YOU

Eating Flowers. While you rarely see someone munching on a rose, flowers do creep into our diets in unexpected ways. Did you know, for example, that vegetables such as broccoli, cauliflower, and artichoke are actually the flowers of the plants? Spices such as cloves and capers are from flowers—indeed, the most expensive of all spices, saffron, consists of dried stigmas of a crocus! How about eating fresh flowers? Yep. Hundreds of fresh flowers are edible and nutritious, although few are widely used as food by humans—at least not directly. Marigold flowers are fed to chickens to give their egg yokes a golden yellow color. We ourselves sometimes put fresh flower petals into salads to add color and unusual flavors. Edible flowers you can put into your salad include nasturtium, chrysanthemum, carnation, honeysuckle, and sunflower. Dried flowers are more often made into herbal teas. The dried flowers of chrysanthemum, rose, jasmine, or chamomile are mixed with tea leaves or infused into tea for the fragrance they add.

TABLE 17.1	Plant Phyla			
Phylum	Typical Examples		Key Characteristics	Approximate Number of Living Species
Nonvascular Plants				
Hepaticophyta (liverworts)	Liverwort		Without true vascular tissues; lack true roots and leaves; live in moist habitats and require water for fertilization; gametophyte is the dominant structure in the life cycle	15,600
Anthocerophyta (hornworts)	Hornwort			
Bryophyta (mosses)	Peat moss			
Seedless Vascular Plants				
Lycophyta (lycopods)	Club mosses		Seedless vascular plants, similar in appearance to mosses but diploid; require water for fertilization; sporophyte is dominant structure in life cycle; found in moist woodland habitats	1,150
Pterophyta (ferns)	Tree ferns Horsetails Whisk ferns		Seedless vascular plants; require water for fertilization; sporophytes diverse in form and dominate the life cycle	11,000
Seed Plants				
Coniferophyta (conifers)	Pines, spruce, fir, redwood, cedar		Gymnosperms; wind pollinated; ovules partially exposed at time of pollination; flowerless; seeds are dispersed by the wind; sperm lack flagella; leaves are needlelike or scalelike; most species are evergreens and live in dense stands; among the most common trees on earth	601
Cycadophyta (cycads)	Cycads		Gymnosperms; wind pollination or possibly insect pollination; very slow growing, palmlike trees; sperm have flagella; trees are either male or female	206
Gnetophyta (shrub teas)	Mormon tea, *Welwitschia*		Gymnosperms; nonmotile sperm; shrubs and vines; wind pollination and possibly insect pollination; plants are either male or female; sporophyte is dominant in the life cycle	65
Ginkgophyta (ginkgo)	Ginkgo trees		Gymnosperms; fanlike leaves that are dropped in winter (deciduous); seeds fleshy and ill-scented; motile sperm; trees are either male or female	1
Anthophyta (flowering plants, also called angiosperms)	Oak trees, corn, wheat, roses		Flowering; pollination by wind, animal, and water; characterized by ovules that are fully enclosed by the carpel; fertilization involves two sperm nuclei: one forms the embryo, the other fuses with the polar body to form endosperm for the seed; after fertilization, carpels and the fertilized ovules (now seeds) mature to become fruit	250,000

Seedless Plants

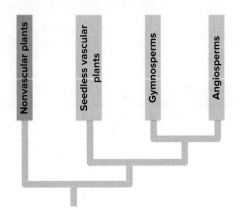

17.3 Nonvascular Plants

> **LEARNING OBJECTIVE 17.3.1** Compare the water-conducting abilities of mosses and hornworts.

The earliest plants to evolve lacked vascular tissue.

Liverworts and Hornworts

The first successful land plants had no vascular system—no tubes or pipes to transport water and nutrients throughout the plant. This greatly limited the maximum size of the plant body because all materials had to be transported by osmosis and diffusion (see section 4.9). However, these simple plants are highly adapted to a variety of terrestrial environments. Only two phyla of living plants, the **liverworts** (phylum Hepaticophyta) and the **hornworts** (phylum Anthocerophyta), completely lack a vascular system. The word *wort* meant *herb* in medieval Anglo-Saxon when these plants were named. Liverworts are the simplest of all living plants. About 6,000 species of liverworts and 100 species of hornworts survive today, usually growing in moist and shady places.

Mosses Use Primitive Conducting Systems

Another phylum of plants, the **mosses** (phylum Bryophyta), were the first plants to evolve strands of specialized cells that conduct water and carbohydrates up the stem of the gametophyte. The conducting cells do not have specialized wall thickenings; instead they are like nonrigid pipes. Because they cannot carry water very high, mosses are usually grouped by botanists with the liverworts and hornworts as "nonvascular" plants.

Today about 9,500 species of mosses grow in moist places all over the world. In the Arctic and Antarctic, mosses are the most abundant plants. Peat moss (genus *Sphagnum*) can be used as a fuel or a soil conditioner. The moss *Physcomitrella patens,* whose genome was sequenced in 2006, has been the subject of many genetic studies. The life cycle of a moss is illustrated in **figure 17.6**. Notice that the majority of the life cycle consists of the haploid gametophyte generation (the green part of the plant), which exists as male or female plants. The diploid sporophyte (the brown stalk with the swollen head) grows out of the gametophyte of the female plant ❸, from the zygote, which is the fertilized egg. Cells within the sporophyte undergo meiosis to produce haploid spores ❹ that grow into gametophytes ❺.

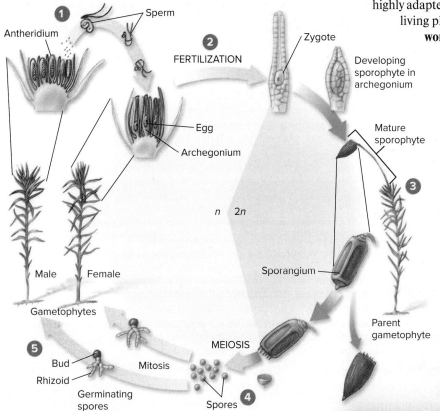

Figure 17.6 The life cycle of a moss.

On the haploid gametophytes, sperm are released from each antheridium (sperm-producing structure) ❶. They then swim through free water to an archegonium (egg-producing structure) and down to the egg. Fertilization takes place there ❷; the resulting zygote develops into a diploid sporophyte. The sporophyte grows out of the archegonium, forming a structure called a sporangium at its apex ❸. The sporophyte grows on the gametophyte, as shown in the photo, and eventually produces spores as a result of meiosis. The spores are shed from the sporangium ❹. The spores germinate, giving rise to gametophytes ❺.

A hair-cup moss, *Polytrichum*
blickwinkel/Alamy Stock Photo

> **Putting the Concept to Work**
> Is the moss that you see growing on a rock haploid or diploid?

17.4 The Evolution of Vascular Tissue

> **LEARNING OBJECTIVE 17.4.1** Differentiate between xylem and phloem and between primary and secondary growth.

The remaining seven phyla of plants, which have efficient vascular systems made of highly specialized cells, are called **vascular plants.** The earliest vascular plants for which we have relatively complete fossils, the extinct phylum Rhyniophyta, lived 410 million years ago. Among them is the oldest known vascular plant, *Cooksonia*. The fossil of *Cooksonia* in figure 17.7 clearly shows that the plant had branched, leafless shoots that formed spores at the tips in structures called *sporangia*.

The Advent of Plumbing

Cooksonia and the other early plants that followed became successful colonizers of the land through the development of efficient water- and food-conducting systems known as **vascular tissues** (Latin, *vasculum,* vessel or duct). These tissues consist of strands of specialized cylindrical or elongated cells that form a network of tubelike structures throughout a plant. This network extends from near the tips of the roots (when present), through the stems, and into the leaves (when present; figure 17.8). One type of vascular tissue, *xylem,* conducts water and dissolved minerals upward from the roots; another type, *phloem,* conducts carbohydrates throughout the plant.

Growing Taller

Most early vascular plants seem to have grown by cell division at the tips of the stem and roots. Imagine stacking dishes—the stack can get taller but not wider! This sort of growth is called **primary growth** and was quite successful. During the so-called Coal Age (between 350 and 290 million years ago, when much of the world's fossil fuel was formed), the lowland swamps that covered Europe and North America were dominated by an early form of seedless tree called a lycophyte. Lycophyte trees grew to heights of 10 to 35 meters (33 to 115 ft), and their trunks did not branch until they attained most of their total height. The pace of evolution was rapid during this period, for the world's climate was changing, growing dryer and colder. As the world's swamplands began to dry up, the lycophyte trees vanished, disappearing abruptly from the fossil record. They were replaced by tree-sized ferns, a form of vascular plant that will be described in detail in section 17.5. Tree ferns grew to heights of more than 20 meters (66 ft) with trunks 30 centimeters (12 in) thick. Like the lycophytes, the trunks of tree ferns were formed entirely by primary growth.

Growing Thicker

About 380 million years ago, vascular plants developed a new pattern of growth in which a cylinder of cells beneath the bark divides, producing new cells around the plant's periphery. This growth is called **secondary growth.** Secondary growth makes it possible for a plant stem to increase in diameter. Only after the evolution of secondary growth could vascular plants become thick-trunked and therefore considerably taller. Redwood trees (see figure 17.14a) today reach heights of up to 117 meters (384 ft) and trunk diameters in excess of 11 meters (36 ft). This evolutionary advance made possible the dominance of the tall forests that today cover northern North America. You are familiar with the product of plant secondary growth as **wood.**

> **Putting the Concept to Work**
> Does a moss show secondary growth?

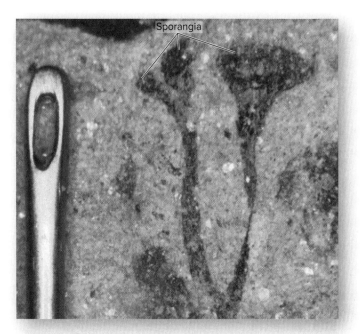

Figure 17.7 The earliest vascular plant.

The earliest vascular plant of which we have complete fossils is *Cooksonia*. This fossil shows a plant that lived some 410 million years ago; its upright branched stems terminated in spore-producing sporangia at the tips.

The Natural History Museum/Alamy Stock Photo

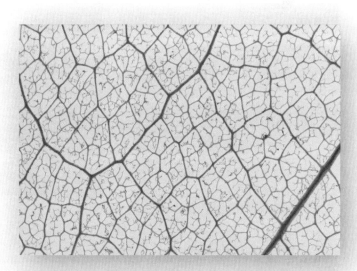

Figure 17.8 The vascular system of a leaf (×3).

The veins of a vascular plant contain strands of specialized cells for conducting food and water. The veins from a leaf are shown in this magnified view of a leaf with the rest of the tissue removed.

Ray Simons/Science Source

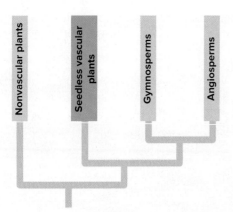

17.5 Seedless Vascular Plants

LEARNING OBJECTIVE 17.5.1 Compare the gametophyte and sporophyte generations of mosses with those of ferns.

The earliest vascular plants lacked seeds, and two of the seven phyla of modern-day vascular plants do not have them. The two phyla of living seedless vascular plants include the ferns, phylum Pterophyta, and the club mosses, phylum Lycophyta. In addition to the typical ferns seen growing

Figure 17.9 Seedless vascular plants.

(a) A tree fern in the forests of Malaysia (phylum Pterophyta). The ferns are by far the largest group of spore-producing vascular plants. (b) Ferns on the floor of a redwood forest. (c) A whisk fern. Whisk ferns have no roots or leaves. (d) A horsetail, *Equisetum telmateia*. This species forms two kinds of erect stems; one is green and photosynthetic, and the other, which terminates in a spore-producing "cone," is mostly light brown. (e) A club moss (phylum Lycophyta). Although superficially similar to the gametophytes of mosses, the conspicuous club moss plants shown here are sporophytes.

(a) Natphotos/Digital Vision/Getty Images; (b) The Jon B. Lovelace Collection of California Photographs in Carol M. Highsmith's America Project, LOC Prints & Photo Div.; (c) Kingsley R. Stern; (d) Stuart Wilson/Science Source; (e) Dieter Hopf/imageBROKER/Alamy Stock Photo

in forests, like those shown in figure 17.9*a, b*, the phylum Pterophyta also contains the whisk ferns (figure 17.9*c*) and the horsetails (figure 17.9*d*). The other phylum, Lycophyta, contains the club mosses (figure 17.9*e*). Both phyla have free-swimming sperm that require the presence of free water for fertilization.

By far the most abundant of seedless vascular plants are the **ferns,** with about 11,000 living species. Ferns are found throughout the world, although they are much more abundant in the tropics than elsewhere. Many are small, with stems only a few centimeters in diameter, but some of the largest plants that live today are also ferns. Descendants of ancient tree ferns can have trunks more than 20 meters (66 ft) tall and leaves up to 5 meters (16 ft) long!

The Life of a Fern

In ferns, the life cycle of plants begins a revolutionary change that culminates later with seed plants. Nonvascular plants such as mosses are made largely of gametophyte (haploid) tissue. Vascular seedless plants such as ferns have both gametophyte and sporophyte individuals, each independent and self-sufficient. The haploid gametophyte is a small structure, 1 to 2 cm in diameter (the heart-shaped plant in figure 17.10 ❶, not drawn to scale), which produces eggs and sperm. Rhizoids (anchoring structures) project from their lower surface. After sperm swim through water and fertilize the egg ❷, the zygote grows into a sporophyte. Eventually, the sporophyte becomes much larger than the gametophyte ❸. Most ferns have horizontal stems, called rhizomes, that creep along belowground. The sporophyte bears haploid spores on the underside of its leaves, in the brown clusters called sori (singular, sorus ❹). The spores are released from the sorus and float to the ground where they germinate, growing into haploid gametophytes. The fern gametophytes are small, thin photosynthetic plants that live in moist places.

The fern sporophytes are much larger and more complex, with long vertical leaves called **fronds.** When you see a fern, you are almost always looking at the sporophyte.

Figure 17.10 **The life cycle of a fern.**

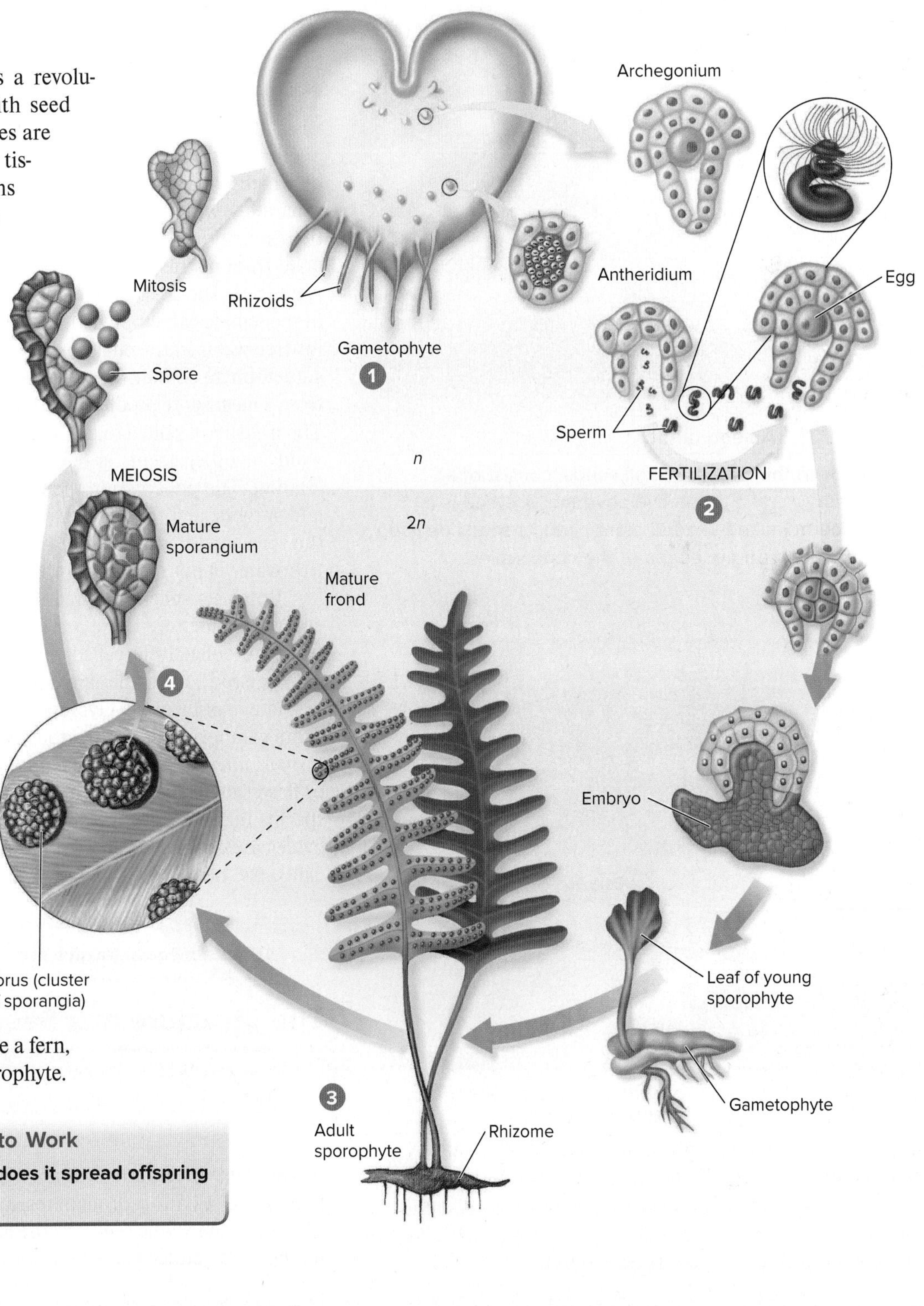

> **Putting the Concept to Work**
> If a fern doesn't have seeds, how does it spread offspring to distant locations?

The Advent of Seeds

Figure 17.11 A seed plant.

The seeds of this cycad, like all seeds, consist of a plant embryo and a protective covering. A cycad is a gymnosperm (naked-seeded plant), and its seeds develop out in the open on the edges of the cone scales.

Kingsley R. Stern

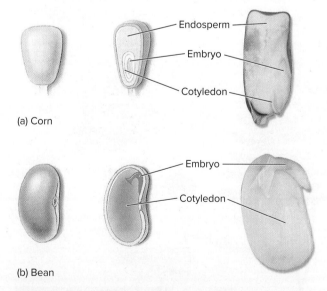

(a) Corn

(b) Bean

Figure 17.12 Basic structure of seeds.

A seed contains a sporophyte (diploid) embryo and a source of food, either endosperm (a) or food stored in the cotyledons (b). A seed coat surrounds the seed and protects the embryo. The seed structures are visible in the cutaway photos of seeds on the *right*.

17.6 Evolution of Seed Plants

> **LEARNING OBJECTIVE 17.6.1** Discriminate between gymnosperms and angiosperms and between a microspore and a megaspore.

When we think of plants, we often think of seed plants, but none of the plants discussed up to this point has seeds. Seeds were a very important evolutionary adaptation in the history of plants. The **seed** is a crucial adaptation to life on land because it protects the embryonic plant when it is at its most vulnerable stage.

Seeds

The dominance of the sporophyte (diploid) generation in the life cycle of vascular plants reaches its full force with the advent of the seed plants. Seed plants produce two kinds of gametophytes—male and female, each of which consists of just a few cells. Both kinds of gametophytes develop separately within the sporophyte and are completely dependent on it for their nutrition. Male gametophytes, commonly referred to as **pollen grains,** arise from **microspores.** The pollen grains become mature when sperm are produced. The sperm are carried to the egg in the female gametophyte without the need for free water in the environment. A female gametophyte contains the egg and develops from a **megaspore** produced within an **ovule.** The transfer of pollen to an ovule by insects, wind, or other agents is referred to as **pollination.** The pollen grain then cracks open and sprouts, and the *pollen tube,* containing the sperm cells, grows out, transporting the sperm directly to the egg. Thus, free water is not required in the pollination and fertilization process.

> Pollination, fertilization, and seed formation are discussed in detail in section 32.1, but it is important to note here that unlike haploid spores in the seedless vascular plants that develop into gametophytes, seeds contain diploid cells that develop into sporophytes.

Botanists generally agree that all seed plants are derived from a single common ancestor. There are five phyla with living representatives. In four of them, collectively called the **gymnosperms** (Greek, *gymnos,* naked, and *sperma,* seed), the ovules are not completely enclosed by sporophyte tissue at the time of pollination. Gymnosperms were the first seed plants (figure 17.11). From gymnosperms evolved the fifth group of seed plants, called **angiosperms** (Greek, *angion,* vessel, and *sperma,* seed), phylum Anthophyta. Angiosperms, or flowering plants, are the most recently evolved of all the plant phyla. Angiosperms differ from all gymnosperms in that at the time of pollination, their ovules are completely enclosed by a vessel of sporophyte tissue in the flower, called the **carpel.**

> **Putting the Concept to Work**
> Does a megaspore give rise to male or female gametes, or to both?

The Structure of a Seed

> **LEARNING OBJECTIVE 17.6.2** Discuss how the structure of seeds better adapts plants to life on land.

A seed has three parts that are visible in the corn and bean seeds shown in figure 17.12: (1) a drought-resistant protective cover called the seed coat (visible in the whole seeds on the left), (2) a source of food for the developing embryo (either endosperm or cotyledons), and (3) a sporophyte plant

Figure 17.13 Seeds allow plants to bypass the dry season.

Seeds can remain dormant until conditions are favorable for growth. For example, when it rains, seeds can germinate, and plants can grow rapidly to take advantage of the relatively short periods when water is available. This palo verde desert tree (*Cercidium floridum*) has tough seeds (*inset*) that germinate only after they are cracked. Rains leach out the chemicals in the seed coats that inhibit germination, and the hard coats of the seeds may be cracked when they are washed down gullies in temporary floods.

(top) Ron and Patty Thomas/Getty Images; (bottom) Dan Suzio/Science Source

embryo (visible in the cutaway views in the middle and on the right). The **endosperm** makes up most of the seed in corn and is the white part of popcorn, but in the seeds of some plants such as beans, the endosperm is stored as food by the embryo in thick "leaflike" structures called **cotyledons.**

Why Seeds Are Important

Seeds are one way in which plants, anchored by their roots to one place in the ground, are able to disperse their progeny to new locations. The hard cover of the seed protects the seed while it travels to a new location. The seed travels by many means such as by air, water, and animals. Seed dispersal will be discussed in chapter 32.

Once a seed has fallen to the ground, it may lie there, dormant, for many years. When conditions are favorable, however, and particularly when moisture is present, the seed germinates and begins to grow into a young plant (**figure 17.13**). Most seeds have abundant food stored in them to provide a ready source of energy for the new plant as it starts its growth.

The advent of seeds had an enormous influence on the evolution of plants. Seeds are particularly adapted to life on land in at least four respects:

1. **Dispersal.** Most important, seeds facilitate the migration and dispersal of plant offspring into new habitats.
2. **Dormancy.** Seeds permit plants to postpone development when conditions are unfavorable, as during a drought, keeping the embryonic plant protected until conditions improve.
3. **Germination.** By making the reinitiation of development dependent upon environmental factors such as temperature, seeds permit the course of embryonic development to be synchronized with critical aspects of the plant's habitat, such as the season of the year.
4. **Nourishment.** Seeds offer nourishment during the critical period just after germination, when the seedling must establish itself.

> Seed germination, as discussed in section 32.4, occurs before the plant is able to carry out photosynthesis. However, the embryo must have energy to start growing. The food stored in the seed provides a source of energy for this early growth.

Putting the Concept to Work
Is the plant embryo within a seed a gametophyte or a sporophyte?

17.7 Gymnosperms

LEARNING OBJECTIVE 17.7.1 Describe the life cycle of the most widespread phylum of gymnosperms.

Four phyla constitute the gymnosperms (figure 17.14): (1) the conifers (Coniferophyta), (2) the cycads (Cycadophyta), (3) the gnetophytes (Gnetophyta), and (4) the ginkgo (Ginkgophyta). The conifers are the most familiar of the four phyla of gymnosperms and include pine, spruce, hemlock, cedar, redwood, yew, cypress, and fir trees (figure 17.14a). Conifers are trees that produce their seeds in cones. The seeds (ovules) of conifers develop on scales within the cones and are exposed at the time of pollination. Most of the conifers have needlelike leaves, an evolutionary adaptation for retarding water loss. Conifers are often found growing in moderately dry regions of the world, including the vast taiga forests of the northern latitudes. Many are very important as sources of timber and pulp.

Kinds of Gymnosperms

There are about 600 living species of conifers. The tallest living vascular plant, the coastal sequoia (*Sequoia sempervirens*), found in coastal California and Oregon, is a conifer and reaches over 100 meters (328 ft). The biggest redwood, however, is the mountain sequoia redwood species (*Sequoiadendron gigantea*) of the Sierra Nevadas. The largest individual tree is nicknamed after General Sherman of the Civil War—it stands more than 83 meters (274 ft) tall and measures 31 meters (102 ft) around its base. Another much smaller type of conifer, the bristlecone pines in Nevada, may be the oldest trees in the world—about 5,000 years old.

The other three gymnosperm phyla are much less widespread. Cycads (figures 17.11 and 17.14b), the predominant land plant in the golden age of dinosaurs, the Jurassic period (213–144 million years ago), have short stems

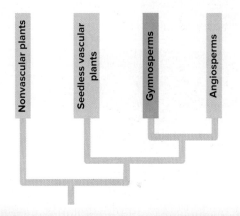

Figure 17.14 Gymnosperms.
(a) This coast redwood tree (*Sequoia sempervirens*), a long-lived type of conifer, typically grows in dense stands.
(b) An African cycad, *Encephalartos transvenosus,* phylum Cycadophyta. The cycads have fernlike leaves and seed-forming cones, like the ones shown here.
(c) Maidenhair tree, *Ginkgo biloba,* the only living representative of the phylum Ginkgophyta, a group of plants that was abundant 200 million years ago. Among living seed plants, only the cycads and ginkgo have swimming sperm.

(a) Fuse/Getty Images; (b) Images Etc Ltd/Getty Images; (c) Nancy Nehring/Getty Images

and palmlike leaves. They are still widespread throughout the tropics. The gnetophytes, phylum Gnetophyta, contains only three kinds of plants, all unusual. One of them is perhaps the most bizarre of all plants, *Welwitschia*, which grows on the exposed sands of the harsh Namibian Desert of southwestern Africa. *Welwitschia* acts like a plant standing on its head! Its two beltlike, leathery leaves are generated continuously from their base, splitting as they grow out over the desert sands. There is only one living species of ginkgo, which has fan-shaped leaves (shown in figure 17.14c) that are shed in the autumn. Because ginkgos are resistant to air pollution, they are commonly planted along city streets.

The fossil record indicates that members of the ginkgo phylum were once widely distributed, particularly in the Northern Hemisphere; today, only one living species, the maidenhair tree (*Ginkgo biloba*), remains. The reproductive structures of ginkgos are produced on separate trees. The fleshy outer coverings of the seeds of female ginkgo plants exude the foul smell of rancid butter caused by butyric and isobutyric acids.

The Life of a Gymnosperm

We will examine conifers as a typical gymnosperm. The conifer life cycle is illustrated in figure 17.15. Conifer trees form two kinds of cones. Pollen cones ❶ contain the male gametophytes, pollen grains; seed cones ❸ contain the female gametophytes, with their egg cells. Conifer pollen grains ❷ are small and light and are carried by the wind to seed cones. Because it is very unlikely that any particular pollen grain will succeed in being carried to a seed cone (the wind can take it anywhere), a great many pollen grains are produced to be sure that at least a few succeed in pollinating seed cones. For this reason, pollen grains are shed from their cones in huge quantities, often appearing as a sticky yellow layer on the surfaces of ponds and lakes—and even on windshields.

When a grain of pollen settles down on a scale of a female cone, a slender tube grows out of the pollen cell up into the scale, delivering the male gamete to the female gametophyte ❹ containing the egg, or ovum. Fertilization occurs when the sperm cell fuses with the egg ❺, forming a zygote that develops into an embryo. This zygote is the beginning of the sporophyte generation. What happens next is the essential improvement in reproduction achieved by seed plants. Instead of the zygote growing immediately into an adult sporophyte—just as you grew directly into an adult from a fertilized zygote—the fertilized ovule forms a seed ❻. The pine seed contains a saillike structure that helps carry the seed by the wind. The seed can then be dispersed into new habitats. If conditions are favorable where the seed lands, it will germinate and begin to grow, forming a new sporophyte plant ❼.

> **Putting the Concept to Work**
>
> If needlelike leaves are an adaptation to dry conditions, why are North America's great conifer forests found in Canada and not Mexico?

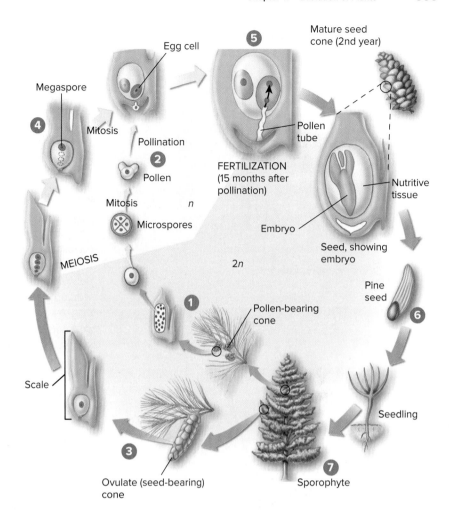

Figure 17.15 The life cycle of a conifer.

In all seed plants, the haploid gametophyte generation (yellow sector) is greatly reduced.

> **IMPLICATION FOR YOU** Each year, Americans spend some $1.5 billion on 36 million Christmas trees grown on 22,000 tree farms occupying half a million acres. Almost all American Christmas trees are Douglas or balsam firs because fir trees don't shed their leaves when they dry out, unlike spruce, hemlock, pine, and other evergreens. American trees are sheared during growth to achieve a more conical shape, whereas Europeans favor the open aspect of unsheared trees. Which do you prefer?

The Evolution of Flowers

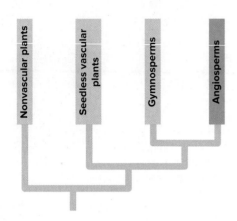

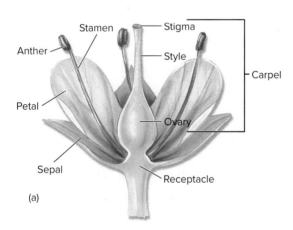

Figure 17.16 An angiosperm flower.

(a) The basic structure of a flower is a series of four concentric circles, or whorls: the sepals, petals, stamens, and the carpel (ovary, style, and stigma), also seen in (b) the wild woodland plant *Geranium*.

(b) Nick Kurzenko/Alamy Stock Photo

17.8 Rise of the Angiosperms

> **LEARNING OBJECTIVE 17.8.1** Name, locate, and describe the four whorls of a flower.

Angiosperms are plants in which the ovule is completely enclosed by sporophyte tissue when it is fertilized. Ninety percent of all living plants are angiosperms, about 250,000 species, including many trees, shrubs, herbs, grasses, vegetables, and grains—in short, nearly all of the plants that we see every day.

Angiosperms successfully meet the last difficult challenge posed by terrestrial living: the need to find mates of the same species. The wind pollination of gymnosperms is inherently inefficient. Angiosperms are able to deliver their pollen *directly*, as if in an addressed envelope, from one individual of a species to another. How? *By inducing insects and other animals to carry it for them!* The tool that makes this animal-dictated pollination possible, the great advance of the angiosperms, is the flower. Though some very successful later-evolving angiosperms such as grasses have reverted to wind pollination of their flowers, the directed pollination of flowering plants has led to phenomenal evolutionary success.

The Flower

Flowers are the reproductive organs of angiosperm plants. A flower is a sophisticated pollination machine. It may employ bright colors to attract the attention of insects (or birds or small mammals), nectar to induce the insect to enter the flower, and structures that coat the insect with pollen grains while it is visiting. Then, when the insect visits another flower, it carries the pollen with it. Parts of the flower also develop into the seed and the fruit after fertilization.

The basic structure of a flower consists of four concentric circles, or whorls, connected to a base called the receptacle:

1. The outermost whorl, called the **sepals** of the flower, typically serves to protect the flower from physical damage. These are the green leaflike structures in **figure 17.16a** and are in effect modified leaves that protect the flower while it is a bud.
2. The second whorl, called the **petals** of the flower, serves to attract particular pollinators. Petals have particular pigments, often vividly colored like the light purple color in **figure 17.16**.
3. The third whorl, called the **stamens** of the flower, contains the "male" parts that produce the pollen grains. Stamens are the slender, threadlike filaments in **figure 17.16** with a pollen-rich **anther** at the tip.
4. The fourth and innermost whorl, called the **carpel** of the flower, contains the "female" parts that produce eggs. The carpel is the vase-shaped structure in **figure 17.16**. The carpel is sporophyte tissue that completely encases the ovules within which the egg cell develops. The ovules occur in the bulging lower portion of the carpel, called the **ovary;** usually there is a slender stalk rising from the ovary called the **style,** with a sticky tip called a **stigma,** which receives pollen. When the flower is pollinated, a pollen tube grows down from the pollen grain on the stigma through the style to the ovary to fertilize the egg.

> **Putting the Concept to Work**
> Which of the four whorls of a flower tend to be brightly colored? Why?

Inquiry and Analysis

How Does Arrowgrass Tolerate Salt?

Plants grow almost everywhere on earth, thriving in many places where exposure, drought, and other severe conditions challenge their survival. In deserts, a common stress is the presence of high levels of salt in the soils. Soil salinity is also a problem for millions of acres of abandoned farmland because the accumulation of salt from irrigation water restricts growth. Why does excess salt in the soil present a problem for a plant? For one thing, high levels of sodium ions that are taken up by the roots are toxic. For another, a plant's roots cannot obtain water when growing in salty soil. Osmosis (the movement of water molecules to areas of higher solute concentrations, see section 4.9) causes water to move in the opposite direction, drawn out of the roots by the soil's high levels of salt. And yet plants do grow in these soils. How do they manage?

To investigate this, researchers have studied seaside arrowgrass (*Triglochin maritima*), the plant you see to the right. Arrowgrass plants are able to grow in very salty seashore soils, where few other plants survive. How are they able to survive? Researchers found that their roots do not take up salt and so do not accumulate toxic levels of salt.

However, this still leaves the arrowgrass plant the challenge of preventing its root cells from losing water to the surrounding salty soil. How then do the roots achieve osmotic balance? In an attempt to find out, researchers grew arrowgrass plants in nonsalty soil for two weeks, then transferred them to one of several soils that differed in salt level. After 10 days, shoots were harvested and analyzed for amino acids because accumulating amino acids could be one way that the cells maintain osmotic balance. Results are presented in the graph.

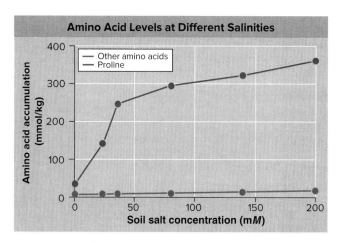

blickwinkel/Alamy Stock Photo

Analysis

1. **Applying Concepts** What do the abbreviations "mM" and "mmol/kg" mean?
2. **Interpreting Data**
   a. In salt-free soil (that is, the mM soil salt concentration = 0), how much proline has accumulated in the roots after 10 days? How much of other amino acids?
   b. In salty beach soils with salt levels of 35 mM, how much proline has accumulated in the roots after 10 days? How much of other amino acids?
3. **Making Inferences**
   a. In general, what is the effect of soil salt concentration on an arrowgrass plant's accumulation of the amino acid proline? Of other amino acids?
   b. Is the effect of salt on proline accumulation the same at lower salt levels (below 50 mM) as at higher salt levels (above 50 mM)?
4. **Drawing Conclusions** Are these results consistent with the hypothesis that arrowgrass accumulates proline to achieve osmotic balance with salty soils?

18 Evolution of Animals

LEARNING PATH ▼

Introduction to the Animals
1. General Features of Animals
2. Six Key Transitions in Body Plan

Evolution of the Animal Phyla
3. The Simplest Animals
4. Advent of Bilateral Symmetry
5. Changes in the Body Cavity
6. Redesigning the Embryo

The Parade of Vertebrates
7. Overview of Vertebrate Evolution
8. Fishes Dominate the Sea
9. Amphibians and Reptiles Invade the Land
10. Birds Master the Air
11. Mammals Adapt to Colder Times
12. Human Evolution

TODAY'S PEOPLE CARRY the genes of several human species.
Christian Darkin/Science Source

Meet Our Hobbit Cousin

Science is by its nature a feisty enterprise, with researchers evaluating each other's work—not always kindly—and constantly seeking better ways to test ideas against the hard reality of data. When a scientific field is active, it is alive with controversy, with the slam-around give-and-take of aggressive inquiring minds. This is nowhere more true than in the field of paleoanthropology, the study of human evolution. For the last several years, the field has been stood on its head by a tiny fossil from Indonesia.

Discovery of the Hobbit People

In 2003, Australian paleoanthropologists hit paydirt at Liang Bua ("cool cave") on Flores island 300 miles south of Borneo in Indonesia. Six meters under the floor of the cave you see in the photo on the next page, in 50,000-year-old sediment, the research team found the nearly complete fossil skeleton of a hominid, an early human. The teeth are worn, and the skull bones are knitted together in an adult way, so the fossilized individual was an adult—but an adult hominid only 1 meter tall, with a brain of only 380 cubic centimeters! This is the stature and brain size of a chimpanzee. The fossil foot, essentially complete, is extremely large, leading the researchers to nickname the newly discovered hominid "the hobbit" after characters in *Lord of the Rings*. In 2004, they found a total of eight more tiny individuals.

This *Homo floresiensis*, as they dubbed the fossil, makes no sense when viewed against the detailed picture paleoanthropologists have pieced together of human evolution. It is too short, with too small a brain. This tiny human simply cannot be. And yet there it is, hard data. Only 50,000 years old, it was still alive on Flores when our sister species *Homo neanderthalensis* arrived nearby in Australia.

A Different Evolutionary Path

What are we to make of it? Large mammals colonizing remote small islands often tend to evolve into isolated

dwarf species. Many examples are known, including pigmy hippos, ground sloths, deer, and even dinosaurs. Indeed, pigmy elephants no bigger than cattle lived on Flores island at that very time! Why do dwarf species evolve? The theory is that on islands with few or no predators, there is no advantage to being big, and with limited food, there is great advantage to being small. On Flores island, the human evolutionary story seems to have played out differently than it did in Africa, evolution favoring the small.

So where did *Homo floresiensis* come from? Careful comparisons of upper arm bones suggest the hobbits of Flores are descendants of *H. erectus*, the ancestor of both our species, *H. sapiens,* and our sister species, the Neanderthals. Seven-hundred-thousand-year-old fossil teeth found nearby are intermediate. It now looks as if a group of *H. erectus* arrived on Flores about 1 million years ago.

We Carry the Genes of Three Species

With modern DNA technology, we are learning a lot about our early ancestors. In 2010, Swedish researcher Svante Paabo sequenced the complete nuclear genome of Neanderthals from the bones of three females who lived in Croatia more than 38,000 years ago. The sequence reveals that Neanderthals and modern humans interbred, although not a lot: 1% to 4% of European and Asian (but not African) nuclear DNA is of Neanderthal origin. Modern humans arose in Africa but interbred with Neanderthals (the descendants of *Homo erectus*) when they spread out into Europe and Asia. Seven months later, Paabo published the complete genome of a girl who lived more than 50,000 years ago in southern Siberia. Using powerful technology, he got the complete sequence from one pinky finger bone found in Denisova Cave. Neither Neanderthal nor modern human, she was an entirely new species of human, never before known to science! Comparing genomes suggests that Denisovans split from Neanderthals about 200,000 years ago, and 17% of their genome is Neanderthal. Modern humans interbred with Neanderthals when they left Africa 100,000 years ago. Some of these modern humans, carrying Neanderthal DNA, migrated to Melanesia and interbred with Denisovans on the way. As much as 8% of today's Melanesian DNA comes from archaic people. Modern humans today carry the genes of three different species. Nor does the story end there: 4% of the Denisovan genome appears to come from another ancient human, a ghost human as yet unknown to science. Today's people are a patchwork, woven through time.

Would the Hobbits Have Carried Denisovan Genes?

When Neanderthals first migrated to the Indonesian subcontinent from China, they interbred with Denisovan humans along the way. If *H. floresiensis* evolved their pigmy size due to prolonged isolation on an island, they might be descendants of these early Indonesians and thus would be expected to carry Denisovan genes. If they are far more ancient and evolved from an isolated *H. erectus* population as early as 1.3 million years ago, then of course they would not carry Denisovan genes.

Do they? This is not an easy question to answer as the relatively few *H. floresiensis* fossils available for study have been preserved for many thousands of years at tropical temperatures. DNA breaks down far more easily at warmer temperatures (the Denisovan genome was sequenced from DNA in a fingerbone preserved in Siberia, quite a cold climate). Several attempts have been made to extract DNA from a *H. floresiensis* molar tooth unearthed in 2003, so far without success.

Introduction to the Animals

18.1 General Features of Animals

> **LEARNING OBJECTIVE 18.1.1** List nine features all animals share.

Key Animal Features

From protist ancestors, an astonishing diversity of animals has evolved. All animals, however, have certain key features in common (table 18.1): (1) Animals are heterotrophs. (2) All animals are multicellular, but unlike protists, fungi, and plants, animal cells lack cell walls. (3) Animals are able to move from place to place. (4) Animals are very diverse in form and habitat. (5) Most animals reproduce sexually. (6) Animals have characteristic tissues and patterns of development.

> **Putting the Concept to Work**
> Do most animal phyla occur in the sea or on land?

TABLE 18.1 General Features of Animals

Heterotrophs. Unlike autotrophic plants and algae, animals cannot construct organic molecules from inorganic chemicals. All animals are heterotrophs—that is, they obtain energy and organic molecules by ingesting other organisms. Some animals (herbivores) consume autotrophs, other animals (carnivores) consume heterotrophs; others, such as the bear to the right, are omnivores that eat both autotrophs and heterotrophs, and still others (detritivores) consume decomposing organisms.

Ron Crabtree/Getty Images

Multicellular. All animals are multicellular, often with complex bodies like that of this brittlestar (*right*). The unicellular heterotrophic organisms called protozoa, which were at one time regarded as simple animals, are now considered members of the large and diverse kingdom Protista, discussed in chapter 16.

Purestock/SuperStock

No Cell Walls. Animal cells are distinct among those of multicellular organisms because they lack rigid cell walls and are usually quite flexible, like these cancer cells. The many cells of animal bodies are held together by extracellular lattices of structural proteins such as collagen. Other proteins form a collection of unique intercellular junctions between animal cells.

Science Photo Library/Alamy Stock Photo

TABLE 18.1	General Features of Animals *(continued)*

Active Movement. The ability of animals to move more rapidly and in more complex ways than members of other kingdoms is perhaps their most striking characteristic, one that is directly related to the flexibility of their cells and the evolution of nerve and muscle tissues. A remarkable form of movement unique to animals is flying, an ability that is well developed among vertebrates (animals with backbones) and insects such as this butterfly. The only terrestrial vertebrate group never to have evolved flight is amphibians.

Wild Horizon/Getty Images

Diverse in Form. Almost all animals (99%) are invertebrates, which, such as this millipede, lack a backbone. Of the estimated 10 million living animal species, only 42,500 have a backbone and are referred to as vertebrates. Animals are very diverse in form, ranging in size from organisms too small to see with the unaided eye to enormous whales and giant squids.

George Grall/National Geographic/Getty Images

Diverse in Habitat. The animal kingdom includes about 35 phyla, most of which, such as these jellyfish (phylum Cnidaria), occur in the sea. Far fewer phyla occur in freshwater and fewer still occur on land. Members of three successful marine phyla, Arthropoda (insects), Mollusca (snails), and Chordata (vertebrates), dominate animal life on land.

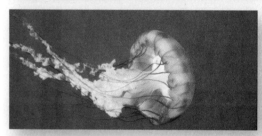
Pavel Vakhrushev/Shutterstock

Sexual Reproduction. Most animals reproduce sexually, as these tortoises are doing. Animal eggs, which are nonmotile, are much larger than the small, usually flagellated sperm. In animals, cells formed in meiosis function directly as gametes. The haploid cells do not divide by mitosis first, as they do in plants and fungi, but rather fuse directly with each other to form the zygote. Consequently, with a few exceptions, there is no counterpart among animals to the alternation of haploid (gametophyte) and diploid (sporophyte) generations characteristic of plants.

M&G Therin-Weise/Getty Images

Embryonic Development. Most animals have a similar pattern of embryonic development. The zygote first undergoes a series of mitotic divisions, called *cleavage*, and, like this dividing frog egg, becomes a solid ball of cells, the morula, then a hollow ball of cells, the blastula. In most animals, the blastula folds inward at one point to form a hollow sac with an opening at one end called the blastopore. An embryo at this stage is called a gastrula. The subsequent growth and movement of the cells of the gastrula differ widely from one phylum of animals to another.

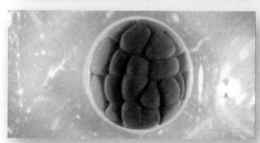
Dr. Keith Wheeler/Science Source

Unique Tissues. The cells of all animals except sponges are organized into structural and functional units called tissues, collections of cells that have joined together and are specialized to perform a specific function. Animals are unique in having two tissues associated with movement: (1) muscle tissue, which powers animal movement, and (2) nervous tissue, which conducts signals among cells. Neuromuscular junctions, where nerves connect with muscle tissue, are shown here.

Ed Reschke/Getty Images

18.2 Six Key Transitions in Body Plan

> **LEARNING OBJECTIVE 18.2.1** List and describe six key transitions in animal body design.

The evolution of animals is marked by six key body transitions: (1) tissues, (2) bilateral symmetry, (3) a body cavity, (4) deuterostome development, (5) molting, and (6) segmentation. These six body transitions are indicated at the branchpoints of the animal evolutionary tree in **figure 18.1**.

1. Evolution of Tissues

The simplest animals, the Parazoa, lack both defined tissues and organs. Characterized by the sponges, these animals exist as aggregates of cells with minimal intercellular coordination.

Radial symmetry
Russell Illig/Photodisc/Getty Images

2. Evolution of Bilateral Symmetry

Sponges also lack any definite symmetry, growing as irregular masses. Virtually all other animals have a definite shape and symmetry that can be defined along an imaginary axis through the animal's body. **Radial symmetry** is where the body parts are arranged around a central axis, whereas **bilateral symmetry** is where the body has a right and a left half that are mirror images of each other.

3. Evolution of a Body Cavity

A third key transition in the evolution of the animal body plan was the evolution of the body cavity. The evolution of efficient organ systems within the animal body was not possible until a body cavity evolved for supporting organs, distributing materials, and fostering complex developmental interactions.

4. Evolution of Deuterostome Development

All but two of the major phyla of bilateral animals are members of a group called the **protostomes** (from the Greek words *protos,* first, and *stoma,* mouth) that undergo similar embryonic development. Two outwardly dissimilar groups, the echinoderms and the chordates, comprise a second group, the **deuterostomes** (Greek, *deuteros,* second, and *stoma,* mouth), with key differences in their pattern of embryonic development.

5. Evolution of Molting

Most coelomate animals grow by gradually adding mass to their body. However, this creates a serious problem for animals with a hard exoskeleton, (external skeleton) which can hold only so much tissue. To grow further, the individual must shed its hard exoskeleton, a process called molting (or, more formally, **ecdysis**). Ecdysis occurs among both roundworms and arthropods.

Bilateral symmetry
Gary W. Carter/Corbis Super/Alamy Stock Photo

6. Evolution of Segmentation

The sixth key transition in the animal body plan involves the subdivision of the body into **segments**. Just as it is efficient for workers to construct a tunnel from a series of identical prefabricated parts, so segmented animals are assembled from a succession of identical segments. Segmentation has evolved in many groups.

Segmentation
IT Stock Free/Alamy Stock Photo

> **Putting the Concept to Work**
> Starfish are far along the bilaterally symmetrical branch of figure 18.1. Are they bilaterally symmetrical? Explain.

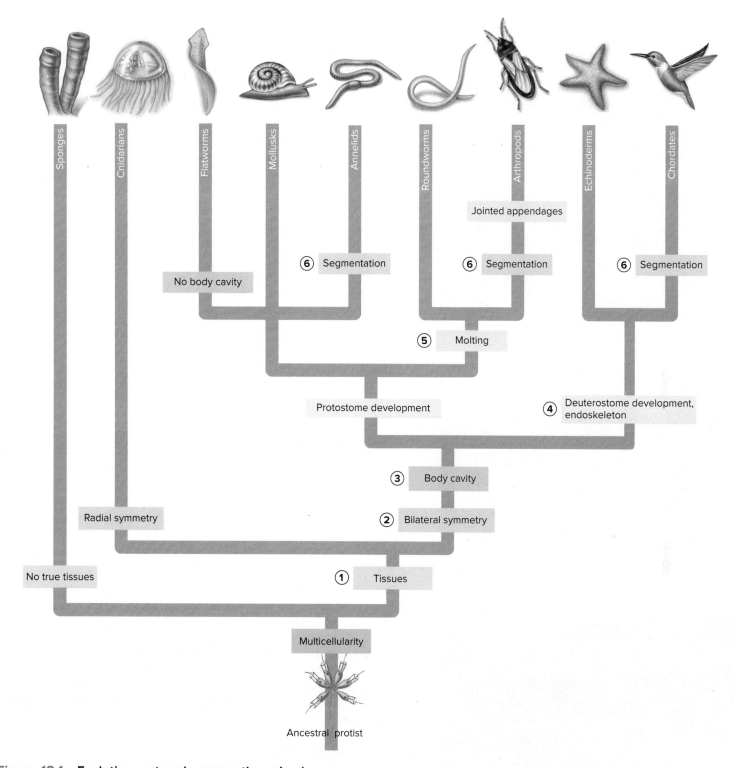

Figure 18.1 **Evolutionary trends among the animals.**

In this chapter, we examine a series of key evolutionary innovations in the animal body plan, shown here along the branches. Some of the major animal phyla are shown on this tree.

Evolution of the Animal Phyla

18.3 The Simplest Animals

The early history of animals involved quite different kinds of body plan.

Sponges: Multicellularity

> **LEARNING OBJECTIVE 18.3.1** Identify the key body innovation of sponges.

Sponges, members of the phylum Porifera, are the simplest animals. Most sponges completely lack symmetry, and although some of their cells are highly specialized, they are not organized into tissues. The bodies of sponges consist of little more than masses of specialized cells embedded in a gel-like matrix, like chopped fruit in Jell-O. However, sponge cells do possess a key property of animal cells: cell recognition. For example, when a sponge is passed through a fine silk mesh, individual cells separate and then reaggregate on the other side to re-form the sponge. Clumps of cells disassociated from a sponge can give rise to entirely new sponges.

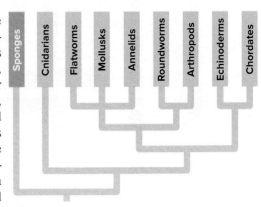

About 5,000 species exist, almost all in the sea. Some are tiny, and others are more than 2 meters in diameter (**figure 18.2a**). The body of an adult sponge is anchored in place on the seafloor and is shaped like a vase (**figure 18.2b**). The outside of the sponge is covered with a skin of flattened cells called epithelial cells that protect the sponge.

Choanocytes. The body of the sponge is perforated by tiny holes. The name of the phylum, Porifera, refers to this system of pores. Unique flagellated cells called **choanocytes,** or collar cells, line the body cavity of the sponge. The beating of the flagella of the many choanocytes draws water in through the pores. Why all this moving of water? The sponge is a "filter-feeder." The beating of each choanocyte's flagellum draws water through its collar, made of small hairlike projections resembling a picket fence. Any food particles in the water, such as protists and tiny animals, are trapped in the fence.

The choanocytes of sponges very closely resemble a kind of protist called choanoflagellates, which seem almost certain to have been the ancestors of sponges and thus of all animals.

> **Putting the Concept to Work**
> When a sponge is passed through a fine silk mesh, and the sponge re-forms, how do the choanocytes "know" to orient outward?

(a)

(b)

Figure 18.2 Diversity in sponges.

These two marine sponges are barrel sponges. They are among the largest of sponges, with well-organized forms. Many are more than 2 meters in diameter (a), while others are smaller (b).

(a) Twilight Zone Expedition Team 2007, NOAA-OE;
(b) Stephen Frink/Getty Images

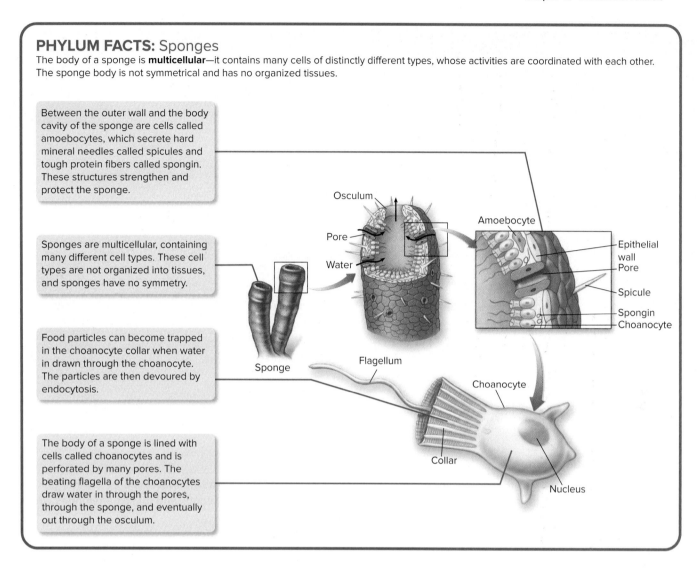

PHYLUM FACTS: Sponges
The body of a sponge is **multicellular**—it contains many cells of distinctly different types, whose activities are coordinated with each other. The sponge body is not symmetrical and has no organized tissues.

Between the outer wall and the body cavity of the sponge are cells called amoebocytes, which secrete hard mineral needles called spicules and tough protein fibers called spongin. These structures strengthen and protect the sponge.

Sponges are multicellular, containing many different cell types. These cell types are not organized into tissues, and sponges have no symmetry.

Food particles can become trapped in the choanocyte collar when water in drawn through the choanocyte. The particles are then devoured by endocytosis.

The body of a sponge is lined with cells called choanocytes and is perforated by many pores. The beating flagella of the choanocytes draw water in through the pores, through the sponge, and eventually out through the osculum.

Cnidarians: Symmetry and Tissues

LEARNING OBJECTIVE 18.3.2 Identify the key body innovations of cnidarians, and contrast their two basic body forms.

All animals other than sponges have both symmetry and tissues. The most primitive are organized around a central axis like the petals of a daisy. Radial symmetry offers advantages to these animals, as their bodies—attached to the surface or free-floating—don't pass through the environment, but rather interact with it on all sides. Called Cnidaria (pronounced ni-DAH-ree-ah), this phylum includes hydra (figure 18.3*a*), jellyfish (figure 18.3*b*), corals, and sea anemones. The bodies of all other animals with tissues are marked by a fundamental bilateral symmetry.

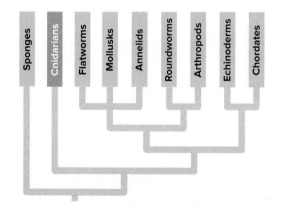

Harpoon Hunters. Cnidarians (phylum Cnidaria) are carnivores that capture prey such as fishes and shellfish with tentacles that ring their mouths. These tentacles, and sometimes the body surface, bear unique stinging cells called **cnidocytes** that give the phylum its name. Within each cnidocyte is a small but powerful harpoon called a **nematocyst** that cnidarians use to spear their prey and then draw the harpooned prey back. The cnidocyte builds up a very high internal osmotic pressure and uses it to push the nematocyst outward so explosively that the barb can penetrate the hard shell of a crab.

Figure 18.3 **Cnidarian body forms.**

(a) Hydroids are a group of cnidarians that are mostly marine and colonial. However, Hydra, shown above, is a freshwater genus whose members exist as solitary polyps. (b) Jellyfish are translucent, marine cnidarians. Together, (c) corals and (d) sea anemones comprise the largest group of cnidarians.

(a) Ted Kinsman/Science Source; (b) Fuse/Getty Images; (c) Darryl Leniuk/Getty Images; (d) Allan Bergmann Jensen/Alamy Stock Photo

Eating Big. A major evolutionary innovation that arose in the Radiata is *extracellular digestion* of food. In sponges, food is taken directly into cells by endocytosis and digested. In cnidarians, food enters a cavity, the *gastrovascular cavity,* where digestive enzymes break it down; the products of this digestion are then absorbed by cells that line the cavity. For the first time, it became possible to digest an animal larger than oneself.

Two Body Forms. Cnidarians have two basic body forms. **Medusae** are free-floating, gelatinous, umbrella-shaped forms (see figure 18.3*b*). Their mouths point downward, with a ring of tentacles hanging down around the edges. Medusae are commonly called "jellyfish" because of their gelatinous interior. **Polyps** are cylindrical, pipe-shaped forms that usually attach to a rock. *Hydra* (see figure 18.3*a*), sea anemones, and corals are examples of polyps. For shelter and protection, corals deposit an external "skeleton" of calcium carbonate within which they live. This is the structure usually identified as coral. Many cnidarians exist only as medusae, others only as polyps, and still others alternate between these two phases during the course of their life cycles.

> **Putting the Concept to Work**
> How do you suppose cnidarians manage to avoid stinging themselves?

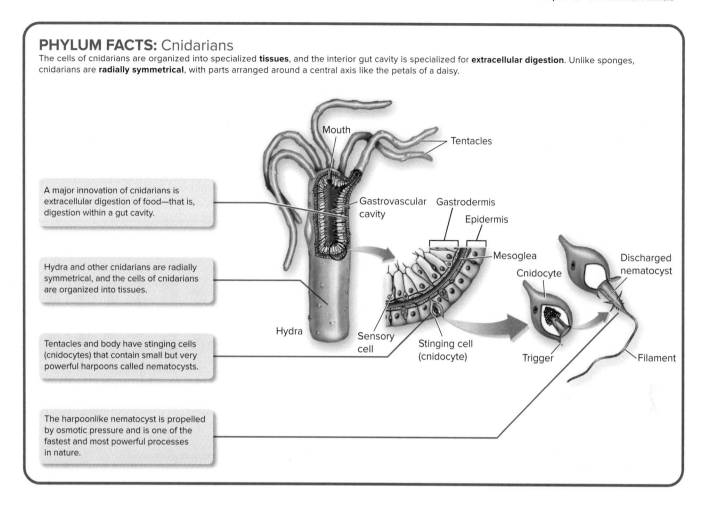

18.4 Advent of Bilateral Symmetry

> **LEARNING OBJECTIVE 18.4.1** Contrast radial and bilateral symmetry.

All animals with tissues, other than cnidarians and ctenophores, are **bilaterally symmetrical**—that is, they have a right half and a left half that are mirror images of each other. This is apparent when you compare the radially symmetrical sea anemone, as shown in **figure 18.4** with the bilaterally symmetrical squirrel. Any of the three planes that cut the sea anemone in half produce mirror images, but only one plane, the green sagittal plane, produces mirror images of the squirrel. In looking at a bilaterally symmetrical animal, you refer to the top half of the animal as **dorsal** and the bottom half as **ventral.** The front is called **anterior** and the back **posterior.** Bilateral symmetry was a major evolutionary advance among the animals because it allows different parts of the body to become specialized in different ways. For example, most bilaterally symmetrical animals have evolved a definite head end. Animals that have heads are often active and mobile, moving through their environment headfirst, with sensory organs concentrated in front so the animal can test for food, danger, and mates, as it enters new surroundings.

The bilaterally symmetrical animals produce three embryonic layers that develop into the tissues of the body: an outer **ectoderm** (colored blue in the drawing of a solid worm in **figure 18.5**), an inner **endoderm** (colored yellow), and a third layer, the **mesoderm** (colored red), between the ectoderm and endoderm. In general, the outer coverings of the body and the nervous system develop from the ectoderm, the digestive organs and intestines develop from the endoderm, and the skeleton and muscles develop from the mesoderm.

> **Putting the Concept to Work**
> What embryonic layer produces the skin of all bilaterally symmetrical animals?

Flatworms

> **LEARNING OBJECTIVE 18.4.2** Describe the key body innovations of flatworms.

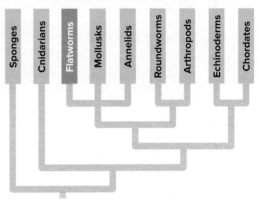

The simplest of all bilaterally symmetrical animals are the solid worms. By far the largest phylum of these, with about 20,000 species, is Platyhelminthes (pronounced plat-ee-hel-MIN-theeze), which includes the **flatworms** (figure 18.7). Flatworms are the simplest animals in which organs occur. An organ is a collection of different tissues that function as a unit.

A Solid Body. Flatworms lack any internal cavity other than the digestive tract, and they have soft bodies, flattened from top to bottom like a piece of tape or ribbon. If you were to cut a flatworm in half across its body, as in figure 18.5, you would see that the gut is completely surrounded by tissues and organs. This solid body construction is called **acoelomate,** meaning without a body cavity.

Although flatworms have a simple body design, they do have a definite head at the anterior end, and they possess organs. Flatworms range in size from a millimeter or less to many meters long, as in some tapeworms.

Most species of flatworms are parasitic, occurring within the bodies of many other kinds of animals. Other flatworms are free-living, occurring in a wide variety of marine and freshwater habitats (figure 18.6), as well as moist places on land. Free-living flatworms eat various small animals and bits of organic debris. They move from place to place by means of ciliated epithelial cells concentrated on their ventral surfaces.

How a Flatworm Eats. Flatworms that have a digestive cavity have a gut with only one opening, a muscular tube called the pharynx. Through this opening, food is taken in as well as waste material expelled. Thus, flatworms cannot feed continuously, as more advanced animals can. The gut is branched and extends throughout the body, functioning in both digestion and the transport of food. Cells that line the gut engulf most of the food particles by phagocytosis and digest them. Parasitic flatworms lack digestive systems and absorb their food through their body walls.

Unlike cnidarians, flatworms have an excretory system, which consists of a network of fine tubules that runs throughout the body. Cilia line the hollow centers of bulblike **flame cells,** which are located on the side branches of the tubules. Cilia in the flame cells move water and excretory substances into the

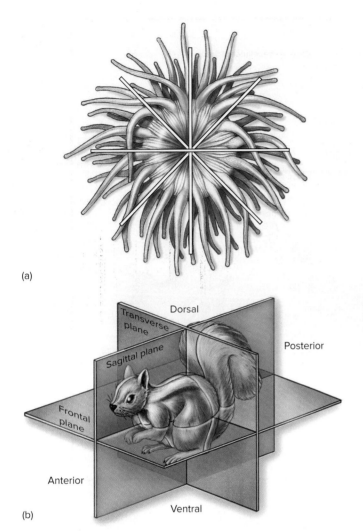

Figure 18.4 Radial and bilateral symmetry.

(a) Radial symmetry is the regular arrangement of parts around a central axis. (b) Bilateral symmetry is reflected in a body form that has a left and right half.

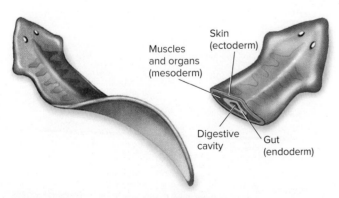

Figure 18.5 Body plan of a solid worm.

All bilaterally symmetrical animals produce three layers during embryonic development: an outer ectoderm, a middle mesoderm, and an inner endoderm. These layers differentiate to form the skin, muscles and organs, and gut, respectively, in the adult animal.

tubules and then out through exit pores located between cells on the epidermis (outer body layer). More metabolic wastes diffuse directly into the gut. Flame cells were named because of the flickering movements of the tuft of cilia within them. They also play a major role in regulating the body's water balance.

Body Organization. Like sponges, cnidarians, and ctenophores, flatworms lack a circulatory system. Instead, flatworms have thin bodies and highly branched digestive cavities, so that all flatworm cells are within diffusion distance of oxygen and food. The nervous system of flatworms is very simple, a longitudinal nerve cord that acts as a simple central nervous system. Between two longitudinal cords are cross connections, so the flatworm nervous system resembles a ladder. Free-living flatworms also have eyespots on their heads. These are inverted, pigmented cups containing light-sensitive cells connected to the nervous system. These eyespots enable the worms to distinguish light from dark. Flatworms are far more active than cnidarians or ctenophores. Such activity is characteristic of bilaterally symmetrical animals.

Unusual Reproduction. The reproductive systems of flatworms are complex. Most flatworms are **hermaphroditic,** with each individual containing both male and female sexual structures. In some parasitic flatworms, there is a complex succession of distinct larval forms. Some genera of flatworms are also capable of asexual regeneration; when a single individual is divided into two or more parts, each part can regenerate an entirely new flatworm.

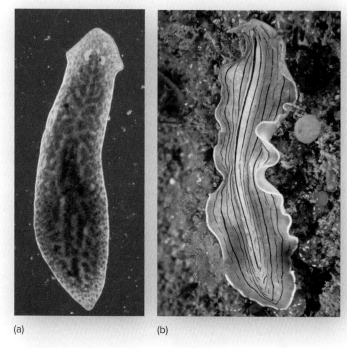

Figure 18.6 Flatworms.
(a) A common flatworm, *Planaria*. (b) A marine free-living flatworm.
(a) M. I. Walker/NHPA/Photoshot; (b) Charles Stirling (Diving)/Alamy Stock Photo

Putting the Concept to Work
Why do you suppose the bodies of flatworms that are capable of asexual regeneration possess male and female sexual structures?

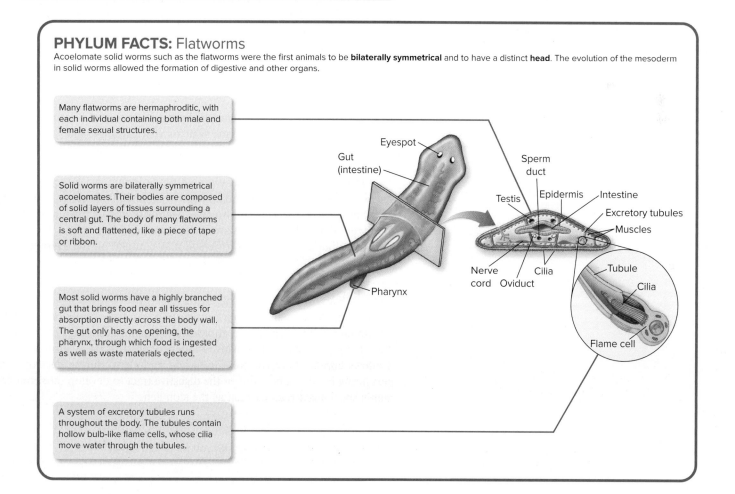

PHYLUM FACTS: Flatworms
Acoelomate solid worms such as the flatworms were the first animals to be **bilaterally symmetrical** and to have a distinct **head**. The evolution of the mesoderm in solid worms allowed the formation of digestive and other organs.

Many flatworms are hermaphroditic, with each individual containing both male and female sexual structures.

Solid worms are bilaterally symmetrical acoelomates. Their bodies are composed of solid layers of tissues surrounding a central gut. The body of many flatworms is soft and flattened, like a piece of tape or ribbon.

Most solid worms have a highly branched gut that brings food near all tissues for absorption directly across the body wall. The gut only has one opening, the pharynx, through which food is ingested as well as waste materials ejected.

A system of excretory tubules runs throughout the body. The tubules contain hollow bulb-like flame cells, whose cilia move water through the tubules.

18.5 Changes in the Body Cavity

> **LEARNING OBJECTIVE 18.5.1** Distinguish acoelomate, pseudocoelomate, and coelomate.

A key innovation in the evolution of the animal body plan was the evolution of the body cavity. All bilaterally symmetrical animals other than solid worms have a cavity within their body. The evolution of an internal body cavity was an important improvement in body design for three reasons:

1. **Circulation.** Fluids that move within the body cavity can serve as a circulatory system, permitting the rapid passage of materials from one part of the body to another and opening the way to larger bodies.
2. **Movement.** Fluid in the cavity makes the animal's body rigid, permitting resistance to muscle contraction and thus opening the way to muscle-driven body movement.
3. **Organ function.** In a fluid-filled enclosure, body organs can function without being deformed by surrounding muscles. For example, food can pass freely through a gut suspended within a cavity, at a rate not controlled by when the animal moves.

Kinds of Body Cavities

There are three basic kinds of body plans found in bilaterally symmetrical animals (**figure 18.7**): (1) **Acoelomates**, such as the solid flatworms that we discussed in section 18.4, have no body cavity. (2) **Pseudocoelomates** have a body cavity called the *pseudocoel* located between the mesoderm and endoderm. (3) **Coelomates** have a fluid-filled body cavity that develops entirely within the mesoderm. Such a body cavity is called a *coelom*.

Pseudocoelomates

Several animal phyla are characterized by the possession of a pseudocoel. The pseudocoel serves as a hydrostatic skeleton that gains its rigidity from being filled with fluid under pressure. The animal's muscles can work against this "skeleton," making its movements more efficient than those of acoelomates.

Only one of the pseudocoelomate phyla includes a large number of species. This phylum, Nematoda, includes some 20,000 recognized species of nematodes and other roundworms. Members of this phylum are found everywhere. Many nematodes are microscopic and live in soil. It has been estimated that a spadeful of fertile soil may contain, on the average, a million nematodes.

Coelomates

The bulk of the animal kingdom consists of coelomates (mollusks, annelids, arthropods, echinoderms, and chordates). A major advantage of the coelomate body plan, which has evolved independently in many groups, is that it allows contact between mesoderm and endoderm during development, permitting localized portions of the digestive tract to develop into complex, highly specialized regions such as the stomach.

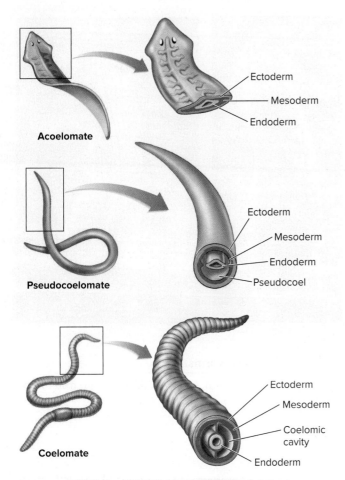

Figure 18.7 Three body plans for bilaterally symmetrical animals.

Acoelomates, such as flatworms, have no body cavity between the digestive tract (endoderm) and the outer body layer (ectoderm). Pseudocoelomates have a body cavity, the pseudocoel, between the endoderm and the mesoderm. Coelomates have a body cavity, the coelom, that develops entirely within the mesoderm and so is lined on both sides by mesoderm tissue.

> **Putting the Concept to Work**
> What sort of body cavity do you think you have?

Mollusks

> **LEARNING OBJECTIVE 18.5.2** Identify the key body innovation of mollusks, and describe the three parts of the mollusk body.

The **mollusks** (phylum Mollusca) are the second largest animal phylum, with over 110,000 species. Mollusks, while mostly marine, are found nearly everywhere.

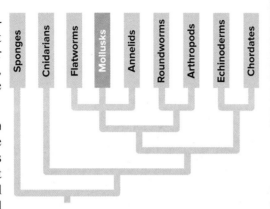

The Complex Body of a Mollusk. Mollusks are coelomates with bodies composed of three distinct parts: a head-foot, a central section called the visceral mass that contains the body's organs, and a mantle. The foot of a mollusk is muscular and may be adapted for locomotion, attachment, food capture, or various combinations of these functions. The **mantle** is a heavy fold of tissue wrapped around the visceral mass like a cape, with the respiratory organs (gills or lungs) positioned on its inner surface like the lining of a coat. The three major groups of mollusks are the *gastropods* (snails and slugs), *bivalves* (clams, oysters, scallops, and mussels), and *cephalopods* (octopuses, squids, and nautiluses) (**figure 18.8**). All terrestrial mollusks are gastropods, in which the mantle often secretes a hard shell. Bivalves secrete a two-part shell with a hinge and filter-feed by drawing water into their shell. Cephalopods have a large brain and a modified mantle cavity that creates a jet propulsion system.

Nephridia. Mollusks were among the first animals to develop an efficient excretory system. Tubular structures called *nephridia* (a type of kidney) gather wastes from the coelom to discharge into the mantle cavity. Mollusks and all other coelomates also have a *circulatory system*, a network of vessels that carries fluids, oxygen, and food molecules to all parts of the body. The circulating fluid is usually pushed through the circulatory system by contraction of one or more heart-like muscular pumps.

The Radula. One of the most characteristic features of gastropods and cephalopods is the **radula**, a rasping, tongue-like organ. With rows of pointed, backward-curving teeth, the radula is used by some snails to scrape algae off rocks. The small holes often seen in oyster shells are produced by gastropods that have bored holes to kill the oyster and extract its body.

Pearls. Pearls are formed when a foreign object, such as a grain of sand, becomes lodged between the mantle and the inner shell layer of a bivalve, including clams and oysters. The mantle coats the foreign object with layer upon layer of shell material to reduce irritation. The shell serves primarily for protection, with some mollusks withdrawing into their shells when threatened.

> **Putting the Concept to Work**
> Explain how an octopus and a clam have the same basic body design.

(a)

(b)

(c)

Figure 18.8 Mollusks.

(a) A gastropod. (b) A bivalve. (c) A cephalopod.

(a) janpietruszk/123RF; (b) Comstock Images/Getty Images; (c) Juniors Bildarchiv GmbH/Alamy Stock Photo

PHYLUM FACTS: Mollusks

The body cavity of a mollusk is a **coelom**, completely enclosed within the mesoderm. This allows physical contact between the mesoderm and the endoderm, permitting interactions that lead to the development of highly specialized organs.

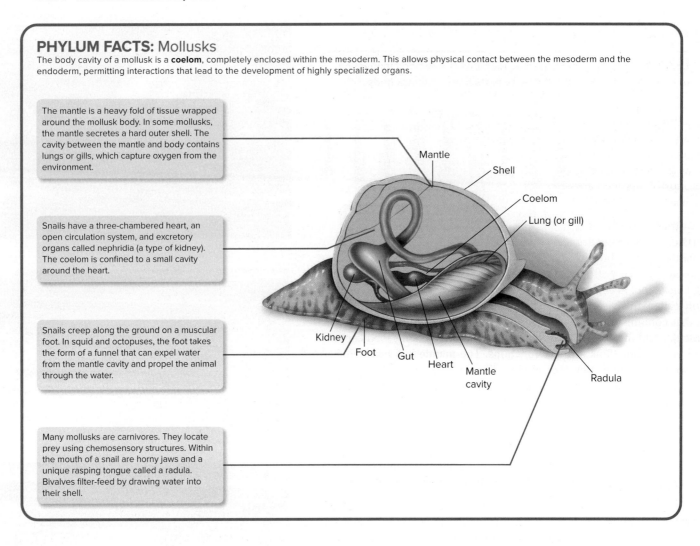

The mantle is a heavy fold of tissue wrapped around the mollusk body. In some mollusks, the mantle secretes a hard outer shell. The cavity between the mantle and body contains lungs or gills, which capture oxygen from the environment.

Snails have a three-chambered heart, an open circulation system, and excretory organs called nephridia (a type of kidney). The coelom is confined to a small cavity around the heart.

Snails creep along the ground on a muscular foot. In squid and octopuses, the foot takes the form of a funnel that can expel water from the mantle cavity and propel the animal through the water.

Many mollusks are carnivores. They locate prey using chemosensory structures. Within the mouth of a snail are horny jaws and a unique rasping tongue called a radula. Bivalves filter-feed by drawing water into their shell.

Annelids

LEARNING OBJECTIVE 18.5.3 Describe the key body innovation of annelids.

A key innovation in body plan that arose in many groups was **segmentation**, the building of a body from a series of similar segments. The first segmented animals to evolve were worms called **annelids**, phylum Annelida (figure 18.9). They are assembled as a chain of nearly identical segments, like the boxcars of a train. The great advantage of such segmentation is the evolutionary flexibility it offers—a small change in an existing segment can produce a new kind of segment with a different function. Thus, segments are modified for reproduction, feeding, and eliminating wastes.

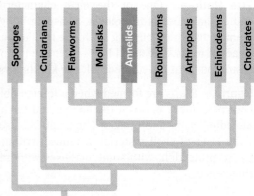

The Segmented Body of an Annelid. Two-thirds of all annelids live in the sea (about 8,000 species), but some live in freshwater, and most of the rest—some 3,100 species—are earthworms. The basic body plan of an annelid is a tube within a tube: The digestive tract is suspended within the coelom, which is itself a tube running from mouth to anus. The body segments of an annelid are visible as a series of ring-like structures running the length of the body, looking like a stack of doughnuts. The segments are divided internally from one another by partitions. In each of the cylindrical segments, the excretory and locomotor organs are repeated. The body fluid within the coelom of each segment creates a hydrostatic (liquid-supported) skeleton that gives the segment rigidity, like an inflated balloon. Because each segment is separate, each is able to expand or contract independently. This lets the worm's body move in ways that are quite complex.

Organizing the Body's Segments. The anterior (front) segments of annelids contain the sensory organs. Elaborate eyes with lenses and retinas have evolved in some annelids. One anterior segment contains a well-developed cerebral ganglion, or brain. The digestive tract, circulatory system, and nervous system are connected between segments. In annelids, a closed circulatory system carries blood from one segment to another through a network of blood vessels. Nerve cords connect the nerve centers located in each segment with each other and the brain. The brain can then coordinate the worm's activities.

Figure 18.9 Annelids.
(a) Earthworms, such as this *Lumbricus terrestris*, are terrestrial annelids. (b) This bristle worm is an aquatic annelid, a polychaete.
(a) Colin Varndell/Getty Images;
(b) Darlyne A. Murawski/Getty Images

Putting the Concept to Work
Most annelids are marine. How do you imagine they move through the water?

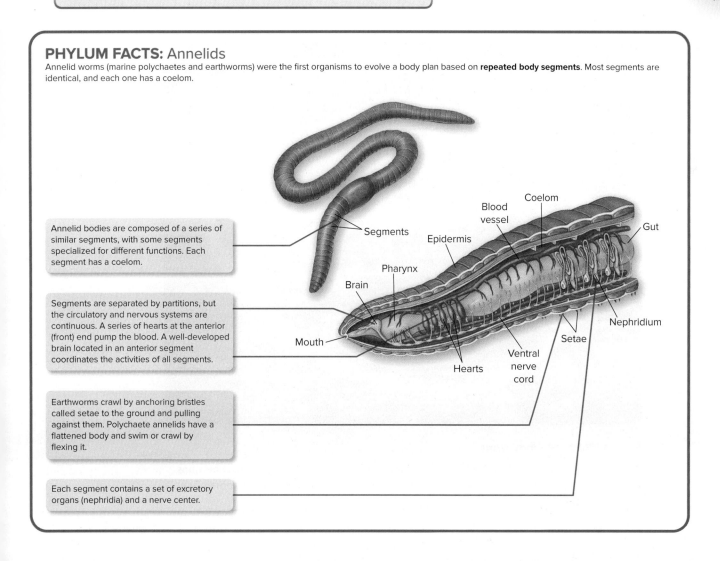

PHYLUM FACTS: Annelids
Annelid worms (marine polychaetes and earthworms) were the first organisms to evolve a body plan based on **repeated body segments**. Most segments are identical, and each one has a coelom.

- Annelid bodies are composed of a series of similar segments, with some segments specialized for different functions. Each segment has a coelom.

- Segments are separated by partitions, but the circulatory and nervous systems are continuous. A series of hearts at the anterior (front) end pump the blood. A well-developed brain located in an anterior segment coordinates the activities of all segments.

- Earthworms crawl by anchoring bristles called setae to the ground and pulling against them. Polychaete annelids have a flattened body and swim or crawl by flexing it.

- Each segment contains a set of excretory organs (nephridia) and a nerve center.

The Roundworms

LEARNING OBJECTIVE 18.5.4 Identify the key body innovation of roundworms.

Roundworms are bilaterally symmetrical, cylindrical, and unsegmented worms (figure 18.10). They are covered by a flexible, thick cuticle, which they shed as they grow, a process called *molting*. Their muscles constitute a layer beneath the epidermis and extend along the length of the worm, rather than encircling its body. These longitudinal muscles attach to the outer layer of the body (figure 18.11) and pull against the cuticle and the pseudocoel, which forms a type of fluid skeleton. When roundworms move, their bodies whip about from side to side.

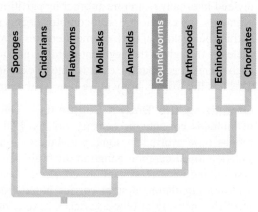

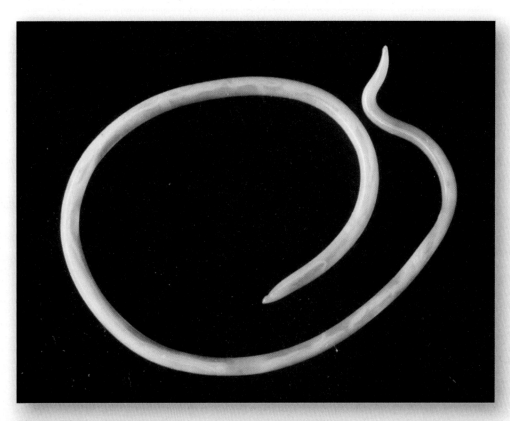

Figure 18.10 Roundworms.

These roundworms (phylum Nematoda) are intestinal roundworms that infect humans and some other animals. Their fertilized eggs pass out with feces and can remain viable in soil for years.

London Scientific Films/Getty Images

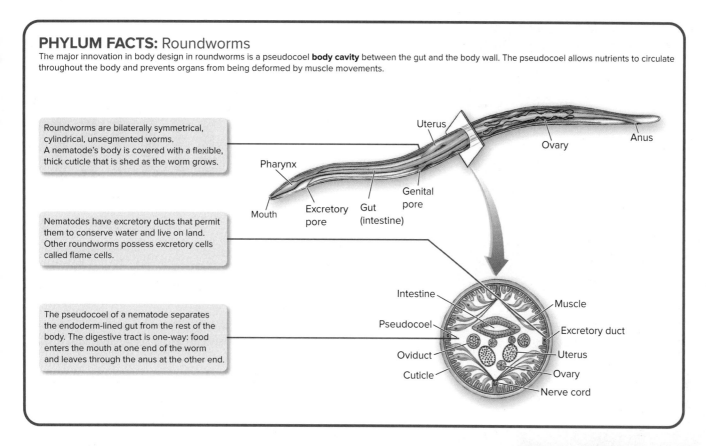

PHYLUM FACTS: Roundworms

The major innovation in body design in roundworms is a pseudocoel **body cavity** between the gut and the body wall. The pseudocoel allows nutrients to circulate throughout the body and prevents organs from being deformed by muscle movements.

Roundworms are bilaterally symmetrical, cylindrical, unsegmented worms. A nematode's body is covered with a flexible, thick cuticle that is shed as the worm grows.

Nematodes have excretory ducts that permit them to conserve water and live on land. Other roundworms possess excretory cells called flame cells.

The pseudocoel of a nematode separates the endoderm-lined gut from the rest of the body. The digestive tract is one-way: food enters the mouth at one end of the worm and leaves through the anus at the other end.

The Body of a Roundworm. Like all pseudocoelomates, roundworms lack a defined circulatory system; this role is performed by the fluids that move within the pseudocoel. Most pseudocoelomates have a complete, one-way digestive tract that acts like an assembly line. Food is broken down, absorbed, and then treated and stored.

How a Roundworm Eats. Near the mouth of a roundworm, at its anterior end, are usually 16 raised, hair-like sensory organs. The mouth is often equipped with piercing organs called *stylets*. Food passes through the mouth as a result of the sucking action of the pharynx. After passing through a short corridor into the pharynx, food continues through the other portions of the digestive tract, where it is broken down and then digested. Some of the water with which the food has been mixed is reabsorbed near the end of the digestive tract, and material that has not been digested is eliminated through the anus.

Roundworm Reproduction. Roundworms completely lack flagella or cilia, even on sperm cells. Reproduction is sexual, with sexes usually separate. Their development is simple, and the adults consist of very few cells. For this reason, the roundworms called nematodes have become extremely important subjects for genetic and developmental studies. The 1-millimeter-long *Caenorhabditis elegans* matures in only three days, its body is transparent, and it has only 959 cells. It is the only animal whose complete developmental cellular anatomy is known and the first animal whose genome (97 million DNA bases encoding over 21,000 different genes) was fully sequenced.

Some roundworms are parasitic in humans, cats, dogs, and other animals. Trichinosis is a roundworm-caused disease in temperate regions. A more prevalent human parasitic roundworm is *Ascaris lumbricoides*.

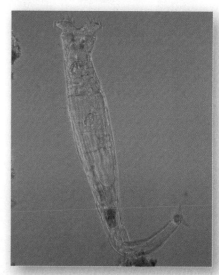

Figure 18.11 Pseudocoelomates: rotifers.

Rotifers (phylum Rotifera) are small, basically aquatic pseudocoelomate animals that have a crown of cilia at their heads; they range from 0.04 to 2 millimeters long.

MELBA PHOTO AGENCY/Alamy Stock Photo

Putting the Concept to Work
Why do you think roundworms periodically shed their cuticle?

(a)

(b) (c)

Figure 18.12 Crustaceans.
(a) Dark-fingered coral crab. (b) Sowbugs, *Porcellio scaber*. (c) Barnacles are sessile animals that permanently attach themselves to a hard substrate.

(a) Comstock/Getty Images; (b) George Grall/Getty Images; (c) MJ Photography/Alamy Stock Photo

Arthropods

> **LEARNING OBJECTIVE 18.5.5** Identify two key body innovations of arthropods.

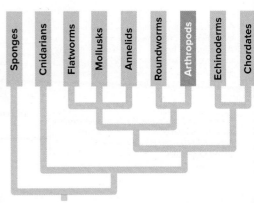

The most successful of all animal groups are the **arthropods,** phylum Arthropoda. Arthropod bodies are segmented like those of annelids. Two-thirds of all named animal species on earth are arthropods, which include the crustaceans (lobsters, crabs, shrimp, and barnacles) (figure 18.12), arachnids (spiders, ticks, mites, scorpions, and daddy longlegs) (figure 18.13), centipedes and millipedes, and insects. About 80% of all arthropod species are insects, and about half of these are beetles!

Jointed Appendages. Arthropods exhibit a profound innovation in body design: jointed appendages. Indeed, the name *arthropod* comes from two Greek words, *arthros,* jointed, and *podes,* feet. All arthropods have jointed appendages. Some are legs, and others may be modified for other uses. To gain some idea of the importance of jointed appendages, imagine yourself without them—no hips, knees, ankles, shoulders, elbows, wrists, or knuckles. Without jointed appendages, you could not walk or grasp an object. Arthropods use jointed appendages as legs for moving, as antennae to sense their environment, as mouthparts for sucking, ripping, and chewing prey, and, very importantly, as wings to fly. Other than vertebrates, insects are the only animals able to fly, an ability that has played a major role in their great evolutionary success.

(a)

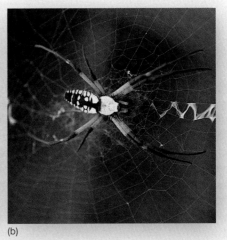

(b)

Figure 18.13 Arachnids.

Spiders are the most ancient and among the most familiar arachnids. (a) One of the poisonous spiders in the United States and Canada is the black widow spider, *Latrodectus mactans*. (b) Another of the poisonous spiders in this area is the brown recluse, *Loxosceles reclusa*. Both species are common throughout temperate and subtropical North America, but they rarely bite humans.

(a) Mark Kostich/Getty Images; (b) dmvphotos/Shutterstock

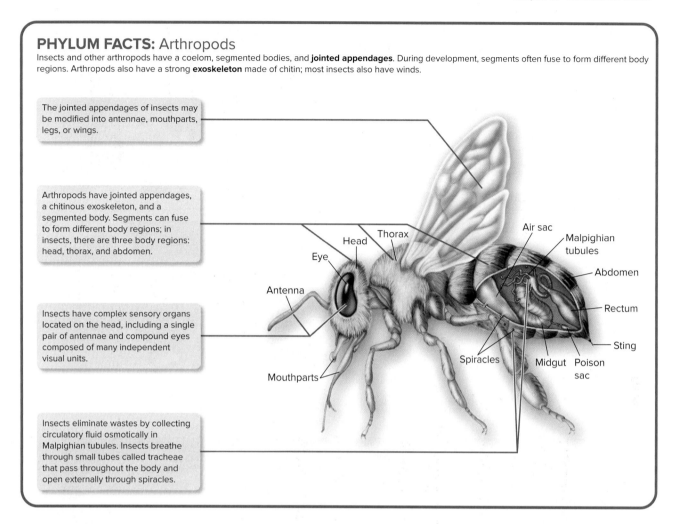

PHYLUM FACTS: Arthropods

Insects and other arthropods have a coelom, segmented bodies, and **jointed appendages**. During development, segments often fuse to form different body regions. Arthropods also have a strong **exoskeleton** made of chitin; most insects also have wings.

- The jointed appendages of insects may be modified into antennae, mouthparts, legs, or wings.

- Arthropods have jointed appendages, a chitinous exoskeleton, and a segmented body. Segments can fuse to form different body regions; in insects, there are three body regions: head, thorax, and abdomen.

- Insects have complex sensory organs located on the head, including a single pair of antennae and compound eyes composed of many independent visual units.

- Insects eliminate wastes by collecting circulatory fluid osmotically in Malpighian tubules. Insects breathe through small tubes called tracheae that pass throughout the body and open externally through spiracles.

Rigid Exoskeleton. The arthropod body plan has a second great innovation: a rigid external skeleton, or **exoskeleton,** made of chitin. As the body grows, its hard exoskeleton is discarded in the process of molting, to be replaced by a new larger one. In arthropods, the muscles attach to the interior surface of this hard chitin shell, which also protects the animal from predators and impedes water loss.

However, though chitin is hard and tough, it is also brittle and cannot support great weight. As a result, the exoskeleton must be much thicker to bear the pull of the muscles in large insects than in small ones, so there is a limit to how big an arthropod body can be. The great majority of arthropod species consist of small animals—mostly about a millimeter in length—but members of the phylum range in adult size from about 80 micrometers long (some parasitic mites) to 3.6 meters across (a gigantic crab found in the sea off Japan).

> **Putting the Concept to Work**
>
> **Is there anything in your body design that would preclude your being able to fly?**

Figure 18.14 Insect diversity.

(a) Some insects have a tough exoskeleton, such as this stag beetle (order Coleoptera). There are more kinds of beetles than all other insects put together. (b) A flea (order Siphonaptera). Fleas are flattened laterally, slipping easily through hair. (c) The honeybee, *Apis mellifera* (order Hymenoptera), is a widely domesticated and efficient pollinator of flowering plants. (d) This dragonfly (order Odonata) has a fragile exoskeleton. (e) A true bug, *Edessa rufomarginata* (order Hemiptera), in Panama. (f) Copulating grasshoppers (order Orthoptera). (g) Luna moth, *Actias luna*, in Virginia. Luna moths and their relatives the butterflies are among the most spectacular of insects (order Lepidoptera).

(a) Simon Murrell/IT Stock Free/Alamy Stock Photo; (b) Cosmin Manci/Shutterstock; (c) IT Stock/age fotostock; (d) ChatchawalPhumkaew/iStock/Getty Images; (e) Paul Harcourt Davies/Science Source; (f) NHPA/James Carmichael, Jr.; (g) StevenRussellSmithPhotos/Shutterstock

18.6 Redesigning the Embryo

> **LEARNING OBJECTIVE 18.6.1** Contrast the embryonic development of protostomes and deuterostomes.

There are two major kinds of bilaterally symmetrical animals representing two distinct evolutionary lines.

Protostomes

All of the animals we have met so far have essentially the same kind of embryonic development. Cell divisions of the fertilized egg produce a hollow ball of cells, a blastula, which indents to form a two-layer-thick ball with a blastopore opening to the outside. In mollusks, annelids, and arthropods, the mouth (stoma) develops from or near the blastopore. An animal whose mouth develops in this way is called a **protostome** (figure 18.15, *column on left*). If such an animal has a distinct anus or anal pore, it develops later in another region of the embryo.

Deuterostomes

A second distinct pattern of embryological development occurs in the echinoderms and the chordates. In these animals, the anus forms from or near the blastopore, and the mouth forms subsequently on another part of the blastula. This group of phyla consists of animals that are called the **deuterostomes** (figure 18.15, *column on right*).

Deuterostomes represent a revolution in embryonic development. In addition to the fate of the blastopore, deuterostomes differ from protostomes in two other features:

1. The progressive division of cells during embryonic growth is called *cleavage*. The cleavage pattern relative to the embryo's polar axis determines how the cells array. In nearly all protostomes, each new cell buds off at an angle oblique to the polar axis. As a result, a new cell nestles into the space between the older ones in a closely packed array. This pattern is called spiral cleavage because a line drawn through a sequence of dividing cells spirals outward from the polar axis, (indicated by the curving blue arrow at the 32-cell stage in figure 18.15).

    In deuterostomes, the cells divide parallel to and at right angles to the polar axis. As a result, the pairs of cells from each division are positioned directly above and below one another; this process gives rise to a loosely packed array of cells. This pattern is called radial cleavage because a line drawn through a sequence of dividing cells describes a radius outward from the polar axis (indicated by the straight blue arrow at the 32-cell stage).

2. In protostomes, the developmental fate of each cell in the embryo is fixed when that cell first appears. Even at the four-celled stage, each cell is different, containing different chemical developmental signals, and no one cell, if separated from the others, can develop into a complete animal. In deuterostomes, on the other hand, the first cleavage divisions of the fertilized egg produce identical daughter cells, and any single cell, if separated, can develop into a complete organism.

> **Putting the Concept to Work**
> Are you a protostome or a deuterostome? Can any cell of your body, if separated, develop into a complete person? Discuss.

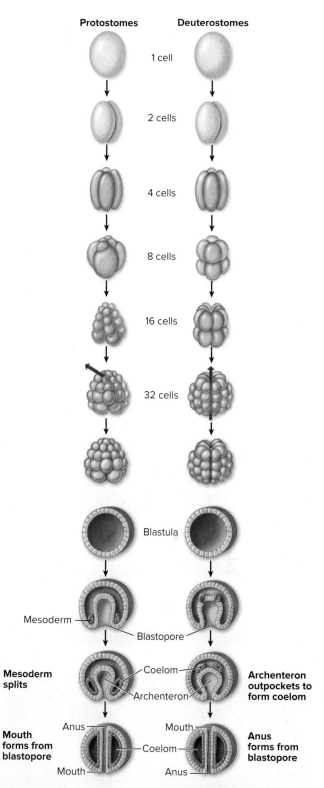

Figure 18.15 Embryonic development in protostomes and deuterostomes.

Cleavage of the egg produces a hollow ball of cells called the blastula. Invagination, or infolding, of the blastula produces the blastopore. In protostomes, embryonic cells cleave in a spiral pattern and become tightly packed. The blastopore becomes the animal's mouth. In deuterostomes, embryonic cells cleave radially and form a loosely packed array. The blastopore becomes the animal's anus.

Echinoderms

> **LEARNING OBJECTIVE 18.6.2** Describe two key body innovations of echinoderms.

The first deuterostomes, marine animals called **echinoderms** in the phylum Echinodermata, appeared more than 650 million years ago.

Advent of the Endoskeleton. The term *echinoderm* means "spiny skin" and refers to an **endoskeleton** composed of hard, calcium-rich plates called 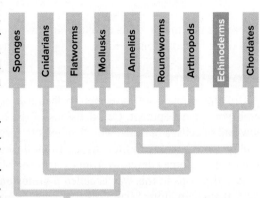 ossicles that lie just beneath the delicate skin. When they are first formed, the plates are enclosed in living tissue, and so are truly an endoskeleton, although in adults they fuse, forming a hard shell. About 6,000 species of echinoderms are living today, almost all of them on the ocean bottom (figure 18.16). Many of the most familiar animals seen along the seashore are echinoderms, including sea stars (starfish), sea urchins, sand dollars, and sea cucumbers.

Figure 18.16 Echinoderms.

(a) Sea star, *Oreaster occidentalis*. (b) Sand dollar, *Echinarachnius parma*. (c) Giant red sea urchin, *Strongylocentrotus franciscanus*.

(a) Ian Cartwright/Getty Images; (b) Jeffrey L. Rotman/Getty Images; (c) NatureDiver/Shutterstock

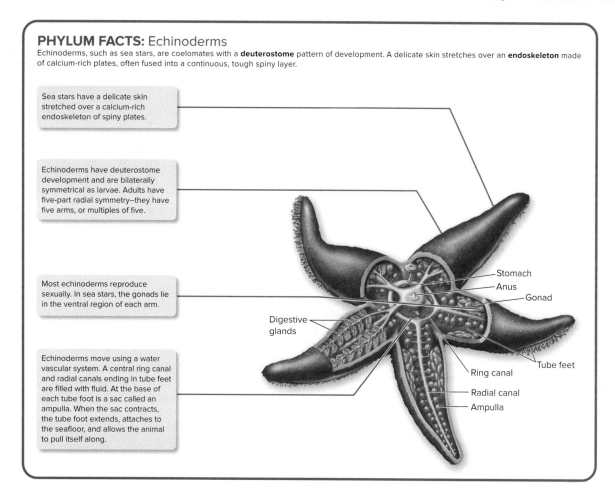

PHYLUM FACTS: Echinoderms
Echinoderms, such as sea stars, are coelomates with a **deuterostome** pattern of development. A delicate skin stretches over an **endoskeleton** made of calcium-rich plates, often fused into a continuous, tough spiny layer.

- Sea stars have a delicate skin stretched over a calcium-rich endoskeleton of spiny plates.
- Echinoderms have deuterostome development and are bilaterally symmetrical as larvae. Adults have five-part radial symmetry—they have five arms, or multiples of five.
- Most echinoderms reproduce sexually. In sea stars, the gonads lie in the ventral region of each arm.
- Echinoderms move using a water vascular system. A central ring canal and radial canals ending in tube feet are filled with fluid. At the base of each tube foot is a sac called an ampulla. When the sac contracts, the tube foot extends, attaches to the seafloor, and allows the animal to pull itself along.

A Shift in Symmetry. The body plan of echinoderms undergoes a fundamental shift during development: All echinoderms are bilaterally symmetrical as larvae but become radially symmetrical as adults. Many biologists believe that early echinoderms were sessile and evolved adult radial symmetry as an adaptation to the sessile existence. Adult echinoderms have a five-part body plan, easily seen in the five arms of a sea star. Some echinoderms such as feather stars have 10 or 15 arms, but always multiples of five. The nervous system consists of a central ring of nerves from which radial branches arise—while the animal is capable of complex response patterns, there is no centralization of function, no "brain." Most echinoderms reproduce sexually, but they have the ability to regenerate lost parts, which can lead to asexual reproduction.

Using Hydraulics to Move. A key evolutionary innovation of echinoderms is the development of a hydraulic system to aid movement. Called a **water vascular system,** this fluid-filled system is composed of a central ring canal from which radial canals extend out. From each radial canal, tiny vessels extend through short side branches into thousands of tiny, hollow tube feet. At the base of each tube foot is a fluid-filled muscular sac, or ampulla, that acts as a valve. When a sac contracts, its fluid is prevented from reentering the radial canal and instead is forced into the tube foot, thus extending it. When extended, the tube foot attaches itself to the ocean bottom, often aided by suckers. An echinoderm can then pull against these tube feet and so haul itself over the seafloor.

> **Putting the Concept to Work**
> Sea stars haul themselves over the ocean floor. Why do you think evolution has not favored their being bilaterally symmetrical?

Chordates

LEARNING OBJECTIVE 18.6.3 Identify the key body innovations of chordates.

Chordates (phylum Chordata) evolved an endoskeleton that is truly internal, based initially on a flexible rod called a **notochord** that develops along the back of the embryo. Muscles attached to this rod allowed early chordates to swing their bodies back and forth, swimming through the water. This key evolutionary innovation, attaching muscles to an internal element, started chordates along an evolutionary path that leads to the vertebrates and, for the first time, to truly large animals.

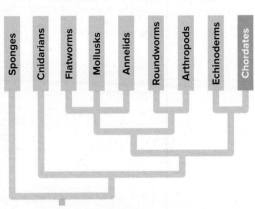

Four Features of All Chordates. The approximately 56,000 species of chordates are distinguished by four principal features, which all chordates have at some time in their lives (figure 18.17):

1. **Notochord.** A stiff but flexible rod that forms beneath the nerve cord in the early human embryo.
2. **Nerve cord.** A single hollow dorsal nerve cord to which the nerves that reach the different parts of the body are attached.

Figure 18.17 Nonvertebrate chordates.

(a) Beautiful blue and gold tunicates. (b) A lancelet, *Branchiostoma lanceolatum*, partly buried in shell gravel, with its anterior end protruding. The muscle segments are clearly visible in this photograph.

(a) Jung Hsuan/Shutterstock; (b) D.P. Wilson/FLPA/Science Source

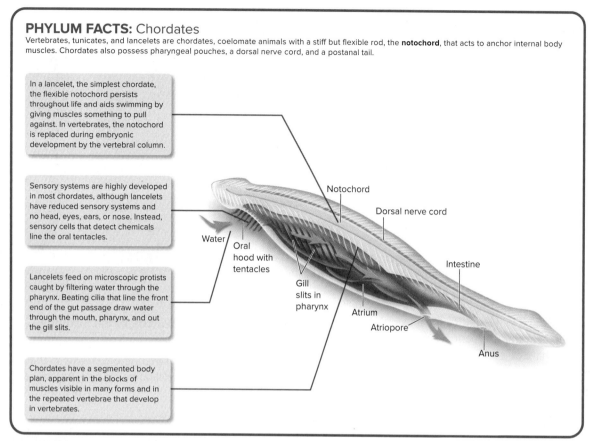

PHYLUM FACTS: Chordates

Vertebrates, tunicates, and lancelets are chordates, coelomate animals with a stiff but flexible rod, the **notochord**, that acts to anchor internal body muscles. Chordates also possess pharyngeal pouches, a dorsal nerve cord, and a postanal tail.

In a lancelet, the simplest chordate, the flexible notochord persists throughout life and aids swimming by giving muscles something to pull against. In vertebrates, the notochord is replaced during embryonic development by the vertebral column.

Sensory systems are highly developed in most chordates, although lancelets have reduced sensory systems and no head, eyes, ears, or nose. Instead, sensory cells that detect chemicals line the oral tentacles.

Lancelets feed on microscopic protists caught by filtering water through the pharynx. Beating cilia that line the front end of the gut passage draw water through the mouth, pharynx, and out the gill slits.

Chordates have a segmented body plan, apparent in the blocks of muscles visible in many forms and in the repeated vertebrae that develop in vertebrates.

Rene Frederic/Pixtal/age fotostock

3. **Pharyngeal pouches.** A series of pouches behind the mouth that develop into slits in some animals.
4. **Postanal tail.** A tail that extends beyond the anus. The pharyngeal pouches and postanal tail disappear during human development, and the notochord is replaced with the vertebral column.

Vertebrates

With the exception of tunicates and lancelets, all chordates are **vertebrates** (figure 18.18). Vertebrates differ from tunicates and lancelets in two important respects:

1. **Backbone.** The notochord becomes surrounded and then replaced during the course of development by a bony vertebral column. The vertebral column is a stack of hollow bones called *vertebrae* that encloses the dorsal nerve cord like a sleeve and protects it.
2. **Head.** All vertebrates except the earliest fishes have a distinct and well-differentiated head, with a skull and brain.

All vertebrates have an internal skeleton made of bone or cartilage against which the muscles work. This endoskeleton makes possible the great size of vertebrates.

> **Putting the Concept to Work**
> What precisely is the difference between a notochord and a nerve cord?

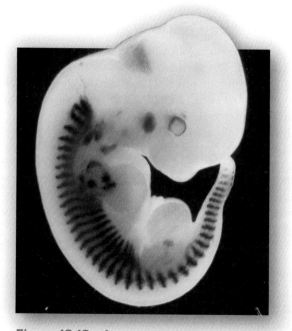

Figure 18.18 A mouse embryo.

At 11.5 days of development, the muscle is already divided into segments called somites (stained dark blue in this photo), reflecting the fundamentally segmented nature of all chordates.

Eric N. Olson/The University of Texas MD Anderson Cancer Center

The Parade of Vertebrates

18.7 Overview of Vertebrate Evolution

> **LEARNING OBJECTIVE 18.7.1** Name the blocks of time scientists use to describe the earth's past, stating in which of them the major animal groups appeared.

The Paleozoic

Virtually all of today's major groups of animals originated in the sea at the beginning of the **Paleozoic era** (545–250 million years ago—M.Y.A.), during or soon after the Cambrian period (545–490 M.Y.A.). Thus, the major diversification of animal life on earth occurred largely in the sea, and fossils from the early Paleozoic era are found in the marine fossil record. Many of the animal phyla that appeared in the Cambrian period, such as the bizarre trilobites that you see in figure 18.19, have no surviving close relatives. Ammonites, shelled, cephalopod mollusks that originated in the Paleozoic era, were among the most abundant creatures on earth 100 million years ago (figure 18.20).

The first vertebrates evolved about 500 million years ago in the oceans—fishes without jaws. They didn't have paired fins either—many of them looked something like a flattened hotdog with a hole at one end and a fin at the other. For over 100 million years, a parade of different kinds of fishes were the only vertebrates on earth. They became the dominant creatures in the sea, some as large as 10 meters, longer than the length of two cars.

Invasion of the Land

The first organisms to colonize the land were fungi and plants, over 500 million years ago. The ancestors of plants were specialized members of a group of photosynthetic protists known as the green algae. It seems probable that plants first occupied the land in symbiotic association with fungi, as discussed in chapter 17.

Only a few of the animal phyla that evolved in the Cambrian seas have invaded the land successfully. The first invasion of the land by animals, and perhaps the most successful invasion of the land, was accomplished by the arthropods, a phylum of hard-shelled animals with jointed legs and a segmented body. This invasion occurred about 410 million years ago.

Vertebrates invaded the land during the Carboniferous period (360–280 M.Y.A.). The first vertebrates to live on land were the amphibians, represented today by frogs, toads, salamanders, and caecilians (legless amphibians). The earliest amphibians known are from the Devonian period. Then about 300 million years ago, the first reptiles appeared. Within 50 million years, the reptiles, better suited than amphibians to living out of water, replaced them as the dominant land animal on earth. The pelycosaur, the fan-backed animal pictured in figure 18.21, was an early reptile. By the Permian period (280 M.Y.A.), the major evolutionary lines on land had been established and were expanding.

The **Mesozoic era** (248–66 M.Y.A.) followed the Paleozoic era and has traditionally been divided into three periods: (1) the Triassic, (2) the Jurassic, and (3) the Cretaceous. The Mesozoic era was a time of intensive evolution of terrestrial plants and animals. With the success of the reptiles, vertebrates truly came to dominate the surface of the earth. Many kinds of reptiles evolved,

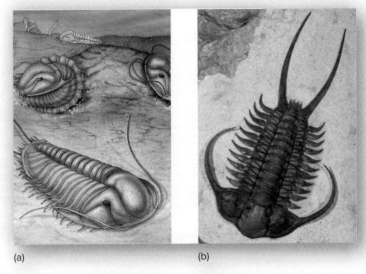

(a) (b)

Figure 18.19 Life in the Cambrian.

(a) Trilobites are shown in this reconstruction of a community of marine organisms in the Cambrian period, 545 to 490 million years ago. (b) A fossil trilobite.

(a) Laurie O'Keefe/Science Source; (b) Aneese/iStock/Getty Images

Figure 18.20 Ammonite fossil.

This shelled mollusk evolved during the Paleozoic era and became the most abundant creature on earth until the end of the Mesozoic era (65 M.Y.A.).

Blackbeck/iStock/Getty Images

Figure 18.21 An early reptile: the pelycosaur.

Leonello Calvetti/Stocktrek Images/Getty Images

from those smaller than a chicken to others even larger than a semitrailer truck (figure 18.22). Dinosaurs dominated the face of the earth for over 150 million years. And then, 66 million years ago, dinosaurs disappeared, along with the flying reptiles (figure 18.23) and the great marine reptiles. This extinction marks the end of the Mesozoic era. During the **Cenozoic era** (66 M.Y.A. to present), new forms of life were able to invade new habitats. Most present-day orders of mammals appeared at this time, a period of great diversity.

Mass Extinctions

The history of life on earth has been marked by occasional profound catastrophes, sudden planetwide exterminations wiping out more than half of all species on earth. Five of these **mass extinctions** have occurred, each ending a major geological era.

1. End-Ordovician extinction, 445 million years ago
2. End-Devonian extinction, 374 million years ago
3. End-Permian extinction, 252 million years ago
4. End-Triassic extinction, 201 million years ago
5. End-Cretaceous extinction, 66 million years ago

In each case, the impact on life was profound. The worst, the end-Permian mass extinction, killed 96% of marine life in the oceans and 75% of life on land—it nearly sterilized life on earth.

What could cause such a mass extinction? The most obvious candidate is collision of the earth with a large asteroid. Just such an asteroid slammed into the earth 66 million years ago, when the dinosaurs disappeared. But this cannot be the whole story, as even larger asteroids hitting earth 214 and 35 million years ago did not lead to mass extinctions. A growing body of evidence suggests that the culprit is large-scale volcanism. Occurring at the same time as each of the last four mass extinctions, massive volcanic eruptions in each case covered millions of square miles with lava, emitting massive amounts of carbon dioxide into the air. This flush of carbon dioxide produced both intense global warming and ocean acidification (CO_2 dissolving in seawater forms carbonic acid).

How close is the correlation of volcanism with mass extinctions? The great end-Permian extinction occurred 251.9 million years ago. The age of the Permian lava, dated with new highly precise uranium radioactive decay techniques, is 252.3 million years old—spot on. Global temperatures are thought to have soared by as much as 7°C per year at the height of the end-Permian eruptions, which blanketed lava over an area of Siberia larger than Western Europe.

The most famous and well-studied extinction, although not as drastic, occurred at the end of the Cretaceous period 66 million years ago, when the dinosaurs went extinct. Here, collision with a huge asteroid 10 kilometers in diameter does seem to have been the trigger, both injecting massive amounts of material into the atmosphere and, more importantly, jarring the earth's crust so extremely as to unlock massive volcanism in far-away India—the lava flows there cover 2 million square kilometers! While volcanic activity in India began several million years before the asteroid hit, new accurate dating reveals that most of India's lava erupted in one stupendous squirt of molten lava ejected from deep inside the earth over less than a million years' time. This lava dates to precisely 66 million years ago.

Figure 18.22 Some dinosaurs were truly enormous.

This young girl is peering into the mouth of the fossil skeleton of a large carnivorous dinosaur. Each tooth is bigger than her hand. Bipedal dinosaurs such as *Tyrannosaurus, Allosaurus,* and *Gigantosaurus* are the largest terrestrial carnivores that have ever lived on earth.

Stephen Wilkes/Getty Images

Figure 18.23 An extinct flying reptile.

Pterosaurs, such as the fossil pictured here, became extinct with the dinosaurs 66 million years ago. Flight has evolved three separate times among vertebrates; however, birds and bats are the only representatives still in existence.

Kevin Schafer/Photolibrary/Getty Images

Putting the Concept to Work

As many as one-fourth of all species will become extinct in the near future, largely due to habitat destruction and global warming. Do you think we are currently experiencing a sixth mass extinction event?

Today's Biology

Fish Out of Water

Vertebrates evolved in the sea, and most vertebrates still live there. The oldest tetrapod (critter with four legs) of which we have any knowledge lived on ocean edges 375 million years ago. With the awkward name *Tiktaalik,* it looked a lot like other fish, but unlike other fish, this guy was a tetrapod, with strong hind legs that let it "walk" in muddy shallows. Early amphibians evolved from it, becoming the first vertebrates to invade land from the sea.

Why Leave the Water?

Why do you suppose *Tiktaalik* ventured out of the water? It could be there was more to eat in the shallows—delicious algae grow on moist rocks. It also might be safer where there are no other fish to eat you. Whatever its reason, 375 million years ago, the choice it made to leave water behind began the long evolutionary journey that has led to us.

But why only once? If this was such a good idea, why aren't fish still trying to do it?

Meet the Blenny

They are.

The speckled, sausage-shaped fish in the photo is the Pacific leaping blenny (*Alticus arnoldorum*). Measuring just a few inches long, this enterprising fish spends its entire life out of water, flopping from rock to rock on the shores of Pacific islands. Blennies have gills like any other fish, but gills don't work very well out of water. Solution? Blennies breathe largely through their skin, much as a frog does today, absorbing oxygen directly from the air. They don't go far from the water's edge, of course, for if they did, they would dry out and die. But they spend their days on land, out of the open water.

At low tide, they can be found swimming in rock pools in the splash zone at water's edge or shuffling along just above it, pausing occasionally to scrape off and eat the algae and bacteria that cover the rocks there.

Rene Frederic/Pixtal/age fotostock

As the tide rises, the blennies use their strong tail fins to "leap" farther up onto higher rocks, avoiding the incoming water. It is as if these fish were afraid of the water.

Testing Why the Leaping Blenny Leaps

They *are* afraid of the water, it seems. Biologists have long suspected that the blenny leaps out of water to escape being eaten. You see, seawater teams with lionfish, flounders, and other predators of small fish, whereas on land a fish need contend only with the occasional hungry gull. To test this idea, researchers went to Rarotonga, a small South Pacific island located about midway between New Zealand and Tahiti. There they found several species of blenny happily living on the rocks by the island's shore. The team fashioned 250 fake plastic blennies, each about 2.5 inches long, painted to look just like the Rarotonga blennies. They put half the mimics on the rocks and half in the water, then watched to see which mimics were more often attacked by predators. After eight days, the research team retrieved all the mimics and for each counted the number of puncture wounds and missing "body parts." While the blennies on land were occasionally hacked to death by hungry seagulls, they were much safer than the blenny mimics placed in the sea. "There were at least three times more attacks on our model blennies placed in the water than there were on the models positioned on land," reported the researchers.

With life far less hostile on the rocks, the Pacific leaping blenny seems to have started once again on the long evolutionary journey onto land. It's still a long way from being a full-fledged land dweller, but the first tentative steps are being taken in the South Pacific as you read this. Who knows how far they will get? We must wait a few million years to see.

18.8 Fishes Dominate the Sea

> **LEARNING OBJECTIVE 18.8.1** Describe three major adaptations of fishes to the challenge of being a predator in water.

A series of key evolutionary advances allowed vertebrates to first conquer the sea and then the land. Figure 18.24 shows a phylogeny of the vertebrates. Branch points in the family tree indicate key adaptations. About half of all vertebrates are **fishes**. The most diverse and successful vertebrate group, they provided the evolutionary base for the invasion of land by amphibians.

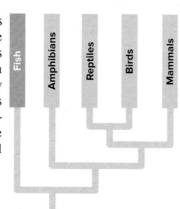

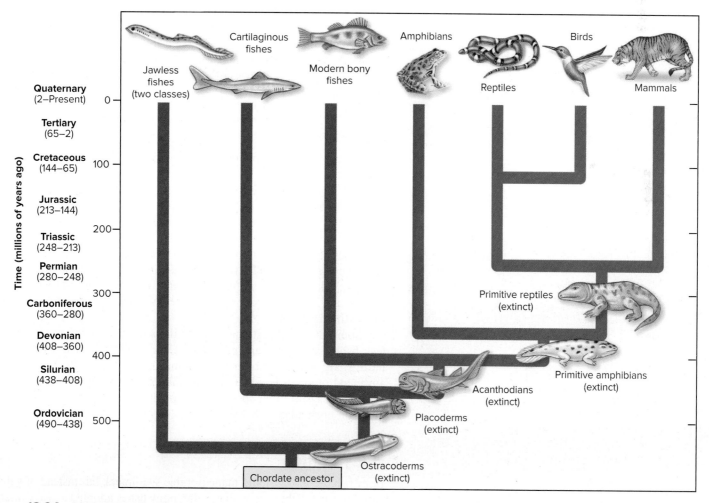

Figure 18.24 **Vertebrate family tree.**

Primitive amphibians arose from lobe-finned fishes. Primitive reptiles arose from amphibians and gave rise in turn to mammals and to dinosaurs, which are the ancestors of today's birds.

Figure 18.25 Chondrichthyes.
The Galápagos shark is a member of the class Chondrichthyes, which are mainly predators and spend most of their time in graceful motion. As they move, they create a flow of water past their gills from which they extract oxygen.

Stephen Frink Collection/Alamy Stock Photo

Characteristics of Fishes

From whale sharks that are 12 meters long to tiny cichlids no larger than your fingernail, fishes vary considerably in size, shape, color, and appearance. However varied, all fishes have four important characteristics in common:

1. **Gills.** Fish are water-dwelling creatures, and they use gills to extract dissolved oxygen gas from the water around them. Gills are fine filaments of tissue rich in blood vessels. When water passes over the gills at the back of the mouth, oxygen gas diffuses from the water into the fish's blood.
2. **Vertebral column.** All fishes have an internal skeleton made of bone or cartilage with a spine surrounding the dorsal nerve cord. The brain is fully encased within a protective case, called the skull or cranium.
3. **Single-loop blood circulation.** Blood is pumped from the heart to the gills. From the gills, the oxygenated blood passes to the rest of the body and then returns to the heart.
4. **Nutritional deficiencies.** Fishes are unable to synthesize the aromatic amino acids and must consume them in their diet. This inability has been inherited by all their vertebrate descendants.

History of Fishes

The first backboned animals were jawless fishes that appeared in the sea about 500 million years ago. Jawless and toothless, these fishes sucked up small food particles from the ocean floor. Early forms of jawed fishes were eventually replaced by fishes that moved through the water faster: the sharks and the bony fishes.

Sharks

Sharks are fast and maneuverable swimmers due to their light and flexible cartilaginous skeleton. Members of this group, the class Chondrichthyes, consist of sharks, skates, and rays. Sharks are very powerful swimmers, with a back fin, a tail fin, and two sets of paired side fins for controlled thrusting through the water (**figure 18.25**). Skates and rays are flattened sharks that are bottom-dwellers. 19 million years ago, sharks underwent a mass extinction with over 90% of sharks disappearing. What happened is obscure. There was no asteroid impact around that time, and bony fishes were unaffected. Sharks never recovered. Today, there are about 750 species of sharks, skates and rays, far fewer than before this extinction event.

Some of the largest sharks filter their food from the water like jawless fishes, but most are predators, their mouths armed with rows of hard, sharp teeth. Reproduction among the Chondrichthyes is the most advanced of any fish. About 40% of sharks, skates, and rays lay fertilized eggs. The eggs of other species develop within the female's body, and the pups are born alive.

Bony Fishes

Bony fish are also fast and maneuverable swimmers, but instead of gaining speed through lightness, as sharks did, bony fishes adopted a heavy internal skeleton made completely of bone. This internal skeleton is very strong, providing a base against which very strong muscles can pull, but it is also heavy. Bony fishes are buoyant because they possess a **swim bladder.** The swim bladder (figure 18.26) is a gas-filled sac that allows fish to regulate their buoyant density and so remain effortlessly suspended at any depth in the water.

Bony fish are the most successful of all fishes, indeed of all vertebrates. Bony fishes (class Osteichthyes) consist of about 30,000 species. The remarkable success of the bony fishes has resulted from a series of significant adaptations. In addition to the swim bladder, they have a highly developed **lateral line system,** a system of sensory cells that run the length of the body on each side that enables them to detect changes in water pressure and thus the movement of predators and prey in the water. Also, most bony fishes have a hard plate called the **operculum** that covers the gills on each side of the head. Flexing the operculum permits bony fishes to pump water over their gills.

> **Putting the Concept to Work**
> Why do you suppose sharks and bony fishes have not one but *two* sets of paired side fins?

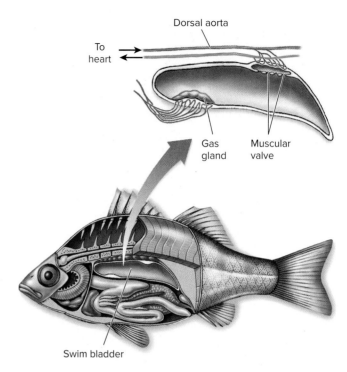

Figure 18.26 Diagram of a swim bladder.

The bony fishes use a swim bladder to control their buoyancy in water. The swim bladder can be filled or drained of gas taken from the blood. The gas gland secretes the gases into the swim bladder; gas is released from the bladder by a muscular valve.

18.9 Amphibians and Reptiles Invade the Land

All land vertebrates have four limbs, which evolved from the fins of ancestors in the sea.

Amphibians

> **LEARNING OBJECTIVE 18.9.1** Describe the five characteristics of amphibians that allowed them to successfully invade land.

Frogs, salamanders, and caecilians, the damp-skinned vertebrates, are direct descendants of fishes. They are the sole survivors of a very successful group, the **amphibians,** the first vertebrates to walk on land. Amphibians likely evolved from a type of bony fish called the lobe-finned fishes, fish with paired fins that consist of a long fleshy muscular lobe supported by a central core of bones that form fully articulated joints with one another. In 2006, a new fossil fish (genus *Tiktaalik*) was discovered that exhibited the transition between fish and amphibians (figure 18.27c).

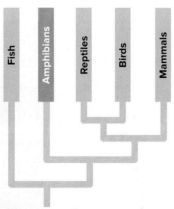

Chordate ancestor

Characteristics of Amphibians. Amphibians have five key characteristics that allowed them to successfully invade the land:

1. **Legs.** Frogs and salamanders have four legs and can move about on land quite well. The way in which legs are thought to have evolved from fins is illustrated in figure 18.27.

2. **Lungs.** Most amphibians possess a pair of lungs, necessary because the delicate structure of fish gills requires the buoyancy of water for support.
3. **Cutaneous respiration.** Frogs, salamanders, and caecilians all supplement the use of lungs by respiring directly across their skin, which is kept moist and provides an extensive surface area (figure 18.28).
4. **Pulmonary veins.** After blood is pumped through the lungs, two large veins called pulmonary veins return the aerated blood to the heart for repumping. This allows aerated blood to be pumped to tissues at a much higher pressure than when it leaves the lungs.
5. **Partially divided heart.** Greater amounts of oxygen are required by muscles for movement and support on land. The chambers of the amphibian heart are separated by a dividing wall that helps prevent aerated blood from the lungs from mixing with nonaerated blood being returned to the heart from the rest of the body. The separation is incomplete, however, and some mixing does occur.

History of Amphibians. Amphibians were the dominant land vertebrates for 100 million years, when 40 families existed. Sixty percent of them were fully terrestrial, with bony plates and armor covering their bodies. Many of these terrestrial amphibians grew to be very large—some as big as a pony! Only a tiny fraction of these have living descendants, all descended from three aquatic families of amphibians that survived competition with reptiles by reinvading the water. Today's 4,850 species of amphibians are classified in three orders: (1) Anura (frogs and toads), (2) Urodela (salamanders and newts), and (3) Apoda (caecilians). Most of today's amphibians are tied to moist if not aquatic environments and must reproduce in water. Many have a tadpole or intermediate larval phase in water, when development pauses for growth before adult limbs develop. In moist habitats, particularly in the tropics, amphibians are often the most abundant vertebrates.

> **Putting the Concept to Work**
> What do you think would be the effect if the separation of amphibian heart chambers were complete?

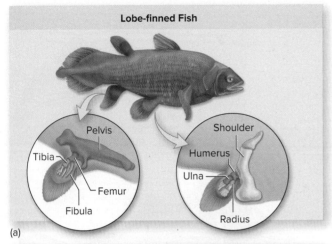
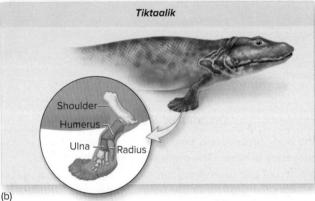
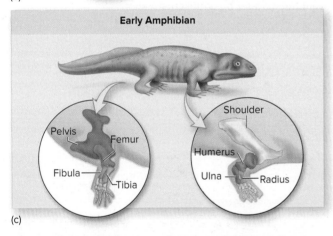

Figure 18.27 A key adaptation of amphibians: the evolution of legs.

In ray-finned fish, the fins contain only bony rays. (a) In lobe-finned fish, the fins have a central core of bones (inside a fleshy lobe) in addition to rays. Some lobe-finned fishes could move out onto land. (b) In the incomplete fossil of *Tiktaalik* (which did not contain the hindlimbs), the shoulder, forearm, and wrist bones were like those of amphibians, but the end of the limb was like that of lobe-finned fishes (importantly, the *Tiktaalik* shoulder bones are separated from the skull—*Tiktaalik* had a neck, which every amphibian and no fish has). (c) In primitive amphibians, the positions of the limb bones are shifted, and bony "toes" are present.

Figure 18.28 Cutaneous respiration.

This red-eyed tree frog, *Agalychnis callidryas*, is a member of the group of amphibians that includes frogs and toads (order Anura). Frogs live in moist environments, allowing for cutaneous respiration.

Paul Souders/Digital Vision/Getty Images

Reptiles

> **LEARNING OBJECTIVE 18.9.2** Describe the features of reptiles that allowed them to replace amphibians as the dominant land animal.

If we think of amphibians as the "first draft" of a manuscript about survival on land, then **reptiles** were the published book.

Characteristics of Reptiles. All living reptiles share certain fundamental characteristics. Among the most important are:

1. **Amniotic egg.** Amphibians never succeeded in becoming fully terrestrial because amphibian eggs must be laid in water to avoid drying out. Most reptiles lay watertight, leathery amniotic eggs that offer various layers of protection from drying out (figure 18.29).
2. **Dry skin.** Many of today's amphibians have a moist skin and must remain in moist places to avoid drying out. In all reptiles, a layer of scales or armor covers their bodies, preventing water loss.
3. **Thoracic breathing.** Amphibians breathe by squeezing their throat to pump air into their lungs; this limits their breathing capacity to the volume of their mouth. Reptiles developed thoracic breathing, expanding the rib cage to suck air into the lungs and then force it out.

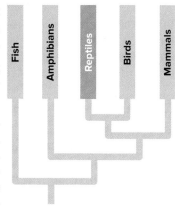

Chordate ancestor

History of Reptiles. Reptiles first evolved about 300 million years ago when the world was entering a long, dry period. Early reptiles rose to prominence and also gave rise to the ancestor of mammals. Reptiles continued their diversification over millions of years, giving rise to many different forms, including marine reptiles (plesiosaurs and ichthyosaurs), flying reptiles (pterosaurs), dinosaurs, and their descendants the birds (figure 18.30). Dinosaurs became the most successful of all land vertebrates until their abrupt extinction around 66 million years ago.

Today some 7,000 species in the class are found in practically every wet and dry habitat on earth. Modern reptiles include four groups: (1) turtles and tortoises, (2) crocodiles and alligators, (3) snakes and lizards, and (4) tuataras.

> **Putting the Concept to Work**
> Construct an argument that reptiles are basically desert animals.

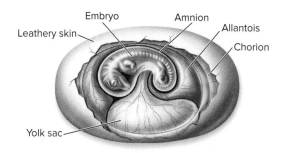

Figure 18.29 A key adaptation of reptiles: watertight eggs.

The watertight amniotic egg allows reptiles to live in a wide variety of terrestrial habitats.

Figure 18.30 An extinct ancestor of dinosaurs.

Euparkeria, a thecodont reptile, had rows of bony plates along the sides of the backbone, as seen in modern crocodiles and alligators. Thecodonts gave rise to crocodiles, dinosaurs, pterosaurs (flying reptiles), and birds.

The Natural History Museum/Alamy Stock Photo

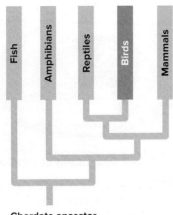

18.10 Birds Master the Air

> **LEARNING OBJECTIVE 18.10.1** Describe three characteristics that distinguish birds from living reptiles.

Birds evolved from small bipedal dinosaurs about 150 million years ago. Birds are so structurally similar to dinosaurs that some biologists consider them "feathered dinosaurs." Most biologists, however, continue to place birds in their own class, Aves, rather than lumping them in with the reptiles.

Characteristics of Birds

Modern birds lack teeth and have only vestigial tails, but they still retain many reptilian characteristics. For instance, birds lay amniotic eggs, although the shells of bird eggs are hard rather than leathery. Also, reptilian scales are present on the feet and lower legs of birds. What makes birds unique? What distinguishes them from living reptiles?

1. **Feathers.** Derived from reptilian scales, feathers serve two functions: providing lift for flight and conserving heat. Feathers consist of a center shaft with barbs extending out (**figure 18.31**). The barbs are held together with secondary branches called barbules that hook over each other. This reinforces the structure of the feather without adding much weight to it. Like scales, feathers can be replaced. Among living animals, feathers are unique to birds. Several types of dinosaurs also possessed feathers; however, these feathers were most likely used for insulation or display.

2. **Flight skeleton.** The bones of birds are thin and hollow. Many of the bones are fused, making the bird skeleton more rigid than a reptilian skeleton and forming a sturdy frame that anchors muscles during flight. The power for active flight comes from large breast muscles that can make up 30% of a bird's total body weight. They stretch down from the wing and attach to the breastbone, which is greatly enlarged and bears a prominent keel for muscle attachment. They also attach to the fused collarbones that form the so-called wishbone. No other living vertebrates have a fused collarbone or a keeled breastbone.

Birds are endothermic. Like mammals, they generate heat through metabolism to create and maintain a high body temperature. Birds maintain body temperatures significantly higher than mammals. This speeds metabolism, necessary to satisfy the large energy requirements of flight.

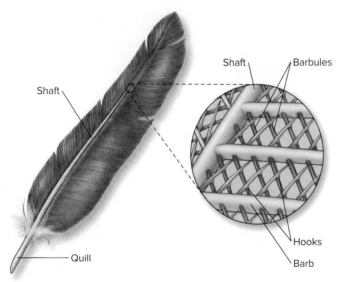

Figure 18.31 A key adaptation of birds: feathers.
The barbs off the main shaft of a feather have secondary branches called barbules that attach to one another by microscopic hooks.

History of Birds

The oldest bird of which there is a clear fossil is *Archaeopteryx* (meaning "ancient wing"; **figure 18.32**), which was about the size of a crow and shared many features with small, bipedal, carnivorous dinosaurs. For example, it had teeth and a long reptilian tail. Its bones were solid, unlike the hollow bones of today's birds, and it had feathers (recently discovered fossils in China show that many species of dinosaurs possessed feathers). By the early Cretaceous, a diverse array of birds had evolved, and they shared the skies with the flying reptiles called pterosaurs for 70 million years. Today about 8,600 species of birds (class Aves) occupy a variety of habitats all over the world.

Figure 18.32 *Archaeopteryx*.
Archaeopteryx lived about 150 million years ago and is the oldest fossil bird. Its fossil feathers contain black pigment granules (melanocytes), indicating the bird was black.
Paul D. Stewart/Science Source

> **Putting the Concept to Work**
> If birds are basically feathered dinosaurs, why do you suppose predatory birds such as hawks have beaks rather than teeth?

18.11 Mammals Adapt to Colder Times

> **LEARNING OBJECTIVE 18.11.1** Describe three characteristics shared by all mammals.

Mammals were the last of the great groups of vertebrates to evolve.

Characteristics of Mammals

The **mammals** (class Mammalia) that evolved about 220 million years ago side by side with the dinosaurs would look strange to you, not at all like modern-day lions and tigers and bears. They shared three key characteristics with mammals today:

1. **Mammary glands.** Female mammals have mammary glands, which produce milk to nurse the newborns. Even baby whales are nursed by their mother's milk. Milk is a very-high-calorie food (human milk has 750 kcal per liter), important because of the high energy needs of a rapidly growing newborn mammal.
2. **Hair.** Among living vertebrates, only mammals have hair (even whales and dolphins have a few sensitive bristles on their snout). A hair is a filament composed of dead cells filled with the protein keratin. The primary function of hair is insulation. The insulation provided by fur may have ensured the survival of mammals when the dinosaurs perished.
3. **Middle ear.** All mammals have three middle-ear bones, which evolved from bones in the reptile jaw. These bones play a key role in hearing by amplifying vibrations created by sound waves that beat upon the eardrum.

History of Mammals

The first mammals arose from therapsids in the late Triassic about 220 million years ago, just as the first dinosaurs evolved from thecodonts. Tiny, shrewlike creatures that ate insects, most mammals were only a minor element in a land that quickly came to be dominated by dinosaurs. Fossils reveal that these early mammals had large eye sockets, evidence that they may have been active at night. At the end of the Cretaceous period, 66 million years ago, when dinosaurs and many other land and marine animals became extinct, mammals rapidly diversified, taking over areas vacated by the extinction of dinosaurs. Mammals reached their maximum diversity about 15 million years ago.

Today, over 4,500 species of mammals occupy all the areas that dinosaurs once claimed, among many others (**figure 18.33**). Present-day mammals have heterodont dentition (different types of teeth that are highly specialized to match particular eating habits), and today's mammals range in size from 1.5-gram shrews to 100-ton whales. Almost half of all mammals are rodents—mice and their relatives. Almost one-quarter of all mammals are bats! Mammals have even invaded the seas, as plesiosaur and ichthyosaur reptiles did so successfully millions of years earlier—79 species of whales and dolphins live in today's oceans. Mammals with binocular vision and opposable thumbs arose about 65 million years ago and eventually gave rise to today's primates, the group to which humans such as ourselves belong.

> **Putting the Concept to Work**
> Which are more like reptiles, mammals or birds?

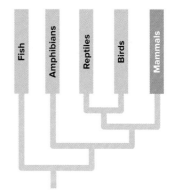

Figure 18.33 Today's mammals.

Today, mammals include the *monotremes* (egg-laying mammals), the *marsupials* (pouched mammals), and the placental mammals (mammals that produce a true placenta). (a) Monotremes include the duck-billed platypus and the echidna, or spiny anteater, such as this *Tachyglossus aculeatus*. (b) Marsupials include kangaroos, such as this adult with young in its pouch. In marsupials, the tiny embryo is born and crawls into the marsupial pouch, where it completes its development. (c) This female African lion, *Panthera leo* (order Carnivora), is a placental mammal. In placental mammals, a specialized organ called the placenta nourishes the embryo throughout its development. Most species of mammals living today, including humans, are placental mammals.

(a) Sir Francis Canker Photography/Getty Images; (b) kjuuurs/Getty Images; (c) Mogens Trolle/Shutterstock

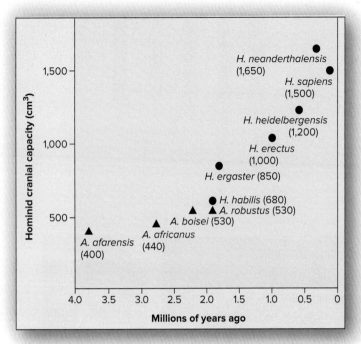

Figure 18.34 Brain size has increased as hominids evolved.

Brain size increased only slowly as *Australopithecus* evolved. The first members of the genus *Homo* had brains similar but slightly larger than those of the australopithecines with whom they shared East African grasslands. Subsequently, brain size increased at a more rapid rate as the genus *Homo* evolved.

18.12 Human Evolution

> **LEARNING OBJECTIVE 18.12.1** Compare and contrast the species of humans.

The first humans (genus *Homo*) evolved from more primitive hominid ancestors of the older, smaller-brained ape-like genus *Australopithecus* about 2 million years ago in Africa.

Homo habilis

In the early 1960s, stone tools were found scattered among hominid bones. Painstaking reconstruction of the many pieces suggested a skull with a brain volume of about 680 cubic centimeters (cc), larger than the australopithecine range of 400 to 550 cubic centimeters (figure 18.34). Because of its association with tools, this early human was called *Homo habilis*, meaning "handy man." Partial skeletons discovered in 1986 indicate that *H. habilis* was small in stature, with arms longer than legs and a skeleton so similar to *Australopithecus* that many researchers at first questioned whether this fossil was human. There is no such doubt about the species that replaced it, *Homo erectus*. Many specimens have been found, and *H. erectus* was without any doubt a true human.

Homo erectus

Homo erectus was a lot larger than *H. habilis*—about 1.5 meters tall. It walked erect, as did *H. habilis,* but had a larger brain, about 1,000 cubic centimeters (see figure 18.34). The cranial capacity of *H. erectus* was about halfway between that of *Australopithecus* and *Homo sapiens,* modern humans. Its skull had prominent brow ridges and, like modern humans, a rounded jaw (figure 18.35). Most interesting of all, the shape of the skull interior suggests that *H. erectus* was able to talk.

Far more successful than *H. habilis*, *H. erectus* quickly became widespread and abundant in Africa and within 1 million years had migrated

Figure 18.35 A hominid evolutionary tree.

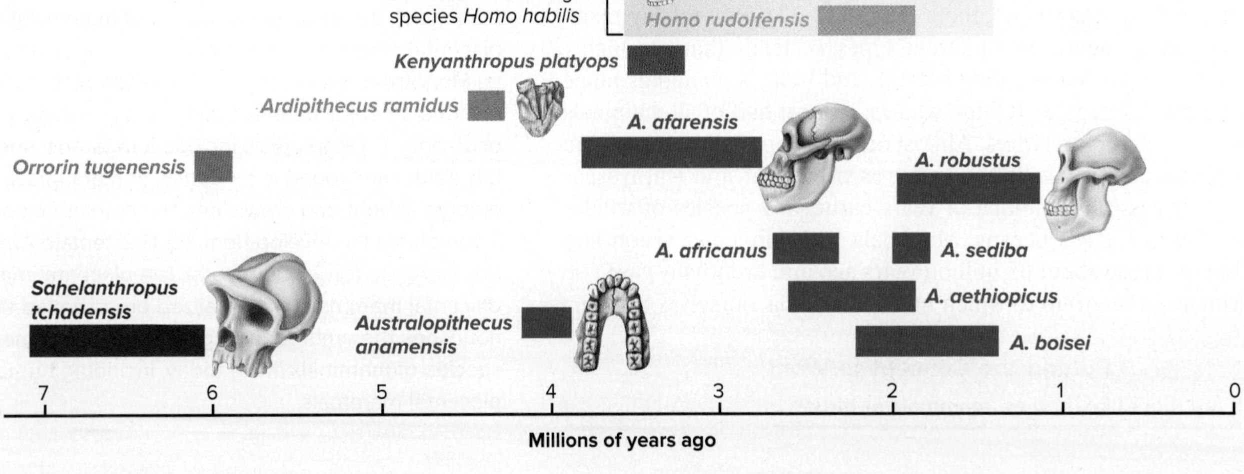

into Asia and Europe. A social species, *H. erectus* lived in tribes of 20 to 50 people, often dwelling in caves and hunting large animals. *H. erectus* survived for over a million years, longer than any other species of human.

Neanderthals

As *Homo erectus* was becoming rarer, about 130,000 years ago, a new species of human, *H. neanderthalensis,* appeared in Europe. Compared to modern humans, Neanderthals were short, stocky, and powerfully built. Their skulls were massive, with protruding faces, heavy, bony ridges over the brows, and larger braincases. The closest evolutionary relatives of present-day humans, Neanderthals were common in large parts of Europe and western Asia 70,000 years ago. The Neanderthals made diverse tools, including scrapers, spearheads, and hand axes. They lived in huts or caves.

In 2010, the bulk of the Neanderthal genome was determined using fragments of DNA extracted from fossil bones of three Neanderthal females who lived in Croatia more than 38,000 years ago. The genome sequence confirms that the Neanderthals have quite different DNA than *Homo sapiens* and were a separate species. Interestingly, comparing this composite Neanderthal genome with the complete genomes of five living humans from different parts of the world, both European and Asian *H. sapiens* today contain about 1% to 4% of sequences inherited from Neanderthals. This offers tantalizing evidence that early modern humans may have interbred with Neanderthals! Although the degree of gene sharing is slight, this suggestion that each of us carries a few Neanderthal genes is sure to provoke controversy.

Homo sapiens

Fossils of *H. neanderthalensis* abruptly disappear from the fossil record about 34,000 years ago and are replaced by fossils of *H. sapiens.* We can only speculate why this sudden replacement occurred, but it was complete all over Europe in a short period. The oldest fossil known of *Homo sapiens,* our own species, is from Africa and is about 130,000 years old. Outside of Africa and the Middle East, there are no clearly dated *H. sapiens* fossils older than roughly 40,000 years of age. The implication is that *H. sapiens* evolved in Africa and then migrated to Europe and Asia (figure 18.36).

Early *H. sapiens* used sophisticated stone tools, had a complex social organization, and are thought to have had full language capabilities (figure 18.37). They lived by hunting large herds of grazing animals.

Humans of modern appearance eventually spread across Siberia to North America, which they reached at least 13,000 years ago, after the ice had begun to retreat and a land bridge still connected Siberia and Alaska. By 10,000 years ago, about 5 million people inhabited the entire world (compared with over 7 *billion* today).

Homo floresiensis

Evidence has begun to accumulate suggesting that until as recently as 50,000 years ago, a fifth species of human existed, hidden away from us on a tiny island in Indonesia. Only 1 meter tall, *Homo floresiensis* has been a startling discovery, as discussed in this chapter's opening.

> **Putting the Concept to Work**
> If scientists are able to extract DNA from *H. floresiensis* bones, which other species of human (or australopithecine) do you think its genome will most closely resemble?

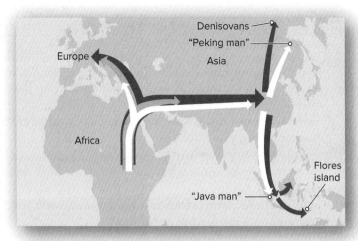

Figure 18.36 Out of Africa—many times.

Several lines of evidence indicate that *Homo* spread from Africa to Europe and Asia repeatedly. First, *H. erectus* (*white arrow*) spread as far as Java and China. Later, *H. heidelbergensis* (*orange arrow*) migrated out of Africa and into Europe and western Asia. And the pattern is repeated again still later by *H. sapiens* (*red arrow*), who migrated out of Africa and into Europe and Asia and eventually into Australia and North America. Lowering of sea levels due to Ice Age coolings made it possible to migrate between landmasses that are currently separated by water. Fossils of *H. neanderthalensis* are known only from Europe and Mediterranean regions, implying that this species may have evolved in Europe. Fossils of Denisovans are known only from Siberia.

Figure 18.37 Europeans, then and now.

The cranial volume of *Homo neanderthalensis* (on the *left*) is typically 1,650 cubic centimeters (cc), and some skulls are as large as 1,740 cc. A typical *H. sapiens* skull (on the *right*) has a cranial volume of 1,500 cc.

John Reader/Science Source

Putting the Chapter to Work

1 In your biology lab class, your instructor shows you a video of a sponge in which water colored with a dye is being drawn through the sponge body cavity. The purpose of this video is to show the filter-feeding ability of the sponge.

What cells are being used by the sponge to propel the water through its filter?

2 During your visit to an aquarium, you observe a cnidarian in the medusae form. It is free-floating, has radial symmetry, and has extracellular digestion.

Out of these characteristics, which one allowed you to differentiate it from the polyp form?

3 Your dog is due for its annual vaccinations. When you bring it to the veterinarian, she positions the dog so that the tail end is facing her to administer the shots.

Because the dog has bilateral symmetry, what plane of symmetry is she viewing?

4 *The developmental stage of which structure differentiates those organisms classified as protostomes and deuterostomes?*

5 In 1974, "Lucy," the first *Australopithecus afarensis* skeleton, was found. Scientists have been able to trace the evolution from the *Australopithecus* genus to our human genus *Homo*.

What difference was observed from the Australopithecus genus to the Homo genus?

Retracing the Learning Path

Introduction to the Animals

18.1 General Features of Animals
1. Animals are complex multicellular heterotrophs. They are mobile and reproduce sexually.

18.2 Six Key Transitions in Body Plan
1. The large diversity of animals can be traced to six key evolutionary changes in body plan: (1) the evolution of tissues, (2) bilateral symmetry, (3) a body cavity, (4) molting, (5) a deuterostome pattern of development, and (6) segmentation.

Evolution of the Animal Phyla

18.3 The Simplest Animals
1. Sponges, in the subkingdom Parazoa, are aquatic and have specialized cells, but lack definite symmetry and organized tissues.
2. Cnidarians have radially symmetrical bodies and tissues. They are carnivores, capturing their prey using harpoonlike nematocysts that are launched from cells called cnidocytes.

18.4 Advent of Bilateral Symmetry
1. Bilaterally symmetrical animals have a right half and a left half that are mirror images of each other.
2. The simplest bilaterally symmetrical animals are the solid worms, which have three embryonic tissue layers but lack a body cavity, and so are called acoelomates.

18.5 Changes in the Body Cavity
1. The evolution of a body cavity improved circulation, movement, and organ function.
2. Roundworms have a body cavity, but it is not a true body cavity, so they are called pseudocoelomates.
3. Mollusks have a true body cavity and so they are called coelomates. Although a diverse group, they all contain a head-foot, a visceral mass, and a mantle.
4. Segmentation first evolved in annelid worms, which allowed for specialization of body parts.
5. Arthropods are the most successful animal phylum. Jointed appendages first evolved in this group, as well as the evolution of a rigid exoskeleton.

18.6 Redesigning the Embryo
1. Deuterostome eggs cleave radially, and the blastopore becomes the animal's anus. A deuterostome developmental pattern evolved in echinoderms and chordates. All other coelomates are protostomes.
2. Echinoderms have an endoskeleton made up of bony plates that lie under the skin. The adults are radially symmetrical, but that appears to be an adaptation to their environment.
3. The chordates have a truly internal endoskeleton and are distinguished by the presence of a notochord, a dorsal nerve cord, pharyngeal pouches, and a postanal tail. The notochord is replaced with the backbone in vertebrates.

The Parade of Vertebrates

18.7 Overview of Vertebrate Evolution
1. Animal life began in the sea primarily during the Paleozoic era. Mass extinctions, where the loss of species outpaces new species formation, have occurred five times throughout earth's history.

18.8 Fishes Dominate the Sea
1. Fishes, the ancestors of all vertebrates, have certain characteristics in common, such as gills, a vertebral column, a single-loop circulatory system, and nutritional deficiencies. Bony fish, the most successful group of vertebrates, have swim bladders to control buoyancy in the water.

18.9 Amphibians and Reptiles Invade the Land
1. Amphibians, the first vertebrates to invade the land, evolved from lobe-finned fishes. Adaptations to a terrestrial environment included the development of legs, lungs, cutaneous respiration, a pulmonary circulatory system, and a partially divided heart.
2. The key adaptations found in reptiles that made them better suited to a terrestrial environment were the evolution of a watertight amniotic egg, watertight skin, and thoracic breathing.

18.10 Birds Master the Air
1. Birds evolved directly from dinosaurs but were a minor group until the mass extinction of dinosaurs. Two key adaptations allowed birds to dominate the skies: feathers and a skeleton modified for flight with its sturdy but lightweight, hollow bones.

18.11 Mammals Adapt to Colder Times
1. Mammals expanded and diversified after the mass extinction of the dinosaurs, as did birds. Mammals are distinguished by the presence of mammary glands for feeding their young, hair for insulation, and bones in the middle ear that amplify sound.

18.12 Human Evolution
1. Our genus, *Homo*, evolved in Africa and then spread worldwide. Our species, *H. sapiens*, appears to have interbred slightly with *H. neanderthalensis*. There were several other species of the genus *Homo*, all now extinct.

Inquiry and Analysis

Is Atmospheric Carbon Dioxide Dissolving the World's Coral Reefs?

From 1991 to 1993, scientists began a unique experiment in the Arizona desert: They built a three-acre totally enclosed world, complete with people, then sealed it tight and watched what happened! With future prolonged space missions in mind, scientists wanted to explore the viability of a closed ecological system to support and maintain human life. Could it? Not really, it turned out. But much was learned. One of the mesocosms in this enclosed world was a large water tank, full of marine life. Studying the coral reef in this tiny ocean, researchers were able to ask a very interesting question. A coral reef is an underwater ecosystem, a massive calcium carbonate exoskeleton secreted by coral polyp colonies. All over the world, coral reefs are dying as ocean pH falls due to global warming (increasing CO_2 levels in the air cause more CO_2 to dissolve into seawater, where it combines with water to form carbonic acid; the lower pH kills the algae living in the reef). But might the dying of coral reefs worldwide be due to more than just the killing of the algae? Maybe something else is going on, too. Perhaps the added CO_2 in ocean water is actually dissolving the reefs themselves, leeching out their calcium from the calcium carbonate exoskeleton.

Using the 9,100-sq-ft ocean in Biosphere 2, scientists set out to explore that possibility. Using hydrochloric acid and sodium hydroxide, scientists were able to vary the levels of dissolved carbon dioxide in the Biosphere 2 ocean, measured as the partial pressure of CO_2 in micro atmospheres (pCO_2 μatm), and at the same time measure changes in the calcification rate of the corals. Scientists conducted the experiment using three different treatments: 200 pCO_2 μatm, 350 pCO_2 μatm, and 700 pCO_2 μatm. This cycle of treatments was repeated several times and the amount of calcium carbonate, as an indicator of coral calcification, was measured. The data are shown on the graph below.

Analysis

1. **Applying Concepts** Variable: In the graph, which is the dependent variable?
2. **Interpreting Data** How does the rate of calcification compare between coral when the water has a dissolved CO_2 concentration of 350 pCO_2 μatm versus 200 pCO_2 μatm?
3. **Making Inferences**
   a. Based on the data presented here, what general statement can you make regarding the growth of corals as the levels of dissolved CO_2 increase?
   b. What would you predict would happen to the growth of the coral if the levels of dissolved CO_2 were reduced in half?
4. **Drawing Conclusions** Does this data support the hypothesis that increases in atmospheric CO_2 levels have a negative impact on the calcification of coral?

5. **Further Analysis**
   a. In the oceans today, the amount of dissolved CO_2 measures approximately 400 pCO_2 μatm. Why was it important for the scientists to try to replicate this CO_2 level in their experiments?
   b. Since the 1980s, approximately 90% of the coral found in the Caribbean Sea has died off. Scientists have captured coral spawn and eggs, grown them into larva, and transplanted the larva into the damaged reef. The transplanted coral had great success and thrived. Why do you think the implanted coral survived better in the ecosystem than the existing coral that died off?

ClassicStock/Alamy Stock Photo

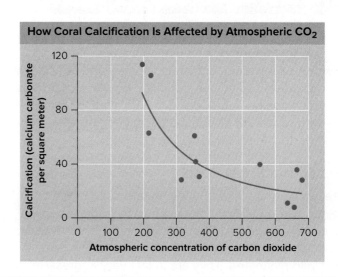

PART 5 The Living Environment

19 Populations and Communities

LEARNING PATH ▼

Ecology
1. What Is Ecology?

Populations
2. Population Growth
3. The Influence of Population Density
4. Life History Adaptations
5. Population Demography

How Competition Shapes Communities
6. Communities
7. The Niche and Competition

Species Interact in Many Ways
8. Coevolution and Symbiosis
9. Predation

Community Stability
10. Ecological Succession

THIS AFRICANIZED HONEYBEE has an attitude.
Darwin Dale/age fotostock

Invasion of the Killer Bees

One of the harshest lessons of environmental biology is that the unexpected does happen. Precisely because science operates at the edge of what we know, scientists sometimes stumble over the unexpected. This lesson has been brought clearly to mind in recent years with reports of killer bees being discovered east of the Mississippi River.

Bees with a Short Fuse

In the continental United States, we are used to the mild-mannered European honeybee, a subspecies called *Apis mellifera mellifera*. The African variety, subspecies *A. m. scutela,* looks very much like them, but they are hardly mild mannered. They are in fact very aggressive critters with a chip on their shoulder, and when swarming, they do not have to be provoked to start trouble. A few individuals may see you at a distance, "lose it," and lead thousands of bees in a concerted attack. The only thing you can do is run—fast. They will keep after you for up to a mile.

They are nicknamed "killer" bees not because any one sting is worse than the European kind, but rather because so many of the bees try to sting you. An average human can survive no more than 300 stings. A total of more than 8,000 stings is not unusual for a killer bee attack. An American graduate student attacked and killed in a Costa Rican jungle had 10,000 stings.

A Queen from Tanzania

Killer bees were brought from Africa to Brazil over 50 years ago by a prominent Brazilian scientist who was attempting to establish tropical bees in Brazil to expand the commercial bee industry (bees pollinate crops and make commercial honey in the process). African bees seemed ideal candidates, better adapted to the tropics than European bees and more prolific honey producers. In 1957, he established a quarantined colony of African

382

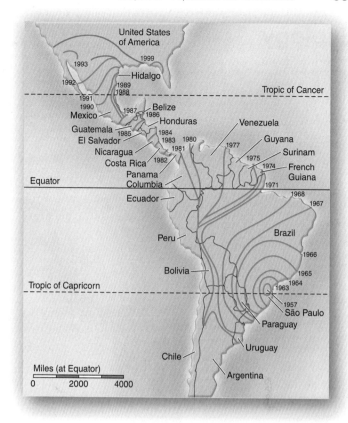

bees at a remote field station several hundred miles from his university at São Paulo. Although he had brought back many queens from South Africa, the colony came to be dominated by the offspring of a single very productive queen from Tanzania. He noted at the time that she appeared unusually aggressive.

Accidents Happen

That fall, the field station was visited by a beekeeper. As no one else was around that day, the visitor performed the routine courtesy of tending the hives. A hive with a queen in it has a set of bars across the door so the queen can't get out. Called a "queen excluder," the bars are far enough apart that the smaller worker bees can squeeze through. Once a queen starts to lay eggs, she never leaves the hive, so there is little point in slowing down entry of workers to the hive, and the queen excluder is routinely removed. On that day, the visitor saw the Tanzanian queen laying eggs in the African colony hive and so removed the queen excluder.

One and a half days later, the Tanzanian queen and 26 of her daughter queens had decamped. Out into the neighboring forest they went. And that's what was unexpected. European queen bees never leave the hive after they have started to lay eggs. No one could have guessed that the Tanzanian queen would behave differently. But she did.

Invasion

By 1970, the superaggressive African bees had blanketed Brazil, totally replacing local colonies. Their queen had evolved in a very different and far more competitive community, a jungle world in which rapid action and fierce aggression were rewarded with survival. Like inviting a fierce Viking warrior to a Texas neighborhood social, the arrival of Africanized honeybees into the community of local honeybees led to a predictable outcome: soon only one kind of honeybee was living there.

The Africanized bees had reached Central America by 1980, Mexico by 1986, and Texas by 1990, having conquered 5 million square miles in 33 years. In the process, they killed an estimated 1,000 people and over 100,000 cows. The first to be killed in the United States, a rancher named Lino Lopez, died of multiple stings in Texas in 1993. All during the 1990s, the bees continued their invasion of the United States. All of Arizona and much of Texas were quickly occupied. A few years later, Los Angeles County was officially declared colonized, and African bees moved on, eventually halfway up California.

Soon after, however, the invasion ceased, on a line roughly from San Francisco, California, to Richmond, Virginia. Winter cold has limited any further northward advance of the invading hoards. For the states below this line, sure as the sun rises, the bees are coming, not caring one bit that they were unanticipated. The deep lesson—that unexpected things do happen—is being driven home by millions of tiny aggressive teachers.

Ecology

Figure 19.1 A redwood community.

(a) The redwood forest of coastal California and southwestern Oregon is dominated by the population of redwoods (Sequoia sempervirens). Other species in the redwood community include (b) redwood sorrel (Oxalis oregana), (c) sword ferns (Polystichum munitum), and (d) ground beetles (Pterostichus lama).

(a) keldridge/Shutterstock; (b) komezo/Getty Images; (c) age fotostock/Alamy Stock Photo; (d) Stuart Wilson/Science Source

19.1 What Is Ecology?

Ecology is the study of how organisms interact with each other and with their environment. Ecology also encompasses the study of the distribution and abundance of organisms, which includes population growth and the limits and influences on population growth. The word *ecology* comes from the Greek words *oikos* (house, place where one lives) and *logos* (study of). Our study of ecology, then, is a study of the house in which we live. Do not forget this simple analogy built into the word *ecology*—most of our environmental problems could be avoided if we treated the world in which we live the same way we treat our own homes. Would you pollute your own house?

Levels of Ecological Organization

> **LEARNING OBJECTIVE 19.1.1** Describe the six levels of organization of organisms.

Ecologists consider groups of organisms at six progressively more encompassing levels of organization. As mentioned in chapter 1, new characteristics called *emergent properties* arise at each higher level, resulting from the way components of each level interact.

1. **Populations.** Individuals of the same species that live together are members of a **population.** They potentially interbreed with one another, share the same habitat, and use the same pool of resources the habitat provides.
2. **Species.** All populations of a particular kind of organism form a species. Populations of the species can interact and affect the ecological characteristics of the species as a whole.
3. **Communities.** Populations of different species that live together in the same place are called communities. Different species typically use different resources within the habitat they share (figure 19.1).
4. **Ecosystems.** A community and the nonliving factors with which it interacts is called an ecosystem. An ecosystem is affected by the flow of energy, ultimately derived from the sun, and the cycling of the essential elements on which the lives of its constituent organisms depend. The redwood forest community pictured in figure 19.1 is part of an ecosystem in which the giant trees and other organisms interact with each other and with their physical surroundings.
5. **Biomes.** Biomes are major terrestrial assemblages of plants, animals, and microorganisms that occur over wide geographical areas that have distinct physical characteristics. Examples include deserts, tropical rain forests, and grasslands. Similar types of groupings occur in marine and freshwater habitats.
6. **The biosphere.** All the world's biomes, along with its marine and freshwater assemblages, together constitute an interactive system we call the biosphere. Changes in one biome can have profound consequences for others.

> **Putting the Concept to Work**
> What is the difference between a community and an ecosystem?

Populations

19.2 Population Growth

> **LEARNING OBJECTIVE 19.2.1** Differentiate population size, population density, and population growth.

A species typically is found living in local populations, separated to at least some extent from other populations of that species. In this section, we will focus on the factors that influence whether a population will grow or shrink and at what rate. These factors are also important when considering our own population. Although we humans picture ourselves as different from populations of animals living in the wild, factors that affect wild populations affect human populations in similar ways.

How Populations Grow

One of the critical properties of any population is its **population size**—the number of individuals in the population. For example, if an entire species consists of only one or a few small populations, that species is likely to become extinct, especially if it occurs in areas that have been or are being radically changed. In addition to population size, **population density**—the number of individuals that occur in a unit area, such as per square kilometer—is often an important characteristic. The density of a population, how closely individuals associate with each other, is an indication of how they live. For example, animals that live in large groups, such as herds of wildebeests, may find safety in numbers (**figure 19.2**). In addition to size and density, another key characteristic of any population is its capacity to grow. To understand populations, we must consider what factors in nature promote or limit **population growth**.

The simplest model of population growth defines a population's growth rate (the change in its numbers over time) as the difference between the birth rate and the death rate, corrected for any movement of individuals into (immigration) or out of (emigration) the population. Movements of individuals can have a major impact on population growth rates. For example, the increase in human population in the United States during the closing decades of the 20th century was mostly due to immigrants. Less than half of the increase came from the reproduction of the people already living there.

> **Putting the Concept to Work**
> Explain how the density of a population may affect its size.

The Exponential Growth Model

> **LEARNING OBJECTIVE 19.2.2** Contrast exponential and logistic growth.

The innate capacity for growth of any population is exponential, which can result in *exponential growth* (see **figure 19.4**). Even when a population's growth rate remains constant, the actual increase in the *number* of individuals in the population accelerates rapidly; the larger a population is, the faster it grows. This sort of growth pattern is similar to that obtained by compounding interest on an investment. In practice, such patterns prevail

(a)

(b)

Figure 19.2 Population density.

(a) Siberian tigers occupy enormous territories (typically 60–100 km² for an adult male) because of the relative lack of prey, especially in winter. (b) This Serengeti wildebeest herd numbers over 1 million animals.

(a) Alan and Sandy Carey/Getty Images; (b) Gavin Hellier/Jon Arnold Images Ltd/Alamy Stock Photo

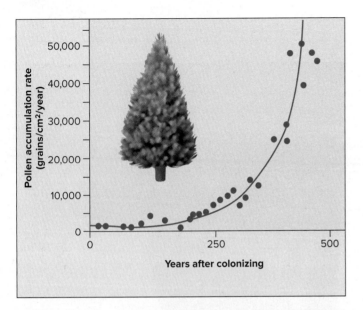

Figure 19.3 Exponential growth in a tree population.
These data present the exponential growth of Scotch pine trees after glaciers receded from the Norfolk region of Great Britain. The size of the population was estimated based on the rate of pollen accumulation in lake sediments.

only for short periods, usually when an organism reaches a new habitat with abundant resources. Figure 19.3 shows the establishment of tree populations in the British Isles after the glaciers receded in the Northern Hemisphere 20,000 years ago. The tree population grew at an exponential rate.

Carrying Capacity

No matter how rapidly populations grow, they eventually reach a limit imposed by shortages of important environmental factors such as space, light, water, or nutrients, and will experience *logistic growth*. A population usually ultimately stabilizes at a certain size, called the **carrying capacity** of the particular place where it lives, and the size of the population levels off, like the flat portion of the blue line in figure 19.4. The carrying capacity is the maximum number of individuals that an area can support.

The Logistic Growth Model

As a population approaches its carrying capacity, its rate of growth slows greatly as fewer and fewer resources remain for each new individual to use, until the population size levels off at or near the carrying capacity. The growth curve of such a population, which is always limited by one or more factors in the environment, is approximated by an S-shaped sigmoid growth curve, the blue line in figure 19.4. The curve is called "sigmoid" because its shape has a double curve like the letter *S*. As the size of a population stabilizes at the carrying capacity, its rate of growth slows down, eventually coming to a halt. The fur seal population in figure 19.5 has a carrying capacity of about 10,000 breeding male seals.

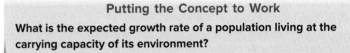

Putting the Concept to Work
What is the expected growth rate of a population living at the carrying capacity of its environment?

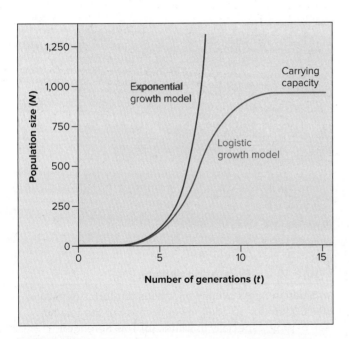

Figure 19.4 Two models of population growth.
The *red line* illustrates the exponential growth model for a population. The *blue line* illustrates the logistic growth model. At first, logistic growth accelerates exponentially, and then, as resources become limiting, the birth rate decreases or the death rate increases, and growth slows. Growth ceases when the death rate equals the birthrate. The carrying capacity ultimately depends on the resources available in the environment.

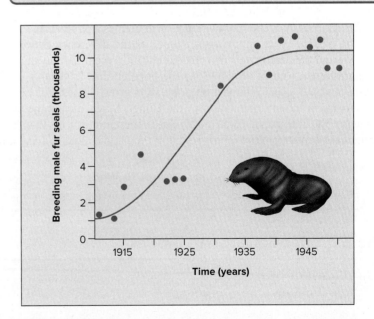

Figure 19.5 Most natural populations exhibit logistic growth.
These data present the history of a fur seal (*Callorhinus ursinus*) population on St. Paul Island, Alaska. Driven almost to extinction by hunting in the late 1800s, the fur seal made a comeback after hunting was banned in 1911. Today the number of breeding males with "harems" oscillates around 10,000 individuals, presumably the carrying capacity of the island.

Today's Biology

The War Against Urban Deer

Early in the 20th century, white-tailed deer were rare in the United States. Uncontrolled hunting had reduced their numbers to about 500,000 nationwide, and some states had no deer at all. To protect the remaining deer, laws were passed in the 1920s and 1930s to restrict hunting, particularly of does (females). This course of action seemed sensible but ignored a fundamental fact about deer: Deer reproduce quickly. A doe matures at 2 or 3 years and then typically gives birth to twins each year for 10 or more years.

Exponential Growth in Action

Any student of biology could have predicted what would happen, what would *have* to happen, if hunting were severely restricted. Since Malthus and Darwin, biologists have recognized that populations that grow unchecked do not increase in a linear fashion but exponentially. That is to say, the population does not increase the same amount each year but rather grows by ever-greater amounts as the babies have babies, like compounding interest. A deer herd that has plenty to eat and is not hunted by humans or other predators will double in size every three years!

With the restrictions on hunting, deer populations indeed began to grow. Numbers of white-tailed deer rebounded slowly at first, then more and more quickly. In the last few decades, deer numbers have literally exploded, and now exceed 30 million nationwide. There are more deer today than at any time in our nation's history.

Suburban Living

Much of the growth in the national deer population has occurred near urban areas. Deer have adapted well to encroaching suburbia, for two reasons:

1. Growth of suburbs. Paradoxically, land development tends to improve deer food supplies. Because the reproduction and survival of deer depend directly upon the quality of the food available to them, the improved food has led to more deer. Deer browse on leaves and require large quantities of new growth with high nutritional content to maintain normal reproduction. Deer populations 40 years ago rarely grew large, for the simple reason that most of the trees in an undisturbed forest are old and only the undergrowth provides suitable food. It is because land development usually involves clearing land that urban development leads to increased deer food supplies. There are far fewer trees, but the trees are new growth and very munchable. So are garden shrubs. Deer eat very, very well in suburbia.
2. Restriction of hunting. Nobody wants someone shooting at deer near their kids. Not surprisingly, then, most of the areas in which deer populations are growing most rapidly have been declared off limits to hunters. Removing their only significant predator—hunters—allows deer populations to grow unchecked.

Anyone who lives in the suburbs knows the result of providing ample food and no hunting: lots of deer. "A deer in the backyard is wonderful," says wildlife biologist William Porter. "Twenty-five deer in the backyard is a problem." Nancy Hoffstetter of the St. Louis county suburb Town and Country has been quoted as saying, "I consider them long-legged rats."

Raymond Gehman/Getty Images

The Problem: Unwanted Guests

What should we do to respond to this plague of deer? You can't just "remove" the deer to some forest far from the suburb. Why won't this humane approach work? Other deer from surrounding areas just take their place! Imagine trying to empty people from a prime section of a new baseball stadium by physically removing individuals one at a time. You would never get anywhere because other people would just crowd in. For every deer removed, there are two eager to get in the chow line and have a good meal.

There were only two real options open to lower deer numbers: decrease the birth rate or increase the death rate.

Solution 1: Birth Control

Decreasing the birth rate is certainly the most ethically palatable approach. However, deer birth control has proven impractical. Every female deer must be captured for the first dose and redarted for each subsequent booster shot. Only in very small isolated populations is this practical. Nor is there any effective oral contraceptive for deer.

Solution 2: Hunting

This leaves increasing the death rate. On more remote forest land, laws can be passed that extend the hunting season, increase the bag limit, and encourage the shooting of antlerless females. But there is no hunting in suburbs. In suburbs, the only realistic approach is to thin out the local herds. How? In the summer of 2009, a 76-year-old housewife from a Cleveland suburb discovered a fawn nestled in her garden, picked up a shovel, and beat it to death. She was arrested and fined for cruelty—no one wants this type of "bambicide." In suburbs throughout the country, the thinning is being done with professional sharpshooters. Deer are still safe in people's backyards, but the herds are being culled by taking deer from nearby open fields.

Killing the cute deer in your backyard is not ideal, but fawns and other deer starving to death is every bit as unpleasant an option, and if local deer herds continue to grow, that is what is going to happen. A lot more deer will die, a lot more miserably. Can you think of any better solution to this haunting problem?

19.3 The Influence of Population Density

> **LEARNING OBJECTIVE 19.3.1** Differentiate between density-dependent and density-independent effects on population growth.

Many factors act to regulate the growth of populations in nature. Some act independently of the size of the population; others do not.

Density-Independent Effects

Effects that are independent of the size of a population and act to regulate its growth are called *density-independent effects*. A variety of factors may affect populations in a density-independent manner. Most of these are aspects of the external environment, such as weather (extremely cold winters, droughts, storms, floods) and physical disruptions (volcanic eruptions and fire). Individuals often will be affected by these activities regardless of the size of the population. Populations that occur in areas in which such events occur relatively frequently will display erratic population growth patterns, increasing rapidly when conditions are good but suffering extreme reductions whenever the environment turns hostile.

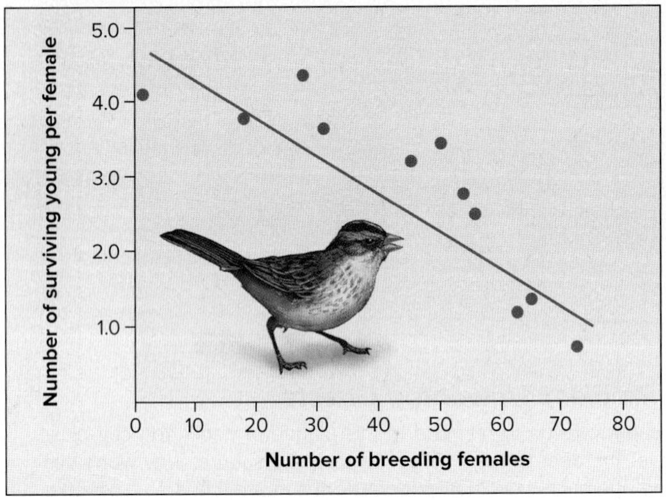

Density-Dependent Effects

Effects that are dependent on the size of the population and act to regulate its growth are called *density-dependent effects*. Among animals, these effects may be accompanied by hormonal changes that can alter behavior that will directly affect the ultimate size of the population. One striking example occurs in migratory locusts. When they become crowded, the locusts produce hormones that cause them to enter a migratory phase; the locusts take off as a swarm and fly long distances to new habitats. Density-dependent effects, in general, have an increasing effect as population size increases. For example, as the population of song sparrows in **figure 19.6** grows, the individuals in the population compete with increasing intensity for limited resources.

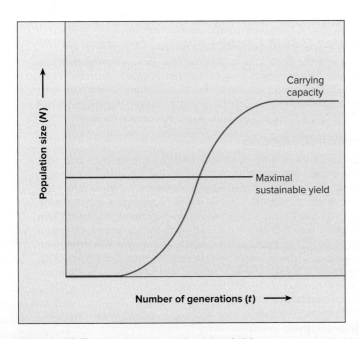

Figure 19.6 Density-dependent effects.

Reproductive success of the song sparrow (*Melospiza melodia*) decreases as population size increases.

Maximizing Population Productivity

In natural systems that are exploited by humans, such as fisheries, the aim is to maximize productivity by exploiting the population early in the rising portion of its sigmoid growth curve. At such times, populations and individuals are growing rapidly, and net productivity—in terms of the amount of material incorporated into the bodies of these organisms—is highest. The point of *maximal sustainable yield* (the red line in **figure 19.7**) lies partway up the sigmoid curve. Harvesting the population of an economically desirable species near this point will result in the best sustained yields. Overharvesting a population that is smaller than this critical size can destroy its productivity for many years or even drive it to extinction.

Figure 19.7 Maximal sustainable yield.

Harvesting the organisms when the population is in the rapid growth phase of the sigmoidal curve, but not overharvesting, will result in sustained yields.

> **Putting the Concept to Work**
> Are density-dependent effects more or less pronounced as population size increases?

19.4 Life History Adaptations

> **LEARNING OBJECTIVE 19.4.1** Contrast *r*-selected and *K*-selected adaptations.

Maximum Growth Rate

Populations of many species, including annual plants, some insects, and most bacteria, can have very fast rates of growth when not limited by dwindling environmental resources, rates which often approximate the exponential growth model discussed earlier. In mathematical formulas used to calculate population growth rates, the maximum growth rate is indicated by *r*.

Populations of most animals have much slower rates of growth when numbers are limited by available resources. Producing a sigmoid growth curve approximating the logistic growth model discussed earlier. Habitats with limited resources lead to more intense competition for resources and favor individuals that can survive and successfully reproduce more efficiently. The number of individuals that can survive at this limit is the carrying capacity of the environment. In mathematical formulas used to calculate population growth rates, the carrying capacity is indicated by *K*.

Life History Strategies

The complete life cycle of an organism constitutes its life history. Some life history adaptations favor very rapid growth in a habitat with abundant resources or in unpredictable or volatile environments in which organisms have to take advantage of the resources when they are available. In these situations, reproducing early, producing many small offspring that mature quickly, and engaging in other aspects of "big bang" reproduction are favored. Using the terms of the exponential model, these adaptations that favor a high rate of increase, approaching the maximum growth rate or *r*, are called ***r*-selected adaptations**. Examples of organisms displaying *r*-selected life history adaptations include dandelions, aphids, mice, and cockroaches (figure 19.8).

Other life history adaptations favor survival in an environment in which individuals are competing for limited resources. These features include reproducing late, having small numbers of larger-sized offspring that mature slowly and receive intensive parental care, and other aspects of "carrying capacity" reproduction. In terms of the logistic model, these adaptations favoring reproduction near the carrying capacity of the environment or *K* are called ***K*-selected adaptations**. Examples of organisms displaying *K*-selected life history adaptations include coconut palms, whooping cranes, and whales.

Most natural populations show life history adaptations that exist along a continuum, ranging from completely *r*-selected traits in rapidly changing habitats to completely *K*-selected traits in very stable competitive habitats. Table 19.1 outlines the adaptations at the extreme ends of the continuum.

> **Putting the Concept to Work**
> Would you expect a mosquito to exhibit more *r*-selected or more *K*-selected adaptations? Explain why.

Figure 19.8 The consequences of exponential growth.
All organisms have the potential to produce populations larger than those that actually occur in nature. The German cockroach *(Blatella germanica)*, a major household pest, produces 80 young every six months. If every cockroach that hatched survived for three generations, kitchens might look like this theoretical culinary nightmare concocted by the Smithsonian Museum of Natural History.

Courtesy National Museum of Natural History, Smithsonian Institution

TABLE 19.1 *r*-Selected and *K*-Selected Life History Adaptations

Adaptation	*r*-Selected Populations	*K*-Selected Populations
Age at first reproduction	Early	Late
Homeostatic capability	Limited	Often extensive
Life span	Short	Long
Maturation time	Short	Long
Mortality rate	Often high	Usually low
Number of offspring produced per reproductive episode	Many	Few
Number of reproductions per lifetime	Usually one	Often several
Parental care	None	Often extensive
Size of offspring or eggs	Small	Large

19.5 Population Demography

LEARNING OBJECTIVE 19.5.1 Explain how the growth rate of a population is influenced by its age structure, fecundity, and mortality.

Demography is the statistical study of how big populations are. Populations grow if births outnumber deaths and shrink if deaths outnumber births. Because birth and death rates also depend on age and sex, the future size of a population depends on its present age structure and sex ratio.

Age Structure

Many annual plants and insects time their reproduction to particular seasons of the year and then die. All members of these populations are the same age. Perennial plants and longer-lived animals contain individuals of more than one generation, so that in any given year, individuals of different ages are reproducing within the population (figure 19.9). A group of individuals of the same age is referred to as a *cohort*.

Within a population, every cohort has a characteristic birthrate, or *fecundity*, defined as the number of offspring produced in a standard time (for example, per year), and a characteristic death rate, or *mortality*, the number of individuals that die in that period. The rate of a population's growth depends on the difference between these two rates.

The relative number of individuals in each cohort defines a population's age structure. A population with a large proportion of young individuals tends to grow rapidly because an increasing proportion of its individuals are reproductive. A population with more females usually also tends to grow more rapidly. The proportion of males to females in a population is its *sex ratio*.

Mortality and Survivorship Curves

A population's intrinsic rate of increase depends on the ages of the organisms in it and the reproductive performance of the individuals in the various age groups. When a population lives in a constant environment for a few generations, its **age distribution**—the proportion of individuals in different age categories—tends to stabilize. A population whose size remains fairly constant through time is called a stable population. In such a population, births plus immigration must balance deaths plus emigration.

One way to express the age distribution characteristics of populations is through a *survivorship curve*. Survivorship is defined as the percentage of an original population that survives to a given age. Examples of different kinds of survivorship curves are shown in figure 19.10. By convention, survival (the vertical axis) is plotted on a log scale. In hydra, animals related to jellyfish, individuals are equally likely to die at any age, as indicated by the straight survivorship curve (the blue line, type II). Oysters, like plants, produce vast numbers of offspring, only a few of which live to reproduce. However, once they become established as reproductive individuals, their mortality is extremely low (red line, type III survivorship curve). Even though human babies are susceptible to death at relatively high rates, mortality in humans, as in many animals and protists, rises in the postreproductive years (green line, type I survivorship curve).

Figure 19.9 Age structure in different types of populations.

The grass you see in this photo is an annual plant. The individual plants are all the same age because they die every year. By contrast, the herd of bison contains individuals of varying ages because they can survive from year to year.

Glen Allison/Photodisc/Getty Images

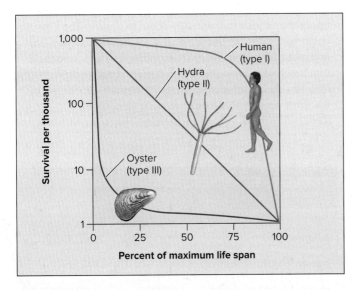

Figure 19.10 Survivorship curves.

> **Putting the Concept to Work**
> If fecundity increases, does the population growth rate go up or down?

How Competition Shapes Communities

19.6 Communities

> **LEARNING OBJECTIVE 19.6.1** Contrast individualistic and holistic concepts of community.

The term **community** refers to the mix of species that occur at any particular locality. Some communities contain many species (figure 19.11), whereas others contain only a few species, such as in the near-boiling waters of Yellowstone's geysers (where a limited number of microbial species live). Communities can be characterized either by their constituent species, a list of all species present in the community, or by their properties, such as species richness (the number of different species present) or primary productivity (the amount of solar energy captured through photosynthesis and stored as organic compounds).

Interactions among community members govern many ecological and evolutionary processes. These interactions, such as predation (figure 19.11*b*) and competition (figure 19.11*c*), affect the population biology of particular species, as well as the ways in which energy and nutrients are used in the ecosystem. Interactions among different species are often influenced by particular keystone species (chapter 22, section 22.6). As discussed further in chapter 20, an *ecosystem* includes a community of living organisms and the nonliving components that surround them.

Studying How Species in a Community Interact

Scientists study biological communities in many ways, ranging from detailed observations to elaborate, large-scale experiments. In some cases, such studies focus on the entire community, whereas in other cases, only a subset of species that are likely to interact with each other are studied. Regardless of how they are studied, two views exist on the makeup and functioning of communities.

The *individualistic concept* of communities holds that a community is nothing more than an aggregation of species that happen to coexist in one place. By contrast, the *holistic concept* of communities views communities as an integrated unit. In this sense, the community could be viewed as a superorganism whose constituent species have coevolved to the extent that they function as a part of a greater whole, just as the kidneys, heart, and lungs all function together within an animal's body. In this view, then, a community would amount to more than the sum of its parts.

Most ecologists today favor the individualistic concept. For the most part, species seem to respond independently to changing environmental conditions.

> **Putting the Concept to Work**
> Why is the individualistic concept of communities supported by the fact that species respond independently to changing environmental conditions?

(a)

(b)

(c)

Figure 19.11 A Tanzanian savanna community.

A community consists of all the species—plants, animals, fungi, protists, and prokaryotes—that occur at a locality. (a) A savanna community in Lake Manyara National Park in Tanzania. Species within a community interact with each other, such as through (b) predation or (c) competition for a resource.

(a) Tim Davis/Science Source; (b) Rostislav Stach/Shutterstock; (c) Cathy Withers-Clarke/Shutterstock

19.7 The Niche and Competition

> **LEARNING OBJECTIVE 19.7.1** Contrast the fundamental niche and the realized niche.

Within a community, each organism occupies a particular biological role, or **niche.** The niche an organism occupies is the sum total of all the ways it uses the resources of its environment, including space, food, and many other factors of the environment. A niche may be described in terms of space utilization, food consumption, temperature range, appropriate conditions for mating, requirements for moisture, and other factors. *Niche* is not synonymous with **habitat,** the place where an organism lives. *Habitat* is a place, and *niche* is a pattern of living. Many species can share a habitat, but as we shall see, no two species can long occupy exactly the same niche.

Sometimes species are not able to occupy their entire niche because of the presence or absence of other species. **Competition** describes the interaction when two organisms attempt to use the same resource when there is not enough of the resource to satisfy both. Competition between individuals of different species is called *interspecific competition*. Interspecific competition is often greatest between organisms that obtain their food in similar ways and between organisms that are more similar. Another type of competition, called *intraspecific competition,* is competition between individuals of the same species.

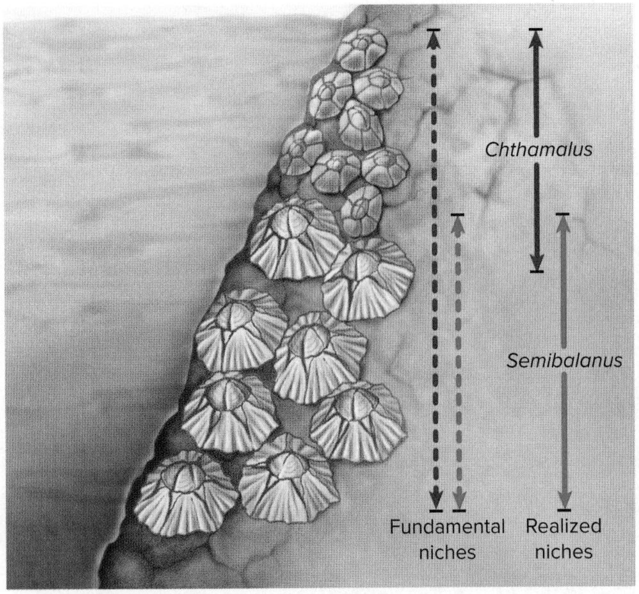

The Realized Niche

Because of competition, organisms may not be able to occupy the entire niche they are theoretically capable of using, called the *fundamental niche* (or theoretical niche). The actual niche the organism is able to occupy in the presence of competitors is called its *realized niche.*

A Classic Study. Barnacles are marine animals (crustaceans) that have free-swimming larvae. The larvae eventually settle down, cementing themselves to rocks and remaining attached for the rest of their lives. Two species of barnacles grow together on rocks along the coast of Scotland. *Chthamalus stellatus* (the smaller barnacle in figure 19.12) lives in shallower water, where tidal action often exposes it to air, and *Semibalanus balanoides* (the larger barnacle) lives at lower depths, where it is rarely exposed to the atmosphere. In the deeper zone, *Semibalanus* could always outcompete *Chthamalus* by crowding it off the rocks, undercutting it, and replacing it even where it had begun to grow. When a researcher removed *Semibalanus* from the area, however, *Chthamalus* was easily able to occupy the deeper zone, indicating that no physiological or other general obstacles prevented it from becoming established there. In contrast, *Semibalanus* could not survive in the shallow-water habitats where *Chthamalus* normally occurs; it evidently does not have the special physiological and morphological adaptations that

Figure 19.12 Competition among two species of barnacles limits niche use.

Chthamalus can live in both deep and shallow zones (its fundamental niche), but *Semibalanus* forces *Chthamalus* out of the part of its fundamental niche that overlaps the realized niche of *Semibalanus*.

allow *Chthamalus* to occupy this zone. Thus, the fundamental niche of the barnacle *Chthamalus* in these experiments included that of *Semibalanus* (the red dashed arrow), but its realized niche was much narrower (the red solid arrow) because *Chthamalus* was outcompeted by *Semibalanus* in its fundamental niche.

> **Putting the Concept to Work**
> What happens to the unoccupied part of a fundamental niche?

Competitive Exclusion

> **LEARNING OBJECTIVE 19.7.2** State the principle of competitive exclusion.

In classic experiments carried out between 1934 and 1935, Russian ecologist G. F. Gause studied competition among three species of *Paramecium,* a tiny protist. All three species grew well alone in culture tubes (figure 19.13a), preying on bacteria and yeasts that fed on oatmeal suspended in the culture fluid. However, when Gause grew *P. aurelia* together with *P. caudatum* in the same culture tube (figure 19.13b), the numbers of *P. caudatum* (the green line) always declined to extinction, leaving *P. aurelia* the only survivor. Why? Gause found *P. aurelia* was able to grow six times faster than its competitor, *P. caudatum,* because it was able to better use the limited available resources.

From experiments such as this, Gause formulated what is now called the *principle of competitive exclusion*. This principle states that if two species are competing for a resource, the species that uses the resource more efficiently will eventually eliminate the other locally—no two species with the same niche can coexist.

> **Putting the Concept to Work**
> If two species attempt to occupy the same niche, how can you tell which one will be eliminated?

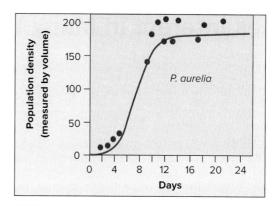

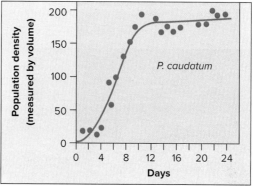

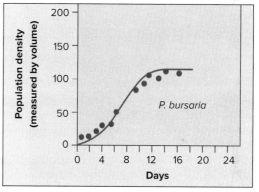

(a)

Figure 19.13 **Competitive exclusion among three species of *Paramecium*.**

In the microscopic world, *Paramecium* is a ferocious predator. Paramecia eat by ingesting their prey; their plasma membranes surround bacterial or yeast cells, forming a food vacuole containing the prey cell. In his experiments, (a) Gause found that three species of *Paramecium* grew well alone in culture tubes. (b) However, *P. caudatum* declined to extinction when grown with *P. aurelia* because they shared the same realized niche, and *P. aurelia* outcompeted *P. caudatum* for food resources.

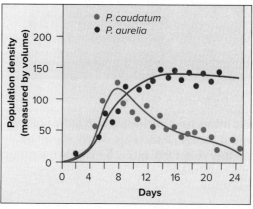

(b)

Species Interact in Many Ways

19.8 Coevolution and Symbiosis

> **LEARNING OBJECTIVE 19.8.1** Explain the difference between evolution and coevolution.

The previous section described the "winner take all" results of competition between two species whose niches overlap. Other relationships in nature are less competitive and more cooperative.

Coevolution

The plants, animals, protists, fungi, and prokaryotes that live together in communities have changed and adjusted to one another continually over millions of years. For example, many features of flowering plants have evolved in relation to the dispersal of the plant's gametes by animals (figure 19.14). These animals, in turn, have evolved a number of special traits that enable them to obtain food or other resources efficiently from the plants they visit, often from their flowers. In addition, the seeds of many flowering plants have features that make them more likely to be dispersed to new areas of favorable habitat.

Such interactions, which involve the long-term, mutual evolutionary adjustment of the characteristics of the members of biological communities, are examples of **coevolution.** Coevolution is the adaptation of two or more species to each other. In this section, we consider the many ways species interact, some of which involve coevolution.

> **Putting the Concept to Work**
> What is the minimal number of species that may be involved in a coevolutionary relationship?

Figure 19.14 Pollination by bat.
Many flowering plants have coevolved with other species to facilitate pollen transfer. Insects are widely known as pollinators, but they're not the only ones. Notice the cargo of pollen on the bat's snout.
Dr. Merlin D. Tuttle/Science Source

Symbiosis Is Widespread

> **LEARNING OBJECTIVE 19.8.2** Describe the three major kinds of symbiotic relationships.

In symbiotic relationships, two or more kinds of organisms live together in often elaborate and more or less permanent relationships. All symbiotic relationships carry the potential for coevolution between the organisms involved, and in many instances, the results of this coevolution are fascinating. Examples of symbiosis include lichens, which are associations of certain fungi with green algae or cyanobacteria. Other important examples are mycorrhizae, the associations between fungi and the roots of most kinds of plants. The fungi expedite the plant's absorption of certain nutrients, and the plants in turn provide the fungi with carbohydrates. Similarly, root nodules that occur in legumes and certain other kinds of plants contain bacteria that fix atmospheric nitrogen and make it available to their host plants (see chapter 20).

> As you will recall from the discussion in section 16.8, mycorrhizae are symbiotic relationships between fungi and the roots of plants; lichens are symbiotic relationships between fungi and a photosynthetic partner such as cyanobacteria or green algae.

In the tropics, leaf-cutter ants are often so abundant that they can remove a quarter or more of the total leaf surface of the plants in a given area. They do not eat these leaves directly; rather, they take them to underground nests, where they chew them up and inoculate them with the spores of particular fungi. These fungi are cultivated by the ants and brought from one specially prepared bed to another, where they grow and reproduce. In turn, the fungi constitute the primary food of the ants and their larvae. The relationship between leaf-cutter ants and these fungi is an excellent example of symbiosis.

The major kinds of symbiotic relationships include (1) **mutualism,** in which both participating species benefit; (2) **commensalism,** in which one species benefits while the other neither benefits nor is harmed; and (3) **parasitism,** in which one species benefits but the other is harmed. Parasitism can also be viewed as a form of predation, although the organism that is preyed upon does not necessarily die.

Mutualism

Mutualism is a symbiotic relationship in which both species benefit. The pistol shrimp patrolling the surface of the coral in figure 19.15 is defending its homestead from sea stars, which prey on coral. When it encounters a sea star on the coral, the shrimp attacks, pinching the sea star's spines and tube feet and making loud snapping sounds with its enlarged pincers. The loud popping sounds, which have given the shrimp its name, are so intense they stun small fish. Protection from being eaten by sea stars certainly provides a benefit to the coral. The shrimp also clearly benefit, obtaining food and shelter from the coral. Because both parties benefit, their relationship is an example of mutualism.

Commensalism

Commensalism is a symbiotic relationship that benefits one species and neither hurts nor helps the other. The best-known examples of commensalism involve the relationships between certain small tropical fishes and sea anemones, marine animals that have stinging tentacles (see chapter 18). These fish have evolved the ability to live among the tentacles of sea anemones, even though these tentacles would quickly paralyze other fishes that touched them. The fishes feed on the detritus left from the meals of the host anemone, remaining uninjured under remarkable circumstances. In another example, birds called oxpeckers eat ticks and other insects off of grazing animals (figure 19.16). In this symbiotic relationship, the oxpeckers receive a clear benefit in the form of nutrition. If the removal of the ticks benefits the impala, then the relationship is mutually beneficial and the relationship would be considered a form of mutualism. However, there is no evidence that this is so. In this instance, as in most examples of commensalism, it is difficult to be certain whether the partner receives a benefit or not.

Parasitism

Parasitism is a symbiotic relationship which benefits one species at the expense of the other. Typically, the parasite is much smaller than its host and remains closely associated with it. Parasitism is sometimes considered a special kind of predator-prey relationship in which the predator is much smaller than the prey, but unlike a predator, the parasite typically does not kill its host. Parasites are very common among animals. The head louse you see in figure 19.17, for example, is one of two types of sucking lice that parasitize humans.

> **Putting the Concept to Work**
> Ants protecting aphids but also eating the "honeydew" the aphids excrete is an example of what type of symbiotic relationship?

Figure 19.15 The pistol shrimp defends the coral, which he calls home—an example of mutualism.

WaterFrame/Alamy Stock Photo

Figure 19.16 Oxpeckers eat insects off an impala—an example of commensalism.

StuPorts/iStock/Getty Images

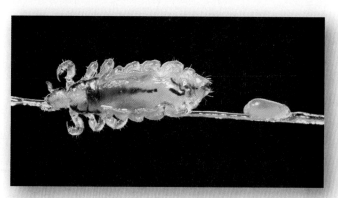

Figure 19.17 The head louse, shown here with its egg, feeds on its host and is an example of parasitism.

Denis Crawford/Alamy Stock Photo

19.9 Predation

LEARNING OBJECTIVE 19.9.1 Discuss the ways predators can affect prey populations.

Species that live together in a community interact in many ways, one of which is to eat one another. **Predation** is the consuming of one organism by another. The organism doing the eating is called the *predator*, while the organism being eaten is the *prey*. Examples of predation include a leopard capturing and eating an antelope, a whale grazing on millions of microscopic ocean plankton, and locusts eating the leaves of plants.

In nature, predators often have large effects on prey populations. Some of the most dramatic examples involve situations in which humans have either added or eliminated predators from an area. For example, the elimination of large carnivores from much of the eastern United States has led to population explosions of white-tailed deer, which strip the habitat of all edible plant life within their reach. Similarly, when sea otters were hunted to near extinction on the western coast of the United States, populations of sea urchins, a principal prey item of the otters, exploded. Appearances, however, sometimes can be deceiving. On Isle Royale in Lake Superior, moose reached the island by crossing over ice in an unusually cold winter and multiplied freely there in isolation. When wolves later reached the island by crossing over the ice, naturalists widely assumed that the wolves were playing a key role in controlling the moose population. More careful studies have demonstrated that this is not in fact the case. The moose that the wolves eat are, for the most part, old or diseased animals that would not survive long anyway. In general, the moose are controlled by food availability, disease, and other factors rather than by the wolves (figure 19.18).

Figure 19.18 **Wolves chasing a moose—what will the outcome be?**

On Isle Royale, Michigan, a large pack of wolves pursue a moose. They chased this moose for almost 2 kilometers; it then turned and faced the wolves, which by that time were exhausted from running through chest-deep snow. The wolves laid down, and the moose walked away.

Rolf O. Peterson

Putting the Concept to Work

Explain why deer are far more common in New Jersey than in Montana.

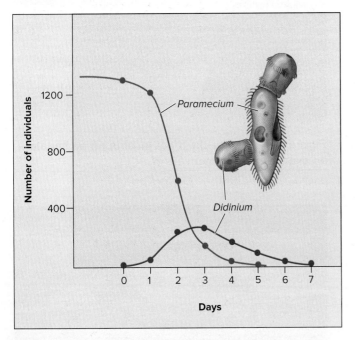

Figure 19.19 **Predator-prey in the microscopic world.**

When the predatory *Didinium* is added to a *Paramecium* population, the numbers of *Didinium* initially rise, while the numbers of *Paramecium* steadily fall. As the *Paramecium* population is depleted, however, the *Didinium* individuals also die.

Predator/Prey Cycles

> **LEARNING OBJECTIVE 19.9.2** Evaluate which factors are responsible for predator-prey oscillations.

Why doesn't a predator exterminate its prey and then become extinct itself, having nothing left to eat? This is just what happens in laboratory experiments, such as the experiment shown in **figure 19.19**, where the predator, *Didinium* (red line) often exterminates its prey, *Paramecium* (blue line). However, if refuges are provided for the prey, its population drops to low levels but not to extinction. Low prey population levels then provide inadequate food for the predators, causing the predator population to decrease. When this occurs, the prey population can recover. In this way, predator and prey populations cycle in their abundance.

Population cycles are characteristic of some species of small mammals, such as lemmings, and they appear to be stimulated, at least in some situations, by their predators. Ecologists have studied cycles in hare populations since the 1920s. They have found that the North American snowshoe hare, *Lepus americanus*, follows a "10-year cycle" (in reality, it varies from 8 to 11 years). Its numbers fall 10-fold to 30-fold in a typical cycle, and 100-fold changes can occur. Two factors appear to be generating the cycle: food plants and predators.

1. **Food plants.** The preferred foods of snowshoe hares are willow and birch twigs. As hare density increases, the quantity of these twigs decreases, leading to a precipitous decline in willow and birch twig abundance. The result is that hares are forced to feed on high-fiber (low-quality) food, causing lower birthrates, low juvenile survivorship, low growth rates, and a corresponding fall in hare abundance.
2. **Predators.** A key predator of the snowshoe hare is the Canada lynx, *Lynx canadensis*. The Canada lynx shows a "10-year cycle" of abundance that seems remarkably entrained to the hare abundance cycle (**figure 19.20**). As hare numbers increase, lynx numbers do, too, rising in response to the increased availability of lynx food. When hare numbers fall, so do lynx numbers, their food supply depleted.

Which factor is responsible for the predator-prey oscillations? Do increasing numbers of hares lead to overharvesting of plants (a hare-plant cycle), or do increasing numbers of lynx lead to overharvesting of hares (a hare-lynx cycle)? Field experiments have shown that both factors can affect the cycle, which, in practice, seems to be generated by the interaction between the two factors.

> **Putting the Concept to Work**
> What do you expect will happen to hare numbers in an experimental plot if predators are excluded from the plot?

(a)

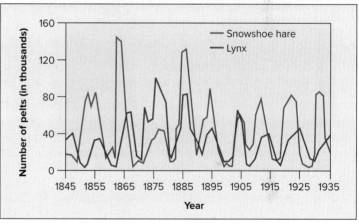
(b)

Figure 19.20 A predator-prey cycle.

(a) A snowshoe hare being chased by a lynx. (b) The numbers of lynxes and snowshoe hares oscillate in tune with each other in northern Canada. The data are based on numbers of animal pelts from 1845 to 1935. As the number of hares grows, so does the number of lynxes, with the cycle repeating about every 10 years. Both predators (lynxes) and available food resources control the number of hares. The number of lynxes is controlled by the availability of prey (snowshoe hares).

(a) Alan & Sandy Carey/Science Source

Community Stability

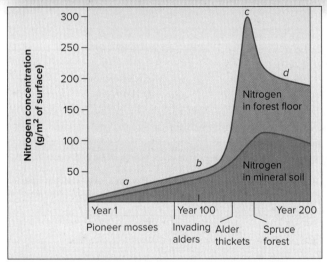

Figure 19.21 Plant succession at Glacier Bay, Alaska, produces progressive changes in the soil.

Source: USGS/Bruce F. Molnia

> **IMPLICATION FOR YOU** Plant succession occurs wherever habitats are disturbed, even in your own backyard. If you are wildly interested in this, you can demonstrate it for yourself. In the fall, dig up an area of your yard about the size of a dining room table, turning all the soil over and removing as many plants as possible. The following summer, lay out a 1-meter-wide path across the disturbed area and onward an equal distance into the untouched yard; this is called a transect line. Inventory the plants you find within the transect, classifying them as either abundant, occasional, or absent. How do the disturbed and untouched areas compare?

19.10 Ecological Succession

> **LEARNING OBJECTIVE 19.10.1** Explain why succession happens.

Marked changes in habitat can result in the orderly replacement of one community with another, from simple to complex, in a process known as **succession**. This process is familiar to anyone who has seen a vacant lot or cleared woods slowly become occupied by an increasing number of plants or a pond become dry land as it is filled with vegetation encroaching from the sides.

Secondary Succession

If a wooded area is cleared and left alone, plants slowly reclaim the area. Eventually, traces of the clearing disappear, and the area is again woods. Similarly, intense flooding may clear a stream bed of many organisms, leaving mostly sand and rock; afterward, the bed is progressively reinhabited by protists, invertebrates, and other aquatic organisms. This kind of succession, which occurs in areas where an existing community has been disturbed, is called **secondary succession**. Humans are often responsible for initiating secondary succession, such as when a fire has burned off an area or in abandoned agricultural fields.

Primary Succession

In contrast, **primary succession** occurs on bare, lifeless substrate, such as rocks. Primary succession occurs in lakes left behind after the retreat of glaciers, on volcanic islands that rise above the sea, and on land exposed by retreating glaciers. Primary succession on glacial moraines provides an example. The graph in **figure 19.21** shows how nitrogen concentrations change in the soil as primary succession occurs. On bare, mineral-poor soil, lichens grow first, forming small pockets of soil. Acidic secretions from the lichens help to break down the substrate and add to the accumulation of soil. Mosses then colonize these pockets of soil (**figure 19.21a**), eventually building up enough nutrients in the soil for alder shrubs to take hold (**figure 19.21b**). These first plants to appear form a *pioneering community*. Over 100 years, the alders (**figure 19.21c** and **photo**) build up the soil nitrogen levels until spruce are able to thrive, eventually crowding out the alder and forming a dense spruce forest (**figure 19.21d**).

Primary successions end with a community called a *climax community*, whose populations remain relatively stable and are characteristic of the region as a whole.

Why Succession Happens

Succession happens because species alter the habitat and the resources available in it, often in ways that favor other species. As ecosystems mature, and more *K*-selected species replace *r*-selected ones, species richness and total biomass increase, but net productivity decreases.

> **Putting the Concept to Work**
> Does step (d) in figure 19.21 indicate a pioneering or a climax community? What is the effect of succession on net productivity?

Today's Biology

How to Stop Cars from Hitting Deer? Add Wolves!

One of the most feared of all North American predators is the wolf. In the early 20th century, the gray wolf population was close to 500,000 strong and they roamed nearly two-thirds of the United States. A single wolf could eat up to 20 pounds of meat a day. Ranchers were afraid of losing their livestock to wolf attacks, and families were concerned for the safety of the unprotected children. Solution: get rid of the wolves. By the 1960s the gray wolf had been all but eradicated from the continental United States.

In 1978, the gray wolf became protected under the Endangered Species Act. Soon after, their population began bouncing back. Canadian gray wolves were deliberately reintroduced into Yellowstone National Park, while others came in on their own, recolonizing states like Wisconsin that border Canada. Everywhere wolves came, they thrived. Wisconsin now has a thousand or more wolves. While conservationists cheered at their success, ranchers complained bitterly of lost stock.

But in Wisconsin, these predators have provided a great benefit, one no one expected: wolves have saved the lives of people.

It all has to do with deer. There are a lot of deer in Wisconsin, so many that an average of 19,757 drivers collide with deer every year, leading to some 477 serious injuries and four deaths. But since wolves have become more common in the last few decades, they have reduced the deer-vehicle collisions by a quarter. This not only saves the state $10.9 million in losses each year (a figure 63 times greater than the compensation paid for loss of livestock or pets to wolves), but by reducing collisions, the wolf population has saved human lives.

How do wolves stop cars from hitting deer? These predators tend to prowl along human-made corridors like roads. Cleared roadways like the one you see below provide wolves a way to move quickly through the area they are hunting. Any deer near the road become the first prey attacked. It didn't take very many years for deer to learn to fear roads. By pushing deer away from roads, wolves reduced the chance one would be hit by a car!

Attempting to decrease auto-deer collisions, states like Wisconsin have tried road signs warning drivers of "deer crossing" wherever collisions have occurred, but these relatively cheap measures don't work, as the high number of collisions attest. Overpasses for deer are a multi-million dollar solution too expensive to use often. As a civil engineering problem, it seems that deer-vehicle collisions are a problem best solved by adding wolves.

Colleen Gara/Getty Images

Putting the Chapter to Work

 Read the following passage and address the questions below:

By the end of 2017, wildfires had blazed through more than 280,000 acres of southern California. During the fire, the number of redwood trees, ponderosa pines, and Coulter pines decreased, but they were not eliminated.

   a. Were the wildfires a density-independent or density-dependent effect on the tree populations?

   b. Ecologists are recording the number of trees per square kilometer each year as a way to document the regrowth of these forests. Which property of the tree populations are they recording?

   c. Eventually, after slowly increasing for many years, the population numbers of these trees will reach a maximum. Ecologists use what term to refer to this number?

 Elephants have an average life span of 65 years. Females meet sexual maturity at 14 years of age. Once pregnant, their gestation period is approximately two years, and they bear one calf per pregnancy.

Based on this information, what type of life history adaptations are observed in elephants?

Retracing the Learning Path

Ecology

19.1 What Is Ecology?

1. Ecology is the study of how the organisms that live in a place interact with each other and with their physical environment.
   - An ecosystem is a dynamic ecological system composed of a community and the nonliving factors with which it interacts.

Populations

19.2 Population Growth

1. Population size, density, and growth are other key characteristics of populations.
2. Exponential growth occurs in a population when no factors are limiting its growth. When resources are limiting, a population increases in size to the carrying capacity of the environment.

19.3 The Influence of Population Density

1. Factors such as weather and physical disruptions are density-independent, acting on population growth regardless of population size. Density-dependent effects are factors affected by increases in population size.

19.4 Life History Adaptations

1. Populations whose resources are abundant experience little competition and reproduce rapidly; these organisms favor near-exponential growth and exhibit r-selected adaptations. Populations that experience competition over limited resources favor logistic growth and exhibit K-selected adaptations.

19.5 Population Demography

1. The growth rate of a population is a sensitive function of its age structure. Survivorship curves illustrate the impact of mortality rates among different age groups in a population.

How Competition Shapes Communities

19.6 Communities

1. The array of organisms that live together in an area is called a community. These individuals compete and cooperate with each other to make the community stable.

19.7 The Niche and Competition

1. A niche is defined as the way an organism uses the available resources in its environment. Competition limits an organism from using its entire niche.
2. When two species try to use the same niche, one can outcompete the other, driving it to extinction (competitive exclusion).

Species Interact in Many Ways

19.8 Coevolution and Symbiosis

1. Coevolution is the long-term evolutionary adjustment of two or more species to each other.
2. Symbiotic relationships involve organisms of different species that live together. The major kinds of symbioses are mutualism, commensalism, and parasitism.

19.9 Predation

1. In predator–prey relationships, the predator kills and consumes the prey.
2. Predator and prey populations often exhibit cycles, the prey population being hunted to a low number, which then begins to negatively affect predator population size.

Community Stability

19.10 Ecological Succession

1. Succession is the replacement of one community with another. Secondary succession occurs following the disturbance of an existing community, and primary succession is the emergence of a pioneering community where no life existed before.

Inquiry and Analysis

Are Island Populations of Song Sparrows Density Dependent?

When island populations are isolated, receiving no visitors from other populations, they provide an attractive opportunity to test the degree to which a population's growth rate is affected by its size. A population's size can influence the rate at which it grows because increased numbers of individuals within a population tend to deplete available resources, leading to an increased risk of death by deprivation. Also, predators tend to focus their attention on common prey, resulting in increasing rates of mortality as populations grow. However, simply knowing that a population is decreasing in numbers does not tell you that the decrease has been caused by the size of the population. Many factors such as severe weather, volcanic eruption, and human disturbance can influence island population sizes too.

The graph to the right displays data collected from 13 song sparrow populations on Mandarte Island (see map below). In an attempt to gauge the impact of population size on the evolutionary success of these populations, each population was counted, and its juvenile mortality rate estimated. On the graph, these juvenile mortality rates have been plotted against the number of breeding adults in each population. Although the data appear scattered, the "best-fit" regression line is statistically significant (statistically significant means that there is a less than 5% chance that there is in fact no correlation between dependent and independent variables).

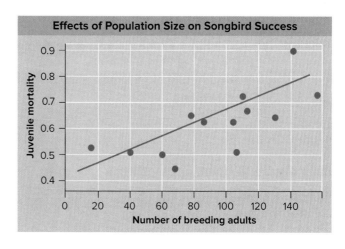

Analysis

1. **Analyzing Scattered Data** What is the size of the song sparrow population (based on breeding adults) with the lowest juvenile mortality? With the greatest?

2. **Interpreting Data**
   a. What is the average juvenile mortality of all 13 populations, estimated from the 13 points on the graph?
   b. How many populations were observed to have juvenile mortality rates *below* this average value? What is the average size of these populations?
   c. How many populations were observed to have juvenile mortality rates *above* this average value? What is the average size of these populations?

3. **Making Inferences** Are the populations with lower juvenile mortality bigger or smaller than the populations with higher juvenile mortality?

4. **Drawing Conclusions** Do the population sizes of these song sparrows appear to exhibit density dependence?

Tim Zurowski/All Canada Photos/Getty Images

20 Ecosystems

LEARNING PATH ▼

The Energy in Ecosystems
1. Ecosystems
2. Ecological Pyramids

Materials Cycle Within Ecosystems
3. The Water Cycle
4. The Carbon Cycle
5. The Nitrogen and Phosphorus Cycles

How Weather Shapes Ecosystems
6. The Sun and Atmospheric Circulation
7. Latitude and Elevation

Major Kinds of Ecosystems
8. Ocean Ecosystems
9. Freshwater Ecosystems
10. Land Ecosystems

THIS ARKANSAS SWAMP may be home to a ghost.

baluzek/Getty Images

Ghostbusters

Have you ever seen a ghost? A living being that you thought was dead and gone, but then you see it there in front of you? Floating along the Cache River that flows through the Cache River National Wildlife Refuge in the Big Woods region of Arkansas during winter seems the perfect environment to encounter such a ghost. It is cold and damp in this forest wetland. In the above photo of it, you can see the bald cypress, tupelo, and swamp maples towering above the swampy ground. The river runs slow here with an eerie calm that makes you jump with each sound. Gene Sparling of Hot Springs, Arkansas, was kayaking down the river on February 11, 2004, when something caught his eye. A huge red-crested woodpecker flew toward him and landed on a nearby tree. The red head and white-and-black color pattern suggested it was an ivory-billed woodpecker (*Campephilus principalis*). Sparling had seen a ghost.

Extinct

The ivory-billed woodpecker is extinct, you see. An unpaired female, the last of its kind, was seen in 1944 flying in a dense stand of trees in the Singer Tract, an isolated remnant of virgin timber along northeastern Louisiana's Tensas River. Then, for 60 years, silence. There had been sporadic unconfirmed sightings over the years, but the species, found only in the southeastern United States and Cuba, was regrettably consigned to the same category as the dodo and the passenger pigeon: extinct.

Why?

What happened to this wonderful woodpecker? It lost its home. The ivory-billed woodpecker was a lover of the deep forest. It made nests in cavities of dying trees and ate beetle larvae that crawled beneath the bark. The ivory-billed woodpecker thrived deep in the great expanses of virgin timber that covered much of the South before the Civil War.

After the Civil War, that changed. A rapidly rebuilding nation developed a huge appetite for lumber, and humans began logging the forests of the southeastern United States. The destruction of the ivory-billed woodpecker's ecosystem continued for almost a century, until by the

SHIRA/SIPA/Newscom

1940s, there was no more timber left to cut. Twenty million acres of bottomland hardwood forests that once covered the Mississippi River delta region were replaced by swamp and scrub. The ivory bills were forced into small bits of fragmented forestland such as the Singer tract. Through the late 1800s and early 1900s, their numbers dwindled, until that last lonely bird was seen in 1944.

But Maybe Not

But are they really extinct? Did Sparling see one? The bird dodged his video camera's lens, leaving only his memories as evidence. Harder evidence was needed to convince a skeptical public. What followed was a year-long search by a research team of renowned ornithologists. Their search generated five additional sightings, numerous audio identifications, and four seconds of a video clip. The video image is blurred and pixilated because of the bird's distance and speed with which it flew, but field markings led the research team to conclude that it was indeed an ivory-billed woodpecker. And then—silence.

An Ongoing Search

The search for the ghostbird has continued for a decade now in the Arkansas Big Woods, as well as similar areas in Mississippi, Louisiana, South Carolina, and Florida, using robotic digital video to monitor areas 24/7. Research teams are in hot pursuit of that one clear picture or video, undisputable proof that the ivory-billed woodpecker still lives. But nothing yet. . . .

As you will learn in this chapter, each piece of an ecosystem affects other pieces. When one piece is removed, the balance is disrupted. That is what has happened with the ivory-billed woodpecker. Is it possible that a small population of ivory-billed woodpecker has survived, descendants perhaps of the one pictured above? One of the last ivory-bills ever photographed, this bird's image was captured by ornithologist Arthur Allen on an expedition to the Singer Tract in 1935. Maybe a population of its descendants still lives in the swampy bottomland forests of Louisiana or Arkansas. In 2021 the U.S. Fish and Wildlife Service gave up hope, and officially declaired the ivory-billed woodpecker extinct. Hope is not yet extinct among researchers, however, who still continue their search for the ghost.

The Energy in Ecosystems

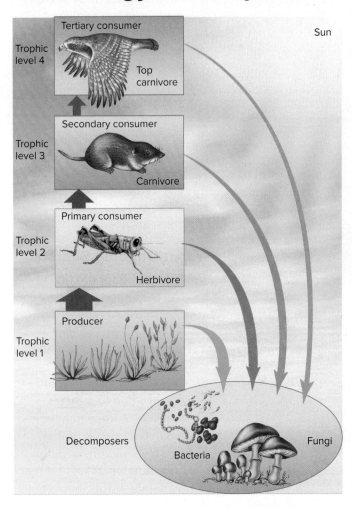

Figure 20.1 Trophic levels within an ecosystem.
Ecologists assign all the members of a community to various trophic levels based on feeding relationships.

20.1 Ecosystems

Life on earth is organized into ecological systems, groups of animals, plants, and other organisms sharing an environment.

What Is an Ecosystem?

> **LEARNING OBJECTIVE 20.1.1** Distinguish among community, habitat, and ecosystem.

Ecologists, the scientists who study ecology, view the world as a patchwork quilt of different environments, all bordering on and interacting with one another. Consider for a moment a patch of forest, the sort of place a deer might live. Ecologists call the collection of creatures that live in a particular place a **community**—all the animals, plants, fungi, and microorganisms that live together in a forest, for example, are the forest community. Ecologists call the place where a community lives its **habitat**—the soil and the water flowing through it are key components of the forest habitat. The sum of these two, community and habitat, is an ecological system, or **ecosystem.** An ecosystem is a largely self-sustaining collection of organisms and their physical environment. An ecosystem can be as large as a forest or as small as a tide pool.

The ecosystem is the most complex level of biological organization. The **biosphere** includes all the ecosystems on earth. The earth is a closed system with respect to chemicals but an open system in terms of energy. The organisms in ecosystems regulate the capture and expenditure of energy and the cycling of chemicals. All organisms depend on the ability of photosynthetic organisms to recycle the basic components of life.

> **Putting the Concept to Work**
> Is your family a community and your home its habitat? Discuss.

Energy Flows Through Ecosystems

> **LEARNING OBJECTIVE 20.1.2** Trace the path of energy through the trophic levels of an ecosystem.

Energy flows into the biological world from the sun, which shines a constant beam of light on our earth. Life exists on earth because some of that light energy can be captured and transformed into chemical energy through the process of photosynthesis and used to make organic molecules such as carbohydrates, nucleic acids, proteins, and fats. These organic molecules are what we call food. Living organisms use the energy in food to make new materials for growth, to repair damaged tissues, to reproduce, and to do myriad other things requiring energy.

You can think of all the organisms in an ecosystem as chemical machines fueled by energy captured in photosynthesis. The organisms that first capture the energy, the **producers,** are plants, algae, and some bacteria, which produce their own energy-storing molecules by carrying out photosynthesis. They are also referred to as *autotrophs*. All other organisms in an ecosystem are **consumers,** obtaining energy-storing molecules by consuming plants or other animals, and are referred to as *heterotrophs*. Ecologists assign every organism in an ecosystem to a trophic (or feeding) level, depending on the source of its energy. A **trophic level** is composed of those organisms within an ecosystem whose source of energy is the same number of consumption "steps" away from the sun. Thus, as shown in figure 20.1, a plant's trophic level is 1, whereas *herbivores* (animals that graze on plants) are in trophic level 2, and *carnivores* (animals that eat these grazers) are in trophic level 3. Higher trophic levels exist for carnivores that eat other carnivores (trophic level 4 in figure 20.1). Food energy passes through an ecosystem from one trophic level to another. When the path is a simple linear progression, like the links of a chain, it is called a **food chain.** The chain ends with *decomposers,* who break down dead organisms or their excretions, and return the organic matter to the soil.

(a) Producers and herbivores

Producers

The lowest trophic level of any ecosystem is occupied by the producers (figure 20.2a)—green plants in most land ecosystems (and, usually, algae in aquatic ecosystems). Plants use the energy of the sun to build energy-rich sugar molecules. They also absorb carbon dioxide from the air and nitrogen and other key substances from the soil, and use them to build biological molecules. It is important to realize that plants consume as well as produce. The roots of a plant, for example, do not carry out photosynthesis—there is no sunlight underground. Roots obtain their energy the same way you do, by using energy-storing molecules produced elsewhere (in this case, in the leaves of the plant).

(b) Carnivores

Herbivores

At the second trophic level are **herbivores,** animals that eat plants (figure 20.2a). They are the *primary consumers* of ecosystems. Deer and horses are herbivores, and so are rhinoceroses, chickens (primarily herbivores), and caterpillars. Most herbivores rely on "helpers" to aid in the digestion of cellulose, a structural material found in plants. A cow, for instance, has a thriving colony of bacteria in its gut that digests cellulose. So does a termite. Humans cannot digest cellulose because we lack these bacteria—that is why a cow can live on a diet of grass and you cannot.

(c) Omnivore

Figure 20.2 Members of the food chain.

(a) The East African grasslands are covered by a dense growth of grasses, the primary producers. Grazing herbivores like these zebras obtain their food from plants.
(b) These wolves are carnivores that live in North American forests. (c) This grizzly bear is an omnivore, this one fishing for salmon.
(d) This crab is a detritivore. (e) Fungi, such as this basidiomycete growing through the soil, and bacteria are the primary decomposers of terrestrial ecosystems.

(a) imagebroker/Alamy Stock Photo; (b) NPS photo; (c) Ron Crabtree/Getty Images; (d) aodaodaod/Getty Images; (e) Emmanuel LATTES/Alamy Stock Photo

(d) Detritivore

(e) Decomposer

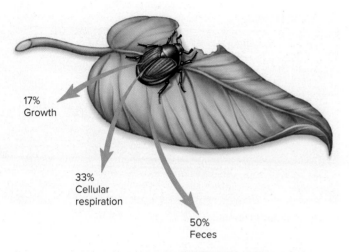

Figure 20.3 How heterotrophs use food energy.

A heterotroph assimilates only a fraction of the energy it consumes. For example, if a "bite" is composed of 500 joules of energy (1 joule = 0.239 calories), about 50%, 250 J, is lost in feces, about 33%, 165 J, is used to fuel cellular respiration, and about 17%, 85 J, is converted into consumer biomass. Only this 85 J (or roughly 20 calories) is available to the next trophic level.

Carnivores

At the third trophic level are animals that eat herbivores, called **carnivores** (meat-eaters). They are the *secondary consumers* of ecosystems. Tigers and wolves are carnivores (figure 20.2*b*), and so are mosquitoes and blue jays. Some animals, like bears and humans, eat both plants and animals and are called **omnivores** (figure 20.2*c*). They use the simple sugars and starches stored in plants as food but not the cellulose. Many complex ecosystems contain a fourth trophic level, composed of animals that consume other carnivores. They are called *tertiary consumers,* or *top carnivores.* A weasel that eats a blue jay is a tertiary consumer. Only rarely do ecosystems contain more than four trophic levels, for reasons we will discuss later.

Detritivores and Decomposers

In every ecosystem, there is a special class of consumers that include **detritivores** (figure 20.2*d*), organisms that eat dead organisms (also referred to as scavengers) and **decomposers,** organisms that break down organic substances making the nutrients available to other organisms (figure 20.2*e*). Worms and vultures are examples of detritivores. Bacteria and fungi are the principal decomposers in land ecosystems.

Energy Flows Through Trophic Levels

How much energy passes through an ecosystem? *Primary productivity* is the total amount of light energy converted by photosynthetic organisms into organic compounds in a given area per unit of time. An ecosystem's *net primary productivity* is the total amount of energy fixed by photosynthesis per unit of time minus that which is expended by photosynthetic organisms to fuel metabolic activities. In short, it is the energy stored in organic compounds that is available to heterotrophs. The total weight of all an ecosystem's organisms, called its **biomass,** increases as a result of the ecosystem's net productivity.

Food Chains. When a plant uses the energy from sunlight to make structural molecules such as cellulose, it loses a lot of energy as heat. In fact, only about half of the energy captured by the plant ends up stored in its molecules. The other half of the energy is lost. This is the first of many such losses as the energy passes through the ecosystem. When the energy flow through an ecosystem is measured at each trophic level, we find that 80% to 95% of the energy available at one trophic level is not transferred to the next. In other words, only 5% to 20% of the available energy passes from one trophic level to the next. For example, the amount of energy that ends up in the beetle's body in figure 20.3 is approximately only 17% of the energy present in the plant molecules it eats. Similarly, when a carnivore eats the herbivore, a comparable amount of energy is lost from the amount of energy present in the herbivore's molecules. This is why food chains generally consist of only three or four steps. So much energy is lost at each step that little usable energy remains after it has been incorporated into the bodies of organisms at four successive trophic levels.

Food Webs. In most ecosystems, the path of energy is not a simple linear one, because individual animals often feed at several trophic levels. This creates a more complicated path of energy flow called a **food web** (figure 20.4).

> **Putting the Concept to Work**
> What trophic level do you occupy in a food chain? Explain.

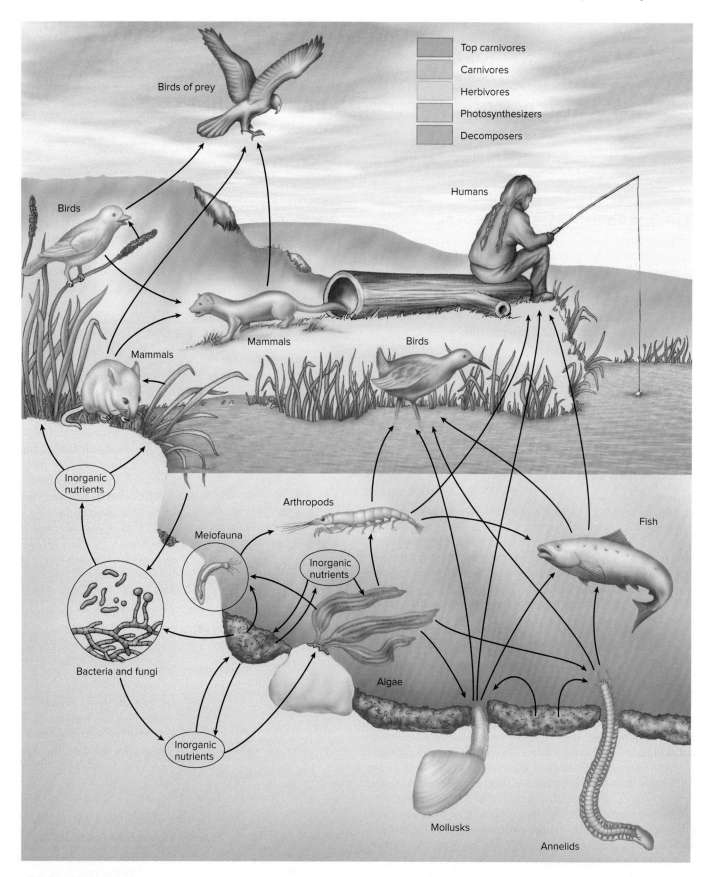

Figure 20.4 **A food web.**

A food web is much more complicated than a linear food chain. The path of energy passes from one trophic level to another and back again in complex ways.

A Closer Look

Metabolic Efficiency and the Length of Food Chains

In the earth's ecosystems, the organisms that carry out photosynthesis are often consumed as food by other organisms. We call these "organism-eaters" *heterotrophs*. Humans are heterotrophs, because no human photosynthesizes.

When Glycolysis Was All There Was

It is thought that the first heterotrophs were ancient bacteria living in a world where photosynthesis had not yet introduced much oxygen into the oceans or atmosphere. The only mechanism they possessed to harvest chemical energy from their food was glycolysis. It has been estimated that a heterotroph limited to glycolysis, as these ancient bacteria were, captures only 3.5% of the energy in the food it consumes. Hence, if such a heterotroph preserves 3.5% of the energy in the photosynthesizers it consumes, then any other heterotrophs that consume the first heterotroph will capture through glycolysis 3.5% of the energy in it, or 0.12% of the energy available in the original photosynthetic organisms. A very large base of photosynthesizers would thus be needed to support a small number of heterotrophs.

With Oxygen Available

When organisms became able to extract energy from organic molecules by oxidative cellular respiration, which we discussed in chapter 7, this constraint became far less severe, because the efficiency of oxidative respiration is estimated to be about 32%. This increased efficiency results in the transmission of much more energy from one trophic level to another than does glycolysis. The efficiency of oxidative cellular respiration has made possible the evolution of food chains, in which photosynthesizers are consumed by heterotrophs, which are consumed by other heterotrophs, and so on.

Eating Wastes Energy

Even with this very efficient oxidative metabolism, approximately two-thirds of the available energy is lost at each trophic level, and that puts a limit on how long a food chain can be. Most food chains, such as the East African grassland ecosystem illustrated here, involve only three or rarely four trophic levels. Too much energy is lost at each transfer to allow chains to be much longer than that. For example, it would be impossible for a large human population to subsist by eating lions captured from the grasslands of East Africa; the amount of grass available there would not support enough zebras and other herbivores to maintain the number of lions needed to feed the human population. Thus, the ecological complexity of our world is fixed in a fundamental way by the chemistry of oxidative cellular respiration.

Photosynthesizers. The grass under this yellow fever tree grows actively during the hot, rainy season, capturing the energy of the sun and storing it in molecules of glucose, which are then converted into starch and stored in the grass.
Ralph A. Clevenger/Getty Images

Herbivores. These zebras consume the grass and transfer some of its stored energy into their own bodies.
Alan and Sandy Carey/Getty Images

Carnivores. The lion feeds on zebras and other animals, capturing part of their stored energy and storing it in its own body.
Patrick Gijsbers/iStock/Getty Images

Scavengers. This hyena and the vultures occupy the same stage in the food chain as the lion. They also consume the body of the kill, after it has been abandoned by the lion.
Peter Miller/Getty Images

Refuse utilizers. These butterflies are feeding on the material left in the hyena's dung after the food the hyena consumed had passed through its digestive tract.
Kittisak Chysree/Shutterstock

A food chain in the open grasslands of East Africa.

At each of these levels in the food chain, only about a third or less of the energy present is used by the recipient.

20.2 Ecological Pyramids

LEARNING OBJECTIVE 20.2.1 Explain why herbivore biomass tends to be greater than that of the carnivores that consume them.

Pyramids

A plant fixes about 1% of the sun's energy that falls on its green parts. The successive members of a food chain, in turn, process into their own bodies, on average, about 10% of the energy available in the organisms on which they feed. For this reason, there are generally far more individuals at the lower trophic levels of any ecosystem than at the higher levels. Similarly, the biomass of the primary producers present in a given ecosystem is greater than the biomass of the primary consumers, with successive trophic levels having a lower and lower biomass and so less and less potential energy.

These key ecological relationships appear as pyramids when expressed as diagrams. Ecologists speak of "pyramids of numbers," where the sizes of the blocks reflect the number of individuals at each trophic level. The "pyramid of energy" in figure 20.5 shows plankton (the producers) as the largest block.

Top Carnivores

The loss of energy that occurs at each trophic level places a limit on how many top-level carnivores a community can support. As we have seen, only about one-thousandth of the energy captured by photosynthesis passes all the way through a three-stage food chain to a tertiary consumer, such as a snake or hawk. This explains why there are no predators that subsist on lions—the biomass of these animals is simply insufficient to support another trophic level.

Rostislav Stach/Shutterstock

In the pyramid of numbers, top-level predators tend to be fairly large animals. Thus, the small residual biomass available at the top of the pyramid is concentrated in a relatively small number of individuals.

Putting the Concept to Work
Explain why the pyramids of numbers, biomass, and energy resemble one another.

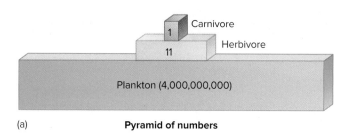

(a) **Pyramid of numbers**

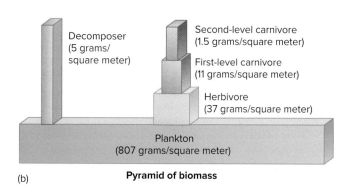

(b) **Pyramid of biomass**

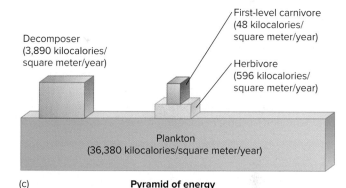

(c) **Pyramid of energy**

Figure 20.5 Ecological pyramids measure different characteristics of each trophic level. In these aquatic ecosystems, plankton are the primary producers. (a) Pyramid of numbers. (b) Pyramid of biomass. (c) Pyramid of energy.

Materials Cycle Within Ecosystems

20.3 The Water Cycle

> **LEARNING OBJECTIVE 20.3.1** Contrast the environmental and organismic water cycles.

Unlike energy, which flows through the earth's ecosystems in one direction (from the sun to producers to consumers), the physical components of ecosystems are passed around and reused within ecosystems. Ecologists speak of such constant reuse as recycling or, more commonly, cycling. Materials that are constantly recycled include all the chemicals that make up the soil, water, and air. While many are important, the proper cycling of four materials is particularly critical to the health of any ecosystem: (1) water, (2) carbon, and (3) the soil nutrients nitrogen and phosphorus.

Geochemical Cycles

The paths of water, carbon, and soil nutrients as they pass from the environment to living organisms and back form closed circles, or cycles. In each cycle, the chemical resides for a time in an organism and then returns to the nonliving environment, often referred to as a *biogeochemical cycle*.

Of all the nonliving components of an ecosystem, water has the greatest influence on the living portion. The availability of water and the way in which it cycles in an ecosystem in large measure determines the biological richness of that ecosystem—the kinds of creatures that live there and how many of each.

Water cycles within an ecosystem in two ways: the environmental water cycle and the organismic water cycle. Both cycles are shown in **figure 20.6**.

BIOLOGY & YOU

You and the Water Cycle. How many gallons of water do you personally consume in a day? Take a guess. Here are some numbers to help you make an estimate: *bath*: 40 gallons; *shower*: 2 gallons per minute; *teeth brushing*: 1 gallon; *hands/face washing*: 1 gallon; *dishwasher*: 15 gallons/load; *clothes machine load*: 25 gallons/load; *toilet flush*: 3 gallons; *glasses of water drunk*: 1/16 gallon/glass. Your per-capita water use (*per* is Latin for "by" and *capita* is Latin for "head") can rapidly become quite alarming—and the number you get from a calculation like this is quite likely an underestimation. It does not take into account everything you use water for, such as cooking, washing your dog, or cleaning muddy shoes. You might water your lawn, wash your car, or leave the water running while you brush your teeth. It all adds up. Even using water conservation measures such as low-flow shower heads and toilets, few of us have a per-capita water use less than 100 gallons a day.

Figure 20.6 The water cycle.

Precipitation on land eventually makes its way to the ocean via groundwater, lakes, and rivers. Solar energy causes evaporation, adding water to the atmosphere. Plants give off excess water through transpiration, also adding water to the atmosphere. Atmospheric water falls as rain or snow over land and oceans, completing the water cycle.

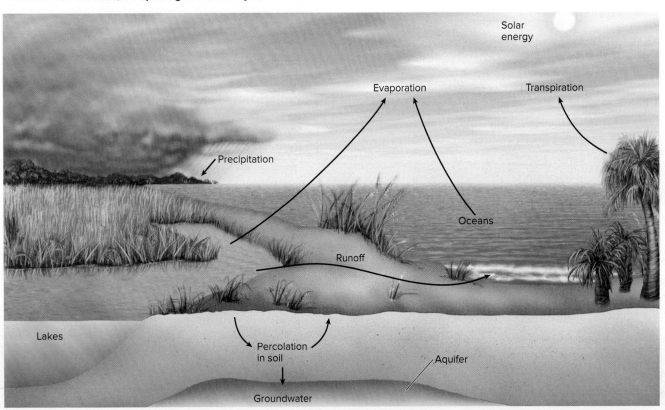

The Environmental Water Cycle

In the environmental water cycle, water vapor in the atmosphere condenses and falls to the earth's surface as rain or snow (called precipitation in figure 20.6). Heated there by the sun, it reenters the atmosphere by **evaporation** from lakes, rivers, and oceans, where it condenses and falls to the earth again.

The Organismic Water Cycle

In the organismic water cycle, surface water does not return directly to the atmosphere. Instead, it is taken up by the roots of plants. After passing through the plant, the water reenters the atmosphere through tiny openings (stomata) in the leaves, evaporating from their surface. This evaporation from leaf surfaces is called **transpiration.** Transpiration is driven by the sun: Its heat creates wind currents that draw moisture from the plant by passing air over the leaves.

Breaking the Cycle

In very dense forest ecosystems, such as tropical rain forests, more than 90% of the moisture in the ecosystem is taken up by plants and then transpired back into the air. Because so many plants in a rain forest are doing this, the vegetation is the primary source of local rainfall. In a very real sense, these plants create their own rain: The moisture that travels up from the plants into the atmosphere falls back to earth as rain.

Where forests are cut down, the organismic water cycle is broken, and moisture is not returned to the atmosphere. Water drains off to the sea instead of rising to the clouds and falling again on the forest. During his expeditions from 1799 to 1805, the great German explorer Alexander von Humboldt reported that stripping the trees from a tropical rain forest in Colombia prevented water from returning to the atmosphere and created a semiarid desert. Once the water cycle is broken, it does not spontaneously repair itself. After 200 years, that part of Colombia is still a desert. This is a clear early instance of local "climate change" induced by human activity. We will have much more to say about climate change in chapter 22. It is a tragedy of our time that this continues today. In many tropical areas, rain forests are being clear-cut or burned in the name of "development" (figure 20.7).

Groundwater

Much less obvious than the surface waters seen in streams, lakes, and ponds is the groundwater, which occurs in permeable, saturated, underground layers of rock, sand, and gravel called *aquifers.* In many areas, groundwater is the most important water reservoir; for example, in the United States, more than 96% of all freshwater is groundwater, which provides about 50% of the population with its drinking water.

Because of the greater rate at which groundwater is being used, the increasing chemical pollution of groundwater is a very serious problem. Pesticides, herbicides, and fertilizers are key sources of groundwater pollution. Because of the large volume of water, its slow rate of turnover, and its inaccessibility, removing pollutants from aquifers is virtually impossible.

Figure 20.7 Burning or clear-cutting forests breaks the water cycle.

The high density and large size of plants in a forest translate into great quantities of water being transpired to the atmosphere, creating rain over the forests. In this way, rain forests perpetuate the wet climate that supports them. Tropical deforestation permanently alters the climate in these areas, creating arid zones.

luoman/E+/Getty Images

> **Putting the Concept to Work**
> How does clear-cutting break the environmental water cycle?

20.4 The Carbon Cycle

LEARNING OBJECTIVE 20.4.1 Contrast the effects of respiration, erosion, and combustion on the carbon cycle.

The earth's atmosphere contains plentiful carbon, present as carbon dioxide (CO_2) gas. This carbon cycles between the atmosphere and living organisms, often being locked up for long periods of time in organisms or deep underground. The cycle is begun by plants that use CO_2 in photosynthesis to build organic molecules—in effect, they trap the carbon atoms of CO_2 within the living world. The carbon atoms are returned to the atmosphere's pool of CO_2 through respiration, combustion, and erosion, as shown in **figure 20.8**.

Respiration

All organisms in ecosystems respire—that is, they extract energy from organic food molecules, which involves stripping away the carbon atoms and combining them with oxygen to form CO_2. This end product of respiration is released into the atmosphere.

Combustion

Plants that become buried in sediment may be transformed by pressure into coal or oil. The carbon originally trapped by these plants is only released back into the atmosphere when the coal or oil (called *fossil fuels*) is burned.

Erosion

Very large amounts of carbon are present in limestone, formed from calcium carbonate shells of marine organisms. When the limestone becomes exposed to weather and erodes, the carbon washes back and is dissolved again in oceans.

> **Putting the Concept to Work**
>
> How is carbon released into the atmosphere from plants?

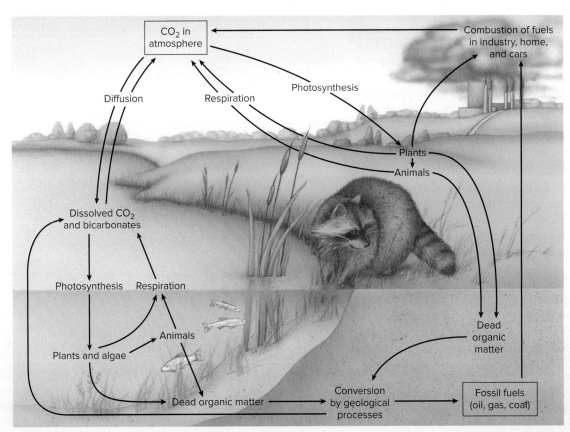

Figure 20.8 The carbon cycle.

Carbon from the atmosphere and from water is fixed by photosynthetic organisms and returned through respiration, combustion, and erosion.

20.5 The Nitrogen and Phosphorus Cycles

LEARNING OBJECTIVE 20.5.1 Compare the nitrogen and phosphorus cycles.

The Nitrogen Cycle

Earth's atmosphere is 78.08% nitrogen gas (N_2), but most living organisms are unable to use the N_2 so plentifully available in the air surrounding them. The two nitrogen atoms of N_2 are bound together by a particularly strong "triple" covalent bond that is very difficult to break. Luckily, a few kinds of bacteria can break this triple bond and bind nitrogen atoms to hydrogen, forming "fixed" nitrogen, ammonia (NH_3), in a process called **nitrogen fixation.**

Bacteria evolved the ability to fix nitrogen early in the history of life, before photosynthesis had introduced oxygen gas into the earth's atmosphere, and that is still the only way the bacteria are able to do it—even a trace of oxygen poisons the process. In today's world, awash with oxygen, these bacteria live encased within bubbles called cysts that admit no oxygen, or within special airtight cells in nodules of tissue on the roots of beans, aspen trees, and a few other plants. Figure 20.9 shows the workings of the nitrogen cycle.

The growth of plants in ecosystems is often severely limited by the availability of "fixed" nitrogen in the soil, which is why farmers fertilize fields. Today, most fixed nitrogen added to soils by farmers is produced in factories by industrial rather than bacterial nitrogen fixation. This industrial process today accounts for a prodigious 30% of the entire nitrogen cycle.

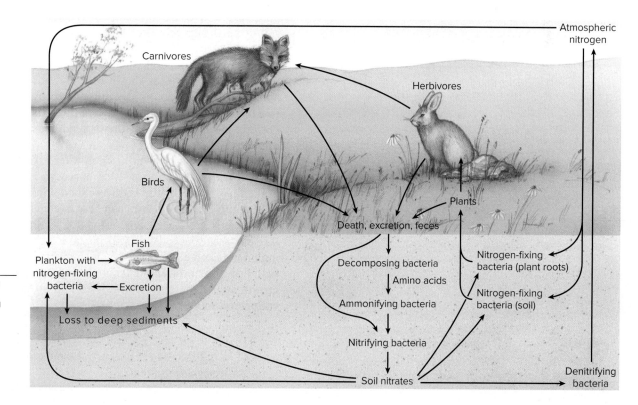

Figure 20.9 The nitrogen cycle. Relatively few kinds of organisms—all of them bacteria—can convert atmospheric nitrogen into forms that can be used for biological processes.

Figure 20.10
The phosphorus cycle.

Phosphorus plays a critical role in plant nutrition. Next to nitrogen, phosphorus is the element most likely to be so scarce that it limits plant growth.

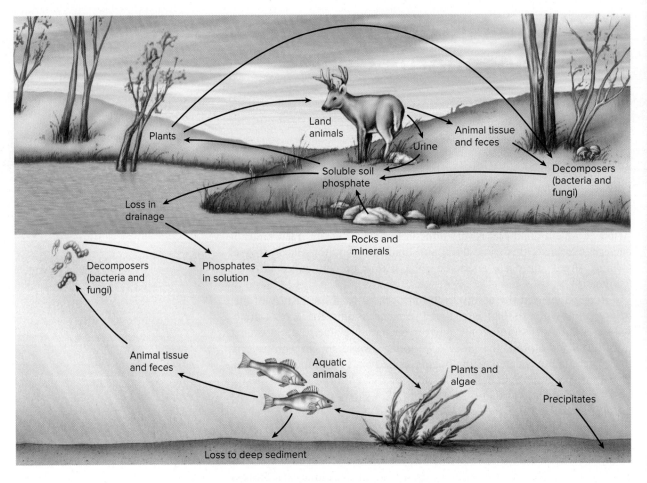

The Phosphorus Cycle

Phosphorus is an essential element in all living organisms, a key part of both ATP and DNA. Phosphorus is often in very limited supply in the soil of particular ecosystems, and because phosphorus does not form a gas, none is available in the atmosphere. Most phosphorus exists in soil and rock as the mineral calcium phosphate, which dissolves in water to form phosphate ions, as shown in figure 20.10. These phosphate ions are absorbed by the roots of plants and used by them to build organic molecules such as ATP and DNA. When the plants and animals die and decay, bacteria in the soil convert the organic phosphorus back into phosphate ions, completing the cycle.

The phosphorus level in freshwater lake ecosystems is often quite low, preventing much growth of photosynthetic algae in these systems. Pollution of a lake by the inadvertent addition of phosphorus to its waters (agricultural fertilizers and many commercial detergents are rich in phosphorus) first produces a green scum of algal growth on the surface of the lake and then proceeds to "kill" the lake: As aging algae die, bacteria feeding on the dead algae cells use up so much of the lake's dissolved oxygen that fish and invertebrate animals suffocate. Such rapid, uncontrolled growth caused by excessive nutrients in an aquatic ecosystem is called **eutrophication.**

> **Putting the Concept to Work**
> **If air is 78% nitrogen gas, why is plant growth often severely limited by available nitrogen?**

How Weather Shapes Ecosystems

20.6 The Sun and Atmospheric Circulation

> **LEARNING OBJECTIVE 20.6.1** Describe how the sun drives circulation of the atmosphere.

The world contains a wide variety of ecosystems because its climate varies a great deal from place to place. On a given day, Miami and Boston often have very different weather. There is no mystery about this. The tropics are warmer than the temperate regions because the sun's rays arrive almost perpendicular (that is, dead on) at regions near the equator. As you move from the equator into temperate latitudes, sunlight strikes the earth at more oblique angles, which spreads it out over a much greater area, thus providing less energy per unit of area (figure 20.11). This simple fact—because the earth is a sphere, some parts of it receive more energy from the sun than others—is responsible for much of the earth's different climates and thus, indirectly, for much of the diversity of its ecosystems.

Seasons

The earth's annual orbit around the sun and its daily rotation on its own axis are also both important in determining world climate. Because of the daily cycle, the climate at a given latitude is relatively constant. Because of the annual cycle and the inclination of the earth's axis, all parts away from the equator experience a progression of seasons. In summer in the Southern Hemisphere, the earth is tilted toward the sun, as shown in figure 20.11, and rays hit more directly, leading to higher temperatures. As the earth reaches the opposite position in its annual orbit, the Northern Hemisphere receives more direct rays from the sun and experiences summer.

How Latitude Determines Climate

Near the equator, warm air rises and flows toward the poles (indicated in figure 20.12 by arrows at the equator that rise and circle). As it rises and cools, this air loses most of its moisture because cool air holds less water vapor than warm air. (This explains why it rains so much in the tropics where the air is warm.) When this air has traveled to about 30 degrees N (north) and S (south) latitudes, the cool, dry air sinks and becomes reheated, soaking up water like a sponge as it warms, producing a broad zone of low rainfall. It is no accident that all of the great deserts of the world lie near 30 degrees N or 30 degrees S latitude. Air at these latitudes is still warmer than it is in the polar regions, and thus it continues to flow toward the poles. At about 60 degrees N and S latitudes, air rises and cools and sheds its moisture, and such are the locations of the great temperate forests of the world. Finally, this rising air descends near the poles, producing zones of very low precipitation.

> **Putting the Concept to Work**
> Why do all the earth's great deserts lie near 30 degrees N or 30 degrees S latitude?

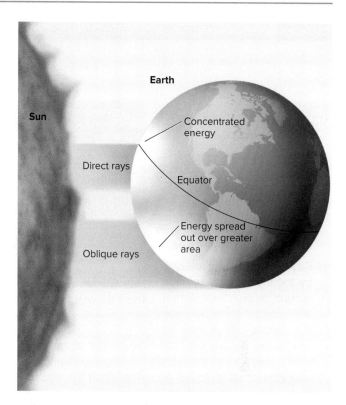

Figure 20.11 Latitude affects climate.

The relationship between the earth and sun is critical in determining the nature and distribution of life on earth. The tropics are warmer than the temperate regions because the sun's rays strike at a direct angle, providing more energy per unit of area.

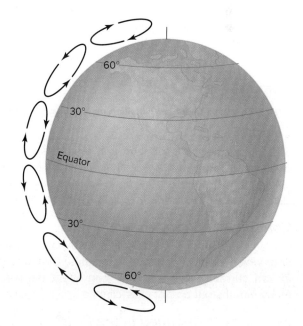

Figure 20.12 Air rises at the equator and then falls.

The pattern of air movement out from and back to the earth's surface forms three pairs of great cycles.

Answering Your Questions About Weather and Climate

Weather

What Is Weather?

Weather is the state of the atmosphere at a place and time—basically, how hot, wet, and still the air is.

What Causes Changes in Weather?

Day and night are caused by the earth's daily rotation around its own axis. Winter and summer occur because the earth's axis is inclined with respect to the sun: Over the course of a rotation (that is, over a year's time), the weather changes seasonally, with summer in the United States occurring when the earth's Northern Hemisphere is tilted toward the sun.

Muir Glacier, Alaska—These two photos taken on August 13, 1941 (left) and August 31, 2004 (right) show 63 years of the glacier's retreat.

left: U.S. Geological Survey (USGS); **right:** Bruce Molnia/U.S. Geological Survey (USGS)

Is Our Weather Getting More Extreme?

Hurricanes, droughts, and other extreme weather events are not more frequent than in the past but do seem to be getting stronger. In a single recent year, 2017, three major hurricanes battered Texas (Hurricane Harvey), Florida (Hurricane Irma), and Puerto Rico (Hurricane Maria). In that same year in Australia, a heat wave literally cooked the continent, causing many deaths, while lack of rainfall in California led to wildfires that torched Napa Valley. Further south, in Ventura County, California, wildfires denuded the hills, leading to mud slides that killed 20.

Climate

What Is the Difference Between Weather and Climate?

Basically, weather is short-term and climate is long-term. In a given locality, the prevailing weather conditions are fairly predictable over long periods of time. The tropics have a warm, wet climate, for example, whereas polar regions are dry and cold.

Can Climate Change?

Not easily. Climate is largely dictated by latitude and elevation, which change only over centuries. Changes in Pacific Ocean currents called El Niño can affect the weather for many months but do not alter climate.

Has Climate Changed in the Past?

Yes. The earth has experienced several ice ages in the distant past and is currently in the midst of one that started about 2.6 million years ago. So why isn't it icy? Because the earth's orbit wobbles a bit, creating interglacial cycles when the wobble tilts earth toward the sun a bit. For the last 10,000 years, we have been in an interglacial period, predicted to last for another 40,000 years. Then the ice comes back.

Climate Change and Global Warming

Is Climate Changing Today?

Yes. In the last century, the earth has seen an increase in the mean global temperature of over two degrees. The 6 hottest years ever recorded have all occurred since 2015. The consequences of this global warming to life on earth will be, and indeed already are, significant (see section 22.2).

How Can We Know If This Change Is Caused by Human Activity?

If wobble in the earth's orbit can mitigate an ice age, might it also be responsible for the global warming happening today, rather than elevated CO_2 levels? No. The earth's orbit has not changed in the last thousand years. Instead, many lines of evidence confirm CO_2 as the culprit. One very strong argument: Our measurements of CO_2 levels always vary a bit from month to month in a random fashion, as do our measures of global mean temperatures—but variation within the two data sets is not independent. Instead, they vary hand-in-hand (a statistician would say the covariance is high), which is only possible if the underlying cause of the variation is the same for both.

If Earth Is Warming, Why Was Last Winter So Cold?

Weather does not equal climate.

20.7 Latitude and Elevation

LEARNING OBJECTIVE 20.7.1 Explain why changes in latitude and elevation have similar effects on ecosystems.

On earth, the climate you experience depends a great deal on where you are.

Why the Tropics Are Warmer

Temperatures are higher in tropical ecosystems for a simple reason: more sunlight per unit area falls on tropical latitudes (see figure 20.11). Solar radiation is most intense when the sun is directly overhead, and this occurs only in the tropics, where sunlight strikes the equator perpendicularly. Temperature also varies with elevation, with higher altitudes becoming progressively colder. At any given latitude, air temperature falls about 6°C for every 1,000-meter increase in elevation. The ecological consequences of temperature varying with elevation are the same as temperature varying with latitude. Figure 20.13 illustrates this principle, comparing changes in ecosystems that occur with increasing latitudes in North America with the ecosystem changes that occur with increasing elevation at the tropics. A 1,000-meter increase in elevation on a mountain in southern Mexico (figure 20.13b) results in a temperature drop equal to that of an 880-kilometer increase in latitude on the North American continent (figure 20.13a). This is why the timberline (the elevation above which trees do not grow) occurs at progressively lower elevations as one moves farther from the equator.

Rain Shadows

When a moving body of air encounters a mountain (figure 20.14), it is forced upward, and as it is cooled at higher elevations, the air's moisture-holding capacity decreases, producing the rain you see on the windward side of the mountains—the side from which the wind is blowing. The effect on the other side of the mountain—the leeward side—is quite different. As the air passes the peak and descends on the far side of the mountains, it is warmed, so its moisture-holding capacity increases. Sucking up all available moisture, the air dries the surrounding landscape, often producing a desert. This effect, called a rain shadow, is responsible for deserts such as Death Valley, which is in the rain shadow of Mount Whitney, the tallest mountain in the Sierra Nevada.

> **Putting the Concept to Work**
> Death Valley, within 160 miles of the Pacific Ocean, is one of the driest places on earth. How could it be so dry, if it is so close to so much water?

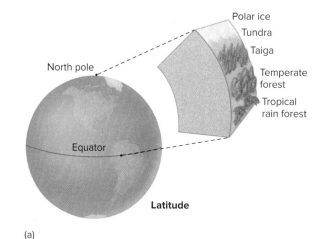

(a)

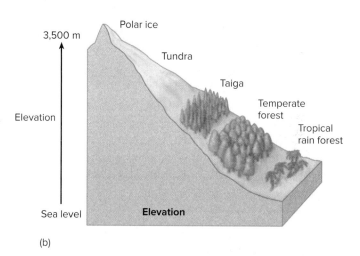

(b)

Figure 20.13 How elevation affects ecosystems.
The same land ecosystems that normally occur as latitude increases north and south of the equator at sea level (a) can occur in the tropics as elevation increases (b).

Figure 20.14 The rain shadow effect.
Moisture-laden winds from the Pacific Ocean rise and are cooled when they encounter the Sierra Nevada. As they cool, their moisture-holding capacity decreases, and precipitation occurs. As the air descends on the east side of the range, it warms, its moisture-holding capacity increases, and the air picks up moisture from its surroundings. As a result, arid conditions prevail on the east side of these mountains.

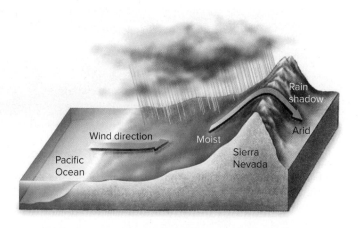

Major Kinds of Ecosystems

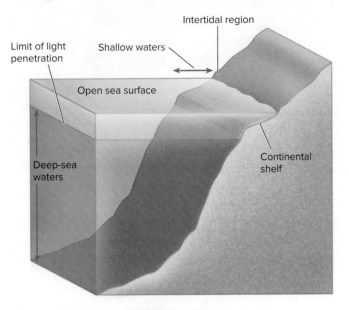

Figure 20.15 Ocean ecosystems.

There are three primary ecosystems found in the earth's oceans. Shallow-water ecosystems occur along the shoreline and at areas of coral reefs. Open-sea surface ecosystems occur in the upper 100 meters where light can penetrate. Finally, deep-sea water ecosystems are bottom areas below 300 meters.

20.8 Ocean Ecosystems

> **LEARNING OBJECTIVE 20.8.1** Compare the communities that occupy shallow waters, open-sea surface, and deep-sea waters.

Most of the earth's surface—nearly three-quarters—is covered by water. The seas have an average depth of more than 3 kilometers, and they are, for the most part, cold and dark. Photosynthetic organisms are confined to the upper few hundred meters because light does not penetrate any deeper. Almost all organisms that live below this level feed on organic debris that rains downward. The three main kinds of marine ecosystems are shallow waters, open-sea surface, and deep-sea waters (figure 20.15).

Shallow Waters

Very little of the earth's ocean surface is shallow—mostly that along the shoreline—but this small area contains many more species than other parts of the ocean (figure 20.16*a*). The world's great commercial fisheries occur on banks in the coastal zones, where nutrients derived from the land are more abundant than in the open ocean. Part of this zone consists of the intertidal region, which is exposed to the air whenever the tides recede. Partly enclosed bodies of water, such as those that often form at river mouths and in coastal bays, where the salinity is intermediate between that of seawater and freshwater, are called **estuaries.** Estuaries are among the most naturally fertile areas in the world, often containing rich stands of submerged and emergent plants, algae, and microscopic organisms. They provide the breeding grounds for most of the coastal fish and shellfish that are harvested both in the estuaries and in open water.

Open-Sea Surface

Drifting freely in the upper, better-illuminated waters of the ocean is a diverse biological community of microscopic organisms. Most of the plankton occurs in the top 100 meters of the sea. Many fish swim in these waters

(a) (b)

Figure 20.16 Shallow waters and open sea surface.

(a) Fish and many other kinds of animals find food and shelter among the coral in the coastal waters of some regions. (b) The upper layers of the open ocean contain plankton and large schools of fish, such as these bigeye snappers.

(a) Stephen Frink/Getty Images; (b) Jeff Hunter/Photographer's Choice RF/Getty Images

as well, feeding on the plankton and one another (figure 20.16b). Some members of the plankton, including algae and some bacteria, are photosynthetic and are called phytoplankton. Collectively, these organisms are responsible for about 40% of all photosynthesis that takes place on earth. Over half of this is carried out by organisms less than 10 micrometers in diameter—at the lower limits of size for organisms—and almost all of it near the surface of the sea, in the zone into which light from the surface penetrates freely.

Deep-Sea Waters

In the deep waters of the sea, below 300 meters, little light penetrates. Very few organisms live there, compared to the rest of the ocean, but those that do include some of the most bizarre organisms found anywhere on earth. Many deep-sea inhabitants have bioluminescent (light-producing) body parts that they use to communicate or to attract prey (figure 20.17a).

Frigid and bare, the floors of the deep sea have long been considered a biological desert. Recent close-up looks taken by marine biologists, however, paint a different picture (figure 20.17b). The ocean floor is teeming with life. Often kilometers deep, thriving in pitch darkness under enormous pressure, crowds of marine invertebrates have been found in hundreds of deep samples from the Atlantic and Pacific. Rough estimates of deep-sea diversity have soared to hundreds of thousands of species. Many appear endemic (local). The diversity of species is so high, it may rival that of tropical rain forests! This profusion is unexpected. New species usually require some kind of barrier to diverge (see chapter 14), and the ocean floor seems boringly uniform. However, little migration occurs among deep populations, and this lack of movement may encourage local specialization and species formation. A patchy environment may also contribute to species formation there; deep-sea ecologists find evidence that fine but nonetheless formidable resource barriers arise in the deep sea.

No light falls in the deep ocean. From where do deep-sea organisms obtain their energy? While some utilize energy falling to the ocean floor as debris from above, other deep-sea organisms are autotrophic, gaining their energy from hydrothermal vent systems, areas in which seawater circulates through porous rock surrounding fissures where molten material from beneath the earth's crust comes close to the surface. Hydrothermal vent systems, also called deep-sea vents, support a broad array of heterotrophic life (figure 20.17c).

> **Putting the Concept to Work**
> How can bacteria living near deep-sea vents be autotrophs, when no light penetrates so deep?

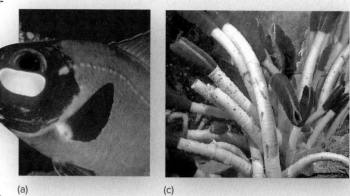
(a) (c)

(b)

Figure 20.17 Deep-sea waters.

(a) The luminous spot below the eye of this deep-sea fish results from the presence of a symbiotic colony of luminous bacteria. (b) Looking for all the world like some undersea sunflower, these two sea anemones (actually animals) use a glass-sponge stalk to catch "marine snow," food particles raining down on the ocean floor from the ocean surface several kilometers above. (c) These giant beardworms live along vents where water jets from fissures at 350°C and then cools to the 2°C of the surrounding water.

(a) Ron and Valerie Tay/age fotostock; (b) Kenneth L. Smith; (c) NOAA Okeanos Explorer Program, Galapagos Rift

20.9 Freshwater Ecosystems

LEARNING OBJECTIVE 20.9.1 Differentiate the ecological zones of a freshwater lake, explaining the cause of the lake's spring and fall overturns.

Freshwater ecosystems (lakes, ponds, rivers, and wetlands) are distinct from both ocean and land ecosystems, and they are very limited in area. Inland lakes cover about 1.8% of the earth's surface, and rivers, streams, and wetlands about 0.4%. All freshwater habitats are strongly connected to land habitats, with marshes and swamps (wetlands) constituting intermediate habitats. In addition, a large amount of organic and inorganic material continually enters bodies of freshwater from communities growing on the land nearby. Many kinds of organisms are restricted to freshwater habitats (figure 20.18). When they occur in rivers and streams, they must be able to attach themselves in such a way as to resist or avoid the effects of current or risk being swept away.

Figure 20.18 Freshwater organism.
Some organisms, such as this giant waterbug with eggs on its back, can only live in freshwater habitats.
John Mitchell/Science Source

Lakes and Ponds

Like the ocean, ponds and lakes have three zones in which organisms live (figure 20.19a): a shallow "edge" zone (the littoral zone), an open-water surface zone (the limnetic zone), and a deep-water zone where light does not penetrate (the profundal zone). Also, lakes can be divided into two

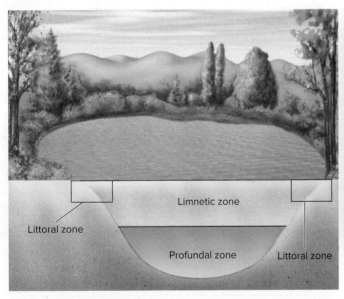

(a)

(b) Oligotrophic lake

(c) Eutrophic lake

Figure 20.19 Characteristics of ponds and lakes.
(a) Ponds and lakes can be divided into three zones based on the types of organisms that live in each. A shallow "edge" (littoral) zone lines the periphery of the lake where attached algae and their insect herbivores live. An open-water surface (limnetic) zone lies across the entire lake and is inhabited by floating algae, zooplankton, and fish. A dark, deep-water (profundal) zone overlies the sediments at the bottom of the lake. The profundal zone contains numerous bacteria and wormlike organisms that consume dead debris settling at the bottom of the lake. Lakes can be oligotrophic (b), containing scarce amounts of organic material, or eutrophic (c), containing abundant amounts of organic material.
(b) Jeff R. Clow/Moment Open/Getty Images; (c) Martin Shields/Alamy Stock Photo

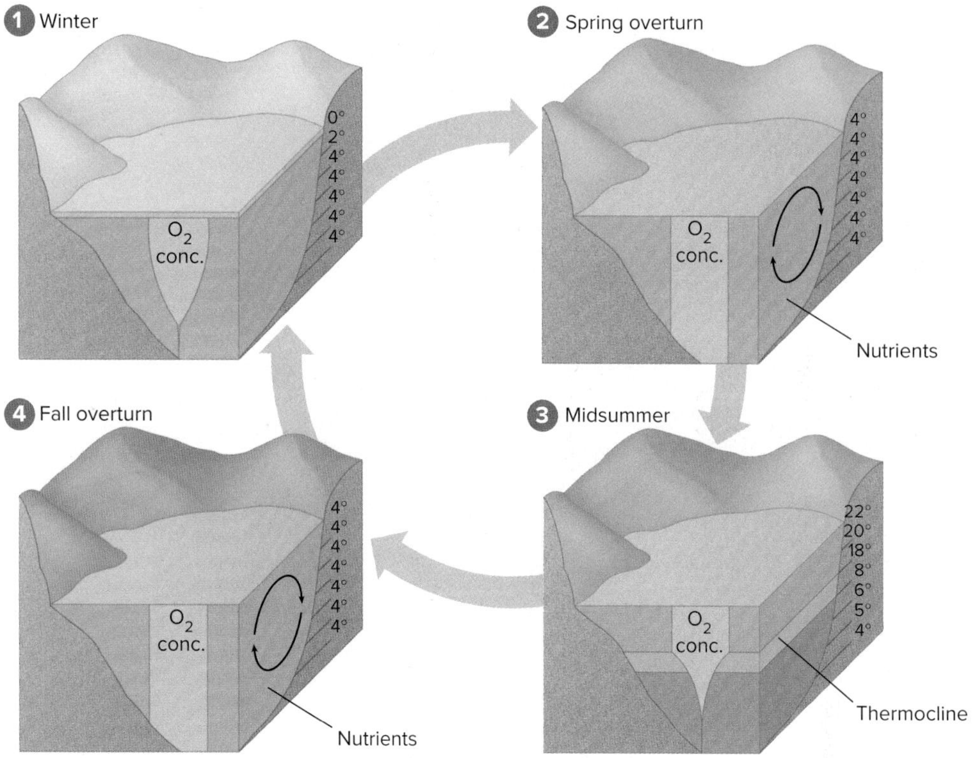

Figure 20.20 Spring and fall overturns in freshwater ponds or lakes.

The pattern of stratification in a large pond or lake in temperate regions is upset in the spring and fall overturns. Of the three layers of water shown in midsummer (*lower right*), the densest water occurs at 4°C. The warmer water at the surface is less dense. The thermocline is the zone of abrupt change in temperature that lies between them. In summer and winter, oxygen concentrations are lower at greater depths, whereas in the spring and fall, they are more similar at all depths.

categories, based on their production of organic material. In **oligotrophic lakes** (figure 20.19*b*), organic matter and nutrients are relatively scarce. Such lakes are often deep, and their deep waters are always rich in oxygen. Oligotrophic lakes are highly susceptible to pollution from excess phosphorus from sources such as fertilizer runoff, sewage, and detergents. **Eutrophic lakes,** on the other hand, have an abundant supply of minerals and organic matter (figure 20.19*c*). Oxygen is depleted at the lower depths in the summer because of the abundant organic material and high rate at which aerobic decomposers in the lower layer use oxygen. These stagnant waters circulate to the surface in the fall (during the fall overturn, as discussed below) and are then infused with more oxygen.

Thermal Layers

Thermal stratification, characteristic of the larger lakes in temperate regions, is the process whereby water at a temperature of 4°C (which is when water is most dense) sinks beneath water that is either warmer or cooler. Follow through the changes in a large lake in figure 20.20 beginning in winter ❶, where water at 4°C sinks beneath cooler water that freezes at the surface at 0°C. Below the ice, the water remains between 0° and 4°C, and plants and animals survive there. In spring ❷, as the ice melts, the surface water is warmed to 4°C and sinks below the cooler water, bringing the cooler water to the top with nutrients from the lake's lower regions. This process is known as the *spring overturn.*

In summer ❸, warmer water forms a layer over the cooler water that lies below. In the area between these two layers, called the thermocline, temperature changes abruptly. You may have experienced the existence of these layers if you have dived into a pond in temperate regions in the summer. Depending on the climate of the particular area, the warm upper layer may become as much as 20 meters thick during the summer. In autumn ❹, its surface temperature drops until it reaches that of the cooler layer underneath—4°C. When this occurs, the upper and lower layers mix—a process called the fall overturn. Therefore, colder waters reach the surfaces of lakes in the spring and fall, bringing up fresh supplies of dissolved nutrients.

> **Putting the Concept to Work**
>
> If you break through the ice on a large lake in winter and dive in, does it get warmer or colder as you swim deeper?

Figure 20.21 Distribution of the earth's major land ecosystems.

The seven primary types of terrestrial ecosystems, discussed in detail in the next section of this chapter, are tropical rain forest, savanna, desert, temperate grassland, temperate deciduous forest, taiga, and tundra. In addition, seven less widespread biomes are chaparral; polar ice; mountain zone (photo *above*); temperate evergreen forest; warm, moist evergreen forest; tropical monsoon forest; and semidesert.

Felix Behnke/Cultura Creative RF/Alamy Stock Photo

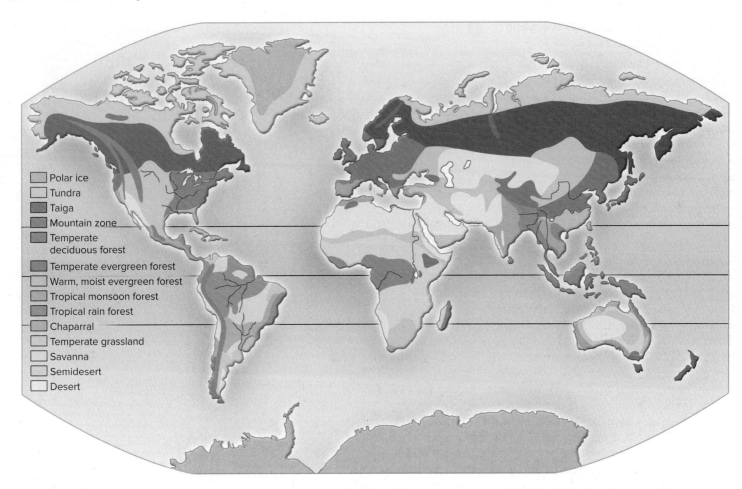

20.10 Land Ecosystems

LEARNING OBJECTIVE 20.10.1 Identify and describe the seven most widespread terrestrial biomes.

Biomes are major types of ecosystems that occur on land. The seven most widely occurring biomes (figure 20.21) are (1) tropical rain forest, (2) savanna, (3) desert, (4) temperate grassland, (5) temperate deciduous forest, (6) taiga, and (7) tundra. There are seven other less widespread biomes also shown in figure 20.21.

Lush Tropical Rain Forests

Rain forests, which experience over 250 centimeters of rain a year, are the richest ecosystems on earth (figure 20.22). They contain at least half of the earth's species of terrestrial plants and animals—more than 2 million species! In a single square mile of tropical forest in Rondonia, Brazil, there are 1,200 species of butterflies—twice the total number found in the United States and Canada combined. The communities that make up tropical rain forests are diverse in that each kind of animal, plant, or microorganism is often represented in a given area by very few individuals. There are extensive tropical rain forests in South America, Africa, and Southeast Asia.

Figure 20.22 **Tropical rain forest.**

Salparadis/Shutterstock

Savannas: Dry Tropical Grasslands

In the dry climates that border the tropics are found the world's great grasslands, called savannas. Landscapes are open, often with widely spaced trees, and rainfall (75 to 125 cm annually) is seasonal. Many of the animals and plants are active only during the rainy season. The huge herds of grazing animals that inhabit the African savanna are familiar to all of us (figure 20.23). Such animal communities occurred in the temperate grasslands of North America during the Pleistocene epoch but have persisted mainly in Africa. On a global scale, the savanna biome is transitional between tropical rain forest and desert. As these savannas are increasingly converted to agricultural use to feed rapidly expanding human populations in subtropical areas, their inhabitants are finding it difficult to survive. The elephant, rhino, and cheetah are now endangered species; the lion and giraffe will soon follow them.

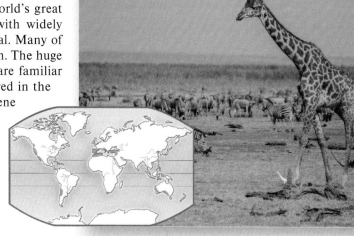

Figure 20.23 Savanna.

David Min/Getty Images

Deserts: Burning Hot Sands

In the interior of continents are found the world's great deserts, especially in Africa (the Sahara), Asia (the Gobi), and Australia (the Great Sandy Desert). **Deserts** are dry places where less than 25 centimeters of rain falls in a year—an amount so low that vegetation is sparse, and survival depends on water conservation (figure 20.24). One quarter of the world's land surface is desert. The plants and animals that live in deserts may restrict their activity to favorable times of the year, when water is present. To avoid high temperatures, most desert vertebrates live in deep, cool, and sometimes even somewhat moist burrows. Those that are active over a greater portion of the year emerge only at night, when temperatures are relatively cool. Some, such as camels, can drink large quantities of water when it is available and then survive long, dry periods. Many animals simply migrate to or through the desert, where they exploit food that may be abundant seasonally.

Figure 20.24 Desert.

Comstock/Stockbyte/Getty Images

Grasslands: Seas of Grass

Halfway between the equator and the poles are temperate regions where rich **grasslands** grow. These grasslands once covered much of the interior of North America, and they were widespread in Eurasia and South America as well. Such grasslands are often highly productive when converted to agriculture. Many of the rich agricultural lands in the United States and southern Canada were originally occupied by **prairies,** another name for temperate grasslands. These natural temperate grasslands are one of the biomes adapted to periodic fire.

The roots of perennial grasses characteristically penetrate far into the soil, and grassland soils tend to be deep and fertile. Temperate grasslands are often populated by herds of grazing mammals. In North America, the prairies were once inhabited by huge herds of bison and pronghorns (figure 20.25). The herds are almost all gone now, with most of the prairies having been converted to the richest agricultural region on earth.

Figure 20.25 Temperate grassland.

Finn O'Hara/Stockbyte/Getty Images

Deciduous Forests: Rich Hardwood Forests

Mild climates (warm summers and cool winters) and plentiful rains promote the growth of **deciduous** ("hardwood") **forests** in Eurasia, the northeastern United States, and eastern Canada (figure 20.26). A deciduous tree is one that drops its leaves in the winter. Deer, bears, beavers, and raccoons are the familiar animals of the temperate regions. Because the temperate deciduous forests represent the remnants of more extensive forests that stretched across North America and Eurasia several million years ago, these remaining areas—especially those in eastern Asia and eastern North America—share animals and plants that were once more widespread. Alligators, for example, are found only in China and in the southeastern United States. The deciduous forest in eastern Asia is rich in species because climatic conditions have remained constant there.

Figure 20.26 Temperate deciduous forest.
Bereczki Barna/Alamy Stock Photo

Taiga: Trackless Conifer Forests

A great ring of northern forests of coniferous trees (spruce, hemlock, larch, and fir) extends across vast areas of Asia and North America. Coniferous trees are ones with leaves like needles that are kept all year long. This ecosystem, called **taiga,** is one of the largest on earth (figure 20.27). Here, the winters are long and cold. Rain, often as little as in hot deserts, falls in the summer. Because it has too short a growing season for farming, few people live there. Many large mammals, including elk, moose, deer, and such carnivores as wolves, bears, lynx, and wolverines, live in the taiga. Traditionally, fur trapping has been extensive in this region. Lumber production is also important. Marshes, lakes, and ponds are common and are often fringed by willows or birches. Most of the trees occur in dense stands of one or a few species.

Figure 20.27 Taiga.
Valerii_M/Shutterstock

Tundra: Cold, Boggy Plains

In the far north, above the great coniferous forests and below the polar ice, there are few trees. There the grassland, called **tundra,** is open, windswept, and often boggy (figure 20.28). Enormous in extent, this ecosystem covers one-fifth of the earth's land surface. Very little rain or snow falls. When rain does fall during the brief Arctic summer, it sits on frozen ground, creating a sea of boggy ground. *Permafrost,* or permanent ice, usually exists within a meter of the surface. Trees are small and are mostly confined to the margins of streams and lakes. Large grazing mammals, including musk-oxen, caribou, reindeer, and carnivores such as wolves, foxes, and lynx live in the tundra. Lemming populations rise and fall on a long-term cycle, with important effects on the animals that prey on them.

Figure 20.28 Tundra.
Cliff LeSergent/Alamy Stock Photo

Chaparral

Chaparral consists of evergreen, often spiny shrubs and low trees that form communities in regions with what is called a "Mediterranean" dry summer climate. These regions include California, central Chile, the cape region of South Africa, southwestern Australia, and the Mediterranean area itself (figure 20.29). Many plant species found in chaparral can germinate only when they have been exposed to the hot temperatures generated during a fire. The chaparral of California and adjacent regions is historically derived from deciduous forests.

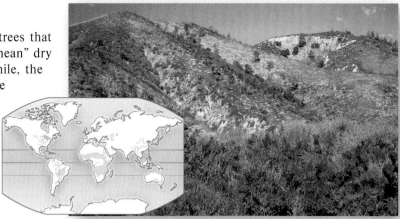

Figure 20.29 **Chaparral.**
Steven P. Lynch

Polar Ice Caps

Polar ice caps lie over the Arctic Ocean in the north and Antarctica in the south (figure 20.30). The poles receive almost no precipitation, so although ice is abundant, freshwater is scarce. The sun barely rises in the winter months. Life in Antarctica is largely limited to the coasts. Because the Antarctic ice cap lies over a landmass, it is not warmed by the latent heat of circulating ocean water and becomes very cold, so only prokaryotes, algae, and some small insects inhabit the vast Antarctic interior.

Figure 20.30 **Polar ice.**
Frank Krahmer/Getty Images

Tropical Monsoon Forest

Tropical upland forests (figure 20.31) occur in the tropics and semitropics at slightly higher latitudes than rain forests or where local climates are drier. Most trees in these forests are deciduous, losing many of their leaves during the dry season. This loss of leaves allows sunlight to penetrate to the understory and ground levels of the forest, where a dense layer of shrubs and small trees grows rapidly. Rainfall is typically very seasonal, measuring several inches daily in the monsoon season and approaching drought conditions in the dry season, particularly in locations far from oceans, such as in central India.

Figure 20.31 **Tropical monsoon forest.**
H Lansdown/Alamy Stock Photo

> **Putting the Concept to Work**
> In what biome do you live? How many others have you visited?

Putting the Chapter to Work

1. You are on a class trip visiting a local forest. Your instructor identifies a species of alga growing on a rock in a stream and points out fish that feed on the algae as their sole source of nutrients.

   a. What is the primary producer in this forest stream ecosystem?

   b. Based on the given information, in what trophic level would you place the fish?

2. More than 1.8 million people in the United States rely on the Ogallala Aquifer for the water they drink.

   An aquifer is an example of what part of the water cycle?

3. Bacteria are used in water treatment systems in sewage plants to break down the nitrogen present in the wastes. Microbiologists at the plants monitor the levels of oxygen in the process to make sure none is present. If oxygen is present, they shut down the process and readjust the parameters.

   What process does the presences of oxygen interfere with?

4. Your instructor is describing the largest land ecosystem that consists of mainly gymnosperms. This ecosystem also has a wide variety of hooved animals and carnivores.

   Which ecosystem is being described?

Retracing the Learning Path

The Energy in Ecosystems

20.1 Ecosystems

1. An ecosystem includes the community and the habitat present in a particular area.
2. Energy constantly flows into an ecosystem from the sun and is passed among organisms in a food chain. Energy from the sun is captured by photosynthetic producers, which are eaten by herbivores, which are in turn eaten by carnivores. Organisms at all trophic levels die and are consumed by detritivores and decomposers. A food chain is organized linearly, but in nature, the flow of energy is often more complex and is called a food web.

- An ecosystem's net primary productivity is the total amount of energy that is captured by producers. Energy is lost at every level in a food chain, such that only about 5% to 20% of available energy is passed on to the next trophic level.

20.2 Ecological Pyramids

1. Because energy is lost at every step as it passes up through the trophic levels of the food chain, there tends to be more individuals at the lower trophic levels. Similarly, the amount of biomass is also less at the higher trophic levels, as is the amount of energy. Ecological pyramids illustrate this distribution of number of individuals, biomass, and energy.

Materials Cycle Within Ecosystems

20.3 The Water Cycle

1. Physical components of the ecosystem cycle through the ecosystem, being used, then recycled and reused, a process called a biogeochemical cycle.

- In the environmental cycle, water cycles from the atmosphere as precipitation, where it falls to the earth and reenters the atmosphere through evaporation. In the organismic cycle, water cycles through plants, entering through the roots and leaving as water vapor by transpiration. Groundwater held in underground aquifers cycles more slowly through the water cycle.

20.4 The Carbon Cycle

1. Carbon cycles from the atmosphere through plants via carbon fixation of CO_2 in photosynthesis. Carbon then returns to the atmosphere as CO_2 via cellular respiration, but some carbon is also stored in the tissues of the organisms. Eventually, this carbon reenters the atmosphere by the burning of fossil fuels and by diffusion after erosion.

20.5 The Nitrogen and Phosphorus Cycles

1. Nitrogen gas in the atmosphere cannot be readily used by organisms and needs to be fixed by certain types of bacteria into ammonia and nitrate that is then used by plants. Animals eat plants that have taken up the fixed nitrogen. Nitrogen reenters the ecosystem through decomposition and animal excretions.

- Phosphorus also cycles through the ecosystem and may limit growth when not available. Phosphorus is not a gas and so does not cycle through the atmosphere. Instead, it cycles from soil and rock through plants and back to the soil.

How Weather Shapes Ecosystems

20.6 The Sun and Atmospheric Circulation

1. The heating power of the sun drives circulation of the atmosphere, causing certain parts of the globe, such as the tropics, to have larger amounts of precipitation.

20.7 Latitude and Elevation

1. Temperature and precipitation are similarly affected by elevation and latitude. Changes in ecosystems from the equator to the poles are similarly reflected in the changes in ecosystems from sea level to mountaintops. Changes in temperature cause the rain shadow effect, in which precipitation is deposited on the windward side of mountains and deserts form on the lee side.

Major Kinds of Ecosystems

20.8 Ocean Ecosystems

1. There are three primary ocean ecosystems: shallow waters, open-sea surfaces, and deep-sea waters.

20.9 Freshwater Ecosystems

1. Freshwater ecosystems cover 2% of the earth's surface and are closely tied to the terrestrial environments that surround them. The penetration of light divides a freshwater lake into three zones with varying amounts of light. Temperature variations in a lake, called thermal stratification, also bring about an overturning of the lake that distributes nutrients.

20.10 Land Ecosystems

1. Biomes are terrestrial communities found throughout the world. Each biome contains its own characteristic group of organisms based on temperature and rainfall patterns.

Inquiry and Analysis

Does Clear-Cutting Forests Cause Permanent Damage?

The lumber industry practice called clear-cutting has been common in many states. Loggers find it more efficient to simply remove all trees from a watershed and sort the logs out later than to selectively cut only the most desirable mature trees. While the open cuts seem a desolation to the casual observer, the loggers claim that new forests can become established more readily in the open cut as sunlight now more easily reaches seedlings at ground level. Ecologists counter that clear-cutting fundamentally changes the forest in ways which cannot be easily reversed.

Who is right? The most direct way to find out is to try it: Clear-cut an area and watch it very carefully. Just this sort of massive field test was carried out in a now-classic experiment at the Hubbard Brook Experimental Forest in New Hampshire. Hubbard Brook is the central stream of a large watershed that drains a region of temperate deciduous forest in northern New Hampshire. The research team, led by then-Dartmouth College professors Herbert Bormann and Gene Likens, first gathered a great deal of information about the forest watershed. Starting in 1963, they censused the trees, measured the flow of water through the watershed, and carefully documented the levels of minerals and other nutrients in the water leaving the ecosystem via Hubbard Brook. To keep track, they constructed concrete dams across each of the six streams that drain the forest and monitored the runoff, chemically analyzing samples. The undisturbed forest proved very efficient at retaining nitrogen and other nutrients. The small amounts of nutrients that entered the ecosystem in rain and snow were approximately equal to the amounts of nutrients that ran out of the valleys into Hubbard Brook.

Now came the test. In the winter of 1965, the investigators felled all the trees and shrubs in 48 acres drained by one stream (as shown in the photo) and examined the water running off. The immediate effect was dramatic: The amount of water running out of the valley increased by 40%. Water that otherwise would have been

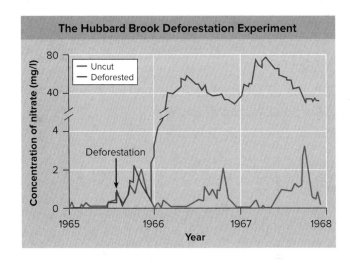

U.S. Forest Service Archives

taken up by vegetation and released into the atmosphere through evaporation was now simply running off. It was clear that the forest was not retaining water as well, but what about the soil nutrients, the key to future forest fertility?

The red line in the graph above shows nitrogen minerals leaving the ecosystem in the runoff water of the stream draining the clear-cut area; the blue line shows the nitrogen runoff in a neighboring stream draining an adjacent uncut portion of the forest.

Analysis

1. **Applying Concepts**
   Scale. What is the significance of the break in the vertical axis between 4 and 40?
2. **Interpreting Data**
   a. What is the approximate concentration of nitrogen in the runoff of the uncut valley before cutting? Of the cut valley before cutting?
   b. What is the approximate concentration of nitrogen in the runoff of the uncut valley one year after cutting? Of the clear-cut valley one year after cutting?
3. **Making Inferences**
   a. Is there any yearly pattern to the nitrogen runoff in the uncut forest? Can you explain it?
   b. How does the loss of nitrogen from the ecosystem in the clear-cut forest compare with nitrogen loss from the uncut forest?
4. **Drawing Conclusions**
   a. What is the impact of this forest's trees upon its ability to retain nitrogen?
   b. Has clear-cutting harmed this ecosystem? Explain.

21 Behavior and the Environment

LEARNING PATH ▼

The Study of Behavior
1. Instinctive Behavioral Patterns
2. Genetic Effects on Behavior

Behavior Can Be Influenced by Learning
3. How Animals Learn
4. Animal Cognition

Evolutionary Forces Shape Behavior
5. A Cost-Benefit Analysis of Behavior
6. Migratory Behavior

Social Behavior
7. Animal Societies
8. Human Social Behavior

The Search for Love Potion #9

> "I told her that I was a flop with chicks.
> I've been disgraced since 1956.
> She looked at my palm and she made a magic sign.
> She said what you need is Love Potion number 9."

This song, made famous by the Searchers in 1965, tells of a man's search for a love potion—a search that has been ongoing in the research community for nearly half a century. Not a love potion per se, but a chemical that triggers a sexual response—a human sex pheromone.

Pheromones

As you will discover in this chapter, pheromones are chemical compounds released by animals that function as signals in communication. Most major animal groups, from insects and sea urchins to mammals, use pheromones to communicate.

ARE CHEMICALS INCREASING their sexual attraction to one another?
Marcus Castillo

In one familiar example, pheromones released in urine are used by dogs to mark territories. However, while the presence and actions of pheromones in other animal groups are indisputable, their presence and action in humans are highly debatable, especially sex pheromones—although some manufacturers of perfume and cologne would have you believe otherwise. Let's look at the research.

428

Do Humans Use Pheromones?

For a long time, researchers have tried to identify a substance that was thought to function as a sex pheromone in humans. Research in other species has shown that secretions found in sweat and that contribute to general body odor are known to be involved in determining sexual attractiveness. This seemed a good place to start looking. Two chemicals were identified in humans that might function as sex pheromones: androstadienone (AND), found in men's underarm sweat, and estratetraenol (EST), found in women's urine.

Sweat and Urine

Is there a quantifiable physiological response in humans to the proposed sex pheromones AND and EST? A research group in Sweden has been studying changes in brain activity in response to AND and EST. In a series of experiments, they measured brain activity with positron emission tomography (PET) scans following the exposure of men and women to the potential sex pheromones. The participants were asked to breathe concentrated samples of AND or EST, as well as odorless air as controls. PET scans revealed that the area of the brain called the anterior hypothalamus showed a spike of activity in women after inhaling AND, while no such activity occurred with EST. Conversely, the anterior hypothalamus was active in males when inhaling EST, but not when inhaling AND.

Other researchers at the University of California at Berkeley have recently measured the effects of AND on hormone levels in women. Women were asked to sniff a solution containing AND, and over a two-hour period, saliva samples were taken and analyzed for levels of the stress hormone, cortisol. The women reported improved mood and significantly higher sexual arousal. These changes coincided with increased blood pressure, heart rate, and breathing. Researchers also determined that cortisol levels increased within 15 minutes and remained elevated for over an hour. Cortisol primes the body for the "fight or flight" response, which could explain the elevated physiological responses. However, because responses are influenced by past experiences, context, and many other more dominant sensory stimuli, the action of a human sex pheromone is not clear-cut.

SolStock/Getty Images

Better Sex Through Chemistry?

While the research on human sex pheromones is inconclusive as yet, the search for human sex pheromone goes on. There is mounting evidence that humans do respond to certain chemical cues that could be considered pheromones, if not necessarily sex pheromones. Research by Martha McClintock while an undergraduate at Wellesley College noted that women in her dorm were menstruating at the same time, suggesting to her that human pheromones involved in reproduction indeed exist. In subsequent studies, 30% of women living together for the first time in a summer school dorm had started to cycle with each other by the end of the summer.

Is menstrual synchrony caused by EST? EST is found in high concentration in underarm sweat, so a simple test would be to place extracts from women's underarm sweat on the upper lip of female volunteers. When done, this caused the menstrual cycles of the volunteer subjects to shorten or lengthen, becoming synchronized with the timing of the menstrual cycle of the sweat-donor's cycle. Application of underarm sweat extracts obtained from males had no effect. This is perhaps the strongest evidence that exists for pheromone action in humans.

The Study of Behavior

21.1 Instinctive Behavioral Patterns

LEARNING OBJECTIVE 21.1.1 Define innate releasing mechanism and fixed action pattern.

Some behaviors seem hardwired, determined by genes.

A Controversial Field of Biology

Biologists define behavior (figure 21.1) as the way an animal responds to stimuli in its environment. The study of behavior has had a long history of controversy. A key argument has centered on the question of whether an animal's behavior is determined more by its genes or by its experiences—is behavior the result of nature (genes) or nurture (learning)?

Early research in the field of animal behavior focused on behavioral patterns that appeared to be instinctive (or "innate"). Because behavior in animals is often stereotyped (that is, appearing in the same way in different individuals of a species), behavioral scientists argued that the behaviors must be based on preset paths in the nervous system. In their view, these neural paths are structured from genetic blueprints and cause animals to show essentially the same behavior from the first time it is produced throughout their lives.

This study of the instinctive nature of animal behavior was typically carried out in the field rather than on laboratory animals. The study of animal behavior in natural conditions is called **ethology.**

An Example of Innate Behavior

The study by Konrad Lorenz of the egg retrieval behavior of geese provides a clear example of what ethologists mean by innate behavior. Figure 21.2 illustrates how when a goose is incubating its eggs in a nest and it notices that an egg has been knocked out of the nest, it will extend its neck toward the egg, get up, and roll the egg back into the nest with a side-to-side motion of its neck while the egg is tucked beneath its bill. Even if the egg is removed during retrieval, the goose completes the behavior, as if driven by a program released by the initial sight of the egg outside the nest.

According to ethologist Lorenz, egg retrieval behavior is triggered by a *sign stimulus* (also called a key stimulus), which in this case is the appearance of an egg outside the nest. The way in which nerves are connected in the goose's brain, the *innate releasing mechanism,* responds to the sign stimulus by providing the neural instructions for the motor program, or *fixed action pattern,* which causes the goose to carry out the intricate egg retrieval behavior. More generally, the sign stimulus is a "signal" in the environment that triggers a behavior. The innate releasing mechanism is the hardwired element of the brain, and the fixed action pattern is the stereotyped act.

Figure 21.1 **Two ways to look at a behavior.**
This male songbird can be said to be singing because his levels of testosterone are elevated, triggering innate "song" programs in his brain. Viewed in a different light, he is singing to defend his territory and attract a mate, behaviors that have evolved to increase his reproductive fitness.
photography by JHWilliams/iStock/Getty Images

> **Putting the Concept to Work**
> What are the three components of the instinctive nesting behavior of geese?

21.2 Genetic Effects on Behavior

> **LEARNING OBJECTIVE 21.2.1** Defend the conclusion that genes play a key role in many behaviors.

Although most animal behaviors are not hardwired instincts, animal behaviorists have clearly demonstrated that many of them are strongly influenced by genes passed from parent to offspring. It appears "nature" plays a key role.

If genes determine behavior, then it should be possible to study their inheritance, much as Mendel studied the inheritance of flower color in garden peas. This sort of investigation is called behavioral genetics.

Studies of Twins

The influence of genes on behavior can be clearly seen in humans. Identical twins are genetically identical. In some instances, identical twins have been separated at birth and raised apart. A recent study of 50 such sets of twins revealed many similarities in personality, temperament, and even leisure-time activities, even though the twins were raised in different circumstances. These results show that genes can play a key role in determining human behavior.

A Detailed Look at How One Gene Affects a Behavior

One well-studied gene mutation in mice provides a clear look at how a particular gene influences a behavior. A gene, *fosB,* determines whether or not female mice will nurture their young. Females with both *fosB* alleles knocked out (experimentally removed) will initially investigate their newborn babies, but then ignore them, in stark contrast to the maternal care provided by unaffected females (figure 21.3).

How does *fosB* work? When mothers of new babies initially inspect them, *fosB* alleles are activated in an area of the brain called the hypothalamus, producing a protein signal that activates neural circuitry within the hypothalamus directing the female to react maternally toward her offspring. In a general way, the information gained from inspecting the newborn babies can be viewed as acting like a sign stimulus, the *fosB* gene as an innate releasing mechanism, and the maternal behavior as the resulting action pattern.

> **Putting the Concept to Work**
> How does the *fosB* gene influence the maternal behavior in mice?

Figure 21.2 Sign stimulus and fixed action pattern.
The series of movements used by a goose to retrieve an egg is a fixed action pattern. Once it detects the sign stimulus (in this case, an egg outside the nest), the goose goes through the entire set of movements: It will extend its neck toward the egg, get up, and roll the egg back into the nest with a side-to-side motion of its neck while the egg is tucked underneath its bill.

Alexander von Düren/imageBROKER/Alamy Stock Photo

> **IMPLICATION FOR YOU** Like the goose egg retrieval described above, human laughter is also considered to be a reflex action, the same fixed action pattern in every human. But is it really the same? Congenitally deaf people laugh out loud, but do they produce the same sounds as those who hear normally? How would you go about determining if they do?

Figure 21.3 A gene alters maternal care.
In mice, normal mothers take very good care of their offspring, retrieving them if they move away and crouching over them. Mothers with the mutant *fosB* allele perform neither of these behaviors, leaving their pups exposed.

Tom McHugh/Science Source

Behavior Can Be Influenced by Learning

21.3 How Animals Learn

> **LEARNING OBJECTIVE 21.3.1** Contrast classical versus operant conditioning and associative versus nonassociative learning.

Many of the behavioral patterns displayed by animals are not the result solely of instinct. In many cases, animals alter their behavior as a result of previous experiences, a process termed *learning*. The simplest type of learning, **nonassociative learning,** does not require an animal to form an association between two stimuli, or between a stimulus and a response. One form of nonassociative learning is *habituation,* learning not to respond to a repeated stimulus. As an everyday example, are you still conscious of the chair you are sitting on?

Classical Conditioning

A change in behavior that involves an association between two stimuli or between a stimulus and a response is called **associative learning.** The behavior is modified, or *conditioned,* through the association. In **classical conditioning,** the paired presentation of two kinds of stimuli causes the animal to form an association between the stimuli. When the Russian psychologist Ivan Pavlov presented meat powder, an *unconditioned stimulus,* to a dog, the dog responded by salivating. If an unrelated stimulus, such as the ringing of a bell, was present at the same time as the meat powder, then over repeated trials the dog would come to salivate in response to the sound of the bell alone. The dog had learned to associate the unrelated sound stimulus with the meat powder stimulus. Its response to the sound stimulus had become conditioned, and the bell was now a *conditioned stimulus.*

Operant Conditioning

In **operant conditioning,** an animal learns to associate its behavioral response with a reward or punishment. Psychologist B. F. Skinner studied operant conditioning in rats by placing them in an experimental cage nicknamed a "Skinner box." As the rat explored the box, it would occasionally press a lever by accident, causing a pellet of food to appear. At first, the rat would ignore the lever, eat the food pellet, and continue to move about. Soon, however, it would learn to associate pressing the lever (the behavioral response) with food (the reward). When a conditioned rat was hungry, it would spend all its time pressing the lever.

Imprinting

As an animal matures, it may form preferences or social attachments to other individuals that will profoundly influence behavior later in life, a process called **imprinting.** In *filial imprinting,* social attachments form between parents and offspring. For example, young birds of some species begin to follow their mother within a few hours after hatching, forming a strong bond between mother and young. Ethologist Konrad Lorenz raised geese from eggs, and when he offered himself as a model for imprinting, the goslings treated him as if he were their parent, following him dutifully (**figure 21.4**).

Figure 21.4 An unlikely parent.
The eager goslings follow ethologist Konrad Lorenz as if he were their mother. He is the first object they saw when they hatched, and they used him as a model for imprinting.
Thomas D. McAvoy/Getty Images

> **IMPLICATION FOR YOU** It appears that humans also exhibit imprinting. One of the most widely studied examples of human imprinting is reverse sexual imprinting, also known as the Westermarck effect. When people live in close proximity during the first few years of their lives, they are desensitized to later sexual attraction. In Israeli kibbutz collective farms, children of both sexes were reared together. Of 3,000 marriages of couples both raised during their first six years in kibbutz, none of the couples had been reared in the same kibbutz. Why do you think evolution might have favored the development of this form of imprinting in humans?

> **Putting the Concept to Work**
> Is filial imprinting a form of operant or classical conditioning?

21.4 Animal Cognition

> **LEARNING OBJECTIVE 21.4.1** Defend the proposition that animals can reason.

In recent years, serious attention has been given by researchers to the topic of animal awareness. The central question is whether animals other than humans show cognitive behavior—that is, do they process information and respond in a manner that suggests thinking?

Evidence of Conscious Planning

What kinds of behavior would demonstrate cognition? Some birds remove the foil caps from nonhomogenized milk bottles to get at the cream beneath. Japanese macaques (a kind of monkey) learn to float grain on water to separate it from sand and teach other macaques to do it. A chimpanzee pulls the leaves off of a tree branch and uses the stick to probe the entrance to a termite nest to gather termites, suggesting that the ape is consciously planning ahead, with full knowledge of what it intends to do. A sea otter will use a rock as an "anvil" against which it bashes a clam to break it open, often keeping a favorite rock for a long time, as though it had a clear idea of its future use (figure 21.5).

Problem Solving

Some instances of problem solving by animals are hard to explain in any other way than as a result of some sort of cognitive process—what, if we were doing it, we would call reasoning. For example, in a series of classic experiments conducted in the 1920s, a chimpanzee was left in a room with a banana hanging from the ceiling out of reach. Also in the room were several boxes lying on the floor. After unsuccessful attempts to jump up and grab the banana, the chimpanzee suddenly looked at the boxes and immediately proceeded to move them underneath the banana, placing one on top of another, and climbed up the boxes to claim its prize. Would you have solved the problem as well?

It is not surprising to find obvious intelligence in animals so closely related to us as chimpanzees. Perhaps more surprising, however, are recent studies finding that other animals also show evidence of cognition. Ravens have always been considered among the most intelligent of birds. Researchers placed a piece of meat on the end of a string and hung it from a branch in an outdoor aviary containing six ravens. The birds like to eat meat but had never seen string before and were unable to get at the meat. After several hours, during which time the birds periodically looked at the meat but did nothing else, one bird flew to the branch, reached down, grabbed the string with its beak, pulled it up, and placed it under his foot. He then reached down and pulled up another length of the string, repeating this action over and over, each time bringing the meat closer (figure 21.6). Eventually, the meat was within reach and was grasped and eaten by the bird. The raven, presented with a completely novel problem, had devised a solution. Eventually, three of the other five ravens also figured out how to get the meat. This result can leave little doubt that ravens have advanced cognitive abilities.

> **Putting the Concept to Work**
> Devise an experiment to compare the reasoning ability of dogs and cats.

Figure 21.5 Conscious planning.
This sea otter is using a rock as an "anvil" to break open a clam.
Malcolm Schuyl/Alamy Stock Photo

Figure 21.6 Problem solving by a raven.
Confronted with a problem it had never previously encountered, the raven figures out how to get the meat at the end of the string by repeatedly pulling up a bit of string and stepping on it.
Courtesy of Bernd Heinrich

Evolutionary Forces Shape Behavior

21.5 A Cost-Benefit Analysis of Behavior

> **LEARNING OBJECTIVE 21.5.1** Explain the evidence in favor of optimal foraging theory.

Darwin tells us that evolution should favor traits whose net benefit increases survival—traits that help more than they hurt.

Foraging Behavior

One important way to examine the evolutionary advantage of a behavior is to ask if it provides an evolutionary benefit greater than its cost. For many animals, the food that they eat can be found in many sizes and in many places. An animal must choose what food to select and how far to go in search of it. These choices are called the animal's **foraging behavior.** Each choice involves benefits and associated costs. Thus, although a larger food may contain more energy, it may be harder to capture and less abundant. In addition, more desirable foods may be farther away than other types. Hence, an animal's foraging behavior involves a trade-off between a food's energy content and the cost of obtaining it.

The net energy (in calories) gained by feeding on each kind of food available to a foraging animal is simply the energy content of the food minus the energy costs of pursuing and handling it. One might expect that evolution would favor foraging behaviors that are as energetically efficient as possible. This sort of reasoning has led to what is known as the **optimal foraging theory,** which predicts that animals will select food items that maximize their net energy intake per unit of foraging time (figure 21.7).

Is the optimal foraging theory correct? Many foragers do preferentially use food items that maximize the energy return per unit time. In some cases, however, the costs of foraging seem to outweigh the benefits. An animal in danger of being eaten itself is often better off minimizing the amount of time it spends foraging. Many animals alter their foraging behavior when predators are present, reflecting this trade-off.

Figure 21.7 Optimal foraging.
Optimal foraging in this golden mantled ground squirrel pays off with increased net energy gain, which can lead to increased reproductive success.

David R. Frazier Photolibrary, Inc./Alamy Stock Photo

Territorial Behavior

Animals often move over a large area, or *home range.* In many species, the home range of several individuals overlap, but each individual uses a portion of its home range exclusively. This behavior is called **territoriality.**

Territories are defended by displays that advertise that the territories are occupied and by overt aggression. A bird sings from its perch within its territory to prevent invasion of its territory by a neighboring bird. If the intruding bird is not deterred by the song, the territory owner may attack and attempt to drive the invader away.

Why aren't all animals territorial? The answer involves a cost-benefit analysis. Territoriality offers clear benefits, including increased food intake from nearby resources (figure 21.8), access to refuges from predators, and exclusive access to mates. The costs of territorial behavior, however, may also be significant. The singing of a bird, for example, is energetically expensive, and advertisement through song or visual display can reveal one's location to a predator. In many instances, particularly when food sources are abundant, defending easily obtained resources is simply not worth the cost.

Figure 21.8 The benefit of territoriality.
Sunbirds, found in Africa and ecologically similar to hummingbirds, increase nectar availability by defending flowers. A sunbird will expend 3,000 calories per hour chasing intruders away.

Hoberman/age fotostock

> **Putting the Concept to Work**
> Why aren't all animals territorial?

TABLE 21.1 Animal Behaviors

Foraging Behavior	Territorial Behavior	Migratory Behavior
Select, obtain, and consume food	Defend portion of home range and use it exclusively	Move to a new location for part of the year

Oystercatchers forage for food by stabbing the ground or rocks in search of arthropods or mollusks.

Eric and David Hosking/Getty Images

Male elephant seals fight with each other for possession of territories. Only the largest males can hold territories, which contain many females.

worldswildlifewonders/Shutterstock

Wildebeests undergo an annual migration in search of new pastures and sources of water. Migratory herds can contain up to a million individuals and can stretch thousands of miles.

Jonathan Ryan/Alamy Stock Photo

Courtship	Parental Care	Social Behavior
Attract and communicate with potential mates	Produce and rear offspring	Interact with members of a social group

This male frog is vocalizing, producing a call that attracts females.

stanley45/Getty Images

Female lions share the responsibility of raising the pride's young, increasing the probability that the young will survive into adulthood.

Thomas Retterath/Shutterstock

This dog's bow—tail-end shoved high in the air, forelegs flat on the ground—is its way of saying to another dog, "Let's play." Every dog invites play in exactly this way—a retriever like this, a terrier, even a wolf.

Mary H. Swift/Alamy Stock Photo

21.6 Migratory Behavior

> **LEARNING OBJECTIVE 21.6.1** Distinguish between the compass sense and the map sense.

Many animals breed in one part of the world and spend the rest of the year in another. Long-range, two-way annual movements like this are called **migrations.** Migratory behavior is particularly common in birds. Ducks and geese migrate southward along flyways from northern Canada across the United States each fall, overwinter in the south, and then return northward each spring to nest. Warblers and many other songbirds winter in the tropics and breed in the United States and Canada in spring and summer, when insects are plentiful.

In studying how animals are able to navigate accurately over such long distances, biologists differentiate between *compass sense* (an innate ability to move in a particular direction, called following a bearing) and *map sense* (a learned ability to adjust a bearing depending on the animal's location). Experiments on starlings shown in **figure 21.9** indicate that inexperienced birds migrate using a compass sense, and older birds that have migrated previously also employ a map sense to help them navigate—in essence, they learn the route. Migrating birds were captured in Holland, the halfway point of their migration, and were taken to Switzerland where they were released. Inexperienced birds (the red arrows) kept flying in their original direction, while experienced birds (the blue arrow) were able to adjust course and reached their normal wintering grounds.

The Compass Sense

We now have a good understanding of how the compass sense is achieved in birds. Many migrating birds have the ability to detect the earth's magnetic field and to orient themselves with respect to it. In a closed indoor cage, they will attempt to move in the correct geographical direction, even though there are no visible external clues. However, the placement of a powerful magnet near the cage can alter the direction in which the birds attempt to move.

The first migration of young birds appears to be innately guided by the earth's magnetic field. Inexperienced birds also use the sun and particularly the stars to orient themselves (migrating birds fly mainly at night).

The Map Sense

Much less is known about how migrating birds and other animals acquire their map sense. During their first migration, young birds move with a flock of experienced older birds that know the route, and during the course of the journey, they appear to learn to recognize certain cues, such as the position of mountains and coastline.

Animals that migrate through featureless terrain present more of a puzzle. Consider the green sea turtle. Every year, great numbers of these large turtles migrate with incredible precision from Brazil halfway across the Atlantic to Ascension Island, over 1,400 miles of open ocean, where females lay their eggs. Plowing head down through the waves, how do they find this tiny rocky island? No one knows for sure, although recent studies suggest that the direction of wave action provides an important navigational clue.

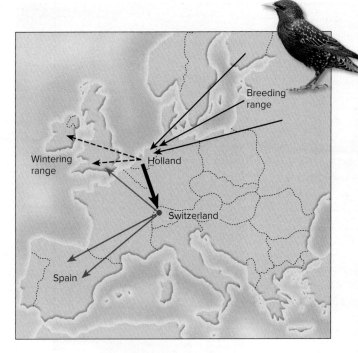

Figure 21.9 Starlings learn how to navigate.
The navigational abilities of inexperienced birds differ from those of adults that have made the migratory journey before. Starlings were captured in Holland, halfway along their full migratory route from Baltic breeding grounds to wintering grounds in the British Isles. These captured birds were transported to Switzerland and released. Experienced older birds compensated for the displacement and flew toward the normal wintering grounds (*blue* arrow). Inexperienced young birds kept flying in the same direction as before, on a course that took them toward Spain (*red* arrows). These observations indicate that inexperienced birds migrate with an innate compass sense, while experienced birds utilize a learned map sense to aid their navigation.
Susan Walker/Getty Images

> **Putting the Concept to Work**
> Do you think humans have a compass sense? How would you test this proposition?

A Closer Look

The Dance Language of the Honeybee

The European honeybee, *Apis mellifera*, lives in hives consisting of 30,000 to 40,000 individuals whose behaviors are integrated into a complex colony. Worker bees may forage miles from the hive, collecting nectar and pollen from a variety of plants on the basis of how energetically rewarding their food is. The food sources used by bees tend to occur in patches, and each patch offers much more food than a single bee can transport to the hive. A colony is able to exploit the resources of a patch because of the behavior of scout bees, which locate patches and communicate their location to hivemates through a *dance language*. Over many years, Nobel laureate Karl von Frisch was able to unravel the details of this communication system.

The Waggle Dance

After a successful scout bee returns to the hive, she performs a remarkable behavior pattern called a *waggle dance* on a vertical comb (**figure**). The path of the bee during the dance resembles a figure-eight. On the straight part of the path, the bee vibrates or waggles her abdomen while producing bursts of sound. She may stop periodically to give her hivemates a sample of the nectar she has carried back to the hive in her crop. As she dances, she is followed closely by other bees, which soon appear as foragers at the new food source.

Von Frisch and his colleagues claimed that the other bees use information in the waggle dance to locate the food source. According to their explanation, the scout bee indicates the *direction* of the food source by representing the angle between the food source, the hive, and the sun as the deviation from vertical of the straight part of the dance performed on the hive wall (that is, if the bee moved straight up, then the food source would be in the direction of the sun, but if the food was at a 30-degree angle relative to the sun's position, then the bee would move upward at a 30-degree angle from vertical). The *distance* to the food source is indicated by the tempo, or degree of vigor, of the dance.

Challenging von Frisch

Adrian Wenner, a scientist at the University of California, did not believe that the dance language communicated anything about the location of food, and he challenged von Frisch's explanation. Wenner maintained that flower odor was the most important cue allowing recruited bees to arrive at a new food source. A heated controversy ensued as the two groups of researchers published articles supporting their positions.

A Test

Such controversies can be very beneficial, because they often generate innovative experiments. In this case, the "dance language controversy" was resolved (in the minds of most scientists) in the mid-1970s by the creative research of James L. Gould. Gould devised an experiment in which hive members were tricked into misinterpreting the directions given by the scout bee's dance. As a result, Gould was able to manipulate where the hive members would go if they were using visual signals. If odor were the cue they were using, hive members would have appeared at the food source, but instead they appeared exactly where Gould predicted. This confirmed von Frisch's ideas. Recently, researchers have expanded the study of the honeybee dance language by building robot bees whose dances can be completely controlled. The use of robot bees has allowed scientists to determine precisely which cues direct hivemates to food sources.

(a)

The waggle dance of honeybees.

(a) The angle between the food source, the nest, and the sun is represented by a dancing bee as the angle between the straight part of the dance and vertical. Here the food is 20 degrees to the right of the sun, and the straight part of the bee's dance on the hive is 20 degrees to the right of vertical. (b) A scout bee dancing in the hive.

(b)

(b) Paul Starosta/Getty Images

Social Behavior

BIOLOGY & YOU

Lord of the Flies. Is the ordered human society in which we live the predictable product of evolutionary forces or just order imposed by the force of convention? This is a very good biological question, although not an easy one to approach scientifically. Novelists have not been so shy, however. Many important works of literature have attempted to address this question head-on, none more powerfully than *Lord of the Flies* by Nobel Prize-winning novelist William Golding. In Golding's story, a group of boys from a British school are shipwrecked on an isolated, unpopulated island. The book portrays their descent into savagery. Left to themselves in an island Eden far from modern civilization, these well-educated children regress, casting off one societal rule after another. By the time they are rescued, all vestiges of civilized society are gone. An allegory, this story contrasts the impulse toward civilization (living peacefully by rules) with the individual impulse for power and survival. Golding's view is that there is nothing inevitable about civil society, nothing biologically driven. Other major novelists have come to the opposite conclusion. Jules Verne's novels argue for the inevitable rise of a harmonious society driven by human nature, which Verne viewed as ingrained, part of what it means to be human. Science cannot judge which approach is correct, which novelist is more insightful. But as you study animal societies, it is useful, and interesting, to keep these questions in mind.

Figure 21.10 Savanna-dwelling African weaver birds form colonial nests.

David Hosking/Science Source

21.7 Animal Societies

LEARNING OBJECTIVE 21.7.1 Contrast how insect and vertebrate societies are organized.

Organisms as diverse as bacteria, cnidarians, insects, fish, birds, prairie dogs, whales, and chimpanzees exist in social groups. To encompass the wide variety of social phenomena, we can broadly define a **society** as a group of organisms of the same species that are organized in a cooperative manner.

Insect Societies

In insects, sociality has chiefly evolved in two orders, the Hymenoptera (ants, bees, and wasps) and the Isoptera (termites), although a few other insect groups include social species. These social insect colonies are composed of different **castes,** groups of individuals that differ in size and morphology and that perform different tasks, such as workers and soldiers.

Honeybees. In honeybees, the queen maintains her dominance in the hive by secreting a pheromone, called "queen substance," that suppresses development of the ovaries in other females, turning them into sterile workers. Drones (male bees) are produced only for purposes of mating. When the colony grows larger in the spring, some members do not receive a sufficient quantity of queen substance, and the colony begins preparations for swarming. Workers make several new queen chambers, in which new queens begin to develop. Scout workers look for a new nest site and communicate its location to the colony. The old queen and a swarm of female workers then move to the new site. Left behind, a new queen emerges, kills the other potential queens, mates, and assumes "rule" of the hive.

Leaf-Cutter Ants. The leaf-cutter ants provide another fascinating example of the remarkable lifestyles of social insects. Leaf-cutters live in colonies of up to several million individuals, their moundlike nests covering more than 100 square meters, with hundreds of entrances and chambers as deep as 5 meters beneath the ground. The division of labor among the worker ants is related to their size. Every day, workers travel along trails from the nest to a tree or a bush, cut its leaves into small pieces, and carry the pieces back to the nest. Smaller workers chew the leaf fragments into a mulch, which they spread like a carpet in the underground fungus chambers. Even smaller workers implant fungal hyphae in the mulch (recent molecular studies suggest that ants have been cultivating these fungi for more than 50 million years!). Soon a luxuriant garden of fungi is growing. While other workers weed out undesirable kinds of fungi, nurse ants carry the larvae of the nest to choice spots in the garden, where the larvae graze. Some of these larvae grow into reproductive queens that start new colonies.

Vertebrate Societies

In contrast to the highly structured and integrated insect societies and their remarkable forms of altruism, vertebrate social groups are usually less rigidly organized and cohesive. Each social group of vertebrates has a certain size, stability of members, number of breeding males and females, and type of mating system. Behavioral ecologists have learned that the way a vertebrate group such as the African weaver birds seen in figure 21.10 is organized is influenced most often by ecological factors such as food type and predation.

Putting the Concept to Work
Why do different vertebrates have differently organized societies?

21.8 Human Social Behavior

> **LEARNING OBJECTIVE 21.8.1** State and evaluate the key prediction of sociobiology.

One of the most profound lessons of biology is that we human beings are animals, quite close relatives of the chimpanzee, and not some special form of life set apart from the earth's other creatures. This biological view of human life raises an important issue: To what degree are the animal behaviors we have described in this chapter characteristic of the social behavior of humans?

Genes and Human Behavior

As we have seen repeatedly, genes play a key role in determining many aspects of animal behavior. From maternal behavior in mice to migratory behavior in songbirds, changes in genes have been shown to have a profound impact. This leads to an important prediction. If behaviors have a genetic basis and if behaviors affect an animal's ability to survive and raise young, then surely behaviors must be subject to natural selection, and the complex social behaviors of animals, including humans, should evolve. The study of this facet of animal behavior, sometimes called **sociobiology,** has proven highly controversial.

Genes certainly affect human behavior in important ways. Human facial expressions are similar the world over, no matter the culture or language, arguing that they have a deep genetic basis. Infants born blind still smile and frown, even though they have never seen these expressions on another face.

It is also true, however, that learning has a huge impact on how humans behave. Human babies in the United States quickly lose the ability to make the consonant sounds they do not hear. Just as birds learn songs from experienced adults, so human babies learn to speak from listening to adults around them.

Diversity Is the Hallmark of Human Culture

When a group of biologists that studied social behavior in chimpanzees among seven widely separated areas compared notes, they identified 39 behaviors involving such things as social behavior, courtship, and tool use that were common in some groups but absent in others. Each population seemed to have a distinct repertoire of behaviors. It seems that each population has its own collection of customary behaviors, which each generation teaches its children. In a word, chimpanzees have culture.

No animal, not even our close relative the chimpanzee, exhibits cultural differences to the degree seen in human populations (figure 21.11). To what degree human culture is the product of evolution acting on behavior-determining genes is a question that is hotly debated by behavioral biologists. However one chooses to answer this question, it is perfectly clear that much of our behavior is molded by experience. Human cultures and the behaviors that produce them change very rapidly, far too rapidly to reflect genetic evolution. Human cultural diversity, a hallmark of our species, is certainly strongly influenced, if not largely determined, by learning and experience.

> **Putting the Concept to Work**
> What sort of social controversy would you expect sociobiology to engender? Might these disagreements be resolved by experiment? How?

Figure 21.11 A New York City street scene.

Glow Images/Getty Images

Putting the Chapter to Work

1 Over the past few weeks, you have been training your puppy. You have instructed it to sit and rewarded it with a dog biscuit when it obeyed your "sit" command. Now when the puppy sees a biscuit, it automatically sits.

What type of learning and conditioning has occurred?

2 At an animal shelter, a person brings in a baby squirrel that requires care. The employees give the squirrel to a cat who is feeding kittens. She begins feeding it and caring for it. Weeks later, the baby squirrel demonstrates behaviors like those of the kittens.

What type of learning has occurred?

3 Without being taught, female flatback sea turtles navigate across thousands of miles of ocean to land on the same shore where they were born, travel up onto the beach, and dig a burrow to lay their eggs. Each nest of eggs, called a clutch, contains approximately 50 eggs. After laying the eggs, the female sea turtle returns to the sea. After the eggs hatch, the newly hatched sea turtles will proceed to move out to the sea without any assistance.

Which type of behavior is exhibited by the female adult and her newly hatched sea turtles?

4 A male lion will lead a pride of approximately nine females and their young. The pride of lions will forage in a defined area of land, the male marking the area with his scent glands.

The male lion is exhibiting which type of behavior?

Retracing the Learning Path

The Study of Behavior

21.1 Instinctive Behavioral Patterns

1. Instinctive, or innate, behaviors are those that are the same in all individuals of a species and appear to be controlled by preset pathways in the nervous system. A sign stimulus triggers the behavior, called a fixed action pattern, such as egg retrieval in geese.

21.2 Genetic Effects on Behavior

1. Most behaviors are not hardwired instincts, but many are strongly influenced by genes and can be studied as inherited traits. Hybrids, twins, and genetically altered animals are used to study genetically influenced behaviors.

Behavior Can Be Influenced by Learning

21.3 How Animals Learn

1. Many behaviors are learned, having been formed or altered based on previous experiences. Classical conditioning results when two stimuli are paired such that the animal learns to associate the two stimuli. Operant conditioning results when an animal associates a behavior with a reward or punishment. Imprinting is when an animal forms social attachments, usually during a critical window of time.

- Behavior is often both genetically determined (instinctive) and modified by learning. Genes can limit the extent to which a behavior can be modified through learning. Ecology has a lot to do with behavior, and knowing an animal's ecological niche can reveal much about its behavior.

21.4 Animal Cognition

1. While humans have evolved a great capacity for cognitive thought, studies show that other animals possess varying degrees of cognitive abilities. Some behaviors in animals show conscious planning ahead. Still other animals show problem-solving abilities. When presented with a novel situation, such as a piece of meat dangling out of reach of a raven, these animals respond to the situation in a problem-solving manner.

Evolutionary Forces Shape Behavior

21.5 A Cost-Benefit Analysis of Behavior

1. For every behavior that offers an individual an advantage for survival, there is usually an associated cost. For example, foraging and territorial behaviors offer benefits by providing food and shelter for the individual and their offspring but may endanger the parents through predation or expenditure of energy. The benefits have to outweigh the costs in order for the behaviors to be favored by natural selection.

21.6 Migratory Behavior

1. Many animals migrate in predictable ways, navigating by looking at the sun and stars and in some cases by detecting magnetic fields. Often, young individuals learn the route by following experienced adults.

Social Behavior

21.7 Animal Societies

1. Many types of animals live in social groups, or societies. Some insect societies are highly structured.

21.8 Human Social Behavior

1. Both genetics and learning play key roles in human behaviors and establishing cultures, but the extent of each is hotly debated.

Inquiry and Analysis

Do Crabs Eat Sensibly?

Many behavioral ecologists claim that animals exhibit so-called optimal foraging behavior. The idea is that because an animal's choice in seeking food involves a trade-off between the food's energy content and the cost of obtaining it, evolution should favor foraging behaviors that optimize the trade-off.

While this all makes sense, it is not at all clear that this is what animals would actually do. This optimal foraging approach makes a key assumption, that maximizing the amount of energy acquired will lead to increased reproductive success. In some cases, this is clearly true. In ground squirrels, zebra finches, and orb-weaving spiders, researchers have found a direct relationship between net energy intake and the number of offspring raised successfully.

However, animals have other needs besides energy, and sometimes these needs conflict. One obvious "other need," important to many animals, is to avoid predators. It makes little sense for you to eat a little more food if doing so greatly increases the probability that you will yourself be eaten. Often the behavior that maximizes energy intake increases predation risk. A shore crab foraging for mussels on a beach exposes itself to predatory gulls and other shore birds with each foray. Thus, the behavior that maximizes fitness may reflect a trade-off, obtaining the most energy with the least risk of being eaten. Not surprisingly, a wide variety of animals use a more cautious foraging behavior when predators are present—becoming less active and staying nearer to cover.

So what does a shore crab do? To find out, an investigator looked to see if shore crabs in fact feed on those mussels that provide the most energy, as the theory predicts. He found that the mussels on the beach he studied come in a range of sizes, from small ones less than 10 millimeters in length that are easy for a crab to open but yield the least amount of energy, to large mussels over 30 millimeters in length that yield the most energy but also take considerably more energy to pry open. To obtain the most net energy, the optimal approach, described by the blue curve in the graph above, would be for shore crabs to feed primarily on intermediate-sized mussels about 22 millimeters in length. Is this in fact what shore crabs do? To find out, the researcher carefully monitored the size of the mussels eaten each day by the beach's population of shore crabs. The results he obtained—the numbers of mussels of each size actually eaten—are presented in the red histogram.

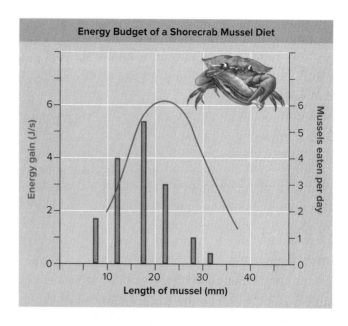

Analysis

1. **Applying Concepts** What variable do the curve and the histogram have in common?
2. **Making Inferences**
   a. What is the most energetically optimal mussel size for the crabs to eat, in mm?
   b. What size mussel is most frequently eaten by crabs, in mm?
3. **Drawing Conclusions** Do shore crabs tend to feed on those mussels that provide the most energy?
4. **Further Analysis** What factors might be responsible for the slight difference in peak prey length relative to the length optimal for maximal energy gain?

Kevin Schafer/Alamy Stock Photo

22 Human Influences on the Living World

LEARNING PATH ▼

Global Change
1. Pollution
2. Global Warming
3. Loss of Biodiversity

Saving Our Environment
4. Preserving Nonreplaceable Resources
5. Curbing Population Growth

Solving Environmental Problems
6. Preserving Endangered Species
7. Individuals Can Make the Difference

AS THE SEA ice melts, this polar bear has a lot to be anxious about.

Paul Souders/Corbis/Getty images

Polar Bears Lose Their Home

Who doesn't love polar bears? The sight of a mother polar bear with her cub makes you just want to cuddle them. Not a good idea. For people who live near the Arctic Circle, their encounters with polar bears are often deadly. Luckily, such meetings are not common.

Hunting Seals

Polar bears are hunters, traveling up to 15 miles a day in an endless search for food. They feed primarily on seals that spend most of their time in the water. In search of a meal, a polar bear will sit perched over a hole in the ice. When a seal surfaces to breathe, the polar bear will grab it and drag it out. Because of its limited diet and method of hunting, polar bears only feed during the winter when the ice pack is largest and thickest. The bears move onshore and fast during the ice-free summer months and migrate back to the ice when it re-forms.

Bears Need Ice

The Arctic ice is the only place where polar bears can successfully hunt. Without these ice platforms, they cannot feed. Although they are strong swimmers, which allows them to travel long distances between large patches of ice, they don't have the speed and agility of a seal in water, so they cannot catch and kill a seal in open water. Without the Arctic sea ice, polar bears would starve—and this vital habitat is melting away. Why? Air and water temperatures are rising because of global warming.

But the Ice Is Disappearing

As global temperatures increase, the outer edges of the Arctic ice cap are retreating earlier in the season and re-forming later. The ice cap is also covering a smaller area, about 20% smaller. This causes the polar bears to migrate to land earlier and remain there longer. The result: Polar bears have a shorter hunting season and a longer fasting season.

So Polar Bears Are Endangered

There are actually more polar bears now than there were in the 1950s. Overhunting in the first half of the 20th century had polar bear numbers reduced to 5,000. In response to this killoff, the Marine Mammal Protection Act was passed to protect them from overhunting and poaching, bringing their numbers up to over 22,000 today.

franzfoto.com/Alamy Stock Photo

However, it is their future that is endangered. The "threatened species" classification of polar bears sought by environmentalists under the Endangered Species Act dictates that protecting the habitat is paramount to saving the species, something the current laws don't require.

Those who object to the threatened species classification point out that true protection of the polar bear's habitat would require putting a stop to global warming. As you will discover in this chapter, global warming is a complex problem with no easy solutions. Cutting the emission of greenhouse gases is a first major step, but because these gases remain in the atmosphere for a long time, any reduction of greenhouse gases that is achieved now may not have any measurable effects for years and years.

They Face an Uncertain Future

What do the polar bears do in the meantime? The polar ice cap continues to shrink. As the ice melts in summer,

(left) NOAA; (right) NASA/Goddard Space Flight Center Scientific Visualization Studio

the polar bear is trapped on land, not able to hunt. If it takes longer for the ice to re-form, the polar bears may starve waiting out the delayed freezing. If the border of the ice cap is farther away from land and they have to swim farther to get to it, they may drown in their weakened state. Global warming is a challenge the polar bears are facing every day. Their only hope is that we successfully address that challenge.

Global Change

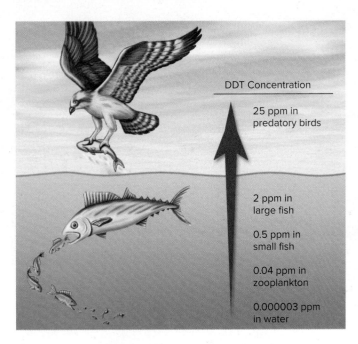

Figure 22.1 Biological magnification of DDT.
Because DDT accumulates in animal fat, the compound becomes increasingly concentrated in higher levels of the food chain, leading to thin, fragile eggshells in many bird species. In the 1960s, DDT was banned from the United States, but is still widely used in the tropics.

BIOLOGY & YOU

Do You Use Bottled Water? One sort of chemical pollution that many of us contribute to without thinking about it is drinking bottled water packaged in plastic. Bottled water is purified water or spring water, popular in the United States both because water is a healthy choice instead of soft drinks and because bottles can be purchased conveniently in stores or vending machines. Americans buy some 28 billion water bottles a year, consuming more bottled water than beer, even though in the United States, tap water quality is as high as bottled water and an awful lot cheaper. Bottled water typically costs $1 to $2 per bottle (90% of the cost is making the bottle, label, and cap), whereas tap water is essentially free. Why do the Sierra Club, the World Wildlife Fund, and other conservation groups urge you to consume less bottled water? Because the bottles, made of plastic, do not degrade, and 80% are not recycled—they simply accumulate by the billions in refuse dumps, landfills, and the world's oceans. Adding to the environmental damage, producing the bottles creates more than 2.5 million tons of CO_2 each year, an estimated 250 grams for each bottle. That's like filling a quarter of every bottle with oil and burning it. If you want to do your bit to curb global warming and pollution, drink more tap water.

22.1 Pollution

> **LEARNING OBJECTIVE 22.1.1** Explain how modern industry and agriculture are leading to higher levels of chemical pollution.

Our world is one highly interactive biosphere, and damage done to any one ecosystem can have ill effects on many others. Burning high-sulfur coal in Illinois kills trees in Vermont, while dumping refrigerator coolants in New York destroys atmospheric ozone over Antarctica and leads to increased skin cancer in Madrid. Biologists call such widespread effects on the worldwide ecosystem *global change*. The pattern of global change that has become evident within recent years, is one of the most serious problems facing humanity's future.

Chemical Pollution

The problem posed by chemical pollution has grown very serious in recent years, both because of the growth of heavy industry and because of an overly casual attitude in industrialized countries. For example, a chemical widely used in air conditioners and aerosol dispensers had been depleting the ozone in earth's upper atmosphere, an important shield against the sun's cancer-causing ultraviolet radiation. International agreements signed by more than 180 countries have led to a steep reduction in the use of these ozone-killing chemicals.

Air Pollution. Air pollution is a major problem in the world's cities. In Mexico City, oxygen is sold routinely on street corners for patrons to inhale. Laws regulating automobile exhaust have largely eliminated serious city air pollution in the United States, but smog is having a major impact on health in rapidly industrializing countries such as China and India. On a fall day in 2017, all 4,000 schools in New Delhi, India, were closed for a week because air pollution was making the children sick.

Water Pollution. Disposing of chemicals into the water, a "flushing it down the sink" approach, doesn't work in today's crowded world. There is simply not enough water available to dilute the many substances that the enormous human population produces continuously. Lakes and rivers throughout the world are becoming increasingly polluted with sewage, fertilizers, and insecticides that wash from the land to the water in great quantities. Even rainwater is being polluted with acids released from coal-burning power plants, as recounted in the Today's Biology feature, *Acid Rain,* in chapter 2.

Agricultural Chemicals

The spread of "modern" agriculture has caused very large amounts of many kinds of new chemicals to be introduced into the global ecosystem, particularly pesticides, herbicides, and fertilizers. Chlorinated hydrocarbons such as DDT accumulate in animal fat tissue, so as they pass through a food chain, they become increasingly concentrated, a process called **biological magnification.** **Figure 22.1** shows how a minute concentration of DDT in plankton increases to significant levels as it is passed up through this aquatic food chain.

> **Putting the Concept to Work**
> Why is DDT, widely used in the tropics, banned in the United States?

22.2 Global Warming

> **LEARNING OBJECTIVE 22.2.1** Assess the proposition that global warming is the consequence of increased CO_2 in the atmosphere.

Few issues facing society are more critical than our growing impact on earth's climate.

Burning Fossil Fuels

For nearly 200 years, much of the growth of our industrial society has been fueled by burning fossil fuels—coal, oil, and gas. Coal, oil, and gas are the remains of ancient plants, transformed by pressure and time into carbon-rich "fossil fuels." When such fossil fuels are burned, this carbon is combined with oxygen atoms, producing carbon dioxide (CO_2). Industrial society's burning of fossil fuels has released huge amounts of carbon dioxide into the atmosphere. No one paid any attention to this because the carbon dioxide was thought to be harmless and because the atmosphere was thought to be a limitless reservoir, able to absorb and disperse any amount. It turns out neither assumption is true, and in recent decades, the levels of carbon dioxide in the atmosphere have risen sharply and are continuing to rise.

What is alarming is that the carbon dioxide doesn't just sit in the air doing nothing. The chemical bonds in carbon dioxide molecules transmit radiant energy from the sun but trap the longer wavelengths of infrared light, or heat, that are reflected off the earth's surface, and prevent them from radiating back into space. This creates what is known as the **greenhouse effect.** Planets that lack this type of "trapping" atmosphere are much colder than those that possess one. If the earth did not have a "trapping" atmosphere, the average earth temperature would be about $-20°C$, instead of the actual $+15°C$.

Global Warming Due to Greenhouse Gases

The rise in average global temperatures during recent decades, a profound change in the earth's climate referred to as **global warming**, is correlated with increased carbon dioxide concentrations in the atmosphere (**figure 22.2**). The suggestion that global warming might in fact be caused by the accumulation of greenhouse gases (carbon dioxide, chlorofluorocarbons, nitrogen oxides, and methane) in the atmosphere has been controversial. However, after serious examination of the evidence, the overwhelming consensus among scientists is that, indeed, greenhouse gases are causing global warming.

Increases in the amounts of greenhouse gases raise average global temperatures from 1° to 4°C, which could have a serious impact on rain patterns, prime agricultural lands, and sea levels.

Effects on Rain Patterns. Global warming is predicted to have a major effect on rainfall patterns. Areas that have already been experiencing droughts may see even less rain, contributing to even greater water shortages and increased danger of forest fires.

Effects on Agriculture. Warmer temperatures and increased levels of carbon dioxide in the atmosphere would be expected to increase the yields of some crops, while having a negative impact on others. Droughts that may result from global warming will also negatively affect crops.

Effects on Sea Level. Much of the water on earth is locked into ice in glaciers and polar ice caps. As global temperatures increase, these large stores of ice have begun to melt. Most of the water from the melted glaciers

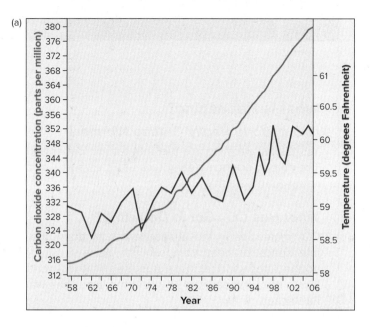

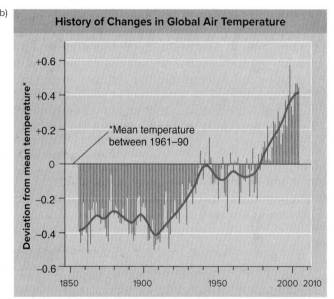

Figure 22.2 The grenhouse effect.

(a) The concentration of carbon dioxide in the atmosphere has shown a steady increase for many years (*blue line*). The *red line* shows the average global temperature for the same period of time. Note the general increase in temperature since the 1950s and, specifically, the sharp rise beginning in the 1980s.
(b) The effect of global warming can be seen clearly by comparing earth temperatures in any given year to the average: things are getting hotter.

(a) Source: The National Center for Atmospheric Research and other sources.
(b) Source: NASA/GISS.

Answering Your Questions About Global Warming

What Is Happening?

Is Our World Really Getting Warmer?

Yes. Global warming is a fact that can no longer be ignored. The culprit, science clearly tells us, is carbon dioxide (CO_2) released into earth's atmosphere by burning fossil fuels.

What Has CO_2 Got to Do with It?

CO_2 molecules in the atmosphere slow down the release into space of heat from sunlight. As a result, the heat accumulates—and temperatures rise. In the last century, CO_2 levels have increased 43% while the earth has warmed 2° Fahrenheit.

How Do We Know CO_2 Is Responsible?

Science has established that global warming is a direct consequence of rising CO_2 levels. One example of proof from among many is that the variability (technically, the "variance") of global temperature over the last century exactly matches the variability in CO_2 levels in the earth's atmosphere: When one goes up or down, so does the other in lockstep. This is only possible if global warming and levels of CO_2 are responding to the same underlying cause.

Could Natural Factors Be the Cause?

Science has further established that humans are responsible for the rise in atmospheric CO_2 levels: Hard evidence using radioactivity to distinguish industrial emissions of CO_2 from natural emissions shows without any doubt that the extra gas is coming from human activity.

Is the Science of Global Warming Disputed?

Not among scientists. There is overwhelming consensus that CO_2 emissions from burning fossil fuels are the root cause of the rapid warming of earth's climate. The U.S. National Academy of Science and the national academies of science of 33 other countries have all made formal declarations to this effect. In a recent survey of 10,306 climate scientists worldwide, over 97% agreed that global warming is real and largely caused by humans. This level of consensus is equivalent to the level of agreement among scientists that smoking causes cancer.

What Could Happen?

How Much Trouble Is the World In?

A lot. The global warming of 2°F we have already experienced may not seem much, but as an average over the surface of the entire planet, this amount of warming is actually enormous, causing much of the world's land ice to start to melt and permafrost to thaw. As a consequence, earth's oceans are rising at an accelerating pace. If CO_2 emissions continue

Hurricane Michael destruction in the Florida Panhandle 2018
Pool/Pool/Getty Images

Flooding in New Orleans caused by Hurricane Katrina 2005
Department of Commerce/NOAA

unchecked, scientists predict global warming will exceed 8°F, with disastrous consequences.

Are Recent Hurricanes and Heat Waves Tied to Global Warming?

Hurricanes such as Sandy (New York and New Jersey), Harvey (Texas), Irma (Florida), and Maria (Puerto Rico) all were stronger because of global warming, their wind speed accelerated by greater energy from warmer ocean waters. They dumped more rain because of simple physics: Warmer air holds more water. The heat waves of 2017, made worse by global warming, caused many deaths worldwide. 2020 tied for the hottest year on record, with the six hottest years occurring since 2015.

Am I Going to Get Flooded?

The world's oceans are going to rise. The key question is how fast. The current rate of a foot per century is manageable with massive efforts to fight coastal erosion. However, most experts believe that even if CO_2 emissions stopped tomorrow, 15 to 20 feet of more sea-level rise is inevitable, enough to flood many of earth's largest cities. If emissions continue unabated, the ultimate rise could be 80 or 100 feet, placing much of Florida and the East Coast under water.

What Can We Do?

What Can the World Do?

Stop burning so much fossil fuel. Governments can help by imposing limits: fuel economy standards for cars, emission limits for coal-powered power plants, tax breaks for renewable energy, and the like. Virtually every country in the world agreed in Paris in December 2015 to join together in such efforts. In the United States, states such as California are going even further. To offset the worst effects of climate change, even more will have to be done soon. Of utmost importance is the development of new "negative emissions" technologies to remove carbon dioxide from the atmosphere. Current attempts, such as pumping CO_2 into the ground or combining it with calcium to form limestone, can have only limited impact. Entirely new technologies are needed that can be applied on a massive scale. If we are to reverse global warming, the task we face is to remove some 40 billion tons of CO_2 from the atmosphere yearly.

Can Renewable Energy Help?

In a word, YES. New technology has greatly increased the energy recovery of wind turbines; wind farms of hundreds of massive turbine towers are producing increasing amounts of renewable energy. Solar panels, harvesting energy from sunlight with no emissions, are becoming widely used and are now common on housetops. Power plants burning natural gas produce fewer emissions than those burning coal. However, nuclear power plants, while emission-free, pose other problems and do not seem to be the road forward.

What About Fracking and Clean Coal?

Hydraulic fracturing, or fracking, is a drilling technology that releases natural gas from deposits in shale rocks. Burning this gas in power plants instead of coal reduces but does not eliminate CO_2 emissions. So-called "clean coal" is a proposed approach in which CO_2 emissions from coal-burning power plants are captured and pumped underground. Quite difficult to do economically, the approach is still largely a slogan of coal power advocates.

Will Electric Cars Make a Difference?

Yes, potentially a huge difference. While sales are still small, they are rising fast worldwide. Some countries such as France have banned the sale of all gasoline- and diesel-fueled cars after year 2040. Although e-cars are powered by electricity that is today generated largely by burning coal and natural gas, the electric grid itself is becoming increasingly green through the use of renewable power such as solar and wind.

What Can I Personally Do?

Reduce your carbon footprint by using efficient light bulbs and smart thermostats, turning off unused lights, and driving fewer miles. Here's a big one: Consider taking one or two fewer plane rides per year—that will save as much in emissions as all your other efforts combined! The most important thing you can do is to inform yourself, then speak up and demand change. The political policies needed to address the problem will only come about as the result of large-scale collective action.

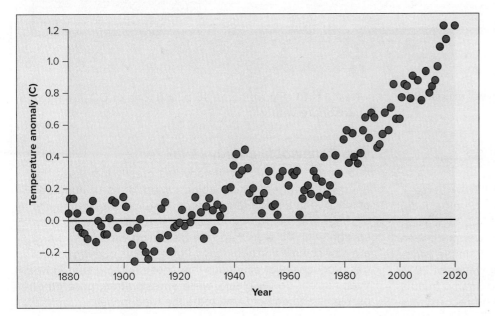

Figure 22.3 Annual global surface temperature.

The black line indicates the average temperature in the late 19th century. Since the 1950s, temperatures have been rising steadily.

ends up in the oceans, causing water levels to rise. Higher water levels can be expected to cause increased flooding of low-lying lands.

Effects on Extreme Weather. Warmer ocean waters will fuel stronger hurricanes, as a hurricane's power is determined in large degree by water temperature. And it gets worse: Because warmer air holds more water, the rainfall associated with these storms will be far more intense.

Combating Global Warming

Atmospheric carbon dioxide levels are at a 2-million-year high, and 2020 tied with 2016 for the hottest year on historical record (**Figure 22.3**). Of the 20 hottest years recorded, 19 have occurred since 2000. If nothing is done to reverse this trend, earth's oceans are going to rise as polar ice melts and flood coastal cities such as Miami, Florida, and Norfolk, Virginia.

Taking the danger seriously, the industrialized nations of the world met in Paris in 2015 and agreed to attempt better control of carbon emissions. Long-range targets were set and focused on reducing the burning of coal and other fossil fuels. Every industrial country on earth has signed on to the effort. Although the United States briefly withdrew from the agreement, California and several other U.S. states and cities have pledged to meet the emission standards set in the agreement, regardless of the federal government's actions. Each of you reading this text can make a real contribution to solving the problem of global warming by limiting the size of your carbon footprint.

Your carbon footprint is the amount of greenhouse gases (carbon dioxide and methane) you cause to be emitted at your standard of living through the course of your life. For most Americans, their carbon footprint is 10 times that of someone living in India. There are many ways to reduce your footprint. If you are like many people, flying is responsible for a large portion of your carbon footprint. Your share of one round-trip flight between New York and California generates about 20% of the greenhouse gases that your car emits over an entire year. So the single most impactful way to reduce your carbon footprint is to fly less often. If you want to take a more detailed look at reducing your carbon footprint, the Nature Conservancy provides a free carbon footprint estimator at www.nature.org/greenliving/carboncalculator.

Antarctica hit an all-time high temperature in 2020, and the American Pacific Northwest had the hottest summer in recent memory in 2021. Global warming is not a future danger. The threat is now.

> **Putting the Concept to Work**
> Do you expect the melting of the polar ice cap covering the North Pole to raise global sea levels significantly? Explain.

22.3 Loss of Biodiversity

LEARNING OBJECTIVE 22.3.1 Identify the three main causes of today's loss of biodiversity.

Just as death is as necessary to a normal life cycle as reproduction, so extinction is as normal and necessary to a stable world ecosystem as species formation. Most species, probably all, go extinct eventually. More than 99% of species known to science (most from the fossil record) are now extinct. However, current rates of extinctions are alarmingly high. The extinction rate for birds and mammals was about one species every decade from 1600 to 1700, but it rose to one species every year during the period from 1850 to 1950 and four species per year between 1986 and 1990. It is this increase in the rate of extinction that is the heart of the **biodiversity** crisis.

Factors Responsible for Extinction

What factors are responsible for extinction? Studying a wide array of recorded extinctions, biologists have identified three factors that seem to play a key role in many extinctions: habitat loss, species overexploitation, and introduced species (figure 22.4).

Habitat Loss. Habitat loss is the single most important cause of extinction. Given the tremendous amounts of ongoing destruction of all types of habitat, from rain forest to ocean floor, this should come as no surprise. Natural habitats may be adversely affected by human influences in four ways: (1) destruction, (2) pollution, (3) human disruption, and (4) habitat fragmentation (dividing up the habitat into small isolated areas). For example, habitat destruction in Madagascar is rapidly endangering species that live in the rain forest (figure 22.5).

Species Overexploitation. Species that are hunted or harvested by humans have historically been at grave risk of extinction, even when the species populations are initially very abundant. There are many examples in our recent history of overexploitation: passenger pigeons, bison, many species of whales, and mahogany trees are but a few.

Introduced Species. Occasionally, a new species will enter a habitat and colonize it, usually at the expense of native species. Colonization occurs in nature, but it is rare; however, humans have made this process more common, with devastating ecological consequence. The introduction of exotic species has wiped out or threatened many native populations. Species introductions occur in many ways, usually unintentionally. Plants and animals can be transported in nursery plants, in the ballast of large ocean vessels, as stowaways in boats, cars, and planes, and as beetle larvae within wood products. These species enter new environments where they have no native predators to keep their population sizes in check. Free to populate the habitat, they crowd out native species.

> **Putting the Concept to Work**
> Name one species of animal or plant driven extinct in your lifetime.

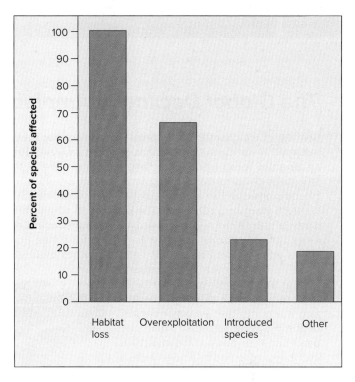

Figure 22.4 Factors responsible for animal extinction.

These data represent known extinctions of mammals in Australia, Asia, and the Americas. Some extinctions have more than one cause.

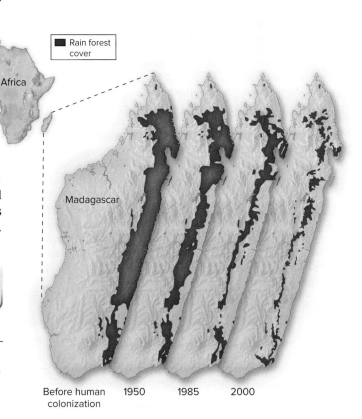

Figure 22.5 Extinction and habitat destruction.

The rain forest covering the eastern coast of Madagascar, an island off the coast of East Africa, has been progressively destroyed as the island's human population has grown. Ninety percent of the original forest cover is now gone. Many species have become extinct, and many others are threatened, including 16 of Madagascar's 31 primate species.

Today's Biology

The Global Decline in Amphibians

Sometimes important things happen right before our eyes, without anyone noticing. That thought occurred to David Bradford as he stood looking at a quiet lake high in the Sierra Nevada Mountains of California in the summer of 1988. Bradford, a biologist, had hiked all day to get to the lake, and when he got there, his worst fears were confirmed. The lake was on a list of mountain lakes that Bradford had been visiting that summer in Sequoia-Kings Canyon National Parks while looking for a little frog with yellow legs. The frog's scientific name was *Rana muscosa,* and it had lived in the lakes of the parks for as long as anyone had kept records. But this silent summer evening, the little frog was gone. The last major census of the frog's populations within the parks had been taken in the mid-1970s, and *R. muscosa* had been everywhere, a common inhabitant of the many freshwater ponds and lakes within the parks. Now, for some reason Bradford did not understand, the frogs had disappeared from 98% of the ponds that had been their homes.

A Wave of Extinction

After Bradford reported this puzzling disappearance to other biologists, an alarming pattern soon became evident. Throughout the world, local populations of amphibians (frogs, toads, and salamanders) were becoming extinct. Waves of extinction have swept through high-elevation amphibian populations in the western United States and the frog populations of Central America and coastal Australia.

Alfred Schauhuber/Getty Images

Amphibians have been around for 350 million years, since long before the dinosaurs. Their sudden disappearance from so many of their natural homes sounded an alarm among biologists. What are we doing to our world? If amphibians cannot survive the world we are making, can we?

In 2004, the first worldwide assessment of amphibian decline found that 32% of species were threatened globally and that as many as 122 species had become extinct since 1980.

In 2007, scientists worldwide met in Atlanta, Georgia, to form a group called the Amphibian Ark. They started captive breeding programs in an attempt to save more than 6,000 species from disappearing.

As of 2010, 486 amphibian species have been listed as "critically endangered" on the International Union for Conservation of Nature Red List.

What's Going On?

After years of intensive investigation, biologists have begun to sort out the reasons for the global decline in amphibians. Five factors seem to be contributing in a major way to the worldwide amphibian decline: (1) habitat deterioration and destruction, particularly clear-cutting of forests, which drastically lower the humidity (water in the air) that amphibians require; (2) the introduction of exotic species that outcompete local amphibian populations; (3) chemical pollutants that are toxic to amphibians; (4) fatal infections by pathogens; and (5) global warming, which is making some habitats unsuitable.

A Fish Virus

Infection by parasites appears to have played a particularly important role in the western United States and coastal Australia. Amphibian ecology expert James Collins of Arizona State University has reported one clear instance of infection leading to amphibian decline. When Collins examined populations of salamanders living on the Kaibab Plateau along the Grand Canyon rim, he found many sick salamanders. Their skin was covered with white pustules, and most infected ones died, their hearts and spleens collapsed. The infectious agent proved to be a virus common in fish called a ranavirus. Ranavirus isolated by Collins from one sick salamander would cause the disease in a healthy salamander, so there was no doubt that ranavirus was the culprit responsible for the salamander decline on the Kaibab Plateau.

A Fungus

A second kind of infection, very common in Australia but also seen in the United States, is having more widespread effects. Populations infected with this microbe, a kind of fungus called a chytrid (pronounced "kit-rid," see chapter 16), do not recover. Usually a harmless soil fungus that decomposes plant material, this particular chytrid (with the Latin name of *Batrachochytrium dendrobatidis*) is far from harmless to amphibians. It dissolves and absorbs the keratinous mouthparts of amphibian larvae, killing them.

This killer chytrid appeared in Australia near Melbourne in the early 1980s. Now almost all Australia is affected. How did the disease spread so rapidly? Apparently, it traveled by truck. Infected frogs moved across Australia in wooden boxes with bunches of bananas. In one year, 5,000 frogs were collected from banana crates in one Melbourne market alone.

Habitat Loss

In other parts of the world, infection does not seem to play as important a role as acid precipitation, habitat loss, and introduction of exotic species. Worldwide amphibian decline has no one culprit. Instead, all five factors play important roles. It is their total impact that has shifted the worldwide balance toward extinction.

Saving Our Environment

22.4 Preserving Nonreplaceable Resources

> **LEARNING OBJECTIVE 22.4.1** Evaluate the importance of three nonreplaceable resources.

Among the many ways ecosystems are being damaged, one problem stands out as more serious than all the rest: consuming or destroying resources that we all share in common but cannot replace (figure 22.6).

Topsoil

Soil is composed of a mixture of rocks and minerals with partially decayed organic matter called humus. Plant growth is strongly affected by soil composition. Minerals such as nitrogen and phosphorus are critical to plant growth and are abundant in humus-rich soils. Our midwestern farm belt sits astride what was once a great prairie. The **topsoil** (upper layer of soil) accumulated bit by bit from countless generations of animals and plants until, by the time humans came to plow, the humus-rich soil extended down several feet.

We cannot replace this rich topsoil, yet we are allowing it to be lost at a rate of centimeters every decade. Our country has lost one-quarter of its topsoil since 1950! By repeatedly tilling (turning the soil over) to eliminate weeds, we permit rain to wash more and more of the topsoil away, into rivers and eventually out to sea. New approaches are desperately needed to lessen

Figure 22.6 The tragedy of the commons.

In a now-famous essay, ecologist Garrett Hardin argued that destruction of the environment is driven by freedom without responsibility.

Martial Colomb/Photographer's Choice/Getty Images

Tragedy of the Commons

An 18th-century European settler cleared a tract of land, planted crops, and considered that land his own property. The Native Americans he drove away from this land never understood this. *How could one person own land?* they asked. No one Native American could conceive of owning the land we all walk on, any more than the air we all breathe. The grazing pasture you see in the photo is owned as property, its owner making sure it is not overgrazed by too large a herd. But what if the residents of the township owned the pasture in common? What if any resident could graze as many cows there as that resident wanted? Put yourself in that town, making a living by raising cows. To maximize your profit, how many cows should *you* graze in the common pasture? Do you see the problem? The answer is always "one more." You get the entire profit from that extra cow, you see, while any damage to the grass from overgrazing caused by that extra cow is shared equally by all the owners, your share being only a tiny fraction. The Native Americans had it right all along. When a natural resource is used in common by all the people, the use by any one individual cannot be unlimited because natural resources are limited. There is only so much grass, so much topsoil, so much groundwater—so much ocean water, so much air. We do not own the world. We live in it. Just like the Native Americans said.

the reliance on intensive cultivation. Some possible solutions include using genetic engineering to make crops resistant to weed-killing herbicides and terracing to recapture lost topsoil.

Groundwater

A second resource that we cannot replace is **groundwater,** water trapped beneath the soil within porous rock reservoirs called aquifers. This water seeped into its underground reservoir very slowly during the last ice age over 12,000 years ago. We should not waste it, for we cannot replace it.

In most areas of the United States, local governments exert relatively little control over the use of groundwater. As a result, a very large portion is wasted watering lawns, washing cars, and running fountains. A great deal more is inadvertently being polluted by poor disposal of chemical wastes—and once pollution enters the groundwater, there is no effective means of removing it. Some cities, such as Phoenix and Las Vegas, may completely deplete their groundwater within several decades.

Biodiversity

The number of species in danger of extinction during your lifetime is far greater than the number that became extinct with the dinosaurs. This disastrous loss of biodiversity is important because as these species disappear, so does our chance to learn about them and their possible benefits for ourselves. The fact that our entire supply of food is based on 20 kinds of plants, out of the 250,000 available, should give us pause. Like burning a library without reading the books, we don't know what it is we waste. All we can be sure of is that we cannot retrieve it. Extinction is forever.

Over the last 30 years, about half of the world's tropical rain forests have been either burned to make pasture land or cut for timber (**figure 22.7**). Over 6 million square kilometers have been destroyed. Every year, the rate of loss increases as the human population in the tropics grows. About 160,000 square kilometers were cut each year in the 1990s, a rate greater than 0.6 hectares (1.5 acres) per second! At this rate, all the rain forests of the world will be gone in your lifetime. In the process, it is estimated that one-fifth or more of the world's species of animals and plants will become extinct—more than a million species. This would be an extinction event unparalleled since the Age of Dinosaurs.

You should not be lulled into thinking that loss of biodiversity is a problem limited to the tropics. The ancient forests of the Pacific Northwest are being cut at a ferocious rate today. At the current rate, very little will remain in a decade. Nor is the problem restricted to one area. Throughout our country, natural forests are being clear-cut, replaced by pure stands of lumber trees planted in rows like so many lines of corn. It is difficult to scold those living in the tropics when we do such a poor job of preserving our own country's biodiversity.

(a)

(b)

(c)

Figure 22.7 Tropical rain forest destruction.
(a) These fires are destroying the rain forest in Brazil, which is being cleared for cattle pasture. (b) The flames are so widespread and so high that their smoke can be viewed from space. (c) The consequences of deforestation can be seen on these middle-elevation slopes in Brazil. The slopes now support only low-grade pastures where they used to support highly productive forest.

(a) Digital Vision/Getty Images; (b) NASA; (c) Mattias Klum/Getty Images

> **Putting the Concept to Work**
> Do you think it is bad to lose a species such as wild horses whose overbreeding harms the environment? Explain.

22.5 Curbing Population Growth

> **LEARNING OBJECTIVE 22.5.1** Describe the growth of the human population over the last 10,000 years.

If we were to solve all the problems mentioned in this chapter, we would merely buy time to address the fundamental problem: There are getting to be too many of us.

Exponential Growth

Humans first reached North America at least 12,000 to 13,000 years ago, crossing the narrow straits between Siberia and Alaska and moving swiftly to the southern tip of South America. By 10,000 years ago, when the continental ice sheets withdrew and agriculture first developed, about 5 million people lived on earth, distributed over all the continents except Antarctica. With the new and much more dependable sources of food that became available through agriculture, the human population began to grow more rapidly. By about 2000 years ago, an estimated 130 million people lived on earth. By the year 1650, the world's population had doubled and doubled again, reaching 500 million. Starting in the early 1700s, changes in technology have given humans more control over their food supply, led to the development of cures for many diseases, and produced improvements in shelter and storage capabilities that make humans less vulnerable to climatic uncertainties. Recall from chapter 19 that populations grow exponentially until they reach the limits of their environment, called the carrying capacity. These changes that occurred since the 1700s allowed humans to expand the carrying capacity of the habitats in which they lived and thus to escape the confines of logistic growth and reenter the exponential phase of the sigmoidal growth curve, shown by the explosive growth in figure 22.8.

Although the human population has grown explosively for the last 300 years, the average human birthrate has stabilized at about 18 births per year per 1,000 people worldwide. However, with the spread of better sanitation and improved medical techniques, the death rate has fallen steadily, to its present level of 2 per 1,000 per year. The difference between birth and death rates amounts to a population growth rate of 1.1% per year, which seems like a small number, but it is not, given our large population size (figure 22.9).

Approaching Earth's Limit

The world population reached 7.8 billion people in 2020, and the annual increase now amounts to about 83 million people, which leads to a doubling of the world population in about 62 years. Put another way, about 219,000 people are added to the world population each day, or almost 152 every minute. At this rate, the world's population will continue to grow and perhaps stabilize at a figure around 10 billion. Such growth cannot continue because our world cannot support it. Just as a cancer cannot grow unabated in your body without eventually killing you, so humanity cannot continue to grow unchecked in the biosphere without killing it.

Most countries are devoting considerable attention to slowing the growth rate of their populations, and there are genuine signs of progress, but the world population may still gain another 1 to 4 billion people before it stabilizes. No one knows whether the world can support so many people indefinitely. Finding a way to do so is the greatest task facing humanity. The quality of life that will be available for your children in this new century will depend to a large extent on our success.

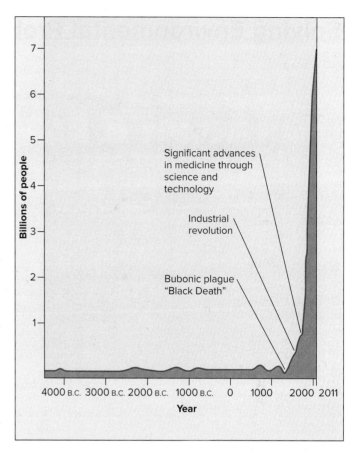

Figure 22.8 Growth curve of the human population.
Over the past 300 years, the world population has been growing steadily. Currently, there are about 7.6 billion people on the earth.

Figure 22.9 Mexico City has a population of about 21 million people.

Dan Fairchild Photography/Getty Images

Solving Environmental Problems

22.6 Preserving Endangered Species

Figure 22.10 Habitat restoration.
The University of Wisconsin-Madison Arboretum has pioneered restoration ecology. (a) The restoration of the prairie was at an early stage in November 1935. (b) The prairie as it looks today. This picture was taken at approximately the same location as the 1935 photograph.
(a, b) Courtesy University of Wisconsin-Madison Arboretum

LEARNING OBJECTIVE 22.6.1 Explain how the design of recovery plans for endangered species is related in each instance to the cause of species loss.

Once you understand the reasons a particular species is endangered, it becomes possible to think of designing a recovery plan. If the cause is commercial overharvesting, regulations can be designed to lessen the impact and protect the threatened species. If the cause is habitat loss, plans can be instituted to restore lost habitat. Loss of genetic variability in isolated subpopulations can be countered by transplanting individuals from genetically different populations. Populations in immediate danger of extinction can be captured, introduced into a captive breeding program, and later reintroduced to other suitable habitats.

Habitat Restoration

Conservation biology typically concerns itself with preserving populations and species in danger of decline or extinction. However, in many situations, conservation is no longer an option. The clear-cutting of the temperate forests of Washington state leaves little behind to conserve, nor does converting a piece of land into a wheat field or an asphalt parking lot. Redeeming these situations requires restoration rather than conservation. Three quite different sorts of habitat restoration programs might be undertaken, depending very much on the cause of the habitat loss.

Pristine Restoration. In situations where all species have been effectively removed, one might attempt to restore the plants and animals that are believed to be the natural inhabitants of the area, when such information is available. Restoring abandoned farmland to prairie (figure 22.10) requires that you know the identity of all of the original inhabitants and the ecologies of each of the species. We rarely have this much information, so no restoration is truly pristine.

Removing Introduced Species. Sometimes the habitat of a species has been destroyed by a single introduced species. In such a case, habitat restoration involves removal of the introduced species. For example, Lake Victoria, Africa, was home to over 300 species of cichlid fishes, small perchlike fishes that display incredible diversity. However, in 1954, the Nile perch, a commercial fish with a voracious appetite, was introduced into Lake Victoria. This resulted in the loss of over 70% of cichlid species, including all open-water species.

Restoration of the once-diverse cichlid fishes to Lake Victoria will require that the Nile perch population (as well as other introduced species such as water hyacinth plants) be brought under control or removed, as well as breeding and restocking the endangered species.

Cleanup and Rehabilitation. Habitats seriously degraded by chemical pollution cannot be restored until the pollution is cleaned up. The successful restoration of the Nashua River in New England, discussed in section 22.7, is one example of how a concerted effort can succeed in restoring a heavily polluted habitat to a relatively pristine condition.

Captive Propagation

Recovery programs, particularly those focused on one or a few species, often must involve direct intervention in natural populations to avoid an immediate threat of extinction. Introducing wild-caught individuals into captive breeding programs is being used in an attempt to save the black-footed ferret and California condor populations in immediate danger of disappearing. Several other such captive propagation programs have had success.

Case History: The Peregrine Falcon. U.S. populations of birds of prey such as the peregrine falcon (*Falco peregrinus*) began an abrupt decline shortly after World War II. Of the approximately 350 breeding pairs east of the Mississippi River in 1942, all of these falcons had disappeared by 1960. The culprit proved to be the chemical pesticide DDT and related organochlorine pesticides. Birds of prey are particularly vulnerable to DDT because they feed at the top of the food chain, where DDT becomes concentrated (see the discussion of biological magnification in section 22.1). DDT interferes with the deposition of calcium in the bird's eggshells, causing most eggs to break before they hatch.

The use of DDT was banned by federal law in 1972, causing levels of DDT in the eastern United States to fall quickly. Because there were no peregrine falcons left in the eastern United States to reestablish a natural population, falcons from other parts of the country were used to establish a captive breeding program at Cornell University in 1970. By the end of 1986, over 850 birds had been released in 13 eastern states, producing an astonishingly strong recovery (figure 22.11).

Figure 22.11 Captive propagation.

The reestablishment of peregrine falcon populations in the eastern United States is a success story for captive propagation programs.

U.S. Fish & Wildlife Service/Craig Koppie

Sustaining Genetic Diversity

One of the chief obstacles to a successful species recovery program is that a species is generally in serious trouble by the time a recovery program is instituted. When populations become very small, much of their genetic diversity is lost. If a program is to have any chance of success, it must sustain as much genetic diversity as possible.

Case History: The Black Rhino. Three of the five species of rhinoceros are critically endangered. The three Asian species live in a forest habitat that is rapidly being destroyed, while the two African species are illegally killed for their horns. Fewer than 22,000 individuals of all five species survive today. The problem is intensified by the fact that many of the remaining animals live in very small, isolated populations. The 4,000 wild-living individuals of the black rhino, *Diceros bicornis* (figure 22.12), live in approximately 75 small, widely separated groups that are adapted to local conditions throughout the species' range. The West African subspecies of black rhino was recently declared extinct, and the three surviving subspecies appear to have low genetic variability. Analysis of mitochondrial DNA suggests that in these populations, most individuals are genetically very similar.

> Small populations are at greater risk of extinction. Genetic drift, as discussed in section 14.9, is caused by random events that eliminate individuals, and therefore alleles, from a population and can be devastating if that population is small to begin with.

This lack of genetic variability represents one of the greatest challenges to the future of the species. Much of the range of the black rhino is still open and not yet subject to human encroachment. To have any significant chance of success, a species recovery program will have to find a way to sustain the genetic diversity that remains in this species. Heterozygosity could be best maintained by bringing all black rhinos together

Figure 22.12 Sustaining genetic diversity.

The black rhino is highly endangered, living in 75 small, widely separated populations. Only about 4,000 individuals survive in the wild. Conservation biologists have the difficult job of finding ways to preserve genetic diversity in small, isolated populations.

Franz Aberham/Getty Images

IMPLICATION FOR YOU A basic problem faced by black rhinos is that the species has lost much of its genetic diversity as its numbers have decreased. Do you think that expanding the size of today's small rhino populations would fix this problem?

in a single breeding population, but this is not a practical possibility. A more feasible solution would be to move individuals between populations. Managing the black rhino populations for genetic diversity could prevent the fatal loss of genetic variation.

Preserving Keystone Species

Keystone species are species that exert a particularly strong influence on the structure and functioning of their ecosystem. Their removal can have disastrous consequences.

Case History: Flying Foxes. The severe decline of many species of pteropodid bats, or "flying foxes," in the Old World tropics is an example of how the loss of a keystone species can have dramatic effects on the other species living within an ecosystem, sometimes even leading to a cascade of further extinctions (figure 22.13). These bats have very close relationships with important plant species on the islands of the Pacific and Indian Oceans. Widespread on the islands of the South Pacific, flying foxes are the most important—and often the only—pollinators and seed dispersers. A study in Samoa found that 80% to 100% of the seeds landing on the ground during the dry season were deposited by flying foxes. Many species are entirely dependent on these bats for pollination.

Flying foxes are being driven to extinction by human hunting. They are hunted for food and by orchard farmers, who consider them pests. Flying foxes are particularly vulnerable because they live in large, easily seen groups of up to a million individuals. Because they move in predictable patterns and can be tracked to their home roost, hunters can easily bag thousands at a time.

In Guam, where the two local species of flying fox have recently been driven extinct or nearly so, the impact on the ecosystem appears to be substantial. Many plant species are not fruiting or are doing so only marginally, with fewer fruits than normal. Attempts to preserve flying fox populations are underway. In each instance, legal protection, captive breeding programs, and habitat preservation are key ingredients to successful programs.

Figure 22.13 Preserving keystone species.
The flying fox is a keystone species in many Old World tropical islands. It pollinates many of the plants and is a key disperser of seeds. Its elimination by hunting and habitat loss is having a devastating effect on the ecosystems of many South Pacific islands.
Dr. Merlin D. Tuttle/Science Source

Conservation of Ecosystems

Habitat fragmentation is one of the most pervasive enemies of biodiversity conservation efforts. Some species simply require large patches of habitat to thrive, and conservation efforts that cannot provide suitable habitat of such a size are doomed to failure. It has become clear that species placed in isolated patches of habitat are lost far more rapidly than species placed in large preserves. As a result, conservation biologists have promoted the creation, particularly in the tropics, of so-called mega-reserves, large areas of land containing a core of one or more undisturbed habitats (figure 22.14).

However, ample land to form mega-reserves may not be available. In these cases, the linking together of habitat islands through the use of corridors is proving an option. Corridors are areas with suitable vegetation and topography between habitats that allow animals to wander outside of their home ranges. Corridors allow for seasonal migrations, the dispersal of juveniles into new territories, and the expansion of growing populations.

Figure 22.14 A mega-reserve.
The Baviaanskloof mega-reserve in South Africa's Eastern Cape Province includes 270,000 hectares of unspoiled terrain, habitat for many species of plants and animals.
Shaen Adey/Getty Images

> **Putting the Concept to Work**
> Explain the importance of keystone species to biodiversity.

22.7 Individuals Can Make the Difference

LEARNING OBJECTIVE 22.7.1 Recount how Lake Washington and the Nashua River were restored through individual action.

The development of appropriate solutions to the world's environmental problems must rest partly on the shoulders of politicians, economists, bankers, and engineers; many kinds of public and commercial activity will be required. However, it is important not to lose sight of the key role often played by informed individuals in solving environmental problems. Often, one person has made the difference; two examples serve to illustrate the point.

The Nashua River

Running through the heart of New England, the Nashua River was severely polluted by mills established in Massachusetts in the early 1900s. By the 1960s, the river was clogged with pollution and declared ecologically dead. When Marion Stoddart moved to a town along the river in 1962, she was appalled. She approached the state about setting aside a "greenway" (trees running the length of the river on both sides), but the state wasn't interested in buying land along a filthy river. So Stoddart organized the Nashua River Cleanup Committee and began a campaign to ban the dumping of chemicals and wastes into the river. The committee presented bottles of dirty river water to politicians, spoke at town meetings, recruited businesspeople to help finance a waste treatment plant, and began to clean garbage from the Nashua's banks. This citizen's campaign, coordinated by Stoddart, greatly aided passage of the Massachusetts Clean Water Act of 1966. Industrial dumping into the river is now banned, and the river has largely recovered (figure 22.15).

Figure 22.15 Cleaning up the Nashua River.
The Nashua River, seen on the left in the 1960s, was severely polluted because factories set up along its banks dumped their wastes directly into the river. Seen on the right today, the river is mostly clean.

Courtesy Nashua River Watershed Association

Lake Washington

A large, 86-square-kilometer freshwater lake east of Seattle, Lake Washington, became surrounded by Seattle suburbs in the building boom following the Second World War. Between 1940 and 1953, a ring of 10 municipal sewage plants discharged their treated effluent into the lake. Safe enough to drink, the effluent was believed "harmless." By the mid-1950s, a great deal of effluent had been dumped into the lake (try multiplying 80 million liters/day × 365 days/year × 10 years). In 1954, an ecology professor at the University of Washington in Seattle, W. T. Edmondson, noted that his research students were reporting filamentous blue-green algae growing in the lake. Such algae require plentiful nutrients, which deep freshwater lakes usually lack. The sewage had been fertilizing the lake! Edmondson, alarmed, began a campaign in 1956 to educate public officials to the danger: Bacteria decomposing dead algae would soon so deplete the lake's oxygen that the lake would die. After five years, joint municipal taxes were levied that financed the building of a sewer to carry the effluent out to sea. The lake is now clean (figure 22.16).

Figure 22.16 Lake Washington, Seattle.
Lake Washington in Seattle is surrounded by residences, businesses, and industries. By the 1950s, the dumping of sewage and the runoff of fertilizers had caused an algal bloom in the lake, which would eventually deplete the lake's oxygen. Efforts to reverse this effect and clean up the lake were started by W. T. Edmondson of the University of Washington in 1956. The lake is now clean.

Neil Rabinowitz/Getty Images

> **Putting the Concept to Work**
> Why didn't Professor Edmondson simply recommend that the municipalities around Lake Washington poison the algae growing in the lake with chemicals that don't harm humans, in order to return the lake's water to its pristine state?

Putting the Chapter to Work

1. The northern white rhino, *Ceratotherium* simum, is targeted by poachers for its horns. In 2018, the last male, named Sudan, died, bringing the species to the brink of extinction. Only two females are left alive, one Sudan's daughter and the other his granddaughter. Before Sudan's death, sperm was collected in hopes of preserving the species.

Even if the two surviving females were to receive Sudan's sperm in an effort to save the species from extinction and their offspring were protected from poachers, what might be an obstacle in sustaining species recovery of the northern white rhino?

2. Sea otters, *Enhydra lutris,* live along the northern Pacific coast and prey on a variety of shellfish and sea urchins. This predation provides sustainability to the kelp ecosystem, because sea urchins love to eat kelp. When sea otters are removed, the kelp ecosystem is quickly overgrazed by sea urchins and soon dies out.

Based on this information, the sea otter is considered what type of species?

Retracing the Learning Path

Global Change

22.1 Pollution

1. All over the world, increasing industrialization is leading to higher levels of pollution. Pollution leads to global change because its effects can spread far from the source. Air and water become polluted when chemicals that are harmful to organisms are released into the ecosystem. The use of agricultural chemicals, such as pesticides, herbicides, and fertilizers, has been widespread with devastating effects on animals. Biological magnification occurs when harmful chemicals become more concentrated as they pass up through the food chain.

22.2 Global Warming

1. Humanity's burning of fossil fuels has greatly increased levels of carbon dioxide in the atmosphere, where it traps infrared light (heat) from the sun, a phenomenon known as the greenhouse effect. As a result, the average global temperatures have been steadily increasing as CO_2 levels increase in the atmosphere, a process known as global warming. Global warming is predicted to have major impacts on global rain patterns, agriculture, and sea levels.

22.3 Loss of Biodiversity

1. The current rate of species extinction is alarmingly high. The three factors mostly responsible include loss of habitat, overexploitation, and the introduction of new species. Of these, loss of habitat is the most devastating.

Saving Our Environment

22.4 Preserving Nonreplaceable Resources

1. The consumption or destruction of nonreplaceable resources is perhaps the most serious problem humans face. Topsoil, necessary for agriculture, is being depleted rapidly. Groundwater, which percolates through the soil to underground reservoirs, is our primary source of drinking water, but it is being wasted and polluted. Biodiversity is being reduced through extinctions, due primarily to loss of habitat, such as rain forests.

22.5 Curbing Population Growth

1. The problem at the core of our environmental concerns is the rapid growth of the world's human population. Technology has allowed the human population to grow exponentially for the last 300 years to a current population of about 7.6 billion.

Solving Environmental Problems

22.6 Preserving Endangered Species

1. In an attempt to slow the loss of biodiversity, recovery programs are under way designed to save endangered species. These programs include habitat restoration, breeding in captivity, sustaining genetic diversity, preserving keystone species, and conservation of ecosystems.

22.7 Individuals Can Make the Difference

1. There are environmental success stories where one or a few individuals made a difference and reversed an ecological disaster.

Inquiry and Analysis

Whole Lotta Shakin' Goin' On

Mario Tama/Getty Images

The COVID-19 pandemic has been shaking our world—or rather, de-shaking it. The graph you see is a seismograph measuring minute movements in the earth's crust. While earthquakes cause such a graph to gyrate violently, everyday changes due to daily warming of the crust by the sun cause the graph to slide up and down a bit. These tiny movements are considered the daily background noise. The seismograph, recorded near Los Angeles, California between February and April 2020, surprised investigators very much: starting around March 1, the power of the daily shaking—the background noise—suddenly began to lessen, a 33% reduction in background noise. The earth was vibrating less!

What might cause this? You could look back over the entire last year and not see such a dramatic effect. Seasonal changes are far more gradual than this abrupt drop. It was as if something had changed the earth's very vibration, like tightening the fret on a guitar string.

A possible solution to this mystery is suggested by the photo you see on the top. It is of downtown Los Angeles at noon on Thursday, April 23, 2020. Early in the coronavirus pandemic, at the beginning of March, stay-at-home orders were put in place throughout the United States—except for emergencies and to pick up curbside food and groceries, you were instructed not to leave your house. What do you see in the photo of Los Angeles? No cars! 320 million Americans, staying at home, were not driving their cars after the first week of March.

To understand how not driving the family car might affect a seismograph, it helps to view the coronavirus pandemic from the highway's point of view. At a microscopic level, a roadway is not perfectly smooth, but a series of bumps and ridges—a tire travelling over it slams against these tiny bumps like so many minute hammer blows: bam, bam, bam . . . Each very tiny, but could all these bams add up enough to move the earth?

Analysis

1. **Applying Concepts** How long does each up-and-down inflection of the graph persist? How many occur within a week?
2. **Interpreting Data**
   a. Measured as relative power, what is the maximum power recorded in the second week of February (February 9 to 15)? The third week of February? The fourth week of February?
   b. What is the maximum power recorded in the first week of March? The second week of March? The third week of March? The fourth week of March?
3. **Making Inferences**
   a. Do you see a trend in the weekly maximum power recordings of February? What, if any, is it?
   b. Do you see a trend in the weekly maximum power recordings of March? What, if any, is it?
   c. Do you see a difference in the trends for February and March? What is it?
4. **Drawing Conclusions** Does the lack of vibration correspond to the date at which Americans stopped driving cars?
5. **Further Analysis** Botanists have long known that mechanical vibration of a growing plant's body results in reduced plant height. So, might COVID-19's reduction in man-made vibrations of the earth's crust have caused the grass on our lawns to grow taller? How would you test this?

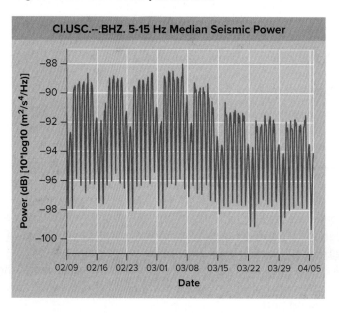

PART 6 Animal Life

23 The Animal Body and How It Moves

LEARNING PATH ▼

The Animal Body Plan
1. Organization of the Vertebrate Body

Tissues of the Vertebrate Body
2. Epithelium Is Protective Tissue
3. Connective Tissue Carries Out Various Functions
4. Muscle Tissue Lets the Body Move
5. Nerve Tissue Conducts Signals Rapidly

The Skeletal and Muscular Systems
6. Types of Skeletons
7. Muscles and How They Work

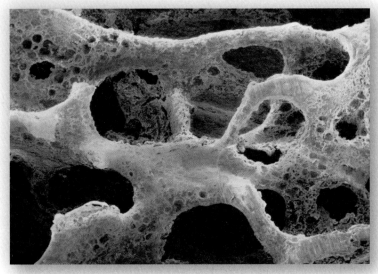

RAVAGED BY OSTEOPORISIS, this bone is greatly weakened.
Professor Pietro M. Motta/Science Source

Why Osteoporosis Is a Woman's Problem

Running, jumping, dancing, swimming, even squirming through mud—all are things that you can do when you want to. As you will discover in this chapter, your body accomplishes these feats through the action of muscles pulling against an internal skeleton made of bone.

Bones Are Not Made of Stone

The bones that make up your skeleton are not lifeless, like the scaffolding of a building, but are living tissue that continues to grow and change throughout your life. Just because you reach your full height does not mean that your bones stop growing. They stop elongating, but that is merely one aspect of bone growth. Every day of your life, your bones are adding new tissue and breaking down old tissue, reshaping your skeleton to meet the needs of your body. If you break a bone, your body heals the fracture by adding new bone tissue at the site of the break. However, when the process of forming new bone is impaired, the bone becomes weak and brittle.

Osteoporosis

As you age, your backbone, hips, and other bones tend to lose bone mass much faster than new bone mass is produced. Excessive bone loss is a condition called *osteoporosis*. The photograph of bone you see here is from someone who has had osteoporosis. You can see how much bone tissue has been lost. You can see how much less bone tissue is present with osteoporosis.

Unfortunately for the person suffering from osteoporosis, even everyday activities such as bending or lifting can result in a broken bone. In a person with osteoporosis, just the stress of the body's weight on the weakened bones causes small fractures up and down the spine.

Osteoporosis is not a rare condition. Over 10 million people in the United States today have osteoporosis, with twice that many at risk. Osteoporosis causes about 1.5 million fractures a year, with spine and hip fractures making up the majority of these.

More Common in Women

Importantly, osteoporosis is more common in women than in men. One out of every two women over 50 years of age experiences osteoporosis to some degree. The incidence

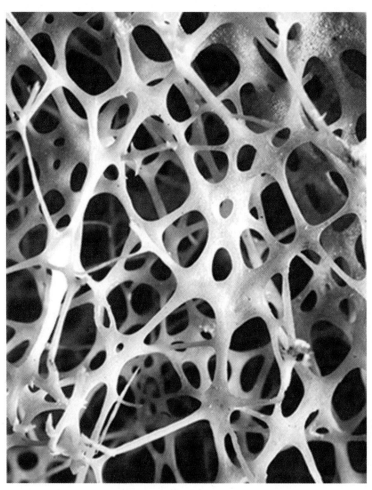
Science Source

Uwe Bumann/Getty Images

of osteoporosis is considerably less in men. Of the 10 million Americans affected by osteoporosis, over 80% are women over the age of 50. Why so prevalent in women and not men? The process of building up bone and breaking it down is controlled by two key hormones: (1) parathyroid hormone, which activates release of calcium from bone, and (2) calcitonin, which acts in reverse to inhibit the release of calcium from bone. The balance of these hormones in a woman's body shifts after puberty, the cyclic building up and breaking down of bone shifting toward breakdown as her body makes more demands for calcium. Osteoporosis develops progressively with age, bone loss exceeding the replacement of bone year after year to the point when the bone becomes weak and brittle.

Planning Ahead

Although osteoporosis is a disease expressed most profoundly in the elderly, it can be prevented by actions taken when a person is younger. The best treatment for osteoporosis is to avoid getting it by building up your bones when you're young. The "sculpting" of bone occurs throughout your lifetime, but it is in your childhood and teenage years when new bone is added at a faster rate than it is lost by the breakdown of bone tissue. This net addition of bone tissue continues until a person reaches about 30 years of age, when the balance begins to tip toward the direction of bone loss. As discussed in this chapter, bone forms when calcium phosphate hardens around collagen fibers, and so an adequate supply of calcium is important to healthy bone development. Because of this, it is important that you have enough calcium in your diet to fulfill the needs of developing bone, as well as other body activities that require calcium. Building up strong, healthy, dense bones before age 30 means that you have a larger pool of calcium in your bones to offset bone loss in your older years. Foods high in calcium include low-fat dairy products, dark green leafy vegetables, and calcium-fortified orange juice.

Physical activity is also key to preventing osteoporosis. New bone tissue is added to areas of your bones that are stressed to make those areas stronger. Weight-bearing exercises help to strengthen areas of your bones that might be most susceptible to fractures.

The Animal Body Plan

23.1 Organization of the Vertebrate Body

> **LEARNING OBJECTIVE 23.1.1** Describe the four general classes of tissues and how they are organized into organs.

All vertebrates and other coelomates have the same general architecture: a long internal tube (the gut or digestive system) that extends from mouth to anus, which is suspended within an internal body cavity called the *coelom*. The coelom of many terrestrial vertebrates is divided into two parts: (1) the *thoracic cavity*, which contains the heart and lungs, and (2) the *abdominal cavity*, which contains the stomach, intestines, and liver. The vertebrate body is supported by an internal scaffold, or skeleton, made up of jointed bones or cartilage.

Like all animals, the vertebrate body is composed of cells—over 10 to 100 trillion of them in your body. It's difficult to picture how large this number actually is. A line of 10 trillion cars would stretch from the earth to the sun and back 100 times! Not all of these cells in your body are the same, of course. If they were, we would not be bodies but amorphous blobs. Vertebrate bodies contain over 100 different kinds of cells.

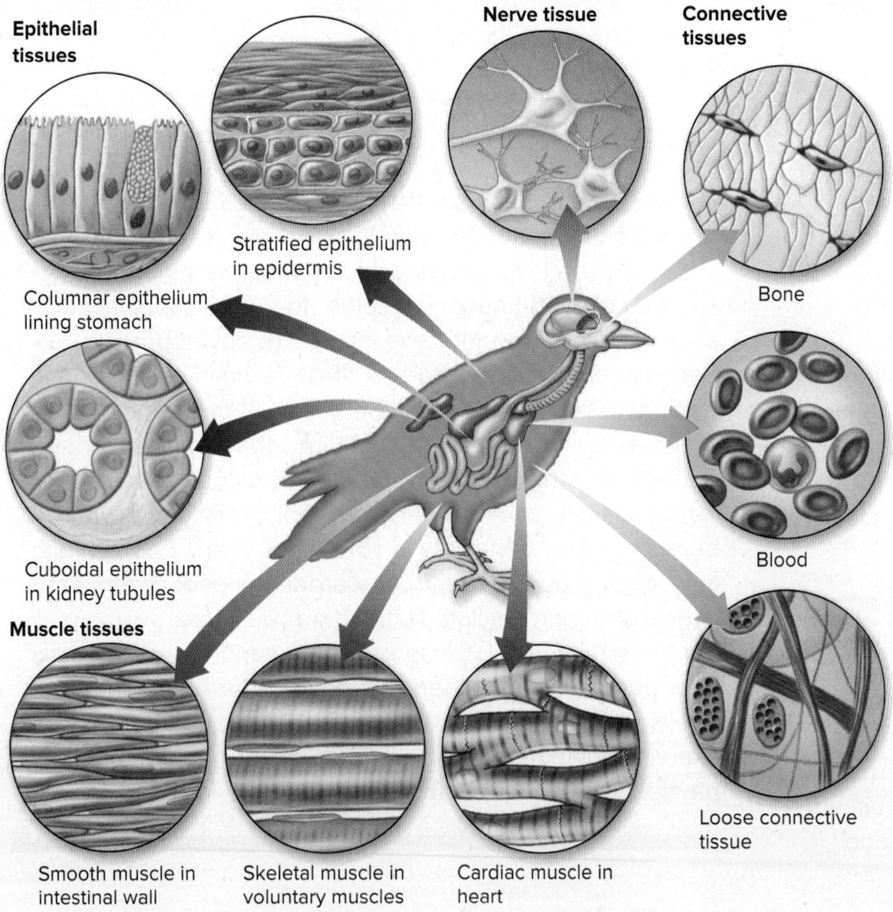

Tissues

Groups of cells of the same type are organized within the body into **tissues**, which are the structural and functional units of the vertebrate body. A tissue is a group of cells of the same type that performs a particular function. It is possible to assemble many different kinds of tissue from 100 cell types, but biologists have traditionally grouped adult tissues into four general classes: *epithelial, connective, muscle,* and *nerve tissue* (figure 23.1).

Organs

Organs are body structures composed of several different tissues grouped together into a larger structural and functional unit, just as a factory is a group of people with different jobs who work together to make something. The heart is an organ. It contains cardiac muscle tissue wrapped in connective tissue and joined to many nerves. All of these tissues work together to pump blood through the body: The cardiac muscles contract, which squeezes the heart to push the blood; the connective tissues act as a bag to hold the heart in the proper shape and ensure that the different chambers of the heart squeeze in the proper order; and the nerves control the rate at which the heart beats. No single tissue can do the job of the heart, any more than one piston can do the job of an automobile engine.

Figure 23.1 Vertebrate tissue types.
The four basic classes of tissue are epithelial, nerve, connective, and muscle.

Epithelial tissues: Columnar epithelium lining stomach; Stratified epithelium in epidermis; Cuboidal epithelium in kidney tubules
Nerve tissue
Connective tissues: Bone; Blood; Loose connective tissue
Muscle tissues: Smooth muscle in intestinal wall; Skeletal muscle in voluntary muscles; Cardiac muscle in heart

> **Putting the Concept to Work**
> Name a human organ other than the heart. What tissues is it made of?

Organ Systems

> **LEARNING OBJECTIVE 23.1.2** Describe the principal organ systems of the vertebrate body.

An **organ system** is a group of organs that work together to carry out an important function (figure 23.2). For example, the vertebrate digestive system is an organ system composed of individual organs that break up food (beaks or teeth), pass the food to the stomach (esophagus), break down the food (stomach and intestine), absorb the food (intestine), and expel the solid residue (rectum). If all of these organs do their job right, the body obtains energy and necessary building materials from food. The digestive system is a particularly complex organ system with many different organs consisting of many different types of cells, all working together to carry out a complex function.

The vertebrate body contains 11 principal organ systems:

1. **Skeletal.** The most distinguishing feature of the vertebrate body is its internal skeleton made of cartilage or bone. The skeletal system protects the body and provides support for locomotion and movement. Its principal components are bones, cartilage, and ligaments. Like arthropods, vertebrates have jointed appendages—arms, hands, legs, and feet.
2. **Circulatory.** The circulatory system transports oxygen, nutrients, and chemical signals to the cells of the body, and removes carbon dioxide, chemical wastes, and water. Its principal components are the heart, blood vessels, and blood.
3. **Endocrine.** The endocrine system coordinates and integrates the activities of the body through the release of hormones. Its principal components are the pituitary, adrenal, thyroid, and other ductless glands.
4. **Nervous.** The activities of the body are coordinated by the nervous system. Its principal components are the nerves, sense organs, brain, and spinal cord.
5. **Respiratory.** The respiratory system captures oxygen and exchanges gases, and is composed of the lungs, trachea, and other air passageways.
6. **Immune.** The immune system removes foreign bodies from the bloodstream using special cells, such as lymphocytes and macrophages.
7. **Digestive.** The digestive system captures soluble nutrients from ingested food. Its principal components are the mouth, esophagus, stomach, intestines, liver, and pancreas.
8. **Urinary.** The urinary system removes metabolic wastes from the bloodstream. Its principal components are the kidneys, bladder, and associated ducts.
9. **Muscular.** The muscular system produces movement, both within the body and of its limbs. Its principal components are skeletal muscle, cardiac muscle, and smooth muscle.
10. **Reproductive.** The reproductive system carries out reproduction. Its principal components are the testes in males, ovaries in females, and associated reproductive structures.
11. **Integumentary.** The integumentary system covers and protects the body. Its principal components are the skin, hair, nails, and sweat glands.

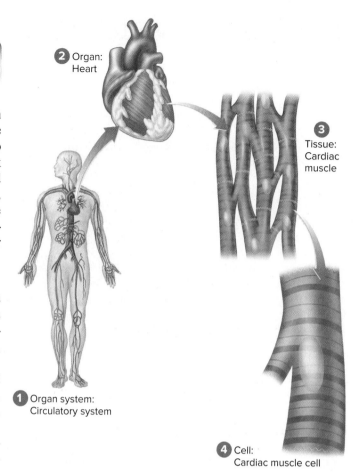

Figure 23.2 Levels of organization within the vertebrate body.

❶ The circulatory system you see illustrated here is an example of an organ system. ❷ The heart is one of several organs that work together to carry out the function of circulation for the body. ❸ Within an organ such as the heart, individual tissues such as cardiac muscle function together with other tissues. ❹ Particular cell types, such as cardiac muscle cells, operate together with other similar types of cells to form each kind of tissue.

> **IMPLICATION FOR YOU** Ten of the human body parts are spelled with three letters. Can you name all 10? (No slang names such as "a__.")

> **Putting the Concept to Work**
> Which organ systems extend throughout the major parts of the human body?

Tissues of the Vertebrate Body

23.2 Epithelium Is Protective Tissue

LEARNING OBJECTIVE 23.2.1 Describe the three types of vertebrate epithelial cells and how they function.

Tissues are the basic building blocks of the animal body. We begin our discussion of tissues with one located on the body's surfaces, epithelium. Epithelial cells are the guards and protectors of the body. **Epithelial tissue** covers both internal and external surfaces of the body, and determines which substances enter the body and which do not. The body's epithelial layers function in three ways:

1. They *protect underlying tissues* from water loss and mechanical damage. Because epithelium encases all the body's surfaces, every substance that enters or leaves the body must cross an epithelial layer, even one as thick as the gila monster's in figure 23.3.
2. They *provide sensory surfaces.* Many of a vertebrate's sense organs are modified epithelial cells.
3. They *secrete materials.* Most secretory glands are derived from pockets of epithelial cells that pinch together during embryonic development.

Types of Epithelial Cells and Epithelial Tissues

Epithelial cells are classified into three types according to their shapes: squamous, cuboidal, and columnar. Layers of epithelial tissue are usually only one or a few cells thick. Individual epithelial cells possess only a small amount of cytoplasm and have a relatively low metabolic rate. A characteristic of all epithelia is that sheets of cells are tightly bound together, with very little space between them. This forms the barrier that is key to their functioning.

> Like animals, plants also have a surface layer of protective epithelial cells, some of which are specialized for different functions, as described in section 31.2.

The cells of epithelial layers are constantly being replaced throughout the life of the organism. The cells lining the digestive tract, for example, are continuously replaced every few days. The epidermis, the epithelium that forms the skin, is renewed every two weeks.

There are two general kinds of epithelial tissue. First, the membranes that line the lungs and the major cavities of the body are a **simple epithelium** only a single cell layer thick. These single-celled layers are surfaces across which many materials must pass, entering and leaving the body's compartments. Second, the skin, or epidermis, is a **stratified epithelium** composed of more complex epithelial cells several layers thick. Multilayers are necessary to provide adequate cushioning and protection, and to enable the skin to continuously replace its cells.

A type of simple epithelial tissue that has a secretory function is cuboidal epithelium, which is found in the **glands** of the body. Endocrine glands secrete hormones into the blood. Exocrine glands (those with ducts that open to the body's outside) secrete sweat, milk, saliva, and digestive enzymes.

Figure 23.3 The epithelium prevents dehydration.
The tough, scaly skin of this gila monster provides a layer of protection against dehydration and injury. For all land-dwelling vertebrates, the relative impermeability of the surface epithelium (the epidermis) to water offers essential protection from dehydration and airborne pathogens (disease-causing organisms).
Tim Flach/Getty Images

BIOLOGY & YOU

Dandruff. Have you ever had dandruff? You know, the white flaky stuff on your collar and shoulders. If so, you are not alone. Many, if not most, Americans experience dandruff at one time or another. Dandruff is the shedding of excessive dead skin cells from the scalp. It is normal for skin cells to die and flake off, producing small flakes too tiny to see; the epidermal layer continually replaces itself from its inside, pushing cells outward, where they eventually die and flake off. However, in people with dandruff, this process is greatly accelerated, with skin cells maturing and shedding in two to seven days, as opposed to around a month in people without dandruff. Dead skin cells build up so fast that they are shed in large, oily clumps that appear as white flakes. What causes dandruff? A scalp fungus called *Malassezia globosa*. It metabolizes skin oils, producing a by-product that triggers an inflammatory response in susceptible persons, leading to accelerated division of skin cells. Shampoos control dandruff in different ways: Sebulex uses salicylic acid to remove dead skin cells from the scalp and slow cell division; Head and Shoulders uses zinc pyrithione to kill the *Malassezia* fungus; Selsun Blue uses selenium sulfide to achieve the results of both salicylic acid and zinc pyrithione. All reduce flaking dramatically.

> **Putting the Concept to Work**
> In what way is your skin like the bark of a tree? In what way is it different?

23.3 Connective Tissue Carries Out Various Functions

LEARNING OBJECTIVE 23.3.1 Describe the three functional categories of connective tissue and their structural features.

The cells of **connective tissue** provide the vertebrate body with its structural building blocks and with its most potent defenses. These cells are sometimes densely packed together and sometimes loosely arrayed. Although very diverse, all connective tissues share a key common structural feature: They all have abundant extracellular **matrix** material between widely spaced cells.

Immune Connective Tissue Defends the Body

The cells of the immune system, the many kinds of "white blood cells," roam the body within the bloodstream. They are mobile hunters of invading microorganisms and cancer cells. The two principal kinds of immune system cells are **macrophages,** which engulf and digest invading microorganisms, and **lymphocytes,** which attack virus-infected cells or make antibodies that tag cells for destruction. Immune cells are carried through the body in a fluid matrix, called *plasma*.

Skeletal Connective Tissue Supports the Body

Three kinds of connective tissue are the principal components of the skeletal system: fibrous connective tissue, cartilage, and bone. Although composed of similar cells, they differ in the nature of the matrix that is laid down between individual cells.

1. **Fibrous connective tissue.** The most common kind of connective tissue in the vertebrate body is fibrous connective tissue. It is composed of flat, irregularly branching cells called **fibroblasts** that secrete structurally strong proteins such as collagen into the spaces between the cells. Fibroblasts are active in wound healing; scar tissue, for example, possesses a collagen matrix.
2. **Cartilage.** In cartilage, a collagen matrix between cartilage cells (technically called *chondrocytes*) forms in long parallel arrays along the lines of mechanical stress. What results is a firm and flexible tissue of great strength (**figure 23.4**), just as strands of nylon molecules laid down in long, parallel arrays produce strong, flexible ropes.
3. **Bone.** Bone is similar to cartilage, except the collagen fibers are coated with a calcium phosphate salt, making the tissue rigid.

Blood and Fat Cells: Transport and Storage

The third general class of connective tissue is made up of cells that are specialized to accumulate and transport particular molecules. They include the fat-accumulating cells of **adipose tissue** and red blood cells, called **erythrocytes,** that function in transport and storage. About 5 billion erythrocytes are present in every milliliter of your blood.

> **Putting the Concept to Work**
> What is the common structural feature of all connective tissues?

BIOLOGY & YOU

Getting Fat. It will come as no surprise to you that overeating can make you fat. Is this because overeating triggers the production of more fat cells or because the fat cells get larger? It turns out, both situations occur. You are born with a predetermined number of fat cells, with women generally inheriting more than men. The number then grows through childhood until late puberty, after which it is pretty much set. It was generally thought that the only difference between obese people and nonobese people was that obese people's fat cells were filled up to capacity. When overweight people lose weight, they empty their fat cells, but the fat cells don't go away. They are still there, like little sponges waiting to refill. We now know that we can—and do—"grow" more fat cells in adulthood and that overweight and obese people do in fact have more fat cells. It turns out that when fat cells expand to their maximum size, this triggers cell division, producing an increase in the actual number of fat cells. It is one of the curses of dieters that these new cells do not go away when you lose weight. Dieting may shrink the cells, but dieters tend to regain lost weight rapidly. If you wind up with a lot of extra fat cells, you have to work harder than lean people to keep excess weight off.

Figure 23.4 Cartilage: skeletal connective tissue.

The entire skeleton of this Caribbean reef shark is made of cartilage, a strong but flexible type of connective tissue. In other vertebrates, cartilage is found in joints, noses, and external ears.

Richard Carey/Alamy Stock Photo

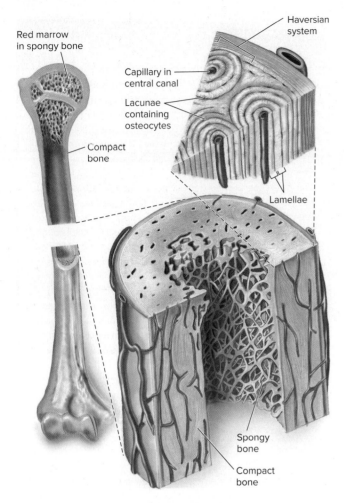

Figure 23.5 The structure of bone.

Some parts of bones are dense and compact, giving the bone strength. Other parts are spongy, with a more open lattice; red blood cells form in the bone marrow. Osteocytes, which are mature osteoblasts, lie in tight spaces called lacunae.

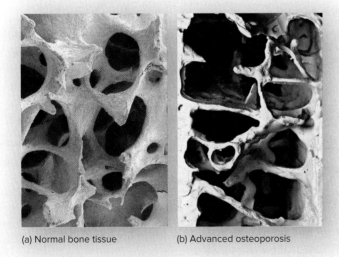

(a) Normal bone tissue (b) Advanced osteoporosis

Figure 23.6 Osteoporosis.

Common in older women, osteoporosis is a bone disorder in which bones progressively lose minerals.

(a) Science Photo Library/Alamy Stock Photo; (b) Pasieka/Science Source

A Closer Look at Bone

> **LEARNING OBJECTIVE 23.3.2** Describe the dynamic structure of bone, contrasting osteoblasts and osteoclasts.

Bone consists of living bone cells embedded within an inert matrix composed of the structural protein collagen coated with a calcium phosphate salt called *hydroxyapatite*. Why coat the collagen fibers with calcium salts? The construction of bone is similar to that of fiberglass, a composite composed of glass fibers embedded in epoxy glue: Small, needle-shaped crystals of hydroxyapatite surround and impregnate collagen fibers within bone. No crack can penetrate far into bone because any stress that breaks a hard hydroxyapatite crystal passes into the collagenous matrix, which dissipates the stress before it encounters another crystal.

A Living Tissue. Most of us think of bones as solid and rocklike. But actually, bone is a dynamic tissue that is constantly being reconstructed. The cross section through a bone in **figure 23.5** shows that the outer layer of bone is very dense and compact, and so is called **compact bone.** The interior is less compact, with a more open lattice structure, and is called **spongy bone.** Red blood cells form in the red marrow of spongy bone. New bone is formed in two stages: First, collagen is secreted by cells called **osteoblasts,** which lay down a matrix of fibers along lines of stress. Then calcium minerals impregnate the fibers. Bone is laid down in thin, concentric layers, like layers of paint on an old pipe. The layers form as a series of tubes around a narrow channel called a *central canal,* also called a *Haversian canal,* which runs parallel to the length of the bone (**figure 23.5**). The many central canals within a bone are all interconnected and contain blood vessels and nerves that provide a lifeline to its living, bone-forming cells.

Bone Remodeling. When bone is first formed in the embryo, osteoblasts use a cartilage skeleton as a template for bone formation. During childhood, bones grow actively. The total bone mass in a healthy young adult, by contrast, does not increase much from one year to the next. This does not mean change is not occurring. Large amounts of calcium and thousands of **osteocytes** (mature osteoblasts) are constantly being removed and replaced, but total bone mass does not change because deposit and removal take place at about the same rate.

Two cell types are responsible for this dynamic bone "remodeling": *Osteoblasts* deposit bone, and **osteoclasts** secrete enzymes that digest the organic matrix of bone, liberating calcium for reabsorption by the bloodstream. The dynamic remodeling of bone adjusts bone strength to workload, new bone being formed along lines of stress. When a bone is subjected to compression, mineral deposition by osteoblasts exceeds withdrawals by osteoclasts. That is why long-distance runners slowly increase the distances they attempt, to allow their bones to strengthen along lines of stress, lest stress fractures cripple them.

Osteoporosis. As a person ages, the backbone and other bones tend to decline in mass. Excessive bone loss is a condition called **osteoporosis.** After the onset of osteoporosis, the replacement of calcium and other minerals lags behind withdrawal, causing the bone tissue to gradually erode. Compare the normal bone in **figure 23.6a** with bone of a person with osteoporosis in **figure 23.6b**. Eventually, the bones become brittle and easily broken. Women are four times more likely to develop osteoporosis than men are.

> **Putting the Concept to Work**
> How do the bones of a runner's legs adjust strength to workload?

23.4 Muscle Tissue Lets the Body Move

> **LEARNING OBJECTIVE 23.4.1** Compare smooth, skeletal, and cardiac muscle.

Muscle cells are the motors of the vertebrate body. The distinguishing characteristic of muscle cells, the thing that makes them unique, is the abundance of contractible protein fibers within them. These fibers, called **myofilaments,** are made of the proteins actin and myosin. Crammed in like the fibers of a rope, they take up practically the entire volume of the muscle cell. When actin and myosin slide past each other, the muscle contracts. Like slamming a spring-loaded door, the shortening of all of these fibers together within a muscle cell can produce considerable force. The process of muscle contraction will be discussed later in this chapter.

Smooth Muscle

Smooth muscle cells are long and spindle-shaped, each containing a single nucleus. Smooth muscle tissue is organized into sheets of cells. In some tissues, smooth muscle cells contract only when they are stimulated by a nerve or hormone. Examples are the muscles that line the walls of many blood vessels and those that make up the iris of the vertebrate eye. In other smooth muscle tissue, such as that found in the wall of the gut, the individual cells contract spontaneously, leading to a slow, steady contraction of the tissue.

Skeletal Muscle

Skeletal muscles are attached to and move the bones of the skeleton when they contract during body movement (figure 23.7). Skeletal muscle cells are produced during development by the fusion of several cells at their ends to form a very long fiber still containing all the original nuclei. Each **muscle fiber** consists of many elongated **myofibrils** composed of myofilaments of the proteins actin and myosin (figure 23.8). The arrangement of actin and myosin gives the muscle fibers a banded appearance, called *striations*.

Cardiac Muscle

The vertebrate heart is composed of striated **cardiac muscle** in which the fibers are arranged very differently from the fibers of skeletal muscle. Instead of very long, multinucleate cells running the length of the muscle, heart muscle is composed of chains of single cells, organized into fibers that branch and interconnect, forming a latticework. This lattice structure is critical to the way heart muscle functions, as electrical impulses pass from cell to cell through small openings called gap junctions, causing the heart to contract in an orderly pulsation.

> **Putting the Concept to Work**
> What is the difference between a myofilament and a myofibril?

Figure 23.7 Locomotion.

Animals are unrivaled among the inhabitants of the living world in their ability to move about from one place to another, swimming, burrowing, crawling, slithering, sliding, walking, jumping, running, gliding, soaring, and flying. This sidewinder rattlesnake can move surprisingly rapidly over the desert sand by a coordinated series of muscle contractions, literally throwing its long body into a series of sinuous curves.

Yvette Cardozo/Alamy Stock Photo

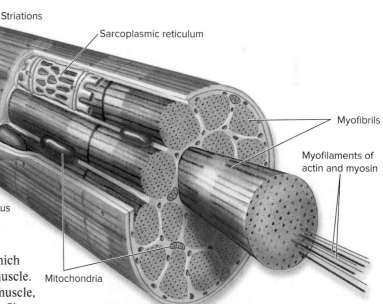

Figure 23.8 A skeletal muscle fiber, or muscle cell.

Each muscle is composed of bundles of muscle cells, or fibers. Each fiber is composed of many myofibrils, which are each, in turn, composed of myofilaments. Muscle cells have a modified endoplasmic reticulum called the sarcoplasmic reticulum that is involved in the regulation of calcium ions in muscles.

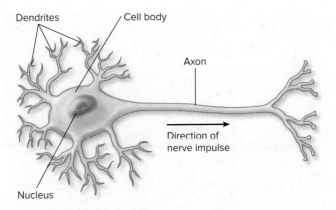

Figure 23.9 Neurons carry nerve impulses.

Neurons carry nerve impulses, which are electrical signals, from their initiation in dendrites to the cell body and down the length of the axon, where they may pass the signal to a neighboring cell.

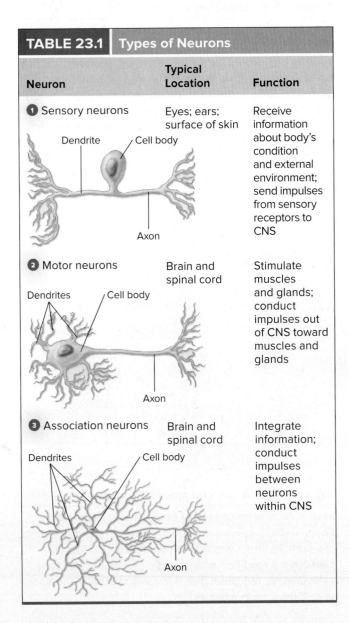

23.5 Nerve Tissue Conducts Signals Rapidly

> **LEARNING OBJECTIVE 23.5.1** Describe the three-part body of a typical neuron, and outline the functions of the three general categories of neuron.

Nerve tissue, the fourth major class of vertebrate tissue, is composed of two kinds of cells: (1) **neurons,** which are specialized for the rapid transmission of nerve impulses from one organ to another, and (2) supporting **glial cells,** which supply the neurons with nutrients, support, and insulation.

Neurons have a highly specialized cell architecture that enables them to conduct signals rapidly throughout the body. Their plasma membranes are rich in ion-selective channels that maintain a voltage difference between the interior and the exterior of the cell, the equivalent of a battery. When ion channels in a local area of the membrane open, ions flood in from the exterior, temporarily wiping out the charge difference. This process, called depolarization, tends to open nearby voltage-sensitive channels in the neuron membrane, resulting in a wave of electrical activity that travels down the entire length of the neuron as a nerve impulse.

Neurons

Structure of a Neuron. Each neuron is composed of three parts, as illustrated in **figure 23.9**: (1) a **cell body,** which contains the nucleus; (2) threadlike extensions called **dendrites** extending from the cell body, which act as antennae, bringing nerve impulses to the cell body from other cells or sensory systems; and (3) a single, long extension called an **axon,** which carries nerve impulses away from the cell body. Axons often carry nerve impulses for considerable distances: The axons that extend from the skull to the pelvis in a giraffe are about 3 meters long!

Kinds of Neurons. The body contains neurons of various sizes and shapes. Some are tiny and have only a few projections, others are bushy and have more projections, and still others have extensions that are meters long. However, all fit into one of three general categories of neurons as shown in **table 23.1.** *Sensory neurons* ❶ generally carry electrical impulses from the body to the central nervous system (CNS), the brain and spinal cord. *Motor neurons* ❷ generally carry electrical impulses from the central nervous system to the muscles. *Association neurons* ❸ occur within the central nervous system and act as a "connector" between sensory and motor neurons. These will be discussed in more detail in chapter 28.

The Synapse. Neurons are not normally in direct contact with one another. Instead, a tiny gap called a **synapse** separates them. Neurons communicate with other neurons by passing chemical signals called **neurotransmitters** across the gap.

Nerves

Vertebrate nerves appear as fine white threads when viewed with the naked eye, but they are actually composed of bundles of axons. Like a telephone trunk cable, nerves include large numbers of independent communication channels—bundles composed of hundreds of axons, each connecting a nerve cell to a muscle fiber or other type of cell. It is important not to confuse a nerve with a neuron. A nerve is made up of the axons of many neurons, just as a cable is made of many wires.

> **Putting the Concept to Work**
> If you were to slice through a slender nerve thread within muscle tissue, what part or parts of its neurons would you damage?

The Skeletal and Muscular Systems

23.6 Types of Skeletons

> **LEARNING OBJECTIVE 23.6.1** Describe and contrast the three principal types of animal skeletal systems.

With muscles alone, the animal body could not move. It would simply pulsate as its muscles contracted and relaxed in futile cycles. For a muscle to produce movement, it must direct its force against another object. Animals are able to move because the opposite ends of their muscles are attached to a rigid scaffold, or *skeleton*, so that the muscles have something to pull against. There are three types of skeletal systems in the animal kingdom: hydraulic skeletons, exoskeletons, and endoskeletons.

Hydraulic skeletons are found in soft-bodied invertebrates such as earthworms and jellyfish. In this case, a fluid-filled cavity is encircled by muscle fibers that raise the pressure of the fluid when they contract. The earthworm in figure 23.10 moves forward by a wave of contractions of circular muscles that begins anteriorly and compresses the body, so that the fluid pressure pushes it forward. Contractions of longitudinal muscles then pull the rest of the body along.

Exoskeletons surround the body as a rigid hard case to which muscles attach internally. When a muscle contracts, it moves the section of exoskeleton to which it is attached. Arthropods, such as crustaceans (figure 23.11) and insects, have exoskeletons made of the polysaccharide *chitin*. An animal with an exoskeleton cannot grow too large because its exoskeleton would have to become thicker and heavier to prevent collapse. If an insect were the size of an elephant, its exoskeleton would have to be so thick and heavy, it would hardly be able to move.

Endoskeletons, found in vertebrates and echinoderms, are rigid internal skeletons to which muscles are attached. Vertebrates have a soft, flexible exterior that stretches to accommodate the movements of their skeleton. The endoskeleton of some vertebrates is made of cartilage, but in many others, it is composed of bone (figure 23.12). Unlike chitin, bone is a cellular, living tissue capable of growth, self-repair, and remodeling in response to physical stresses.

> **Putting the Concept to Work**
> In what way would your athletic training be different if your skeleton were made of chitin rather than bone?

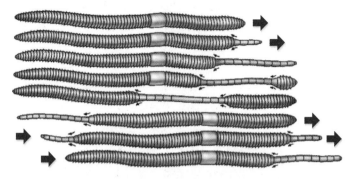

Figure 23.10 Earthworms have a hydraulic skeleton.
When an earthworm's circular muscles contract, the internal fluid presses on the longitudinal muscles, which then stretch to elongate segments of the earthworm. A wave of contractions down the body of the earthworm produces forward movement.

Figure 23.11 Crustaceans have an exoskeleton.
The exoskeleton of this rock crab is bright orange.
Maskot/Image Source

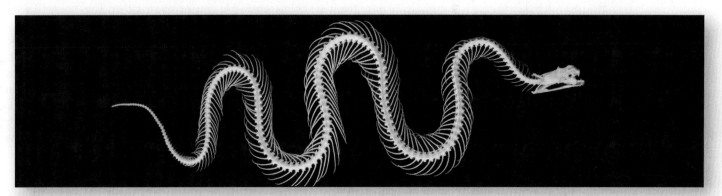

Figure 23.12 Snakes have an endoskeleton.
The endoskeleton of most vertebrates is made of bone. A snake's skeleton is specialized for quick lateral movement.
Garry Gay/Getty Images

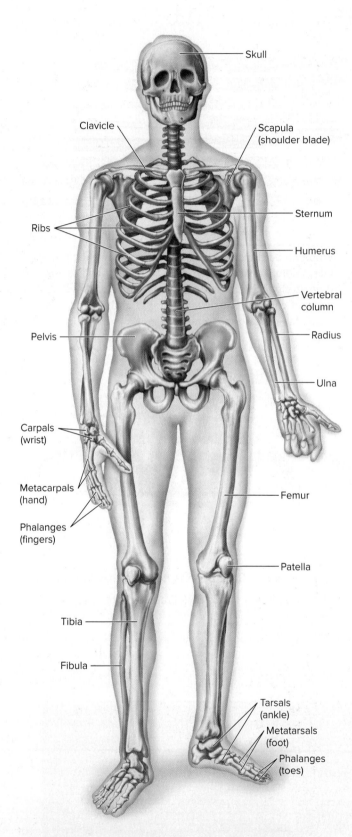

Figure 23.13 Axial and appendicular skeletons.
The axial skeleton is shown in *purple,* and the appendicular skeleton is shown in *tan.* Some of the joints are shown in *green.*

A Vertebrate Endoskeleton: The Human Skeleton

> **LEARNING OBJECTIVE 23.6.2** Contrast the axial and appendicular human skeletons and their main classes of joints.

The human skeleton is made up of 206 individual bones. If you saw them as a pile of bones jumbled together, it would be hard to make any sense of them. To understand the skeleton, it is necessary to group the 206 bones according to their function and position in the body. The 80 bones of the **axial skeleton,** the purple-colored bones in figure 23.13, support the main body axis, while the remaining 126 bones of the **appendicular skeleton,** the tan-colored bones, support the arms and legs. These two skeletons function more or less independently—that is, the muscles controlling the axial skeleton (postural muscles) are managed by the brain separately from those controlling the appendages (manipulatory muscles).

The Axial Skeleton

The axial skeleton is made up of the skull, vertebral column (backbone), and rib cage, which includes the sternum. Of the skull's 28 bones, only 8 form the cranium, which encases the brain; the rest are facial bones and middle ear bones.

The skull is attached to the upper end of the vertebral column, which is also called the spine. The spine is made up of 26 individual bones called vertebrae, stacked one on top of the other to provide a flexible column surrounding and protecting the spinal cord. Curving forward from the vertebrae are 12 pairs of ribs, attached at the front to the sternum, or breastbone, and forming a protective cage around the heart and lungs.

The Appendicular Skeleton

The 126 bones of the appendicular skeleton are attached to the axial skeleton at the shoulders and hips. The shoulder, or *pectoral girdle,* is composed of two large, flat shoulder blades, each connected to the top of the sternum by a slender, curved collarbone (clavicle). The arms are attached to the pectoral girdle; each arm and hand contain 30 bones. The clavicle is the most frequently broken bone of the body. Can you guess why? Because if you fall on an outstretched arm, a large component of the force is transmitted to the clavicle.

The *pelvic girdle* forms a bowl that provides strong connections for the legs, which must bear the weight of the body. Each leg and foot contain a total of 30 bones.

Joints are points where two bones come together. They confer flexibility to the rigid endoskeleton, allowing a range of motion determined by the type of joint. There are three main classes of joints, based on mobility. (1) *Immovable joints,* such as the sutures of the skull, are capable of little to no movement. (2) *Slightly movable joints,* such as the joints between vertebrae in the spine, allow the bones some movement. (3) And *freely movable joints* allow a range of motion and are seen in the limbs (for example, the shoulder, elbow, hip, knee), the jaw, and fingers and toes. Depending on the type of joint, bones are held together at the joint by cartilage, fibrous connective tissue, or a fibrous capsule filled with a lubricating fluid.

> **Putting the Concept to Work**
> If skull sutures are incapable of movement, why have them?

23.7 Muscles and How They Work

> **LEARNING OBJECTIVE 23.7.1** Describe the structure and functioning of a skeletal muscle and its filaments.

Three kinds of muscle together form the vertebrate muscular system. As we have discussed, the vertebrate body is able to move because *skeletal muscles* pull the bones with considerable force. The heart pumps because of the contraction of *cardiac muscle*. Food moves through the intestines because of the rhythmic contractions of *smooth muscle*.

Actions of Skeletal Muscle

Skeletal muscles move the bones of the skeleton. Some of the major human muscles are labeled on the right in figure 23.14. Muscles are attached to bones by straps of dense connective tissue called **tendons.** Bones pivot about flexible connections called joints, pulled back and forth by the muscles attached to them. Each muscle pulls on a specific bone. One end of the muscle, the *origin,* is attached by a tendon to a bone that remains stationary during a contraction. This provides an object against which the muscle can pull. The other end of the muscle, the *insertion,* is attached to a bone that moves if the muscle contracts. For example, the origin and insertion for the sartorius muscle are labeled on the left in figure 23.14. This muscle helps bend the leg at the hip, bringing the knee to the chest. The origin of the muscle is at the hip and stays stationary. The insertion is just below the knee, such that when the muscle contracts (gets shorter) the knee is pulled up toward the chest.

Muscles can only pull, not push, because myofibrils contract rather than expand. For this reason, the muscles that extend across movable joints of vertebrates are attached in opposing pairs, called flexors and extensors, which, when contracted, move the bones in different directions. As you can see in figure 23.15, when the *flexor* muscle at the back of your upper leg contracts, the lower leg is moved closer to the thigh. When the *extensor* muscle at the front of your upper leg contracts, the lower leg is moved in the opposite direction, away from the thigh.

Muscle Filaments

Recall from figure 23.8 that myofibrils are composed of bundles of myofilaments. Far too fine to see with the naked eye, the individual myofilaments of vertebrate muscles are only 8 to 12 nanometers thick. Each is a long, threadlike filament of the proteins actin or myosin. An **actin filament** consists of two strings of actin molecules wrapped around one another, like two strands of pearls loosely wound together. A **myosin filament** is about twice as long as an actin filament and is composed of bundles of myosin molecules. A myosin molecule consists of two polypeptide chains wound about each other. Each polypeptide chain has a shape somewhat like that of a golf club: a very long rod with a globular region, or "head." This odd structure is the key to how muscles work.

> **Putting the Concept to Work**
> Why can a muscle pull but not push? If your arm muscles cannot push, how do your arms manage to do so?

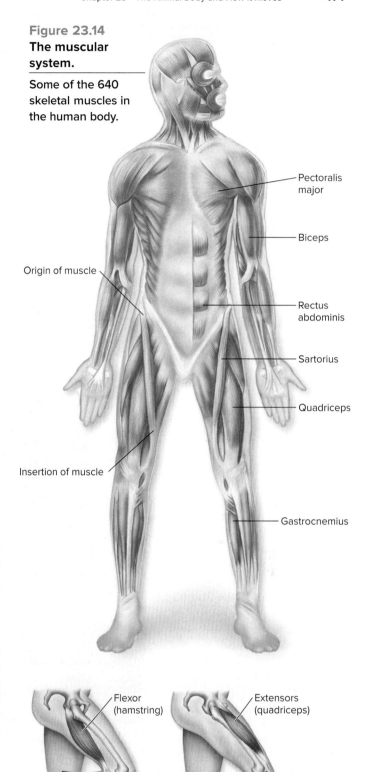

Figure 23.14
The muscular system.
Some of the 640 skeletal muscles in the human body.

Figure 23.15 Flexor and extensor muscles.
Limb movement is always the result of muscle contraction, never muscle extension. Muscles that retract limbs are called flexors; those that extend limbs are called extensors.

A Closer Look

A Day in the Life of Your Body

Your heart will beat more than 80,000 times today, pumping out about 2,000 gallons (7,570 liters) of blood. In your lifetime, your heart will beat 2 billion times without once stopping for a rest.

Red blood cells (RBCs) race around your body, taking less than 60 seconds to complete a full circuit. Each RBC makes 1,440 trips around every day! Worn down by all this tearing about, an individual RBC cell lives only about 40 days, having made 60,000 trips around your body before being replaced by a younger one.

Your lungs take in about 17,000 breaths a day, on average. That's a lot of air. In your lifetime, you will breathe 180 million liters of air. That's about the volume of the Hindenberg (see figure 2.8).

Your digestive system extracts energy from the food you eat. It takes about six to eight hours for food to pass through your stomach and small intestine, and two more days to complete the digestion process in your large intestine. You will eat over 50 tons of food in your lifetime, all of which will eventually leave your body as CO_2, urine, or feces.

To power your body's many activities, you use massive amounts of ATP. At any one time, your body contains around 250 grams of this molecule, but when you exercise vigorously, the ATP is so rapidly recycled that the entire stock is put to use twice in a minute! Just sitting around doing nothing, you turn your entire stock of ATP over around 160 times each day.

Tom Grill/Getty Images

Each of your two kidneys contains 1 million tiny filters that clean an average of 2.2 pints (1.3 liters) of blood every minute—that's 500 gallons of blood a day passing through your kidneys, expelling about 2.5 pints of urine. In your lifetime, your kidneys will filter 40 million liters of blood, and you will excrete 8,000 gallons of urine.

Your hair grows slowly but steadily all your life, each hair extending about half a millimeter each day. A full scalp has around 100,000 hairs, so that's 50 meters of hair growing on your head every day. In your lifetime, you will grow 670 miles of hair—and 72 feet of fingernails!

Your body cools itself by secreting water (sweat) onto its surface that cools as it evaporates. There are 3 million sweat glands on your body, enough to pump out over a liter of sweat in an hour when you exercise strenuously. Just hanging around, you sweat over 10 liters each day. In your lifetime, you will sweat 57,000 gallons.

Your eyes blink to keep their surface clean and moist. As this is a tenth-of-a-second involuntary reflex, you may not be aware of how often you do it—about 28,000 times every day. In your lifetime, you will blink 622 million times.

All your life, your tissues regenerate themselves. You have a new set of taste buds every 10 days, new nails every 6 to 10 months, new bones every 10 years—even a new heart every 20 years. In your lifetime, you will die only once. When you do, all your tissues will stop growing, even your nails and hair (they only appear to grow longer on your corpse as its skin shrinks and they do not).

How Myofilaments Contract

> **LEARNING OBJECTIVE 23.7.2** Describe the sliding filament model of muscle contraction.

The Sliding Filament Model. The **sliding filament model** of muscle contraction, illustrated in **figure 23.16**, describes how actin and myosin cause muscles to contract. Focus on the knob-shaped myosin head in **panel 1**. When a muscle contraction begins, the heads of the myosin filaments, which are attached to actin, move first. Like flexing your hand downward at the wrist, the heads bend backward and inward, as shown in **panel 2**. This moves the myosin heads closer to their rodlike backbones in the direction of the flex. In itself, this myosin head-flex accomplishes nothing, but because the myosin head is attached to the actin filament, the actin filament is pulled along with the myosin head as it flexes. This causes the actin filament to slide by the myosin filament in the direction of the flex (indicated by the arrows in **panel 2**). As one myosin head after another flexes, the myosin in effect "walks" step by step along the actin.

The Role of ATP. Each step uses a molecule of ATP to recock the myosin head (shown in **panel 3**) before it attaches to the actin again, ready for the next flex. It is important to note that ATP hydrolysis does not occur at the same time as the power stroke in the cycle. The hydrolysis of ATP to cock the myosin head is like your thumb cocking the spring-loaded hammer of an old-fashioned revolver, contributing the energy for when you pull the trigger later.

How a Sarcomere Contracts. How does this sliding of actin past myosin lead to myofibril contraction and muscle cell movement? The actin filament is anchored at one end, at a position in striated muscle called the Z line, indicated by the lavender-colored bars toward the edges in **figure 23.17**. Two Z lines with the actin and myosin filaments in between make up a contractile unit called a **sarcomere**. Because it is tethered like this, the actin cannot simply move off. Instead, the actin pulls the anchor with it! As actin moves past myosin, it drags the Z line toward the myosin. The secret of muscle contraction is that each myosin is interposed between two pairs of actin filaments, which are anchored at both ends to Z lines, as shown in **panel 1** of **figure 23.17**. One moving to the left and the other to the right, the two pairs of actin molecules drag the Z lines toward each other as they slide past the myosin core, shown in **panel 2**. As the Z lines are pulled closer together, the plasma membranes to which they are attached move toward one another, and the cell contracts.

> **Putting the Concept to Work**
> To which molecule of a muscle fiber does ATP contribute its energy and to do precisely what?

Figure 23.17 The sliding filament model of muscle contraction.

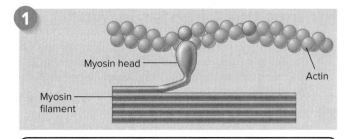
The myosin head is attached to actin.

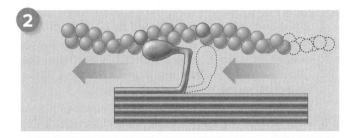
The myosin head flexes, advancing the actin filament.

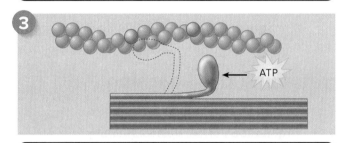
The myosin head releases and unflexes, powered by ATP. The myosin head is then able to reattach to actin, farther along the fiber.

Figure 23.16 How myofilament contraction works.

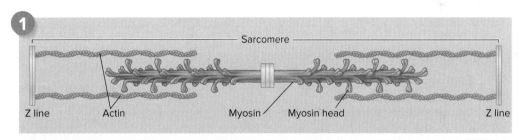
The heads on the two ends of the myosin filament are oriented in opposite directions.

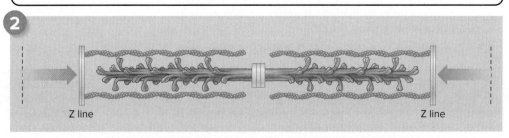

Thus, as the right-hand end of the myosin filament "walks" along the actin filaments, pulling them and their attached Z line leftward toward the center, the left-hand end of the same myosin filament "walks" along the actin filaments, pulling them and their attached Z line rightward toward the center. The result is that both Z lines move toward the center—and contraction occurs.

Putting the Chapter to Work

1. During a dissection demonstration, your instructor cuts open a pig stomach and points to the lining tissue. They describes its function as a passage for nutrients and fluid through the stomach wall.

What type of epithelium tissue are they pointing to?

2. Athletes may develop nerve damage due to continuous injuries. One of the most common injuries is to the brachial plexus, which is a nerve bundle around the shoulder. If an injury is severe, a nerve impulse may travel through the injured neuron but not away from its cell body. This produces a feeling of pain in the shoulder.

What part of the neuron would be damaged?

3. Lobsters grow quickly in their first five to seven years. They molt, shedding their exterior, approximately 25 times during that growth stage. This shedding allows them to grow a new exterior of chitin that fits their increased body size.

What type of skeleton does a lobster have?

4. Using a human skeleton model, your instructor points to the joint that connects the femur to the tibia and fibula. They then describe the straps that assist in attaching muscle to the bones for movement of the joint.

What structure are they referring to as "straps"?

Retracing the Learning Path

The Animal Body Plan

23.1 Organization of the Vertebrate Body

1. All vertebrates and some invertebrates have the same general body architecture: a tube (the gut or digestive system) suspended in a cavity (the coelom) that in many terrestrial vertebrates is divided into a thoracic cavity and an abdominal cavity.

2. Cells that group together into tissues act as functional units. Organs of the body are composed of several different kinds of tissues that act together to perform a higher level of function. Organs work together in an organ system to perform larger-scale body functions.

Tissues of the Vertebrate Body

23.2 Epithelium Is Protective Tissue

1. Epithelial tissue is composed of different types of epithelial cells. It covers internal and external surfaces of the body and provides protection.

- The structure of the epithelium determines its function. Some types of epithelium are a single layer of cells through which substances can pass. Stratified epithelium provides protection. Cuboidal epithelium lines glands in the body and has a secretory function, producing and releasing hormones into the blood.

23.3 Connective Tissue Carries Out Various Functions

1. The connective tissues of the body are very diverse in structure and function, but all are composed of cells embedded in an extracellular matrix. The matrix may be hard as in bone, flexible as in fibrous connective tissue, adipose tissue, and cartilage, or fluid as in blood.

- Immune connective tissue contains white blood cells that float in the blood plasma and protect the body from infection.

- Skeletal connective tissues—such as fibrous connective tissue, cartilage (which makes up the skeleton of sharks), and bone—provide structural support of the body.

- Adipose tissue stores fat deposits, and red blood cells, erythrocytes, transport substances throughout the body.

2. Bone is a living tissue. Bone cells called osteoblasts lay down a fibrous matrix. Calcium minerals then impregnate the fibers, causing the matrix to harden into compact bone.

23.4 Muscle Tissue Lets the Body Move

1. There are three types of muscle tissue: smooth, skeletal, and cardiac muscle. All three types of muscle contain actin and myosin myofilaments but differ in the organization of the myofilaments.

- Smooth muscle cells are long, spindle-shaped cells organized into sheets. Smooth muscle is found in the walls of blood vessels and in the walls of the digestive system.

- Skeletal muscle cells are fused into long fibers. Skeletal muscle is attached to the skeleton, so when the muscle contracts, the skeleton moves.

- Cells of cardiac muscle found in the heart are interconnected so that they contract together in an orderly pulsation.

23.5 Nerve Tissue Conducts Signals Rapidly

1. Nerve tissue is composed of neurons and supporting glial cells. Neurons have three parts: branching dendrites, the cell body that contains the nucleus, and a long axon. Neurons carry electrical impulses from one area of the body to another. The electrical impulse travels down the length of the neuron's axon by the movement of ions across the plasma membrane.

The Skeletal and Muscular Systems

23.6 Types of Skeletons

1. The skeletal system provides a framework on which muscles act to move the body. Soft-bodied invertebrates have hydraulic skeletons, where muscles act on a fluid-filled cavity.

- Arthropods have exoskeletons, where muscles attach from within to the hard outer covering of the body.

2. Vertebrates and echinoderms have endoskeletons, where muscles attach to bones or cartilage inside the body. The human skeleton has 206 individual bones that make up the axial and appendicular skeletons.

23.7 Muscles and How They Work

1. Skeletal muscles attach to the skeleton at two points. The end that attaches to the stationary bone is called the origin. The muscle passes over a joint and attaches to another bone at a point called the insertion. As the muscle contracts, the insertion is brought closer to the origin and the joint flexes. Muscles act in opposing pairs to flex or extend a joint.

2. Myofibrils are composed of bundles of actin and myosin. During muscle contraction, myosin attaches to actin.

- A flexing myosin filament pulls the actin along its length, causing the myofilaments to slide past each other. Actin myofilaments are anchored at each end to a structure called the Z line. As the actin slides along the myosin, the anchor points are brought closer together, resulting in a shortening of the muscle. Energy from ATP causes the myofilaments to dissociate and reset, triggering the sliding motion again.

Inquiry and Analysis

Which Mode of Locomotion Is the Most Efficient?

Running, flying, and swimming require more energy than sitting still, but how do they compare? The greatest differences between moving on land, in the air, and in water result from the differences in support and resistance to movement provided by water and air. The weight of a swimming animal is fully supported by the surrounding water, and no effort goes into supporting the body, whereas running and flying animals must support the full weight of their bodies. On the other hand, water presents considerable resistance to movement, air much less, so that flying and running require less energy to push the medium out of the way.

A simple way to compare the costs of moving for different animals is to determine how much energy it takes to move. The energy cost to run, fly, or swim is in each case the energy required to move one unit of body mass over one unit of distance with that mode of locomotion. (Energy is measured in the **metric system** as a **kilocalorie** [kcal] or technically 4.184 kilojoules [note that the Calorie measured in food diets and written with a capital C is equivalent to 1 kcal]; body mass is measured in kilograms, where 1 kilogram [kg] is 2.2 pounds; distance is measured in kilometers, where 1 kilometer [km] is 0.62 miles). The graph to the right displays three such "cost-of-motion" studies. The blue squares are running, the red circles are flying, and the green triangles are swimming. In each study, the line is drawn as the statistical "best fit" for the points. Some animals such as humans have data in two lines, as they both run (well) and swim (poorly). Ducks have data in all three lines, as they not only fly (very well), but also run and swim (poorly).

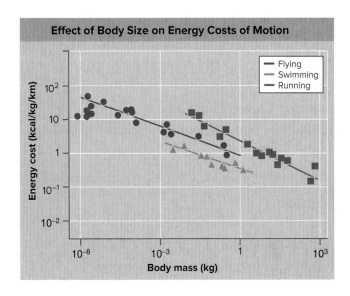

Analysis

1. **Applying Concepts** Do the three modes of locomotion have the same or different costs?
2. **Interpreting Data**
   a. For any given mode of locomotion, what is the impact of body mass on cost of moving?
   b. Is the impact of body mass the same for all three modes of locomotion? If not, which mode's cost is least affected by body mass? Why do you think this is so?
3. **Making Inferences**
   a. Comparing the energy costs of running versus flying for animals of the same body mass, which mode of locomotion is the most expensive? Why would you expect this to be so?
   b. Comparing the energy costs of swimming to flying, which uses the least energy? Why would you expect this to be so?
4. **Drawing Conclusions** In general, which mode of locomotion is the most efficient? The least efficient? Why do you think this is so?

24 Circulation and Respiration

LEARNING PATH ▼

Circulation
1. Open and Closed Circulatory Systems
2. Architecture of the Vertebrate Circulatory System
3. Blood
4. Human Circulatory System

Respiration
5. Types of Respiratory Systems
6. The Human Respiratory System
7. How Respiration Works: Gas Exchange

A SCENE FROM the iconic 1922 film *Nosferatu*.
World History Archive/Alamy Stock Photo

Vampire

There are few images more shivery to students today than that of a vampire, an undead creature of myth that feeds on the blood of living victims. From *Nosferatu* and the immortal Transylvanian Count Dracula to today's *True Blood, Twilight,* and *Buffy the Vampire Slayer,* vampires seem intriguing to many.

There Really Are Vampires

However, there are no such things as undead vampires. But what about vampire bats? Bats are flying mammals—the only mammals that fly, actually—and some of them really are vampires. Not many, though. Whereas there are lots of kinds of bats—over 1,000 species, one quarter of all the kinds of living mammals—there is little danger you will encounter a vampire bat that would suck your blood. Only three species of bat, found in Central and South America, are vampire bats, feeding on blood.

Nor need you fear them. Vampire bats, made famous by horror movies, are not the blood-sucking predators that they are made out to be in the movies. They feed on large birds, cattle, and other large animals, and in only a few instances have human bites been verified. The common vampire bat shown in the photo on the next page, *Desmodus rotundus*, has specialized thermoreceptors in its nose for locating areas of skin on its prey where the blood flows close to the surface.

Hunting for Blood

Vampire bats hunt only when it is fully dark. The common vampire bat feeds mostly on the blood of sleeping mammals, while the other two species feed on the blood of birds. Vampire bats don't bite and drain the blood from their victims, but rather, with their sharp teeth, they make small cuts in their prey as it sleeps and lap up blood as it oozes out of the cut—usually less than two tablespoons worth. Their saliva contains a chemical that acts as a local anesthetic, numbing the area of the bite to keep the animal from waking up as they feed, and a second chemical that keeps the blood from clotting.

Family Folks

Vampire bats have a strong social sense. The neocortex of their brain, the part linked to social intelligence and

Samuel Betkowski/Moment/Getty Images

behavior, is about twice the average size of all other bats. Vampire bats are believed to be the only species of bat in the world to "adopt" another young bat if something happens to orphan it. And they share. A vampire bat can only survive about two days without a meal of blood, and yet on many nights, a bat cannot find food. When a bat fails to find food, it will "beg" another bat in its colony, which may regurgitate a small amount of blood to sustain it.

Warm and fuzzy is not what we were told that bats are like when we were growing up. As a kid, you were probably warned to cover your head at night if bats were flying around. Bats, it was thought, were essentially blind and would swoop down and get tangled up in your hair. They have been characterized as dirty, blood-sucking, rabies-carrying flying rodents—no wonder we grew up afraid of them!

You needn't be. Bats are not, in fact, rodents, nor are they filthy—they groom themselves constantly when not sleeping or feeding. It is true that bats, like dogs and many other mammals, can carry rabies, but a bat is no more likely to have rabies than the raccoon that might invade your garbage cans at night. Bats are not blind, either, despite the saying "blind as a bat." Bats actually have better eyesight than most people—on par with military night-vision scopes.

W. Perry Conway/Aerie Nature Series, Inc./Corbis/Getty Images

Insect Eaters

Happily, most bats eat insects, not blood. Bats are the only consistent predators of nocturnal flying insects, playing a major role in controlling insect pests. For example, every night, bats emerge by the millions from Bracken Cave in Texas, as shown in the photo here. The estimated 20 million bats that inhabit this one cave cover thousands of square kilometers nightly and will eat more than 200 tons of insects before they return to the cave at dawn.

Circulation

24.1 Open and Closed Circulatory Systems

> **LEARNING OBJECTIVE 24.1.1** Contrast open and closed circulatory systems.

Among the unicellular protists, oxygen and nutrients are obtained directly by simple diffusion from the aqueous external environment. Animals such as flatworms have cells that are directly exposed either to the external environment or to a body cavity called the gastrovascular cavity (figure 24.1*a*), which branches extensively to supply every cell of the flatworm with the oxygen and nourishment obtained by digestion. Larger animals, however, have tissues that are several cell layers thick, so many cells are too far away from the body surface or digestive cavity to exchange materials directly with the environment. Instead, oxygen and nutrients are transported from the environment and digestive cavity to the body cells by an internal fluid within a **circulatory system.**

Open Circulatory Systems

There are two main types of circulatory systems: *open* and *closed.* In an **open circulatory system,** such as that found in arthropods, there is no distinction between the circulating fluid (called **hemolymph**) and the extracellular fluid of the body tissues. Insects, such as the fly in figure 24.1*b*, have a muscular tube that serves as a heart to pump the hemolymph through a network of open-ended channels that empty into the cavities in the body (downward-pointing arrows). There, the hemolymph delivers nutrients to the cells of the body. It then reenters the circulatory system through pores in the heart (upward-pointing arrows).

Closed Circulatory Systems

In a **closed circulatory system,** the circulating fluid, or *blood,* is always enclosed within blood vessels that transport blood away from and back to the heart. Annelids and all vertebrates have a closed circulatory system. In annelids, such as the earthworm in figure 24.1*c*, a dorsal blood vessel and five lateral hearts contract rhythmically to function as pumps.

In vertebrates, blood vessels form a tubular network that permits blood to flow from the heart to all the cells of the body and then back to the heart. *Arteries* carry blood away from the heart, whereas *veins* return blood to the heart. Blood passes from the arterial to the venous system in *capillaries,* which are the thinnest and most numerous of the blood vessels.

> **Putting the Concept to Work**
> If you had an open rather than a closed circulatory system, would your body have to operate differently? Discuss.

(a) *Planaria:* gastrovascular cavity

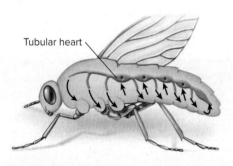

(b) Insect: open circulation

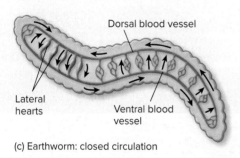

(c) Earthworm: closed circulation

Figure 24.1 Three types of circulatory systems found in the animal kingdom.

The Functions of Vertebrate Circulatory Systems

> **LEARNING OBJECTIVE 24.1.2** Discuss the three principal functions of the vertebrate circulatory system.

The functions of the circulatory system can be divided into three areas: transportation, regulation, and protection.

1. *Transportation.* Substances essential for cellular functions are transported by the circulatory system and can be categorized as follows:
   Respiratory. Red blood cells, or erythrocytes, transport oxygen to the tissue cells. In the capillaries of the lungs or gills, oxygen attaches to hemoglobin molecules within the erythrocytes and is transported to the cells for aerobic respiration. Carbon dioxide produced by cell respiration is carried by the blood to lungs or gills for elimination.
   Nutritive. The digestive system is responsible for the breakdown of food so that nutrients can be absorbed through the intestinal wall and into the blood vessels of the circulatory system. The blood then carries these absorbed products of digestion to the cells of the body.
   Excretory. Metabolic wastes, excessive water and ions, and other molecules in plasma (the fluid portion of blood) are filtered through the capillaries of the kidneys and excreted in urine.
   Endocrine. The blood carries hormones from the endocrine glands, where they are secreted, to the distant target organs they regulate.
2. *Regulation.* The cardiovascular system regulates body temperature.
   Temperature regulation. In warm-blooded vertebrates, or homeotherms, a constant body temperature is maintained, regardless of the surrounding temperature. This is accomplished in part by blood vessels located just under the epidermis. When the ambient temperature is cold, the superficial vessels constrict to divert the warm blood to deeper vessels, reducing heat loss. When the ambient temperature is warm, the superficial vessels dilate so that the warmth of the blood can be lost by radiation. Some vertebrates also retain heat in a cold environment by using a **countercurrent heat exchange.** Figure 24.2 shows how a countercurrent heat exchange system works in the flipper of a killer whale. In this process, the warm blood going out heats the cold blood returning from the body surface, helping to maintain a stable core body temperature.
3. *Protection.* The circulatory system protects against injury and foreign microbes or toxins introduced into the body.
   Blood clotting. The clotting mechanism protects against blood loss when vessels are damaged. This clotting mechanism involves both proteins from the blood plasma and blood cell structures called platelets.
   Immune defense. The blood contains proteins and white blood cells, or leukocytes, that provide immunity against many disease-causing agents. Some white blood cells are phagocytic, some produce antibodies, and some act by other mechanisms to protect the body.

> **Putting the Concept to Work**
> If the heart were to stop so that blood stops circulating, which of the functions discussed in this section would cease to operate?

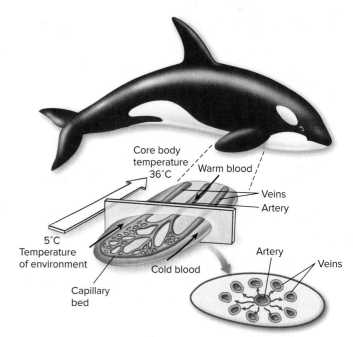

Figure 24.2 Countercurrent heat exchange.
Many marine mammals, such as this killer whale, limit heat loss in cold water by countercurrent flow that allows heat exchange between arteries and veins. The warm blood pumped from within the body in arteries warms the cold blood returning from the skin in veins, so that the core body temperature can remain constant in cold water. The cut-away portion in the figure shows how the veins surround the artery, maximizing the heat exchange between the artery and the veins.

24.2 Architecture of the Vertebrate Circulatory System

LEARNING OBJECTIVE 24.2.1 Describe the elements of the vertebrate circulatory system.

The vertebrate circulatory system, also called the **cardiovascular system,** is made up of three elements: (1) the *heart,* a muscular pump that pushes blood through the body; (2) the *blood vessels,* a network of tubes through which the blood moves; and (3) the *blood,* which circulates within these vessels.

The Blood Vessels

Blood moves through the body in a cycle, from the heart, through a system of vessels: from the arteries and arterioles, into the capillaries, and then back to the heart through the venules and veins, as shown in figure 24.3. The designation of a blood vessel as a vein or artery is determined by the direction of blood flow in relation to the heart (veins carry blood toward the heart, arteries carry blood away from the heart), not whether the vessel is carrying oxygenated or deoxygenated blood.

A Circular Path. Blood leaves the heart through vessels known as **arteries.** From the arteries, the blood passes into a network of smaller arteries called **arterioles,** shown in figure 24.4a ❶. From these, it is eventually forced through a capillary bed ❷, a fine latticework of very narrow tubes called **capillaries** (from the Latin, *capillus,* "a hair"). While passing through the capillaries, the blood exchanges gases and metabolites (glucose, vitamins, hormones) with the cells of the body. After traversing the capillaries, the blood passes into venules, or small veins, ❸. A network of venules empties into larger **veins** that collect the circulating blood and carry it back to the heart.

Vessel Size. The capillaries have a much smaller diameter than the other blood vessels of the body. Blood leaves the mammalian heart through a large artery, the aorta, a tube that in your body has a diameter of about 2 centimeters. But when blood reaches the capillaries, it passes through vessels with an average diameter of only 8 micrometers, a reduction in radius of some 1,250 times!

This decrease in size of blood vessels has a very important consequence. Although each capillary is very narrow, there are so many of them that the capillaries have the greatest *total* cross-sectional area of any other type of vessel. Consequently, this allows more time for blood to exchange materials with the surrounding extracellular fluid. By the time the blood reaches the end of a capillary, it has released some of its oxygen and nutrients and picked up carbon dioxide and other waste products. Blood loses most of its pressure and velocity in passing through the vast capillary networks and so is under very low pressure when it enters the veins.

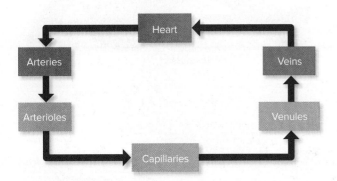

Figure 24.3 The flow of blood through the circulatory system.

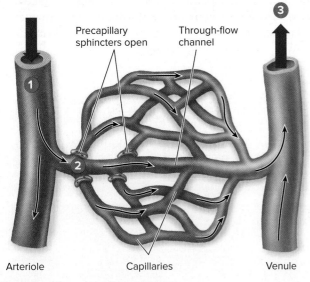

Figure 24.4 The capillary network connects arteries with veins.

Through-flow channels connect arterioles directly to venules. Branching from these through-flow channels are the capillaries. Most of the exchange between the body tissues and red blood cells occurs in this capillary network.

> **Putting the Concept to Work**
> How is a venule different from a vein? Which receives blood first?

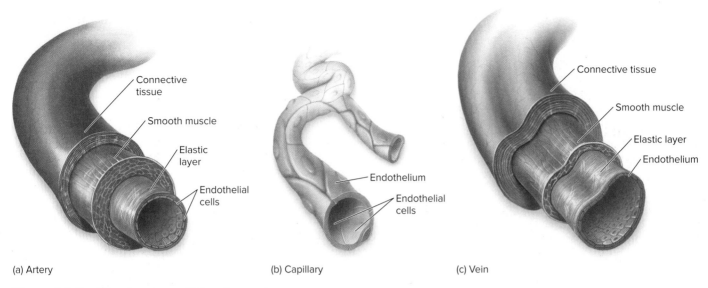

(a) Artery (b) Capillary (c) Vein

Figure 24.5 **The structure of blood vessels.**

(a) Arteries, which carry blood away from the heart, are expandable and are composed of layers of tissue. (b) Capillaries are simple tubes whose thin walls facilitate the exchange of materials between the blood and the cells of the body. (c) Veins, which transport blood back to the heart, do not need to be as sturdy as arteries. The walls of veins have thinner muscle layers than arteries, and they collapse when empty.

Arteries: Highways from the Heart

> **LEARNING OBJECTIVE 24.2.2** Explain why arteries have to expand and how they do it.

The arterial system, composed of arteries and arterioles, carries blood (figure 24.6) away from the heart. An artery is more than simply a pipe. Blood comes from the heart in pulses rather than in a smooth flow, forcing blood into the arteries in big slugs as the heart ejects its contents with each contraction. Arteries have to be able to *expand* to withstand the pressure caused by each contraction of the heart. An artery, then, is an expandable tube, with its walls made up of four layers of tissue: endothelial cells, an elastic layer, a thick layer of smooth muscle, and protective connective tissue (figure 24.5a). The artery is able to expand its volume considerably when the heart contracts, pumping a new volume of blood into the artery—just as a tubular balloon expands when you blow more air into it. The steady contraction of the smooth muscle layer stops the wall of the vessel from overexpanding.

When stimulated by nerves, the muscle lining of arterioles contracts, constricting their diameter. Such a contraction limits the flow of blood to the extremities during periods of low temperature or stress. You turn pale when you are scared or cold because the arterioles in your skin are constricting. You blush for just the opposite reason. When you overheat or are embarrassed, the nerve fibers connected to muscles surrounding the arterioles are inhibited, relaxing the smooth muscle so the arterioles in the skin expand, bringing heat to the surface for escape.

> **Putting the Concept to Work**
> How does an arteriole differ from an artery? Which receives blood first?

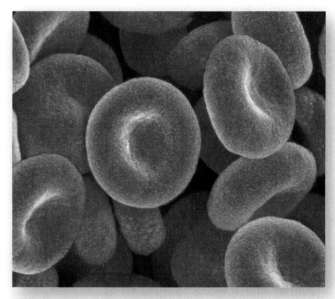

Figure 24.6 **Red blood cells.**

Red blood cells such as those seen above are the oxygen transporters of the vertebrate circulatory system. There are approximately 5 million of them in each microliter (μl) of blood! Each red blood cell is shaped like a rounded cushion, squashed in the center, and is crammed full of a protein called hemoglobin, an iron-containing molecule that gives blood its red color. Oxygen binds easily to the iron in hemoglobin, making red blood cells efficient oxygen carriers. A single red blood cell contains about 250 million hemoglobin molecules, and each red blood cell can carry about 1 billion molecules of oxygen at one time. The average life span of a red blood cell is only 120 days. About 2 million new ones are produced in the bone marrow every second to replace those that die or are worn out.

Susumu Nishinaga/Getty Images

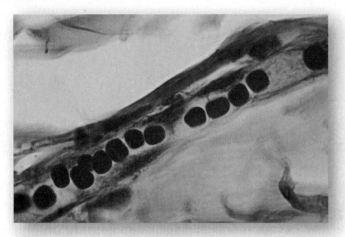

Figure 24.7 Red blood cells within a capillary.
The red blood cells in this capillary pass along in single file. Red blood cells can pass through capillaries even narrower than their own diameter, pushed along by the pressure of the pumping heart.
Ed Reschke/Getty Images

Capillaries: Where Exchange Takes Place

> **LEARNING OBJECTIVE 24.2.3** Explain why the narrow diameter of capillaries is critical to their function.

Capillaries are where oxygen and food molecules are transferred from the blood to the body's cells and where metabolic waste and carbon dioxide are picked up. To facilitate this back-and-forth traffic, capillaries are narrow and have thin walls across which gases and metabolites pass easily. Capillaries have the simplest structure of any element in the cardiovascular system. They are built like a soft-drink straw, simple tubes with walls only one cell thick (see figure 24.5b). The average capillary is about 1 millimeter long and connects an arteriole with a venule. All capillaries are very narrow, with an internal diameter of about 8 micrometers, just bigger than the diameter of a red blood cell (5 to 7 micrometers). This design is critical to the function of capillaries. By bumping against the sides of the vessel as they pass through (like the cells in figure 24.7), the red blood cells are forced into close contact with the capillary walls, making exchange easier.

Almost all cells of the vertebrate body are no more than 100 micrometers from a capillary. At any one moment, about 5% of the circulating blood is in capillaries, a network that amounts to several thousand miles in overall length. If all the capillaries in your body were laid end to end, they would extend across the United States!

> **Putting the Concept to Work**
> If capillaries are so narrow, why do food molecules not pass INTO the bloodstream from the body's tissues?

Veins: Returning Blood to the Heart

> **LEARNING OBJECTIVE 24.2.4** Explain why the diameter of veins is so much larger than that of arteries.

Veins are vessels that return blood to the heart. Veins do not have to accommodate the pulsing pressures that arteries do, because much of the force of the heartbeat is weakened by the high resistance of the capillary network. For this reason, vein walls have much thinner layers of muscle and elastic fiber (figure 24.5c). An empty artery will stay open, like a pipe, but when a vein is empty, its walls collapse like an empty balloon.

Because the pressure of the blood flowing within veins is low, it becomes important to avoid any further resistance to flow, lest there not be enough pressure to get the blood back to the heart. Because a wide tube presents much less resistance to flow than a narrow one, the internal passageway of veins is often quite large. The diameters of the largest veins in the human body, the venae cavae, which lead into the heart, are fully 3 centimeters; this is wider than your thumb! Veins have unidirectional valves (figure 24.8) that ensure the return of blood by preventing it from flowing backward.

> **Putting the Concept to Work**
> Why do empty veins collapse but empty arteries do not?

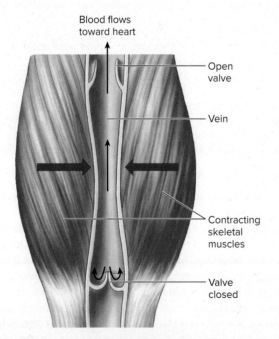

Figure 24.8 Flow of blood through veins.
Venous valves ensure that blood moves through the veins in only one direction back to the heart. This movement of blood is aided by the contraction of skeletal muscles surrounding the veins.

24.3 Blood

About 5% of your body mass is composed of the blood circulating through the arteries, veins, and capillaries of your body. This blood is composed of a fluid called **plasma,** together with materials and cells that circulate within it.

Blood Plasma: The Blood's Fluid

> **LEARNING OBJECTIVE 24.3.1** Describe the substances dissolved in blood plasma.

Blood plasma is a complex solution of water with three very different sorts of substances dissolved within it:

1. **Metabolites and wastes.** If the circulatory system is viewed as a highway of the vertebrate body, the blood contains the traffic traveling on that highway. Dissolved within its plasma are glucose, vitamins, hormones, and wastes that circulate between the cells of the body.
2. **Salts and ions.** Plasma is a dilute salt solution. The chief plasma ions are sodium, chloride, and bicarbonate. In addition, trace amounts of other salts, such as calcium and magnesium, as well as metallic ions, including copper, potassium, and zinc, are present in plasma. The composition of the plasma is not unlike that of seawater.
3. **Proteins.** Blood plasma is 90% water. Passing by all the cells of the body, blood would soon lose most of its water to them by osmosis if it did not contain as high a concentration of proteins as the cells it passes. More than half the amount of protein that is necessary to balance the protein content of the cells of the body consists of a single protein, serum albumin, which circulates in the blood as an osmotic counterforce. Human blood contains 46 grams of serum albumin per liter—that's over half a pound of it in your body. Starvation and protein deficiency result in reduced levels of protein in the blood. This lack of plasma proteins produces swelling of the body because the body's cells, which now have a higher level of solutes than the blood, take up water from the albumin-deficient blood. A symptom of protein deficiency diseases such as kwashiorkor is edema, a swelling of tissues, although other factors can also result in edema.

The liver produces most of the plasma proteins, including both albumin and *fibrinogen,* which is required for blood clotting. When blood in a test tube clots, the fibrinogen is converted into insoluble threads of *fibrin* that become part of the clot. You can see a blood clot forming in **figure 24.9**. Red blood cells, the disk-shaped cells, are becoming trapped in and among the threads of fibrin. The fluid that's left, which lacks fibrinogen and so cannot clot, is called serum.

> **Putting the Concept to Work**
> About a half-pound of serum albumin protein is circulating through your bloodstream at any one moment. What is it doing?

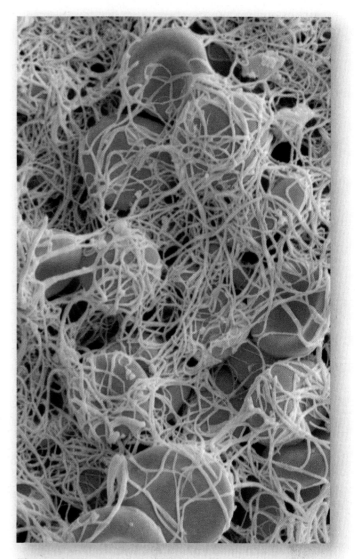

Figure 24.9 Threads of fibrin.

This scanning electron micrograph (×1,430) shows fibrin threads among red blood cells. Fibrin is formed from a soluble protein, fibrinogen, in the plasma to produce a blood clot when a blood vessel is damaged.

Steve Gschmeissner/Science Photo Library/Getty Images

Blood Cells: Cells that Circulate Through the Body

> **LEARNING OBJECTIVE 24.3.2** List the three principal cellular components of blood, and describe their functions.

Although blood is liquid, nearly half of its volume is actually occupied by cells. The three principal cellular components of blood are red blood cells, white blood cells, and cell fragments called platelets. The fraction of the total volume of the blood that is occupied by red blood cells is referred to as the blood's hematocrit. In humans, the hematocrit is usually about 45%.

Red Blood Cells Carry Hemoglobin. Each microliter (1 μl) of blood contains about 5 million **red blood cells.** Each human red blood cell (pictured at the top of figure 24.10) is a flat disk with a central depression on both sides, something like a doughnut with a hole that doesn't go all the way through. Red blood cells carry oxygen to the cells of the body. Almost the entire interior of a red blood cell is packed with hemoglobin, a protein that binds oxygen in the lungs and delivers it to the cells of the body.

> Hemoglobin is a large molecule made up of four subunits. Each subunit, as described in section 24.7, contains a central heme group that binds oxygen; therefore, each hemoglobin molecule can transport four molecules of oxygen.

Mature red blood cells function like boxcars rather than trucks. Like a vehicle without an engine, red blood cells contain neither a nucleus nor the machinery to make proteins. Because they lack a nucleus, these cells are unable to repair themselves and therefore have a rather short life; any one human red blood cell lives only about four months. New red blood cells are constantly being synthesized and released into the blood by cells within the soft interior marrow of bones.

White Blood Cells Defend the Body. Less than 1% of the cells in mammalian blood are **white blood cells.** There are many different kinds of white blood cells, which are shown in figure 24.10. They are the somewhat transparent cells with large or odd-shaped nuclei. Most are larger than red blood cells. They contain no hemoglobin and are essentially colorless. Each type of white blood cell has a different function. Neutrophils are the most numerous, followed in order by lymphocytes, monocytes, eosinophils, and basophils. *Neutrophils* attack in a self-sacrificing way, responding to foreign cells by releasing chemicals that kill all the cells in the neighborhood—including themselves. Monocytes give rise to *macrophages,* which attack and kill foreign cells by ingesting them (the name means "large eater"). White blood cells also include *B cells,* which make antibodies, and *T cells,* which kill infected cells.

Platelets Help Blood to Clot. Certain large cells within the bone marrow, called megakaryocytes, regularly pinch off bits of their cytoplasm. These cell fragments, called **platelets** (shown at the bottom of figure 24.10), contain no nuclei. Entering the bloodstream, they play a key role in blood clotting. In a clot, a gluey mesh of fibrin protein fibers (see figure 24.9) sticks platelets together to form a mass that plugs the rupture in the blood vessel. The clot provides a tight, strong seal, much as the inner lining of a tubeless tire seals punctures. The fibrin that forms the clot is made in a series of reactions that start when circulating platelets first encounter the site of an injury. Responding to chemicals released by the damaged blood vessel, platelets release a protein factor into the blood that starts the clotting process.

Blood cell	Life span in blood	Function
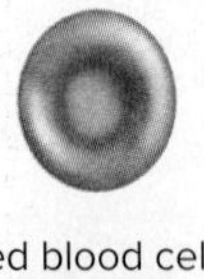 Red blood cell	120 days	O_2 and CO_2 transport
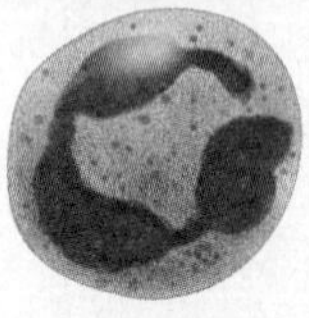 Neutrophil	7 hours	Immune defenses
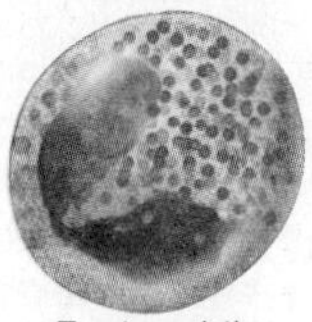 Eosinophil	Unknown	Defense against parasites
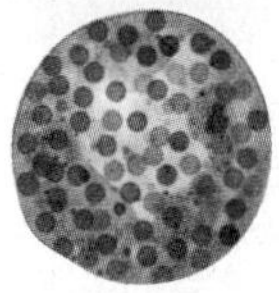 Basophil	Unknown	Inflammatory response
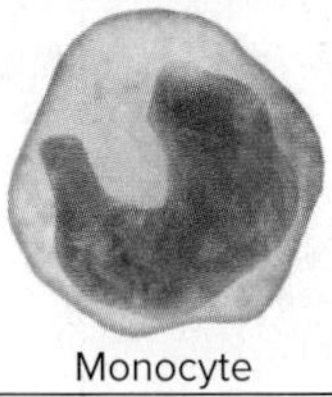 Monocyte	3 days	Immune surveillance (precursor of tissue macrophage)
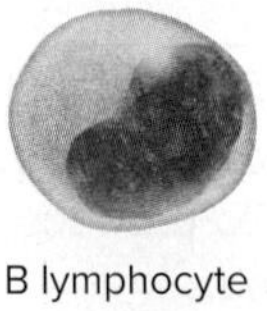 B lymphocyte	Unknown	Antibody production (precursor of plasma cells)
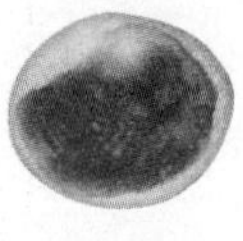 T lymphocyte	Unknown	Cellular immune response
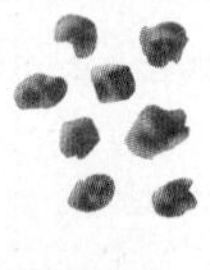 Platelets	7–8 days	Blood clotting

Figure 24.10 Types of blood cells.
Red blood cells, white blood cells (neutrophils, eosinophils, basophils, monocytes, and lymphocytes), and platelets are the three principal cellular components of blood in vertebrates.

> **Putting the Concept to Work**
> What is the difference between red blood cells and white blood cells?

24.4 Human Circulatory System

LEARNING OBJECTIVE 24.4.1 Describe the path of blood through the human heart and how it is monitored.

Humans and other mammals have a four-chambered heart that is really two separate pumping systems operating together within a single unit. One system pumps blood to the lungs, while the other pumps blood to the rest of the body. The left side has two connected chambers, and so does the right, but the two sides are not connected with one another.

Circulation Through the Heart

Let's follow the journey of blood through the human heart in **figure 24.11a**, starting with the entry of oxygen-rich blood into the heart from the lungs. Oxygenated blood (the pink arrows) from the lungs enters the left side of the heart (which is on the right as you look at the figure), emptying directly into the **left atrium** through large vessels called the **pulmonary veins.** From the atrium, blood flows down through an opening into the adjoining chamber, the **left ventricle.** Most of this flow, roughly 70%, occurs while the heart is relaxed. When the heart starts to contract, the atrium contracts first, pushing the remaining 30% of its blood into the ventricle.

After a slight delay, the ventricle contracts. The walls of the ventricle are far more muscular than those of the atrium (as seen in this cross section), and thus this contraction is much stronger. It forces most of the blood out of the ventricle in a single strong pulse. The blood is prevented from going back into the atrium by a large, one-way valve, the *bicuspid (mitral) valve,* or left atrioventricular valve, whose two flaps are pushed shut as the ventricle contracts.

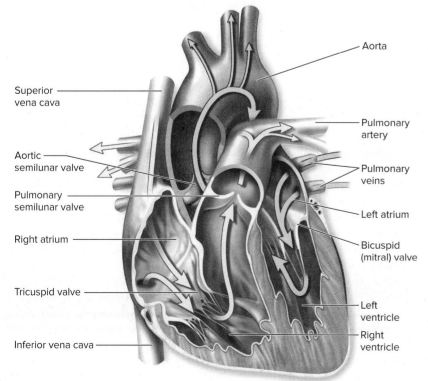

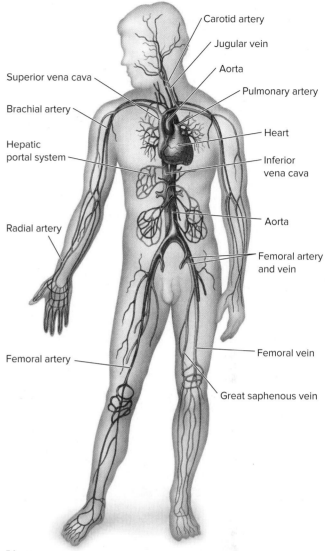

Figure 24.11 The heart and circulation in humans.

(a) In mammals, the heart has a septum dividing the ventricle into left and right ventricles. Oxygenated blood from the lungs enters the left atrium of the heart by way of the pulmonary veins. This blood then enters the left ventricle, from which it passes into the aorta to circulate throughout the body and deliver oxygen to the tissues. When gas exchange has taken place at the tissues, veins return blood to the heart. After entering the right atrium by way of the superior and inferior venae cavae, deoxygenated blood passes into the right ventricle and then through the pulmonary valve to the lungs by way of the pulmonary artery. (b) Some of the major arteries and veins in the human circulatory system are shown.

Prevented from reentering the atrium, the blood within the contracting left ventricle takes the only other passage out, a large blood vessel called the **aorta.** The aorta and its many branches carry oxygen-rich blood to all parts of the body.

Eventually, this blood returns to the heart after delivering its cargo of oxygen to the cells of the body. In returning, it passes through a series of progressively larger veins, ending in two large veins that empty into the right atrium.

The right side of the heart is similar in organization to the left side. Deoxygenated blood (the blue-colored arrows) passes from the **right atrium** into the **right ventricle** through a one-way valve, the *tricuspid valve* or right atrioventricular valve. It passes out of the contracting right ventricle through a second valve, the *pulmonary semilunar valve,* into the **pulmonary arteries,** which carry the deoxygenated blood to the lungs. The blood then returns from the lungs to the left side of the heart with a new cargo of oxygen, which is pumped to the rest of the body. Figure 24.11*b* shows the major veins and arteries in the human body. Veins carry blood to the heart and arteries carry blood away from the heart.

Monitoring the Heart's Performance

The simplest way to monitor the heartbeat is to listen to the heart at work, using a stethoscope. The first sound you hear, a low-pitched *lub,* is the closing of the bicuspid and tricuspid valves at the start of ventricular contraction. A little later, you hear a higher-pitched *dub,* the closing of the pulmonary and aortic valves at the end of ventricular contraction. If the valves are not closing fully or if they open incompletely, a turbulence is created within the heart. This turbulence can be heard as a liquid sloshing sound, called a *heart murmur.*

A second way to examine the events of the heartbeat is to monitor blood pressure (figure 24.12). A cuff wrapped around the upper arm is tightened enough to stop the flow of blood to the lower part of the arm ❶. As the cuff is loosened, blood begins pulsating through the arm's arteries and can be detected using a stethoscope. Two measurements are recorded: The systolic pressure ❷ is recorded when a pulse is heard, and the diastolic pressure ❸ is recorded when the pressure in the cuff is so low that the sound stops.

During the first part of the heartbeat, the atria are filling. At this time, the pressure in the arteries leading from the left side of the heart out to the tissues of the body decreases slightly. This low pressure is referred to as the **diastolic** pressure. During the contraction of the left ventricle, a pulse of blood is forced into the systemic arterial system, immediately raising the blood pressure within these vessels. The high blood pressure produced in this pushing period, which ends with the closing of the aortic valve, is referred to as the **systolic** pressure. Normal blood pressure values are 70 to 80 diastolic and 110 to 120 systolic. When the inner walls of the arteries accumulate fats, as they do in the condition known as *atherosclerosis,* the diameters of the passageways are narrowed. If this occurs, the systolic blood pressure is elevated.

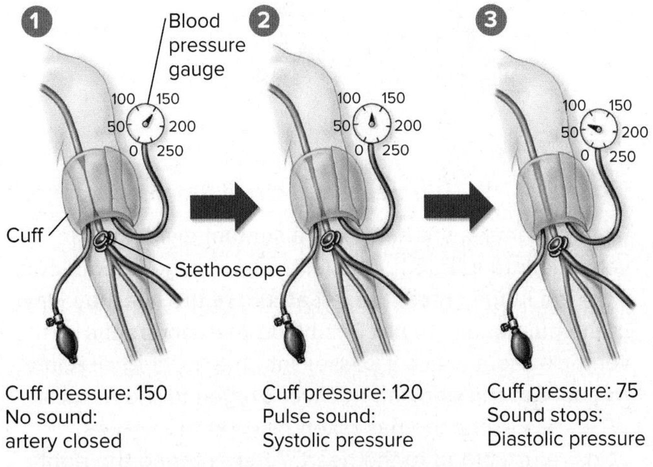

Cuff pressure: 150
No sound:
artery closed

Cuff pressure: 120
Pulse sound:
Systolic pressure

Cuff pressure: 75
Sound stops:
Diastolic pressure

Figure 24.12 Measuring blood pressure.

The blood pressure cuff is tightened to stop the blood flow through the brachial artery ❶. As the cuff is loosened, the systolic pressure is recorded as the pressure at which a pulse is heard through a stethoscope ❷. The diastolic pressure is recorded as the pressure at which a sound is no longer heard ❸.

Putting the Concept to Work
Is blood circulating in the pulmonary vein oxygenated? Explain.

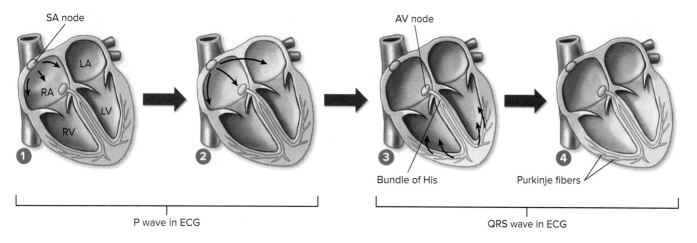

Figure 24.13 **How the mammalian heart contracts.**

How the Heart Contracts

> **LEARNING OBJECTIVE 24.4.2** Describe the characteristic events of a heartbeat, and explain what each signifies.

Contraction of the heart consists of a carefully orchestrated series of muscle contractions. First, both atria contract simultaneously. Then, after a brief time interval, both ventricles contract simultaneously. Contraction is initiated by the sinoatrial (SA) node (figure 24.13 ❶). Its membranes spontaneously depolarize (that is, admit ions that cause it to become more positively charged), creating electrical signals with a regularity that determines the rhythm of the heart's beating. In this way, the SA node acts as a *pacemaker* for the heart. Each electrical signal initiated within this pacemaker passes quickly from one heart muscle cell to another in a wave that envelops the left and the right atria at almost the same time (indicated by the yellow coloring in figure 24.13 ❶ and ❷).

But the electrical signals do not immediately spread to the ventricles. There is a pause before the lower half of the heart starts to contract. The reason for the delay is that the atria of the heart are separated from the ventricles by connective tissue that does not propagate the electrical signal. The signal would not pass to the ventricles at all except for a slender connection of cardiac muscle cells known as the **atrioventricular (AV) node** (labeled in ❸), which connects across the gap to a strand of specialized muscle known as the atrioventricular bundle, or bundle of His. Bundle branches divide into fast-conducting Purkinje fibers, which initiate the almost simultaneous contraction of all the cells of the right and left ventricles about 0.1 seconds after the atria contract. This delay permits the atria to finish emptying their contents into the corresponding ventricles before those ventricles start to contract. The contraction of the ventricles begins at the apex (the bottom of the heart), where depolarization of Purkinje fibers begins the electrical signal. The contraction then spreads up toward the atria (❸ and ❹). This results in a "wringing out" of the ventricles, forcing blood up and out of the heart.

Because the human body contains so much water, it conducts electrical currents rather well. These depolarizations generate an electrical current that can be detected using an electrocardiogram (ECG, more commonly known as an **EKG**).

> **Putting the Concept to Work**
> If your bundle of His were to be severed, what effect would this have on your heartbeat?

Respiration

24.5 Types of Respiratory Systems

LEARNING OBJECTIVE 24.5.1 Compare respiration among the vertebrates, identifying and describing the respiratory organs.

As discussed in chapter 7, animals obtain the energy they need by harvesting energy-rich electrons from organic molecules and then using these electrons to drive the synthesis of ATP and other molecules. This process, called oxidative respiration, requires oxygen because the spent electrons are donated to oxygen gas (O_2) to form water (H_2O). The leftover carbon atoms are released as carbon dioxide (CO_2), a "waste" by-product of the process.

Different Approaches

The uptake of oxygen and the release of carbon dioxide together are called **respiration,** neatly defining one of the principal evolutionary challenges facing all animals—how to obtain oxygen and dispose of carbon dioxide.

Direct Diffusion. Most of the primitive phyla of organisms obtain oxygen by direct diffusion from their aquatic environments, which contains dissolved oxygen. Sponges, cnidarians, many flatworms and roundworms, and some annelid worms obtain their oxygen by diffusion from surrounding water. Oxygen and carbon dioxide diffuse across the surface of the body, as shown in the flatworm in figure 24.14a. Similarly, some members of the vertebrate class Amphibia conduct gas exchange by direct diffusion through their moist skin.

Gills. The more advanced marine invertebrates (mollusks, arthropods, and echinoderms) and fishes possess special respiratory organs called gills. Gills increase the surface area available for diffusion of oxygen. A **gill** is basically a thin sheet of tissue that waves through the water. Gills can be simple or complex, as in the highly convoluted gills of fish (figure 24.14b). In most bony fish, the gills are protected by a covering called an operculum (not shown in figure 24.14b so that the underlying gills are visible). Because of this, their gills do not wave in the water. Instead, water is pumped over the gills, and gas exchange occurs across the walls of capillaries contained in the gills.

Trachea. Insects and other terrestrial arthropods do not have a single major respiratory organ like a gill. Instead, a network of air ducts, beginning with the **trachea** (the purple tubes in figure 24.14c) that branches into smaller and smaller tubes, carry air to every part of the body. The tracheae open to the outside of the body through structures called **spiracles,** which can be closed and opened.

Lungs. Terrestrial vertebrates, except for some amphibians, have a pair of respiratory organs called the lungs. Gas exchange in the lungs occurs across the walls of capillaries and air sacs, which in humans are called *alveoli* (shown in the enlargement in figure 24.14d).

> **Putting the Concept to Work**
> If fish can extract oxygen from seawater, why do we drown in it?

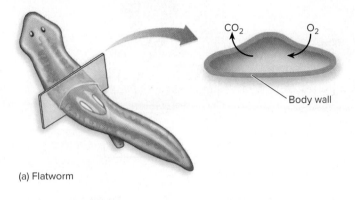
(a) Flatworm

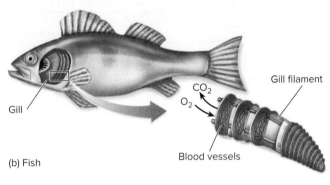
(b) Fish

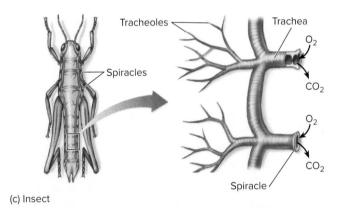
(c) Insect

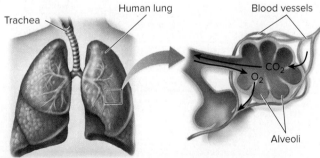
(d) Mammal

Figure 24.14 Gas exchange in animals.
(a) Gases diffuse directly across the body wall in many invertebrates and in some species of amphibians. (b) Fish gills provide a very large respiratory surface. (c) Terrestrial arthropods respire through tracheae, which open to the outside through spiracles. (d) The alveoli of human lungs provide a large respiratory surface area.

A Closer Look

On Being a Whale

Whales, made famous by Herman Melville's *Moby Dick* as man killers, are intelligent, often gentle giants that roam the Earth's oceans. What captivates us all is not so much their awesome size or dangerous disposition, but rather the way they live. Whales are mammals like us—they are warm-blooded, air-breathing creatures that somehow live and thrive in the oceans. They are not part-time ocean-dwellers, like the sea lions, seals, and walruses that feed in the oceans but live on the land. Whales, along with their close relatives the dolphins, live all their lives in the sea and cannot exist for very long out of the water. Descendants of African grazing animals with hooves, their bodies have adapted so well to ocean life that they die when stranded on land—not because they can't breathe like a fish out of water, but rather because their skeletons cannot support their massive weight on land. A whale's rib cage collapses under the weight of its body.

Whales Breathe Air, Just Like You Do

While whales are aquatic animals, they cannot extract oxygen from water as fish do. Like all mammals, they gain their oxygen from air, and so all whales need to surface from time to time in order to breathe.

Barcroft/Contributor/Getty Images

Most land mammals can breathe either through their nose or their mouth, but whales breathe only through their noses. A key adaptation in whales is that the nose is positioned on the top of its head, called a blowhole. This adaptation allows the whale to stay almost completely submerged while it is breathing, like the whales in the photo above. Some whales, including the humpback whale and other baleen whales, have two openings, called nostrils, making up the blowhole, whereas the toothed whales have only one nostril. The blowhole connects to the lungs, the oxygen-extracting organ in the body. A muscle surrounds the blowhole, and when it contracts, the blowhole opens, while relaxing the muscle closes the blowhole. When a whale comes to the surface to breathe, it swims just beneath the surface, arching its back to expose the blowhole to the air. It will first exhale, which is the spout of water that is usually the first sign of a whale's presence. The whale exhales only air. The mist that makes up the spout is water being sprayed away from the covering of the blowhole and the condensation of the moist air coming from the whale's lungs, similar to the way you can "see your breath" when you exhale in the cold air. The whale then inhales air and submerges, where it can stay down for long periods of time.

How a Whale Holds Its Breath

Stephen Frink/Getty Images

A whale's ability to remain underwater for extended periods of time, during which it is able to "hold its breath," is the result of other adaptations of the whale's respiratory system. Most humans can hold their breath for a minute or two. Free-divers can train themselves to hold their breath for about seven minutes and longer if they breathe pure oxygen. The deep-sea-diving sperm whale might take 50 to 60 breaths on the surface but then can stay down for over an hour.

To achieve this breath-holding feat, a whale's body has adapted in two ways: taking in more oxygen, and using it more efficiently. In one breath, a human will extract about 15% of the oxygen available in the air. A whale, on the other hand, can extract about 90% of the oxygen. The whale stores this extra oxygen in its blood and muscles. Whales have more blood per unit of body weight than humans and can store three times as much oxygen in their muscles.

A Whale Never Sleeps

Unlike land mammals, a whale consciously controls its breathing. We breathe even when unconscious during sleep, but a whale has complete voluntary control over its breathing, breathing only when its blowhole is exposed to the surface. Because of this, another adaptation was necessary for a whale's watery existence: Whales never enter a state of total unconsciousness. A whale never falls asleep—it might not wake up in time to breathe. Marine biologists have found that a whale's brain sleeps in shifts, letting only one-half of the brain sleep at a time. The "awake" part of the brain has enough awareness of its surroundings and its body condition to bring the whale to the surface when it needs to breathe.

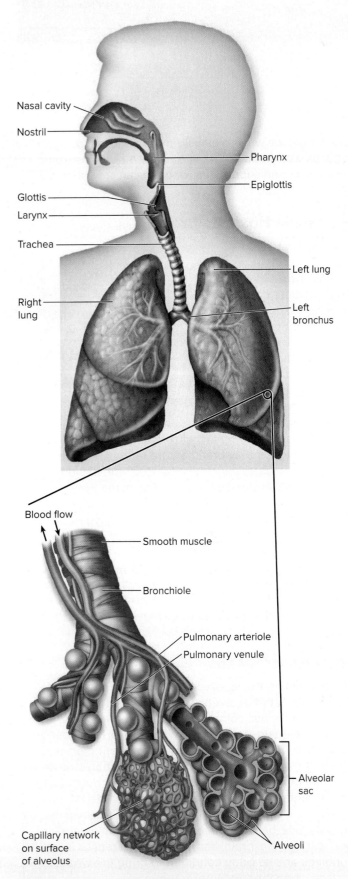

Figure 24.15 The human respiratory system.

The respiratory system consists of the lungs and the passages that lead to them.

24.6 The Human Respiratory System

LEARNING OBJECTIVE 24.6.1 Describe the components of the human respiratory system, explaining the functions of the pleural membrane and the diaphragm.

The oxygen-gathering mechanism of humans and all mammals, although less efficient than fishes' gills, adapts them well to their terrestrial habitat. Humans, like all other terrestrial vertebrates, obtain the oxygen they need for metabolism from air, which is about 21% oxygen gas. A pair of lungs is located in the chest within the **thoracic** cavity. As you can see in figure 24.15, the two lungs hang free within the cavity, connected to the rest of the body only at one position, where the lung's blood vessels and air tube enter. This air tube is called a **bronchus** (plural, bronchi). It connects each lung to a long tube called the **trachea,** which passes upward and opens into the rear of the mouth. The trachea and both the right and left bronchi are supported by C-shaped rings of cartilage.

The Path of Air

Air normally enters through the nostrils into the nasal cavity, where it is moistened and warmed. In addition, the nostrils are lined with hairs that filter out dust and other particles. As the air passes through the nasal cavity, an extensive array of cilia further filters it. The air then passes to the back of the mouth, through the pharynx (the common passage of food and air), and then through the larynx (voice box) and the trachea. Because the air crosses the path of food at the back of the throat, a special flap called the epiglottis covers the trachea whenever food is swallowed, to keep it from "going down the wrong pipe." From the trachea, air passes down through several branchings of bronchi in the lungs and eventually to bronchioles that lead to alveoli. Mucous secretory ciliated cells in the trachea and bronchi also trap foreign particles and carry them upward to the pharynx, where they can be swallowed. The lungs contain millions of alveoli, tiny sacs clustered like grapes. The alveoli are surrounded by an extremely extensive capillary network. All gas exchange between the air and blood takes place across the walls of the capillaries and the alveoli via diffusion.

The Respiratory Apparatus

The human respiratory apparatus is simple in structure. The thoracic cavity is bounded on its sides by the ribs and on the bottom by a thick layer of muscle, the **diaphragm,** which separates the thoracic cavity from the abdominal cavity. Each lung is covered by a very thin, smooth membrane called the pleural membrane. This membrane also folds back on itself to line the interior wall of the thoracic cavity, into which the lungs hang. The space between these two layers of membrane is very small and filled with fluid. This fluid causes the two membranes to adhere to each other in the same way a thin film of water can hold two plates of glass together, effectively coupling the lungs to the walls of the thoracic cavity.

The thin film of fluid on the inner surface of the alveoli could cause these air sacs to collapse. This doesn't occur because the epithelial cells that line the alveoli secrete onto their surface a mixture of lipoprotein molecules called *surfactant*, reducing the surface tension that would otherwise cause the alveoli to collapse.

Lungs operate as simple one-cycle pumps. Air is drawn into the lungs by the creation of negative pressure—that is, pressure in the lungs is less

Essential Biological Process 24A

Breathing

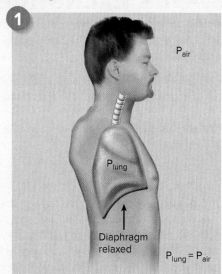
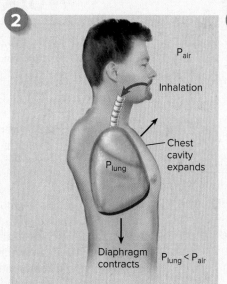
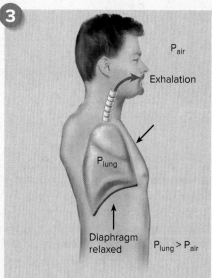

Before inhalation, the air pressure in the lungs (P_{lung}) is equal to the atmospheric pressure (P_{air}).

During inhalation, the diaphragm contracts, and the chest cavity expands downward and outward. This increases the volume of the chest cavity and lungs, which reduces the air pressure inside the lungs, and the air from outside the body flows into the lungs.

During exhalation, the diaphragm relaxes, decreasing the volume of the chest cavity. The pressure increases in the lungs, forcing air out of the lungs.

than atmospheric pressure. The pressure in the lungs is reduced when the volume of the lungs is increased. This is similar to how a bellow pump or accordion works. In both cases, when the bellow is extended, the volume inside increases, causing air to rush in. How does this occur in the lungs? The volume in the lungs increases when muscles that surround the thoracic cavity contract, causing the thoracic cavity to increase in size.

The Mechanics of Breathing

The active pumping of air in and out of the lungs is called breathing. During *inhalation,* muscular contractions cause the rib cage to move outward and upward. The diaphragm, the red-colored lower border of the lung in *Essential Biological Process 24A*, is dome-shaped when relaxed (**panel 1**) but moves downward and flattens during contraction (**panel 2**). These rib cage movements expand the chest cavity, which decreases the air pressure in the lungs compared to the atmosphere ($P_{lung} < P_{air}$). When this happens, air flows into the lungs. In effect, inhalation sucks air into the lungs.

During *exhalation* (**panel 3**), the ribs and diaphragm return to their original resting position. In doing so, they exert pressure on the lungs. This pressure is transmitted uniformly over the entire surface of the lung, forcing air from the inner cavity back out to the atmosphere. Because the lungs fill with a mixture of fresh and partly deoxygenated air, the respiratory efficiency of mammalian lungs is far from maximal.

Putting the Concept to Work
Why is water in the space between pleural membrane layers?

BIOLOGY & YOU

Asthma. Over 9% of American children are affected by asthma, a chronic disease of the respiratory system in which the airways occasionally constrict, causing shortness of breath, coughing, and in some instances a life-threatening difficulty in breathing. Asthma attacks can be triggered by a variety of events, including viral illnesses such as colds, stress, airborne allergens, smoke, and exercise. Researchers do not yet fully understand the cause of asthma or why it seems to have become more common. Many genetic and environmental factors appear to interact in complex ways to produce this common disorder. At least 25 different genes have been implicated, some of which may cause asthma only when combined with specific environmental triggers. The symptoms can be treated by inhaling chemicals called bronchodilators that expand the constricted passageways. Longer-term preventive medications such as inhaled corticosteroids reduce inflammation of the airway lining and so suppress attacks. The most effective treatment for asthma sufferers is to identify their triggers, such as pets or aspirin, and limit exposure to them. As children mature into adults, half of asthma sufferers (54%) lose their sensitivity to asthma triggers; the less fortunate suffer from asthma all their lives.

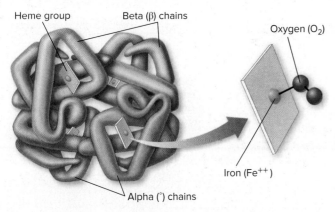

Figure 24.16 The hemoglobin molecule.

The hemoglobin molecule is actually composed of four protein chain subunits: two copies of the "alpha chain" and two copies of the "beta chain." Each chain is associated with a heme group, and each heme group has a central iron atom, which can bind to a molecule of oxygen.

24.7 How Respiration Works: Gas Exchange

LEARNING OBJECTIVE 24.7.1 Describe how the hemoglobin molecule functions in mammalian respiration.

When oxygen has diffused from the air into the moist cells lining the inner surface of the lung, its journey has just begun. Passing from these cells into the bloodstream, the oxygen travels throughout the body in the circulatory system, described earlier in this chapter.

O_2 Transport

Oxygen (O_2) moves within the circulatory system carried piggyback on the protein **hemoglobin**. A hemoglobin molecule contains four heme groups, each with a central atom of iron, which binds oxygen, as shown in figure 24.16. Hemoglobin molecules act like little oxygen sponges, soaking up oxygen within red blood cells and causing more to diffuse in from the blood plasma. The oxygen binds in a reversible way, which is necessary so that the oxygen can unload when it reaches the tissues of the body. Hemoglobin is manufactured within red blood cells and never leaves these cells, which circulate in the bloodstream like ships bearing cargo.

At the high O_2 levels that occur in the lung, most hemoglobin molecules carry a full load of oxygen atoms. In tissue, the presence of carbon dioxide (CO_2) causes the hemoglobin molecule to assume a different shape, one that gives up its oxygen more easily. The effect of CO_2 on oxygen unloading is important because CO_2 is produced by the tissues through cell metabolism. For this reason, the blood unloads oxygen more readily within tissues undergoing metabolism.

CO_2 Transport

At the same time the red blood cells are unloading oxygen, they are also absorbing CO_2 from the tissue. About 8% of the CO_2 in blood is simply dissolved in plasma. Another 20% is bound to hemoglobin; however, this CO_2 binds, not to the iron-containing heme group, but instead to another site on the hemoglobin molecule, and so it does not compete with oxygen binding. The remaining 72% of the CO_2 diffuses into the red blood cells. Inside the lungs, gaseous CO_2 diffuses outward from the blood into the alveoli. With the next exhalation, this CO_2 leaves the body. The diffusion of CO_2 out from the red blood cells causes the hemoglobin within these cells to release its bound CO_2 and take up O_2 instead. The cells with a new load of O_2 then start their next respiratory journey.

Putting the Concept to Work

What stops CO_2 from diffusing out of red blood cells into plasma?

BIOLOGY & YOU

Hiccups. A hiccup is a spasmodic contraction of the diaphragm that repeats several times a minute. Increasing respired CO_2 by breathing into a paper bag can stop hiccups for the interesting reason that hiccups may be an evolutionary remnant of earlier amphibian respiration. Frogs and other amphibians don't have a diaphragm and instead gulp air via a simple motor reflex much like your hiccuping reflex. Air gulping in frogs is inhibited by CO_2, just as your hiccuping is. In humans, the hiccuping motor pathway forms early in fetal development, well before the motor pathways that drive normal breathing. Premature infants born before their lung motor pathways are fully functional spend 2.5% of their time hiccuping, gulping air just like amphibians.

Answering Your Questions *About Vaping*

What Is an E-Cigarette?

Cigarettes have been an immense commercial success because even though the tars in tobacco may lead to lung cancer, the nicotine in tobacco leads to addiction. Said simply, smokers keep buying cigarettes, even if tobacco is dangerous, because they quickly become addicted to the nicotine. Eighty-five percent of people who try to quit smoking on their own relapse. However, as the evidence connecting cigarette smoking to lung cancer became impossible to avoid, public acceptance of smoking has faded. As you might expect, cigarette sales rapidly fell as fewer people took up smoking. While 40% of Americans smoked in 1965, only 16% did in 2017. The commercial answer? Sell the addictive nicotine without the cancer-causing tars! Rather than burning tobacco to deliver nicotine in smoke, simply deliver the nicotine to a customer directly—an e-cigarette. First patented back in the 1960s, e-cigarettes really didn't come onto the scene in the United States until 2007. In the years since their introduction, e-cigarettes, have become as widely used as tobacco cigarettes.

What Is Vaping?

E-cigarettes do not burn tobacco products. Rather they contain nicotine solutions that are heated to make a vapor (hence the term *vape* or *vaping*). Typically, the nicotine in an e-cigarette is combined with chemicals to vaporize the nicotine, chiefly propylene glycol (closely related to ethylene glycol, a key ingredient in antifreeze) and glycerin, which together comprise about 95% of the e-liquid. Flavorings may also be added, either natural or artificial. Stored in a tube that looks like a cigarette or small flashlight, the e-liquid is heated by a battery to about 250°C, producing a nicotine-laced aerosol vapor. The amount of nicotine delivered by an e-cigarette into a smoker's bloodstream is comparable to that delivered by smoking a tobacco cigarette.

Steve Heap/Shutterstock

What Is a Juul Vape Pen?

The Juul (pronounced "jewel") vape pen is an e-cigarette with a lot of nicotine. Smoking a Juul delivers far more nicotine than most commercially available e-cigarettes. How does it do this? Each Juul vaporizes, rather than heats, a liquid that is 5% nicotine. Packaged in a sleek, small, flat stick, a Juul looks a bit like a USB computer thumb drive. Juuls are not particularly expensive, at $35 for the pen and $16 for a four-pack of prefilled cartridges (a "Juul pod"). The hit of nicotine is powerful: One pod of Juuls contains as much nicotine as a pack of cigarettes. A Juul is thus an ideal commercial product to get its users addicted to its use, in just the same way tobacco hooked so many cigarette smokers. Juuls represent half of the $2 billion e-cigarettes sold annually today and are particularly trendy among high school students and young adults.

How Widespread Is Vaping Among Young Adults?

While manufacturers claim that e-cigarettes are not aimed at kids, there are over 7,000 flavors on the market, including gummy bear, bubble gum, and cotton candy. Is the marketing working? Government statistics found millions of teens have tried vaping: Nearly 36% of 12th graders had tried some form of e-cigarette in 2017, 17% within the last 30 days (that's more than smoked cigarettes). Because they are tobacco-free, e-cigarettes are currently unregulated in the United States, although that may change. Many states are considering restrictions.

Is Vaping Safer than Smoking Cigarettes?

Yes. The DNA-mutating tars responsible for lung cancer are not present in e-cigarette vapor. Nor do Juul pens produce secondhand smoke. If e-cigarettes were used only as replacement products for tobacco smokers who have been unable to quit smoking, they would certainly be a harm-reducing product. However, e-cigarette use among young adults is not primarily for tobacco use cessation, for few of the high school kids being introduced to vaping are tobacco users. The safety of vaping is much disputed. Manufacturers tout their safety, while public health officials point out that while e-cigarettes do not produce carcinogenic smoke, there is accumulating evidence that vaping may carry negative health effects that should not be ignored.

Are E-Cigarettes a Gateway Drug?

This is, of course, the key question. Does vaping lead to cigarette smoking? Researchers report that teens who vape are six times more likely as those who have never vaped to later begin smoking. Once addicted to nicotine, a vaping young adult can as easily satisfy a craving for nicotine with a cigarette as a Juul. The danger is very real.

Putting the Chapter to Work

1 Your instructor shows you a diagram of the human circulatory system. She points to weblike structures and describes their function as a delivery system of oxygen and a pickup system of carbon dioxide.

What structures is she referring to?

2 Varicose veins are swollen veins in the legs that result in enlarged, visible venous networks. These can result from long periods of standing that exert pressure on the legs. Because of the pressure, the blood in the veins can flow backward.

What structure in the veins should prevent this backflow under normal pressure conditions?

3 You bring your dog to the veterinarian because it is not acting like itself. The veterinarian recommends blood work to check the levels of the different kinds of blood cells. Based on the results, the veterinarian wants to treat the dog for parasites.

What blood cell type was elevated to point the veterinarian to this treatment?

Retracing the Learning Path

Circulation

24.1 Open and Closed Circulatory Systems

1. Animal cells acquire oxygen from the environment and nutrients from the food they eat, but not all animals do this in the same way. Flatworms have a gastrovascular cavity that circulates the products of gas exchange and digestion.
- Mollusks and arthropods have open circulatory systems. Hemolymph is pumped by tubular hearts through blood vessels that open into a body cavity.
- Annelids and all vertebrates have closed circulatory systems. The blood is propelled through the closed vessels by the pumping of a heart.

2. The vertebrate circulatory system functions in the transportation of substances throughout the body, which includes gases in respiration, nutrients in digestion, metabolic wastes in excretion, and hormones in endocrine functions. It also functions in the regulation of body temperature and in the protection of the body through blood clotting and immunity.

24.2 Architecture of the Vertebrate Circulatory System

1. In vertebrates, blood circulates from the heart through arteries and arterioles to capillaries. Blood flows from the capillaries back to the heart through venules and larger veins.
2. Arteries have a thick layer of smooth muscle that absorbs the force generated by the strong contraction of the heart. A layer of elastic tissue allows the artery to expand with the force of the blood flow.
3. The walls of capillaries are only a single cell layer thick and small in diameter, which aids in the exchange of gases and nutrients with surrounding tissue.
4. Veins bring blood back to the heart. The blood moves through the veins with the help of contractions of skeletal muscles and one-way valves inside the veins that keep the blood from backing up in the vessel.

24.3 Blood

1. Blood consists of a fluid called plasma that contains metabolites, waste products, salts and ions, and proteins.
2. Various cells also circulate in the plasma, including red blood cells that are involved in gas exchange, and various types of white blood cells that defend the body against infection. Platelets are fragments of cells that circulate in the blood and are involved in blood clotting.

24.4 Human Circulatory System

1. The human heart has four chambers that function as a two-cycle pump, pumping blood from the heart to the lungs and oxygenated blood from the heart to the body.
- Oxygenated blood from the lungs enters the left side of the heart through the pulmonary vein, emptying into the left atrium. From there, the blood passes into the left ventricle, where it is pumped out to the body. After delivering oxygen to the tissues of the body, the deoxygenated blood flows back to the heart and enters the right atrium. The blood then passes into the right ventricle, and when it contracts, the blood is pumped to the lungs.

2. A pacemaker, called the sinoatrial (SA) node, controls heartbeat rate. The electrical impulses of the heart can be detected and recorded as an electrocardiogram (EKG).

Respiration

24.5 Types of Respiratory Systems

1. To acquire energy, all animals need oxygen to fuel cellular respiration and need to expel carbon dioxide, a waste product produced during cellular respiration. However, animals carry out gas exchange in different ways.
- Aquatic animals extract oxygen dissolved in water. Some do so directly across the body surface; others use gills, which greatly increase the surface area for gas exchange.
- Terrestrial animals extract oxygen from the air with tracheae or lungs. The tracheae system in arthropods is a network of air ducts with openings to the outside, called spiracles. Other terrestrial vertebrates use lungs where gas exchange occurs across the walls of capillaries and air sacs called alveoli.

24.6 The Human Respiratory System

1. Human lungs are positioned within an internal cavity, called the thoracic cavity. A bronchus connects each lung to the trachea.
- Air is filtered as it enters through the nostrils and passes through the nasal cavity and down the trachea. At the back of the throat, the air crosses the path of food. The epiglottis will close over the trachea when food is swallowed so that the food doesn't go down the trachea.
- The airways end in alveoli, which are surrounded by capillaries. Gas exchange occurs across the single cell layers of the alveoli and capillary walls.
- Pleural membranes line the outside of the lungs and the inner layer of the thoracic cavity. Contraction of the muscles that line the thoracic cavity expands the space in the lungs, causing air to rush into the lungs. Relaxation of the muscles causes air to be exhaled.

24.7 How Respiration Works: Gas Exchange

1. In the lungs where oxygen concentrations are high, oxygen diffuses into the red blood cells in the blood. Oxygen binds to iron atoms contained within hemoglobin. The oxygen is then carried to areas of the body that are low in oxygen. The oxygen diffuses down its concentration gradient, entering cells in the surrounding tissues.

Inquiry and Analysis

Do Big Hearts Beat Faster?

Small animals live at a much faster pace than large animals. They reproduce more quickly and live shorter lives. As a rule, they tend to move about more quickly and so consume more oxygen per unit body weight. Interestingly, small and large mammals have about the same size heart, relative to body size (about 0.6% of body mass). It is interesting to ask whether all mammalian hearts beat at the same rate. The heart of a 7,000-kilogram (a kilogram is 1,000 grams) African bull elephant must push a far greater volume of blood through its body than the heart of a 20-gram mouse, but the elephant is able to do it through much-larger-diameter arteries, which impose far less resistance to the blood's flow. Does the elephant's heart beat faster? Or does the mouse's, in order to deliver more oxygen to its muscles? Or perhaps the mouse's heart beats more slowly because of increased resistance to flow through narrower blood vessels?

The graph shown here displays the pulse rate of a number of mammals of different body sizes (the **pulse rate** is the number of heartbeats counted per minute, a measure of how rapidly the heart is beating). Note that both the *x* and *y* axes use log scales (see figure 0.15). For comparison, the pulse rate of an adult human at rest is about 70 beats per minute. The largest mammal is the blue whale, as big as a supersized moving van with a body mass as great as 136,000 kilograms; the smallest is the pygmy shrew, smaller than a cockroach with a body mass of a few hundredths of a gram.

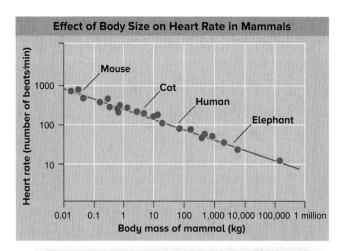

James Gathany/CDC

johan63/Getty Images

Analysis

1. **Applying Concepts** All mammals have the same size hearts relative to their body size. Do their hearts beat at the same rate?
2. **Interpreting Data**
   a. What is the resting pulse rate of a 7,000-kilogram African bull elephant?
   b. What is the resting pulse rate of a 20-gram mouse? How many complete heartbeats is that per second?
   c. What general statement can be made regarding the effect of body size on heart rate in mammals?
3. **Making Inferences**
   a. The data in the graph, plotted on logarithmic coordinates (that is, the scale rises in powers of 10), fall nicely upon a straight line. How would you expect them to look plotted on linear coordinates?
   b. As you walk through the graph from left to right, the line slopes down. This is called a negative slope. What does the negative slope of the line signify?
4. **Drawing Conclusions** If you plotted data for an experiment measuring body mass versus resting oxygen consumption, you would get exactly the same slope of the line as shown in this graph. What does this tell us about why body size affects heart rate in mammals as it does?

25 The Path of Food Through the Animal Body

LEARNING PATH ▼

Food Energy and Essential Nutrients
1. Food for Energy and Growth

Digestion
2. Types of Digestive Systems
3. The Human Digestive System
4. The Mouth and Teeth
5. The Esophagus and Stomach
6. The Small and Large Intestines
7. Accessory Digestive Organs

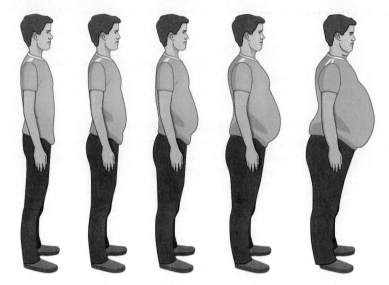

IF HIS WAIST measurement divided by his hip measurement is above 0.9, he is obese.

Supersizing Ourselves

The number of overweight and obese people in the United States is skyrocketing. Thirty years ago, about 47% of adults were considered overweight, with 15% of them being obese. By 2020, those statistics increased to 73% of the population being overweight, with half of them (42.4%) tipping the scales to obesity. In 2020, 120 million adult Americans were obese.

An Unsuspected Culprit

What is fueling this trend to obesity? It doesn't take a scientist to hypothesize that a change in lifestyle is contributing to an increase in weight. Today's "super-size it" fast foods, "big gulp" drinks, and "all-you-can-eat" buffets, coupled with lack of physical exercise, leaves Americans consuming more and more calories without the physical activity needed to burn off those calories. Unexpectedly, scientists have recently stumbled over another potential cause of today's obesity outbreak. The culprit? Microorganisms that live in your gut.

The human digestive system is teeming with 10 to 100 trillion microorganisms, primarily bacteria. These bacteria outnumber the cells in the human body by at least 10 times! This is a symbiotic microbe–human relationship, with both partners benefiting. The bacteria have a moist, warm place to live and ample food. The human host also benefits because the bacteria help in the digestion of the food you eat.

Two Intestinal Residents Compete

Researchers at the Washington University School of Medicine in St. Louis recently found that although humans can harbor thousands of different types of intestinal microbes, almost all (more than 90%) are bacteria that belong to one of two groups: the Firmicutes and the Bacteroidetes. Unexpectedly, the distribution of these two groups is different in obese people compared to their leaner counterparts! Obese people have a higher proportion of Firmicutes than lean people do, with the lean patients having a higher proportion of Bacteroidetes.

While this is an interesting observation, it might in principle be a result of factors other than just weight. To test this possibility, the obese patients in this study were put on diets to lose weight. Did the mix of microbes change? Yes. The balance of the two kinds of bacteria in their guts changed to reflect more closely that of lean patients: decreases in the proportion of Firmicutes and increases in Bacteroidetes.

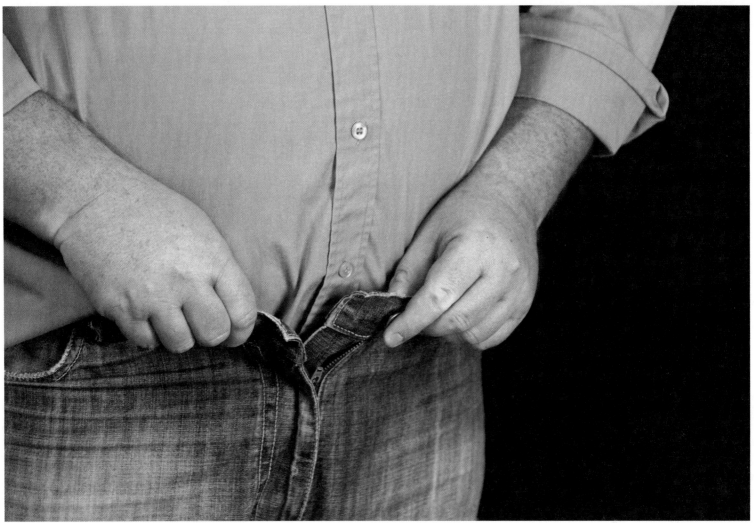

shutterupeire/Shutterstock

The Role Your Genes Play

What role might host genes play? To look into this, the study also examined the intestinal microbes of genetically obese mice. Like humans, genetically obese mice also contain a higher proportion of Firmicutes in their guts, whereas the guts of normal lean mice are dominated by the Bacteroidetes. But are the Firmicutes bacteria in genetically obese mice a cause or result of obesity? To address this very key question, the investigators "infected" the guts of germ-free mice (mice that have never been exposed to bacteria) with microbial samples taken from the intestines of lean and obese mice. Both experimental groups gained weight, but the mice that received bacteria from obese donors gained significantly more weight without increased food consumption. The "obese bacteria" mice gained weight not because they ate more food, but because they extracted more energy from their food.

Cause or Effect?

But a human is not a mouse. While fat and lean people have the same differences in gut bacteria that mice do, perhaps in humans this is an effect of obesity rather than a cause. To eliminate this possibility once and for all, the Washington University research team searched out pairs of female human twins, ranging in age from 21 to 32, in which one twin is obese and the other lean. This allowed the researchers to cancel out much of the effect of genetics and environment. They gave fecal material isolated from the twins to mice reared in a sterile environment. After five weeks, the mice with bacteria from the fat twins had much more body fat than those with bacteria from the thin twins! When the two sorts of mice were placed in the same cage, where they eat each others' droppings, bacteria from the lean twins took over the guts of the fat mice, who became thin. Are fecal transplants in our overweight futures?

Food Energy and Essential Nutrients

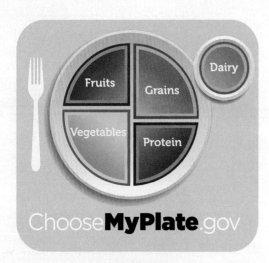

Figure 25.1 The nutrition plate.
The U.S. Department of Agriculture guidelines for a healthy diet utilize a "food plate" icon, organized the way people eat. It is recommended that one-half of your dietary plate be filled with fruits and vegetables, the other half with whole grains and lean protein. Fat is to be avoided in all choices. Dairy, indicated on the side, should favor low-fat items as well, such as skim milk and yogurt. Filling half the plate with fruit and vegetables reduces calories. The plate emphasizes whole grains rather than refined grains such as white rice or white bread, which are stripped of nutrients, such as vitamins, fiber, and iron.
Source: U.S. Department of Agriculture

25.1 Food for Energy and Growth

> **LEARNING OBJECTIVE 25.1.1** Discuss how a balanced diet influences the BMI and why this is important.

The food animals eat provides both a source of energy and essential molecules such as certain amino acids and fats that the animal body is not able to manufacture for itself. An optimal diet contains a balance of fruits, vegetables, grains, protein, and dairy, as recommended by the federal government's "nutrition plate" in figure 25.1. The plate is intended as a general guideline of what a person should eat. About half of a person's diet should be fruits and vegetables, and the other half should contain proteins, whole grains, and dairy. Fats are recommended in small amounts because they have a far greater number of energy-rich carbon–hydrogen bonds and thus a much higher energy content per gram than carbohydrates or proteins.

Carbohydrates are obtained primarily from grains (the brown section), fruits (the red section), and vegetables (the green section). On average, carbohydrates contain 4.1 calories per gram; fats, by comparison, contain 9.3 calories per gram, over twice as much. Dietary fats are obtained from oils, margarine, and butter and are abundant in fried foods, meats, and processed snack foods, such as potato chips and crackers. Like carbohydrates, proteins have 4.1 calories per gram and can be obtained from many foods, including dairy products, poultry, meat (the blue and purple sections), and grains.

Body Mass Index. Are you overweight? As a rough measure, run a tape measure around your middle: "waist inches/hip inches" greater than 0.9 is considered overweight. The international standard measure of appropriate body weight is the body mass index (BMI), estimated as your body weight in kilograms, divided by your height in meters squared. A BMI chart is presented in figure 25.2. To determine your BMI, find your height in the left-hand column (in feet and inches) and trace it across to the column with your weight (in pounds). A BMI value of 25 (dark blue boxes) and above is considered overweight and 30 or over is considered obese. In the United States, the National Institutes of Health estimated in 2016 that 70.7% of adults, 160 million Americans, were overweight, with a BMI of 25 or more. More than half of these, an alarming 39.6% of American adults, were considered obese with a BMI of 30 or greater. Being overweight is highly correlated with coronary heart disease, diabetes, and many other disorders. However, a BMI of less than 18.5, often resulting from eating disorders such as anorexia nervosa, is also unhealthy.

Basal Metabolic Rate. All animals must eat. Even an animal that is completely at rest requires energy to support its metabolism. This minimum rate of energy consumption, called the *basal metabolic rate* (BMR), is relatively constant for a given individual. Exercise raises the metabolic rate above the basal levels, so the amount of energy the body requires per day is determined not only by the BMR but also by the level of physical activity. Energy that is not used for metabolism or exercise is stored as fat. Therefore, energy needs can be altered by the choice of diet (caloric intake) and the amount of energy expended in exercise.

> **Putting the Concept to Work**
> Calculate your personal body mass index. Are you overweight?

	25 OVERWEIGHT LIMIT			OVERWEIGHT																		
WEIGHT	100	105	110	115	120	125	130	135	140	145	150	155	160	165	170	175	180	185	190	195	200	205
HEIGHT																						
5' 0"	20	21	21	22	23	24	25	26	27	28	29	30	31	32	33	34	35	36	37	38	39	40
5' 1"	19	20	21	22	23	24	25	26	26	27	28	29	30	31	32	33	34	35	36	37	38	39
5' 2"	18	19	20	21	22	23	24	25	26	27	27	28	29	30	31	32	33	34	35	36	37	37
5' 3"	18	19	19	20	21	22	23	24	25	26	27	27	28	29	30	31	32	33	34	35	35	36
5' 4"	17	18	19	20	21	21	22	23	24	25	26	27	27	28	29	30	31	32	33	33	34	35
5' 5"	17	17	18	19	20	21	22	22	23	24	25	26	27	27	28	29	30	31	32	32	33	34
5' 6"	16	17	18	19	19	20	21	22	23	23	24	25	26	27	27	28	29	30	31	31	32	33
5' 7"	16	16	17	18	19	20	20	21	22	23	23	24	25	26	27	27	28	29	30	31	31	32
5' 8"	15	16	17	17	18	19	20	21	21	22	23	24	24	25	26	27	27	28	29	30	30	31
5' 9"	15	16	16	17	18	18	19	20	21	21	22	23	24	24	25	26	27	27	28	29	30	30
5' 10"	14	15	16	17	17	18	19	19	20	21	22	22	23	24	24	25	26	27	27	28	29	29
5' 11"	14	15	15	16	17	17	18	19	20	20	21	22	22	23	24	24	25	26	26	27	28	29
6' 0"	14	14	15	16	16	17	18	18	19	20	20	21	22	22	23	24	24	25	26	26	27	28
6' 1"	13	14	15	15	16	16	17	18	18	19	20	21	21	22	22	23	24	24	25	26	26	27
6' 2"	13	13	14	15	15	16	17	17	18	19	19	20	21	21	22	22	23	24	24	25	26	26
6' 3"	12	13	14	14	15	16	16	17	17	18	19	19	20	21	21	22	22	23	24	24	25	26
6' 4"	12	13	13	14	15	15	16	16	17	18	18	19	19	20	21	21	22	23	23	24	24	25

Figure 25.2 Are you overweight?
This chart presents the body mass index (BMI) values used by federal health authorities to determine who is overweight. Your body mass index is at the intersection of your height and weight.

Essential Substances for Growth

> **LEARNING OBJECTIVE 25.1.2** Explain why we require essential amino acids, trace elements, and vitamins.

Over the course of their evolution, many animals have lost the ability to manufacture certain substances they need. Mosquitoes, for example, cannot manufacture cholesterol and must obtain it in their diet—human blood is rich in cholesterol. Humans are unable to manufacture eight of the 20 amino acids used to make proteins: lysine, tryptophan, threonine, methionine, phenylalanine, leucine, isoleucine, and valine. These amino acids, called **essential amino acids,** must therefore be obtained from proteins in the food we eat.

Trace Elements. In addition to supplying energy, food must also supply the body with a wide variety of **trace elements,** which are minerals required in very small amounts. Among the trace elements are iodine (a component of thyroid hormone), cobalt (a component of vitamin B_{12}), zinc and molybdenum (components of enzymes), manganese, and selenium.

Vitamins. Essential organic substances that are used in trace amounts are called **vitamins.** Humans require at least 13 different vitamins. Humans, monkeys, and guinea pigs, for example, lack the ability to synthesize ascorbic acid (vitamin C) and will develop the potentially fatal disease called scurvy—characterized by weakness, spongy gums, and bleeding of the skin and mucous membranes—if vitamin C is not supplied in their diets.

> **Putting the Concept to Work**
> Why are only eight of the 20 amino acids called essential?

BIOLOGY & YOU

Vitamin Supplements: Nutrition in a Pill? It seems like wherever we go these days, we are bombarded with advertisements touting the health benefits of food supplements. Megadoses of vitamin C are said to aid in avoiding colds, and supplements of antioxidant vitamins (A, C, and E) are said to help prevent heart attacks and cancer. Is any of this true? Do dietary supplements make us healthier? According to most health professionals, the answer is no. The American Heart Association states that healthy people get adequate nutrients by eating a healthy diet and recommends no supplements. Clinical trials are under way to see if increased vitamin antioxidant intake may have an overall benefit, but in early results, a large placebo-controlled, randomized study failed to show any benefit from vitamin E on heart disease. While dietary supplements may be necessary if you are a vegetarian or consume fewer than 1,600 calories a day, it appears they are wasted on most of us. The lone exception: omega-3 fatty acid supplements, which have been associated with decreased risk of heart disease. If you don't eat fish (salmon, herring, and trout are rich in omega-3), the American Heart Association suggests this supplement may be of value.

Answering Your Questions About Food Labels

What Is a Calorie?
A calorie measures the amount of energy release by food as it is digested by the human body. On a food label, the Calories per serving indicate the energy available to your body from eating the food in the package. In general, 1 gram of fat contains 9,000 calories (9 kilocalories), whereas carbohydrates and proteins contain about 4 kcal; alcohol contains 7 kcal per gram. It is important to pay attention to the calorie count on a food label, as consuming more food energy than your body uses will cause your body to store the excess calories for a rainy day as body fat—and this will lead quickly to an increase in your body weight. As a good rule of thumb in selecting foods with a low calorie density, look to see that the Calories per serving on the label do not exceed the grams per serving.

Why Is Cholesterol Important?
Cholesterol, a type of lipid molecule, is an essential structural component of cell membranes. Its amount in a membrane—about 30%—determines how fluid the membrane will be. A typical 150-pound individual contains about 35 g of cholesterol, mostly within the body's cell membranes. Your body makes about 1,000 mg of cholesterol a day, and you eat about 300 mg more. It is a good idea to minimize your consumption, because cholesterol tends to lead to plaque deposits within your arteries, a direct cause of heart attacks.

What Are Trans Fats?
Trans fats are fats that have been altered during food production so that their hydrogen atoms are oriented differently. Rare in nature, trans fats are produced as an undesirable by-product when preparing food from vegetable fats. Stick margarine, packaged baked goods, and deep-fried fast food all can be rich in trans fats. Trans fats are dangerous because they are consistently associated with coronary heart disease, a leading cause of death in the United States. On a food label, look to confirm that the amount of trans fat is zero.

What Are Added Sugars?
High levels of sugar are sometimes added to make the flavor of a packaged food more appealing and to extend its shelf life. Notorious among sugar additives is high-fructose corn syrup, often added in high amounts to many products to sweeten the taste. To avoid excessive added sugars, select foods with a food label indicating an amount of added sugars that does not exceed 5% Daily Value.

Why Is Sodium Important?
Americans consume an average of 3,400 milligrams of sodium each day, much more than the 2,300 mg considered healthy. Only a small amount of the sodium you consume comes from the table salt shaker. Most of your daily sodium intake comes from salt added to packaged, processed foods to improve their flavor. And it can be a surprisingly large amount. A single tablespoon of ketchup contains 160 mg of sodium, for example, whereas the two slices of white bread in a sandwich contain 180 mg. What is the problem with eating all this sodium? High salt reduces the ability of your kidneys to remove water; the result is a higher blood pressure due to the extra fluid, a major factor in heart disease. To avoid excess sodium, select foods for which the milligrams of sodium per serving indicated on the food label do not exceed the number of calories per serving indicated on the label.

Source: U.S. Food and Drug Administration

What Does the Label "USDA Organic" Mean?
The U.S. Department of Agriculture (USDA) certifies crops as organic if they were grown without the use of pesticides, petroleum-based fertilizers, or sewage sludge. In addition, chicken, beef, and pork cannot be labeled organic if the birds, cows, or pigs received growth hormones or antibiotics, were genetically modified, or fed genetically modified food. It is important to understand that the food molecules themselves are the same in organic and nonorganic foodstuffs. The "USDA Organic" label is intended to help you identify food whose production does not involve chemicals potentially harmful to the environment or you.

What Does the Label "Free Range" Mean?
The label "free range" is regulated by the USDA only for poultry produced for meat—not for egg-producing chickens or for cattle or pigs. While the label signifies that the individual chicken was not raised in a cage, the requirements are not very high: The label may be used if the chicken had any access to the outdoors each day, even if only for a few minutes. In fact, the label does not ensure that the animal was ever able to roam freely or actually went outdoors at all.

Digestion

25.2 Types of Digestive Systems

LEARNING OBJECTIVE 25.2.1 Contrast intracellular and extracellular digestion, and the specializations of the digestive tract.

Heterotrophs are divided into three groups on the basis of their food sources. Animals that eat plants exclusively are classified as **herbivores**; common examples include cows, horses, rabbits, and sparrows. Animals that are meat eaters, such as cats, eagles, trout, and frogs, are **carnivores**. **Omnivores** are animals that eat both plants and other animals. We humans are omnivores, as are pigs, bears, and crows.

Simple Digestion

Single-celled organisms and lower animals, such as sponges, digest their food intracellularly, breaking down food particles with digestive enzymes inside their cells. Other animals digest their food extracellularly, within a digestive cavity. In this case, the digestive enzymes are released into a cavity that is continuous with the animal's external environment. In flatworms (such as *Planaria*) and cnidarians, such as the hydra in figure 25.3, the digestive cavity has only one opening at the top that serves as both mouth (the red arrow) and anus (the blue arrow). There can be no specialization within this type of digestive system, called a *gastrovascular cavity*, because every cell is exposed to all stages of food digestion.

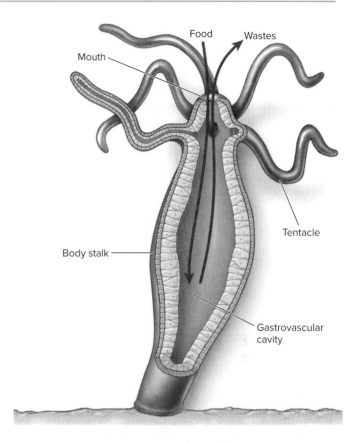

Figure 25.3 The gastrovascular cavity of *Hydra*.

A Digestive Tract

Specialization occurs when the digestive tract, or alimentary canal, has a separate mouth and anus, so that transport of food is one way. Three examples are shown in figure 25.4. The most primitive digestive tract is seen in nematodes (phylum Nematoda), where it is simply a tubular *gut* lined by an epithelial membrane. Earthworms (phylum Annelida) have a digestive tract specialized in different regions for the ingestion, storage (crop), fragmentation (gizzard), digestion, and absorption of food (intestine). All higher animals, such as the salamander and humans like you, show similar specializations.

Chemical Digestion. The process of chemical digestion occurs primarily in the intestine, breaking down the larger food molecules of polysaccharides, fats, and proteins into smaller subunits. Chemical digestion involves hydrolysis reactions that liberate the subunits—primarily monosaccharides, amino acids, and fatty acids—from the food. These products of chemical digestion pass through the epithelial lining of the gut and ultimately into the blood, in a process known as absorption. Any molecules in the food that are not absorbed cannot be used by the animal. These waste products are excreted from the body through the anus.

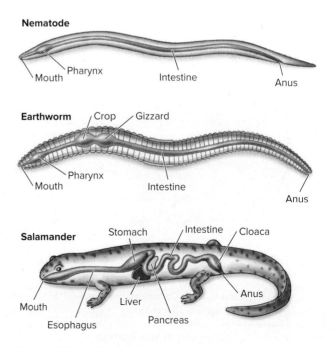

Figure 25.4 One-way digestive tracts.

One-way movement through the digestive tract allows different regions of the digestive system to become specialized for different functions.

> **Putting the Concept to Work**
> Why does specialization not occur within the digestive tract of flatworms (refer to figure 24.1), when it does in earthworms?

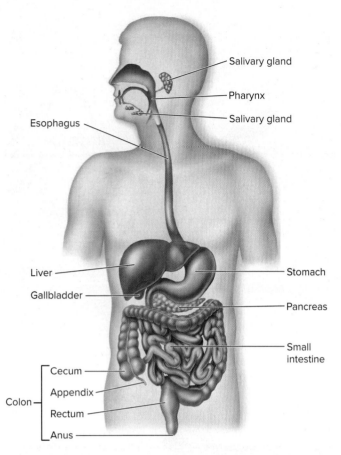

Figure 25.5 The human digestive system.

The tubular gastrointestinal tract and accessory digestive organs are shown. The colon extends from the cecum to the anus.

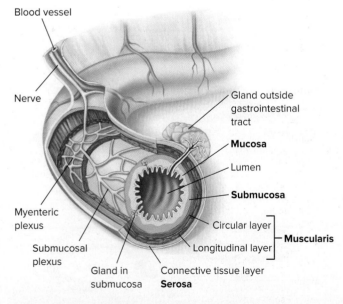

Figure 25.6 The layers of the human gastrointestinal tract.

The mucosa contains a lining epithelium, the submucosa is composed of connective tissue (as is the outer serosa layer), and the muscularis consists of smooth muscles.

25.3 The Human Digestive System

LEARNING OBJECTIVE 25.3.1 Describe the elements of the human digestive system and the layers of the digestive tract.

In humans and other vertebrates, the digestive system consists of a tubular gastrointestinal tract and accessory digestive organs (figure 25.5). Working through the figure from the top down, the initial components of the gastrointestinal tract are the mouth and the pharynx, which is the common passage of the oral and nasal cavities. The pharynx leads to the esophagus, a muscular tube that delivers food to the stomach. From the stomach, where some preliminary digestion occurs, food passes to the first part of the small intestine, where a battery of digestive enzymes continues the digestive process. Accessory digestive organs, such as the liver, gallbladder, and pancreas aid in digestion. The products of digestion pass across the wall of the small intestine into the bloodstream. The small intestine empties what remains into the large intestine, also called the colon, where water and minerals continue to be absorbed.

The Human Intestinal Tube

The tubular gastrointestinal tract of humans and other vertebrates has a characteristic layered structure (figure 25.6). Working from the inside (the lumen) outward, the innermost layer is the mucosa, an epithelium that lines the lumen. The next major tissue layer, composed of connective tissue, is called the submucosa. Just outside the submucosa is the muscularis, which consists of a double layer of smooth muscles. The muscles in the inner layer have a circular orientation, and those in the outer layer are arranged longitudinally. An outer connective tissue layer, the serosa, covers the external surface of the tract. Nerves, intertwined in regions called plexuses, are located in the submucosa and help regulate the gastrointestinal activities.

> **Putting the Concept to Work**
> How is the function of the small intestine different from that of the large intestine?

25.4 The Mouth and Teeth

Specializations of the digestive systems in different kinds of vertebrates reflect differences in the way these animals live. Many vertebrates have teeth (figure 25.7), and chewing (mastication) breaks up food into small particles and mixes it with fluid secretions. Birds, which lack teeth, break up food in a stomach chamber called the gizzard. The gizzard contains small pebbles ingested by the bird that are churned together with the food by muscular action. This churning grinds up the seeds and other hard plant material into smaller bits that can be digested more easily in the intestine.

Teeth

> **LEARNING OBJECTIVE 25.4.1** Describe the interior structure of a tooth and the functions of the four kinds of teeth.

Reptiles and fish have homodont dentition (teeth that are all the same). However, humans and most other mammals have heterodont dentition (teeth of different specialized types): Incisors are chisel-shaped teeth used for nipping and biting; canines are sharp, pointed teeth used for tearing food; and premolars (bicuspids) and molars usually have flattened, ridged surfaces used for grinding and crushing food. The front teeth in the upper and lower jaws of mammals are incisors. On each side of the incisors are the canines. Behind the canines are premolars and then molars.

Diet Determines Teeth. This general pattern of heterodont dentition is modified in different mammals depending on their diet (figure 25.8). For example, in carnivorous mammals, the canines are prominent, and the premolars and molars are more bladelike, with sharp edges adapted for cutting and shearing. Carnivores often tear off pieces of their prey but have little need to chew them because digestive enzymes can act directly on animal cells. (Have you ever noticed how a cat or dog gulps down its food?) By contrast, grass-eating herbivores, such as cows and horses, must pulverize the cellulose cell walls of plant tissue before digesting it. In these mammals, the incisors are used to cut grass and other plants, the canines are reduced or absent, and the premolars and molars are large, flat teeth with complex ridges for grinding.

Human Teeth. Humans are omnivores, and human teeth are adapted for eating both plant and animal food. Viewed simply, humans are carnivores in the front of the mouth and herbivores in the back. Children have only 20 teeth, but these deciduous teeth are lost during childhood and are replaced by 32 adult teeth. The third molars are the wisdom teeth, which usually grow in during the late teens or early twenties, when a person is assumed to have gained some "wisdom."

The Living Tooth. As you can see in figure 25.9, the tooth is a living organ, composed of connective tissue, nerves, and blood vessels, held in place by cementum, a bonelike substance that anchors the tooth in the jaw. The interior of the tooth contains connective tissue called pulp that extends into the root canals and contains nerves and blood vessels. A layer of calcified tissue called dentin surrounds the pulp cavity. The portion of the tooth that projects above the gums is called the crown and is covered with an extremely hard, nonliving substance called enamel. Enamel protects the tooth against abrasion and acids that are produced by bacteria living in the mouth. Cavities form when bacterial acids break down the enamel, allowing bacteria to infect the inner tissues of the tooth.

> **Putting the Concept to Work**
> Why does brushing your teeth prevent tooth decay?

Figure 25.7 Beginning the digestive journey.

There are no photosynthetic animals. All animals must continuously consume plants or other animals in order to live. The passage of food on its journey into and through a mammal begins with the teeth. The grass in this prairie dog's mouth will be chewed before it is ingested and converted within the prairie dog's cells to body tissue, energy, and refuse.

Corbis/VCG/Getty Images

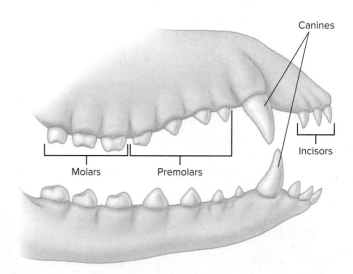

Figure 25.8 Diagram of heterodont dentition.

Different mammals have specific variations of heterodont dentition, depending on whether the mammal is an herbivore, carnivore, or omnivore. In this carnivore, the canines are prominent, and the premolars and molars are pointed—adaptations for tearing and ripping food.

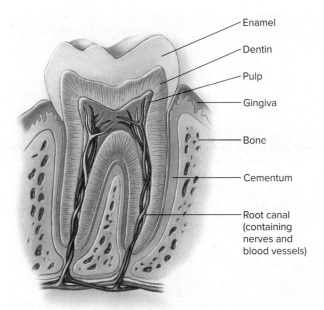

Figure 25.9 Human teeth.

Each vertebrate tooth is alive, with a central pulp containing nerves and blood vessels. The actual chewing surface is a hard enamel layered over the softer dentin, which forms the body of the tooth.

Processing Food in the Mouth

> **LEARNING OBJECTIVE 25.4.2** Describe the role of saliva and the stages of the swallowing process.

Inside the mouth, the tongue mixes food with a mucous solution, called *saliva*. In humans, three pairs of salivary glands secrete saliva into the mouth through ducts in the mouth's mucosal lining. Saliva moistens and lubricates the food so that it is easier to swallow and does not abrade the tissue it passes on its way through the esophagus. Saliva also contains the hydrolytic enzyme *salivary amylase*, which initiates the breakdown of the polysaccharide starch into the disaccharide maltose. This digestion is usually minimal in humans, however, because most people don't chew their food very long.

The secretions of the salivary glands are controlled by the nervous system, which in humans maintains a constant flow of about half a milliliter of saliva per minute when the mouth is empty of food. This continuous secretion keeps the mouth moist. The presence of food in the mouth triggers an increased rate of saliva secretion, as taste-sensitive neurons in the mouth send impulses to the brain, which responds by stimulating the salivary glands. The most potent stimuli are acidic solutions; lemon juice, for example, can increase the rate of salivation eightfold. The sight, sound, or smell of food can stimulate salivation markedly in dogs, but in humans, these stimuli are much less effective than thinking or talking about food.

Swallowing

When food is ready to be swallowed, the tongue moves it to the back of the mouth. In mammals, the process of swallowing begins when the soft palate elevates, pushing against the back wall of the pharynx (figure 25.10). Elevation of the soft palate seals off the nasal cavity and prevents food from entering it ❶. Pressure against the pharynx stimulates neurons within its walls, which send impulses to the swallowing center in the brain. In response, muscles are stimulated to contract and raise the *larynx* (voice box). This pushes the *glottis*, the opening from the larynx into the trachea (windpipe), against a flap of tissue called the *epiglottis* ❷. These actions keep food out of the respiratory tract, directing it instead into the esophagus ❸.

> **Putting the Concept to Work**
> When you breathe through your mouth, air travels down your throat to your lungs. However, when you drink through your mouth, the liquid moving down your throat does *not* travel to your lungs. Why not?

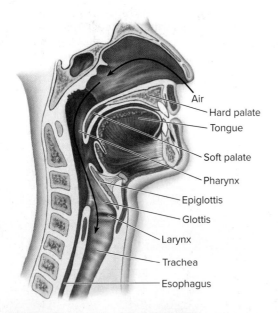

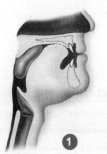
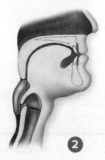
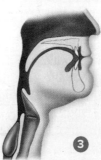

Figure 25.10 The human pharynx, palate, and larynx.

Swallowing triggers the closing of the epiglottis over the trachea, which keeps food and liquids from going down the windpipe.

25.5 The Esophagus and Stomach

LEARNING OBJECTIVE 25.5.1 Outline the structure and function of the esophagus and stomach.

Structure and Function of the Esophagus

Swallowed food enters a muscular tube called the **esophagus,** which connects the pharynx to the stomach. In adult humans, the esophagus is about 25 centimeters long; the upper third is enveloped in skeletal muscle, for voluntary control of swallowing, whereas the lower two-thirds is surrounded by involuntary smooth muscle. The swallowing center stimulates successive waves of contraction in these muscles that move food along the esophagus to the stomach. The muscles relax ahead of the food, allowing it to pass freely, and contract behind the food to push it along, as shown in **figure 25.11**. These rhythmic waves of muscular contraction are called **peristalsis;** they enable humans and other vertebrates to swallow even if they are upside down.

Stomach contents can be brought back up during vomiting, when the **sphincter** (a ring of circular smooth muscle between the stomach and esophagus) is relaxed and the contents of the stomach are forcefully expelled through the mouth. The relaxing of this sphincter can also result in the movement of stomach acid into the esophagus, causing an irritation called *heartburn.* Chronic and severe heartburn is a condition known as *acid reflux.*

Structure and Function of the Stomach

The **stomach** is a saclike portion of the digestive tract (**figure 25.12**). Its inner surface is highly convoluted, enabling it to fold up when empty and open out like an expanding balloon as it fills with food. Thus, while the human stomach has a volume of only about 50 milliliters when empty, it may expand to contain 2 to 4 liters of food when full.

The stomach contains an extra layer of smooth muscle for churning food and mixing it with *gastric juice,* an acidic secretion of the tubular gastric glands of the mucosa. The gastric glands lie at the bottom of deep depressions, the gastric pits shown in the enlargement in **figure 25.12**. These exocrine glands contain two kinds of secretory cells: *parietal cells,* which secrete hydrochloric acid (HCl); and *chief cells,* which secrete pepsinogen, a weak protease (protein-digesting enzyme) that requires a very low pH to be active. This low pH is provided by the HCl. Activated pepsinogen molecules then cleave each other at specific sites, producing a much more active protease, pepsin. This process of secreting a relatively inactive enzyme that is then converted into a more active enzyme outside the cell prevents the chief cells from digesting themselves. It should be noted that only proteins are partially digested in the stomach; there is no significant digestion of carbohydrates or fats there.

Action of Acid

The human stomach produces about 2 liters of HCl and other gastric secretions every day, creating a very acidic solution inside the stomach. The concentration of HCl in this solution is about 10 millimolar, corresponding to a pH of 2. Thus, gastric juice is about 250,000 times more acidic than blood, whose normal pH is 7.4. The low pH in the stomach disrupts hydrogen bonds and so denatures (opens up) food proteins, making them easier to digest, and keeps pepsin maximally active. Active pepsin hydrolyzes

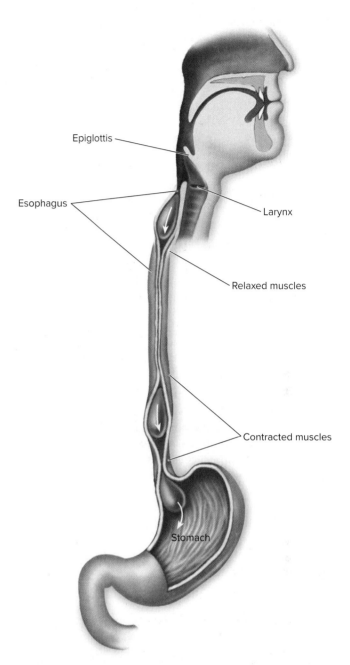

Figure 25.11 The esophagus and peristalsis.

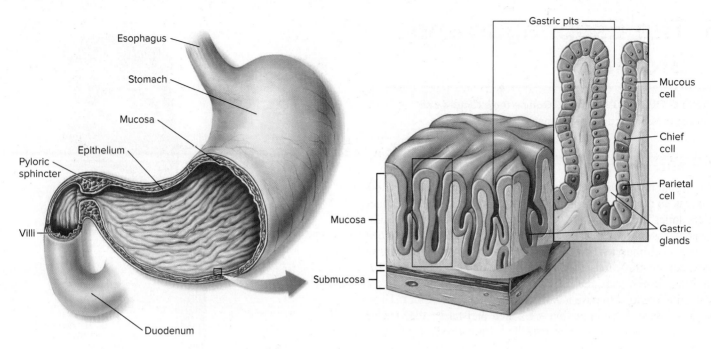

Figure 25.12 The stomach and gastric glands.
Food enters the stomach from the esophagus. The epithelial walls of the stomach are dotted with gastric pits, which contain glands that secrete hydrochloric acid (HCl) and the enzyme pepsinogen. The gastric glands consist of mucous cells, chief cells that secrete pepsinogen, and parietal cells that secrete HCl. Gastric pits are the openings of the gastric glands.

food proteins into shorter chains of polypeptides that are not fully digested until the mixture enters the small intestine. The mixture of partially digested food and gastric juice is called **chyme.**

Ulcers

It is important that the stomach not produce too much acid. If it did, the body could not neutralize the acid later in the small intestine, a step essential for the final stage of digestion. Production of acid is controlled by hormones. These hormones are produced by endocrine cells scattered within the walls of the stomach. The hormone gastrin regulates the synthesis of HCl by the parietal cells of the gastric pits, permitting HCl to be made only when the pH of the stomach is higher than about 1.5.

Overproduction of gastric acid can occasionally eat a hole through the wall of the stomach. Such **gastric ulcers** are rare, however, because epithelial cells in the mucosa of the stomach are protected by a layer of alkaline mucus and because those cells are rapidly replaced by cell division if they become damaged (gastric epithelial cells are replaced every two to three days). Over 90% of gastrointestinal ulcers are **duodenal ulcers,** which are ulcers of the small intestine. These may be produced when the mucosal barriers to self-digestion are weakened by an infection of the bacterium *Helicobacter pylori.* Modern antibiotic treatments can reduce symptoms and often cure the ulcer.

Leaving the Stomach

Chyme leaves the stomach at its base through the *pyloric sphincter* and enters the small intestine. This is where all terminal digestion of carbohydrates, fats, and proteins occurs and where the products of digestion—amino acids, glucose, and fatty acids—are absorbed into the blood.

> **Putting the Concept to Work**
> **Acid causes both ulcers and heartburn. Why do antibiotics cure one and not the other?**

25.6 The Small and Large Intestines

LEARNING OBJECTIVE 25.6.1 Contrast the locations and roles of the duodenum, jejunum, ileum, and large intestine.

Digestion and Absorption: The Small Intestine

The digestive tract exits from the stomach into the **small intestine** (figure 25.13), where large molecules are broken down into small ones. Only relatively small portions of food are introduced into the small intestine at one time to allow time for acid to be neutralized (using bicarbonate released from the pancreas and discussed later) and enzymes to act. The small intestine is the primary digestive organ of the body. Within it, carbohydrates are broken down into simple sugars, proteins into amino acids, and fats into fatty acids. Once these small molecules have been produced, they pass across the epithelial wall of the small intestine into the bloodstream.

Digesting Proteins and Carbohydrates. Some of the enzymes necessary to digest proteins and carbohydrates are secreted by the cells of the intestinal wall. Most, however, are made in a large gland called the *pancreas* (discussed in section 25.7), situated near the junction of the stomach and the small intestine. It is one of the body's major exocrine glands (secreting through ducts). The pancreas sends its secretions into the small intestine through a duct that empties into its initial segment, the **duodenum.** Your small intestine is approximately 6 meters long—unwound and stood on its end, it would be far taller than you are! Only the first 25 centimeters, about 4% of the total length, is the duodenum. It is within this initial segment, where the pancreatic enzymes enter the small intestine, that the majority of digestion occurs.

Digesting Fats. Much of the food energy the vertebrate body harvests is obtained from fats. The digestion of fats is carried out by a collection of molecules known as *bile salts* secreted into the duodenum from the *liver* (discussed in section 25.7). Because fats are insoluble in water, they enter the intestine as drops within the watery chyme. The bile salts, which are partly lipid-soluble and partly water-soluble, work like detergents. They combine with fats to form microscopic droplets in a process called emulsification. These tiny droplets have greater surface areas upon which the enzyme that breaks down fats, called *lipase*, can work. This allows the digestion of fats to proceed more rapidly. The digested fats are first absorbed into lymphatic vessels called lacteals before they later enter the bloodstream.

Reclaiming Water. Two areas make up the rest of the small intestine (96% of its length): the **jejunum** and the **ileum.** Digestion continues into the jejunum, but the ileum is devoted to absorbing water and the products of digestion into the bloodstream. The lining of the small intestine is folded into ridges, as shown in figure 25.13. The ridges are covered with fine fingerlike projections called **villi** (singular, **villus**), each too small to see with the naked eye. In turn, each of the cells covering a villus is covered on its outer surface by a field of cytoplasmic projections called **microvilli.** The enlargement of the villus shows epithelial cells lining the villus, and the further enlargement of these cells shows the microvilli on the surface side of the cells. Scanning and transmission electron micrographs in

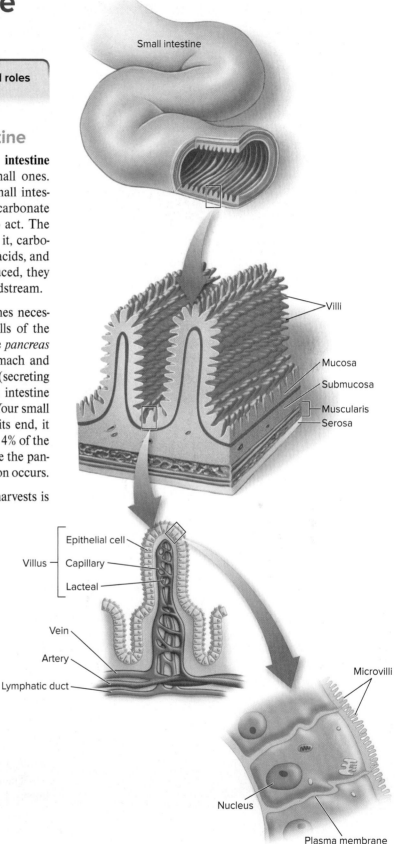

Figure 25.13 The small intestine.

A cross section of the small intestine shows the structure of the villi and microvilli.

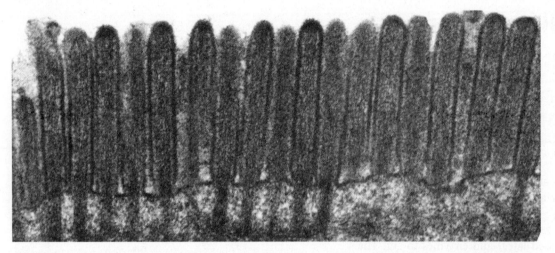

Figure 25.14 Microvilli in the small intestine.

Microvilli, shown in a transmission electron micrograph, are very densely clustered, giving the small intestine an enormous surface area, which is very important for efficient absorption.

Steve Gschmeissner/SPL/Science Source

figure 25.14 give you different perspectives of the microvilli. Both villi and microvilli greatly increase the absorptive surface of the lining of the small intestine. The average surface area of the small intestine of an adult human is about 300 square meters, more than the surface of many swimming pools!

An Active Passageway. The amount of material passing through the small intestine is startlingly large. An average human consumes about 800 grams of solid food and 1,200 milliliters of water per day, for a total volume of about 2 liters. To this amount is added about 1.5 liters of fluid from the salivary glands, 2 liters from the gastric secretions of the stomach, 1.5 liters from the pancreas, 0.5 liters from the liver, and 1.5 liters of intestinal secretions. The total adds up to a remarkable 9 liters—more than 10% of the total volume of your body! However, although the flux is great, the *net* passage is small. Almost all these fluids and solids are reabsorbed during their passage through the small intestine—about 8.5 liters across the walls of the small intestine and 0.35 liters across the wall of the large intestine. Of the 800 grams of solids and 9 liters of liquids that enter the digestive tract each day, only about 50 grams of solids and 100 milliliters of liquids leave the body as feces.

Concentration of Solids: The Large Intestine

The **large intestine,** or **colon,** is much shorter than the small intestine, approximately 1 meter long, but it is called the large intestine because of its larger diameter. The small intestine empties directly into the large intestine at a junction where the cecum and the appendix are located, which are two structures no longer actively used in humans (see figure 25.5). No digestion takes place within the large intestine, and only about 6% to 7% of fluid absorption occurs there. The large intestine is not convoluted, lying instead in three relatively straight segments, and its inner surface does not possess villi. As a consequence, the large intestine has only one-thirtieth the absorptive surface area of the small intestine. Although some water, sodium, and vitamin K are absorbed across its walls, the primary function of the large intestine is to act as a refuse dump. Within it, undigested material, including large amounts of plant fiber and cellulose, is compacted and stored. Many bacteria live and actively divide within the large intestine, where they play a role in the processing of undigested material into the final excretory product, *feces.* Bacterial fermentation produces gas within the human colon at a rate of about 500 milliliters per day. This rate increases greatly after the consumption of beans or vegetable matter because the passage of undigested plant material (fiber) into the large intestine provides substrates for fermentation.

The final segment of the digestive tract is a short extension of the large intestine called the **rectum.** Compact solids within the colon pass through the rectum as a result of the peristaltic contractions of the muscles encasing the large intestine and then out of the body through the **anus.**

> **Putting the Concept to Work**
> If you ingest 800 grams (1.8 lb) of solid food each day and excrete only 50 grams (0.1 lb), why don't you gain a *lot* of weight?

25.7 Accessory Digestive Organs

> **LEARNING OBJECTIVE 25.7.1** Describe the functions of the pancreas, liver, and gallbladder.

The Pancreas

The **pancreas**, a large gland situated near the junction of the stomach and the small intestine (see **figure 25.5**), is one of the accessory organs that contribute secretions to the digestive tract. Fluid from the pancreas is secreted into the duodenum through the *pancreatic duct*, shown in **figure 25.15**. This fluid contains a host of enzymes that digest proteins, starch, and fats. Pancreatic enzymes digest proteins into smaller polypeptides, polysaccharides into shorter chains of sugars, and fats into free fatty acids and other products.

Pancreatic fluid also contains bicarbonate, which neutralizes the HCl from the stomach and gives the chyme in the duodenum a slightly alkaline pH. In addition to its exocrine role in digestion, the pancreas also functions as an endocrine gland, secreting several hormones into the blood that control the blood levels of glucose and other nutrients. These hormones are produced in the **islets of Langerhans,** clusters of endocrine cells scattered throughout the pancreas and shown in the enlarged view in **figure 25.15**. Two pancreatic hormones, insulin and glucagon, are discussed in chapters 26 and 29.

> Endocrine glands release their products, hormones, directly into the bloodstream, as described in section 29.1, whereas exocrine glands release their products into ducts that connect to nearby structures. The pancreatic hormones are discussed in section 29.4.

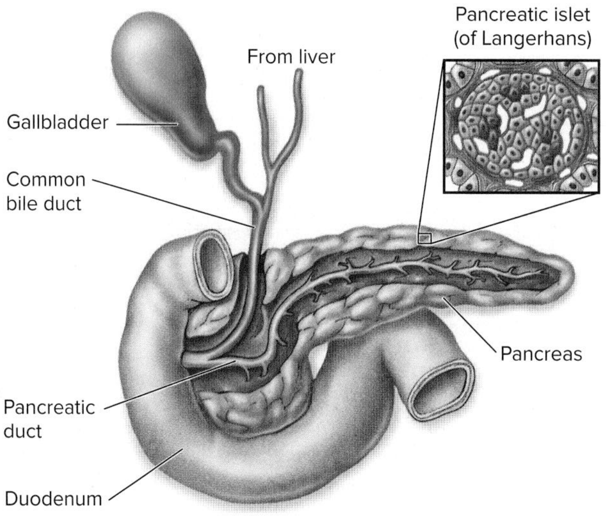

Figure 25.15 The pancreatic and bile ducts empty into the duodenum.

The Liver and Gallbladder

The **liver** is the largest internal organ of the body. In an adult human, the liver weighs about 1.5 kilograms and is the size of a football. The main exocrine secretion of the liver is **bile,** a fluid mixture consisting of *bile pigments* and *bile salts* that is delivered into the duodenum during the digestion of a meal. The bile salts work like detergents, dispersing large drops of fat into a fine suspension of smaller droplets. This breaking up, or emulsification, of the fat into droplets produces a greater surface area of fat upon which the lipase enzymes can act and thus allows the digestion of fat to proceed more rapidly.

After it is produced in the liver, bile is stored and concentrated in the gallbladder (the green organ in **figure 25.15**). The arrival of fatty food in the duodenum stimulates the gallbladder to contract, causing bile to be injected into the duodenum through the common bile duct.

Regulatory Functions of the Liver

A large vein carries blood from the stomach and intestine directly to the liver, which is the first stop for substances absorbed from the gastrointestinal tract. The liver acts as a kind of purification plant. Ingested alcohol and other drugs are taken into liver cells and metabolized; this is why the liver is often damaged as a result of alcohol and drug abuse. The liver also removes toxins, pesticides, carcinogens, and other poisons, converting them into less toxic forms. The liver and all of the digestive organs work together (**figure 25.16**).

> **Putting the Concept to Work**
> If your liver ceases to function ("liver failure"), what happens that makes you seriously ill?

BIOLOGY & YOU

Vegans. We humans are omnivores, meaning we can eat a broad range of plant and animal tissues—but not all of us choose to do so. Some people don't like spinach and love steak, whereas others, called vegetarians, choose not to eat meat. Some become vegetarians because they judge it a more healthy diet—plants are low in saturated fats linked to heart disease. Others make the choice for ethical reasons, sensitive to the animal rights issues associated with livestock agriculture. Still others simply don't like meat. The most extreme form of vegetarian diet is the vegan diet. Vegans avoid all animal proteins. They don't eat red meat, poultry, fish, eggs, or milk. Instead, they obtain all protein and nutrients from grains, vegetables, fruits, legumes, nuts, and seeds. The vegan diet, mirroring that of our early human ancestors, is very challenging because no single fruit, vegetable, or grain contains all the essential amino acids that humans require in their diet. Vegetal foods must be eaten in particular combinations to provide this necessary balance. Beans and rice together provide a balanced diet, but neither food does so when eaten alone. For calcium, which is usually obtained from milk, vegans must eat green leafy vegetables such as broccoli or spinach. In practice, it is not difficult to achieve this balance if a vegan eats a variety of plants.

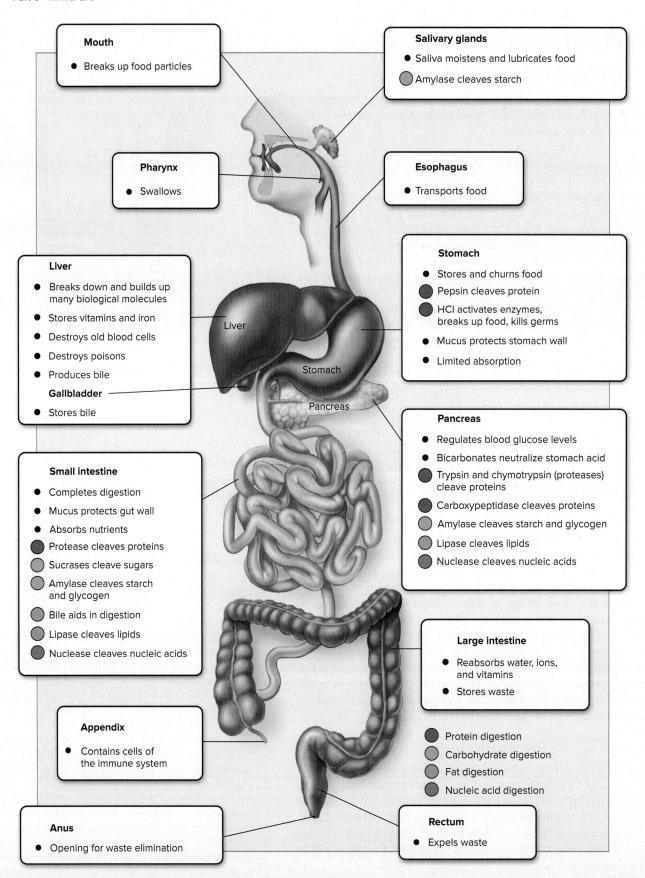

Figure 25.16 The organs of the digestive system and their functions.

The digestive system contains some dozen different organs that act on the food that is consumed, starting with the mouth and ending with the anus. All of these organs must work properly for the body to effectively obtain nutrients.

Today's Biology

Test-Tube Hamburgers?

Progress sometimes takes us to unexpected places, none more surprising than changes in the food we eat. Few things have had as big an impact on our human diet as fast food, and in particular the standardized hamburger. It has been said that today's cow is a biological machine invented by humans to turn grass into hamburgers. On average, Americans eat three hamburgers a week—that's a total of nearly 50 billion burgers per year, a lot of cows!

But what if you didn't need cows to make the burgers? For many years, scientists have been learning how to grow animal cells in tissue culture. The *in vitro* cultivation of muscle cells was performed as early as 1971. It doesn't take a genius to see where this was going: artificial meat! The National Aeronautics and Space Administration (NASA) has been conducting experiments on lab-grown test-tube meat from turkey cells since 2001: In 2002, NASA produced fish filets grown from goldfish cells! Soon laboratories around the world were working on cultured meat research. In 2009, scientists from the Netherlands announced lab-grown pig meat, the first artificial meat patties.

The First Lab-Grown Burger

In August 2013, the world's first test-tube hamburger was cooked and eaten at a news conference in London. Scientists had first harvested muscle stem cells from an adult cow. Called myoblasts, these cells grow rapidly, all of their descendant cells destined to become adult muscle cells. It was necessary to use stem cells because adult muscle cells divide hardly at all. The harvested stem cells are treated by applying a special protein that triggers cell division, then allowed to grow and divide in a liquid culture medium, swirling in a flask where they divide, divide again, and again and again. Each eventually grows into a tiny myofibril of muscle tissue. To create three-dimensional strips of meat, the cells are grown on a scaffold, itself edible so the meat does not have to be removed from it. Gathering them up, the researchers combined 20,000 thin strips of this muscle tissue to make the burger.

PA Photos/ABACA/Newscom

Tasty?

So what was the test-tube burger like? Well, for one thing, it violated the "burger golden ratio" of 80% lean beef to 20% fat, said to make the best burgers luscious and aromatic. These test-tube burgers are 100% protein, 0% fat, making them more chewy and less tasty—professional chefs compared the taste to "meatloaf without any salt and pepper." The first food writer to taste one called the test-tube burger an animal protein cake, tasting like the scientists had mixed soy-based meat substitute with a Big Mac.

Impossible Burgers

Impossible Burgers are vegetable-based patties that look, taste, and even smell like beef. Soy leghemoglobin, which is structurally like the hemoglobin found in your blood, gives the Impossible Burger the color and flavor of beef, and even mimics the juiciness of a hamburger when it is cut.

The Benefits of Eating Meat Grown in Test Tubes

The idea of raising meat in a test tube is known as the clean meat movement. The goal of this movement is to provide meat products for vegetarians and other people who have moral reservations about killing an animal for food. Nor is beef production an efficient use of earth's resources, movement proponents point out. The traditional way of raising animals for meat production consumes approximately one-third of the world's grain supply and requires massive amounts of land for grazing. Growing one cow can expend up to 11,000 gallons of water a year.

Test-tube meat makers hope to be able to raise meat in a way that decreases the environmental impact and animal welfare issues that come with traditional meat production. Their meat technology produces 90% fewer greenhouse emissions than traditional meat agriculture (worldwide, livestock are thought to be responsible for 15% of greenhouse emissions) and does not use antibiotics or the additives common in normal meat production. Overall, they claim, meat grown in test tubes is healthier and more environmentally friendly than traditionally produced meat.

Coming Soon to a Store Near You

Don't hold your breath, however, waiting to taste a test-tube burger. Lab-grown beef and chicken are not at the market yet. Perhaps the biggest hurdle preventing test-tube meats from becoming a staple at cookouts and barbeques is their price tag. Large-scale commercial development is going to require major improvements in scale. Few in the food industry doubt they are coming. The makers of test-tube chicken plan on cutting costs dramatically and making their meat products available to consumers in 2021.

Putting the Chapter to Work

1 What is the BMI of a person who is 5 foot, 11 inches tall and weighs 195 pounds?

Are they overweight?

2 Your instructor gives you a mammalian skull to examine. When observing the teeth, you notice that the premolars and molars are flat and large. There is also an absence of canines.

Based on these findings, what type of diet did this mammal have?

3 Approximately 65% of stroke patients develop a swallowing difficulty called dysphagia. As a result, they are more prone to aspiration of food particles and water. This can cause pneumonia to develop.

What part of the respiratory system that would normally prevent particles from entering the trachea has been damaged?

4 All the organs in the digestive system work together to break down food to provide molecules for energy. If any of these organs are damaged, the digestive system may not function well.

If a person is having difficulty neutralizing the stomach contents before they enter the small intestines, what organ has possibly been damaged?

Retracing the Learning Path

Food Energy and Essential Nutrients

25.1 Food for Energy and Growth

1. Animals consume food as a source of energy and of essential molecules and minerals. For humans, a balanced diet of fruits, vegetables, grains, proteins, and dairy is recommended.

- The energy from food is either used up through metabolic activity or is stored as fat in fat cells. If a person consumes a lot of calories, especially in the form of fats, but doesn't burn off those calories through exercise, the energy is stored as fat. A measurement called a body mass index (BMI) is an easy guide to determining whether a person is overweight or obese.

2. Many animals must consume foods that contain essential amino acids, minerals, and vitamins that the body needs but cannot produce itself.

Digestion

25.2 Types of Digestive Systems

1. All animals are heterotrophs. Herbivores eat exclusively plants, carnivores are meat eaters, and omnivores have diets that contain both plants and meat.

- Single-celled organisms and lower animal phyla, such as sponges, digest food intracellularly. Food is taken up by individual cells and is broken down inside the cells.

- All other animals digest food extracellularly. Digestive enzymes are released into a cavity or tract where they break down food. The products of digestion are then absorbed by cells in the body.

- The evolution of a one-way digestive tract has allowed specialization of the digestive tract, where different regions of the tract are involved in different digestive functions.

25.3 The Human Digestive System

1. Vertebrate digestion occurs in a tubular gastrointestinal tract with specialized areas for different digestive functions. The size or structure of specialized digestive areas varies in different animal groups, depending on the animal's diet.

25.4 The Mouth and Teeth

1. In vertebrates, food is first brought into the mouth. Reptiles and mammals tear and grind food with teeth. Birds break up food in the gizzard, a compartment of their digestive system. In the gizzard, food is churned and ground up with pebbles that the bird has swallowed. In other vertebrates, teeth are used to chew up food, breaking it into smaller pieces.

2. The chewed food mixes with saliva in the mouth. Saliva moistens and lubricates the food, and it contains the enzyme salivary amylase that begins the digestion of starches. The moistened food is then swallowed, passing from the mouth into the esophagus. A flap called the epiglottis closes over the trachea so that the food passes down the esophagus and not into the windpipe.

25.5 The Esophagus and Stomach

1. Food is moved along the esophagus to the stomach by peristaltic waves of muscle contractions. A ring of smooth muscle, called a sphincter, closes off the esophagus from the stomach, keeping food from coming back up.

- In the stomach, muscle contractions churn up the food with gastric juice, which contains hydrochloric acid and pepsin, a protein-digesting enzyme activated by HCl. Gastric juice is produced by two gastric glands.

- Proteins are partially digested in the stomach by pepsin. The acidic conditions in the stomach help denature proteins so they are easier to digest in the small intestine. The partially digested food and gastric juice that leaves the stomach is called chyme.

25.6 The Small and Large Intestines

1. Most digestion occurs in the initial upper portion of the small intestine, called the duodenum. The acidic chyme from the stomach passes into the duodenum, where it is neutralized and mixed with other digestive enzymes. Some enzymes are secreted by the cells that line the walls of the intestine, but most enzymes and other digestive substances are produced in the pancreas or other accessory organs. The rest of the small intestine is involved in absorption of food molecules and water. The lining of the small intestine is folded into ridges that are covered with fingerlike projections called villi. The surface of the cells that line the villi are themselves covered with cytoplasmic projections called microvilli. Villi and microvilli increase the surface area for absorption.

- The large intestine collects and compacts solid waste, releasing it through the rectum and anus.

25.7 Accessory Digestive Organs

1. The pancreas produces protein-digesting enzymes, a starch-digesting enzyme, and a fat-digesting enzyme. The pancreatic fluid also contains bicarbonate, which neutralizes the acidic chyme.

- The liver produces bile (a mixture of bile pigments and bile salts), which breaks down fats. Bile is stored in the gallbladder and released into the small intestine. All of the organs of digestion work together.

Inquiry and Analysis

Why Do Diabetics Excrete Glucose in Their Urine?

Late-onset diabetes is a serious and increasingly common disorder in which the body's cells lose their ability to respond to insulin, a hormone that is needed to trigger their uptake of glucose. As illustrated below, the binding of insulin to a receptor in the plasma membrane causes the rapid insertion of glucose transporter channels into the plasma membrane, allowing the cell to take up glucose. In diabetics, however, glucose molecules accumulate in the blood while the body's cells starve for the lack of them. In mild cases, blood glucose levels rise to several times the normal value of 4 mM; in severe untreated cases, blood glucose levels may become enormously elevated, up to 25 times the normal value. A characteristic symptom of even mild diabetes is the excretion of large amounts of glucose in the urine. The name of the disorder, *diabetes mellitus*, means "excessive secretion of sweet urine." In normal individuals, by contrast, only trace amounts of glucose are excreted. The kidney very efficiently reabsorbs glucose molecules from the fluid passing through it. Why doesn't it do so in diabetic individuals?

The graph displays so-called glucose tolerance curves for a normal person (*blue line*) and a diabetic (*red line*). After a night without food, each individual drank a test dose of 100 grams of glucose dissolved in water. Blood glucose levels were then monitored at 30-minute and 1-hour intervals. The dotted line indicates the kidney threshold, the maximum concentration of blood glucose molecules (about 10 mM) that the kidney is able to retrieve from the fluid passing through it when all of its glucose-transporting channels are being utilized full-bore.

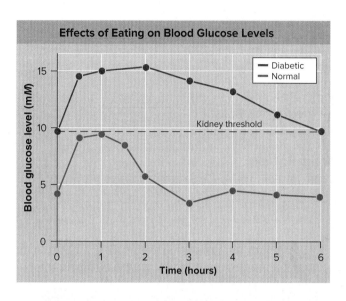

Analysis

1. **Applying Concepts**
   Reading a Curve. What is the immediate impact on the normal individual's blood glucose levels of consuming the test dose of glucose? How long does it take for the normal person's blood glucose level to return to the level before the test dose?

   Comparing Curves. Is the impact any different for the diabetic person? How long does it take for the diabetic person's blood glucose levels to return to the level before the test dose?

2. **Interpreting Data**
   a. Is there any point at which the normal individual's blood glucose levels exceed the kidney threshold?
   b. Is there any point at which the diabetic individual's blood glucose levels do *not* exceed the kidney threshold?

3. **Making Inferences**
   a. Why do you suppose the diabetic individual took so long to recover from the test dose?
   b. Would you expect the normal individual to excrete glucose? Explain. The diabetic individual? Explain.

4. **Drawing Conclusions** Why do diabetic individuals secrete sweet urine?

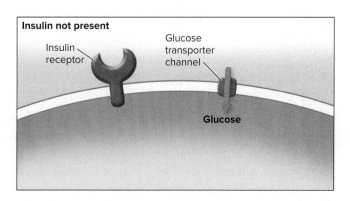

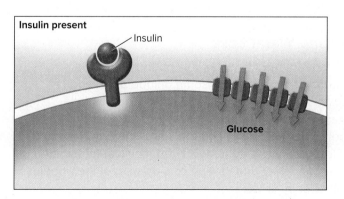

26 Maintaining the Internal Environment

LEARNING PATH ▼

Homeostasis
1. How the Animal Body Maintains Homeostasis

Osmoregulation
2. Regulating the Body's Water Content
3. Eliminating Nitrogenous Wastes

A Kangaroo Rat Never Drinks

The kangaroo rat (genus *Dipodomys*) is native to the deserts of North America. Like its namesake, it hops around on its long hind legs. It uses the long tail as a rudder to help balance and control its direction while hopping—the kangaroo rat can make a 90-degree turn in midflight using its long, bushy tail. However, under this cute and cuddly exterior is a water-conservation machine. A kangaroo rat obtains and conserves water so efficiently that it doesn't have to drink at all. An adult kangaroo rat may never have a single sip of water its whole life and wouldn't know what to do if confronted with a bowl of water, but its body has about the same water content as other desert animals. What is its secret? Water management: reducing water loss, squeezing out every drop of water that is available, and recycling water in their bodies.

Reducing Water Loss

Kangaroo rats reduce water loss both by modifying their behavior and through some unique physiological adaptations. Like many desert animals, they escape the heat by burrowing deep underground where the temperatures are cooler and the soil or sand is more humid. They forage only at night, which reduces water loss through evaporation. They don't sweat or pant to cool their bodies like other animals do.

IN ALL HIS life, this kangaroo rat has had not one drink of water, and it never will.
Rick & Nora Bowers/Alamy Stock Photo

Body adaptations help to further reduce water loss. Kangaroo rats have external cheek pouches that they use to store food during foraging. They will stuff seeds in these external pouches and then empty them upon returning to their dens. Most other rodents have internal cheek pouches, which require opening up the mouth to fill and empty the pouches; every time the mouth opens, valuable water escapes from the moist lining of the mouth. This is water that the kangaroo rat can't afford to lose.

Where Does the Water Come From?

Reducing water loss cannot be the whole story, however. Where does the kangaroo rat's body water come from in the first place, if it never drinks? From the food it eats. Kangaroo rats literally eat their moisture, extracting water from the food molecules themselves.

Tom McHugh/Science Source

Tom McHugh/Science Source

Design Pics Inc/Alamy Stock Photo

Most of a kangaroo rat's diet consists of dry seeds. They usually store these dry seeds for a while in their relatively humid dens before eating them. The seeds act like sponges, absorbing moisture from the humid air in the den.

The Secret: Metabolic Water

The real source of water, however, and the key to a kangaroo rat's ability to live without drinking, is embedded within the carbohydrates, fats, and proteins of the seeds it eats. Metabolizing these molecules produces a constant amount of water through oxidative respiration. Recall from chapter 7 that water is a chemical by-product of cellular respiration. This "metabolic water" is expelled from your body when you breathe, the water vapor expelled from the lungs with the exhaled air. Not so with the kangaroo rat. The surface area of its nasal passages is cooler than the air being exhaled. Water condenses on these surfaces where it is collected and returned to the body through swallowing.

Recycling

The kangaroo rat will also recycle water from its body by recapturing much of the water in the kidneys before it is expelled in the urine. To fully understand this final elegant adaptation, you are going to have to first look closely in this chapter at how a mammal's kidneys function. Like the kangaroo rat, you avoid losing water by recapturing it as it passes through your kidneys.

Homeostasis

26.1 How the Animal Body Maintains Homeostasis

> **LEARNING OBJECTIVE 26.1.1** Define homeostasis, and explain how negative feedback loops help maintain it.

The animal body is a sophisticated machine, its many kinds of cells finely tuned to carry out precise roles. Such specialization of cell function is possible only when extracellular conditions are kept within narrow limits. Temperature, pH, the concentration of glucose and oxygen, and many other factors must be held constant for cells to function efficiently.

Homeostasis may be defined as the dynamic constancy of the internal environment. The term *dynamic* is used because conditions are never absolutely constant but fluctuate continuously within narrow limits. Homeostasis is essential for life.

Negative Feedback Loops

To maintain internal constancy, the vertebrate body must have sensors that are able to measure each condition of the internal environment (indicated by the light green box in **figure 26.1**). Sensors constantly monitor the extracellular conditions and relay this information (usually via nerve signals) to an integrating center that contains the *set point,* which is the proper value for that condition. This set point is analogous to the temperature setting on a house thermostat. In the body, there are set points for temperature, blood glucose concentration, the tension on a tendon, and so on. The integrating center is often a particular region of the brain or spinal cord, but in some cases, it can also be cells of endocrine glands. It receives messages from several sensors, weighs the relative strengths of each sensor input, and then determines whether the value of the condition is deviating from the set point. When a deviation occurs, the integrating center sends a message to increase or decrease the activity of particular effectors. Effectors (the blue box in the figure) are generally muscle or glands and can change the value of the condition back toward the set point value, which is "the response."

For example, if the human body temperature exceeds the set point of 37°C, sensors in the brain relay this information to an integrating center (also in the brain), which then stimulates effectors, such as sweat glands, that act to lower body temperature. This response feeds back to the integrating center, which then stops stimulating the effectors. Because the activity of the effectors is influenced by the effects they produce, this type of control system is known as a negative feedback loop (**figure 26.1**).

The nature of the negative feedback loop becomes clear when we again refer to the analogy of the thermostat in a house. After an air conditioner has been on for some time, the room temperature may fall significantly below the set point of the thermostat. When this occurs, the air conditioner will be turned off. The effector (air conditioner) is turned on by high temperature; and when activated, it produces a negative change (lowering of the temperature) that ultimately causes the effector to be turned off.

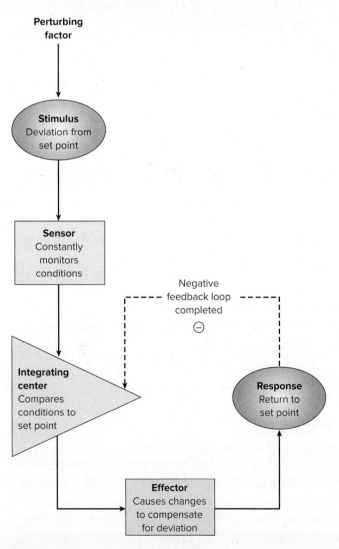

Figure 26.1 A generalized diagram of a negative feedback loop.

Negative feedback loops maintain a state of homeostasis, or dynamic constancy of the internal environment, by correcting deviations from a set point.

> **Putting the Concept to Work**
> Diagram how a negative feedback loop might be used in winter to maintain warm temperatures in a house heated by hot-water radiators.

26.3 Eliminating Nitrogenous Wastes

> **LEARNING OBJECTIVE 26.3.1** Contrast the ways in which fish, reptiles, and mammals eliminate nitrogenous wastes from their bodies.

Amino acids and nucleic acids are nitrogen-containing molecules. When animals catabolize these molecules for energy or convert them into carbohydrates or lipids, they produce nitrogen containing by-products called *nitrogenous wastes* that must be eliminated from the body.

Ammonia

The first step in the metabolism of amino acids and nucleic acids is the removal of the amino (–NH$_2$) group and its combination with H$^+$ to form **ammonia** (NH$_3$) in the liver, ❶ in figure 26.6. Ammonia is quite toxic to cells and therefore is safe only in very dilute concentrations. The excretion of ammonia is not a problem for the bony fish and tadpoles, which eliminate most of it by diffusion through the gills and the rest by excretion in very dilute urine ❷.

Urea

In sharks, adult amphibians, and mammals, the nitrogenous wastes are eliminated in the far less toxic form of **urea** ❸. Urea is water-soluble and so can be excreted in large amounts in the urine. It is carried in the bloodstream from its place of synthesis in the liver to the kidneys, where it is excreted in the urine.

Uric Acid

Reptiles, birds, and insects excrete nitrogenous wastes in the form of **uric acid** ❹, which is only slightly soluble in water. As a result of its low solubility, uric acid precipitates and thus can be excreted using very little water. Uric acid forms the pasty white material in bird droppings.

> **Putting the Concept to Work**
> What do birds gain by not eliminating their nitrogenous wastes as urea, as mammals do?

Figure 26.6 Nitrogenous wastes.

When amino acids and nucleic acids are metabolized, the immediate nitrogen by-product is ammonia ❶, which is quite toxic but can be eliminated through the gills of bony fish ❷. Mammals convert ammonia into urea ❸, which is less toxic. Birds and terrestrial reptiles convert it instead into uric acid ❹, which is insoluble in water.

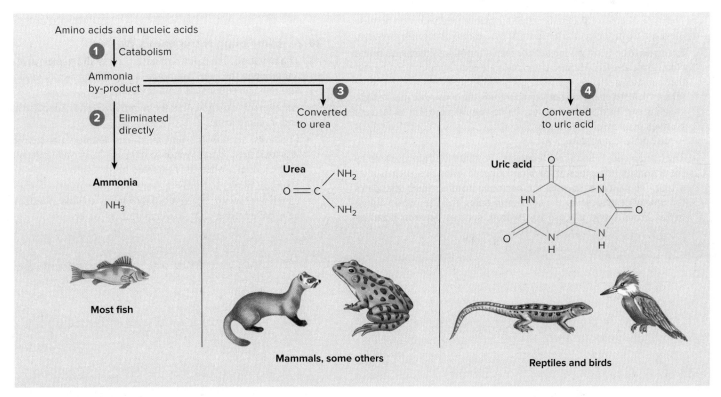

Putting the Chapter to Work

1 Type 1 diabetes is also known as insulin-dependent diabetes. Individuals with type 1 diabetes are not able to secrete insulin from the pancreas.

If insulin is not substituted by injections, will the patients' blood sugar be high or low?

2 Kidney failure is often caused by diseases such as diabetes or high-blood pressure. The most severe stage of kidney failure is called chronic kidney disease. In some cases of chronic kidney disease, urine can back up from the bladder and into the ureters and kidneys.

Chronic kidney failure is often caused by type 2 diabetes, your body progressively loosing its ability to respond to insulin.

Untreated, what would be the effect of kidney failure on your body's functioning?

3 Dialysis is an option for patients who have chronic kidney disease. A dialysis treatment is designed to purify the blood by removing waste and excess fluid.

What part of the nephron is dialysis taking the place of?

4 To change the pH in soil, some farmers add chicken or bird droppings.

What component in the droppings will change the pH of the soil?

Retracing the Learning Path

Homeostasis

26.1 How the Animal Body Maintains Homeostasis

1. Animals maintain relatively constant internal conditions, a process called homeostasis, a dynamic constancy of the internal environment within narrow ranges.
   - The body uses sensors to measure internal conditions. If the measurement is deviating from the brain's set point, the body will make adjustments through muscles or glands to bring the body back to the set point.
   - This type of control system is called a negative feedback loop because the response to correct the deviation is in a reverse or negative direction.
2. Examples of homeostasis include the regulation of body temperature and blood glucose levels. Mammals and birds are endothermic, meaning that they maintain a relatively constant body temperature. When temperature sensors detect a change in body temperature, the body responds through sweating and the dilation of surface blood vessels in the skin to decrease body temperature or shivering and a constriction of surface blood vessels to raise body temperature.
   - The human body maintains a relatively constant blood glucose level by the actions of hormones. If the blood glucose levels rise, such as after eating, the pancreas releases the hormone insulin, which stimulates the uptake of glucose by the cells in muscles, the liver, and adipose tissue. When blood glucose levels drop, such as between meals or during exercise, the hormone glucagon is released from the pancreas or adrenaline from the adrenal glands. These hormones trigger the release of glucose stores from the liver.

Osmoregulation

26.2 Regulating the Body's Water Content

1. Kidneys are the excretory organs in vertebrates. They filter fluid under pressure and then reabsorb molecules from the filtrate that are important to the body.
2. Each kidney is composed of roughly 1 million nephrons. Each nephron contains a Bowman's capsule, a loop of Henle, and a collecting duct. The function of the nephron is to collect the filtrate from the blood, selectively reabsorb ions and water, and expel the waste from the body.
   - The blood is first filtered from the glomerulus, a network of blood capillaries encapsulated by a structure called Bowman's capsule. Water, small molecules, ions, and urea pass into the capsule. The Bowman's capsule is connected to a long tubule where selected molecules and ions are reabsorbed by the body. The tubule forms a hairpin loop called the loop of Henle. The tubule empties into a collecting duct, which recovers additional water from the filtrate.
   - Different areas of the loop of Henle are permeable to different substances. In the proximal tubule, much water and important molecules are reclaimed. In the descending arm of the loop of Henle, water passes back into the surrounding tissue by osmosis, but the loop at this point is impermeable to either salts or urea. At the turn of the loop, the walls become permeable to salts and other molecules, which pass out of the tubules and into the surrounding tissue. This results in a higher concentration of solutes in the tissue surrounding the nephron tubule.
   - As the filtrate passes into the collecting duct, whose walls are permeable to water, water is absorbed back into the tissues through osmosis. Osmosis occurs because the surrounding tissue is highly concentrated with solutes such as ions and other molecules that were reabsorbed from the ascending loop.
   - The fluid that remains in the collecting duct is a highly hypertonic urine that passes to the ureter and is excreted from the body.

26.3 Eliminating Nitrogenous Wastes

1. The metabolic breakdown of amino acids and nucleic acids produces ammonia in the liver. Ammonia, a nitrogenous waste product, is toxic and must be eliminated from the body.
   - Ammonia is excreted directly in bony fish and tadpoles through gills and in dilute urine.
   - In sharks, amphibians, and mammals, ammonia is converted to the less toxic urea. Urea is water-soluble and is carried in the bloodstream to the kidneys, where it is excreted in the urine.
   - Reptiles, birds, and insects excrete nitrogenous wastes as uric acid, which has low toxicity and is only slightly soluble in water and so can be excreted using only small amounts of water.

Inquiry and Analysis

How Do Sleeping Birds Stay Warm?

Mammals and birds are endothermic: They maintain relatively constant body temperatures regardless of the temperature of their surroundings. This lets them reliably run their metabolism even when external temperatures fall—the rates of most enzyme-catalyzed reactions slow two- to threefold for every 10°C temperature drop. Your body keeps its temperature within narrow bounds at 37°C (98.6°F), and birds maintain even higher temperatures. To stay warm like this, mammals and birds continuously carry out oxidative metabolism, which generates heat. This requires a several-fold increase in metabolic rate, which is expensive, particularly when the animal is not active. The logical solution is to give up the struggle to keep warm and let the body temperature drop during sleep, a condition known as *torpor*. Humans don't adopt this approach, but many other mammals and birds do. This raises an interesting question: What prevents a sleeping bird in torpor from freezing? Does its body simply adopt the temperature of its surroundings, or is there a body temperature below which metabolic heating kicks in to avoid freezing?

Glenn Bartley/All Canada Photos/Alamy Stock Photo

The graph displays an experiment examining this issue in the tropical hummingbird *Eulampis*. The study examines the effect on metabolic rate (measured as oxygen consumption) of decreasing air temperature. Oxygen consumption was assessed over a range of air temperatures from 3° to 37°C for two contrasting physiological states: The blue data were collected from birds that were awake, the red data from sleeping torpid birds. The blue and red lines, called regression lines, were plotted using curve-fitting statistics that provide the best fit to the data.

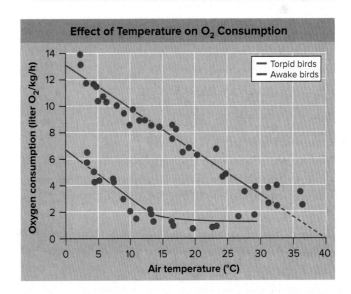

Analysis

1. **Applying Concepts**
   Comparing Two Data Sets. Do awake hummingbirds maintain the same metabolic rate at all air temperatures? Sleeping torpid ones? At a given temperature, which has the higher metabolic rate, an awake bird or a sleeping one?

2. **Interpreting Data**
   a. How does the oxygen consumption of awake hummingbirds change as air temperature falls? Why do you think this is so? Is the change consistent over the entire range of air temperatures examined?
   b. How does the oxygen consumption of sleeping torpid birds change as air temperature falls? Is this change consistent over the entire range of air temperatures examined? Explain any difference you detect.
   c. Are there any significant differences in the slope of the two regression lines below 15°C? What does this suggest to you?

3. **Making Inferences**
   a. For each five-degree air temperature interval, estimate the average oxygen consumption for awake and for sleeping birds, and plot the difference as a function of air temperature.
   b. Based on this curve, what would you expect to happen to a sleeping bird's body temperature as air temperatures fall from 30° to 20°C? From 15° to 5°C?

4. **Drawing Conclusions** How do *Eulampis* hummingbirds avoid becoming chilled while sleeping on cold nights?

27 How the Animal Body Defends Itself

LEARNING PATH ▼

Three Lines of Defense
1. Skin: The First Line of Defense
2. Cellular Counterattack: The Second Line of Defense
3. Specific Immunity: The Third Line of Defense

The Immune Response
4. Initiating the Immune Response
5. T Cells: The Cellular Response
6. B Cells: The Humoral Response
7. Active Immunity Through Clonal Selection
8. Vaccination
9. Antibodies in Medical Diagnosis

Defeat of the Immune System
10. Overactive Immune System
11. AIDS: Immune System Collapse

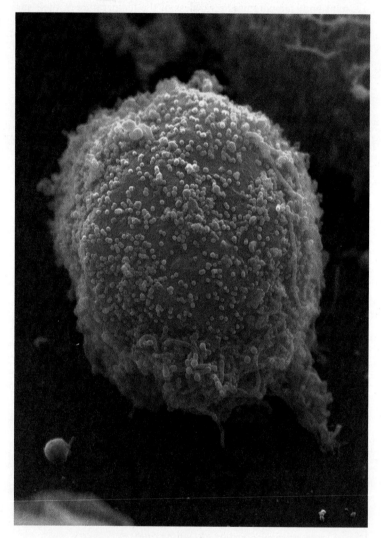

HIV PARTICLES EMERGING from an infected human blood cell.
Eye of Science/Science Source

AIDS: Shooting at a Moving Target

Since the AIDS epidemic burst upon us in 1981, scientists have feverishly sought a vaccine to protect people from this deadly and incurable disease. But the path to a vaccine has not been easy. Nearly 40 years and 1 million American AIDS cases later, an effective AIDS vaccine still eludes the best efforts of researchers.

The Need for an AIDS Vaccine

What is a vaccine, and why is an AIDS vaccine proving so difficult to develop? As you will discover in this chapter, the body wages an all-out war against bacteria and viruses when they invade the body, using cells to attack them and chemicals to kill them. This defense is much stronger when the body knows the enemy. It can elicit a stronger and swifter response to a returning invader. A vaccine uses dead or weakened microbes, ones that can't cause the disease but can alert the body to a future invasion. The body identifies what the microbe "looks like" and defends itself with cells and proteins called antibodies that attack only that microbe. Later, if the real thing should enter the body, looking the same as the microbe in the

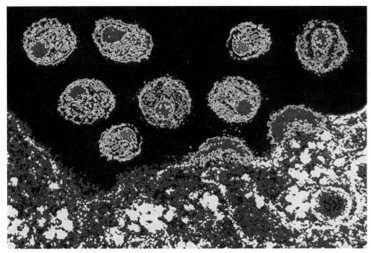

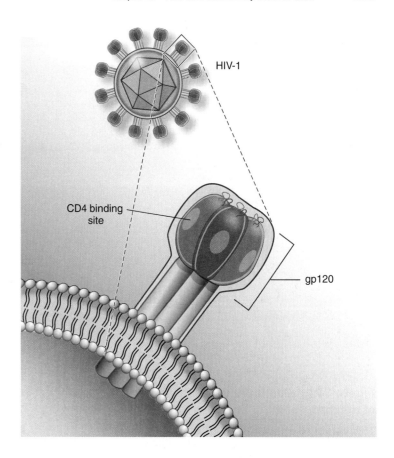

vaccine, the body is ready to renew the war, eliminating the invading microbe before the infection can gain a foothold in the body.

Vaccinations have been highly effective in wiping out certain diseases. Smallpox was eradicated worldwide through an aggressive use of vaccinations. Many illnesses such as polio that were commonplace 25 to 50 years ago are now rare in the United States. If these vaccinations are so effective and researchers have succeeded in combating many different diseases using them, why not AIDS?

A Difficult Challenge

One problem has thwarted all efforts: The HIV virus that causes AIDS has an unusually mistake-prone DNA copying enzyme, making mistakes here and there as it produces offspring virus particles. HIV generates mutations at a prodigious rate. The rate of change is so great that the HIV viruses you see budding from a single white blood cell in the photo above may not all be the same. A vaccine targeted against one version of HIV may be ineffective against others. Like a thief of many disguises, at least a few of the HIV particles in an infection are able to dodge any antibody a vaccine throws at them. Shooting at a moving target, AIDS researchers have been unable to develop a vaccine that is effective.

A Possible Solution

For nearly 40 years, this problem has seemed insurmountable; but recently, researchers are closing in on a solution, seeking out antibodies that neutralize a broad range of HIV strains. The search for broadly neutralizing antibodies was revolutionized in 2009 by the introduction of techniques for isolating antivirus antibodies from single cells—so-called monoclonal antibodies. Using these new approaches, researchers have discovered certain sites on HIV's surface that must be conserved for the virus to be infective. Antibodies directed against these sites target a broad range of HIV strains.

Recipe for an AIDS Vaccine

The first of these hidden weaknesses to be discovered was on a key HIV surface protein: gp120. Despite the great diversity in sequence and shape that the gp120 envelope protein presents, the *Env* gene that encodes it must conserve certain elements of a basic protein shape, or gp120 would lose its ability to bind to the CD4 receptor, and HIV could no longer dock on human cells. Wedged between two arms of the V-shaped gp120 protein is a recessed cavity into which a portion of the CD4 receptor protein fits when first binding to gp120. This section of gp120 acts as an "on" switch for the binding event and appears to be always conserved in successfully attaching HIV viruses. It presents a unique site of HIV vulnerability and offers the exciting possibility that vaccines using antibodies directed against it may successfully neutralize HIV. The moving target is in our sights.

Three Lines of Defense

Figure 27.1 Skin is the vertebrate body's first line of defense.

This young elephant has tough, leathery skin thicker than a belt, allowing it to follow the herd through dense thickets without injury.

Don Farrall/Photodisc/Getty Images

BIOLOGY & YOU

Pimples. Many of us as teenagers develop acne, our face pocked with blemishes called pimples. A pimple is the result of a blockage of one of the skin's pores. What causes them? Not stress, despite what you might read in magazines. Inside the pores of your face's skin are sebaceous glands that become active around puberty, stimulated by male hormones from the adrenal glands of both boys and girls. Activated, these glands begin to produce an oil called sebum, which protects your aging skin. Unfortunately, when the outer layers of your skin shed (as they do continuously), sometimes the dead skin cells left behind become "glued" together by the sebum, causing a blockage in the pore. When this happens, the sebum continually being produced by the sebaceous gland builds up behind the blockage, causing a pimple to form. Bacteria grow in the oil within the pimple, causing the surrounding tissues to become inflamed. If the oil breaks through to the surface, the result is a "whitehead." If the oil accumulates melanin pigment or becomes oxidized, the oil changes from white to black, and the result is a "blackhead." Blackheads are not dirt and do not reflect poor hygiene. How do you treat acne? Lots of scrubbing to keep the affected area clean, plus over-the-counter medications such as benzoyl peroxide and salicylic acid that help the skin slough off more easily, plus patience. In time, acne tends to fade, although severe cases may require consulting a physician.

27.1 Skin: The First Line of Defense

> **LEARNING OBJECTIVE 27.1.1** Describe how the two layers of skin protect the body from infection.

Multicellular bodies offer a feast of nutrients for tiny, single-celled creatures, as well as a sheltered environment in which they can grow and reproduce. We live in a world awash with bacteria and viruses (microbes). Some are **pathogens,** causing disease and wreaking havoc in the body; no animal can long withstand their onslaught unprotected. Animals survive because they have a variety of very effective defenses against this constant attack.

Overview of the Three Lines of Defense

The vertebrate body is defended from infection the same way knights defended medieval cities. "Walls and moats" make entry difficult; "roaming patrols" attack strangers; and "guards" challenge anyone wandering about and signal for an attack if a proper "ID" is not presented.

1. **Walls and moats.** The outermost layer of the vertebrate body, the skin, is the first barrier to penetration by microbes. Mucous membranes in the respiratory and digestive tracts are also important barriers that protect the body from invasion.
2. **Roaming patrols.** If the first line of defense is penetrated, the response of the body is to mount a *cellular counterattack,* using a battery of cells and chemicals that kill microbes. These defenses act very rapidly after the onset of infection.
3. **Guards.** Lastly, the body is also policed by cells that circulate in the bloodstream and scan the surfaces of every cell they encounter. They are part of the *specific immune response.* One kind of immune cell aggressively attacks and kills any cell identified as foreign or infected with viruses, whereas the other type marks the foreign cell or virus for elimination by the roaming patrols.

The Skin

Skin, like the thick, tough skin of the elephants in figure 27.1, is the outermost layer of the vertebrate body and provides the first defense against invasion by microbes. Skin is our largest organ, comprising some 15% of our total weight. One square centimeter of skin from your forearm (about the size of a dime) contains 200 nerve endings, 10 hairs and muscles, 100 sweat glands, 15 oil glands, 3 blood vessels, 12 heat-sensing organs, 2 cold-sensing organs, and 25 pressure-sensing organs.

A Barrier to Infection. The section of skin you see in figure 27.2 has two distinct layers: an outer epidermis and a lower dermis. A subcutaneous layer lies underneath the **dermis.** Cells of the outer **epidermis** are continually being worn away and replaced by cells moving up from below—in one hour, your body loses and replaces approximately 1.5 million skin cells!

Chemical Defenses. The skin not only defends the body by providing a nearly impermeable barrier, but also reinforces this defense with

Answering Your Questions About Plastic Microbeads

What Are Plastic Microbeads?

Plastic microbeads are exactly what their name says: tiny balls of plastic. Colorful and harmless-looking, each sphere is almost too small to see (less than a millimeter in diameter). But they are not in fact harmless. The harm is indirect (you could eat the stuff) but serious and derives from the nature of the material used to make the tiny balls—plastic. Plastic is not found in nature. Invented in 1907 and widely used after the Second World War, plastic is a synthetic material derived from petrochemicals. It is cheap and easy to make, hard, and impervious to water. It is also very stable and not usually biodegradable. And therein lies the problem.

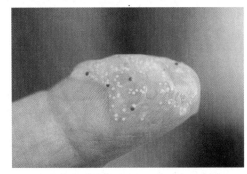
imagegallery2/Alamy Stock Photo

What Products Contain Plastic Microbeads?

Plastic microbeads are a relatively new product, added as an exfoliating agent to cosmetic and personal care products such as facial scrubs, face soaps, and toothpastes. Exfoliation is the removal of the oldest dead skin cells on the skin's surface, said by its proponents you see in TV ads to restore a healthy glow and skin tone. Plastic microbeads are also often included in "age-defying" makeup, as well as lip gloss and nail polish. Being tiny and spherical, plastic microbeads create a ball-bearing effect on creams and lotions, resulting in a silky texture and ease of spreading.

So What's the Problem?

The problem is that the tiny plastic beads are nearly indestructible. Practically every plastic microbead ever made still exists—somewhere. Washed off from your face or hands, they travel down the drain, so small they pass unfiltered through sewage treatment plants, making their way into rivers and lakes, and eventually into the earth's oceans. A single shower can flush as many as 100,000 microbeads. When they are being used by millions of people daily, that adds up pretty quickly.

Isn't Plastic Safe?

About 8 million tons of plastic enter the world's oceans each year. By 2020, there will be more plastic in seawater than fish! But plastic is inert, isn't it? Messy but not dangerous, right? Wrong. The problem is the tiny size of plastic microbeads. In lakes and rivers, the little beads absorb and concentrate pollutants and pesticides. Researchers have found that fish in the Great Lakes eat these polluted plastic microbeads instead of real food such as zooplankton, which are the same size. This unnatural diet is not only toxic to the fish, but also to other animals farther up the food chain. Like you. Plastic microbeads have reached high concentrations in the Great Lakes. A 2017 study found as many as 1.1 million microbeads per square mile on the surface of the Great Lakes. That's 94,250 square miles at 1.1 million microbeads each. A lot of microbeads.

How Do I Know If I Am Contributing to This Problem?

You have probably seen many products containing plastic microbeads on store shelves. It's not always obvious, of course. The labels don't shout "MICROBEADS!" However, the Food and Drug Administration requires that all products list their ingredients. If you see any of the following ingredients on a face-cleansing product—polyethylene, polypropylene, polyethylene terephthalate, or polymethyl methacrylate—you would be cleaning your face with plastic microbeads. As governments have begun banning microbeads, cosmetic companies are phasing out microbeads in their products.

Is Anything Being Done to Stop the Use of Plastic Microbeads?

Yes. In 2015, Congress enacted the Microbead-Free Waters Act, prohibiting the manufacture, packaging, and distribution of rinse-off cosmetics containing plastic microbeads, to take effect in 2017–2018. Great Britain, Canada, Australia, and New Zealand soon followed suit. Other countries have not been as quick to act.

What About All the Plastic Microbeads Already Released into the Environment?

Untreated, microbeads will remain a serious form of environmental pollution for years to come. Fortunately, scientists have recently stumbled onto a way to potentially reduce existing plastic pollution—an enzyme that will eat plastic! They made the discovery in 2018 while examining microbes found in a Japanese waste recycling center. Studying a microbe that seemed to be able to live on plastic, they isolated the enzyme in the microbe that interacted with plastic. Attempting to understand the enzyme's structure, the researchers accidentally engineered an enzyme that is very, very good at breaking down plastic! While much remains to be done, this discovery offers hope of the removal of millions of tons of plastic beads, bottles, and other wastes that would otherwise persist for hundreds of years in the ocean environment.

chemical weapons housed within structures in the dermis layer of the skin. For example, the oil glands that occur along the shaft of the hair and sweat glands secrete chemicals that make the skin's surface very acidic (pH of between 4 and 5.5), which inhibits the growth of many microbes. Sweat also contains the enzyme lysozyme, which attacks and digests the cell walls of many bacteria.

In addition to the skin, other surfaces, such as the eyes, are exposed to the outside. Like sweat, the tears that bathe the eyes contain lysozyme to fight bacterial infections. The digestive tract and the respiratory tract are two other potential routes of entry by bacteria and viruses. Microbes present in food are killed by saliva (which contains lysozyme), the acid environment in the stomach (pH of 2), and digestive enzymes in the intestines. Microbes present in inhaled air are trapped by a layer of sticky mucus secreted by cells that line the bronchi and bronchioles before they can reach the lungs. Other cells that line these passages have cilia that continually sweep the mucus up to the throat, like an escalator, where they can be swallowed and killed.

The surface defense provided by skin is very effective, but it is occasionally breached. Through breathing, eating, or cuts and nicks, bacteria and viruses now and then enter our bodies. When these invaders reach deeper tissue, a second line of defense comes into play, a cellular counterattack.

> **Putting the Concept to Work**
> Why does acidic skin protect against bacterial infection?

Figure 27.2 A section of human skin.

The skin defends the body by providing a barrier with sweat and oil glands, whose secretions make the skin's surface acidic enough to inhibit the growth of microorganisms.

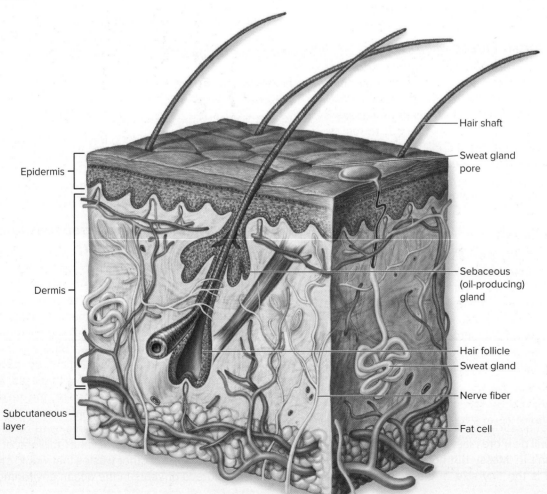

27.2 Cellular Counterattack: The Second Line of Defense

When the body's interior is invaded, a host of cellular and chemical defenses swing into action. Four are of particular importance: (1) cells that kill invading microbes; (2) proteins that kill invading microbes; (3) the temperature response, which elevates body temperature to slow the growth of invading bacteria; and (4) the inflammatory response, which speeds defending cells to the point of infection. Although these cells and proteins roam throughout the body, they are stored in and distributed from the **lymphatic system.** The lymphatic system consists of the structures shown in figure 27.3.

Cells that Kill Invading Microbes

> **LEARNING OBJECTIVE 27.2.1** Describe the cellular defenses activated by infection.

The most important counterattack to infection is mounted by white blood cells, which attack invading microbes. These cells patrol the bloodstream and await invaders within the tissues. The three basic kinds of killing cells are macrophages and neutrophils, which are phagocytes, and natural killer cells. Natural killer cells can distinguish the body's cells (self) from foreign cells (nonself).

Macrophages. White blood cells called **macrophages** (Greek, "big eaters") kill bacteria by ingesting them, much as an amoeba ingests a food particle. The macrophage in figure 27.4 is sending out long, sticky cytoplasmic extensions that catch the sausage-shaped bacteria and draw them back to the cell where they are engulfed. Although some macrophages are anchored within particular organs, most patrol the byways of the body, circulating as precursor cells called **monocytes** in the blood, lymph, and fluid between cells. Macrophages are among the most actively mobile cells of the body.

Neutrophils. Other white blood cells called **neutrophils** act like kamikazes. In addition to ingesting microbes, they release chemicals (identical to household bleach) to "neutralize" the entire area, killing any bacteria in the neighborhood—and themselves in the process. A neutrophil is like a grenade tossed into an infection. It kills everything in the vicinity. Macrophages, by contrast, kill only one or a few invading cells at a time but live to keep on doing it.

Natural Killer Cells. A third kind of white blood cell, called a **natural killer cell,** does not attack invading microbes but instead attacks the body cells that are infected by them. Natural killer cells puncture the membrane of the infected target cell with a molecule called *perforin*. The natural killer cell in figure 27.5 is releasing several perforin molecules that insert in the membrane of the infected cell, like boards on a picket fence, forming a pore that allows water to rush in, causing the cell to swell and burst. Natural killer cells are particularly effective at detecting and attacking body cells that have been infected with viruses. They are also one of the body's most potent defenses against cancer.

> **Putting the Concept to Work**
> Why aren't macrophages able to defend against virus infection?

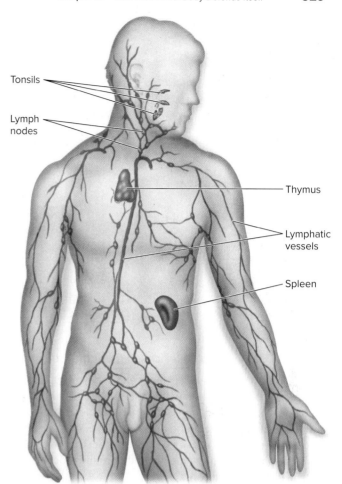

Figure 27.3 The lymphatic system.

Immune system cells and proteins roam through the body and are stored and distributed by the lymphatic system. The lymphatic system consists of lymph nodes, lymphatic organs, and a network of lymphatic capillaries that drain into lymphatic vessels emptying back into the circulatory system.

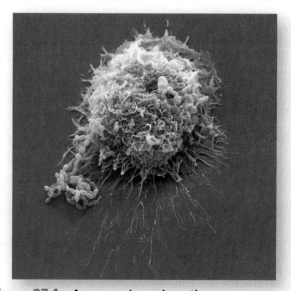

Figure 27.4 A macrophage in action.

In this scanning electron micrograph, a macrophage is "fishing" for rod-shaped bacteria.

Eye of Science/Science Source

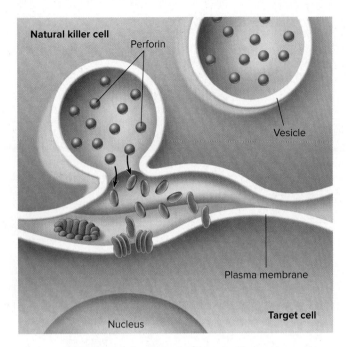

Figure 27.5 Natural killer cells attack target cells.
Tight binding of the natural killer cell to the target cell initiates a chain of events within the natural killer cell in which vesicles loaded with perforin molecules move to the outer plasma membrane and expel their contents into the intercellular space over the target. The perforin molecules insert into the membrane, forming a pore that admits water and ruptures the cell.

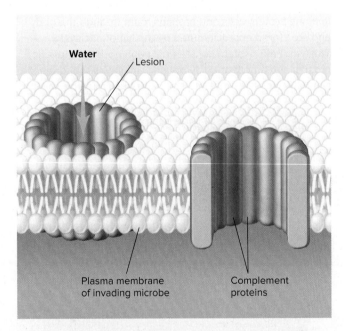

Figure 27.6 Complement proteins attack invaders.
Complement proteins form a transmembrane channel resembling the perforin-lined lesion made by natural killer cells, but complement proteins are free-floating, and they attach to the invading microbe directly, whereas perforin molecules insert into infected body cells.

Proteins that Kill Invading Microbes

> **LEARNING OBJECTIVE 27.2.2** Describe the chemical defenses activated by infection.

The cellular defenses of vertebrates are complemented by a very effective chemical defense called the **complement system.** This system consists of approximately 20 different proteins that circulate freely in the blood plasma in an inactive state. Their defensive activity is triggered when they encounter the cell walls of bacteria or fungi. The complement proteins then aggregate to form a *membrane attack complex* that inserts itself into the foreign cell's plasma membrane, forming a pore like that produced by natural killer cells. Like the perforin pore, the membrane attack complex in figure 27.6 allows water to enter the foreign cell, causing the cell to swell and burst. The difference between the two is that perforin released from natural killer cells attacks host cells that have become infected, whereas free-floating complement proteins attack the foreign cell directly. Aggregation of the complement proteins is also triggered by the binding of antibodies to invading microbes, as we'll see in section 27.6.

The Temperature Response

Human pathogenic bacteria do not grow well at high temperatures. Thus, when macrophages initiate their counterattack, they increase the odds in their favor by sending a message to the brain to raise the body's temperature. The cluster of brain cells that serves as the body's thermostat responds to the chemical signal by boosting the body temperature several degrees above the normal 37°C (98.6°F). The higher-than-normal temperature that results is called a *fever*. Although a fever is quite effective at inhibiting microbial growth, very high fevers are dangerous because excessive heat can inactivate critical cellular enzymes. In general, temperatures greater than 39.4°C (103°F) are dangerous; those greater than 40.6°C (105°F) are often fatal.

The Inflammatory Response

The aggressive cellular and chemical counterattacks to infection are made more effective by the **inflammatory response.** The inflammatory response can be broken down into three stages:

1. First, infected or injured cells release chemical alarm signals, most notably histamine and prostaglandins.
2. These chemical alarm signals cause blood vessels to expand, both increasing the flow of blood to the site of infection or injury and, by stretching their thin walls, making the capillaries more permeable. This produces the redness and swelling so often associated with infection.
3. Now the increased blood flow through larger, leakier capillaries promotes the migration of phagocytes from the blood to the site of infection, squeezing between cells in the capillary walls. Neutrophils arrive first, spilling out chemicals that kill the microbes (as well as tissue cells in the vicinity and themselves). Monocytes follow and become macrophages, which engulf the pathogens as well as the remains of all the dead cells. This counterattack takes a considerable toll; the pus associated with infections is a mixture of dead neutrophils, cells, and pathogens.

> **Putting the Concept to Work**
> Compare and contrast perforin and complement.

27.3 Specific Immunity: The Third Line of Defense

> **LEARNING OBJECTIVE 27.3.1** Contrast the origins of T cells and B cells, describing how each responds to infection.

The second line of defense, with both chemical and cellular weapons, provides a sophisticated defense against microbial infection. Only occasionally do bacteria or viruses overwhelm this defense. When this happens, they face yet a third line of defense, more difficult to evade than any they have encountered. It is the specific immune response, the most elaborate of the body's defenses.

Specific immune defense mechanisms of the body involve the actions of white blood cells, or leukocytes. They are very numerous—of the 10 to 100 trillion cells of your body, two in every 100 are white blood cells! Macrophages are white blood cells, as are neutrophils and natural killer cells. In addition, there are T cells, B cells, plasma cells, mast cells, and monocytes (**table 27.1**). T cells and B cells are called **lymphocytes** and are critical to the specific immune response.

TABLE 27.1 Cells of the Immune System

Cell Type	Function	Cell Type	Function
Helper T cell	Commander of the immune response; detects infection and sounds the alarm, initiating both T cell and B cell responses	Plasma cell	Biochemical factory devoted to the production of antibodies directed against specific foreign antigens
Memory T cell	Provides a quick and effective response to an antigen previously encountered by the body	Mast cell	Initiator of the inflammatory response, which aids the arrival of leukocytes at a site of infection; secretes histamine, and is important in allergic responses
Cytotoxic T cell	Detects and kills infected body cells; recruited by helper T cells		
Suppressor T cell	Dampens the activity of T and B cells, scaling back the defense after the infection has been checked	Monocyte	Precursor of macrophage
B cell	Precursor of plasma and memory cells; specialized to recognize specific foreign antigens	Macrophage	The body's first cellular line of defense; also serves as antigen-presenting cell to B and T cells and engulfs antibody-covered cells
Memory B cell	Provides a quick and effective response to an antigen previously encountered by the body		
Neutrophil	Engulfs invading bacteria and releases chemicals that kill neighboring bacteria	Natural killer cell	Recognizes and kills infected body cells; natural killer cell detects and kills cells infected by a broad range of invaders

T Cells

After their origin in the bone marrow, **T cells** migrate to the thymus (hence the designation "T"), a gland just above the heart (see figure 27.3). There they develop the ability to identify cells that have been infected by microorganisms or viruses. These infected cells display antigens on their surfaces that are produced by the infecting microorganism or virus. An **antigen** is a molecule that provokes a specific immune response. Antigens are large, complex molecules, such as proteins, and they are generally foreign to the body, usually belonging to bacteria and viruses. Tens of millions of different T cells are made, each specializing in the recognition of one particular antigen. No invader can escape being recognized by at least a few T cells. There are four principal kinds of T cells: *Helper T cells* initiate the immune response; *memory T cells* provide a quick response to a previously encountered antigen; *cytotoxic T cells* lyse cells that have been infected by microbes; and *suppressor T cells* terminate the immune response.

In 2018, the Nobel Prize for Physiology or Medicine was awarded for research employing T cells as a new kind of blockbuster cancer-fighting drug. American cancer researcher James Allison and Japanese immunologist Tasuku Honjo discovered and were studying a "braking system" the body uses to restrain its T cells, preventing T cells from attacking the body's own cells. They reasoned that T cells might attack cancer cells if released from these brakes, called "check points." To see if this was true, they developed drugs called check point inhibitors that release the brakes on T cells. Did the drugs work? Yes! They proved to be powerful weapons against many cancers, slowing and often curing deadly cancers like melanomas. One of these drugs cured U.S. President Carter of potentially fatal melanoma that had spread to many parts of his body. While they have many side-effects that can prove harmful, these cancer immunotherapy drugs offer great promise.

B Cells

Unlike T cells, **B cells** do not travel to the thymus; they complete their maturation in the bone marrow. From the bone marrow, B cells are released to circulate in the blood and lymph. Individual B cells, like T cells, are specialized to recognize particular foreign antigens. When a B cell encounters the antigen to which it is targeted, it begins to divide rapidly, and its progeny differentiate into *plasma cells* and *memory B cells*. Each plasma cell is a miniature factory producing markers called *antibodies*. These antibodies are large proteins that stick like flags to that antigen wherever it occurs in the body, marking any cell bearing the antigen for destruction. So, B cells don't kill foreign invaders directly, but they mark these cells so that they are more easily recognized by the other white blood cells that do the dirty work.

> **Putting the Concept to Work**
> Which cells provide specific immunity in the body?

The Immune Response

27.4 Initiating the Immune Response

> **LEARNING OBJECTIVE 27.4.1** Identify the cells that initiate the immune response, explaining the roles of helper T cells and interleukin-1.

To understand how this third line of defense works, imagine you have just come down with the flu. Influenza viruses enter your body in small water droplets inhaled into your respiratory system. If they are able to pass across the epithelium lining the respiratory system (first line of defense) and avoid consumption by macrophages (second line of defense), the viruses infect mucous membrane cells.

The Alarm Signal

At this point, T cells and macrophages initiate the immune response. Macrophages that encounter pathogens—either a foreign cell such as a bacterial cell or a virus-infected body cell with telltale viral proteins stuck to its surface—respond by secreting a chemical alarm signal. The alarm signal is a protein called *interleukin-1* (Latin for "between white blood cells"). This protein stimulates **helper T cells**. The helper T cells respond to the interleukin-1 alarm by simultaneously initiating two different parallel lines of immune system defense: the cellular immune response carried out by T cells and the antibody or humoral response carried out by B cells. The immune response carried out by T cells is called the *cellular response* because the T lymphocytes attack the cells that carry antigens (figure 27.7). The B cell response is called the *humoral response* because antibodies are secreted into the blood and body fluids (*humor* refers to a body fluid).

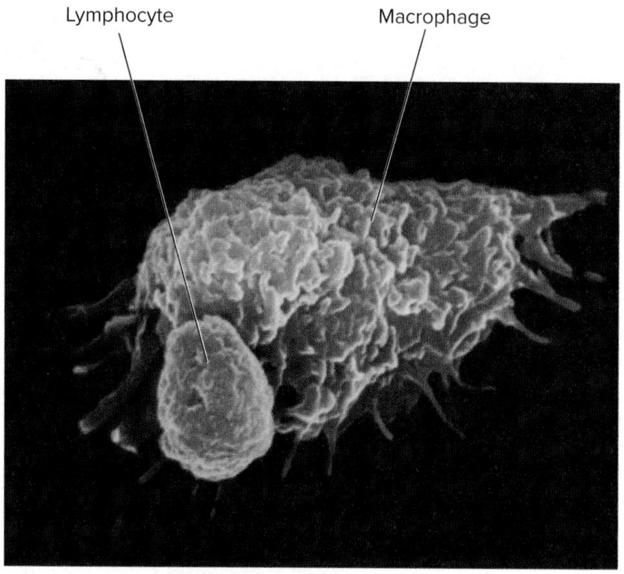

Figure 27.7 Why your antibodies don't attack your own body's cells.

Cells of the body have proteins on their surfaces that identify them as "self" cells. Immune system cells do not attack these cells. In this scanning electron micrograph, a macrophage is beginning to engulf a smaller lymphocyte.

CNRI/Science Source

> **Putting the Concept to Work**
> How does a macrophage "know" when to release interleukins and so initiate the immune response?

27.5 T Cells: The Cellular Response

> **LEARNING OBJECTIVE 27.5.1** Describe the five stages of the cellular immune response.

When macrophages process foreign antigens, they trigger the **cellular immune response**, illustrated in **figure 27.8**. Macrophages secrete interleukin-1 ❶, which stimulates cell division and proliferation of T cells. Helper T cells become activated by macrophages. The helper T cells then secrete

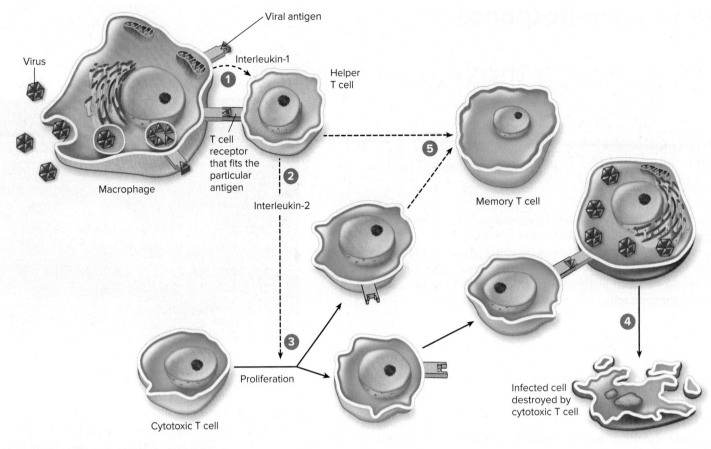

Figure 27.8 The T cell immune defense.

A macrophage releases interleukin-1, signaling helper T cells to bind to the antigen on the infected cell surface. This triggers the helper T cell to release interleukin-2, which stimulates the multiplication of cytotoxic T cells. Body cells that have been infected by the antigen are destroyed by the cytotoxic T cells.

interleukin-2 ❷, which stimulates the proliferation of cytotoxic T cells ❸, which recognize and destroy infected body cells displaying the foreign antigen ❹.

The body makes millions of different versions of T cells. Each version bears a single, unique kind of receptor protein on its exterior, a receptor able to bind to a particular antigen on the surface of an invading bacterium or virus-infected body cell. Any cytotoxic T cell whose receptor fits the particular antigen begins to multiply rapidly. Large numbers of infected cells can be quickly eliminated because the single T cell able to recognize the invading virus is amplified in number to form a large clone of identical T cells, all able to carry out the attack. Any of the body's cells that bear traces of viral infection are destroyed. The method used by cytotoxic T cells to kill infected body cells is similar to that used by natural killer cells and complement proteins: They puncture the plasma membrane of the infected cell. Following an infection, some activated T cells give rise to memory T cells ❺ that remain in the body, ready to mount an attack quickly if the antigen is encountered again.

> **Putting the Concept to Work**
> Contrast how T cells respond to interleukin-1 and interleukin-2.

27.6 B Cells: The Humoral Response

LEARNING OBJECTIVE 27.6.1 Describe how antibodies label pathogens for destruction.

B cells also respond to helper T cells activated by interleukin-1. Like cytotoxic T cells, B cells have receptor proteins on their surfaces, a different receptor for each version of B cell. B cells recognize invading microbes much as cytotoxic T cells recognize infected cells, but unlike cytotoxic T cells, they do not go on the attack themselves. Rather, they mark the pathogen for destruction by mechanisms that have no "ID check" system of their own. Early in the immune response known as the **humoral immune response,** the markers placed by B cells alert complement proteins to attack the cells carrying them. Later in the response, the markers placed by B cells activate macrophages and natural killer cells.

B Cells

The way B cells do their marking is simple and foolproof. B cell receptors bind to foreign antigens, shown in step ❶ of figure 27.9. When a B cell encounters such an antigen, antigen particles enter the B cell by endocytosis and get processed and placed on the surface. Helper T cells recognize the specific antigen, bind to the antigen B cell complex ❷, and release interleukin-2, which stimulates the B cell to divide. In addition, free, unprocessed antigens stick to antibodies (the green Y-shaped structures) on the B cell surface. This antigen exposure triggers even more B cell proliferation. B cells divide to produce plasma cells that serve as short-lived antibody factories ❸ and long-lived

BIOLOGY & YOU

Immunity from Your Mother. Your humoral immune response protects you all your life—except at its beginning. As a newborn infant, you had no prior exposure to microbes and so were particularly vulnerable to infection. Luckily for you, several layers of passive protection were provided by your mother. During pregnancy, a particular type of antibody, called IgG, was transported from your mother to you across the placenta, so that you had high levels of protective antibodies even at birth, offering you the same protection that your mom's humoral immune response offered her. After birth, this protection is continued with breast milk, which also contains antibodies that protected you as a newborn until you could synthesize your own antibodies. During the first several months of your dangerous entrance to this world, you did not actually make any protective antibodies—you only borrowed them.

Figure 27.9 The B cell immune defense.

Invading particles are bound by B cells, which interact with helper T cells and are activated to divide. The multiplying B cells produce memory B cells or plasma cells that secrete antibodies that bind to invading microbes and tag them for destruction by macrophages.

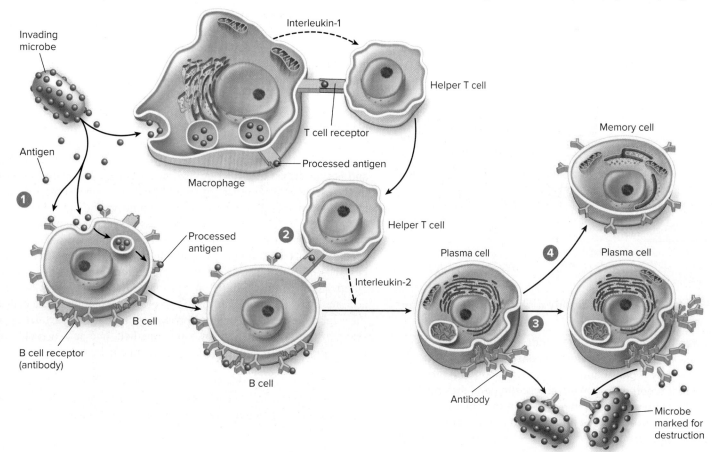

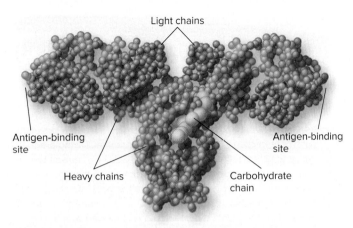

Figure 27.10 An antibody molecule.

In this molecular model of an antibody molecule, each amino acid is represented by a small sphere. Each molecule consists of four protein chains, two "light" (*red*) and two "heavy" (*blue*). The four protein chains wind around one another to form a Y shape. Foreign molecules, called antigens, bind to the arms of the Y.

memory B cells ❹ that remain in the body after the initial infection and are available to mount a quick attack if the antigen enters the body again.

Plasma Cells

The **plasma cells** that are derived from B cells produce lots of the same specific antibody that was able to bind the antigen in the initial immune response. **Antibodies** are proteins that consist of two short sections called light chains and two longer sections called heavy chains, all held together to form a Y-shaped molecule (figure 27.10). Flooding through the bloodstream, these antibody proteins are able to stick to antigens on any cells or microbes that present them, flagging those cells and microbes for destruction. Complement proteins, macrophages, or natural killer cells then destroy the antibody-tagged cells and microbes.

The B cell defense is very powerful because it amplifies the reaction to an initial pathogen encounter a millionfold. It is also a very long-lived defense in that a few of the multiplying B cells do not become antibody producers. Instead, they become a line of **memory B cells** that continue to patrol your body's tissues, circulating through your blood and lymph for a long time—sometimes for the rest of your life.

> **Putting the Concept to Work**
> How is an invading cell killed after being tagged with an antibody?

Antibody Diversity

> **LEARNING OBJECTIVE 27.6.2** Describe the process responsible for generating antibody diversity.

The vertebrate immune system is capable of recognizing as foreign practically any nonself molecule presented to it—literally millions of different antigens. Although vertebrate chromosomes contain only a few hundred receptor-encoding genes, it is estimated that human B cells can make between 10^6 and 10^9 different antibody molecules. How do vertebrates generate millions of different antigen receptors when their chromosomes contain only a few hundred versions of the genes encoding those receptors?

The answer to this question is that the millions of immune receptor genes do not exist as single sequences of nucleotides. Rather, they are assembled by stitching together three or four DNA segments that code for different parts of the receptor molecule. One representative is selected at random from a collection of similar but slightly different versions of the segment, and is transcribed into mRNA. The mRNA is then joined to other bits of mRNA selected from other segment libraries to form a composite gene that codes for the four different regions of each heavy and light chain in an antibody molecule (figure 27.11). This process is called **somatic rearrangement**.

Because a cell may end up with any heavy chain gene and any light chain gene during its maturation, the total number of different antibodies possible is staggering: 16,000 heavy chain combinations × 1,200 light chain combinations = 19 million different possible antibodies. If one also takes into account that additional changes are created by mutation as the cells divide, the total can exceed 200 million! Imagine going into a department store and selecting one shirt, one pair of pants, one belt, one pair of socks, and one pair of shoes. Think of the number of combinations that might be possible!

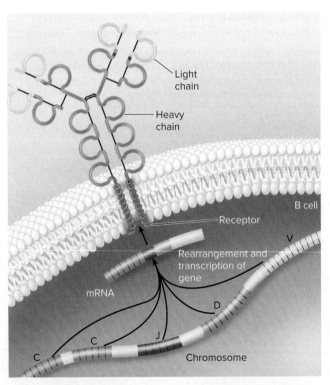

Figure 27.11 The antibody molecule is produced by a composite gene.

Different regions of the DNA code for different regions of the antibody (C, constant regions; J, joining regions; D, diversity regions; and V, variable regions) and are brought together to make a composite gene that codes for the antibody. By combining bits from different segments, an enormous number of different antibodies can be produced.

> **Putting the Concept to Work**
> What is the minimum number of versions of the four heavy chain regions required to produce 16,000 combinations?

27.7 Active Immunity Through Clonal Selection

> **LEARNING OBJECTIVE 27.7.1** Define clonal selection, and differentiate between the primary and secondary immune response.

As we discussed earlier, B cells and T cells have receptors on their cell surfaces that recognize and bind to specific antigens. When a particular antigen enters the body, it must, by chance, encounter the specific lymphocyte with the appropriate receptor to provoke an immune response.

Primary Immune Response

The first time a pathogen invades the body, there are only a few B cells or T cells that may have the receptors that can recognize the invader's antigens. Binding of the antigen to its receptor on the lymphocyte surface, however, stimulates cell division and produces a *clone* (a population of genetically identical cells). This process is known as **clonal selection**. For example, in the first encounter with a chicken pox virus in figure 27.12, there are only a few cells that can mount an immune response, and the response is relatively weak. This is called a **primary immune response** and is indicated by the first curve, which shows the initial amount of antibody produced upon exposure to the virus.

Secondary Immune Response

The primary immune response involves B cells, and so some of the cells become plasma cells that secrete antibodies (taking 10 to 14 days to clear the chicken pox virus from the system), and some become memory cells. Some of the T cells involved in the primary response also become memory cells. Because a clone of memory cells specific for that antigen develops after the primary response, the immune response to a second infection by the same pathogen is swifter and stronger, as shown by the second curve in the figure. Many memory cells can be produced following the primary response, providing a jump-start for the production of antibodies should a second exposure occur. The next time the body is invaded by the same pathogen, the immune system is ready. As a result of the first infection, there is now a large clone of lymphocytes that can recognize that pathogen. This more effective response, elicited by subsequent exposures to an antigen, is called a **secondary immune response**. The "Inquiry & Analysis" feature at the end of this chapter further explores the nature of the secondary immune response.

Memory cells can survive for several decades, which is why people rarely contract chicken pox a second time after they have had it once. Memory cells are also the reason vaccinations are effective. The viruses causing childhood diseases have surface antigens that change little from year to year, so the same antibody is effective for decades. Other diseases, such as influenza, are caused by viruses whose genes that encode surface proteins mutate rapidly. This rapid genetic change causes new strains to appear every year or so that are not recognized by memory cells from previous infections.

Overall, the cellular and humoral immune responses occur simultaneously in the body and work together to produce the body's primary and secondary immune responses.

> **Putting the Concept to Work**
> Why is a secondary immune response greater than a primary one?

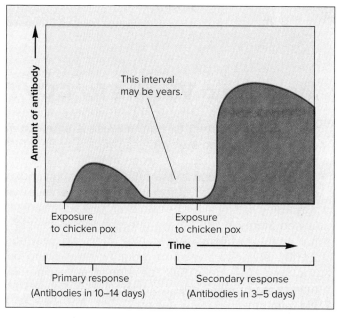

Figure 27.12 The development of active immunity.
Immunity to chicken pox occurs because the first exposure stimulated the development of lymphocyte clones with receptors for the chicken pox virus. As a result of clonal selection, a second exposure stimulates the immune system to produce large amounts of the antibody more rapidly than before, keeping the person from getting sick again.

Today's Biology

Inventing a Vaccine for COVID-19

Vaccination to protect us from disease is so much a part of today's American life that most of us give little thought to it. A generation ago, life as an American was considerably more dangerous. In the 1950s, for example, polio, the disease caused by a virus that infects the nervous system, was killing half a million people each year worldwide. In the United States in 1952, polio infected 57,628 people, over half of them children five to nine years old, leaving over 24,000 dead, paralyzed, or requiring help just to breath. That year Dr. Jonas Salk developed a vaccine. Rather than trying to administer a similar virus as Edward Jenner did with smallpox, Salk simply "killed" the polio virus by heating it, and then administered the inactivated virus. Did this work? Yes! Several doses administered to young children protected them completely. Starting in February 1954, large-scale inoculation of American schoolchildren eradicated the disease in this country. The United States has been polio-free since 1979.

Today we have vaccines for many serious bacterial and viral diseases, including influenza, cholera, diphtheria, Ebola, yellow fever, rubella, measles, chickenpox, tetanus, malaria, mumps, rabies, shingles, human papilloma virus, and tuberculosis. Some vaccinations like measles last a lifetime; others like influenza (flu) must be renewed yearly to combat newly-arisen variants of the virus.

Not all of these vaccines are made the way Salk made his polio vaccine, by administering heat-killed microbes to induce the patient to produce antibodies directed against the disease. All, however, took many years to develop. Thus while Ebola virus outbreaks have been known since 1976, the Ebola vaccine was not approved until 2019. Trying since 1981, we have yet to devise a successful vaccine against AIDS.

At the heart of every vaccine is an antigen, a molecule which provokes the body's immune system to generate antibodies against the virus—like poking a sleeping wasp nest with a stick. There are five general ways to present COVID-19 antigen to a patient, four that have been used successfully in the past and a new approach are described below:

- **Recombinant Vector Vaccines.** One way to present a COVID-19 antigen to a human immune system is to insert the COVID-19 gene for spike protein into a harmless

Everett Collection/Shutterstock

virus. A common choice is adenovirus, which is very good at getting into cells. Scientists remove one of its genes, so it cannot replicate and so is safe, and replace it with the COVID-19 spike protein gene. When injected into a person, this so-called "recombinant vector" virus will induce the cells it infects to make COVID-19 spike proteins, which these cells will then display as antigens on their surfaces, waving them around like little flags to draw the attention of the immune system. The immune system responds to this warning by making antibodies directed against COVID-19 spike proteins, and so protects the vaccinated person against the virus. This is the approach taken recently and very successfully against Ebola. Piggyback vaccines of this sort were developed against COVID-19 by *Johnson&Johnson* and by *Oxford/AstraZenica*. Both have had about 60% effectiveness.

- **Live Attenuated Vaccines.** Another way to expose a patient to COVID-19 antigens is to damage COVID-19 virus particles in a way that renders them harmless without killing them (the virus is said to have been "attenuated"), and then inject this live but hobbled virus into people. The spike proteins are still present on the attenuated virus, so antibodies to COVID-19 are produced, but the vaccinated person doesn't get sick because the virus has been disabled. This is the approach taken in the past with many commonly-used vaccines such as those directed against chickenpox, measles, mumps, and rubella. The *Sinovac* COVID-19 vaccine developed in China and widely used in Brazil and Indonesia has been about 50% effective.

- **Inactivated Virus Vaccines.** A third way is to simply "kill" the virus and inject it. All you have to do is heat the virus solution. The spike proteins on the heat-inactivated virus can still induce antibody formation. This is the way the Salk polio vaccine was produced. Your seasonal flu shots are another example of this very direct approach. The *Sinopharm* COVID-19 vaccine made in this way has been widely used in China.
- **Subunit Vaccines.** In a fourth kind of vaccine, the patient may be presented with the antigen itself, spike proteins manufactured by cells growing in tissue culture, harvested, and attached to nanoparticles to create a vaccine. Injected, these particles direct a patient's immune system to produce antibodies against COVID-19 spike proteins. This approach is the one taken to make the vaccines for HPV and hepatitis B. Although very successful in clinical trials, manufacturing difficulties delayed the introduction of such nanoparticle COVID-19 vaccines by *Novavax* and *GlaxoSmithKline*.
- **Nucleic Acid Vaccines.** In principle, it should be possible to simply make jillions of RNA copies of the COVID-19 gene encoding the spike protein, and inject this mRNA directly into patients. Their cells would be expected to take up the mRNA and begin producing the spike proteins, inducing the patient to manufacture antibodies directed against it. These antibodies would then protect the patient from any future infection by the COVID-19 virus. While no RNA vaccine had ever been licensed for use anywhere in the world, this approach proved the most successful of all, and produced a vaccine within a single year, an unheard of speed-of-development. The *Pfizer* and *Moderna* vaccines both encapsulate the mRNA within tiny fat bubbles to create lipid nanoparticles. The lipid coverings both stabilize the RNA and protect it from a patient's immune system until it can enter the relative safety of a human cell. Widely used in the United States, these two mRNA COVID-19 vaccines have proven very effective, blocking 95% of illness.

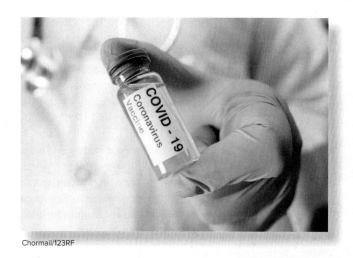

Chormail/123RF

Figure 27.13 The birth of immunology.
This famous painting shows Edward Jenner inoculating patients with cowpox in the 1790s and thus protecting them from smallpox.
Bettmann/Corbis/Getty Images

Figure 27.14 The flu epidemic of 1918 killed over 20 million people in 18 months.
With 25 million Americans alone infected during the influenza epidemic, it was hard to provide care for everyone. The Red Cross often worked around the clock.
PhotoQuest/Archive Photos/Getty Images

27.8 Vaccination

> **LEARNING OBJECTIVE 27.8.1** Explain how vaccination protects against infection.

In 1796, an English country doctor named Edward Jenner carried out an experiment that marks the beginning of the study of immunology. Smallpox was a common and deadly disease in those days. Jenner observed, however, that milkmaids who had caught a much milder form of "the pox" called cowpox (presumably from cows) rarely caught smallpox. Jenner set out to test the idea that cowpox conferred protection against smallpox by infecting people with mild cowpox (figure 27.13). As he predicted, many of them became immune to smallpox.

Defense Against Multiple Enemies. We now know that smallpox and cowpox are caused by two different but similar viruses. Jenner's patients who were injected with the cowpox virus mounted a defense that was also effective against a later infection of the smallpox virus. Jenner's procedure of injecting a harmless microbe to confer resistance to a dangerous one is called vaccination. **Vaccination** is the introduction into your body of a dead or disabled pathogen or, more commonly these days, of a harmless microbe with pathogen proteins displayed on its surface. The vaccination triggers an immune response against the pathogen, without an infection ever occurring. Afterward, the vaccinated person has memory cells (B and T cells) directed against that specific pathogen. The vaccinated person is said to have been "immunized" against the disease.

Immunity Isn't Forever. If the activities of memory cells provide such an effective defense against future infection, why can you catch some diseases such as flu more than once? The reason you don't stay immune to flu is that the flu virus has evolved a way to evade the immune system—it changes. The genes encoding the surface proteins of the flu virus mutate very rapidly. Thus, the shapes of these surface proteins alter swiftly. Your memory cells do not recognize viruses with altered surface proteins as being the same viruses they have already successfully defeated or been vaccinated against, because the memory cells' receptors no longer "fit" the new shape of the flu surface proteins.

Some new versions of the flu can be quite deadly. When mutations arose in a bird flu in 1918 that allowed this flu virus to pass easily from one infected human to another, over 20 million Americans and Europeans died in 18 months (see pages 305 and 306, and figure 27.14). Less profound changes in flu virus surface proteins occur periodically, resulting in new strains of flu for which we are not immune. The annual flu shots are vaccines against these new strains. Researchers are able to predict the current year's strain of the flu by examining preseason flu reports from across the globe and to prepare a vaccine against the year's strain.

> **Putting the Concept to Work**
> How can you come down with the flu if you have had it before?

27.9 Antibodies in Medical Diagnosis

> **LEARNING OBJECTIVE 27.9.1** Describe the antigens of the ABO and Rh systems and their potential impact on blood transfusions.

Blood Typing

A person's blood type indicates the class of antigens found on the red blood cell surface. Red blood cell antigens are clinically important because their types must be matched between donors and recipients for blood transfusions. There are several groups of red blood cell antigens, but the major group is known as the **ABO system.** In terms of the antigens present on the red blood cell surface, a person may be *type A* (with only A antigens), *type B* (with only B antigens), *type AB* (with both A and B antigens), or *type O* (with neither A nor B antigens).

> Human ABO blood types are genetically determined, as discussed in section 10.6. The inheritance of the ABO blood type is an example of codominance. A person who inherits both the A and B genes expresses both antigens on their red blood cells.

The immune system is tolerant to its own red blood cell antigens but produces antibodies against nonself blood antigens. For example, a person who is type A does not produce anti-A antibodies but does make antibodies against the B antigen. People who are type AB develop tolerance to both antigens and thus do not produce either anti-A or anti-B antibodies. Those who are type O make both anti-A and anti-B antibodies.

If type A blood is mixed on a glass slide with serum from a person with type B blood, the anti-A antibodies in the serum cause the type A blood cells to clump together, or agglutinate (this is shown in the upper-right panel of **figure 27.15**). These tests allow the blood types to be matched prior to transfusions, so that agglutination will not occur in the blood vessels, where it could lead to inflammation and organ damage.

Rh Factor. Another group of antigens found on most red blood cells is the *Rh factor*. People who have these antigens are said to be Rh-positive, whereas those who do not are Rh-negative. The Rh-negative allele is less common in the human population. The Rh factor is of particular significance when Rh-negative mothers give birth to Rh-positive babies.

Erythroblastosis. Because fetal and maternal blood are normally kept separate across the placenta (see chapter 30), the Rh-negative mother is not usually exposed to the Rh antigen of an Rh-positive fetus during pregnancy. At the time of birth, however, a varying degree of exposure may occur, and the Rh-negative mother's immune system may become sensitized and produce antibodies against the Rh antigen. If so, these antibodies can cross the placenta in subsequent pregnancies and cause hemolysis of the Rh-positive red blood cells of the fetus, a condition called erythroblastosis fetalis, or hemolytic disease of the newborn. Erythroblastosis fetalis can be prevented by injecting the Rh-negative mother with an antibody preparation against the Rh factor soon after the birth of each Rh-positive baby. The injected antibodies inactivate the Rh antigens and thus prevent the mother from becoming actively immunized to them.

> **Putting the Concept to Work**
> Does a transfusion of type O blood elicit more or fewer antibodies than a transfusion of type AB blood?

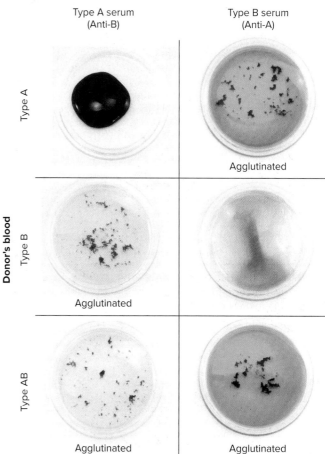

Figure 27.15 Blood typing.

Agglutination of the red blood cells is seen when blood types are mixed with sera containing antibodies against the A and B antigens. Note that no agglutination would be seen if type O donor blood (not shown) was used.

Ivan Ivanov/iStock/Getty Images

Defeat of the Immune System

Figure 27.16 The house dust mite.
This tiny animal, *Dermatophagoides*, causes an allergic reaction in many people. The dust mite lives in mattresses and pillows and feeds on the large quantities of dead skin cells that we all shed.
Sebastian Kaulitzki/Shutterstock

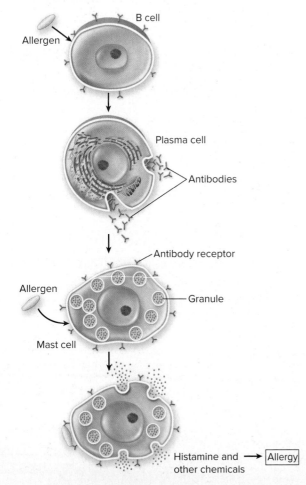

Figure 27.17 An allergic reaction.
In an allergic response, B cells secrete antibodies that attach to the plasma membranes of mast cells, which secrete histamine in response to antigen–antibody binding.

27.10 Overactive Immune System

> **LEARNING OBJECTIVE 27.10.1** Compare autoimmune diseases to allergies, explaining the role of mast cells and histamines.

Although the immune system is one of the most sophisticated systems of the vertebrate body, it is still not perfect. Several major diseases we face, and some minor irritations as well, reflect an overactive immune system.

Autoimmune Diseases

The ability of T cells and B cells to distinguish cells of your own body—"self" cells—from nonself cells is the key ability of the immune system that makes your body's third line of defense so effective. In certain diseases, this ability breaks down, and the body attacks its own tissues. Such diseases are called **autoimmune diseases.** For reasons we don't understand, autoimmune diseases are far more common among women than men.

Multiple sclerosis is an autoimmune disease that usually strikes people between the ages of 20 and 40. In multiple sclerosis, the immune system attacks and destroys a sheath of fatty material called myelin that insulates motor nerves (like the rubber covering electrical wires). Degeneration of the myelin sheath interferes with transmission of nerve impulses until eventually they cannot travel at all. Voluntary functions, such as the movement of limbs, and involuntary functions, such as bladder control, are lost, leading finally to paralysis and death. Scientists do not know what stimulates the immune system to attack myelin.

Another autoimmune disease is type I diabetes, thought to result from an immune attack on the insulin-manufacturing cells of the pancreas. Also, rheumatoid arthritis is an immune system attack on the tissues of the joints, lupus is an autoimmune attack on connective tissue and kidneys, and in Graves' disease, the thyroid is attacked.

Allergies

Although your immune system provides very effective protection against pathogens, sometimes it does its job too well, mounting an immune response that is greater than necessary to eliminate an antigen. The antigen in this case is called an *allergen,* and such an immune response is called an **allergy.** Hay fever, sensitivity to plant pollen, is a familiar example of an allergy. Many people are also allergic to proteins released from the feces of a minute mite that lives on grains of house dust (figure 27.16).

What makes an allergic reaction uncomfortable, and sometimes dangerous, is the involvement of antibodies attached to a kind of white blood cell called a *mast cell.* It is the job of the mast cells in an immune response to initiate an inflammatory response (figure 27.17). When mast cells encounter something that matches their antibodies, they release histamines and other chemicals that cause capillaries to swell. Histamines also increase mucus production by cells of the mucous membranes, resulting in runny noses and nasal congestion. Most allergy medicines relieve these symptoms with antihistamines, chemicals that block the action of histamines.

> **Putting the Concept to Work**
> If mast cells make us so uncomfortable, why do we have them?

27.11 AIDS: Immune System Collapse

> **LEARNING OBJECTIVE 27.11.1** Describe how HIV cripples the immune defense.

AIDS (acquired immunodeficiency syndrome) was first recognized as a disease in 1981. By the end of 2020 in the United States, 700,000 individuals had died of AIDS, and 1.2 million others were thought to be infected with **HIV** (human immunodeficiency virus), the virus that causes the disease (**figure 27.18**). Worldwide, over 37 million are infected, and 35 million have died. HIV apparently evolved from a very similar virus that infects chimpanzees in Africa when a mutation arose that allowed the virus to recognize a human cell surface receptor called CD4. This receptor is present in the human body on certain immune system cells, notably macrophages and helper T cells. It is the identity of these immune system cells that leads to the devastating nature of the disease.

How HIV Attacks the Immune System

HIV attacks and cripples the immune system by targeting cells that have CD4 receptors (CD4⁺ cells), including helper T cells. The significance of killing helper T cells is that it leaves the immune system unable to mount a response to *any* foreign antigen. AIDS is a deadly disease for just this reason. With little defense against infection, any of a variety of commonplace infections proves fatal. Also, with no ability to recognize and destroy cancer cells when they arise, death due to cancer becomes far more likely. More AIDS victims die of cancer than from any other cause.

The fatality rate of AIDS is 100%; no patient exhibiting the symptoms of AIDS has recovered. The disease is *not* highly contagious because it is only transmitted from one individual to another through the transfer of internal body fluids, typically in semen or vaginal fluid during sexual intercourse and in blood transmitted by needles during drug use. However, symptoms of AIDS do not usually show up for several years after infection with HIV, apparent in the delay of the onset of AIDS in the United States (the red line in **figure 27.19**) after initial infection with HIV (the green line). Because of this symptomless delay, infected individuals often unknowingly spread the virus to others. Awareness campaigns have helped reduce the numbers of new cases.

A variety of drugs inhibit HIV in the test tube. These include AZT and its analogs (which inhibit virus nucleic acid replication) and protease inhibitors (which inhibit the production of functional viral proteins). A combination of a protease inhibitor and two AZT analog drugs entirely eliminates the HIV virus from many patients' bloodstreams. Widespread use of this *combination therapy* has cut the U.S. AIDS death rate by almost two-thirds since its introduction in the mid-1990s, from 51,414 AIDS deaths in 1995 to 38,074 in 1996; 24 years later, in 2020, deaths were even lower, at 15,815.

> **Putting the Concept to Work**
> How is HIV spread from one person to another?

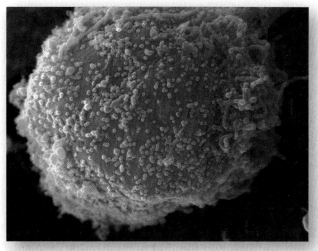

Figure 27.18 Human cell infected with HIV.
The cell you see here is a human immune system cell called a macrophage that has been infected by HIV, the virus that causes AIDS. Progeny HIV viruses (artificially colored blue here) are being released from the infected cell, budding out from its surface.
Eye of Science/Science Source

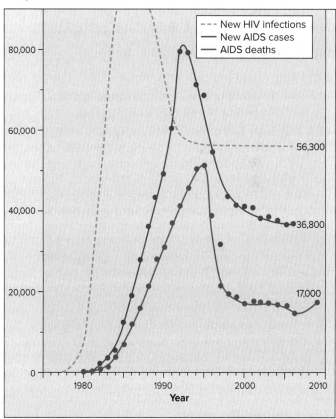

Figure 27.19 History of the AIDS epidemic in the United States.
The U.S. Centers for Disease Control and Prevention reports that 40,040 new AIDS cases were reported in 2015 in the United States, with a total of 675,000 deaths since the beginning of the epidemic. Over 1 million other individuals are thought to be infected with the HIV virus in the United States and over 37 million worldwide.

Putting the Chapter to Work

1. A dangerous form of skin cancer known as melanoma is often deadly because it is resistant to current chemotherapy and radiation treatments. The immune system participates in early recognition of melanoma cells.

What type of immune system cell in the second line of defense would be involved in defending the body against these cells, once they have been recognized?

2. A healthy person has approximately 1,200 helper T cells per cubic millimeter of blood. If a person is exposed to HIV, the virus will target, infect, and kill T cells. A person is diagnosed with AIDS when the helper T cell count is below 200 per cubic millimeter of blood.

This decrease hinders the communication between T cells and what other type of cell in the third line of defense?

3. The pneumococcal vaccine protects the patient from 13 types of bacteria that can cause pneumonia. The vaccine is made from polysaccharides that would be found on the bacteria's cell surface. By exposing the body to these polysaccharides, the immune system can develop memory cells to be on alert to a real infection.

What do these memory cells do when exposed to the polysaccharides?

Retracing the Learning Path

Three Lines of Defense

27.1 Skin: The First Line of Defense

1. The body has three lines of defense against infection, the first being skin. Skin forms a barrier to pathogens.

27.2 Cellular Counterattack: The Second Line of Defense

1. The second line of defense is a nonspecific cellular attack. The cells and chemicals used in this line of defense attack all foreign agents they encounter.
- Macrophages and neutrophils attack the invading pathogen by engulfing them before they can infect cells of the body. Natural killer cells kill infected cells before the pathogen can spread to other cells by inserting perforin proteins into the plasma membrane, causing the cell to burst.
2. Free-floating proteins in the blood, called complement proteins, insert into the membranes of foreign cells, killing them.
- Increasing body temperature slows bacterial growth, and inflammation brings immune cells to the site of infection.

27.3 Specific Immunity: The Third Line of Defense

1. The third line of defense is a specific immune response and involves lymphocytes called T cells and B cells "programmed" by exposure to specific antigens. Once programmed, they seek out the antigens or cells carrying the antigens and destroy them.

The Immune Response

27.4 Initiating the Immune Response

1. A bacteria, cancer cell, or virus that exhibits "nonself" proteins is engulfed by a macrophage, and antigens from the microbe are inserted into the surface membrane of the macrophage. This macrophage presents these antigens to T cells, activating the T cell response. This cell also secretes interleukin-1. Interleukin-1 stimulates helper T cells that trigger B cell immune responses.

27.5 T Cells: The Cellular Response

1. The cellular response is carried out by T cells. Antigen-presenting cells activate helper T cells, which respond by releasing interleukin-2. Interleukin-2 stimulates the cloning of cytotoxic T cells that recognize and kill infected cells that display the specific antigen. Following the infection, memory T cells form and remain in the body to fight subsequent infections.

27.6 B Cells: The Humoral Response

1. Interleukin-2 also activates B cells. In the humoral immune response, B cells label antigens on pathogens, virus-infected cells, or cancer cells with antibodies that tag these cells for destruction by complement proteins, natural killer cells, and macrophages.
2. The vertebrate immune system can produce some 10^9 different antibodies from only a few hundred versions of the genes encoding the antibody receptor proteins. This is accomplished by somatic rearrangement of the genes encoding different parts of the antibody molecule. The various antibodies are produced from composite genes.

27.7 Active Immunity Through Clonal Selection

1. The initial immune response triggered by infection is called the primary response. B and T cells are stimulated to begin dividing, producing a clone of cells. This is called clonal selection. The primary immune response is a delayed and somewhat weak response. But through clonal selection, a large population of memory cells is present in the body, such that a second infection by the same antigen will trigger a quicker and larger response, called the secondary response.
- Memory cells can survive for several decades, conferring disease immunity to a person carrying the memory cells for that particular pathogen.

27.8 Vaccination

1. Vaccination takes advantage of the mechanism of the primary and secondary immune responses. Vaccines introduce pathogenic antigens into the body in a way that doesn't cause disease. This triggers the primary immune response. A later actual infection elicits a swift and large secondary immune response.

27.9 Antibodies in Medical Diagnosis

1. Antibodies are keen detectors of antigens and so are used in many medical diagnostic applications such as blood typing.

Defeat of the Immune System

27.10 Overactive Immune System

1. Sometimes the immune system attacks antigens that are not foreign or pathogenic. In autoimmune responses, the body attacks its own cells. In allergic reactions, the body mounts an attack against a harmless substance, called an allergen. The allergic response includes activating mast cells that release histamines. Histamines trigger the inflammatory response that results in allergy symptoms.

27.11 AIDS: Immune System Collapse

1. AIDS is a fatal disease caused by infection with the HIV virus. HIV attacks macrophages and helper T cells, eventually destroying T cells that protect the body from other infections.

Inquiry and Analysis

Is Immunity Antigen-Specific?

The immune response provides a valuable protection against infection because it can remember prior experiences. We develop lifelong immunity to many infectious diseases after one childhood exposure. This long-term immunity is why vaccines work. A key question about immune protection is whether or not it is specific. Does exposure to one pathogen confer immunity to only that one, or is the immunity you acquire a more general response, protecting you from a range of infections?

The graph displays the results of an experiment designed to answer this question. A colony of rabbits is immunized once with the antigen to Severe Acute Respiratory Syndrome (SARS), and the level of antibody directed against this antigen is monitored in each individual. After 40 days, each rabbit is reinjected, some with the SARS antigen and others with the antigen to COVID-19, and the level of antibody directed against the reinjected antigens is monitored. The red line is typical of results for SARS antigen, the blue line for COVID-19 antigen.

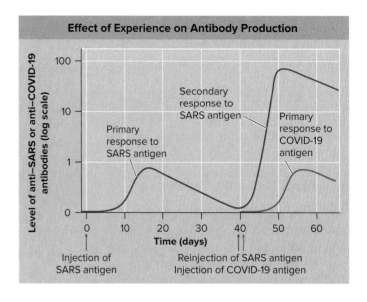

Analysis

1. **Applying Concepts**
   Reading a Continuous Curve. Does each injection of the SARS antigen result in detectable antibody production? Antigen B?

2. **Interpreting Data**
   a. The initial response to the SARS antigen is called the "primary" response, and the second response to the SARS antigen administered 40 days later is called the "secondary" response. Compare the speed of the primary and secondary responses. Which reaches maximal antibody response quicker?
   b. Compare the magnitude of the primary and secondary immune responses to the SARS antigen. Are they similar, or is one response of greater magnitude?

3. **Making Inferences**
   a. Why is the secondary response induced by a second exposure to the SARS antigen different from the primary response?
   b. Is the response to the COVID-19 antigen more similar to the primary or secondary response of the SARS antigen?

4. **Drawing Conclusions**
   a. Does the prior exposure to the SARS antigen have any impact on the speed or magnitude of the primary response to the COVID-19 antigen?
   b. Is the immune response to these antigens antigen-specific?

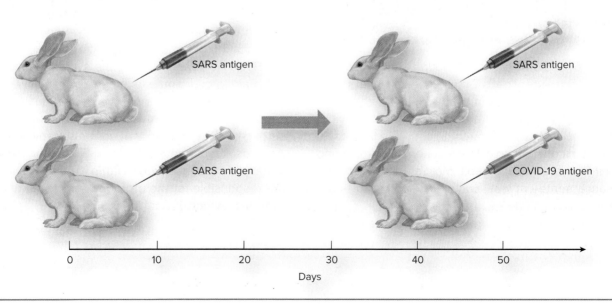

28 The Nervous System

LEARNING PATH ▼

Neurons and How They Work
1. The Nervous System
2. Nerve Impulses
3. The Synapse

The Central Nervous System
4. How the Brain Works
5. The Spinal Cord

The Peripheral Nervous System
6. The Voluntary and Autonomic Nervous Systems

The Sensory Nervous System
7. Sensing the Internal Environment
8. Sensing Gravity and Motion
9. Sensing Chemicals: Taste and Smell
10. Sensing Sounds: Hearing
11. Sensing Light: Vision

Smoking an e-cigarette delivers vapor laced with highly addictive nicotine to the lungs.

Martina Paraninfi/Moment/Getty Images

Dancing with Death: Our Love Affair with Nicotine

Each day, over 5,000 teens and young adults smoke their first cigarette. That's 2 million new smokers annually. As you learned in chapter 24, if they continue to smoke, lots of them are going to get lung cancer—but many find, when they try to quit, that they can't. Addicted to the nicotine in cigarette smoke, they simply find it too difficult to overcome the habit. Most studies indicate a "quitting" success rate—at least two years' abstinence—of about 20%.

Nicotine

The chemical nature of nicotine addiction is a tragedy but not a mystery. Scientists understand nicotine addiction quite well. Individual nicotine molecules attach themselves to a key protein on the surface of brain nerve cells that act to "fine tune" the sensitivity of a wide variety of other brain activities to chemical signals within the brain. Adjusting particular sensitivities up or down to slow some activities and speed others, this protein is responsible for overall coordination of the brain's activities.

By artificially stimulating the system normally used by the brain to coordinate its many activities, nicotine alters the pattern of release by nerve cells of many chemical signals, called neurotransmitters—like turning up the setting on a TV remote that controls many television sets. As a result, changes in the level of activity occur in a wide variety of nerve pathways within the brain. These changes are responsible for the profound effect smoking has on the brain's activities.

piksel/123RF

How Nicotine Causes Addiction

Addiction occurs because the nervous system responds piecemeal to nicotine's fiddling with its central control. The brain attempts to "turn the volume back down" by readjusting the sensitivities of each kind of activity to its signals individually, eventually restoring an appropriate balance of activities. Unfortunately, these readjustments may occur after only a few cigarettes.

Now what happens if you stop smoking? Everything is out of whack! The newly coordinated system now *requires* nicotine to achieve an appropriate balance of nerve pathway activities. The body's physiological response is profound and unavoidable. You are addicted to nicotine. There is no way to prevent this addiction with willpower, any more than willpower can stop a bullet when playing Russian roulette with a loaded gun. If you smoke cigarettes for long, you will become addicted.

Quitting Isn't Easy

Many people attempting to quit smoking use gum or a patch containing nicotine to help them, the idea being that providing nicotine removes the craving for cigarettes. This is true, it does—so long as you keep chewing the gum or using the patch. Actually, using them simply substitutes one (admittedly less dangerous) nicotine source for another.

Nicotine Without Tobacco

Another approach being trumpeted by tobacco companies in recent years is the "smokeless" electronic cigarette, or *e-cigarette*. A tube made to look like a rather large cigarette, an e-cigarette contains no tobacco, delivering to the smoker a measured dose of nicotine within water vapor. Because no tars are introduced into the lung, e-cigarettes do not cause lung cancer. However, they are highly addictive—all that nicotine. When you start smoking e-cigarettes, you are signing on to contribute significant amounts of money to the tobacco industry for years to come. No wonder the tobacco industry promotes them. Instead of curing your nicotine addiction, they chain you to it.

The Only Way to Quit

If you are going to quit smoking, there is no way to avoid the necessity of eliminating the drug to which you are addicted: nicotine. Hard as it is to hear the bad news, there is no easy way out. The only way to quit is to quit.

If all tobacco use were to be replaced with e-cigarettes, e-cigars, and e-pipes, would this reduce lung cancer rates to pre-1900 levels, when very few smoked? Unfortunately, the answer is no.

Nonsmokers represent fully 15% of newly diagnosed lung cancers. Importantly, two-thirds of these nonsmoking lung cancer patients are women! In 2007, lung cancer among nonsmoking women killed three times as many women as did cervical cancer.

Why so many more women than men? Very little research has been done to explore why, but what little has been learned suggests that the 5% of lung cancers affecting nonsmoking men are due to some as-yet-unidentified generalized mutagen such as smoke particles in the air we all breathe; an equal portion of women are affected. In addition, 5% of lung cancers may be caused by an endocrine disruptor such as BPA (see chapter 29, Why Don't Men Get Breast Cancer?) that mimics estrogen and so affects only women—even if they don't smoke.

Neurons and How They Work

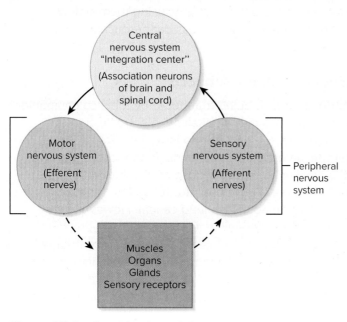

Figure 28.1 Organization of the vertebrate nervous system.

The central nervous system, consisting of the brain and spinal cord, issues commands via the motor nervous system and receives information from the sensory nervous system. The motor and sensory nervous systems together make up the peripheral nervous system.

28.1 The Nervous System

LEARNING OBJECTIVE 28.1.1 Identify the three types of neurons in the nervous system, and describe their functions.

To live and survive, an animal must be able to respond to environmental stimuli. To do this, it must have sensory receptors that can detect the stimulus and motor effectors that can respond to it. In humans and all vertebrates, sensory receptors and motor effectors are linked by way of the **nervous system** (figure 28.1).

Neurons

The basic structural unit of the nervous system, whether central, motor, or sensory, is the nerve cell, or **neuron**. All neurons have the same basic structure, as you can see by comparing the three cell types in figure 28.2. The *cell body* in figure 28.2 is the flat region of the neuron containing the nucleus. Short, slender branches called *dendrites* extend from one end of a neuron's cell body. Dendrites are input channels. Nerve impulses travel inward along them, toward the cell body. Motor and association neurons possess a profusion of highly branched dendrites, enabling those cells to receive information from many different sources simultaneously. Projecting out from the other end of the cell body is a single, long, tubelike extension called an *axon*. Axons are output channels. Nerve impulses travel outward along them, away from the cell body, toward muscles or glands, or to other neurons. The axons of many neurons are enveloped at intervals by a myelin sheath, an insulating cover consisting of multiple layers of membrane.

There are three types of neurons. **Sensory** (or *afferent*) **neurons** (❶ in figure 28.2) carry impulses from sensory receptors to the **central nervous system (CNS),** which consists of the brain and spinal cord. **Motor** (or *efferent*) **neurons** ❸ carry impulses away from the CNS to effectors—muscles

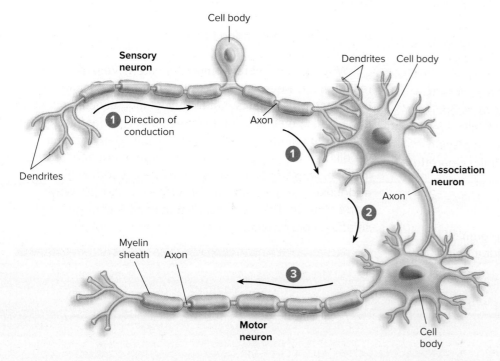

Figure 28.2 Three types of neurons.

Sensory neurons carry information about the environment to the brain and spinal cord. *Association neurons* are found in the brain and spinal cord, and often provide links between sensory and motor neurons. *Motor neurons* carry impulses to muscles and glands (effectors).

and glands. Association neurons (or *interneurons*) ❷ link these two types of neurons together in the CNS. Together, motor and sensory neurons constitute the **peripheral nervous system (PNS)** of vertebrates (the bracketed tan circles in **figure 28.1**).

> **Putting the Concept to Work**
> Do sensory nerves carry impulses toward or away from the CNS?

28.2 Nerve Impulses

> **LEARNING OBJECTIVE 28.2.1** Explain how ion movements create an action potential and propagate a nerve impulse.

The key function of neurons is to transmit information throughout the body, often over long distances. Your sciatic nerve from toe to hip is almost 4-feet long! A nerve impulse must pass this entire length. How? As an electric impulse.

❶ **The Resting Potential.** When a neuron is "at rest," not carrying an impulse, active transport proteins in the neuron's plasma membrane transport sodium ions (Na^+) out of the cell and potassium ions (K^+) in. This sodium-potassium pump, which was described in chapter 4, requires an expenditure of energy to function. Sodium ions cannot easily move back into the cell once they are pumped out, so the concentration of sodium ions builds up outside the cell. Similarly, potassium ions accumulate inside the cell, although they are not as highly concentrated because many potassium ions are able to diffuse out through open channels. The result is to make the outside of the neuron more positive than the inside, a condition called the *resting membrane potential*, indicated by the yellow coloring in **panel 1** of **figure 28.3**. The resting plasma membrane is said to be "polarized."

Neurons are constantly expending energy to pump sodium ions out of the cell in order to maintain the resting membrane potential. The resting membrane potential is the starting point for a nerve impulse.

❷ **Starting a Nerve Impulse.** The impulse starts when pressure or other sensory inputs disturb a neuron's plasma membrane, causing sodium channels called **voltage-gated channels** (that is, protein channels in the neuron membrane that open or close in response to an electric voltage) to open (the purple channels in **panel 2**). As a result, sodium ions flood into the neuron from outside, down their concentration gradient, and for a brief moment, a localized area inside of the membrane is "depolarized," becoming more positively charged in that immediate area of the axon (indicated by the pink coloring in **panel 3**).

❸ **Propagating a Nerve Impulse.** The sodium channels in the small patch of depolarized membrane remain open for only about a half a millisecond. However, if the change in voltage becomes large enough, reaching threshold potential, it causes nearby voltage-gated sodium and potassium channels to open (**panel 3**). The sodium channels open first, which starts a wave of depolarization moving down the neuron. The opening of the gated channels causes nearby voltage-gated channels to open, like a chain of falling dominoes.

❹ **The Action Potential.** This local reversal of voltage moving along the axon is called an **action potential**. An action potential follows an all-or-none law: A large enough depolarization produces either a full action potential or none at all, because the voltage-gated Na⁺ channels open completely or not at all. Once they open, an action potential occurs.

The Refractory Period. After a slight delay, potassium voltage-gated channels open, and K⁺ flows out of the cell down its concentration gradient, making the inside of the cell more negative. The increasingly negative membrane potential (colored green in panel 4) causes the voltage-gated sodium channels to snap closed again. This period of time after the action potential has passed and before the resting membrane potential is restored is called the *refractory period*. A second action potential cannot fire during the refractory period; the resting potential must first be restored by the actions of the sodium-potassium pump.

The depolarization and restoration of the resting membrane potential takes only about 5 milliseconds. Fully 100 such cycles could occur, one after another, in the time it takes to say the word *nerve*.

> **Putting the Concept to Work**
>
> What is the difference between a resting potential, a threshold potential, and an action potential?

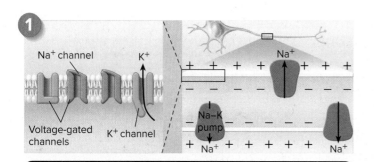

At the resting membrane potential, the inside of the axon is negatively charged because the sodium-potassium pump keeps a higher concentration of Na⁺ outside. Voltage-gated ion channels are closed, but there is some leakage of K⁺.

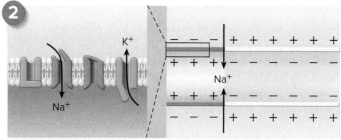

In response to a stimulus, the membrane depolarizes: voltage-gated Na⁺ channels open, Na⁺ flows into the cell, and the inside becomes more positive.

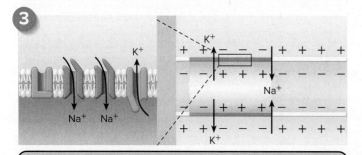

The local change in voltage opens adjacent voltage-gated Na⁺ channels, and an action potential is produced.

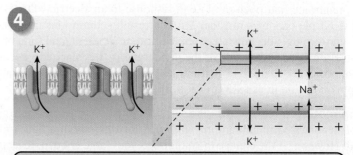

As the action potential travels farther down the axon, voltage-gated Na⁺ channels close and K⁺ channels open, allowing K⁺ to flow out of the cell and restoring the negative charge inside the cell. Ultimately, the sodium-potassium pump restores the resting membrane potential.

Figure 28.3 How an action potential works.

28.3 The Synapse

LEARNING OBJECTIVE 28.3.1 Describe a synapse, and explain how a neurotransmitter transmits a nerve impulse across it.

A nerve impulse travels along a neuron until it reaches the end of the axon, usually positioned very close to another neuron, a muscle cell, or gland. Axons, however, do not actually make direct contact with other cells. Instead, a narrow gap, 10 to 20 nanometers across, called the *synaptic cleft,* separates the axon tip and the target cell. This junction of an axon with another cell is called a **synapse**. The cell on the axon side of a synapse (on the left in figure 28.4) is called the **presynaptic neuron;** the cell on the receiving side of the synapse is called the **postsynaptic cell.**

Neurotransmitters

When a nerve impulse reaches the end of an axon, its message must cross the synapse if it is to continue. Messages do not "jump" across synapses. Instead, they are carried across by chemical messengers called **neurotransmitters.** These chemicals are packaged in tiny sacs, or vesicles, at the tip of the axon. When a nerve impulse arrives at the tip, it causes the sacs to release their contents into the synapse, as shown in figure 28.5a. The neurotransmitters diffuse across the synaptic cleft and bind to receptors (the purple structures) in the postsynaptic membrane. The signal passes to the postsynaptic cell when the binding of the neurotransmitter opens special ion channels, allowing ions to enter the postsynaptic cell and cause a change in electrical charge across its membrane. The enlarged view of figure 28.5a shows how the channel opens and the ion (the yellow ball) enters the cell. Because these channels open when stimulated by a chemical, they are said to be *chemically gated.*

Why go to all this trouble? Why not just wire the neurons directly together? For the same reason that the wires of your house are not all connected but instead are separated by a host of switches. When you turn on one light switch, you don't want every light in the house to go on, the toaster to start heating, and the television to come on! If every neuron in your body were connected to every other neuron, it would be impossible to move your hand without moving every other part of your body at the same time. Synapses are the control switches of the nervous system. However, the control switch must be turned off at some point by getting rid of the neurotransmitter, or the postsynaptic cell would stay stimulated and keep firing action potentials. In some cases, the neurotransmitter molecules diffuse away from the synapse. In other cases, the neurotransmitter molecules are either reabsorbed by the presynaptic cell or degraded in the synaptic cleft.

> **Putting the Concept to Work**
> Why remove neurotransmitters from the synapse if they are transmitting the impulse?

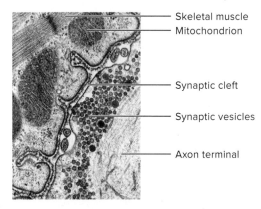

Figure 28.4 A synapse between two neurons.
This micrograph clearly shows the space between the presynaptic and postsynaptic membranes, which is called the synaptic cleft.

Don W. Fawcett/T. Reese/Science Source

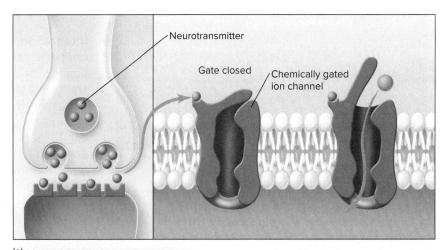

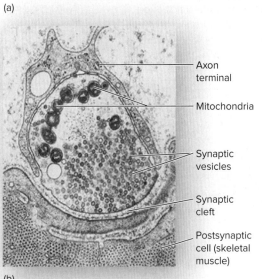

Figure 28.5 Events at the synapse.

(a) When a nerve impulse reaches the end of an axon, it releases a neurotransmitter into the synaptic cleft. The neurotransmitter molecules diffuse across the synapse and bind to receptors on the postsynaptic cell, opening ion channels. (b) A transmission electron micrograph of the tip of an axon filled with synaptic vesicles.

(b) Don W. Fawcett/Science Source

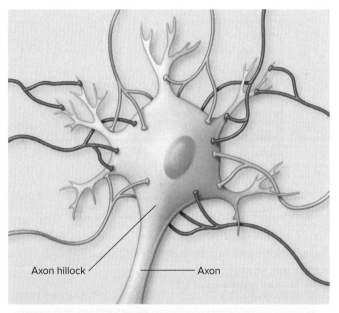

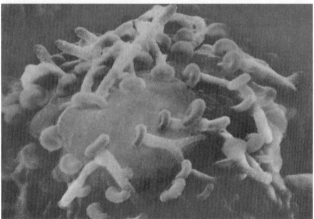

Figure 28.6 Integration.
Many different axons synapse with the cell body and dendrites of the postsynaptic neuron, illustrated here. Excitatory synapses are shown in *red*, and inhibitory synapses are shown in *blue*. The summed influence of their input at the base of the axon, called the axon hillock, determines whether or not a nerve impulse will be sent down the axon extending below. The scanning electron micrograph shows a neuronal cell body with numerous synapses.

(bottom) Omikron/Science Source

Kinds of Synapses

> **LEARNING OBJECTIVE 28.3.2** Contrast excitatory and inhibitory synapses, and explain how they together produce neural integration.

The vertebrate nervous system uses dozens of different kinds of neurotransmitters, each recognized by specific receptors on receiving cells. They fall into two classes, depending on whether they excite or inhibit the postsynaptic cell.

Excitatory Synapses. In an *excitatory synapse,* the receptor protein is usually a chemically gated sodium channel. On binding with a neurotransmitter whose shape fits it, the sodium channel opens, allowing sodium ions to flood inward. If enough sodium ion channels are opened by neurotransmitters, an action potential begins.

Inhibitory Synapses. In an *inhibitory synapse,* the receptor protein is a chemically gated potassium or chloride channel. Binding with its neurotransmitter opens these channels, leading to the exit of positively charged potassium ions or the influx of negatively charged chloride ions, resulting in a more negative interior in the receiving cell. This inhibits the start of an action potential because the negative voltage change inside means that even more sodium ion channels must be opened to get a domino effect started among voltage-gated sodium channels to start an action potential.

Integration. An individual nerve cell, such as the neuron in figure 28.6, can possess both kinds of synaptic connections to other nerve cells. In the drawing, the excitatory synapses are colored red, and the inhibitory synapses are colored blue. When signals from both excitatory and inhibitory synapses reach the cell body of a neuron, the excitatory effects (which cause less internal negative charge) and the inhibitory effects (which cause more internal negative charge) interact with one another. The result is a process of **integration** in which the various excitatory and inhibitory electrical effects tend to cancel or reinforce one another. If the result of the integration is a large enough depolarization (where the inside of the cell becomes more positive), an action potential will fire. Neurons often receive many inputs. A single motor neuron in the spinal cord may have as many as 50,000 synapses on it!

Neurotransmitters and Their Functions

Acetylcholine (ACh) is the neurotransmitter released at the neuromuscular junction, the synapse that forms between a neuron and a muscle fiber. ACh forms an excitatory synapse with skeletal muscle but has the opposite effect on cardiac muscle, causing an inhibitory synapse.

Glycine and *GABA* are inhibitory neurotransmitters. This inhibitory effect is very important for neural control of body movements and other brain functions. Interestingly, the drug diazepam (Valium) causes its sedative and other effects by enhancing the binding of GABA to its receptors.

Biogenic amines are a group of neurotransmitters that include *dopamine*, important in controlling body movements, *norepinephrine* and the hormone *epinephrine*, both involved in the autonomic nervous system, and *serotonin*, which is involved in sleep and emotional states.

> **Putting the Concept to Work**
> Which type of synapse, excitatory or inhibitory, uses sodium channels?

The Central Nervous System

28.4 How the Brain Works

> **LEARNING OBJECTIVE 28.4.1** Describe the principal role of the cerebral cortex, the thalamus, the hypothalamus, the cerebellum, and the brain stem.

The structure and function of the vertebrate brain have long been the subject of scientific inquiry. Despite ongoing research, scientists are still not sure how the brain performs many of its functions. For instance, scientists continue to look for the mechanism the brain employs to store memories, and they do not understand how some memories can be "locked away," only to surface in times of stress. The brain is the most complex vertebrate organ ever to evolve, and it can perform a bewildering variety of complex functions.

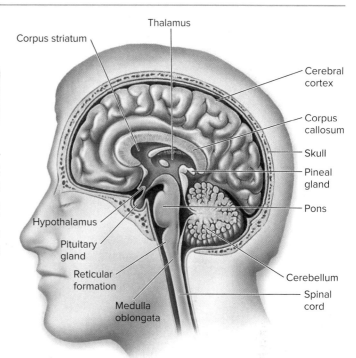

Figure 28.7 A section through the human brain.
The cerebrum occupies most of the brain. Only its outer layer, the cerebral cortex, is visible on the surface.

The Cerebrum Is the Control Center of the Brain

Although vertebrate brains differ in the relative importance of different components, the human brain is a good model of how vertebrate brains function. About 85% of the weight of the human brain is made up of the cerebrum, the tan convoluted area in **figure 28.7**. The cerebrum is a large rounded area of the brain divided by a groove into right and left halves called cerebral hemispheres. The sectioned brain in **figure 28.7** is cut along the center groove, with the left hemisphere removed, showing the right hemisphere. The cerebrum functions in language, conscious thought, memory, personality development, vision, and a host of other activities we call "thinking and feeling." The cerebrum, which looks like a wrinkled mushroom, is positioned over and surrounding the rest of the brain, like a hand holding a fist. Much of the neural activity of the cerebrum occurs within a thin, gray outer layer only a few millimeters thick called the **cerebral cortex** (*cortex* is Latin for "bark of a tree"). This layer is gray because it is densely packed with neuron cell bodies. The human cerebral cortex contains the cell bodies of more than 10 billion nerve cells (**figure 28.8**), roughly 10% of all the neurons in the brain. The wrinkles in the surface of the cerebral cortex increase its surface area (and number of cell bodies) threefold. Underneath the cortex is a solid white region of myelinated nerve fibers that shuttle information between the cortex and the rest of the brain.

Two Brains in One. Researchers have found that the two sides of the cerebrum can operate as two different brains. For instance, in some people the tract between the two hemispheres has been cut by accident or surgery. In laboratory experiments, one eye of an individual with such a "split brain" is covered and a stranger is introduced. If the other eye is then covered instead, the person does not recognize the stranger who was just introduced!

Sometimes blood vessels in the brain are blocked by blood clots, causing a disorder called a *stroke*. During a stroke, circulation to an area in the brain is blocked and the brain tissue dies. A severe stroke in one side of the cerebrum often causes paralysis of the other side of the body.

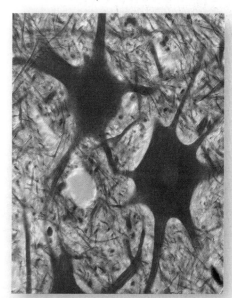

Figure 28.8 The cerebral cortex is a neural network of astonishing complexity.
The network of neurons seen here, magnified over a thousand times, is transmitting signals within the cerebral cortex, a layer of grey matter only a few millimeters thick on the brain's outer surface. Densely packed with neurons and highly convoluted, it is the site of higher mental activities, such as your reading and understanding these words.

Biophoto Associates/Science Source

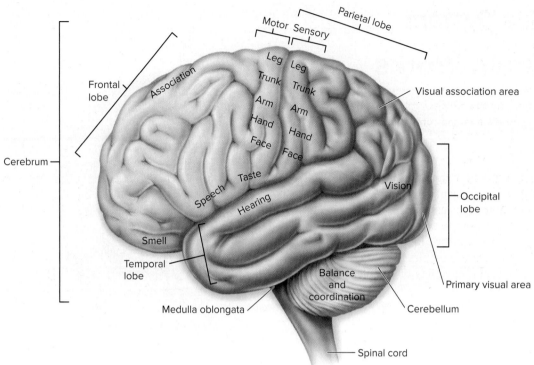

Figure 28.9 **The major functional regions of the human brain.**

Specific areas of the cerebral cortex are associated with different regions and functions of the body.

Figure 28.9 shows the general areas of the brain and the functions they control. The hemispheres of the cerebrum are divided into the frontal, parietal, occipital, and temporal lobes.

The Thalamus and Hypothalamus Process Information

Beneath the cerebrum are the thalamus and hypothalamus, important centers for information processing.

The Thalamus. The **thalamus** is the major site of sensory processing in the brain. Auditory (sound), visual, and other information from sensory receptors enters the thalamus and then is passed to the sensory areas of the cerebral cortex (indicated in figure 28.9). The thalamus also controls balance. Information about posture, derived from the muscles, and information about orientation, derived from sensors within the ear, combine with information from the cerebellum and pass to the thalamus. The thalamus processes the information and channels it to the appropriate motor center on the cerebral cortex.

The Hypothalamus. The **hypothalamus** integrates all the internal activities of the body. It controls centers in the brain stem that in turn regulate body temperature, blood pressure, respiration, and heartbeat. It also directs the secretions of the brain's major hormone-producing gland, the pituitary gland. The hypothalamus is linked by an extensive network of neurons to other areas of the cerebral cortex.

The Limbic System

This network, along with parts of the hypothalamus and areas of the brain called the *hippocampus* and *amygdala,* make up the **limbic system.** The areas highlighted in green in figure 28.10 indicate the components of the limbic

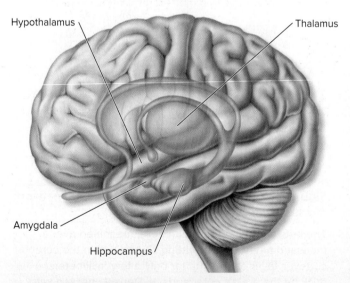

Figure 28.10 **The limbic system.**

The hippocampus and the amygdala are the major components of the limbic system, which controls our most deep-seated drives and emotions.

system. The operations of the limbic system are responsible for many of the most deep-seated drives and emotions of vertebrates, including pain, anger, sex, hunger, thirst, and pleasure, centered in the amygdala. The limbic system is also the area of the brain affected by drugs such as cocaine, and the hippocampus is involved in memory.

The Cerebellum Coordinates Muscle Movements

Extending back from the base of the brain is a structure known as the **cerebellum.** The cerebellum controls balance, posture, and muscular coordination. This small, cauliflower-shaped structure, while well developed in humans and other mammals, is even more developed in birds. Birds perform more complicated feats of balance than we do because they move through the air in three dimensions. Imagine the kind of balance and coordination needed for a bird to land on a branch, stopping at precisely the right moment without crashing into it.

The Brain Stem Controls Vital Body Processes

The **brain stem,** a term used to collectively refer to the midbrain, pons, and medulla oblongata, connects the rest of the brain to the spinal cord. This stalklike structure contains nerves that control your breathing, swallowing, and digestive processes, as well as the beating of your heart and the diameter of your blood vessels. A network of nerves called the *reticular formation* runs through the brain stem and connects to other parts of the brain. These widespread connections make these nerves essential to consciousness, awareness, and sleep. One part of the reticular formation filters sensory input, enabling you to sleep through repetitive noises, such as traffic, yet awaken instantly when a telephone rings.

Language and Memory

Although the two cerebral hemispheres seem structurally similar, they are responsible for different activities. The most thoroughly investigated example of this lateralization of function is language. The left hemisphere is the "dominant" hemisphere for language—the hemisphere in which most neural processing related to language is performed (figure 28.11).

While the dominant hemisphere for language is adept at sequential reasoning, such as that needed to formulate a sentence, the nondominant hemisphere (the right hemisphere in most people) is adept at spatial reasoning, the type of reasoning needed to assemble a puzzle or draw a picture. It is also the hemisphere primarily involved in musical ability—a person with damage to the speech area in the left hemisphere may not be able to speak but may retain the ability to sing!

One of the great mysteries of the brain is the basis of memory and learning. There is no one part of the brain in which all aspects of a memory appear to reside. Although memory is impaired if portions of the brain, particularly the temporal lobes, are removed, it is not lost entirely. Many memories persist in spite of the damage, and the ability to access them gradually recovers with time. Therefore, investigators who have tried to probe the physical mechanisms underlying memory often have felt that they were grasping at a shadow.

> **Putting the Concept to Work**
> Contrast the limbic system with the reticular formation.

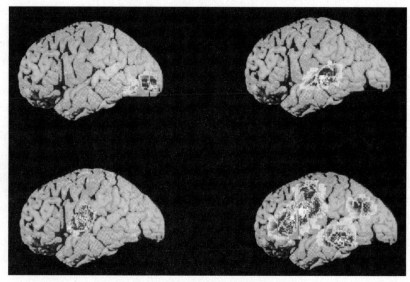

Figure 28.11 **Different brain regions control various activities.**

The colors indicate how the brain reacts in human subjects asked to listen to a spoken word, to read that same word silently, to repeat the word out loud, and then to speak a word related to the first. Regions of white, red, and yellow show the greatest activity.

WELLCOME DEPT. OF COGNITIVE NEUROLOGY/ SCIENCE PHOTO LIBRARY/Getty Images

Answering Your Questions About the Opioid Crisis

What's the Crisis?
A rise of opioid painkiller prescriptions from doctors and aggressive sales promotions from pharmaceutical companies in the 1990s and early 2000s have sparked a wave of opioid drug use. Now people are dying in the streets—half a million Americans have died of drug overdoses since the year 2000. On the day you read this, 91136 more will die. The rate of addiction has shot up 500% in the last 15 years. A total of 64,000 Americans died of opioid overdoses in 2016 alone.

What Is an Opioid?
The rubbery latex juice of the opium poppy plant has been used to alleviate pain for thousands of years—ancient Sumerians referred to poppy plants as "joy plants" in 3400 B.C. In 1806, a German chemist isolated the active ingredient of opium, naming it morphine. Morphine proved to be a powerful pain reliever but even more addictive than opium. Treatment of battle wounds with morphine in the American Civil War led to mass addictions. In an attempt to remedy these addictions, morphine was chemically modified in 1898, but the modified form, heroin, proved to be even more powerful—and more addictive. In 2020, the world produced 6,300 tons of opium, and 70% of that was converted to heroin. In that year, some 17 million people around the globe used heroin, morphine, or opium.

What Is OxyContin?
The production of heroin was made illegal in 1924 because the drug proved so dangerously addictive. In that same year, a powerful pain-relieving drug called oxycodone was synthesized. While addictive and requiring a doctor's prescription, it was marketed as less addictive than heroin and became widely used in the United States in the 1950s. In 1995, a continuous-release form was produced called OxyContin (the "Contin" for continuous release, get it?), which released the opioid over the course of hours rather than all at once. By 2010, OxyContin, widely marketed toward noncancer patients such as those with chronic back pain, reached over $3 billion in annual sales. Oxycodone mixed with aspirin (called Percodan) was even being prescribed by some dentists for toothache!

What Is Fentanyl?
Fentanyl is a synthetic opioid. First made in 1959, fentanyl is 75 times stronger than morphine and was developed for use in patches that relieve the chronic pain of cancer patients. Some more recent chemically modified versions of fentanyl, as much as 10,000 times stronger than morphine, are being added by drug dealers to illegal heroin to increase its strength. There is little margin for error—a tiny bit too much fentanyl, and the drug dose is deadly. Over 20,000 Americans died in 2016 from a fentanyl overdose.

Why Are Opioids So Deadly?
When you take an opioid such as oxycodone, it binds to a particular receptor called the mu opiate receptor. Binding to these receptors creates an inhibitory synapse, switching off the neuron. Binding of opiates to mu receptors located in the brain stem slows respiration, which is why a drug overdose can kill you—your brain stem slows breathing until it stops.

Why Are Opioids So Addictive?
It is the binding of opiates to mu receptors in the midbrain that leads to addiction. Your body inhibits pain by binding endorphins ("endogenous morphine") to these mu receptors, present on a batch of nerves that normally serve as the OFF switches for the brain's pleasure and euphoria networks. But opioids are far more powerful than endorphins. Once opioids shut off these neurons, the pleasure circuits turn ON far more emphatically, flooding the midbrain with a neurotransmitter called dopamine and so triggering a surge of happiness. Communicating with the prefrontal cortex, the brain's decision-making center, the good feelings reinforce what quickly becomes a desired behavior—and soon a habit. The decision by the brain to seek out the drug becomes automatic. As the brain's neurons adapt to the opioid drug, the addiction becomes entrenched, further pushing a desire for opioids. Any withdrawal of the drug now triggers intense anxiety as the brain attempts to force the body to provide more opiates.

How Do Overdose and Replacement Therapies Work?
First responders carry a drug called naloxone that, if administered soon enough, can counteract an overdose. How does it work? The naloxone molecule knocks the opioid off of the mu receptor. Free to act normally, the brain stem neurons resume normal breathing. Curing addiction is a bigger task. Compounds such as methadone can counteract opioid addiction to some degree. One opiate has been replaced by a less dangerous one, reducing withdrawal symptoms as an individual attempts to taper off opioid use.

What Is Being Done?
Clearly, overprescribing opioids has led to today's crisis. In 2010, the number of opioid prescriptions peaked and began to fall. However, in 2020 there were still three times as many opioid prescriptions being written by doctors as there were 20 years earlier, when pharmaceutical companies lit this fire. And many thousands of Americans, already addicted, need help.

28.5 The Spinal Cord

> **LEARNING OBJECTIVE 28.5.1** Describe the structure of the spinal cord, and explain its general function.

The **spinal cord** is a cable of neurons extending from the brain down through the backbone; a section of the spinal cord is shown in **figure 28.12**. The cross section through the spinal cord in **figure 28.13** shows a darker gray area in the center that consists of neuron cell bodies, which form a column down the length of the cord. This column is surrounded by axons and dendrites, which make the outer edges of the cord white because they are coated with myelin. The spinal cord is surrounded and protected by a series of bones called the vertebrae. Spinal nerves pass out to the body from between the vertebrae. Messages between the body and the brain run up and down the spinal cord, like an information highway.

In each segment of the spine, motor nerves extend out of the spinal cord to the muscles. Motor nerves from the spine control most of the muscles below the head. This is why injuries to the spinal cord often paralyze the lower part of the body. A muscle is paralyzed and cannot move if its motor neurons are damaged.

> **Putting the Concept to Work**
> Does a nerve signal pass through the grey or white matter of the spine, or both?

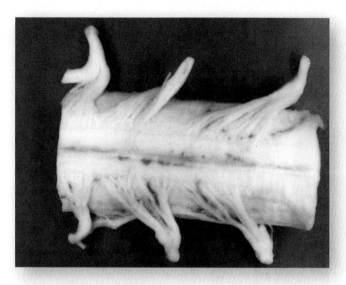

Figure 28.12 A portion of the human spinal cord.
Pairs of spinal nerves can be seen extending out from the spinal cord. Along these nerves, the brain and spinal cord communicate with the body.
Norbert Dr. Lange/Alamy Stock Photo

Figure 28.13 The vertebrate nervous system.
The spinal cord connects to the base of the brain, and nerves extend out from the spinal cord to all parts of the body.

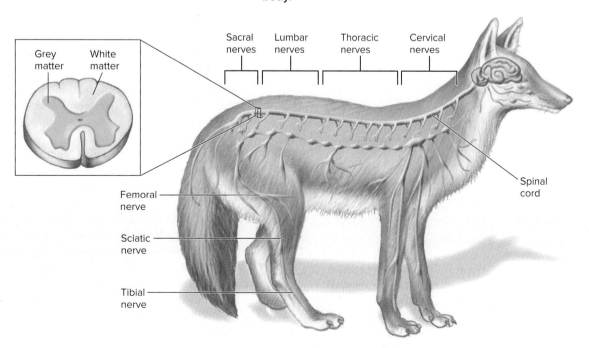

The Peripheral Nervous System

28.6 The Voluntary and Autonomic Nervous Systems

> **LEARNING OBJECTIVE 28.6.1** Contrast the voluntary and autonomic nervous systems, describing the opposing actions of the sympathetic and parasympathetic nervous systems.

As you learned in the opening discussion in section 28.1, the nervous system is divided into two main parts: the central nervous system and the peripheral nervous system. The motor pathways of the peripheral nervous system of a vertebrate can be further subdivided into the **somatic (voluntary) nervous system,** which relays commands to skeletal muscles, and the **autonomic (involuntary) nervous system,** which stimulates glands and relays commands to the smooth muscles of the body and to cardiac muscle. The voluntary nervous system can be controlled by conscious thought. You can, for example, command your hand to move. The autonomic nervous system, by contrast, cannot be controlled by conscious thought. You cannot, for example, tell the smooth muscles in your digestive tract to speed up their action. The central nervous system issues commands over both voluntary and autonomic systems, but you are conscious of only the voluntary commands.

Voluntary Nervous System

Motor neurons of the voluntary nervous system stimulate skeletal muscles to contract in two ways. First, motor neurons may stimulate the skeletal muscles of the body to contract in response to conscious commands. For example, if you want to bounce a basketball, your CNS sends messages through motor neurons to the muscles in your arms and hands. However, skeletal muscle can also be stimulated as a part of reflexes that do not require conscious control.

Autonomic Nervous System

Some motor neurons are active all the time. These neurons carry messages from the CNS that keep the body going even when it is not active. These neurons make up the autonomic nervous system. The autonomic nervous system carries messages to muscles and glands that work without the animal noticing.

The autonomic nervous system is the command network used by the CNS to maintain the body's homeostasis. Using it, the CNS regulates heartbeat and controls muscle contractions in the walls of the blood vessels. It directs the muscles that control blood pressure, breathing, and the movement of food through the digestive system. It also carries messages that help stimulate glands to secrete tears, mucus, and digestive enzymes.

The autonomic nervous system is composed of two divisions that act in opposition to one another. One division, the **sympathetic nervous system,** dominates in times of stress. It controls the "fight-or-flight" reaction, increasing blood pressure, heart rate, breathing rate, and blood flow to the muscles. The sympathetic nervous system is colored pink in figure 28.14, with neurons extending from the middle section of the spinal cord. Long motor

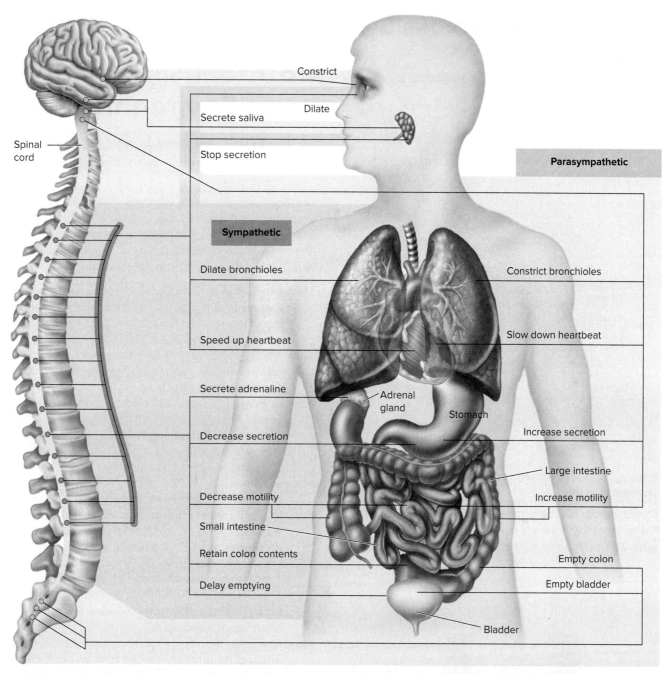

Figure 28.14 How the sympathetic and parasympathetic nervous systems interact.

A nerve path runs from both systems to every organ indicated except the adrenal gland, which is only innervated by the sympathetic nervous system.

neurons extend from the ganglia directly to each target organ. Another division, the **parasympathetic nervous system,** has the opposite effect. It conserves energy by slowing the heartbeat and breathing rate and by promoting digestion and elimination. The parasympathetic nervous system is colored in blue in the figure, with neurons extending from the upper and lower sections of the spinal cord.

Most glands, smooth muscles, and cardiac muscles get constant input from *both* systems. The CNS controls activity by varying the ratio of the two signals to either stimulate or inhibit the organ.

> **Putting the Concept to Work**
> Which element of the autonomic nervous system—the sympathetic or parasympathetic—acts to slow the heartbeat?

The Sensory Nervous System

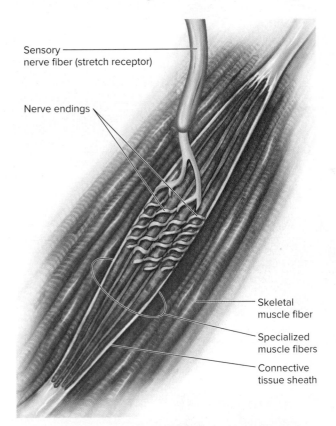

Figure 28.15 A stretch receptor embedded within skeletal muscle.

Stretching the muscle elongates the specialized muscle fibers, which deforms the nerve endings, causing them to send a nerve impulse out along the nerve fiber.

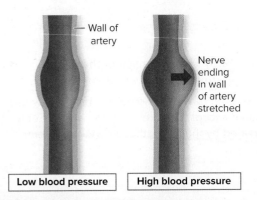

Figure 28.16 How a baroreceptor works.

A network of nerve endings covers a region where the wall of the artery is thin. High blood pressure causes the wall to balloon out there, stretching the nerve endings and causing them to fire impulses.

28.7 Sensing the Internal Environment

> **LEARNING OBJECTIVE 28.7.1** Describe how the body's sensory receptors respond to aspects of its internal environment.

Sensory receptors inside the body called **interoceptors** inform the CNS about the condition of the body. Much of this information passes to the hypothalamus, the part of the brain responsible for the body's homeostasis. The vertebrate body uses a variety of different sensory receptors to respond to different aspects of its internal environment.

Temperature change. Two kinds of nerve endings in the skin are sensitive to changes in temperature, one stimulated by cold, the other by warmth. By comparing information from the two, the CNS can learn what the temperature is and if it is changing.

Blood chemistry. Receptors in the walls of arteries sense CO_2 levels in the blood. The brain uses this information to regulate the body's respiration rate, increasing it when CO_2 levels rise above normal.

Pain. Damage to tissue is detected by special nerve endings within tissues, usually near the surface, where damage is most likely to occur. When these nerve endings are physically damaged or deformed, the CNS responds by reflexively withdrawing the body part and often by changing heartbeat and blood pressure.

Muscle contraction. Buried deep within muscles are sensory receptors called stretch receptors. In each, the end of a sensory neuron is wrapped around a muscle fiber, like the receptor shown in **figure 28.15**. When the muscle is stretched, the fiber elongates, stretching the spiral nerve ending (like stretching a spring) and causing repeated nerve impulses to be sent to the brain. From these signals, the brain can determine the rate of change of muscle length at any given moment. The CNS uses this information to control movements that require the combined action of several muscles, such as those that carry out breathing or locomotion.

Blood pressure. Blood pressure is sensed by neurons called baroreceptors with highly branched nerve endings within the walls of major arteries. When blood pressure increases, the stretching of the arterial wall, like the expansion of the artery in **figure 28.16**, causes the sensory neuron to increase the rate at which it sends nerve impulses to the CNS. When the wall of the artery is not stretched, the rate of firing of the sensory neuron goes down. Thus, the frequency of impulses provides the CNS with a continuous measure of blood pressure.

Touch. Touch is sensed by pressure receptors buried below the surface of the skin. There are a variety of different types, some specialized to detect rapid changes in pressure, others to measure the duration and extent to which pressure is applied, and still others sensitive to vibrations.

> **Putting the Concept to Work**
> How does the brain know whether an incoming sensory impulse is temperature, pressure, or pain?

28.8 Sensing Gravity and Motion

LEARNING OBJECTIVE 28.8.1
Describe how the inner ear senses gravity and acceleration.

Two types of receptors in the ear inform the brain where the body is in three dimensions. This knowledge enables an animal to move freely and maintain its balance. **Figure 28.17** shows the anatomy of the inner ear and the locations of these receptors.

Balance

To keep the body's balance, the brain needs a frame of reference, and the reference point it uses is gravity. The sensory receptors that detect gravity are hair cells within the utricle and the saccule of the inner ear (**figure 28.17 ❶**). The tips of the hair cells project into a gelatinous matrix with embedded particles called **otoliths** ❷. To illustrate how these receptors work, imagine a pencil standing in a glass. No matter which way you tip the glass, the pencil rolls along the rim due to the pull of gravity, applying pressure to the lip of the glass. If you want to know the direction the glass is tipped, you need only ask where on the rim pressure is being applied. Similarly, otoliths in the utricle and saccule will shift in the matrix in response to the pull of gravity and stimulate hair cells. The brain uses information from the hair cells to determine vertical positioning.

Motion

The brain senses motion by employing a receptor in which fluid deflects cilia of hair cells in a direction opposite that of the motion. Within the inner ear are three fluid-filled **semicircular canals,** each oriented in a different plane at right angles to the other two (❶) so that motion in any direction can be detected. Protruding into the canal are groups of cilia from sensory hair cells. The cilia from each cell are arranged in a tentlike assembly called a *cupula,* shown in ❸, which is pushed when fluid in the semicircular canals moves in a direction opposite that of the head's movement. Because the three canals are oriented in all three planes, movement in any plane is sensed by at least one of them, and the brain is able to analyze body movements by comparing the sensory inputs from each canal.

The semicircular canals do not react if the body moves in a straight line because the fluid in the canals does not move. That is why traveling in a car or airplane at a constant speed in one direction gives no sense of motion.

Putting the Concept to Work
Why does someone falling from a high window sense movement, whereas someone descending in an elevator does not?

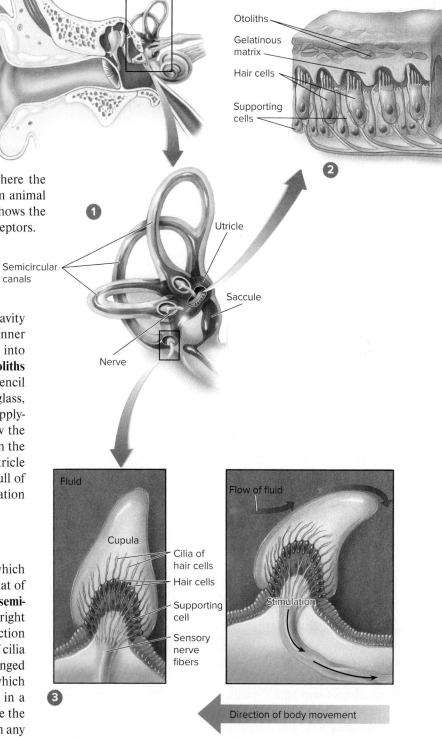

Figure 28.17 How the inner ear senses gravity and motion.

❶ The semicircular canals are part of the inner ear. ❷ Enlargement of a section of the utricle or saccule. Otoliths embedded in the gelatinous matrix move in response to the pull of gravity. ❸ The cupulae within the semicircular canals are surrounded by fluid and contain hair cells. Movement in a particular direction causes fluid in the semicircular canal of that plane to move; the cupula is displaced, thereby stimulating the hair cells.

A Closer Look

A Sense of Where You Are

LeBron James Is Able to Sink a Jump Shot Without Looking at the Basket

Maddie Meyer/Staff/Getty Images

It is June 2017 and you are watching the NBA championship game. LeBron James has the ball and is moving past the basket, closely guarded. Without looking back, he unexpectedly tosses the ball back over his shoulder. The ball rises up in a tight arc and drops smoothly right through the net, behind James, who is still moving away from the basket. Two more points among the 41 he scores that night.

But wait a minute. LeBron James wasn't even looking at the basket! How did he DO that? To those who watch a lot of basketball, the shot does not come as a great surprise. The over-the-shoulder basket was a specialty of Bill Russell, a great of a past generation, and of Bill Bradley, one of the most accurate of all shooters from out on the floor in the days before you got three points for doing it. A reporter for *The New Yorker* magazine who interviewed Bradley about his over-the-shoulder shot wrote that Bradley tossed a ball over his shoulder and into the basket while he was talking and looking the reporter in the eye. The reporter retrieved the ball and handed it back to him. "When you have played basketball for a while, you don't need to look at the basket when you are in close like this," Bradley said, throwing it over his shoulder again and right through the hoop. "You develop a sense of where you are."

A sense of where you are. That is what LeBron James had that June night. But still—how did he DO that?

Place Cells

To answer this question, we must look back to 1948, when an experimental psychologist in London named Ed Tolman began studying how animals learn to navigate. He came to the conclusion that animals form some sort of cognitive "map" in their brain that allows them to move sensibly through their environment. He had no idea, of course, of what a cognitive map might look like or where it might be located in the brain. The prevailing view among behavioral psychologists was that Tolman was being overly simplistic and that his so-called "map" was not a thing, just a metaphor.

There the problem sat until a young researcher, John O'Keefe, began to study animal behavior in the late 1960s. Working carefully, O'Keefe was able to implant a tiny electrode into a single cell within the brain of a rat and record when that cell emitted a signal (in researcher lingo, the cell is said to have "fired"). He chose cells in the hippocampus, a small region of the brain that appears to be involved with balance, and after inserting the electrode, he let the experimental rat run about in a box. As the rat explored the box, O'Keefe recorded exactly where the rat went and when the cell fired. The firing pattern was unlike anything he might have expected! Individual cells fired only when the rat was at a particular place on the floor of the box and nowhere else. Different hippocampus cells fired when the rat was at different places in the box.

O'Keefe called the individual hippocampus cells "place cells" and concluded that an animal's memory of a place is stored as a specific combination of place cells in the hippocampus. Tolman's map was real after all. For his discovery of place cells, O'Keefe was awarded half of the 2014 Nobel Prize in Physiology or Medicine.

Grid Cells

Now we must fast-forward to the 1990s. Tolman's exciting finding had sparked a large number of scientists to study how place cells create a map of the environment within the brain. A young couple, Edvard and May-Britt Moser, set out to address a simple question about O'Keefe's hippocampus maps: Might the firing of place cells be triggered by activity in other parts of the brain outside the hippocampus? A simple question, it led right off an intellectual cliff. At first, studies of cells in a brain structure adjacent to the hippocampus seemed to show place cells similar to those discovered by O'Keefe in the hippocampus. Then the young investigators did something only young investigators tend to do—they thought "outside the box."

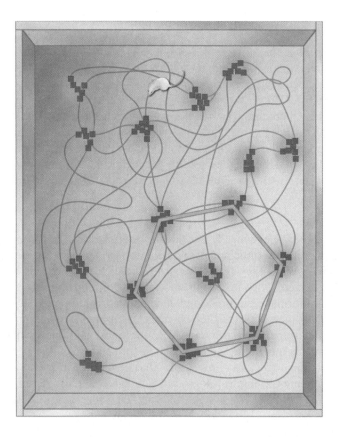

The way the Mosers changed the standard O'Keefe experiment was the way an inquisitive child might—they made a bigger box. When a rat was allowed to run about in this far larger environment, the firing of the place cells they had found showed an astonishing pattern: Individual cells were active in multiple places! These firing locations were not at all random. Instead, a cell's firing locations form the nodes of a grid, interlacing hexagons like the chambers of a beehive. In recognition of this repeating pattern, the Mosers called these cells "grid cells." For their discovery of grid cells, they shared the other half of the 2014 Nobel Prize.

How LeBron James Does It

This was a very exciting result because a grid system allows the brain to measure how far a body moves and so put scale to the place map in the hippocampus. Here's how (you will need to look at the diagram to get this): Imagine the firing pattern of a particular grid cell is the red grid of boxes on the diagram. The pattern of red boxes forms a hexagon.

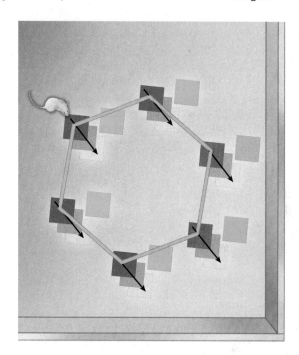

Now imagine a second grid cell has a firing pattern represented by the orange boxes and a third grid cell, a pattern represented by the yellow boxes. Now look what happens when the animal moves from where it is on top of the red node, traveling toward the lower right: the orange grid cells fire, then the yellow grid cells. This gives the brain information about direction, speed, and distance! As the rat moves, the grid cell system is constantly updating the place map of the hippocampus with information about distance and direction of movement.

It is important to remember that humans are not rats. Often, experiments carried out on rats produce results that differ from what happens when the same experiment is carried out in humans. Who says a human like LeBron locates himself in space the same way a rat does? Actually, it seems likely. Human grid cells were reported in 2013, and while the experiments are more indirect (you can't go around inserting electrodes into someone's brain and asking the person to walk around while you record), human grid cells seem to work just like rat grid cells.

So that's how LeBron James does it. He's got great grid cells!

28.9 Sensing Chemicals: Taste and Smell

LEARNING OBJECTIVE 28.9.1 Contrast the senses of taste and smell.

Vertebrates are able to detect many of the chemicals in air and in food.

Taste

Embedded within the surface of the tongue are *taste buds* located within *papillae*, which are the raised areas on the tongue in **figure 28.18**. Taste buds (the onion-shaped structures in the figure) contain many taste receptor cells, each of which has fingerlike microvilli that project into an opening called the taste pore. Chemicals from food dissolve in saliva and contact the taste cells through the taste pore. Salty, sour, sweet, bitter, and umami (a "meaty" taste) are perceived because chemicals in food are detected in different ways by taste buds. When the tongue encounters a chemical, information from the taste cells passes to sensory neurons, which transmit the signals to the brain.

Smell

The nose contains chemically sensitive neurons whose cell bodies are embedded within the epithelium of the nasal passage, shown in cross section in **figure 28.19**. When they detect chemicals, these sensory neurons (the red cells in the enlarged view) transmit information to a location in the brain where smell information is processed and analyzed. Although humans can sense only five different tastes, they can detect thousands of different smells. It appears that as many as a thousand different genes may code for different receptor proteins for smell. The particular set of neurons that respond to a given odor might serve as a "fingerprint" that the brain can use to identify the odor. In many vertebrates (dogs are a familiar example), these neurons are far more sensitive than in humans.

Smell and taste are very important senses in telling an animal about its food. That is why when you have a bad cold and your nose is stuffed up, your food has little taste. Other receptors also play a role. For example, the "hot" sensation of foods such as chili peppers is detected by pain receptors, not chemical receptors.

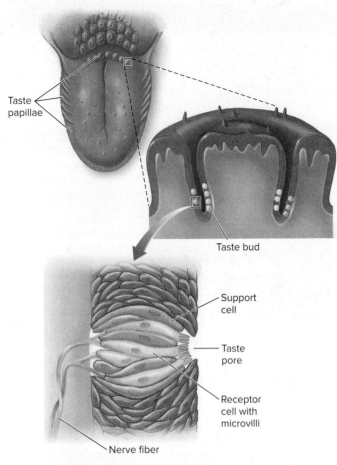

Figure 28.18 Taste.
Taste buds on the human tongue are typically grouped into projections called papillae. Individual taste buds are bulb-shaped collections of taste receptor cells that open out into the mouth through a taste pore.

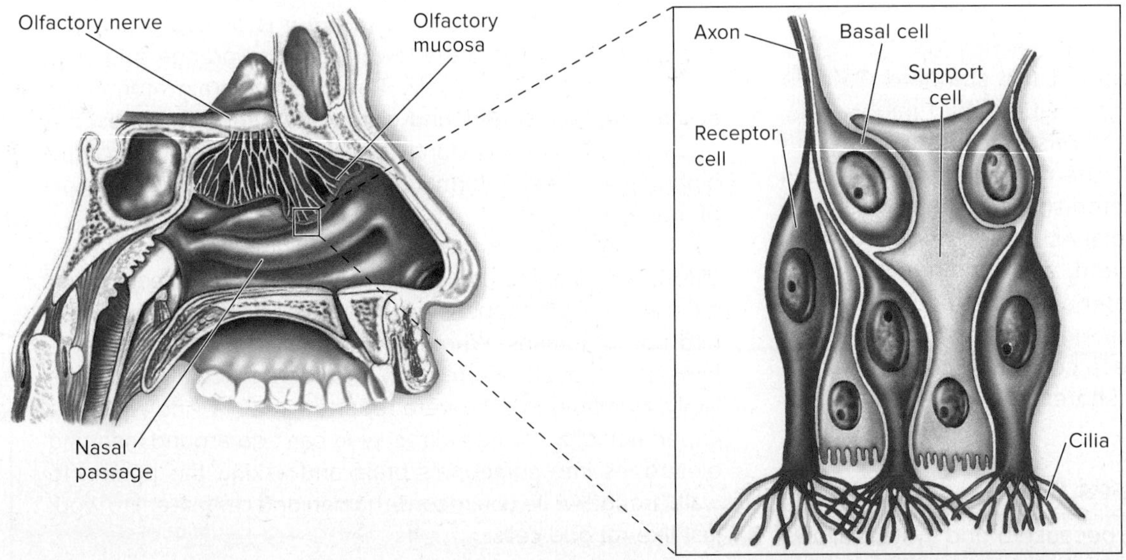

Figure 28.19 Smell.
Humans smell by using receptor cells located in the lining of the nasal passage. The receptor cells are neurons. Axons from these sensory neurons project back through the olfactory nerve directly to the brain.

> **Putting the Concept to Work**
> Why is your dog able to smell better than you can? Is it just because it has a bigger nose?

28.10 Sensing Sounds: Hearing

> **LEARNING OBJECTIVE 28.10.1** Explain how sound receptors within the ear's cochlea differentiate between sounds of different frequencies and different intensities.

When you hear a sound, you are detecting the air vibrating—waves of pressure in the air beating against your ear, pushing a membrane called the eardrum in and out.

Vibrating Air

As you can see in **figure 28.20**, on the inner side of the eardrum are three small bones, called ossicles, that act as a lever system to increase the force of the vibration. They transfer the amplified vibration across a second membrane to fluid within the inner ear. The fluid-filled chamber of the inner ear is shaped like a tightly coiled snail shell and is called the cochlea, from the Latin name for "snail." The middle ear, where the ossicles are located, is connected to the throat by the eustachian tube in such a way that there is no difference in air pressure between the middle ear and the outside. That is why your ears sometimes "pop" when landing in an airplane—the pressure is equalizing between the two sides of the eardrum. This equalized pressure is necessary for the eardrum to work.

Hair Cells. The sound receptors within the cochlea are hair cells that rest on a membrane that runs up and down the middle of the spiraling chamber, separating it into two halves, the upper and lower fluid-filled canals in the enlarged view of the figure. The hair cells do not project into the fluid-filled canals of the cochlea; instead, they are covered by a second membrane (the darker blue membrane in the figure). When a sound wave enters the cochlea, it causes the fluid in the chambers to move. The moving fluid causes this membrane "sandwich" to vibrate, bending the hairs pressed against the upper membrane and causing them to send nerve impulses to sensory neurons that travel to the brain.

Detecting Sound Frequency. Sounds of different frequencies travel different distances down the length of the cochlea and cause different parts of the membrane to vibrate. Each area of the membrane fires a different set of sensory neurons; the identity of the sensory neuron being fired tells the CNS the frequency of the sound. Sound waves of higher frequencies, about 20,000 vibrations (or cycles) per second, also called hertz (Hz), don't travel very far into the cochlea and move the membrane in the area closest to the middle ear. Medium-length frequencies, about 2,000 Hz, travel farther and move the membrane in the area about midway down the length of the cochlea. The lowest-frequency sound waves, about 500 Hz, move the membrane near the tip of the cochlea.

Detecting Sound Intensity. The intensity of the sound is determined by how *often* the neurons fire. Our ability to hear depends upon the flexibility of the membranes within the cochlea. Humans cannot hear low-pitched

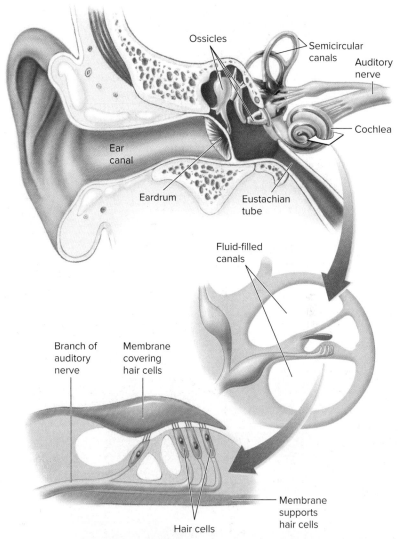

Figure 28.20 Structure and function of the human ear.

Sound waves passing through the ear canal beat on the eardrum, pushing a set of three small bones, or ossicles, against an inner membrane. This sets up a wave motion in the fluid filling the canals within the cochlea. The wave causes the membrane covering the hair cells to move back and forth against the hair cells, which causes associated neurons to fire impulses.

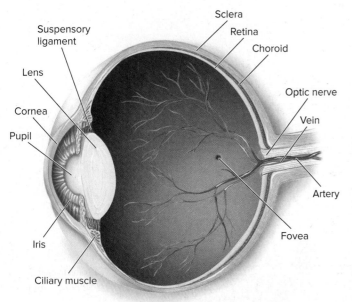

Figure 28.21 The structure of the human eye.
Light passes through the transparent cornea and is focused by the lens on the rear surface of the eye, the retina. The retina is rich in photoreceptors, with a high concentration in an area called the fovea.

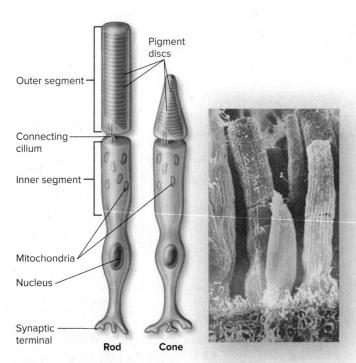

Figure 28.22 Rods and cones.
The broad tubular cell on the *left* is a rod. The shorter, tapered cell next to it is a cone. An electron micrograph of rods and cones is also shown.
Omokron/Science Source

sounds, below 20 Hz, although some vertebrates can. As children, we can hear high-pitched sounds, up to 20,000 cycles per second, but this ability decreases as we get older. Other vertebrates can hear sounds at far higher frequencies. Dogs readily hear sounds of 40,000 cycles per second and so respond to a high-pitched dog whistle that seems silent to a human.

> **Putting the Concept to Work**
> How does the ear distinguish a sound's frequency?

28.11 Sensing Light: Vision
Structure of the Vertebrate Eye

> **LEARNING OBJECTIVE 28.11.1** Diagram the structure of the human eye, explaining the functions of its elements.

The vertebrate eye works like a lens-focused camera. Light first passes through a transparent protective covering, the **cornea** (the light blue layer in figure 28.21), which begins to focus the light onto the rear of the eye. The beam of light then passes through the **lens,** which completes the focusing. The lens is attached by stringlike *suspensory ligaments* to **ciliary muscles.** When these muscles contract and relax, they change the shape of the lens and thus allow the eye to view objects that are far and near. The amount of light entering the eye is controlled by a shutter, called the **iris** (the colored part of your eye), between the cornea and the lens. The transparent zone in the middle of the iris, the **pupil,** gets larger in dim light and smaller in bright light. The pupil also gets smaller when the eye is viewing close objects.

The Retina. The light that passes through the pupil is focused by the lens onto the back of the eye. An array of light-sensitive receptor cells lines the back surface of the eye, called the **retina.** The retina is the light-sensing portion of the eye. The vertebrate retina contains two kinds of photoreceptors, called **rods** and **cones,** which, when stimulated by light, generate nerve impulses that travel to the brain along a short, thick nerve pathway called the optic nerve. Rods, the taller, flat-topped cells in figure 28.22, are receptor cells that are extremely sensitive to light, and they can detect various shades of gray even in dim light. However, they cannot distinguish colors, and because they do not detect edges well, they produce poorly defined images. Cones, the pointed-topped cells, are receptor cells that detect color and are sensitive to edges so that they produce sharp images. The center of the vertebrate retina contains a tiny pit, called the **fovea,** densely packed with some 3 million cones. This area produces the sharpest image, which is why we tend to move our eyes so that the image of an object we want to see clearly falls on this area.

> **Putting the Concept to Work**
> How do ciliary muscles focus the eye?

How the Eye Senses Light

> **LEARNING OBJECTIVE 28.11.2** Explain how a photon of light initiates a sensory nerve impulse.

A rod or cone cell in the eye is able to detect a single photon of light. How can it be so sensitive? The primary sensing event of vision is the absorption of a photon of light by a pigment. The pigments in rods and cones are made from plant pigments called carotenoids. That is why eating carrots is said to be good for night vision—the orange color of carrots is due to the presence of carotenoids called carotenes. The visual pigment in the human eye is a fragment of carotene called *cis*-retinal. The pigment is attached to a protein called opsin to form a light-detecting complex called **rhodopsin**.

When it receives a photon of light, the pigment undergoes a change in shape. This change in shape must be large enough to alter the shape of the opsin protein attached to it. When light is absorbed by the *cis*-retinal pigment (the upper molecule in figure 28.23), the linear end of the molecule rotates sharply upward, straightening out that end of the molecule. The new form of the pigment is referred to as *trans*-retinal, and the dashed outline in the figure shows the shape before it was stimulated by light. This radical change in the pigment's shape induces a change in the shape of the protein opsin to which the pigment is bound, initiating a chain of events that leads to the generation of a nerve impulse.

Each rhodopsin activates several hundred molecules of a protein called transducin. Each of these activates several hundred molecules of an enzyme whose product stimulates sodium channels in the photoreceptor membrane at a rate of about 1,000 per second. This cascade of events allows a single photon to have a large effect on the receptor.

> **Putting the Concept to Work**
> Why doesn't the eye run out of *cis*-retinal pigment?

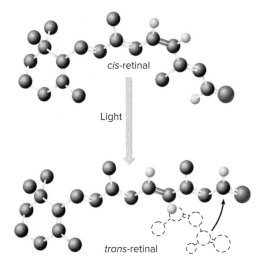

Figure 28.23 Absorption of light.

When light is absorbed by *cis*-retinal, the pigment undergoes a change in shape and becomes *trans*-retinal.

Color Vision

> **LEARNING OBJECTIVE 28.11.3** Distinguish between rods and cones, and explain how cones produce color vision.

Three kinds of cone cells provide us with color vision. Each possesses a different version of the opsin protein (that is, one with a distinctive amino acid sequence and thus a different shape). These differences in shape affect the flexibility of the attached retinal pigment, shifting the wavelength at which it absorbs light. The absorption spectrum in figure 28.24 shows the wavelength of light that is absorbed by each cone and rod cell. In rods, light is absorbed at 500 nanometers. In cones, the three versions of opsin absorb light at 420 nanometers (blue-absorbing), 530 nanometers (green-absorbing), or 560 nanometers (red-absorbing). By comparing the relative intensities of the signals from the three types of cones, the brain can calculate the intensity of other colors.

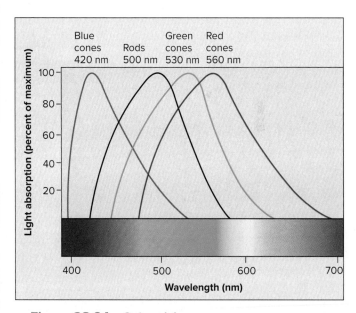

Figure 28.24 Color vision.

The absorption spectrum of *cis*-retinal is shifted in cone cells from the 500 nanometers characteristic of rod cells. The amount of the shift determines what color the cone absorbs: 420 nanometers yields blue absorption; 530 nanometers yields green absorption; and 560 nanometers yields red absorption. Red cones do not peak in the red part of the spectrum, but they are the only cones that absorb light there.

Figure 28.25 Structure of the retina.

The rods and cones are at the rear of the retina. Light passes over four other types of cells in the retina before it reaches the rods and cones. Nerve impulses then travel through the bipolar cells to the ganglion cells and on to the optic nerve (as indicated by the black arrows).

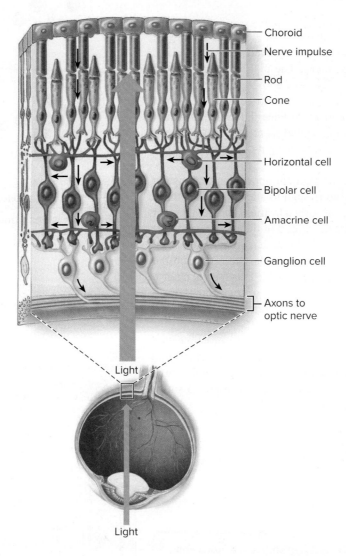

The path of light through each eye is the reverse of what you might expect. The rods and cones are at the rear of the retina, not the front. If you track the path that light would take in figure 28.25, you will see that light passes through several layers of ganglion and bipolar cells before it reaches the rods and cones. Once the photoreceptors are activated, they stimulate bipolar cells, which in turn stimulate ganglion cells. The direction of nerve impulses in the retina is thus opposite to the direction of light.

Putting the Concept to Work
Does light pass through a cone receptor from its pointed tip to its base or up from its base and out through its tip?

A Closer Look

How the Platypus Sees with Its Eyes Shut

The duck-billed platypus (*Ornithorhynchus anatinus*) is abundant in freshwater streams of eastern Australia. These mammals have a unique mixture of traits—in 1799, British scientists were convinced that the platypus skin they received from Australia was a hoax. The platypus is covered in soft fur and has mammary glands, but in other ways it seems very reptilian. Females lay eggs as reptiles do, and like reptilian eggs, the yolk of the fertilized egg does not divide. In addition, the platypus has a tail not unlike that of a beaver, a bill not unlike that of a duck, and webbed feet!

It turns out that platypuses also have some very unique behaviors. Until recently, few scientists had studied the platypus in its natural habitat—it is elusive, spending its days in burrows it constructs on the banks of waterways. Also, a platypus is active mostly at night, diving in streams and lagoons to capture bottom-dwelling invertebrates such as shrimps and insect larvae. Interestingly, unlike whales and other marine mammals, a platypus cannot stay under water long. Its dives typically last a minute and a half. (Try holding your breath that long!)

When scientists began to study the platypus's diving behavior, they noticed something curious: The eyes and ears of a platypus are located within a muscular groove, and when a platypus dives, the sides of these grooves close over tightly. Imagine pulling your eyebrows down to your cheeks: Effectively blindfolded, you wouldn't be able to see a thing! To complete its isolation, the nostrils at the end of the snout also close. So how in the world does the animal find its prey?

For over a century, biologists have known that the soft surface of the platypus bill is pierced by hundreds of tiny openings. In recent years, Australian neuroscientists (scientists that study the brain and nervous system) have learned that these pores contain sensitive nerve endings. Nestled in an interior cavity, the nerves are protected from damage by the bill, but are linked to the outside streamwater via the pore. The nerve endings act as sensory receptors, communicating to the brain information about the animal's surroundings. These pores in the platypus bill are its diving "eyes."

Platypuses have two types of sensory cells in these pores. Clustered in the front are so-called mechanoreceptors, which act like tiny pushrods. Anything pushing against them triggers a signal. Your ears work the same way, sound waves pushing against tiny mechanoreceptors within your ears. These pushrods evoke a response over a much larger area of the platypus brain than does stimulation from the eyes and ears—for the diving platypus, the bill is the primary sense organ. What responses do the pushrod receptors evoke? Touching the bill with a fine glass probe reveals the answer—a lightning-fast, snapping movement of its jaws. When the platypus contacts its prey, the pushrod receptors are stimulated, and the jaws rapidly snap and seize the prey.

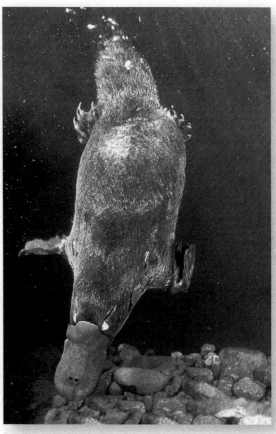

Dave Watts/Alamy Stock Photo

But how does the platypus locate its prey at a distance, in murky water with its eyes shut? That is where the other sort of sensory receptor comes in. When a platypus feeds, it swims along, steadily wagging its bill from side to side, two or three sweeps per second, until it detects and homes in on prey. How does the platypus detect the prey individual and orient itself to it? The platypus does not emit sounds like a bat, which rules out the possibility of sonar as an explanation. Instead, electroreceptors in its bill sense the tiny electrical currents generated by the muscle movements of its prey as the shrimp or insect larva moves to evade the approaching platypus!

It is easy to demonstrate this, once you know what is going on. Just drop a small 1.5-volt battery into the stream. A platypus will immediately orient to it and attack it, from as far away as 30 centimeters. Some sharks and fishes have the same sort of sensory system. In muddy murky waters, sensing the muscle movements of prey is far superior to trying to see the prey's body or hear it move—which is why the platypus you see in the photo above is hunting with its eyes shut.

Putting the Chapter to Work

1 Your instructor shows you a video on how an action potential passes along a neuron's axon. She stops the video at a part that shows the initial entry of Na⁺ ions into the cell in response to a stimulus.

What has happened to the ion potential of the membrane at this point?

2 In 1848, a young man named Phineas Gage was working on a railroad when an explosion occurred, causing an iron rod to pierce his left cheek, pass through the roof of his mouth, into and through his brain, and out the top of his head. He recovered from surgery and was released after 10 weeks in the hospital. While most of his brain functions were not impaired, he did suffer epileptic seizures, uncontrollable outbursts in communication, and disorders related to his ability to form associations.

What part of his brain was permanently injured?

3 You are home alone, and you hear your home alarm system go off. You frantically jump out of bed and feel your heart beating quickly. You are sweating and scared. You peer out the window to see a cat jumping down from the eaves. Relieved, you realize that it was not an intruder who activated the alarm—just a cat.

Which of your autonomic nervous systems responded to this stressful situation?

Retracing the Learning Path

Neurons and How They Work

28.1 The Nervous System

1. The nervous system is the communication network in the body. The animal nervous system is organized into a central nervous system that includes a brain and spinal cord and contains associative neurons, and a peripheral nervous system that contains sensory and motor neurons.

28.2 Nerve Impulses

1. Neurons are cells that conduct electrical impulses but are supported by neuroglial cells. These cells closely associate with the axons, wrapping them in a fatty material called myelin.

2. Electrical signals begin in dendrites and travel down an axon. Nerve impulses result from the movement of Na⁺ and K⁺ ions across the plasma membranes through voltage-gated channels. The movement of Na⁺ in one area of the membrane causes a change in electrical properties, called depolarization, which causes the opening of adjacent ion channels. If the depolarization is large enough, it will trigger an action potential that will spread down the axon. The action of the Na⁺/K⁺ pumps restores the resting membrane potential.

28.3 The Synapse

1. A synapse is the junction of an axon with another cell. When a nerve impulse reaches the end of an axon, it triggers the release of neurotransmitters that pass across a small gap, called a synaptic cleft, between the axon and the postsynaptic cell. Neurotransmitter molecules bind to receptors on the postsynaptic cell, causing chemically gated ion channels to open. Ions flow across the plasma membrane, creating electrical impulses in the postsynaptic cell.

2. Depending on the type of ion that flows into the cell, the synapse is excitatory or inhibitory. All neural inputs are integrated in the postsynaptic cell, producing an overall positive or negative change in membrane potential.

The Central Nervous System

28.4 How the Brain Works

1. The associative activity of the brain is centered in the cerebral cortex, which lies over the cerebrum.
   - The thalamus and hypothalamus, which lie underneath the cerebrum, process information and integrate bodily functions.
   - The cerebellum controls balance, posture, and muscular coordination. The brain stem controls vital functions, such as breathing, swallowing, heartbeat, and digestion. Language, memory, and learning are localized in the cerebrum.

28.5 The Spinal Cord

1. The spinal cord is a cable of neurons that extends from the brain down the back and is encased in the bony vertebrae of the backbone. Motor nerves carry impulses from the brain and spinal cord out to the body, and sensory nerves carry impulses from the body to the brain and spinal cord.

The Peripheral Nervous System

28.6 The Voluntary and Autonomic Nervous Systems

1. The voluntary nervous system relays commands between the CNS and skeletal muscles, and can be consciously controlled. The autonomic nervous system consists of opposing sympathetic and parasympathetic divisions that unconsciously relay commands between the CNS and muscles and glands.

The Sensory Nervous System

28.7 Sensing the Internal Environment

1. Interoceptors inform the CNS about the internal condition of the body. Baroreceptors in blood vessels, for example, provide the CNS with a continuous measure of blood pressure.

28.8 Sensing Gravity and Motion

1. Sensory receptors in the inner ear sense gravity and acceleration. The otolith sensory receptors detect gravity by the deflection of hair cells caused by the movement of otoliths in a gelatin-like matrix. Motion is detected by the deflection of hair cells in the cupulae of the semicircular canals.

28.9 Sensing Chemicals: Taste and Smell

1. Chemicals are detected through the sense of taste, using taste buds on the tongue, and the sense of smell, using olfactory receptors that line the nasal passages.

28.10 Sensing Sounds: Hearing

1. Sound receptors detect vibrations of air through the deflection of hair cells in the inner ear. Sound waves cause the eardrum to vibrate. That vibration is amplified by bones in the middle ear, which displace fluid in the inner ear. This movement of fluid causes the deflection of hair cells.

28.11 Sensing Light: Vision

1. Sensory receptors in the eye use a lens to focus light on pigment-containing receptors.

2. Light striking the visual pigment retinal initiates a cascade of reactions that open sodium channels in the photoreceptor.

3. Rod cells detect the intensity of light, while cone cells detect different colors of light.

4. Light passes through layers of cells before reaching the rods and cones at the rear of the retina.

Inquiry and Analysis

Do Birds Use Magnetic Particles as Compass Needles?

Some migrating birds use infrasound to orient themselves. Others may use visual cues, like the angle of polarizing light or the direction of a sunset. Many birds that migrate long distances use the earth's geomagnetic field as a source of compass information. If the magnetic field of a blind "orientation cage" (see image) is deflected 120 degrees clockwise by an artificial magnet, a bird that normally orients to the north will orient toward the southeast instead.

The sensory system underlying the magnetic compass of these birds is one of the great mysteries of sensory biology. There are two competing hypotheses.

The Magnetite Hypothesis One hypothesis is that crystals of the magnetic mineral magnetite within brain cells of migrating birds act as miniature compass needles. While trace amounts of magnetite are indeed present in some brain cells, intensive research has failed to confirm that information about the orientation of magnetite particles within these cells is transmitted to any other cells of the brain.

The Photoreceptor Hypothesis An alternative hypothesis is that the primary process underlying the compass is instead a magnetically sensitive chemical reaction within the photoreceptors of the bird's eyes. The alignment of photopigment molecules with the earth's magnetic field might alter the visual pattern in a way that could be used to obtain directional information.

Which hypothesis is correct? Experiments have shown that the magnetic detector used by birds in blind cages is light sensitive, as a photoreceptor compass should be—but this would also be true of a light-activated magnetite compass.

In 2004, University of California, Irvine researchers devised a clever experiment to distinguish between the two hypotheses. They studied the way in which migrating European robins held in orientation cages use the magnetic field as a source of compass information to hop in the appropriate migratory direction. They found that the robins oriented 16 degrees north during the spring migration, the appropriate direction. To distinguish between the magnetite and photoreceptor hypotheses, the robins in the cages were exposed to oscillating, low-level radio frequencies (7 MHz) that would disrupt the energy state of any light-absorbing photoreceptor molecules involved in sensing the magnetic field, but would not affect the alignment of magnetite particles.

The chart presents the results of this study. Each data entry is the mean of three recordings. For each recording, the bird was placed in a 35-inch conical orientation cage lined with coated paper, and the vector (the directional position of first contact with the paper) recorded relative to magnetic north (north = 360 degrees; east = 90 degrees; south = 180 degrees; west = 270 degrees).

Effect of Radio-Disruption on Orientation

Bird	Mean Heading (degrees)	
	Geomagnetic field only	Radio-disrupted
1	26	110
2	20	126
3	4	86
4	350	17
5	15	162
6	1	330
7	18	297
8	20	220
9	354	58
10	24	261
11	358	278
12	37	3

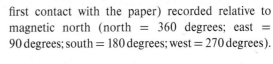

Analysis

1. **Applying Concepts** In the chart, is there a dependent variable? If so, what is it? Discuss.
2. **Interpreting Data** Plot each column on a circle. For birds orienting to the geomagnetic field without radio interference, what is the greatest difference (expressed in degrees) between recorded vectors and the mean vector of 16 degrees? For birds orienting with 7 MHz radio interference?
3. **Making Inferences**
   a. For birds orienting to the geomagnetic field without radio interference, how many of the 12 birds oriented with an accuracy of +/–30 degrees relative to the mean vector of 16 degrees? For birds orienting with 7 MHz radio interference, how many?
   b. If you were to select a bird at random, what is the probability that it would orient within +/–30 degrees of the appropriate migration direction (16 degrees north) without radio interference? With radio interference?
4. **Drawing Conclusions** Is the ability of European robins to orient correctly with respect to geomagnetic fields disrupted by 7 MHz radio frequencies? Is it fair to conclude that the birds' compass sense involves a molecule sensitive to radio disruption, such as a photoreceptor? That it does not involve particles not sensitive to radio disruption?

29 Chemical Signaling Within the Animal Body

LEARNING PATH ▼

The Endocrine System
1. Hormones
2. How Hormones Target Cells

The Major Endocrine Glands
3. The Pituitary
4. The Pancreas
5. The Thyroid, Parathyroid, and Adrenal Glands

Eating Our Way Out of Whack

We Americans love to eat, but recently the Centers for Disease Control and Prevention released a report warning we are eating ourselves into a diabetes epidemic. Diabetes affected 7 million Americans in 1991. By the end of 2020, the number was over 34 million, more than 9% of all Americans, an alarming increase in just 29 years! The same explosion of diabetes is being seen worldwide. Diabetes now affects 463 million people and kills almost 4 million each year. Every 10 seconds, one person dies of diabetes. In the same 10 seconds, two more people develop the disease.

Type II Diabetes

Diabetes is a disorder in which the body's cells fail to take up glucose from the blood. Tissues waste away as glucose-starved cells are forced to consume their own proteins. Diabetes is the leading cause of kidney failure, blindness, and amputation in adults. Almost all the increase

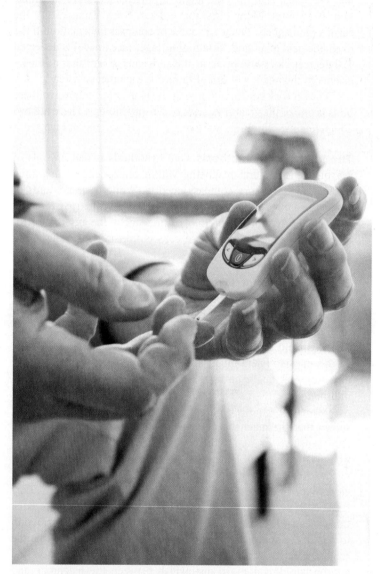

THIS DIABETES PATIENT is monitoring his blood glucose levels with a pinprick of his finger.

Garry Wade/The Image Bank/Getty Images

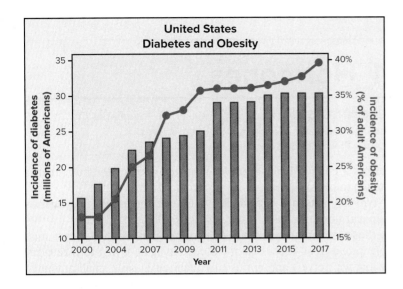

in diabetes in the last decade is in the 90% of diabetics who suffer from type II, or "adult-onset," diabetes. These individuals lack the ability to use the hormone insulin.

Your body manufactures insulin after a meal as a way to alert cells that higher levels of glucose are coming soon. The insulin signal attaches to special receptors on the cell surfaces, which respond by causing the cell to turn on its glucose-transporting machinery. Most individuals who suffer from type II diabetes have normal or even elevated levels of insulin in their blood and normal insulin receptors, but for some reason, the binding of insulin to their cell receptors does not turn on the glucose-transporting machinery as it is supposed to do. For 30 years, researchers have been trying to figure out why not.

Diabetes and Obesity Are Both Exploding

Over that same period that diabetes has exploded among Americans, the obesity rate has increased from 13% of the adult population in 1999 to 42.4% in 2020—an increase of 29% in 21 years! Is there a connection between type II diabetes and obesity? An estimated 80% of those who develop type II diabetes are obese, a tantalizing clue. Look at the graph. During the same time that diabetes has undergone its explosive increase (red bars), the obesity rate increased from 6% of the U.S. population to over 40% (blue line). In addition, type II diabetes, which was once considered a disease of adults, is also on the rise in children. Type II diabetes now makes up one-half of all new cases of diabetes in children under 18 years of age, and most of these children are overweight or obese.

What Is the Link?

What is the link between type II diabetes and obesity? Recent research suggests an answer to this key question. A team of scientists at the University of Pennsylvania School of Medicine had been investigating why a class of drugs called thiazolidinediones (TZDs) helped combat diabetes. They found that TZDs cause the body's cells to use insulin more effectively, suggesting that the TZD drug might be targeting a hormone.

A Hormone Called Resistin

The researchers then set out to see if they could find such a hormone in mice. Looking to see which mouse genes were activated or deactivated by TZD, they were able to zero in on the hormone they sought. Dubbed resistin, the hormone is produced by fat cells and prompts tissues to resist insulin. The same resistin gene is present in humans too. The researchers speculate that resistin may have evolved to help the body deal with periods of famine.

Mice given resistin by the researchers lost much of their ability to take up blood sugar. When given a drug that lowers resistin levels, these mice recovered the lost glucose-transporting ability. Importantly, dramatically high levels of the hormone were found in mice that were obese from overeating. Finding this sort of result is like ringing a dinner bell to diabetes researchers. If obesity is causing high resistin levels in humans, leading to type II diabetes, then resistin-lowering drugs might offer a diabetes cure!

On the scent of something important, research into resistin has been intense. While many studies support a link between resistin and type II diabetes, some researchers have reported conflicting results. Still, the excitement is tangible. With diabetes levels approaching epidemic proportions, a clearer understanding of the link to obesity is essential. Until a cure is found, the best thing for you to do to avoid developing type II diabetes is to eliminate the risk of obesity. Other factors also contribute to the risk, but this one is a no-brainer.

The Endocrine System

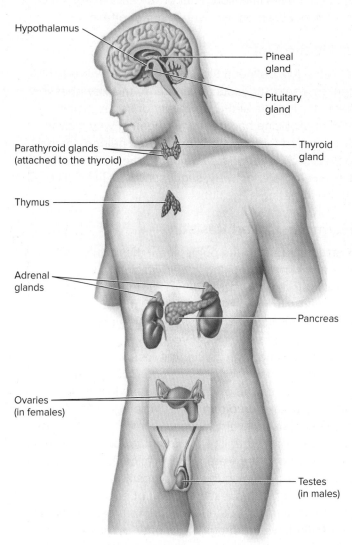

Figure 29.1 Major glands of the human endocrine system.

Hormone-secreting cells are clustered in endocrine glands. The pituitary and adrenal glands are each composed of two glands.

29.1 Hormones

> **LEARNING OBJECTIVE 29.1.1** Explain the advantages of communication by hormones rather than nerves.

A **hormone** is a chemical signal produced in one part of the body that is stable enough to be transported in active form far from where it is produced and that typically acts at a distant site. There are three big advantages to using chemical hormones as messengers rather than speedy electrical signals (like those used in nerves) to control body organs. First, chemical molecules can spread to all tissues via the blood (imagine trying to wire every cell with its own nerve!) and are usually required in only small amounts. Second, chemical signals can persist much longer than electrical ones, a great advantage for hormones controlling slow processes such as growth and development. Third, many different kinds of chemicals can act as hormones, so different hormone molecules can be targeted at different tissues. For all these reasons, hormones are excellent messengers for signaling widespread, slow-onset, long-duration responses.

The Endocrine System

Hormones, in general, are produced by glands, most of which are controlled by the central nervous system. Because these glands are completely enclosed in tissue rather than having ducts that empty to the outside, they are called **endocrine glands** (from the Greek, *endon*, within). Hormones are secreted from them directly into the bloodstream (this is in contrast to **exocrine glands,** such as sweat glands, which have ducts). Your body has a dozen principal endocrine glands (figure 29.1) that together make up the endocrine system.

Hormones secreted by endocrine glands belong to four different chemical categories:

1. **Polypeptides** are composed of chains of amino acids that are shorter than 100 amino acids. Examples include insulin and antidiuretic hormone (ADH).
2. **Glycoproteins** are composed of polypeptides longer than 100 amino acids with attached carbohydrates. Examples include follicle-stimulating hormone (FSH) and luteinizing hormone (LH).
3. **Amines**, derived from the amino acids tyrosine and tryptophan, include a wide variety of hormones secreted by the adrenal and thyroid glands.
4. **Steroids** are lipids derived from cholesterol. Examples include the hormones testosterone, estrogen, and cortisol.

> **Putting the Concept to Work**
> What is the difference between endocrine and exocrine glands?

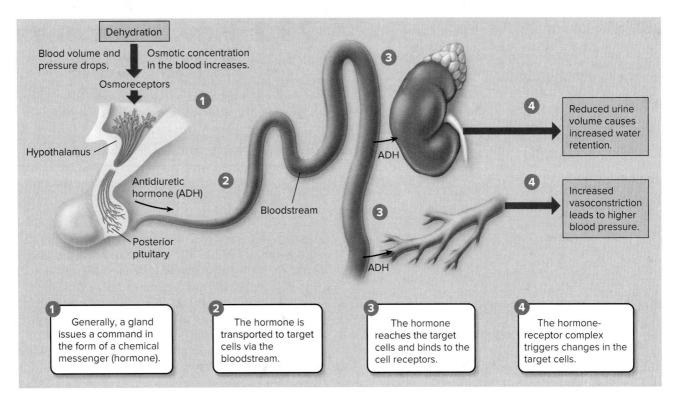

Figure 29.2 **How hormonal communication works.**

How Hormones Work

> **LEARNING OBJECTIVE 29.1.2** Delineate the four stages of hormonal signaling.

Hormones are effective messengers within the body mainly because a particular hormone can influence a specific target cell. How does the target cell recognize that hormone, ignoring all others? Embedded in the plasma membrane or within the target cell are receptor proteins that match the shape of the potential signal hormone like a hand fits a glove. The system is highly specific because cells that the body has targeted to respond to a particular hormone have receptor proteins shaped to fit that hormone and no other. Thus, chemical communication within the body involves *two* elements: a molecular signal (the hormone) and a protein receptor on or in target cells.

The path of communication taken by a hormonal signal can be visualized as the series of simple steps shown in **figure 29.2**:

1. **Issuing the command.** Some hormones produced in glands are released into the bloodstream in response to a signal from the brain.
2. **Transporting the signal.** The hormones travel through the bloodstream to the target cells.
3. **Hitting the target.** When a hormone encounters a cell with a matching receptor, called a target cell, the hormone binds to that receptor.
4. **Having an effect.** When the hormone binds to the receptor protein, the protein responds by changing shape, triggering a cell change.

> **Putting the Concept to Work**
> How do you suppose hormones made in the hypothalamus get to the pituitary to be released?

29.2 How Hormones Target Cells

The chemical nature of hormones determines how they enter the cells of your body. There are two sorts: steroids and peptides.

Steroid Hormones Enter Cells

> **LEARNING OBJECTIVE 29.2.1** Describe the structure and mode of action of steroid hormones.

Some protein receptors designed to recognize hormones are located in the cytoplasm or nucleus of the target cell. The hormones in these cases are typically lipid-soluble **steroid hormones.** All steroid hormones are manufactured from cholesterol, a complex molecule composed of four rings. Steroids include the hormones that promote the development of secondary sexual characteristics such as testosterone, estrogen, and progesterone, discussed in detail in chapter 30.

Because lipids are hydrophobic, steroid hormones such as estrogen, "E" in **figure 29.3**, can pass across the lipid bilayer of the cell plasma membrane ❶ and bind to receptors within the cell and often, as in the case with estrogen, within the nucleus. The hormone-receptor complex then binds to the DNA in the nucleus ❷ and activates the gene for a progesterone receptor protein, which is transcribed ❸. The protein is synthesized ❹, and the receptor is available to bind progesterone when it enters the cell ❺, which itself activates another set of genes.

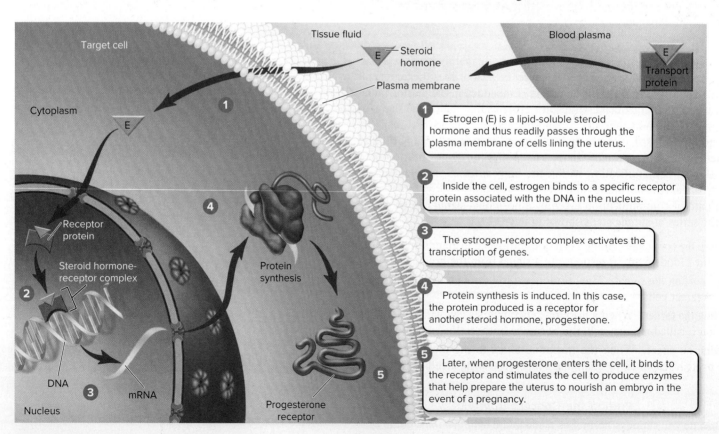

Figure 29.3 How steroid hormones work.

Biology and Staying Healthy

Why Don't Men Get Breast Cancer?

In 2020, an estimated 281,550 new cases of breast cancer were expected among women in the United States and 2650 new cases among men—over 99% of the new breast cancer victims were women. Similarly, of 43,600 breast cancer deaths anticipated that year, all but 530 of them were women. It is impossible not to wonder: Why so few men? An obvious answer would be that men don't have breasts, but men do have breast tissue. It just isn't as developed as a woman's. So why so few men?

The Cause of Breast Cancer Is a Mystery

While 5% of breast cancers are due to inherited genetic mutations (*BRCA1* and *BRCA2*), the cause of the other 95%—the overwhelming majority—remains a mystery. Hormone differences seem the most promising place to start because the female sex hormone estrogen controls breast development in women. Men, by contrast, lack physiologically significant amounts of estrogen. A logical suggestion is that, in breast cancer patients, the effect of estrogen on breast cells is being altered by exposure to a so-called endocrine disrupter. Endocrine disrupters are man-made chemicals that mimic hormones. By sheer chance, their molecules are perfectly shaped to fit particular hormone receptors. In this case, the culprit would be a chemical mimic of estrogen that promotes cancerous growth in breast cells.

BPA

One candidate is bisphenol A (BPA), a molecule that is structurally similar to estrogen, with carbon rings at each end tipped with OH groups. Used to form the plastic packaging of many foods and drinks, as well as the clear plastic liners of metal food and beverage cans, BPA is a chemical to which all of us are exposed daily. Six billion pounds are produced worldwide each year.

It has been known since 1938 that BPA promotes excess estrogen production in rats. Alarmingly, in 1993, BPA was shown to have the same effects on human breast cancer cells growing in culture. What was disturbing was that the effect could be measured at concentrations as low as 2 parts per billion, near the levels to which we humans are routinely exposed.

Does BPA Induce Breast Cancer?

Does BPA in fact induce breast cancer? In an early test of this possibility, Dr. Ana Soto of Tufts University School of Medicine and her research team in 2006 tested its effects on rats, which get breast cancer in much the same way as humans do. They exposed pregnant female rats to a range of BPA concentrations and after 50 days sacrificed them for examination of their breast tissue. The researchers looked in particular for aberrant cell growth patterns in breast tissue called ductal hyperplasias, which in both rats and people are considered to be the precursors of breast cancer. BPA was administered to four groups of rats. Some received low doses not unlike what humans are exposed to, while others received much higher doses. In a fifth group, which served as a control, no BPA was administered.

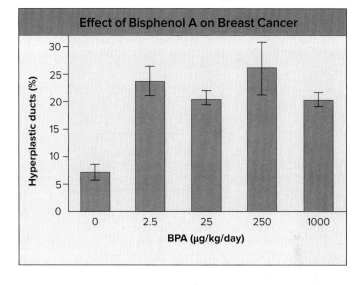

Bisphenol A

Estrogen

BPA Is a Carcinogen

The histogram above shows what the researchers found. Did any of the four doses of BPA they administered result in a percent of hyperplastic ducts significantly higher than that seen in the BPA control group? (Hint: If their error bars overlap with the control's, the difference is not significant.) The answer: Yes, all four doses are significant. Do higher doses of BPA increase the incidence of hyperplastic ducts? Do the error bars of the four doses fail to overlap? No, all doses yielded similar results. It is difficult to avoid the conclusion that exposure to even low levels of bisphenol A induces ductal hyperplasias in laboratory rats. These results suggest the rather alarming conclusion that BPA, a chemical to which we are all exposed every day, may cause breast cancer.

Other Effects

In a much larger study reported in 2012, pregnant mice were exposed to low doses of BPA, analogous to the chronic levels of BPA to which humans are typically exposed. Researchers observed profound behavioral effects on both mothers and offspring. The very low doses of BPA appear to have acted epigenetically to modify how hormone-encoding DNA is regulated, interfering with the expression of hormones such as oxytocin and vasopressin in a way that mimics autism and attention deficit disorder.

A mouse or a rat is not a human, but the possibility that BPA is a carcinogen and an epigenetic endocrine disruptor certainly merits further investigation. Most government-funded breast cancer research has focused on the search for more effective breast cancer treatments; far less money is spent on searching for the causes of breast cancer. The most encouraging aspect of this study is its suggestion that such a search, if it became a priority, might prove fruitful.

Anabolic steroids are synthetic compounds that resemble the male sex hormone testosterone. The injection of anabolic steroids into muscles activates growth genes and causes the muscle cells to produce more protein, resulting in bigger muscles and increased strength. However, anabolic steroids have many dangerous side effects, including liver damage, heart disease, and high blood pressure. Anabolic steroids are illegal, and athletes in many sports are tested for their use.

> **Putting the Concept to Work**
> What does a steroid hormone do after it enters a cell?

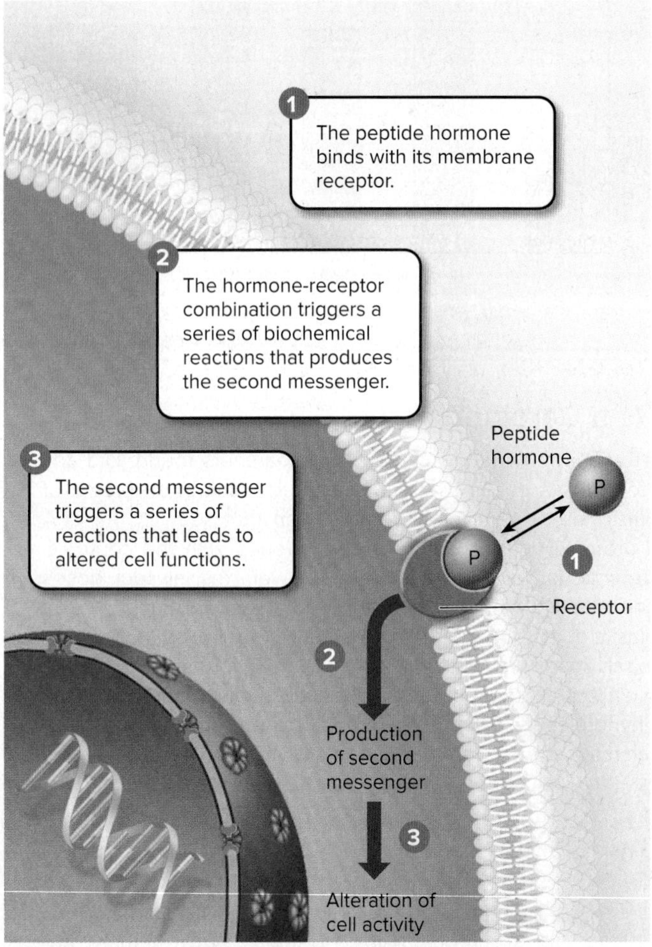

Peptide Hormones Act at the Cell Surface

LEARNING OBJECTIVE 29.2.2 Describe the structure and mode of action of peptide hormones.

Other hormone receptors are embedded within the plasma membrane, with their recognition regions directed outward from the cell surface. **Peptide hormones**, like the one binding to the receptor in **figure 29.4** ❶, are typically short peptide chains (although some are full-sized proteins). The binding of the peptide hormone to the receptor triggers a change in the cytoplasmic end of the receptor protein. This change then triggers events within the cell cytoplasm, usually through intermediate within-cell signals called **second messengers** ❷, which greatly amplify the original signal and result in changes in the cell ❸.

How does a second messenger amplify a hormone's signal? Second messengers activate enzymes. One of the most common second messengers is cyclic AMP (cAMP). A single hormone molecule binding to a receptor in the plasma membrane can result in the formation of many second messengers in the cytoplasm. Each second messenger can activate many molecules of a certain enzyme, and sometimes each of these enzymes can in turn activate many other enzymes. Thus, second messengers enable each hormone molecule to have a tremendous effect inside the cell.

> **Putting the Concept to Work**
> How do second messengers amplify hormone signals?

Figure 29.4 How peptide hormones work.

① The peptide hormone binds with its membrane receptor.

② The hormone-receptor combination triggers a series of biochemical reactions that produces the second messenger.

③ The second messenger triggers a series of reactions that leads to altered cell functions.

The Major Endocrine Glands

29.3 The Pituitary

> **LEARNING OBJECTIVE 29.3.1** Contrast the posterior and anterior pituitary glands.

The **pituitary gland,** located in a bony recess in the brain just below the hypothalamus, produces hormones that influence the body's other endocrine glands. Hormones produced by the back portion of the pituitary, or *posterior lobe,* regulate water conservation, as well as milk letdown and uterine contraction in women; hormones produced by the front portion, or *anterior lobe,* regulate the other endocrine glands.

The Posterior Pituitary

The posterior pituitary contains axons that originate in cell bodies within the hypothalamus. The hormones released from the posterior pituitary are actually produced by neuron cell bodies located in the hypothalamus. The hormones are transported to the posterior pituitary through axon tracts and are stored and released from the posterior pituitary (figure 29.5).

ADH. The role of the posterior pituitary first became evident in 1912, when a remarkable medical case was reported: A man who had been shot in the head developed a surprising disorder: He began to urinate every 30 minutes, unceasingly. The bullet had lodged in his pituitary gland, and subsequent research demonstrated that surgical removal of the pituitary also produces these unusual symptoms. Pituitary extracts were shown to contain a substance that makes the kidneys conserve water, and in the early 1950s, the peptide hormone vasopressin (now more commonly called **antidiuretic hormone, ADH**) was isolated. As you learned in chapter 26, ADH regulates the kidney's retention of water. When ADH is missing, the kidneys cannot retain water, which is why the bullet caused excessive urination. Excessive consumption of alcohol and caffeine, which inhibit ADH secretion, has a similar effect.

Oxytocin. The posterior pituitary also releases a second hormone, **oxytocin,** of very similar structure; both ADH and oxytocin are short peptides composed of nine amino acids. However, oxytocin has a very different function; it initiates uterine contraction during childbirth and milk release in mothers. Here is how milk release works: Sensory receptors in the mother's nipples, when stimulated by sucking, send messages to the hypothalamus, causing the hypothalamus to stimulate the release of oxytocin from the posterior pituitary. The oxytocin travels in the bloodstream to the breasts, where it stimulates contraction of the muscles around the ducts into which the mammary glands secrete milk.

> **Putting the Concept to Work**
> Why does pituitary injury lead to excessive urination?

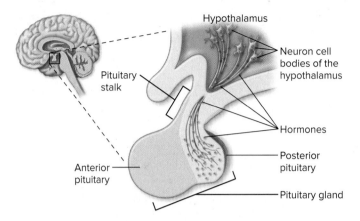

Figure 29.5 The posterior pituitary contains cells that originate in the hypothalamus.

A tract of nerve cells originates in the hypothalamus, extends down along the pituitary stalk, and ends in the posterior pituitary. The cell bodies of the neurons produce hormones, which travel down the axons and are stored in the posterior pituitary. Thus, the hormones released from the posterior pituitary are actually synthesized in neurons in the hypothalamus.

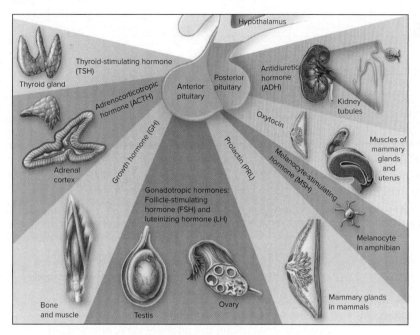

Figure 29.6 The role of the pituitary.

The Anterior Pituitary

> **LEARNING OBJECTIVE 29.3.2** List seven peptide hormones that the anterior pituitary produces, and explain how the hypothalamus regulates their release.

The anterior pituitary gland produces seven major peptide hormones (figure 29.6), each controlled by a particular releasing signal secreted from the hypothalamus:

1. **Thyroid-stimulating hormone (TSH).** TSH stimulates the thyroid gland to produce the thyroid hormone thyroxine, which in turn stimulates oxidative respiration.
2. **Adrenocorticotropic hormone (ACTH).** ACTH stimulates the adrenal gland to produce a variety of steroid hormones. Some regulate the production of glucose; others regulate the balance of sodium and potassium ions in the blood.
3. **Growth hormone (GH).** GH stimulates the growth of muscle and bone throughout the body.
4. **Follicle-stimulating hormone (FSH).** FSH is significant in the female menstrual cycle by triggering the maturation of egg cells and stimulating the release of estrogen. In males, it stimulates cells in the testes, regulating development of the sperm.
5. **Luteinizing hormone (LH).** LH plays an important role in the female menstrual cycle by triggering ovulation, which is the release of a mature egg. It also stimulates the male gonads to produce testosterone.
6. **Prolactin (PRL).** Prolactin stimulates the breasts to produce milk, which is released in response to oxytocin.
7. **Melanocyte-stimulating hormone (MSH).** In reptiles and amphibians, MSH stimulates color changes in the epidermis (figure 29.7). The function of this hormone in humans is still poorly understood.

> Follicle-stimulating hormone and luteinizing hormone (along with estrogen and progesterone that are produced elsewhere in the body) coordinate the female reproductive cycle, discussed in more detail in section 30.4.

> **Putting the Concept to Work**
> Why doesn't the hypothalamus just release hormones directly into the bloodstream? Why employ the pituitary gland?

Figure 29.7 A hormone in action.

Melanocyte-stimulating hormone stimulates color changes in reptiles and amphibians. The green anole (*Anolis punctatus*) shown here in the upper photo has changed to a tan color in the lower photo, in response to an environmental cue.

William H. Mullins/Science Source

29.4 The Pancreas

> **LEARNING OBJECTIVE 29.4.1** Identify the hormones produced in the pancreas, and describe how they interact to regulate blood glucose levels.

The **pancreas** gland is located behind the stomach and is connected to the front end of the small intestine by a narrow tube. It secretes a variety of digestive enzymes into the digestive tract through this tube, and for a long time it was thought to be solely an exocrine gland. In 1869, however, a German medical student named Paul Langerhans described some unusual clusters of cells scattered throughout the pancreas. In 1893, doctors concluded that these clusters of cells, which came to be called islets of Langerhans, produced a substance that prevented diabetes mellitus. *Diabetes mellitus* is a serious disorder in which affected individuals' cells are unable to take up glucose from the blood, even though their levels of blood glucose become very high. Some individuals lose weight and literally starve; others develop poor circulation, sometimes resulting in amputation of limbs with restricted circulation. Diabetes is the leading cause of blindness among adults, and it accounts for one-third of all kidney failures. It is the seventh leading cause of death in the United States.

Insulin and Glucagon

The islets of Langerhans in the pancreas produce two hormones that interact to govern the levels of glucose in the blood. These hormones are *insulin* and *glucagon*. Insulin is a storage hormone, designed to put away nutrients for leaner times. It promotes the accumulation of glycogen in the liver and triglycerides in fat cells. When food is consumed (left side of figure 29.8), beta cells in the islets of Langerhans secrete insulin, causing the cells of the body to take up and store glucose as glycogen and triglycerides to be used later. When body activity causes the level of glucose in the blood to fall as it is used as fuel (right side of figure 29.8), other cells in the islets of Langerhans, called alpha cells, secrete glucagon, which causes liver cells to release stored glucose and fat cells to break down triglycerides for energy use. The two hormones work together to keep glucose levels in the blood within a narrow range.

Type I Diabetes. Over 34.2 million people in the United States and over 422 million people worldwide have **diabetes.** There are two kinds of diabetes mellitus. About 5% to 10% of affected individuals suffer from type I diabetes, an autoimmune disease in which the immune system attacks the islets of Langerhans, resulting in abnormally low insulin secretion. Called juvenile-onset diabetes, this type usually develops before age 20. Affected individuals can be treated by daily injections of insulin. Active research on the possibility of transplanting islets of Langerhans holds promise as a lasting treatment for type I diabetes.

Type II Diabetes. In type II diabetes, the level of insulin in the blood is often higher than normal, but cells don't respond to insulin. This form of diabetes usually develops in people over 40 years of age. It is almost always a consequence of excessive weight; in the United States, 80% of those who develop type II diabetes are obese. The cells of some type II diabetics, overwhelmed with food, adjust their appetite for glucose downward, reducing their sensitivity to insulin by reducing their number of insulin receptors. To compensate, the pancreas pumps out ever-more insulin. Type II diabetes is usually treatable with diet and exercise, and most affected individuals do not need daily injections of insulin.

> **Putting the Concept to Work**
> Distinguish between type I and type II diabetes.

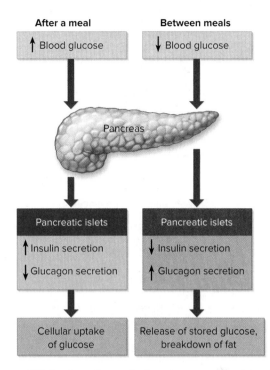

Figure 29.8 Insulin and glucagon secreted by the pancreas regulate blood glucose levels.

After a meal, an increased secretion of insulin by the beta cells of the pancreatic islets of Langerhans promotes the movement of glucose from blood into tissue cells. Between meals, an increased secretion of glucagon by the alpha cells of the pancreatic islets and decreased secretion of insulin cause the release of stored glucose and the breakdown of fat.

BIOLOGY & YOU

Pancreatic Cancer. Among the most deadly of all cancers are malignant tumors of the pancreas. In 2020 in the United States, about 57,600 individuals were diagnosed with this condition, and 47,050 died from the disease. Less than 7% of those diagnosed with pancreatic cancer are still alive five years after diagnosis; 50% were dead within six months. Pancreatic cancer is sometimes called a "silent killer" because early pancreatic cancer often does not cause symptoms, so a tumor is often not diagnosed until it is far advanced. Unfortunately, we don't have much of an idea of what causes pancreatic cancer. It is more common in men over 60 and slightly more common in cigarette smokers, diabetics, and obese people, but there are no established guidelines for preventing pancreatic cancer.

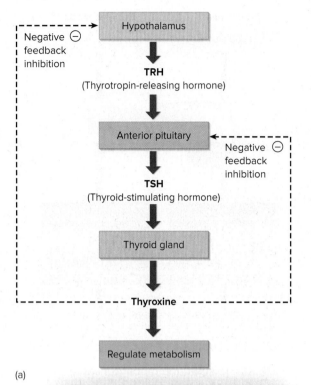

Figure 29.9 The thyroid gland secretes thyroxine.
(a) Thyroxine exerts negative feedback control of the hypothalamus and anterior pituitary. (b) A goiter is caused by a lack of iodine in the diet, which causes thyroxine secretion to decrease. As a result, there is less negative feedback, TSH is not inhibited from stimulating the thyroid gland, and the thyroid gland becomes enlarged.
(b) Chris Hellier/Science Source

29.5 The Thyroid, Parathyroid, and Adrenal Glands

> **LEARNING OBJECTIVE 29.5.1** Locate the thyroid, parathyroid, and adrenal glands, identify the hormones they produce, and discuss how these hormones function.

The Thyroid: A Metabolic Thermostat

The **thyroid gland** is shaped like a shield (its name comes from *thyros,* the Greek word for "shield") and lies just below the Adam's apple in the front of the neck. The thyroid makes several hormones, the two most important of which are *thyroxine,* which increases metabolic rate and promotes growth, and *calcitonin,* which inhibits the release of calcium from bones.

Thyroxine regulates the level of metabolism in the body in several important ways. Without adequate thyroxine, growth is retarded. For example, children with underactive thyroid glands are not able to carry out carbohydrate breakdown and protein synthesis at normal rates, a condition called cretinism, which results in stunted growth. Mental retardation can also result, because thyroxine is needed for normal development of the central nervous system.

The thyroid is stimulated to produce thyroxine by the hypothalamus, which is inhibited by thyroxine via negative feedback. The dashed lines in figure 29.9*a* illustrate how thyroxine inhibits the release of TRH (thyrotropin-releasing hormone) and TSH (thyroid-stimulating hormone) from the hypothalamus and anterior pituitary, respectively.

Thyroxine contains iodine, and if the amount of iodine in the diet is too low, the thyroid cannot make adequate amounts of thyroxine to keep the hypothalamus inhibited. The hypothalamus will then continue to stimulate the thyroid, which will grow larger in a futile attempt to manufacture more thyroxine. The greatly enlarged thyroid gland that results is called a goiter (figure 29.9*b*). This need for iodine in the diet is why iodine is added to table salt.

Calcitonin, which is also produced by the thyroid and will be discussed later, plays a key role in maintaining proper calcium levels in the body.

The Parathyroids: Regulating Calcium

The **parathyroid glands** are four small glands attached to the thyroid. Small and unobtrusive, they were ignored by researchers until well into the last century. The first suggestion that the parathyroids produce a hormone came from experiments in which they were removed from dogs: The concentration of calcium in the dogs' blood plummeted to less than half the normal level. However, if an extract of the parathyroid gland was administered, calcium levels returned to normal. If an excess was administered, calcium levels in the blood became *too* high, and the bones of the dogs were literally dismantled by the extract. It was clear that the parathyroid glands were producing a hormone that acted on calcium, both its uptake into bones and its release from bones.

The hormone produced by the parathyroids is **parathyroid hormone (PTH).** It is one of only two hormones in the body that is absolutely essential for survival (the other is aldosterone, a hormone produced by the adrenal glands, discussed on the next page). PTH regulates the level of calcium in blood. Calcium ions are needed in muscle contraction—by initiating calcium release, nerve impulses cause muscles to contract. A vertebrate cannot live without the muscles that pump the heart and drive the body, and these muscles cannot function if calcium levels are not kept within narrow limits.

PTH acts as a fail-safe to make sure calcium levels never fall too low. If they do, as in figure 29.10a, PTH is released into the bloodstream, travels to the bones, and acts on the osteoclast cells (the blue cells) within bones, stimulating them to dismantle bone tissue and release calcium into the bloodstream. PTH also acts on the kidneys to reabsorb calcium ions from the filtrate and leads to activation of vitamin D, necessary for calcium absorption by the intestine. A diet deficient in vitamin D leads to poor bone formation, a condition called rickets. The hormone calcitonin is released from the thyroid gland and acts in reverse of PTH. When calcium levels in the blood rise (figure 29.10b), calcitonin activates osteoblast cells (the orange cells) to take up calcium and rebuild bone.

The Adrenals: Two Glands in One

Mammals have two **adrenal glands,** one located just above each kidney (see figure 29.1). Each adrenal gland is composed of two parts: (1) an inner core, the **medulla,** which produces the hormones epinephrine (also called adrenaline) and norepinephrine; and (2) an outer shell, the **cortex,** which produces the steroid hormones cortisol and aldosterone.

The Adrenal Medulla: Emergency Warning Siren. The adrenal medulla releases *epinephrine* and *norepinephrine* in times of stress. These hormones act as emergency signals that stimulate rapid deployment of body fuel. The "alarm" response these hormones produce throughout the body is identical to the individual effects achieved by the sympathetic nervous system, but it is much longer lasting. Among their effects are accelerated heartbeat, increased blood pressure, higher levels of blood sugar, and increased blood flow to the heart.

The Adrenal Cortex: Inflammation, Stress, and Maintaining the Proper Amount of Salt. The adrenal cortex produces the steroid hormone cortisol. *Cortisol* (also called hydrocortisone) acts on many different cells in the body to maintain nutritional well-being. It stimulates carbohydrate metabolism and reduces inflammation. Synthetic derivatives of this hormone, such as prednisone, have widespread medical use as anti-inflammatory agents. Cortisol is also called the *stress hormone*, released in times of stress to help the body deal with acute stress. Problems arise when the body experiences chronic stress and cortisol levels remain high in the body. This can lead to problems with high blood pressure, reduced immune function, fat accumulation, and maintaining blood sugar, among others. These chronic effects of cortisol are unhealthy.

The adrenal cortex also produces *aldosterone*. Aldosterone acts primarily in the kidney to promote the uptake of sodium and other salts from the urine, which also increases the reabsorption of water. Aldosterone is, with PTH, one of the two endocrine hormones essential for survival. Removal of the adrenal glands is invariably fatal.

> **Putting the Concept to Work**
> **Which two hormones are so essential for human survival that you would die if they ceased to be produced? Why are they so important?**

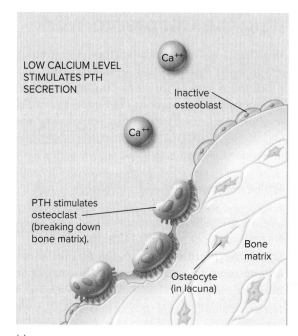

(a)

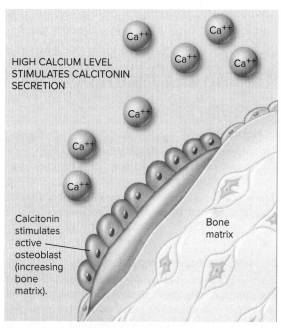

(b)

Figure 29.10 Maintenance of proper calcium levels in the blood.

(a) When calcium levels in the blood become too low, the parathyroid gland produces additional amounts of PTH, which stimulates the breakdown of bone, releasing calcium. (b) Conversely, abnormally high levels of calcium in the blood trigger the thyroid gland to secrete calcitonin, which inhibits the release of calcium from bone, and promotes the activity of osteoblasts to remove calcium from the blood and deposit it in bone.

Putting the Chapter to Work

1 Hypothyroidism, low production of the thyroid-stimulating hormone, was previously suggested to be the cause of obesity in many people.

If this were true, what sort of weight would you expect to see in a patient with hyperthyroidism (overproduction of the thyroid-stimulating hormone)?

2 Your instructor shows you a model of a kidney and discusses the hormones associated with water uptake and release. She focuses her discussion on a key hormone that is used to remove sodium from the urine.

Which hormone is she referring to?

3 Imagine you are watching a scary movie in a theater. It is dark and quiet. Suddenly, you are startled by a loud noise, and you "jump" in your seat.

Which two hormones would be released into your bloodstream by your brain to warn you of potential danger?

4 During a traumatic and/or stressful time in your life, many hormones can fluctuate.

Which hormone, if maintained at a chronically high level, would result in your gaining weight and having difficulty maintaining healthy blood sugar?

Retracing the Learning Path

The Endocrine System

29.1 Hormones

1. Hormones are chemical signals produced in glands or other endocrine tissues and transported to distant sites in the body. Endocrine glands produce hormones and release them into the bloodstream.
2. Many hormones are effective signals because only cells that have receptors for that particular hormone respond to it. The hormone binds to the receptor on such a "target cell" and elicits a response in that cell, often a change in cellular activity or genetic expression.

29.2 How Hormones Target Cells

1. Steroid hormones are lipid-soluble molecules. They pass through the plasma membrane of the target cell and bind to receptors in the cytoplasm or nucleus. The hormone-receptor complex binds to DNA, causing a change in gene expression that alters cell function.
2. Peptide hormones are unable to pass through the plasma membrane. Instead, they bind to membrane protein receptors. The binding of the hormone causes a change in the internal side of the receptor that activates a second messenger. Second messengers, such as cyclic AMP, activate enzymes in the cell. The enzymes then trigger a change in cellular activity. The second messenger system is a cascade of reactions that amplifies the signal and facilitates the change in cellular activity.

The Major Endocrine Glands

29.3 The Pituitary

1. The pituitary gland is actually two glands: the posterior and anterior pituitary glands. The posterior pituitary develops as an extension of the hypothalamus and contains axons that extend from cell bodies in the hypothalamus.

- The hormones released from the posterior pituitary are actually produced in the hypothalamus and transported by the axons to the posterior pituitary for storage and release. The hormones of the posterior pituitary include antidiuretic hormone (ADH), which regulates water retention in the kidneys, and oxytocin, which initiates uterine contractions during childbirth and milk release in the mother.

2. The anterior pituitary originates from epithelial tissue and produces the hormones it releases. Seven hormones are produced in the anterior pituitary. They are thyroid-stimulating hormone (TSH), adrenocorticotropic hormone (ACTH), growth hormone (GH), follicle-stimulating hormone (FSH), luteinizing hormone (LH), prolactin (PRL), and melanocyte-stimulating hormone (MSH). The hypothalamus controls the anterior pituitary. The hypothalamus produces hormones that are released into blood capillaries that surround the pituitary stalk. They travel a short distance to the anterior pituitary.

29.4 The Pancreas

1. The pancreas secretes two hormones—insulin and glucagon—into the blood. These hormones interact to maintain stable blood glucose levels. Insulin stimulates cell uptake of glucose from the blood. Glucagon stimulates the breakdown of glycogen to glucose. Two different types of cells in the islets of Langerhans produce insulin and glucagon.

- These hormones work opposite to each other. An increase in blood glucose levels triggers the release of insulin, and a decrease in blood glucose levels triggers the release of glucagon. When insulin is not available or cells fail to respond to insulin, diabetes mellitus can result.

29.5 The Thyroid, Parathyroid, and Adrenal Glands

1. The thyroid gland is a shield-shaped organ that lies just beneath the Adam's apple in the front of the neck. The thyroid makes several hormones, but the two most important hormones produced by the thyroid are thyroxine, which increases metabolism and growth, and calcitonin, which stimulates calcium uptake by bones. Thyroxine is controlled by negative feedback. When enough of the hormone has been released, it feeds back to inhibit the hormone-production process.

- Underproduction of thyroxine can lead to serious health problems. The underproduction of thyroxine in children can stunt growth and lead to mental retardation. A lack of iodine inhibits the production of thyroxine. In the absence of adequate levels of thyroxine, the hypothalamus will continue to stimulate the thyroid gland, leading to the formation of a goiter.

- The parathyroid glands are four small glands that are attached to the thyroid. The parathyroids produce parathyroid hormone. PTH regulates the levels of calcium in the blood. Low calcium ion concentrations stimulate the release of PTH from the parathyroid glands. PTH acts on the bones to dismantle bone tissue, releasing Ca^{++} into the blood. When Ca^{++} levels again increase, calcitonin is released from the thyroid and stimulates the uptake of Ca^{++} by bone cells and the synthesis of new bone tissue.

- The adrenal gland is actually two glands: The adrenal medulla is the inner core, and the adrenal cortex is the outer shell. The adrenal medulla secretes epinephrine and norepinephrine that stimulate an emergency response in the body. The adrenal cortex secretes cortisol, which is involved in inflammation and the regulation of glucose and stress responses. It also secretes aldosterone, which promotes the uptake of sodium and water from urine.

Inquiry and Analysis

Is the Bisphenol A in Your Plastic Water Bottle Dangerous?

Since the mid 1960s, plastic manufacturers have been using a chemical called bisphenol A (BPA) to produce a kind of clear plastic. Nowadays, these so-called polycarbonate plastics are found everywhere: in plastic water bottles, lining the insides of every can that populates our kitchens, and even in dental sealants. Leeching out of all this plastic in minute amounts, BPA has slowly accumulated in the bodies of over 90% of adult Americans. While the concentration of BPA is low (only about 2 micrograms per liter of blood), the effects of exposure to trace amounts of BPA on humans over a long period of time are unknown. Researchers have seen a link between exposure to low levels of BPA and cancer in mice (see page 577). Might prolonged exposure to BPA be harmful to humans?

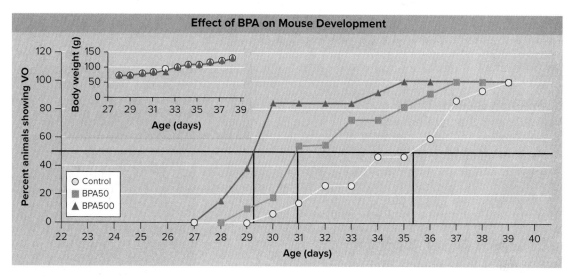

Because BPA has some structural similarity to the hormone estrogen, and tiny amounts of hormones can have major effects on human development, scientists looked to see if BPA affects the rate of embryonic development. Because you cannot ethically experiment on humans, researchers chose to look at how BPA affects the speed of sexual maturity in rats. Maturity of a female rat is reached at the first appearance of her vaginal opening, which normally occurs after the newborn rat is 35 days old. For this study, scientists housed pregnant rats individually until their pups were born. After birth, all of the male newborn pups were removed. The female neonates were divided into three treatment groups and left with the mother. Starting when the newborn rats were 1 day old, researchers subcutaneously injected the female rats daily, for 10 days, with different treatments: 50 µg/50 µl BPA (BPA 50), 500 µg/50 µl BPA (BPA 500), or a castor oil control. Following weaning, which occurs around 14 days after birth, the young rats were weighed daily and examined for the appearance of the vaginal opening. The age and weight of the rats at the time of the appearance of the vaginal opening are presented in the graph here.

Analysis

1. **Applying Concepts** Variable: In the graph, what is the dependent variable?
2. **Interpreting Data**
   a. When performing research on live animals, researchers often look at the data when 50% of the population have responded to the treatment. What reason do you think scientists use for looking at a 50% response instead of a 100% response?
   b. How does the rate of the appearance of the vaginal opening compare between rats treated with 50 µg/50 µl BPA versus 500 µg/50 µl BPA? How does this differ from the castor oil treatment?
   c. What is the purpose of the castor oil control?
   d. Why was it important to measure the body weight of the rats over the 38-day period of the experiment? Does it change? What is the effect of BPA?
3. **Making Inferences**
   a. Based on the data presented here, what general statement can you make regarding the appearance of the vaginal opening as the levels of BPA exposure increase?
   b. If the concentration of BPA in the injections were doubled to 1000 µg/µl, what would you predict would happen to the trend line for the appearance of the vaginal opening?
   c. What would you predict would happen if the injections for each treatment group were discontinued after 5 days as opposed to 10 days?
4. **Drawing Conclusions** Do these data support the hypothesis that exposure to BPA impacts the sexual maturity rate of rats?

30 Reproduction and Development

LEARNING PATH ▼

Modes of Reproduction
1. Asexual and Sexual Reproduction

The Human Reproductive System
2. Males
3. Females
4. Hormones Coordinate the Reproductive Cycle

The Course of Development
5. Embryonic Development
6. Fetal Development

Birth Control and Sexually Transmitted Diseases
7. Contraception and Sexually Transmitted Diseases

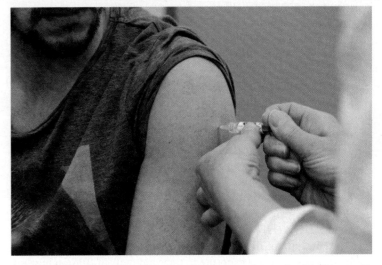

THIS YOUNG MAN is being vaccinated with Gardasil, a vaccine against HPV virus.

MediaforMedical/Andrea Aubert/Alamy Stock Photo

Should Boys Get a Vaccine for Cervical Cancer?

Death is always unfortunate, but it is tragic when avoidable. Every year in the United States, about 3,700 women die of cervical cancer, and most of these deaths are completely avoidable. How can a cancer that is the second leading cause of cancer deaths among women around the world be avoidable? Because unlike most cancers, cervical cancer is not caused by radiation or carcinogenic chemicals like those in cigarette smoke but by a virus—a virus for which an effective vaccine has been developed. Women vaccinated before they are exposed to the virus do not get cervical cancer.

HPV Virus

Cervical cancer is caused by human papilloma virus (HPV). HPV, the spherical particle shown on the next page, is sexually transmitted among humans, and is the most common sexually transmitted virus in the United States. Approximately 20 million people in the United States are infected, with about 6 million more getting infected each year. Most of the infections don't cause symptoms and die out on their own. More than 50% of sexually active men and women in the United States are infected with HPV at some point in their lives, most in their late teens and early 20s, few of them knowing it. The Centers for Disease Control and Prevention (CDC) estimates that in 2014, fully 44% of women aged 20 to 24 were infected.

HPV Causes Cancer

The problem is that while most infections pass unnoticed, some of the infections are not successfully fought off by the body's immune system and lead to serious health problems. In both men and women, HPV can cause genital warts. And in about 17,000 women and 9,000 men, HPV causes cancer. In both women and men, HPV is a leading cause of anal and mouth/throat cancer. Women are even more at risk because HPV is the principal cause of cervical cancer.

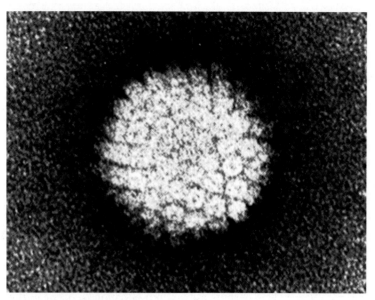

Vaccination Prevents It

In the last decade, vaccines have been developed against the two types of HPV that cause 70% of cervical cancers in women and 80% of anal cancers in men and women. The first of these to be introduced, Gardasil, is a preventive vaccine. It does not treat infection or cure cancer. What it does do is prevent HPV infection if it has not already occurred.

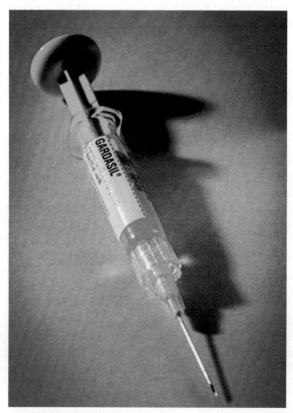

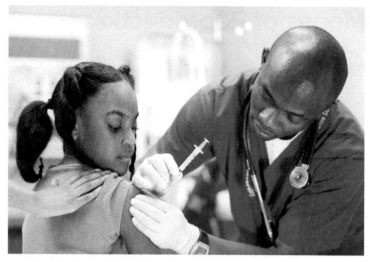

Broadly used, these vaccines have the potential to largely eliminate cervical cancer in the United States. Vaccination is recommended for women who are 9 to 25 years old who have not yet been exposed to HPV. However, because HPV is so widespread among young adults, early vaccination is strongly recommended, preferably before sexual activity commences.

Early Vaccination Raises Concerns

Therein lies the problem. Vaccinating schoolgirls for a sexually transmitted disease (STD), however sensible, offends many parents because they feel it encourages promiscuity. Aren't people who catch an STD promiscuous? Are you saying my little girl is going to be promiscuous? Won't giving her this vaccine be like giving her a license to be sexually active?

Still, serious progress is being made. Several states are requiring school vaccinations of girls, and the program of extensive early vaccination is working. Since introduction of the vaccine in 2006, the prevalence of HPV among girls ages 14 to 19 had fallen fully 56% by 2014—and only a third of girls 13 to 17 had received vaccination.

And Boys?

But what about the boys? Men have higher rates of HPV than women, the CDC says. And although males do not have a cervix, every transmission of HPV requires *two* partners. Vaccination of boys, not yet widely done, would greatly amplify the impact of the nation's vaccination program on cervical cancer rates. And remember: Men do get anal cancer, and genital warts. So should schoolboys be vaccinated for HPV? You bet.

Modes of Reproduction

30.1 Asexual and Sexual Reproduction

> **LEARNING OBJECTIVE 30.1.1** Contrast asexual and sexual reproduction, discriminating between parthenogenesis and hermaphroditism.

Not all reproduction involves two parents. Asexual reproduction, in which the offspring are genetically identical to the parent, is the primary means of reproduction among protists, cnidarians, and tunicates, and occurs in some more complex animals.

Through mitosis, genetically identical cells are produced from a single parent cell. This permits asexual reproduction to occur in the *Euglena* in figure 30.1 by division of the organism, or **fission**. The DNA replicates and cell structures, such as the flagellum, duplicate. The nucleus divides with identical nuclei going to each daughter cell. Cnidaria commonly reproduce by **budding**, in which a part of the parent's body becomes separated from the rest and differentiates into a new individual. The new individual may become an independent animal or may remain attached to the parent, forming a colony.

Unlike asexual reproduction, sexual reproduction occurs when a new individual is formed by the union of *two* cells. These cells are called **gametes**, and the two kinds that combine are generally called *sperm* and *eggs* (or ova). The union of a sperm and an egg produces a fertilized egg, or **zygote**, that develops by mitotic division into a new multicellular organism. The zygote and the cells it forms by mitosis are diploid; they contain both members of each homologous pair of chromosomes. The gametes, formed by meiosis in the sex organs, or **gonads**—the *testes* and *ovaries*—are haploid (see chapter 9). The processes of spermatogenesis (sperm formation) and oogenesis (egg formation) are described in later sections.

Different Approaches to Sex

Parthenogenesis, a type of reproduction in which offspring are produced from unfertilized eggs, is common in many species of arthropods. Some species are exclusively parthenogenic, whereas others switch between sexual reproduction and parthenogenesis in different generations. In honeybees, for example, a queen bee mates only once and stores the sperm. She then can control the release of sperm. If no sperm are released, the eggs develop parthenogenetically into drones, which are males; if sperm are allowed to fertilize the eggs, the fertilized eggs develop into other queens or worker bees, which are female. Parthenogenesis also occurs among populations of some lizard genera.

Hermaphroditism, another variation in reproductive strategy, is when one individual has both testes and ovaries and so can produce both sperm and eggs. A tapeworm is hermaphroditic and can fertilize itself as well as cross-fertilize, a useful strategy because it is unlikely that there will be another tapeworm living inside the host.

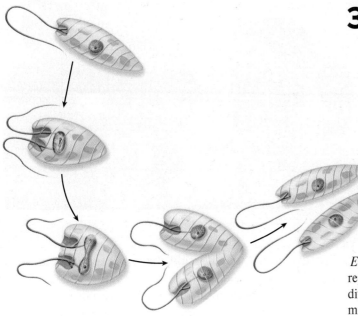

Figure 30.1 Asexual reproduction in protists.
The protist *Euglena* reproduces asexually: A mature individual divides by fission, and two complete individuals result.

> **Putting the Concept to Work**
> Give an example of a parthenogenic animal.

Sex Determination

> **LEARNING OBJECTIVE 30.1.2** Discuss sex determination in mammals.

In mammals, the sex is determined early in embryonic development. The reproductive systems of human males and females appear similar for the first 40 days after conception. During this time, the cells that will give rise to ova or sperm migrate to the embryonic gonads, which have the potential to become either ovaries in females or testes in males. If the embryo is XY, it is a male and will carry a gene on the Y chromosome whose protein product converts the gonads into testes (as on the left in figure 30.2). In females, who are XX, this Y chromosome gene and the protein it encodes are absent, and the gonads become ovaries (as on the right). Recent evidence suggests that the sex-determining gene may be one known as ***SRY*** (for "*s*ex-determining *r*egion of the *Y* chromosome"). The *SRY* gene appears to have been highly conserved during the evolution of different vertebrate groups.

Once testes form in the embryo, they secrete testosterone and other hormones that promote the development of the male external genitalia and accessory reproductive organs (indicated in the blue box in the figure). If testes do not form, the embryo develops female external genitalia and accessory reproductive organs. The ovaries do not promote this development of female organs because the ovaries are nonfunctional at this stage. In other words, all mammalian embryos will develop female sex accessory organs and external genitalia by default unless they are masculinized by the secretions of the testes.

Gender Identity and Sexual Orientation

In humans, the sex of the child—its assigned sex—is based on the physical appearance of the newborn child: Does it have male or female genitalia? However, not everyone has a gender identity that matches their assigned sex. Gender identity is the expression of the innermost sense people have about who they are. This sense is shaped early in development, long before birth, by a combination of developmental events turning some genes ON and others OFF. Sexual orientation is how a person views others as potential romantic and sexual partners. Sexual orientation exists on a continuum, from attraction to someone of the same gender, to someone of a different gender, to any gender, and possibly not having any sexual attraction to someone. Gender identity is who you are, sexual orientation is who you want to be with. Like being left-handed, this is something we are all born with, not something that can be "cured"—it is simply who the person is.

> **Putting the Concept to Work**
> If genetic engineering could be used to transfer the *SRY* gene into an X chromosome, would a zygote carrying a normal X and this *SRY*-X be a female or a male? Explain.

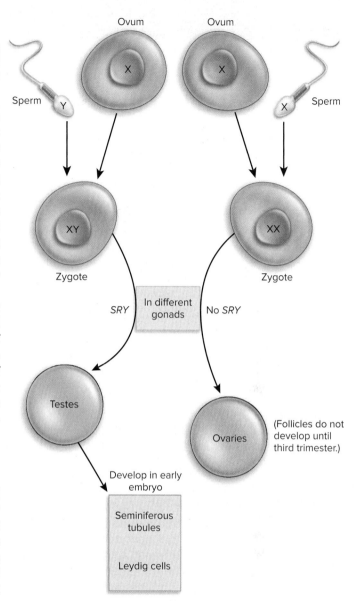

Figure 30.2 Sex determination.

Sex determination in mammals is made by a gene on the Y chromosome designated *SRY*. Testes are formed when the Y chromosome and *SRY* are present; ovaries are formed when they are absent.

A Closer Look

A Closer Look at the Science Behind Sexual Orientation

Classic experiments in the 1950s and '60s showed us how the sex chromosomes of a fetus determine whether that individual will grow up behaving as a male or a female.

The Key Impact of Testosterone

One key to adult sexual behavior, it turns out, is the activation early in development of fetal genes located on the Y chromosome that lead the developing testis to secrete the hormone testosterone. This testosterone acts on the brain of the fetus in a way that determines that individual's future sexual behavior.

Early experiments were carried out on rats. In a baby rat, the release of testosterone does not occur until after birth. When experimenters gave testosterone to newborn female rats, their sexual behavior was malelike when they grew up. Conversely, when experimenters removed the testes of newborn males, the result was femalelike behavior. Similar experiments have since been carried out in species whose babies are born much more mature than rats, such as monkeys. The same results were obtained as with rats, indicating that the same processes occur in the brain of these species, but earlier, during pregnancy.

Epigenetics Is the Tool

But a monkey is not a human. Does this process of determining sexual preference during pregnancy occur in humans, too? Researchers think so. Male human embryos form testes very early in pregnancy (at about 10 weeks), and these testes immediately start to secrete testosterone. So the male brain is exposed to testosterone early in development. What does the testosterone do in the male brain? Recent experiments give a hint. In the hypothalamus (a portion of the brain that plays a major role in sexual behavior), a female fetus suppresses malelike behavior by chemically modifying certain key genes. The process is an example of epigenetics (see section 13.6), the female fetus adding methyl groups to the genes to silence them. Giving females testosterone reduces this, removing some of the methyl groups and releasing these genes to become active. Such females behave more like males. In much the same fashion, giving a drug to newborn males that blocks demethylation causes these males to behave more like females. It seems that the default human sexual behavior is female, with testosterone in the male fetus releasing genes that determine future malelike behavior.

So, in a nutshell, how we behave sexually—our sexual orientation—is determined by what **genes** we have, by **hormones** that influence these genes in developing males, and by **environmental factors** that may modify the rate at which these genes are epigenetically activated.

LGBTQ

As a rough measure, your sexual orientation (your enduring pattern of sexual or romantic attraction) can be to a person or persons of the opposite gender (heterosexual), to the same gender (homosexual), or to both genders (bisexual). Nonheterosexual individuals commonly refer to themselves as **LGBTQ**, a term that stands for **l**esbian, **g**ay, **b**isexual, **t**ransgender, and **q**ueer. The CDC estimates the number of nonheterosexual individuals as 5% of the total U.S. population in 2014—some 13 million LGBTQ individuals.

In the CDC study, 1.5% of adult American women self-identify as lesbian, or homosexual women, while about 1.8% of adult men self-identify as gay, or homosexual men. About 1.3% of adult Americans self-identify as bisexual.

There is another possibility, of course—individuals whose gender identity differs from their anatomical sex at birth. These people are said to be transgender. The CDC estimates that transgender individuals, an often-not-discussed but not rare minority, represent 0.6% of the U.S. population.

So what about the Q in LGBTQ? While for decades a demeaning label for any LGBT individual, the term *queer* has been embraced by younger members of the LGBTQ community as signifying "questioning"—someone who is in the process of figuring out their sexual identity.

Conversion Therapy

Can the sexual orientation of an adult be changed? No. No major mental health professional organization sanctions efforts to change a person's sexual orientation (a discredited process its advocates call "conversion therapy"), and virtually all of these organizations have adopted policy statements cautioning against conversion therapy. The underlying assumption of conversion therapy—that homosexuality is the result of troubled family dynamics and so can be reversed by intensive psychotherapy—is now uniformly understood to have been based on misinformation and prejudice. There is no substantive evidence that the nature of parenting or early childhood experience plays any role in the formation of a person's sexual orientation.

The developmental decisions that determine sexual orientation act at the gene level during pregnancy. There is much we do not yet know about which genes are involved and the factors influencing the timing of their epigenetic activation, but this much is clear: The events determining sexual orientation, once made in a fetus, are irrevocable and are not subject to later change. Just as left-handed children, about 10% of all kids, can be trained to write with their right hand but continue for all their lives to be left-handed, so it is with the 5% of children born LGBTQ. Their sexual preference, like left-handedness, is not voluntary or "curable," although the overt expression of their sexuality is certainly subject to conscious choice. You are born with your sexuality but choose how to express it.

The Human Reproductive System

30.2 Males

> **LEARNING OBJECTIVE 30.2.1** Describe the male reproductive system, including where and how sperm are formed and delivered.

The human male gamete, or **sperm,** is highly specialized for its role as a carrier of genetic information. Produced by meiosis, sperm cells have 23 chromosomes instead of the 46 found in other cells of the male body. Sperm do not successfully complete their development at 37°C (98.6°F), the normal human body temperature. The sperm-producing organs, the **testes** (singular, **testis**), move during the course of fetal development into a sac called the scrotum (figure 30.3), which hangs between the legs of the male, maintaining the two testes at a temperature about 3°C cooler than the rest of the body. The testes contain cells that secrete the male sex hormone *testosterone*.

Male Gametes Are Formed in the Testes

An internal view of a testis in figure 30.4 ❶ shows that it is composed of several hundred compartments, each packed with large numbers of tightly coiled tubes called *seminiferous tubules* (seen in cross section in ❷). Sperm production, *spermatogenesis,* takes place inside the tubules. The process of spermatogenesis

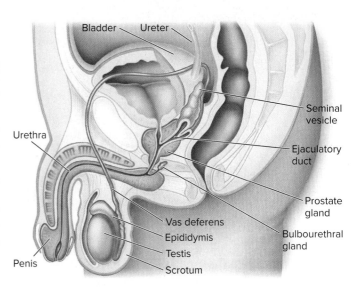

Figure 30.3 The male reproductive organs.
The testis is where sperm are formed. Cupped above the testis is the epididymis, a highly coiled passageway within which sperm complete their maturation. Extending away from the epididymis is a long tube, the vas deferens.

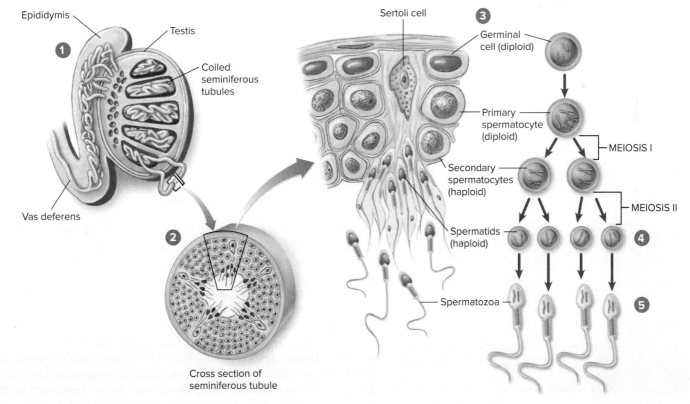

Figure 30.4 The testis and formation of sperm.
Inside the testis ❶, the seminiferous tubules ❷ are the sites of sperm formation. Germinal cells in the seminiferous tubules ❸ give rise to primary spermatocytes (diploid), which undergo meiosis to form haploid spermatids ❹. Spermatids develop into mobile spermatozoa, or sperm ❺. Sertoli cells are nongerminal cells within the walls of the seminiferous tubules. They assist spermatogenesis in several ways, such as helping to convert spermatids into spermatozoa.

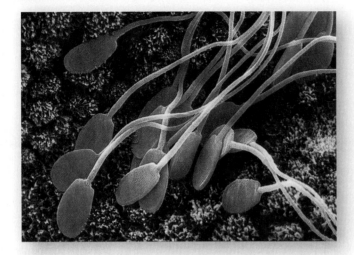

begins in germinal cells toward the outside of the tubule (shown in the enlarged view in ❸). As the cells undergo meiosis, they move toward the lumen of the tubule ❹, with spermatozoa being released into the tubule ❺. The number of sperm produced is truly incredible. A typical adult male produces several hundred million sperm each day of his life! Those that are not ejaculated from the body are broken down, and their materials are reabsorbed and recycled.

Sperm. After a sperm cell is manufactured within the testes through intermediate stages of meiosis, it is delivered to a long, coiled tube called the **epididymis** (see figure 30.4), where it matures. A sperm cell is not motile when it arrives in the epididymis, and it must remain there for at least 18 hours before its motility develops. Mature sperm are relatively simple cells, consisting of a head, body, and tail (figure 30.5). The head encloses a compact nucleus and is capped by a vesicle called an *acrosome*. The acrosome contains enzymes that aid in the penetration of the protective layers surrounding the egg. The body and tail provide a propulsive mechanism: Within the tail is a flagellum, and inside the body are centrioles, which act as a basal body for the flagellum, and mitochondria, which generate the energy needed for flagellar movement.

Semen. From the epididymis, the sperm is delivered to another long tube, the **vas deferens.** When sperm are released, they travel through a tube from the vas deferens to the **urethra,** where the reproductive and urinary tracts join, emptying through the penis. Sperm is released in a fluid called **semen,** which also contains secretions mostly from the seminal vesicles and the prostate gland that provide metabolic energy sources for the sperm. Benign enlargement of the prostate occurs in 90% of men by age 70, but it can be cancerous. Prostate cancer is the second most common cancer in men and can be treated effectively if detected early during physical examinations, before it spreads.

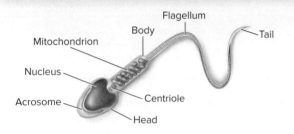

Figure 30.5 Human sperm cells.

Each sperm possesses a long tail that propels the sperm, and a head that contains the nucleus. The tip, or acrosome, contains enzymes to help the sperm cell digest a passageway into the egg for fertilization.

(Top) David M. Phillips/Science Source

Male Gametes Are Delivered by the Penis

In the case of humans and some other mammals, the **penis** is an external tube containing two long cylinders of spongy tissue side by side (figure 30.6). A third cylinder of spongy tissue contains in its center the urethra, through which both semen (during ejaculation) and urine (during urination) pass. Why this unusual design? The penis is designed to inflate. The spongy tissues that make up the three cylinders are riddled with small spaces between the cells, and when nerve impulses from the CNS cause the arterioles leading into this tissue to expand, blood collects within these spaces. Like blowing up a balloon, this causes the penis to become erect and rigid.

Erection can be achieved without any physical stimulation of the penis, but physical stimulation is required for semen to be delivered. Stimulation of the penis, as by repeated thrusts into the vagina of a female, leads first to the mobilization of the sperm. In this process, muscles encircling the vas deferens contract, moving the sperm along the vas deferens into the urethra. The bulbourethral glands also secrete a clear, slippery fluid that neutralizes the acidity of any residual urine and lubricates the head of the penis. Further penis stimulation then leads to **ejaculation,** the forceful ejection of 2 to 5 milliliters of semen. Within this small 5-milliliter volume are several hundred million sperm. Because there are extraordinarily high odds against any one individual sperm cell successfully completing the long journey to the egg and fertilizing it, successful fertilization requires a high sperm count. Males with fewer than 20 million sperm per milliliter are generally considered sterile.

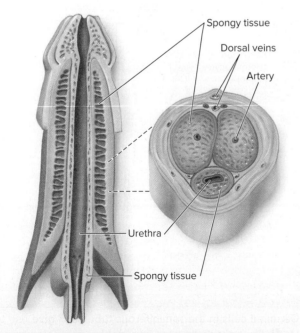

Figure 30.6 Structure of the penis.

(*Left*) longitudinal section; (*right*) cross section.

> **Putting the Concept to Work**
> What is the function of the prostate gland?

30.3 Females

> **LEARNING OBJECTIVE 30.3.1** Diagram the female reproductive system, describing where and how ova are formed and fertilized.

In females, eggs develop from cells called **oocytes,** located in the outer layer of compact masses of cells called **ovaries** within the abdominal cavity (**figure 30.7**). Recall that in males, the gamete-producing cells are constantly dividing. In females, all of the oocytes needed for a lifetime are already present at birth. During each reproductive cycle, one or a few of these oocytes are initiated to continue their development in a process called **ovulation;** the others remain in developmental holding patterns.

Usually Only One Female Gamete Matures Each Month

At birth, a female's ovaries contain some 2 million oocytes, all of which have begun the first meiotic division. At this stage, they are called *primary oocytes* (❶ in **figure 30.8**). Each primary oocyte waits to receive the proper developmental "go" signal before continuing on with meiosis. Until then, its meiosis remains arrested in prophase of the first meiotic

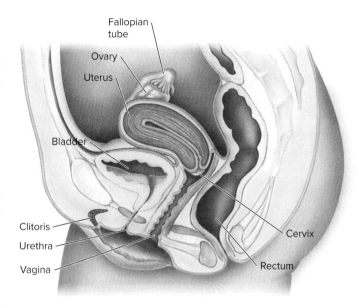

Figure 30.7 The female reproductive system.

The organs of the female reproductive system are specialized to produce gametes and to provide a site for embryonic development if a gamete is fertilized.

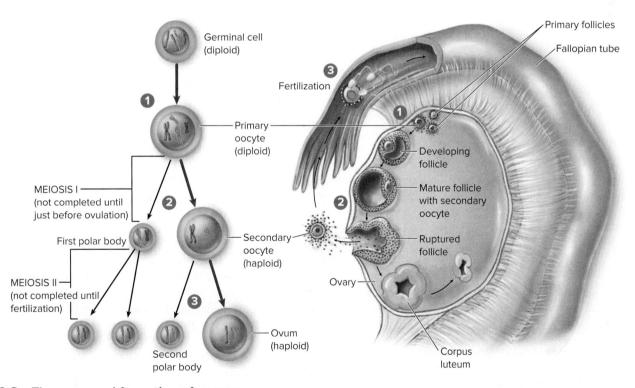

Figure 30.8 The ovary and formation of an ovum.

In this figure, the maturation of the ovum through meiosis is shown on the left, and the developmental journey of the ovum is on the right, with corresponding stages numbered on each. At birth, a human female's ovaries contain about 2 million egg-forming cells called oocytes, which have begun the first meiotic division and stopped. At this stage, they are called primary oocytes ❶, and their further development is halted until they receive the proper developmental signals, which are the hormones FSH and LH. Beginning at puberty, a monthly cycle of hormone secretion is established. When the hormones FSH and LH are released, meiosis resumes in a few oocytes, but only one oocyte usually continues to mature while the others regress. The primary oocyte (diploid) completes the first meiotic division, and one division product becomes a nonfunctional polar body. The other product, the secondary oocyte, is released during ovulation ❷, along with the polar body. The secondary oocyte does not complete the second meiotic division until fertilization ❸; that division yields two more nonfunctional polar bodies and a single haploid egg, or ovum. Fusion of the haploid egg with a haploid sperm during fertilization produces a diploid zygote.

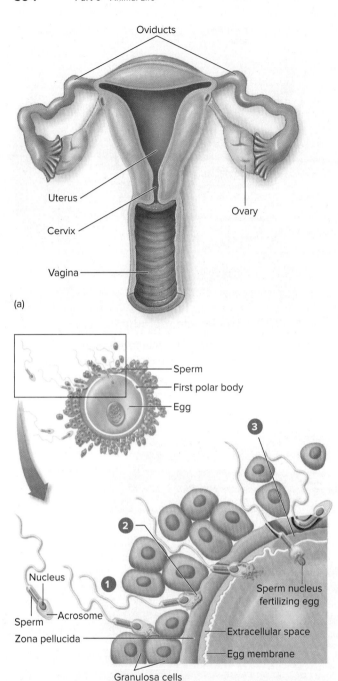

Figure 30.9 Fertilization occurs in the oviducts.

(a) The oviducts extend out from the uterus. Sperm are deposited in the vagina and travel to the oviducts. (b) Fertilization occurs in the oviduct when a sperm cell penetrates the outer layers of the egg cell.

division. Very few primary oocytes ever receive the awaited signal, which turns out to be the pituitary hormones FSH and LH, which were discussed in chapter 29.

With the onset of puberty, females mature sexually. At this time, the release of FSH and LH initiates the resumption of the first meiotic division in a few oocytes. The first meiotic division produces the *secondary oocyte* and a nonfunctional *polar body* ❷. In humans, usually only a single oocyte is ovulated, and the others regress. In some instances, more than one oocyte develops; if both are fertilized, they become fraternal twins. Approximately every 28 days after that, another oocyte matures and is ovulated, although the exact timing may vary from month to month. Only about 400 of the approximately 2 million oocytes a woman is born with mature and are ovulated during her lifetime.

Fertilization Occurs in the Oviducts

The **oviducts** (also called **fallopian tubes** or uterine tubes) transport eggs from the ovaries to the **uterus**. In humans, the uterus is a muscular, pear-shaped organ about the size of a fist that narrows to a muscular ring called the **cervix**, which leads to the vagina (figure 30.9a). The uterus is lined with a stratified epithelial membrane, called the **endometrium**. The surface of the endometrium is shed approximately once a month during menstruation, while the underlying portion remains to generate a new surface during the next cycle.

The Egg's Journey. After ovulation, smooth muscles lining the fallopian tubes contract rhythmically, moving the egg down the tube to the uterus in much the same way that food is moved down through your digestive system (see chapter 25), pushing it along by squeezing the tube behind it. The journey of the egg through the fallopian tube is a slow one, taking from five to seven days to complete. However, if the egg is not fertilized within 24 hours of ovulation, it loses its capacity to develop.

Fertilization. During sexual intercourse, sperm are deposited within the vagina, a thin-walled muscular tube about 7 centimeters long that leads to the mouth of the uterus. Using their flagella, sperm entering the uterus swim up to and enter the fallopian tubes. Sperm can remain viable within the female reproductive tract for up to six days. If sexual intercourse takes place five days before ovulation or one day after, a viable egg will be present high up in the fallopian tubes. Of the several hundred million sperm that are ejaculated, only a few dozen make it to the egg. Once they reach the egg, the sperm must penetrate through two protective layers that surround the secondary oocyte (figure 30.9b ❶): a layer of granulosa cells and a protein layer called the zona pellucida. Enzymes within the acrosome cap of the sperm help digest the second of these layers ❷. The first sperm to make it through the second layer stimulates the oocyte to block the entry of other sperm ❸ and to complete meiosis II. Meiosis II produces the **ovum** (plural, **ova**) and two more nonfunctional polar bodies (see figure 30.8 ❺). When the female haploid nucleus within the ovum combines with the male haploid nucleus, the egg is fertilized and becomes a zygote. The zygote then begins a series of cell divisions while traveling down the fallopian tube. After about six days, it reaches the uterus, attaches itself to the endometrial lining, and continues the long developmental journey that eventually leads to the birth of a child.

> **Putting the Concept to Work**
> Where is the endometrium found? What does it do? When?

30.4 Hormones Coordinate the Reproductive Cycle

> **LEARNING OBJECTIVE 30.4.1** Describe the two phases of the menstrual cycle, and explain how four hormones regulate them.

The female reproductive cycle, called a **menstrual cycle,** is composed of two distinct phases, the *follicular phase,* in which an egg reaches maturation and is ovulated, and the *luteal phase,* where the body continues to prepare for pregnancy. These phases are coordinated by a family of hormones. Hormones play many roles in human reproduction. Sexual development is initiated by hormones, released from the anterior pituitary and ovary, that coordinate simultaneous sexual development in many kinds of tissues. The production of gametes is another closely orchestrated process, involving a series of carefully timed developmental events. Successful fertilization initiates yet another developmental "program," in which the female body continues its preparation for the many changes of pregnancy.

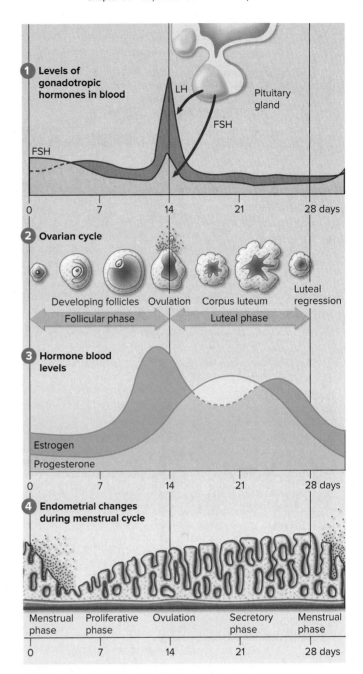

Figure 30.10 **The human menstrual cycle.**

Ovulation and the preparation of the uterine lining for implantation are controlled by a group of four hormones during the menstrual cycle.

Triggering the Maturation of an Egg: The Follicular Phase

The first phase of the menstrual cycle, the follicular phase, corresponds to days 0 through 14 in figure 30.10. During this time, several follicles (an oocyte and its surrounding tissue is called a *follicle*) are stimulated to develop. This development is carefully regulated by hormones. The anterior pituitary starts the cycle by secreting small amounts of *follicle-stimulating hormone (FSH)* and *luteinizing hormone (LH)* ❶. These hormones stimulate follicular growth ❷ and the secretion of the female sex hormone *estrogen* ❸, more technically known as *estradiol,* from the developing follicles. Several follicles are stimulated to grow under FSH stimulation.

> Estrogen is a steroid hormone, and as shown in figure 29.3, steroid hormones are able to pass through the plasma membrane and bind to receptors inside the cell. The binding of estrogen to its receptor activates the transcription of a receptor protein for progesterone.

Initially, the relatively low levels of estrogen have a negative-feedback effect on FSH and LH secretion. The low but rising levels of estrogen in the bloodstream feed back to the hypothalamus, which responds to the rising estrogen by commanding the anterior pituitary to decrease production of FSH and LH. As FSH levels fall, usually only one follicle achieves maturity. Late in the follicular phase, estrogen levels in the blood have increased drastically, and these higher levels of estrogen begin to have a positive-feedback effect on FSH and LH secretion. The rise in estrogen levels signals the completion of the follicular phase of the menstrual cycle.

Preparing the Body for Fertilization: The Luteal Phase

The second phase of the cycle, the luteal phase (days 14 through 28), follows smoothly from the first. In a positive-feedback response to high levels of estrogen, the hypothalamus causes the anterior pituitary to rapidly secrete large amounts of LH and FSH (see figure 30.10 ❶). The surge of

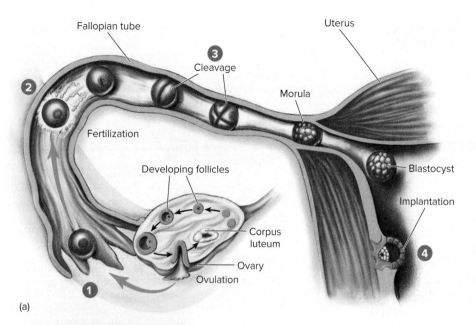

LH is larger than the surge of FSH and can last up to 24 hours. The peak in LH secretion triggers ovulation: LH causes the wall of the follicle to burst, and the egg within the follicle is released into one of the fallopian tubes extending from the ovary to the uterus (see ❶ in figure 30.11).

After the egg's release and departure, estrogen levels decrease, and LH directs the repair of the ruptured follicle, which fills in and becomes yellowish. In this condition, it is called the *corpus luteum,* which is simply the Latin phrase for "yellow body." The corpus luteum soon begins to secrete the hormone *progesterone* (the light green curve in figure 30.10 ❸), in addition to small levels of estrogen. Increased levels of progesterone and estrogen have a negative-feedback effect on the secretion of FSH and LH, preventing further ovulations. Progesterone completes the body's preparation of the uterus for fertilization, including the thickening of the endometrium (figure 30.10 ❹).

If fertilization does *not* occur soon after ovulation, however, production of progesterone slows and eventually ceases, marking the end of the luteal phase. The decreasing levels of progesterone cause the thickened layer of blood-rich tissue to be sloughed off, a process that results in the bleeding associated with menstruation. **Menstruation,** or "having a period," usually occurs about midway between successive ovulations (shown in figure 30.10 at 28 days).

Restarting the Cycle

At the end of the luteal phase, neither estrogen nor progesterone is being produced. In their absence, the anterior pituitary can again initiate production of FSH and LH, thus starting another reproductive cycle. Each cycle begins immediately after the preceding one ends. A cycle usually occurs every 28 days, or a little more frequently than once a month, although this varies in individual cases. The Latin word for "month" is *mens,* which is why the reproductive cycle is called the menstrual cycle, or monthly cycle.

If fertilization does occur high in the fallopian tube (❷ in figure 30.11a), the zygote undergoes a series of cell divisions called cleavage ❸, while traveling toward the uterus. At the blastocyst stage, it implants in the lining of the uterus ❹. The tiny embryo secretes human chorionic gonadotropin (hCG), an LH-like hormone, which maintains the corpus luteum. By maintaining the corpus luteum, hCG keeps the levels of estrogen and progesterone high, thereby preventing menstruation, which would terminate the pregnancy. Because hCG comes from the embryo and not from the mother, it is hCG that is tested in all pregnancy tests.

> **Putting the Concept to Work**
> When in the menstrual cycle does luteinizing hormone peak?

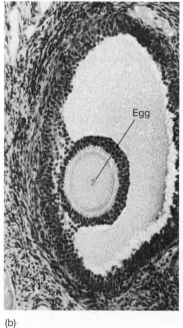

Figure 30.11 The journey of an ovum.

(a) Produced within a follicle and released at ovulation, an ovum is swept up into a fallopian tube ❶ and carried down by waves of contraction of the tube walls. Fertilization occurs within the tube ❷ by sperm journeying upward. Several mitotic divisions occur while the fertilized ovum undergoes cleavage and continues its journey down the fallopian tube ❸, becoming first a morula, then a blastocyst. The blastocyst implants itself within the wall of the uterus ❹, where it continues its development. (b) A mature egg within an ovarian follicle. In each menstrual cycle, a few follicles are stimulated to grow under the influence of FSH and LH, but usually only one achieves full maturity and ovulation.

(b) Ed Reschke/Getty Images

The Course of Development

30.5 Embryonic Development

> **LEARNING OBJECTIVE 30.5.1** Outline how the events of cleavage and embryonic development determine body architecture.

Cleavage: Setting the Stage for Development

The first major event in human embryonic development is the rapid division of the zygote into a larger and larger number of smaller and smaller cells, becoming first 2 cells, then 4, then 8, and so on. The first of these divisions occurs about 30 hours after union of the egg and the sperm, and the second, 30 hours later. During this period of division, called **cleavage,** the overall size does not increase from that of the zygote. The cells continue to divide, each cell secreting a fluid into the center of the cell mass. Eventually, a hollow ball of 500 to 2,000 cells is formed. This is the **blastocyst,** which contains a fluid-filled cavity called the **blastocoel** (figure 30.12). Within the ball is an *inner cell mass* concentrated at one pole that goes on to form the developing embryo. The outer sphere of cells, called the *trophoblast,* releases the hCG hormone, discussed in section 30.4.

During cleavage, the developing embryo journeys down the mother's fallopian tube and reaches the uterus on about the sixth day; it attaches to the uterine lining and penetrates into the tissue of the lining. The blastocyst now begins to grow rapidly, initiating the formation of the membranes that will later surround, protect, and nourish it. One of these membranes, the **amnion,** will enclose the developing embryo, whereas another, the **chorion,** which forms from the trophoblast, will interact with uterine tissue to form the **placenta,** which will nourish the growing embryo (see figure 30.14). The placenta connects the developing embryo to the blood supply of the mother. Fully 61 of the cells at the 64-celled stage develop into the trophoblast and only 3 into the embryo proper.

Determination of Body Architecture

In the third week of embryonic development, the embryo begins its development into the tissues and organs of the body.

The first characteristic vertebrate feature to form is the **notochord,** a flexible rod that forms along the midline of the embryo, below its dorsal surface. The **neural tube** forms above the notochord and later differentiates into the spinal cord and brain.

While the neural tube is forming, the rest of the basic architecture of the human body is being rapidly determined. On either side of the developing notochord, segmented blocks of tissue form. Ultimately, these blocks, or **somites,** give rise to the muscles, vertebrae, and connective tissues. As development continues, more and more somites are formed. Within another strip of tissue that runs alongside the somites, many of the significant glands of the body, including the kidneys, adrenal glands, and gonads, develop.

By the end of the third week, over a dozen somites are evident, and the blood vessels and gut have begun to develop. At this point, the embryo is about 2 millimeters (less than a tenth of an inch) long.

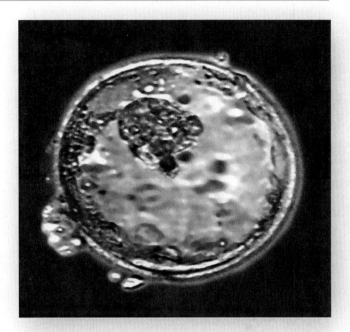

Figure 30.12 The beginnings of human development.

A human blastocyst. The formation of the blastocyst occurs when the zygote undergoes cleavage, producing a hollow ball of cells. An inner cell mass will later differentiate into the different tissues of the embryo.
BSIP/Newscom

> **Putting the Concept to Work**
> What tissues do trophoblast cells give rise to in the adult body?

The First and Second Months: Organogenesis

> **LEARNING OBJECTIVE 30.5.2** Describe how the embryo takes shape.

In the fourth week of pregnancy, the body organs begin to form, a process called **organogenesis** (figure 30.13a). The eyes form, and the heart begins a rhythmic beating and develops four chambers. At 70 beats per minute, the little heart is destined to beat more than 2.5 billion times during a lifetime of about 70 years. More than 30 pairs of somites are visible by the end of the fourth week, and the arm and leg buds have begun to form. The embryo more than doubles in length during this week, reaching about 5 millimeters. By the end of the fourth week, the developmental scenario is far advanced, although most women are not yet aware that they are pregnant.

During the second month of pregnancy, great changes in morphology occur as the embryo takes shape (figure 30.13b). The miniature limbs of the embryo assume their adult shapes. The arms, legs, knees, elbows, fingers, and toes can all be seen, as well

Figure 30.13 The developing human.

Photos (*top*) and illustrations (*bottom*) of a developing human show a four-week embryo (a), seven-week embryo (b), 12-week fetus (c), and four-month fetus (d).

(a, b - *top*) Bradley Smith; (c - *top*) Dopamine/Science Source; (d - *top*) Tissuepix/Science Source

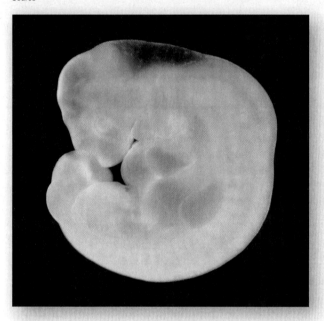
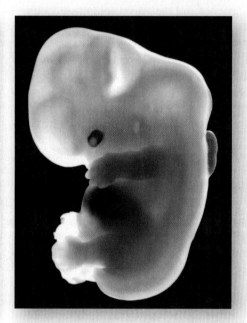
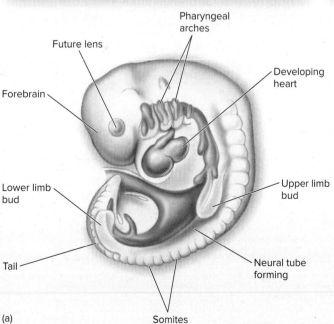
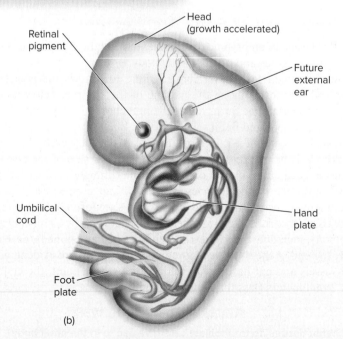

(a) (b)

as a short, bony tail. The bones of the embryonic tail, an evolutionary reminder of our past, later fuse to form the coccyx, or tailbone. Within the body cavity, the major internal organs are evident, including the liver and pancreas. By the end of the second month, the embryo has grown to about 25 millimeters in length—it is 1-inch long. It weighs perhaps a gram and is beginning to look distinctly human.

> **Putting the Concept to Work**
> What happened to the tail you had as a two-month embryo?

BIOLOGY & YOU

Fetal Alcohol Syndrome. The first eight weeks is a crucial time in human development because the proper course of events can be interrupted easily. For example, alcohol use by pregnant women during the first months of pregnancy is one of the leading causes of birth defects, producing fetal alcohol syndrome, in which the baby is born with facial abnormalities and often severe intellectual disabilities. One in 250 newborns in the United States is affected with fetal alcohol syndrome.

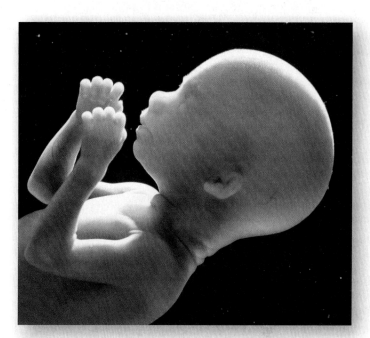

Development is essentially complete.

The developing human is now referred to as a fetus.

Facial expressions and primitive reflexes are carried out.

All of the major body organs have been established.

Arms and legs begin to move.

Bones actively enlarge.

Mother can feel the fetus kicking.

Following a period of rapid growth, the fetus is born.

Neurological growth continues after birth.

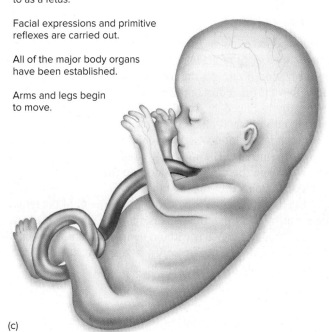

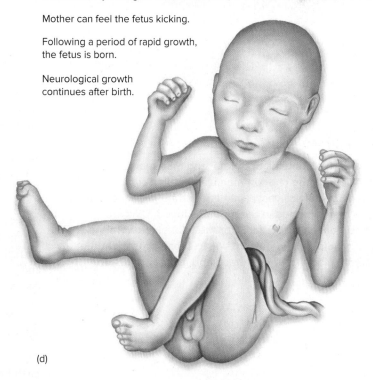

(c)

(d)

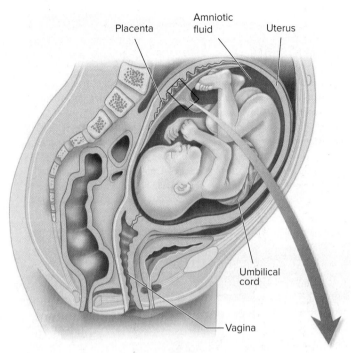

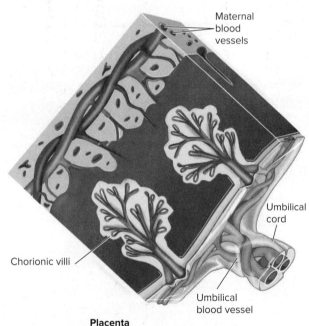

Figure 30.14 Structure of the placenta.

The placenta contains both fetal and maternal tissues. Extensions of the chorion membrane called chorionic villi contain fetal blood vessels and penetrate into maternal tissue that is bathed in the mother's blood. Oxygen and nutrients are able to enter the fetal blood from the maternal blood by diffusion. Waste substances enter the maternal blood from the fetal blood, also by diffusion.

30.6 Fetal Development

LEARNING OBJECTIVE 30.6.1 Outline the events that take place between the end of the second month and birth.

By the end of the second month, most of the key developmental steps have been taken. What remains is growth.

The Third Month: Completion of Development

Development of the embryo is essentially complete except for the lungs and brain. The lungs don't complete development until the third trimester, and the brain continues to develop even after birth. From this point on, the developing human is referred to as a **fetus** rather than an embryo. What remains is essentially growth. The nervous system and sense organs develop during the third month. The fetus begins to show facial expressions and carries out primitive reflexes such as the startle reflex and sucking. By the end of the third month, all of the major organs of the body have been established, and the arms and legs begin to move (figure 30.13c).

The Second Trimester: The Fetus Grows in Earnest

The second trimester is a time of growth. In the fourth (figure 30.13d) and fifth months of pregnancy, the fetus grows to about 175 millimeters in length (almost 7 in long), with a body weight of about 225 grams. Bone formation occurs actively during the fourth month. During the fifth month, the head and body become covered with fine hair. This downy body hair, called *lanugo,* is another evolutionary relic and is lost later in development. By the end of the fourth month, the mother can feel the baby kicking; by the end of the fifth month, they can hear its rapid heartbeat with a stethoscope. In the sixth month, growth accelerates. By the end of the sixth month, the fetus is over 0.3 meter (1 ft) long and weighs 0.6 kilograms (about 1.5 lb)—and most of its prebirth growth is still to come. At this stage, the fetus cannot yet survive outside the uterus without special medical intervention.

The Third Trimester: The Pace of Growth Accelerates

The third trimester is a period of rapid growth. In the seventh, eighth, and ninth months of pregnancy, the weight of the fetus more than doubles. This increase in bulk is not the only kind of growth that occurs. Most of the major nerve tracts are formed within the brain during this period, as are new brain cells.

All of this growth is fueled by nutrients provided by the mother's bloodstream, passing into the fetal blood supply within the placenta. In the placenta (figure 30.14), fetal blood vessels extend from the umbilical cord into tissues that line the uterus. The mother's blood bathes this tissue so that nutrients can pass from the mother's blood into the fetal blood vessels without the two blood systems ever mixing blood.

By the end of the third trimester, the neurological growth of the fetus is far from complete, but by this time, the fetus is about as large as it can get and still be delivered through the pelvis without damage to mother or child. As any woman who has had a baby can testify, it is a tight fit. Birth takes place as soon as the probability of survival is high.

Birth

At approximately 40 weeks from the last menstrual cycle, the process of birth begins as hormonal changes in the mother initiate the onset of **labor.**

During labor and delivery, the cervix gradually dilates (the opening becomes larger), the amnion ruptures, causing amniotic fluid to flow out through the vagina (sometimes referred to as the "water breaking"), and uterine contractions become strong and regular, usually resulting in the expulsion of the fetus from the uterus. Hormones called *oxytocin* and *prostaglandins* work in a positive-feedback mechanism to stimulate and increase uterine contractions. The fetus is usually in a head-down position near the end of pregnancy. In a vaginal birth, the fetus is pushed down through the cervix and out through the vagina (figure 30.15). The umbilical cord is still attached to the baby, and is clamped and cut after birth. The baby transitions from living in a fluid environment to a gaseous one, and many of its organ systems undergo major changes. After the birth of the fetus, continuing uterine contractions expel the placenta and associated membranes, collectively called the "afterbirth." In some cases, such as when a vaginal delivery would cause harm to the fetus or the mother, the fetus and placenta are surgically removed from the uterus in a procedure called a *caesarian section (C-section)*.

In the mother, hormones during late pregnancy prepare the *mammary glands* for nourishing the baby after birth. For the first couple of days after childbirth, the mammary glands produce a fluid called *colostrum,* which contains protein (including antibodies) and lactose but little fat. Then, milk production is stimulated by the anterior pituitary hormone *prolactin,* usually by the third day after delivery. When the infant suckles at the breast (figure 30.16), the posterior pituitary hormone oxytocin is released, initiating milk release, or milk "letdown."

Postnatal Development

Growth continues rapidly after birth. Babies typically double their birth weight within a few months. Different organs grow at different rates, however, and the body proportions of infants are different than that of adults. The head, for example, is disproportionately large in newborns, but after birth, it grows more slowly than the rest of the body. Such a pattern of growth, in which different components grow at different rates, is referred to as *allometric growth*.

The fact that the human brain continues to grow significantly for the first few years of postnatal life means that adequate nutrition and a safe environment are particularly crucial during this period.

> **Putting the Concept to Work**
> Distinguish between an embryo and a fetus.

Figure 30.16 A mother nursing her child.

It is difficult not to feel warmed by the look of wonder and delight seen on the face of a nursing mother.

JGI/Blend Images/Blend Images/Getty Images

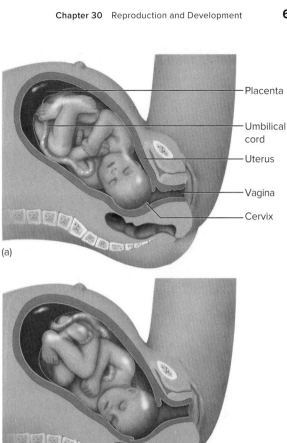

(a)

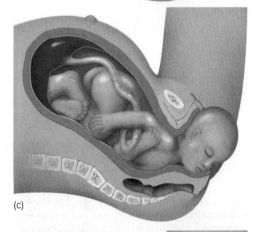

(b)

(c)

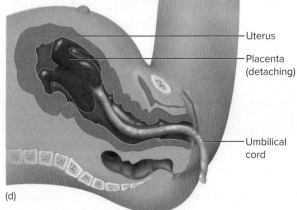

(d)

Figure 30.15 Stages of childbirth.

Birth Control and Sexually Transmitted Diseases

30.7 Contraception and Sexually Transmitted Diseases

> **LEARNING OBJECTIVE 30.7.1** Evaluate the effectiveness of five methods of contraception.

Some of the most important lessons to be learned about human reproduction involve how not to succeed in reproducing.

Contraception

Not all couples want to initiate a pregnancy every time they have sex, yet sexual intercourse may be a necessary and important part of their emotional lives together. The solution to this dilemma is to find a way to avoid reproduction without avoiding sexual intercourse, an approach that is commonly called **birth control**, or **contraception**.

Abstinence. The simplest and most reliable way to avoid pregnancy is not to have sex at all. Of all methods of birth control, this is the most certain—and the most limiting because it denies a couple the emotional support of a sexual relationship. A variant of this approach is to avoid sex only on the days when successful fertilization is likely to occur. The rest of the sexual cycle is considered relatively "safe" for intercourse. This approach, called the rhythm method, or natural family planning when other indicators are also monitored, is satisfactory in principle but difficult in application because ovulation is not easy to predict and may occur unexpectedly. Failure rates are as high as 20% to 30%.

Prevention of Egg Maturation. A widespread form of birth control in the United States has been the daily ingestion of hormones, or **birth control pills** (figure 30.17a). These pills contain estrogen and progesterone, which shut down production of the pituitary hormones FSH and LH. The ovarian follicles do not ripen in the absence of FSH, and ovulation does not occur in the absence of LH. Other methods of hormone delivery include medroxy progesterone (Depo-Provera), which is injected every one to three months, the weekly birth control patch, which releases the hormones through the skin, and surgically implanted capsules that release hormones. Failure rates are less than 2%.

Emergency contraception, called *Plan B* or the "morning after pill," is a high-dose progesterone pill that can block ovulation if taken soon after unprotected sex. Its failure rate varies, and it should not be used as a primary method of birth control.

Prevention of Embryo Implantation. The insertion of a coil or other irregularly shaped object into the uterus is an effective means of birth control. The irritation in the uterus prevents the implantation of the descending embryo within the uterine wall. Such **intrauterine devices (IUDs)** are very effective because once inserted, they can be forgotten. They have a failure rate of less than 2%. A chemical means of preventing embryo implantation or ending an early pregnancy is *RU-486*. This pill blocks the action of progesterone, causing the endometrium to slough off. RU-486 must be administered under a doctor's care because of potentially serious side effects.

BIOLOGY & YOU

RU-486. Emergency contraception such as the *"morning after pill" (Plan B)* should not be confused with the highly publicized RU-486 (Mifepristone). RU-486 is not an emergency contraceptive—it is not a contraceptive at all. It ends an unwanted pregnancy after conception. To be effective, RU-486 has to be administered within 49 days (or seven weeks) following the first day of the last menstrual period. A pregnancy terminated by RU-486 involves two drugs given two days apart. The first drug is RU-486 (Mifepristone), a synthetic steroid compound that works by blocking progesterone receptors in the uterus. Without progesterone stimulation, the endometrium degenerates and is sloughed off, ejecting the embryo with the shed lining of the uterus. Because RU-486 by itself is effective only 60% of the time in inducing removal of the embryo, a second medication, a prostaglandin called misoprostol, is given two days later to induce strong contractions of the uterus, forcing the embryo's expulsion. Because RU-486 has numerous side effects, some potentially severe (including sepsis, a severe infection of the bloodstream, and excessive bleeding), it must be administered by a doctor.

Sperm Blockage. Fertilization cannot occur without sperm. One way to prevent the delivery of sperm is to encase the penis within a thin rubber bag, or **condom** (figure 30.17b). In principle, this method is easy to apply, but in practice, it proves to be less effective due to incorrect use, with a failure rate of up to 15%. A second way to prevent the entry of sperm is to cover the cervix with a rubber dome called a **diaphragm** (figure 30.17c), inserted immediately before intercourse. Because the dimensions of individual cervices vary, diaphragms must be fitted by a physician. Failure rates average 20%.

Sperm Destruction. Another approach to birth control is to destroy the sperm within the vagina. Sperm can be destroyed with **spermicidal jellies, suppositories,** and **foams** applied immediately before intercourse. The failure rate varies widely, from 10% to 25%. The use of a spermicide with a condom or diaphragm increases the effectiveness over each method used independently.

> **Putting the Concept to Work**
> What is the failure rate of condoms? Of birth control pills?

Sexually Transmitted Diseases

> **LEARNING OBJECTIVE 30.7.2** Describe six significant STDs.

Sexually transmitted diseases (STDs) are diseases that spread from one person to another through sexual contact. AIDS, discussed in chapter 27, is a deadly viral STD. Other significant STD include:

Gonorrhea. The primary symptom of this disease, which is caused by the bacterium *Neisseria gonorrhoeae*, is discharge from the penis or vagina. It can be treated with antibiotics. If left untreated in women, gonorrhea can cause pelvic inflammatory disease (PID), a condition in which the fallopian tubes become scarred and blocked. PID can eventually lead to sterility.

Chlamydia. Caused by the bacterium *Chlamydia trachomatis,* this disease is sometimes called the "silent STD" because women usually experience no symptoms until after the infection has become established. Like gonorrhea, chlamydia can cause PID in women if left untreated.

Syphilis. Caused by the bacterium *Treponema pallidum,* this disease is one of the most potentially devastating STDs. Left untreated, the disease progresses to heart disease, memory loss, and nerve damage that may include loss of motor function or blindness.

Genital herpes. Caused by the herpes simplex virus type 2 (HSV-2), this disease is the most common STD in the United States. The virus causes red blisters on the penis or on the labia, vagina, or cervix that scab over.

Cervical cancer. About 70% of cervical cancer is caused by HPV (human papillomavirus), a sexually transmitted virus. Gardasil, a vaccine that blocks HPV in women not yet exposed to the virus, could cut worldwide deaths by about 290,000 women each year.

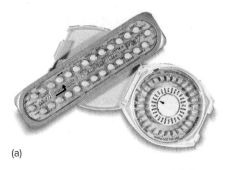

(a)

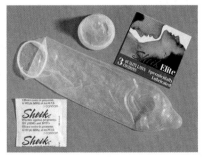

(b)

(c)

Figure 30.17 Three common birth control methods.

(a) Oral contraceptives; (b) condom; and (c) diaphragm and spermicidal jelly.

Bob Coyle/McGraw Hill

> **Putting the Concept to Work**
> Which STD is the most common in the United States?

Putting the Chapter to Work

1. Your biology instructor shows the class a diagram of a sea star. They discuss the reproductive organs and point to the testes and ovaries. They inform the class that sea stars can reproduce without a partner.

   What mechanism of reproduction do sea stars have?

2. In your biology lab, you have been instructed to identify labeled structures on a model of the male reproductive system. There is an arrow pointing at a coiled tube in which sperm matures.

   What is this part?

3. In a human reproduction video, you are shown the steps of ovulation. The video shows the signaling carried out by the anterior pituitary hormones FSH and LH, and the ovulation of two oocytes. In most instances, only one oocyte is ovulated.

   If both are fertilized, what will be the result?

4. Home pregnancy tests can be very reliable if they are used correctly. Most require a waiting period of one week after a missed menstrual cycle before use.

   During this time, the body begins to accumulate which hormone produced by the fetus?

5. You are shown a picture of a human embryo. It has all its limbs, and major internal organs, and is approximately 25 millimeters in length.

   By which month in pregnancy have these changes taken place?

Retracing the Learning Path

Modes of Reproduction

30.1 Asexual and Sexual Reproduction

1. Asexual reproduction through fission or budding is the primary means of reproduction among protists and some animals, but most animals reproduce sexually.

   - In animals, the most common mode of reproduction is sexual, but parthenogenesis and hermaphroditism are two variations. In parthenogenesis, offspring are produced from unfertilized eggs. In hermaphroditism, an individual has both testes and ovaries, producing both sperm and eggs. Some hermaphrodites can self-fertilize.

2. In mammals, sex is genetically determined and appears during embryonic development. An embryo that is XY develops into a male, and an XX embryo develops into a female.

The Human Reproductive System

30.2 Males

1. Male testes continuously produce large numbers of male gametes, sperm. The testes contain a large number of tightly coiled tubes called seminiferous tubules, where sperm develop in a process called spermatogenesis. As sperm develop and undergo meiosis, they move toward the lumen of the tubules, and from there, they pass into the epididymis. Once matured, they are stored in the vas deferens. During sexual intercourse, sperm are delivered through the penis into the female.

30.3 Females

1. Female gametes, eggs or ova, develop in the ovary from oocytes. The ovaries are located in the lower abdomen. A female is born with some 2 million oocytes, all arrested during the first meiotic division. The hormones FSH and LH initiate the resumption of meiosis I in a few oocytes, but usually only one oocyte completes development in each monthly cycle.

   - The egg ruptures from the ovary, called ovulation, and enters the fallopian tube. Sperm deposited in the vagina travel up through the cervix and uterus and into the fallopian tube, or oviduct, to reach the egg. Usually only one sperm cell penetrates the egg's protective layers. At this point, the oocyte completes meiosis II, and fertilization occurs. The fertilized egg, called a zygote, is transported to the uterus through the oviduct. The arriving zygote attaches to the endometrial lining of the uterus, where it completes development.

30.4 Hormones Coordinate the Reproductive Cycle

1. The human reproductive cycle, called a menstrual cycle, is divided into two phases: the follicular and luteal phases. The follicular phase begins with the secretion of FSH and LH, which stimulates the resumption of oocyte development and the secretion of estrogen. The estrogen acts as a negative-feedback signal to stop the secretion of FSH from the anterior pituitary.

   - As the level of estrogen increases, it begins to have a positive-feedback effect on FSH and LH. The luteal phase begins with a surge of LH, which causes ovulation and the formation of the corpus luteum. The corpus luteum begins secreting progesterone, which acts to prepare the uterus for implantation of the zygote.

   - If fertilization does not occur, estrogen and progesterone levels drop, and the endometrial lining of the uterus sloughs off, a process called menstruation, and a new cycle begins.

   - If a zygote implants in the lining of the uterus, estrogen and progesterone levels remain high due to the release of human chorionic gonadotropin (hCG), and the uterus is maintained.

The Course of Development

30.5 Embryonic Development

1. The vertebrate embryo develops over hundreds of cell divisions that eventually produce a hollow ball of cells called a blastocyst.

2. Organs begin forming by the fourth week. By the end of the third month, all major organs except the brain and lungs are developed, and the fetus looks distinctly human.

30.6 Fetal Development

1. Most of the key events in human development occur early in the first trimester. The second and third trimesters are periods of considerable growth.

   - During labor and delivery, the fetus and placenta are expelled from the uterus. Hormones coordinate the production of milk in the mother for nourishing the newborn.

Birth Control and Sexually Transmitted Diseases

30.7 Contraception and Sexually Transmitted Diseases

1. Various birth control methods are available and work by preventing egg maturation, preventing embryo implantation, and blocking or killing sperm.

2. Sexually transmitted diseases are spread through sexual contact. AIDS is a deadly STD. Other STDs may not be as fatal as AIDS but are quite destructive, especially if left untreated.

Inquiry and Analysis

Why Do STDs Vary in Frequency?

As a general rule, the incidence of a sexually transmitted disease (STD) is expected to increase with increasing frequencies of unprotected sexual contact. With the emergence of AIDS in the early 1980s, intense publicity and education have lessened such dangerous behavior. Both the number of sexual partners and the frequency of unprotected sex have fallen significantly in the United States in the last decade. It would follow, then, that the frequencies of STDs such as syphilis, gonorrhea, and chlamydia should also be falling—but they are not. The Centers for Disease Control and Prevention reports 2 million new cases of these three STDs in 2019! And yet not all of their frequencies are rising: Detailed yearly statistics are reported in the graph to the right. The results are a little surprising: While the level of chlamydia infection has risen steadily since 1984, the level of gonorrhea infection has fallen steeply over the same period! What are we to make of this?

The simplest explanation of such a difference is that the two STDs are occurring in different populations, and one population has rising levels of sexual activity, while the other has falling levels. However, nationwide statistics encompass all population subgroups, and each major subgroup contains all three major STDs mentioned above. So this would seem an unlikely explanation for the rise in frequency of chlamydia.

A second possible explanation would be a change in the infectivity of one of the STDs. A less infective STD would tend to fall in frequency in the population for the simple reason that fewer sexual contacts result in infection.

Syphilis is most infective in its initial stage, but this stage lasts only about a month. Most transmissions occur during the much longer second stage, marked by a pink rash and sores in the mouth. The bacteria can be transmitted at this stage by kissing or shared liquids. Any drop in infectivity of this STD would be expected to shorten this stage—but no such shortening has been observed.

Gonorrhea can be transmitted by various forms of sexual contact with an infected individual at any time during the infection. Symptoms are typically experienced soon after infection. Just as with syphilis, no drop in gonorrhea infectivity per sexual contact has been reported.

Chlamydia offers the most interesting possibility of changes in infectivity because of its unusual nature. *Chlamydia trachomatis* is genetically a bacterium but is an obligate intracellular parasite, much like a virus in this respect—it can reproduce only inside human cells. The red structures in the photo are chlamydia bacteria inside human cells. Like gonorrhea, chlamydia is transmitted through vaginal, anal, or oral intercourse with an infected person. As with the other two STDs, there does not appear to have been a drop in infectivity.

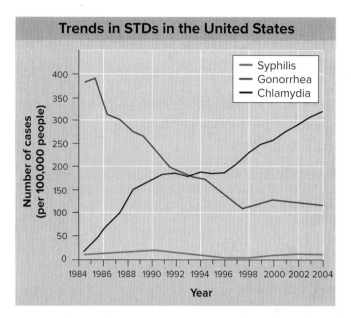

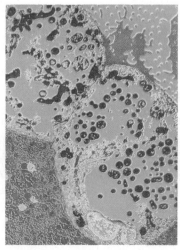

R. Dourmashkin, MD/Science Source

There is, however, a third possible explanation for why the frequency of one STD in a population might rise while the frequency of another STD in that same population falls. To grasp this third possible explanation, we need to focus on the experience of infected individuals. Unlike someone infected with syphilis and gonorrhea, who soon knows they are infected, a person infected with chlamydia may show no symptoms. Programs to raise public awareness of STDs might lower the sexual activity of individuals aware of their infection, while not influencing the sexual behavior of individuals NOT aware of their infection.

To assess this possibility, examine carefully the trends in the incidence in the United States of gonorrhea, chlamydia, and syphilis seen in the graph.

Analysis

1. **Applying Concepts** What is the dependent variable?
2. **Making Inferences**
   a. Gonorrhea: What is the incidence in 1985? In 1995? Has the frequency declined or increased? In general, are individuals aware they are infected when they transmit the STD?
   b. Chlamydia: What is the incidence in 1985? In 1995? Has the frequency declined or increased? In general, are individuals aware they are infected when they transmit the STD?
3. **Drawing Conclusions** Might heightened public awareness explain why the trend in levels of gonorrhea differs from that of chlamydia? Explain.

PART 7 Plant Life

31 Plant Form and Function

LEARNING PATH ▼

Structure and Function of Plant Tissues
1. Organization of a Vascular Plant
2. Plant Tissue Types

The Plant Body
3. Roots
4. Stems
5. Leaves

Plant Transport and Nutrition
6. Water Movement
7. Carbohydrate Transport

THIS SQUIRREL IS licking maple sap, a tasty and nutritious treat on a cold, late-winter day.
Judy Freilicher/Alamy Stock Photo

Squirrel Candy

Pancakes, flapjacks, or whatever name you give them, they are an American breakfast favorite—stacked high with melting butter and dripping with golden maple syrup. Truth be told, the fried batter that is the pancake wouldn't be all that appealing without the warm maple syrup, but did you ever wonder where maple syrup came from? Most people know that it comes from maple trees, but how did it first make its way from the inside of a tree to our breakfast table? The responsible party, oddly, seems to have been a squirrel.

Native Americans Learn from Squirrels

Early Native Americans were pure scientists, keen observers of the natural world. Tribes of eastern Canada and the northeastern United States discovered the delicious sweet liquid that graces our breakfast table by observing the antics of squirrels. Legend has it that a young Iroquois boy watched a squirrel scamper up a tree and gnaw on a branch. The squirrel then appeared to lick the branch, as the squirrel in the photo is doing. Why would the squirrel do this? Being a curious young lad, the boy did the same (probably using a knife rather than his teeth to cut the tree!) and discovered a sweet liquid dripping from the wound of the tree. The sap of the maple tree becomes even sweeter when the water evaporates from it. Squirrels will score the tree and allow the sap to seep out, dripping down the tree. After a while, when water has evaporated from the sap and it begins to harden, the squirrels will return and lick the congealing sap off the tree. Native Americans improved on the squirrels' discovery. They found that by boiling the sap to a thick syrup, it became even sweeter. They used it to

Don Johnston/age fotostock/Alamy Stock Photo

Anna Sirotina/iStock/Getty Images

flavor foods and by the mid-1600s were using maple syrup for bartering with early European settlers.

Sweet Sap from Maple Trees

The typical sugar maple tree (*Acer saccharum*) produces sap that is about 2% sugar. This, when boiled down, produces a syrup that is approximately 66% sugar (no wonder we like it so much!). Sugar maples are found primarily in northeastern and north central areas of the United States and in eastern Canada—Vermont, New Hampshire, and Quebec are particularly well known for their maple syrup production. Sweet sap only flows during late winter to early spring. In the chapter opening photo, the squirrel is collecting sap from ice-covered branches. The sap is in fact a key source of energy for squirrels in late winter, when their stash of nuts runs low.

Making Maple Syrup

Maple sap is harvested now in much the same way as it was by the Native Americans over 400 years ago. A three-inch-deep hole is drilled into the trunk of the tree. Up to three taps can be put into larger trees. A spout is placed in the hole through which the sap drains, and a container of some kind collects the sap as it drips out. The sap is taken to a sugar house where it is boiled so that the water evaporates, which increases the sugar content and maple flavor. The clear sap darkens as it boils, giving maple syrup its golden brown color. Each tap can produce about 10 gallons of sap per year, which is converted into about 1 quart of maple syrup. If done correctly, tapping does not damage the tree, and the same tree can be retapped year after year; some maple trees have been producing sap for over 100 years.

Why Do Maple Trees Go to All This Trouble?

Why does the maple tree produce sweet-tasting sap? Surely it is not for the palatable enjoyment of squirrels? As temperatures begin to drop during autumn, maples stop growing but not photosynthesizing. The tree stores this newly manufactured carbohydrate in specialized ray cells within the wood of the tree. The increase in temperatures during late winter and early spring triggers enzymes located in the ray cells to convert the starches into sugars, primarily sucrose, and also creates a buildup of pressure within the tissues of the tree, causing the sap to flow. This flowing maple sap distributes sugars throughout the tree to be used as a source of energy when the trees begin to emerge from their winter dormancy. This early start provides maple trees with a key evolutionary advantage in the competitive environment of forest life.

Structure and Function of Plant Tissues

31.1 Organization of a Vascular Plant

LEARNING OBJECTIVE 31.1.1 Diagram the basic body plan of a plant, distinguishing between apical and lateral meristems.

While the plants you encountered in chapter 17 are remarkably diverse, most possess the same fundamental architecture and the same three major groups of organs—roots, stems, and leaves. The organs of vascular plants have an outer covering of protective tissue and an inner matrix of tissue. The cells and tissues of vascular plants and how the plant body carries out the functions of living are the focus of this chapter.

A vascular plant is organized along a vertical axis (**figure 31.1**). The part belowground is called the **root,** and the part aboveground is called the **shoot,** made up of the stem and leaves (however, in some instances, roots may extend above the ground and stems can extend below it). Although roots and shoots differ in their basic structure, growth at the tips throughout the life of the individual is characteristic of both. The root penetrates the soil and absorbs water and various minerals, which are crucial for plant nutrition. It also anchors the plant. The **stem** serves as a framework for the positioning of the **leaves,** where most photosynthesis takes place. The arrangement, size, and other characteristics of the leaves are critically important in the plant's production of food.

Meristems

When an animal grows taller, all parts of its body lengthen—when you grew taller as a child, your arms and legs lengthened, and so did your torso. A plant doesn't grow this way. Instead, it adds tissues to the tips of its roots and shoots. If you grew like this, your legs would get longer, and your head taller, while the central portion of your body would not change.

Why do plants grow in this way? The plant body contains growth zones of unspecialized cells called **meristems.** Meristems are areas with actively dividing cells that result in plant growth but also continually replenish themselves. That is, one cell divides to give rise to two cells. One remains meristematic, while the other is free to differentiate and contribute to the plant body, resulting in plant growth. In this way, meristem cells function much like "stem cells" in animals (see chapter 13).

Apical Meristems. In plants, **primary growth** is initiated at the tips by the **apical meristems,** regions of active cell division that occur at the tips of roots and stems, colored lime green in **figure 31.1**. The growth of these meristems extends the plant body. As the body tip elongates, it forms what is known as the primary plant body, which is made up of the primary tissues.

Lateral Meristems. Growth in thickness, **secondary growth,** involves the activity of the **lateral meristems,** which are cylinders of meristematic tissue, colored yellow in **figure 31.1**. The continued division of their cells results primarily in the thickening of the plant body. There are two kinds of lateral meristems: the *vascular cambium,* which ultimately gives rise to thick accumulations of secondary xylem and phloem, and the *cork cambium,* from which arise the outer layers of bark on both roots and stems.

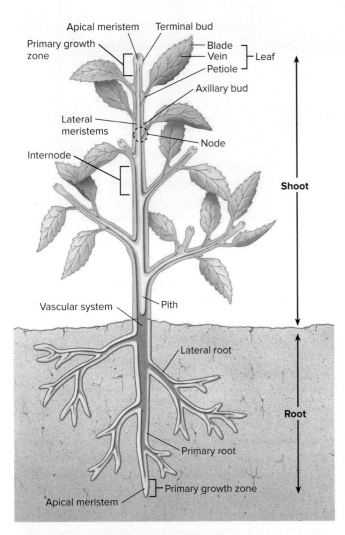

Figure 31.1 The body of a plant.

The body of this eudicot plant consists of an aboveground portion called the shoot (stems and leaves) and a belowground portion called the root. Elongation of the plant, so-called primary growth, takes place when clusters of cells called the apical meristems (*lime green areas*) divide at the ends of the roots and the stems. Thickening of the plant, so-called secondary growth, takes place in the lateral meristems (*yellow areas*) of the stem, allowing the plant to increase in girth like letting out a belt.

> **Putting the Concept to Work**
> Distinguish between primary and secondary growth.

31.2 Plant Tissue Types

The organs of a plant—the roots, stem, leaves, and in some cases, flowers and fruits—are composed of different combinations of tissues. A tissue is a group of similar cells—cells that are specialized in the same way and are organized into a structural and functional unit. Each major tissue type is composed of distinctive kinds of cells whose structures are related to the functions of the tissues in which they occur. Most plants have three major tissue types: (1) *ground tissue*, in which the vascular tissue is embedded; (2) *dermal tissue*, the outer protective covering of the plant; and (3) *vascular tissue*, which conducts water and dissolved minerals up the plant and conducts the products of photosynthesis throughout.

Ground Tissue

> **LEARNING OBJECTIVE 31.2.1** Describe the functioning of ground tissue, differentiating between parenchyma, collenchyma, and sclerenchyma.

Parenchyma. **Parenchyma cells** are the least specialized and the most common of all plant cell types; they form masses in leaves, stems, and roots. Parenchyma cells, unlike some other cell types, are characteristically alive at maturity, with fully functional cytoplasm and a nucleus. They are the cells that carry out the basic functions of living, including photosynthesis, cellular respiration, and food and water storage. The edible parts of most fruits and vegetables are composed of parenchyma cells. They are capable of cell division and are important in cell regeneration and wound healing. Most parenchyma cells have only thin cell walls, as seen in **figure 31.2**, called **primary cell walls,** which are mostly cellulose that is laid down while the cells are still growing.

Collenchyma. **Collenchyma cells,** which are also living at maturity, form strands or continuous cylinders beneath the epidermis of stems or leaf stalks and along veins in leaves. They are usually elongated, with unevenly thickened primary walls, which are clearly visible in **figure 31.3**, and are their distinguishing feature. Strands of collenchyma provide much of the support for plant organs in which secondary growth has not occurred.

Sclerenchyma. **Sclerenchyma cells,** in contrast to parenchyma and collenchyma cells, have tough, thick cell walls called **secondary cell walls;** they usually do not contain living cytoplasm when mature. The secondary cell wall is laid down inside of the primary cell wall after the cell has stopped growing and expanding in size. The secondary cell wall provides cells with strength and rigidity. There are two types of sclerenchyma: *fibers,* which are long, slender cells that usually form strands, and *sclereids,* which are variable in shape but often branched. Sclereids, the reddish-colored cells in **figure 31.4**, are sometimes called "stone cells." Clusters of sclereids form the gritty texture you feel in the flesh of pears. Both fibers and sclereids are thick-walled and strengthen the tissues in which they occur. Compare the thickness of the cell walls in the sclereid cells in **figure 31.4** with the blue-stained parenchyma cells that surround them.

> **Putting the Concept to Work**
> How are the cell walls of parenchyma, collenchyma, and sclerenchyma different?

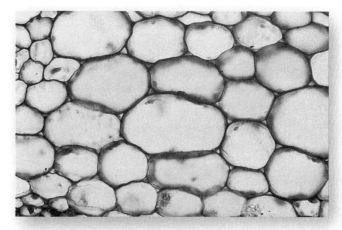

Figure 31.2 Parenchyma cells.

Cross section of parenchyma cells from grass. Only thin primary cell walls are seen in this living tissue.

Biophoto Associates/Science Source

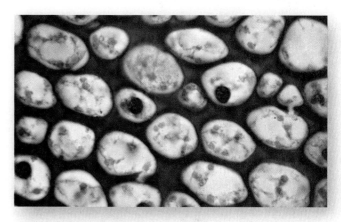

Figure 31.3 Collenchyma cells.

Cross section of collenchyma cells, with thickened side walls, from a young branch. In some collenchyma cells, the thickened areas may occur at the corners of the cells; in others, as strands.

Ed Reschke/Getty Images

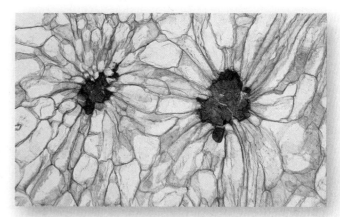

Figure 31.4 Sclerenchyma cells in sclereids.

Clusters of sclereids ("stone cells"), stained *red* in this preparation, in the pulp of a pear. These sclereid clusters give pears their gritty texture. The surrounding thin-walled cells, stained *blue,* are parenchyma cells.

Garry DeLong/Getty Images

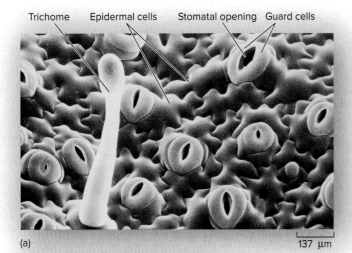

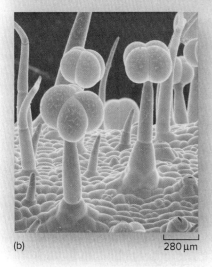

Figure 31.5 Guard cells and trichomes.

(a) Numerous stomata occur among the leaf epidermal cells of this tobacco (*Nicotiana tabacum*) leaf. A trichome is also visible. (b) Trichomes on a tomato plant (*Lycopersicon lycopersicum*).

(a) Dr Jeremy Burgess/Science Source; (b) Andrew Syred/Science Source;

Dermal Tissue

> **LEARNING OBJECTIVE 31.2.2** Describe the functioning of dermal tissue, including trichomes, stomata, root hairs, and the cuticle.

All parts of the outer layer of a primary plant body are covered by flattened epidermal cells. These are the most abundant cells in the plant *epidermis*, or outer covering. The epidermis is one cell layer thick and is a protective layer that provides an effective barrier against water loss. The epidermis is often covered with a thick, waxy layer called the **cuticle,** which protects against ultraviolet light damage and water loss. In some cases, the dermal tissue is more extensive and forms the bark of trees.

Guard Cells. Specialized cells and outgrowths can occur in the epidermis. **Guard cells** (figure 31.5*a*) are paired cells with an opening called a **stoma** (plural, **stomata**) that lies between them. Guard cells and stomata occur frequently in the epidermis of leaves and occasionally on other parts of the shoot, such as on stems or fruits. Oxygen, carbon dioxide, and water vapor pass across the epidermis almost exclusively through the stomata, which open and shut in response to external factors such as moisture, temperature, and light.

Trichomes. **Trichomes** are outgrowths of the epidermis that occur on the surfaces of stems and leaves. Trichomes vary greatly in form in different kinds of plants, from the rounded tip trichome in figure 31.5*a* to the pointed and globular-tipped ones in figure 31.5*b*. Trichomes play an important role in regulating the heat and water balance of the leaf, much as the hairs of an animal's coat provide insulation.

Root Hairs. Other outgrowths of the epidermis occur belowground, on the surface of roots near their tips. Called **root hairs,** these tubular extensions of single epidermal cells keep the root in intimate contact with the particles of soil (see figure 31.24). Root hairs play an important role in the absorption of water and minerals from the soil by increasing the surface area of the root.

> **Putting the Concept to Work**
> How does a change in light cause the stomata of a leaf to open?

Vascular Tissue

> **LEARNING OBJECTIVE 31.2.3** Describe vascular tissue, contrasting the structure and function of xylem and phloem.

Xylem. Xylem is a vascular plant's principal water-conducting tissue, forming a continuous system that runs throughout the plant body. Within this system, water (and minerals dissolved in it) passes from the roots up through the stem in an unbroken stream (figure 31.6). When water reaches the leaves, much of it passes into the air as water vapor, through the stomata.

The two principal types of conducting cells in the xylem are **tracheids** and **vessel elements,** both of which have thick secondary walls that are laid down inside the primary cell wall. These cells are elongated, and they have no living cytoplasm (they are dead) at maturity. Tracheids are elongated cells that overlap at the ends, as shown in figure 31.7a. In conducting elements composed of tracheids, water flows from tracheid to tracheid through openings called *pits* in the secondary walls. In contrast, vessel elements are elongated cells that line up end-on-end, as shown in figure 31.7b,c. The end walls of vessel elements may be almost completely open or may have bars or strips of wall material perforated by pores through which water flows. A linked row of vessel elements forms a vessel. Primitive angiosperms and other vascular plants have only tracheids, but the majority of angiosperms have vessels. Vessels conduct water much more efficiently than do strands of tracheids.

Figure 31.6 Trees are a plumber's dream.

The trees of this hardwood forest capture water and nutrients from the soil with their roots. Connecting the roots of a tree to its leaves is the stem. The stem of a tall tree is an engineering marvel, piping water and nutrients many meters up and down the plant.

FrancescoRizzato/iStockphoto/Getty Images

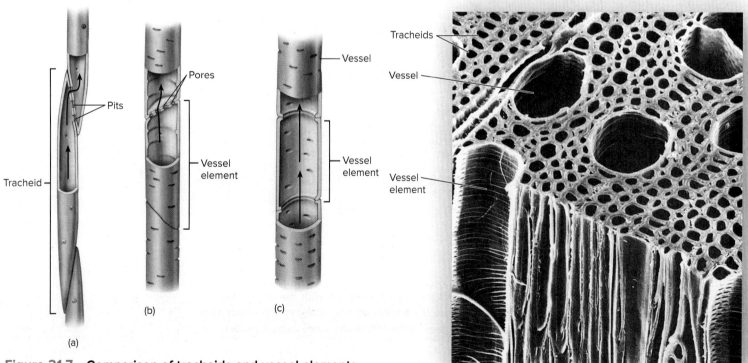

Figure 31.7 Comparison of tracheids and vessel elements.

(a) In tracheids, the water passes from cell to cell by means of pits. (b) In vessel elements, pores in the end walls allow water to move from cell to cell, or (c) the end walls of vessel elements may be almost completely open.

Figure 31.8 A scanning electron micrograph of red maple (*Acer rubrum*) wood, showing the xylem.

Courtesy N.C. Brown Center for Ultrastructure Studies, SUNY-ESF, Syracuse, NY

Phloem. **Phloem** is the principal food-conducting tissue in vascular plants. Food conduction in phloem is carried out through two kinds of elongated cells: **sieve cells** and **sieve-tube members.** Both cell types are living, but their nuclei are lost during maturation. The cells differ in the extent of the perforations between the cells, with the sieve cells having smaller perforations between cells. Seedless vascular plants and gymnosperms have only sieve cells; most angiosperms have sieve-tube members. In sieve-tube members, some sieve areas have larger pores and are called *sieve plates.* Sieve-tube members occur end to end, as shown in figure 31.9, forming longitudinal series called **sieve tubes.** Specialized parenchyma cells known as **companion cells** occur regularly in association with sieve-tube members; their cytoplasms are connected to the sieve-tube members through openings called *plasmodesmata.*

> **Putting the Concept to Work**
> What is the difference between a tracheid and a vessel element?

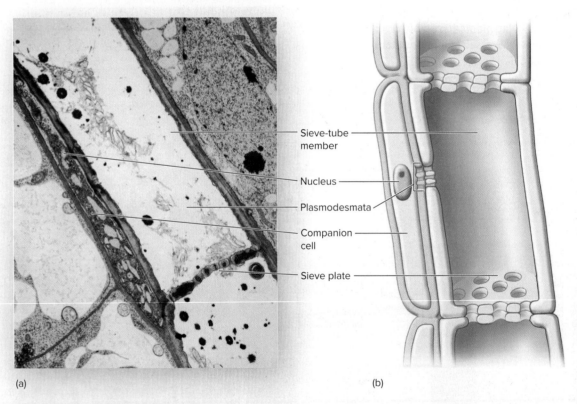

Figure 31.9 Sieve tubes.

(a) Sieve-tube member from the phloem of *Arabidopsis thaliana*, connected with the cells above and below to form a sieve tube. (b) In this diagram, note the thickened end walls, which are at right angles to the sieve tube. The narrow cell with the nucleus at the left of the sieve-tube member is a companion cell.

(a) Biophoto Associates/Science Source

The Plant Body

31.3 Roots

> **LEARNING OBJECTIVE 31.3.1** Diagram the basic structure of a root, differentiating between the three primary root meristems and explaining the functioning of the Casparian strip.

We now consider the three kinds of vegetative organs that form the body of a plant: roots, stems, and leaves. Though we will examine their basic structures, it is important to understand that these organs are often also modified for different functions. For example, roots and stems can be modified for water and food storage, and leaves can be modified for defenses.

Architecture of a Root

Roots have a simpler pattern of organization and development than do stems, and we will examine them first. Although different patterns exist, the kind of root described here and shown in **figure 31.10a** is found in many plants. Roots contain xylem (the pink areas in the figure) and phloem (the light green areas). As you can see, roots have a central column of xylem with radiating arms. Alternating with the radiating arms of xylem are strands of primary phloem. Surrounding the column of vascular tissue and forming its outer boundary is a cylinder of cells one or more cell layers thick called the **pericycle**. Branch, or lateral, roots are formed from cells of the pericycle. The outer layer of the root is the epidermis. The mass of parenchyma in which the root's vascular tissue is located is the cortex. Its innermost layer—the endodermis—consists of cells that regulate the flow of water between the vascular tissues and the root's outer portion. The **endodermis** is a single cell layer (**figure 31.11b**) just outside the pericycle.

Casparian Strip. Endodermis cells are encircled by a thickened, waxy band called the **Casparian strip.** Figure 31.11c is a drawing of endodermal cells showing how the wax substance that makes up the Casparian strip surrounds each cell. As the black arrows indicate, the Casparian strip blocks the movement of water *between* cells and instead directs the movement of water *through* the plasma membrane of the endodermal cells. In this way, the Casparian strip controls the passage of minerals into the xylem because transport through the endodermal cells is regulated by special channels embedded in the plasma membrane.

Root Meristems. The apical meristem of the root (shown as a group of cells at the base of the root in **figure 31.10a, b**) divides and produces cells both inwardly, back toward the body of the plant, and outwardly. The three primary meristems, shown in the cross section of the root in **figure 31.10a** and in the photo of a root in **figure 31.10b**, are the **protoderm,** which becomes the epidermis; the **procambium,** which produces primary vascular tissues (primary xylem and primary phloem); and the **ground meristem,** which differentiates further into ground tissue composed of parenchyma cells. Outward cell division results in the formation of a thimblelike mass of relatively unorganized cells, called the **root cap,** which you can clearly see in **figure 31.10b.** The root cap covers and protects the root's apical meristem as it grows through the soil.

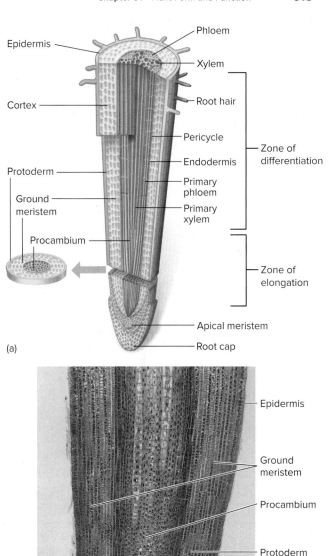

Figure 31.10 Root structure.

(a) Diagram of primary meristems in a plant root, showing their relation to the apical meristem. The three primary meristems are the protoderm, which differentiates further into epidermis; the procambium, which differentiates further into primary vascular strands; and the ground meristem, which differentiates further into ground tissue. (b) Median longitudinal section of a root tip in corn, *Zea mays,* showing the differentiation of protoderm, procambium, and ground meristem.

(b) Garry DeLong/Science Source

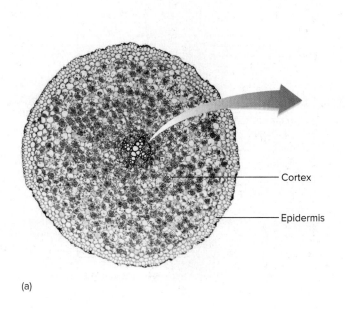
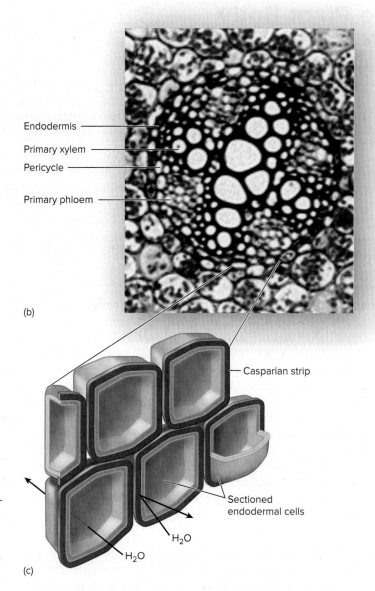

Figure 31.11 A root cross section.

(a) Cross section through a root of a buttercup (Ranunculus), a eudicot (×10). (b) The enlargement shows the various tissues present. (c) Endodermal cells are surrounded by a waterproofing, waxy band, called a Casparian strip, that forces water and minerals to pass through the cell rather than between two cells.

(a, b) Ed Reschke/Getty Images

Root Elongation

The root elongates relatively rapidly just behind its tip in the area called the *zone of elongation*. Abundant root hairs (see figure 31.24), extensions of single epidermal cells, form above that zone, in the area called the *zone of differentiation*. Virtually all water and minerals are absorbed from the soil through the root hairs, which greatly increase the root's surface area and absorptive powers. In plants with symbiotic mycorrhizae, the root hairs are often greatly reduced in number, and the fungal filaments of the mycorrhizae play a role similar to that of the root hairs, increasing the surface area for absorption. Another symbiotic relationship involving plant roots is often key to the health of ecosystems. The roots of some plants, specifically plants of the pea family, called legumes, form symbiotic relationships with bacteria that are able to break down atmospheric nitrogen into a source that can be taken up and used by plants. These plants are a key component of the nitrogen cycle, cycling nitrogen back into the ecosystem in a usable form.

Root Branching. One of the fundamental differences between roots and stems has to do with the nature of their branching. In stems, branching occurs from buds on the stem surface; in roots, branching is initiated well back of the root tip as a result of cell divisions in the pericycle. The developing lateral roots (the red-stained mass of cells in figure 31.12) grow out through the cortex toward the surface of the root, eventually breaking through and becoming established as lateral roots. In some plants, roots may arise along a stem or in some place other than the root of the plants. These roots are called *adventitious roots*. Adventitious roots occur in ivy, bulb plants such as onions, perennial grasses, and other plants that produce rhizomes, which are horizontal stems that grow underground.

> **Putting the Concept to Work**
> If you were able to make the Casparian strip of a root nonwaxy, what would happen to water movement through the root?

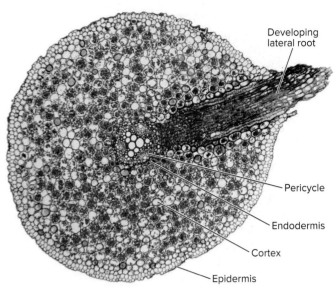

Figure 31.12 Lateral roots.

A lateral root growing out through the cortex of the meadow buttercup, *Ranunculus acris*. Lateral roots originate beneath the surface of the main root, whereas lateral stems originate at the surface.

Patrick J. Lynch/Science Source

31.4 Stems

> **LEARNING OBJECTIVE 31.4.1** Differentiate between primary and secondary stem growth, contrasting the arrangement of vascular bundles in monocots and eudicots.

Stems serve as the main structural support of the plant and the framework for the positioning of the leaves. Often experiencing both primary and secondary growth, stems are the source of an economically important product—wood.

Primary Growth

In the primary growth of a stem, leaves first appear as leaf primordia (singular, primordium), rudimentary young leaves that cluster around the apical meristem, unfolding and growing as the stem itself elongates. The places on the stem at which leaves form are called nodes (indicated by the small bracket in figure 31.13). The portions of the stem between these attachment points are called the internodes (the larger bracket). As the leaves expand to maturity, a bud, a tiny undeveloped side shoot, develops in the **axil** of each leaf, the angle between a leaf and the stem from which it arises. These buds, which have their own immature leaves (shown in figure 31.13), may elongate and form lateral branches, or they may remain small and dormant. A hormone moving downward from the terminal bud of the stem continually suppresses the expansion of the lateral buds in the upper portions of the stem. Lateral buds begin forming lower down when the stem lengthens to the point where the amount of hormone reaching the lower portion of the stem is reduced or if the terminal bud is removed, such as when you prune a plant.

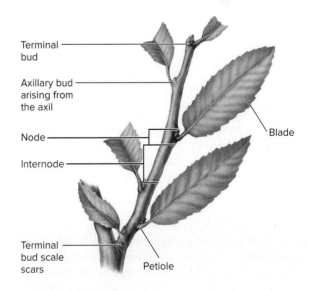

Figure 31.13 A woody twig.

This twig shows key stem structures, including the node and internode areas, the axillary bud in the axil, and leaves.

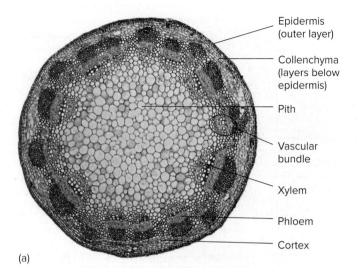

Eudicots Versus Monocots.
Within the soft, young stems, the strands of vascular tissue, xylem and phloem, are arranged differently in eudicots versus monocots. In eudicots, the vascular bundles are arranged around the outside of the stem as a cylinder (figure 31.14a). In monocots, the vascular bundles are scattered throughout the stem (figure 31.14b). This difference in vascular tissue organization is one of several differences between these two major groups of angiosperms (see the discussion of monocots and eudicots in chapter 17). The vascular bundles contain both primary xylem and primary phloem. At the stage when only primary growth has occurred, the inner portion of the ground tissue of a eudicot stem is called the **pith** (the center cells in figure 31.14a), and the outer portion is the **cortex** (the cells located toward the outside).

Secondary Growth

Vascular Cambium.
In stems, secondary growth (the thickening of the stem) is initiated by the differentiation of a lateral meristem called the **vascular cambium,** a thin cylinder of actively dividing cells located between the bark and the interior region of the stem in woody plants. The vascular cambium develops from cells within the vascular bundles of the stem, between the xylem (the purple-colored areas in figure 31.15) and the phloem (the light green area). The cylindrical form of the vascular cambium is completed by the differentiation of some of the parenchyma cells that lie between the bundles. Once established, the vascular cambium consists of elongated and somewhat flattened cells with large vacuoles. The cells that divide from the vascular cambium outwardly, toward the bark, become secondary phloem; those that divide from it inwardly become secondary xylem.

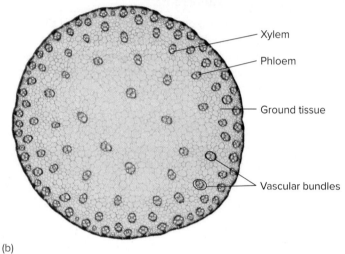

Figure 31.14 A comparison of eudicot and monocot stems.

(a) Transection of a young stem of a eudicot, the common sunflower, *Helianthus annuus*, in which the vascular bundles are arranged around the outside of the stem. (b) Transection of a monocot stem, corn, *Zea mays*, with the scattered vascular bundles characteristic of the group.

(a, b) Ed Reschke/Getty Images

Figure 31.15 Vascular cambium and secondary growth.

The vascular cambium and cork cambium (lateral meristems) produce secondary tissues, causing the stem's girth to increase. Each year, a new layer of secondary tissue is laid down, forming rings in the wood.

> **IMPLICATION FOR YOU** Musical instruments such as violins, guitars, clarinets, and recorders are made mostly or entirely of wood. The choice of wood makes a significant difference to the tone of the instrument. The bodies of clarinets are made from hard, dense African blackwood, whereas light spruce and sycamore are used in violins. How do you think the nature of the wood affects the sound of the instrument?

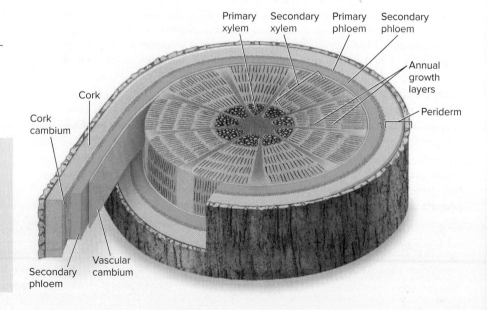

Cork Cambium. While the vascular cambium is becoming established, a second kind of lateral meristem, the **cork cambium,** develops in the stem's outer layers. The cork cambium usually consists of plates of dividing cells that move deeper and deeper into the stem as they divide. Outwardly, the cork cambium splits off densely packed cork cells; they contain a fatty substance and are nearly impermeable to water. Cork cells are dead at maturity. Inwardly, the cork cambium divides to produce a layer of parenchyma cells. The cork, the cork cambium that produces it, and this layer of parenchyma cells make up a layer called the **periderm** (see figure 31.15), which is the plant's outer protective covering.

Bark. Cork covers the surfaces of mature stems or roots. The term **bark** refers to all of the tissues of a mature stem or root outside of the vascular cambium. Because the vascular cambium has the thinnest-walled cells that occur anywhere in a secondary plant body, it is the layer at which bark breaks away from the accumulated secondary xylem.

Wood. Wood is one of the most useful, economically important, and beautiful products obtained from plants. Anatomically, wood is accumulated secondary xylem (the light-purple pie-shaped areas in figure 31.15). As the secondary xylem ages, its cells become infiltrated with gums and resins, and the wood may become darker. For this reason, the wood located nearer the central regions of a given trunk, called heartwood, can be darker and denser than the wood nearer the vascular cambium, called sapwood, which is still actively involved in water transport within the plant.

Because of the way it is accumulated, wood often displays rings (figure 31.16). The rings reflect the fact that the vascular cambium of trees divides more actively in the spring and summer, when water is plentiful and temperatures are suitable for growth, than in the fall and winter, when water is scarce and the weather is cold. As a result, layers of larger, thinner-walled cells formed during the growing season (the lighter rings) alternate with the smaller, darker layers of thick-walled cells formed during the rest of the year. New rings are laid down each year toward the outer edge of the stem. A count of such annual rings in a tree trunk can be used to calculate the tree's age, and the width of rings can reveal information about environmental factors. Can you estimate the age of the pine tree shown in figure 31.16?

Figure 31.16 Annual rings in a section of pine.

To test your understanding of how they form, answer this question: Are the inner rings older or younger than the outer rings?

malerapaso/Getty Images

> **Putting the Concept to Work**
> Of what tissue is wood composed?

31.5 Leaves

> **LEARNING OBJECTIVE 31.5.1** Diagram a leaf in cross section, explaining how it grows in size.

Leaves are usually the most prominent shoot organs and are structurally diverse (figure 31.17). As outgrowths of the stem apex, leaves are the major light-capturing organs of most plants. Most of the chloroplast-containing cells of a plant are within its leaves, and it is there that the bulk of photosynthesis occurs (see chapter 6). Exceptions to this are found in some plants, such as cacti, whose green stems have largely taken over the function of photosynthesis for the plant. Photosynthesis is conducted mainly by the "greener" parts of plants because they contain more chlorophyll, the major photosynthetic pigment.

Parts of a Leaf

The apical meristems of stems and roots are capable of growing indefinitely under appropriate conditions. Leaves, in contrast, grow by means of **marginal meristems,** which flank their thick central portions. These marginal meristems grow outward and ultimately form the **blade** (flattened portion) of the leaf, while the central portion becomes the midrib. Once a leaf is fully expanded, its marginal meristems cease to grow.

In addition to the flattened blade, most leaves have a slender stalk, the **petiole.** Two leaflike organs, the **stipules,** may flank the base of the petiole where it joins the stem. Veins, consisting of both xylem and phloem, run through the leaves. In most eudicots, the pattern is net, or reticulate, venation—as you can see in figure 31.18a. In many monocots, the veins are parallel, like the parallel veins that pass vertically up through the monocot leaf in figure 31.18b.

Leaf Shape. Leaf blades come in a variety of forms from oval to deeply lobed to having separate leaflets (the blade being divided but attached to a single petiole, such as the black walnut leaf in figure 31.17c). In *simple leaves* (see figure 31.17a, b), such as those of birch or maple trees, there is a single blade, undivided, but some simple leaves may have teeth, indentations, or lobes, such as the leaves of maples and oaks. In *compound leaves,* such as those of ashes, box elders, and walnuts, the blade is divided into leaflets. If the leaflets are arranged in pairs along a common axis—the equivalent of the main central vein, or *midrib,* in simple leaves—the leaf is *pinnately compound,* such as in the black walnut (see figure 31.17c). If, however, the leaflets radiate out from a common point at the blade end of the petiole, the leaf is *palmately compound,* such as in buckeyes, horse chestnuts (figure 31.17d), and Virginia creepers. Leaves may be alternately arranged (alternate leaves usually spiral around a stem), or they may be in opposite

Figure 31.17 Leaves.

Leaves are stunningly variable. (a) *Simple leaves* from a birch, in which there is a single blade. (b) A simple leaf, its margin lobed, from the vine maple. (c) A *pinnately compound* leaf of a black walnut tree, where leaflets occur in pairs along the central axis of the main vein. (d) *Palmately compound* leaves of a horse chestnut tree, in which the leaflets radiate out from a single point. (e) The leaves of pine trees are tough and needlelike. (f) Many unusual types of modified leaves occur in different kinds of plants. For example, some plants produce floral leaves or bracts; the most conspicuous parts of this poinsettia flower are the red bracts, which are not petals but rather are modified leaves that surround the small yellowish true flowers in the center.

(a) VIDOK/Getty Images; (b) Robert Glusic/Corbis/Getty Images; (c) DEA/S. MONTANARI/Getty Images; (d) Frank Krahmer/Getty Images; (e) David Chapman/Design Pics Inc./Alamy Stock Photo; (f) Don Hammond/Design Pics

pairs. Less often, three or more leaves may be in a whorl, a circle of leaves at the same level at a node (figure 31.19).

Interior of a Leaf. A typical leaf contains masses of parenchyma, called **mesophyll** ("middle leaf"), through which the vascular bundles, or veins, run. Beneath the upper epidermis of a leaf are one or more layers of closely packed, columnlike parenchyma cells called **palisade mesophyll** (the red-stained cells in the photo in figure 31.20). These cells contain more chloroplasts than other cells in the leaf and so are more capable of carrying out photosynthesis. This makes sense when you consider that the cells on the surface receive more sun. The rest of the leaf interior, except for the veins, consists of a tissue called **spongy mesophyll.** Between the spongy mesophyll cells are large intercellular spaces that function in gas exchange and particularly in the passage of carbon dioxide from the atmosphere to the mesophyll cells. You can see the spongy mesophyll in the photo of figure 31.20, but the air spaces that are the basis of this tissue's function might be easier to see in the drawing. These intercellular spaces are connected, directly or indirectly, with the stomata in the lower epidermis.

> **Putting the Concept to Work**
> Where on a plant would you expect to find marginal meristems?

Figure 31.18 Eudicot and monocot leaves.

(a) The leaves of eudicots have netted, or reticulate, veins; (b) those of monocots have parallel veins.

(a) irin-k/Shutterstock; (b) sot/Getty Images

Alternate (spiral): Ivy Opposite: Periwinkle Whorled: Sweet woodruff

Figure 31.19 Types of leaf arrangements.

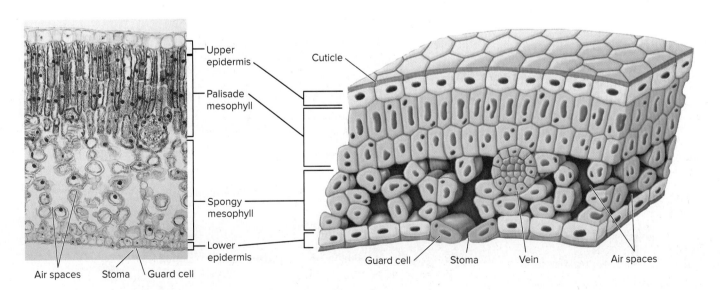

Figure 31.20 A leaf in cross section.

Cross section of a leaf, showing the arrangement of palisade and spongy mesophyll, a vascular bundle or vein, and the epidermis, with paired guard cells flanking the stoma.

Ed Reschke/Getty Images

Plant Transport and Nutrition

31.6 Water Movement

> **LEARNING OBJECTIVE 31.6.1** Contrast how minerals and carbohydrates move in stems, tracing the movement of a water molecule through a plant from soil to atmosphere.

Vascular plants have a conducting system, as humans do, for transporting fluids and nutrients from one part to another. Functionally, a plant is essentially a bundle of tubes with its base embedded in the ground. At the base of the tubes are roots, and at their tops are leaves. For a plant to function, two kinds of transport processes must occur: First, the carbohydrate molecules produced in the leaves by photosynthesis must be carried to all of the other living cells in the plant. To accomplish this, liquid with dissolved carbohydrate molecules must move both up and down the tubes. Second, minerals and water in the soil must be taken up by the roots and ferried to the leaves and other plant cells. In this process, liquid moves up the tubes. Plants accomplish these two processes by using chains of specialized cells. Cells of the phloem transport photosynthetically produced carbohydrates up and down the plant (red arrows in figure 31.21), and those of the xylem carry water and minerals upward (blue arrows in figure 31.21).

Cohesion-Adhesion-Tension Theory

Many of the leaves of a large tree may be more than 10 stories off the ground. How does a tree manage to raise water so high? Several factors are at work to move water up the height of a plant.

Osmotic Pressure. The initial movement of water into the roots of a plant involves osmosis. Water moves into the cells of the root because the fluid in

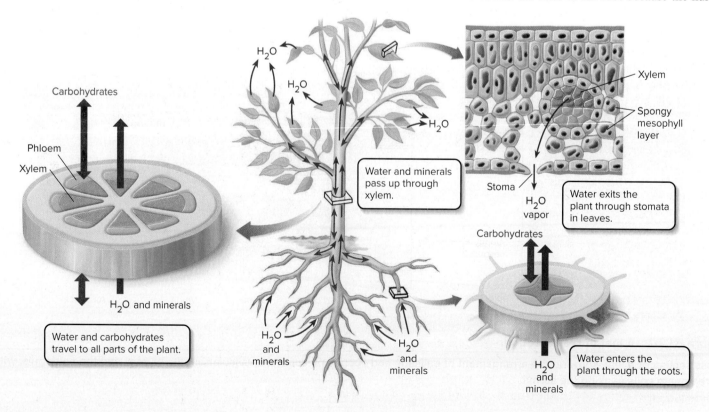

Figure 31.21 The flow of materials into, out of, and within a plant.

Water and minerals enter through the roots of a plant and are transported through the xylem to all parts of the plant body (*blue arrows*). Water leaves the plant through the stomata in the leaves. Carbohydrates synthesized in the leaves are circulated throughout the plant by the phloem (*red arrows*).

the xylem contains more solutes than the surroundings—recall from chapter 4 that water will move across a membrane from an area of lower solute concentration to an area of higher solute concentration. However, this force, called *root pressure*, is not by itself strong enough to "push" water up a plant's stem.

Adhesion. Capillary action adds a "pull" to the process. *Capillary action* results from the tiny electrical attractions of polar water molecules to surfaces that carry an electrical charge, a process called *adhesion*. In the laboratory, a column of water rises up a tube of glass because the attraction of the water molecules to the charged molecules on the interior surface of the glass tube "pull" the water up in the tube. In figure 31.22, which illustrates this process, why does the water travel higher up in the narrower tube? The water molecules are attracted to the glass molecules, and the water travels up farther in the narrower tube because the amount of surface area available for adhesion is greater than in the larger-diameter tube.

> Adhesion and cohesion are properties of water, as discussed in section 2.4. Water molecules form hydrogen bonds with other polar surfaces (adhesion) and with each other (cohesion). These properties allow water to move up the stem of a plant.

Evaporation. However, although capillary action can produce enough force to raise water a meter or two, it cannot account for the movement of water to the tops of tall trees. A second very strong "pull" accomplishes this, provided by **transpiration,** the process by which water leaves a plant. Blowing air across the upper end of the tube in figure 31.22 demonstrates how transpiration draws water up a plant stem. The stream of relatively dry air causes water molecules at the water column's exposed top surface to evaporate from the tube. The water level in the tube does not fall because as water molecules are drawn from the top through evaporation, they are replenished by new water molecules pulled up from the bottom. This, in essence, is what happens in plants. The passage of air across leaf surfaces results in the loss of water by evaporation, creating a "pull" at the open upper end of the plant. New water molecules that enter the roots are pulled up the plant. Adhesion of water molecules to the walls of the narrow vessels in plants also helps to maintain water flow to the tops of plants.

Cohesion. A column of water in a tall tree does not collapse under its weight because water molecules have an inherent strength that arises from their tendency to form hydrogen bonds with one another. These hydrogen bonds cause *cohesion* of the water molecules; in other words, a column of water resists separation. The beading of water droplets illustrates the property of cohesion. This resistance, called *tensile strength,* varies inversely with the diameter of the column; that is, the smaller the diameter of the column, the greater the tensile strength. Therefore, plants must have very narrow transporting vessels to take advantage of tensile strength.

How the combination of osmosis, adhesion, and tensile strength due to cohesion affects water movement in plants is called the **cohesion-adhesion-tension theory.** It is important to note that the movement of water up through a plant is a passive process and requires no expenditure of energy on the part of the plant.

Transpiration

The process by which water leaves a plant is called **transpiration.** More than 90% of the water taken in by plant roots is ultimately lost to the atmosphere, almost all of it from the leaves. It passes out primarily through the stomata in the evaporation of water vapor, as you can see in **panel 1** of *Essential Biological Process 31A*. On its journey from the plant's interior to the outside, a molecule of water first diffuses from the xylem into the spongy mesophyll

Figure 31.22 Capillary action.

The attraction of water molecules to the glass surface of narrow tubes draws water upward, while the force of gravity tends to draw it down. The narrower the tube, the greater the surface area available for adhesion for a given volume of water and the higher the water rises in the tube.

Essential Biological Process 31A

Transpiration

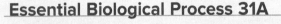

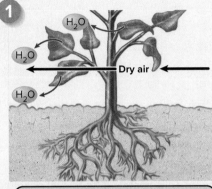

Dry air passes across the leaves and causes water vapor to evaporate out of the stomata.

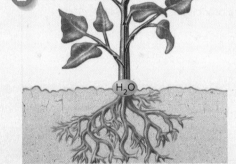

The loss of water from the leaves creates a type of "suction" that draws water up the stem through the xylem.

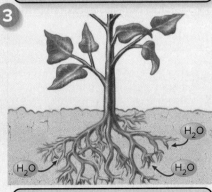

New water enters the plant through the roots to replace the water moving up the stem.

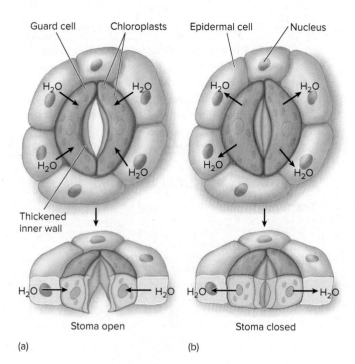

Figure 31.23 How guard cells regulate the opening and closing of stomata.

(a) When guard cells contain a high level of solutes, water enters the guard cells by osmosis, causing them to swell and bow outward. This bowing opens the stoma. (b) When guard cells contain a low level of solutes, water leaves the guard cells, causing them to become flaccid. This flaccidity closes the stoma.

Figure 31.24 Root hairs.

Abundant fine root hairs can be seen behind the root apex of this germinating radish seedling, *Raphanus sativus*.

Kingsley R. Stern

cells of the leaf (see figure 31.20). Then, water passes into the pockets of air within the leaf by evaporating from the walls of the spongy mesophyll that line the intercellular spaces. These intercellular spaces open to the outside of the leaf by way of the stomata. The water that evaporates from these surfaces of the spongy mesophyll cells is continuously replenished from the xylem in the leaves. Because the strands of xylem conduct water within the plant in an unbroken stream all the way from the roots to the leaves, when a portion of the water vapor in the intercellular spaces passes out through the stomata, the supply of water vapor in these spaces is continually renewed from lower down in the column (panel 2) and ultimately from the roots (panel 3). Because the process of transpiration is dependent upon evaporation, factors that affect evaporation also affect transpiration. In addition to the movement of air across the stomata, mentioned earlier, humidity levels in the air will affect the rate of evaporation—high humidity reducing it and low humidity increasing it. Temperature will also affect the rate of evaporation—high temperatures increasing it and lower temperatures reducing it. This temperature effect is especially important because evaporation also acts to cool plant tissues.

Regulation of Transpiration: Open and Closed Stomata

The only way plants can control water loss on a short-term basis is to close their stomata. Many plants can do this when subjected to water stress. But the stomata must be open at least part of the time so that carbon dioxide, which is necessary for photosynthesis, can enter the plant. In its pattern of opening or closing its stomata, a plant must respond to both the need to conserve water and the need to admit carbon dioxide. A number of environmental factors affect the opening and closing of stomata. The most important is water loss.

The stomata open and close because of changes in the water pressure of their guard cells. Stomatal guard cells are long, sausage-shaped cells attached at their ends. These are the green cells in figure 31.23. The cellulose microfibrils of their cell wall wrap around the cell such that when the guard cells are **turgid** (plump and swollen with water), they expand in length. This causes the cells to bow, opening the stomata, shown on the left side of the figure. Turgor in guard cells results from the active uptake of ions, causing water to enter osmotically as a consequence. Loss of water, as in a wilted plant, causes the guard cells to become flaccid and the stomata close.

Water Absorption by Roots

Most of the water absorbed by plants comes in through the root hairs, extensions of epidermal cells. These give a root the feathery appearance, as shown in figure 31.24. These root hairs greatly increase the surface area and therefore the absorptive powers of the roots. Root hairs are turgid—plump and swollen with water—because they contain a higher concentration of dissolved minerals and other solutes than does the water in the soil solution; water, therefore, tends to move into them steadily. Once inside the roots, water passes inward to the conducting elements of the xylem.

Water is not the only substance that enters the roots by passing into the cells of root hairs. Minerals also enter the root. Membranes of root hair cells contain a variety of ion transport channels that actively pump specific ions into the plant, even against large concentration gradients. These ions, many of which are plant nutrients, are then transported throughout the plant as a component of the water flowing through the xylem.

> **Putting the Concept to Work**
> What is the difference between cohesion and adhesion?

31.7 Carbohydrate Transport

> **LEARNING OBJECTIVE 31.7.1** Describe how carbohydrates move through the stem without the expenditure of energy by the plant.

Most of the carbohydrates manufactured in plant leaves and other green parts move through the phloem to other parts of the plant.

Translocation

The movement of carbohydrates up and down the stem is critical to the life of the plant. This process of **translocation** makes carbohydrate building blocks available at the plant's actively growing regions. The carbohydrates are concentrated in storage organs such as underground stems (potatoes), roots (carrots), and leaves (onions and cabbage), often in the form of starch. The starch is converted into transportable molecules such as sucrose and moves through the plant.

The pathway by which sugars and other substances travel within the plant has been demonstrated precisely by using radioactive isotopes and aphids, a group of insects that suck the sap of plants. Aphids thrust their piercing mouthparts into the phloem cells of leaves and stems to obtain the abundant sugars there. When the aphids are cut off of the leaf, the liquid continues to flow from the detached mouthparts protruding from the plant tissue and is thus available in pure form for analysis. The liquid in the phloem contains 10% to 25% sugar.

The harvesting of sap from maple trees uses a similar process. A hole is drilled in the tree, and the sugar-rich fluid is drained from the tree into buckets. The sap is then processed into maple syrup.

Mass Flow

Using aphids to obtain the critical samples and radioactive tracers to mark them, researchers have learned that movement of substances in the phloem can be remarkably fast—rates of 50 to 100 centimeters per hour have been measured. This translocation movement is a passive process that does not require the expenditure of energy by the plant. The **mass flow** of materials transported in the phloem occurs because of water pressure, which develops as a result of osmosis. *Essential Biological Process 31B* walks you through the process of translocation. Sugar produced as a result of photosynthesis is actively "loaded" into the sieve tubes (or sieve cells) of the vascular bundles (**panel 1**). This loading increases the solute concentration of the sieve tubes, so water passes into them by osmosis (**panel 2**). The area where the sugar is made is called a *source*, and the area where sugar is delivered from the sieve tubes is called a *sink*. Sinks include the roots and other regions of the plant that are not photosynthetic, such as young leaves and fruits. Water flowing into the phloem forces the sugary substance in the phloem to flow down the plant (**panel 3**). The sugar is unloaded and stored in sink areas (**panel 4**). There the solute concentration of the sieve tubes is decreased as the sugar is removed. As a result of these processes, water moves through the sieve tubes from the areas where sucrose is being added into those areas where it is being withdrawn, and the sugar moves passively with the water. This is called the *pressure-flow hypothesis*.

> **Putting the Concept to Work**
> How does sugar made in a leaf reach the cells of a root?

Essential Biological Process 31B

Translocation

1. Sugar created in the leaves by photosynthesis ("source") enters the phloem by active transport.

2. When the sugar concentration in the phloem increases, water is drawn into phloem cells from the xylem by osmosis.

3. The addition of water from the xylem causes pressure to build up inside the phloem and pushes the sugar down.

4. Sugar from the phloem enters the root cells ("sink") by active transport.

Putting the Chapter to Work

1. You are observing a plant tissue with a microscope. This tissue appears globular and tightly packed together. Your instructor tells you that this tissue is involved in photosynthesis and cellular respiration.

   What kind of plant cells are you observing?

2. Plants do not have layers of fat or hair to insulate the body to retain warmth, as your human body does. Instead, plants rely on the small structures that extend out from the surfaces of leaves to help regulate temperature and reduce water loss.

   What are these structures called?

3. You place a freshly-cut piece of celery in blue-colored water. Over the next 24 hours you observe if the blue water moves up through the interior of the celery stalk or not.

   a. *Does the blue water move up the stalk?*
   b. *What type of vascular tissue could be used to transport the water up through the celery stalk?*

4. Different sections of wood are used in building materials. Some areas of the tree are softer than others. One such soft area, often densely layered, is found directly under the bark layer.

   What is it called?

   (Hint: It is often used as a porous cap to seal wine bottles.)

Retracing the Learning Path

Structure and Function of Plant Tissues

31.1 Organization of a Vascular Plant

1. Most plants possess roots, stems, and leaves, although they may not always look the same. Vascular tissue extends throughout the plant, connecting roots, stems, and leaves.
- Growth occurs in regions called meristems. The tips of the roots and shoots contain apical meristems, which are the sites of primary growth that extends the plant body lengthwise. Secondary growth, extending the thickness or girth of the plant, occurs at the lateral meristems, which are cylinders of meristematic tissue.

31.2 Plant Tissue Types

1. Parenchyma cells are the most common type of cell in plants. They have thin cell walls and are alive at maturity. They carry out functions such as photosynthesis and food and water storage. The edible parts of fruits and vegetables are primarily parenchyma cells.
- Collenchyma cells form strands that provide support, especially for plants that do not have secondary growth. They are elongated cells with unevenly thickened primary cell walls.
- Sclerenchyma cells have thick secondary cell walls that provide strength and rigidity. The secondary cell wall is laid down after the cell has stopped growing. They form long fibers or branched structures called sclereids or "stone cells."

2. Vascular tissue is composed of xylem and phloem. Xylem contains water-conducting cells, such as the tracheids and vessel elements. They have thick secondary cell walls that provide structural support. They form long strands that are connected by pits and pores through which water passes.
- Phloem contains food-conduction cells, sieve cells, and sieve-tube members. The cells fit end-to-end, and food is conducted through pores between the cells.

The Plant Body

31.3 Roots

1. Roots are organs adapted to absorb water and minerals from the soil. Vascular tissue forms the core of the root.
- A single layer of cells, the endodermis, surrounds the vascular tissue. A waxy Casparian strip encircles the endodermal cells and blocks the passage of water between the cells.
- The root grows at the tip, where new cells are added by the apical meristem. Sometimes roots will emerge from the stems of a plant and are called adventitious roots.

31.4 Stems

1. The stem serves as a framework for positioning the leaves. Primary growth occurs at the apical meristem. Leaves grow out of the stems at node areas. Secondary growth occurs at the lateral meristems with the differentiation of vascular cambium into xylem and phloem, and the cork cambium into the layers of cork inside bark. Wood forms from the accumulation of secondary xylem, which is thicker in spring and summer, resulting in rings in the wood.

31.5 Leaves

1. Leaves are the primary site for photosynthesis. They grow out from the stem by means of marginal meristems that form the blade. They vary in size, shape, and arrangement. Photosynthetic palisade mesophyll cells lie toward the surface. An underlying spongy mesophyll cell layer has large intercellular spaces that function in gas exchange.

Plant Transport and Nutrition

31.6 Water Movement

1. Carbohydrates and water are transported by phloem and xylem respectively.
- Water enters the roots by osmosis. Root pressure and capillary action cause the water to pass up into the tissues. However, for water to travel up the length of the stem, it requires a stronger force, the combination of cohesion and adhesion; this is known as the cohesion-adhesion-tension theory. Transpiration, the evaporation of water vapor from the leaves, creates the "pull" that raises water through the xylem.
- The guard cells flanking stomata will swell up when water is plentiful. This turgid pressure opens stomata, letting water vapor out. Under conditions of water stress, water leaves the guard cells and stomata close, reducing water loss.
- Minerals enter the plant along with water through ion channels in the cells of root hairs.

31.7 Carbohydrate Transport

1. Carbohydrates produced in the leaves travel throughout the plant in phloem tissue. Translocation involves osmotic movement of water into the phloem cells, forcing the sugars to "sinks" for carbohydrate storage until needed.

Inquiry and Analysis

Does Water Move Up a Tree Through Phloem or Xylem?

Before reading this chapter, you may have wondered how water gets to the top of a tree, 10 stories above its roots. Earlier scientists also wondered about this. A column of water that tall weighs an awful lot. If you were to make a tube of drinking straws that tall and fill it with water, you would not be able to lift it. The answer to this puzzle was first proposed by biologist Otto Renner in Germany in 1911. He suggested that dry air moving across the tree's leaves captured water molecules by evaporation and that this water was replaced with other water molecules coming in from the roots. Renner's idea, which was essentially correct, forms the core of the cohesion-adhesion-tension theory described in this chapter. Essential to the theory is that there is an unbroken water column from leaves to roots, a "pipe" from top to bottom through which the water can move freely.

There are two candidates for the role of water pipe, each a long series of narrow vessels that runs the length of the stem of a tree. As you have learned earlier, these two vessel systems are called xylem and phloem. In principle, either xylem or phloem could provide the plumbing through which water moves up a tree trunk or other stem. Which one is it?

An elegant experiment demonstrates which of these vessel systems carries water up a tree stem. A section of a stem was placed in water containing the radioactive potassium isotope ^{42}K. A piece of wax paper was carefully inserted between the xylem and the phloem in a 23-centimeter (cm) section of the stem to prevent any lateral transport of water between xylem and phloem.

After enough time had elapsed to allow water movement up the stem, the 23-cm section of the stem was removed, cut into six segments, and the amounts of ^{42}K measured both in the xylem and in the phloem of each segment, as well as in the stem immediately above and below the 23-cm section. The amount of radioactivity recorded provides a

Richard Rowan/Science Source

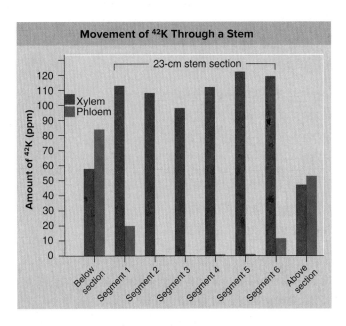

direct measure of the amount of water that has moved up from the bottom of the stem through either the xylem or phloem.

The results are presented in the graph.

Analysis

1. **Interpreting Data**
   a. In the portion of the stem below where the 23-cm section was removed, do xylem and phloem both contain radioactivity? How about in the portion above where the 23-cm section was removed?
   b. In the central portion of the 23-cm segment of the stem (segments 2, 3, 4, 5), do both xylem and phloem contain radioactivity?
2. **Making Inferences**
   a. In the 23-cm section, is more ^{42}K found in xylem or phloem? What might you conclude from this?
   b. Above and below the 23-cm section, is more ^{42}K found in xylem or phloem? How would you account for this? [Hint: These sections did not contain the wax paper barrier that prevents lateral transport between xylem and phloem.]
   c. Within the 23-cm section, the phloem in segments 1 and 6 contains more ^{42}K than interior segments. What best accounts for this?
   d. Is it fair to infer that water could move through either xylem or phloem vessel systems?
3. **Drawing Conclusions** Does water move up a stem through phloem or xylem? Explain.

32 Plant Reproduction and Growth

LEARNING PATH ▼

Flowering Plant Reproduction
1. Angiosperm Reproduction
2. Seeds
3. Fruit
4. Germination

Regulating Plant Growth
5. Plant Hormones
6. Auxin

Plant Responses to Environmental Stimuli
7. Photoperiodism and Dormancy
8. Tropisms

How Tall Is Too Tall?

Plants can grow at surprising rates and for a very long time. Bamboo plants can grow as fast as three feet per day—that's 1.5 inches per hour! At that rate, you can almost watch it grow. If plants can grow fast and for a long time, it should come as no surprise that some plants become very tall.

The World's Tallest Trees

Three types of trees vie for the title of the world's tallest tree: (1) the coastal redwood (*Sequoia sempervirens*), (2) the Douglas fir (*Pseudotsuga menziesii*), and (3) the Australia mountain ash (*Eucalyptus regnans*). Up until the summer of 2006, the title of the "tallest tree" went to a coastal redwood called the Stratosphere Giant that stood a breathtaking 370 feet (112.8 meters) tall. This towering giant lost its title to not one but three other coastal redwoods in Redwood National Park. The tallest of these, and now owner of the title "The World's Tallest Tree," is a coastal redwood called Hyperion that stands at a height of 379.1 feet

VERY LARGE TREES like this one can grow to surprising heights.
Fuse/Getty Images

(115.5 meters). Laid on its side, this tree would span just over the full length of a football field, including the endzones. The stand of trees that contains Hyperion is in a remote location of the forest, but its exact position is not being released for fear that a flood of visitors would damage the delicate forest ecosystem that contains this tree.

The "General Sherman" giant sequoia.
P.Burghardt/Shutterstock

The World's Largest Tree

As tall as Hyperion is, it is not the tallest tree ever found. The tallest tree was an Australia mountain ash, called the Robinson Tree, that was felled in Mount Baw Baw, Victoria, Australia, in 1889. The tree measured 470 feet. Other trees have impressive statistics; for example, the tree with the largest canopy (that is the leafy part of the tree) is a great banyan in Calcutta's Indian Botanical Garden that covers 3 acres. But the grandparent of all trees, the largest tree in the world and considered by many to be the largest organism on earth, is a giant sequoia named General Sherman, shown above. While not the tallest tree, it is the largest in overall size. Named after the American Civil War leader, General Sherman stands at 274.9 feet (83.8 meters (m)) and has a maximum base diameter of 36.5 feet (11.1 m) and a 102.6 feet (31.1 m) circumference. It would take over 60 people standing shoulder-to-shoulder to encircle its base. In January 2006, the largest branch on the tree broke off. The branch alone was the size of a tree, with a diameter of over 6 feet and 140 feet (30 m) in length.

General Sherman is located in the Giant Forest of Sequoia National Park in California. Dating of the tree's rings suggest that it is 2,200 years old. It is large not because it has been growing for so long but because it grows so fast, adding more wood each year. The trunk of General Sherman is very unified in size from the base up to the first branch, giving it a total trunk volume of 52,508 cubic feet, about 4,000 cubic feet more than its nearest giant sequoia competitor.

How Tall Trees Survive

It is important to remember that the primary reason why General Sherman, Hyperion, Stratosphere Giant, and other sequoia and redwood trees remain standing is because they are growing in national forests, protected from logging. More than likely, if these trees were located outside of the national forests, they would have met with the same fate as the Robinson Tree in Australia, cut down for its lumber.

How Tall Can a Tree Grow?

When you consider the massive size of General Sherman and Hyperion, it is impossible to avoid asking "How tall can these trees grow?" Can we expect them to grow another 50, 100, 200 feet—or more? In this chapter, you will learn that a tree grows tall in a different way than you do and can grow broader without becoming taller. How tall can a tree grow? As long as the tree is living, the apical meristem will continue to add cells, so is "the sky the limit"?

We know that General Sherman will not grow any taller than its 274.9 feet because the top of the tree is dead. This means that there is no more primary growth adding to the height of the tree. Hyperion, however, is still growing taller. How tall can it realistically be expected to grow? Researchers climbed to the tops of towering trees in Humboldt Redwoods State Park and discovered that although water is readily available to their roots, the water has a difficult time reaching the tops of these trees. The pull of gravity and friction that builds up between the water molecules and the walls of the xylem vessels in the tree's trunk limits how high a column of water can be supported. The high end for redwoods appears to be about 427 feet (130 meters).

So, how tall will Hyperion get? According to this research, maybe only another 50 feet, but that is still taller than a 30-story building—a long way up indeed!

Flowering Plant Reproduction

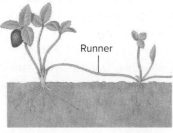

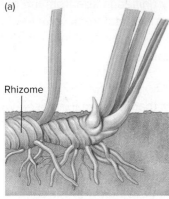

Figure 32.1 Vegetative reproduction.
(a) Runners are slender stems that grow along the ground, sending out roots and shoots at the nodes. (b) Rhizomes are underground horizontal stems that give rise to new shoots. (c) Small plants arise from notches along the leaves of the house plant *Kalanchoë daigremontiana*.

(c) Jerome Wexler/Science Source

32.1 Angiosperm Reproduction

Although reproduction varies greatly among the members of the plant kingdom, we focus in this chapter on reproduction among flowering plants. The evolution of their unique sexual reproductive features, flowers and fruits, have contributed to their success, but angiosperms also reproduce asexually.

Asexual Reproduction

> **LEARNING OBJECTIVE 32.1.1** Contrast asexual and sexual plant reproduction, describing four forms of vegetative reproduction.

In **asexual reproduction,** an individual inherits all of its chromosomes from a single parent and is, therefore, genetically identical to that parent. Asexual reproduction produces a "clone" of the parent.

In a stable environment, asexual reproduction may prove more advantageous than sexual reproduction because it allows individuals to reproduce with a lower investment of energy and to maintain successful traits. A common type of asexual reproduction, called *vegetative reproduction,* results when new individuals are simply cloned from parts of the parent. Vegetative reproduction in plants varies and includes:

Runners. Some plants reproduce by means of runners—long, slender stems that grow along the surface of the soil. The strawberry plant shown in figure 32.1a reproduces by runners. Notice that at node regions on the stem, adventitious roots form, extending into the soil. Leaves and flowers form, and a new stem is sent out, continuing the runner.

Rhizomes. Rhizomes are underground horizontal stems that create a network underground. As in runners, nodes give rise to new flowering shoots (figure 32.1b). The noxious character of many weeds results from this type of growth pattern, but so do grasses and many garden plants such as irises. Other specialized stems, called tubers, function in food storage and reproduction. White potatoes are specialized underground stems that store food, and the "eyes" give rise to new plants.

Suckers. The roots of some plants produce "suckers," or sprouts, which give rise to new plants, such as found in cherry, apple, raspberry, and blackberry plants. When the root of a dandelion is broken, which may occur if one tries to pull it from the ground, each root fragment may give rise to a new plant.

Adventitious Plantlets. In a few species, even the leaves are reproductive. One example is the house plant *Kalanchoë daigremontiana,* familiar to many people as the "maternity plant," or "mother of thousands." The common names of this plant are based on the fact that numerous plantlets arise from meristematic tissue located in notches along the leaves. You can see the numerous little plantlets in figure 32.1c.

> **Putting the Concept to Work**
> How does a runner differ from a rhizome?

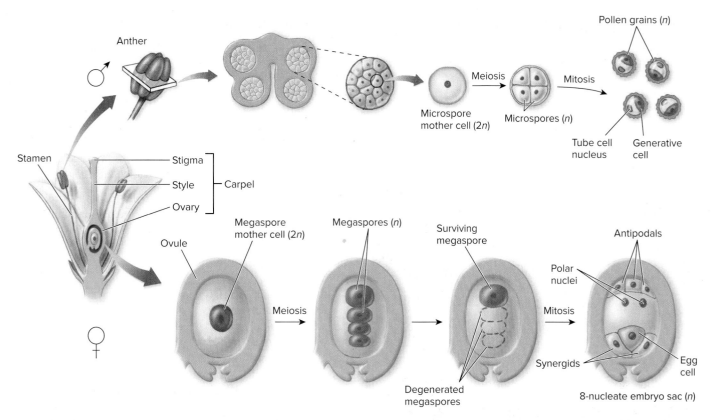

Figure 32.2 Formation of pollen and egg.

Sexual Reproduction

> **LEARNING OBJECTIVE 32.1.2** Describe the basic structure of a flower and the processes of gamete formation and pollination.

Plant sexual life cycles are characterized by an alternation of generations, in which a diploid *sporophyte generation* gives rise to a haploid *gametophyte generation,* as described in chapter 17. In angiosperms, the developing gametophyte generation is completely enclosed within the tissues of the parent sporophyte (see **figure 17.17**). The male gametophytes are **pollen grains,** and they develop from *microspores.* The female gametophyte is the **embryo sac,** which develops from a *megaspore.* Pollen grains and the embryo sac both are produced in separate, specialized structures of the angiosperm flower.

Structure of the Flower. Flowers contain the organs for sexual reproduction in angiosperms. However, angiosperm reproductive structures are not permanent parts of the adult individual but develop seasonally. Most flowers contain male and female parts. The male parts, called *stamens,* are the long, filament structures you see in the cutaway flower in **figure 32.2**. At the tip of each filament is a swollen portion, called the *anther,* that contains pollen. The female part, called the *carpel,* is the vase-shaped structure in **figure 32.2**. The carpel consists of a lower bulging portion called the *ovary,* a slender stalk called the *style,* and a sticky tip called the *stigma* that receives pollen. In some plants, there are separate male and female flowers, but they occur on the same plant. These plants are called *monoecious,* meaning "one house." In monoecious plants, the male and female flowers may mature at different times, which keeps the plant from pollinating itself.

Pollen Formation. If you were to cut an anther in half, you would see four pollen sacs (see **figure 32.2**). Each pollen sac contains microspore mother cells, which undergo meiosis to form four haploid microspores. Subsequently, mitotic divisions form pollen grains that contain a generative cell and a tube cell nucleus. The tube cell nucleus forms the pollen tube; the generative cell will later divide to form two sperm cells.

Egg Formation. Eggs develop in the **ovule** of the angiosperm flower, which is contained within the ovary at the base of the carpel. Within each ovule is a megaspore mother cell, which undergoes meiosis to produce four haploid megaspores. In most plants, only one of these megaspores survives and undergoes repeated mitotic divisions to produce eight haploid nuclei, which are enclosed within a structure called an *embryo sac*.

Pollination. The process by which pollen is transferred from the anther to the stigma (the top of the carpel) is called **pollination.** The pollen may be carried to the flower by wind or by animals, or it may originate within the individual flower itself. When pollen from a flower's anther pollinates the same flower's stigma, the process is called *self-pollination,* which can lead to *self-fertilization.*

In many angiosperms, the pollen grains are carried from flower to flower by insects and other animals that visit the flowers for food or other rewards or are deceived into doing so because the flower's characteristics suggest such rewards. A liquid called **nectar,** which is rich in sugar as well as amino acids and other substances, is often the reward sought by animals.

For pollination by animals to be effective, a particular insect or other animal must visit plant individuals of the same species. A flower's color and form have been shaped by evolution to promote such specialization. Yellow and blue flowers are particularly attractive to bees (**figure 32.3a**), whereas red flowers attract birds but are not particularly noticed by most insects. Some flowers have very long floral tubes with the nectar produced deep within them; only the long, slender beaks of hummingbirds or the long, coiled proboscis of moths or butterflies (**figure 32.3b**) can reach such nectar supplies.

In certain angiosperms and all gymnosperms, pollen is blown about by the wind and reaches the stigmas passively. For such a system to operate efficiently, the individuals of a given plant species must grow relatively close together because wind does not carry pollen very far compared to transport by insects or other animals. Because gymnosperms, such as spruces or pines, grow in dense stands, wind pollination is very effective. Wind-pollinated angiosperms, such as birches, grasses, and ragweed, also tend to grow in dense stands.

Fertilization. Once a pollen grain has been spread by wind, an animal, or self-pollination, it adheres to the sticky, sugary substance that covers the stigma and begins to grow a **pollen tube,** which pierces the style. The pollen tube, nourished by the sugary substance, grows until it reaches the ovule in the ovary (as shown in **figure 17.17**).

Figure 32.4 traces the steps from fertilization through seed formation. When the pollen tube reaches the entry to the embryo sac in the ovule ❶, the tip of the tube bursts and releases the two sperm cells that form from the generative cell. One of the sperm cells fertilizes the egg cell, forming a zygote. The other sperm cell fuses with the two polar nuclei located at the center of the embryo sac, forming the triploid ($3n$) primary endosperm nucleus ❷. This process of fertilization in angiosperms in which two sperm cells are used is called **double fertilization.** Once fertilization is complete, the cells of the zygote divide numerous times. The primary endosperm nucleus eventually develops into the endosperm, shown in ❹, which nourishes the embryo.

Figure 32.3 **Insect pollination.**

(a) Bees are usually attracted to yellow flowers. (b) This alfalfa butterfly (*Colias eurytheme*) has a long proboscis that allows it to feed on nectar deep in the flower.

(a) PictureNet/Getty Images; (b) manfredxy/Shutterstock

Putting the Concept to Work
Why do you think there are more yellow flowers than red flowers?

32.2 Seeds

LEARNING OBJECTIVE 32.2.1 Trace the development of a seed.

After fertilization, active cell division forms an organized mass of cells, the embryo shown in **figure 32.4** ❻. By the fifth day, the principal tissue systems can be detected within the embryo mass ❼. The developing embryo is first nourished by the endosperm and then later in some plants by the seed leaves, thick leaflike food storage structures called **cotyledons.**

Early in the development of an angiosperm embryo, a profoundly significant event occurs: The embryo simply stops developing and becomes dormant as a result of drying. In many plants, embryo development is arrested soon after apical meristems and the cotyledons are differentiated ❽. The ovule of the plant has now matured into a **seed,** which includes the dormant embryo and a source of stored food, both surrounded by a protective and relatively impermeable seed coat that develops from the outer cover of the ovule. Once the seed coat fully develops, most of the embryo's metabolic activities cease.

Germination, or the resumption of metabolic activities that leads to the growth of a mature plant, cannot take place until water and oxygen reach the embryo, a process that sometimes involves cracking the seed. Seeds of some plants have been known to remain viable for hundreds, and in some cases thousands, of years.

> **BIOLOGY & YOU**
>
> **Deadly Seeds.** Whereas many seeds are edible, providing humans with the bulk of their nutrition, some seeds are poisonous, containing deadly chemicals that discourage herbivores and seed predators. Some of these compounds (such as those in mustard plants) simply taste bad to insects, but other compounds are quite toxic to humans. One of the most famous is ricin, a chemical found in the seeds of the castor bean—two to eight seeds provide enough ricin for a lethal dose. Another deadly seed chemical is cyanide, released by the breakdown of a defensive chemical called amygdalin that is found in the seeds of bitter almond, apricot, peach, and others. The seeds of many legumes, including the common bean and soybeans, contain proteins called lectins that can cause gastric distress if the beans are eaten without cooking, which degrades lectins to a harmless form.

> **Putting the Concept to Work**
> What structure(s) found in seeds is(are) produced through double fertilization?

Figure 32.4 Development in an angiosperm embryo.

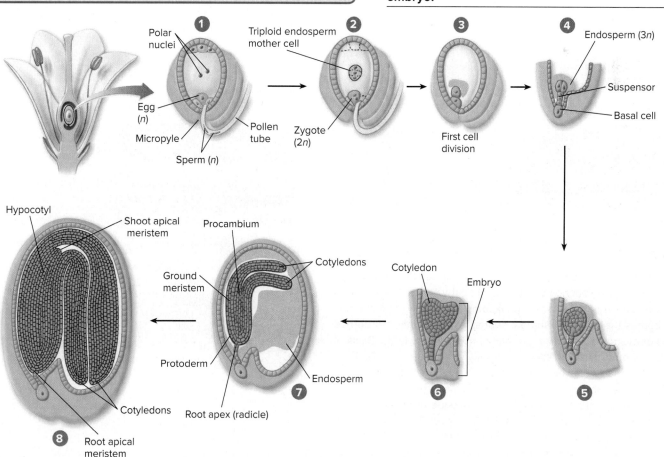

32.3 Fruit

> **LEARNING OBJECTIVE 32.3.1** Define fruit, and compare the three main kinds of fleshy fruits.

During seed formation, the flower ovary begins to develop into fruit. The evolution of flowers was key to the success and diversification of the angiosperms. But of equal importance to angiosperm success has been the evolution of these fruits in response to their modes of dispersal.

Fleshy Fruits

There are three main kinds of fleshy fruits: (1) berries, (2) drupes, and (3) pomes. In *berries*—such as grapes, tomatoes (**figure 32.5a**), and peppers—which are typically many-seeded, the inner layers of the ovary wall are fleshy. In *drupes*—such as peaches (**figure 32.5b**), olives, plums, and cherries—the inner layer of the fruit is stony and adheres tightly to a single seed. In *pomes*—such as apples (**figure 32.5c**) and pears—the fleshy part of the fruit forms from the portion of the flower that is embedded in the receptacle (the swollen end of the flower stem that holds the petals and sepals). The inner layer of the ovary is a leathery membrane that encloses the seeds.

Fruits that have fleshy coverings, often black, bright blue, or red (like those you see in **figure 32.5d**), are normally dispersed by birds and other vertebrates. By feeding on these fruits, the animals carry seeds from place to place before excreting the seeds as solid waste. The seeds, not harmed by the animal's digestive system, are carried to another habitat.

(a) Berries

(b) Drupes

(c) Pomes

(d) Eaten by animals

(e) Dispersed by wind

(f) Dispersed by water

Figure 32.5 Types of fruits and common modes of dispersion.

(a) Tomatoes are a type of fleshy fruit called berries that have multiple seeds. (b) Peaches are a type of fleshy fruit called drupes that contain a single large seed. (c) Apples are a type of fleshy fruit called pomes that contain multiple seeds. (d) The bright red berries of this honeysuckle, *Lonicera,* are highly attractive to birds. Birds may carry the berry seeds either internally or stuck to their feet for great distances. (e) The seeds of this dandelion, *Taraxacum officinale,* are enclosed in a dry fruit with a "parachute" structure that aids their dispersal by wind. (f) This coconut, the seed of the palm tree *Cocos nucifera* carried on ocean waves from a far island, has sprouted and will soon form a new palm tree.

(a) Emilio Ereza/Alamy Stock Photo; (b) Jack Dykinga/USDA; (c) Roman Samokhin/Shutterstock; (d) Eric Crichton/Getty Images; (e) image100/Corbis/Getty Images; (f) izanbar/Getty Images

Dry Fruits

Fruits that are dispersed by wind or by attaching themselves to the fur of mammals or the feathers of birds are called dry fruits because they lack the fleshy tissue of edible fruits. Dry fruits can have structures that aid in their dispersion, as seen by the fluffy "parachute" structure of the wind-dispersed dandelion seed (figure 32.5e). Still other fruits, such as those of mangroves, coconuts (figure 32.5f), and certain other plants that characteristically occur on or near beaches or swamps, are spread from place to place by water.

> **Putting the Concept to Work**
> Why are many fruits fleshy?

32.4 Germination

> **LEARNING OBJECTIVE 32.4.1** Discuss how plants meet the two essential requirements for initiating germination.

What happens to a seed when it encounters conditions suitable for its germination? First, it absorbs water. Seed tissues are so dry at the start of germination that the seed takes up water with great force, after which metabolism resumes. Initially, the metabolism may be anaerobic, but when the seed coat ruptures, aerobic metabolism takes over. At this point, oxygen must be available to the developing embryo because plants require oxygen for active growth (see chapter 7), and plants can "drown" for the same reason people do if submersed in water. Few plants produce seeds that germinate successfully underwater, although some, such as rice, have evolved a tolerance of anaerobic conditions and can initially respire anaerobically. The first stage in both cases is the emergence of the roots. Following that, in **eudicots,** the cotyledons emerge from underground along with the stem. The cotyledons eventually wither, and the first leaves begin the process of photosynthesis. In **monocots,** the cotyledon doesn't emerge from underground; instead, a structure called the *coleoptile* (a sheath wrapped around the emerging shoot) pushes through to the surface where the first leaves emerge and begin photosynthesis.

> **Putting the Concept to Work**
> Why won't grass seed germinate if placed in a glass of water?

Regulating Plant Growth

32.5 Plant Hormones

> **LEARNING OBJECTIVE 32.5.1** Describe the evidence that the cells of adult plants contain all the genes necessary to develop an adult plant.

Differentiation in Plants Is Reversible

After a seed germinates, the pattern of growth and differentiation that was established in the embryo is repeated indefinitely until the plant dies. But differentiation in plants, unlike that in animals, is largely reversible. Botanists first demonstrated in the 1950s that individual differentiated cells isolated from mature individuals could give rise to entire individuals. For example, botanist F. C. Steward was able to induce isolated bits of phloem tissue taken from carrots to form new plants, plants that were normal in appearance and fully fertile (figure 32.6). Regeneration of entire plants from differentiated tissue has since been carried out in many plants, including cotton, tomatoes, and cherries. These experiments clearly demonstrate that the original differentiated phloem tissue still contains cells that retain all of the genetic potential needed for the differentiation of entire plants.

Once a seed has germinated, the plant's further development depends on the activities of the meristematic tissues, which interact with the environment through hormones (discussed below). The shoot and root apical meristems give rise to all of the other cells of the adult plant.

The tissue regeneration experiments of Steward and many others have led to the general conclusion that some nucleated cells in differentiated plant tissue are capable of expressing their hidden genetic information when provided with suitable environmental signals. What halts the expression of genetic potential when the same kinds of cells are incorporated into normal, growing plants? The expression of some of these genes is controlled by plant hormones.

Figure 32.6 **How Steward regenerated a plant from differentiated tissue.**

Hormones Control Plant Growth and Development

At least five major kinds of hormones are found in plants: auxin, gibberellins, cytokinins, ethylene, and abscisic acid. Other kinds of plant hormones exist but are less well understood. Hormones have multiple functions in the plant; the same hormone may work differently in different parts of the plant, at different times, and interact with other hormones in different ways. The study of plant hormones, especially how hormones produce their effects, is today an active and important field of research.

> **Putting the Concept to Work**
> What do the five kinds of plant hormones do?

32.6 Auxin

LEARNING OBJECTIVE 32.6.1 Explain how auxin causes plants to bend toward light.

In his later years, the great evolutionist Charles Darwin and his son Francis published a book called *The Power of Movement in Plants,* in which they reported their systematic experiments concerning the way in which growing plants bend toward light, a phenomenon known as **phototropism.**

Phototropism

After conducting a series of experiments, shown in **figure 32.7**, they observed that plants grew toward light. If the tip of the seedling was covered, the plant didn't bend toward the light. A control experiment showed that the cap was not interfering with the directional growth pattern. Another control showed that covering the lower portions of the plant did not block the directional growth. The Darwins hypothesized that when plant shoots were illuminated from one side, an "influence" that arose in the uppermost part of the shoot was then transmitted downward, causing the shoot to bend. Later, several botanists conducted experiments that demonstrated that the substance causing the shoots to bend was a chemical we call **auxin.**

How Auxin Controls Plant Growth

How auxin controls plant growth was discovered in 1926 by Frits Went, a Dutch plant physiologist, in the course of studies for his doctoral dissertation. From his experiments Went was able to show that a substance flowed into agar from the tips of light-grown grass seedlings, and that this substance enhanced cell elongation. Phototropism occurs because this chemical messenger caused the tissues on the side of the seedling into which it flowed to grow more than those on the opposite side, and thus causes the plant to bend toward the light. He named the substance that he had discovered auxin, from the Greek word *auxin,* meaning "to increase."

How Auxin Works. Auxin appears to act by increasing the "stretchability" of the plant cell wall within minutes of its application. Researchers speculate that the covalent bonds linking the polysaccharides of the cell wall to one another change extensively in response to auxin, allowing the cells to take up water and thus enlarge.

Synthetic Auxins. Synthetic auxins are routinely used to control weeds. One of the most important of the synthetic auxins used in this way is 2,4-dichlorophenoxyacetic acid, usually known as 2,4-D. It kills weeds in lawns without harming the grass because 2,4-D affects only broad-leaved eudicots. When treated, the weeds literally "grow to death," rapidly reducing ATP production so that no energy remains for transport or other essential functions.

> **Putting the Concept to Work**
> How can an herbicide cause a weed to grow to death?

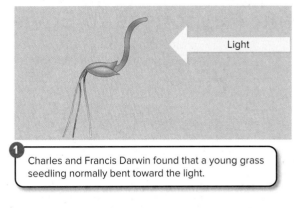
① Charles and Francis Darwin found that a young grass seedling normally bent toward the light.

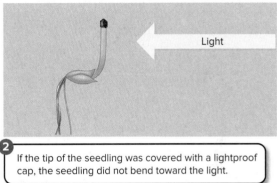
② If the tip of the seedling was covered with a lightproof cap, the seedling did not bend toward the light.

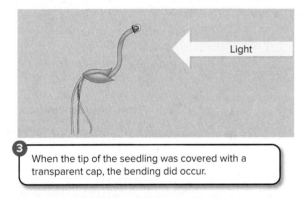
③ When the tip of the seedling was covered with a transparent cap, the bending did occur.

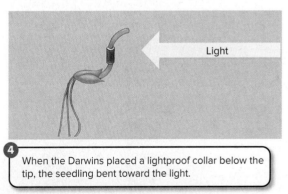
④ When the Darwins placed a lightproof collar below the tip, the seedling bent toward the light.

Figure 32.7 The Darwins' experiment with phototropism.

From these experiments, the Darwins concluded that, in response to light, an "influence" that causes bending was transmitted from the tip of the seedling to the area below the tip, where bending usually occurs.

Plant Responses to Environmental Stimuli

32.7 Photoperiodism and Dormancy

LEARNING OBJECTIVE 32.7.1 Describe how day length influences flowering time.

Photoperiodism

Essentially all eukaryotic organisms are affected by the cycle of night and day, and many features of plant growth and development are keyed to changes in the proportions of light and dark in the daily 24-hour cycle. Such responses constitute **photoperiodism,** a mechanism by which organisms measure seasonal changes in relative day and night length. One of the most obvious of these photoperiodic reactions concerns angiosperm flower production.

Day length changes with the seasons; the farther from the equator you are, the greater the variation. Plants' flowering responses fall into three basic categories in relation to day length: long-day plants, short-day plants, and day-neutral plants. Long-day plants such as the iris in figure 32.8 initiate flowers in the summer, when nights become shorter than a certain length (and days become longer). Short-day plants, on the other hand, begin to form flowers when nights become longer than a critical length (and days become shorter); the goldenrod in figure 32.8 doesn't flower in summer but instead flowers in fall. Thus, many spring and early summer flowers are long-day

Figure 32.8 How photoperiodism works in plants.

① *Early summer.* Short periods of darkness induce flowering in long-day plants, such as iris, but not in short-day plants, such as goldenrod.

② *Late fall.* Long periods of darkness induce flowering in short-day plants, such as goldenrod, but not in long-day plants, such as iris.

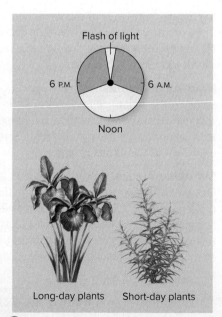
③ *Interrupted night.* If the long night of winter is artificially interrupted by a flash of light, the goldenrod will not bloom and the iris will.

plants, and many fall flowers are short-day plants. An "interrupted night" experiment makes it clear that it is actually the length of uninterrupted dark that is the flowering trigger. In addition, a number of plants are described as day-neutral and produce flowers without regard to day length.

Dormancy

Plants respond to their external environment largely by changes in growth rate. A plant's ability to stop growing altogether when conditions are not favorable—to become dormant—is critical to its survival. In temperate regions, dormancy is generally associated with winter, when low temperatures and the unavailability of water because of freezing make it impossible for plants to grow. During this season, the buds of deciduous trees and shrubs remain dormant, and the apical meristems remain well protected inside enfolding scales. Perennial herbs spend the winter underground as stout stems or roots packed with stored food. Many other kinds of plants, including most annuals, pass the winter as seeds.

> **Putting the Concept to Work**
> What controls flowering in plants, the length of day or of night?

32.8 Tropisms

> **LEARNING OBJECTIVE 32.8.1** Contrast phototropism, gravitropism, and thigmotropism.

Tropisms are directional and irreversible growth responses of plants to external stimuli. They control patterns of plant growth and thus plant appearance. Three major classes of plant tropisms are phototropism (figure 32.9a, and discussed earlier), gravitropism, and thigmotropism.

Gravitropism

Gravitropism causes stems to grow upward and roots downward, in response to gravity. Both of these responses clearly have adaptive significance. Stems, like the one growing from a tipped-over flower pot in figure 32.9b, grow upward and are apt to receive more light than those that do not; roots that grow downward are more apt to encounter a more favorable environment than those that do not.

Thigmotropism

Still another commonly observed response of plants is **thigmotropism**, a name derived from the Greek root *thigma*, meaning "touch." Thigmotropism is defined as the response of plants to touch. Examples include plant tendrils, which rapidly curl around and cling to stems or other objects, and twining plants, such as bindweed, which also coil around objects (figure 32.9c). These behaviors result from rapid growth responses to touch.

> **Putting the Concept to Work**
> Are growth patterns called tropisms reversible? Explain.

Figure 32.9 Tropism guides plant growth.

(a) Phototropism is exhibited by this *Coriandrum* plant growing toward light. (b) Gravitropism is exhibited by this plant, *Phaseolus vulgaris*. Note the negative gravitational response of the shoot. (c) The thigmotropic response of these twining stems causes them to coil around the object with which they have come in contact.

(a) Maryann Frazier/Science Source; (b) Martin Shields/Science Source; (c) PhotoAlto/Odilon Dimier/Getty Images

Putting the Chapter to Work

1 You are visiting a botanical garden and trying to differentiate the types of reproductive strategies employed by the plants you encounter. The first one you confront is a mint plant. It has node regions on various sections of the stem and adventitious roots that travel down to and along the soil.

What type of asexual reproduction does this plant employ?

2 You are designing a flower bed to attract hummingbirds. You purchase plants that will produce blue flowers with ample nectar. However, after these flowers bloom, you do not observe any hummingbirds in your garden.

What should you have planted instead?

3 *If a researcher studying the effects of hormones on plants were interested in learning how plants could increase bud formation, which hormone would she alter?*

4 If you were to place a potted plant in your window sill and position the plant away from the sun in the morning, at the end of day the plant stem may have repositioned itself to face the sun. This phenomenon is known as phototropism.

What chemical is produced by plants to promote this bending?

Retracing the Learning Path

Flowering Plant Reproduction

32.1 Angiosperm Reproduction

1. Angiosperms reproduce sexually and asexually. In asexual reproduction, offspring are genetically identical to the parent; this often involves vegetative reproduction. Plants use many forms of vegetative reproduction, including runners, rhizomes, suckers, and adventitious plantlets.
2. Sexual reproduction in plants involves an alternation of generations. A diploid sporophyte gives rise to a haploid gametophyte that produces gametes—egg and pollen. Pollination and fertilization bring the gametes together.
- Flowers have male and female parts. The male parts include the stamen and the pollen-producing anthers. Pollen grains, the male gametophyte, are produced in the anthers. The female parts include the stigma, style, and ovary, which make up the carpel. The embryo sac is the female gametophyte, and it forms in the ovule, within the ovary.
- Pollen grains are carried to flowers by wind or animals. A pollen grain lands on the stigma and extends a pollen tube through the style to the base of the ovule. Two sperm cells travel down the pollen tube. One sperm fertilizes the egg, and the other fuses with the two polar nuclei, a process called double fertilization.

32.2 Seeds

1. The fertilized egg begins dividing, forming the embryo. After the shoot and root apical meristems form, the embryo stops growing and becomes dormant in a structure called a seed.
- A seed contains the dormant embryo and substantial food reserves in the form of endosperm, all encased within a tough, drought-resistant coat.
- The endosperm provides a source of food for the plant embryo. In some cases, the endosperm is consumed in the seed and stored in structures called cotyledons (seed leaves). The outer layer of the ovule becomes the seed coat. Seed development resumes when conditions are favorable.

32.3 Fruit

1. During seed formation, the flower's ovary begins to develop into fruit that surrounds the seed. The walls of the ovary can develop differently, which accounts for the variety of fruits. Fruits are dispersed in different ways. Fleshy fruits are eaten by animals and then dispersed through their feces. Dry fruits are usually dispersed by wind, water, or animals. Dry fruits have structures that aid in their dispersal.

32.4 Germination

1. Germination is the resumption of a seed's growth and development, typically triggered by water. The seed absorbs water and uses the endosperm or cotyledons as a food source. The seed coat cracks open, and the plant begins to grow. The overall process is similar in monocots and eudicots but there are differences in the structures involved.

Regulating Plant Growth

32.5 Plant Hormones

1. Differentiation in plants is largely reversible. New plants can be grown from parts of adult plants. The shoot and root meristems differentiate early in development and give rise to all of the other cell types.
- Unlike animals, which use genes within cells to control the cell's development, the development of plant tissues is controlled by hormones that interact with the environment to regulate the expression of key genes. There are at least five major kinds of plant hormones: auxin, gibberellins, cytokinins, ethylene, and abscisic acid.

32.6 Auxin

1. The primary growth-promoting hormone of plants is auxin. Charles Darwin and his son described a process called phototropism. In a series of experiments, they showed that the tip of a plant will grow toward light.
- Frits Went identified a chemical he called auxin as the hormone involved in Darwin's phototropism. When a plant is exposed to light on one side, auxin is released from the tip and causes cells on the shady side of the plant to elongate. This causes the plant to bend toward the light as it grows.

Plant Responses to Environmental Stimuli

32.7 Photoperiodism and Dormancy

1. Flowering is affected by the length of daylight, a process called photoperiodism. A plant pigment, phytochrome, exists in two forms converted by darkness. In short-day plants, the active form of phytochrome inhibits flowering. Darkness converts the active form to the inactive form, allowing flowering to occur.
- Plants survive unfavorable conditions by entering a phase of dormancy, in which they stop growing.

32.8 Tropisms

1. Tropisms are irreversible growth patterns in response to external stimuli. Phototropism is a growth response toward light. Gravitropism is growth in response to the pull of gravity; this causes stems to grow upward and roots to grow downward. Thigmotropism is growth in response to touch.

Inquiry and Analysis

Are Pollinators Responsible for the Evolution of Flower Color?

Evolution results from many types of interactions among organisms, including predator–prey relationships, competition, and mate selection. An important type of coevolution among plants and animals involves flowering plants and their pollinators. Pollinators need flowers for food, and plants need pollinators for reproduction. It is logical, then, to hypothesize that evolutionary changes in flower shape, size, odor, and color are driven to a large extent by pollinators such as bees. We know that insects respond to variation in flower traits by visiting flowers with certain features, but few studies have been carried out to predict and then evaluate the response of plant populations to selection by pollinators. In wild radish populations, honeybees preferentially visit yellow and white flowers, whereas syrphid flies prefer pink flowers. Rebecca Irwin and Sharon Strauss at the University of California, Davis, studied the response of wild radish flower color to selection by pollinators in natural populations. They compared the frequency of four flower colors (yellow, pink, white, and bronze) in two populations of wild radishes. Bees created the first population by visiting flowers based on their color preferences. The scientists produced the second population by hand pollinating wild radish flowers with no discrimination due to flower color. However, the pollen used for artificial pollinations contained varying proportions of each type of plant, based on the frequencies of each in the wild. In the graph shown here, you can see the distribution of flower colors in these two populations. The blue bars indicate the number of plants with yellow, white, pink, and bronze flowers in the population generated by the bee pollination. The red bars show the number of each type of flower in the population generated by researcher pollination, with no selection for flower color.

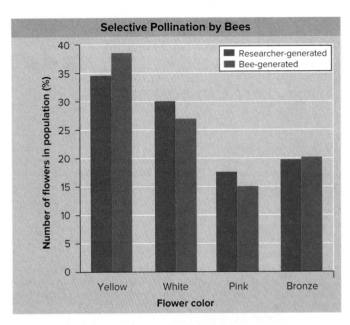

An uncommon visitor: Here a syrphid fly is visiting a yellow wild radish flower, usually preferred by bees.

Jeff K. Conner

Analysis

1. **Applying Concepts**
   a. Variable. In the graph, what is the dependent variable?
   b. Reading a bar graph. Does this graph reflect data on flower colors from syrphid fly pollinations?
2. **Interpreting Data**
   a. Which flower color was most common?
   b. Which flower color(s) did the bees preferentially appear to visit?
3. **Making Inferences**
   a. Which type of insect do you suppose was most abundant in the region where this study was carried out and why?
   b. Because there were pink flowers in these populations, can you say that syrphid flies *had* to be present in the area?
4. **Drawing Conclusions**
   a. Does it appear that insects are influencing the evolution of flower color in wild radish populations?
   b. Why did the researcher-pollinated population of flowers exhibit a pattern of flower colors similar to that of the bee-pollinated population? Were the researchers showing some experimental bias in their color selections?
5. **Further Analysis** Assume the wild radish population in this study is visited again in 10 years and, although a slight increase in yellow-flowered plants is observed, the proportion is not as high as would be predicted by this study. Provide some explanation for a slower-than-expected increase in the yellow-flowered plants.

Glossary

Terms and Concepts

A

ABO blood groups A set of four phenotypes produced by different combinations of three alleles at a single locus; blood types are A, B, AB, and O, depending on which alleles are expressed as antigens on the red blood cell surface.

absorption (L. *absorbere*, to swallow down) The movement of water and substances dissolved in water into a cell, tissue, or organism.

acetyl-CoA The product of the transition reaction between glycolysis and the Krebs cycle. Pyruvate is oxidized to acetyl-CoA by NAD^+, also producing CO_2, and NADH.

acid Any substance that dissociates to form H^1 ions when dissolved in water. Having a pH value less than 7.

acoelomate (Gr. *a*, not + *koiloma*, cavity) A bilaterally symmetrical animal not possessing a body cavity, such as a flatworm.

actin (Gr. *actis*, ray) One of the two major proteins that make up myofilaments (the other is myosin). It provides the cell with mechanical support and plays major roles in determining cell shape and cell movement.

action potential A single nerve impulse. A transient all-or-none reversal of the electrical potential across a neuron membrane. Because it can activate nearby voltage-sensitive channels, an action potential propagates along a nerve cell.

activation energy The energy a molecule must acquire to undergo a specific chemical reaction.

activator A regulatory protein that binds to the DNA and makes it more accessible for transcription.

active site The region of an enzyme surface to which a specific set of substrates binds, lowering the activation energy required for a particular chemical reaction and so facilitating it.

active transport The transport of a solute across a membrane by protein carrier molecules to a region of higher concentration by the expenditure of chemical energy. One of the most important functions of any cell.

adaptation (L. *adaptare*, to fit) Any peculiarity of structure, physiology, or behavior that promotes the likelihood of an organism's survival and reproduction in a particular environment.

adenosine triphosphate (ATP) A molecule composed of ribose, adenine, and a triphosphate group. ATP is the chief energy currency of all cells. Cells focus all of their energy resources on the manufacture of ATP from ADP and phosphate, which requires the cell to supply 7 kilocalories of energy obtained from photosynthesis or from electrons stripped from foodstuffs to form 1 mole of ATP. Cells then use this ATP to drive endergonic reactions.

adhesion (L. *adhaerere*, to stick to) The molecular attraction exerted between the surfaces of unlike bodies in contact, as water molecules to the walls of the narrow tubes that occur in plants.

aerobic (Gr. *aer*, air + *bios*, life) Oxygen-requiring.

allele (Gr. *allelon*, of one another) One of two or more alternative forms of a gene.

allele frequency The relative proportion of a particular allele among individuals of a population. Not equivalent to gene frequency, although the two terms are sometimes confused.

allosteric interaction (Gr. *allos*, other + *stereos*, shape) The change in shape that occurs when an activator or repressor binds to an enzyme. These changes result when specific, small molecules bind to the enzyme, molecules that are not substrates of that enzyme.

alternation of generations A reproductive life cycle in which the multicellular diploid phase produces spores that give rise to the multicellular haploid phase, and the multicellular haploid phase produces gametes that fuse to give rise to the zygote. The zygote is the first cell of the multicellular diploid phase.

alternative splicing In eukaryotes, the production of different mRNAs from a single primary transcript by including different sets of exons.

alveolus, *pl*. alveoli (L. *alveus*, a small cavity) One of the many small, thin-walled air sacs within the lungs in which the bronchioles terminate.

amino acid The subunit structure from which proteins are produced, consisting of a central carbon atom with a carboxyl group (—COOH), an amino group (—NH_2), a hydrogen, and a side group (R group); only the side group differs from one amino acid to another.

amnion The innermost of the extraembryonic membranes; the amnion forms a fluid-filled sac around the embryo in amniotic eggs.

amniotic egg An egg that is isolated and protected from the environment by a more or less impervious shell. The shell protects the embryo from drying out, nourishes it, and enables it to develop outside of water.

anabolic steroids Synthetic hormones derived from testosterone that facilitate the growth of muscle tissue. Their use in sports is illegal and dangerous.

anaerobic (Gr. *an*, without + *aer*, air + *bios*, life) Any process that can occur without oxygen. Includes glycolysis and fermentation. Anaerobic organisms can live without free oxygen.

analogous Structures that are similar in function but different in evolutionary origin, such as the wing of a bat and the wing of a butterfly.

anaphase In mitosis and meiosis II, the stage initiated by the separation of sister chromatids, during which the daughter chromosomes move to opposite poles of the cell; in meiosis I, marked by separation of replicated homologous chromosomes.

aneuploidy An abnormal number of chromosomes.

angiosperms The flowering plants, one of five phyla of seed plants. In angiosperms, the ovules at the time of pollination are completely enclosed by tissues.

anterior (L. *ante*, before) Located before or toward the front. In animals, the head end of an organism.

anther (Gr. *anthos*, flower) The part of the stamen of a flower that bears the pollen.

antibody (Gr. *anti*, against) A protein "label" placed on foreign cells that marks them for destruction by the immune system. Produced by B cells that circulate in the blood, your body has at least a few that will bind to any conceivable invading bacterium or virus.

anticodon The three-nucleotide sequence of a tRNA molecule that is complementary to, and base pairs with, an amino acid-specifying codon in mRNA.

antigen (Gr. *anti*, against + *genos*, origin) A foreign substance. When your body's immune defenses circulating in your bloodstream encounter an antigen (typically a protein found on the surface of a microbe or virus) they react by proliferating and secreting antibodies that bind to the antigen, marking it for destruction.

anus The terminal opening of the gut; the solid residues of digestion are eliminated through the anus.

aorta (Gr. *aeirein*, to lift) The major artery of vertebrate systemic blood circulation; in mammals, carries oxygenated blood away from the heart to all regions of the body except the lungs.

apical meristem (L. *apex*, top + Gr. *meristos*, divided) In vascular plants, the growing point at the tip of the root or stem.

archaea A group of prokaryotes that are among the most primitive still in existence, characterized by the absence of peptidoglycan in their cell walls, a feature that distinguishes them from bacteria.

arteriole A smaller artery, leading from the arteries to the capillaries.

artery (L. *arteria*, artery) Any blood vessel that carries blood away from the heart.

arthropod (Gr. *arthros*, jointed + *podes*, feet) An animal of the phylum Arthropoda, which includes the arachnids, crustaceans, centipedes, millipedes, and insects.

artificial selection Change in the genetic structure of populations due to selective breeding by humans. Many domestic animal breeds and crop varieties have been produced through artificial selection.

asexual reproduction Reproducing without forming gametes. Asexual reproduction does not involve sex. Its outstanding characteristic is that an individual offspring is genetically identical to its parent.

atom (Gr. *atomos*, indivisible) A core (nucleus) of protons and neutrons surrounded by an orbiting cloud of electrons. The chemical behavior of an atom is largely determined by the distribution of its electrons, particularly the number of electrons in its outermost level.

atomic number The number of protons in the nucleus of an atom. In an atom that does not bear an electric charge (that is, one that is not an ion), the atomic number is also equal to the number of electrons.

atrioventricular (AV) node A slender connection of cardiac muscle cells that receives the heartbeat impulses from the sinoatrial node and conducts them by way of the bundle of His.

atrium An antechamber; in the heart, a thin-walled chamber that receives venous blood and passes it on to the thick-walled ventricle; in the ear, the tympanic cavity.

Australopithicus The genus of primitive hominid ancestors of humans that lived in Africa from 2 to 6 million years ago.

autonomic nervous system (Gr. *autos*, self + *nomos*, law) The motor pathways that carry commands from the central nervous system to regulate the glands and nonskeletal muscles of the body. Also called the involuntary nervous system.

autosome (Gr. *autos*, self + *soma*, body) Any of the 22 pairs of human chromosomes that are similar in size and morphology in both males and females.

autotroph (Gr. *autos*, self + *trophos*, feeder) An organism that can harvest light energy from the sun or from the oxidation of inorganic compounds to make organic molecules.

auxin (Gr. *auxein*, to increase) A plant hormone that controls cell elongation, among other effects.

axial skeleton The skeleton of the head and trunk of the human body containing 80 bones.

axil In plants, the angle between a leaf's petiole and the stem to which it is attached.

axon (Gr., axle) A process extending out from a neuron that conducts impulses away from the cell body.

B

B cell The circulating cells of your immune system that produce antibodies. Each B cell is able to manufacture only one specific antibody (there are LOTS of types of B cells). When activated by a pathogen encounter, a B cell begins to churn out its antibody, which mark any other pathogens for destruction.

bacteriophage A virus that infects bacterial cells; also called a *phage*.

bacterium, *pl.* **bacteria** (Gr. *bakterion*, dim. of *baktron*, a staff) The most common type of prokaryotic organism. Cell walls contain peptidoglycan. Play many important ecological roles.

basal body In eukaryotic cells that contain flagella or cilia, a form of centriole that anchors each flagellum.

base Any substance that combines with H^+ ions, thereby reducing the H^+ ion concentration of a solution. Having a pH value above 7.

bilateral symmetry (L. *bi*, two + *lateris*, side; Gr. *symmetria*, symmetry) A body form in which the right and left halves of an organism are approximate mirror images of each other.

binary fission (L. *binarius*, consisting of two things or parts + *fissus*, split) Asexual reproduction of a cell by division into two equal or nearly equal parts. Bacteria divide by binary fission.

binding site The site on a substrate or reactant that binds to an enzyme.

binomial system (L. *bi*, twice, two + Gr. *nomos*, usage, law) A system of nomenclature that uses two words. The first names the genus, and the second designates the species.

biochemical pathway A sequence of chemical reactions in which the product of one reaction becomes the substrate of the next reaction. The Krebs cycle is a biochemical pathway.

biodiversity The number of species and their range of behavioral, ecological, physiological, and other adaptations in an area.

biological species concept The concept that defines species as groups of populations that have the potential to interbreed and that are reproductively isolated from other groups.

biomass (Gr. *bios*, life + *maza*, lump or mass) The total weight of all of the organisms living in an ecosystem.

biome (Gr. *bios*, life + *-oma*, mass, group) A major terrestrial assemblage of plants, animals, and microorganisms that occur over wide geographical areas and have distinct characteristics. The largest ecological unit.

blade The broad, expanded part of a leaf; also called the lamina.

blastocoel The central cavity of the blastula stage of vertebrate embryos.

blastocyst (Gr. *blastos*, germ + *kystis*, bladder) Embryonic stage in mammals that consists of a hollow ball of cells and inner cell mass surrounding a fluid-filled cavity.

Bowman's capsule In the vertebrate kidney, the bulbous unit of the nephron, which surrounds the glomerulus.

bronchus, *pl.* **bronchi** One of a pair of respiratory tubes branching from the lower end of the trachea (windpipe) into either lung.

bud An asexually produced outgrowth that develops into a new individual. In plants, an embryonic shoot, often protected by young leaves; buds may give rise to branch shoots.

buffer A substance that takes up or releases hydrogen ions (H^+) to maintain the pH within a certain range.

C

C_3 photosynthesis The main cycle of the dark reactions of photosynthesis, in which CO_2 binds to ribulose 1,5-bisphosphate (RuBP) to form two 3-carbon phosphoglycerate (PGA) molecules.

C_4 photosynthesis A process of CO_2 fixation in photosynthesis by which the first product is the 4-carbon oxaloacetate molecule.

calorie (L. *calor*, heat) The amount of energy in the form of heat required to raise the temperature of 1 gram of water 1 degree Celsius.

Calvin cycle The dark reactions of C_3 photosynthesis; also called the Calvin–Benson cycle.

calyx (Gr. *kalyx*, a husk, cup) The sepals collectively. The outermost flower whorl.

cancer Unrestrained invasive cell growth. A tumor or cell mass resulting from uncontrollable cell division.

capillary (L. *capillaris*, hairlike) A blood vessel with a very small diameter. Blood exchanges gases and metabolites across capillary walls. Capillaries join the end of an arteriole (small artery) to the beginning of a venule (small vein).

carbohydrate (L. *carbo*, charcoal + *hydro*, water) An organic compound consisting of a chain or ring of carbon atoms to which hydrogen and oxygen atoms are attached in a ratio of approximately 1:2:1. A compound of carbon, hydrogen, and oxygen having the generalized formula $(CH_2O)_n$, where n is the number of carbon atoms.

carbon fixation The conversion of CO_2 into organic compounds during photosynthesis; the first stage of the dark reactions of photosynthesis in which carbon dioxide from the air is combined with ribulose 1,5-bisphosphate.

carcinogen (Gr. *karkinos*, cancer + *-gen*) Any cancer-causing agent.

cardiovascular system (Gr. *kardia*, heart + L. *vasculum*, vessel) The blood circulatory system and the heart that pumps it. Collectively, the blood, heart, and blood vessels.

Coronavirus A kind of respiratory virus named for their crown (corona) of surface "spike" proteins protruding outward from its surface–seen in cross-section, it is like a crown. Some 20%–30% of common colds are caused bny coronaviruses. Three are far more deadly: SARS, MERS, and COVID-19. All three are bat viruses that have passed to humans.

carpel (Gr. *karpos*, fruit) A leaflike organ in angiosperms that encloses one or more ovules.

carrying capacity The maximum population size that a habitat can support.

Casparian strip In plants, a band that encircles the cell wall of root endodermal cells. Adjacent cells' strips connect, forming a layer through which water cannot pass; therefore, all water entering roots must pass through cell membranes and cytoplasm.

catabolism (Gr. *katabole*, throwing down) A process in which complex molecules are broken down into simpler ones.

catalysis (Gr. *katalysis*, dissolution + *lyein*, to loosen) The enzyme-mediated process in which the subunits of polymers are positioned so that their bonds undergo chemical reactions.

catalyst (Gr. *kata*, down + *lysis*, a loosening) A general term for a substance that speeds up a specific chemical reaction by lowering the energy required to activate or start the reaction. An enzyme is a biological catalyst.

cell (L. *cella*, a chamber or small room) The smallest unit of life. The basic organizational unit of all organisms. Composed of a nuclear region containing the hereditary apparatus within a larger volume called the cytoplasm bounded by a lipid membrane.

cell cycle The repeating sequence of growth and division through which cells pass each generation.

cellular respiration The process in which the energy stored in a glucose molecule is released by oxidation. Hydrogen atoms are lost by glucose and gained by oxygen.

central vacuole A large, membrane-bounded sac found in plant cells that stores proteins, pigments, and waste materials, and is involved in water balance.

centriole (Gr. *kentron*, center of a circle + L. *ola*, small) A cytoplasmic organelle located outside the nuclear membrane, identical in structure to a basal body; found in animal cells and in the flagellated cells of other groups; divides and organizes spindle fibers during mitosis and meiosis.

centromere (Gr. *kentron*, center + *meros*, a part) A constricted region of the chromosome joining two sister chromatids, to which the kinetochore is attached.

cerebellum The hindbrain region of the vertebrate brain that lies above the medulla (brainstem) and behind the forebrain; it integrates information about body position and motion, coordinates muscular activities, and maintains equilibrium.

cerebral cortex The thin surface layer of neurons and glial cells covering the cerebrum; well developed only in mammals, and particularly prominent in humans. The cerebral cortex is the seat of conscious sensations and voluntary muscular activity.

chemical bond The force holding two atoms together. The force can result from the attraction of opposite charges (ionic bond) or from the sharing of one or more pairs of electrons (a covalent bond).

chemiosmosis The cellular process responsible for almost all of the adenosine triphosphate (ATP) harvested from food and for all the ATP produced by photosynthesis.

chemoautotroph An autotrophic bacterium that uses chemical energy released by specific inorganic reactions to power its life processes, including the synthesis of organic molecules.

chiasma, *pl.* **chiasmata** (Gr. a cross) In meiosis, the points of crossing over where portions of chromosomes have been exchanged during synapsis. A chiasma appears as an X-shaped structure under a light microscope.

chlorophyll (Gr. *chloros*, green + *phyllon*, leaf) The primary type of light-absorbing pigment in photosynthesis. Chlorophyll *a* absorbs light in the violet-blue and the red ranges of the visible light spectrum; chlorophyll *b* is an accessory pigment to chlorophyll *a*, absorbing light in the blue and red-orange ranges. Neither pigment absorbs light in the green range, 500–600 nm.

chloroplast (Gr. *chloros*, green + *plastos*, molded) A cell-like organelle present in algae and plants that contains chlorophyll (and usually other pigments) and is the site of photosynthesis.

choanocyte (Gr. *choane*, funnel + *kytos*, hollow vessel) A type of flagellated cell that lines the body cavity of a sponge.

chorion The outer member of the double membrane that surrounds the embryo of reptiles, birds, and mammals; in placental mammals, it contributes to the structure of the placenta.

chromatid (Gr. *chroma*, color + L. *-id*, daughters of) One of two daughter strands of a duplicated chromosome that is joined by a single centromere.

chromatin (Gr. *chroma*, color) The complex of DNA and proteins of which eukaryotic chromosomes are composed.

chromosome (Gr. *chroma*, color + *soma*, body) The vehicle by which hereditary information is physically transmitted from one generation to the next. In a eukaryotic cell, long threads of DNA that are associated with protein and that contain hereditary information.

cilium, *pl.* **cilia** (L. eyelash) Refers to flagella that are numerous and organized in dense rows. Cilia propel cells through water. In human tissue, they move water or mucus over the tissue surface.

circulatory system A network of vessels in coelomate animals that carries fluids to and from different areas of the body.

clade A taxonomic group composed of an ancestor and all its descendents.

cladistics A taxonomic technique used for creating hierarchies of organisms based on derived characters that represent true phylogenetic relationship and descent.

cladogram A branching diagram that represents the phylogeny of an organism.

class A taxonomic category ranking below a phylum (division) and above an order.

classical conditioning The repeated presentation of a stimulus in association with a response that causes the brain to form an association between the stimulus and the response, even if they have never been associated before.

cleavage In vertebrates, a rapid series of successive cell divisions of a fertilized egg, forming a hollow sphere of cells, the blastula.

clonal selection Amplification of a clone of immune cells initiated by antigen recognition.

clone (Gr. *klon*, twig) A line of cells, all of which have arisen from the same single cell by mitotic division. One of a population of individuals derived by asexual reproduction from a single ancestor. One of a population of genetically identical individuals.

closed circulatory system A circulatory system in which the blood is physically separated from other body fluids.

cnidarian (Gr. *knide*, nettle) An animal of the phylum Cnidaria, which includes hydra, jellyfish, corals, and sea anemones.

cnidocyte (Gr. *knide*, nettle + *kytos*, hollow vessel) Modified cell of cnidarians that holds the nematocyst.

codominance In genetics, a situation in which the effects of both alleles at a particular locus are apparent in the phenotype of the heterozygote.

codon (L. code) The basic unit of the genetic code. A sequence of three adjacent nucleotides in DNA or mRNA that codes for one amino acid or for polypeptide termination.

coelom (Gr. *koilos*, a hollow) A body cavity formed between layers of mesoderm and in which the digestive tract and other internal organs are suspended.

coenzyme A cofactor of an enzyme that is a nonprotein organic molecule.

coevolution (L. *co-*, together + *e-*, out + *volvere*, to fill) A term that describes the long-term evolutionary adjustment of one group of organisms to another.

cohesion (L. *cohaes*, to cohere) The molecular attraction between the surfaces of like bodies in contact, such as that between water molecules and other water molecules.

collenchyma cell In plants, the cells that form a supporting tissue called collenchyma; often found in regions of primary growth in stems and in some leaves.

commensalism (L. *cum*, together with + *mensa*, table) A symbiotic relationship in which one species benefits, while the other neither benefits nor is harmed.

community (L. *communitas*, community, fellowship) The populations of different species that live together and interact in a particular place.

Community transmission Transmission of a virus like COVID-19 from one member of a community to another member of that same community. Typically, rates of community transmission are high only when many members of a community are already infected.

companion cell A specialized parenchyma cell that is associated with each sieve-tube member in the phloem of a plant.

competition Interaction between individuals for the same scarce resources. Intraspecific competition is competition between individuals of a single species. Interspecific competition is competition between individuals of different species.

competitive exclusion The hypothesis that if two species are competing with one another for the same limited resource in the same place, one will be able to use that resource more efficiently than the other and eventually will drive that second species to extinction locally.

complement system The chemical defense of a vertebrate body that consists of a battery of proteins that insert in bacterial and fungal cells, causing holes that destroy the cells.

complementary DNA (cDNA) A DNA copy of an mRNA transcript; produced by the action of the enzyme reverse transcriptase.

concentration gradient The concentration difference of a substance as a function of distance. In a cell, a greater concentration of its molecules in one region than in another.

condensation The coiling of the chromosomes into more and more tightly compacted bodies begun during the G_2 phase of the cell cycle.

cone (1) In plants, the reproductive structure of a conifer. (2) In vertebrates, a type of light-sensitive neuron in the retina concerned with the perception of color and with the most acute discrimination of detail.

conjugation (L. *conjugare*, to yoke together) An unusual mode of reproduction in unicellular organisms in which genetic material is exchanged between individuals through tubes connecting them during conjugation.

consumer In ecology, a heterotroph that derives its energy from living or freshly killed organisms or parts thereof. Primary consumers are herbivores; secondary consumers are carnivores or parasites.

continuous variation Variation in a trait that occurs along a continuum, such as the trait of height in human beings; often occurs when a trait is determined by more than one gene.

cork cambium The lateral meristem that forms the periderm, producing cork (phellem) toward the surface (outside) of the plant and phelloderm toward the inside.

coronavirus A kind of respiratory virus named for their crown (corona) of surface "spike" proteins protruding outward from its surface—seen in cross section, it is like a crown. Some 20%–30% of common colds are caused by coronaviruses. Three are far more deadly: SARS, MERS, and COVID-19.

cornea The transparent outer layer of the vertebrate eye.

cortex (L. bark) In vascular plants, the primary ground tissue of a stem or root, bounded externally by the epidermis and internally by the central cylinder of vascular tissue. In animals, the outer, as opposed to the inner, part of an organ, as in the adrenal, kidney, and cerebral cortexes.

cotyledon (Gr. *kotyledon,* a cup-shaped hollow) Seed leaf. Monocot embryos have one cotyledon, and dicots have two.

countercurrent flow In organisms, the passage of heat or of molecules (such as oxygen, water, or sodium ions) from one circulation path to another moving in the opposite direction. Because the flow of the two paths is in opposite directions, a concentration difference always exists between the two channels, facilitating transfer.

covalent bond (L. *co-,* together + *valare,* to be strong) A chemical bond formed by the sharing of one or more pairs of electrons.

COVID-19 Corona Virus Disease. The number **19** refers to the year this virus first appeared. A disease caused by the SARS-CoV-2 virus that first appeared in 2019.

COVID-19 Genome The COVID-19 genome is large for a virus, with 30,000 nucleotides encoding 29 different proteins. Four of these proteins make up the protein coat of the virus, including the spike, while the rest stabilize the viral genes and helps them replicate.

crassulacean acid metabolism (CAM) A mode of carbon dioxide fixation by which CO_2 enters open leaf stomata at night and is used in photosynthesis during the day, when stomata are closed to prevent water loss.

CRISPR A molecular tool used to edit genes. The tool is composed of two molecules that work together. One part is the nucleotide sequence manufactured by the investigator to guide the CRISPR tool to a specific gene sequence; the other part is a DNA cutting enzyme. When a cut is made at the target sequence, the hole that results is filled in with whatever sequence the investigator floods into the reaction.

crista, *pl.* **cristae** (L. *crista,* crest) A folded extension of the inner membrane of a mitochondrion. Mitochondria contain numerous cristae.

crossing over An essential element of meiosis occurring during prophase when nonsister chromatids exchange portions of DNA strands.

cuticle (L. *cutis,* skin) A very thin film covering the outer skin of many plants.

cytokinesis (Gr. *kytos,* hollow vessel + *kinesis,* movement) The C phase of cell division in which the cell itself divides, creating two daughter cells.

cytoplasm (Gr. *kytos,* hollow vessel + *plasma,* anything molded) A semifluid matrix that occupies the volume between the nuclear region and the cell membrane. It contains the sugars, amino acids, proteins, and organelles (in eukaryotes) with which the cell carries out its everyday activities of growth and reproduction.

cytoskeleton (Gr. *kytos,* hollow vessel + *skeleton,* a dried body) In the cytoplasm of all eukaryotic cells, a network of protein fibers that supports the shape of the cell and anchors organelles, such as the nucleus, to fixed locations.

D

deciduous (L. *decidere,* to fall off) In vascular plants, shedding all the leaves at a certain season.

dehydration reaction Water-losing. The process in which a hydroxyl (OH) group is removed from one subunit of a polymer, and a hydrogen (H) group is removed from the other subunit, linking the subunits together and forming a water molecule as a by-product.

Delta variant A highly infectious form of COVID-19 that can evade the antibodies produced by COVID-19 vaccination, resulting in many "breakthrough" infections of fully vaccinated individuals. Because the Delta variant does not evade T cells, such breakthrough infections do not usually lead to serious illness.

demography (Gr. *demos,* people + *graphein,* to draw) The statistical study of population. The measurement of people or, by extension, of the characteristics of people.

dendrite A process extending from the cell body of a neuron, typically branched, that conducts impulses toward the cell body.

deoxyribonucleic acid (DNA) The basic storage vehicle or central plan of heredity information. It is stored as a sequence of nucleotides in a linear nucleotide polymer. Two of the polymers wind around each other like the outside and inside rails of a circular staircase.

depolarization The movement of ions across a cell membrane that wipes out locally an electrical potential difference.

derived character A characteristic used in taxonomic analysis representing a departure from the primitive form.

detritivore (L. *detri,* a rubbing away + *vorare,* to devour) An organism that feeds on or breaks down dead organisms.

deuterostome (Gr. *deuteros,* second + *stoma,* mouth) An animal in whose embryonic development the anus forms from or near the blastopore, and the mouth forms later on another part of the blastula. Also characterized by radial cleavage.

diaphragm (1) In mammals, a sheet of muscle tissue that separates the abdominal and thoracic cavities and functions in breathing. (2) A contraceptive device used to block the entrance to the uterus temporarily and thus prevent sperm from entering during sexual intercourse.

diastolic pressure In the measurement of human blood pressure, the minimum pressure between heartbeats (repolarization of the ventricles). *Compare with* systolic pressure.

diffusion (L. *diffundere,* to pour out) The net movement of molecules to regions of lower concentration as a result of random, spontaneous molecular motions. The process tends to distribute molecules uniformly.

dihybrid (Gr. *dis,* twice + L. *hibrida,* mixed offspring) An individual heterozygous for two genes.

dikaryotic In fungi, having pairs of nuclei within each cell.

dioecious (Gr. *di,* two + *eikos,* house) Having male and female flowers on separate plants of the same species.

diploid (Gr. *diploos,* double + *eidos,* form) Having two sets of chromosomes ($2n$), in contrast to haploid (n).

directional selection A form of selection in which selection acts to eliminate one extreme from an array of phenotypes. Thus, the genes promoting this extreme become less frequent in the population.

disaccharide (Gr. *dis,* twice + *sakcharon,* sugar) A sugar formed by linking two monosaccharide molecules together. Sucrose (table sugar) is a disaccharide formed by linking a molecule of glucose to a molecule of fructose.

disruptive selection A form of selection in which selection acts to eliminate rather than favor the intermediate type.

diurnal (L. *diurnalis,* day) Active during the day.

DNA Deoxyribo Nucleic Acid, the molecules that make up genes. Each DNA molecule is composed of two long chains of subunits called nucleotides. The two chains wind around each other to form a double helix. There are four kinds of subunits (A,T,G,C). The sequence of these four letters encodes the message of a gene.

DNA fingerprinting An identification technique that makes use of a variety of molecular techniques to identify differences in the DNA of individuals.

DNA vaccine A type of vaccine that uses DNA from a virus or bacterium that stimulates the cellular immune response.

domain In taxonomy, the level higher than kingdom. The three domains currently recognized are Bacteria, Archaea, and Eukarya.

dominant allele An allele that dictates the appearance of heterozygotes. One allele is said to be dominant over another if an individual heterozygous for that allele has the same appearance as an individual homozygous for it.

dorsal (L. *dorsum,* the back) Toward the back, or upper surface. Opposite of ventral.

double fertilization A process unique to the angiosperms, in which one sperm nucleus fertilizes the egg and the second one fuses with the polar nuclei. These two events result in the formation of the zygote and the primary endosperm nucleus, respectively.

double helix The structure of DNA, in which two complementary polynucleotide strands coil around a common helical axis.

duodenum In vertebrates, the upper portion of the small intestine.

E

Ebola A lethal filamentous virus arising in Central African fruit bats that attacks human connective tissue. Typically, half of infected individuals die.

ecdysis (Gr. *ekdysis,* stripping off) The shedding of the outer covering or skin of certain animals. Especially the shedding of the exoskeleton by arthropods.

echinoderm (Gr. *echinos,* sea urchin, hedgehog + *derma,* skin) An animal of the phylum Echinodermata, which includes the sea stars, sea urchins, sea cucumbers, and sand dollars.

ecology (Gr. *oikos,* house + *logos,* word) The study of the relationships of organisms with one another and with their environment.

ecosystem (Gr. *oikos,* house + *systema,* that which is put together) A community, together with the nonliving factors with which it interacts.

ectoderm (Gr. *ecto,* outside + *derma,* skin) One of three embryonic germ layers that forms in the gastrula; giving rise to the outer epithelium and to nerve tissue.

ectothermic Referring to animals whose body temperature is regulated by their behavior or their surroundings.

electron A subatomic particle with a negative electrical charge. The negative charge of one electron exactly balances the positive charge of one proton. Electrons orbit the atom's positively charged nucleus and determine its chemical properties.

electron transport chain A collective term describing the series of membrane-associated electron carriers embedded in the inner mitochondrial membrane. It puts the electrons harvested from the oxidation of glucose to work driving proton-pumping channels.

electron transport system A collective term describing the series of electron carriers embedded in the thylakoid membrane of the chloroplast. It puts the electrons harvested from water molecules and energized by photons of light to work driving proton-pumping channels.

element A substance that cannot be separated into different substances by ordinary chemical methods.

embryonic stem cell A cell that is derived from the inner cell mass of an early embryo and is able to develop into any tissue or give rise to an adult organism when injected into a blastocyst.

emergent properties Novel properties in the hierarchy of life that were not present at the simpler levels of organization.

endemic Native to a place. A disease is endemic when it occurs constantly in a geographic area. Influenza ("flu") is endemic in the United States because this virus infects large numbers of people every year; malaria is not endemic because it does not.

endergonic (Gr. *endon*, within + *ergon*, work) Reactions in which the products contain more energy than the reactants and require an input of usable energy from an outside source before they can proceed. These reactions are not spontaneous.

endocrine gland (Gr. *endon*, within + *krinein*, to separate) A ductless gland producing hormonal secretions that pass directly into the bloodstream or lymph.

endocytosis (Gr. *endon*, within + *kytos*, cell) The process by which the edges of plasma membranes fuse together and form an enclosed chamber called a vesicle. It involves the incorporation of a portion of an exterior medium into the cytoplasm of the cell by capturing it within the vesicle. The way an animal cell is able to "gobble up" a nearby molecule or virus. The cell's membrane surrounds the object, then the membrane's edges fuse together to enclose the object within the cell.

endoderm (Gr. *endon*, outside + *derma*, skin) One of three embryonic germ layers that forms in the gastrula; giving rise to the epithelium that lines internal organs and most of the digestive and respiratory tracts.

endodermis In vascular plants, a layer of cells forming the innermost layer of the cortex in roots and some stems.

endomembrane system A system of connected membranous compartments found in eukaryotic cells.

endometrium The lining of the uterus in mammals; thickens in response to secretion of estrogens and progesterone and is sloughed off in menstruation.

endoplasmic reticulum (ER) (L. *endoplasmic*, within the cytoplasm + *reticulum*, little net) An extensive network of membrane compartments within a eukaryotic cell; attached ribosomes synthesize proteins to be exported.

endoskeleton (Gr. *endon*, within + *skeletos*, hard) In vertebrates, an internal scaffold of bone or cartilage to which muscles are attached.

endosperm (Gr. *endon*, within + *sperma*, seed) A nutritive tissue characteristic of the seeds of angiosperms that develops from the union of a male nucleus and the polar nuclei of the embryo sac. The endosperm is either digested by the growing embryo or retained in the mature seed to nourish the germinating seedling.

endospore A highly resistant, thick-walled bacterial spore that can survive harsh environmental stress, such as heat or desiccation, and then germinate when conditions become favorable.

endosymbiotic (Gr. *endon*, within + *bios*, life) **theory** A theory that proposes how eukaryotic cells arose from large prokaryotic cells that engulfed smaller ones of a different species. The smaller cells were not consumed but continued to live and function within the larger host cell. Organelles that are believed to have entered larger cells in this way are mitochondria and chloroplasts.

endothermic The ability of animals to maintain an elevated body temperature using their metabolism.

energy The capacity to bring about change, to do work.

enhancer A site of regulatory protein binding on the DNA molecule distant from the promoter and start site for a gene's transcription.

entropy (Gr. *en*, in + *tropos*, change in manner) A measure of the disorder of a system. A measure of energy that has become so randomized and uniform in a system that the energy is no longer available to do work.

enzyme (Gr. *enzymos*, leavened; *from en*, in + *zyme*, leaven) A protein capable of speeding up specific chemical reactions by lowering the energy required to activate or start the reaction but that remains unaltered in the process.

Epidemic - a rapid spread of an infectious agent in a localized population in a given period of time.

epidermis (Gr. *epi*, on or over + *derma*, skin) The outermost layer of cells. In vertebrates, the nonvascular external layer of skin of ectodermal origin; in invertebrates, a single layer of ectodermal epithelium; in plants, the flattened, skinlike outer layer of cells.

epididymis A sperm storage vessel; a coiled part of the sperm duct that lies near the testis.

epigenetic modification Change in the packaging of DNA, often due to chemical alteration of histones. Epigenetic modifications can be passed on to offspring. As DNA packaging affects gene expression, these changes are inherited, even though no gene is changed.

epigenetics The conditioning of gametic DNA by the parent; also called gene reprogramming.

epistasis (Gr. *epistasis*, a standing still) An interaction between the products of two genes in which one modifies the phenotypic expression produced by the other.

epithelium (Gr. *epi*, on + *thele*, nipple) A thin layer of cells forming a tissue that covers the internal and external surfaces of the body. Simple epithelium consists of the membranes that line the lungs and major body cavities and that are a single cell layer thick. Stratified epithelium (the skin or epidermis) is composed of more complex epithelial cells that are several cell layers thick.

erythrocyte (Gr. *erythros*, red + *kytos*, hollow vessel) A red blood cell, the carrier of hemoglobin. Erythrocytes act as the transporters of oxygen in the vertebrate body. During the process of their maturation in mammals, they lose their nuclei and mitochondria, and their endoplasmic reticulum is reabsorbed.

estrus (L. *oestrus*, frenzy) The period of maximum female sexual receptivity. Associated with ovulation of the egg. Being "in heat."

estuary (L. *aestus*, tide) A partly enclosed body of water, such as those that often form at river mouths and in coastal bays, where the salinity is intermediate between that of saltwater and freshwater.

ethology (Gr. *ethos*, habit or custom + *logos*, discourse) The study of patterns of animal behavior in nature.

eudicot Short for eudicotyledon; a class of flowering plants generally characterized by having two cotyledons, netlike veins, and flower parts in fours or fives.

eukaryote (Gr. *eu*, true + *karyon*, kernel) A cell that possesses membrane-bounded organelles, most notably a cell nucleus, and chromosomes whose DNA is associated with proteins; an organism composed of such cells. The appearance of eukaryotes marks a major event in the evolution of life, as all organisms on earth other than bacteria and archaea are eukaryotes.

eumetazoan (Gr. *eu*, true + *meta*, with + *zoion*, animal) A "true animal." An animal with a definite shape and symmetry and nearly always distinct tissues.

eutrophic (Gr. *eutrophos*, thriving) Refers to a lake in which an abundant supply of minerals and organic matter exists.

evaporation The escape of water molecules from the liquid to the gas phase at the surface of a body of water.

evolution (L. *evolvere*, to unfold) Genetic change in a population of organisms over time (generations). Darwin proposed that natural selection was the mechanism of evolution.

exergonic (L. *ex*, out + Gr. *ergon*, work) Any reaction that produces products that contain less free energy than that possessed by the original reactants and that tends to proceed spontaneously.

exocrine gland A type of gland that releases its secretion through a duct, such as a digestive gland or a sweat gland.

exocytosis (Gr. *ex*, out of + *kytos*, cell) The extrusion of material from a cell by discharging it from vesicles at the cell surface. The reverse of endocytosis.

exon (Gr. *exo*, outside) A segment of DNA that is both transcribed into RNA and translated into protein.

exoskeleton (Gr. *exo*, outside + *skeletos*, hard) An external hard shell that encases a body. In arthropods, comprised mainly of chitin.

experiment The test of a hypothesis. An experiment that tests one or more alternative hypotheses and those that are demonstrated to be inconsistent with experimental observation are rejected.

F

facilitated diffusion The transport of molecules across a membrane by a carrier protein in the direction of lowest concentration.

family A taxonomic group ranking below an order and above a genus.

feedback inhibition A regulatory mechanism in which a biochemical pathway is regulated by the amount of the product that the pathway produces.

fermentation (L. *fermentum*, ferment) A catabolic process in which the final electron acceptor is an organic molecule.

fertilization (L. *ferre*, to bear) The union of male and female gametes to form a zygote.

fibroblast A flat, irregularly branching cell of connective tissue that secretes structurally strong proteins into the matrix between the cells.

first law of thermodynamics Energy cannot be created or destroyed but can only undergo conversion from one form to another; thus, the amount of energy in the universe is unchangeable.

fitness The genetic contribution of an individual to succeeding generations, relative to the contributions of other individuals in the population.

flagellum, *pl.* **flagella** (L. *flagellum*, whip) A fine, long, threadlike organelle protruding from the surface of a cell. In bacteria, a single protein fiber capable of rotary motion that propels the cell through the water. In eukaryotes, an array of microtubules with a characteristic internal 9 + 2 microtubule structure that is capable of vibratory but not rotary motion. Used in locomotion and feeding. Common in protists and motile gametes. A cilium is a short flagellum.

flame cell A specialized cell found in the network of tubules inside flatworms that assists in water regulation and some waste excretion.

food web The food relationships within a community. A diagram of who eats whom.

foraging behavior A collective term for the many complex, evolved behaviors that influence what an animal eats and how the food is obtained.

founder effect The effect by which rare alleles and combinations of alleles may be enhanced in new populations.

fovea A small depression in the center of the retina with a high concentration of cones; the area of sharpest vision.

frameshift mutation A mutation in which a base is added or deleted from the DNA sequence. These changes alter the reading frame downstream of the mutation.

frequency In statistics, defined as the proportion of individuals in a certain category, relative to the total number of individuals being considered.

fruit In angiosperms, a mature, ripened ovary (or group of ovaries) containing the seeds.

G

gamete (Gr. *wife*) A haploid reproductive cell. Upon fertilization, its nucleus fuses with that of another gamete of the opposite sex. The resulting diploid cell (zygote) may develop into a new diploid individual, or in some protists and fungi, may undergo meiosis to form haploid somatic cells.

gametophyte (Gr. *gamete,* wife + *phyton,* plant) In plants, the haploid (n), gamete-producing generation, which alternates with the diploid ($2n$) sporophyte.

ganglion, *pl.* **ganglia** (Gr. a swelling) A group of nerve cells forming a nerve center in the peripheral nervous system.

gardasil A vaccine for cervical cancer.

gene (Gr. *genos,* birth, race) The basic unit of heredity. A sequence of DNA nucleotides on a chromosome that encodes a polypeptide or RNA molecule and so determines the nature of an individual's inherited traits.

gene expression The process in which an RNA copy of each active gene is made, and the RNA copy directs the sequential assembly of a chain of amino acids at a ribosome.

gene frequency The frequency with which individuals in a population possess a particular gene. Often confused with allele frequency.

gene silencing See "RNA interference."

genetic code The "language" of the genes. The mRNA codons specific for the 20 common amino acids constitute the genetic code.

genetic counseling The process of evaluating the risk of genetic defects occurring in offspring, testing for these defects in unborn children, and providing the parents with information about these risks and conditions.

genetic drift Random fluctuations in allele frequencies in a small population over time.

genetic map A diagram showing the relative positions of genes.

genetics (Gr. *genos,* birth, race) The study of the way in which an individual's traits are transmitted from one generation to the next.

genome (Gr. *genos,* offspring + L. *oma,* abstract group) The genetic information of an organism.

genomics The study of genomes as opposed to individual genes.

genotype (Gr. *genos,* offspring + *typos,* form) The total set of genes present in the cells of an organism. Also used to refer to the set of alleles at a single gene locus.

genus, *pl.* **genera** (L. *race*) A taxonomic group that ranks below a family and above a species.

germination (L. *germinare,* to sprout) The resumption of growth and development by a spore or seed.

germ-line cells During zygote development, cells that are set aside from the somatic cells and that will eventually undergo meiosis to produce gametes.

gill (1) In aquatic animals, a respiratory organ, usually a thin-walled projection from some part of the external body surface, endowed with a rich capillary bed and having a large surface area. (2) In basidiomycete fungi, the plates on the underside of the cap.

gland (L. *glandis,* acorn) Any of several organs in the body, such as exocrine or endocrine, that secrete substances for use in the body. Glands are composed of epithelial tissue.

glomerulus (L. a little ball) A network of capillaries in a vertebrate kidney, whose walls act as a filtration device.

glycolysis (Gr. *glykys,* sweet + *lyein,* to loosen) The anaerobic breakdown of glucose; this enzyme-catalyzed process yields two molecules of pyruvate with a net of two molecules of ATP.

Golgi complex Flattened stacks of membrane compartments that collect, package, and distribute molecules made in the endoplasmic reticulum.

granum, *pl.* **grana** A stacked column of flattened, interconnected disks (thylakoids) that are part of the thylakoid membrane system in chloroplasts.

gravitropism (L. *gravis,* heavy + *tropes,* turning) The response of a plant to gravity, which generally causes shoots to grow up and roots to grow down.

greenhouse effect The process in which carbon dioxide and certain other gases, such as methane, that occur in the earth's atmosphere transmit radiant energy from the sun but trap the longer wavelengths of infrared light, or heat, and prevent them from radiating into space.

ground meristem The primary meristem, or meristematic tissue, that gives rise to the plant body (except for the epidermis and vascular tissues).

guard cells Pairs of specialized epidermal cells that surround a stoma. When the guard cells are turgid, the stoma is open; when they are flaccid, it is closed.

gymnosperm (Gr. *gymnos,* naked + *sperma,* seed) A seed plant with seeds not enclosed in an ovary. The conifers are the most familiar group.

H

habitat (L. *habitare,* to inhabit) The place where individuals of a species live.

half-life The length of time it takes for half of a radioactive substance to decay.

haploid (Gr. *haploos,* single + *eidos,* form) Having only one set of chromosomes (n), in contrast to diploid ($2n$).

Hardy-Weinberg equilibrium After G. H. Hardy, English mathematician, and G. Weinberg, German physician. A mathematical description of the fact that the relative frequencies of two or more alleles in a population do not change because of Mendelian segregation. Allele and genotype frequencies remain constant in a random-mating population in the absence of inbreeding, selection, or other evolutionary forces. Usually stated as: If the frequency of allele A is p and the frequency of allele a is q, then the genotype frequencies after one generation of random mating will always be $(p + q)^2 + p^2 + 2pq + q^2$.

Haversian canal After Clopton Havers, English anatomist. Narrow channels that run parallel to the length of a bone and contain blood vessels and nerve cells.

helper T cell A class of white blood cells that initiates both the cell-mediated immune response and the humoral immune response; helper T cells are the targets of the AIDS virus (HIV).

hemoglobin (Gr. *haima,* blood + L. *globus,* a ball) A globular protein in vertebrate red blood cells and in the plasma of many invertebrates that carries oxygen and carbon dioxide.

hemolymph (Gr. *haima,* blood + L. *lympha,* water) Circulating fluid in the coelom of some invertebrates, such as insects.

herbivore (L. *herba,* grass + *vorare,* to devour) Any organism that eats only plants.

Herd immunity The protection gained when enough members of a population are immune from infection (say, because they are vaccinated against the infecting agent). The vaccinated individuals (the "herd") form a protective wall around each still-vulnerable (not-yet-vaccinated) individual, preventing the virus from reaching that person.

heredity (L. *heredis,* heir) The transmission of characteristics from parent to offspring.

hermaphroditism Condition in which an organism has both male and female functional reproductive organs.

heterokaryon (Gr. *heteros,* other + *karyon,* kernel) A fungal hypha that has two or more genetically distinct types of nuclei.

heterotroph (Gr. *heteros,* other + *trophos,* feeder) An organism that does not have the ability to produce its own food. *See also* autotroph.

heterozygote (Gr. *heteros,* other + *zygotos,* a pair) A diploid individual carrying two different alleles of a gene on its two homologous chromosomes.

heterozygote advantage The situation in which individuals heterozygous for a trait have a selective advantage over those who are homozygous; an example is sickle-cell anemia.

histone (Gr. *histos,* tissue) A complex of small, very basic polypeptides rich in the amino acids arginine and lysine. A basic part of chromosomes, histones form the core around which DNA is wrapped.

homeostasis (Gr. *homeos,* similar + *stasis,* standing) The maintaining of a relatively stable internal physiological environment in an organism or steady-state equilibrium in a population or ecosystem.

homeotherm (Gr. *homeo,* similar + *therme,* heat) An organism, such as a bird or mammal, capable of maintaining a stable body temperature.

hominid (L. *homo*, man) Human beings and their direct ancestors. A member of the family Hominidae. *Homo sapiens* is the only living member.

Homo The genus of as many as eight human species, only one of which (*H. sapiens*) survives today.

homologous chromosome (Gr. *homologia*, agreement) One of the two nearly identical versions of each chromosome. Chromosomes that associate in pairs in the first stage of meiosis. In diploid cells, one chromosome of a pair that carries equivalent genes.

homology (Gr. *homologia*, agreement) A condition in which the similarity between two structures or functions is indicative of a common evolutionary origin.

homozygote (Gr. *homos*, same or similar + *zygotos*, a pair) A diploid individual whose two copies of a gene are the same. An individual carrying identical alleles on both homologous chromosomes is said to be homozygous for that gene.

hormone (Gr. *hormaein*, to excite) A chemical messenger, often a steroid or peptide, produced in a small quantity in one part of an organism and then transported to another part of the organism, where it brings about a physiological response.

humoral immunity Arm of the adaptive immune system involving B cells that produce soluble antibodies specific for foreign antigens.

hybrid (L. *hybrida*, **the offspring of a tame sow and a wild boar**) A plant or animal that results from the crossing of dissimilar parents.

hybridization The mating of unlike parents of different taxa.

hydrogen bond A molecular force formed by the attraction of the partial positive charge of one hydrogen atom of a water molecule with the partial negative charge of the oxygen atom of another.

hydrolysis reaction (Gr. *hydro*, water + *lyse*, break) The process of tearing down a polymer by adding a molecule of water. A hydrogen is attached to one subunit and a hydroxyl to the other, which breaks the covalent bond. Essentially the reverse of a dehydration reaction.

hydrophilic (Gr. *hydro*, water + *philic*, loving) Describes polar molecules, which form hydrogen bonds with water and therefore are soluble in water.

hydrophobic (Gr. *hydro*, water + *phobos*, hating) Describes nonpolar molecules, which do not form hydrogen bonds with water and therefore are not soluble in water.

hydroskeleton (Gr. *hydro*, water + *skeletos*, hard) The skeleton of most soft-bodied invertebrates that have neither an internal nor an external skeleton. They use the relative incompressibility of the water within their bodies as a kind of skeleton.

hypertonic (Gr. *hyper*, above + *tonos*, tension) A solution surrounding a cell that contains a higher concentration of solutes than does the cell.

hypha, *pl.* **hyphae** (Gr. *hyphe*, web) A filament of a fungus. A mass of hyphae comprises a mycelium.

hypothalamus (Gr. *hypo*, under + *thalamos*, inner room) The region of the brain under the thalamus that controls temperature, hunger, and thirst and that produces hormones that influence the pituitary gland.

hypothesis (Gr. *hypo*, under + *tithenai*, to put) A proposal that might be true. No hypothesis is ever proven correct. All hypotheses are provisional—proposals that are retained for the time being as useful but that may be rejected in the future if found to be inconsistent with new information. A hypothesis that stands the test of time—often tested and never rejected—is called a theory.

hypotonic (Gr. *hypo*, under + *tonos*, tension) A solution surrounding a cell that has a lower concentration of solutes than does the cell.

I

inbreeding The breeding of genetically related plants or animals. In plants, inbreeding results from self-pollination. In animals, inbreeding results from matings between relatives. Inbreeding tends to increase homozygosity.

incomplete dominance The ability of two alleles to produce a heterozygous phenotype that is different from either homozygous phenotype.

independent assortment Mendel's second law: The principle that segregation of alternative alleles at one locus into gametes is independent of the segregation of alleles at other loci. Only true for gene loci located on different chromosomes or those so far apart on one chromosome that crossing over is very frequent between the loci.

industrial melanism (Gr. *melas*, black) The evolutionary process in which a population of initially light-colored organisms becomes a population of dark organisms as a result of natural selection.

inflammatory response (L. *inflammare*, to flame) A generalized nonspecific response to infection that acts to clear an infected area of infecting microbes and dead tissue cells so that tissue repair can begin.

integument (L. *integumentum*, covering) The natural outer covering layers of an animal. Develops from the ectoderm.

intermembrane space The outer compartment of a mitochondrion that lies between the two membranes.

interoceptor A receptor that senses information related to the body itself, its internal condition, and its position.

interphase That portion of the cell cycle preceding mitosis. It includes the G_1 phase, when cells grow, the S phase, when a replica of the genome is synthesized, and a G_2 phase, when preparations are made for genomic separation.

intron (L. *intra*, within) A segment of DNA transcribed into mRNA but removed before translation. These untranslated regions make up the bulk of most eukaryotic genes.

ion An atom in which the number of electrons does not equal the number of protons. An ion carries an electrical charge.

ionic bond A chemical bond formed between ions as a result of the attraction of opposite electrical charges.

isolating mechanisms Mechanisms that prevent genetic exchange between individuals of different populations or species.

isotonic (Gr. *isos*, equal + *tonos*, tension) A solution having the same concentration of solutes as the cell.

isotope (Gr. *isos*, equal + *topos*, place) An atom that has the same number of protons but different numbers of neutrons.

J

joint The part of a vertebrate where one bone meets and moves on another.

K

karyotype (Gr. *karyon*, kernel + *typos*, stamp or print) The particular array of chromosomes that an individual possesses.

kidney In vertebrates, the organ that filters the blood to remove nitrogenous wastes and regulates the balance of water and solutes in blood plasma.

kinetic energy The energy of motion.

kinetochore (Gr. *kinetikos*, putting in motion + *choros*, chorus) A disk of protein bound to the centromere to which microtubules attach during cell division, linking chromatids to the spindle.

kingdom The chief taxonomic category. This book recognizes six kingdoms: Archaea, Bacteria, Protista, Fungi, Animalia, and Plantae.

Krebs cycle Another name for the citric acid cycle; also called the tricarboxylic acid (TCA) cycle.

L

lagging strand The DNA strand that must be synthesized discontinuously because of the 5'-to-3' directionality of DNA polymerase during replication and the antiparallel nature of DNA. Compare *leading strand*.

lateral line system A sensory system encountered in fish through which mechanoreceptors in a line down the side of the fish are sensitive to motion.

lateral meristems In vascular plants, the meristems that give rise to secondary tissue; the vascular cambium and cork cambium.

law of independent assortment Mendel's second law of heredity, stating that genes located on nonhomologous chromosomes assort independently of one another.

law of segregation Mendel's first law of heredity, stating that alternative alleles for the same gene segregate from each other in production of gametes.

leading strand The DNA strand that can be synthesized continuously from the origin of replication. Compare *lagging strand*.

ligament (L. *ligare*, to bind) A band or sheet of connective tissue that links bone to bone.

light-dependent reactions In photosynthesis, the reactions in which light energy is captured and used in production of ATP and NADPH. In plants, this involves the action of two linked photosystems.

light-independent reactions In photosynthesis, the reactions of the Calvin cycle in which ATP and NADPH from the light-dependent reactions are used to reduce CO_2 and produce organic compounds such as glucose. This involves the process of carbon fixation, or the conversion of inorganic carbon (CO_2) to organic carbon (ultimately carbohydrates).

limbic system The hypothalamus, together with the network of neurons that link the hypothalamus to some areas of the cerebral cortex. Responsible for many of the most deep-seated drives and emotions of vertebrates, including pain, anger, sex, hunger, thirst, and pleasure.

lipid (Gr. *lipos*, fat) A loosely defined group of molecules that are insoluble in water but soluble in oil. Oils such as olive, corn, and coconut are lipids, as well as waxes, such as beeswax and earwax.

lipid bilayer The basic foundation of all biological membranes. In such a layer, the nonpolar tails of phospholipid molecules point inward, forming a nonpolar zone in the interior of the bilayers. Lipid bilayers are selectively permeable and do not permit the diffusion of water-soluble molecules into the cell.

littoral (L. *litus*, shore) Referring to the shoreline zone of a lake or pond or the ocean that is exposed to the air whenever water recedes.

locus, *pl.* **loci (L. place)** The position on a chromosome where a gene is located.

loop of Henle After F. G. J. Henle, German anatomist. A hairpin loop formed by a urine-conveying tubule when it enters the inner layer of the kidney and then turns around to pass up again into the outer layer of the kidney.

lymph (L. *lympha*, clear water) In animals, a colorless fluid derived from blood by filtration through capillary walls in the tissues.

lymphatic system An open circulatory system composed of a network of vessels that function to collect the water within blood plasma forced out during passage through the capillaries and to return it to the bloodstream. The lymphatic system also returns proteins to the circulation, transports fats absorbed from the intestine, and carries bacteria and dead blood cells to the lymph nodes and spleen for destruction.

lymphocyte (Gr. *lympha*, water + Gr. *kytos*, hollow vessel) A type of white blood cell. A cell of the immune system that either synthesizes antibodies (B cells) or attacks virus-infected cells (T cells).

lyse (Gr. *lysis*, loosening) To disintegrate a cell by rupturing its plasma membrane.

lysosome (Gr. *lysis*, loosening + *soma*, body) A membrane-bounded vesicle containing digestive enzymes that is produced by the Golgi apparatus in eukaryotic cells.

M

macromolecule (Gr. *makros*, large + L. *moliculus*, a little mass) An extremely large molecule. Refers specifically to carbohydrates, lipids, proteins, and nucleic acids.

macrophage (Gr. *makros*, large + *phage*, eat) A phagocytic cell of the immune system able to engulf and digest invading bacteria, fungi, and other microorganisms, as well as cellular debris.

mantle The soft, outermost layer of the body wall in mollusks; the mantle secretes the shell.

marrow The soft tissue that fills the cavities of most bones and is the source of red blood cells.

marsupial A mammal in which the young are born early in their development, sometimes as soon as eight days after fertilization, and are retained in a pouch.

mass extinction A relatively sudden, sharp decline in the number of species; for example, the extinction at the end of the Cretaceous period in which the dinosaurs and a variety of other organisms disappeared.

mass flow The overall process by which materials move in the phloem of plants.

mass number The mass number of an atom consists of the combined mass of all of its protons and neutrons.

matrix (L. *mater*, mother) In mitochondria, the fluid in the interior space surrounded by the cristae that contains the enzymes and other molecules involved in oxidative respiration; more generally, that part of a tissue within which an organ or process is embedded.

medusa A free-floating, often umbrella-shaped body form found in cnidarian animals, such as jellyfish.

megaspore (Gr. *megas*, large + *spora*, seed) In plants, a haploid reproductive cell that develops into a female gametophyte; in most groups, megaspores are larger than microspores.

meiosis (Gr. *meioun*, to make smaller) A special form of nuclear division that precedes gamete formation in sexually reproducing eukaryotes. It results in four haploid daughter cells.

Mendelian ratio After Gregor Mendel, Austrian monk. Refers to the characteristic 3:1 segregation ratio that Mendel observed, in which pairs of alternative traits are expressed in the F_2 generation in the ratio of three-fourths dominant to one-fourth recessive.

menstruation (L. *mens*, month) Periodic sloughing off of the blood-enriched lining of the uterus when pregnancy does not occur.

meristem (Gr. *merizein*, to divide) In plants, a zone of unspecialized cells whose only function is to divide.

MERS **M**iddle **E**ast **R**espiratory **S**yndrome (MERS), a coronaviruses that originated in bats that spread to humans via camels (No one knows how the batto-camel transfer happened!). Some 2000 cases of MERS were reported over two years before the outbreak subsided; 36% of them died of the infection.

mesoderm (Gr. *mesos*, middle + *derma*, skin) One of the three embryonic germ layers that form in the gastrula. Gives rise to muscle, bone, and other connective tissue; the peritoneum; the circulatory system; and most of the excretory and reproductive systems.

mesophyll (Gr. *mesos*, middle + *phyllon*, leaf) The photosynthetic parenchyma of a leaf, located within the epidermis. The vascular strands (veins) run through the mesophyll.

messenger RNA (mRNA) The RNA transcribed from structural genes; RNA molecules complementary to a portion of one strand of DNA, which are translated by the ribosomes to form protein.

metabolism (Gr. *metabole*, change) The process by which all living things assimilate energy and use it to grow.

metaphase (Gr. *meta*, middle + *phasis*, form) The stage of mitosis characterized by the alignment of the chromosomes on a plane in the center of the cell.

metastasis, *pl.* **metastases (Gr. to place in another way)** The spread of cancerous cells to other parts of the body, forming new tumors at distant sites.

microevolution (Gr. *mikros*, small + L. *evolvere*, to unfold) Refers to the evolutionary process itself. Evolution within a species. Also called adaptation.

microspore (Gr. *mikros*, small + *spora*, seed) In plants, a haploid reproductive cell that develops into a male gametophyte.

microtubule (Gr. *mikros*, small + L. *tubulus*, little pipe) In eukaryotic cells, a long, hollow cylinder about 25 nanometers in diameter and composed of the protein tubulin. Microtubules influence cell shape, move the chromosomes in cell division, and provide the functional internal structure of cilia and flagella.

mimicry (Gr. *mimos*, mime) The resemblance in form, color, or behavior of certain organisms (mimics) to other more powerful or more protected ones (models), which results in the mimics being protected in some way.

mitochondrion, *pl.* **mitochondria (Gr. *mitos*, thread + *chondrion*, small grain)** A tubular or sausage-shaped organelle 1 to 3 micrometers long. Bounded by two membranes, mitochondria closely resemble the aerobic bacteria from which they were originally derived. As chemical furnaces of the cell, they carry out its oxidative metabolism.

mitosis (Gr. *mitos*, thread) The M phase of cell division in which the microtubular apparatus is assembled, binds to the chromosomes, and moves them apart. This phase is the essential step in the separation of the two daughter cell genomes.

mole (L. *moles*, mass) The atomic weight of a substance, expressed in grams. One mole is defined as the mass of $6.0222 + 10^{23}$ atoms.

molecule (L. *moliculus*, a small mass) The smallest unit of a compound that displays the properties of that compound.

mollusk (L. *molluscus*, soft) An animal of the phylum Mollusca, which includes snails and slugs, bivalves (such as clams and oysters), and cephalopods (octopuses, squids, and nautiluses).

monocot Short for monocotyledon; flowering plant in which the embryos have only one cotyledon, the flower parts are often in threes, and the leaves typically are parallel-veined.

monocyte A type of leukocyte that becomes a phagocytic cell (macrophage) after moving into tissues.

monomers (Gr. *mono*, single + *meris*, part) Simple molecules that can join together to form polymers.

monosaccharide (Gr. *monos*, one + *sakcharon*, sugar) A simple sugar.

monosomic Describes the condition in which a chromosome has been lost due to nondisjunction during meiosis, producing a diploid embryo with only one of these autosomes.

monotreme An egg-laying mammal.

morphogenesis (Gr. *morphe*, form + *genesis*, origin) The formation of shape. The growth and differentiation of cells and tissues during development.

morula Solid ball of cells in the early stage of embryonic development.

motor (efferent) neuron Neuron that transmits nerve impulses from the central nervous system to an effector, which is typically a muscle or gland.

multicellularity A condition in which the activities of the individual cells are coordinated and the cells themselves are in contact. A property of eukaryotes alone and one of their major characteristics.

muscle (L. *musculus*, mouse) The tissue in the body of humans and animals that can be contracted and relaxed to make the body move.

muscle cell A long, cylindrical, multinucleated cell that contains numerous myofibrils and is capable of contraction when stimulated.

muscle fiber A long, cylindrical, multinucleated cell containing numerous myofibrils, which is capable of contraction when stimulated.

muscle spindle A sensory organ that is attached to a muscle and sensitive to stretching.

mutagen (L. *mutare*, to change) A chemical capable of damaging DNA.

mutation (L. *mutare*, to change) A change in a gene sequence. Often a mutation results from a copying mistake during DNA replication, the wrong nucleotide being inserted in the message, changing its meaning.

mutualism (L. *mutuus*, lent, borrowed) A symbiotic relationship in which both participating species benefit.

mycelium, *pl.* **mycelia** (Gr. *mykes*, fungus) In fungi, a mass of hyphae.

mycology (Gr. *mykes*, fungus) The study of fungi. A person who studies fungi is called a mycologist.

mycorrhiza, *pl.* **mycorrhizae** (Gr. *mykes*, fungus + *rhiza*, root) A symbiotic association between fungi and plant roots.

myofibril (Gr. *myos*, muscle + L. *fibrilla*, little fiber) An elongated structure in a muscle fiber, composed of myosin and actin.

myofilament A contractile microfilament, composed largely of actin and myosin, within muscle.

myosin (Gr. *myos*, muscle + *in*, belonging to) One of two protein components of myofilaments. (The other is actin.)

N

natural killer cell A cell that does not kill invading microbes but rather the cells infected by them.

natural selection The differential reproduction of genotypes caused by factors in the environment. Leads to evolutionary change.

nematocyst (Gr. *nema*, thread + *kystos*, bladder) A coiled, threadlike stinging structure of cnidarians that is discharged to capture prey and for defense.

nephron (Gr. *nephros*, kidney) The functional unit of the vertebrate kidney. A human kidney has more than 1 million nephrons that filter waste matter from the blood. Each nephron consists of a Bowman's capsule, glomerulus, and tubule.

nerve A bundle of axons with accompanying supportive cells, held together by connective tissue.

nerve impulse A rapid, transient, self-propagating reversal in electrical potential that travels along the membrane of a neuron.

neural crest A special strip of cells that develops just before the neural groove closes over to form the neural tube in embryonic development.

neural tube The dorsal tube, formed from the neural plate, that differentiates into the brain and spinal cord.

neuromodulator A chemical transmitter that mediates effects that are slow and longer lasting and that typically involve second messengers within the cell.

neuromuscular junction The structure formed when the tips of axons contact (innervate) a muscle fiber.

neuron (Gr. nerve) A nerve cell specialized for signal transmission.

neurotransmitter (Gr. *neuron*, nerve + L. *trans*, across + *mitere*, to send) A chemical released at an axon tip that travels across the synapse and binds a specific receptor protein in the membrane on the far side.

neurulation (Gr. *neuron*, nerve) The elaboration of a notochord and a dorsal nerve cord that marks the evolution of the chordates.

neutron (L. *neuter*, neither) A subatomic particle located within the nucleus of an atom. Similar to a proton in mass, but as its name implies, a neutron is neutral and possesses no charge.

neutrophil An abundant type of white blood cell capable of engulfing microorganisms and other foreign particles.

niche (L. *nidus*, nest) The role an organism plays in the environment; realized niche is the niche that an organism occupies under natural circumstances; fundamental niche is the niche an organism would occupy if competitors were not present.

nitrogen fixation The incorporation of atmospheric nitrogen into nitrogen compounds, a process that can be carried out only by certain microorganisms.

nocturnal (L. *nocturnus*, night) Active primarily at night.

node of Ranvier After L. A. Ranvier, French histologist. A gap formed at the point where two Schwann cells meet and where the axon is in direct contact with the surrounding intercellular fluid.

nondisjunction The failure of homologous chromosomes to separate in meiosis I. The cause of Down syndrome.

nonrandom mating A phenomenon in which individuals with certain genotypes sometimes mate with one another more commonly than would be expected on a random basis.

notochord (Gr. *noto*, back + L. *chorda*, cord) In chordates, a dorsal rod of cartilage that forms between the nerve cord and the developing gut in the early embryo.

nuclear envelope The bounding structure of the eukaryotic nucleus. Composed of two phospholipid bilayers with the outer one connected to the endoplasmic reticulum.

nuclear pore One of a multitude of tiny but complex openings in the nuclear envelope that allow selective passage of proteins and nucleic acids into and out of the nucleus.

nucleic acid A nucleotide polymer. A long chain of nucleotides. Chief types are deoxyribonucleic acid (DNA), which is double-stranded, and ribonucleic acid (RNA), which is typically single-stranded.

nucleolus A region inside the nucleus where rRNA and ribosomes are produced.

nucleosome (L. *nucleus*, kernel + *soma*, body) The basic packaging unit of eukaryotic chromosomes, in which the DNA molecule is wound around a ball of histone proteins. Chromatin is composed of long strings of nucleosomes, like beads on a string.

nucleotide A single unit of a nucleic acid, composed of a phosphate, a 5-carbon sugar (either ribose or deoxyribose), and a purine or a pyrimidine.

nucleus (L. kernel, dim. Fr. *nux*, nut) A spherical organelle (structure) characteristic of eukaryotic cells. The repository of the genetic information that directs all activities of a living cell. In atoms, the central core, containing positively charged protons and (in all but hydrogen) electrically neutral neutrons.

O

oligotrophic (Gr. *oligo*, little, few + *trophein*, to nourish) Containing a scarcity of organic material and nutrients, and having a high oxygen content; in reference to a body of water, such as a lake.

oncogene (Gr. *onkos*, protuberance, tumor + *genos*, decent) Any of a number of genes that when inappropriately activated can cause unrestrained cell growth (cancer).

oocyte (Gr. *oion*, egg + *kytos*, vessel) A cell in the outer layer of the ovary that gives rise to an ovum. A primary oocyte is any of the 2 million oocytes a female is born with, all of which have begun the first meiotic division.

open circulatory system A circulatory system in which the blood flows into sinuses in which it mixes with body fluid and then reenters the vessels in another location.

operant conditioning A learning mechanism in which the reward follows only after the correct behavioral response.

operculum (L. cover) A flat, bony, external protective covering over the gill chamber in fish.

operon (L. *operis*, work) A cluster of functionally related genes transcribed onto a single mRNA molecule. A common mode of gene regulation in prokaryotes; it is rare in eukaryotes other than fungi.

order A taxonomic category ranking below a class and above a family.

organ (L. *organon*, tool) A complex body structure composed of several different kinds of tissue grouped together in a structural and functional unit.

organ system A group of organs that function together to carry out the principal activities of the body.

organelle (Gr. *organella*, little tool) A specialized compartment of a cell. Mitochondria are organelles.

organism Any individual living creature, either unicellular or multicellular.

osmoregulation The maintenance of a constant internal solute concentration by an organism, regardless of the environment in which it lives.

osmosis (Gr. *osmos*, act of pushing, thrust) The diffusion of water across a membrane that permits the free passage of water but not that of one or more solutes. Water moves from an area of low solute concentration to an area with higher solute concentration.

osmotic pressure The increase of hydrostatic water pressure within a cell as a result of water molecules that continue to diffuse inward toward the area of lower water concentration (the water concentration is lower inside than outside the cell because of the dissolved solutes in the cell).

osteoblast (Gr. *osteon*, bone + *blastos*, bud) A bone-forming cell.

osteoclast (Gr. *osteon*, bone + *klan*, to break) A bone-dissolving cell.

osteocyte (Gr. *osteon*, bone + *kytos*, hollow vessel) A mature osteoblast.

outcross A term used to describe species that interbreed with individuals other than those like themselves.

ovary (L. *ovum*, egg) (1) In animals, the organ in which eggs are produced. (2) In flowering plants, the enlarged basal portion of a carpel that contains the ovule(s); the ovary matures to become the fruit.

oviduct In vertebrates, the passageway through which ova (eggs) travel from the ovary to the uterus.

oviparous (L. *ovum*, egg + *parere*, to bring forth) Refers to reproduction in which the eggs are developed after leaving the body of the mother, as in reptiles.

ovulation The successful development and release of an egg by the ovary.

ovule (L. *ovulum*, **a little egg**) A structure in a seed plant that becomes a seed when mature.

ovum, *pl.* **ova** (L. **egg**) A mature egg cell. A female gamete.

oxidation (Fr. *oxider*, **to oxidize**) The loss of an electron during a chemical reaction from one atom to another. Occurs simultaneously with reduction. Is the second stage of the 10 reactions of glycolysis.

oxidation-reduction reaction A type of paired reaction in living systems in which electrons lost from one atom (oxidation) are gained by another atom (reduction). Termed a *redox reaction* for short.

oxidative respiration Respiration in which the final electron acceptor is molecular oxygen.

oxytocin A hormone of the posterior pituitary gland that affects uterine contractions during childbirth and stimulates lactation.

P

p53 gene The gene that produces the p53 protein that monitors DNA integrity and halts cell division if DNA damage is detected. Many types of cancer are associated with a damaged or absent *p53 gene*.

pancreas (Gr. *pan*, **all** + *kreas*, **flesh**) A gland located near the stomach that secretes digestive enzymes into the small intestine and hormones into the blood.

pandemic A world-wide outbreak of a deadly disease.

parasitism (Gr. *para*, **beside** + *sitos*, **food**) A symbiotic relationship in which one organism benefits and the other is harmed.

parenchyma cell The most common type of plant cell; characterized by large vacuoles, thin walls, and functional nuclei.

parthenogenesis (Gr. *parthenos*, **virgin** + Eng. *genesis*, **beginning**) The development of an adult from an unfertilized egg. A common form of reproduction in insects.

partial pressures (P) The components of each individual gas—such as nitrogen, oxygen, and carbon dioxide—that together constitute the total air pressure.

pathogen (Gr. *pathos*, **suffering** + Eng. *genesis*, **beginning**) A disease-causing organism.

PCR The **P**olymerase **C**hain **R**eaction a laboratory technique used to amplify DNA. Short DNA sequences called primers are used to target the part of the genome to be amplified. The temperature of the sample is repeatedly raised and lowered to help DNA polymerase copy the target DNA sequence. The technique can produce billions of copies of the target sequence in just a few hours.

pedigree (L. *pes*, **foot** + *grus*, **crane**) A family tree. The patterns of inheritance observed in family histories. Used to determine the mode of inheritance of a particular trait.

peptide (Gr. *peptein*, **to soften, digest**) Two or more amino acids linked by peptide bonds.

peptide bond A covalent bond linking two amino acids. Formed when the positive (amino, or NH_2) group at one end and a negative (carboxyl, or COOH) group at the other end undergo a chemical reaction and lose a molecule of water.

pericycle In vascular plants, one or more cell layers surrounding the vascular tissues of the root, bounded externally by the endodermis and internally by the phloem.

periderm Outer protective tissue in vascular plants that is produced by the cork cambium and functionally replaces epidermis when it is destroyed during secondary growth; the periderm includes the cork, cork cambium, and phelloderm.

peristalsis (Gr. *peri*, **around** + *stellein*, **to wrap**) The rhythmic sequences of waves of muscular contraction in the walls of a tube.

petal A flower part, usually conspicuously colored; one of the units of the corolla.

petiole The stalk of a leaf.

pH Refers to the concentration of H^+ ions in a solution. The numerical value of the pH is the negative of the exponent of the molar concentration. Low pH values indicate high concentrations of H^+ ions (acids), and high pH values indicate low concentrations (bases).

phagocyte (Gr. *phagein*, **to eat** + *kytos*, **hollow vessel**) A cell that kills invading cells by engulfing them. Includes neutrophils and macrophages.

phagocytosis (Gr. *phagein*, **to eat** + *kytos*, **hollow vessel**) A form of endocytosis in which cells engulf organisms or fragments of organisms.

phenotype (Gr. *phainein*, **to show** + *typos*, **stamp or print**) The realized expression of the genotype. The observable expression of a trait (affecting an individual's structure, physiology, or behavior) that results from the biological activity of proteins or RNA molecules transcribed from the DNA.

pheromone (Gr. *pherein*, **to carry** + [hor]mone) A chemical signal emitted by certain animals as a means of communication.

phloem (Gr. *phloos*, **bark**) In vascular plants, a food-conducting tissue basically composed of sieve elements, various kinds of parenchyma cells, fibers, and sclereids.

phosphodiester bond The bond that results from the formation of a nucleic acid chain in which individual sugars are linked together in a line by the phosphate groups. The phosphate group of one sugar binds to the hydroxyl group of another, forming an —O—P—O bond.

phospholipid A macromolecule similar in structure to a fat, but having only two fatty acids attached to the glycerol backbone, with the third space linked to a phosphorylated molecule; contains a polar hydrophilic "head" end (phosphate group) and a nonpolar hydrophobic "tail" end (fatty acids).

photon (Gr. *photos*, **light**) The unit of light energy.

photoperiodism (Gr. *photos*, **light** + *periodos*, **a period**) A mechanism that organisms use to measure seasonal changes in relative day and night length.

photorespiration A process in which carbon dioxide is released without the production of ATP or NADPH. Because it produces neither ATP nor NADPH, photorespiration acts to undo the work of photosynthesis.

photosynthesis (Gr. *photos*, **light** + *-syn*, **together** + *tithenai*, **to place**) The process by which plants, algae, and some bacteria use the energy of sunlight to create from carbon dioxide (CO_2) and water (H_2O) the more complicated molecules that make up living organisms.

photosystem An organized complex of chlorophyll, other pigments, and proteins that traps light energy as excited electrons. Plants have two linked photosystems in the thylakoid membrane of chloroplasts.

phototropism (Gr. *photos*, **light** + *trope*, **turning to light**) A plant's growth response to a unidirectional light source.

phylogenetic tree A pattern of descent generated by analysis of similarities and differences among organisms. Modern gene-sequencing techniques have produced phylogenetic trees showing the evolutionary history of individual genes.

phylogeny (Gr. *phylon*, **race, tribe**) The evolutionary relationships among any group of organisms.

phylum, *pl.* **phyla** *(Gr. phylon*, **race, tribe**) A major taxonomic category, ranking above a class.

physiology (Gr. *physis*, **nature** + *logos*, **a discourse**) The study of the function of cells, tissues, and organs.

pigment (L. *pigmentum*, **paint**) A molecule that absorbs light.

pili (**pilus**) Short flagella that occur on the cell surface of some prokaryotes.

pinocytosis (Gr. *pinein*, **to drink** + *kytos*, **cell**) A form of endocytosis in which the material brought into the cell is a liquid containing dissolved molecules.

pistil (L. *pistillum*, **pestle**) Central organ of flowers, typically consisting of ovary, style, and stigma; a pistil may consist of one or more fused carpels and is more technically and better known as the gynoecium.

pith The ground tissue occupying the center of the stem or root within the vascular cylinder.

pituitary gland (L. *pituitarius*, **mucus**) Endocrine gland at the base of the hypothalamus, composed of anterior and posterior lobes. Pituitary hormones affect a wide variety of processes in vertebrates.

placebo A substance designed to have no therapeutic value. Thus in clinical trials of a vaccine, some patients are administered a salt solution instead of the vaccine as a control.

placenta, *pl.* **placentae** (1) In flowering plants, the part of the ovary wall to which the ovules or seeds are attached. (2) In mammals, a tissue formed in part from the inner lining of the uterus and in part from other membranes, through which the embryo (later the fetus) is nourished while in the uterus and through which wastes are carried away.

plankton (Gr. *planktos*, **wandering**) The small organisms that float or drift in water, especially at or near the surface.

plasma (Gr. **form**) The fluid of vertebrate blood. Contains dissolved salts, metabolic wastes, hormones, and a variety of proteins, including antibodies and albumin. Blood minus the blood cells.

plasma cell An antibody-producing cell resulting from the multiplication and differentiation of a B lymphocyte that has interacted with an antigen.

plasma membrane A lipid bilayer with embedded proteins that control the cell's permeability to water and dissolved substances.

plasmid (Gr. *plasma,* a form or something molded) A small fragment of DNA that replicates independently of the bacterial chromosome.

plasmodesmata (Gr. *plasma,* form + *desma,* bond) In plants, cytoplasmic connections between adjacent cells.

platelet (Gr. dim of *plattus,* flat) In mammals, a fragment of a white blood cell that circulates in the blood and functions in the formation of blood clots at sites of injury.

pleiotropy (Gr. *pleros,* more + *trope,* a turning) A gene that produces more than one phenotypic effect.

point mutation An alteration of one nucleotide in a chromosomal DNA molecule.

polar molecule A molecule with positively and negatively charged ends. One portion of a polar molecule attracts electrons more strongly than another portion, with the result that the molecule has electron-rich (−) and electron-poor (+) regions, giving it magnetlike positive and negative poles. Water is one of the most polar molecules known.

polarization The charge difference of a neuron so that the interior of the cell is negative with respect to the exterior.

pollen (L. fine dust) A fine, yellowish powder consisting of grains or microspores, each of which contains a mature or immature male gametophyte. In flowering plants, pollen is released from the anthers of flowers and fertilizes the pistils.

pollen tube A tube that grows from a pollen grain. Male reproductive cells move through the pollen tube into the ovule.

pollination The transfer of pollen from the anthers to the stigmas of flowers for fertilization, as by insects or the wind.

polygenic inheritance Describes a mode of inheritance in which more than one gene affects a trait, such as height in human beings; polygenic inheritance may produce a continuous range of phenotypic values, rather than discrete either–or values.

polygyny (Gr. *poly,* many + *gyne,* woman, wife) A mating system in which a male mates with more than one female.

polymer (Gr. *polus,* many + *meris,* part) A large molecule formed of long chains of similar molecules called monomers.

polymerase chain reaction (PCR) A process by which DNA polymerase is used to copy a sequence of DNA repeatedly, making millions of copies of the same DNA.

polymorphism (Gr. *polys,* many + *morphe,* form) The presence in a population of more than one allele of a gene at a frequency greater than that of newly arising mutations.

polyp A cylindrical, pipe-shaped cnidarian usually attached to a rock with the mouth facing away from the rock on which it is growing. Coral is made up of polyps.

polypeptide (Gr. *polys,* many + *peptein,* to digest) A general term for a long chain of amino acids linked end to end by peptide bonds. A protein is a long, complex polypeptide.

polysaccharide (Gr. *polys,* many + *sakcharon,* sugar) A sugar polymer. A carbohydrate composed of many monosaccharide sugar subunits linked together in a long chain.

population (L. *populus,* the people) Any group of individuals of a single species, occupying a given area at the same time.

posterior (L. *post,* after) Situated behind or farther back.

potential difference A difference in electrical charge on two sides of a membrane caused by an unequal distribution of ions.

potential energy Energy with the potential to do work. Stored energy.

predation (L. *praeda,* prey) The eating of other organisms. The one doing the eating is called a predator, and the one being consumed is called the prey.

primary growth In vascular plants, growth originating in the apical meristems of shoots and roots, as contrasted with secondary growth; results in an increase in length.

primary immune response The first response of an immune system to a foreign antigen. If the system is challenged again with the same antigen, the memory cells created during the primary response will respond more quickly.

primary plant body The part of a plant that arises from the apical meristems.

primary producers Photosynthetic organisms, including plants, algae, and photosynthetic bacteria.

primary structure of a protein The sequence of amino acids that makes up a particular polypeptide chain.

primordium, *pl.* **primordia** *(L. primus,* first + *ordiri,* begin) The first cells in the earliest stages of the development of an organ or structure.

prions Infectious proteinaceous particles.

procambium In vascular plants, a primary meristematic tissue that gives rise to primary vascular tissues.

productivity The total amount of energy of an ecosystem fixed by photosynthesis per unit of time. Net productivity is productivity minus that which is expended by the metabolic activity of the organisms in the community.

prokaryote (Gr. *pro,* before + *karyon,* kernel) A simple organism that is small, single-celled, and has little evidence of internal structure.

promoter An RNA polymerase binding site. The nucleotide sequence at the end of a gene to which RNA polymerase attaches to initiate transcription of mRNA.

prophase (Gr. *pro,* before + *phasis,* form) The first stage of mitosis during which the chromosomes become more condensed, the nuclear envelope is reabsorbed, and a network of microtubules (called the spindle) forms between opposite poles of the cell.

protein (Gr. *proteios,* primary) A long chain of amino acids linked end to end by peptide bonds. Because the 20 amino acids that occur in proteins have side groups with very different chemical properties, the function and shape of a protein is critically affected by its particular sequence of amino acids.

protist (Gr. *protos,* first) A member of the kingdom Protista, which includes unicellular eukaryotic organisms and some multicellular lines derived from them.

protoderm The primary meristem that gives rise to the dermal tissue.

proton A subatomic particle in the nucleus of an atom that carries a positive charge. The number of protons determines the chemical character of the atom because it dictates the number of electrons orbiting the nucleus and available for chemical activity.

proto-oncogene A normal cellular gene that can act as an oncogene when mutated.

protostome (Gr. *protos,* first + *stoma,* mouth) An animal in whose embryonic development the mouth forms at or near the blastopore. Also characterized by spiral cleavage.

pseudocoel (Gr. *pseudos,* false + *koiloma,* cavity) A body cavity similar to the coelom except that it forms between the mesoderm and endoderm.

punctuated equilibrium A hypothesis of the mechanism of evolutionary change that proposes that long periods of little or no change are punctuated by periods of rapid evolution.

Punnett square A diagrammatic way of showing the possible genotypes and phenotypes of genetic crosses.

purine (L. *purus,* pure + *urina,* urine) The larger of the two general kinds of nucleotide base found in DNA and RNA; a nitrogenous base with a double-ring structure, such as adenine or guanine.

pyrimidine (alter. of pyridine, from Gr. *pyr,* fire + *id,* adj. suffix + *ine*) The smaller of two general kinds of nucleotide base found in DNA and RNA; a nitrogenous base with a single-ring structure, such as cytosine, thymine, or uracil.

Q

quaternary structure of a protein A term to describe the way multiple protein subunits are assembled into a whole.

R

r_0 The number of individuals an infected person is likely to infect. Any value of r_0 greater than 1.0 will cause the infection to spread.

radial cleavage The embryonic cleavage pattern of deuterostome animals in which cells divide parallel to and at right angles to the polar axis of the embryo.

radial symmetry (L. *radius,* a spoke of a wheel + Gr. *summetros,* symmetry) The regular arrangement of parts around a central axis so that any plane passing through the central axis divides the organism into halves that are approximate mirror images.

radioactivity The emission of nuclear particles and rays by unstable atoms as they decay into more stable forms. Measured in curies, with 1 curie equal to 37 billion disintegrations a second.

radula (L. scraper) A rasping, tonguelike organ characteristic of most mollusks.

Receptor The door molecules use to enter animal cells. Typically a protein on the cell surface that acts like the trip on a trap - when a molecule encounters the receptor protein and fits correctly, the receptor protein responds by altering its shape, triggering receptor activation or endocytosis.

recessive allele An allele whose phenotype effects are masked in heterozygotes by the presence of a dominant allele.

recombination The formation of new gene combinations. In bacteria, it is accomplished by the transfer of genes into cells, often in association with viruses. In eukaryotes, it is accomplished by reassortment of chromosomes during meiosis and by crossing over.

reducing power The use of light energy to extract hydrogen atoms from water.

reduction (L. *reductio,* a bringing back; originally, "bringing back" a metal from its oxide) The gain of an electron during a chemical reaction from one atom to another. Occurs simultaneously with oxidation.

refractory period The recovery period after membrane depolarization during which the membrane is unable to respond to additional stimulation.

renal (L. *renes,* kidneys) Pertaining to the kidney.

repression (L. *reprimere,* **to press back, keep back)** The process of blocking transcription by the placement of the regulatory protein between the polymerase and the gene, thus blocking movement of the polymerase to the gene.

repressor (L. *reprimere,* **to press back, keep back)** A protein that regulates transcription of mRNA from DNA by binding to the operator and so preventing RNA polymerase from attaching to the promoter.

reproductive isolating mechanism Any barrier that prevents genetic exchange between species.

resolving power The ability of a microscope to distinguish two points as separate.

respiration (L. *respirare,* **to breathe)** The utilization of oxygen. In terrestrial vertebrates, the inhalation of oxygen and the exhalation of carbon dioxide.

resting membrane potential The charge difference that exists across a neuron's membrane at rest (about 70 millivolts).

restriction enzyme (restriction endonuclease) A special kind of enzyme that can recognize and cleave DNA molecules into fragments. One of the basic tools of genetic engineering.

retina The photosensitive layer of the vertebrate eye; contains several layers of neurons and light receptors (rods and cones); receives the image formed by the lens and transmits it to the brain via the optic nerve.

retrovirus (L. *retro,* **turning back)** A virus whose genetic material is RNA rather than DNA. When a retrovirus infects a cell, it makes a DNA copy of itself, which it can then insert into the cellular DNA as if it were a cellular gene.

ribonucleic acid (RNA) A class of nucleic acids characterized by the presence of the sugar ribose and the pyrimidine uracil; includes mRNA, tRNA, rRNA, and siRNA.

ribose A 5-carbon sugar.

ribosomal RNA (rRNA) A class of RNA molecules found, together with characteristic proteins, in ribosomes; transcribed from the DNA of the nucleolus.

ribosome A cell structure composed of protein and RNA that translates RNA copies of genes into protein.

RNA interference A type of gene silencing in which the mRNA transcript is prevented from being translated; small interfering RNAs (siRNAs) have been found to bind to mRNA and target its degradation prior to its translation.

RNA polymerase The enzyme that transcribes RNA from DNA.

rod Light-sensitive nerve cell found in the vertebrate retina; sensitive to very dim light; responsible for "night vision."

root The usually descending axis of a plant, normally belowground, which anchors the plant and serves as the major point of entry for water and minerals.

root cap In plants, a tissue structure at the growing tips of roots that protects the root apical meristem as the root pushes through the soil; cells of the root cap are continually lost and replaced.

root hair In plants, a tubular extension from an epidermal cell located just behind the root tip; root hairs greatly increase the surface area for absorption.

S

saltatory conduction A very fast form of nerve impulse conduction in which the impulses leap from node to node over insulated portions.

sarcoma (Gr. *sarx,* **flesh)** A cancerous tumor that involves connective or hard tissue, such as muscle.

sarcomere (Gr. *sarx,* **flesh** + *meris,* **part of)** The fundamental unit of contraction in skeletal muscle. The repeating bands of actin and myosin that appear between two Z lines.

sarcoplasmic reticulum (Gr. *sarx,* **flesh** + *plassein,* **to form, mold; L.** *reticulum,* **network)** The endoplasmic reticulum of a muscle cell. A sleeve of membrane that wraps around each myofilament.

SARS The deadly coronavirus Severe Acute Respiratory Syndrome (SARS) appeared in China in 2002. The outbreak infected 8,098 people, killing just under 10% of them. Because the virus made people very sick quickly, it didn't spread before symptoms appeared. Immediately quarantining infected people stopped the spread in its tracks, and within a few months it was eliminated from the human population. No known transmission of SARS has occurred since 2004.

scientific creationism A view that the biblical account of the origin of the earth is literally true, that the earth is much younger than most scientists believe, and that all species of organisms were individually created just as they are today.

sclerenchyma cell Tough, thick-walled cells that strengthen plant tissues.

second law of thermodynamics A statement concerning the transformation of potential energy into heat; it says that disorder (entropy) is continually increasing in the universe as energy changes occur, so disorder is more likely than order.

second messenger An intermediary compound that couples extracellular signals to intracellular processes and also amplifies a hormonal signal.

secondary cell wall In plants, the innermost layer of the cell wall. Secondary walls have a highly organized microfibrillar structure and are often impregnated with lignin.

secondary growth In vascular plants, growth that results from the division of a cylinder of cells around the plant's periphery. Secondary growth causes a plant to grow in diameter.

secondary immune response The swifter response of the body the second time it is invaded by the same pathogen because of the presence of memory cells, which quickly become antibody-producing plasma cells.

secondary structure of a protein The folding and bending of a polypeptide chain, which is held in place by hydrogen bonds.

seed A structure that develops from the mature ovule of a seed plant. Contains an embryo and a food source surrounded by a protective coat.

segmentation The division of the developing animal body into repeated units; segmentation allows for redundant systems and more efficient locomotion.

selection The process by which some organisms leave more offspring than competing ones and their genetic traits tend to appear in greater proportions among members of succeeding generations than the traits of those individuals that leave fewer offspring.

selective permeability Condition in which a membrane is permeable to some substances but not to others.

self-fertilization The transfer of pollen from an anther to a stigma in the same flower or to another flower of the same plant.

semen In reptiles and mammals, sperm-bearing fluid expelled from the penis during male orgasm.

semicircular canal Any of three fluid-filled canals in the inner ear that help to maintain balance.

sepal (L. *sepalum,* **a covering)** A member of the outermost whorl of a flowering plant. Collectively, the sepals constitute the calyx.

septum, *pl.* **septa (L.** *saeptum,* **a fence)** A partition or cross-wall, such as those that divide fungal hyphae into cells.

sex chromosomes In humans, the X and Y chromosomes, which are different in the two sexes and are involved in sex determination.

sex-linked characteristic A genetic characteristic that is determined by genes located on the sex chromosomes.

sexual reproduction Reproduction that involves the regular alternation between syngamy and meiosis. Its outstanding characteristic is that an individual offspring inherits genes from two parent individuals.

sexual selection A type of differential reproduction that results from variable success in obtaining mates.

shoot In vascular plants, the aboveground parts, such as the stem and leaves.

sieve cell In the phloem (food-conducting tissue) of vascular plants, a long, slender sieve element with relatively unspecialized sieve areas and with tapering end walls that lack sieve plates. Found in all vascular plants except angiosperms, which have sieve-tube members.

sister chromatid One of two identical copies of each chromosome, still linked at the centromere, produced as the chromosomes duplicate for mitotic division; similarly, one of two identical copies of each homologous chromosome present in a tetrad at meiosis.

soluble Refers to polar molecules that dissolve in water and are surrounded by a hydration shell.

solute The molecules dissolved in a solution. *See also* solution, solvent.

solution A mixture of molecules, such as sugars, amino acids, and ions, dissolved in water.

solvent The most common of the molecules in a solution. Usually a liquid, commonly water.

somatic cells (Gr. *soma,* **body)** All the diploid body cells of an animal that are not involved in gamete formation.

somatic nervous system In vertebrates, the neurons of the peripheral nervous system that control skeletal muscle.

somite A segmented block of tissue on either side of a developing notochord.

speciation The process by which new species arise, either by transformation of one species into another or by the splitting of one ancestral species into two descendant species.

species, *pl.* **species (L. kind, sort)** A group of interbreeding organisms that are reproductively isolated from all other such groups; a taxonomic unit ranking below a genus and designated by a two-part scientific name consisting of its genus and the species name.

sperm (Gr. *sperma*, sperm, seed) A sperm cell. The male gamete.

sphincter In vertebrate animals, a ring-shaped muscle capable of closing a tubular opening by constriction (e.g., between stomach and small intestine or between anus and exterior).

spike protein Protruding out from the COVID-19 virus particle's "skin" are a host of protein spikes. When a COVID-19 particle encounters a human cell, the head of the spike protein interacts with a protein on the human cell surface called ACE2. This binding initiates entry into the cell, where the virus particle disassembles and begins to reproduce.

spindle The mitotic assembly that carries out the separation of chromosomes during cell division. Composed of microtubules and assembled during prophase at the centrioles of the dividing cell.

spiral cleavage The embryonic cleavage pattern of some protostome animals in which cells divide at an angle oblique to the polar axis of the embryo; a line drawn through the sequence of dividing cells forms a spiral.

spore (Gr. *spora*, seed) A haploid reproductive cell, usually unicellular, that is capable of developing into an adult without fusion with another cell. Spores result from meiosis, as do gametes, but gametes fuse immediately to produce a new diploid cell.

sporophyte (Gr. *spora*, seed + *phyton*, plant) The spore-producing, diploid ($2n$) phase in the life cycle of a plant having alternation of generations.

stabilizing selection A form of selection in which selection acts to eliminate both extremes from a range of phenotypes.

stamen (L. thread) The part of the flower that contains the pollen. Consists of a slender filament that supports the anther. A flower that produces only pollen is called staminate and is functionally male.

steroid (Gr. *stereos*, solid + L. *ol*, from oleum, oil) A kind of lipid. Many of the molecules that function as messengers and pass across cell membranes are steroids, such as the male and female sex hormones and cholesterol.

steroid hormone A hormone derived from cholesterol. Those that promote the development of the secondary sexual characteristics are steroids.

stigma (Gr. mark) A specialized area of the carpel of a flowering plant that receives the pollen.

stipules Leaflike appendages that occur at the base of some flowering plant leaves or stems.

stoma, *pl.* **stomata** (Gr. mouth) A specialized opening in the leaves of some plants that allows carbon dioxide to pass into the plant body and allows water vapor and oxygen to pass out of them.

stroma (Gr. *stroma*, anything spread out) In chloroplasts, the semiliquid substance that surrounds the thylakoids and that contains the enzymes needed to assemble organic molecules from CO_2.

style (Gr. *stylos*, column) In flowers, the slender column of carpel tissue that arises from the top of the ovary and through which the pollen tube grows.

substrate (L. *substratus*, strewn under) A molecule on which an enzyme acts.

substrate-level phosphorylation The generation of ATP by coupling its synthesis to a strongly exergonic (energy-yielding) reaction.

subunit vaccine A type of vaccine created by using a subunit of a viral protein coat to elicit an immune response; may be useful in preventing viral diseases such as hepatitis B.

succession In ecology, the slow, orderly progression of changes in community composition that takes place through time. Primary succession occurs in nature on bare substrates, over long periods of time. Secondary succession occurs when a climax community has been disturbed.

sugar Any monosaccharide or disaccharide.

surface tension A tautness of the surface of a liquid, caused by the cohesion of the liquid molecules. Water has an extremely high surface tension.

surface-to-volume ratio Describes cell size increases. Cell volume grows much more rapidly than surface area.

swim bladder An organ encountered only in the bony fish that helps the fish regulate its buoyancy by increasing or decreasing the amount of gas in the bladder via the esophagus or a specialized network of capillaries.

symbiosis (Gr. *syn*, together with + *bios*, life) The condition in which two or more dissimilar organisms live together in close association; includes parasitism, commensalism, and mutualism.

synapse (Gr. *synapsis*, a union) A junction between a neuron and another neuron or muscle cell. The two cells do not touch. Instead, neurotransmitters cross the narrow space between them.

synapsis (Gr. *synapsis*, contact, union) The close pairing of homologous chromosomes that occurs early in prophase I of meiosis. With the genes of the chromosomes thus aligned, a DNA strand of one homologue can pair with the complementary DNA strand of the other.

syngamy (Gr. *syn*, together with + *gamos*, marriage) Fertilization. The union of male and female gametes.

systolic pressure A measurement of how hard the heart is contracting. When measured during a blood pressure reading, ventricular systole (contraction) is what is being monitored.

T

T cell A type of lymphocyte involved in cell-mediated immune responses and interactions with B cells. Also called a T lymphocyte.

taxon, *pl.* **taxa** (Gr. *taxis*, arrangement) A group of organisms at a particular level in a classification system.

taxonomy (Gr. *taxis*, arrangement + *nomos*, law) The science of the classification of organisms.

telophase (Gr. *telos*, end + *phasis*, form) The phase of cell division during which the spindle breaks down, the nuclear envelope of each daughter cell forms, and the chromosomes uncoil and decondense.

tendon (Gr. *tenon*, stretch) A strap of connective tissue that attaches muscle to bone.

tertiary structure of a protein The three-dimensional shape of a protein. Primarily the result of hydrophobic interactions of amino acid side groups and, to a lesser extent, of hydrogen bonds between them. Forms spontaneously.

testcross A cross between an individual of unknown genotype and a recessive homozygote. A procedure used to determine a dominant individual's genotype.

testis, *pl.* **testes** (L. witness) In animals, the sperm-producing organ.

theory (Gr. *theorein*, to look at) A well-tested hypothesis supported by a great deal of evidence.

thigmotropism (Gr. *thigma*, touch + *trope*, a turning) The growth response of a plant to touch.

thorax (Gr. a breastplate) The part of the body between the head and the abdomen.

threshold The minimum amount of stimulus required for a nerve to fire (depolarize).

thylakoid (Gr. *thylakos*, sac + *-oides*, like) A flattened, saclike membrane in the chloroplast of a eukaryote. Thylakoids are stacked on top of one another in arrangements called grana and are the sites of photosystem reactions.

tissue (L. *texere*, to weave) A group of similar cells organized into a structural and functional unit.

totipotent (L. *totus*, entire + *potent*, to be able, have power) The condition in which a cell has the potential to form any body tissue or an entire organism.

trachea, *pl.* **tracheae** (L. windpipe) In vertebrates, the windpipe.

tracheid (Gr. *tracheia*, rough) An elongated cell with thick, perforated walls that carries water and dissolved minerals through a plant and provides support. Tracheids form an essential element of the xylem of vascular plants.

transcription (L. *trans*, across + *scribere*, to write) The first stage of gene expression in which the RNA polymerase enzyme synthesizes an mRNA molecule whose sequence is complementary to the DNA.

transfer RNA (tRNA) A class of small RNAs (about 80 nucleotides) with two functional sites; at one site, an "activating enzyme" adds a specific amino acid, while the other site carries the nucleotide triplet (anticodon) specific for that amino acid.

translation (L. *trans*, across + *latus*, that which is carried) The second stage of gene expression in which a ribosome assembles a polypeptide, using the mRNA to specify the amino acids.

translocation (L. *trans*, across + *locare*, to put or place) In plants, the process in which most of the carbohydrates manufactured in the leaves and other green parts of the plant are moved through the phloem to other parts of the plant.

transpiration (L. *trans*, across + *spirare*, to breathe) The loss of water vapor by plant parts, primarily through the stomata.

transposon (L. *transponere*, to change the position of) A DNA sequence carrying one or more genes and flanked by insertion sequences that confer the ability to move from one DNA molecule to another. An element capable of transposition (the changing of chromosomal location).

trichome In plants, a hairlike outgrowth from an epidermal cell; glandular trichomes secrete oils or other substances that deter insects.

trisomic Describes the condition in which an additional chromosome has been gained due to nondisjunction during meiosis, and the diploid embryo therefore has three of these autosomes. In humans, trisomic individuals may survive if the autosome is small; Down syndrome individuals are trisomic for chromosome 21.

trophic level (Gr. *trophos*, feeder) A step in the movement of energy through an ecosystem.

tropism (Gr. *trop*, turning) A plant's response to external stimuli. A positive tropism is one in which the movement or reaction is in the direction of the source of the stimulus. A negative tropism is one in which the movement or growth is in the opposite direction.

tumor-suppressor gene A gene that normally functions to inhibit cell division; mutated forms can lead to the unrestrained cell division of cancer, but only when both copies of the gene are mutant.

turgor pressure (L. *turgor*, a swelling) The pressure within a cell that results from the movement of water into the cell. A cell with high turgor pressure is said to be turgid.

U

unicellular Composed of a single cell.

urea (Gr. *ouron*, urine) An organic molecule formed in the vertebrate liver. The principal form of disposal of nitrogenous wastes by mammals.

urethra The tube carrying urine from the bladder to the exterior of mammals.

uric acid Insoluble nitrogenous waste products produced largely by reptiles, birds, and insects.

urine (Gr. *ouron*, urine) The liquid waste filtered from the blood by the kidneys.

V

vaccination The injection of a harmless microbe into a person or animal to confer resistance to a dangerous microbe.

vaccine A substance injected into the body that is designed to illicit an immune response and to provide protection against a virus. There are five general ways to present antigens to patients:

- *Recombinant vector vaccines* have genes of the infectious virus inserted into a modified cold virus.
- *Live attenuated virus vaccines* are made with weakened forms of the virus that are unable to cause an infection but can still illicit an immune response.
- *Inactivated virus vaccines* are made of heat-killed virus particles.
- *Subunit vaccines* use a piece of the virus, like it's capsid or surface proteins, to induce a response.
- *Nucleic acid vaccines* use DNA or mRNA from the infectious agent. The body uses the templates to form viral proteins which illicit the response.

vacuole (L. *vacuus*, empty) A cavity in the cytoplasm of a cell that is bound by a single membrane and contains water and waste products of cell metabolism. Typically found in plant cells.

variable Any factor that influences a process. In evaluating alternative hypotheses about one variable, all other variables are held constant so that the investigator is not misled or confused by other influences.

vas deferens In mammals, the tube carrying sperm from the testes to the urethra.

vascular bundle In vascular plants, a strand of tissue containing primary xylem and primary phloem. These bundles of elongated cells conduct water with dissolved minerals and carbohydrates throughout the plant body.

vascular cambium In vascular plants, the meristematic layer of cells that gives rise to secondary phloem and secondary xylem. The activity of the vascular cambium increases stem or root diameter.

vascular tissue Containing or concerning vessels that conduct fluid.

vein (L. *vena*, a vein) (1) In plants, a vascular bundle forming a part of the framework of the conducting and supporting tissue of a stem or leaf. (2) In animals, any blood vessel that carries blood toward the heart.

ventral (L. *venter*, belly) Refers to the bottom portion of an animal. Opposite of dorsal.

ventricle A muscular chamber of the heart that receives blood from an atrium and pumps blood out to either the lungs or the body tissues.

vertebrate An animal having a backbone made of bony segments called vertebrae.

vesicle (L. *vesicula*, a little ladder) Membrane-enclosed sacs within eukaryotic cells.

vessel element In vascular plants, a typically elongated cell, dead at maturity, that conducts water and solutes in the xylem.

villus, *pl.* **villi (L. a tuft of hair)** In vertebrates, fine, microscopic, fingerlike projections on epithelial cells lining the small intestine that serve to increase the absorptive surface area of the intestine.

virus (L. slimy liquid, poison) Tiny infectious particles. Although referred to as live or dead, in fact, no virus is alive. A virus is a combination of two chemicals, a nucleic acid (DNA or RNA) encased within a protective protein coat. Viruses can only reproduce by injecting their genetic material into the cells of living creatures.

vitamin (L. *vita*, life + *amine*, of chemical origin) An organic substance that the organism cannot synthesize but is required in minute quantities by an organism for growth and activity.

viviparous (L. *vivus*, alive + *parere*, to bring forth) Refers to reproduction in which eggs develop within the mother's body and young are born free-living.

voltage-gated channel A transmembrane pathway for an ion that is opened or closed by a change in the voltage, or charge difference, across the cell membrane.

W

water vascular system The system of water-filled canals connecting the tube feet of echinoderms.

wood Accumulated secondary xylem. Heartwood is the central, nonliving wood in the trunk of a tree. Hardwood is the wood of dicots, regardless of how hard or soft it actually is. Softwood is the wood of conifers.

X

xylem (Gr. *xylon*, wood) In vascular plants, a specialized tissue, composed primarily of elongate, thick-walled conducting cells, that transports water and solutes through the plant body.

Y

yolk (O.E. *geolu*, yellow) The stored substance in egg cells that provides the embryo's primary food supply.

Z

zygote (Gr. *zygotos*, paired together) The diploid ($2n$) cell resulting from the fusion of male and female gametes (fertilization).

Index

A

AAV (adeno-associated virus) vector, 246f, 247, 247f
Abdominal cavity, 462
ABO blood groups, 180, 180f, 541, 541f
Abstinence, 602
Accessory digestive organs, 502, 502f, 502f, 509, 509f, 510f
Acetylcholine (ACh), 552
Acetyl-CoA, 123, 123f, 124f
Acid rain, 45, 45f, 444
Acid reflux, 44, 505
Acids, 44. *See also specific types*
Acid, stomach, 505–506
Acne, 526
Acoelomates, 352, 353f, 354, 354f
Acquired immunodeficiency syndrome. *See* HIV/AIDS
Acrosome, 592
ACTH (adrenocorticotropic hormone), 580, 580f
Actin, 80, 473, 473f
Actin filaments, 80–81, 471
Actinopoda, 315t
Action potentials, 549–550, 550f
Activating enzymes, 213
Activation energy, 94, 94f
Activators, 97, 97f, 219, 219f, 220, 220f
Active immunity, 537, 537f
Active sites of enzymes, 95, 95f
Active transport, 87, 100t
Adaptation
    life history, 389, 389t
    of plants to terrestrial life, 324–325
    within populations, 269–272
Adaptive zone, 293
Added sugars, 500
Addiction, 546–547, 556
Adenine, 56, 56f, 57, 192, 192f
Adeno-associated virus (AAV) vector, 246f, 247, 247f
Adenosine diphosphate (ADP), 99
Adenosine triphosphate (ATP),
    photosynthesis production of, 109, 112f
    cellular activities power by, 87, 100t
    cellular respiration producing, 99, 120–124, 121f
    chemiosmosis production of, 112, 126
    electrons for production of, 125–126, 125f
    muscle contraction and, 473, 473f
    photosynthesis production of, 99
    structure of, 99, 99f
Adenovirus, 246, 246f
ADH (antidiuretic hormone), 520, 520f, 574, 580
Adhesion, 43, 621, 621f
Adipose tissue, 465
ADP (adenosine diphosphate), 99
Adrenal cortex, 583
Adrenal glands, 583, 583f

Adrenaline, 517, 517f
Adrenal medulla, 583
Adrenocorticotropic hormone (ACTH), 580, 580f
Adult stem cells, 244
Adventitious plantlets, 628, 628f
Adventitious roots, 615
Afferent neurons. *See* Sensory neurons
African Americans, sickle-cell disease in, 176, 187, 270
African elephants, 291f, 495, 495f
African weaver birds, 438, 438f
Agalychnis callidryas (red-eyed tree frog), 374f
Agar, 635
Agaricus bisporus (button mushrooms), 318f
Age
    of Earth, 254
    maternal, 182–183, 182f
    paternal, 197, 197f
    of trees, 617, 617f
Age distribution, of populations, 390, 390f
Age structure, of populations, 390, 390f
Agglutination, 541, 541f
Aging, 151
Agricultural Improvement Act, 339
Agriculture
    chemical pollution from, 414, 444, 444f
    diet and, 130–131
    genetic engineering and, 226–227, 226f, 227f, 230, 230f, 242–244, 242f, 243t, 244f, 253, 253f
    global warming and, 445
    herbicides and, 635
    organic, 500
    topsoil in, 324, 451–452
Agrostis gigantea, 249
Agrostis stolonifera, 249
AIDS. *See* HIV/AIDS
Air pollution, 45, 271–272, 444
Albinism, 172t
Albumin, 483
Alcohol
    energy drinks mixed with, 127
    fetal alcohol syndrome and, 599
Alcoholic fermentation, 304
Aldosterone, 520, 582, 583
Ale, 304
Algae, 420f
    colonies of, 313, 313f
    impact on humans, 314
    life cycle of, 325
    multicellular, 313
    in ocean fertilization, 105
    in ponds, 64–65
Alkaptonuria, 172t
Alleles, 172–173
    environmental effects on, 179, 179f
    frequencies of, 263–264

Allen, Arthur, 403
Allergen, 542, 542f
Allergies, 542, 542f
Alligators, 424
Allison, James, 532
Allometric growth, 601
Allosteric enzymes, 97, 97f
Alper, T., 58
Alternation of generations, 325, 325f, 326, 326f
Alternative hypothesis, 25
Alternative splicing, 216, 228
Alticus arnoldorum (blenny), 370
Alveolata, 315t
Alveoli, 488, 488f, 490
Amanita, 316f
Amanita muscaria, 318f
Ambulocetus (walking whale), 258, 258f
American Cancer Society, 11
American Heart Association, 119, 499
Amines, 574
Amino acids, 53, 53f
    aromatic, 238
    essential, 499
Amish, 265f
Ammonia, 521, 521f
Ammonites, 368, 368f
Amnion, 597
Amniotic eggs, of reptiles, 375, 375f
Amoebas, 314, 314f
Amoebic parasites, 312
Amoebozoa, 315t
Amphibians
    characteristics of, 373–374
    evolution of, 368, 373–374, 374f
    extinctions and global decline of, 450
    invasion of land by, 368
    nitrogenous waste elimination by, 521, 521f
    respiration of, 488, 488f
Amygdala, 554–555, 554f
Amygdalin, 631
Amylases, 304
Anabolic steroids, 48–49, 578
Analogous structures, 259
Anaphase I, in meiosis I, 157, 157f, 181, 181f
Anaphase II, meiosis II, 159, 159f
Andro, 49
Androstadienone (AND), 429
Aneuploidy, 181
Angiosperms, 327t
    embryonic development of, 631, 631f
    evolution of, 326, 326f, 332, 336, 338
    life cycle of, 338, 338f
    photoperiodism in, 636–637, 636f

Angiosperms (*Cont.*)
   pollination of, 630
   reproduction by, 169, 628-633, 628*f*-632*f*
   structure of, 336, 336*f*
Angiotensin converting enzyme 2 (ACE2), 310
Animal evolution
   bilateral symmetry in, 346, 351-353, 352*f*
   body cavity in, 354-362
   embryonic development in, 344, 344*t*, 363-367, 363*f*
   key transitions in body plan, 346, 347*f*
   simplest animals in, 348-351, 348*f*-350*f*
   of vertebrates, 259, 259*f*, 368-379
Animals (Animalia)
   body plan of, 462-463
   body plan of, key transitions in, 346, 347*f*
   cells of, 70*f*, 344, 344*t*
   cloning of, 238-239, 238*f*-239*f*
   cognition by, 433, 433*f*
   diversity of, 344, 345*t*
   extinction factors for, 449, 449*f*
   general features of, 344, 344*t*-345*t*
   kingdom of, 18*f*, 288, 289*t*
   learning by, 432
   multicellularity of, 344, 344*t*
   oldest, 337
   sexual life cycles in, 155, 155*f*
   viruses infecting, 308
Anions, 38, 38*f*
Annelids (Annelida), 356-357, 357*f*
   circulatory system of, 478, 478*f*
   digestive system of, 501, 501*f*
Antarctica, 27
Anteaters, 377*f*
Antenna complex, 110-112, 110*f*
Anterior, definition of, 351
Anterior pituitary, 580, 580*f*
Anthers, 336, 336*f*, 629, 629*f*
Anthocerophyta, 327*t*, 328
Anthophyta, 327*t*, 332
Anthracotheres, 258, 258*f*
Antibiotics, 97, 223, 300, 301
Antibodies, 532
   definition of, 536
   diversity of, 536, 536*f*
   maternal, 535
   in medical diagnosis, 541
   monoclonal, 525
   production of, 535-536, 535*f*
   structure of, 536, 536*f*
Anticodons, 213, 213*f*
Antidiuretic hormone (ADH), 520, 520*f*, 574, 580
Antigen-presenting cells, 533-534, 533*f*, 534*f*
Antigens, 532
   in allergies, 542, 542*f*
   in blood typing, 541, 541*f*
   cellular immune response and, 533-534, 534*f*
   humoral immune response and, 535-536, 535*f*
   immune specificity for, 545
Antihistamines, 542
Ants
   cooperation in, 23
   social behavior of, 438
   symbiosis in, 395
Anura, 374
Anus, 508
Aorta, 480, 485*f*, 486
Apes, 256
Apical meristems, 608, 608*f*, 613, 613*f*

Apicomplexans, 315*t*
Apis mellifera mellifera (killer bees), 382-383, 382*f*, 383*f*, 437, 437*f*
Apoda, 374
Appendicular skeleton, 470, 470*f*
Appendix, 259, 510*f*
Apples, 632, 632*f*
Aquaporins, 82, 83
Aquifers, 411, 452
Arachnids, 360*f*
Archaea
   cells of, 69
   domain of, 30, 30*f*, 288-289, 289*t*, 290, 290*f*
   kingdom of, 18*f*, 288-289, 289*t*
Archaeopteryx, 376, 376*f*
Archaeplastida, 315*t*
Arctic fox, 179, 179*f*
Arctic ice, 442-443
Aristotle, 280
Arithmetic progression, 255, 255*f*
Armadillo, 254, 254*f*
Armstrong, Lance, 49
Aromatic amino acids, 238
Arrowgrass (*Triglochin maritima*), 341
Arrows, in textbooks, 5, 5*f*, 18
Arteries, 478, 480, 480*f*, 481, 481*f*
Arterioles, 480, 480*f*
Arthritis, 542
Arthropods (Arthropoda), 360*f*, 361
   circulatory system of, 478, 478*f*
   exoskeletons of, 360, 361*f*, 469, 469*f*
   invasion of land by, 368
   reproduction by, 152, 588
Artificial selection, 22, 265
Ascaris lumbricoide, 359
Ascomycota, 318*t*
Asexual reproduction, 154*f*
   by angiosperms, 628, 628*f*
   definition of, 154
   by echinoderms, 365
   by fungi, 317
   by invertebrates, 152
   mutations in sexual reproduction compared to, 153
   by protists, 313, 313*f*, 588, 588*f*
Aspirin, 556
Assigned reading, 7
Association neurons, 468, 468*t*, 549
Associative learning, 432
Asthma, 491
Atherosclerosis, 486
Athletes
   anabolic steroid use by, 48-49
   exercise and, 119, 475
Atkins diet, 118-119
Atlantic bluefin tuna, 389
Atmosphere, history of, 296-297
Atmospheric circulation, 415, 415*f*
Atomic numbers, 36, 37*t*
Atoms
   as atoms and isotopes, 38, 38*f*
   energy in, 37, 37*f*
   in hierarchy of complexity, 20, 20*f*
   structure of, 36, 36*f*
ATP. *See* Adenosine triphosphate
ATP-ADP cycle, 99, 99*f*
ATP synthase, 112, 126
Atrioventricular (AV)

   nodes, 487, 487*f*
   valves, 485*f*, 486
Atrium, 485-487, 485*f*
Attendance, learning and, 6
Australia mountain ash (*Eucalyptus regnans*), 626
Australopithecus, 378
Autism, paternal age and, 197, 197*f*
Auto-deer collisions, 399
Autoimmune diseases, 542
Autonomic (involuntary) nervous system, 558-559, 559*f*
Autosomal DNA testing, 182
Autosomes, 181
Autotrophs, 324, 404
Auxin, 635, 635*f*
AV (atrioventricular) nodes, 487, 487*f*
AV (atrioventricular) valves, 485*f*, 486
Axial skeleton, 470, 470*f*
Axils, in leaves, 615, 615*f*
Axon hillocks, 552, 552*f*
Axons
   junction with cells (synapse), 551-552, 551*f*, 552*f*
   nerve impulse along, 468, 468*f*, 548, 548*f*, 549-550, 550*f*
   size of, 67, 468
Axopodia, 312
AZT, 543

B

Bacillus thuringiensis (Bt), 238
Backbone, vertebrate, 367
Bacteria
   cells of, 67, 67*f*, 300-302
   as decomposers, 405*f*
   domain of, 30, 30*f*, 289, 289*t*, 290, 290*f*
   flagella of, 302
   genetic engineering of, 234
   gram-negative, 300
   gram-positive, 300
   in human gut, 496-497
   intestinal, 508
   kingdom of, 18*f*, 288-289, 289*t*
   in nitrogen cycle, 414
   obesity and, 496-497
   in ocean ecosystems, 419, 419*f*
   pathogenic, 526
   restriction enzymes in, 231
   viruses infecting, 305, 308, 308*f*
Bacteriophages, 305, 305*f*, 308, 308*f*
Balance, 561, 561*f*
Baldness, 172*t*
Bar graphs, 14, 14*f*
Bark, 617
Barnacles, 360, 361*f*, 392-393, 392*f*
Baroreceptors, 560, 560*f*
Barrel sponges, 348*f*
Basal bodies, 81
Basal metabolic rate (BMR), 498
Baseball, anabolic steroid use in, 48-49
Bases, 44
Base substitution mutations, 196*f*, 198*t*
Basidiomycota, 319*t*
Basilosaurus, 258, 258*f*
Basophils, 484, 484*f*
Batrachochytrium dendrobatidis, 450
Bats
   keystone species of, 456, 456*f*
   pollination by, 394*f*

Bats (*Cont.*)
   vampire, 476-477, 477*f*
B cells, 484, 484*f*, 531-532, 531*t*, 535-537, 535*f*, 536*f*
Beagle, HMS, 252-254, 252*f*, 253*f*
Bean seeds, 332-333, 332*f*
Bears, 405*f*, 406, 442-443, 442*f*, 443*f*
Beason, William, 237
Beer making, 128, 304
Bees
   flower color and pollination by, 630, 630*f*, 639, 639*f*
   killer, invasion of, 382-383, 382*f*, 383*f*
   parthenogenesis in, 152, 588
   pheromones of, 438
   social behavior of, 438
   societies of, 438
   waggle dance of, 437, 437*f*
Beetles, 384*f*
Behavior
   courtship, 435*t*
   definition of, 430
   evolutionary forces shaping, 434-436
   foraging, 434, 434*f*, 435*t*, 439
   genetic effects on, 431, 431*f*
   instinctive (innate), 431, 431*f*
   learning influencing, 432-433
   migratory, 435*t*, 436, 436*f*
   parental care, 435*t*
   social, 429*f*, 435*t*, 438-439, 439*f*
   territorial, 434, 434*f*, 435*t*
Behavioral genetics, 431
Behavioral isolation, 273*t*, 274
Beneden, Pierre-Joseph van, 154
Benzo(a)pyrene, 147
Berries, 632, 632*f*
Best-fit lines, 14, 277
Beta-hemoglobin, 176, 269
β-Oxidation, 129
Bicarbonate, 483, 509
Bicuspid (mitral) valve, 485-486, 485*f*
Bilateral symmetry, 346, 351-353, 352*f*
Bile salts, 507
Binary fission, 136, 136*f*, 302
Binding sites of enzymes, 95, 95*f*
Binomial system of classification, 280-281, 280*f*
Biochemical pathways, 95-96, 96*f*
Biodiversity, 18, 18*f*
   ecosystems and, 288
   loss of, 449, 449*f*, 452, 452*f*
   in medicine, 288
   value of, 452
Biogenic amines, 552
Biogeochemical cycle, 410
Biological magnification, 444, 444*f*
Biological processes, illustrations of, 8, 8*f*
Biological species concept, 273, 283
Biology
   definition of, 18
   major theories of, 28, 28*f*-31*f*, 30
   media coverage of, 11, 11*f*
   scientific processes in, 24-26, 24*f*, 25*f*
   themes of, 22-23
Bioluminescence, 419, 419*f*
Biomass
   of ecosystems, 406
   pyramids of, 409, 409*f*
Biomes, 384, 421*f*, 422-425, 422*f*
Biosphere, 384
Biosynthesis, ATP for, 100*t*

Bird flu, 309, 540
Birds. *See also specific types*
   body temperature of, 517, 523
   characteristics of, 376
   classification of, 285, 285*f*
   evolution of, 376, 376*f*
   feathers of, 376, 376*f*
   flower color and pollination by, 630, 630*f*
   gizzard of, 503
   instinctive behavior and, 430, 431*f*
   magnetic compass of, 571
   migration of, 436, 436*f*
   nitrogenous waste elimination by, 521, 521*f*
   problem solving by, 433, 433*f*
   singing of, 430, 430*f*, 434, 434*f*
   societies of, 438, 438*f*
   territorial behavior of, 434, 434*f*
   viral diseases in, 306-307, 306*f*, 307*f*
Birth, 600-601, 601*f*
   control, 387, 602-603, 603*f*
   control pills, 602, 603*f*
   defects, 156
   rates, 390
Bison, 423, 423*f*
Bisphenol A (BPA), 577
Biston betularia (peppered moth), 271-272
Bithorax mutations, 196*f*
Bivalves, 355, 355*f*
Black-bellied seedcracker finch (Pyrenestes ostrinus), 268, 268*f*
Black cherry trees, 34-35
Blackheads, 526
Black rhinoceroses, 455-456, 455*f*
Blades of leaves, 618
Blastocladiomycota, 319*t*
Blastocoel, 597
Blastocysts, 244, 596, 596*f*, 597, 597*f*
Blastopores, 363, 363*f*
Blastula, 363, 363*f*
Blatella germanica (cockroaches), 389*f*
Blenny *(Alticus arnoldorum),* 370
Blinking, 472
BLM protein, 158
Blood, 465
   cells, 484, 484*f*
   composition of, 481*f*
   CO_2 transport in, 479
   fetal, 600, 600*f*
   flow of, from heart, 480-482
   functions of, 479, 479*f*
   glucose, regulation of, 513, 513*f*, 517, 517*f*
   hemoglobin in, 479, 481*f*, 484, 492
   hormone transport in, 479
   oxygen transport in, 479, 481*f*, 485-486, 485*f*
   pH of, 44
   pH of fish, 133, 133*f*
   plasma of, 481*f*, 483, 483*f*
   vampire bats feeding on, 476-477
Blood chemistry, sensory receptors for, 560
Blood clotting, 262, 262*f*, 479, 483, 483*f*, 484
Blood groups, ABO, 180, 180*f*, 541, 541*f*
Blood pressure, 486, 486*f*, 560, 560*f*
Blood typing, 541, 541*f*
Blood vessels, 478, 480, 480*f*, 481*f*
Bloom's syndrome, 158
Blue jays, 406
BMI (body mass index), 498, 499*f*
BMR (basal metabolic rate), 498

Body cavity, 346, 354-362, 462
Body mass index (BMI), 498, 499*f*
Body temperature
   homeostasis in, 516-517, 517*f*
   immune response of, 528, 529
   regulation of, 479, 516-517, 517*f*, 523
   sweat and, 517, 517*f*
Bonds, Barry, 49
Bone
   calcium and health of, 461
   composition of, 465, 466
   fetal formation of, 600
   osteoporosis and, 460-461, 460*f*, 466, 466*f*
   remodeling of, 466
   structure of, 466, 466*f*
   types of, 466, 466*f*
Bony fishes, 293, 293*f*, 372-373, 373*f*
Bormann, Herbert, 426
Bottled water, 444
Bovine spongiform encephalopathy (BSE), 58
Bowman's capsule, 519, 519*f*
BPA (bisphenol A), 577
Brachydactyly, 172*t*
Bracts, 618*f*
Bradford, David, 450
Bradley, Bill, 562
Brain
   addiction and, 556
   anatomy of, 553-555, 553*f*-555*f*
   during sleep, 2-3
   functional areas of, 554*f*, 554*f*
   functioning of, 553-555
   hominid, size of, 342, 378, 378*f*
   language, learning, memory and, 555, 555*f*
   navigation and, 562-563
   stem, 554-556
Branchiostoma lanceolatum, 366*f*
BRCA1, 577
BRCA2, 577
Bread making, 128
Breast cancer, 143, 577, 577*f*
Breastfeeding, 601, 601*f*
Breast milk, 579-580, 601
Breathing, mechanics of, 491, 491*f*
Brewer's yeast, 304
Bristlecone pine trees, 337
Bristle worms, 357*f*
Bronchioles, 490
Bronchus, 490, 490*f*
Brown algae, 313, 315*t*
Bryophyta, 327*t*
BSE (bovine spongiform encephalopathy), 58
Bubble links, in illustrations, 8
Bubble model, 297, 297*f*
Budding, 588
Buds, 615, 615*f*
Buffers, 44
Build-a-baby process, 208-209
Bumpus, H. C., 267
Bundle of His, 487
Bundle-sheath cells, 115, 115*f*
Burt, Austin, 203
Buttercups *(Ranunculus),* 614*f*, 615*f*
Butterflies
   body temperature of, 517
   flower color and pollination by, 630, 630*f*
   in food chain, 408
Button mushrooms *(Agaricus bisporus),* 318*f*

C

Cactus finches, 254f
Caecilians, 373–374
Caenorhabditis elegans, 221, 359
Caesarian section (C-section), 601
Caffeine, 127
Calcitonin, 582–583, 583f
Calcium
    atomic number and mass number of, 37t
    bone health and, 461
    muscles and, 582–583
    plant absorption of, 324
    PTH and, 582–583, 583f
Callorhinus ursinus (fur seals), 386, 386f
Calment, Jeanne, 337
Calories, 119, 500
CAM (crassulacean acid metabolism), 115
Cambium
    cork, 608, 616f, 617
    procambium, 613, 613f
    vascular, 608, 616, 616f
Cambrian period, 368
Camels, 423
cAMP (cyclic AMP), 578
Campbell, Keith, 242
Campephilus principalis (ivory-billed woodpecker), 402–403, 403f
Camptodactyly, 172t
Canada lynx *(Lynx canadensis),* 397, 397f
Cancer. *See also* Lung cancer
    adenovirus and, 246
    breast, 143, 577, 577f
    cell cycle in, 143–147
    cervical, 586–587, 603
    and COVID, 148
    CRISPR treatment for, 202
    e-cigarettes and, 149
    environmental factors in, 146
    HPV causing, 586
    incidence of, 146, 146f
    metastases of, 143
    mutations causing, 143, 146
    pancreatic, 581
    prevention of, 144–145, 147
    prostate, 592
    radioactive tracers and, 38, 38f
    skin, 134–135, 135f
    vaccines against, 235
    viruses in, 143
Canids, 250–251
Canines (teeth), 503, 503f
Canis, 250
Cannabis, 114
Capillaries, 478, 480, 480f, 482, 482f
Capsids, 305f, 306
Capsule
    bacterial, 300
    prokaryotic, 69
Captive propagation programs, 455, 455f
Carbohydrates, 50
    blood glucose and, 517
    diets low in, 118–119
    energy from, 98
    as food source, 498
    formation of, 51f
    functions of, 59, 60t
    structure of, 59, 59f
    transportation in plants, 623, 623f
    types of, 59, 59f, 60t
Carbon
    atomic number and mass number of, 36, 37t
    in covalent bonds, 40
    isotopes of, 38, 38f, 47
Carbon-12, 47
Carbon cycle, 410, 412, 412f
Carbon dioxide (CO_2), 117, 381
    blood transport of, 479
    coral calcification, 381f
    in carbon cycle, 412, 412f
    in cellular respiration, 123–124, 412
    as fermentation product, 128
    in global warming, 12, 12f, 104, 412, 416, 445, 445f, 446
    in mass extinctions, 369
    respiration transport of, 492
    sensory receptors and, 560
    sequestration of, 104
Carbon fixation, 113, 115, 115f
Carbon footprint, 412, 447, 448
Carboniferous period, 368
Carcinogens, 577
Carcinomas, 143, 143f
Cardiac muscles, 467, 471
Cardiovascular disease, 72
Cardiovascular systems, 480–482
Caribbean reef sharks, 465f
Carnivores, 405f, 406, 408, 501, 503
    top, 409
Carpels, 332, 336, 336f, 629, 629f
Carrier proteins, 84, 84f
Carroll, Lewis, 153
Carrying capacity, 386, 386f, 389, 453
Carter, Jimmy, 532
Cartilage, 465, 465f
Casparian strip, 613, 614f
Castes, 438
Catalysis, 94
Caterpillars, 35, 405
Cations, 38, 38f
Cats
    cloning of, 241
    family tree of, 285, 286f
    genetic change in populations of, 263, 263f
Causation, correlation and, 13
Cavities, tooth, 503
cDNA (complimentary DNA), 231, 231f
CD4 receptor protein, 525, 525f
Cedar trees, 327t
Cell body of neurons, 468, 468f, 468t, 548, 548f
Cell cycle
    in cancer, 143, 147
    prokaryotic, 136, 136f
Cell division. *See also* Meiosis; Mitosis
    in cancer, 143, 147
    in eukaryotes, 140–142, 140f–141f, 300t
    in prokaryotes, 136, 136f, 300t
Cell plate, 142
Cells. *See also* Eukaryotic cells
    aging of, 151
    of animals, 70f
    ATP as energy for, 87, 99, 100t
    B, 484, 484f, 531–532, 531t, 535–537, 535f, 536f
    blood, 484, 484f
    bundle-sheath, 115, 115f
    in cell theory, 28, 28f, 66–67
    chief, 505, 506f
    collenchyma, 609, 609f
    companion, 612
    in culture, 165
    daughter, 136, 136f, 160
    definition of, 19
    diploid, 138, 154, 154f
    discovery of, 28, 66
    enzyme regulation by, 96–97, 97f
    epithelial, 464
    fat, 465
    germ-line, 155, 160
    glial, 468
    grid, 563
    guard, 324f, 610, 610f, 621, 622f
    haploid, 154, 154f
    in hierarchy of complexity, 20, 20f
    mast, 531, 531t, 542, 542f
    mesophyll, 107, 115, 115f, 621–622
    movement of, 81
    organization of, 19, 19f, 20, 20f
    origin of first, 296–297
    origin of term for, 66
    parenchyma, 609, 609f, 613
    parietal, 505, 506f
    place, 562–563
    of plants, 71f
    plasma, 531, 531t, 532, 535–536, 535f
    prokaryotic, 69, 69f, 300–302, 300t
    protein recycling by, 89
    sclerenchyma, 609, 609f
    Sertoli, 591f
    sieve, 612
    size of, 66, 66f, 67, 300t
    somatic, 155
    structure of, 66, 66f
    T, 484, 484f, 531–537, 531t, 534f, 535f, 543
    visualizing, 68, 68f
Cell surface antigens, 180
Cell surface proteins, 73
Cell theory, 28, 28f, 66–67
Cellular counterattack, 526, 529–530, 529f, 530f
Cellular immune response, 533–534, 534f
Cellular respiration, 78
    ATP production by, 99, 120–124, 121f
    CO_2 in, 123–124, 412
    energy and, 98
    of fats and lipids, 129, 129f
    of glucose, 120–126
    glycolysis in, 120, 121f, 122, 122f
    overview of, 120–121, 121f
    oxidative, 123–125, 408, 488
    of proteins, 129, 129f
Cellular slime molds, 315t
Cellulose, 59, 59f, 60t
Cell walls
    of animals, 344, 344t
    of bacteria, 300–301
    collar, 348
    eukaryotic, 71, 71f, 75t
    flame, 352–353
    of fungi, 316
    of plants, 609, 609f
Cenozoic era, 369
Centers for Disease Control and Prevention, 572, 586, 605
Centipedes, 360

Central canal of bone, 466
Central dogma, 210, 210f, 215, 215f
Central nervous system (CNS), 548-549, 553-557
Central vacuoles, 71, 71f, 77, 77f
Centrioles, 70f, 71
Centromeres, 138, 138f
Centrosomes, 165
Cephalopods, 355, 355f
Cercidium floridum (Palo verde trees), 232, 333f
Cercozoans, 315t
Cerebellum, 553f, 555
Cerebral cortex, 553, 553f, 554f
Cerebral hemispheres, 553-555
Cerebrum, 553-554, 553f, 554f
Certainty, 25, 26
Cervical cancer, 586-587, 603
Cervix, 593f, 594, 594f
CF. *See* Cystic fibrosis
cf gene, 86, 178f, 246
CFCs. *See* Chlorofluorocarbons CFTR protein, 86
Chaparral, 425, 425f
Chargaff, Erwin, 192
Charophytes, 315t
Checkered whiptail lizard, 152, 153f
Chemical activation, ATP for, 100t
Chemical bonds, 123-124. *See also specific types*
Chemical digestion, 501
Chemically gated ion channels, 551-552
Chemical pollution, 411, 414, 444, 444f
Chemical reactions, 92, 94, 94f
Chemiosmosis, 112, 126
Chemistry
    atoms in, 36-37, 36f, 37f
    cyanide and, 34-35
    definition of, 36
    molecules in, 39-41
Chemoautotrophs, 300t
Cherry trees, 34-35
Chewing, 503
Chicken pox virus, 537, 537f
Chickens, 405
Chief cells, 505, 506f
Childbirth, 600-601, 601f
Chimpanzees, 306, 306f, 433
Chipmunks, 124f
Chitin, 59, 60t, 316, 360, 361f, 469
Chlamydia, 603, 605, 605f
Chlamydia trachomatis, 603, 605
Chlamydomonas, 111f, 315t
Chloride, 483, 552
Chlorine, 37t
Chlorofluorocarbons (CFCs), 12, 24-26
Chlorophyll, 108-112, 117
Chlorophyta, 313, 315t
Chloroplasts
    endosymbiosis and, 79
    eukaryotic, 71, 71f, 75t, 78-79, 79f
    in photosynthesis, 107-109
Choanocytes, 348
Choanoflagellates, 315t, 348
Choanoflagellida, 315t
Cholesterol, 61, 72, 500
Chondrocytes, 465
Chordates (Chordata), 366-367, 366f
Chorion, 597
Chromalveolata, 315t
Chromatin, 74, 74f, 139, 219-220, 220f
    chromosomes, 75t

coiling of, 139, 139f
discovery of, 138
in DNA repair hypothesis, 153
in gametes, 154
homologous, 138, 138f, 157, 157f, 160
human, 138, 139, 139f, 181-184
number of, 138
RNA in, 74, 139
structure of, 74, 139, 139f
in chromosomal theory of inheritance, 30, 30f
DNA replication and, 195
eukaryotic compared to prokaryotic, 300t
gene distribution among, 228-229
sex, 181-184
Chrysophyta, 315t
Chthamalus stellatus, 392-393, 392f
Chyme, 506, 509
Chytridiomycota, 319t
Chytrids, 450
Cichlids, 454
Cigarette smoking. *See* Smoking
Cilia
    ear, 561, 561f
    eukaryotic, 70f, 75t, 81, 81f
    immune function of, 528
    protist, 312
Ciliates, 315t
Circulatory systems
    closed, 478, 478f
    daily activity of, 472
    functions of, 479, 479f
    of humans, 485-487, 485f-487f
    mollusk, 355
    open, 478, 478f
    types of, 478, 478f
    of vertebrates, 463, 463f, 478-487
cis-retinal, 567, 567f
Cisternae, 77
Clades, 284
Cladistics, 284-285
Cladograms, 284-285, 284f, 285f
Clams, 355
Class (taxonomic rank), 282
Classical conditioning, 432
Classification of organisms, 280-281
Clean Air Act of 1963, 272
"Clean coal", 447
Clean meat movement, 511
Clear-cutting, 411, 411f, 426, 426f
Cleavage, 363, 596, 596f, 597, 597f
Cleavage furrow, 142, 142f
Climate, 416
    latitude and, 415, 415f
    Mediterranean, 417
    weather compared to, 416
Climate change, 117, 416. *See also* Global warming
Climax communities, 398
Cloaca, 278
Clonal selection, 537, 537f
Cloning
    ethics and, 246
    reproductive, 245-244, 245f-246f
Club mosses, 327t, 330, 330f
Cnidarians (Cnidaria), 349-351, 350f
    gastrovascular cavity of, 478, 478f, 501, 501f
    reproduction by, 588
Cnidocytes, 349, 351
CNS (central nervous system), 548-549, 553-557

CO_2. *See* Carbon dioxide
Coal, "clean,", 447
Cochlea, 565, 565f
Cockroaches (*Blatella germanica*), 389f
Coconuts, 632f
Codominance, 180, 180f
Codons, 212, 212f
Coelom, 354, 354f, 462
Coelomates, 354, 354f
Coevolution, 394-395, 394f, 395f
Coffee, 127
Cognition, animals and, 433, 433f
Cohesin, 160
Cohesion, 42t, 43, 43f, 621, 621f
Cohesion-adhesion-tension theory, 620-621, 621f, 625
Cohorts, 390
Coleoptile, 633
Colias eurytheme, 630f
Collagen, 52, 52f
Collar cells, 348
Collecting duct, 519, 519f
Collenchyma cells, 609, 609f
Collins, James, 450
Colon, 502, 502f, 502f, 508, 510f
Colonial organisms, 313, 313t
Color blindness, 172t
Colorectal cancer, 143
Color vision, 567-568, 567f
Colostrum, 601
Columnar epithelium, 464
Combination therapy, for HIV/AIDS, 543
Combustion, in carbon cycle, 412, 412f
Commensalism, 395, 395f
Common names, 281, 281f
Communities
    climax, 398
    coevolution in, 394-395, 394f, 395f
    competition in, 391-393
    definition of, 384, 391, 404
    in ecosystems, 391, 404
    of forests, 384, 384f
    in hierarchy of complexity, 21, 21f, 384
    individualistic compared to holistic concept of, 391
    niches in, 392-393, 392f
    pioneering, 398
    species interactions in, 394-397
    stability in, 398
    succession in, 398, 398f
    symbiosis in, 394-395, 395f
Compact bone, 466, 466f
Companion cells, 612
Compass sense, 436, 436f, 571, 571f
Competition
    in communities, 391-393
    interspecific, 392
    intraspecific, 392
    niches and, 392-393, 392f
Competitive exclusion, principle of, 393, 393f
Competitive inhibition, 97, 97f
Complement system, 530, 530f
Complex carbohydrates, 59, 59f
Complexity
    as argument against evolution, 261, 262
    hierarchy of, 20-21, 20f-21f, 384
    as property of life, 19
Complimentary DNA (cDNA), 231, 231f
Compound leaves, 618, 618f
Compound microscopes, 68

Concentration gradient, 82, 82
Concentration, in studying, 6
Conclusions, 24f, 25
Condensation, 140, 140f
Conditioned stimulus, 432
Condoms, 603, 603f
Cones, 566-568, 566f, 568f
Conifers (Coniferophyta), 327t, 334-335, 334f, 335f, 424
Conjugation, 302, 302f, 313f
Conjugation bridge, 302, 302f
Connecting, in learning, 5
Connective tissue, 462, 462f, 465-466, 465f
Connect program, 9
Consciousness, as emergent property of life, 21
Conscious planning, 433, 433f
Conservation efforts
    ecosystems and, 456, 456f
    endangered species in, 454-456
    human population growth and, 453
    of individuals, 457, 457f
    resource preservation and, 451-453
Consumers, 405, 406
Continuous data, 14
Continuous variation, 178, 178f
Contraception, 602-603, 603f
Contractile proteins, 52f
Contraction of muscle, 100t, 473, 473f, 560, 560f
Control experiments, 25
Controls, 24f, 25
Convergent evolution, 259
Conversion therapy, 163, 590
Cooksonia, 324, 329, 329f
Cooperation, 23
Corals, 350f
Cork cambium, 608, 616f, 617
Corn
    classification of, 327t
    genetically modified, 239, 239t
    root structure of, 613f
    seeds of, 332-333, 332f
Coronavirus
    spread of, 311
Corpus callosum, 553f
Corpus luteum, 595f, 596, 596f
Corpus striatum, 553f
Correlation, causation and, 13
Corridors, 456
Cortex, stem, 616, 616f
Cortisol, 574, 583
Cotton, 230f, 242, 243, 243t
Cotyledons, 333, 338, 631, 631f, 633
Countercurrent heat exchange, 479, 479f
Coupled reactions, 99, 122
Coupled transport proteins, 87
Courtship, 435t
Covalent bonds, 39-41, 40f
Covariance, 166-167
COVID-19, 17, 148
    attenuated vaccines, 538-539
    inactivated virus vaccines, 539
    nucleic acid vaccines, 539
    outbreak spreads, 17
    pandemic, 459
    recombinant vector vaccines, 538
    seismograph, 459
    "Stay-At-Home" Orders, 33
    subunit vaccines, 539
Cowpox, 540, 540f

Cows, 405
    cloning of, 241
Coyotes, 250-251
C_3 photosynthesis (Calvin cycle), 109, 113, 113f
C_4 photosynthesis, 115, 115f
Crabs, 360, 361f, 405f
Cramming, 2
Crassulacean acid metabolism (CAM), 115
Crawling, cell, 100t
Creationism, 93
Cretaceous period, 368-369
Creutzfeldt-Jakob disease, 58
Crichton, Michael, 241
Crick, Francis, 192, 192f
Criminal cases, DNA in, 190-191, 232f, 236-237
CRISPR, 86
    discovery of, 200
    for disease treatment, 201-202, 205, 208-209, 270
    -edited human babies, 205
    for gene drives, 203-204, 203f
    gene editing and, 208-209
    mutation corrections with, 202
Cristae, 78, 78f
Crohn's disease, 98
Crop plants
    genetic engineering of, 226-227, 242-244, 242f, 243f, 244t
    genetic variation in, 288
    herbicide and, 635
Crossing over, in meiosis, 157, 157f, 162
Crustaceans, 360, 361f, 392-393, 392f
    exoskeletons of, 469, 469f
Cryptosporidium, 312
C-section (caesarian section), 601
Cuboidal epithelium, 464
Cuenot, Lucien, 178
Cupula, 561, 561f
Cuticles, of plants, 324-325, 324f, 610
Cyanide, 34-35, 631
Cyanobacteria, 64
Cycads (Cycadophyta), 327t, 332f, 334-335, 334f
Cyclic AMP (cAMP), 578
Cystic fibrosis (CF)
    gene therapy for, 86, 246, 246f
    pleiotropic effects in, 178f, 179
Cysts, protist formation of, 312
Cytoplasm, 69, 70, 70f, 142, 317
Cytoplasmic transport, 100t
Cytosine, 56, 56f, 57, 192, 192f
Cytoskeleton, 70, 70f, 71f, 75t, 80-81
Cytotoxic T cells, 531t, 532, 534, 534f, 535, 535f

D

Daddy longlegs, 360
Dance language, of honeybees, 437, 437f
Dandelions, 632f, 633
Dandruff, 464
Darwin, Charles, 22, 30. *See also* Evolution, theory of Natural selection
    Beagle voyage of, 252-254, 252f, 253f
    on canids, 250-251
    critics of, 256, 256f, 261
    finch research of, 31f
    on phototropism, 635f
Darwin, Francis, 635, 635f
Date palms, 333

Daughter cells, 136, 136f, 158, 160
DDT, 444, 444f, 455
Deamination, 129, 129f
Deciduous forests, 422, 422f, 424, 424f
Decomposers, 319, 405f, 406
Deep sea ecosystems, 419, 419f
Deep-sea vents, 419
Deer, 387, 396
Defenses. *See also* Immune responses
    cellular, 526, 529-530, 529f, 530f
    as circulatory system function, 479
    digestive tract as, 526
    respiratory tract as, 526
    skin as first line of, 526, 526f, 528, 528f
Defensive proteins, 52f
Dehydration, 128, 520
    epithelial protection from, 464f
    metabolic water and, 515
Dehydration synthesis, 50-51, 51f
Delbruck, Max, 207
Deletion in DNA, 198, 198t
Demography, 390
Denaturation, 55, 55f
Dendrites, 468, 468f, 468t, 548, 548f
Denisova Cave, 343
De novo mutations, 197
Density-dependent effects, 388, 388f
Density-independent effects, 388
Deoxyribonucleic acid (DNA)
    autosomal testing of, 182
    base substitution in, 198, 198t
    in central dogma, 210, 210f
    in chloroplasts, 79
    in chromosomes, 74, 139, 195
    complimentary, 231, 231f
    CRISPR gene editing and, 86, 200-204, 203f
    definition of, 19, 28, 192
    deletion in, 198, 198t
    discovering shape of, 192-193, 193f
    in DNA repair hypothesis, 153
    double helix of, 57, 57f, 192-193, 193f
    eukaryotic, 70, 74
    evolution and, 260, 260f
    family history and, 182-183
    first cells and, 297
    forensic science using, 190-191, 236-237
    functions of, 57
    in gene theory, 28, 28f, 29f
    in heredity theory, 30
    Innocence Project and, 236-237
    insertion in, 198, 198t
    mitochondrial, 78, 183
    noncoding, 229, 229t
    in phenotypes, 176
    prokaryotic, 69
    in prokaryotic cell cycle, 136, 136f
    proofreading and, 195
    replication of, 136, 136f, 194, 194f, 195f
    sequencing of, 228, 228f, 229t
    structural, 229, 229t
    structure of, 28, 28f, 57, 57f, 192-193, 192f
    Y chromosomes in testing of, 182-183
Department of Agriculture (USDA), 500
Dependent variables, 13
Depolarization, 549-550, 550f
Depo-Provera, 602
Derived characters, 284
Dermal tissue of plants, 609, 610, 610f

Dermatophagoides (dust mites), 542, 542*f*
Dermis, 526, 528, 528*f*
The Descent of Man (Darwin, C.), 256
Deserts, 422, 422*f*, 423, 423*f*
Desmids, 64
Detritivores, 405*f*, 406
Deuterostomes, 346, 363, 363*f*
Development. *See also* Embryonic development
  fetal, 600-601, 600*f*
  postnatal, 601
Devonian period, 368, 369
Diabetes mellitus
  genetic engineering and, 230, 234
  global incidence of, 581
  glucose excretion and, 513
  insulin for, 230, 234
  juvenile-onset, 581
  obesity and, 12, 12*f*, 572*f*, 573
  type I, 542, 581
  type II, 12, 12*f*, 572-573, 581
Diaphragm (contraceptive), 603, 603*f*
Diaphragm (muscle), 490
Diastolic pressure, 486
Diatoms, 314, 315*t*
Diazepam (Valium), 552
Dicambra, 226-227
Didinium, 396, 396*f*
Diet. *See also* Food
  agriculture and, 130-131
  fads in, 118-119, 130-131
  flowers in, 326
  food labels and, 500
  nutritional labels and, 50, 50*f*
  teeth and, 503, 503*f*
  vegetarian and vegan, 90-91, 90*f*, 91*f*, 509
Differentiation, 614, 634, 634*f*
Diffusion, 82, 82*f*, 84, 84*f*
Digestion, 502-509
  extracellular, of cnidarians, 350
  of fats, 507
  fungal external, 317
  roundworm, 359
Digestive systems
  daily activity of, 472
  defensive function of, 526
  esophagus in, 502, 502*f*, 502*f*, 505, 505*f*, 510*f*
  gallbladder in, 502, 502*f*, 502*f*, 509, 509*f*, 510*f*
  of humans, 502, 502*f*
  large intestine in, 502, 502*f*, 502*f*, 508, 510*f*
  liver in, 502, 502*f*, 502*f*, 509, 509*f*, 510*f*
  mouth in, 502, 502*f*, 502*f*, 504, 504*f*, 510*f*
  organs of, 510*f*
  pancreas in, 502, 502*f*, 502*f*, 507, 509, 509*f*, 510*f*
  small intestine in, 502, 502*f*, 502*f*, 507-508, 507*f*, 507*f*, 508*f*, 508*f*, 510*f*
  stomach in, 502, 502*f*, 502*f*, 505-506, 505*f*, 506*f*, 510*f*
  teeth in, 503, 503*f*, 504*f*
  types of, 501, 501*f*
  of vertebrates, 463, 502, 502*f*
Digestive tract, 501, 501*f*, 508, 526
Dihybrid individuals, 175, 175*f*
Dikaryotic fungal hyphae, 317
Dileptus, 64-65, 64*f*
Dinoflagellates, 314, 315*t*
Dinosaurs
  cloning of, 241
  evolution of, 368-369

extinction of, 369
fossils of, 369*f*
Diploid cells, 138, 154, 154*f*
Diplomonads (Giardia), 315*t*
Dipodmys, 514-515, 514*f*, 515*f*
Direct diffusion, 488, 488*f*
Directional selection, 266*f*, 268
Disaccharides, 59, 60*t*
Diseases and disorders. *See also specific types*
  CRISPR for treatment of, 208-209, 270
  gene silencing for, 221
  genetic engineering for, 230, 230*f*, 234, 234*f*, 238, 238*f*
  sexually transmitted, 586-587, 603, 605, 605*f* stem cell treatment for, 244-245, 244*f*, 245*f*
  viral, 306-307, 306*f*, 307*f*
Dispersal, 333
Disruptive selection, 266*f*, 268, 268*f*
Diversity. *See also* Biodiversity
  of animals, 344, 345*t*
  genetic, 162, 455-456, 455*f*
  human social behavior and, 439, 439*f*
  of insects, 362*f*
DNA. *See* Deoxyribonucleic acid
DNA fingerprinting, 232, 232*f*
DNA ligase, 195, 195*f*, 230
DNA-mutating tars, 493
DNA particle guns, 238
DNA polymerase, 194, 194*f*
DNA repair hypothesis, 153
DNA vaccines, 235
Dogs
  cloning of, 243, 243*f*, 244
  evolution of, 250-251, 250*f*, 251*f*
  gene therapy in, 247*f*
  heredity in coat color of, 174
  learning by, 432
  social behavior of, 435*t*
Doll, Richard, 199
Dolly (sheep), 246
Domains, 30, 30*f*, 282, 288-291, 289*t*, 290*f*
Dominant traits, 170-171, 170*t*, 171*f*, 172*t*
Donkeys, 283, 283*f*
Dopamine, 552
Doping, 49
Dormancy, 333, 333*f*, 637
Dorsal, definition of, 351
Double covalent bonds, 40, 40*f*
Double fertilization, 338, 338*f*, 630
Double helix, 57, 57*f*, 192-193, 193*f*
Doublings, 151, 151*f*
Douglas fir *(Pseudotsuga menziesii)*, 626
Down, J. Langdon, 181
Down syndrome, 181-182, 181*f*, 182*f*
Dragonflies, 362*f*
Drosophila, 196*f*, 268, 268*f*
Drug addiction, 556
Drugs. *See also specific types*
  addiction and, 546-547, 556
  marijuana, 114
  overdosing from, 556
  overprescription of, 556
  superbugs and resistance to, 223, 301
DrugsFromDirt.org, 223, 301
Drupes, 632, 632*f*
Dry fruits, 632*f*, 633
Duck-billed platypus, 278-279, 278*f*, 279*f*, 569
Ducks, 306, 306*f*

Duodenal ulcers, 506
Duodenum, 506, 507
Duplicated sequences, 229, 229*t*
Dust mites *(Dermatophagoides)*, 542, 542*f*
Dysentery, amoebic, 312

E

Ear
  hearing function of, 565-566, 565*f*
  of humans, 565-566, 565*f*
  inner, 561, 561*f*, 565
  middle, 377
  motion/balance of, 561, 561*f*
Earth
  age of, 254
  population limit of, 453
Earthworms, 357*f*
  circulatory system of, 478, 478*f*
  digestive system of, 501, 501*f*
  hydraulic skeletons of, 469, 469*f*
Eastern gray squirrel, 282, 282*f*
Eastern tent caterpillar, 35
Ebola virus, 303, 306, 306*f*
eBook, 9
Ecdysis, 346
ECG. *See* Electrocardiogram
Echinarachnius parma, 364*f*
Echinoderms (Echinodermata), 364-365, 364*f*
E-Cigarette, 493
  gateway drug, 493
  Juul Vape Pen, 493
  vaping, 493
E-cigarettes, 149, 547
Ecological isolation, 273*t*, 274
Ecological organization, 384
Ecological pyramids, 409, 409*f*
Ecological succession, 398, 398*f*
Ecology, 384
Ecosystems. *See also specific types*
  biodiversity and, 288
  biomass of, 406
  communities in, 384, 404
  conservation of, 456, 456*f*
  definition of, 384, 404
  Emergent properties, of life, 21
  energy flows in, 404-405, 404*f*, 408
  energy in, 404-409
  habitats in, 404
  in hierarchy of complexity, 21, 21*f*, 384
  materials cycles within, 410-414
  niches in, 392-393, 392*f*
  types of, 418-425
  weather in, 415-417, 415*f*, 417*f*
Ecosystem services, 288
Ectoderm, 351-352, 352*f*, 597
Ectotherms, 517
Edema, 483
Edible mushrooms, 318*f*
Edmondson, W. T., 457, 457*f*
Effectors, 516, 516*f*
Eggs
  formation of, 588, 593-594, 593*f*, 630
  instinctive behavior and, 430, 431*f*
Ejaculation, 592
EKG (electrocardiogram), 487
Electric vehicles, 447
Electrocardiogram (EKG), 487

Electromagnetic spectrum, 110, 110f
Electron microscopes, 68
Electrons, 36, 36f
    ATP production by, 125-126, 125f
    in chemical bonds, 39-41
    harvest from chemical bonds, 123-124
    potential energy of, 37, 37f
Electron shells, 37
Electron transport chain, 125-126, 125f
Electron transport system, 111-112
Elements, 36, 37t
Elephants, 291f, 495, 495f, 526, 526f
Elephant seals, 435t
Elevation, weather and, 417, 417f
Ellis-van Creveld syndrome, 265f
Elongation, zone of, 614
Embryonic development
    of angiosperms, 631, 631f
    in animals, 344, 345t, 363-367, 363f
    body architecture determination in, 597
    cleavage in, 597, 597f
    evolution of, 346
    human, 597-601
    organogenesis in, 598-599, 598f-599f
    sex determination in, 589, 589f
    sexual orientation and, 590
    vertebrate, 259, 259f
Embryonic stem cells, 244
Embryo sac, 629
Emergency contraception, 602
Emergent properties, of life, 384
Emerging viruses, 306-307
Emigration, 264
Enamel, tooth, 503, 504f, 504f
Encephalartos transvenosus, 334f
Endangered species, 454-456
Endangered Species Act, 443
Endergonic reactions, 94, 94f
Endocrine glands, 464, 574, 574f, 579-583, 634
Endocrine systems
    anatomy of, 574, 574f
    hormone secretion in, 509, 574-578, 581, 581f, 581f
    major glands of, 579-583
    of vertebrates, 463
Endocytosis, 79, 84, 84f
Endoderm, 352, 352f
Endodermis, of roots, 613, 613f
Endomembrane systems, 70f, 71, 71f, 76-77, 76f, 77f
Endometrium, 594, 596
Endonucleases, restriction, 230, 231f
Endoskeletons, 469, 469f
Endosperm, 333, 338, 338f
Endospores, 302
Endosymbiosis, 79, 79f, 291, 291f
Endosymbiotic theory, 79
Endothelial cells, 481f
Endothelium, 481f
Endotherms, 517
Energy
    activation, 94, 94f
    from carbohydrates, 98
    carried by electrons, 37, 37f
    cellular respiration and, 98
    in chemical bonds, 39
    in chemical reactions, 92, 94, 94f
    definition of, 37, 92
    in ecosystems, 404-409
    enzymes and, 95-97
    flow of, 92
    food for, 498-499
    forms of, 92
    heat, 92
    kinetic, 92, 92f
    laws of, 92-93
    locomotion and, 475, 475f
    measurement of, 92
    potential, 37, 37f, 92, 92f
    in protein recycling, 89
    pyramids of, 409, 409f
    in thermodynamics, 92-93, 93f
Energy drinks, 127
Energy flows
    as biological theme, 22
    in ecosystems, 404-405, 404f, 408
    in living things, 92
    sunlight in, 92, 106, 404
    through trophic levels, 406
Enhancers, 220, 220f
Entamoeba histolytica, 312
Entropy, 93, 93f
Envelope, 305f, 306
Environment
    alleles affected by, 179, 179f
    cancer factors in, 146
    GM crops and, 226-227, 242-244, 243t, 253, 253f
    in human intelligence, 166-167
    pigment and, 179
    plant responses to stimuli from, 636-637, 636f, 637f
    prokaryotes and, 302
Environmental water cycle, 410, 410f, 411
Enzyme polymorphism, 277
Enzymes, 23
    activating, 213
    active sites of, 95, 95f
    allosteric, 97, 97f
    binding sites of, 95, 95f
    in biochemical pathways, 95-96, 96f
    cell regulation of, 96-97, 97f
    definition of, 51, 95
    energy and, 95-97
    restriction, 230, 231, 231f
    shape of, 95, 95f, 97, 97f
    structure and function of, 52, 52f, 55, 55f, 95, 95f
    temperature and pH affecting, 96, 96f
Eosinophils, 484, 484f
Epidermis, 526, 528, 528f
    plant, 107, 610, 613, 613f
Epididymis, 591f, 592
Epigenetic modifications, 179
Epigenetics, 163, 246, 590
Epiglottis, 504, 504f
Epinephrine, 552, 583
Epithelial cells, 464
Epithelial tissues, 462, 462f, 464, 464f
Epithelium, 464, 464f
Equisetum telmateia (horsetail), 327t, 330f, 331
ER (endoplasmic reticulum), 75t, 76-77, 76f, 77f
Erections, 592
Erosion, in carbon cycle, 412, 412f
Erythroblastosis fetalis, 541
Erythrocytes. See Red blood cells
Escherichia coli, 207, 219
Esophagus, 44, 502, 502f, 505, 505f, 505f, 510f
Essay on the Principle of Population (Malthus), 255
Essential amino acids, 499
EST (estratetraenol), 429

Estradiol, 61
Estratetraenol (EST), 429
Estrogen, 574, 576, 576f, 595-596, 595f, 602
Estuaries, 418
Ethanol fermentation, 128, 128f
Ethics
    cloning and, 244
    stem cell therapy and, 244-245
Ethology, 430
Eucalyptus regnans (Australia mountain ash), 626
Eudicots (Eudicotyledons), 338, 608f, 616, 616f, 618, 619f, 633
Euglena (euglenozoa), 315t, 588, 588f
Eukaryotes (Eukarya)
    domain of, 30, 30f, 289, 289t, 291, 291f
    evolution of, 300
    genomes of, 228
    kingdom of, 289, 289t
    origin of name, 70
    prokaryotes compared to, 300, 300t
    protein synthesis in, 216, 217f
    sexual life cycles in, 155, 155f
    transcription in, 216, 217f
    translation in, 216, 217f
Eukaryotic cells
    cytoskeleton of, 80-81
    gene expression in, 215, 215f-217f, 216, 219-220
    nucleus of, 70, 70f, 71f, 74, 74f, 75t
    organelles of, 75t, 78-79
    origin of first, 300
    plasma membrane of, 70, 70f, 72-73, 75t, 82-87
    prokaryotic cells compared to, 300, 300t
    structure of, 70, 70f, 71f
    walls of, 71, 71f, 75t
Eulampis (hummingbird), 523
Euparkeria, 375f
Europa, life on, 298
Eutrophication, 414
Eutrophic lakes, 420f, 421
Evaporation, 410f, 411, 620f, 621
Evidence, analyzing, 12
Evolution. *See also* Animal evolution
    adaptation within populations in, 269-272
    agents of, 198, 264-268, 264t
    of amphibians, 368, 373-374, 374f
    anatomical record of, 259, 259f
    of animal tissues, 346
    behavior shaped by, 434-436
    as biological theme, 22
    of birds, 376, 376f
    of blood clotting, 262, 262f
    coevolution, 394-395, 394f, 395f
    convergent, 259
    definition of, 22, 252
    DNA and, 260, 260f
    dogs, 250-251, 250f, 251f
    of embryonic development, 346
    of eukaryotes, 300
    evidence of, 254, 257-261, 257f, 258f, 261f
    extraterrestrial, 298-299
    of fishes, 293, 293f, 368, 370-373
    of flower color, 639, 639f
    of flowers, 336, 338, 639, 639f
    of hemoglobin, 260, 260f
    of hominids, 342, 378, 378f
    of humans, 342-343, 378-379, 378f
    of mammals, 369, 377
    of meiosis, 152, 162

Evolution (Cont.)
　molecular record of, 260, 260f
　of molting, 346
　origin of life and, 296-297, 296f, 297f
　of plants, 326, 326f
　of populations, 263-272
Evolution. See also Animal evolution
　of prokaryotes, 300
　of reptiles, 368-369, 375, 375f
　of seeds, 332-333
　of segmentation, 346
　of sexual reproduction, 152-153, 162
　species formation in, 273-275
　treadmill, 153
　of vascular plants, 329, 329f
　of vertebrates, 368-379
　of whales, 258, 258f
Evolutionary discordance hypothesis, 131
Evolution, theory of. See also Natural selection
　controversy over, 256
　critics of, 256, 256f, 261-262, 261f, 262f
　Darwin, C.'s, 22, 252
　evidence supporting, 254, 257-261, 257f, 258f, 261f
　thermodynamics and, 93
Exams, reviewing for, 7
Excavata, 315t
Excitatory synapses, 552
Excretion
　as circulatory system function, 479
　of glucose in urine, 513
　systems of, 518-519, 518f, 519f
Exercise, 119, 475
Exergonic reactions, 94, 94f
Exhalation, 491, 491f
Exocrine glands, 509, 574
Exocytosis, 84, 84f
Exons, 216, 216f
Exoplanets, 299
Exoskeletons, 360, 361f, 469, 469f
Experiments, 24f, 25
Exponential growth model, 385-386, 386f, 387
Extensor muscles, 471, 471f
External digestion, fungal, 317
Extinctions
　of amphibians, 450
　causes of, 369
　factors responsible for, 449, 449f
　mass, 369
　rates of, 369
Extracellular digestion, of cnidarians, 350
Extracellular matrix, 465
Extraterrestrial life, 298-299
Extreme weather, 416
　global warming and, 447, 448
Extrinsic pathway, 262, 262f
Eyes
　blinking of, 472
　color vision of, 567-568, 567f
　light sensed by, 567, 567f
　structure of, 261, 261f, 560, 560f, 566, 566f
　of vertebrates, 261, 261f, 566-568, 566f

F

Facial expressions, of humans, 439
Facilitated diffusion, 84, 84f
Factor IX, 247
Fad diets, 118-119, 130-131

$FADH_2$, 124
Falco peregrinus (peregrine falcons), 455, 455f
Fallopian tubes (oviducts), 593f, 594, 594f
Families (taxonomic rank), 282
Family history, DNA and, 182-183
Fat cells, 465
Fats, 61
　cellular respiration of, 129, 129f
　digestion of, 507
　food sources of, 498
Fatty acids, 61, 72, 129
Feathers, of birds, 376, 376f
Feces, 508
Fecundity, 390
Feedback
　negative, 516, 516f, 595-596
　positive, 595-596
Feedback inhibition, 97, 97f
Female reproductive systems, 593-594, 593f, 594f
Fentanyl, 556
Fermentation
　alcoholic, 304
　CO_2 as product of, 128
　definition of, 128
　ethanol, 128, 128f
　lactic acid, 128, 128f
Ferns, 327t, 330, 330f, 384f
　life cycle of, 331, 331f
　sexual life cycles in, 155, 155f
Fertilization
　double, 338, 338f, 630
　gametes and, 154
　human, 594, 594f
　ocean, 104-105, 104f, 105f
　random, 162
　self-, 169, 169f
　in vitro, 156
Fertilizers, chemical, 414
Fetal alcohol syndrome, 599
Fetal development, 600-601, 600f
Fetus, 600
Fever, 530
F_1 generation, 170-171
F_2 generation, 169-171, 169f
F_3 generation, 171, 171f
Fibers, 609
Fibrin, 483, 483f
Fibrinogen, 483
Fibroblasts, 465
Filial imprinting, 432
Finches
　Darwin, C.'s, research on, 31f
　disruptive selection in, 268, 268f
Fingerprinting, DNA, 232, 232f
Fingers, 172t
Fire, Andrew, 221
First law of thermodynamics, 93
Fir trees, 327t, 335
Fishes
　binary, 136, 136f
　blood pH of, 133, 133f
　characteristics of, 372
　evolution of, 293, 293f, 368, 370-373
　nitrogenous waste elimination by, 521, 521f
　nutrition and, 372
　in ocean ecosystems, 418-419, 418f
　respiration of, 488, 488f
　respiratory system of, 372

　symbiosis in, 395
　teeth of, 503
　vertebral columns in, 372
Fission, 588, 588f
　binary, 302
Fittest, survival of, 256
Fixed action patterns, 430
Flagella
　ATP in movements of, 100t
　bacterial, 302
　eukaryotic, 70f, 75t, 81, 81f, 300t
　prokaryotic, 69, 300t, 302
　protist, 312
Flame cells, 352, 353
Flatworms
　body plan of, 352-353, 352f
　digestion of, 352-353
　gastrovascular cavity of, 478, 478f, 501, 501f
　reproduction by, 353
　respiration of, 488, 488f
Fleas, 362f
Fleming, Alexander, 223, 301
Flemming, Walther, 138, 154
Fleshy fruits, 632, 632f
Flexor muscles, 471, 471f
Flooding, 447
Floral leaves, 618f
Flores island, 342-343
Flowers, 169, 332
　color of, evolution of, 639, 639f
　color of, pollination and, 630, 630f, 639, 639f
　in diet, 326
　egg formation in, 630
　evolution of, 326, 336, 338, 639, 639f
　fertilization of, 630, 630f
　life cycle of, 338, 338f
　monoecious, 629
　photoperiodism in, 636-637, 636f
　pigment forms in, 637
　pollen formation by, 629-630
　structure of, 336, 336f, 629, 629f
Flu. See Influenza
Fluctuation tests, 207, 207f
Fluid mosaic model, 72
Fluorescent dye, 165
Fly, circulatory system of, 478, 478f
Flying foxes, 303, 456, 456f
Foams, 603
Follicle-stimulating hormone (FSH), 574, 580, 580f, 593f, 594-596, 595f, 596f, 602
Follicular phase of menstrual cycle, 595, 595f
Food. See also Diet
　chewing, 503
　for energy, 498-499
　for growth, 498-499
　labels on, 500
　swallowing, 504, 504f
　test-tube, 511
Food and Drug Administration, 234, 527
Food chains, 406, 406f, 408
Food vacuoles, 313
Food webs, 406, 407
Foraging behavior, 434, 434f, 435t, 439
Foraminifera, 315t
Forams, 315t
Forelimbs, 259, 259f

Forensic science, DNA in, 190–191, 232, 232f, 236–237
Forests
    acid rain and, 45, 45f
    clear-cutting of, 411, 411f, 426, 426f
    communities of, 384, 384f
    logging of, 402–403, 411, 411f, 426, 426f, 452, 452f
    water cycle in, 411, 411f
Forgetting, 4–5
fosB gene, 431, 431f
Fossil fuels, 412, 445
Fossil intermediates, 261, 261f
Fossil record
    as evidence of evolution, 254, 257–261, 257f, 258f, 261f
    hominid, 342, 378
    radioisotopic dating of, 38, 47
    vertebrates in, 368–369
Founder effect, 265, 265f
Foxes, 179, 179f
Fox, Michael J., 244f
Fracking (hydraulic fracturing), 447
Frameshift mutations, 198, 198t
Franklin, Rosalind, 192, 193f
Freely movable joints, 470
Free range labels, 500
Frequencies, allele, 263–264
Frequency, sound, 565
Freshwater ecosystems, 64–65, 414, 420–421, 420f, 421f
Frisch, Karl von, 437
Frogs
    courtship by, 435t
    evolution of, 373–374, 374f
    postzygotic isolating mechanisms in, 275, 275f
    sexual life cycles in, 155, 155f
Fronds, 331, 331f
Frontal lobe, 554, 554f
Fruits, 326
    dry, 632f, 633
    fleshy, 632, 632f
    poisonous, 631
FSH (follicle-stimulating hormone), 593f, 594–596, 595f, 596f, 602
Function, structure determining, 23
Fundamental niches, 392–393
Fundulus heteroclitus, 277, 277f
Fungi
    body of, 316–317, 316f
    commercial uses of, 319
    as decomposers, 319, 405f
    kingdom of, 18f, 288, 289t
    mutualistic associations of, 319, 319f
    nutrition of, 317, 317f
    plant association with, 324
    plants compared to, 316
    population decline from, 450
    reproduction by, 316f, 317, 317f
    spores of, 317, 317f
    symbiosis in, 324
    types of, 318–319, 319t
Fur seals (*Callorhinus ursinus*), 386, 386f

G

GABA, 552
Galápagos Islands, Darwin, C.'s, research on, 253f, 254, 254f
Gallbladder, 502, 502f, 509, 509f, 510f
Gamete fusion, prevention of, 273t, 275
Gametes. *See also* Eggs; Ovum; Sperm
    chromosomes in, 154
    female, 593–594, 593f, 594f
    male, 591–592, 591f
    types of, 588
Gametophytes, 325, 325f, 629
Gardasil, 587, 603
Gar pikes, 293, 293f
Gas exchange, 488, 488f
Gastric juice, 505–506
Gastric ulcers, 506
Gastrin, 506
Gastrointestinal tract, 502, 502f
Gastropods, 355, 355f
Gastrovascular cavity, 350, 478, 478f, 501, 501f
Gause, G. F., 393, 393f
Geese, 430, 431f, 432
Gene drives, CRISPR for, 203–204, 203f
Gene editing, with CRISPR, 86, 200–204, 203f, 205, 208–209
Gene expression
    in central dogma, 210, 210f, 215, 215f
    in eukaryotic cells, 215, 215f, 220–220
    gene architecture and, 216
    overview of, 210, 210f
    in prokaryotic cells, 215, 215f, 216, 216f, 218–219
    regulation of, 218–219
    RNA interference, 221, 221f
    stages of, 210, 214
Gene prospecting, 288
Genera (genus), 280
Generations, alternation of, 325, 325f, 326, 326f
Gene reprogramming, 243
Genes, 163
    architecture of, 216
    in central dogma, 210, 210f, 215, 215f
    cf, 178f, 246
    definition of, 19, 28, 172
    distribution among chromosomes, 228–229
    in evolution theory, 30
    in gene theory, 28, 28f, 29f
    Hb, 176, 177f
    for hemoglobin, 176, 177f
    in heredity theory, 30, 172
    human social behavior and, 439
    noncoding DNA within, 229, 229t
    origin of new, 297
    sexual orientation and, 590
    silencing of, 221, 222f
    SRY, 181, 589, 589f
    traits influenced by, 176, 177f
    tumor-suppressor, 143, 146
Gene silencing, 221, 222f
Gene theory, 28, 28f, 29f
Gene therapy, 246
    for cystic fibrosis, 86, 246, 246f
    vectors for, 246–247, 246f, 247f
Genetically modified (GM) crops, 226–227, 242–244, 243t, 253, 253f
Genetic code, 212, 212f
Genetic disorders
    nondisjunction in, 181–182
    race and, 287
Genetic diversity, 162, 455–456, 455f
Genetic drift, 264t, 265, 265f
Genetic engineering, 151. *See also* CRISPR
    agriculture and, 226–227, 226f, 227f, 230, 230f, 242–244, 242f, 243t, 244f, 253, 253f
    of bacteria, 234
    cDNA formation and, 231
    definition of, 230
    DNA fingerprinting and, 232, 232f
    injection methods in, 238, 238f
    in medicine, 230, 230f, 234, 234f, 238, 238f
    polymerase chain reaction in, 232
    restriction enzymes in, 230
    viruses and, 235
Genetics
    behavior and, 431, 431f
    codominance and, 180, 180f
    continuous variation in, 178, 178f
    environmental effects in, 179, 179f
    incomplete dominance in, 179, 179f
    of intelligence, 166–167
    Mendelian, 168
    non-Mendelian, 178–180
    pleiotropic effects in, 178–179, 178f
Genital herpes, 603
Genomes
    comparison of, 260
    DNA sequences in, 228, 228f, 229t
    human, 226f, 176, 177f, 208, 228–229, 229f, 229t
    nematode, 228
    protist, 228, 314, 314t
Genotypes
    Hardy-Weinberg equilibrium in, 263, 263f
    in Mendel's theory of heredity, 172–174
Genus (genera), 280
Geographical isolation, 273t, 274, 275
Geometric progression, 255, 255f
George, Russ, 105
Germination, 333, 333f, 631, 631f, 633
Germ-line cells, 155
Germ-line tissues, 155, 196
GH (growth hormone), 580, 580f
Giardia (diplomonads), 315t
Gila monsters, 464, 464f
Gills, 372, 488, 488f
Ginkgo biloba, 334f, 335
Ginkgo trees, 327t, 334–335, 334f
Giraffes, 423, 423f
Gizzard, 501, 501f, 503
Glacial moraines, 398, 398f
Glands, 464
Glial cells, 468
Global change, caused by humans, 444–450
Global warming
    agriculture and, 445
    climate change and, 416
    CO_2 in, 12, 12f, 104, 412, 416, 445, 445f, 446
    combating, 447, 448
    evidence of, 446
    extreme weather and, 447, 448
    fossil fuels in, 412, 445
    habitat loss caused by, 443
    impact of, 446–447
    ocean fertilization and, 104–105, 104f, 105f
    polar ice caps and, 443, 443f
    rain and, 445
    sea level and, 445, 448
    sequestration and, 104
    temperature rise and, 448, 448f
Glomeromycota, 319t
Glomerulus, 519, 519f
Glottis, 504, 504f

Glucagon, 517, 517f, 581, 581f
Glucose
　in blood plasma, 483
　blood, regulation of, 513, 513f, 517, 517f
　cellular respiration of, 120-126
　excretion in urine, 513
　monitoring levels of, 572f, 573
　structure of, 59, 59f
Glucose tolerance curves, 513, 513f
Glutamic acid, 269
Glycerol, 61
Glycine, 552
Glycogen, 59, 60t, 517, 517f
Glycolysis, 120, 121f, 122, 122f, 408
Glycoproteins, 574
Glyphosate, 238, 238f
Glyptodonts, 254, 254f
GM (genetically modified) crops, 226-227, 242-244, 243t, 253, 253f
Gnetophytes (Gnetophyta), 327t, 334-335, 334f
Goldilocks planets, 299
Golding, William, 438
Golgi bodies, 76f, 77
Golgi complexes, 75t, 78-79
Gonads, 588
Gonorrhea, 603, 605, 605f
Gould, James L., 437
gp120 protein, 525, 525f
Gram, Hans, 300
Gram-negative bacteria, 300
Gram-positive bacteria, 300
Grams, 14
Grana, 78, 108
Graphs, 13-15, 13f-15f
Grasses, genetically modified, 249
Grasshoppers, 362f
Grasslands, 408, 422, 422f, 423, 423f
Graves' disease, 542
Gravitropism, 637, 637f
Gravity, sensory receptors for, 561, 561f
Gray matter, 553, 557f
Great Ape Project, 256
Green algae, 313, 313f
Greenhouse effect, 445, 445f
Grid cells, 563
Griffith, J., 58
Grizzly bears, 405f
Gropp, John, 236
Ground beetles, 384f
Ground finches, 254, 254f
Ground meristem, 613, 613f
Ground tissue of plants, 609, 609f
Groundwater, 411, 452
Growth
　allometric, 601
　fetal, 600
　food for, 498-499, 600-601
　of plants, 329, 337, 608, 608f, 634-635
　as property of life, 19
　stem, primary, 615-616, 615f, 616f
　stem, secondary, 616-617, 616f, 617f
Growth factors, 143, 146
Growth hormone (GH), 580, 580f
Guanine, 56, 56f, 57, 192, 192f
Guard cells, 324f, 610, 610f, 622, 622f
Gut bacteria, 496-497
Gymnosperms
　evolution of, 326, 326f
　life cycle of, 335, 335f
　pollination of, 630
　types of, 334-335, 334f

H

Habitats
　definition of, 404
　diversity of, 344, 345t
　in ecosystems, 404
　fragmentation of, 456
　loss of, 402-403, 443, 449, 450
　niche compared to, 392
　restoration of, 454, 454f
Habituation, 432
Hair
　dominant and recessive traits in, 172t
　fetal formation of, 600
　growth of, 472
　of mammals, 377
　woolly, 189
Hair cells, 565, 565f
Half-life, 47
Hamburgers, test-tube, 511
Hangovers, 128
Haploid cells, 154, 154f
Haplopappus gracilis, 138
Haplotypes, 287
Hardin, Garrett, 451, 451f
Hardy, G. H., 263-264
Hardy-Weinberg equilibrium, 263-264, 263f
Hares, 397, 397f
Haversian canal of bone, 466
Hayflick, Leonard, 151
Hayflick limit, 151
Hb gene, 176, 177f
hCG (human chorionic gonadotropin), 596, 597
HCl (hydrochloric acid), 44, 505-506, 506f
Head louse, 395f
Head, vertebrate, 367
Hearing, 565-566, 565f
Heart, 472
　blood flow from, 480-482, 485-486, 485f
　in closed circulatory systems, 478, 478f
　contraction of, 487, 487f
　human, 485-486, 485f, 495
　of mammals, 485-486, 485f, 495
　monitoring performance of, 486, 486f
　in open circulatory systems, 478, 478f
　return of blood to, 482, 482f, 485f, 486
　size and rate of, 495, 495f
Heart attacks, 127
Heartburn, 44, 505
Heart murmurs, 486
Heartwood, 617
Heat energy, 92
Heat exchange, countercurrent, 479, 479f
Heat of vaporization, 42t, 43
Heat production, ATP for, 100t
Heat storage, in water, 42, 42t
Heat waves, 447
Helianthus annuus (sunflowers), 616f
Helicase, 194, 194f, 195f
Helicobacter pylori, 506
Helium, 37, 37f
Helper T cells, 531t, 532, 533-534, 534f, 535, 535f
Hemochromatosis, 287
Hemoglobin, 63, 479
　in blood, 481f, 484, 492
　evolution of, 260, 260f
　gene for, 176, 177f
　molecule of, 492, 492f
　in sickle-cell disease, 187, 187f, 269-270, 269f
　structure of, 176, 176f
Hemolymph, 478
Hemolytic disease of newborn, 541
Hemophilia, 184t, 185-186, 186f, 247
Hemp, 114, 339
Henslow, John Stevens, 253
Hepaticophyta, 327t, 329
Hepatitis B, 235
Herbicides, 226-227, 242, 242f
Herbivores, 405, 405f, 408, 501, 503
Heredity
　definition of, 19, 168
　Mendel's experiments on, 168
　Mendel's theory of, 30, 172-174, 173f, 174f
　non-Mendelian, 178-180
　as property of life, 19
Heritability, 166-167
Hermaphroditism, 353, 588
Heroin, 556
Herpes simplex virus, 235, 235f
Herpes simplex virus type 2 (HSV-2), 603
Herrick, James, 176
Heterotrophs, 312, 344, 344t, 501
　in energy flow, 405, 408
Heterozygosity, 172-174, 173f, 174f
Heterozygote advantage, 270
HGH (human growth hormone), 234
Hiccups, 492
Hill, John, 199
Hippocampus, 554-555, 554f, 562-563
Hippopotamus, 23, 258
Histograms, 14, 14f, 15, 15f, 249
Histones, 139, 139f, 220, 220f
Hitting deer, 399
HIV/AIDS
　combination therapy for, 543
　CRISPR for treatment of, 201
　history of, 543, 543f
　immune system attack in, 543, 543f
　mutations and, 525
　structure of virus, 305f, 306
　vaccine against, 247, 524-525
H1N1 flu virus, 309
Hoffstetter, Nancy, 387
Holistic community concept, 391
Homeostasis
　as biological theme, 23
　in blood glucose, 517, 517f
　in body temperature, 516-517, 517f
　definition of, 516
　hypothalamus and, 517, 554
　negative feedback loops in, 516, 516f
　as property of life, 19
　sensory nervous system and, 560
Home range, 434
Hominids, 342, 378, 378f
Homo, 378
Homodont dentition, 503
Homo erectus, 343, 378-379, 378f
Homo floresiensis, 342-343, 379
Homogentisic acid, 172t
Homo habilis, 378, 378f

Homologous chromosomes (homologues), 138, 138f, 139f, 157, 157f, 160
Homologous structures, 259, 259f
Homo neanderthalensis, 342-343, 378f, 379, 379f
Homo sapiens, 378f, 379, 379f
Honeybees, 362f
    dance language of, 437, 437f
    flower color and pollination by, 639, 639f
    killer, invasion of, 382-383, 382f, 383f
    parthenogenesis by, 588
    pheromones of, 438
    societies of, 438
Honeysuckles, 632f
Hong Kong flu, 309
Honjo, Tasuku, 532
Hooke, Robert, 28, 66
Hops, 304
Hormones, 163. *See also specific types*
    of adrenal glands, 583, 583f
    anabolic steroids in sports and, 48-49
    for birth control, 602, 603f
    in blood plasma, 483
    blood transport of, 479
    communication of, 575, 575f
    definition of, 574
    endocrine secretion of, 574-578, 581, 581f
    kidney secretion of, 520, 520f
    of pancreas, 509, 581, 581f
    parathyroid, 582-583, 583f
    peptide, 578, 578f
    of plants, 634-635
    reproductive cycle coordination by, 595-596, 595f-596f
    sexual orientation and, 590
    steroid, 576, 576f, 578
    stress, 583
    thyroid, 582, 582f
    types of, 574
Hornworts, 327t, 328
Horses
    classification of, 283, 283f
    cyanide and, 34-35
    roan pattern in, 180, 180f
Horsetail *(Equisetum telmateia)*, 327t, 330f, 331
HPV (human papilloma virus), 586-587, 586f-587f, 603
HSV-2 (herpes simplex virus type 2), 603
Human chorionic gonadotropin (hCG), 596, 597
Human diet. *See* Diet
Human Genome Project, 182
Human growth hormone (HGH), 234, 234f
Human immunodeficiency virus. *See* HIV/AIDS
Human papilloma virus (HPV), 586-587, 586f-587f, 603
Human reproductive systems
    female, 593-594, 593f, 594f
    fertilization in, 594, 594f
    hormone coordination of, 595-596, 595f-596f
    male, 591-592, 591f, 592f
Humans, 314
    bacteria in gut of, 496-497
    birth weight in, 267, 267f
    body temperature of, 517
    brain of, anatomy of, 553-555, 553f-555f
    cells of, 151
    chromosomes of, 138, 139, 139f, 181-184
    circulatory system of, 485-487, 485f-487f
    cloning of, 244
    daily activity of, 472
    digestive system of, 502, 502f
    dominant and recessive traits in, 172t
    ear of, 565-566, 565f
    embryonic development of, 597-601
    evolution of, 342-343, 378-379, 378f
    extinctions due to activity of, 449, 449f
    eyes of, 68
    facial expressions of, 439
    female reproductive system of, 593-594, 593f, 594f
    genomes of, 176, 177f, 208, 226f, 228-229, 229f, 229t
    global change caused by, 444-450
    heart of, 485-486, 485f, 495
    height of, 178, 178f
    IQ scores of, 166-167
    kidneys of, 518, 518f
    language of, 439
    male reproductive system of, 591-592, 591f, 592f
    in nitrogen cycle, 414
    oldest, 337
    as omnivores, 501, 503
    pedigrees of, 185-186, 185f, 186f
    pheromones in, 428-429
    in phosphorus cycle, 414
    population growth of, 385, 453, 453f
    predation by, 396
    protist's impact on, 314
    respiratory system of, 490-491, 490f
    in secondary succession, 398
    sexual orientation and, 590
    similar facial features in families of, 168f, 172
    skeleton of, 470, 470f
    social behavior of, 439, 439f
    survivorship curves for, 390, 390f
    teeth of, 503, 504f
    urine production by, 519
    water use by, 410, 452
Humboldt, Alexander von, 411
Hummingbird *(Eulampis)*, 523
Humoral immune response, 533, 535-536, 535f
Hunter-gatherers, 130
Hunter, John, 278
Hunting, of deer, 387
Hurricanes, 447, 448
Hybrids
    dihybrid individuals and, 175, 175f
    reproductive isolation and, 273, 273t, 274f, 275, 283
Hydra, 350, 350f
    digestive system of, 501, 501f
    survivorship curves for, 390, 390f
Hydration shells, 43, 43f
Hydraulic fracturing (fracking), 447
Hydraulic skeletons, 469, 469f
Hydrochloric acid (HCl), 44, 505-506, 506f
Hydrocortisone, 583
Hydrogen
    atomic number and mass number of, 37t
    in covalent bonds, 40, 40f
    in hydrogen bonds, 39, 41-43, 41f-43f
Hydrogenated fats, 61
Hydrogen bonds, 39, 41-43, 41f-43f
Hydrogen ions, 44, 45
Hydrogen peroxide (H_2O_2)
    antiseptic properties of, 103
    decomposition curves, 103, 103f
    tissue samples, 103
Hydroids, 350f
Hydrolysis, 51, 51f
Hydrophilic molecules, 43
Hydrophobic molecules, 43
Hydrothermal vents, 419
Hydroxide ions, 44
Hydroxyapatite, 466
Hydroxyurea, 270
Hyenas, 408
Hypertonic solutions, 83, 83f
Hyphae, of fungi, 316-317
Hypogonadism, 49
Hypothalamus, 553f, 554f
    anterior pituitary regulation by, 580, 580f
    body temperature regulated by, 517
    control function of, 554
    homeostasis and, 517, 554
    neuroendocrine function of, 554
    posterior pituitary regulation by, 579, 579f
    reproductive regulation by, 595-596
Hypotheses, 24-25, 24f
Hypotonic solutions, 83, 83f

I

Ice, formation of, 42, 42f, 42t
Ice caps. *See* Polar ice caps
ID (intelligent design), 26, 261
IgG, 535
Ileum, 507
Illustrations, in textbooks, 8, 8f
Immigration, 264
Immovable joints, 470
Immune connective tissue, 465
Immune responses
    active, 537, 537f
    antigen specificity of, 545
    blood cells and, 479
    cellular, 533-534, 534f
    cellular counterattack in, 526, 529-530, 529f, 530f
    humoral, 533, 535-536, 535f
    inflammatory, 530
    to influenza, 533
    initiation of, 533, 533f
    macrophages in, 529, 529f, 531t, 533, 533f, 534f, 543, 543f
    natural killer cells in, 529-530, 530f, 531t
    neutrophils in, 529-530, 531t
    primary, 537, 537f, 545f
    proteins in, 529, 530, 530f
    secondary, 537, 545f
    skin as first line of defense in, 526, 526f, 528, 528f
    specific immunity in, 526, 531-532, 531t
    temperature, 529, 530
    vaccination and, 525
Immune systems
    defeat of, 542-543
    HIV attack on, 543, 543f
    overactive, 542, 542f
    of vertebrates, 463
Imprinting, 432, 432f
Inbreeding, 265
Incisors, 503, 503f
Incomplete dominance, 179, 179f
Independent assortment, 157, 158f, 162, 162f, 175, 175f
Independent variables, 13
Individualistic community concept, 391
Industrial melanism, 271-272, 271f, 272f

Inflammatory response, 530
Influenza, 306
　deadly cases of, 540, 540f
　immune response to, 533
　pandemics, 309
　vaccine for, 235, 540
Ingroups, 284
Inhalation, 491, 491f
Inheritance. *See also* Genes; Genetics; Heredity
　chromosomal theory of, 30, 30f
　polygenic, 178
Inhibitory synapses, 552
Initiation complex, 220
Innate (instinctive) behavior, 430, 431f
Innate releasing mechanism, 430, 431f
Inner cell mass, 597
Inner ear, 561, 561f, 565
Innocence Project, 236-237
Inquiry and Analysis questions, 10, 15
Insects. *See also specific types*
　body temperature of, 517
　circulatory system of, 478, 478f
　cooperation in, 23
　diversity of, 362f
　respiration of, 488, 488f
　societies of, 438
Insertion in DNA, 198, 198t
Instinctive (innate) behavior, 430
Insulin, 573, 574, 581, 581f
　genetically engineered production of, 230, 234
　response to, 513, 517, 517f
Integrating center, 516, 516f
Integration, synaptic, 552, 552f
Integumentary systems, of vertebrates, 463
Intelligence, poverty and, 167
Intelligence quotient (IQ) scores, 166-167
Intelligent design (ID), 26, 261
Intensity, sound, 565-566
Interleukins, 533-534, 534f
Intermediate filaments, 80-81
Intermembrane space, 78, 78f, 125
Internal compartmentalization, 300
Internet resources, in textbooks, 9-10, 9f, 10f
Interneurons. *See* Association neurons
Interoreceptors, 560
Interspecific competition, 392
Intervals, studying at, 6
Intestines, 502, 502f, 502f, 507-508, 507f, 508f, 510f
Intraspecific competition, 392
Intrauterine devices (IUDs), 602
Intrinsic pathway, 262, 262f
Introduced species, 449, 449f, 454
Introns, 216, 216f
Invertebrates. *See also specific types*
　asexual reproduction by, 152
　body temperature of, 517
　respiratory system of, 489, 489f
　societies of, 438
Inverted biomass pyramids, 409, 409f
in vitro fertilization, 156
Involuntary (autonomic) nervous system, 558-559, 559f
Ion channels, 84, 549-552, 550f
Ionic bonds, 39, 39f
Ionization, 44
Ions, 38, 38f
　in blood plasma, 483
IQ (intelligence quotient) scores, 166-167

Iron
　atomic number and mass number of, 37t
　deficiency, 239
　Isotopes, 38, 38f, 47
　in oceans, 104-105, 104f, 105f, 117, 117f
Irons, Ernest, 176, 269f
Irwin, Rebecca, 639, 639f
Island ecosystems, 322-323, 322f, 323f, 342-343, 401
Islets of Langerhans, 509, 517, 581
Isotonic solutions, 83, 83f
IUDs (intrauterine devices), 602
Ivory-billed woodpecker *(Campephilus principalis)*, 402-403, 403f
Ivy, 619f

J

Jackals, 250-251
James, LeBron, 562-563
Japanese four o'clocks, 179, 179f
Jejunum, 507
Jellyfish, 350, 350f
　hydraulic skeletons of, 469
Jenner, Edward, 235, 540, 540f
Johnson, Eldred, 237
Joint appendages, in arthropods, 360
Joints, 470
Jones, Marion, 49
Journal of the American Heart Association, 127
Jurassic Park (Crichton), 241
Jurassic period, 368-369
Juul Vape Pen, 493
Juvenile-onset diabetes, 581

K

Kalanchoë daigremontiana, 628, 628f
Kangaroo rats, 23, 514-515, 514f, 515f
Kangaroos, 377f
Karyotypes, 138, 139, 139f
Kelp, 315t
Keratin, 52, 52f, 55
Kettlewell, Bernard, 271-272
Key stimulus, 430
Keystone species, 391, 456, 456f
Kidneys, 472
　function of, 519, 519f
　hormones secreted by, 520, 520f
　mammalian, 518, 518f
Kidney stones, 518
Killer bees *(Apis mellifera mellifera),* 382-383, 382f, 383f, 437, 437f
Killer whales, 479f
Kilocalories, 475
Kinetic energy, 92, 92f
Kinetoplastids (Trypanosoma), 315t
Kingdoms, 18, 18f, 30, 282, 288-289, 288f, 289t
Klinefelter syndrome, 181, 183f, 184
Krebs cycle, 120-121, 123-126, 124f
K-selected adaptation, 389, 389t
Kuru, 58

L

Labels, food, 500
Labor, 600-601
Lab simulations, 10

lac operon, 218, 219f
Lactate dehydrogenase, 128, 277
Lacteals, 507, 507f
Lactic acid, 128, 128f, 133, 133f
Lactose, 60t, 98
Lactose intolerance, 98, 131
Lactuca canadensis, 274
Lactuca graminifolia, 274
Lager, 304
Lagging strands, 194f, 195, 195f
Lake ecosystems, 414, 420-421, 420f, 421f
Lake Washington, Seattle, 457, 457f
Lancelets, 366f
Land ecosystems, 422-425
Langerhans, Paul, 581
Language
　brain and, 555, 555f
　of honeybees, 437, 437f
　of humans, 439
Lanugo, 600
Large intestine, 502, 502f, 508, 510f
Larynx, 490, 490f, 504, 504f
Lateral line system, 373
Lateral meristems, 608, 608f, 616, 616f
Latitude, 415, 415f, 417, 417f
Law of independent assortment, 175
Law of segregation, 175
Laws
　antipollution, 272
　on endangered species, 443
　on marine mammals, 442
Leading strands, 194, 194f, 195f
Leaf-cutter ants, 438
Leaflets, 618, 618f
Learning
　in animals, 432
　associative and nonassociative, 432
　behavior influenced by, 432-433
　brain and, 555, 555f
　definition of, 4
　memory in, 2-5
　sleep and, 2-3
　stimulation in, 432
　by students, 2-10, 4f-6f
　visual, 9
Learning objectives, in textbooks, 7
LearnSmart Questions, 8
Leaves
　axils in, 615, 615f
　interior of, 619, 619f
　parts of, 618-619, 619f
　in photosynthesis, 106-107, 617
　shape of, 618-619, 619f
　types of, 617-618, 618f
　vascular, 329, 329f, 608, 608f
Lectins, 631
Leeuwenhoek, Anton van, 28
Left atrium, 485-486, 485f
Left ventricle, 485-486, 485f
Lemmings, 424
Leopard frogs *(Rana genus),* 275, 275f
Lepus americanus (snowshoe hare), 397, 397f
Lerman, Louis, 297f
Lesbian, gay, bisexual, transgender, and queer (LGBTQ), 590
Lettuce, wild, 274
Leukocytes (white blood cells), 484, 484f, 529-533, 530f, 531t

LGBTQ (lesbian, gay, bisexual, transgender, and queer), 590, 163
LH (luteinizing hormone), 574, 580, 580f, 593f, 594-596, 595f, 596f, 602
Lice, 395f
Lichens, 271-272, 319, 319f, 323
Life
    domains of, 30, 30f
    emergent properties of, 21, 384
    extraterrestrial, 298-299
    kingdoms of, 18, 18f, 30
    organization of, 20-21, 20f-21f
    origin of, 296-297, 296f, 297f
    properties of, 19, 19f
Life history adaptations, 389, 389t
Ligers, 274, 274f
Light. *See also* Sunlight
    eyes sensing, 567, 567f
    visible, 110
Light-dependent reactions, 109, 110-112, 111f
Light-independent reactions, 109
Light microscopes, 68, 68f
Likens, Gene, 426
Limbic systems, 554-555, 554f
Limestone, 412
Limnetic zone, 420, 420f
Linear scales, 13-14, 13f
Line graphs, 14, 14f, 15, 15f
Linking, in learning, 5, 5f, 7
Linnaean classification, 280-281, 280f
Linnaeus, Carolus, 280, 280f
Lions, 377f, 408
    ecological isolation of, 274, 274f
    parental care and, 435t
Lipase, 507
Lipid bilayer, 72
Lipids, 50
    cellular respiration of, 129, 129f
    formation of, 51f
    in plasma membranes, 61, 61f
Littoral zone, 420, 420f
Liver, 502, 502f, 510f
    digestive functions of, 509, 509f
    glucose metabolism in, 517, 517f
    plasma protein production in, 483
    regulatory functions of, 509
    secretions of, 507, 509
Liverworts, 325, 327t, 328
Lizards
    body temperature of, 517
    epithelium of, 464, 464f
    parthenogenesis by, 152, 588
Lobsters, 360
Locomotion, 467, 467f, 475, 475f
Locomotor organelles, protist, 312
Locusts, 388
Logging, 402-403, 411, 411f, 426, 426f
Log (logarithmic) scales, 13-14, 13f
Long-term memory, 4-5
Look-alike meter, 172
Loop of Henle, 519, 519f
Lord of the Flies (Golding), 438
Lorenz, Konrad, 251, 430, 432, 432f
Lumbricus terrestris, 357f
Lumen, 502, 502f
Luna moths, 362f
Lung cancer
    e-cigarettes and, 149
    incidence of, in United States, 585, 585f
    nature of, 146
    prevention of, 147
    smoking and, 11, 11f, 147, 147f, 199, 547, 585, 585f
Lungs, 488, 488f
    daily activity of, 472
    human, 490-491, 490f
Lupus, 542
Luria, Salvadore, 207
Luteal phase of menstrual cycle, 595-596, 595f
Luteinizing hormone (LH), 574, 580, 580f
Luteinizing hormone (LH), 593f, 594-596, 595f, 596f, 602
Lycopersicon lycopersicum (tomato plant), 610
Lycophyta, 327t, 330f, 331
Lycophyte trees, 329
Lycopods, 327t
Lyell, Charles, 254
Lymphatic systems, 529, 529f
Lymphocytes, 465, 484, 484f
Lynx canadensis (Canada lynx), 397, 397f
Lysosomes, 71, 75t, 77, 78f, 313
Lysozyme, 95, 95f, 528

M

Macroevolution, 273
Macromolecules. *See also* Carbohydrates; Lipids; Nucleic acids; Proteins
    anabolic steroids and, 48-49
    definition of, 50
    formation of, 50-51, 51f
    in hierarchy of complexity, 20, 20f
    origin of first, 297
    types of, 52-55
Macronucleus, 65
Macrophages, 465, 484
    as antigen-presenting cells, 533-534, 533f, 534f
    in immune response, 529, 529f, 531t, 533, 533f, 534f, 543, 543f
Mad cow disease, 58
Magnesium, plant absorption of, 324
Magnetic field, in compass sense, 571
Magnetic hypothesis, 571
Malacidin, 223, 301
Malaria, 187, 187f, 270, 270f, 315t
    CRISPR gene drive for treatment of, 203-204, 203f
Malate, 115, 115f
Male reproductive systems, 591-592, 591f, 592f
Malignant melanoma, 134-135, 135f
Malthus, Thomas, 255, 256
Maltose, 504
Mammals. *See also specific types*
    body temperature of, 517
    in Cenozoic era, 369
    characteristics of, 377
    evolution of, 369, 377
    hair of, 377
    heart of, 485-486, 485f, 495
    kidneys of, 518, 518f
    mammary glands of, 377
    in Mesozoic era, 369
    middle-ear bones of, 377
    nitrogenous waste elimination by, 521, 521f
    respiration of, 488, 488f
    respiratory system of, 489-492
    sex determination in, 487, 487f, 589, 589f
    teeth of, 503, 503f
Mammary glands, 377, 601
Mantle of mollusks, 355
Maple syrup, 607, 607f, 623, 623f
Maple trees, 606-607, 606f, 607f, 623, 623f
Map sense, 436, 436f
Marburg virus, 306
Mare reproductive loss syndrome, 34
Marginal meristems, 618
Marijuana, 114, 339
Marine Mammal Protection Act, 442
Mars, life on, 298
Marsupials, 377, 377f
Martin, John, 104-105
Mash, 304
Masks, wearing, 311
Mass, 36
Mass extinctions, 369
Mass flow, 623, 623f
Mass numbers, 36, 37t
Mast cells, 531, 531t, 542, 542f
Maternal age, 182-183, 182f
Maternal antibodies, 535
Mating, nonrandom, 264t, 265
Matrix, mitochondrial, 78, 78f, 125
Matter, 36
Maximal sustainable yield, 388f, 389
McClintock, Martha, 429
McGwire, Mark, 49
McKormick, Robert, 253
Measurement, units of, 14
Mechanical isolation, 273t, 275
Mechanoreceptors, 569
Media coverage, of biology, 11, 11f
Medical marijuana, 114
Medicine
    biodiversity in, 288
    genetic engineering in, 230, 230f, 234, 234f, 238, 238f
    stem cell treatment in, 244-245, 244f, 245f
Mediterranean climate, 417
Medulla oblongata, 553f, 554f, 555
Medusae, 350, 350f
Mega-reserves, 456, 456f
Megaspores, 332, 629, 629f
Meiosis
    crossing over in, 157, 157f, 162
    definition of, 154
    discovery of, 154
    evolution of, 152, 162
    in human female reproduction, 593-594, 593f
    in human male reproduction, 591f, 592
    mitosis compared to, 160, 160f, 161f
    nondisjunction in, 181-182, 181f
    in plants, 325
    in sexual life cycle, 155, 155f
    stages of, 157-159, 158f, 159f
Meiosis I, 156-158, 158f, 181, 181f
Meiosis II, 157-159, 159f, 191, 594
Melanin, 172t
Melanism, industrial, 271-272, 271f, 272f
Melanocyte-stimulating hormone (MSH), 580, 580f
Melanoma, 134-135, 135f
Mello, Craig, 221
Melospiza melodia (song sparrow), 388, 388f, 401, 401f
Melville, Herman, 258
Membrane attack complex, 530
Membrane potentials, 549-550, 550f
Membrane proteins, 72-73

Memorization, 5
Memory
　brain and, 555, 555f
　in learning, 2-5
Memory B cells, 531t, 536, 537
Memory RNA, 536, 536f
Memory T cells, 531t, 532
Mendel, Gregor
　chromosomal theory of inheritance and, 30
　experiments of, 168-171, 169f-171f, 170t
　laws of, 175
　theory of heredity proposed by, 30, 172-174, 173f, 174f
Mendelian segregation, 30
Mendelian traits, 175
Menstrual cycle, 595-596, 595f-596f
Menstrual synchrony, 429
Menstruation, 595f, 596
Meristems
　apical, 608, 608f, 613, 613f
　ground, 613, 613f
　lateral, 608, 608f, 616, 616f
　marginal, 618
Mesoderm, 352, 352f
Mesophyll cells, 107, 621-622
Mesophylls, 619, 619f
Mesozoic era, 368-369
Messenger RNA (mRNA), 176, 177f, 211, 221, 221f
Metabolic water, 515
Metabolism
　crassulacean acid, 115
　as emergent property of life, 21
　eukaryotic compared to prokaryotic, 300t
　oxidative, 78, 523
　as property of life, 19, 19f
Metabolites, 100t, 483
Metaphase II, meiosis II, 159, 159f
Metaphase I, in meiosis I, 157, 157f
Metastases, 143
Methadone, 556
Methane, 40, 40f
Metric system, 14, 475
Mice
　behavioral genetics in, 431, 431f
　BPA tests on, 577
　embryo of, 367f
　natural selection for melanism in, 272
Microbead-Free Waters Act, 527
Microbeads, plastic, 527
Microevolution, 273
Microfilaments, 80-81
Micrographs, 165, 165f
Micro ribonucleic acid (miRNA), 221, 221f
Microscopes, 28, 66, 68, 68f
Microspheres, 297
Microspores, 332, 629, 629f
Microsporidia, 319t
Microtubules, 80-81, 165, 165f
Microvilli
　eukaryotic, 70f
　intestinal, 507-508, 507f, 508f
　taste bud, 564, 564f
Midbrain, 555
Mid-digital hair, 172t
Middle-ear bones, of mammals, 377
Mifepristone (RU-486), 602
Migration
　as agent of evolution, 264, 264t
　behavior and, 435t, 436, 436f

　of birds, 436, 436f
　definition of, 264
　of *Homo sapiens*, 379, 379f
Migratory behavior, 435t, 436, 436f
Milk, breast, 579-580, 601
Miller, Stanley, 296-297
Millipedes, 360
Minerals, plant absorption of, 620, 620f
miRNA (micro ribonucleic acid), 221, 221f
Mites, 360
Mitochondria
　endosymbiosis and, 79
　eukaryotic, 70, 70f, 71f, 75t, 78, 78f
　matrix of, 78, 78f, 125
　in vitro fertilization problems with, 156
Mitochondrial DNA (mtDNA), 78, 183
Mitochondrial medicine, 78
Mitosis
　in fungi, 316
　meiosis compared to, 160, 160f, 161f
　in protists, 313
Mitral (bicuspid) valve, 485-486, 485f
Mobb Deep, 176
Molars, 503, 503f
Molecular record, of evolution, 260, 260f
Molecules, 39-41. *See also specific types*
　of hemoglobin, 492, 492f
　in hierarchy of complexity, 20, 20f
Mollusks, 355-356, 355f
Molting, 358
　evolution of, 346
Monoclonal antibodies, 525
Monocots (monocotyledons), 338, 616, 616f, 618, 619f
Monocytes, 484, 484f, 529, 530, 531, 531t
Monoecious plants, 629
Monomers, 50, 51f
Monosaccharides, 59, 59f
Monosomy, 138, 181
Monotremes, 377, 377f
Moore, Mary Tyler, 244f
Moose, 396, 396f
Morels, 316f
Mormon tea, 327t
"Morning after pill" (Plan B), 602
Morphine, 556
Mortality rates, 390
Mose, Edvard, 562-563
Mose, May-Britt, 562-563
Mosquitos, 406
　malaria and, 203-204, 203f
　West Nile virus and, 307
Mosses
　life cycle of, 328, 328f
　in succession, 398
Moths
　body temperature of, 517
　industrial melanism in, 271-272, 271f, 272f
　structure and function in, 23
Motion receptors, 561, 561f
Motor neurons, 468, 468t, 548-549, 548f, 557
Mountains, 417
Mouth, 502, 502f, 504, 504f, 510f
Movement
　of animals, 344, 345t
　motion receptors and, 561, 561f
　navigation and, 562-563
　as property of life, 19
　of protists, 312

mRNA (messenger RNA), 176, 177f, 221, 221f
MSH (melanocyte-stimulating hormone), 580, 580f
mtDNA (mitochondrial DNA), 78, 183
Mucosa, 502, 502f
Mules, 283, 283f
Muller, Herman, 153
Muller's Ratchet, 152
Multicellularity
　of animals, 344, 344t
　of protists, 313
　of sponges, 348-349
Multiple sclerosis, 542
Mu opiate, 556
Muscle fibers, 467, 467f
Muscle filaments, 471
Muscles
　anabolic steroids and, 48-49, 578
　calcium and, 582-583
　cardiac, 467, 471
　contraction of, 100t, 473, 473f, 560, 560f
　skeletal, 467, 467f, 471, 471f, 560, 560f
　smooth, 467, 471
　of vertebrates, 467, 467f, 471, 473
Muscle tissues, 462, 462f, 467
Muscularis, 502, 502f
Muscular systems, of vertebrates, 463, 471, 471f
Mushrooms, 316-317, 316f, 317f, 318f
Musical instruments, from wood, 616
Mussels, 355, 439
Mutations
　as agent of evolution, 198, 264, 264t
　in cancer, 143, 146
　CRISPR correcting, 202
　definition of, 143, 176, 196, 264
　HIV/AIDS and, 525
　paternal age and, 197, 197f
　phenotype alteration by, 176
　protecting against, 199
　as random compared to directed, 207, 207f
　in sexual compared to asexual reproduction, 153
　types of, 196, 196f, 198, 198t
Mutualism, 395, 395f
Mycelium, 316-317, 316f
Mycorrhizae, 319, 319f, 324, 614
Myelin sheath, 542, 548, 548f
Myoblasts, 511
Myofibrils, 467, 467f, 471
Myofilaments, 467, 473, 473f
Myosin filaments, 471, 473, 473f
MyPlate, 498, 498f
Myrmecia, 138

N

NAD$^+$, 123-124, 128, 128f
NADH, 123-124, 128, 128f
NADPH (nicotinamide adenine dinucleotide phosphate), 109, 112
Na$^+$-K$^+$ (sodium-potassium) pump, 87, 87f, 100t, 549-550, 550f
Naloxone, 556
Nashua River, 457, 457f
National Academy of Medicine, 204
National Academy of Science, 204
National Collaborative Prenatal Project, 167
National Institutes of Health, 167
Natural killer cells, 529-530, 530f, 531t

Natural selection
    adaptation within populations and, 269-272
    Darwin, C.'s, theory of, 252, 255-256
    definition of, 22, 252, 255, 265
    directional, 266f, 268
    disruptive, 266f, 268, 268f
    enzyme polymorphism and, 277
    industrial melanism in, 271-272, 271f, 272f
    in speciation, 273-275
    stabilizing, 266f, 267, 267f, 270
Natural Theology (Paley), 261
Nautiluses, 355
Navigation, 562-563
Neanderthals, 342-343, 378f, 379, 379f
Negative feedback, 516, 516f, 595-596
Neisseria gonorrhoeae, 603
Nematocysts, 349, 351
Nematodes (Nematoda), 354, 358-359, 358f, 359f
    digestive system of, 501, 501f
    gene expression in, 221
    genome of, 228
Neocallimastigomycota, 319t
Nephridia, 355
Nephron, 518, 519f
Nerve cords, of chordates, 366
Nerve impulses, 468, 468f, 548-550, 548f, 550f
Nerve stem cells, 244
Nerve tissues, 462, 462f, 468, 468f, 468t
Nervous systems
    autonomic (involuntary), 558-559, 559f
    cells of, 67
    central, 548-549, 553-557
    organization of, 548, 548f
    parasympathetic, 559, 559f
    peripheral, 549, 558-559
    sensory, 560-568
    somatic (voluntary), 558
    sympathetic, 558-559, 559f
    of vertebrates, 463, 548, 548f, 557, 557f
Neufeld, Peter, 237
Neural tubes, 597
Neurons
    nerve impulses generated by, 549-550, 550f
    postsynaptic, 551, 551f
    presynaptic, 551, 551f
    structure of, 468, 468f, 548-549, 548f
    types of, 468, 468t, 548-549, 548f
Neurotransmitters, 468, 551-552, 551f
Neutrons, 36, 36f
Neutrophils, 484, 484f, 529-530, 531t
Newts, 141f
Niches, 392-393, 392f
Nicotiana tabacum (tobacco leaf), 610f
Nicotinamide adenine dinucleotide phosphate (NADPH), 109, 112
Nicotine, 149, 199, 546-547
Nile perch, 454
Nitrogen
    atomic number and mass number of, 37t
    plant absorption of, 324
    structure of, 37, 37f
Nitrogen cycle, 410, 413, 413f, 614
Nitrogen fixation, 413, 413f
Nitrogenous wastes, 521, 521f
Nodes, stem, 615, 615f
Noel, Walter Clement, 269f
Nonassociative learning, 432

Noncoding DNA, 229, 229t
Noncompetitive inhibition, 97, 97f
Nondisjunction, 181-184, 181f, 183f
Nonpolar molecules, 43
Nonrandom mating, 264t, 265
Nonvascular plants, 326, 326f, 328
Norepinephrine, 552, 583
Note taking, 4, 6, 7
Notochords, 366
Nuclear envelope, 74, 74f
Nucleariida, 315t
Nuclear membrane, 74, 74f
Nuclear pores, 74, 74f
Nucleic acids, 50, 51f
Nucleoid region, 69
Nucleolus, eukaryotic, 74, 74f, 75t
Nucleosomes, 139, 139f
Nucleotides, 56, 56f, 192, 192f
Nucleus
    eukaryotic, 70, 70f, 71f, 74, 74f, 75t
    prokaryotic, 69
Numbers, pyramids of, 409, 409f
Nutrition, 498-499, 498f, 499f
    as circulatory system function, 479
    fish and, 372
    of fungi, 317, 317f
    GM crops and, 239
    of plants, 620-623
    of protists, 312-313
    vegans and, 509
Nutritional labels, 50, 50f
Nutrition plate, 498, 498f

O

Oak trees, 280, 280f, 327t
Obesity
    bacteria and, 496-497
    BMI and, 498, 499f
    fat cells and, 465
    food labels and, 500
    measurement of, 498, 499f
    type II diabetes and, 12, 12f, 573, 573f
Observation, 24, 24f
Occipital lobe, 554, 554f
Oceans
    ecosystems of, 418-419, 418f, 419f
    fertilization of, 104-105, 104f, 105f
    iron in, 104-105, 104f, 105f
    origin of life in, 296-297, 297f
    rising sea levels in, 445, 448
Octopuses, 355
Oil glands, 528
O'Keefe, John, 562-563
Oligotrophic lakes, 420f, 421
Olympics, anabolic steroid use in, 49
Omega-3 fatty acid supplements, 499
Omnivores, 90-91, 405f, 406, 501, 503
Oncogenes, 143
On the Origin of Species (Darwin, C.), 252, 256, 257
Oocytes, 593-594, 593f
Oogenesis, 588
Oomycetes, 315t
Open circulatory system, 478, 478f
Open-sea ecosystems, 418-419, 418f
Operant conditioning, 432
Operculum, 373

Operons, 219, 219f
Opioids, 556
Optimal foraging theory, 434, 434f, 439
Oral contraceptives, 602, 603f
Orbitals, 37, 37f
Order (taxonomic rank), 282
Ordovician period, 369
Oregon newt *(Taricha granulosa)*, 141f
Organelles. See also *specific types*
    eukaryotic, 70-71, 70f, 71f, 75t, 78-79
    in hierarchy of complexity, 20, 20f
    prokaryotic, 69
    protist, 312
Organic food labels, 500
Organic molecules, 50. See also *specific types*
Organismic water cycle, 410, 410f, 411
Organisms
    classification of, 280-291
    in hierarchy of complexity, 20f, 21
Organization, in learning, 5
Organogenesis, 598-599, 598f-599f
Organs. See also *specific organs*
    of digestive system, 510f
    in hierarchy of complexity, 20f, 21
    of vertebrates, 462, 463f
Organ systems, 20f, 21
    of vertebrates, 463, 463f
Organ transplants, from pigs, 201
Orientation, magnetic field and, 571
Ornithorhynchus anatinus, 278, 569
Osmoregulation, 518-521, 518f, 519f
Osmosis, 82-83, 83f, 620-621
Osmotic concentration, 83
Osmotic pressure, 83, 83f
Osteichthyes, 293
Osteoblasts, 466
Osteoclasts, 466
Osteocytes, 466
Osteoporosis, 460-461, 460f, 466, 466f
Otoliths, 561, 561f
Outgroups, 284
Ovalbumin, 216f
Ovarian follicles, 595-596, 595f
Ovaries (flower), 336, 336f, 629, 629f
Ovaries (human), 588, 589, 593-596, 593f-596f
Overdoses, drug, 556
Overexploitation of species, 449, 449f
Overharvesting, 389
Overpasses, for deer, 399
Overweight
    bacteria and, 496-497
    BMI and, 498, 499f
    fat cells and, 465
    food labels and, 500
    measurement of, 498, 499f
Oviducts (fallopian tubes), 593f, 594, 594f
Ovulation, 593-596, 593f, 595f, 596f
Ovules, of flowers, 332, 630
Ovum, 593f, 594, 596, 596f
Oxalis oregana, 384f
Oxaloacetate, 115, 115f
Oxidation, 120-121, 120f
Oxidation-reduction (redox) reactions, 120, 120f
Oxidative metabolism, 78, 523
Oxidative respiration, 123-125, 488
    efficiency of, 408
Oxycodone, 556
OxyContin, 556

Oxygen
    in atmosphere, 296–297
    atomic number and mass number of, 37t
    blood transport of, 479, 481f, 485–486, 485f
    in cellular respiration, 123–125, 408, 488
    in covalent bonds, 40f
    in freshwater ecosystems, 421, 421f
    in hydrogen bonds, 41, 41f
    memory and, 4
    respiration transport of, 492
Oxygen concentration, 63
Oxygen loading curves, 63, 63f
Oxytocin, 580, 601
Oystercatchers, 435t
Oyster mushroom (Pleurotus ostreatus), 318f
Oysters, 355
    survivorship curves for, 390, 390f
Ozone hole
    causes of, 12, 12f
    graphs of, 14, 14f, 15, 15f
    scientific processes applied to, 25, 25f

P

P generation, 169
p53 proteins, 146
Paabo, Svante, 251, 343
Pacemakers, 487
Pain, sensory receptors for, 560
Paleoanthropology, 342
Paleo diet, 118–119, 130–131
Paleozoic era, 368, 368f
Paley, William, 261, 262
Palisade mesophyll, 619, 619f
Palmately compound leaves, 618, 618f
Palmer amaranth (waterhemp), 226
Palm trees, 632f
Palo verde trees (Cercidium floridum), 232, 333f
Pancreas
    blood glucose regulated by, 517, 511f
    digestive function of, 502, 502f, 507, 509, 509f, 510f
    endocrine functions of, 509, 581
    exocrine functions of, 509
    hormones of, 509, 581, 581f
Pancreatic cancer, 581, 581f
Pancreatic duct, 509, 509f
Pancreatic islets, 517
Pandemic, 303
    COVID-19, 17
    influenza, 309
    outbreak spreads, 17
    SARS, 17
Papillae, of tongue, 564, 564f
Parabasalids (Trichomonas), 315t
Paramecium
    characteristics of, 315t
    competition among, 393, 393f
    predation on, 396
    reproduction by, 313f
Paraphrasing, 4
Parasaurolophus, 257f
Parasites, 312
Parasitism, 395, 395f
Parasympathetic nervous systems, 559, 559f
Parathyroid glands, 582–583, 583f
Parathyroid hormone (PTH), 582–583, 583f
Parazoa, 346
Parenchyma cells, 609, 609f, 613

Parental care, 435t
Parietal cells, 505, 506f
Parietal lobe, 554, 554f
Paris Climate Accord, 27
Parkinson's disease, 244, 244f
Parthenogenesis, 152, 588
Partial charges, 40, 41
Paternal age, mutations and, 197, 197f
Pathogens, 464f, 526
Pavlov, Ivan, 432
PCR (polymerase chain reaction), 232
Peaches, 632, 632f
Peanuts, 239t
Pea plants, 168–171, 168f–171f, 170t
Pearls, 355
Peat moss, 328
Pectoral girdle, 470
Pedigrees, 185–186, 185f, 186f
Pelvic girdle, 470
Pelvic inflammatory disease (PID), 603
Pelycosaurs, 368, 368f
Penicillin, 97, 223, 300, 301
Penicillium, 138
Penis, 592, 592f
Peppered moth (Biston betularia), 271–272, 271f
Pepsin, 96, 96f, 505–506
Peptide bonds, 53, 53f
Peptide hormones, 578, 578f
Perch, 293, 293f, 454
Percodan, 556
Peregrine falcons (Falco peregrinus), 455, 455f
Perforin, 529, 530f
Pericycle, 613, 613f
Periderm, 616f, 617
Peripheral nervous system (PNS), 549, 558–559
Peristalsis, 505
Periwinkle, 619f
Permafrost, 424
Permeability, selective, 84, 84f
Permian period, 369
Pest resistance, 230f, 238, 242
PET (positron-emission tomography), 38, 429
PET/CT (positron-emission tomography/computed tomography), 38
Petals, 336, 336f
Petioles of leaves, 618
Petrochemicals, 527
Petunia plants, 238f
pH
    of acid rain, 45
    of blood, 44
    definition of, 44
    in enzyme activity, 96, 96f
    of fish blood, 133, 133f
    scale of, 44, 44f
    of stomach, 528
Phaeophyta, 313, 315t
Phagocytes, 529, 530
Phagocytosis, 84, 84f
Phagosomes, 313
Phagotrophs, 313
Pharyngeal pouches, 259, 259f, 367
Pharynx, 490, 490f, 502, 502f, 510f
Phenotypes
    definition of, 172
    DNA in, 176, 177f
    Hardy-Weinberg equilibrium in, 263, 263f
    in Mendel's theory of heredity, 172–174

    mutations altering, 176
    proteins determining, 176, 177f
Phenylthiocarbamide (PTC) sensitivity, 172t
Pheromones
    of honeybees, 438
    in humans, 428–429
Phloem, 329, 612, 612f, 613, 613f, 616, 616f, 623, 623f, 625
Phospholipids, 61, 61f, 72
Phosphorus, 37t
    plant absorption of, 324
Phosphorus cycle, 410, 414, 414f
Phosphorylation, substrate-level, 122
Photons, 110, 110f
Photoperiodism, 636–637, 636f
Photoreceptor hypothesis, 571
Photorespiration, 115, 115f
Photosynthesis
    ATP production by, 99
    C_3 (Calvin cycle), 109, 113, 113f
    C_4, 115, 115f
    chloroplasts in, 107–109
    energy capture from sunlight in, 110–111, 110f, 111f
    in flow of energy, 92, 404
    leaves in, 106–107, 617
    light-dependent reactions of, 109, 110–112, 111f
    light-independent reactions of, 109
    in ocean ecosystems, 418–419
    overview of, 106–109
    photorespiration and, 115, 115f
    photosystems and, 108–112, 110f–112f
    pigments in, 110–111
    stomata in, 115, 115f
    trees in, 106
Photosynthesizers, 408
Photosystems, 108–112, 110f–112f
Phototrophs, 312
Phototropism, 268, 268f, 635f, 637, 637f
Phylogenetic trees, 284–285
    protist, 314
    of vertebrates, 371f
Phylogenies, 284
Phylum (phyla), 282, 326, 327t
Phylum Facts illustrations, 8, 8f
Physcomitrella patens, 328
Phytate, 239
Phytochrome, 637
Phytoplankton, in ocean ecosystems, 418–419
Pickling, 55
PID (pelvic inflammatory disease), 603
Pie charts, 14, 14f, 277, 277f
Piggyback (subunit) vaccines, 235
Pigments
    albinism and lack of, 172t
    definition of, 110
    environment and, 179
    flower forms of, 637
    in photosynthesis, 110–111
Pigs, organ transplants from, 201
Pili, 69, 302
Pimples, 526
Pineal gland, 553f
Pine trees
    leaves of, 618f
    life cycle of, 325, 325f
    phyla of, 327t
    population growth in, 386, 386f
    rings and age of, 617, 617f

Pinnately compound leaves, 618, 618f
Pinocytosis, 84, 84f
Pioneering communities, 398
Pistol shrimp, 395, 395f
Pisum sativum, 168f
Pith, 616, 616f
Pituitary gland, 553f, 554, 579-580, 579f, 580f
Place cells, 562-563
Placenta, 597, 600, 600f
Planaria, 353f
    circulatory system of, 478, 478f
    digestive system of, 501, 501f
Plan B ("morning after pill"), 602
Plankton, 409, 409f, 418-419, 418f
Plant cells
    osmotic pressure in, 83, 83f
    structure of, 71f
    walls of, 609, 609f
Plants (Plantae). *See also* Leaves; Roots; Stems; Vascular plants
    adaptation to terrestrial life, 324-325
    body of, 608, 608f, 613-619
    carbohydrate transportation in, 623, 623f
    cooperation in, 23
    cuticles of, 324-325, 324f, 610
    differentiation stages of, 634, 634f
    DNA fingerprinting of, 232
    dormancy of, 637
    ecological role of, 324f
    environmental stimuli responses of, 636-637, 636f, 637f
    epidermis of, 107, 610, 613, 613f
    evolution of, 326, 326f
    extinction factors for, 449, 449f
    fungal association with, 324
    fungi compared to, 316
    growth of, 329, 337, 634-635
    hormones of, 634-635
    kingdom of, 18f, 288, 289t
    life cycle of, 325, 325f
    marijuana, 114
    meiosis in, 325
    mineral absorption by, 324, 620, 620f
    nonvascular, 326, 326f, 328
    nutrition of, 620-623
    oldest, 337
    photoperiodism in, 636-637, 636f
    phyla of, 326, 327t
    as producers and consumers, 405
    regeneration in, 634, 634f
    reproduction by, 325, 628-633, 628f-632f
    sexual life cycles in, 155, 155f
    soil salinity and, 341
    spores of, 325
    in succession, 398
    symbiosis in, 324
    tissues, structure and function of, 608-612
    in trophic levels, 406
    water conservation in, 324-325, 614
    in water cycle, 411
    water movement in, 611, 611f, 620-622, 620f-622f, 625
Plasma, 465, 481f, 483, 483f
Plasma cells, 531, 531t, 532, 535-536, 535f
Plasma membranes, 19
    bulk passage across, 84, 84f
    diffusion across, 82, 82f
    eukaryotic, 70, 70f, 72-73, 75t
    lipids in, 61, 61f
    osmosis and, 82-83, 83f
    prokaryotic, 69, 300t
    selective permeability of, 84, 84f
    structure of, 72-73
    transport across, 82-87
Plasmids, 302, 302f
Plasmodesmata, 71, 71f, 612
Plasmodium, 314, 315t
Plastic microbeads, 527
Platelets, 484, 484f
Platyhelminthes, 352. *See also* Flatworms
Platypus, 278-279, 278f, 279f, 569
Pleiotropic effects, 178-179, 178f
Pleurotus ostreatus (oyster mushroom), 318f
PNS (peripheral nervous system), 549
Point mutations, 198, 198t
Poisonous mushrooms, 318f
Polar bears, 442-443, 442f, 443f
Polar bodies, 593f, 594
Polar ice, 27
Polar ice caps, 425, 425f
    global warming and, 443, 443f
Polarity, of water, 42t, 43
Polar molecules, 40, 41, 41f
Polio, 525
Pollen, 332, 629-630
Pollen cones, 335, 335f
Pollen grains, 332, 335, 335f, 629, 629f
Pollen tubes, 630
Pollination, 332
    by bats, 394f
    flower color and, 630, 630f, 639, 639f, 639f
    self-, 630
    wind, 630
Pollution
    acid rain, 45, 444
    air, 45, 271-272, 444
    bottled water and, 444
    chemical, 411, 414, 444, 444f
    clean up of, 454
    industrial melanism and, 271-272
    plastic microbeads and, 527
    water, 411, 414, 444
Polydactyly, 172t
Polygenic inheritance, 178
Polymerase, 151
Polymerase chain reaction (PCR), 232
Polymers, 50, 51f
Polynucleotide chains, 56
Polypeptides, 53, 574
Polyps, 350, 350f
Polysaccharides, 59, 59f, 60t
Polystichum munitum, 384f
Pomes, 632, 632f
Ponds, 64-65, 420-421, 420f, 421f
Pons, 553f, 555
Populations
    adaptation within, 269-272
    age distribution of, 390, 390f
    age structure of, 390, 390f
    definition of, 384
    Earth's limit for, 453
    evolution of, 263-272
    fungi causing declines in, 450
    geometric progression of, 255
    in hierarchy of complexity, 21, 21f, 384
    life history adaptations in, 389, 389t
    sex ratio of, 390
    viruses causing declining, 450
Population demography, 390
Population density, 385, 385f, 388, 388f, 401, 401f
Population growth
    carrying capacity and, 386, 386f, 389, 453
    curbing, 453, 453f
    definition of, 385
    human, 385, 453, 453f
    life history adaptations and, 389, 389t
    models of, 385-387, 386f
    population size influencing, 401, 401f
Population productivity, 388f, 389
Population size, 385, 401, 401f
Porifera. *See* Sponges
Porter, William, 387
Positive feedback, 595-596
Positron-emission tomography (PET), 429
Postanal tails, of chordates, 367
Posterior, definition of, 351
Posterior pituitary, 579, 579f
Postnatal development, 601
Postsynaptic neurons, 551, 551f
Postzygotic isolating mechanisms, 273, 273t, 275, 275f
Pot, 114
Potassium
    atomic number and mass number of, 37t
    in nerve impulse, 549-550, 550f
    plant absorption of, 324
Potatoes, 239t
Potential energy, 37, 37f, 92, 92f
Pott, Percivall, 199
Poverty, intelligence and, 167
Prairie dogs, 503f
Prairies, 423, 423f, 454
Precapillary sphincters, 480, 480f
Precipitation
    acid rain and, 45, 45f, 444
    global warming and, 445
    in water cycle, 410f
Predation
    cycles of, 396-397, 396f, 397f
    definition of, 396
    parasitism as, 395
    social behavior and, 438
Predictions, 24f, 25
Pregnancies. *See also* Embryonic development
    birth and, 600-601, 601f
    fetal alcohol syndrome and, 599
    fetal development in, 600-601, 600f
    maternal age in, 182-183, 182f
    organogenesis in, 598-599, 598f-599f
    paternal age and, 197, 197f
    testing for, 541
Premolars, 503, 503f
Pressure-flow hypothesis, 623
Pressure receptors, 560
Presynaptic neurons, 551, 551f
Prey, 396-397, 396f, 397f
Prezygotic isolating mechanisms, 273-275, 273t
Primary carbon footprint, 412
Primary cell walls, 609, 609f
Primary consumers, 405
Primary growth of plants, 329, 608, 608f
Primary immune response, 537, 537f, 545f
Primary oocytes, 593-594, 593f
Primary phloem, 616, 616f

Primary RNA transcript, 216
Primary structure of protein, 55, 55f
Primary succession, 398, 398f
Primary xylem, 616, 616f
Primers, 194, 194f
Primordial soup hypothesis, 296-297
Principles of Geology (Lyell), 254
Prions, 58
PRL (prolactin), 580, 580f, 601
Problem solving, 433, 433f
Procambium, 613, 613f
Prodigy, of Mobb Deep, 176
Producers, 404, 405, 405f
Products of reactions, 94, 94f
Profundal zone, 420, 420f
Progesterone, 576
Prokaryotes. *See also* Archaea; Bacteria
    cell cycle of, 136, 136f
    cell division in, 136, 136f, 300t
    cells of, 69, 69f, 300-301, 300t
    definition of, 300
    environment and, 302
    eukaryotes compared to, 300, 300t
    evolution of, 300
    gene expression in, 215, 215f, 216, 216f, 218-219
    genomes of, 228
    kingdoms of, 288, 289t
    protein synthesis in, 216, 216f
    reproduction by, 302, 302f
    as simple organisms, 69
    structure of, 69, 69f, 300
    synthesis of, 216, 216f, 217f
    transcription in, 216, 216f
    translation in, 216, 216f
Prolactin (PRL), 580, 580f, 601
Promoters, 210, 218-220, 219f, 220f
Proofreading, DNA and, 195
Prophase II, meiosis II, 158, 159f
Prophase I, in meiosis I, 157, 158f
Prostaglandins, 601
Prostate cancer, 592
Prostate gland, 591f, 592
Protease inhibitors, 543
Proteasome, 89
Proteins. *See also specific types*
    BLM, 158
    in blood plasma, 483
    carrier, 84, 84f
    CD4 receptor, 525, 525f
    cell surface, 73
    cellular respiration of, 129, 129f
    in central dogma, 210, 210f
    CFTR, 86
    complement, 530, 530f
    coupled transport, 87
    in cytoskeletons, 80
    deficiency of, 483
    definition of, 52
    folding of, 55, 55f
    food sources of, 498
    functions of, 52, 52f, 55, 55f
    gp120, 525, 525f
    in immune responses, 529, 530, 530f
    membrane, 72-73
    p53, 146
    phenotype determination by, 176, 177f
    structure of, 51f, 54, 54f, 55, 55f
    synthesis of, 225
    transmembrane, 73
    types of, 52, 52f
Protein's function, 63
Protists (Protista). *See also specific types; specific types*
    biology of, 312
    classification of, 314, 314f
    flagella of, 312
    genomes of, 228, 314, 314f
    impact on humans, 314
    kingdoms of, 18f, 288, 289t
    multicellularity of, 313
    phylogenetic tree for, 314, 314f
    reproduction by, 313, 313f, 588, 588f
    sexual life cycles in, 155, 155f
    types of, 314, 315t
Protoderm, 613, 613f
Protons, 36, 36f
Proto-oncogenes, 143
Protostomes, 346, 365, 365f
Proximal tubule, 519, 519f
Prusiner, Stanley, 58
Pseudocoel, 354, 354f
Pseudocoelomates, 354, 354f
Pseudopodia, 312, 314f
Pseudotsuga menziesii (Douglas fir), 626
PTC (phenylthiocarbamide) sensitivity, 172t
Pterophyta, 327t, 330, 330f
Pterosaurs, 369f
Pterostichus lama, 384f
PTH (parathyroid hormone), 582-583, 583f
Puffballs, 317, 317f
Pulmonary arteries, 485f, 486
Pulmonary semilunar valve, 485f, 486
Pulmonary veins, 485, 485f
Pulse rate, 495
Punctuality, learning and, 6
Punnett squares, 173, 173f
Purines, 192, 192f
Purkinje fibers, 487
Putting Concepts to Work, 8
Pygmy shrew, 495
Pyloric sphincter, 506
Pyrenestes ostrinus (black-bellied seedcracker finch), 268, 268f
Pyrimidines, 192, 192f
Pyruvate, 123, 123f, 128f
Pyruvate dehydrogenase, 123

Q

Quaternary structure of protein, 55, 55f
Quercus phellos (willow oak), 280, 280f
Quercus rubra (red oak), 280, 280f

R

Race, genetic disorders and, 287
Radial symmetry, 346, 352f
Radioactive decay, 38, 47
Radioactive isotopes, 38, 38f
Radioisotopic dating, 38, 47
Radiolarians, 315t
Radish *(Raphanus sativus),* 622f
Radula, 355
Rain
    acid, 45, 45f, 444
    global warming and, 445
    in water cycle, 410f, 411
Rain forests, 422, 422f
    loss of, 452, 452f
    water cycle in, 411, 411f
Rain shadows, 417, 417f
Rana genus (leopard frogs), 275, 275f
Rana muscosa, 450
Ranavirus, 450
Random fertilization, 162
Ranunculus (buttercups), 614f, 615f
Raphanus sativus (radish), 622f
Rapid-eye-movement (REM) sleep, 3
Rarotonga island, 370
Ratios, 47
Rats, 432, 577
Ravens, 433, 433f
Ray, John, 283
Reactants, 94, 94f
Reading, assigned, 7
Realized niches, 392-393
Recessive traits, 170-171, 170t, 171f
    in human hair, 172t
Recombination, in influenza, 309
Rectum, 502, 502f, 508, 510f
Recycling
    of materials within ecosystems, 410-414
    of proteins by cells, 89
Red algae, 313, 314, 315t
Red blood cells (erythrocytes), 465, 472, 484, 484f
    agglutination in blood typing, 541, 541f
    in capillaries, 482, 482f
    carbon dioxide transport in, 479
    oxygen transport in, 479, 481f
    in sickle-cell disease, 269, 269f
Red-eyed tree frog *(Agalychnis callidryas),* 374f
Red oak *(Quercus rubra),* 280, 280f
Redox (oxidation-reduction) reactions, 120, 120f
Red Queen hypothesis, 153
Red tides, 314, 315t
Reduction, 120
Reduction division, 160, 160f
Redwood National Park, 626-627
Redwood sorrel, 384f
Redwood trees, 327t, 330f, 334, 334f, 626-627, 626f, 627f
Refractory period, 550, 550f
Regeneration
    in plants, 634, 634f
    of tissues, 472
Regression analysis, 277
Regression lines, 277
Rehearsal, in learning, 5, 5f
Remdesivir, 310
Remembering, 4-5
REM (rapid-eye-movement) sleep, 3
Renal cortex, 518, 519f
Renal medulla, 518, 519f
Renal tubule, 519, 519f
Renewable energy, 447
Renner, Otto, 625
Repeated sequences, 229, 229f, 229t
Repetition, in learning, 5
Replication, DNA, 136, 136f, 194-195, 194f, 195f, 215f
Replication forks, 194, 194f
Repressors, 97, 97f, 219, 219f

Reproduction. *See also* Asexual reproduction; Sexual reproduction
    by cnidarians, 588
    contraception and, 602-603, 603f
    in flatworms, 353
    in fungi, 316f, 317, 317f
    modes of, 588-589
    in plants, 325, 628-633, 628f-632f
    by protists, 313, 313f, 588, 588f
    roundworm, 359
    sex determination and, 589, 589f
    STDs and, 586-587, 603, 605, 605f
    vegetative, 628, 628f
Reproductive cloning, 242-243, 242f-243f
Reproductive isolation, 273-275, 273t, 283
Reproductive systems
    female, 593-594, 593f, 594f
    fertilization in, 594, 594f
    male, 591-592, 591f, 592f
    of vertebrates, 463, 463f
Reprogramming, gene, 246
Reptiles. *See also* specific types
    amniotic eggs of, 375, 375f
    characteristics of, 375
    evolution of, 368-369, 375, 375f
    in Mesozoic era, 368-369
    nitrogenous waste elimination by, 521, 521f
    respiratory system of, 375
    teeth of, 503
Resistin, 573
Resolution, 68
Resources, preservation of, 451-453
Respiration. *See also* Cellular respiration
    in carbon cycle, 412, 412f
    as circulatory system function, 479
    CO_2 transport in, 492
    definition of, 488
    oxidative, 123-125, 488
    oxygen transport in, 492
    types of, 488, 488f
Respiratory systems
    defensive function of, 526
    of fishes, 372
    of humans, 490-491, 490f
    of mammals, 489-492
    of reptiles, 375
    types of, 488, 488f
    of vertebrates, 463, 463f, 488-492
    of whales, 489
Respiratory tract, 504, 526
Resting membrane potential, 549, 550f
Restriction endonucleases, 230, 231f
Restriction enzymes, 230, 231, 231f
Reticular formation, 553f, 555
Retinal, 110
Retinal degeneration, 247f
Reverse transcriptase, 231, 231f
Reviewing, for exams, 7
Revising, notes, 4, 6
Rheumatoid arthritis, 542
Rh factor, 541
Rhinoceroses, 405, 455-456, 455f
Rhizoids, 331, 331f
Rhizomes, 628, 628f
Rhodophyta, 313, 315t
Rhodopsin, 567
Rhynia, 325
Rhyniophyta, 329

Rhythm method, 602
Ribonucleic acid (RNA)
    in central dogma, 210, 210f
    in chromosomes, 74, 139
    first cells and, 297
    function of, 58
    in gene expression, 221, 221f
    in gene theory, 28, 29f
    interference, 221, 221f
    memory, 536, 536f
    messenger, 176, 177f, 211, 221, 221f
    micro, 221, 221f
    ribosomal, 74, 213
    small, 225, 225f
    transfer, 213, 213f, 221
    types of, 210
Ribose, 57
Ribosomal RNA (rRNA), 74, 213
Ribosomes
    eukaryotic, 74, 74f, 75t
    prokaryotic, 69
    in translation, 213, 213f, 214, 214f
Ribs, 470, 470f
Rice
    flood-tolerant, 633
    iron deficiency and, 239
Ricin, 631
Right atrium, 485f, 486
Right ventricle, 485f, 486
Rings, in wood, 617, 617f
RNA. *See* Ribonucleic acid
RNA-dependent RNA polymerase (RdRP), 310
RNA interference, 221, 221f
RNA polymerase, 210, 210f, 211, 220, 220f
Roan pattern, in horses, 180, 180f
Robins, 571
Rodents. *See* Mice; Rats
Rodhocetus, 258, 258f
Rodriguez, Alex, 48f, 49
Root canals, 503, 504f
Root cap, 613, 613f
Root hairs, 610, 614, 622, 622f
Roots
    adventitious, 615
    branching of, 615, 615f
    elongation of, 614-615, 615f
    in energy flow, 405
    mineral absorption by, 324
    structure of, 613, 613f, 614f
    of vascular plants, 608, 608f
    water absorption by, 325, 614, 622, 622f
Rotifers (Rotifera), 359f
Rough endoplasmic reticulum, 76-77, 76f, 77f
Roundup, 226, 226f, 227f
Roundworms, 358-359, 358f, 359f
rRNA (ribosomal RNA), 74
r-selected adaptation, 389, 389t
RU-486 (Mifepristone), 602
Rubisco, 115
Runners, 628, 628f
Russell, Bill, 562

S

Saccharomyces bayanus, 304
Saccharomyces uvarum, 304
Salamanders
    digestive system of, 501, 501f
    evolution of, 373-374
    global decline in, 450
Salinity, soil, 341
Saliva, 504, 528
Salivary amylase, 504
Salivary glands, 504, 510f
Salmon, 230f
Salt
    adrenal cortex and, 583
    in blood plasma, 483
    dissolution in water, 43, 43f
    ionic bond in, 39, 39f
    plants tolerating, 341
Sand dollars, 364-365, 364f
SA (sinoatrial) nodes, 487, 487f
Saprozoic feeders, 312
Sapwood, 617
Sarcomeres, 473, 473f
Sarcoplasmic reticulum, 467f
SARS (severe acute respiratory syndrome), 307
SARS-CoV-2 life cycle, 310
Saturated fats, 61, 61f
Savannas, 391f, 408, 422, 422f, 423, 423f
Scale, in graphs, 13-14, 13f
Scallops, 355
Scanning electron microscope (SEM), 68, 68f
Scavengers, 408
Scheck, Barry, 237
Schizophrenia, 197, 197f
Schleiden, Matthias, 28, 66
Schwann, Theodor, 28
SCID (severe combined immune deficiency), 246
Science
    limitations of, 26
    uncertainty in, 25, 26
    as way of thinking, 11-12
Scientific method, 26
Scientific names, 281
Scientific processes, 24-26, 24f, 25f
Sclereids, 609, 609f
Sclerenchyma cells, 609, 609f
Scorpions, 360
Scotch pine trees, 386f
Scrapie, 58
Scurvy, 499
Sea anemones, 350f, 395
Sea cucumbers, 364-365
Sea level, global warming and, 445, 448
Sea otters, 396, 433, 433f
Seasons, 415, 415f
Sea stars, 364-365, 364f
Sea turtles, 255, 255f, 436
Sea urchins, 364-365, 364f
Sebaceous glands, 526, 528f
Sebum, 526
Secondary carbon footprint, 412
Secondary cell walls, 609, 609f
Secondary consumers, 406
Secondary growth of plants, 329, 608, 608f
Secondary immune response, 537, 545f
Secondary oocytes, 593f, 594
Secondary phloem, 616, 616f
Secondary structure of protein, 55, 55f
Secondary succession, 398
Secondary xylem, 616, 616f, 617
Second law of thermodynamics, 93, 261
Second messengers, 578, 578f

Seeds
- dispersal of, 333, 632-633, 632f
- double fertilization of, 338, 338f
- evolution of, 332-333
- germination of, 333, 333f, 631, 631f, 633
- poisonous, 631
- structure of, 332-333, 332f

Seed coats, 326, 327f
Seedless plants
- evolution of, 326, 326f
- nonvascular, 326
- vascular, 326, 330-331, 330f

Segmentation, 356-357, 357f
- evolution of, 346

Segments, 346
Segregation, law of, 175
Selection. *See also* Natural selection
- as agent of evolution, 264t, 265
- artificial, 22, 265
- definition of, 265
- directional, 266f, 268
- disruptive, 266f, 268, 268f
- stabilizing, 266f, 267, 267f, 270

Selective diffusion, 84
Selective permeability, 84, 84f
Self-fertilization, 169, 169f
Self-pollination, 630
SEM (scanning electron microscope), 68, 68f
Semen, 592
Semibalanus balanoide, 392-393, 392f
Semicircular canals, 561, 561f
Seminal vesicles, 591f, 592
Seminiferous tubules, 591, 591f
Sensors, 516, 516f
Sensory nervous systems, 560-568
Sensory neurons, 468, 468t, 548-549, 548f
Sensory receptors
- for gravity and motion, 561, 561f
- for hearing, 565-566, 565f
- internal, 560, 560f
- for taste and smell, 564, 564f

Sepals, 336, 336f
Septa, of fungal hyphae, 317
Sequence questions, 9
Sequoia National Park, 627
Sequoia sempervirens, 334, 334f
Serosa, 502, 502f
Serotonin, 552
Sertoli cells, 591f
Serum, 483
Set point, 516, 516f
Severe acute respiratory syndrome (SARS), 307
Severe combined immune deficiency (SCID), 246
Sex chromosomes, 181-184
Sex determination, 589, 589f
Sex-linked traits, 185-186, 185f
Sex pheromones, 428-429
Sex ratio of populations, 390
Sexual life cycles, 155, 155f
Sexually transmitted diseases (STDs), 586-587, 603, 605, 605f. *See also* HIV/AIDS
Sexual orientation, 163, 590
Sexual reproduction, 152f
- by angiosperms, 629-630, 629f, 630f
- by animals, 155, 155f, 344, 345t
- definition of, 154
- by echinoderms, 365
- evolution of, 152-153, 162
- in fungi, 316f, 317, 317f
- *in vitro* fertilization and, 156
- mutations in asexual reproduction compared to, 153
- by protists, 313, 313f

Shallow water ecosystems, 418, 418f
Sharks, 372, 372f, 465f
Sheep, cloning of, 246
Shivering, 517
Shoots, of vascular plants, 608, 608f
Short-term memory, 4-5
Shoulder girdle, 470
Shrimp, 360
Shrub teas, 327t
Siberian tigers, 385f
Sickle-cell disease
- adaptation within populations and, 269-270, 269f, 270f
- in African Americans, 176, 187, 270
- CRISPR for treatment of, 270
- hemoglobin in, 187, 187f
- inheritance of, 186-187, 187f

Sieve cells, 612
Sieve plates, 612
Sieve-tube members, 612, 612f
Sieve-tubes, 612, 612f
Sigmoid growth curve, 386
Sign stimulus, 430, 431f
Simple carbohydrates, 59, 59f
Simple digestion, 501, 501f
Simple epithelium, 464
Simple leaves, 618, 618f
Simple sugars, 59, 59f
Simpson, George, 293
Single covalent bonds, 40, 40f
Sinoatrial (SA) nodes, 487, 487f
Sister chromatids, 138, 138f, 140-141
Skeletal connective tissue, 465
Skeletal muscles, 467, 467f, 471, 471f, 560, 560f
Skeletal systems, vertebrate, 463
Skeletons. *See also* Bone
- cytoskeleton, 70, 70f, 71f, 75t, 80-81
- endoskeletons, 469, 469f
- exoskeletons, 360, 361f, 469, 469f
- of humans, 470, 470f, 557, 557f
- types of, 469-470, 469f, 470f
- of vertebrates, 469-470

Skin
- cells of, 66
- as first line of defense, 526, 526f, 528, 528f
- sensory receptors in, 560
- structure of, 526, 528, 528f

Skin cancer, 134-135, 135f
Skinner, B. F., 432
Skull, 470, 470f
Sleeping sickness, 314
Sleep, learning and, 2-3
Sliding filament model, 473, 473f
Slightly movable joints, 470
Slime molds, 315t
Slow-wave sleep, 3
Slugs, 355
Small intestine, 502, 502f, 507-508, 507f, 508f, 510f
Smallpox vaccine, 235, 525, 540, 540f
Small RNA, 225, 225f
SmartBook, 10f

SmartBook program, 10, 10f
Smell sense, 564, 564f
Smoking
- addiction and, 546-547
- e-cigarettes and, 149, 547
- lung cancer and, 11, 11f, 147, 147f, 199, 547, 585, 585f
- quitting, 547

Smooth endoplasmic reticulum, 76-77, 76f, 77f
Smooth muscles, 467, 471
Snails, 355
Snakes, 469f
Snowshoe hare (*Lepus americanus*), 397, 397f
Social behavior, 429f
Societies, 438, 438f
Sociobiology, 439
Sodium
- adrenal cortex and, 583
- atomic number and mass number of, 37t
- in blood plasma, 483
- in diet, 500
- in nerve impulse, 549-550, 550f

Sodium chloride. *See* Salt
Sodium ions, 38, 38f
Sodium-potassium (Na^+-K^+) pump, 87, 87f, 100t, 549-550, 550f
Soft palate, 504, 504f
Soils
- acid rain and, 45
- cycling of nutrients in, 410
- in plant absorption of minerals, 324
- salinity of, 341
- in succession, 398
- topsoil, 324, 451-452

Solar energy, 447
Solenoids, 139
Solubility, 43
Solutes, 83
Somatic cells, 155
Somatic (voluntary) nervous system, 558
Somatic rearrangement, 536
Somatic tissue mutations, 198, 199
Somites, 597
Song sparrow (*Melospiza melodia*), 388, 388f, 401, 401f
Sorus (sori), 331, 331f
Soto, Ana, 577
Sound receptors, 565-566, 565f
Sparling, Gene, 402-403
Sparrows
- population density of, 388, 388f
- population size and growth of, 401, 401f
- stabilizing selection in, 267

Speciation, 273-275
Species. *See also* Populations
- biological species concept and, 273
- in classification of organisms, 280-281
- coevolution of, 394-395, 394f, 395f
- definition of, 252, 273, 283, 384
- formation of, 273-275
- in hierarchy of complexity, 21, 21f, 384
- interactions among, 394-397
- introduced, 449, 449f, 454
- keystone, 391, 456, 456f
- names of, 281, 281f
- number of, 283
- overexploitation of, 449, 449f

Specific epithet, 281
Specific immunity, 526, 531–532, 531t
Spemann, Hans, 242
Sperm
    contraception blocking and destroying, 603, 603f
    formation of, 588, 591–592, 591f
    of fungi, 316
    structure of, 592, 592f
Spermatocytes, 591f
Spermatogenesis, 588
Spermicidal jelly, 603, 603f
Sperm whale, 258
Sphincter, 505
Spiders, 360, 360f
Spinal cord, 557, 557f
Spinal nerves, 557, 557f
Spindle fibers, 141
Spindles, 141, 165
Spiracles, 488, 488f
Spiral cleavage, 363
Splicing, alternative, 216, 228
"Split brain", 553
Sponges (Porifera), 348, 348f
    digestive system of, 501
Spongy bone, 466, 466f
Spongy mesophyll, 619, 619f
Spores
    of fungi, 317, 317f
    of plants, 325
Sporophyte generation, 629
Sporophytes, 325, 325f
Sports, anabolic steroid use in, 48–49
Spring overturn, 421, 421f
Spruce trees, 327t
Squamous epithelium, 464
Squids, 355
Squirrels
    classification of, 282, 282f
    foraging behavior of, 434, 434f
    maple sap and, 606–607, 606f, 607f
SRY gene, 181, 589, 589f
Stabilizing selection, 266f, 267, 267f, 270
Stag beetles, 362f
Stamens, 336, 336f, 629, 629f
Starch, 59, 60t, 504, 623
Starlings, 436, 436f
Statistical significance, 401
STDs (sexually transmitted diseases), 586–587, 603, 605, 605f. See also HIV/AIDS
Stem cell therapy, 244–245, 244f, 245, 245f
Stems
    primary growth of, 615–616, 615f, 616f
    secondary growth of, 616–617, 616f, 617f
    of vascular plants, 608, 608f
Sternum, 470, 470f
Steroids, 48–49, 574, 578
Steroid hormones, 576, 576f, 578
Steward, F. C., 634, 634f
Stigma, 336, 336f, 629, 629f
Stimulation
    in learning, 432
    response to, as property of life, 19
    sign, 430, 431f
Stimulus, in homeostasis, 516, 516f
Stipules of leaves, 618
Stoddart, Marion, 457

Stomach, 502, 502f, 510f
    acid in, 505–506
    chyme leaving, 506
    pH of, 528
    structure and function of, 505, 506f
    ulcers in, 506
Stomata, 324f, 325, 610, 610f
    in photosynthesis, 115, 115f
    transpiration regulated by, 621f, 622
Storage polysaccharides, 60t
Stratified epithelium, 464
Stratum corneum, 526, 528, 528f
Strauss, Sharon, 639
Strawberry plants, 154f
Streptomyces, 69f
Streptophytes, 315t
Stress hormone, 583
Stretch receptors, 560, 560f
Striations, of muscle, 467, 467f
Stroke, 553
Stroma, 78, 79f, 108
Structural DNA, 229, 229t
Structural polysaccharides, 60t
Structural proteins, 52, 52f
Structure, function determined by, 23
Studying, 2–10
Sturgeons, 293, 293f
Style, 336, 336f, 629, 629f
Stylets, 359
Submucosa, 502, 502f
Substrates, 94, 94f, 102, 102f
Substrate-level phosphorylation, 122
Subunit (piggyback) vaccines, 235, 235f
Suburbs, deer in, 387
Succession, 398, 398f
Suckers, 628
Sucking lice, 395f
Sucrose, 59, 60t, 623
Sugars, 59, 59f, 623
    added, 500
Sulfur
    acid rain and, 45
    atomic number and mass number of, 37t
    plant absorption of, 324
Sulfur dioxide, 45
Sunbirds, 434, 434f
Sunflowers (*Helianthus annuus*), 616f
Sunlight. See also Photosynthesis
    atmospheric circulation and, 415, 415f
    in flow of energy, 92, 106, 404, 404f
    latitude and, 415, 417
    in skin cancer, 135
Superbugs, 223, 301
Suppositories, 603
Suppressor T cells, 531t, 532
Surface tension of water, 43, 43f
Surfactant, 490
Surtsey (island), 322–323, 322f
Survival of the fittest, 256
Survivorship curves, 390, 390f
Suspensory ligaments, 566, 566f
Sustainable yield, maximal, 388f, 389
Swallowing, 504, 504f
Sweat
    body temperature and, 517, 517f
    daily, 472
    lysozyme in, 528

    pheromones in, 429
Sweat glands, 528, 528f
Sweet woodruff, 619f
Swim bladder, 372–373, 373f
Swine flu, 309
Sword ferns, 384f
Symbiosis
    as biological theme, 23
    commensalism and, 395, 395f
    endosymbiosis and, 79, 79f, 291, 291f
    fungi and plants in, 324, 394–395
    mutualism and, 394, 394f
    parasitism and, 394, 394f
Symmetry
    bilateral, 346, 351–353, 352f
    radial, 346, 352f
Sympathetic nervous systems, 558–559, 559f
Synapses, 160, 160f, 468, 551–552, 551f
Synaptic cleft, 551, 551f
Syngamy, 154
Synthetic auxins, 635
Syphilis, 603, 605, 605f
Systolic pressure, 486

T

T cells, 484, 484f, 531–537, 531t, 534f, 535f, 543
T1 viruses, 207, 207f
$t_{1/2}$, 165
Taiga, 422, 422f, 424, 424f
Tails
    postanal, 367f
    vertebrate embryo, 259, 259f
Tanning, 134–135, 134f
Tapeworms, 588
Taricha granulosa (Oregon newt), 141f
Taste buds, 564, 564f
Taste sense, 564, 564f
Taurine, 127
Taxon (taxa), 281
Taxonomy, 281
    traditional, 285, 285f
Tears, 528
Teeth, 503, 503f, 504f
Telomerase, 151, 151f
Telomeres, 148, 148f
Telophase I, in meiosis I, 157, 158f
Telophase II, meiosis II, 159, 159f
TEM (transmission electron microscope), 68, 68f
Temperature. See also Body temperature
    alleles affected by, 179, 179f
    C_4 photosynthesis and, 115, 115f
    in enzyme activity, 96, 96f
    global warming and rising, 448, 448f
    seasons and, 421, 421f
    sensory receptors for, 560
Temperature response, 529, 530
Temporal isolation, 273, 273t, 274, 274f
Temporal lobe, 554–555, 554f
Tendons, 471
Tensile strength, 621
Terminal buds, 615, 615f
Territorial behavior, 434, 434f, 435t
Territoriality, 434, 434f
Tertiary consumers, 406
Tertiary structure of protein, 55, 55f
TESS (Transiting Exoplanet Survey Satellite), 299

Testcrosses, 174, 174f
Testes, 588, 589, 591–592, 591f
Testing, 24f, 25
Testosterone, 48, 61, 163, 574, 591
Test-tube hamburgers, 511
Tetrahydrocannabinol (THC), 114
Tetrahymena, 81f
Tetrapods, 370
Textbooks
 arrows in, 5, 5f, 18
 Internet resources in, 9, 9f, 10f
 using, 7–8, 8f
Thalamus, 554, 554f
THC (tetrahydrocannabinol), 114
Theoretical niches, 392–393
Theories. *See also* Evolution, theory of
 of biology, 28, 28f–31f, 30
 cell theory, 28, 28f, 66–67
 chromosomal theory of inheritance, 30, 30f
 cohesion-adhesion-tension theory, 620–621, 621f, 625
 critics of evolution as, 261
 endosymbiotic theory, 79
 gene, 28, 28f, 29f
 Mendel's theory of heredity, 30, 172–174, 173f, 174f
 optimal foraging theory, 434, 434f, 439
 rejection of, 26, 26f
 in scientific processes, 25, 26
The Power of Movement in Plants (Darwin, C., and Darwin, F.), 635
Thermal stratification, 421, 421f
Thermocline, 421f
Thermodynamics, 92–93, 93f, 261
Thiazolidinediones (TZDs), 573
Thigmotropism, 637, 637f
Thoracic cavity, 462, 490, 490f
Threshold potentials, 549
Through the Looking Glass (Carroll), 153
Thylakoids, 78, 79f, 108
Thymine, 56, 56f, 57, 192, 192f
Thyroid gland, 582, 582f
Thyroid-stimulating hormone (TSH), 580, 580f
Thyroxine, 582, 582f
Ticks, 360
Tigers, 274, 274f, 385f, 406
Tigons, 274
Tiktaalik, 261f, 370, 373–374, 374f
Timberline, 417
Time, for studying, 6
Tissues
 animal, characteristics of, 344, 345t
 animal, evolution of, 346
 definition of, 462
 germ-line, 155, 196
 in hierarchy of complexity, 20f, 21
 mutations of somatic, 198, 199
 plant, structure and function of, 608–612
 regeneration of, 472
 of vertebrates, 462, 462f, 464–468
TMV (tobacco mosaic virus), 305–306, 305f
Tobacco, 149, 547
Tobacco leaf *(Nicotiana tabacum)*, 610f
Tobacco mosaic virus (TMV), 305–306, 305f
Toes, 172t
Tolman, Ed, 562
Tomatoes, 632, 632f
Tomato plants *(Lycopersicon lycopersicum)*, 238, 610f

Tongue, 504, 504f, 564, 564f
Top carnivores, 406, 409
Topsoil, 324, 451–452
Torpor, 523
Tortoises, 337
Totipotency, 244
Touch, pressure receptors and, 560
Trace elements, 499
Tracers, 38, 38f
Trachea, 488, 488f, 488f, 490, 490f, 504, 504f
Tracheids, 611, 611f
Traditional taxonomy, 285, 285f
"Tragedy of the Commons" (Hardin), 451, 451f
Traits
 definition of, 168
 dominant, 170–171, 170t, 171f, 172t
 genes influencing, 176, 177f
 Mendelian inheritance of, 168–175
 non-Mendelian inheritance of, 178–180
 recessive, 170–171, 170t, 171f, 172t, 184–187, 184t
 sex-linked, 185–186, 185f
Transcription
 in eukaryotes, 216, 217f
 process of, 210–211, 210f, 211f
 in prokaryotes, 216, 216f
 regulation of, 218–220, 222f
Transcription factors, 220, 220f
Trans fats, 61, 500
Transfer RNA (tRNA), 213, 213f, 221
Transiting Exoplanet Survey Satellite (TESS), 299
Translation
 in eukaryotes, 216, 217f
 genetic code in, 212, 212f
 process of, 210, 212–214 211f, 214f, 215f
 in prokaryotes, 216, 216f
Translocation, 623, 623f
Transmembrane proteins, 73
Transmissible spongiform encephalopathies (TSEs), 58
Transmission electron microscope (TEM), 68, 68f
Transpiration, 410f, 411, 620f, 621–622, 621f
Transport
 across plasma membranes, 82–87
 active, 87, 100t
 cytoplasmic, 100t
Transport disaccharides, 60t
Transport proteins, 52f
Transposable elements, 229, 229t
trans-retinal, 567, 567f
Traveler's diarrhea, 312
Treadmill evolution, 153
Trees. *See also* Forests
 DNA fingerprinting of, 232
 height of, 626–627, 626f, 627f
 oldest, 337
 in photosynthesis, 106
 rings and age of, 617, 617f
 water movement in, 611, 611f, 625
Treponema, 69f
Treponema pallidum, 603
Triassic period, 368, 369
Trichomes, 610, 610f
Trichomonas (parabasalids), 315t
Tricuspid valve, 485f, 486
Triglochin maritima (arrowgrass), 341
Trilobites, 368, 368f
Triple covalent bonds, 40
Trisomy, 138, 181

tRNA (transfer RNA), 213, 213f, 221
Trophic levels, 404f, 405–409
Trophoblast, 597
Tropical monsoon forests, 425, 425f
Tropical rain forests, 411, 411f, 422, 422f, 452, 452f
Tropisms, 637, 637f
Trout, 133, 133f
True-breeding, 169, 171, 171f
True bugs, 362f
Trypanosomes, 314
Trypsin, 96, 96f
Tryptophan, 574
TSEs (transmissible spongiform encephalopathies), 58
TSH (thyroid-stimulating hormone), 580, 580f
Tubers, 628
Tubular gastrointestinal tract, 502, 502f
Tubulin, 80
Tumors, 143, 143f
Tumor-suppressor genes, 143, 146
Tuna, 389
Tundra, 422, 422f, 424, 424f
Tunicates, 366f
Turgidity, 622
Turgor pressure, 83
Turner syndrome, 183f, 184
Turtles
 migration by, 436
 natural selection in, 255, 255f
Tutt, J. W., 271
Twin studies, 166–167, 431
Tyrosine, 574
TZDs (thiazolidinediones), 573

U

Ubiquitin, 89, 89f
Ulcers, 506
Unconditioned stimulus, 432
Unicellularity, 300t
United Nations World Food Program, 240
Units of measurement, 14
Unsaturated fats, 61, 61f
Uracil, 56, 56f, 57
Urea, 521, 521f
Ureter, 518, 518f
Urethra, 518, 518f, 592
Urey, Harold, 296–297
Uric acid, 521, 521f
Urinary bladder, 518, 518f
Urinary systems, of vertebrates, 463
Urine, 472
 glucose excretion in, 513
 human production of, 519
 pheromones in, 429
 water in, 520
Urodela, 374
USDA (Department of Agriculture), 500
USDA Organic label, 500
Uterus, 593f, 594, 594f

V

Vaccines
 cancer, 235
 for cervical cancer, 603
 controversy over, 587

Vaccines (*Cont.*)
　DNA, 235
　history of, 540, 540*f*
　for HIV/AIDS, 247, 524–525
　for HPV, 587, 603
　immune responses and, 525
　influenza, 235, 540
　smallpox, 235, 525, 540, 540*f*
　subunit (piggyback), 235, 235*f*
Vacuoles
　central, 71, 71*f*, 77, 77*f*
　food, 313
Vagina, 593*f*, 594, 594*f*
Valium (diazepam), 552
Vampire bats, 476–477, 477*f*
Vaping, 149
　nicotine-laced aerosol vapor, 493
　propylene glycol, 493
　young adults, 493
Vaporization, heat of, 42*t*, 43
Variables, 13, 25
Variant Creutzfeldt-Jakob disease (vCJD), 58
Vascular cambium, 608, 616, 616*f*
Vascular plants
　body of, 608, 608*f*
　carbohydrate transportation in, 623, 623*f*
　evolution of, 326, 326*f*, 329, 329*f*
　organization of, 608, 608*f*
　phyla of, 327*t*
　seedless, 326, 326, 330–331, 330*f*
　tissues of, structure and function of, 608–612
　water movement in, 620–622
Vascular systems, of echinoderms, 365
Vascular tissue of plants, 326, 609, 611–612, 611*f*, 612*f*
Vas deferens, 591*f*
Vasopressin, 579 See also Antidiuretic hormone
vCJD (variant Creutzfeldt-Jakob disease), 58
Vectors, for gene therapy, 246–247, 246*f*, 247*f*
Vegan diet, 90–91, 90*f*, 91*f*, 509
Vegetarian diet, 90–91, 90*f*, 91*f*, 509
Vegetarian finches, 254*f*
Vegetative reproduction, 628, 628*f*
Veins, 478, 480, 480*f*, 481*f*, 482, 482*f*
Veins of leaves, 618–619, 619*f*
Ventral, definition of, 351
Ventricles, 485–487, 485*f*
Venules, 480, 480*f*
Verne, Jules, 438
Vertebrae, 470, 557
Vertebral columns
　fish, 372
　human, 470, 470*f*, 557, 557*f*
Vertebrates. *See also specific types*
　body temperature of, 517, 517*f*
　brain of, anatomy of, 553–555, 553*f*–555*f*
　as chordates, 367
　circulatory systems of, 463, 463*f*, 478–487
　classification of, 284, 284*f*, 371, 371*f*
　digestive function of, 502, 502*f*
　digestive systems of, 463, 463*f*
　evolution of, 259, 259*f*, 368–379
　eyes of, 261, 261*f*
　in fossil record, 368–369
　invasion of land by, 368–369
　muscles of, 467, 467*f*, 471, 473
　nervous systems of, 463, 463*f*, 548, 548*f*, 557, 557*f*
　organs of, 462, 463*f*
　organ systems of, 463, 463*f*
　osmoregulation by, 518–521
　phylogenetic tree of, 371*f*
　respiratory systems of, 463, 463*f*, 488–492
　skeletons of, 469–470
　societies of, 438, 438*f*
　tissues of, 462, 462*f*, 464–468
Vesicles, 70*f*, 71, 71*f*, 76
Vessel elements, 611, 611*f*
Villi, 507–508, 507*f*
Viral diseases, 306–307, 306*f*, 307*f*. *See also specific types*
Virtual Lab Simulations, 10*f*
Viruses. *See also specific types; specific types*
　animal, 308
　bacterial, 305, 308, 308*f*
　in cancer, 143
　chicken pox, 537, 537*f*
　emerging, 306–307, 306*f*, 307*f*
　genetic engineering and, 235
　immune response to, 533
　population decline from, 450
　structure of, 305–307, 305*f*
　T1, 207, 207*f*
Vision, 566–568, 566*f*–568*f*. *See also* Eyes
Visual learning, 9
Vitalographs, 86
Vitamins, 483, 499
Vitamin A, 499
Vitamin C, 499
Vitamin E, 499
Vitamin K, 508
Vitamin supplements, 499
Voltage-gated channels, 549–550, 550*f*
Voluntary (somatic) nervous systems, 558
Volvox, 313, 313*f*, 315*t*
Vorticella, 312, 312*f*
Vultures, 406

W

Waggle dance, of honeybees, 437, 437*f*
Walking whale (*Ambulocetus*), 258, 258*f*
Wallace, Alfred Russel, 256
Walnut trees, 618, 618*f*
Warney, Douglas, 236–237
Wastes
　in blood plasma, 483
　nitrogenous, 521, 521*f*
Water
　bottled, 444
　human use of, 410, 452
　hydrogen bonds in, 39, 41–43, 41*f*–43*f*
　ionization of, 44
　metabolic, 515
　movement across plasma membrane, 82–83, 83*f*
　in origin of life, 296–297, 297*f*
　osmoregulation of, 518–521, 518*f*, 519*f*
　plant conservation of, 324–325, 614
　plant movement of, 611, 611*f*, 620–622, 620*f*–622*f*, 625
　properties of, 42–43, 42*t*
　root absorption of, 325, 614, 622, 622*f*
　small intestine reclaiming, 507–508
　tree movement of, 611, 611*f*, 625
　in urine, 520
Water balance, 23
Water cycle, 410–411, 410*f*, 411*f*
Waterhemp (Palmer amaranth), 226
Water molds, 315*t*
Water pollution, 411, 414, 444
Water vascular systems, 365
Watson, James, 192, 193*f*
Weasels, 406
Weather, 415–417, 415*f*, 416, 417*f*, 447, 448
Weeds, 238
Weight. *See also* Overweight
　mass compared to, 36
　units of measurement for, 14
Weight loss, 118–119, 465
Weinberg, W., 263–264
Weissmann, Charles, 58
Welwitschia, 327*t*, 335
Wenner, Adrian, 437
Went, Frits, 635
Westermarck effect, 432
West Nile virus, 307
Whales
　countercurrent heat exchange in, 479, 479*f*
　evolution of, 258, 258*f*
　heart size of, 495
　respiratory system of, 489
Wheat, 327*t*
Whisk ferns, 327*t*, 330*f*
White blood cells (leukocytes), 484, 484*f*, 529–533, 530*f*, 531*t*
Whiteheads, 526
White matter, 553, 557*f*
White-tailed deer, 387, 396
Wildebeests, 385*f*, 435*t*
Wilkins, Maurice, 192, 193*f*
Willow oak (*Quercus phellos*), 280, 280*f*
Wilmut, Ian, 242*f*–243*f*, 243
Wind energy, 447
Winder, Ernst, 199
Wind-pollination, 630
Wine making, 128, 304
Wisdom teeth, 503
Wolves
　as carnivores, 405*f*, 406
　dog evolution and, 250–251, 251*f*
　predation by, 396, 396*f*
Wood, 329
　musical instruments from, 616
　rings in, 617, 617*f*
　uses of, 617
Woodpecker
　finches, 254*f*
　ivory-billed, 402–403, 403*f*
Woolly hair, 189
Worms, 400. *See also* Earthworms; Flatworms
　bristle worms, 357*f*
　roundworms, 358–359, 358*f*, 359*f*
　tapeworms, 588
Wort, 304

X

x axis, 13, 13*f*
X chromosomes
　nondisjunction involving, 183, 183*f*

sex determination and, 589, 589f
Xylem, 329, 611, 611f, 613, 613f, 616, 616f, 617, 623, 623f, 625

Y

y axis, 13, 13f
Yamagiwa, Katsusaburo, 199
Yamanaka, Shinya, 245
Y chromosomes
    in DNA testing, 182–183
    sex determination and, 589, 589f
Yeasts, 128, 304
Yellow morel, 316f

Z

Z lines, 473, 473f
Zamecnik, Paul, 225
Zebras, 405f, 408
Zika virus, 307
Zona pellucida, 594, 594f
Zone of differentiation, 614, 622, 622f
Zone of elongation, 614, 622, 622f
Zooplankton, 420f
Zygomycota, 319t
Zygotes, 154, 588, 594
    in vitro fertilization and, 154